Advanced Techniques in Tracking and Sensor Management

IEEE Press
445 Hoes Lane
Piscataway, NJ 08854

Advanced Techniques in Tracking and Sensor Management

Theory and Applications

Edited by

Mahendra Mallick

Independent Consultant
Cambria, CA, USA

Ba-Ngu Vo

Curtin University
Perth, WA, Australia

Ratnasingham Tharmarasa

McMaster University
Hamilton, ON, Canada

IEEE*Press*
WILEY

Published by John Wiley & Sons, Inc., Hoboken, New Jersey.

Library of Congress Cataloging-in-Publication Data:

Names: Mallick, Mahendra editor | Vo, Ba-Ngu editor | Tharmarasa, R. editor | John Wiley & Sons publisher
Title: Advanced techniques in tracking and sensor management : theory and applications / edited by Mahendra Mallick, Ba-Ngu Vo, Ratnasingham Tharmarasa.
Description: Hoboken, New Jersey : Wiley, [2026] | Includes bibliographical references and index.
Identifiers: LCCN 2026021288 | ISBN 9781394332014 (hardback) | ISBN 9781394332038 pdf | ISBN 9781394332021 epub
Subjects: LCSH: Sensor networks–Data processing | Tracking (Engineering)–Data processing | Signal processing–Digital techniques–Data processing | Multisensor data fusion | Tracking radar
Classification: LCC TK7872.D48 A38 2026
LC record available at https://lccn.loc.gov/2026021288

Cover Design: Wiley
Cover Image: BazziBa/stock.adobe.com

This book is dedicated to my loving family, Emma, Asha, Sumita, Jessica, Ashish, Aditya, and Anand, in gratitude for their enduring support.

— Mahendra Mallick

I dedicate this to my family Ivy, Ba-Nang and Ba-Huu, for the sacrifices, and for forgiving me when the dinner table turned into "optimal assignments." Thank you for still treating me as family after all that!

— Ba-Ngu Vo

To my wife, Shanthy, and my children, Thivya and Theepak, who are my constant source of strength and joy.

— Ratnasingham Tharmarasa

Contents

List of Contributors

Martin Adams
Department of Electrical Engineering and
Advanced Mining Technology Center
Universidad de Chile
Santiago
Chile

Aranee Balachandran
Department of Electrical and Computer
Engineering
McMaster University
Hamilton
ON
Canada

Giorgio Battistelli
Dipartimento di Ingegneria
dell'Informazione (DINFO)
Università degli Studi di Firenze
Firenze
Italy

Aidan Blair
School of Computing Technologies
RMIT University
Melbourne
Victoria
Australia

Henk A.P. Blom
Aerospace Engineering Department
Delft University of Technology
Delft
The Netherlands

Yvo Boers
Department of Applied Mathematics
University of Twente
The Netherlands

Leonardo Cament
Department of Electrical Engineering and
Advanced Mining Technology Center
Universidad de Chile
Santiago
Chile

Xiaomeng Cao
Electrical and Control Engineering
Shaanxi University of Science and Technology
Xi'an
China

Abhijit Chatterjee
Department of Computer Science
University of Exeter
Exeter
UK

Luigi Chisci
Dipartimento di Ingegneria
dell'Informazione (DINFO)
Università degli Studi di Firenze
Firenze
Italy

Stefano Coraluppi
Systems and Technology Research
Woburn
MA
USA

Javier Correa
Galgo
Chile
Santiago
Chile

Kutluyil Dogancay
UniSA STEM
University of South Australia
Mawson Lakes
SA
Australia

Jindřich Duník
Department of Cybernetics
University of West Bohemia
Pilsen
Czech Republic

Alfonso Farina
Independent Researcher
Rome
Italy

Nicola Forti
Dipartimento di Ingegneria
dell'Informazione (DINFO)
Università degli Studi di Firenze
Firenze
Italy

Lin Gao
School of Information and Communication
Engineering
University of Electronic Science and
Technology of China
Chengdu
China

Amirali K. Gostar
School of Engineering
RMIT University
Carlton
Victoria
Australia

Marcel Hernandez
Department of Electrical Engineering and
Electronics
University of Liverpool
Liverpool
UK

Hatem Hmam
Defence Science & Technology Group
Sensors and Effectors Division
Edinburgh
SA
Australia

Reza Hoseinnezhad
School of Engineering
RMIT University
Melbourne
Victoria
Australia

Felipe Inostroza
Advanced Mining Technology Center
Universidad de Chile
Santiago
Chile

Hao Jiao
National Key Laboratory of Radar Signal
Processing
Xidian University
Xi'an
China

Brandon A. Jones
Department of Aerospace Engineering and
Engineering Mechanics
University of Texas at Austin
Austin
TX
USA

Du Yong Kim
School of Engineering
RMIT University
Melbourne
Victoria
Australia

Thiagalingam Kirubarajan
Department of Electrical and Computer Engineering
McMaster University
Hamilton
ON
Canada

Keith Leung
Applanix Corporation (Trimble)
Richmond Hill
ON
Canada

Keyi Li
School of Electronics and Information Engineering
Harbin Institute of Technology
Harbin
China

Hongwei Liu
National Key Laboratory of Radar Signal Processing
Xidian University
Xi'an
China

Jing Liu
School of Automation Science and Engineering
Xi'an Jiaotong University
Xi'an
China

Weifeng Liu
Electrical and Control Engineering
Shaanxi University of Science and Technology
Xi'an
China

Ronald Mahler
Independent Researcher
Eagan
MN
USA

Mahendra Mallick
Independent Consultant
Anacortes
WA
USA

Pranab K. Mandal
Department of Mathematics and Computer Science
Eindhoven University of Technology
The Netherlands

Jakub Matoušek
Department of Cybernetics
University of West Bohemia
Pilsen
Czech Republic

Hoa Van Nguyen
School of Electrical Engineering, Computing and Mathematical Sciences
Curtin University
Perth
WA
Australia

Tran Thien Dat Nguyen
School of Electrical Engineering, Computing and Mathematical Sciences
Curtin University
Perth
WA
Australia

Wenqiang Pu
Shenzhen Research Institute of Big Data
The Chinese University of Hong Kong (Shenzhen)
Shenzhen
China

Ivo Punčochář
Department of Cybernetics
University of West Bohemia
Pilsen
Czech Republic

Benjamin L. Reifler
Department of Aerospace Engineering and Engineering Mechanics
University of Texas at Austin
Austin
TX
USA

Branko Ristic
School of Engineering
RMIT University
Melbourne
Victoria
Australia

Luke Rosenberg
Advanced Systems & Technologies
Lockheed Martin Australia
Adelaide
Australia

Jifeng Ru
Magna International Inc.
Lowell
MA
USA

C. Shim
School of Electrical Engineering, Computing and Mathematical Sciences
Curtin University
Perth
WA
Australia

Ondřej Straka
Department of Cybernetics
University of West Bohemia
Pilsen
Czech Republic

Ehsan Taghavi
Axon
Toronto
ON
Canada

Ratnasingham Tharmarasa
Department of Electrical and Computer Engineering
McMaster University
Hamilton
ON
Canada

Xiaoqing Tian
School of Electrical and Control Engineering
Xi'an University of Science and Technology
Xi'an
China

Pramod K. Varshney
Department of Electrical Engineering & Computer Science
Syracuse University
Syracuse
NY
USA

Ba-Ngu Vo
School of Electrical Engineering, Computing and Mathematical Sciences
Curtin University
Perth
WA
Australia

Ba-Tuong Vo
School of Electrical Engineering, Computing and Mathematical Sciences
Curtin University
Perth
WA
Australia

Linfeng Xu
School of Automation
Northwestern Polytechnical University
Xi'an
China

Junkun Yan
National Key Laboratory of Radar Signal Processing
Xidian University
Xi'an
China

Peijie Yang
School of Automation
Northwestern Polytechnical University
Xi'an
China

Wei Yi
School of Electronic Engineering
University of Electronic Science and Technology of China
Chengdu
Sichuan
China

Ye Yuan
School of Electronic Engineering
University of Electronic Science and Technology of China
Chengdu
Sichuan
China

Peng Zhang
National Key Laboratory of Radar Signal Processing
Xidian University
Xi'an
China

Gongjian Zhou
School of Electronics and Information Engineering
Harbin Institute of Technology
Harbin
China

Yun Zhu
School of Computer Science
Shaanxi Normal University
Xi'an
China

About the Editors

Mahendra Mallick, PhD, is an international expert in filtering, multitarget tracking, and data fusion. A Life Senior Member of the IEEE, he served as an Associate Editor for *IEEE Transactions on Aerospace and Electronic Systems* and co-edited *Integrated Tracking, Classification, and Sensor Management: Theory and Applications.*

Ba-Ngu Vo, PhD, is a Professor and Signals & Systems Chair at Curtin University, Australia. He is a Fellow of the IEEE. He was an Associate Editor for *IEEE Transactions on Aerospace and Electronic Systems* and *IEEE Transactions on Signal Processing*. He is currently a Senior Area Editor of *IEEE Transactions on Signal Processing*. He co-edited *Integrated Tracking, Classification, and Sensor Management: Theory and Applications.*

Ratnasingham Tharmarasa, PhD, is an Associate Professor of Electrical and Computer Engineering at McMaster University, Canada. He serves as an Associate Editor of *Signal Processing* (Elsevier) and specializes in target tracking, sensor management, and data fusion.

Notes on Contributors

Martin Adams is Professor of Electrical Engineering of the Department of Electrical Engineering, University of Chile and his research work focuses on autonomous robot navigation, sensing and multi-target tracking. He is also a principal investigator in the Advanced Mining Technology Centre (AMTC). He received his D.Phil. at the Robotics Research Group, University of Oxford in 1992 and continued his post-doctoral studies at the ETH, Zurich, Switzerland. From 1996 to 2000, he served as a senior research scientist at the European Semiconductor Equipment Centre (ESEC), Switzerland. From 2000 to 2010, he was Associate Professor at Nanyang Technological University (NTU), Singapore. (98 words)

Aranee Balachandran received her B.Sc. Eng. in Electrical and Electronic Engineering from the University of Peradeniya, Sri Lanka, in 2019, and her Ph.D. in Electrical and Computer Engineering from McMaster University, Canada, in 2025. She is currently a Postdoctoral Fellow in the Department of Electrical and Computer Engineering at McMaster University. Her research focuses on enhancing automotive radar performance through multipath analysis, advanced multi-target tracking, and machine-learning-based tracking methods. She has contributed to research in sensing, information fusion, and intelligent transportation systems across academic and engineering environments.

Giorgio Battistelli received the Laurea degree in electronic engineering and the Ph.D. degree in robotics from the University of Genoa, Italy, in 2000 and 2004. Since 2006, he has been with the University of Florence, Italy, where he is Full Professor of Automatic Control and Head of the Department of Information Engineering. His research interests include adaptive and learning systems, linear and nonlinear estimation, sensor networks, and information fusion. He is an Associate Editor of *Automatica* and a former Associate editor of *IEEE Transactions on Neural Networks and Learning Systems* and *IEEE Control Systems Letters*.

Aidan Blair received the B.Eng. degree (Hons.) in advanced manufacturing and mechatronics, and the Ph.D. degree in mechanical engineering from RMIT University, Australia, in 2019 and 2024, respectively. He is currently working as a Research Assistant at the School of Computing Technologies at RMIT University. His main research interests include multi-target tracking, multi-sensor control, distributed sensor networks, attack-resilient information fusion, computer vision, and machine learning applications.

Henk A.P. Blom is Full Professor with Delft University of Technology and has been Principal Scientist with National Aerospace Laboratory NLR, The Netherlands. He has in-depth experience in exploiting the theory of stochastic hybrid processes for multi-sensor data fusion, and for safety risk assessment of novel aircraft operations. In 2007, he has been elected by IEEE to the grade of Fellow. In 2023, he received (jointly with Dr. Yaakov Bar-Shalom) the 2022 Pioneer Award from the IEEE Aerospace and Electronic Systems Society for the development of the Interacting Multiple Model (IMM) approach to multi-model estimation and maneuvering target tracking.

Yvo Boers (M.Sc. in Applied Mathematics 1994 and Ph.d.in Electrical Engineering 1999) From 1999 to 2013 Dr. Boers has been employed at THALES Nederland as an R&D expert on sensor signal and data processing. Since 2013, Yvo has retired from his regular job due to permanent disability (leave). Yvo remains interested in technical problems with a mathematical component. He is also affiliated with the Applied Mathematics Department of the University of Twente as a volunteer/guest. Yvo has also served as an associate editor for several scientific journals in the field.

Xiaomeng Cao received the B.S. degree in Information Engineering and the Ph.D. degree in Control Science and Engineering from Xi'an Jiaotong University, Xi'an, China, in 2012 and 2021, respectively. Since 2023, she has been an associate professor with the School of Electrical and Control Engineering, Shaanxi University of Science and Technology. Her research interests include extended object tracking and maneuvering target tracking.

Leonardo Cament holds a Ph.D. in Electrical Engineering from Universidad de Chile, with a doctoral specialization in biometric face recognition. Since 2016, he has been a lead researcher at the University's Computer Vision Laboratory. His research focuses on the application of random finite sets (RFS) for object trajectory modeling, bridging the gap between advanced computer vision and practical applications in robotics and space situational awareness (SSA). He currently oversees multiple projects dedicated to improving tracking accuracy and orbital safety through robust mathematical modelling.

Abhijit Chatterjee is a Lecturer in Computer Science at the University of Exeter. He holds a Ph.D. in Electrical and Computer Engineering from McMaster University, Canada, where he developed models for satellite scheduling under uncertainty. His research spans optimization and autonomous decision-making for mission planning, with interests in satellite tasking, resource management, and UAV path-planning problems. He develops scalable algorithms that blend operations research, probabilistic modelling, and machine learning for safety-critical systems. He publishes in aerospace and electronic-systems venues and collaborates across engineering and data science communities. Abhijit also teaches and supervises projects in AI, Prompt Engineering, and intelligent systems.

Luigi Chisci received the Laurea degree in electronic engineering from the University of Florence, Italy, in 1984 and the Ph.D. degree in Systems Engineering from the University of Bologna, Italy, in 1989. Since 2004, he is Full Professor of control engineering with the University of Florence. His current interests concern networked estimation, multi-target multi-sensor tracking, multi-agent systems, and sensor data fusion. Dr. Chisci is currently in the editorial board of the International Journal of Adaptive Control and Signal Processing and Associate Editor of the IEEE Transactions on Aerospace and Electronic Systems.

Stefano Coraluppi is an STR Chief Scientist and IEEE Fellow. He received the BS in Electrical Engineering and Mathematics from Carnegie Mellon University (1990) and MS and Ph.D. in Electrical Engineering from University of Maryland (1992, 1997). His research spans estimation, tracking and control. He has worked at ALPHATECH (1997–2002), the NATO Undersea Research Centre (2002–2010), Compunetix (2010–2014), and STR (since 2014). Service includes IEEE AESS

Board of Governors, ISIF Board of Directors, IEEE T-AES Associate Editor-in-Chief, ISIF JAIF Editor-in-Chief, FUSION 2006 General Co-Chair (Florence, Italy), FUSION 2024 General Co-Chair (Venice, Italy), NATO STO Lecturer, and IEEE AESS Distinguished Lecturer.

Javier Correa is currently working at Galgo Chile as a Data Engineer. His research interests are in multitarget tracking and mobile robotics with a specialization in social environments. He holds a Ph.D. from Imperial College London and both bachelors and a masters degrees in Computing Engineering from the Pontificia Universidad Católica de Chile. His work focuses on developing advanced algorithms for real-time tracking and autonomous navigation in complex, human-centric spaces.

Kutluyil Dogancay is a Professorial Lead in the UniSA STEM Unit at University of South Australia. He received an M.Sc. in communications and signal processing from Imperial College, and a Ph.D. in telecommunications engineering from the Australian National University. His research spans statistical and adaptive signal processing, emitter geolocation, target tracking, and autonomous systems. He has authored over 200 publications and two books, and his work has been supported by major defence and industry partners. He serves as Editor-in-Chief of Digital Signal Processing and is a Fellow of the IEEE.

Jindřich Duník is an associate professor at the Department of Cybernetics, University of West Bohemia in Pilsen, and a senior scientist with Honeywell International, Aerospace Advanced Technology Europe. He focuses on state estimation, system identification, inertial and GNSS/GNSS-denied navigation systems, and integrity monitoring methods. He is a co-author of more than 100 papers and patents. He is a member of the IEEE AESS Navigation System Panel and serves as Associate Editor of IEEE Transactions on Aerospace and Electronic Systems. He received the Honeywell Technology Achievement Awards, the Werner von Siemens Excellence Award, and the IEEE AESS Harry Rowe Mimno Award.

Alfonso Farina Dr. Ing. degree EE, University of Rome Sapienza, Italy; Ph.D. HC in ICT, University of Palermo. From 1973 to 2014, he worked at Selex ES until he became Senior Vice President (SVP) and Chief Technology Officer (CTO). Currently, he is President of the Leonardo Radar & Sensors Academy of the Leonardo Electronics Division. Fellow of several professional societies and international academia. Numerous best paper awards and over 1000 peer reviewed published papers. A multifaceted engineer, scientist, and university professor, he is among the 2% top world scientists and one of the world's leading experts on radar systems.

Nicola Forti received the Laurea degree in automation engineering and the Ph.D. degree in information engineering from the University of Florence, Italy, in 2013 and 2017, respectively. From 2018 to 2023, he was a Scientist with the NATO Centre for Maritime Research and Experimentation, La Spezia, Italy. He is currently Assistant Professor at the University of Florence. His research interests include state estimation, networked control systems, information fusion, multi-agent systems, and adaptive and learning systems. He is an Editor of the *International Journal of Adaptive Control and Signal Processing*.

Lin Gao received the B.S., M.S., and Ph.D. degrees in information and communication engineering from the University of Electronic Science and Technology of China (UESTC). From 2018 to 2020, he served as a Postdoctoral Research Fellow with the Department of Information Engineering, University of Florence. He joined the School of Information and Communication Engineering of UESTC as a Lecturer, from 2021 to 2023, and currently he is an Associate Professor. His research focuses on multisensor multitarget tracking, information fusion, and multivehicle SLAM. He has delivered tutorials at the International Conference on Information Fusion, where he also chaired sessions.

Amirali K. Gostar is a Senior Lecturer in the School of Engineering at RMIT University (Melbourne, Australia). His research focuses on multi-object tracking and multi-sensor information fusion for autonomous and intelligent systems, with particular expertise in Random Finite Set (RFS)-based estimation and distributed fusion. He works at the intersection of statistical signal processing and applied AI, translating algorithms into deployable sensing and analytics pipelines for applications such as intelligent transport systems and distributed sensor networks.

Marcel Hernandez is currently a Senior Data Scientist in the Department of Electrical Engineering and Electronics, University of Liverpool, UK. He holds B.Sc. and Ph.D. degrees in Mathematics and is a Fellow of the Institute of Mathematics and its Applications. His research interests include Bayesian inference, information theory, data fusion, autonomous sensor control and optimization theory. He has co-authored over 100 scientific publications, and has received numerous prizes and distinctions, including the best paper award at Fusion 2005, the second-best paper award at Fusion 2021, and was runner-up for the IEEE M. Barry Carlton award for 2008.

Hatem Hmam received his MS degree from University of Houston (1988) and his Ph.D. degree from University of Cincinnati (1992), both in electrical engineering. He briefly worked as a Post-Doc at Queensland University of Technology (QUT) before joining the Defence Science and Technology Group, Australia, in late 1996. There he worked on a number of projects including missile engagement modeling and analysis, Non-GNSS based positioning and navigation, optimal sensor placement, as well as other defense research topics. He made several contributions to the literature of Alternative Navigation and Positioning, including the scan-based localization technique of adversary radars.

Reza Hoseinnezhad is a Professor of Autonomous Systems in the School of Engineering at RMIT University (Melbourne, Australia). His research is broadly concentrated on statistical information processing and sensor fusion with a particular focus on machine vision and Random Finite Set theory applications for situational awareness in autonomous systems. He was recognized as the Australian lead in the field of Signal Processing in 2025.

Felipe Inostroza is an Electrical Engineer with a Ph.D. from Universidad de Chile, where he currently works as a Research Associate at the Advanced Mining Technology Center (AMTC). His work focuses on autonomous navigation, SLAM, and sensor data processing for mining applications with recent contributions in applying random finite set theory to robotic mapping and localization, and applied localization systems for underground mines. He currently leads a project developing autonomous navigation and loading systems for Load-Haul-Dump vehicles in underground mines.

Brandon A. Jones is an Associate Professor in the Department of Aerospace Engineering and Engineering Mechanics at The University of Texas at Austin (UT). He completed his Ph.D. in 2010 at the University of Colorado Boulder as a NASA Graduate Student Research Program (GSRP) fellow. He started at UT in January 2016, and his research program develops methods of uncertainty quantification, Bayesian estimation, multi-target tracking, and information fusion in the field of astronautics. Dr. Jones has designed multiple algorithms for space object tracking that have been transitioned to industry and government organizations for operational use.

Du Yong Kim received the B.E. degree in electrical and electronics engineering from Ajou University, Korea, in 2005, and the M.S. and Ph.D. degrees from the Gwangju Institute of Science and Technology (GIST) in 2006 and 2011. He held postdoctoral positions at GIST, the University of Western Australia, and Curtin University, focusing on statistical signal and image processing. He was a Vice-Chancellor's Research Fellow at RMIT University (2018–2022) and is currently a Senior Lecturer. His research interests include Bayesian filtering, machine learning, computer vision, sensor networks, and control.

Thiagalingam Kirubarajan received M.S. and Ph.D. degrees in electrical engineering from the University of Connecticut, Storrs, CT, USA, in 1995, and 1998, respectively. He is currently a Professor with the Department of Electrical and Computer Engineering, McMaster University, Canada. He has authored or coauthored more than 360 articles, in addition to one book on estimation, tracking, and navigation, four edited volumes, and 12 book chapters. His research interests include estimation, target tracking, multisource information fusion, sensor resource management, signal detection, and fault diagnosis. He received IEEE Barry Award and Ontario Premiers Research Excellence Award.

Keith Leung received his Ph.D. in 2012 from the University of Toronto, where his research focused on state estimation and simultaneous localization and mapping (SLAM) with applications to mobile robotic applications. After obtaining his Ph.D., Keith became a postdoctoral researcher at the Advanced Mining Technology Center (AMTC), Universidad de Chile. In 2016, Keith joined Trimble, where he currently leads a research and development team applying SLAM and perception technologies to a wide range of civil construction and geospatial field applications.

Keyi Li (Member, IEEE) was born in China in 1991. He received the B.E. degree in communication engineering and the M.E. and Ph.D. degrees in information and communication engineering from Harbin Institute of Technology, Harbin, China, in 2014, 2016, and 2020, respectively. He is currently a Lecturer with the School of Electronics and Information Engineering, Harbin Institute of Technology. From September 2018 to September 2019, supported by China Scholarship Council, he was a visiting Ph.D. student with McMaster University, Hamilton, ON, Canada. His research interests include estimation, target tracking, and information fusion.

Weifeng Liu, currently a Professor at Shaanxi University of Science & Technology Xi'an, Shaanxi. He received his Ph.D. in Control Science and Engineering from Xi'an Jiaotong University in 2009, Xi'an. He worked at Hangzhou Dianzi University (2009–2020) before joining his current institution. His research focuses on multi-source information fusion and big data analysis, multi-sensor configuration and optimal control, as well as semantic understanding and cognitive computing. He has published over 100 academic papers.

Ronald Mahler has a mathematics Ph.D. from Brandeis U. and was a 35-year research scientist at Lockheed Martin Corp. (LMC). He originated the stochastic geometry approach to multitarget tracking and information fusion. He is recipient of the Joseph Mignogna Data Fusion Award, the IEEE AESS Harry Rowe Mimno Award, the IEEE AESS Barry Carlton Award, and two LMC Author of the Year Awards. He was co-principal investigator of the DARPA/IXO project "Group Tracking Techniques for Foliage Penetration Ground Moving Target Indication Radars" (2005–2007). His most influential publication is *Statistical Multisource-Multitarget Information Fusion* (2007), an Artech House best-seller.

Mahendra Mallick, Ph.D., is an international expert in filtering, multitarget tracking, and data fusion. A Life Senior Member of the IEEE, he served as an Associate Editor for *IEEE Transactions* *on Aerospace and Electronic Systems*. He co-edited *Integrated Tracking, Classification, and Sensor Management: Theory and Applications* (Wiley/IEEE, 2012) and *Advanced Techniques in Tracking and Sensor Management: Theory and Applications* (Wiley/IEEE, 2026), and was Guest Editor of the *Sensors* 2021–2022 Special Issue on Angle-Only Filtering and Tracking in Two and Three Dimensions, and the June 2013 *IEEE Journal of Selected Topics in Signal Processing* Special Issue on Multitarget Tracking.

Pranab K. Mandal received his Master's degree in statistics from the Indian Statistical Institute, Kolkata, in 1992 and the Ph.D. degree in statistics from the University of North Carolina at Chapel Hill in 1997. Afterwards, he has held various positions – visiting scientist at MSU, East Lansing; postdoctoral researcher at EURANDOM in The Netherlands (NL) and assistant/associate professor at University of Twente, NL. He is currently a senior lecturer at Eindhoven University of Technology, NL. His research interests include mathematical statistics and stochastic (nonlinear) filtering, and in particular, particle filtering with applications to statistical signal processing.

Jakub Matoušek received his Ph.D. in cybernetics from the University of West Bohemia in Pilsen, Czech Republic, in 2025. He is currently a postdoctoral researcher at The University of Texas at Austin. His research focuses on nonlinear Baye.

Tran Thien Dat Nguyen received the B.E. degree in electrical power engineering in 2018 and the Ph.D. degree in electrical and computer engineering in 2022. He is currently a Research Fellow with the Intelligent Sensing & Perception Laboratory (ISPL), Curtin University, Australia. His research interests include multi-object systems, Bayesian estimation, and random sets.

Benjamin L. Reifler is a postdoctoral fellow in the Oden Institute for Computational Engineering and Sciences at the University of Texas at Austin (UT). He completed his Ph.D. in aerospace engineering at UT in 2025, where his dissertation focused on multisensor multitarget tracking and initial orbit determination following satellite breakups and deployments. Dr. Reifler's research interests include multitarget tracking, hard/soft information fusion, and autonomous navigation.

Branko Ristic received all degrees in electrical engineering: Ph.D. in 1995 from Queensland University of Technology, M.Sc. in 1991 from Belgrade University, and B. Eng. In 1984 from The University of Novi Sad. Dr Ristic held various research/engineering positions in former Yugoslavia and Australia, until he joined in 1996 the Defence Science and Technology Organisation (DSTO), where he stayed until 2015. His role in DSTO, was to carry out research, develop new capabilities and provide technical advice to the Australian Defence Organisation on topics of target tracking, multisensor fusion and reasoning under uncertainty. During 2003/04 he spent a year in IRIDIA (Université libre de Bruxelles, Belgium) on a study leave. Since 2015, Branko Ristic is Principal Research Fellow in the School of Engineering of RMIT University, where his role is to attract funding, conduct research and manage research contracts (over 20 contracts so far, with a total funding over $3.5M). Dr Ristic co-authored two books, "Beyond the Kalman filter: Particle filters for tracking applications" (Artech House, 2004), and "Particle filters for random set models" (Springer, 2013), as well as 10 book chapters and well over 100 journal articles. He received numerous awards for his publications. His research interests are information fusion, Bayesian estimation, target tracking, Monte Carlo methods, search techniques, reasoning and decision making under uncertainty, sensor control.

Luke Rosenberg is the RF Analysis and Sensemaking Team Lead at Advanced Systems & Technologies, Lockheed Martin Australia and an adjunct Associate Professor with the University of Adelaide, Australia. Prior to this, he worked at the Defence Science and Technology Group (DSTG) for 23 years as a research specialist in maritime radar. In 2024, he became an IEEE Fellow for contributions to maritime radar and in 2026 became a Lockheed Martin Associate Fellow. He has over 190 publications including a book on Radar Sea Clutter: Modelling and Detection.

Jifeng Ru received her Ph.D. degree from the University of New Orleans, USA, in 2005. During 2005–2015, she worked as a Senior Research Engineer on air traffic control systems, improving tracking performance and operational safety. She later served as an Advanced Project Leader in several research programs. Since 2015, she has been with Magna, where she is currently a Senior Principal Engineer and Technical Manager, leading tracking and feature development for ADAS systems from concept through production readiness. Dr. Ru is coauthor of more than 20 journal and conference papers and holds multiple patents in ADAS and automotive safety.

C. Shim received the B.E. degree in Electronics and Information Engineering and the Ph.D. in Computer Science and Engineering, from Korea University, South Korea, in 2014 and 2019, respectively. He is currently a postdoctoral research fellow in the School of Electrical Engineering, Computing and Mathematical Sciences (EECMS) at Curtin University, Australia. Prior to joining Curtin, he served as a lecturer at Korea University in 2018 and a visiting researcher at RMIT University in Melbourne, Australia, in 2019. His research interests include spatiotemporal databases, Bayesian estimation, and visual analytics.

Ondřej Straka is an associate professor at the Department of Cybernetics, University of West Bohemia in Pilsen. He heads the Identification and Decision-Making research group. His research interests focus on nonlinear state estimation, including Kalman and particle filtering, fault diagnosis, signal processing, system identification, optimal control, and stochastic processes. He has authored or co-authored more than 150 publications in leading international journals and conferences. His work has received several international awards, including the Jean-Pierre Le Cadre Award and the Werner von Siemens Excellence Award. O. Straka serves as a senior editor of IEEE Transactions on Aerospace and Electronic Systems.

Ehsan Taghavi received the M.Sc. degree in communication engineering from Chalmers University of Technology, Sweden, in 2012, and the Ph.D. degree in computational science and engineering from McMaster University, Canada, in 2016. He is now serving as a senior research scientist in industry focusing on integrating vision, LiDAR, radar, sonar, satellite, GPS, and social data to build robust perception systems for autonomous vehicles and surveillance applications. His research interests span estimation theory, scientific computing, signal processing, parameter estimation, mathematical modeling, algorithm design, and AI-based perception and object detection techniques.

Ratnasingham Tharmarasa received M.A.Sc and Ph.D. degrees in electrical engineering from McMaster University, Canada, in 2003 and 2007, respectively. Currently, he is an associate professor in the Electrical and Computer Engineering Department at McMaster University. He is an Associate Editor for Elsevier Signal Processing, and he was a Guest Editor of a Sensors 2021-2023 Special Issue. His expertise includes target tracking, information fusion, resource management, and performance prediction for surveillance systems and autonomous vehicles. He has published over 75 journal articles, 70 conference papers, and three book chapters, and developed numerous real-world tracking, fusion, and resource management systems.

Pramod K. Varshney received the Ph.D. in electrical engineering from the University of Illinois at Urbana-Champaign in 1976. He is a Distinguished Professor at Syracuse University. A Life Fellow of the IEEE (elected 1997) for contributions to distributed detection and data fusion, he served as

President of the International Society of Information Fusion in 2001. His numerous honors include the IEEE Judith A. Resnik Award (2012), the ISIF Yaakov Bar-Shalom Award (2018), the IEEE Claude Shannon-Harry Nyquist Technical Achievement Award (2021), the IEEE AESS Pioneer Award (2021) and the Syracuse University Lifetime Achievement Award (2023).

Ba-Ngu Vo earned dual Bachelor's degrees in Pure Mathematics and Electrical Engineering with First Class Honours in 1994, followed by a Ph.D. in 1997. He is currently a Professor at Curtin University. Vo is a recipient of the Australian Research Council's inaugural Future Fellowship and the 2010 Eureka Prize for Outstanding Science in Support of Defence or National Security. Widely recognized as a pioneer of the stochastic geometric approach to multi-object systems, his research spans signal processing, systems theory, and stochastic geometry, with applications in target tracking, space situational awareness, robotics, and computer vision.

Ba-Tuong Vo is currently a professor of Signal Processing at Curtin University. He has primary research interests in random sets, control and estimation, and multiple object systems, emphasising both fundamental and applied research and development. He is a recipient of the 2010 DSTO Australian Museum Eureka Prize and a finalist for the 2020 DST Australian Museum Eureka Prize.

Wei Yi received the Ph.D. degree in electronic engineering from the University of Electronic Science and Technology of China (UESTC) in 2012. He is currently a Full Professor at UESTC. He is a Senior Member of the IEEE, and his research focuses on target detection and tracking, radar signal processing, and information fusion. He serves as an Associate Editor for the *IEEE Transactions on Signal Processing* and *IEEE Transactions on Aerospace and Electronic Systems*. He received the First Place Award in the Best Student Paper Competition at the 2012 IEEE Radar Conference and the Best Student Paper Award at the 2012 FUSION Conference.

Ye Yuan received the Ph.D. degree in information and communication engineering from the University of Electronic Science and Technology of China (UESTC), Chengdu, China, in 2022. From 2021 to 2022, he was a Visiting Student with the Wireless Communication and Information Systems Laboratory, Seoul National University. He is currently an Associate Professor with UESTC. His research interests include radar signal processing, target tracking, and resource allocation for joint radar-communication systems.

Gongjian Zhou (Member, IEEE) was born in China, in 1979. He received the B.E., M.E., and Ph.D. degrees in information and communication engineering from Harbin Institute of Technology, China, in 2000, 2002, and 2008, respectively. From April 2011 to May 2012, he was a Visiting Professor with the Department of Electrical and Computer Engineering, McMaster University, Hamilton, ON, Canada. He is currently a Professor with the Department of Electronic Engineering, Harbin Institute of Technology. He is also a Longjiang Young Scholar. His research interests include estimation, tracking, detection, information fusion, and signal processing.

Yun Zhu received her B.S. and Ph.D. degrees from Xidian University in 2012 and 2019, respectively. She is currently an associate professor at Shaanxi Normal University. Her research focuses on multi-target tracking, sensor management, and multi-sensor information fusion.

Preface

The field of target tracking and state estimation has undergone a remarkable transformation over the past several decades, evolving from relatively simple single-target scenarios to complex multi-target, multisensor environments that demand sophisticated mathematical frameworks and computational approaches. This book brings together cutting-edge research and methodologies that address key challenges in modern tracking systems, including civilian air traffic control, ballistic missile tracking, dim target tracking, space situational awareness, distributed sensor networks, and sensor management.

The motivation for this book stems from the recognition that contemporary tracking problems cannot be adequately addressed by classical methods alone. Many applications require solutions that can handle nonlinear dynamics, measurement origin uncertainties, detection ambiguities, and resource constraints—all while operating in real time with limited computational resources. Whether tracking aircraft through holding patterns, monitoring reentry vehicles as they descend through the atmosphere, or coordinating multiple sensors to detect dim targets buried in clutter, practitioners face scenarios of unprecedented complexity that require both theoretical rigor and practical innovation.

This book, with 25 chapters, is organized into four parts, each addressing a critical dimension of modern tracking systems: nonlinear filtering, multitarget multisensor tracking, distributed fusion, and sensor management and control. Together, they provide a comprehensive treatment of the multitarget multisensor tracking and sensor management that defines the current state-of-the-art research.

Part I establishes the theoretical and algorithmic foundations essential for accurate state estimation in nonlinear systems. Chapter 1 considers the 3D trajectory estimation of a civilian aircraft using air traffic control (ATC) 2D radars. The 3D motions analyzed are the nearly constant velocity (NCV) motion at nearly constant altitude (NCAlt), climbing with nearly constant climbing rate, and nearly constant turn (NCT) at NCAlt. The Itô stochastic differential equation (SDE) is used to describe the continuous-time dynamic model during the NCT. Chapter 2 discusses the reentry vehicle (RV) filtering problem using a radar and angle-only passive sensor. The continuous-time dynamic model is represented by an Itô SDE involving accelerations due to a spherical Earth gravity and atmospheric drag. An analytic expression for the discrete-time process noise covariance, valid for a general nonlinear SDE, is derived using a 1.5 order Itô-Taylor expansion (IT-1.5). The accurate continuous-discrete EKF (ACD-EKF) is compared with two approximate discrete filtering algorithms.

In Chapter 3, a complete review of various radar measurements is given, including unresolved targets, passive coherent location, and continuous 2D assignment. The concept of multisensor target tracking is also addressed, including a discussion on centralized, distributed, and decentralized multisensor tracking systems. The radar bias estimation is discussed by introducing some of the advances and new methods to estimate the biases. Finally, simulation results are presented to illustrate the performance of some of the methods discussed in this chapter. The common approach in handling hybrid jumps is to combine a variable structure (VS) approach with interacting multiple model (IMM) estimation. Chapter 4 improves this situation through deriving an exact characterization of the nonlinear filter for a Markov switching diffusion with hybrid jumps, and through showing how this characterization leads to exact hybrid jump extensions of the IMM estimator and the IMM particle filter.

Chapter 5 addresses destination-constrained (DC) systems with unknown arrival times, focusing on modeling DC dynamics and estimating both the state and arrival time. Two DC state estimation algorithms are developed that exploit destination information in prediction, differing in computational cost and accuracy, with one enabling joint state and arrival-time estimation. Numerical simulations demonstrate the effectiveness of the proposed methods for tracking adversarial destination-guided targets. Chapter 6 addresses target tracking with implicit motion constraints arising from partial prior information, such as destinations, trajectory shapes, and guidance laws. Three classes of implicit constraints are modeled using direct, state-augmentation, and parameter-augmentation formulations, with pseudo-measurements incorporated into nonlinear filters to exploit these constraints effectively. Simulations show substantial performance gains over unconstrained methods, near-posterior Cramér–Rao lower bound (PCRLB) accuracy, and robustness to noisy or biased prior information.

In Chapter 7, some fundamental convergence results related to particle filtering (PF) and their implications for practitioners, especially in the field of target tracking, are presented. It is shown that some of the convergence results can be considered versions of the law of large numbers (LLN) for PFs. The chapter aims to inform practitioners about the different types of convergence results for PFs that exist and to provide some guidelines on how to interpret them and use them wisely in various applications. Chapter 8 addresses the computation of the PCRLB in cluttered environments with measurement origin uncertainty and measurement accuracy uncertainty. Novel PCRLB formulations are introduced that condition on sampled measurement sequences and then average the resulting conditional bounds to obtain an unconditional PCRLB. The methods are demonstrated through simulations involving ballistic missile impact point estimation, ground vehicle tracking under radar spoofing, and low-altitude airborne target tracking with specular multipath.

Chapter 9 presents a unifying framework for tensor decomposition–based point mass filters (PMFs) for nonlinear stochastic dynamic systems. The proposed PMF algorithms employ tensor decompositions to approximate both the dynamic model—by factorizing the transient density into current and future state components—and the conditional state probability density function using the tensor-train decomposition. Their performance is demonstrated in a realistic tracking scenario and compared with standard filtering approaches.

Part II brings together a comprehensive collection of theories, algorithms, and applications of multitarget tracking that reflect the remarkable evolution of the field. Beginning with the fundamental problem of jointly estimating target numbers (no s) and trajectories, Chapter 10 introduces the labeled random finite set (RFS) state-space model, a principled Bayesian framework capable of representing full trajectory sets while naturally accommodating crossings, fragmentations, spawning, and trajectory uncertainty. Subsequent chapters examine cutting-edge sensing and estimation paradigms across diverse domains: Bayesian track-before-detect (TBD) methods for maritime radar (Chapter 11), advanced video synesthetic aperture radar (ViSAR) tracking techniques that combine TBD and correlation filtering (Chapter 12), and state-of-the-art space object tracking approaches for space situational awareness (Chapters 13 and 15).

Part II further highlights major advances under difficult sensing conditions (Chapter 14), as well as the growing importance of extended and group target tracking using random matrix and hierarchical models (Chapter 16). Beyond traditional tracking problems, the book showcases the power of the RFS methodology in simultaneous localization and mapping through the joint vector-set simultaneous localization and mapping (SLAM) framework (Chapter 17), which unifies trajectory, map, and cardinality estimation while resolving long-standing data association fragilities. Collectively, the chapters of Part II provide both authoritative references for established researchers and an accessible gateway for newcomers entering the rapidly expanding world of multitarget tracking.

Part III examines distributed multitarget tracking, bringing together advanced methods that unify Bayesian inference, information-theoretic fusion, and resource-constrained sensing for modern multiagent systems. Chapter 18 revisits the well-established multiple-hypothesis tracking (MHT) paradigm, discussing its practical strengths, limitations, and modern considerations such as distributed processing and challenging sensor modalities. Chapter 19 lays the foundation with a Bayesian perspective on multiagent state estimation, in which each agent maintains its own probability density of the state and achieves network-wide consistency through principled fusion. Central to this treatment are the Kullback–Leibler-based fusion rules for correlated information and consensus protocols that guarantee scalable, communication-efficient operation over peer-to-peer networks. These theoretical tools enable the development of distributed recursive Bayesian filters with well-characterized implementation and stability properties in linear-Gaussian settings.

Chapter 20 extends these principles to more complex multiobject and multisensor environments. It examines fusion of multiobject densities—an essential component of cooperative multiagent object detection and tracking—and highlights both Kullback–Leibler divergence-based criteria and complementary-fusion strategies designed to integrate information from agents with limited or non-overlapping fields of view.

Part IV addresses the critical problem of optimally allocating limited sensing resources to achieve tracking objectives. Earth-orbiting satellites provide multiple perspectives for tracking missions but face environmental constraints and limited availability. Chapter 21 introduces mixed open- and closed-loop scheduling approaches for multi-agile/non-agile satellites, and a three-stage optimization model to obtain a suboptimal solution.

Chapter 22 surveys sensor management for multitarget tracking, framing sensing as a sequential decision problem under uncertainty. Sensor control is categorized based on the choice of tracking filter, objective function (task-driven versus information-driven), and purpose (non-selective versus selective). Rewards based on Rényi, Kullback–Leibler, and Cauchy–Schwarz divergences, along with performance bounds like the PCRLB, are reviewed. Optimal sensor placement for angle-of-arrival target localization and tracking with Gaussian priors is studied in Chapter 23. The Fisher information matrix is used to establish A-optimality and D-optimality criteria for information-theoretic optimal sensor placement.

Power allocation in colocated multiple-input multiple-output (MIMO) radar systems for multi-target tracking must balance competing demands with limited resources, employing task utility functions and target-specific weights to optimize resource allocation. A robust power allocation methodology within a quality-of-service framework is proposed in Chapter 24. Optimal resource allocation for fused target tracking in radar sensor networks requires a shift from open-loop signal processing to collaborative optimization that adapts parameters across multiple nodes to improve overall tracking performance. Chapter 25 provides a brief overview of resource allocation methods for radar sensor networks.

This book represents the collective expertise of leading researchers addressing the most challenging problems in target tracking and sensor management. The chapters span theoretical foundations, algorithmic innovations, and practical applications, providing readers with both deep understanding and actionable techniques. From the mathematical rigor of stochastic differential equations and random finite set theory to the pragmatic concerns of sensor bias estimation and computational efficiency, the material reflects the multifaceted nature of modern tracking systems. Whether you are a researcher developing new algorithms, an engineer implementing tracking systems, or a student entering this field, we hope this book serves as both a comprehensive reference and an inspiration for continued innovation.

Acknowledgments

We thank the contributing authors for their high-quality chapters and timely submissions, which were essential to the success of this book.

We are also grateful to the reviewers for their insightful and constructive comments, which substantially improved the technical quality, organization, and clarity of the chapters.

The editors have benefited immensely from long-standing collaborations with many researchers over the years, and we express our sincere thanks to Yaakov Bar-Shalom, X. Rong Li, Mark Morelande, Sanjeev Arulampalam, Barbara La Scala, Kuo-Chu Chang, Yanjun Yan, David Castañon, Robert Washburn, Thomas Kurien, late Samuel Blackman, late Craig Carthel, Mark Coates, Taek Lyul Song, late Darko Mušicki, Lucy Pao, Vikram Krishnamurthy, Fredrik Daum, Roy Streit, Frederic Gustafsson, Simmo Särkkä, Yunmin Zhu, Xiaojing Shen, Neil Gordon, David Salmond, Simon Maskell, Alan Marrs, Huimin Chen, Thuraiappah Sathyan, Abhijit Sinha, Zhansheng Duan, Jian Lan, Subhash Challa, Lyudmila Mihaylova, Steven Rubin, Jean Berger, and Mihai Cristian Florea.

We express our sincere gratitude to the Wiley/IEEE Editorial Board; Editor Mary Hatcher and Senior Editorial Assistant Victoria Bradshaw; and the staff for their professional guidance, responsiveness, and timely support throughout the preparation and publication process. We also thank Managing Editors Gokulakrishnan Sridharan and Jayashree Saishankar at the Wiley Chennai office in India for their prompt and professional support during review and production stages.

Part I

Nonlinear Filtering

1

3D Tracking of an Aircraft Using Air Traffic Control 2D Radars

Mahendra Mallick[1], Linfeng Xu[2], Xiaoqing Tian[3], and Jifeng Ru[4]

[1] *Independent Consultant, Cambria, CA, USA*
[2] *School of Automation, Northwestern Polytechnical University, Xi'an, China*
[3] *School of Electrical and Control Engineering, Xi'an University of Science and Technology, Xi'an, China*
[4] *Magna International Inc., Lowell, MA, USA*

1.1 Introduction

In this chapter, we consider the three-dimensional (3D) trajectory estimation of a civilian aircraft using air traffic control (ATC) two-dimensional (2D) radars used at airports. The 3D motions analyzed are the nearly constant velocity (NCV) motion at nearly constant altitude (NCAlt), climbing with nearly constant climbing rate, and nearly constant turn (NCT) at NCAlt. The NCV motion at NCAlt and climbing with a nearly constant climbing rate represent non-maneuvering motions, whereas the NCT at NCAlt is a maneuvering motion. When an aircraft is in a holding pattern near an airport before landing, it performs an NCT (usually called a coordinated turn in civil aviation) at an NCAlt [1].

An airport surveillance radar (ASR) system has a primary surveillance radar (PSR) and a secondary surveillance radar (SSR). The PSR has a rotating antenna mounted on the radar tower that continuously transmits electromagnetic waves and receives scattered signals from aircraft. A PSR measures the slant range and azimuth of an aircraft. Therefore, a PSR is called a 2D radar. As a result, the 3D position of an aircraft cannot be determined from a single ATC 2D radar measurement. A 2D radar has a maximum sensing range of about 60 nautical miles or 111 kilometers. An SSR transmits signals and receives altitude and identification codes automatically from the onboard Mode C transponder on an aircraft. An SSR has a maximum range of about 300 nautical miles or 556 kilometers. The surveillance of the PSR is noncooperative, whereas that of the SSR is cooperative. The measurements of the PSR and SSR occur at different times. The Federal Aviation Administration (FAA) mandates that all aircraft flying within the controlled airspace (10,000 ft or above for class B airspace) must provide Mode C altitude in response to secondary surveillance radar interrogations. However, an aircraft flying below 10,000 ft is not required to provide an altitude response. A hostile aircraft may not respond to secondary surveillance radar interrogations. Thus, it is important for the ATC system to accurately estimate the 3D position and velocity of an aircraft using ATC 2D radar measurements in this scenario.

One and two radars are used to study the NCV motion with NCAlt. The antenna rotation period for an ATC 2D radar is typically in 4–12 s range. As used in [2], we use 4.6 s in the one-radar scenario and 4.6 s and 12 s in two-radar scenarios. The radars with antenna rotation periods of 4.6 s and 12 s are known as short-range (SR) and long-range (LR) radars, respectively [3, 4]. The measurements are interleaved in two-radar scenarios. Thus, the measurement time interval is half the antenna rotation period due to interleaved measurements. We use two radars to study the climbing motion of the aircraft with an antenna rotation period of 4.6 s and the NCV motion with NCAlt using the antenna rotation periods of 4.6 s and 12 s. An SR radar with an antenna rotation period of 4.6 s is used for the NCT motion at NCAlt.

The NCV model in 3D can handle both the NCV motion at NCAlt and climbing motion. The six-dimensional Cartesian state vector consisting of the 3D Cartesian position and velocity defines the state vector for the NCV model in 3D. For the NCT motion at NCAlt, the state vector comprises the 3D Cartesian position, 2D Cartesian velocity parallel to the XY plane, and nearly constant angular rate. The slant range and azimuth angle are nonlinear functions of the aircraft state for these three types of motions. The dynamic model for the NCV model in 3D is linear. The Itô stochastic differential equation (SDE) can accurately describe the continuous-time dynamic model for the NCT motion, which is nonlinear [3, 4, 5, 6]. Two methods can be used to solve the NCT motion of an aircraft. In the first method, the Itô SDE is used with the discrete-time radar measurement model. This type of filtering problem is called the continuous-discrete filtering (CDF), where the dynamic and measurement models are continuous and discrete in time, respectively [7, 8]. Since the continuous-time dynamic model is nonlinear in aircraft state, it can be discretized only approximately, e.g., by the Itô-Taylor expansion of order 1.5 (IT-1.5) [3]. The second method solves the nonlinear filtering problem by using the approximate discretized dynamic model and discrete-time measurement model.

1.1.1 Introductions to NCV with NCAlt

Gai et al. [9] first proposed the height-parametrized (HP) Cartesian extended Kalman filter (HP-CEKF) for the NCV motion with NCAlt of an aircraft using one 2D radar. Their filter initializations had an error, and the measurement time interval and measurement standard deviations (SDs) did not represent typical values for a 2D radar. In [10], Li et al. use the HP-CEKF with probabilistic data association for aircraft tracking with false alarms and missed detections.

Rakvongthai et al. [2] used two ATC 2D radars assuming constant velocity and constant altitude of the aircraft. They initialized the filter using the first N measurements (e.g., $N = 20$) and computed the maximum likelihood (ML) estimate of the initial 3D position and XY component of the velocity. They did not calculate the associated covariance of the ML estimate and used a heuristic covariance. Then they processed the remaining measurements using a HP-CEKF. Since the ML estimate of the constant altitude is known, there was no need to use the HP-CEKF.

This part of the chapter is based on our previous work in [11], where we developed the HP-CEKF, HP Cartesian UKF (HP-CUKF), HP Cartesian CKF (HP-CCKF), HP modified spherical coordinate-based UKF (HP-MSCUKF), and bias-compensated pseudo-linear estimator (BCPLE) [12]. Due to page limitations, we present the HP-CEKF, HP-CUKF, and HP-CCKF and compared their performance evaluation with the posterior Cramér–Rao lower bound (PCRLB) [13, 14].

1.1.2 Introductions to NCT with NCAlt

Previous researchers have considered the aircraft 3D state estimation for the NCT motion in ATC problem using the slant range, azimuth, and elevation measurements [3, 4, 6, 15]. However, an ATC 2D radar measures the slant range and azimuth only. In the NCT motion considered by the authors, the aircraft moves with the NCT motion in a plane parallel to $XY-$ plane (ground plane) and the motion along the vertical direction ($Z-$axis) is a nearly constant velocity (NCV) motion [16, 17]. The state of the aircraft includes 3D Cartesian position, 3D Cartesian velocity, and the turn rate. An Itô stochastic differential equation (SDE) is used to describe the motion of the aircraft and Brownian motions representing perturbations along the $X-$axis, $Y-$axis, $Z-$axis, and for turn rate are used in the diffusion term.

When an aircraft is in a holding pattern near an airport before landing, it performs an NCT (usually called a coordinated turn in civil aviation) at NCAlt [1]. The International Civil Aviation Organization (ICAO) mandates that all turns should be made, at a bank angle of 25° or at a rate of 3°/s, whichever requires the lesser bank angle [18]. Higher bank angles would create discomfort and injury to passengers. In [3, 4, 6, 15], the horizontal speed of the aircraft is chosen as 150 m/s and turn rates of $(3.0, 4.5, 6.0)^\circ$/s are considered. The bank angles for these parameters are about $(39, 50, 58)^\circ$, respectively. These bank angles are significantly higher than the recommended value of 25°. These high bank angles are acceptable for military aircraft, but not for civilian aircraft moving with the NCT motion near an airport.

When an aircraft moves with an NCT, the height of the aircraft and radius of turn should be such that it avoids mountains and tall buildings. The radius of turn is proportional to the square of horizontal speed and inversely proportional to the tangent of the bank angle [1]. These factors should be taken into account while analyzing the NCT of a civilian aircraft in the ATC scenarios.

We consider the aircraft 3D trajectory estimation during the NCT motion at NCAlt using the ATC 2D radar measurements. Our work is the first work for this challenging ATC filtering problem. We take into account the physics of the NCT motion and civil aviation regulations recommended by the ICAO [18] using the bank angle and radius of turn. These considerations were absent in previous studies. We use the Itô SDE to describe the continuous-time dynamic model during the NCT [3–6, 19]. We present two filtering algorithms. In the first filtering algorithm, the predicted state estimate and predicted covariance are calculated using the continuous-time extended Kalman filter (CEKF), where the differential equations for the conditional mean and conditional covariance are numerically integrated during measurement time interval [4, 20]. In the second filtering algorithm, the continuous-time SDE is discretized using IT-1.5 [21] and the approximate discrete dynamic model is used to calculate the predicted state estimate and predicted covariance using the cubature Kalman filter (CKF) prediction step. For both algorithms, we use the CKF measurement update step.

1.1.3 Notation Convention

For clarity, we use italics to denote scalar quantities and boldface for vectors and matrices. A lower or uppercase Roman letter represents a name (e.g., "s" for "sensor," "RMS" for "root mean square," etc.). We use ":=" to define a quantity and $\mathbf{A}'$ denotes the transpose of the vector or matrix $\mathbf{A}$. The $n-$dimensional identity matrix, $m-$dimensional null matrix, and $m \times n$ null matrix are denoted by $\mathbf{I}_n$, $\mathbf{0}_m$, and $\mathbf{0}_{m\times n}$, respectively.

This chapter is based on our previous work on 3D trajectory estimation of a non-maneuvering aircraft [11]. The contribution for the maneuvering aircraft is new. Motions of the non-maneuvering aircraft include the NCV at NCAlt and climbing motion with nearly constant climbing rate [11]. The second part of the chapter addresses the 3D trajectory estimation of an aircraft executing an NCT at NCAlt.

This work makes the following key contributions:

1. We develop the HP-CUKF and HP-CCKF based on the multiple model-based method for the NCV at NCAlt and climbing motion.

2. We calculate the PCRLB for the NCV at NCAlt and climbing motion.
3. For the first time, we solve the realistic NCT trajectory estimation problem for a civilian aircraft using an ATC 2D radar, where ICAO recommendations on bank angle and turn rate are taken into account. This problem is relatively harder than that considered in previous studies where the range, azimuth, and elevation measurements for a radar were considered [3, 4, 6].
4. We develop an effective filter initialization algorithm for the aircraft NCT motion using the HP-based multiple-model method. Previous work used the two-point difference filter initialization, which is not valid for the CT motion.

The chapter is organized as follows. Section 1.2 presents the measurement model for the ATC 2D radar. Section 1.3 considers the trajectory estimation of a non-maneuvering aircraft. Subsections 1.3.1 and 1.3.3 describe the dynamic model and filter initialization algorithm, respectively. The height parametrization technique and HP-based multiple model-based filters are explained in Subsections 1.3.2 and 1.3.4, respectively. Calculation of the PCRLB for the NCV motion using the 2D radar measurements is presented in Subsection 1.3.5. Section 1.4 presents algorithms for estimating the trajectory of an aircraft moving with NCT at NCAlt. The constant turn (CT) model with reference to the bank angle and NCT at NCAlt are explained in Subsections 1.4.1 and 1.4.2, respectively. Continuous-time and discrete-time NCT models are presented in Subsections 1.4.2.1 and 1.4.2.2, respectively. Subsection 1.4.3 presents the algorithm for determining the 3D position and covariance. The filtering algorithms for the continuous-time and discrete-time NCT models are presented in Subsection 1.4.4. Section 1.5 discusses filter evaluation metrics. Sections 1.6 and 1.7 present the numerical simulations and results for the non-maneuvering and maneuvering aircraft, respectively. Section 1.8 summarizes conclusions and discusses our future work.

1.2 Measurement Model for ATC 2D Radars

The origin of the tracker coordinate frame (T frame) is chosen at a given longitude and latitude on Earth's surface. The X, Y, and Z axes of the T frame are along the local east, north, and upward directions, respectively. Let $\mathbf{p} := [x\, y\, z]'$ denote the 3D position of the aircraft at time t. An ATC 2D radar has a fixed position $\mathbf{s} := [s_x \quad s_y \quad s_z]'$. The slant range r and azimuth α of the aircraft are shown in Figure 1.1.

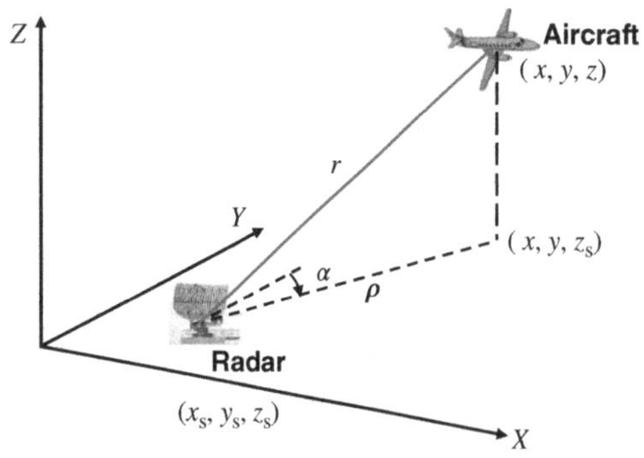

Figure 1.1 An ATC 2D radar and an aircraft in 3D. *Source*: [11] / with permission of IEEE.

Let $\mathbf{z}$ denote the slant range and azimuth measurements of a 2D radar. The nonlinear measurement model for the ATC 2D radar is described by [2, 11, 22]

$$\mathbf{z} = \mathbf{h}(\mathbf{p}, \mathbf{s}) + \mathbf{n}, \tag{1.1}$$

where $\mathbf{h}$ and $\mathbf{n}$ are the nonlinear measurement function and the measurement noise, respectively. The measurement function is defined by [2, 11, 22]

$$\mathbf{h}(\mathbf{p}, \mathbf{s}) := \begin{bmatrix} r \\ \alpha \end{bmatrix} = \begin{bmatrix} [(\mathbf{p} - \mathbf{s})'(\mathbf{p} - \mathbf{s})]^{1/2} \\ \tan^{-1}(x - s_x, y - s_y) \end{bmatrix}, \tag{1.2}$$

where α lies in $[0, 2\pi)$. We assume that $\mathbf{n}$ is zero-mean Gaussian with covariance $\mathbf{R}$

$$\mathbf{n} \sim \mathcal{N}(\mathbf{n}; \mathbf{0}_{2\times1}, \mathbf{R}), \quad \mathbf{R} = \mathrm{diag}(\sigma_r^2, \sigma_\alpha^2), \tag{1.3}$$

where σ_r and σ_α are the measurement error standard deviations for slant range and azimuth, respectively. Let $\mathbf{r}$ denote the range vector of the aircraft relative to the radar

$$\mathbf{r} := \mathbf{p} - \mathbf{s}. \tag{1.4}$$

The ground range of the aircraft is defined by

$$\rho := \sqrt{(x - s_x)^2 + (y - s_y)^2}. \tag{1.5}$$

1.3 Non-maneuvering Aircraft

1.3.1 Dynamic Models

We consider the nearly constant velocity (NCV) motion with nearly constant altitude (NCAlt) and climbing motion with nearly constant climbing rate. These two types of motions are described by the NCV motion in 3D. The Cartesian state vector $\mathbf{x}_k$ for the aircraft at time t_k, is defined by

$$\mathbf{x}_k := [x_k \quad y_k \quad z_k \quad \dot{x}_k \quad \dot{y}_k \quad \dot{z}_k]', \tag{1.6}$$

where x_k, y_k, and z_k are the Cartesian X, Y, and Z position coordinates and $\dot{x}_k$, $\dot{y}_k$, and $\dot{z}_k$ are the corresponding velocity components, respectively, at time t_k. We assume that the measurement time interval T is constant. The 3D NCV motion of the aircraft is described by [16, 11]

$$\mathbf{x}_k = \mathbf{F}_{k-1}\mathbf{x}_{k-1} + \mathbf{w}_{k-1}, \tag{1.7}$$

where $\mathbf{F}_{k-1}$ is the state transition matrix, and $\mathbf{w}_{k-1}$ is zero-mean white Gaussian process noise with covariance matrix $\mathbf{Q}_{k-1}$, during the time interval $[t_{k-1}, t_k)$ [16]. Using the discretized continuous-time NCV model, the state transition matrix and process noise covariance matrix are given by [16]

$$\mathbf{F}_{k-1} = \begin{bmatrix} \mathbf{I}_3 & \mathbf{I}_3 T \\ \mathbf{0}_{3\times3} & \mathbf{I}_3 \end{bmatrix}, \tag{1.8}$$

$$\mathbf{Q}_{k-1} = \begin{bmatrix} \mathrm{diag}(\mathbf{q})T^3/3 & \mathrm{diag}(\mathbf{q})T^2/2 \\ \mathrm{diag}(\mathbf{q})T^2/2 & \mathrm{diag}(\mathbf{q})T \end{bmatrix}, \tag{1.9}$$

where $\mathbf{q} := [q_x\ q_y\ q_z]'$ represents the power spectral densities (PSDs) of the continuous-time acceleration process noise along the X, Y and Z axes, respectively [16].

1.3.2 Height Parametrization

In the HP method, a minimum altitude $a_{\min}$ and a maximum altitude $a_{\max}$ for the aircraft are chosen. Then the interval $[a_{\min}, a_{\max}]$ is divided into a number of sub-intervals, and the mean of a sub-interval is used to denote the initial height in that interval. In our previous work [11], we used the geometric series and uniform sub-intervals. We found that these two methods yield nearly the same results. Therefore, we use uniform sub-intervals in this paper for its simplicity.

Suppose there are N_a sub-intervals each with width Δa. Let $\{a_i\}_{i=0}^{N_a}$ denote the edges of the sub-intervals with $a_0 = a_{\min}$ and $a_{N_a} = a_{\max}$. Then

$$\Delta a = (a_{\max} - a_{\min})/N_a, \tag{1.10}$$

$$a_i = a_0 + i\Delta a, \, i = 0, 1, 2, \ldots, N_a. \tag{1.11}$$

The mean height for the ith interval is given by

$$\bar{a}_i = (a_{i-1} + a_i)/2. \tag{1.12}$$

Assuming the heights to be distributed uniformly, the standard deviation of height in each sub-interval is given by

$$\sigma_h = \frac{\Delta a}{\sqrt{12}}. \tag{1.13}$$

Since each sub-interval has the same width, the initial probabilities or weights $\{w_i\}$ are equal to $1/N_a$.

1.3.3 Filter Initialization

We follow the filter initialization described in [11]. Let $z_{1,r}, z_{1,\alpha}$ be the first range and azimuth measurements from a 2D radar at time t_1. In HP approach, a number of mean heights $\{\bar{a}_i\}_{i=1}^{N_a}$ are chosen for filter initialization, and a bank of N_a filters are used, each filter being associated with a given mean height. Using $z_{1,r}$ and the sensor height $s_{1,z}$ at time t_1, an estimate of the ground range and associated standard deviation using the height $\bar{a}_i$ for the ith filter are given, respectively, by

$$\hat{\rho}_{1,i} = \sqrt{z_{1,r}^2 - (\bar{a}_i - s_{1,z})^2}, \tag{1.14}$$

$$\sigma_{\rho(1,i)} = (\hat{\rho}_{1,i}/z_{1,r})\sigma_r. \tag{1.15}$$

Using the algorithm described in Section 3.3 of [23] as explained below, we calculate the converted unbiased Cartesian XY measurements $(z_{1,x}^i, z_{1,y}^i)$ and associated 2×2 covariance $\mathbf{R}_c^i(1)$ using the first measurement.

1.3.3.1 Filter Initialization for NCV with NCAlt

We set the initial velocity estimate to zero. Then the initial state estimate and position covariance for the ith filter are given, respectively, by [23]

$$\hat{\mathbf{x}}_{1|1}^i = [z_{1,x}^i \quad z_{1,y}^i \quad \bar{a}_i \quad 0 \quad 0 \quad 0]', \tag{1.16}$$

$$\mathbf{P}_{1|1,xy}^i = \mathbf{R}_c^i(1), P_{1|1,z}^i = \sigma_h^2. \tag{1.17}$$

Let $\upsilon_{x,\max}$, $\upsilon_{y,\max}$ and $\upsilon_{z,\max}$ be the maximum possible velocity along the X-axis, Y-axis, and Z-axis, respectively. We assume $\upsilon_{x,\max} = \upsilon_{y,\max}$. Assuming that the velocity along each axis is distributed uniformly [24],

$$P^i_{1|1,\dot{x}} = P^i_{1|1,\dot{y}} = \upsilon^2_{x,\max}/3, P^i_{1|1,\dot{z}} = \upsilon^2_{z,\max}/3. \tag{1.18}$$

1.3.3.2 Filter Initialization for Climbing Motion

The climb rate of an aircraft depends on a number of factors such as the type of aircraft, weight of the aircraft (i.e., same type but different weights), weather, etc. A nearly constant climb rate is usually used for climbing motion. Let $\upsilon_{cz,\min}$ and $\upsilon_{cz,\max}$ be the minimum and maximum climbing rates of an aircraft. Then the initial estimate of the climb rate and associated variance for the ith filter are given, respectively, by

$$\bar{\upsilon}^i_{1,z} = (\upsilon_{cz,\min} + \upsilon_{cz,\max})/2, \tag{1.19}$$

$$P^i_{1|1,\dot{z}} = (\upsilon_{cz,\max} - \upsilon_{cz,\min})^2/12. \tag{1.20}$$

Then the initial state estimate for the ith filter is given by [23]

$$\hat{\mathbf{x}}^i_{1|1} = [z^i_{1,x} \quad z^i_{1,y} \quad \bar{a}_i \quad 0 \quad 0 \quad \bar{\upsilon}^i_{1,z}]'. \tag{1.21}$$

The initial covariance is given by (1.17) and (1.18) where $P^i_{1|1,\dot{z}}$ from (1.20) is used in (1.18).

1.3.4 Multiple Model-based Nonlinear Filters

In the multiple model-based method, a bank of N_a filters are used, each filter being associated with a mean height. The HP-CEKF was used in [2, 9]. For improved trajectory estimation accuracy, we also employ the HP-CUKF and HP-CCKF in addition to the HP-CEKF in this study. Let $\mathbf{Z}_k$ denote the collection of measurements up to time t_k,

$$\mathbf{Z}_k := \{\mathbf{z}_1, \mathbf{z}_2, \ldots, \mathbf{z}_k\}. \tag{1.22}$$

Let $P(j|\mathbf{Z}_k) = w_{k,j}$ denote the probability for the jth model or filter after processing $\mathbf{Z}_k$. Using the Bayes' rule [16], we get

$$P(j|\mathbf{Z}_k) \propto p(\mathbf{z}_k|j, \mathbf{Z}_{k-1})P(j|\mathbf{Z}_{k-1}), \quad j = 1, 2, \ldots, N_a. \tag{1.23}$$

In (1.23), $p(\mathbf{z}_k|j, \mathbf{Z}_{k-1})$ represents the likelihood of the jth model and is given by

$$p(\mathbf{z}_k|j, \mathbf{Z}_{k-1}) = \mathcal{N}(\boldsymbol{\nu}_{k,j}; \mathbf{0}, \mathbf{S}_{k,j}), \quad j = 1, 2, \ldots, N_a, \tag{1.24}$$

where $\boldsymbol{\nu}_{k,j}$ and $\mathbf{S}_{k,j}$ are the innovation and innovation covariance, respectively, for the jth model [16]. Using uniform height intervals,

$$P(j|\mathbf{z}_1) = 1/N_a, \quad j = 1, 2, \ldots, N_a. \tag{1.25}$$

Let $\hat{\mathbf{x}}_{k|k,j}$ and $\mathbf{P}_{k|k,j}$ denote the state estimate and associated covariance in Cartesian coordinates for the jth filter, respectively, at time t_k after sequentially processing $\mathbf{Z}_k$. The state estimate and covariance of the aircraft are calculated, respectively, by

$$\hat{\mathbf{x}}_{k|k} = \sum_{j=1}^{N_a} w_{k,j}\hat{\mathbf{x}}_{k,j}, \tag{1.26}$$

$$\mathbf{P}_{k|k} = \sum_{j=1}^{N_a} w_{k,j}[\mathbf{P}_{k,j} + (\hat{\mathbf{x}}_{k|k,j} - \hat{\mathbf{x}}_{k|k})(\hat{\mathbf{x}}_{k|k,j} - \hat{\mathbf{x}}_{k|k})']. \tag{1.27}$$

1.3.5 Posterior Cramér–Rao Lower Bound

The posterior Cramér–Rao lower bound is a bound on the best achievable accuracy of an unbiased estimator [13] of the state vector. It is therefore a useful metric for evaluating the performance of an estimator. Suppose $\hat{\mathbf{x}}_{k|k}$ is an unbiased estimator of the state vector $\mathbf{x}_k$ based on the measurement sequence $\mathbf{Z}_k$. Then the lower bound for the mean square error matrix $\mathbf{P}^C_{k|k}$ is given by [13, 25]

$$\mathbf{P}^C_{k|k} := E\{(\mathbf{x}_k - \hat{\mathbf{x}}_{k|k})(\mathbf{x}_k - \hat{\mathbf{x}}_{k|k})'\} \geq \mathbf{J}_k^{-1}, \tag{1.28}$$

where $\mathbf{J}_k$ is the Bayesian information matrix [25] and $\mathbf{J}_k^{-1}$ is the PCRLB. Tichavsky et al. [13] developed an elegant recursive algorithm for computing the PCRLB for filtering problems in the state space framework. In [13], it was shown that the Bayesian information matrix can be computed recursively using

$$\mathbf{J}_{k+1} = \mathbf{D}_k^{22} - (\mathbf{D}_k^{12})'\left(\mathbf{J}_k + \mathbf{D}_k^{11}\right)^{-1}\mathbf{D}_k^{12} + \mathbf{J}^z_{k+1}, \tag{1.29}$$

where

$$\mathbf{D}_k^{11} = E\{\left(\nabla_{\mathbf{x}_k} \log p(\mathbf{x}_{k+1}|\mathbf{x}_k)\right)\left(\nabla_{\mathbf{x}_k} \log p(\mathbf{x}_{k+1}|\mathbf{x}_k)\right)'\},$$

$$\mathbf{D}_k^{12} = E\{\left(\nabla_{\mathbf{x}_k} \log p(\mathbf{x}_{k+1}|\mathbf{x}_k)\right)\left(\nabla_{\mathbf{x}_{k+1}} \log p(\mathbf{x}_{k+1}|\mathbf{x}_k)\right)'\}, \tag{1.30}$$

$$\mathbf{D}_k^{22} = E\{\left(\nabla_{\mathbf{x}_{k+1}} \log p(\mathbf{x}_{k+1}|\mathbf{x}_k)\right)\left(\nabla_{\mathbf{x}_{k+1}} \log p(\mathbf{x}_{k+1}|\mathbf{x}_k)\right)'\}, \tag{1.31}$$

$$\mathbf{J}^z_{k+1} = E\{\left(\nabla_{\mathbf{x}_{k+1}} \log p(\mathbf{z}_{k+1}|\mathbf{x}_{k+1})\right)\left(\nabla_{\mathbf{x}_{k+1}} \log p(\mathbf{z}_{k+1}|\mathbf{x}_{k+1})\right)'\}.$$

For the current problem using Cartesian coordinates, the dynamic model is linear with additive Gaussian process noise and the measurement model is nonlinear with additive Gaussian measurement noise. For this type of nonlinear filtering problem the recursion simplifies to [25]

$$\mathbf{J}_{k+1} = \mathbf{Q}^{-1}_{k+1,k} + E\{\mathbf{H}'_{k+1}\mathbf{R}^{-1}_{k+1}\mathbf{H}_{k+1}\} - \mathbf{Q}^{-1}_{k+1,k}\mathbf{F}_{k+1,k}\left(\mathbf{J}_k + \mathbf{F}'_{k+1,k}\mathbf{Q}^{-1}_{k+1,k}\mathbf{F}_{k+1,k}\right)^{-1} \times \mathbf{F}'_{k+1,k}\mathbf{Q}^{-1}_{k+1,k},$$

where $\mathbf{H}_{k+1}$ is the Jacobian of the nonlinear measurement function $\mathbf{h}$ evaluated at the true state $\mathbf{x}_{k+1}$. Using the matrix inversion lemma [16], the above equation reduces to [26]

$$\mathbf{J}_{k+1} = \left(\mathbf{Q}_{k+1,k} + \mathbf{F}_{k+1,k}\mathbf{J}_k^{-1}\mathbf{F}'_{k+1,k}\right)^{-1} + E\{\mathbf{H}'_{k+1}\mathbf{R}^{-1}_{k+1}\mathbf{H}_{k+1}\}. \tag{1.32}$$

The second term on the right-hand side of (1.32) cannot be evaluated analytically due to its nonlinear dependence on the state. We compute it numerically using Monte Carlo simulation.

1.4 Maneuvering Aircraft

1.4.1 Constant Turn Model for an Aircraft

Consider an aircraft of mass m executing a CT at a constant height with speed v. The forces acting on the aircraft are the weight mg acting downward and the lift force **L** [1] as shown in Figure 1.2, where g is the acceleration due to gravity.

The bank angle ϕ is defined as the angle between the airplane wings and the horizontal plane as shown in Figure 1.2. The component of the lift force along the vertical direction is $L\cos\phi$. The total force along the vertical direction is zero. Hence,

$$L\cos\phi = mg. \tag{1.33}$$

The horizontal component of the lift force is $L\sin\phi$. This component provides the centripetal force for the CT motion [27] in a plane parallel to the horizontal plane. The centripetal acceleration is v^2/R_c, where R_c is the radius of the circular turn [27]. Therefore,

$$L\sin\phi = mv^2/R_c. \tag{1.34}$$

Eliminating L from (1.33) and (1.34), we get

$$R_c = v^2/(g\tan\phi). \tag{1.35}$$

Equation (1.35) states that at a given bank angle, the radius of turn is proportional to the square of horizontal speed. If the horizontal speed is doubled, then the radius of the turn increases by four times. The period of CT is given by

$$T_c = 2\pi R_c/v = 2\pi v/(g\tan\phi). \tag{1.36}$$

Hence, the turn rate or angular velocity is

$$\omega = 2\pi/T_c = g\tan\phi/v. \tag{1.37}$$

We observe from (1.37) that for a given bank angle, the turn rate is inversely proportional to the horizontal speed. The bank angle can be expressed by

$$\phi = \arctan(v\omega/g). \tag{1.38}$$

From (1.35), the diameter of the circular turn is $D_c = 2v^2/(g\tan\phi)$. Given the speed and the bank angle, this equation and altitude of the aircraft can be used to choose the area that avoids mountains and buildings during the CT motion.

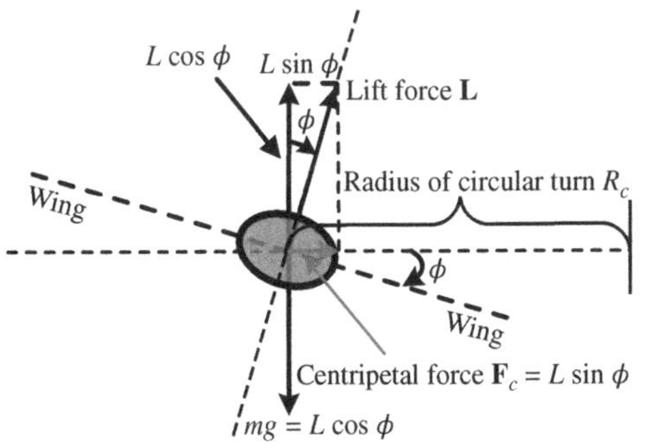

Figure 1.2 Forces acting on an aircraft during the constant turn motion.

In civil aviation, a *standard rate turn* or *rate one turn* is defined as a 3° per second turn [28]. This implies that a complete turn is made in 120 seconds or in two minutes.

1.4.2 NCT Dynamic Model

In the ATC scenario, the aircraft moves with the NCT in a plane parallel to the $XY-$plane [3, 5, 17] at a nearly constant altitude. A nearly constant velocity (NCV) or Wiener velocity motion [16] along the $Z-$axis was first used in [3]. Subsequently, following this scenario in [3], authors in [4, 6, 15] have also used this scenario. In this paper, we use the nearly constant position model along the $Z-$axis to be consistent with the realistic scenario.

Let (x, y, z), $(\dot{x}, \dot{y}, \dot{z})$, and ω denote the 3D Cartesian position, 3D Cartesian velocity, and angular rate of the aircraft, respectively. The state of the aircraft at time t is defined by

$$\mathbf{x}(t) := \begin{bmatrix} x(t) & y(t) & \dot{x}(t) & \dot{y}(t) & \omega(t) & z(t) \end{bmatrix}'. \tag{1.39}$$

1.4.2.1 Continuous-time NCT Model

The Itô stochastic differential equation (SDE) [29, 30] for the aircraft state is described by

$$d\mathbf{x}(t) = \mathbf{f}(\mathbf{x}(t))dt + \mathbf{G}d\boldsymbol{\beta}(t), \tag{1.40}$$

where

$$\mathbf{f}(\mathbf{x}) = \begin{bmatrix} \dot{x} & \dot{y} & -\omega\dot{y} & \omega\dot{x} & 0 & 0 \end{bmatrix}', \tag{1.41}$$

$$d\boldsymbol{\beta}(t) = \begin{bmatrix} d\beta_x(t) & d\beta_y(t) & d\beta_\omega(t) & d\beta_z(t) \end{bmatrix}', \tag{1.42}$$

$$\mathbf{G} = \begin{bmatrix} 0 & 0 & 0 & 0 \\ 0 & 0 & 0 & 0 \\ \sqrt{q_x} & 0 & 0 & 0 \\ 0 & \sqrt{q_y} & 0 & 0 \\ 0 & 0 & \sqrt{q_\omega} & 0 \\ 0 & 0 & 0 & \sqrt{q_z} \end{bmatrix}, \tag{1.43}$$

In (1.40), $\boldsymbol{\beta}(t)$ represents the four-dimensional standard Brownian motion with increments $d\boldsymbol{\beta}(t)$ defined in (1.42). The components of the Brownian motion represent independent perturbations in Cartesian accelerations and angular acceleration with the property

$$E\{d\boldsymbol{\beta}(t)\} = \mathbf{0}_{4\times1}, \quad E\{d\boldsymbol{\beta}(t)d\boldsymbol{\beta}(t)'\} = \mathbf{I}_4 dt. \tag{1.44}$$

The diffusion matrix $\mathbf{Q}$ of the Brownian motion is given by [30]

$$\mathbf{Q}dt = \mathbf{G}E\{d\boldsymbol{\beta}(t)d\boldsymbol{\beta}'(t)\}\mathbf{G}' = \text{diag}(0, 0, q_x, q_y, q_\omega, q_z)dt. \tag{1.45}$$

The continuous extended Kalman filter (CEKF) [4, 20, 29, 31] can be used to obtain the predicted state estimate and covariance.

1.4.2.2 Discrete-time NCT Model

Using IT-1.5 [3], the discretized dynamic model is given by

$$\mathbf{x}_{k+1} = \mathbf{f}_k(\mathbf{x}_k) + \mathbf{w}_k, \tag{1.46}$$

where the noise-free term $\mathbf{f}_k(\mathbf{x}_k)$ is given by

$$\mathbf{f}_k(\mathbf{x}_k) = \mathbf{x}_k + T\mathbf{f}(\mathbf{x}_k) + \frac{T^2}{2}L_0\mathbf{f}(\mathbf{x}_k), \tag{1.47}$$

and the differential operator L_0 is defined by [3, 32]

$$L_0 \; := \frac{\partial}{\partial t} + \sum_{i=1}^{n} f_i \frac{\partial}{\partial x_i} + 1/2 \sum_{l=1}^{p} \sum_{i,j=1}^{n} G_{il}G_{jl} \frac{\partial}{\partial x_i \partial x_j}, \tag{1.48}$$

where n and p are the dimensions of the state and process noise, respectively. For the current problem, $n = 6$ and $p = 4$. It can be shown that [3]

$$\mathbf{f}_k(\mathbf{x}_k) = \begin{bmatrix} x_k + T\dot{x}_k - (T^2/2)\omega_k\dot{y}_k \\ y_k + T\dot{y}_k + (T^2/2)\omega_k\dot{x}_k \\ \dot{x}_k - T\omega_k\dot{y}_k - (T^2/2)\omega_k^2\dot{x}_k \\ \dot{y}_k + T\omega_k\dot{x}_k - (T^2/2)\omega_k^2\dot{y}_k \\ \omega_k \\ z_k \end{bmatrix}. \tag{1.49}$$

The discrete-time process noise for the time interval $[t_k, t_{k+1}]$ is given by [3, 31]

$$\mathbf{w}_k = \sqrt{T}\mathbf{G}\mathbf{u}_1 + (T^{3/2}/2)\mathbf{F}_{\mathbf{x},k}\mathbf{G}(\mathbf{u}_1 + \mathbf{u}_2/\sqrt{3}), \tag{1.50}$$

where $\mathbf{u}_1$ and $\mathbf{u}_2$ are two four-dimensional independent zero-mean standard Gaussian random variables with unit variance,

$$\mathbf{u}_i \sim \mathcal{N}(\mathbf{0}_4, \mathbf{I}_4), \; i = 1, 2. \tag{1.51}$$

Detailed derivation of the discrete-time process noise covariance $\mathbf{Q}_k$ corresponding to $\mathbf{w}_k$ is presented in [31] and has the form

$$\mathbf{Q}_k \; = \mathbf{Q}T + (\mathbf{F}_{\mathbf{x},k}\mathbf{Q} + \mathbf{Q}\mathbf{F}'_{\mathbf{x},k})\frac{T^2}{2} + \mathbf{F}_{\mathbf{x},k}\mathbf{Q}\mathbf{F}'_{\mathbf{x},k}\frac{T^3}{3}, \tag{1.52}$$

$$\mathbf{F}_{\mathbf{x},k} = \frac{\partial \mathbf{f}(\mathbf{x})}{\partial \mathbf{x}}|_{\mathbf{x}=\hat{\mathbf{x}}_{k+1|k}}. \tag{1.53}$$

Remark 1: Multiple sub-intervals can be used in T to improve the prediction accuracy.

The Jacobian of $\mathbf{f}(\mathbf{x})$ with respect to $\mathbf{x}$ is given by

$$\mathbf{F}_{\mathbf{x}} = \frac{\partial \mathbf{f}(\mathbf{x})}{\partial \mathbf{x}} = \begin{bmatrix} 0 & 0 & 1 & 0 & 0 & 0 \\ 0 & 0 & 0 & 1 & 0 & 0 \\ 0 & 0 & 0 & -\omega & -\dot{y} & 0 \\ 0 & 0 & \omega & 0 & \dot{x} & 0 \\ 0 & 0 & 0 & 0 & 0 & 0 \\ 0 & 0 & 0 & 0 & 0 & 0 \end{bmatrix}. \tag{1.54}$$

We can use the discrete cubature Kalman filter (CKF) [21] or unscented Kalman filter (UKF) [33] to calculate the predicted state estimate and covariance using the discretized dynamic model.

1.4.3 3D Position Estimate and Covariance

Given the slant range r, altitude z, and sensor altitude s_z, the ground range ρ (see Figure 1.1) is given by

$$\rho = \sqrt{r^2 - (z - s_z)^2}. \tag{1.55}$$

Using the mean altitude $\bar{a}_i$ in the ith sub-interval, the corresponding ground range is given by

$$\rho_i = \sqrt{r^2 - (\bar{a}_i - s_z)^2}. \tag{1.56}$$

Hence, the 3D Cartesian position using ρ_i, azimuth α, and 3D sensor position $\mathbf{s}$ are given by

$$x_i = s_x + \sqrt{r^2 - (\bar{a}_i - s_z)^2} \sin \alpha, \tag{1.57}$$

$$y_i = s_y + \sqrt{r^2 - (\bar{a}_i - s_z)^2} \cos \alpha. \tag{1.58}$$

Define

$$\mathbf{b}_i := [r \quad \alpha \quad \bar{a}_i]'. \tag{1.59}$$

Then the mean of $\mathbf{b}_i$ is

$$\bar{\mathbf{b}}_i = [\bar{r} \quad \bar{\alpha} \quad \bar{a}_i]' = [z_r \quad z_\alpha \quad \bar{a}_i]', \tag{1.60}$$

where we have set the mean of slant range and azimuth to the corresponding measurements. The covariance of $\mathbf{b}_i$ is given by

$$\mathbf{P}_{\mathbf{b}_i} = \text{diag}(\sigma_r^2, \sigma_\alpha^2, \sigma_h^2). \tag{1.61}$$

Let $\mathbf{p}_i$ denote the 3D Cartesian position corresponding to $\mathbf{b}_i$. Thus, we can represent the nonlinear transformation $\mathbf{g}$ from $\mathbf{b}_i$ to $\mathbf{p}_i$ by

$$\mathbf{p}_i = \mathbf{g}(\mathbf{b}_i), \tag{1.62}$$

where the functional form of $\mathbf{g}$ is known from (1.57) and (1.58). We know the mean and covariance of $\mathbf{b}_i$. We can determine the mean $\bar{\mathbf{p}}_i$ and associated covariance $\mathbf{P}_{\mathbf{p}_i}$ using the cubature transformation [21] or unscented transformation [33].

We estimate the parameters of the constant turn (CT) model at a constant altitude z_0 with constant angular rate ω using a deterministic model. The altitude is decoupled from the parameters of the CT model. Define the state

$$\eta_k := \begin{bmatrix} x_k & y_k & \dot{x}_k & \dot{y}_k & \omega \end{bmatrix}'. \tag{1.63}$$

$$\eta_k = \mathbf{F}^{\text{CT}} \eta_{k-1}, \tag{1.64}$$

$$\mathbf{F}^{\text{CT}} = \begin{bmatrix} 1 & 0 & \dfrac{\sin(\omega T)}{\omega} & -\dfrac{[1 - \cos(\omega T)]}{\omega} & 0 \\ 0 & 1 & \dfrac{[1 - \cos(\omega T)]}{\omega} & \dfrac{\sin(\omega T)}{\omega} & 0 \\ 0 & 0 & \cos(\omega T) & -\sin(\omega T) & 0 \\ 0 & 0 & \sin(\omega T) & \cos(\omega) & 0 \\ 0 & 0 & 0 & 0 & 1 \end{bmatrix}. \tag{1.65}$$

We follow the algorithm in [34] for estimating the parameters of the CT motion in 2D with enhancements for estimating the unknown constant altitude z_0. The position and velocity in the 2D CT motion change with time. It is shown in [34] that the two-point difference (TPD) algorithm [16] commonly used for the constant velocity model (CV) is not appropriate to estimate the velocity in the CT model. Given the slant range and azimuth measurements at times t_1, t_2, and t_3, we used the HP-method, to estimate the 3D position vectors for each of the N_a altitudes. Then we choose the altitude, which has the maximum likelihood. We describe the details of the algorithm next.

$$\mathbf{x}_k^i := \begin{bmatrix} x_k & y_k & \dot{x}_k & \dot{y}_k & \omega & \bar{z}^i \end{bmatrix}'. \tag{1.66}$$

$$\mathbf{X}_3^i := \{\mathbf{x}_1^i, \quad \mathbf{x}_2^i, \quad \mathbf{x}_3^i\}. \tag{1.67}$$

$$p(\mathbf{Z}_k|\mathbf{X}_3^i) = p(\mathbf{z}_1|\mathbf{x}_1^i)p(\mathbf{z}_2|\mathbf{x}_2^i)p(\mathbf{z}_3|\mathbf{x}_3^i). \tag{1.68}$$

Since the measurement errors are Gaussian,

$$p(\mathbf{Z}_k|\mathbf{X}_3^i) \propto \prod_{k=1}^{3} \exp -\left[(\mathbf{z}_k - \mathbf{h}(\mathbf{x}_k^i))' \mathbf{R}^{-1}(\mathbf{z}_k - \mathbf{h}(\mathbf{x}_k^i))/2\right]. \tag{1.69}$$

The cost function for the state $\mathbf{X}_3^i$ is given by

$$J(\mathbf{X}_3^i) = \sum_{k=1}^{3} (\mathbf{z}_k - \mathbf{h}(\mathbf{x}_k^i))' \mathbf{R}^{-1}(\mathbf{z}_k - \mathbf{h}(\mathbf{x}_k^i)). \tag{1.70}$$

We use the altitude $\bar{z}^i = z_0$ for which the likelihood in (1.69) is a maximum or the cost function in (1.70) is a minimum.

1.4.4 NCT Filtering Algorithms

Given $\hat{\mathbf{x}}_{3|3}$ and $\mathbf{P}_{3|3}$ at t_3, our goal is to calculate the state estimates and covariances at $t_k > t_3$, using measurements $\mathbf{z}_k$, $k = 4, 5, \ldots$ For the prediction step, we use two algorithms, Algorithm 1 and Algorithm 2. Algorithm 1 calculates the approximate conditional mean and covariance by numerically integrating two moment differential equations during measurement time interval [4, 20, 29, 31]. Algorithm 2 computes the approximate conditional mean and covariance by using the prediction step of the CKF [21] in the approximate discretized dynamic model described in the Sub-sub-section 1.4.2.2 of Sub-section 1.4.2. We use the measurement update step of the CKF with both Algorithms 1 and 2.

We briefly summarize the prediction step of Algorithm 1 as described in [4, 6, 20, 31]. Suppose we have the conditional mean $\hat{\mathbf{x}}_{k-1|k-1}$ and corresponding covariance $\mathbf{P}_{k-1|k-1}$ after processing $\mathbf{Z}_{k-1}$ at time t_{k-1}, where

$$\hat{\mathbf{x}}_{k-1|k-1} = E\{\mathbf{x}_{k-1}|\mathbf{Z}_{k-1}\}, \tag{1.71}$$

$$\mathbf{P}_{k-1|k-1} = E\{(\mathbf{x}_{k-1} - \hat{\mathbf{x}}_{k-1})(\mathbf{x}_{k-1} - \hat{\mathbf{x}}_{k-1})'|\mathbf{Z}_{k-1}\}. \tag{1.72}$$

Our goal is to propagate $\hat{\mathbf{x}}_{k-1|k-1}$ and $\mathbf{P}_{k-1|k-1}$ from time t_{k-1} to time t_k before processing the measurement $\mathbf{z}_k$. Let $\mathbf{m}(t)$ and $\mathbf{P}(t)$ denote the conditional mean and associated conditional covariance at time t between the observation times t_{k-1} and t_k,

$$\mathbf{m}(t) = E\{\mathbf{x}(t)|\mathbf{Z}_{k-1}\}, \quad t_{k-1} < t \leq t_k, \tag{1.73}$$

$$\mathbf{P}(t) = E\{(\mathbf{x}(t) - \mathbf{m}(t))(\mathbf{x}(t) - \mathbf{m}(t))' | \mathbf{Z}_{k-1}\}, t_{k-1} < t \le t_k. \tag{1.74}$$

Then the approximate moment differential equations to be integrated are given by [4, 6, 20, 29, 31]

$$\frac{d\mathbf{m}}{dt} \approx \mathbf{f}(\mathbf{m}), \quad t_{k-1} < t \le t_k, \tag{1.75}$$

$$\frac{d\mathbf{P}}{dt} \approx \mathbf{P}\mathbf{F}_{\mathbf{x}}'(\mathbf{m}) + \mathbf{F}_{\mathbf{x}}(\mathbf{m})\mathbf{P} + \mathbf{Q}, \quad t_{k-1} < t \le t_k. \tag{1.76}$$

Equations (1.75) and (1.76) represent two ordinary differential equations (ODEs) in $n + n(n+1)/2$ independent variables. They can be integrated by the Runge-Kutta integrator [35] or by the MATLAB function ode45.

1.5 Filter Evaluation Metrics

We use root mean square (RMS) errors for position, velocity, and ω to evaluate performance of the filters in various scenarios. In addition, to determine the overall performance of a filter, we use the root time-averaged mean square (RTAMS) error metric [25]. Let $\mathbf{x}_{k,i}$ and $\hat{\mathbf{x}}_{k,i}$ be the true and estimated states at time t_k in the ith Monte Carlo run, respectively. The RTAMS metric for the XY−position is defined as

$$\text{RTAMS}_{\text{pos},xy} := \sqrt{\frac{\sum_{k=k_1}^{k_2} \sum_{i=1}^{M} \left[(\hat{x}_{k,i} - x_{k,i})^2 + (\hat{y}_{k,i} - y_{k,i})^2 \right]}{(k_2 - k_1 + 1)M}}, \tag{1.77}$$

where M is the number of Monte Carlo runs, and k_1 and k_2 are the initial and final time indices, respectively. The time index k_1 is chosen after the transient phase in filtering and k_2 is chosen as the last time index in filtering. Similar metrics can be defined for other components of the state in both non-maneuvering and NCT motions.

To evaluate the consistency of the filter-calculated covariance and innovation covariance, we use the average normalized estimation error squared (ANEES) and average normalized innovation squared (ANIS) [16, 36], respectively. Let n and m denote the dimension of the state and measurement, respectively. Let $\mathbf{z}_{k,i}$ and $\hat{\mathbf{z}}_{k,i}$ denote the measurement and predicted measurement at time t_k in the ith Monte Carlo run, respectively. The ANEES and ANIS at time t_k are defined, respectively, as

$$\text{ANEES}_k := \frac{\sum_{i=1}^{M} (\hat{\mathbf{x}}_{k,i} - \mathbf{x}_{k,i}) \mathbf{P}_{k,i}^{-1} (\hat{\mathbf{x}}_{k,i} - \mathbf{x}_{k,i})}{nM}, \tag{1.78}$$

$$\text{ANIS}_k := \frac{\sum_{i=1}^{M} (\hat{\mathbf{z}}_{k,i} - \mathbf{z}_{k,i}) \mathbf{S}_{k,i}^{-1} (\hat{\mathbf{z}}_{k,i} - \mathbf{z}_{k,i})}{mM}, \tag{1.79}$$

where $\mathbf{P}_{k,i}$ and $\mathbf{S}_{k,i}$ are the filter-calculated covariance and innovation covariance at time t_k in the ith Monte Carlo run, respectively. The ideal values of ANEES and ANIS are unity.

1.6 Simulations and Results: Non-maneuvering Motions

We consider the following three scenarios using realistic parameter values for aircraft motion and ATC 2D measurements, as presented in Table 1.1:

- NCV and NCAlt using one SR radar in Section 1.6.1,
- NCV and NCAlt using two LR radars in Section 1.6.2, and
- Climbing motion using two SR radars in Section 1.6.3.

Target and radar parameters for true trajectory and sensor measurements are presented in Table 1.2. NCV filter parameters are presented in Table 1.3. We used 500 Monte Carlo runs to compute various metrics. The HP-CEKF, HP-CUKF, and HP-CCKF were used for the one short-range radar scenario. Our numerical results show that in the height-parametrized filtering approach, uniform height intervals and heights based on a geometric series [37, 38] produce nearly the same results. Therefore, all the height-parametrized filtering results in the paper are based on uniform

Table 1.1 Radar and motion parameters of scenarios.

No. of radars	Type of motion	Time interval (s)	No. of measurements
One SR	NCV with NCAlt	4.6	200
Two LR	NCV with NCAlt	6.0	200
Two SR	Climbing motion	2.3	160

Table 1.2 Target and radar parameters for true trajectory and sensor measurements.

	Variable	Value
NCV with NCAlt	Prior mean x-coordinate ($\bar{x}_1$)	−60 km
	Prior mean y-coordinate ($\bar{y}_1$)	(10, 15, 20) km
	Prior mean z-coordinate ($\bar{z}_1$)	9000 ft = 2743.200 m
	Prior mean x-velocity ($\bar{\dot{x}}_1$)	300 knots = 154.333 m/s
	Prior mean y-velocity = z-velocity	0 m/s
	PSD of process noise ($q_x = q_y$)	$0.01\text{m}^2/\text{s}^3$
	PSD of process noise (q_z)	$0.0001\text{m}^2/\text{s}$
	Prior horizontal position covariance	$\text{diag}(50^2, 50^2)\text{m}^2$
	Prior vertical position variance	2^2m^2
	Prior horizontal velocity variance	$\text{diag}(5^2, 5^2)(\text{m/s})^2$
	Prior vertical velocity variance	$(0.1)^2(\text{m/s})^2$
Radar parameters	Range and azimuth SDs (σ_r, σ_α)	(111.12 m, 3.491 millirad)
	Measurement time interval, T	4.6 s for SR, 12 s for LR

Table 1.3 NCV filter parameters.

Variable	Value
NCAlt filter initialization: Maximum speed ($v_{x,\max} = v_{y,\max}$)	400 m/s
NCAlt filter initialization: Maximum speed $v_{z,\max}$	5 m/s
Climbing motion filter initialization: Maximum and minimum speeds ($v_{z,\min}$, $v_{z,\max}$)	(6.096, 10.160) m/s
Minimum and maximum altitudes ($a_{\min}$, $a_{\max}$)	(100, 12000) ft
Number of height intervals N_a	3

height intervals. We experimented with several height intervals and found that three height intervals are sufficient for state estimation accuracy.

Table 1 in [39] presents realistic values for range and azimuth measurement error standard deviations for ATC 2D radars.

Aircraft 1, 2, and 3 with mean y-coordinates 10, 15, and 20 km, respectively, were used in the one SR radar scenario as shown in Figure 1.3. This allows us to analyze the state estimation accuracy and specifically, the height estimation accuracy as a function of distance (y-coordinate) of the aircraft from the radar. In Section 1.6.1, we observe that if the distance of the aircraft from the radar is greater than 20 km in the one SR radar scenario, then the height of the aircraft cannot be estimated accurately.

In the two SR radar scenario, aircraft 4–8 with mean y-coordinates 10, 15, 20, 30, and 40 km, respectively, were used as shown in Figure 1.3. Our results in Section 1.6.2, show that we can estimate the height of the aircraft accurately for distances up to 30 km.

The prior distribution of the initial state in all cases involving the NCV and NCAlt is the same. We have chosen k_1 and k_2 such that the corresponding times t_{k_1} and t_{k_2} are the same or nearly the same in non-climbing scenarios. Therefore, the segments of the true trajectories between t_{k_1} and t_{k_2} will be statistically nearly the same in non-climbing scenarios, and we can compare the RTAMS metrics for these cases fairly. The parameters k_1, k_2, t_{k_1}, and t_{k_2} related to RTAMS computation are presented in Table 1.4.

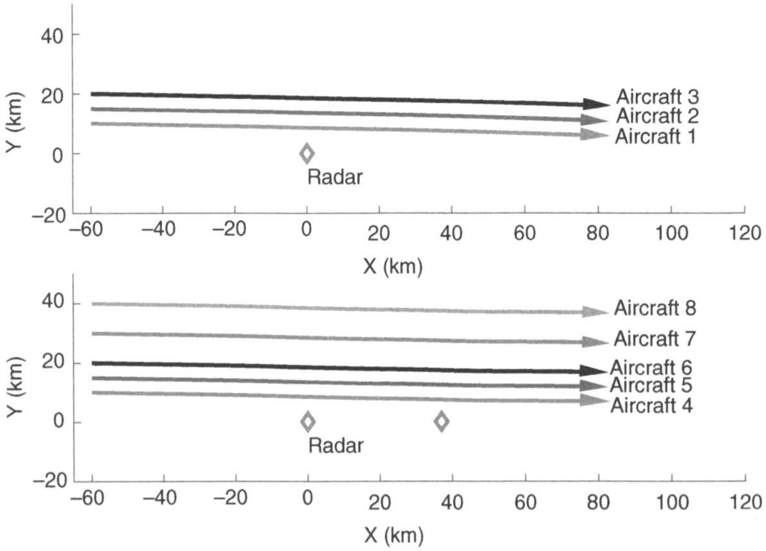

Figure 1.3 True trajectories of aircraft 1, 2, and 3 in one SR radar scenario, and true trajectories of aircraft 4, 5, 6, 7 and 8 in two SR radar scenario from first Monte Carlo run.

Table 1.4 RTAMS parameters of scenarios.

No. of radars	Type of motion	k_1	k_2	t_{k_1} (s)	t_{k_2} (s)
One SR	NCV with NCAlt	101	200	460.0	915.4
Two LR	NCV with NCAlt	77	154	460.0	917.7
Two SR	Climbing	81	160	184.0	356.7

1.6.1 Nearly Constant Velocity and Altitude Using One SR Radar

The radar is at the origin of the coordinate frame. RMS position-XY and altitude errors for aircraft 1, 2, and 3 using the HP-CEKF, HP-CUKF, HP-CCKF, and associated PCRLBs are presented in Figure 1.4. We observe that the RMS position errors for the three filters are nearly the same.

Figure 1.5 shows the variation of the RMS position-XY and altitude errors from HP-CEKF with aircraft y-coordinate. We observe from Figure 1.5 that as the distance (y-coordinate) of the aircraft from the radar increases from 10 to 20 km, the RMS altitude error increases. The minimum and final values of the RMS altitude error for aircraft 1, 2, and 3 are (235.547, 402.551) m, (476.116, 577.690) m, and (735.030, 828.554) m, respectively. Figure 1.5(b) shows that the altitude of the aircraft cannot be estimated accurately for aircraft 3 at a y-coordinate of 20 km. Hence, the altitude of an aircraft cannot be estimated accurately when the distance of the aircraft is more than 20 km in the one SR radar scenario. The variation of the RMS position-XY error with aircraft y−coordinate is smaller than that of the RMS altitude error with aircraft y−coordinate.

The HP-CUKF RTAMS metrics for position-XY and altitude of one SR radar and two LR radars scenarios are compared, and the results as a function of mean y-coordinate are presented in Figure 1.6. Results in Figure 1.6 show that the two LR radars system provides significant improvement over one SR radar in estimating position-XY and altitude. This enhancement can be attributed to the geometry between the radars and the aircraft, which contributes to more precise localization.

The RTAMS metrics are presented in Table 1.5. As observed from Table 1.5, the RTAM metrics for position-XY and velocity-XY from HP-CEKF, HP-CUKF, and HP-CCKF are nearly the same and closely match the corresponding metrics from the PCRLB. However, there is a significant difference between the filter-calculated RTAM metric for altitude and the corresponding RTAM metric from the PCRLB.

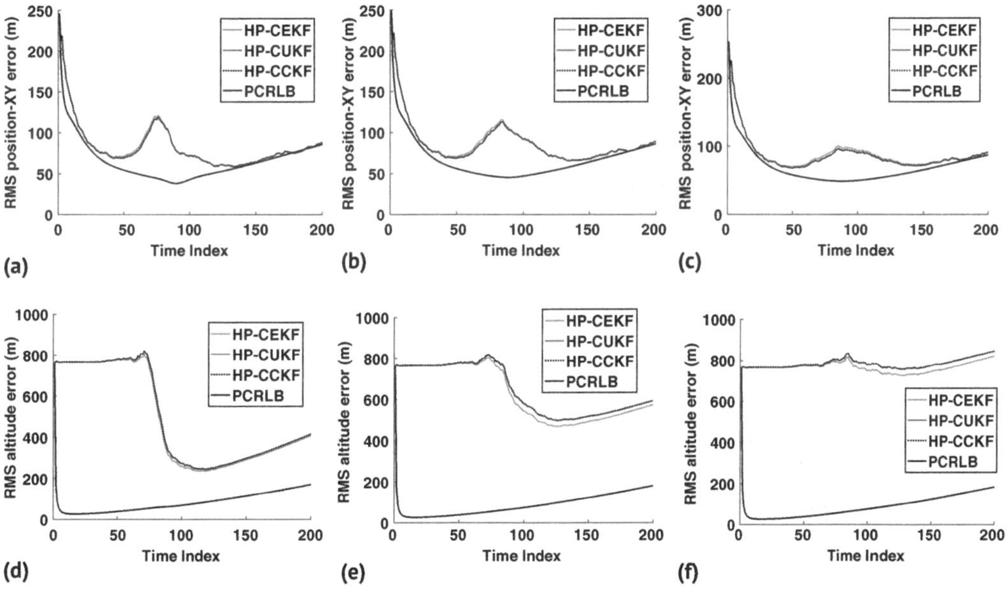

Figure 1.4 RMS position-XY errors and RMS altitude errors for HP-CEKF, HP-CUKF, HP-CCKF, and PCRLB in one SR radar scenario (left to right: Aircraft 1,2,3).

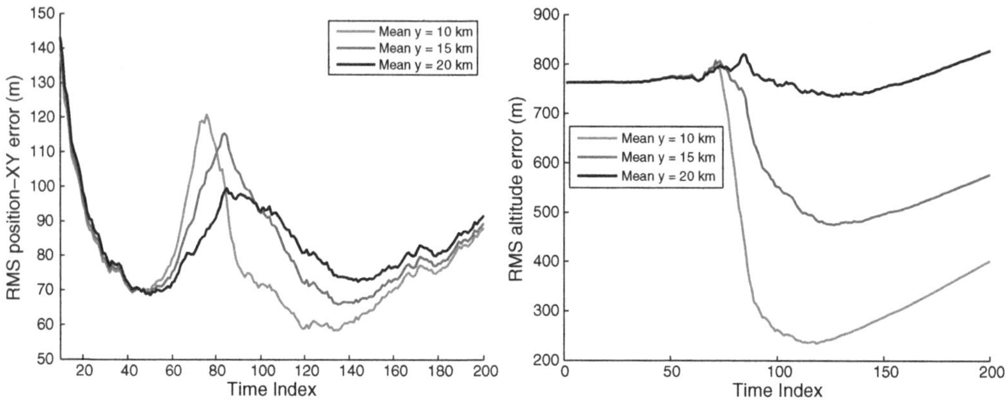

(a) (b)

Figure 1.5 One SR radar scenario: (a) Variation of HP-CEKF RMS position-XY errors for aircraft 1, 2, and 3; (b) Variation of HP-CEKF RMS altitude errors for aircraft 1, 2, and 3.

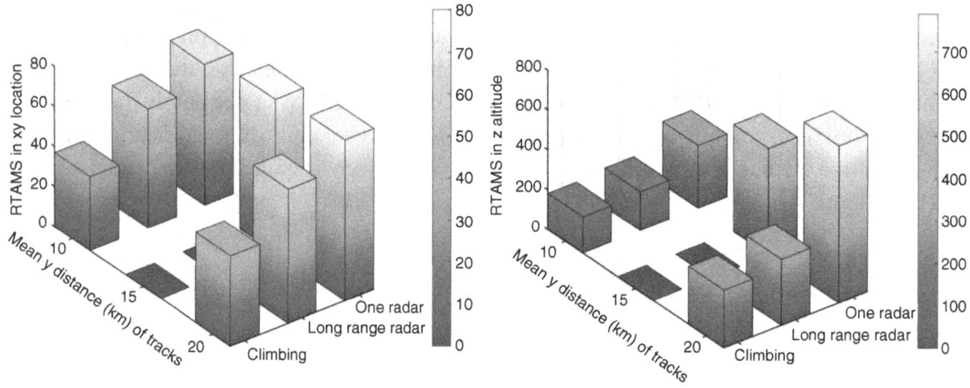

(a) (b)

Figure 1.6 Variation of HP-CUKF position-XY and altitude RTAMS metrics with mean y−coordinate in all simulated scenarios.

1.6.2 Nearly Constant Velocity and Altitude Using Two LR Radars

The antenna rotation period of each radar is 12 s and the radar measurements are interleaved. Hence, the measurement time interval for this scenario is 6 s.

Table 1.6 presents RTAMS metrics for two y-coordinates, 10 and 20 km using HP-CUKF and associated PCRLBs. Results in Table 1.5 show that these three filters have nearly the same accuracy, HP-CUKF being the preferred filter.

Table 1.5 RTAMS metric in one SR radar scenario.

Aircraft	RTAMS	HP-CEKF	HP-CUKF	HP-CCKF	PCRLB
Aircraft 1	Pos-XY (m)	70.079	70.208	70.210	65.172
Aircraft 1	Pos-Z (m)	306.790	314.164	314.541	119.922
Aircraft 1	Vel-XY (m/s)	1.203	1.204	1.204	1.190
Aircraft 1	Vel-Z (m/s)	0.620	0.636	0.637	0.268
Aircraft 2	Pos-XY (m)	76.703	76.411	76.416	66.220
Aircraft 2	Pos-Z (m)	511.618	537.212	537.263	128.320
Aircraft 2	Vel-XY (m/s)	1.223	1.223	1.223	1.199
Aircraft 2	Vel-Z (m/s)	0.577	0.580	0.581	0.277
Aircraft 3	Pos-XY (m)	81.495	80.104	80.119	67.479
Aircraft 3	Pos-Z (m)	760.102	788.376	788.141	130.461
Aircraft 3	Vel-XY (m/s)	1.232	1.230	1.230	1.209
Aircraft 3	Vel-Z (m/s)	0.574	0.561	0.562	0.279

Table 1.6 RTAMS metric in two LR radars scenario.

Aircraft	RTAMS	HP-CUKF	PCRLB
Aircraft mean $y = 10$ km	Pos-XY (m)	59.208	54.986
Aircraft mean $y = 10$ km	Pos-Z (m)	194.955	98.410
Aircraft mean $y = 10$ km	Vel-XY (m/s)	1.168	1.131
Aircraft mean $y = 10$ km	Vel-Z (m/s)	0.635	0.243
Aircraft mean $y = 20$ km	Pos-XY (m)	66.896	62.876
Aircraft mean $y = 20$ km	Pos-Z (m)	330.704	122.007
Aircraft mean $y = 20$ km	Vel-XY (m/s)	1.212	1.193
Aircraft mean $y = 20$ km	Vel-Z (m/s)	0.635	0.269

1.6.3 Climbing Motion Using Two SR Radars

We analyze the climbing motion of an aircraft using a representative climb rate of 1500 ft/min (7.62 m/s) with measurements from two SR ATC radars. The aircraft is assumed to climb in a plane parallel to the $XZ-$plane at a distance (mean y-coordinate) of 10 or 20 km from the $X-$axis. The mean initial velocity components are 300 knots (154.33 m/s) along the $X-$axis, 0 m/s along the $Y-$axis, and 7.62 m/s along the $Z-$axis. The mean initial $X-$coordinate of the aircraft is -10 km. The mean initial height of the aircraft is 200 ft (60.96 m). We use 160 interleaving measurements from two radars with a measurement time interval of 2.3 s. Over the 365.7-s observation period, the aircraft ascends to a final altitude of approximately 9646 ft (2940 m).

RMS errors for aircraft position-XY and altitude estimated by the HP-CUKF are shown in Figure 1.7 for a mean y-coordinate of 10 km. Figure 1.7 shows that in the second half of the trajectory, the RMS position-XY errors for the HP-CUKF are close to the PCRLBs. The RMS altitude error gradually approaches the PCRLB over time.

RTAMS metrics for two y-coordinates, 10 and 20 km, using the HP-CUKF and corresponding PCRLBs are presented in Table 1.7. Results in Table 1.7 indicate that the RTAMS metrics for components in the horizontal plane are close to the corresponding PCRLBs. In contrast, for components along the $Z-$axis, there is a large difference between RTAMS metrics and the corresponding PCRLBs.

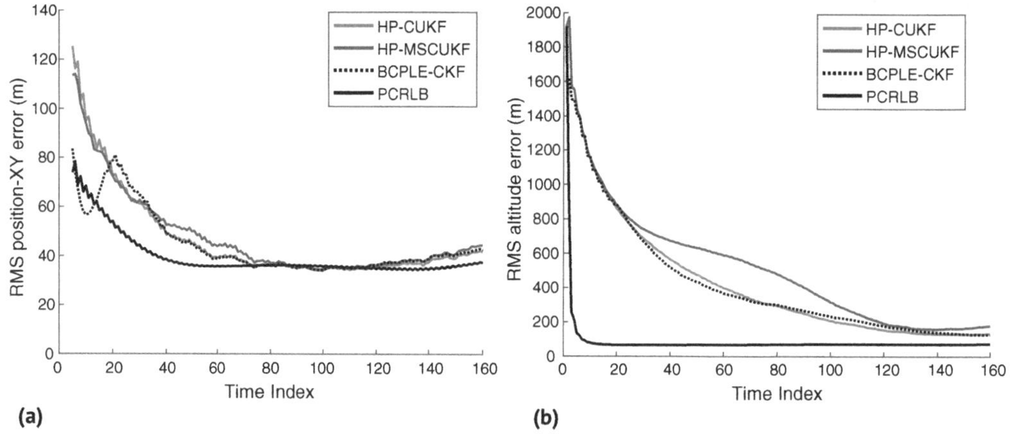

Figure 1.7 Climbing motion with two SR radars: RMS position-XY and RMS altitude errors for mean y-coordinate of 10 km.

Table 1.7 RTAMS metric in climbing scenario using two SR radars.

Aircraft	RTAMS	HP-CUKF	PCRLB
Aircraft mean $y = 10$ km	Pos-XY (m)	36.886	35.641
Aircraft mean $y = 10$ km	Pos-Z (m)	181.326	71.749
Aircraft mean $y = 10$ km	Vel-XY (m/s)	0.989	0.988
Aircraft mean $y = 10$ km	Vel-Z (m/s)	0.873	0.188
Aircraft mean $y = 20$ km	Pos-XY (m)	44.942	41.336
Aircraft mean $y = 20$ km	Pos-Z (m)	289.480	73.877
Aircraft mean $y = 20$ km	Vel-XY (m/s)	1.038	1.046
Aircraft mean $y = 20$ km	Vel-Z (m/s)	0.887	0.191

1.6.4 Computational Cost of Algorithms

Table 1.8 presents average CPU times (averaged over aircraft within each scenario) per Monte Carlo run, as well as relative CPU times for the one SR radar, two LR radars, and climbing (with two SR radars) scenarios. In the one SR radar scenario, The CPU time is averaged over aircraft 1–3. As shown in Table 1.8 in the one SR radar scenario, the HP-CUKF has the lowest CPU time, being about half of those for the HP-CEKF and HP-CCKF. For the two long range-radar and climbing scenarios, CPU times are averaged over two aircraft at $y-$coordinates of 10 and 20 km.

1.7 Simulations and Results: Nearly Constant Turn

We use realistic parameters in our simulation consistent with the NCT motion of an aircraft. An aircraft moves with a speed of 88.330 m/s during the NCT motion. As mentioned in the Introduction, according to ICAO, all turns should be made at a bank angle of 25° or at a rate of 3°/s, whichever requires the lesser bank angle [18]. The height of the NCT motion should be such that the area of the NCT turn avoids mountains and buildings. The perturbation in the atmosphere are higher in the horizontal direction than in the vertical direction. Therefore, we have used a higher diffusion coefficients ($q_x = q_y$) in the horizontal direction than that (q_z) in the vertical direction. Numerical values of the prior mean, PSDs of process noise, and prior covariance are presented in Table 1.9.

Table 1.8 Average CPU times per Monte Carlo run.

Scenario	Algorithm	Avg. per track (s)	Relative
One SR radar	HP-CEKF	0.2494	1.000
One SR radar	HP-CUKF	0.1287	0.516
One SR radar	HP-CCKF	0.2649	1.062
Two LR radars	HP-CUKF	0.1296	—
Climb (two SR radars)	HP-CUKF	0.131	—

Table 1.9 NCT: target initial parameters.

Variable	Value
Prior mean horizontal position $(\bar{x}_1, \bar{y}_1)$	(1667.890, 0.000) m
Prior mean turn rate $(\bar{\omega}_1)$	3.0°/s
Prior mean horizontal velocity $(\bar{\dot{x}}_1, \bar{\dot{y}}_1)$	(0, 87.330) m/s
Prior mean altitude $(\bar{z}_1)$	3000 m
Diffusion coefficients, $q_x = q_y$	$0.1\ m^2/s^3$
Diffusion coefficient, q_ω	$1.0 \times 10^{-8} rad^2/s^3$
Diffusion coefficient, q_z	$0.001\ m^2/s$
Prior horizontal position covariance	$diag(100^2, 100^2)\ m^2$
Prior horizontal velocity covariance	$diag(10^2, 10^2)(m/s)^2$
Prior turn rate variance	$2.5 \times 10^{-5}(rad/s)^2$
Prior vertical position variance	$5^2\ m^2$

For an ATC 2D radar, the antenna rotation period lies in the 4–12 s range. We use a rotation period of $T = 4.6$ s, as used in [2]. This type of radar is called a short-range (SR) radar. The 3D position of the radar is $[0\ 0\ 100]'$ m. Range and azimuth measurement error SDs (σ_r, σ_α) for the ATC 2D radar are 0.06 NM = 111.12 m and 0.2 deg = 3.491 millirad, respectively, [2]. Realistic values for range and azimuth measurement error SDs for ATC 2D radars are presented in Table 1 of [39].

We use the Euler–Maruyama method [30, 32] for generating the true trajectory, where 1000 subintervals are used in the measurement time interval T. The true NCT trajectory parallel to the XY–plane is shown in Figure 1.8 from the first Monte Carlo run. The start and end points are marked with a circle and a diamond, respectively. The XY– position of the sensor is marked with a circle centered at (0.0) m.

We use 500 Monte Carlo runs to calculate RMS errors, ANEES, and ANIS for various simulation scenarios. We consider seven scenarios for analysis of trajectory estimation accuracy and filter evaluation. For these scenarios, we keep the prior mean horizontal velocity ($\bar{\dot{x}}_1, \bar{\dot{y}}_1$), prior mean turn rate ($\bar{\omega}_1$), and measurement time interval (T) constant and vary the range measurement SD (σ_r), prior mean altitude ($\bar{z}_1$), and bank angle (ϕ). The parameters for seven scenarios are presented in Table 1.10.

1.7.1 Variation in Radar Range SD

In Scenario 1, we use the same range and azimuth SDs as used in [11]. The RMS errors for position, velocity, and ω obtained from the CEKF-CKF and CKF-CKF are presented in Figure 1.9. Observing Figure 1.9, we note that the RMS errors do not consistently decrease as measurements are processed. In fact, the RMS position-XY and velocity-XY errors increase sharply after about 80 measurements. This suggests that the 2D radar SDs used in [11] are not appropriate for producing accurate NCT trajectory estimates in our simulations.

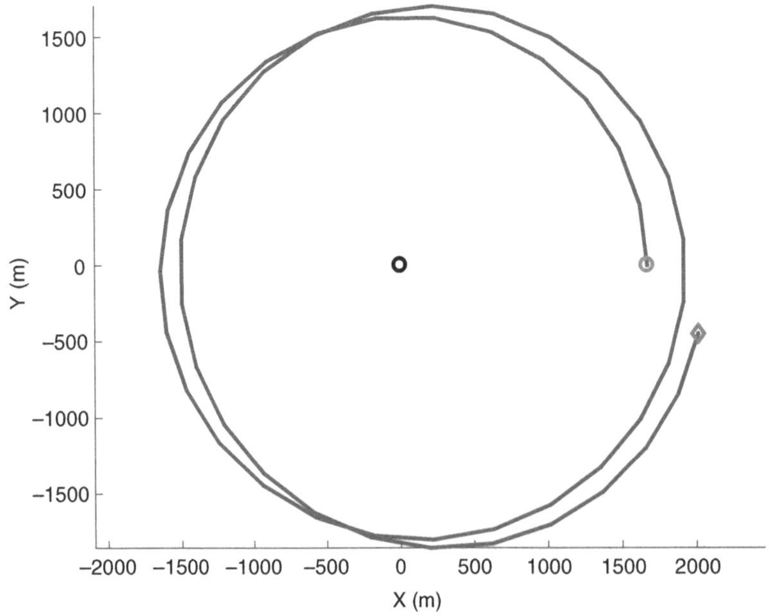

Figure 1.8 True NCT trajectory parallel to the XY–plane from the first Monte Carlo run.

Table 1.10 Parameters used for various scenarios in ATC NCT motion analysis.

Scenario	σ_r (m)	Bank angle (°)	$\bar{z}_1$ (m)
Scenario 1	111.12	25	3000
Scenario 2	50.00	25	3000
Scenario 3	18.52	25	3000
Scenario 4	18.52	25	2500
Scenario 5	18.52	25	2000
Scenario 6	18.52	20	3000
Scenario 7	18.52	30	3000

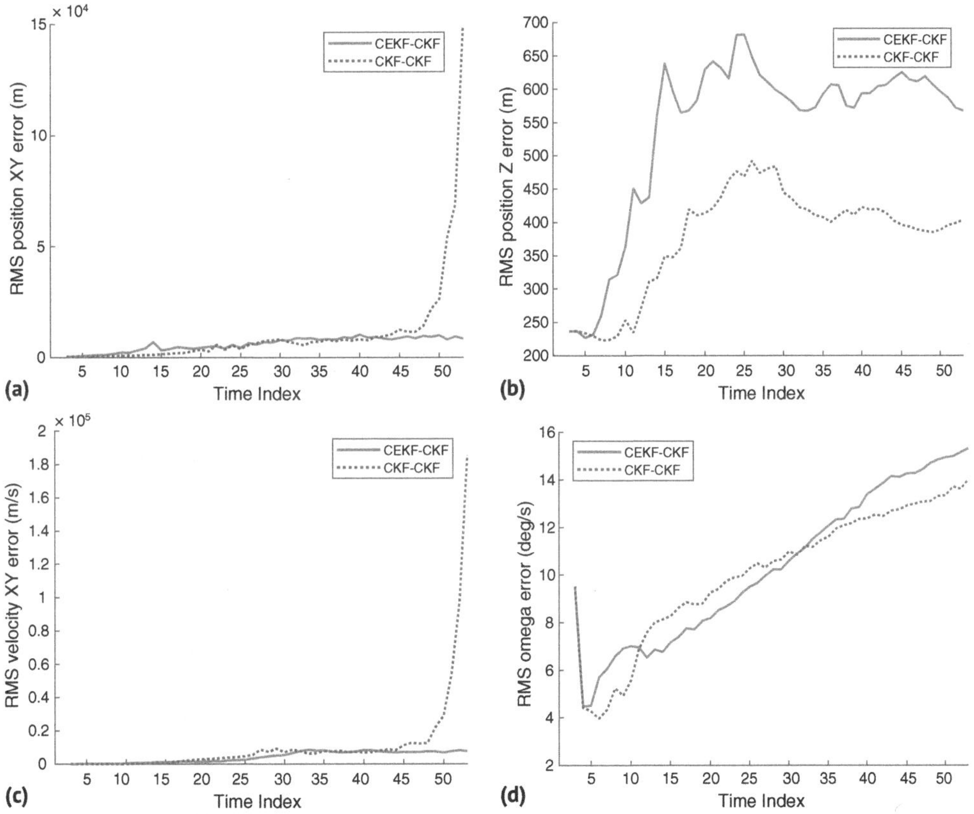

Figure 1.9 (a) RMS position-XY errors, (b) RMS position-Z errors, (c) RMS velocity-XY errors, and (d) RMS ω errors for the CEKF-CKF and CKF-CKF using $\sigma_r = 111.12$ m, bank angle = 25°, and $\bar{z}_1 = 3000$ m.

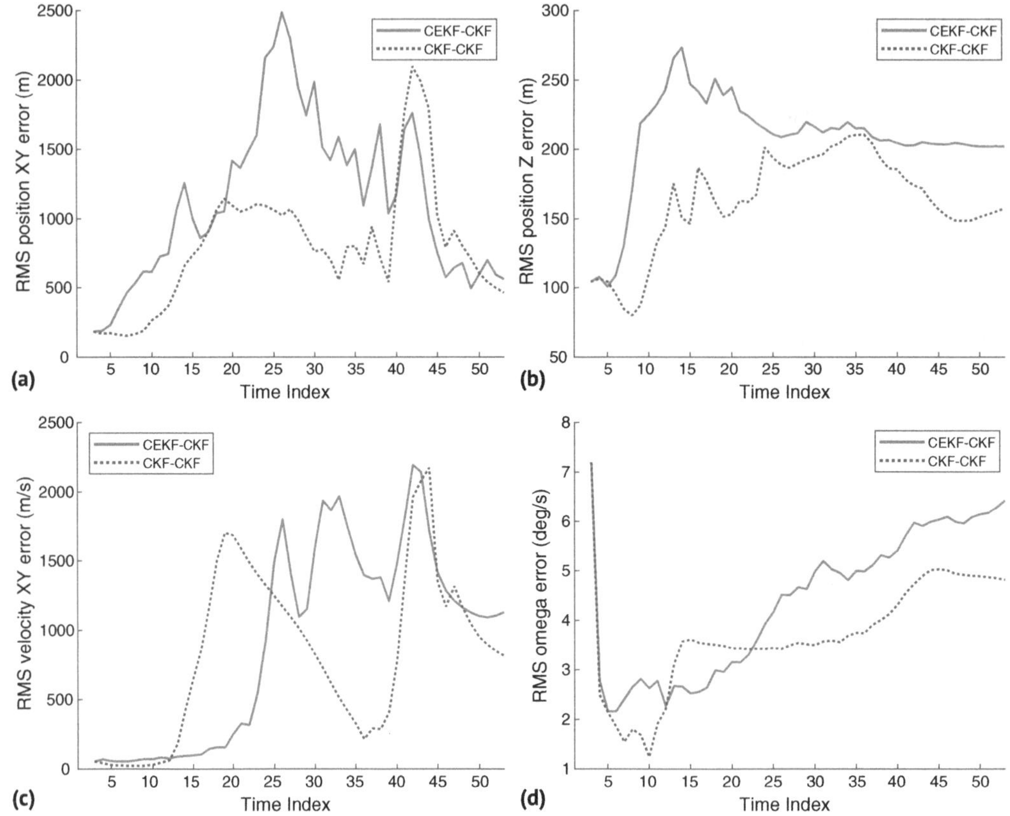

Figure 1.10 (a) RMS position-XY errors, (b) RMS position-Z errors, (c) RMS velocity-XY errors, and (d) RMS ω errors for the CEKF-CKF and CKF-CKF using $\sigma_r = 50$ m, bank angle = 25°, and $\bar{z}_1 = 3000$ m.

To improve the trajectory estimation accuracy, we chose a range SD of 50 m as was used in [3], while keeping other parameters of Scenario 1 intact to obtain Scenario 2 parameters. The RMS errors from the CEKF-CKF and CKF-CKF are shown in Figure 1.10. We observe from Figure 1.10 that the trajectory estimation accuracy is not satisfactory for this scenario too.

From radar hardware perspective, it is easier to achieve a higher range resolution than to achieve a comparable azimuth resolution. Therefore, we have focused on lowering the range SD. We found that using a range SD of 0.01 NM = 18.52 m in Scenario 3, produces a significant improvement in the trajectory estimation accuracy, as shown by the RMS errors in Figure 1.11. After the transient phase, all the RMS errors decrease gradually after about 15 measurements and the RMS position errors of the CEKF-CKF are significantly lower than that of the CKF-CKF. The RMS velocity and ω errors for these two filters are nearly the same.

The ANEES and ANIS for the CEKF-CKF and CKF-CKF with 99% confidence bounds using range SD of 18.52 m, bank angle of 25°, and a prior mean altitude of 3000 m, are presented in Figure 1.12. Plots in Figure 1.12 show that the ANEES and ANIS lie within the confidence bounds

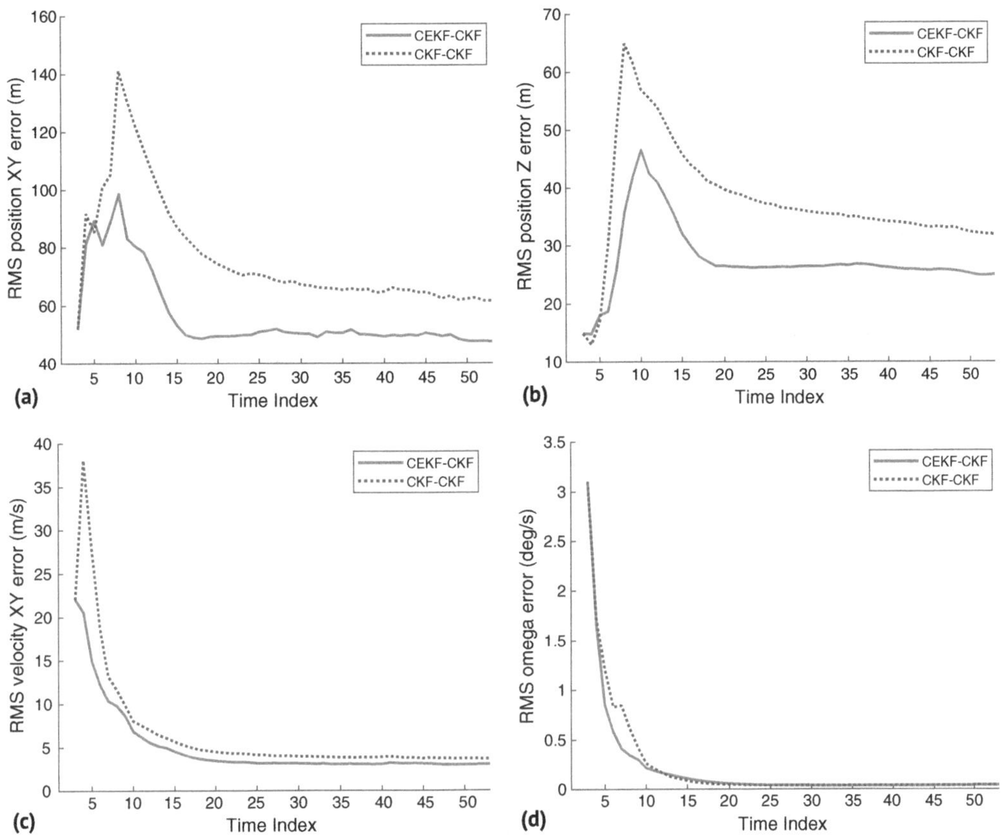

Figure 1.11 (a) RMS position-XY errors, (b) RMS position-Z errors, (c) RMS velocity-XY errors, and (d) RMS ω errors for the CEKF-CKF and CKF-CKF using $\sigma_r = 18.52$ m, bank angle= $25°$ and $\bar{z}_1 = 3000$ m.

most of the time after the transient phase in the filter. We have excluded the AEES and ANIS plots for range SDs of 112.12 m and 50.0 m due to inaccurate state estimation for these values.

1.7.2 Variation in Mean Altitude

To analyze the effect of mean altitude on trajectory estimation accuracy, we consider mean altitudes of 3000 m, 2500 m, and 2000 m while using the radar range SD of 18.52 m and mean bank angle of 25°. To observe the variations of the RMS errors with the mean altitudes clearly, we have plotted the RMS errors together for three altitudes in Figure 1.13. We observe from Figure 1.13 that the RMS XY, Z, and velocity-XY errors increase with decrease in the mean altitude and the CEKF-CKF consistently performs better than the CKF-CKF. We do not observe any variation of the RMS α error with the mean altitude.

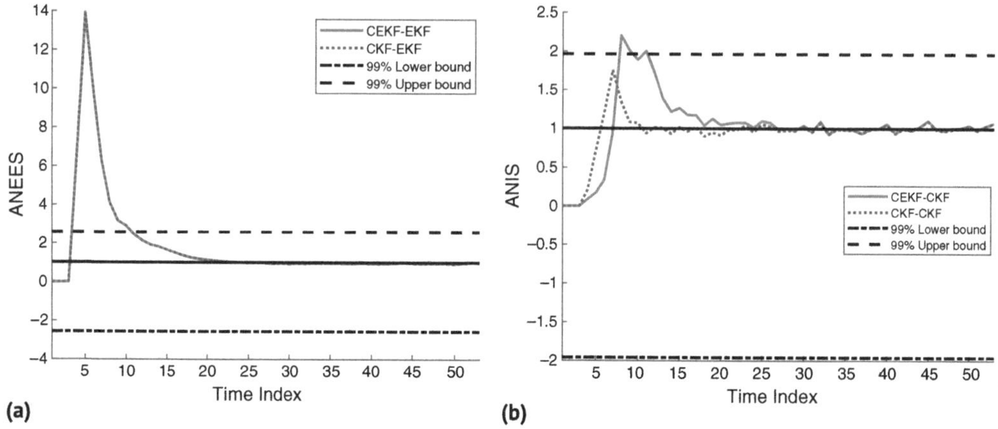

Figure 1.12 ANEES and ANIS for the CEKF-CKF and CKF-CKF using $\sigma_r = 18.52$ m, bank angle= 25°, and $\bar{z}_1 = 3000$ m.

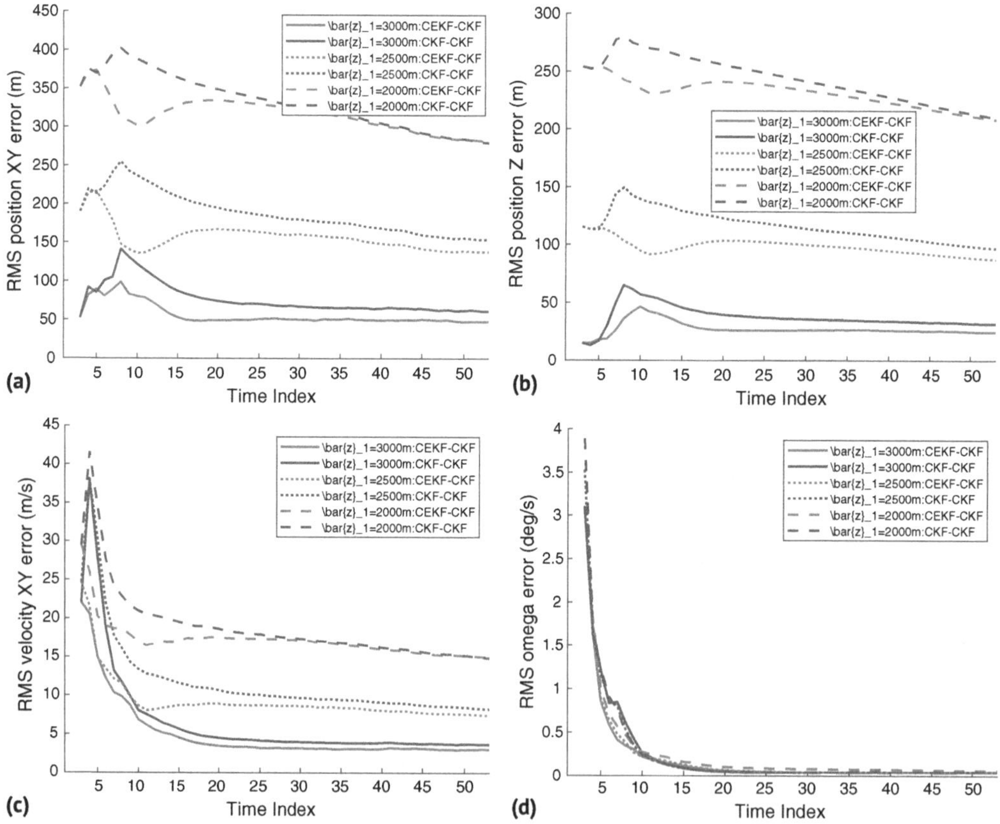

Figure 1.13 (a) RMS position-XY errors, (b) RMS position-Z errors, (c) RMS velocity-XY errors, and (d) RMS ω errors for the CEKF-CKF and CKF-CKF using different $\bar{z}_1$ with $\sigma_r = 18.52$ m, bank angle= 25°.

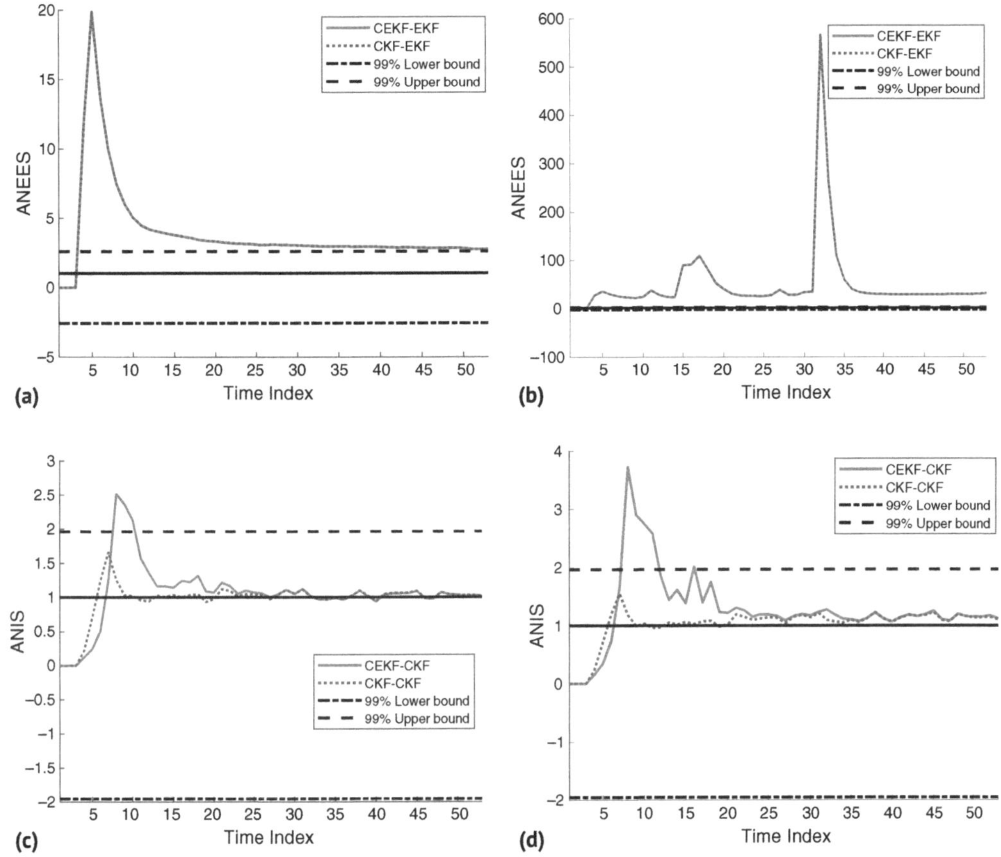

Figure 1.14 ANEES and ANIS for CEKF-CKF and CKF-CKF using $\sigma_r = 18.52$ m, bank angle= 25°, and two values for $\bar{z}_1$.

1.7.3 Variation with Bank Angles

In Figure 1.11, we present RMS errors for the two filters using $\sigma_r = 18.52$ m, $\bar{z}_1 = 3000$ m, and the ICAO recommended bank angle of 25° for Scenario 1. In this subsection, we estimate the trajectories by choosing bank angles of 20° and 30°, which are slightly lower and higher than the ICAO recommended bank angle of 25°. We keep the range SD and mean altitude the same. These two cases represent Scenarios 6 and 7, respectively, as shown in Table 1.10. Figure 1.14 presents ANEES and ANIS for CEKF-CKF and CKF-CKF using $\sigma_r = 18.52$ m, bank angle= 25°, and two values for $\bar{z}_1$. The RMS errors for bank angles of 20°, 25°, and 30° corresponding to Scenarios 6, 3, and 7 are presented in Figure 1.15. We do not observe any systematic pattern in the RMS errors due to

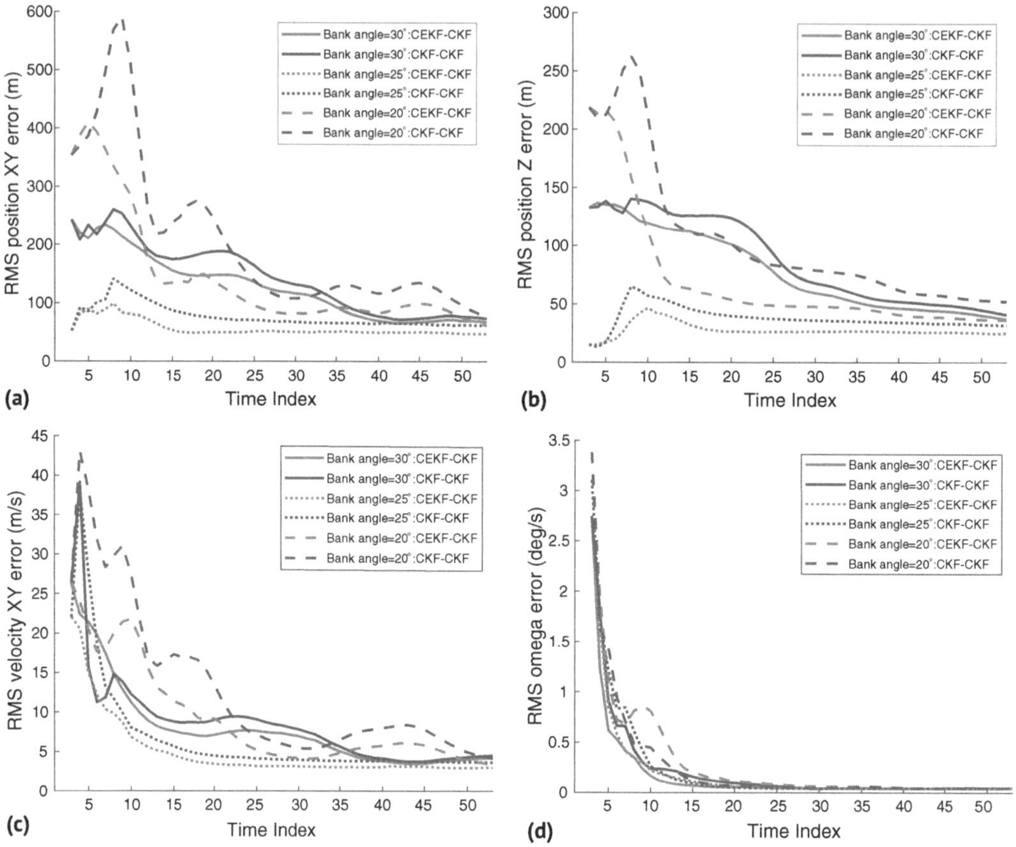

Figure 1.15 (a) RMS position-XY errors, (b) RMS position-Z errors, and (c) RMS velocity-XY errors, and (d) RMS ω errors for the CEKF-CKF and CKF-CKF using different bank angles with $\sigma_r = 18.52$ m, $\bar{z}_1 = 3000$ m.

variation in bank angles. The ANEES and ANIS for bank angles of 20° and 30° corresponding to Scenarios 6 and 7 are presented in Figure 1.16. Results in Figures 1.15 and 1.16 show that the CKF-CKF consistently has lower RMS XY, RMS Z, and velocity-XY errors compared with those from the CEKF-CKF. Both filters are able to estimate the trajectory accurately with varying bank angles.

The RTAMS metrics and average CPU times for the two filters are presented in Tables 1.11 and 1.12, respectively. The CEKF-CKF and CKF-CKF are able to accurately estimate the aircraft trajectory in Scenarios 3–7. The RTAMS metrics for pos-XY, pos-Z, velocity-XY, and ω in Table 1.11 show that the CEKF-CKF consistently performs better than the CKF-CKF. We observe from Table 1.12 that the average CPU time of the CEKF-CKF in Scenario 3, where both filters estimate the aircraft trajectory accurately, is about 1.8 times higher than that of the CKF-CKF.

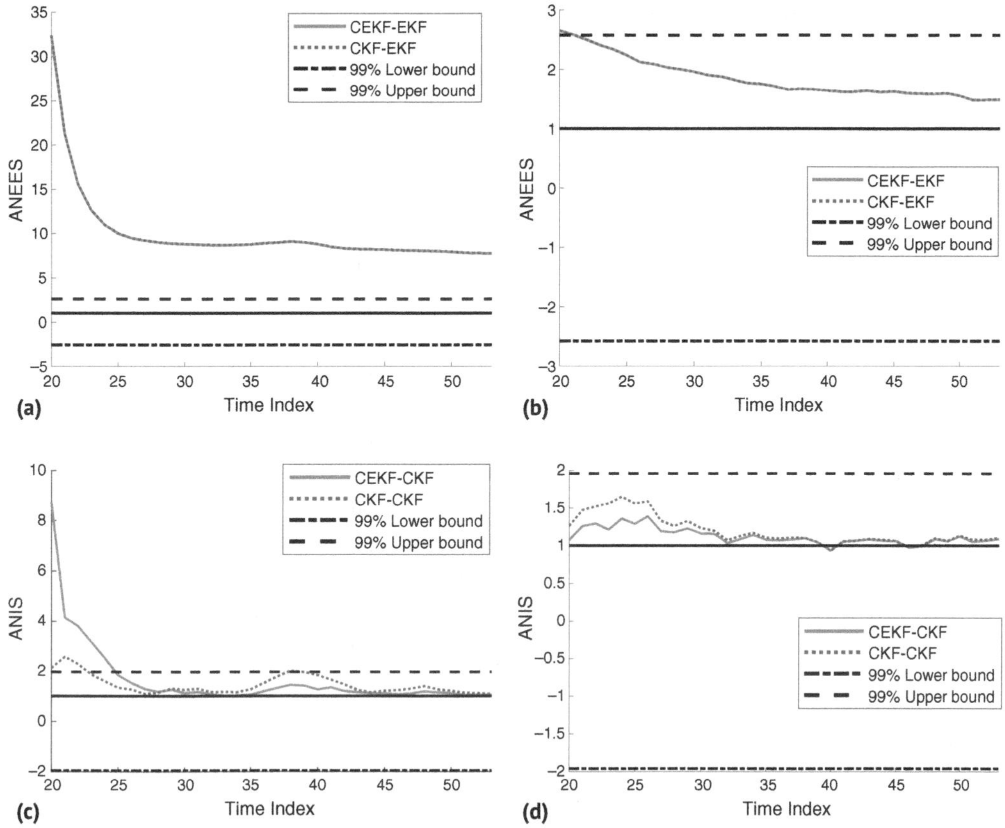

Figure 1.16 ANEES and ANIS for the CEKF-CKF and CKF-CKF using $\sigma_r = 18.52$ m and different bank angles.

Table 1.11 RTAMS metrics.

	σ_r (m)	Bank angle(°)	Mean Alt (m)	RTAMS	CEKF-CKF	CKF-CKF
Scenario 1	111.12	25	3,000	Pos-XY (m)	7920.38	3111.43
				Pos-Z (m)	606.70	424.84
				Vel-XY (m/s)	6491.18	3827.88
				ω (rad/s)	0.2160	0.2065
Scenario 2	50	25	3,000	Pos-XY (m)	1458.15	1047.05
				Pos-Z (m)	210.97	179.28
				Vel-XY (m/s)	1416.60	1158.69
				ω (rad/s)	0.0910	0.0726
Scenario 3	18.52	25	3,000	Pos-XY (m)	49.78	66.33
				Pos-Z (m)	26.10	35.15
				Vel-XY (m/s)	3.14	3.94
				ω (rad/s)	0.0008	0.0007

(Continued)

Table 1.11 RTAMS metrics. (Continued)

	σ_r (m)	Bank angle(°)	Mean Alt (m)	RTAMS	CEKF-CKF	CKF-CKF
Scenario 4	18.52	25	2,500	Pos-XY (m)	152.92	173.04
				Pos-Z (m)	96.66	109.85
				Vel-XY (m/s)	8.24	9.35
				ω (rad/s)	0.0008	0.0008
Scenario 5	18.52	25	2,000	Pos-XY (m)	309.63	313.56
				Pos-Z (m)	226.63	233.27
				Vel-XY (m/s)	16.38	16.68
				ω (rad/s)	0.0014	0.0008
Scenario 6	18.52	20	3,000	Pos-XY (m)	92.16	133.02
				Pos-Z (m)	44.11	71.65
				Vel-XY (m/s)	5.42	7.30
				ω (rad/s)	0.0011	0.0009
Scenario 7	18.52	30	3,000	Pos-XY (m)	100.66	118.23
				Pos-Z (m)	59.59	69.81
				Vel-XY (m/s)	5.59	6.44
				ω (rad/s)	0.0008	0.0009

Table 1.12 Average CPU times per Monte Carlo run in passive sensor scenarios.

Scenario	Averaged CPU Time (s)		
	CEKF-CKF	CKF-CKF	Ratio
Scenario 1	0.0313	0.0204	1.5
Scenario 2	0.0253	0.0140	1.8
Scenario 3	0.0260	0.0141	1.8

1.8 Conclusions

The RMS altitude error increases with mean distance of the aircraft from the radar. In the one SR radar scenario, accurate altitude estimation is feasible when the aircraft is within approximately 15 km of the radar. Our results demonstrate that the trajectory of an aircraft with the NCV and NCAlt can be estimated accurately using two LR radars. Furthermore, we can also estimate the trajectory of a climbing aircraft effectively using two SR 2D radars. In the one SR radar scenario, the HP-CUKF has the lowest computational cost and that for the HP-CEKF and HP-CCKF is about two times higher.

We successfully solved the challenging problem of 3D trajectory estimation of a civilian aircraft executing the NCT using an ATC 2D radar measurements. The Itô stochastic differential equation and guidelines from the ICAO on turn rate and bank angle for the NCT motion were used, which were absent in previous studies. Furthermore, we developed an effective filter initialization algorithm based on the height-parametrized multiple model approach. An important finding of our study is that current ATC 2D radar range and azimuth standard deviations are insufficient for accurate NCT trajectory estimation, necessitating a six-fold improvement in range measurement accuracy. Numerical simulations demonstrate the superior performance of the CEKF-CKF over the CKF-CKF across various scenarios. The computational cost of the CEKF-CKF is about 1.8 times higher than that of the CKF-CKF. Our future research will focus on exploring novel algorithms and systematically analyzing the impact of range and azimuth measurement accuracy on trajectory estimation.

Acknowledgments

This work was supported in part by National Natural Science Foundation of China through Grant 62273283. The authors would like to thank Keyi Li and Gongjian Zhou of the Harbin Institute of Technology, Harbin, China, for valuable explanations regarding their 2D constant turn motion filter initialization algorithm. We also extend our sincere gratitude to Captain Javed Khan, a pilot with IndiGo, for helpful discussions on aircraft turn motion and his assistance in selecting realistic parameters for the aircraft motion.

References

1 M. V. Cook. *Flight Dynamics Principles*. Butterworth-Heinemann, Oxford, UK, 2nd edition, 2007.
2 Y, Rakvongthai, J, Ru, S, and Sivananthan, S. Oraintara, Altitude estimation for 3-D tracking with two 2-D radars. In *Proceedings of the 14th International Conference on Information Fusion*, Chicago, IL, USA, pages 1401–1408,2011.
3 I. Arasaratnam, S. Haykin, and T. R. Hurd. Cubature Kalman filtering for continuous-discrete systems: Theory and simulations. *IEEE Transactions on Signal Processing*, 58(10):4977–4993, 2010.
4 G. Y. Kulikov and M. V. Kulikova. The accurate continuous-discrete extended Kalman filter for radar tracking. *IEEE Transactions on Signal Process*, 64(4):948–958, 2016.
5 M. R. Morelande and N. J. Gordon, Target tracking through a coordinated turn. In *Proceedings of the IEEE International Conference on Acoustics, Speech, and Signal Processing (ICASSP\'05)*, Philadelphia, PA, USA, pages iv/21–iv/24, 2005.
6 S. Särkkä and J. Sarmavuori. Gaussian filtering and smoothing for continuous-discrete dynamic systems. *Signal Processing*, 93(2):500–510, 2013.
7 M. Mallick, M. Morelande, and L. Mihaylova. Continuous-discrete filtering using EKF, UKF, and PF, *In Proceedings of the 2012 International Conference on Information Fusion*, Singapore, pages 1087–1094, 2012.
8 M. Mallick, M. Morelande, L. Mihaylova, S. Arulampalam, and Y. Yan. Angle-only filtering in three dimensions, In M. Mallick, V. Krishnamurthy, and B.-N. Vo, Editors, *Integrated Tracking, Classification, and Sensor Management: Theory and Applications*, Wiley/IEEE, 1, pages 3–42.2012,
9 M-J, Gai, X, Yi, Y. He, and B. Shi. An approach to tracking a 3D-target with 2D-radar. In *Proceedings of IEEE International Radar Conference*, Crystal Gateway Marriott Arlington, VA, pages 763–768, 2005
10 S. Li, Y. Cheng, R. Tharmarasa, M. Efe, R. Jessemi-Zargami, D. Brookes, and T. Kirubarajan. 3-D tracking of air targets using a single 2-D radar. *Signal Processing*, 166: 107241, 2020.
11 M. Mallick, S. Arulampalam, Y. Yan, J. Ru. Three dimensional tracking of an aircraft using 2D radars. *IEEE Transactions on Aerospace Electronic Systems*, 54(2):585–600, 2018.

12 K. Dogançay. Bias compensation for the bearings-only pseudolinear target track estimator. *IEEE Transactions on Signal Processing*, 54(2):59–68, 2006.

13 P. Tichavsky, C. H. Muravchik, and A. Nehorai. Posterior Cramér-Rao bounds for discrete-time nonlinear filtering *IEEE Transactions on Signal Processing*, 46(5):1386–1396, 1998.

14 M. Hernandez. Performance bounds for target tracking: computationally efficient formulations and associated applications. In M. Mallick,V. Krishnamurthy, and B.-N.Vo, editors, *Integrated Tracking, Classification, and Sensor Management: Theory and Applications*, Wiley/IEEE, 7, 255–310, 2012.

15 M. V. Kulikova and G. Y. Kulikov. MATLAB-based general approach for square-root extended-unscented and fifth-degree cubature Kalman filtering methods. *European Journal of Control*, 59: 1–12, 2021.

16 Y. Bar-Shalom, X. R. Li, and T. Kirubarajan. *Estimation with Applications to Tracking and Navigation*. Wiley, New York, 2001.

17 X. R. Li and V. P. Jilkov. Survey of maneuvering target tacking. Part I: Dynamic models. *IEEE Transactions on Aerospace and Electronic Systems*, 39(4):1333–1364, 2003.

18 *Doc 8168 (PANS-OPS)*, International Civil Aviation Organization (ICAO), 5th edition, 2006.

19 M. Roth, G. Hendeby, and F. Gustafsson. EKF/UKF maneuvering target tracking using coordinated turn models with polar/Cartesian velocity Target tracking through a coordinated turn. In *Proceedings of the 17th International Conference on Information Fusion (FUSION)*, pages 1–82014. Salamanca, Spain, 2014.

20 A. Gelb, Editor, *Applied Optimal Estimation*, MIT Press, 1974.

21 I. Arasaratnam and S. Haykin. Cubature Kalman filters. *IEEE Transactions on Automatic Control*, 54(6):1254–1269, 2009.

22 Yan, and J. Ru. Altitude estimation using primary returns from two radars. *Journal of Air Traffic Control*, 54(4):33–37, 2012.

23 M. Mallick and S. Arulampalam. Comparison of nonlinear filtering algorithms in ground moving target indicator (GMTI) target tracking. In *Proceedings of SPIE*, San Diego, CA, 2003.

24 M. Mallick and B. La Scala. Comparison of single-point and two-point difference track initiation algorithms using position measurements. *Acta Automatica Sinica*, 34(3): 258–265, 2008.

25 B. Ristic, S. Arulampalam, and N. Gordon. *Beyond the Kalman Filter*, Artech House, 2004.

26 B. Ristic, S. Zollo, and S. Arulampalam, Performance bounds for manoeuvring target tracking using asynchronous multi-platform angle-only measurements. In *Proceedings of the 4th International Conference on Information Fusion*, Montreal, Qubéc, Canada, 2001.

27 D. Halliday, R. Resnick, and J. Walker. *Fundamentals of Physics Extended*, Wiley, 2010.

28 Wikipedia. Standard rate turn. https://en.wikipedia.org/wiki/Standard_rate_turn, accessed: 2025-05-23.

29 A. H. Jazwinski. *Stochastic Processes and Filtering Theory*, Academic Press, New York, NY, USA, 1970.

30 S. Särkkä and A. Solin. *Applied Stochastic Differential Equations*. Cambridge University Press, 2019.

31 M. Mallick, X. Tian, and L. Xu. Reentry vehicle filtering using a passive angle-only sensor. *IEEE Transactions on Aerospace and Electronic Systems*, 60(6):8539–8554, 2024.

32 P. E. Kloeden and E. Platen. *Numerical Solution of Stochastic Differential Equations*, Springer Verlag, 1999.

33 S. Julier, J. Uhlmann, and H. F. Durrant-Whyte. A new method for the nonlinear transformation of means and covariances in filters and estimators. *IEEE Transactions on Automatic Control*, AC-45(3): 477–482, 2000.

34 K. Li, G. Zhou, and T. Kirubarajan. A general model-based filter initialization approach for linear and nonlinear dynamic systems. *Digital Signal Processing*, 111:102978, 2021.

35 W. H. Press. *Numerical Recipes: The Art of Scientific Computing*. Cambridge University Press, 3rd Edition, 2007.

36 X. R. Li, Z. Zhao, and V. P Jilkov, Estimator's credibility and its measures. *In Proceedings of the IFAC 15th World Congress*, Barcelona, Spain, 2002.

37 T. R. Kronhamn. Bearings-only target motion analysis based on a multihypothesis Kalman filter and adaptive ownship motion control. *IEE Proceedings - Radar, Sonar and Navigation*, 145(4):247–252, 1998.

38 N. Peach. Bearings-only tracking using a set of range-parameterised extended Kalman filters. *IEEE Proceedings-Control Theory and Applications*, 142(1): 73–80, 1995.

39 R. Sittler, J. Ru, S. Sivananthan, and D. Vader. A new algorithm of radar to ADS-B registration *Air Traffic Control Quarterly*, 19(3):191, 2011.

2

Reentry Vehicle Filtering Using Radar and Passive Angle-only Sensors

Mahendra Mallick[1], Xiaoqing Tian[2], and Linfeng Xu[3]

[1] *Independent Consultant, Cambria, CA, USA*
[2] *School of Electrical and Control Engineering, Xi'an University of Science and Technology, Xi'an, China*
[3] *School of Automation, Northwestern Polytechnical University, Xi'an, China*

2.1 Introduction

The ballistic missile tracking problem has been studied extensively over the last few decades [1–12]. Studying the continuous-time dynamic model represented by a stochastic differential equation (SDE) [13–16], presents many theoretical and computational challenges. Second, it has great military importance. Ballistic projectiles [17, 18] and ballistic missiles are examples of ballistic targets. Forces due to Earth's gravity, atmospheric drag, and thrust (in the non-ballistic segment of the trajectory) act on ballistic targets. The dynamic model is highly nonlinear in the target state and is relatively harder compared with the kinematic models commonly used in target tracking [19, 10].

Ballistic projectiles, such as mortars and howitzers [17], travel near Earth's surface characterized by a short range, brief flight duration, and low apogee. Consequently, a constant acceleration due to gravity is sufficient for modeling the gravitational force. In contrast, ballistic missile trajectories involve a long range and extended flight duration, typically comprising three segments: boost, ballistic (or coast), and reentry [1, 20]. Commonly, Newton's second law is integrated in an earth-centered earth-fixed (ECEF) coordinate frame (the E frame). While Newton's second law is strictly valid in an inertial frame where only real forces are considered [21], the E frame is non-inertial due to Earth's rotation. Therefore, when integrating the equations of motion in the E frame, the non-real centrifugal and Coriolis forces [22] are included [1, 5, 6, 20]. For modeling the gravitational force, either a spherical [1] or an ellipsoidal [5, 8] model of Earth is employed. To account for atmospheric disturbances, process noise is introduced into the continuous-time dynamic model [1, 5, 6, 23].

A non-maneuvering ballistic missile during the reentry phase is termed a reentry vehicle (RV). Since a radar measures the range, azimuth, and elevation of a target, the RV filtering problem using a radar is relatively simpler compared with that using a passive angle-only sensor that measures the azimuth and elevation. With range, azimuth, and elevation measurements, we can determine the three-dimensional (3D) Cartesian position at a measurement time. This is not possible for a passive angle-only sensor. Consequently, many papers have been published on RV filtering using a radar [3–6, 8, 23, 9, 10, 12], while only a limited number of publications are available on RV filtering with an angle-only sensor [2, 11, 24]. Furthermore, the state of an RV is observable with a stationary radar, whereas a passive sensor must maneuver to achieve observability [24].

Within the Bayesian framework, the motion of an RV is typically modeled by an SDE [1, 5, 6, 23, 9, 10, 12]. In the RV filtering problem, the dynamic model is continuous in time, whereas the measurement model is discrete in time. This class of problems is known as the continuous discrete filtering (CDF) problem [13–16]. An SDE can be rigorously described by the Itô SDE [14, 16].

In this chapter, we discuss the RV filtering problem using a stationary radar and a maneuvering passive sensor. The work on the passive sensor scenario is based on our previously published paper [24], while the work on the radar scenario is new. The force models include gravity of a spherical-Earth model and atmospheric drag using the 1976 US standard atmospheric density model. For this problem, the centrifugal and Coriolis forces are significantly smaller than the drag and gravity and thus, these two forces are not considered in the equation of motion. The state of the RV is defined by the Cartesian position vector, velocity, and a drag parameter. The equation of motion is integrated in a topographic coordinate frame (TCF), denoted as the T frame fixed to Earth's surface. Although the E frame could have been used, interpreting the simulation and filtering results is simpler in the T frame due to the smaller values of the position vector components compared to thousands of kilometers in the E frame.

Conventional extended Kalman filter (EKF) [19, 25], unscented Kalman filter (UKF) [26], cubature Kalman filter (CKF) [27] and particle filter (PF) [28] use discrete-time dynamic and measurement models. Hence, these algorithms cannot be applied to the current CDF problem. In [4, 8, 12], continuous-time dynamic model and process noise model are not presented and discrete-time dynamic models are not rigorously derived. Given the conditional mean and covariance at time t_{k-1} after processing measurements up to time t_{k-1}, our objective is to propagate these statistics to time t_k.

The continuous white noise acceleration (CWNA) model (or nearly constant velocity (NCV) model with a small power spectral density), continuous Wiener process acceleration (CWPA) model (or nearly constant acceleration (NCA) model), and continuous nearly constant turn (NCT) model [19, 29, 30] are commonly used in filtering and tracking problems. A fundamental problem with the white noise is that it is discontinuous almost everywhere, which violates the requirements for functions within ordinary differential equations (ODEs) [16]. This problem can be resolved by expressing the SDEs for these motion models in the Itô form. The SDEs for these problems are linear and time-invariant. Therefore, the state transition matrix (STM) can be expressed as a matrix exponential [19, 25], which simplifies to closed-form analytic expressions [19, 31, 29]. An exact integral equation for the discrete-time process noise covariance matrix for a linear time-invariant system is provided in [32]. Leveraging the closed-form analytic expressions for the STM, the process noise covariance matrix can be calculated exactly, yielding closed-form analytic expressions for these models. Interestingly, the results obtained using the heuristic white noise-based models in [19, 29, 30] coincide with the corresponding results derived using the Itô formula [16]. However, for nonlinear SDEs, such as those encountered in the current RV filtering problem, the heuristic white noise-based approach will yield incorrect results [16].

Two approaches are prevalent for the prediction step in nonlinear filtering using continuous-time stochastic dynamic models. The first involves deriving approximate ODEs for the conditional mean and covariance over the time interval $[t_{k-1}, t_k)$, which are then numerically integrated from time t_{k-1} to time t_k to obtain the conditional mean and covariance at time t_k [25, 14, 33, 15]. The second approach discretizes the continuous-time dynamic model by dividing the interval $[t_{k-1}, t_k)$ into subintervals and applying a 1.5 order Itô-Taylor expansion (IT-1.5) [13] or second-order stochastic Taylor series approximations [34, 35, 36]. Subsequently, a standard EKF, UKF, CKF, or PF can be employed to obtain the predicted conditional mean and covariance at time t_k. The first method requires calculation of the Jacobian of the drift function in the SDE and the second method requires calculation of the Jacobian and Hessian of the drift function. For complex drift functions, computing these derivatives can be cumbersome and prone to errors.

The accurate continuous-discrete EKF (ACD-EKF) based on the first method [33] and the CD-UKF and CD-CKF based on the IT-1.5 were evaluated for the nearly constant turn (NCT) motion in the challenging air traffic control (ATC) problem, where the target state was seven-dimensional. This evaluation revealed that the accumulated root-mean-square errors for 3D position, 3D velocity, and angular rate were consistently lower for the ACD-EKF compared to the CD-UKF and CD-CKF. Consequently, the authors in [33] concluded that the ACD-EKF exhibits greater flexibility and robustness. Motivated by this observation, we employ a continuous EKF (CEKF) step for the prediction of the conditional mean and covariance from time t_{k-1} to time t_k and the CKF measurement update step is used to calculate the posterior state estimate using the measurement at time t_k. It is important to note that the drift function for the RV filtering problem, considering atmospheric drag and spherical-Earth gravity, is significantly more complex than that for the NCT filtering problem in the ATC scenario [13, 33].

Mehra in [8] used an ODE without process noise and discretized it using the second order Taylor series (TS2) expansion. A discrete-time process noise was then added to the resulting discrete-time equation in an ad hoc manner. This same heuristic approach was also adopted for ballistic projectile filtering in [17]. In [18], an attempt was made to discretize an SDE with continuous-time white noise using the TS2 expansion. However, this derivation contains errors, the specifics of which are detailed in [37]. Despite the different starting points, the final discretized equations presented in [8, 17, 18] are basically the same, and a standard EKF is then applied to calculate the predicted conditional mean and covariance at time t_k. We refer to this overall prediction algorithm as the TS2-EKF. We have shown in [24] that the discrete-time process noise is different from that for the 3D NCV motion.

In [4, 12], a discrete-time dynamic model based on the piece-wise constant acceleration (PCA) approximation is used for RV filtering. However, the discrete-time process noise is for the nearly constant velocity (NCV) model. For filter evaluation in this chapter, we use a standard EKF prediction step, and we refer to the resulting prediction algorithm as the PCA-EKF.

Features of existing filtering algorithms for the RV filtering problem using a radar and a passive sensor are summarized in Tables 2.1 and 2.2, respectively.

Table 2.1 Radar scenario: characteristics of existing estimators and filters.

Feature	Algorithm in [8]	Algorithm in [4]	Our algorithm
Algorithm	EKF	EKF, CADET, UKF, PF	CEKF-KF
RV state	Cartesian	Cartesian	Cartesian
Model of RV	Point mass	Point mass	Sharp-nose projectile
Dynamic model	Stochastic	Stochastic	Stochastic
Atmospheric density model	Not known	1976 US standard atmosphere	1976 US standard atmosphere
Gravity model	Ellipsoidal–Earth	Flat-Earth	Spherical-Earth
Centrifugal and Coriolis forces	Present	Absent	Absent
Process noise	Present	Present	Present
Measurement model	Range, azimuth, elevation	Range, elevation	Range, azimuth, elevation

Table 2.2 Passive sensor scenario: characteristics of existing estimators and filters.

Feature	Algorithm in [2]	Algorithm in [11]	Algorithm in [24]
Algorithm	Iterated LS	MLE	CEKF-CKF
RV state	Cartesian	Cartesian	Cartesian
Model of RV	Point mass	Sharp-nose projectile	Sharp-nose projectile
Dynamic model	Deterministic	Deterministic	Stochastic
Atmospheric density model	Absent	Incorrect model	1976 US standard atmosphere
Gravity model	Spherical-Earth	Flat-Earth	Spherical-Earth
Process noise	Absent	Absent	Present
Measurement model	Azimuth, elevation	Azimuth, elevation	Azimuth, elevation

This chapter contains the following key contributions of our work:

1. The prediction step of the passive sensor RV filtering problem, using the Itô SDE was first solved in [24]. This algorithm is applied to the radar RV filtering problem for the first time in this chapter.
2. In [24], we also derived an analytic expression for the discrete-time process noise covariance using the IT-1.5 [13], which is valid for a general SDE. We demonstrate that the process covariance differs from that derived for the NCV model [19], as used in some RV filtering and related problems [4, 17, 12, 18].
3. For the radar filtering scenario, we present a general filter initialization algorithm using the CA motion.

Remaining sections are organized as follows. Section 2.2 presents the equation of motion of the RV, and Section 2.3 describes the Itô SDE. The measurement models for the radar and passive sensor are presented in Section 2.4. Range-parametrization is explained in Section 2.5. Filter initialization using the radar and passive sensor measurements is described in Sections 2.6 and 2.7, respectively. Three filtering algorithms are explained in Section 2.8. Numerical simulations and results are presented in Section 2.9 and conclusions and future work are discussed in Section 2.10.

Notation Convention:
An italic letter represents a scalar and a boldface letter represents a vector or a matrix. We use lower or uppercase Roman letters for a name (e.g., "g" for "gravity," "d" for "drag," "RV" for "reentry vehicle," etc.). The symbol ":=" is used to define a quantity. For a vector or matrix $\mathbf{A}$, $\mathbf{A}'$ denotes the transpose of the variable. We represent the n-dimensional identity matrix, m-dimensional zero matrix, and $m \times n$ zero matrix by $\mathbf{I}_n$, $\mathbf{0}_m$, and $\mathbf{0}_{m\times n}$, respectively.

2.2 Equation of Motion

In an inertial coordinate frame, the motion of macroscopic objects such as projectiles, missiles, satellites, etc. obeys Newton's second law [21], which states that for fixed mass, mass times acceleration is equal to the total external force acting on the object.

Given the position **r** and velocity **v** of the RV, the equation of motion of the RV with mass m in the E frame is given by [5, 6, 8, 23, 38]

$$m\frac{d\mathbf{v}}{dt} = \mathbf{F}_g + \mathbf{F}_d - m\boldsymbol{\omega} \times (\boldsymbol{\omega} \times \mathbf{r}) - 2m(\boldsymbol{\omega} \times \mathbf{v}), \tag{2.1}$$

where $\mathbf{F}_g$ and $\mathbf{F}_d$ are the forces due to gravity and drag, respectively, and $\boldsymbol{\omega}$ is Earth's angular velocity about the axis of rotation. The pseudo forces (not real) $-m\boldsymbol{\omega} \times (\boldsymbol{\omega} \times \mathbf{r})$ and $-2m(\boldsymbol{\omega} \times \mathbf{v})$ represent the centrifugal and Coriolis forces in the rotating E frame, respectively [22].

Remark 2.1 Although some authors use "centripetal force" in place of "centrifugal force", the correct usage is "centrifugal force" [22]. We assume Earth to be a sphere of constant density with mass M_e and radius R_e. Then

$$\mathbf{F}_g = -m\mu_e \frac{\mathbf{r}}{r^3}, \tag{2.2}$$

where $\mu_e = GM_e$, G denotes the universal gravitational constant, and r denotes the magnitude of **r**. Given the position vector **r**, the altitude of the RV is calculated as $a = r - R_e$. The drag force is given by [6, 38]

$$\mathbf{F}_d = -\frac{1}{2} C_d A_{ref} \rho(a) \mathbf{v} \parallel \mathbf{v} \parallel, \tag{2.3}$$

where C_d, A_{ref}, and $\rho(a)$ are the drag coefficient, reference cross-sectional area of the RV, and atmospheric density at altitude a. The drag coefficient depends on the geometric shape of the RV [39] and its speed. The drag coefficient as a function of the Mach number ($N_M := \parallel \mathbf{v} \parallel / c_0$), where c_0 is the speed of sound, is shown in [39] for a cylinder, sphere, round nose projectile, and a sharp nose projectile. As in [18, 11], we model the RV as a sharp nose projectile. From (2.1) we get

$$\frac{d\mathbf{v}}{dt} = -\mu_e \frac{\mathbf{r}}{r^3} - \frac{1}{2} \alpha_d \rho(a) \mathbf{v} \parallel \mathbf{v} \parallel - \boldsymbol{\omega} \times (\boldsymbol{\omega} \times \mathbf{r}) - 2\boldsymbol{\omega} \times \mathbf{v}, \tag{2.4}$$

where $\alpha_d = C_d A_{ref} / m$ is a drag parameter [6]. The first, second, third, and fourth terms on the right-hand side of (2.4) represent the gravitational, drag, centrifugal and Coriolis accelerations, respectively. The ballistic coefficient $\beta = 1/\alpha_d$ is also used in some papers [6, 23]. We prefer to use the drag parameter, since the drag acceleration is linear in the drag parameter. The drag acceleration can be expressed as

$$\mathbf{a}_d = -\frac{1}{2} \alpha C_d \rho(a) \mathbf{v} \parallel \mathbf{v} \parallel, \tag{2.5}$$

where $\alpha = A_{ref} / m$ is a specific drag coefficient. We model α as a nearly constant random variable and C_d is used from [39] for a sharp nose projectile.

The atmospheric density models used in [4, 23] are nearly the same and are based on the 1976 US standard atmosphere model. This atmospheric density model at an altitude a is described by an exponential model [23]

$$\rho(a) = \rho_0 \exp(-\kappa a), \tag{2.6}$$

where ρ_0 is the density at the surface of the earth and κ is a constant. Numerical values of the parameters are presented in Table 2.3. We refer to this model as Model 1 and use it in this study.

Three scenarios for RV tracking are considered in [11]. Scenarios 2 and 3 have the same initial RV states. To validate our results, we first generated two deterministic true trajectories using the initial state vectors from [11] for Scenarios 1 and 2 with the flat-Earth gravity model and atmospheric drag model based on Model 1. The fourth order Runge–Kutta integrator [40] was used to integrate the equation of motion. We observed that large differences exist between our true trajectories and the corresponding ones in [11]. A close examination showed that large differences are due to differences in the atmospheric density models. An exponential atmospheric density model is also used in [1] derived from [17]. We call this Model 2, whose parameters are presented in Table 2.3. We find from Table 2.3 that ρ_0 for Model 2 has a very low value of 0.002378 kg/m^3. Actually, $\rho_0 = 0.002378\,\text{slug/ft}^3$ [39]. If we convert this to the meter, kilogram, second (MKS) system of units, then we get $\rho_0 = 1.2256\,\text{kg/m}^3$, which close to ρ_0 of Model 1 for altitudes less than 9144 m. We were able to reproduce the incorrect true trajectories in [11] using Model 2.

Let $\mathbf{r}^{\mathrm{E}}$, $\mathbf{v}^{\mathrm{E}}$, and $\boldsymbol{\omega}^{\mathrm{E}}$ denote 3×1 column vectors in the E frame. The direction of the angular velocity is along the Z-axis of the E frame. Then,

$$\mathbf{r}^{\mathrm{E}} := \begin{bmatrix} r_x & r_y & r_z \end{bmatrix}', \tag{2.7}$$

$$\mathbf{v}^{\mathrm{E}} := \begin{bmatrix} \dot{r}_x & \dot{r}_y & \dot{r}_z \end{bmatrix}', \tag{2.8}$$

$$\boldsymbol{\omega}^{\mathrm{E}} = \begin{bmatrix} 0 & 0 & \omega \end{bmatrix}'. \tag{2.9}$$

The centrifugal and Coriolis accelerations in the E frame are expressed, respectively, by

$$-(\boldsymbol{\omega} \times (\boldsymbol{\omega} \times \mathbf{r}))^{\mathrm{E}} = \omega^2 \begin{bmatrix} r_x & r_y & 0 \end{bmatrix}', \tag{2.10}$$

$$-(2\boldsymbol{\omega} \times \mathbf{v})^{\mathrm{E}} = 2\omega \begin{bmatrix} \dot{r}_y & -\dot{r}_x & 0 \end{bmatrix}'. \tag{2.11}$$

Table 2.4 compares the magnitudes of various accelerations at the initial time for two scenarios in [11]. Initial positions and velocities of the RV in Scenarios 2 and 3 are the same. The drag acceleration has the highest magnitude among all accelerations and is opposite to the direction of target velocity. The second strongest acceleration is the acceleration due to gravity, and it is directed

Table 2.3 Atmospheric density model parameters.

Model	a (m)	ρ_0 (kg/m^3)	κ(m^{-1})
Model 1	< 9144 m	1.2256	1/9144.0
Model 1	≥ 9144 m	1.7523	1/6705.6
Model 2	all altitudes	2.3780×10^{-3}	1/9144.0

Table 2.4 Comparison of various accelerations.

Scenario	Speed (Mach)	Drag (g)	Gravity/Drag	Centrifugal/Drag	Coriolis/Drag
1	4.2	15.2	0.06	2.0×10^{-4}	3.6×10^{-4}
2	6.9	33.7	0.03	8.9×10^{-5}	6.6×10^{-4}

towards Earth's center. The magnitude acceleration due to gravity is about 15 and 34 times smaller than that due to drag for scenarios 1 and 2 (or 3). The magnitudes of the centrifugal and Coriolis accelerations are about four orders of magnitude smaller than that of the drag acceleration. Therefore, we have not considered the centrifugal and Coriolis accelerations in this study.

2.3 Itô Stochastic Differential Equation

We perform filtering in a topographic frame (T frame) as shown in Figure 2.1. The origin of the T frame is at a given longitude and latitude on Earth's surface. The X, Y, and Z axes of the T frame represent the local east, north, and upward directions, respectively. Let $\mathbf{p}$ and $\mathbf{v}$ be the 3D position vector and 3D velocity of the RV from the T frame origin, respectively. Since the T frame is fixed relative to the E frame, the velocity of the RV is the same in these two frames. For simplicity in notation, we have omitted the superscript T from the vectors and treat them as 3×1 column vectors. Let $\mathbf{r}$ and $\mathbf{r}_0$ be the position vector of the RV and T frame origin from Earth's center. Then

$$\mathbf{r} = \mathbf{r}_0 + \mathbf{p}, \tag{2.12}$$

$$\mathbf{v} = \dot{\mathbf{r}} = \dot{\mathbf{p}}. \tag{2.13}$$

We estimate the kinematic state of the RV and α in the filter. The RV state at time t is defined by

$$\mathbf{x}(t) := \begin{bmatrix} \mathbf{p}'(t) & \mathbf{v}'(t) & \alpha(t) \end{bmatrix}'. \tag{2.14}$$

We model $\alpha(t)$ as a scalar Brownian motion process with power spectral density (PSD) q_α [14]

$$d\alpha(t) = \sqrt{q_\alpha} d\beta_\alpha(t), \tag{2.15}$$

where $\beta_\alpha(t)$ is a standard Wiener process [14, 16]

$$E\{d\alpha(t)d\alpha(t)\} = q_\alpha dt. \tag{2.16}$$

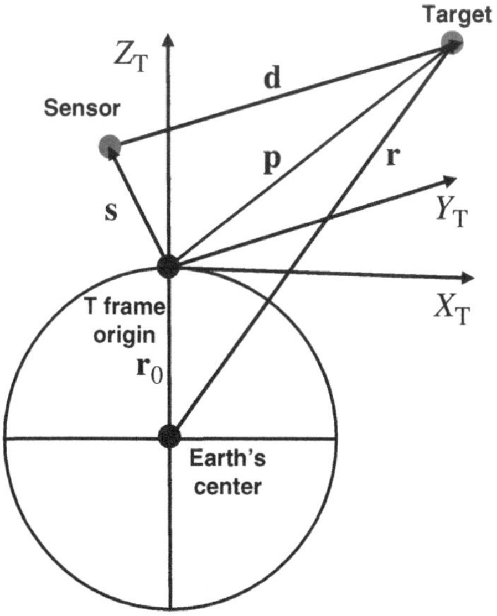

Figure 2.1 Definition of the T frame and target-sensor geometry.

To handle perturbations in Earth's atmosphere, we introduce the acceleration process noise. The Itô SDE [14, 16] for the target state is described by

$$d\mathbf{x}(t) = \mathbf{f}(\mathbf{x}(t))dt + \mathbf{G}d\boldsymbol{\beta}(t), \tag{2.17}$$

where

$$\mathbf{f}(\mathbf{x}) = \begin{bmatrix} \mathbf{v} \\ -\mu_e \frac{\mathbf{r}}{r^3} - \frac{1}{2}\alpha C_d \rho(a)\mathbf{v} \parallel \mathbf{v} \parallel \\ 0 \end{bmatrix}, \tag{2.18}$$

$$d\boldsymbol{\beta}(t) = \begin{bmatrix} d\beta_x(t) & d\beta_y(t) & d\beta_z(t) & d\beta_\alpha(t) \end{bmatrix}', \tag{2.19}$$

$$\mathbf{G} = \begin{bmatrix} \mathbf{0}_{3\times 4} \\ \mathbf{G_q} \end{bmatrix}, \tag{2.20}$$

$$\mathbf{G_q} = \text{diag}(\sqrt{q_x}, \sqrt{q_y}, \sqrt{q_z}, \sqrt{q_\alpha}). \tag{2.21}$$

The terms $\mathbf{f}(\mathbf{x}(t))$ and $\mathbf{G}d\boldsymbol{\beta}(t)$ in (2.17) are known as the drift function and dispersion term, respectively. In (2.17), $d\boldsymbol{\beta}(t)$ represents an increment of the standard Wiener process [14, 16], whose components are defined in (2.19). The increments $d\beta_x(t)$, $d\beta_y(t)$, and $d\beta_z(t)$ correspond to the acceleration process noises along the X, Y, and Z axes respectively. The second row of $\mathbf{f}$ in (2.18) represents the total acceleration $\mathbf{a}$ given by

$$\mathbf{a} = \mathbf{a}_g + \mathbf{a}_d. \tag{2.22}$$

We note that the function $\mathbf{f}$ is differentiable with respect to $\mathbf{x}$. The effective PSD matrix $\mathbf{Q}$ of the process noise is given by

$$\mathbf{Q}dt = \mathbf{G}E\{d\boldsymbol{\beta}(t)d\boldsymbol{\beta}'(t)\}\mathbf{G}'. \tag{2.23}$$

From (2.23) we get

$$\mathbf{Q} = \mathbf{G}\mathbf{G}' = \text{diag}(0, 0, 0, q_x, q_y, q_z, q_\alpha). \tag{2.24}$$

We assume that the prior distribution of $\mathbf{x}$ is Gaussian,

$$\mathbf{x}(0) \sim \mathcal{N}(\bar{\mathbf{x}}_0, \mathbf{P}_0). \tag{2.25}$$

2.4 Sensor Measurement Models

The passive sensor measures the azimuth and elevation, and the radar measures the range, azimuth, and elevation of the RV at discrete times. We assume that the measurement time interval T is constant. The angles are expressed in the T frame. The range vector of the target from the sensor at time t_k is given by

$$\mathbf{d}_k = \mathbf{p}_k - \mathbf{s}_k, \tag{2.26}$$

where $\mathbf{s}_k$ is the 3D sensor position at time t_k. The radar is assumed to be stationary at $\mathbf{s}$. Next, we describe the measurement models of these sensors.

2.4.1 Radar Measurement Model

The measurement model is described by [41]

$$\mathbf{z}_k = \mathbf{h}(\mathbf{p}_k, \mathbf{s}) + \mathbf{n}_k, \tag{2.27}$$

where

$$\mathbf{h}(\mathbf{p}_k, \mathbf{s}) := \begin{bmatrix} d_k \\ \varphi_k \\ \epsilon_k \end{bmatrix} = \begin{bmatrix} \| \mathbf{d}_k \| \\ \tan^{-1}(d_{k,x}, d_{k,y}) \\ \arctan(d_{k,z}/\rho_k) \end{bmatrix}, \tag{2.28}$$
$$d_k > 0, \varphi_k \in [0, 2\pi), \epsilon_k \in (-\pi/2, \pi/2),$$

where d_k, φ_k, and ϵ_k are the range, azimuth, and elevation, respectively, and ρ_k is the ground range given by

$$\rho_k = \sqrt{d_{k,x}^2 + d_{k,y}^2}. \tag{2.29}$$

The measurement noise is assumed as zero-mean Gaussian with covariance $\mathbf{R}$. Let σ_r, σ_φ and σ_ϵ denote the standard deviations (SDs) of range, azimuth, and elevation measurements, respectively. Then

$$\mathbf{n}_k \sim \mathcal{N}(\mathbf{0}, \mathbf{R}), \quad \mathbf{R} = \text{diag}(\sigma_r^2, \sigma_\varphi^2, \sigma_\epsilon^2). \tag{2.30}$$

2.4.2 Passive Sensor Measurement Model

To observe the state of the RV, it is necessary for the passive sensor to perform maneuver. The 3D trajectory of the passive sensor is assumed to be deterministic. The motion of the sensor consists of a sequence of constant velocity (CV) and constant turn (CT) [19]. The motion profile of the sensor is presented in Table 2.9 of Section 2.9.

The state of the passive sensor is composed of the 3D position $\mathbf{s}_k$ and 3D velocity $\dot{\mathbf{s}}_k$,

$$\mathbf{x}_k^{\mathrm{s}} := \begin{bmatrix} \mathbf{s}_k' & \dot{\mathbf{s}}_k' \end{bmatrix}'. \tag{2.31}$$

The CV motion model of the sensor in 3D is modeled as [19]

$$\mathbf{x}_k^{\mathrm{s}} = \mathbf{F}_{k-1}\mathbf{x}_{k-1}^{\mathrm{s}}, \tag{2.32}$$

where the STM for CV, $\mathbf{F}_{k-1}$, is given by,

$$\mathbf{F}_{k-1} = \begin{bmatrix} \mathbf{I}_3 & \mathbf{I}_3 T \\ \mathbf{0}_{3\times3} & \mathbf{I}_3 \end{bmatrix}. \tag{2.33}$$

In our study, the sensor is assumed to move with CV along the $X-$axis and with CT in the $YZ-$plane of the T frame during the CT motion. Let $\Omega_{k-1}^{\mathrm{s}}$ denote angular rate of the sensor during $[t_{k-1}, t_k)$. The dynamic model for the CT motion is [19, 42]

$$\mathbf{x}_k^{\mathrm{s}} = \mathbf{F}^{\mathrm{CT}}(\Omega_{k-1}^{\mathrm{s}})\mathbf{x}_{k-1}^{\mathrm{s}}, \tag{2.34}$$

where the corresponding state transition matrix is specified by [19, 42]

$$\mathbf{F}^{\mathrm{CT}}(\Omega) = \begin{bmatrix} 1 & 0 & 0 & T & 0 & 0 \\ 0 & 1 & 0 & 0 & \sin(\Omega T)/\Omega & -[1-\cos(\Omega T)]/\Omega \\ 0 & 0 & 1 & 0 & [1-\cos(\Omega T)]/\Omega & \sin(\Omega T)/\Omega \\ 0 & 0 & 0 & 1 & 0 & 0 \\ 0 & 0 & 0 & 0 & \cos(\Omega T) & -\sin(\Omega T) \\ 0 & 0 & 0 & 0 & \sin(\Omega T) & \cos(\Omega T) \end{bmatrix}. \tag{2.35}$$

The measurement model is described by [42]

$$\mathbf{z}_k = \mathbf{h}(\mathbf{p}_k, \mathbf{s}_k) + \mathbf{n}_k, \tag{2.36}$$

where

$$\mathbf{h}(\mathbf{p}_k, \mathbf{s}_k) := \begin{bmatrix} \varphi_k \\ \epsilon_k \end{bmatrix} = \begin{bmatrix} \tan^{-1}(d_{k,x}, d_{k,y}) \\ \arctan\left(d_{k,z}/\rho_k\right) \end{bmatrix}, \tag{2.37}$$
$$\varphi_k \in [0, 2\pi),\ \epsilon_k \in (-\pi/2, \pi/2),$$

$$\rho_k = \sqrt{d_{k,x}^2 + d_{k,y}^2}. \tag{2.38}$$

The measurement noise is assumed as zero-mean Gaussian with covariance $\mathbf{R}$. Let σ_φ and σ_ϵ denote the standard deviations (SDs) of azimuth and elevation measurement, respectively. Then

$$\mathbf{n}_k \sim \mathcal{N}(\mathbf{0}, \mathbf{R}), \quad \mathbf{R} = \mathrm{diag}(\sigma_\varphi^2, \sigma_\epsilon^2). \tag{2.39}$$

2.5 Range-parametrization (RP)

The RV state is seven-dimensional and the passive sensor measurement is two-dimensional (2D). Therefore, prior information is used to initialize the filter at initial time t_1 using the first measurement $\mathbf{z}_1$. We use the range-parametrization (RP) approach [43, 44] to determine the initial Cartesian position estimate $\hat{\mathbf{p}}_1$ using the first measurement. Let $d_{\min}$ and $d_{\max}$ denote the minimum and maximum ranges, respectively, and $\{d_i\}_{i=0}^{N_d}$ denote the boundary values of the subintervals with $d_0 = d_{\min}$ and $d_{N_d} = d_{\max}$.

We assume that $\{d_i\}_{i=0}^{N_d}$ form a geometric sequence [43, 44],

$$d_0, d_1 = d_0\eta, d_2 = d_0\eta^2, \ldots, d_{N_d} = d_0\eta^{N_d}, \tag{2.40}$$

where η is the common ratio given by

$$\eta = \left(\frac{d_{N_d}}{d_0}\right)^{1/N_d} = \left(\frac{d_{\max}}{d_{\min}}\right)^{1/N_d}. \tag{2.41}$$

Let $\bar{d}_i$ and Δd_i denote the mean range and range interval for the ith subinterval, respectively,

$$\bar{d}_i = (d_{i-1} + d_i)/2, \quad \Delta d_i = d_i - d_{i-1}, \quad i = 1, 2, \ldots, N_d. \tag{2.42}$$

We assume that the range is normally distributed in a subinterval and the subinterval has a 0.999 probability of containing the range. Thus, a subinterval represents a 6.581 sigma interval. The normal distribution of range in a subinterval is required for polar to Cartesian coordinate conversion

[45, Chapter 2] using the Taylor series expansion [19, 25], unscented transformation [26], or cubature transformation [27]. We use the cubature transformation [27]. Detailed explanations are presented below. Hence, the SD of range in the ith subinterval is given by

$$\sigma_i = \frac{\Delta d_i}{6.581}, \quad i = 1, 2, \ldots, N_d. \tag{2.43}$$

Remark 2.2 Conventionally, the range is assumed to be uniformly distributed in a subinterval and the denominator in (2.43) is $\sqrt{12}$. If the uniform distribution in the range subinterval is used, then the required normal distribution for polar coordinates to Cartesian coordinate is violated.

The initial probabilities or weights are calculated by [44]

$$w_i = \Delta d_i/(d_{\max} - d_{\min}), \quad i = 1, 2, \ldots, N_d. \tag{2.44}$$

For each subinterval, the mean of the azimuth and elevation are set to $z_{1,\varphi}$ and $z_{1,\epsilon}$, respectively, and the azimuth and elevation measurement SDs are chosen as σ_φ and σ_ϵ. Now we have the polar mean and SD for range, azimuth, and elevation for each subinterval. The polar mean and covariance in each subinterval is converted to the 3D Cartesian mean range vector and associated covariance using the cubature transformation [27].

Let $\hat{\mathbf{d}}_{1|1,i}$ and $\mathbf{P}_{1|1,i}$ denote the 3D Cartesian range vector estimate and associated 3×3 covariance for the ith subinterval. Then the 3D Cartesian position vector estimate for the ith interval is given by

$$\hat{\mathbf{p}}_{1|1,i} = \mathbf{s}_1 + \hat{\mathbf{d}}_{1|1,i}. \tag{2.45}$$

The initial 3D position estimate and associated covariance are obtained, respectively, by [19]

$$\hat{\mathbf{p}}_{1|1} = \sum_{i=1}^{N_d} w_i \hat{\mathbf{p}}_{1|1,i}, \tag{2.46}$$

$$\mathbf{P}_{1|1} = \sum_{i=1}^{N_d} w_i [\mathbf{P}_{1|1,i} + (\hat{\mathbf{p}}_{1|1,i} - \hat{\mathbf{p}}_{1|1})(\hat{\mathbf{p}}_{1|1,i} - \hat{\mathbf{p}}_{1|1})']. \tag{2.47}$$

Next we discuss initialization of the Cartesian velocity and α using the minimum and maximum available prior information of a quantity [45, Chapter 6], [46]. The initial Z component of velocity in the T frame is negative. Therefore, we choose negative values for its minimum and maximum values, $v_{z,\min}$ and $v_{z,\max}$, respectively. Then the SD of initial v_z is $\sigma_{v_z} = (v_{z,\max} - v_{z,\min})/6.581$. The initial X and Y components of velocity in the T frame can be positive or negative. Assuming that $\dot{x}$ is uniformly distributed in $[-v_{x,\max}, v_{x,\max}]$, the mean and SD of $\dot{x}$ are zero and $2v_{x,\max}/6.581$. We use $v_{y,\max} = v_{x,\max}$ and calculate the mean and SD of $\dot{y}$. Similarly, we choose an interval $[\alpha_{\min}, \alpha_{\max}]$ for α with mean and SD $(\alpha_{\min} + \alpha_{\max})/2$ and $(\alpha_{\max} - \alpha_{\min})/6.581$, respectively. To make the simulation results robust, in each Monte Carlo run, we draw sample from these distributions and set it to the initial estimate.

2.6 Filter Initialization Using Radar Measurements

The state of the RV state is seven-dimensional. The radar measurement at a single measurement time has three scalar measurements. Thus, radar measurement at time t_1 is not sufficient to initialize the RV state at t_1. We can convert the range, azimuth, and elevation measurements to 3D Cartesian position using the standard conversion (SC) [30], unscented transformation [26], or cubature transformation [27]. Researchers often use the two-point difference (TPD) filter initialization [19] to calculate the position and velocity estimates at t_2. However, this method strictly applies to the constant velocity motion. It is important to know the type of motion of the target and then apply the appropriate filter initialization algorithm as discussed in [47].

The components and magnitude of acceleration of the RV are shown in Figure 2.2. We observe from Figure 2.2 that the acceleration is nearly constant up to 40 s. Thus, a constant acceleration (CA) model can be used to initialize the filter. In the CA motion, the position, velocity, and acceleration are to be estimated at a given time. This requires a minimum of three 3D position measurements. More than three measurements will improve the accuracy of the estimate. It is convenient to estimate the position and velocity at time t_3 and acceleration using converted measurements at t_1, t_2, and t_3. In general, if m measurements are used then the position and velocity at time t_m and acceleration are to be estimated.

Let $\mathbf{p}_3$ and $\mathbf{v}_3$ be the 3D position and velocity at t_3. In [47], an algebraic method is used to obtain estimates of the $\mathbf{p}_3$, $\mathbf{v}_3$, and $\mathbf{a}$ and the associated covariance. We present a simple and direct method to estimate these parameters using the maximum likelihood estimator (MLE). The 3D position at time t for the constant acceleration $\mathbf{a}$ is given by

$$\mathbf{p}(t) = \mathbf{p}_3 + \mathbf{v}_3(t - t_3) + \frac{1}{2}\mathbf{a}(t - t_3)^2. \tag{2.48}$$

At times t_1 and t_2, the above equations can be written as

$$\mathbf{p}_1 = \mathbf{p}_3 + \mathbf{v}_3(t_1 - t_3) + \frac{1}{2}\mathbf{a}(t_1 - t_3)^2, \tag{2.49}$$

$$\mathbf{p}_2 = \mathbf{p}_3 + \mathbf{v}_3(t_2 - t_3) + \frac{1}{2}\mathbf{a}(t_2 - t_3)^2. \tag{2.50}$$

Let θ denote the parameter with components $\mathbf{p}_3$, $\mathbf{v}_3$, and $\mathbf{a}$

$$\theta := [\mathbf{p}_3' \quad \mathbf{v}_3' \quad \mathbf{a}']'. \tag{2.51}$$

We can write the above equations and $\mathbf{p}_3$ in matrix form as

$$\mathbf{y} = \mathbf{H}\theta, \tag{2.52}$$

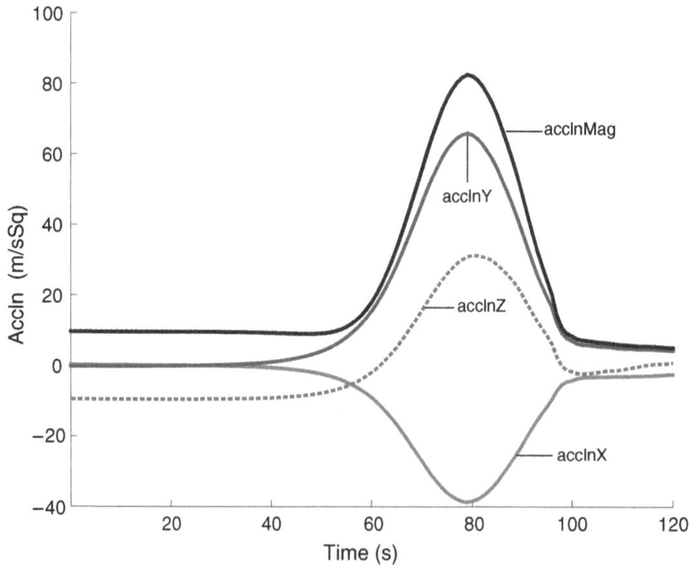

Figure 2.2 Radar Scenario: target acceleration.

where

$$\mathbf{y} = [\mathbf{p}_1' \quad \mathbf{p}_2' \quad \mathbf{p}_3']', \tag{2.53}$$

$$\mathbf{H} = \begin{bmatrix} \mathbf{I}_3 & (t_1 - t_3)\mathbf{I}_3 & \frac{1}{2}\mathbf{a}(t_1 - t_3)^2\mathbf{I}_3 \\ \mathbf{I}_3 & (t_2 - t_3)\mathbf{I}_3 & \frac{1}{2}\mathbf{a}(t_2 - t_3)^2\mathbf{I}_3 \\ \mathbf{I}_3 & \mathbf{0}_3 & \mathbf{0}_3 \end{bmatrix}. \tag{2.54}$$

We have noisy converted position measurements $\hat{\mathbf{p}}_i$, $i = 1, 2, 3$. We note that the errors in $\{\hat{\mathbf{p}}_i\}$ are uncorrelated at different times, but not Gaussian. Using the converted measurements we get

$$\mathbf{z} = \mathbf{H}\theta + \mathbf{e}, \tag{2.55}$$

where

$$\mathbf{z} = [\hat{\mathbf{p}}_1' \quad \hat{\mathbf{p}}_2' \quad \hat{\mathbf{p}}_3']', \tag{2.56}$$

$$\hat{\mathbf{p}}_i = \mathbf{p}_i + \mathbf{e}_i, \quad i = 1, 2, 3. \tag{2.57}$$

$$\mathbf{e}_i \sim \mathcal{N}(\mathbf{0}_3, \boldsymbol{\Sigma}_i), \quad E\{\mathbf{e}_i\mathbf{e}_j'\} = \delta_{ij}\boldsymbol{\Sigma}_i, \quad i, j = 1, 2, 3. \tag{2.58}$$

$$\boldsymbol{\Sigma} = \begin{bmatrix} \boldsymbol{\Sigma}_1 & \mathbf{0}_3 & \mathbf{0}_3 \\ \mathbf{0}_3 & \boldsymbol{\Sigma}_2 & \mathbf{0}_3 \\ \mathbf{0}_3 & \mathbf{0}_3 & \boldsymbol{\Sigma}_3 \end{bmatrix}. \tag{2.59}$$

We note that (2.57) is a static linear measurement model with additive uncorrelated noises at different times [48]. The maximum likelihood estimate $\hat{\theta}$ of θ can be obtained assuming Gaussian noises using the singular value decomposition. Obtaining the estimate of θ by solving the normal equation for (2.57) is numerically inefficient. The covariance of $\hat{\theta}$ is given by [48]

$$\mathbf{P}_{\hat{\theta}} = (\mathbf{H}'\boldsymbol{\Sigma}^{-1}\mathbf{H})^{-1}. \tag{2.60}$$

We choose an interval $[\alpha_{\min}, \alpha_{\max}]$ for α_1. Then the mean and SD are $(\alpha_{\min} + \alpha_{\max})/2$ and $(\alpha_{\max} - \alpha_{\min})/6.581$, respectively. We can calculate the initial state estimate $\hat{\mathbf{x}}_{3|3}$ and associated covariance $\mathbf{P}_{3|3}$ at t_3 from $\hat{\theta}$, $\mathbf{P}_{\hat{\theta}}$, and the mean and SD of α_1.

2.7 Filter Initialization Using Passive Sensor Measurements

The RV state is seven-dimensional and the passive sensor measurement is 2D. Therefore, prior information is used to initialize the filter at time t_1 using the first measurement $\mathbf{z}_1$. We use the range-parametrization (RP) approach [43, 44] to determine the initial Cartesian position estimate $\hat{\mathbf{p}}_1$ using the first measurement. Let $d_{\min}$ and $d_{\max}$ denote the minimum and maximum ranges, respectively, and $\{d_i\}_{i=0}^{N_d}$ denote the boundary values of the subintervals with $d_0 = d_{\min}$ and $d_{N_d} = d_{\max}$.

We assume that $\{d_i\}_{i=0}^{N_d}$ form a geometric sequence [43, 44],

$$d_0, d_1 = d_0\eta, d_2 = d_0\eta^2, \ldots, d_{N_d} = d_0\eta^{N_d}, \tag{2.61}$$

where η is the common ratio given by

$$\eta = \left(\frac{d_{N_d}}{d_0}\right)^{1/N_d} = \left(\frac{d_{\max}}{d_{\min}}\right)^{1/N_d}. \tag{2.62}$$

Let $\bar{d}_i$ and Δd_i denote the mean range and range interval for the ith subinterval, respectively,

$$\bar{d}_i = (d_{i-1} + d_i)/2, \quad \Delta d_i = d_i - d_{i-1}, \quad i = 1, 2, \ldots, N_d. \tag{2.63}$$

We assume that the range is normally distributed in the interval and the interval has a 0.999 probability of containing the range. Then an interval represents a 6.581 sigma interval. Hence, the SD of range in the ith subinterval is given by

$$\sigma_i = \frac{\Delta d_i}{6.581}, \quad i = 1, 2, \ldots, N_d. \tag{2.64}$$

Remark 2.3 Conventionally, the range is assumed to be uniformly distributed in an interval and the denominator in (2.64) is $\sqrt{12}$. The initial probabilities or weights are calculated by [44]

$$w_i = \Delta d_i/(d_{\max} - d_{\min}), \quad i = 1, 2, \ldots, N_d. \tag{2.65}$$

For each subinterval, the mean of the azimuth and elevation are set to $z_{1,\varphi}$ and $z_{1,\epsilon}$, respectively. Now we have the polar mean and SD for each subinterval. The polar mean and SD in each subinterval is converted to the Cartesian mean using the Cubature transformation [27] and SD to obtain $\{\hat{\mathbf{d}}_i\}_{i=1}^{N_d}$. Then

$$\hat{\mathbf{p}}_{1|1,i} = \mathbf{s}_1 + \hat{\mathbf{d}}_i, \tag{2.66}$$

$$\hat{\mathbf{p}}_{1|1} = \sum_{i=1}^{N_d} w_i \hat{\mathbf{p}}_{1|1,i}, \tag{2.67}$$

$$\mathbf{P}_{1|1} = \sum_{i=1}^{N_d} w_i [\mathbf{P}_{1|1,i} + (\hat{\mathbf{p}}_{1|1,i} - \hat{\mathbf{p}}_{1|1})(\hat{\mathbf{p}}_{1|1,i} - \hat{\mathbf{p}}_{1|1})']. \tag{2.68}$$

Next we discuss initialization of the Cartesian velocity and α_1. The initial Z component of velocity in the T frame is negative. Therefore, we choose negative values for its minimum and maximum values, $v_{z,\min}$ and $v_{z,\max}$, respectively. Then the mean of initial $v_{1,z}$ is $\bar{v}_{1,z} = (v_{z,\min} + v_{z,\max})/2$ and the SD is $\sigma_{v_{1,z}} = (v_{z,\max} - v_{z,\min})/6.581$. The initial X and Y components of velocity in the T frame can be positive or negative. Assuming that $\dot{x}_1$ is uniformly distributed in $[-v_{x,\max}, v_{x,\max}]$, the mean and SD of $\dot{x}_1$ are zero and $2v_{x,\max}/6.581$. We set the mean and SD of $\dot{y}_1$ to corresponding values of $\dot{x}_1$. We calculate the initial estimate of α_1 and corresponding SD as was done in the radar scenario.

2.8 Filtering Algorithms

We discuss the CDF filtering algorithm and two approximate filtering algorithms using discretized dynamic models. These three algorithms use the same measurement update step for the radar and passive sensor scenarios.

2.8.1 CDF Filtering Algorithm

The CDF algorithm has two steps: prediction and measurement update. Since the measurement model is discrete in time, a conventional measurement update algorithm can be used. Let $\mathbf{Z}_{k-1}$ denote the collection of measurements up to time t_{k-1},

$$\mathbf{Z}_{k-1} := \{\mathbf{z}_1, \mathbf{z}_2, \dots, \mathbf{z}_{k-1}\}. \tag{2.69}$$

We follow the descriptions in [14, 25, 15] for the prediction step of the CDF. We use the approximation of the Gaussian filter [49, 50] in calculating the conditional mean and covariance in the prediction step,

$$p(\mathbf{x}(t)|\mathbf{Z}_{k-1}) \approx \mathcal{N}(\mathbf{x}(t); \mathbf{m}(t), \mathbf{P}(t)), \quad t_{k-1} < t \leq t_k, \tag{2.70}$$

where $\mathbf{m}(t)$ and $\mathbf{P}(t)$ are the conditional mean and covariance at t between the observation times t_{k-1} and t_k. Suppose we have the conditional mean $\hat{\mathbf{x}}_{k-1|k-1}$ and corresponding covariance $\mathbf{P}_{k-1|k-1}$ after processing $\mathbf{Z}_{k-1}$ at time t_{k-1}, where

$$\hat{\mathbf{x}}_{k-1|k-1} = E\{\mathbf{x}_{k-1}|\mathbf{Z}_{k-1}\}, \tag{2.71}$$

$$\mathbf{P}_{k-1|k-1} = E\{(\mathbf{x}_{k-1} - \hat{\mathbf{x}}_{k-1})(\mathbf{x} - \hat{\mathbf{x}}_{k-1})' |\mathbf{Z}_{k-1}\}. \tag{2.72}$$

Our goal is to propagate $\hat{\mathbf{x}}_{k-1|k-1}$ and $\mathbf{P}_{k-1|k-1}$ from time t_{k-1} to time t_k before processing the measurement $\mathbf{z}_k$.

2.8.2 Prediction Algorithm CEKF

For simplicity in notation we omit the time argument from $\mathbf{m}(t)$, $\mathbf{x}(t)$, and $\mathbf{P}(t)$. Then the differential equations for the prediction step are given by [15]

$$\frac{d\mathbf{m}}{dt} = E\{\mathbf{f}(\mathbf{x})\}, \quad t_{k-1} < t \leq t_k, \tag{2.73}$$

$$\frac{d\mathbf{P}}{dt} = E\{(\mathbf{x} - \mathbf{m})\mathbf{f}'(\mathbf{x})\} + E\{\mathbf{f}(\mathbf{x})(\mathbf{x} - \mathbf{m})'\} + \mathbf{Q}, \quad t_{k-1} < t \leq t_k, \tag{2.74}$$

where the expectation is evaluated with respect to the Gaussian density $\mathcal{N}(\mathbf{x}; \mathbf{m}, \mathbf{P})$. If the function $\mathbf{f}$ is differentiable and $\mathbf{x}$ is a Gaussian random variable, then from [51], we get

$$E\{\mathbf{f}(\mathbf{x})(\mathbf{x} - \mathbf{m})'\} = E\{\mathbf{F}_{\mathbf{x}}(\mathbf{x})\}\mathbf{P}, \tag{2.75}$$

where $\mathbf{F}_{\mathbf{x}}(\mathbf{x})$ is the Jacobian of $\mathbf{f}$ with respect to $\mathbf{x}$ evaluated at $\mathbf{x}$. Then (2.74) can be written as

$$\frac{d\mathbf{P}}{dt} = \mathbf{P}E\{\mathbf{F}'_{\mathbf{x}}(\mathbf{x})\} + E\{\mathbf{F}_{\mathbf{x}}(\mathbf{x})\}\mathbf{P} + \mathbf{Q}, \quad t_{k-1} < t \leq t_k. \tag{2.76}$$

For any nonlinear function $\mathbf{g}$, expected values of functions in (2.73) and (2.76) are evaluated using the Gaussian approximation

$$E\{\mathbf{g}(\mathbf{x})\} \approx \int \mathbf{g}(\mathbf{x})\mathcal{N}(\mathbf{x}; \mathbf{m}, \mathbf{P})d\mathbf{x}. \tag{2.77}$$

Using the first order Taylor series expansion, $\mathbf{g}(\mathbf{x})$ is expressed as

$$\mathbf{g}(\mathbf{x}) \approx \mathbf{g}(\mathbf{m}) + \mathbf{F}_{\mathbf{x}}(\mathbf{x} - \mathbf{m}). \tag{2.78}$$

Then

$$E\{\mathbf{g}(\mathbf{x})\} \approx \mathbf{g}(\mathbf{m}), \tag{2.79}$$

since the second term on the right-hand side vanishes. Similarly,

$$E\{\mathbf{F}_{\mathbf{x}}(\mathbf{x})\} \approx \mathbf{F}_{\mathbf{x}}(\mathbf{m}). \tag{2.80}$$

Using (2.79) and (2.80) in (2.73) and (2.76), we get

$$\frac{d\mathbf{M}}{dt} = \mathbf{f}(\mathbf{m}), \quad t_{k-1} < t \leq t_k, \tag{2.81}$$

$$\frac{d\mathbf{P}}{dt} = \mathbf{P}\mathbf{F}'_{\mathbf{x}}(\mathbf{m}) + \mathbf{F}_{\mathbf{x}}(\mathbf{m})\mathbf{P} + \mathbf{Q}, \quad t_{k-1} < t \leq t_k. \tag{2.82}$$

Equation (2.81) is an ordinary differential equation (ODE) in n independent variables. Since **P** is a symmetric matrix, (2.82) is an ODE in $n(n+1)/2$ independent variables. Thus, (2.81) and (2.82) represent a joint ODE in $n + n(n+1)/2$ independent variables. We can integrate (2.81) and (2.82) jointly using the Runge–Kutta integrator or the Matlab function ode45. This prediction algorithm is called the continuous EKF (CEKF) step.

2.8.3 Approximate Discrete-time Prediction Algorithms

This class of algorithms first use approximate discretization and then use the EKF for the prediction step. We describe these algorithms next.

2.8.3.1 Approximate Prediction Algorithm 1

An ODE representing the continuous-time dynamic model without process noise

$$\frac{d\mathbf{x}(t)}{dt} = \mathbf{f}(\mathbf{x}(t)), \tag{2.83}$$

is first used in [8]. It can be shown that using a Taylor series expansion of $\mathbf{x}(t)$ up to the second order (TS2) gives [8, 37]

$$\mathbf{x}(t+T) = \mathbf{x}(t) + \mathbf{f}(\mathbf{x}(t))T + \frac{1}{2}\mathbf{F}_{\mathbf{x}}(\mathbf{x})\mathbf{f}(\mathbf{x}(t))T^2, \tag{2.84}$$

where $\mathbf{F}_{\mathbf{x}}$ denotes the partial derivative of $\mathbf{f}$ with respect to $\mathbf{x}$

$$\mathbf{F}_{\mathbf{x}}(\mathbf{x}) := \frac{\partial \mathbf{f}(\mathbf{x}(t))}{\partial \mathbf{x}} = \partial_{\mathbf{x}}\mathbf{f}(\mathbf{x}). \tag{2.85}$$

Let $\mathbf{x}(t) = \mathbf{x}_k$ and $\mathbf{x}(t+T) = \mathbf{x}_{k+1}$. Then from (2.79) we get

$$\mathbf{x}_{k+1} = \mathbf{x}_k + \mathbf{f}(\mathbf{x}_k)T + \frac{1}{2}\mathbf{F}_{\mathbf{x}}(\mathbf{x}_k)\mathbf{f}(\mathbf{x}_k)T^2. \tag{2.86}$$

Equation (2.80) represents the discretized dynamic model of the ODE (2.83). Next, a discrete-time zero-mean white Gaussian process noise $\mathbf{w}_k^{\mathrm{a}}$ for the time interval $[t_k, t_{k+1})$ is added to the right-hand side of (2.80) in an ad hoc manner to obtain

$$\mathbf{x}_{k+1} = \mathbf{x}_k + \mathbf{f}(\mathbf{x}_k)T + \frac{1}{2}\mathbf{F}_{\mathbf{x}}(\mathbf{x}_k)\mathbf{f}(\mathbf{x}_k)T^2 + \mathbf{w}_k^{\mathrm{a}}. \tag{2.87}$$

The algorithms in [17, 18] are similar to this algorithm. The prediction step of the EKF is used to calculate the predicted state estimate and covariance in [8]. We call this prediction step the TS2-EKF.

2.8.3.2 Approximate Prediction Algorithm 2

The authors in [4, 12] use an approximate discrete-time dynamic model based on the PCA approximation. However, they use discrete-time process noise for the NCV model [19]. The discretized continuous-time model [19] for process noise is used in [4], whereas the direct discrete model [19] for process noise is used in [12]. Using the IT-1.5 approximation [23], we have shown in [24] that the discrete-time process noise does not arise from the NCV motion. Let $\mathbf{a}_{k-1}$ denote the total acceleration (2.22) at time t_{k-1}. The discretized continuous-time model for position, velocity, and α using the PCA are described, respectively, by

$$\mathbf{p}_k = \mathbf{p}_{k-1} + \mathbf{v}_{k-1}T + \mathbf{a}_{k-1}T^2/2 + \mathbf{w}_{k-1,\mathbf{p}}, \tag{2.88}$$

$$\mathbf{v}_k = \mathbf{v}_{k-1} + \mathbf{a}_{k-1}T + \mathbf{w}_{k-1,\mathbf{v}}, \tag{2.89}$$

$$\alpha_k = \alpha_{k-1} + w_{k-1,\alpha}, \tag{2.90}$$

where $\mathbf{w}_{k-1,\mathbf{p}}$, $\mathbf{w}_{k-1,\mathbf{v}}$, and $w_{k-1,\alpha}$ are the discrete-time process noise for position, velocity, and α, respectively, during the time interval $[t_{k-1}, t_k)$. For a fair comparison of filtering algorithms, we have included α in (2.90). Using $\mathbf{p}_k$, $\mathbf{v}_k$, and α_k, the state vector is defined by

$$\mathbf{x}_k := \begin{bmatrix}\mathbf{p}'_k & \mathbf{v}'_k & \alpha_k\end{bmatrix}'. \tag{2.91}$$

Then the discrete-time dynamic model using the PCA approximation can be written as

$$\mathbf{x}_k = \mathbf{f}^{\mathrm{PCA}}(\mathbf{x}_{k-1}) + \mathbf{w}_{k-1}^{\mathrm{PCA}}, \tag{2.92}$$

where

$$\mathbf{f}^{\mathrm{PCA}}(\mathbf{x}_{k-1}) = \begin{bmatrix}\mathbf{p}_{k-1} + \mathbf{v}_{k-1}T + \mathbf{a}_{k-1}T^2/2 \\ \mathbf{v}_{k-1} + \mathbf{a}_{k-1}T \\ \alpha_{k-1}\end{bmatrix}, \tag{2.93}$$

$$\mathbf{w}_{k-1}^{\mathrm{PCA}} = \begin{bmatrix}\mathbf{w}'_{k-1,\mathbf{p}} \mathbf{w}'_{k-1,\mathbf{v}} w_{k-1,\alpha}\end{bmatrix}', \tag{2.94}$$

$$\mathbf{w}_{k-1}^{\mathrm{PCA}} \sim \mathcal{N}(\mathbf{0}, \mathbf{Q}_{k-1}^{\mathrm{PCA}}), \tag{2.95}$$

$$\mathbf{Q}_{k-1}^{\mathrm{PCA}} = \begin{bmatrix} qT^3/3\mathbf{I}_3 & qT^2/2\mathbf{I}_3 & \mathbf{0}_{3\times1} \\ qT^2/2\mathbf{I}_3 & qT^3/3\mathbf{I}_3 & \mathbf{0}_{3\times1} \\ \mathbf{0}_{1\times3} & \mathbf{0}_{1\times3} & q_\alpha T \end{bmatrix}. \tag{2.96}$$

The function $\mathbf{f}^{\mathrm{PCA}}$ is a nonlinear function of $\mathbf{x}_{k-1}$, since $\mathbf{a}_{k-1}$ is a nonlinear function of $\mathbf{x}_{k-1}$. The prediction step of the discrete-time EKF [19, 25] can be used to calculate the predicted state estimate and associated covariance. This prediction step is called the PCA-EKF.

2.8.4 Measurement Update Step

For the radar scenario, we convert the range, azimuth, and elevation measurements to 3D Cartesian position measurement using the CT [27] as described in Sub-section 2.6 and obtain the associated covariance. Since the measurement model is linear in the Cartesian position measurement, we can use the KF update step to calculate the updated state estimate and covariance [19] after time t_3 while processing measurements $\mathbf{z}_4, \mathbf{z}_5, \ldots$ Strictly speaking the update algorithm is a linear minimum mean square error (LMMSE) estimator [19], since the error in the converted measurement is not Gaussian.

We use the measurement update step of the CKF while processing azimuth and elevation measurements $\mathbf{z}_2, \mathbf{z}_3, \ldots$ after time t_1 in the passive sensor scenario.

2.9 Numerical Simulation and Results

We model the RV as a sharp-nose projectile as used in [11]. The Mach number dependent drag coefficient C_d as a function of Mach number from [39] is plotted in Figure 2.3. We use a spherical-Earth model with constant density where the gravitational constant $\mu_e = 3.986005 \times 10^{14}\,\text{m}^3\text{s}^{-2}$. Assuming an acceleration due to gravity $g_0 = 9.806\,\text{ms}^{-2}$ at Earth's surface, we obtain the radius of Earth $R_e = 6.375628 \times 10^6$ m. Our atmospheric density model is based on the 1976 US standard atmosphere model [23].

Usually, the Euler–Maruyama approximation [34, 16] is used to generate the true trajectory using a large number (e.g., 1000) of subintervals. We observed that this approximation is not suitable for the current problem, possibly due to high degree of nonlinearity of the dynamic model. We integrate the deterministic part of the dynamic model in (2.17) using the Runge–Kutta method for the time interval T and add the process noise using the IT-1.5 [13] approximation. Thus, the discrete-time process noise is given by [13, 24]

$$\mathbf{w}_{k-1} = \sqrt{T}\mathbf{G}\mathbf{u}_1 + (T^{3/2}/2)\mathbf{F}_{\mathbf{x},k-1}\mathbf{G}(\mathbf{u}_1 + \mathbf{u}_2/\sqrt{3}), \tag{2.97}$$

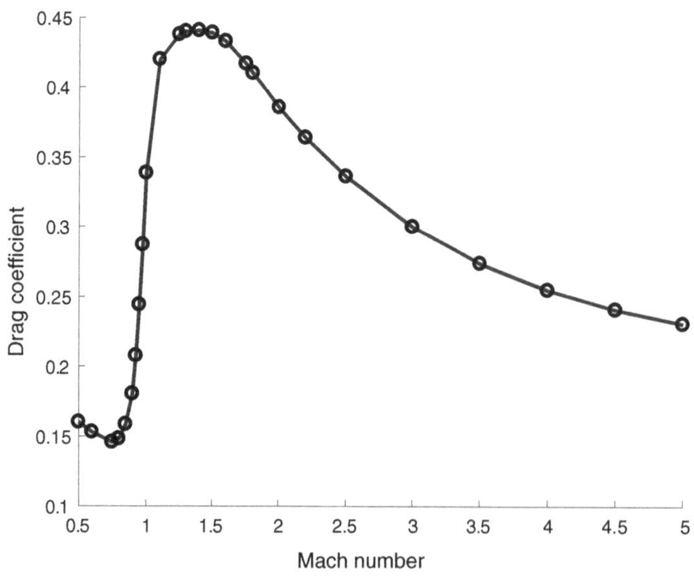

Figure 2.3 Mach number-dependent drag coefficient C_d versus Mach number.

where $\mathbf{u}_1$ and $\mathbf{u}_2$ are two four-dimensional independent zero-mean standard Gaussian random variables with unit variance,

$$\mathbf{u}_i \sim \mathcal{N}(\mathbf{0}_4, \mathbf{I}_4), i = 1, 2. \tag{2.98}$$

Detailed derivation of the discrete-time process noise covariance $\mathbf{Q}_{k-1}$ corresponding to $\mathbf{w}_{k-1}$ is presented in [24] and has the form

$$\mathbf{Q}_{k-1} = \mathbf{Q}T + (\mathbf{F}_{\mathbf{x},k-1}\mathbf{Q} + \mathbf{Q}\mathbf{F}'_{\mathbf{x},k-1})\frac{T^2}{2} + \mathbf{F}_{\mathbf{x},k-1}\mathbf{Q}\mathbf{F}'_{\mathbf{x},k-1}\frac{T^3}{3}. \tag{2.99}$$

Let n and m be the dimension of the state and measurement, respectively. We use the average normalized estimation error squared (ANEES) and average normalized innovation squared (ANIS) [19, 52, 53] to evaluate the consistency of the filter's calculated covariance and innovation covariance, respectively. The ANEES and ANIS at time t_k are defined, respectively, by

$$\text{ANEES}_k = (1/nM) \times \sum_{i=1}^{M} (\widehat{\mathbf{x}}_{k|k,i} - \mathbf{x}_{k,i})\mathbf{P}_{k|k,i}^{-1}(\widehat{\mathbf{x}}_{k|k,i} - \mathbf{x}_{k,i}), \tag{2.100}$$

$$\text{ANIS}_k = (1/mM) \times \sum_{i=1}^{M} (\widehat{\mathbf{z}}_{k|k,i} - \mathbf{z}_{k,i})\mathbf{S}_{k,i}^{-1}(\widehat{\mathbf{z}}_{k|k,i} - \mathbf{z}_{k,i}), \tag{2.101}$$

where M is the number of Monte Carlo runs, and $\mathbf{P}_{k|k,i}$ and $\mathbf{S}_{k,i}$ are the filter's calculated covariance and innovation covariance at time t_k in the ith Monte Carlo run. The ideal values of ANEES and ANIS are unity.

We evaluate the overall performance of the filters using the root time-averaged mean square (RTAMS) error metric [45]. The RTAMS metric for position is defined as

$$\text{RTAMS}_{\text{pos}} = \sqrt{\frac{1}{(k_2 - k_1 + 1)M} \sum_{k=k_1}^{k_2} \sum_{i=1}^{M} \| \widehat{\mathbf{p}}_{k,i} - \mathbf{p}_{k,i} \|^2}, \tag{2.102}$$

where k_1 and k_2 are the initial and final time indices, respectively. Similar metrics are used for velocity and α.

2.9.1 Radar Scenario: Simulation and Results

In [4], a 2D scenario is considered where the RV trajectory lies in the vertical XY−plane and a flat-Earth model for gravity is used. Secondly, the drag parameter α is not estimated. We use a spherical-Earth model for gravity and estimate the drag parameter α. The initial 3D position and velocity of the RV are obtained from [4] by rotating the vertical plane by 30^0 about the Z−axis of the T frame. Numerical values of the prior mean position, velocity, α, power spectral densities (PSDs) of process noise, and prior covariance are presented in Table 2.5. Table 2.5 also presents the radar parameters such as the radar position, measurement SDs, measurement time interval, and number of measurements.

We use 500 Monte Carlo runs to generate true RV trajectories and sensor measurements. We consider Scenarios 1 and 2 with acceleration process noise PSDs of 1.0 m^2/s^3 and 5.0 m^2/s^3, respectively. Other parameters are the same for these scenarios. The true 3D trajectory of the RV from the first Monte Carlo run is shown in Figure 2.4 from Scenario 1. The starting and final points are marked by green and red circles, respectively. We evaluate the filters CEKF-KF, TS2-EKF-KF, and PCA-EKF-KF, where CEKF, TS2-EKF, and PCA-EKF refer to the prediction step. The KF measurement update step is the same for all three filters. The RMS errors for position, velocity, and α for

Table 2.5 Radar scenario: target and sensor parameters.

Variable	Value
Prior mean position $(\bar{x}_1, \bar{y}_1, \bar{z}_1)$	$(-116.000, 200.918, 88.000)$ km
Prior mean velocity $(\bar{\dot{x}}_1, \bar{\dot{y}}_1, \bar{\dot{z}}_1)$	$(1.129, -1.953, -0.398)$ km/s
Prior mean of α $(\bar{\alpha}_1)$	$0.03\ \text{m}^2/\text{kg}$
PSD of acceleration process noise, $q_x = q_y = q_z$	$(1.0, 5.0)\ \text{m}^2/\text{s}^3$
PSD of process noise, q_α	$1.0 \times 10^{-9}\ \text{m}^4/(\text{kg}^2\text{s})$
Prior position covariance	$\text{diag}(200^2, 200^2, 200^2)\ \text{m}^2$
Prior velocity covariance	$\text{diag}(10^2, 10^2, 10^2)\ (\text{m/s})^2$
Prior α variance	$1.0 \times 10^{-6}\ \text{m}^4/\text{kg}^2$
Radar position $(\mathbf{s})$	$(0.0, 0.0, 0.0)$ m
Range, azimuth, elevation measurement SDs $(\sigma_r, \sigma_\varphi, \sigma_\epsilon)$	$(40.0\ \text{m}, 0.1^\circ, 0.1^\circ)$
Measurement time interval (T)	1 s
Number of measurements time interval (N)	121

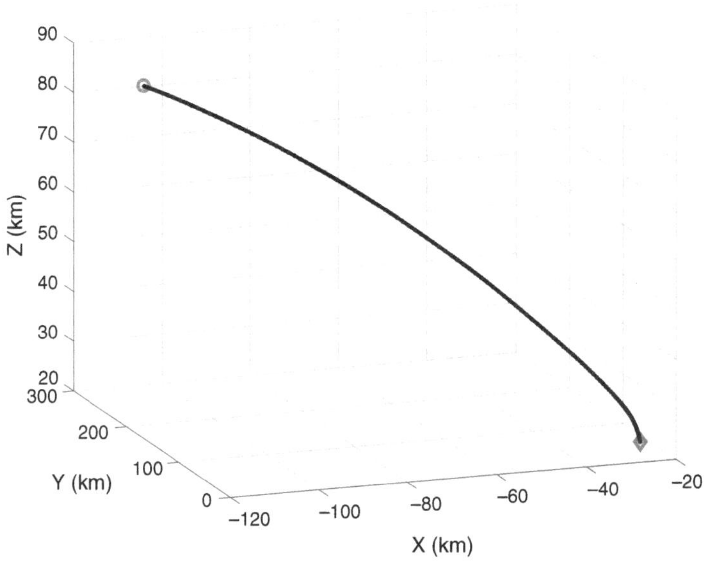

Figure 2.4 Radar Scenario 1. Target true 3D trajectory from the first Monte Carlo run.

these filters are presented in Figures 2.5 and 2.6, respectively, for Scenarios 1 and 2, an acceleration process noise PSD of $1\ \text{m}^2/\text{s}^3$ and $5\ \text{m}^2/\text{s}^3$. We observe that the RMS errors for position and velocity for CEKF and TS2-EKF are nearly the same. The RMS errors for position and velocity of the PCA-EKF are nearly the same as other two filters except for the time interval [75–100] s. From Figure 2.2 we observe that the total acceleration is nearly constant except during the time interval [75–100] s. This explains why the PCA-EKF does not perform well during all time intervals except [75–100] s. The TS2-EKF and PCA-EKF differ from the CEKF-KF due to the wrong model of the process noise.

ANEES and ANIS for radar Scenarios 1 and 2 are presented in Figures 2.7 and 2.8, respectively. Examining Figure 2.8a, we see that the ANEES from CEKF-KF lies within the 99% confidence bounds most of the time whereas the ANEES from the PCA-EKF lies outside the 99% confidence

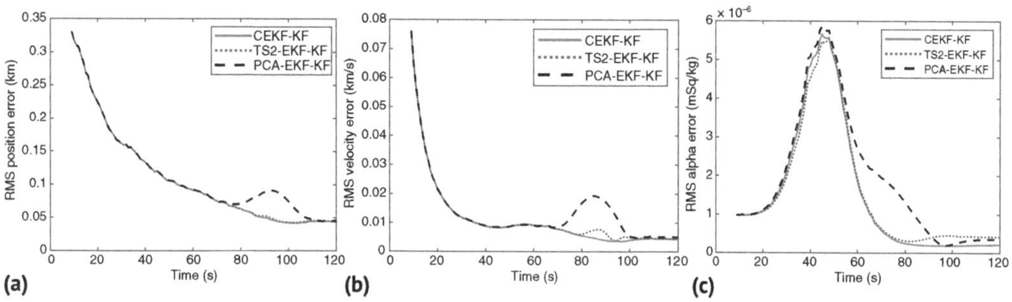

Figure 2.5 Radar Scenario 1. (a) RMS position error, (b) RMS velocity error, and (c) RMS alpha error.

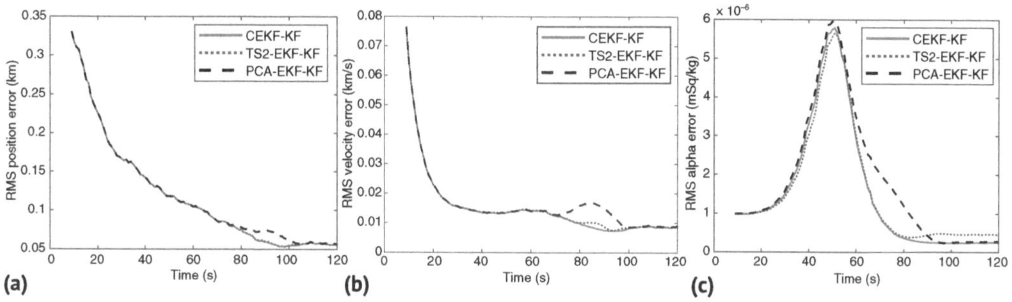

Figure 2.6 Radar Scenario 2. (a) RMS position error, (b) RMS velocity error, and (c) RMS alpha error.

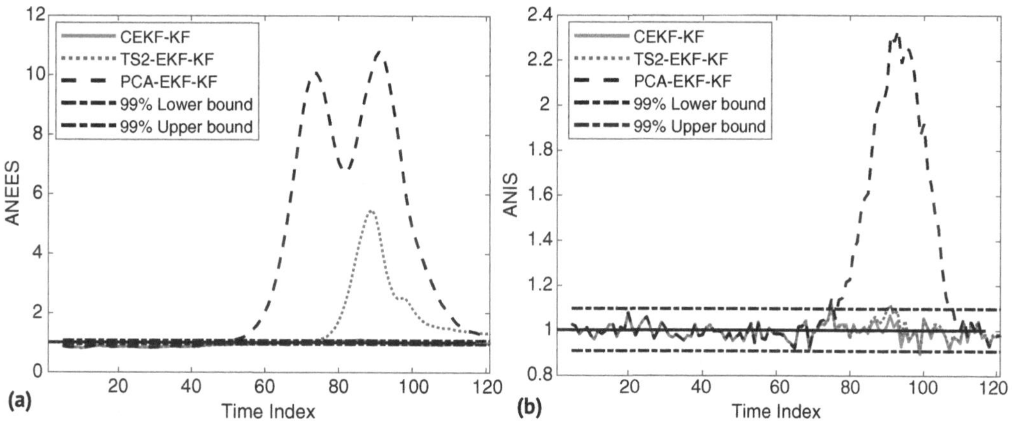

Figure 2.7 Radar Scenario 1. (a) ANEES and (b) ANIS.

bounds during the time interval [60,100]s. Similarly, the ANIS from the CEKF-KF lies within 99% confidence bounds all the time and that from the PCA-EKF lies outside the 99% confidence bounds during the time interval [80,100]s. The non-constant acceleration is the reason for these deviations for the PCA-EKF-KF.

The RTAMS metrics and average CPU times for the three filters in two scenarios are presented in Tables 2.6 and 2.7, respectively. Results in Table 2.6 show that the CEKF-KF has the best state estimation accuracy. However, the average CPU time of the CEKF-KF is the highest among the three filters.

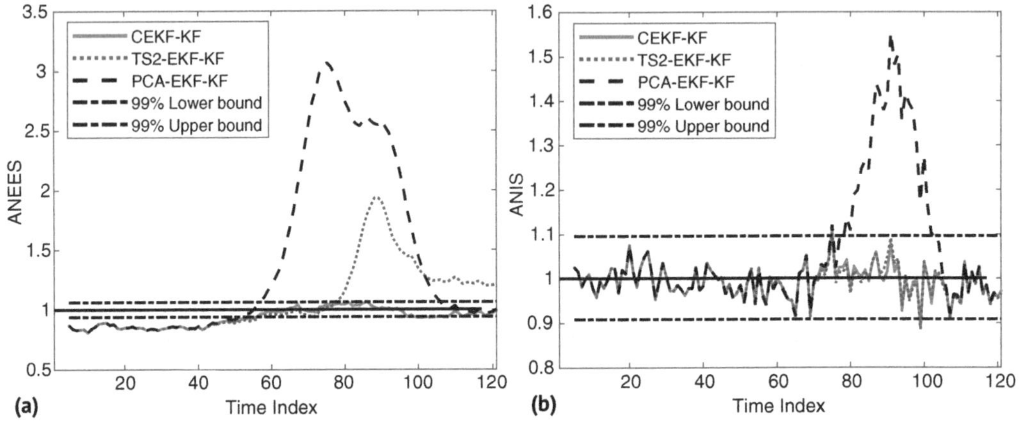

Figure 2.8 Radar Scenario 2. (a) ANEES and (b) ANIS.

Table 2.6 RTAMS metrics in radar scenarios.

	RTAMS	CEKF-KF	TS2-EKF-KF	PAC-EKF-KF
Scenario 1	Pos (m)	88.06	88.32	94.27
	Vel (m/s)	7.21	7.57	10.56
	α (m^2/kg)	0.0025	0.0024	0.0028
Scenario 2	Pos (m)	97.76	97.86	99.53
	Vel (m/s)	11.54	11.72	13.01
	α (m^2/kg)	0.0024	0.0024	0.0027

Table 2.7 Average CPU times per Monte Carlo run in radar scenarios.

Filter	CEKF-KF	TS2-EKF-KF	PAC-EKF-KF
Average CPU Time (s)	0.1760	0.0458	0.0537

2.9.2 Passive Sensor: Simulation and Results

We use two scenarios from [11]. Initial mean Cartesian position, velocity, α, PSDs of the process noise, prior covariances of the RV, and sensor parameters are presented in Table 2.8.

For passive sensing, the sensor must perform maneuver to observe the state of the target. The sensor moves in the YZ-plane and its motion profile is presented in Table 2.9. The sensor trajectory has a sequence of constant velocity (CV) and constant turn (CT) motions. For Scenarios 1 and 2, the sensor and target trajectories in the $YZ-$plane are shown in Figure 2.9.

We use 500 Monte Carlo runs to calculate RMS errors, ANEES, and ANIS. The RMS errors for position, velocity, and α from CEKF-CKF, TS2-EKF-CKF, and PCA-EKF-CKF for Scenario 1 are presented in Figure 2.10a,b, and c, respectively. Results in these figures show that the RMS errors for position, velocity, and α from CEKF-CKF and PCA-EKF-CKF are nearly the same. The corresponding RMS errors for the TS2-EKF-CKF are somewhat higher.

Table 2.8 Passive sensor scenarios: target and sensor parameters.

Variable	Value
Scenario 1: RV initial mean position	(0, 20, 30) km
Scenario 2: RV initial mean position	(−20, 30, 30) km
Scenario 1: RV initial mean velocity	(0.0, −1.0, −1.0) km/s
Scenario 2: RV initial mean velocity	(1.5, −1.5, −1.0) km/s
Prior position covariance	$\text{diag}(500^2, 500^2, 500^2)$ m^2
Prior velocity covariance	$\text{diag}(10^2, 10^2, 10^2)$ (m/s)^2
Prior α variance	0.001 m^4/kg^2
PSD of acceleration process noise, $q_x = q_y = q_z$	(1.0, 3.0) m^2/s^3
PSD of process noise, q_α	1.0×10^{-9} m^4/(kg^{2}s)
Sensor initial position	(5, −10, 15) km
Sensor initial velocity	(0, 0.3, 0) km/s
Azimuth and elevation measurement SD (σ_φ, σ_ϵ)	(0.1, 0.1) mrad
Measurement time interval (T)	1/30 s
Number of measurements for Scenarios 1 and 2 (N)	(907, 977)

Table 2.9 Sensor motion profile in the YZ plane.

Segment	Time interval (s)	Duration (s)	Motion type	Ω (rad/s)
1	[0, 15]	15	CV	0
2	[15, 31]	16	CT	$-\pi/64$
3	[31, 43]	12	CV	0
4	[43, 75]	32	CT	$\pi/64$
5	[75, 86]	11	CV	0
6	[86, 102]	16	CT	$-\pi/64$
7	[102, 122]	20	CV	0
8	[122, 182]	60	CT	$-\pi/60$
9	[182, 218]	36	CV	0

For the passive sensor Scenario 1, the ANEES and ANIS with 99% confidence bounds are shown in Figure 2.11a,b, respectively. The ANEES for the CEKF-CKF lies within the 99% confidence bounds for times greater than 140 s, whereas that for the PCA-EKF-CKF lies within the confidence bounds for a smaller fraction of time. The ANIS for the three filters lie within the 99% confidence bounds most of the time.

The RMS errors for position, velocity, and α, and the ANEES and ANIS with 99% confidence bounds for Scenario 2 are plotted in Figures 2.12a,b, and c, and 2.13a,b, respectively. Similar pattern is observed for Scenario 2 as in Scenario 1.

The RTAMS metrics and average CPU times for the three filters are presented in Tables 2.10 and 2.11, respectively. As in the radar scenarios, RTAMS results in Table 2.10 show that the CEKF-KF has the best state estimation accuracy. Similarly, the average CPU time of the CEKF-KF is the highest among the three filters.

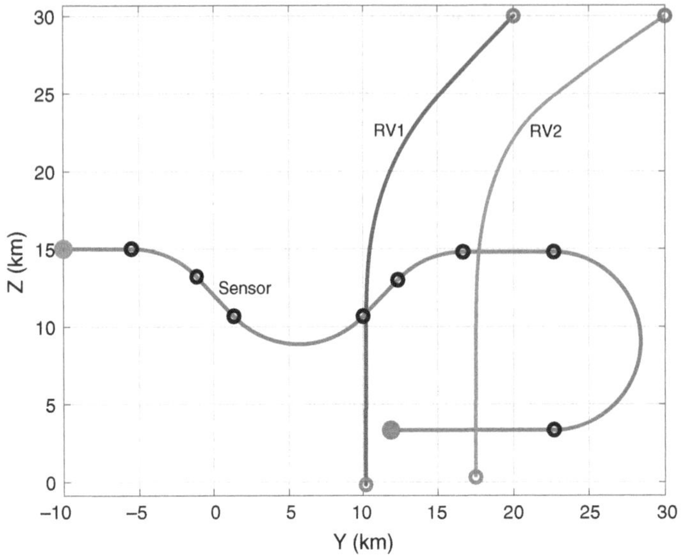

Figure 2.9 Passive Sensor Scenarios 1 and 2. RV and sensor trajectories in the YZ-plane.

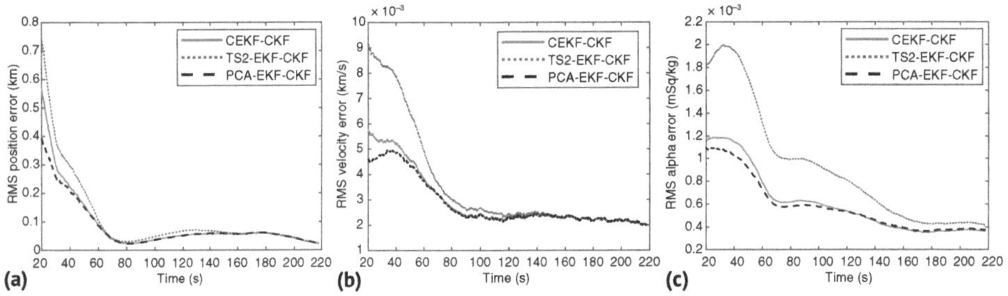

Figure 2.10 Passive Sensor Scenario 1. (a) RMS position error, (b) RMS velocity error, and (c) RMS alpha error.

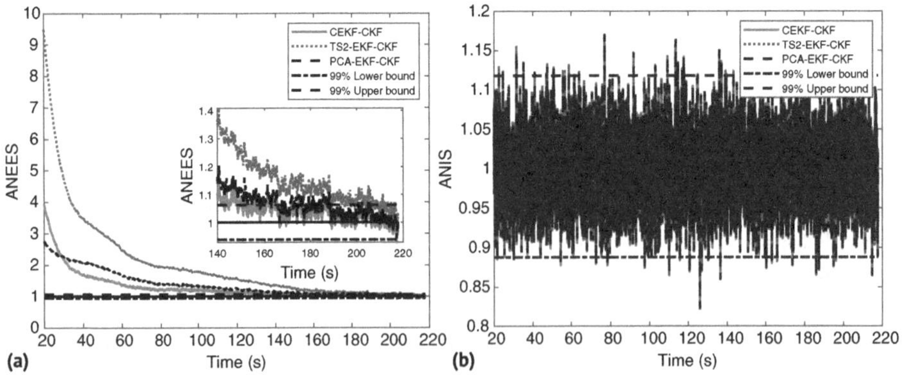

Figure 2.11 Passive Sensor Scenario 1. (a) ANEES and (b) ANIS.

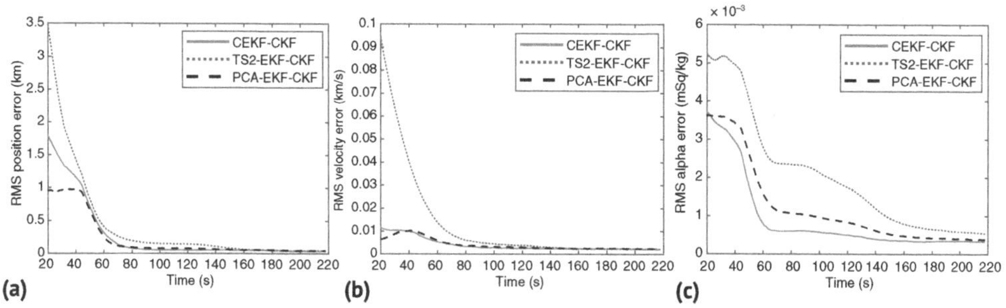

Figure 2.12 Passive Sensor Scenario 2. (a) RMS position error, (b) RMS velocity error, and (c) RMS alpha error.

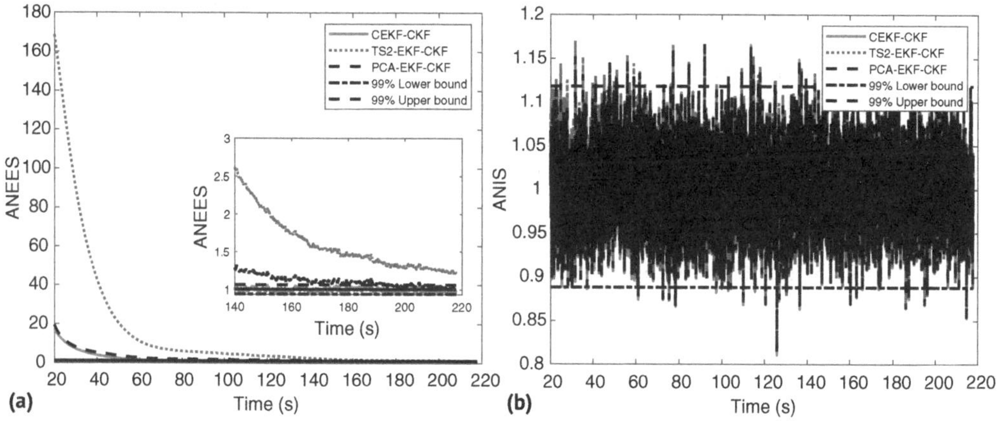

Figure 2.13 Passive Sensor Scenario 2. (a) ANEES. (b) ANIS.

Table 2.10 RTAMS metrics in passive sensor scenarios.

	RTAMS	CEKF-CKF	TS2-EKF-CKF	PCA-EKF-CKF
Scenario 1	Pos (m)	51.49	58.20	51.14
	Vel (m/s)	2.25	2.32	2.24
	α (m^2/kg)	0.0004	0.0006	0.0004
Scenario 2	Pos (m)	52.33	98.44	60.32
	Vel (m/s)	2.38	2.91	2.47
	α (m^2/kg)	0.0004	0.0012	0.0006

Table 2.11 Average CPU times per Monte Carlo run in passive sensor scenarios.

Filter	CEKF-CKF	TS2-EKF-CKF	PCA-EKF-CKF
Averaged CPU time (s)	17.6664	2.3113	5.9256

2.10 Conclusions

This chapter presents algorithms and results for the 3D RV filtering problem using a radar and a passive sensor with angle-only measurements separately. We use a spherical-Earth model and the atmospheric drag model based on the 1976 US Standard Atmosphere, and estimates a drag parameter in both the radar and passive sensor scenarios. Our contribution for the radar scenario is new, whereas that for the passive sensor is based on our recent publication. We present the CEKF algorithm for the prediction step of the 3D RV filtering problem rigorously using the Itô SDE that was first developed in our recent work using a passive angle-only sensor. We successfully apply the CEKF successfully to the RV filtering problem using a radar. For the radar scenario, we present an effective filter initialization algorithm using the constant acceleration model. This algorithm can be used to initialize the filter using three or more radar measurements. We also present and apply our recently derived analytic expression for the discrete-time process noise covariance using the IT-1.5 approximation. This expression is valid for a general SDE. Previous work on RV filtering used ad hoc expressions for the process noise covariance.

We compare our CEKF based filter with two existing filters that use the TS2-EKF and PCA-EKF prediction steps across four radar and passive sensor scenarios. A spherical-Earth model and an atmospheric drag model are used for all three filters, and a seven-dimensional RV state is estimated including a drag parameter. Our numerical results, obtained through Monte Carlo simulations, demonstrate that the CEKF based filter has the best state estimation accuracy compared with the TS2-EKF and PCA-EKF based filters for the scenarios analyzed. However, the CEKF based filter incurs the highest computational cost.

Acknowledgments

This work was supported in part by National Natural Science Foundation of China through Grant 62273283. We would like to thank Gongjian Zhou and Keyi Li of Harbin Institute of Technology, Harbin, China, for providing valuable explanations on their model-based filter initialization paper [47].

References

1 A. Benavoli, L. Chisci, and A. Farina. Tracking of a ballistic missile with A-priori information, *IEEE Transactions on Aerospace and Electronic Systems*, 43(3):1000–1016, 2007.
2 C. B. Chang. Ballistic trajectory estimation with angle-only measurements. *IEEE Transactions on Automatic Control*, 25(3):474–480, 1980.
3 P. Dodin, P. Minvielle, and J.-P.-Le Cadre, Re-entry vehicle tracking observability and theoretical bound, *Proceedings of the 7th International Conference on Information Fusion*, Philadelphia, PA, USA, pages 25–28, 2005.
4 A. Farina, B. Ristic, and D. Benvenuti. Tracking a ballistic target: Comparison of several nonlinear filters. *IEEE Transactions on Aerospace and Electronic Systems*, 38(3): 854–867, 2002.
5 W. J. Farrell. Interacting multiple model filter for tactical ballistic missile tracking. *IEEE Transactions on Aerospace and Electronic Systems*, 44(2):418–426, 2008.
6 S. Ghosh and S. Mukhopadhyay. Tracking reentry ballistic targets using acceleration and Jerk models. *IEEE Transactions on Aerospace and Electronic Systems*, 47(2):666–683, 2011.

7 R. Ji, Y. Liang, L. Xu, and Z. Wei. Trajectory prediction of ballistic missiles using Gaussian process error model. *Chinese Journal of Aeronautics*, 35(1):458–469, 2022.

8 R. K. Mehra. A comparison of several nonlinear filters for reentry vehicle tracking. *IEEE Transactions on Automatic Control*, AC-16(4):307–319, 1971.

9 N. K. Singh, S. Bhaumik, and S. Bhattacharya, A comparison of several nonlinear filters for ballistic missile tracking on re-entry, In *Proceedings of the 2016 IEEE First International Conference on Control, Measurement and Instrumentation (CMI)*, pages 459–463, 2016.

10 G. Siouris, G. Chen, and J. Wang. Tracking an incoming ballistic missile using an extended interval Kalman filter. *IEEE Transactions on Aerospace and Electronic Systems*, 33(1):232–240, 1997.

11 Z. Tian, K. Yang, M. Danino, Y. Bar-Shalom, B. Milgrom. Ballistic object trajectory and impact point estimation in the reentry phase from a moving passive sensor. *IEEE Transactions on Aerospace and Electronic Systems*, 58(5):4540–4540, 2022.

12 M. Yu, L. Gong, H. Oh, W. Chen, and J. Chambers. Multiple model ballistic missile tracking with state-dependent transitions and Gaussian particle filtering. *IEEE Transactions on Aerospace and Electronic Systems*, 54(3):1066–1081, 2018.

13 I. Arasaratnam, S. Haykin, and T. R. Hurd. Cubature Kalman filtering for continuous-discrete systems: Theory and simulations. *IEEE Transactions on Signal Process*, 58(10):4977–4993, 2010.

14 A. H. Jazwinski. *Stochastic Processes and Filtering Theory*. Academic Press, New York, NY, USA, 1970.

15 S. Särkkä and J. Sarmavuori. Gaussian filtering and smoothing for continuous-discrete dynamic systems. *Signal Processing*, 93(2):500–510, 2013.

16 S. Särkkä and A. Solin. *Applied Stochastic Differential Equations*. Cambridge University Press, 2019.

17 V. Ravindra, Y. Bar-Shalom, and P. K. Willett. Projectile identification and impact point prediction. *IEEE Transactions on Aerospace and Electronic Systems*, 46(4):2004–2021, 2010.

18 T. Yuan, Y. Bar-Shalom, P. Willett, E. Mozeson, S. Pollak, and D. Hardiman. A multiple IMM estimation approach with unbiased mixing for thrusting projectiles. *IEEE Transactions on Aerospace and Electronic Systems*, 48(4):3250–3267, 2012.

19 Y. Bar-Shalom, X. R. Li, and T. Kirubarajan. *Estimation with Applications to Tracking and Navigation*, Wiley, New York, 2001.

20 X. R. Li and V. P. Jilkov. Survey of maneuvering target tracking. Part II: Motion Models of Ballistic and Space Targets. *IEEE Transactions on Aerospace and Electronic Systems*, 46(1):96–119, 2010.

21 D. Halliday, R. Resnick, and J. Walker. *Fundamentals of Physics Extended*. Wiley, 2010.

22 J. B. Marion and S. T. Thornton. *Classical Dynamics of Particles and Systems*, Brooks-Cole-Thomson Learning, Belmont, CA, USA, 5th Edition, 2004.

23 K. Moon, T. Kim, and T. Song. Comparison of ballistic-coefficient-based estimation algorithms for precise tracking of a re-entry vehicle and its impact point prediction. *Journal of Astronomy and Space Sciences*, 29(4):363–374, 2012.

24 M. Mallick, X. Tian, and L. Xu. Reentry vehicle filtering using a passive angle-only sensor. *IEEE Transactions on Aerospace and Electronic Systems*, 60(6):8539–8554, 2024.

25 A. Gelb, Editor. *Applied Optimal Estimation*. MIT Press, 1974.

26 S. Julier, J. Uhlmann, and H. F. Durrant-Whyte. A new method for the nonlinear transformation of means and covariances in filters and estimators. *IEEE Transactions on Automatic Control*, AC-45(3):477–482, 2000.

27 I. Arasaratnam and S. Haykin. Cubature Kalman filters. *IEEE Transactions on Automatic Control*, 54(6):1254–1269, 2009.

28 M. S. Arulampalam, S. Maskell, N. Gordon, and T. Clapp. A tutorial on particle filters for online nonlinear/non-Gaussian Bayesian tracking. *IEEE Transactions on Signal Processing*, 50(2):174–188, 2002.

29 X. R. Li and V. P. Jilkov. Survey of maneuvering target tacking. Part I: Dynamic Models, *IEEE Transactions on Aerospace and Electronic Systems*, 39(4):1333–1364, 2003.

30 Y. Bar-Shalom, P. Willett, and X. Tian. *Tracking and Data Fusion: A Handbook of Algorithms*, YBS Publishing, 2011.

31 A. F. García-Fernández and S. Maskell. Continuous-discrete multiple target filtering: PMBM, PHD and CPHD filter implementations. *IEEE Transactions on Signal Processing*, 68:1300–1314, 2020.

32 P. Axelsson and F. Gustafsson. Discrete-time solutions to the continuous-time differential Lyapunov equation with applications to Kalman filtering. *IEEE Transactions on Automatic Control*, 60(3):632–643, 2015.

33 G. Y. Kulikov and M. V. Kulikova. The accurate continuous-discrete extended Kalman filter for radar tracking. *IEEE Transactions on Signal Process*, 64(4):948–958, 2016.

34 P. E. Kloeden and E. Platen. *Numerical Solution of Stochastic Differential Equations*. Springer Verlag, 1999.

35 M. Mallick, M. Morelande, and L. Mihaylova. Continuous-discrete filtering using EKF, UKF, and PF. In *Proceedings of the 2012 International Conference on Information Fusion*, Singapore, 1087–1094, 2012.

36 M. R. Morelande and N. J. Gordon. Target tracking through a coordinated turn, In *Proc. IEEE International Conference on Acoustics, Speech, and Signal Processing (ICASSP'05)*, Philadelphia, PA, USA, 2005.

37 M. Mallick, X. Tian, and L. Xu. Discretization of dynamic model and process noise in ballistic object filtering. In *13th International Conference on Control, Automation and Information Sciences (ICCAIS)*, Ho Chi Minh City, Vietnam, 1–6,2024.

38 F. J. Regan and S. M. Anandakrishnan. *Dynamics of Atmospheric Re-entry*. Washington D.C., American Institute of Aeronautics and Astronautics, 1993.

39 R. A. Braeunig. Rocket and Space Technology: Drag Coefficients, 2012, http://www.braeunig.us/space/cd.htm

40 W. H. Press. *Numerical Recipes: The Art of Scientific Computing*, Cambridge University Press, 3rd Edition, 2007.

41 M. Mallick, K-C. Chang, S. Arulampalam, and Y. Yan. Heterogeneous Track-to-Track fusion in 3D using IRST sensor and air MTI radar. *IEEE Transactions on Aerospace and Electronic Systems*, 55(6):3062–3079, 2019.

42 M. Mallick, M. Morelande, L. Mihaylova, S. Arulampalam, and Y. Yan, Angle-only filtering in three dimensions, In M. Mallick, V. Krishnamurthy, and B.-N. Vo, editors, *Integrated Tracking, Classification, and Sensor Management: Theory and Applications*, Wiley/IEEE, 1, 2012.

43 N. Peach. Bearings-only tracking using a set of range-parameterised extended Kalman filters. *IEE Proceedings-Control Theory Applications*, 142(1):73–80, 1995.

44 T. R. Kronhamn. Bearings-only target motion analysis based on a multihypothesis Kalman filter and adaptive ownship motion control. *IEE Proceedings Radar, Sonar and Navigation*, 145(4):247–252, 1998.

45 B. Ristic, S. Arulampalam, and N. Gordon. *Beyond the Kalman Filter*. Artech House, 2004.

46 M. Mallick, Y. Bar-Shalom, T. Kirubarajan and M. Moreland. An improved single-point track initiation using GMTI measurements. *IEEE Transactions on Aerospace and Electronic Systems*, 51(4):2697–2714, 2015.

47 K. Li, G. Zhou, and T. Kirubarajan. A general model-based filter initialization approach for linear and nonlinear dynamic systems. *Digital Signal Processing*, 111:102978, 2021.

48 J. M. Mendel. *Lessons in Estimation Theory for Signal Processing, Communications, and Control*. Pearson Education, 1995.

49 K. Ito and K. Q. Xiong. Gaussian filters for nonlinear filtering problems. *IEEE Transactions on Automatic Control*, 45(5):910–927, 2000.

50 Y. Wu, D. Hu, M. Wu, and X. Hu. A numerical-integration perspective on Gaussian filters. *IEEE Transactions on Signal Processing*, 54(8):2910–2921, 2006.

51 A. Papoulis. *Probability, Random Variables, and Stochastic Processes*. McGraw-Hill, 1984.

52 X. R. Li, Z. Zhao, and V. P Jilkov. Estimator's credibility and its measures. In *Proceedings of the IFAC 15th World Congress*, Barcelona, Spain, 2002.

53 M. Mallick and S. Arulampalam, Comparison of nonlinear filtering algorithms in ground moving target indicator (GMTI) target tracking. In *Proceedings of SPIE, vol. 5204*, San Diego, CA, 2003.

3

Radar Tracking with Bias Estimation

Ehsan Taghavi[1], Ratnasingham Tharmarasa[2], and Thiagalingam Kirubarajan[2]

[1] *Axon, Toronto, ON, Canada*
[2] *Department of Electrical and Computer Engineering, McMaster University, Hamilton, ON, Canada*

3.1 Introduction

Radar tracking has been one of the major advances of engineering since early 1930s. Radar technology has long evolved and received much attention because of its importance in detecting and tracking objects of interest from meteorological aspects to military applications. There are different methods to deal with radar signals, and each method can result in high-quality object detection and tracking under different conditions. In general tracking algorithms, it is assumed that radar provides a set of point measurements at each scan. The tracking algorithm extracts the desirable parameters by utilizing these point measurements. In general, radars provide more information than only a point measurement report. A practical radar may provide a data image, where each pixel pertains to the power or intensity in a special location (e.g., range bins and azimuth beams) [1]. In this case, one can use a simple thresholding method and possibly interpolation, and apply it to the data image. After this, those cells with intensity or power higher than the threshold can be treated as point measurements.

This chapter begins with an introduction to target dynamics, followed by a comprehensive overview of radar measurement models. It then discusses several radar tracking algorithms, including multisensor fusion, tracking with transmitters of opportunity, and radar bias estimation. Finally, the chapter presents simulation results to demonstrate the performance of some of the mentioned methods.

3.2 Target Motion Models and Clutter Model

In radar surveillance systems, sensors report their measurements to the tracker at regular intervals of time (scans or data frames). However, not all measurements are originated from the targets of interest; some measurements may come from physical background objects such as clutter, and others may be generated by thermal noise. In other words, there is a measurement origin ambiguity. Furthermore, in some scans, the target of interest does not produce any measurements at all, i.e., the probability of detection is less than unity. The objective of a typical tracking system is to first partition all sensors' measurements into sets, such that all observations in one set are produced by the same object, i.e., data association process. Then, the measurements corresponding to the same object are processed in order to estimate the state of the object, i.e., the filtering process [2, 3].

To handle the tracking problem, usually the Bayesian approach is applied. In the Bayesian approach, the final goal is to construct the posterior probability density function (PDF) of the state (in some cases multitarget) given all the received measurements so far. Since this PDF contains all available statistical information, it is the complete solution to the tracking problem. In principle, given a cost function, it is always possible to obtain the optimal estimate under the given cost function from the posterior probability. Note that the state of the system should be a combination of the number of the targets and the state of each target, because, in a real scenario, both the number of the targets and the target states are random and unknown [4, 5].

To distinguish the target-originated measurements from the clutter and estimate the state of the system, the following three models, namely, the target dynamic model, the sensor model, and the clutter model are crucial for all radar tracking systems.

3.2.1 Target Dynamic Models

The target dynamic model, which is also known as the system model, describes the evolution of the state with time, given by

$$x(k+1) = f(k, x(k)) + G(k)v(k), \tag{3.1}$$

where f is, in general, a nonlinear function, $x(k)$ is the state of the target at time step k, and $v(k)$ is the process noise with covariance $q(k)$, which is usually assumed to be Gaussian. The covariance of the process noise multiplied by the gain $G(k)$ is

$$Q(k) = G(k)q(k)G(k)'. \tag{3.2}$$

The nearly constant velocity (NCV) model is widely used by radar tracking systems as the target dynamic model [6]. The state vector of NCV in one generic coordinate is

$$\mathbf{x} = [x\,\dot{x}]'. \tag{3.3}$$

In this case, the state model is linear, i.e., $f(k, x(k)) = F(k)$. The F and G in one generic coordinate are

$$F = \begin{bmatrix} 1 & T \\ 0 & 1 \end{bmatrix}, \tag{3.4}$$

$$G = \begin{bmatrix} T^2/2 \\ T \end{bmatrix}. \tag{3.5}$$

The process noise standard deviation σ_v is

$$\sigma_v = w_v A_{\max}, \tag{3.6}$$

where w_v is the scaling factor, and $A_{\max}$ is the maximum acceleration.

More advanced models like constant acceleration and constant turn models can be found in [6].

3.2.2 A Clutter Model

For a sensor with N resolution cells, sometimes detections will be declared in those cells that are pointed to a region without any targets of interest. The detection, which is not produced by any targets of interest, is known as a false alarm. Assume:

- The events of detection in each cell are independent of each other.
- The probability of false alarms is equal to p_{FA} in each cell and $p_{FA} \ll 1$.
- The number of the resolution cells is large, such that $p_{FA} \cdot N > 1$.

Then, the probability mass function (PMF) of the number of false alarms in these N resolution cells, $\mu_{FA}(m)$, is approximately following the Poisson distribution [2].

$$\mu_{FA}(m) = \exp\{-N \cdot p_{FA}\}\frac{(N \cdot p_{FA})^m}{m!}. \tag{3.7}$$

Furthermore, the spatial distribution of the false alarm is uniform based on the above three assumptions. Thus, if the granularity due to the size of the resolution cells can be neglected, the PDF of a false measurement, i.e., the clutter spatial intensity normalized by the expected number of clutter in the measurement space, is [2]

$$\bar{\lambda}_{FA} = p(z|z \text{ is a false alarm}) = \frac{1}{V}, \tag{3.8}$$

where V represents the volume of the sensor's measurement space. The unnormalized clutter spatial intensity is

$$\lambda_{FA} = \frac{N \cdot p_{FA}}{V}. \tag{3.9}$$

3.3 Tracking with Radar Measurements

In this section, different radar measurements are introduced as the fundamental step to design a complete radar tracking system. Each measurement provides partial information about the targets in the surveillance region, and based on that, a suitable tracking algorithm must be applied to the measurements. There are various types of radars, each of which provides a specific set of measurements to the tracking systems. This section starts with defining all radar measurements and then considers the most important radar tracking models.

3.3.1 Radar Sensor Models

Radar measurement model is a model relating the noisy measurements to the state and is given by

$$z(k) = h(k, \mathbf{x}(k)) + w(k), \tag{3.10}$$

where h is, in general, nonlinear functions, $x(k)$ is the state of the target, $z(k)$ is the measurement vector, and $w(k)$ is the measurement noise at measurement time k, which is usually assumed to be Gaussian. The following measurement models are typical for the radar tracking system:

- Range, r

$$h = \sqrt{(x - x_s)^2 + (y - y_s)^2 + (z - z_s)^2}. \tag{3.11}$$

For 2D tracking, the terms related to z must be deleted.

- Range rate, $\dot{r}$

$$h = ((x - x_s)\dot{x} + (y - y_s)\dot{y} + (z - z_s)\dot{z})/r. \tag{3.12}$$

For 2D tracking, the terms related to z must be deleted.

- Azimuth, θ

$$h = \arctan(y - y_s, x - x_s) \tag{3.13}$$

- Elevation, γ

$$h = \arctan\left(z - z_s, \sqrt{(x - x_s)^2 + (y - y_s)^2}\right) \tag{3.14}$$

- Bistatic range, r_b

$$h = \sqrt{(x - x_r)^2 + (y - y_r)^2} + \sqrt{(x - x_t)^2 + (y - y_t)^2}. \tag{3.15}$$

- Bistatic range rate, $\dot{r}_b$

$$h = \frac{(x - x_r)}{R_r}\dot{x} + \frac{(y - y_r)}{R_r}\dot{y} + \frac{(x - x_t)}{R_t}\dot{x} + \frac{(y - y_t)}{R_t}\dot{y}. \tag{3.16}$$

A combination of these measurements is collected by a specific radar model and further processed to find the targets' states and track them over time.

3.3.2 2D/3D Radar

The basic model for radar measurement consists of range and azimuth for a 2D radar and additional elevation measurement for a 3D radar. As a result, the radar measurement in this case can be written as

$$h(k, \mathbf{x}(k)) = \begin{bmatrix} r(k) \\ \theta(k) \\ \gamma(k) \end{bmatrix}. \tag{3.17}$$

Figure 3.1 shows how the measurements are collected by a 3D radar. By defining $h(k, \mathbf{x}(k))$ and the covariance of $w(k)$ as $\Sigma_{3D} = \mathrm{diag}(\sigma_r^2, \sigma_\theta^2, \sigma_\gamma^2)$, where σ_r, σ_θ, and σ_γ are the error standard deviation of range, azimuth, and elevation, respectively, one can apply EKF or UKF to obtain the state estimates recursively. Moreover, it is possible to convert the measurements to Cartesian coordinates as [7, 8].

$$x = r\cos(\theta)\cos(\gamma), \tag{3.18}$$

$$y = r\cos(\theta)\sin(\gamma), \tag{3.19}$$

$$z = r\sin(\gamma), \tag{3.20}$$

and then use Kalman filter to estimate the states. It is worth noting that, as it is stated in Radar Bias Estimation section, the converted covariance matrix for the new set of measurements must be used instead of $\Sigma_{3D} = \mathrm{diag}(\sigma_r^2, \sigma_\theta^2, \sigma_\gamma^2)$. Further discussion of converted coordinated Kalman filter can be found in [2, 7, 8].

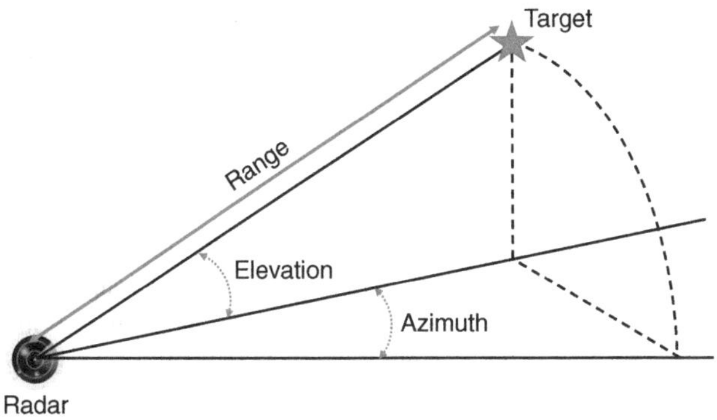

Figure 3.1 To establish the location of the target in space, three quantities must be determined: distance (range), bearing (azimuth), and altitude (elevation). *Source*: [43] / with permission of IEEE.

3.3.3 Radar with Doppler Measurement

In some cases, Doppler or range rate measurements, which provide further information about velocity directly, may be available. Here, the Doppler effect is used to determine the relative velocity of the target [9]. Pulses of radio frequency (RF) energy returning from the target are processed to measure the frequency shift between carrier cycles in each pulse and the original transmitted frequency. Therefore, the Doppler measurement can be incorporated into the tracker to enhance the tracking performance. On the other hand, in a coherent pulsed Doppler radar, ambiguity can occur in both range and radial velocity measurements [10, 11]. The focus here is on the combination of position measurements provided in spherical coordinates and Doppler measurements. The observations are modeled as [6]:

$$h(k, \mathbf{x}(k)) = \begin{bmatrix} r(k) \\ \theta(k) \\ \dot{r}(k) \text{ or } \Delta\dot{r} \end{bmatrix}. \tag{3.21}$$

The noiseless Doppler measurement $\dot{r}$ without ambiguity is defined in (3.12), and $\Delta\dot{r}$ with ambiguity is given by

$$\Delta\dot{r} = \text{mod}(\dot{r}, v_f), \tag{3.22}$$

where v_f is a given constant called the first blind velocity [12], which is determined from the PRF as

$$v_f = \frac{1}{2}\lambda \times \text{PRF}, \tag{3.23}$$

where λ is the wavelength of the radar waves. The measurement noise $w(k)$ is a white, zero-mean Gaussian random variable with covariance

$$R = \begin{bmatrix} \sigma_r^2 & 0 & \rho\sigma_r\sigma_{\dot{r}} \\ 0 & \sigma_\theta^2 & 0 \\ \rho\sigma_r\sigma_{\dot{r}} & 0 & \sigma_{\dot{r}}^2 \end{bmatrix}, \tag{3.24}$$

where σ_r, σ_θ, and $\sigma_r = \sigma_{\dot{r}}$ are the standard deviations of the additive white noise for range, bearing, and Doppler measurements, respectively. According to [13], there is a negative correlation, denoted by ρ in (3.24), between the range measurement error and the Doppler measurement error. However, in this chapter, we assume that they are independent. By having a precise measurement model, one can apply a tracking algorithm for single or multitarget scenarios as in [14].

3.3.4 Unresolved Targets

Targets are called unresolved if more than one target falls in a measurement resolution cell. That is, a single measurement can be originated from multiple targets. To formulate the 2D assignment problem with possibly unresolved measurements, we assume the following notational conventions [15]. At scan k, the M available measurements are to be assigned to the V most likely existing tracks from the previous scans up to scan index $k-1$. Among those tracks, it is assumed that there are U track pairs with possibly unresolved measurements. The track list is augmented to have the V tracks and U track pairs for the assignment. The computational load increases significantly in a dense target environment where the number of potential track pairs is large. This is unavoidable because the number of hypotheses on the measurement origins has increased significantly. We do not consider merged measurements from more than two tracks. The generalization to such a case is conceptually straightforward, but at a larger increase in complexity.

Define the binary assignment variable $\mathcal{X}_{ij}$ as

$$\mathcal{X}_{ij} = \begin{cases} 1 & \text{if measurement } j \text{ is assigned to track } i \text{ or track pair } j \\ 0 & \text{otherwise} \end{cases}. \tag{3.25}$$

Note that $\mathcal{X}_{i0} = 1$ when $i \leq V$ implies that track i is unassociated and has missed a detection at scan k (it is associated with the "dummy" measurement). If the index i corresponds to a pair of tracks, $\mathcal{X}_{i0} = 1$ implies that no (unresolved) measurement is assigned to track pair i. Define c_{ij} as the cost to assign measurement j to track i, i.e., the negative log-likelihood ratio of track i being assigned to measurement $z_j(k)$ at time k is [2, 16]

$$c_{ij} = \frac{1}{2}\left[z_j(k) - \hat{z}_i(k \mid k-1)\right]^{\mathrm{T}} S_i(k)^{-1}\left[z_j(k) - \hat{z}_i(k \mid k-1)\right] + \ln \frac{\lambda_e |2\pi S_i(k)|^{1/2}}{P_D} - \ln P_{\text{res}} \tag{3.26}$$

where $\hat{z}_i(k \mid k-1)$ is the predicted measurement from track i with covariance $S_i(k)$; λ_e is the spatial density of the false alarms (extraneous measurements); P_D is the detection probability of track i; and P_{res} is the probability of obtaining a fully resolved measurement if track i is detected. When no measurement is assigned to track i, i.e., it has a dummy measurement indexed by 0, the cost is

$$c_{i0} = -\ln\left[1 - P_D\right]. \tag{3.27}$$

Similarly, we can define the binary assignment variable $\mathcal{X}_{lj}$ and the cost c_{lj} of assigning measurement j to the track pair l as being an unresolved measurement. Denote the indices of the individual tracks comprising the track pair l as $l_1(l)$ and $l_2(l)$. The prediction of the merged measurement (assuming the same SNR for both tracks) is

$$\hat{z}_l(k \mid k-1) = \frac{1}{2}\left[\hat{z}_{l_1}(k \mid k-1) + \hat{z}_{l_2}(k \mid k-1)\right]. \tag{3.28}$$

The negative log-likelihood ratio of track pair l being assigned to (merged) measurement $z_j(k)$ at time k is

$$c_{lj} = \left[z_j(k) - \hat{z}_l(k \mid k-1)\right]^{\mathrm{T}} S_l(k)^{-1}\left[z_j(k) - \hat{z}_l(k \mid k-1)\right] + \ln \frac{\lambda_e |2\pi S_l(k)|^{1/2}}{P_D} - \ln\left(1 - P_{\text{res}}\right). \tag{3.29}$$

The innovation covariance matrix can be obtained as

$$S_l(k) = H(k)P_l(k \mid k-1)H(k)^{\mathrm{T}} + R_l(k), \tag{3.30}$$

where $H(k)$ is the measurement matrix, and $R_l(k)$ is the covariance of the unresolved measurement at time k. The covariance of state prediction can be approximated by

$$P_l(k \mid k-1) = \frac{1}{2}\left[P_{l_1(l)}(k \mid k-1) + P_{l_2(l)}(k \mid k-1)\right]. \tag{3.31}$$

Note that the probability of having an unresolved measurement is assumed to be independent of P_D. This model simplifies the measurement resolvability uncertainty (single vs. multiple origins) by one quantity Pres and has the advantage of incorporating additional prior (such as target feature information) in a Bayesian framework.

The modified 2D assignment problem is formulated as

$$\min_{\mathcal{X}} \sum_{i=1}^{V+U} \sum_{j=0}^{M} c_{ij}\mathcal{X}_{ij}, \tag{3.32}$$

subject to

$$\sum_{j=0}^{M} \mathcal{X}_{ij} = 1, \quad i = 1, \dots, V+U, \tag{3.33}$$

i.e., each track or track pair gets a measurement or a dummy;

$$\sum_{i=1}^{V+U} \mathcal{X}_{ij} \le 1, \quad j = 1, \dots, M, \tag{3.34}$$

i.e., each measurement can be assigned to at most one track or one track pair;

$$\sum_{j=0}^{M} (\mathcal{X}_{lj} + \mathcal{X}_{l_1(l),j}) = 1, \quad i = 1, \dots, V+U, \tag{3.35}$$

$$\sum_{j=0}^{M} (\mathcal{X}_{lj} + \mathcal{X}_{l_2(l),j}) = 1, \quad i = 1, \dots, V+U, \tag{3.36}$$

i.e., the track pair and the corresponding individual tracks can be assigned to at most one measurement;

$$\mathcal{X}_{ij} \in \{0, 1\}, \quad i = 1, \dots, V+U, j = 1, \dots, M. \tag{3.37}$$

Additional constraints are needed when one considers an unresolved measurement from more than two tracks. For instance, the likelihood function of a measurement from tracks 1, 2, and 3 should be defined and compared with the likelihood functions of a measurement from one of the track pairs (1&2, 1&3, 2&3) with a missed detection of the third track. Only one assignment is allowed for the above four cases, which requires a new constraint. Furthermore, one may want to exclude the case that two track pairs with a common track (e.g., 1&2 and 1&3) being assigned to two different measurements as a feasible solution. However, the number of new constraints to ensure this is proportional to the number of track pairs, which makes the resulting integer programming problem even more complicated. In practice, the overall likelihood of two track pairs with a common track being assigned to two different measurements is relatively small when the measurements have a large separation. Thus, we expand the hypothesis space by ignoring less important constraints to make the problem computationally tractable.

Note that the resulting integer programming problem is not a standard 2D assignment and can not be solved using JVC or auction algorithm. If we relax the integer constraints for all binary assignment variables with the new constraints that they are bounded within [0, 1], the problem becomes a linear program with $(V + U)(M + 1)$ variables and $V + 3U + M$ constraints. In this case, we use interior point linear programming to solve the (relaxed) constrained optimization rather than enumerating all feasible hypotheses as in [2]. The following theorem shows that the linear program yields the same optimal result as the original integer program in 2D problem. In some special cases, such as no common tracks among all track pairs, the 2D program can be converted into a linear cost flow problem with augmented source and destination nodes [17], and more efficient algorithms can be used than solving a linear program.

3.3.5 Passive Coherent Location

Passive coherent location (PCL) is a specific case of bistatic or multistatic radar surveillance systems. Most multistatic PCL systems are based on ellipsoid crossing localization [18]. In general, a simple PCL system consists of one illuminator of opportunity (the transmitter) and a PCL receiver. This bistatic configuration of PCL is illustrated in Figure 3.2. The direct signal and the echo from the target that originate from commercial transmitters like TV and radio stations are received by the PCL receiver.

It is assumed that the PCL system's measurements are bistatic range, bistatic range rate, and bearing (from north). The measurement noise w, is assumed to be a zero-mean Gaussian random variable with covariance $\Sigma = [\sigma_r^2, \sigma_{\dot{r}}^2, \sigma_\theta^2]$. False alarms are assumed to be uniformly distributed throughout the surveillance region.

In a traditional setting, where the locations of the commercial transmitters are known accurately, methods in [19] can be applied to estimate the target's state. If there exists transmitter-origin uncertainty in a multistatic PCL system, i.e., multiple transmitters-of-opportunity in the area of interest transmit the same signal at the same frequency, one cannot use these measurements without resolving transmitter- and measurement-origin uncertainties. Then, the problem is to track multiple time-varying numbers of targets by resolving the measurement origin and reflection point (clutter/target) uncertainties. This adds another level of complexity to the standard data association problem, where the uncertainty is only in measurement origins. That is, there are two uncertainties that need to be resolved in order to track multiple targets. One is the measurement-to-target association, and the other is the measurement-to-transmitter association.

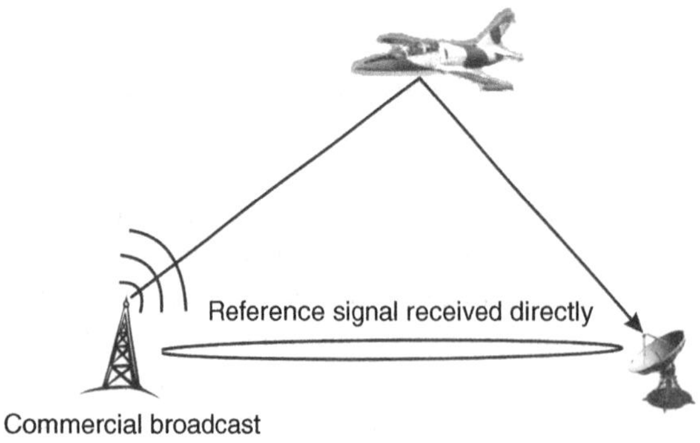

Figure 3.2 PCL system.

The possible approaches to solve the new association problem are as follows [19]:

- Track the targets in the state space and associate the measurements to the targets and transmitters.
- Track the targets in the measurement space by assuming that all the measurements are from a single transmitter. In this case, more than one track may be formed for each target. Hence, tracks need to be fused by finding the correct track-to-transmitter assignment.

To explore more, a brief explanation of the first method is given here. For already established tracks, the problem is to associate the measurements to the targets and the transmitters. This assignment problem is described by the tree diagram in Figure 3.3. In Figure 3.3, the dummy target indicates false alarms and the dummy measurement indicates the miss detections. Note that even though dummy targets and dummy measurements exist, there is no dummy transmitter. To present the assignment, define a binary assignment variable $a(k, m, t, j)$ such that

$$a(k, m, t, j) = \begin{cases} 1 & \text{measurement } z(k, m) \text{ is assigned to track } \mathcal{T}^t(k-1) \\ & \text{and transmitter } j \\ 0 & \text{otherwise} \end{cases} . \tag{3.38}$$

The objective of the assignment is to find the optimal assignment $a^*(k)$, which minimizes the global association cost.

$$C(k \mid \mathbf{a}(k)) = \sum_{m=0}^{M} \sum_{t=0}^{T} \sum_{j=1}^{N} a(k, m, t, j) c(k, m, t, j), \tag{3.39}$$

where $c(k, m, t, j)$ is the cost of assignment $a(k, m, t, j)$. Finally, the constraints are

$$\sum_{m=0}^{M} a(k, m, t, j), \quad t = 1, 2, \dots, T, \quad j = 1, 2, \dots, N, \tag{3.40}$$

$$\sum_{t=0}^{T} a(k, m, t, j), \quad m = 1, 2, \dots, M, \quad j = 1, 2, \dots, N, \tag{3.41}$$

$$\sum_{j=1}^{N} a(k, m, t, j), \quad m = 1, 2, \dots, M, \quad t = 1, 2, \dots, T. \tag{3.42}$$

Despite the simple appearance of the above formulation and the similarity to the standard assignment problem, the objective function (3.39) will result in a suboptimal solution if the $c(k, m, t, j)$

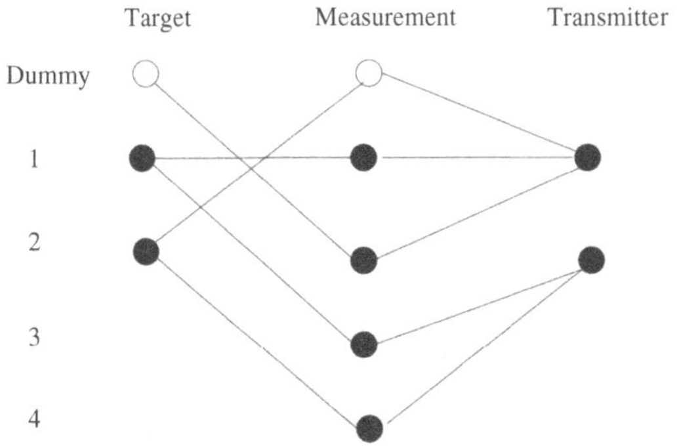

Figure 3.3 Assignment tree.

are calculated independent of any other measurement that can be associated to the same target. In contrast, with the standard 2-D assignment, it is possible to find the optimal solution with polynomial time complexity [20].

3.3.6 Continuous 2D Assignment

Conventional target tracking algorithms rely on receiving all the measurements in a scan or frame to perform measurement-to-track association in order to satisfy the common one measurement per track assumption [2, 3, 21]. For example, multitarget tracking algorithms like the joint probabilistic data association (JPDA) filter [22, 23], the finite set statistics (FISST)-based probability hypothesis density (PHD) filter [4, 24, 25], the multiple hypothesis tracker (MHT) [3], and the multiframe assignment tracker (MFA) [21, 26, 27] require the full set of measurements in a scan in order to proceed with measurement-to-track association and/or filtering. However, most mechanically steered radar systems return measurements continuously, at the instant of each detection or in sectors, while sweeping from one region to another. Waiting for the full set of measurements in a scan (e.g., 360 degrees) to perform data association and filtering results in delayed track estimates. This problem becomes challenging when tracking maneuvering high-speed targets with low scan rate sensors, as in the airborne early warning (AEW) system [28].

The conventional frame-based 2-D assignment algorithm uses the complete set of measurements $Z(k)$ in the scan for measurement-to-track association and state estimation, which requires waiting until the end of the scan. In contrast, in sector-based 2-D assignment algorithm, a partial set of measurements $Z_s(t_{ks}) \in Z(k)$, i.e., measurements in a given sector, are used for the measurement-to-track association and state estimation. One way of implementing the sector-based 2-D assignment is to perform 2-D association at the end of each sector and estimate target states. However, this approach poses a problem whenever the measurement validation region of targets or cluster of targets lies in more than one sector, e.g., targets $\mathcal{T}_5(k)$, $\mathcal{T}_7(k)$ and targets cluster $\mathcal{C}_1(k) = \mathcal{T}_2(k), \mathcal{T}_3(k), \mathcal{T}_4(k)$ in Figure 3.4. For a complete explanation of how to implement a working algorithm, refer [29].

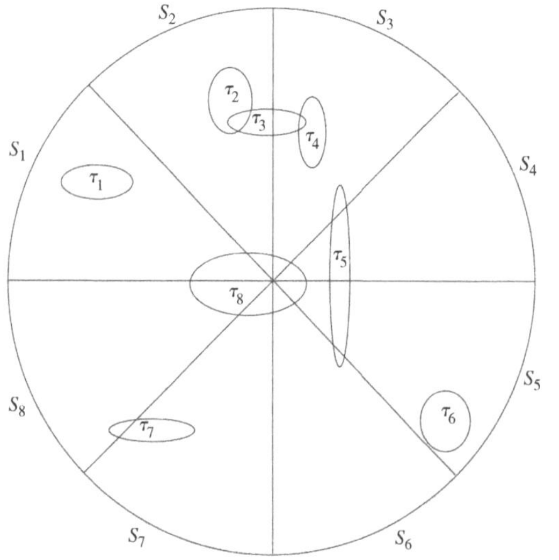

Figure 3.4 Measurement space of rotating radar with eight 45 degrees sectors ($\mathcal{S}_1, \ldots, \mathcal{S}_8$). Measurement validation region for each target ($\mathcal{T}_1, \ldots, \mathcal{T}_8$) is shown by the corresponding ellipse.

3.4 Multisensor Radar Tracking

Three major types of architecture, namely, centralized, distributed and decentralized are commonly used in multisensor–multitarget tracking applications [2, 30, 31]. Here, it is assumed that the data of each sensor has already been converted to a common coordinate system, regardless of whether the sensor measurements are processed in the centralized or distributed architecture.

3.4.1 Centralized Tracking

In the centralized architecture, Figure 3.5a, several sensors monitor the region of interest to detect and track the targets therein. All sensors generate measurements at each revisit time and report those measurements to the central fusion center (CFC). It in turn fuses all the acquired measurements and updates the tracks. This is the optimal architecture in terms of tracking performance.

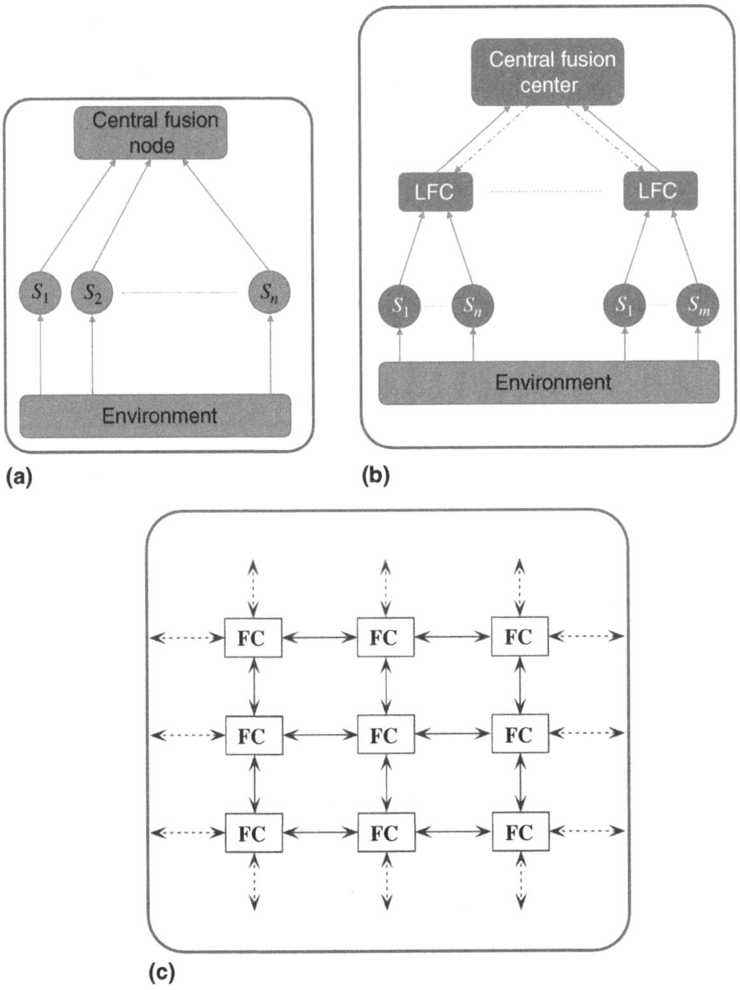

Figure 3.5 Common fusion architectures.

However, in a large surveillance region with many sensors, this architecture may not be feasible because of limited resources, e.g., communication bandwidth and computation power. To further improve the reliability and performance of the centralized architecture, one can use the replicated centralized architecture [32]. In the replicated centralized architecture, there are multiple CFCs operating independently and each sensor reports its measurements to all CFCs. In other words, each CFC processes data from all sensors and there is no communication among CFCs. This architecture has high performance and reliability because the multiple CFCs process the same data. However, it also has higher communication and processing costs.

Two well-known schemes used by the CFC to incorporate measurements from multiple sensors are sequential updating and parallel updating:

- In the sequential updating scheme, the updating is carried out with the measurement of one sensor at a time. The posterior state estimate obtained from the measurements reported by the kth sensor is used as the prior information when the measurements from the $k + 1$th sensor are processed. For one specific sensor, those techniques commonly used in the single sensor–multitarget tracking problem, like one/two point initialization, multiframe data association, and Kalman filtering, are applied.
- In the parallel updating scheme, the measurements from the various sensors are first associated to yield "supermeasurements," then these supermeasurements are processed by the CFC as if they were reported by a single sensor. In other words, after the measurement association procedure, the multisensor–multitarget tracking problem has been converted to a single sensor–multitarget tracking problem in the parallel updating scheme.

In the two schemes mentioned above, it is usually assumed that the sensors are synchronized. In real scenarios, the sensors are seldom perfectly synchronized and too many asynchronous sensors will reduce the optimality of the centralized tracking architecture.

3.4.2 Distributed Tracking

In order to avoid the heavy communication and computational requirements of centralized fusion, distributed or hierarchical architecture, shown in Figure 3.5b, is used alternatively [31]. In this architecture, sensors are connected to local fusion centers (LFCs) and LFCs are in turn connected to a CFC. Each LFC updates its local tracks based on the measurements obtained from the local sensors and sends its tracks information to CFC. Then, the CFC performs the track-to-track fusion and may send back the updated tracks to the LFCs, if the feedback path is available.

There are two crucial questions in this type of architecture. The CFC needs to determine whether local tracks from different LFCs represent the same target, i.e., the problem of track-to-track association. Furthermore, because of the common process noise or the common prior, the local state estimates for the same target from different LFCs are no longer independent anymore. Thus, the dependency between the local tracks for the same target from different LFCs has to be handled optimally in the track–track fusion. First, the cross-covariance between the local tracks has to be estimated and then the covariance of the fused track has to be increased correspondingly. Note that, even if the above two questions are solved optimally, the distributed tracking architecture is still suboptimal compared to the centralized architecture, because the optimal fusion of the state estimates of local tracks can not provide the same performance as the optimal fusion of the entire data set [2].

To optimally handle the dependency between the local tracks for the same target, the computational cost may be large and the required information may take up too much communication bandwidth, especially when there are many LFCs. Thus, several suboptimal methods are proposed to handle the dependency between the local tracks for the same target. One suboptimal way is using the corresponding tracklets of the local tracks in the CFC for the track–track fusion. A tracklet is a track specially calculated from the local track such that its errors are not cross-correlated with the errors of any other data in the system for the same target [33–35]. Another suboptimal method is the covariance intersection technique, which takes a convex combination of the estimate of the mean and the information matrix (i.e., the inverse of the covariance matrix) of the local tracks. The covariance intersection method provides an estimate of the upper bound of the covariance matrix of the optimally fused track [12, 36].

3.4.3 Decentralized Tracking

When there is no CFC that can communicate with all LFCs in a large surveillance region, neither centralized nor distributed tracking is possible. In such cases, an alternative called decentralized architecture, shown in Figure 3.5c, is used. Decentralized architecture is composed of multiple fusion centers (FCs) and no CFC [31]. Here, each FC obtains the measurements from one or more sensors that are connected to it, and uses those measurements to update its tracks. In addition, tracks are also updated whenever an FC gets additional information from its neighbors. Note that even though many FCs are available, each FC can communicate only with its neighbors; the FCs within the communication distance every few measurement time steps. There is no common communication facility in decentralized tracking network. As a result, FC cannot broadcast results and communication must be kept on a strictly neighbor-to-neighbor basis [37]. In decentralized tracking architecture, each FC can be considered as a combination of LFC and CFC, because it updates the track state using measurements from connected sensors, and it also does track–track fusion whenever neighboring FCs report their track states. Usually, the algorithms developed for the distributed tracking architecture can be used here without any major modification.

3.5 Radar Bias Estimation

Bias estimation and compensation are essential steps in distributed tracking systems. The objective of sensor registration is to estimate the biases in sensor measurements, such as scaling and offset biases in range and azimuth measurements of a radar, clock bias and/or uncertainties in sensor positions [2]. In a distributed multisensor tracking scenario, each local tracker provides its own estimates of target states for fusion. Local filters can be, e.g., Kalman filters or interacting multiple model (IMM) estimators with different motion models. These local tracks, i.e., state estimate vectors and associated covariance matrices, are sent to the fusion center for further processing. Next, the fusion center carries out track-to-track fusion. The fusion is done sequentially subsequent to the estimation of biases based on common targets that are tracked by various sensors in different locations.

Usually, bias estimation is considered as a two-sensor problem [38, 39] where a stacked vector is assumed with all unknown biases and states as targets. A drawback of this approach is the computational burden due to the increased dimension of the stacked vector. In addition, most of the algorithms proposed for estimating the biases operate on the measurements directly [40]. That is, such methods perform filtering on the measurements received from sensors, which also include the biases. In many practical tracking systems, access to measurements before tracking at the sensor level is not always feasible. That is, sensors may provide only processed tracks to the user for further processing [2]. Thus, methods that can simultaneously handle track-to-track fusion and bias estimation are needed.

3.5.1 Problem Formulation for a Distributed Tracking System

Assume that there are M sensors reporting range and azimuth measurements[1] in polar coordinates of N targets in the common surveillance region. Note that N is not exactly known to the algorithm and that it could be time-varying. That is, bias estimation is carried out based on time-varying and possibly erroneous numbers of tracks reported by the local trackers. The model for the measurements originating from a target with biases at time k in polar coordinates (denoted by superscript p) for sensor s is [7, 41, 42]

$$z_s^p(k) = \begin{bmatrix} r_s^p(k) \\ \theta_s^p(k) \end{bmatrix} = \begin{bmatrix} [1+\epsilon_s^r(k)]\, r_s(k) + b_s^r(k) + w_s^r(k) \\ [1+\epsilon_s^\theta(k)]\, \theta_s(k) + b_s^\theta(k) + w_s^\theta(k) \end{bmatrix} \qquad s = 1, \ldots, M, \tag{3.43}$$

where $r_s(k)$ and $\theta_s(k)$ are the true range and azimuth, respectively, $b_s^r(k)$ and $b_s^\theta(k)$ are the offset biases in the range and azimuth, respectively, $\epsilon_s^r(k)$ and $\epsilon_s^\theta(k)$ are the scale biases in the range and azimuth, respectively. The measurement noises $w_s^r(k)$ and $w_s^\theta(k)$ in range and bearing are zero-mean with corresponding variances σ_r^2 and σ_θ^2, respectively, and are assumed to be mutually independent.

The bias vector $\beta_s(k) = \begin{bmatrix} b_s^r(k) & b_s^\theta(k) & \epsilon_s^r(k) & \epsilon_s^\theta(k) \end{bmatrix}^{\mathrm{T}}$ can be modeled as an unknown constant over a certain window of scans (nonrandom variable). Consequently, the maximum likelihood (ML) estimator [43] or the weighted least squares (LS) estimator [44] can be used for bias estimation. On the other hand, a Gauss-Markov random model [45] can also be used, in which case a Kalman filter can be adopted for bias estimation. The measurement model can be written as

$$z_s^p(k) = \begin{bmatrix} r_s(k) \\ \theta_s(k) \end{bmatrix} + C_s(k)\beta_s(k) + \begin{bmatrix} w_s^r(k) \\ w_s^\theta(k) \end{bmatrix}, \tag{3.44}$$

where

$$C_s(k) \triangleq \begin{bmatrix} 1 & 0 & r_s(k) & 0 \\ 0 & 1 & 0 & \theta_s(k) \end{bmatrix}. \tag{3.45}$$

Here, the measured azimuth $\theta_s^m(k)$ and range $r_s^m(k)$ can be utilized in (3.45) without any significant loss of performance [7, 41, 42].

After bias estimation, all the biases can be compensated for in the state estimates at the fusion center.

Since target motion is better modeled and most trackers operate in Cartesian coordinates, the polar measurements are converted into Cartesian coordinates. It is assumed that this does not

---

1 While this assumes 2D radars, the extension to 3D radars is straightforward.

introduce biases [6]; this is verified in the simulations. Then, sensor s has the measurement equation (with the same $H_s(k) = H(k)$ for all s).

$$z_s(k) = H(k)\mathbf{x}(k) + B_s(k)C_s(k)\beta_s(k) + w_s(k), \tag{3.46}$$

where the state vector $\mathbf{x}(k) = \begin{bmatrix} x(k) & \dot{x}(k) & y(k) & \dot{y}(k) \end{bmatrix}^{\mathrm{T}}$ and $H(k)$ is the measurement matrix given by

$$H(k) = \begin{bmatrix} 1 & 0 & 0 & 0 \\ 0 & 0 & 1 & 0 \end{bmatrix} \triangleq H. \tag{3.47}$$

Since distributed tracking systems may cover a large geographical area, the earth can no longer be assumed to be flat and coordinate transformations need to include an earth curvature model like WGS-84 [46, 47].

The matrix $B_s(k)$ is a nonlinear function with the true range and azimuth as its arguments. A constant $B_s(k)C_s(k)$ also results in incomplete observability as discussed in [41]. Using the measured azimuth $\theta_s^m(k)$ and range $r_s^m(k)$ from sensor s, $B_s(k)$ can be written as [6].

$$B_s(k) = \begin{bmatrix} \cos\theta_s^m(k) & -r_s^m(k)\sin\theta_s^m(k) \\ \sin\theta_s^m(k) & r_s^m(k)\cos\theta_s^m(k) \end{bmatrix}. \tag{3.48}$$

Finally, the new covariance matrix of the measurement in Cartesian coordinates (omitting index k in the measurements for clarity) is given by

$$R_s(k) = \begin{pmatrix} r_s^2\sigma_\theta^2 \sin^2\theta_s + \sigma_r^2\cos^2\theta_s & (\sigma_r^2 - r_s^2\sigma_\theta^2)\sin\theta_s\cos\theta_s \\ (\sigma_r^2 - r_s^2\sigma_\theta^2)\sin\theta_s\cos\theta_s & r_s^2\sigma_\theta^2\cos^2\theta_s + \sigma_r^2\sin^2\theta_s \end{pmatrix}, \tag{3.49}$$

where one can use the observed range and azimuth as well.

3.5.2 General Radar Bias Estimation Algorithm

Recursive least square bias estimation (RLSB) and optimal MMSE bias estimation (OMBE) [48], are dependent on the Kalman gains provided by the local trackers, in addition to the state estimates and the associated covariance matrices at every time step. Moreover, as the number of sensors increases, the above algorithms face an increase in computational requirements cubic in M.

In order to obtain a general algorithm, first, a simple approach to calculate tracklets based on [35] is obtained. This approach makes it possible to calculate approximate equivalent measurements of the local tracks directly and efficiently without any further processing, and it supports updating the bias estimates whenever a new local track is available at the fusion center. In addition, it can handle asynchronous updates from different local trackers and map them to a common time [33, 35]. Then, a sequential update algorithm is used for the fusion step. Although it is not an optimal approach for fusing the local tracks, it is computationally cheaper than parallel update [2].

3.5.3 Sequential Fusion Algorithm

After calculating the equivalent measurements of the state for each local track at the fusion center, they can be used as new measurements for the estimation of fused state and its covariance matrix. To do this recursively, it should be assumed that the fused state estimate and its covariance matrix at time k' as $\mathbf{x}_f(k' \mid k')$ and $\mathbf{P}_f(k' \mid k')$, respectively, are already computed. For $k' = 1$, the parameters $\mathbf{x}_f(k' \mid k')$ and $\mathbf{P}_f(k' \mid k')$ are initialized with $\mathbf{x}(k' \mid k')$ and $\mathbf{P}(k' \mid k')$, respectively. Then, these two can be updated by following the steps in Algorithm 3.1. Although the sequential update is sub-optimal [2], it has the advantage of being computationally efficient to implement and, in addition, it is not dependent on the previous equivalent measurements at time k'.

Algorithm 3.1: The Sequential Fusion Algorithm with Equivalent Measurements

Input: $\mathbf{x}_f(k' \mid k')$ and $\mathbf{P}_f(k' \mid k')$, $y_{s,k}$ and $R_{s,k}$ for $s = 1, \ldots, M$
Output: $\mathbf{x}_f(k \mid k)$ and $\mathbf{P}_f(k \mid k)$

1: **Compute** $\mathbf{x}_f(k \mid k')$ and $\mathbf{P}_f(k \mid k')$ according to their dynamic model (prediction step)
2: **Assign:**

$$\begin{aligned} \mathbf{x}_{\text{temp}} &= \mathbf{x}_f(k \mid k') \\ \mathbf{P}_{\text{temp}} &= \mathbf{P}_f(k \mid k'). \end{aligned}$$

3: **for** $s = 1, \ldots, M$ **do**
4: Update $\mathbf{x}_{\text{temp}}$ and $\mathbf{P}_{\text{temp}}$ with new measurement and its covariance matrix, i.e., $y_{s,k}$ and $R_{s,k}$ according to

$$\begin{aligned} \mathbf{x}_{\text{temp}} &= \mathbf{x}_{\text{temp}} + W_{\text{temp}}\tilde{\mathbf{y}} \\ W_{\text{temp}} &= \mathbf{P}_{\text{temp}}H^{\mathrm{T}}\left(H\mathbf{P}_{\text{temp}}H^{\mathrm{T}} + R_{s,k}\right)^{-1} \\ \tilde{\mathbf{y}} &= y_{s,k} - H\mathbf{x}_{\text{temp}} \\ \mathbf{P}_{\text{temp}} &= \left(I - W_{\text{temp}}H\right)\mathbf{P}_{\text{temp}} \\ H &= \begin{bmatrix} 1 & 0 & 0 & 0 \\ 0 & 0 & 1 & 0 \end{bmatrix} \end{aligned},$$

5: **end for**
6: **return**

$$\begin{aligned} \mathbf{x}_f(k \mid k) &= \mathbf{x}_{\text{temp}} \\ \mathbf{P}_f(k \mid k) &= \mathbf{P}_{\text{temp}} \end{aligned}.$$

3.5.4 Multisensor Fusion and Track-to-track Bias Estimation

The first step in implementing a general bias estimation algorithm for radar systems is to find the Kalman gains of each local track at the fusion center by using only the state estimates and the associated covariance matrices. To do so, first, one must calculate the equivalent measurement and its covariance matrix, $\mathbf{U}_s(k, k')$, for sensor s, and at time frame k (the target index is omitted for simplicity). Since the measurement model here is linear, one has

$$R_{s,k} = H(k)\mathbf{U}_s(k, k')H(k)^{\mathrm{T}}, \tag{3.50}$$

$$W_{s,k} = \mathbf{P}_s(k \mid k')H(k)^{\mathrm{T}}\left[H(k)\mathbf{P}_s(k \mid k')H(k)^{\mathrm{T}} + R_{s,k}\right]^{-1}, \tag{3.51}$$

$$y_{s,k} = H(k)\mathbf{u}_s(k, k'). \tag{3.52}$$

Note that the reason to keep the position information only is that further in the new bias estimation algorithm, the corrected positions are used as new measurements for sequential fusion. Because the equivalent measurements are used for the fused track, the Kalman gain for it (denoted by subscript f) at time frame k can be recovered as

$$R_{f,k} = H(k)\left[\sum_{i=1}^{M} (\mathbf{U}_i(k,k'))^{-1}\right]^{-1} H(k)^{\mathrm{T}}, \tag{3.53}$$

$$W_{f,k} = \mathbf{P}_f(k \mid k')H(k)^{\mathrm{T}}\left[H(k)\mathbf{P}_f(k \mid k')H(k)^{\mathrm{T}} + R_{f,k}\right]^{-1}. \tag{3.54}$$

Note that the noises/errors in the equivalent measurements are not white, so using a Kalman filter is not optimal. This amounts to the same approximation as in [2, p. 563]. Also in (3.54), a common coordinate system is used for equivalent measurements. As a result, the use of a common measurement matrix $H(k)$ for all the equivalent measurements is feasible.

To be able to apply the bias estimation algorithm, an approximately bias-compensated[2] fused track needs to be found at the fusion center for the set $\{\mathcal{S}\} \setminus s$, where $\mathcal{S} = 1, 2, \ldots, M$. The notation $\{\mathcal{S}\} \setminus s$ stands for the set that contains all those elements of $\mathcal{S}$ excluding element s. Then, the bias estimation problem can be reduced to the case of two sensors. The first one is an equivalent sensor with fusion of bias-corrected measurements of the set $\{\mathcal{S}\} \setminus s$, and the second is sensor s, which has bias. Then, the bias in sensor s can be found with either the RLSB or OMBE algorithm [48].

The next step in this approach is to correct the biases in the measurement domain of the sensors (except for sensor s) and then fuse them together. This can be done by going from the Cartesian coordinate of equivalent measurements to the polar coordinate of the radar and correct with the previously estimated biases. Then by correcting the covariance matrix of the new bias-compensated measurements, they can be fused by sequentially updating the fused track (excluding the track from sensor s) by using the sequential fusion algorithm (SFA). Then, the Kalman gain for the fused and the now-corrected track can be calculated using (3.54).

For the equivalent measurement, define

$$\mathbf{u}_s \triangleq \begin{bmatrix} u_x & \dot{u}_x & u_y & \dot{u}_y \end{bmatrix}^{\mathrm{T}}, \tag{3.55}$$

where time and target indexes are omitted for simplicity. Then, to correct the biases in the measurement domain, assuming that the latest-estimated biases are $\hat{\epsilon}_\theta$, $\hat{\epsilon}_r$, $\hat{b}_\theta$, and $\hat{b}_r$, one has[3]

$$\theta_s^{\text{b-c}} = \frac{\arctan\left(\frac{u_y}{u_x}\right) - \hat{b}_\theta}{(1 + \hat{\epsilon}_\theta)} \tag{3.56}$$

$$r_s^{\text{b-c}} = \frac{\sqrt{(u_x)^2 + (u_y)^2} - \hat{b}_r}{(1 + \hat{\epsilon}_r)}. \tag{3.57}$$

Now that the scale and offset biases are compensated for, one can go back to Cartesian coordinates as follows:

$$u_x^{\text{b-c}} = \lambda_\theta r_s^{\text{b-c}} \cos\left(\theta_s^{\text{b-c}}\right), \tag{3.58}$$

$$u_y^{\text{b-c}} = \lambda_\theta r_s^{\text{b-c}} \sin\left(\theta_s^{\text{b-c}}\right), \tag{3.59}$$

$$\mathcal{Y}^{\text{b-c}} = \begin{bmatrix} u_x^{\text{b-c}} & u_y^{\text{b-c}} \end{bmatrix}^{\mathrm{T}}, \tag{3.60}$$

2 Here, bias-compensated means a fused track, in which the bias-corrected equivalent measurements by using the latest estimated biases are used for fusion.

3 Here, the superscript "b−c" is used to denote the bias-corrected bearing and range.

where

$$\lambda_\theta = \exp\left(-\frac{\sigma_\theta^2}{2}\right), \tag{3.61}$$

is the compensation factor for the bias in coordinate conversion from polar to Cartesian [8]. The method from [8] is used here because of the fact that bias correction and compensation along with the changes in the covariance matrices and uncertainties may violate the assumptions made for the debiased conversion in Section 3.5.1. The next step is to update the covariance matrix of the corrected equivalent measurements. In addition to the term (3.50), the additional uncertainty in the bias estimates, i.e., their associated covariance matrix and the uncertainty in the model of the radar, both in Cartesian coordinates, must be accounted for. The final formula for the covariance matrix with proper conversion from polar to Cartesian coordinate is

$$\begin{aligned} R_s^{\mathrm{b-c}} = {} & H(k)\mathbf{U}_s(k \mid k)H(k)^{\mathrm{T}} + B_s^{\mathrm{b-c}}(k)\begin{bmatrix} \sigma_r^2 & 0 \\ 0 & \sigma_\theta^2 \end{bmatrix} B_s^{\mathrm{b-c}}(k)^{\mathrm{T}} \\ & + K_s^{\mathrm{b-c}}(k)\mathbf{P}_{b,s}(k \mid k)K_s^{\mathrm{b-c}}(k)^{\mathrm{T}} \end{aligned}, \tag{3.62}$$

where

$$K_s^{\mathrm{b-c}}(k) = B_s^{\mathrm{b-c}}(k)C_s^{\mathrm{b-c}}(k), \tag{3.63}$$

and $\mathbf{P}_{b,s}(k \mid k)$ is the latest-updated bias covariance matrix at time k and for sensor s. Now that all the required formulations and variables are available, the new algorithm to find the bias estimates for all the sensors is given in Algorithm 3.2.

Algorithm 3.2: The Fused Bias Estimation Algorithm

Input: inputs of SFA and RLSB (defined within the corresponding algorithms).
Output: $\mathbf{b}_s(k)$ and $\Sigma_s(k)$ for $s = 1, \dots, M$.

At time k
for $s = 1, \dots, M$ **do**
  Compute $W_{s,k}$ as in (3.51) and
  $z_{b,t}^s(k) \triangleq W_{s,k}^{\dagger}\left[\hat{\mathbf{x}}(k \mid k) - (I - W_{s,k}H_s(k))F(k, k-L)\hat{\mathbf{x}}(k-L \mid k-L)\right]$,
  $\bar{s} \in \{1, \dots, M\} \setminus s$
  Call SFA with inputs $\mathbf{x}_f^s(k' \mid k')$ and $\mathbf{P}_f^s(k' \mid k')$, $\mathcal{Y}_{\bar{s}}^{\mathrm{b-c}}(k \mid k)$ and $R_{\bar{s}}^{\mathrm{b-c}}(k \mid k)$.
  Compute $W_{f,k}^s$ as in (3.54) and

$$z_{b,f}^s(k) \triangleq \left(W_{f,k}^s\right)^{\dagger}\left[\hat{\mathbf{x}}_f^s(k \mid k) - (I - W_{f,k}^s H_f(k))F(k, k-L)\hat{\mathbf{x}}_f^s(k-L \mid k-L)\right].$$

  Call RLSB with inputs $z_{b,t}^s(k)$, $z_{b,f}^s(k)$ and the last update of the estimated biases and their associated covariance matrix.
  return $\mathbf{b}_s(k)$ and $\Sigma_s(k)$
end for

As shown in Algorithm 3.2, one only needs to call SFA and RLSB with new input parameters. One of the advantages of this approach is that in each "**for loop**" only a low-dimensional Kalman filter that is *independent* of the size of the stacked bias vector and number of the sensors is needed. In addition, the fusion of local tracks can be done by only adding one sequential update for the latest corrected measurement of sensor s to the previously fused track. It is also important to note that there is no constraint on the rate of receiving local tracks from the individual sensors. To show how well this new algorithm performs, in the next section, the simulation results on two different scenarios are used to compare its performance with those of the previously proposed algorithm in [7, 41, 42] for synchronous sensors.

To better illustrate how the new bias estimation algorithm (FBEA) works, a block-diagram representation of the method is shown in Figure 3.6 for a single time-step estimation of the biases for the first sensor. In Figure 3.6, by receiving the local track estimates from all available sensors at time step k, the first step is to calculate the tracklet for all of them. Then, the equivalent measurement of the first sensor is sent for Kalman gain recovery using (3.50) and (3.51). At the same time, the equivalent measurements of all other sensors are sent to bias correction to first remove the bias from the equivalent measurements by using the previously estimated biases at time step $k-1$ using (3.56)–(3.60) and the Kalman gain recovery in (3.53) and (3.54). The next step is to fuse the tracks by using SFA algorithm. Then, the fused and corrected estimate is sent to the pseudo-measurement calculation block for each individual sensor. At this point, one has a two-sensor problem with only one sensor having biases in the measurement. The output is now sent to the RLSB algorithm along with the previously estimated biases for the first sensor so that the bias estimates can be updated at time step k before proceeding to the next time step.

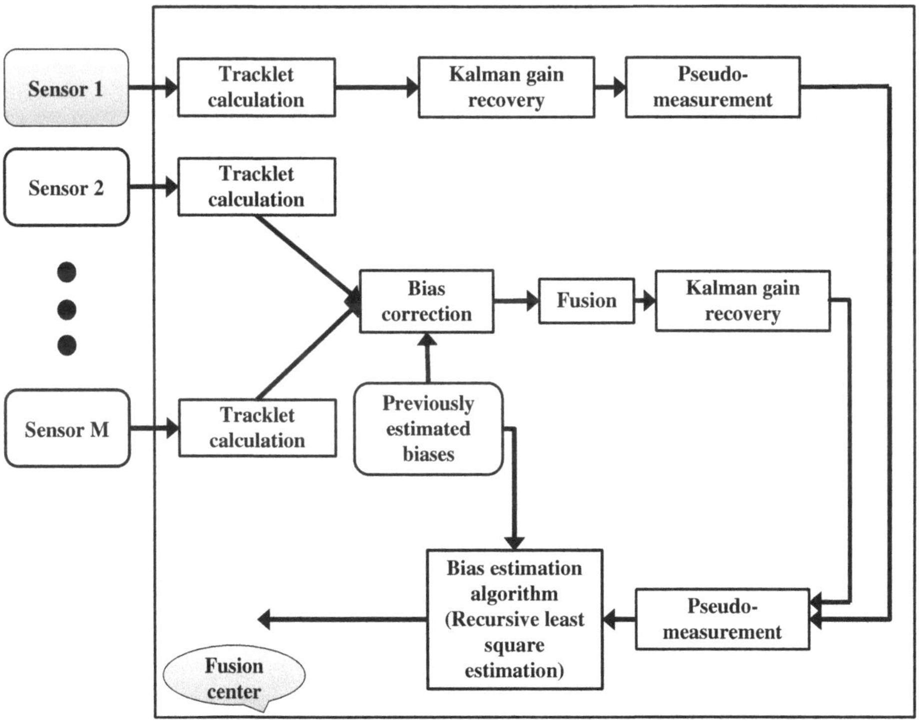

Figure 3.6 Block diagram of the new offset and scaling bias estimation algorithm.

3.6 Simulation Results

3.6.1 Motion Dynamics and Measurement Parameters

Here, a distributed tracking scenario with five sensors and sixteen targets is considered as shown in Figure 3.7. It is assumed that all sensors are synchronized. Without loss of generality and to easily compare the results of bias estimation for different sensors, all the biases are assumed as

$$\beta_s = [20\ \text{m} \quad 1\ \text{mrad}]^T. \tag{3.64}$$

The standard deviations of measurement noises are $\sigma_r = 10$ m and $\sigma_\theta = 1$ mrad for target range and azimuth measurements, respectively. In this problem $r \simeq 20000$. According to [6], $\frac{r\sigma_\theta^2}{\sigma_r} = \frac{2 \times 10^4 \times 10^{-6}}{10} = 2 \times 10^{-3} \ll 0.4$ which is the threshold for polar to Cartesian conversion to be unbiased. Note that this condition holds for $r \simeq 200$ km as well.

The true dynamics of the targets are modeled using the discretized continuous white noise acceleration (DCWNA) or NCV model with $q_x = q_y = 0.1 \frac{\text{m}^2}{\text{s}^3}$ and the constant turn rate model with rate $\omega = 0.1 \frac{\text{deg}}{\text{s}}$. As for the filtering in local trackers, DCWNA and continuous Wiener process acceleration (CWPA) are used with various settings to be able to create different scenarios for the simulation (for details, see [6, p. 268 and p. 467]).

3.6.2 A Five-sensor Problem with NCV Kalman Filter as Local Tracker with Bias

Here, a scenario with five sensors with sixteen targets, $L = k - k' = 10$ and $k = 1, \ldots, 100$ is simulated. The motivation for using 100 time steps is to have enough update steps to demonstrate the convergence results in terms of RMSE. Here, the fused bias estimation algorithm (FBEA, Algorithm 2) is used to estimate the biases at the fusion center. The results of this simulation are shown in Figure 3.8.

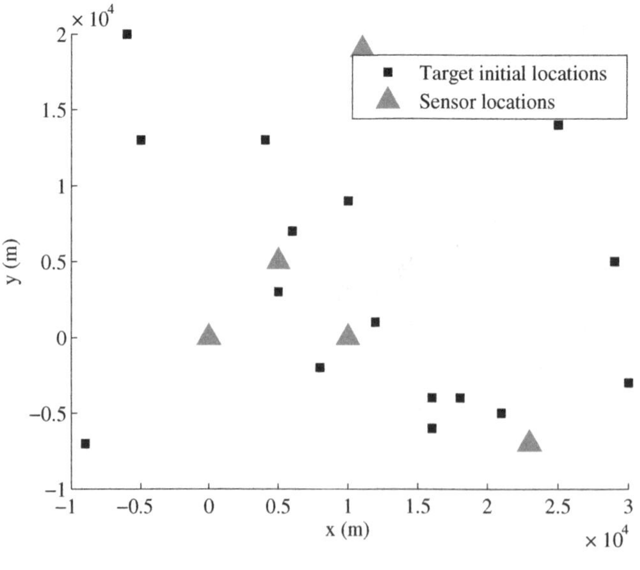

Figure 3.7 Initial locations of the targets and sensors.

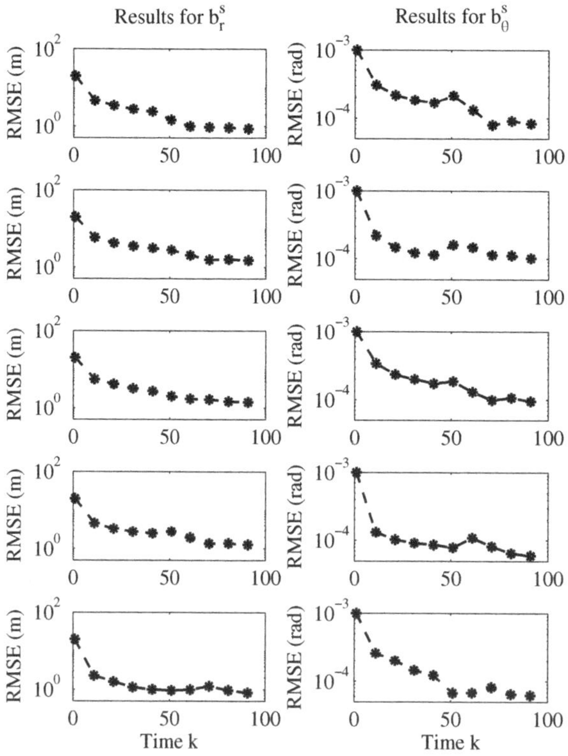

Figure 3.8 RMSE of the bias parameter b_r (left column) and b_θ (right column) for all 5 sensors from sensor 1 (top) to sensor 5 (bottom) in logarithmic scale. Note that residual bias RMSE is an order of magnitude below the measurement noise standard deviations, i.e., it becomes negligible.

From Figure 3.8, it can be seen that the FBEA works well in a scenario with five sensors and with tracklet update at every $L = 10$ steps. Note that the FBEA is using only a two-dimensional state space for the bias estimation step for each sensor. At its core, FBEA is a recursive least square (RLS) estimator. The computational complexity of RLS is of the order of $O(n^2)$, where n is the number of parameters to be estimated.

To show the performance of the FBEA in the fusion step, the results in terms of RMSE of the local track estimates for a specific sensor (sensor 1) and the fused estimates are presented in Figure 3.9 in logarithmic scale. To compare the results, the RMSE values of the local tracks and fused tracks with no biases are also included.

Figure 3.9 shows that the sequential fusion step used in the FBEA is a viable solution to the fusion problem. In this case, there is no need for a feedback channel. This reduces the communication requirements.

In order to further evaluate the performance of the FBEA algorithm, one can assume that all sensors have scaling and offset biases. Let the value of biases be

$$\beta_s = [20\,\text{m} \quad 1\,\text{mrad} \quad 0.001 \quad 0.001]^{\text{T}}. \tag{3.65}$$

The results in terms of the RMSE of the local track estimates for a specific sensor (sensor 1) and the fused estimates are shown in Figure 3.10 in logarithmic scale. Figure 3.10 shows that the FBEA can handle offset and scaling biases at the same time and fuse the corrected tracks in order to achieve better estimates of the targets' state.

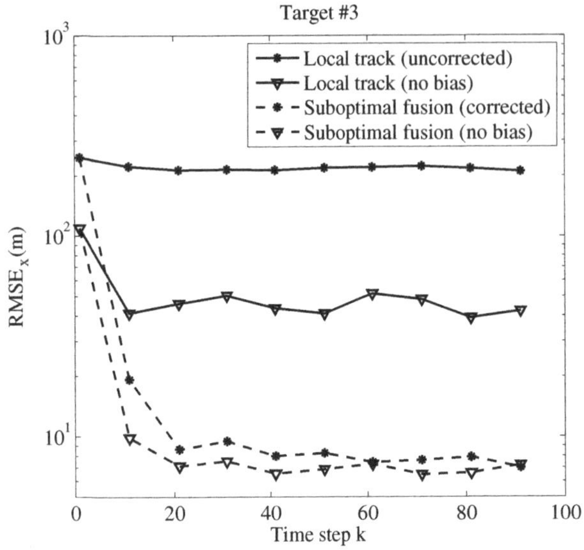

Figure 3.9 RMSE of local track (sensor 1) and the output of the fusion algorithm including offset biases for all sensors in logarithmic scale.

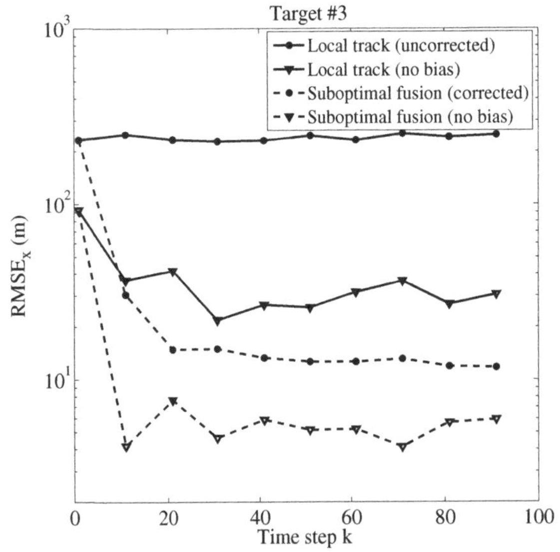

Figure 3.10 RMSE of local track (sensor 1) and the output of the fusion algorithm including scaling and offset biases for all sensors in logarithmic scale.

3.6.3 Continuous 2D Assignment for Multitarget Tracking with Rotating Radars

In this simulation experiment, a surveillance area of 3000 m × 3000 m is considered. Six targets, spanning multiple sectors, are located in the sensor's field of view. The targets follow an NCV motion model [10]. However, in this simulation, the focus is to demonstrate the track update efficiency of the sector-based association in [29] compared with the standard frame-based association. The target trajectories for a duration of 20 scans are shown in Figure 3.11. It can be noted in Figure 3.11 that some of the targets cross consecutive sectors in their lifetime (targets 1,

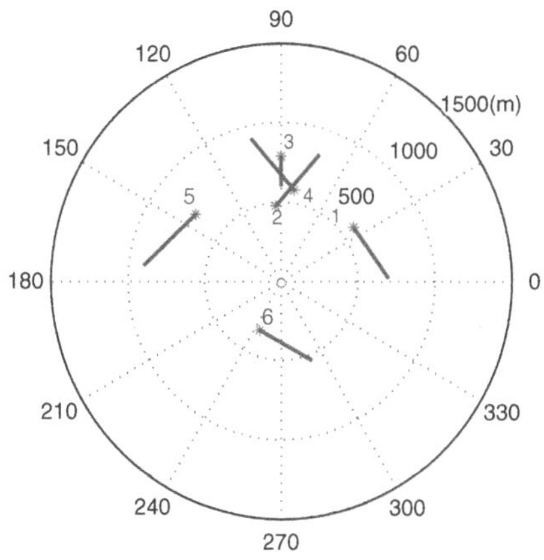

Figure 3.11 Target trajectories.

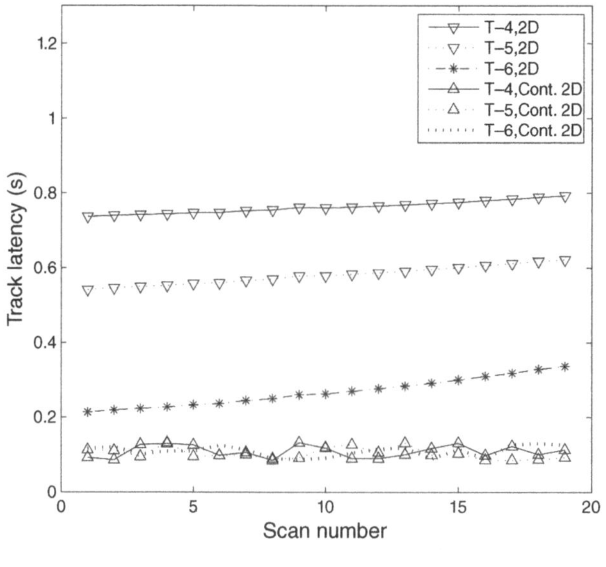

Figure 3.12 Tracks with increasing latency.

2, 4, 5, and 6), some move in the clockwise direction (targets 4, 5, and 6), some move in the counterclockwise direction (targets 1 and 2), and that others move in the radial direction (target 3). In addition, a set of targets is clustered in multiple sectors with an overlapping measurement validation region (targets 2, 3, and 4).

The comparison between the frame-based and dynamic association techniques with respect to track latency based on 100 Monte Carlo runs is shown in Figures 3.12 and 3.13. In 3.12, the track latency for the targets that move in a counterclockwise direction is shown. It can be noted that the track latency with the frame-based approach increases over time for clockwise moving targets. Targets that move mainly in a radial direction have a relatively constant track latency compared with those moving tangentially, as shown in 3.13.

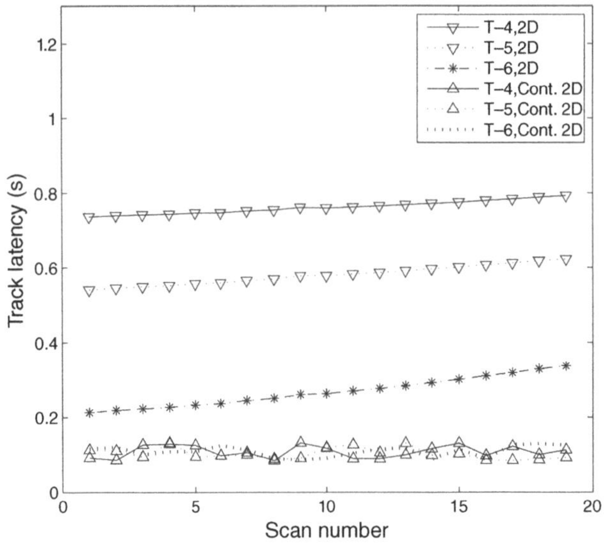

Figure 3.13 Track with constant latency.

3.7 Conclusions

Radar tracking, target detection, and filtering are the main components of today's surveillance systems. In this chapter, we demonstrated the main concepts behind radar tracking systems such as filtering methods, different radar measurements, and issues related to a general radar tracking platform. Furthermore, fusion architectures and radar bias estimation were investigated briefly to help the reader with the current implementation of radar systems.

References

1 S. J. Davey, G. Rutten Mark, and B. Cheung, A comparison of detection performance for several track-before-detect algorithms. *EURASIP Journal on Advances in Signal Processing*, 2008, (1):428036, 2007.

2 Y. Bar-Shalom, P. K. Willett, and X. Tian. *Tracking and Data Fusion: A Handbook of Algorithms*. YBS Publishing, Storr, CT, 2011.

3 S. S. Blackman. Multiple hypothesis tracking for multiple target tracking. *IEEE Aerospace and Electronic Systems Magazine*, 19(1):5–18, 2004.

4 R. Mahler. Multitarget bayes filtering via first-order multitarget moments. *IEEE Transactions on Aerospace and Electronic Systems*, 39(4):1152–1178, 2003.

5 B.-N. Vo, S. Singh, and A. Doucet. Sequential Monte Carlo implementation of the PHD filter for multi–target tracking. In *6th International Conference of Information Fusion*, pages 792–799, 2003.

6 Y. Bar-Shalom, X. R. Li, and T. Kirubarajan. *Estimation with Applications to Tracking and Navigation: Theory, Algorithms and Software*. Wiley, NY, 2001.

7 X. Lin, Y. Bar-Shalom, and T. Kirubarajan. Exact multisensor dynamic bias estimation with local tracks. *IEEE Transactions on Aerospace and Electronic Systems*, 1(40):576–590, 2004.

8 M. Longbin, S. Xiaoquan, Z. Yiyu, S. Z. Kang, and Y. Bar-Shalom. Unbiased converted measurements for tracking. *IEEE Transactions on Aerospace and Electronic Systems*, 34(3):1023–1027, 1998.

9 W. Koch. Effect of Doppler ambiguities on GMTI tracking. In *Proceedings of RADAR,* Edinburgh, UK, pages 153–157, 2002.

10 C. M. Alabaster, E. J. Hughes, and J. H. Matthew. Medium PRF radar PRF selection using evolutionary algorithms. *IEEE Transactions on Aerospace and Electronic Systems,* 39(3):990–1001, 2003.

11 G. Trunk and S. Brockett. Range and velocity ambiguity resolution. In *The Record of the 1993 IEEE National Radar Conference,* Lynnfield, MA, pages 146–149,1993.

12 C. Chong and S. Mori. Convex combination and covariance intersection algorithms in distributed fusion. In *4th International Conference on Information Fusion,* Montreal, QC, Canada, 2001.

13 Y. Bar-Shalom. Negative correlation and optimal tracking with Doppler measurements. *IEEE Transactions on Aerospace and Electronic Systems,* 37(3):1117–1120, 2001.

14 K. Li, B. Habtemariam, R. Tharmarasa, M. Pelletier, and T. Kirubarjan. Multitarget tracking with Doppler ambiguity. *IEEE Transactions on Aerospace and Electronic Systems,* 49(4):2640–2656, 2013.

15 H. Chen, K. Pattipati, T. Kirubarajan, and Y. Bar-Shalom. General data association with possibly unresolved measurements using linear programming. In *2003 Conference on Computer Vision and Pattern Recognition Workshop,* 9, 103–103, 2003.

16 D. P. Bertsekas. *Network Optimization: Continuous and Discrete Models.* Athena Scientific, 1998.

17 R. K. Ahuja, T. L. Magnanti, and J. B. Orlin. *Network Flows: Theory, Algorithms, and Applications.* Prentice Hall, 1993.

18 D. Cristallini, M. Caruso, P. Falcone, D. Langellotti, C. Bongioanni, F. Colone, S. Scafe, and P. Lombardo. Space-based passive radar enabled by the new generation of geostationary broadcast satellites. In *2010 IEEE Aerospace Conference,* Big Sky, MT, pages 1–11, 2010.

19 R. Tharmarasa, M. Subramaniam, N. Nadarajah, T. Kirubarajan, and M. McDonald. Multitarget passive coherent location with transmitter-origin and target-altitude uncertainties. *IEEE Transactions on Aerospace and Electronic Systems,* 48(3):2530–2550, 2012.

20 K. Pattipati, T. Kirubarajan, and R. Popp. Survey of assignment techniques for multitarget tracking. In *Proceedings of the Workshop on Estimation, Tracking, and Fusion: A Tribute to Yaakov Bar-Shalom,* Monterey, CA, pages 77–159, 2001.

21 S. Deb, M. Yeddanapudi, K. Pattipati, and Y. Bar-Shalom. A generalized SD assignment algorithm for multisensor-multitarget state estimation. *IEEE Transactions on Aerospace and Electronic Systems,* 33(2):523–538, 1997.

22 D. Musicki and R. Evans. Joint integrated probabilistic data association: JIPDA. *IEEE transactions on Aerospace and Electronic Systems,* 40(3):1093–1099, 2004.

23 J. A. Roecker. A class of near optimal JPDA algorithms. *IEEE Transactions on Aerospace and Electronic Systems,* 30(2):504–510, 1994.

24 B-N. Vo and W-K. Ma. The Gaussian mixture probability hypothesis density filter. *IEEE Transactions on Signal Processing,* 54(11):4091–4104, 2006.

25 B-N. Vo, S. Singh, and A. Doucet. Sequential Monte Carlo methods for multitarget filtering with random finite sets. *IEEE Transactions on Aerospace and Electronic Systems,* 41(4):1224–1245, 2005.

26 M. R. Chummun, T. Kirubarajan, K. R. Pattipati, and Y. Bar-Shalom. Fast data association using multidimensional assignment with clustering. *IEEE Transactions on Aerospace and Electronic Systems,* 37(3):898–913, 2001.

27 T. Kirubarajan, H. Wang, Y. Bar-Shalom, and K. R. Pattipati. Efficient multisensor fusion using multidimensional data association. *IEEE Transactions on Aerospace and Electronic Systems,* 37(2):386–400, 2001.

28 J. Clarke. Airborne early warning radar. *Proceedings of the IEEE,* 73(2):312–324, 1985.

29 B. Habtemariam, R. Tharmarasa, M. Mcdonald, and T. Kirubarajan. Continuous 2-D assignment for multitarget tracking with rotating radars. *IEEE Transactions on Aerospace and Electronic Systems*, 51(3):2193–2204, 2015.

30 M. E. Liggins, C. Chong, I. Kadar, M. Alford, V. Vannicola, and S. Thomopoulos. Distributed fusion architectures and algorithms for target tracking. *Proceedings of the IEEE*, 85(1):95–107, 1997.

31 N. Xiong and P. Svensson. Multi–sensor management for information fusion: issues and approaches. *Information Fusion*, 3(1):163–186, 2002.

32 C. Chong. Distributed architectures for data fusion. In *Proceedings of the 1st International Conference on Information Fusion,* Las Vegas, NV, 1998.

33 O. E. Drummond. Tracklets and a hybrid fusion with process noise. In *International Society for Optics and Photonics*, 512–524, 1997.

34 O. E. Drummond. A hybrid sensor fusion algorithm architecture and tracklets. In *Proceedings of the SPIE, Signal and Data Processing of Small Targets,* 3613, 1997.

35 O. E. Drummond. Track and tracklet fusion filtering. *Proceedings of the SPIE, Signal and Data Processing of Small Targets*, 4728:176–195, 2002.

36 S. J. Julier and J. K. Uhlmann. A non–divergent estimation algorithm in the presence of unknown correlations. In *Proceedings of the American Control Conference,* Albuquerque, New Mexico, 1997.

37 H. Durrant-Whyte, M. Stevens, and E. Nettleton. Data fusion in decentralised sensing networks. In *Fourth International Conference on Information Fusion*, Montreal, QC, Canada, 302–307, 2001.

38 M. P. Dana. Registration: a prerequisite for multiple sensor tracking. In *Multitarget-Multisensor Tracking: Advanced Applications*, Artech House, Norwood, MA, 155–185, 1990.

39 B. Friedland. Treatment of bias in recursive filtering. *IEEE Transactions on Automatic Control,* 14(4):359–367, 1969.

40 N. Nabaa and R. H. Bishop. Solution to a multisensor tracking problem with sensor registration errors. *IEEE Transactions on Aerospace and Electronic Systems*, 35(1):354–363, 1999.

41 X. Lin, Y. Bar-Shalom, and T. Kirubarajan. Multisensor multitarget bias estimation for general asynchronous sensors. *IEEE Transactions on Aerospace and Electronic Systems*, 41(3):899–921, 2005.

42 X. Lin, Y. Bar-Shalom, and T. Kirubarajan. Multisensor-multitarget bias estimation of asynchronous sensors. *Proceedings of the SPIE*, 5429:105–116, 2004.

43 A. Balleri, A. Nehorai, and J. Wang. Maximum likelihood estimation for compound-Gaussian clutter with inverse gamma texture. *IEEE Transactions on Aerospace and Electronic Systems,* 43(2):775–779, 2007.

44 T. Asparouhov and B. Muthén. Weighted least squares estimation with missing data. *MplusTechnical Appendix, Citeseer*, 2010.

45 H. Rue and L. Held. *Gaussian Markov Random Fields: Theory and Applications.* CRC Press, 2005.

46 I. Li and J. Georganas. Multi-target multi-platform sensor registration in geodetic coordinates. In *International Conference on Information Fusion (FUSION),* Annapolis, MD, pages 366–373, 2002.

47 M. Yeddanapudi, Y. Bar-Shalom, and K. Pattipati. IMM estimation for multitarget-multisensor air traffic surveillance. *Proceedings of the IEEE*, 85(1):80–96, 1997.

48 E. Taghavi, R. Tharmarasa, T. Kirubarajan, Y. Bar-Shalom, and M. Mcdonald. A practical bias estimation algorithm for multisensor–multitarget tracking. *IEEE Transactions on Aerospace and Electronic Systems*, 52(1):2–19, 2016.

4

Exact IMM Estimation of Markov Switching Diffusions with Hybrid Jumps

Henk A.P. Blom

Aerospace Engineering Department, Delft University of Technology, Delft, The Netherlands

4.1 Introduction

Many controlled dynamic systems have a hybrid state that consists of a finite-valued control mode θ_t and a Euclidean-valued state X_t. These controlled dynamic systems typically make hybrid jumps (HJs), in the sense that each time the control mode switches, the Euclidean-valued state makes a simultaneous jump. A typical example of a controlled system is an aircraft where the control mode may switch from a straight flight regime to a coordinated turn (CT) regime, while the rate of turn makes a simultaneous jump. Inside the aircraft, these HJs are well known and under control. However, for an external observer, these HJs can only be estimated from surveillance information.

To handle these HJs in target tracking, the commonly adopted approach is to extend interacting multiple model (IMM) estimation with a variable-structure (VS) IMM approach [1]. To also estimate a simultaneous random jump in a Euclidean-valued state component, the VS-IMM approach runs an extra Euclidean-valued target model for each possible jump value and adds an extra mode to the switching Markov chain for each extra target model. To manage the adding of extra models to the Markov chain, various improvements to the VS-IMM approach have been developed [2–5]. Despite these VS-IMM developments, one of the remaining challenges is that extra target models tend competing each other [6]. The objective of this chapter is to show that proper IMM estimation can manage HJs in an exact way and therefore does not need a VS extension.

As explained in [7, 8], the early continuous-time nonlinear filtering study [9] forms the roots of the original IMM estimator [10, 11]. Another novelty of [9] was to define a hybrid state stochastic process as a solution of the stochastic differential equation (SDE) of [12, pp. 273, 274] on a hybrid state space. Because this SDE is driven by Brownian motion and Poisson random measure, its evolution in a hybrid space involves HJs. As a result of this second novelty, the exact continuous-time filter equations of [9] covered HJs. These early exact results have stimulated the further development of the HJ version of the IMM approach [13–16], under the name of generalized IMM. It also is this HJ version of IMM that successfully has been used in the European air traffic multi-sensor multi-target tracking system [17–20]. Recently, the HJ version of the IMM estimator has also successfully been applied to the more demanding estimation of switching aircraft flight control modes and set points [21].

Having its roots in continuous-time nonlinear filtering, the exact handling of HJs by IMM has largely gone unnoticed in target tracking literature. To improve the situation, this chapter will derive exact IMM estimation characterizations by applying up-to-date nonlinear filtering results [22–24] to continuous-time hybrid state stochastic processes with HJs [25, 26]. These up-to-date derivations make use of a change of measure approach, instead of adopting Kailath's innovation conjecture by [9, 16].

For the derivation of exact IMM estimation equations, this chapter follows three stages. The first stage is to study Markov switching diffusion processes with HJs, i.e., the Euclidean diffusion state may undergo a random jump that occurs simultaneously with a Markov switching moment. It is also illustrated how this setting can model a suddenly maneuvering target. The second stage is to develop exact nonlinear filter equations for the joint probability mass-density of the mode and state, given partial observations of a Markov switching diffusion with HJs. A recent proof of the Feller property for this class of processes [27, 28] allows to use the nonlinear filtering results of [22–24]. The third stage is to show that the resulting exact nonlinear filter equations embed the interaction/mixing equations of IMM.

In addition to showing that continuous-time IMM estimation covers HJs in an exact way, this chapter also develops three HJ estimators for the case that the continuous-time observations are replaced by discrete-time observations. These three are (1) HJ-IMM estimator, (2) HJ version of sampling importance resampling (SIR) particle filter (PF), and (3) HJ version of IMM PF.

The chapter is organized as follows. Section 4.2 characterizes stochastic hybrid systems with HJs. Section 4.3 develops nonlinear filtering equations for stochastic hybrid systems with HJs. Section 4.4 develops for a Markov jump linear system (MJLS) an exact transformation of the nonlinear filter equations of Section 4.3 to a form that shows IMM's unique mixing equation. Section 4.4 also shows how this yields the HJ-IMM estimator for discrete-time observations of a continuous-time MJLS. Section 4.5 develops the HJ-SIR PF and the HJ-IMM PF. Section 4.6 draws conclusions.

4.2 Markov Switching Diffusion with HJs

This section starts with a definition of Markov switching diffusions with HJs, as a pathwise unique solution of an SDE, in Subsection 4.2.1. Subsection 4.2.2 presents a suddenly maneuvering aircraft as an example of a Markov switching diffusion with HJs. Subsection 4.2.3 addresses the strong Markov property of Markov switching systems with HJs.

Throughout the chapter, all processes are defined on a complete stochastic basis $(\Omega, \mathcal{F}, F, P)$, with $(\Omega, \mathcal{F})$ a measurable space, P a probability measure defined on $\mathcal{F}$, and F a filtration that is a collection of increasing right-continuous sub-σ -algebra's $\mathcal{F}_t$, i.e., $F = \{\mathcal{F}_t, t \in [0, \infty)\}$.

4.2.1 Characterization of the Stochastic Process

We consider a continuous-time process $\{\xi_t\} = \{\theta_t, X_t\}$, with $\{\theta_t\}$ the regime switching process component and $\{X_t\}$ the Euclidean-valued state component.

The regime-switching process $\{\theta_t\}$ has cadlag paths (right-continuous with left hand limits), and assumes values in $\Theta = \{e_1, .., e_N\}$, according to transition rates $\lambda_{\theta\eta}$, $\eta \neq \theta$, that do not depend on X_t, i.e.,

$$P\{\theta_{t+\Delta} = \eta | \theta_t = \theta, X_t = x, (X_s, \theta_s; s < t)\} = \begin{cases} \lambda_{\theta\eta}\Delta + o(\Delta), & \eta \neq \theta \\ 1 + \lambda_{\theta\theta}\Delta + o(\Delta), & \eta = \theta \end{cases}, \tag{4.1}$$

with $\lim_{\Delta\to 0} o(\Delta)/|\Delta| = 0$, and $\lambda_{\theta\theta} = -\sum_{\eta\neq\theta} \lambda_{\theta\eta}$.

Let $\{N_t\}$ count the number of regime switchings, i.e.,

$$dN_t = 1\{\theta_t \neq \theta_{t-}\}, \tag{4.2}$$

then the Euclidean-valued process $\{X_t\}$ evolves according to the SDE:

$$dX_t = F(\theta_t, X_t)dt + G(\theta_t, X_t)dB_t + C(\theta_{t-}, \theta_t, X_{t-}, u_{N_t})dN_t, \tag{4.3}$$

where $\{B_t\}$ is standard Brownian motion, and $\{u_i\}$ a sequence of independent random vectors with probability measure $\mu(du)$ on a Borel measurable mark space U. For completeness, we define $C(\theta, \theta, x, u_i) = 0$.

An extra feature is that at a regime switching moment τ_i, there may be a change in the Euclidean dimension $\mathbf{R}^{d(\theta)}$ of $\{X_t\}$. Hence, the full state space of the joint process $\{\xi_t\} = \{\theta_t, X_t\}$ equals:

$$\Xi = \bigcup_{\theta \in \Theta} \{\theta\} \times \mathbf{R}^{d(\theta)}. \tag{4.4}$$

For pathwise uniqueness of a more general stochastic hybrid system with HJs, [28] has derived relaxed sufficient conditions. For the process defined by (4.1–4.4) sufficient conditions for pathwise uniqueness are:

A1. Transition rates $\lambda_{\theta\eta}$ are finite.
A2. $F(\theta, .)$ and $G(\theta, .)$ satisfy local Lipschitz and quadratic growth conditions.
A3. There exists a constant $c_J < \infty$ such that for all θ, η, x:

$$\int_U \left|C(\theta, \eta, x, u)\right|^2 \mu(du) \leq c_J\left(1 + |x|^2\right).$$

4.2.2 Suddenly Maneuvering Aircraft Example

As an illustration of a hybrid system with HJs, we consider an aircraft that undergoes sudden mode switches, e.g., in horizontal direction from straight flight to turn or from uniform motion to acceleration or deceleration, and in vertical direction from level flight to climb or descent. The illustrative example is focused on sudden maneuvers, where an aircraft may switch between three modes: straight flight (S), left turn (L), and right turn (R). Hence, we consider a discrete-valued state space consisting of three horizontal flight modes, i.e., $\Theta = \{S, L, R\}$. The maneuver mode process $\{\theta_t\}$ is assumed to evolve according to nonzero transition rates $\lambda_{SL}, \lambda_{SR}, \lambda_{LS}$, and λ_{RS}, which means that a direct switching from left turn to right turn (or the other way around) is assumed not to happen. The Euclidean-valued state satisfies a CT model with polar velocity components, which is augmented with a rate of turn component, i.e., $X_t = Col\{x_t, y_t, v_t, \varphi_t, \omega_t\}$, with 2-dimensional horizontal position (x_t, y_t), polar velocity (v_t, φ_t), and turn rate ω_t. Because an aircraft undergoes a finite number of discontinuities on a finite time interval, X_t is modeled by (4.3), with:

$$F(\theta, (x, y, v, \varphi, \omega)) = Col\{v \cos \varphi, v \sin \varphi, 0, \omega, 0\}, \tag{4.5a}$$

$$G(\theta, (x, y, v, \varphi, \omega)) = Col\{0, 0, g_v, 0, 0\}, \tag{4.5b}$$

$$C(\eta, \theta, (x, y, v, \varphi, \omega), u) = Col\{0, 0, 0, 0, c_\omega(\theta, \omega, u)\}, \tag{4.5c}$$

$$c_\omega(\theta, \omega, u) = \begin{cases} -\omega, & \text{if } \theta = S \\ -\omega + u, & \text{if } \theta = L, \\ -\omega - u, & \text{if } \theta = R \end{cases} \tag{4.5d}$$

where g_v is a scalar valued model parameter for the amount of Brownian motion influencing the aircraft speed v_t, and u is a random variable according to a strictly positive probability density function with mean ε_ω and standard deviation σ_ω. At the ith mode switching moment τ_i, the rate of turn makes a simultaneous jump, i.e., $\omega_{\tau_i} = \omega_{\tau_i -} + c_\omega(\theta_{\tau_i}, \omega_{\tau_i -}, u_i)$. If $\theta_t = S$, than $\omega_t = 0$, which reduces the effective Euclidean state space from $\mathbf{R}^5$ to $\mathbf{R}^4$.

4.2.3 Probability Mass-density of the Joint Markov Process

We write $p_{\theta_t, X_t}(\theta, x)$ for the probability mass-density of the joint Markov process $\{\theta_t, X_t\}$. To characterize the unconditional evolution of $p_{\theta_t, X_t}(\theta, x)$ (i.e., without receiving observations), the joint process $\{\theta_t, X_t\}$ has to be a strong Markov process that satisfies the following Feller property: for any bounded continuous function $f : \Xi \to \mathbf{R}$, $E\{f(\xi_t)|\xi_0 = \xi\}$ is continuous in $t > 0$, and $\lim_{t\downarrow 0} E\{f(\xi_t)|\xi_0 = \xi\} = f(\xi)$.

Recently Blom [28] has proven that for the process defined by (4.1)–(4.4) sufficient conditions for the Feller property are the pathwise uniqueness assumptions A1–A3 together with the additional assumption:

A4. $C(\theta, \eta, x, u)$ is continuous in x, for each $\theta, \eta \in \Theta$ and $u \in U$.

To prepare the characterization of the evolution of the unconditional probability mass-density $p_{\theta_t, X_t}(\theta, x)$, we define the infinitesimal generator $\mathcal{A}$ and its formal adjoint of Feller process $\{\xi_t\}$.

Definition 4.1 The infinitesimal generator $\mathcal{A}$ of Feller process $\{\xi_t\}$ is defined for functions $f : \Xi \to \mathbf{R}$ as:

$$\mathcal{A}f(\xi) = \lim_{t\downarrow 0} \frac{E\{f(\xi_t)|\xi_0 = \xi\} - f(\xi)}{t}. \tag{4.6}$$

$\mathcal{A}$'s domain $D(\mathcal{A})$ includes all functions $f : \Xi \to \mathbf{R}$ for which this limit exists.

Definition 4.2 The formal adjoint of operator $\mathcal{A}$ is defined as the operator $\mathcal{A}^*$ such that for each twice continuously differentiable and bounded probability mass-density $\rho : \Xi \to \mathbf{R}_+$ for all $f(\xi) \in D(\mathcal{A})$:

$$\int_\Xi \rho(\xi)\mathcal{A}f(\xi)d\xi = \int_\Xi f(\xi)\mathcal{A}^*\rho(\xi)d\xi. \tag{4.7}$$

The domain $D(\mathcal{A}^*)$ includes all measures $\rho : \Xi \to \mathbf{R}$ for which $\mathcal{A}^*\rho(\xi)$ is uniquely defined.

Proposition 4.1 Let assumptions A1–A4 be satisfied. The infinitesimal generator $\mathcal{A}$ of the process solution of (4.1)–(4.3) satisfies for all twice continuously differentiable bounded functions f that vanish at infinity:

$$\mathcal{A}f(\theta, x) = \mathcal{L}^\theta f(\theta, x) + \mathcal{J}f(\theta, x), \tag{4.8}$$

$$\mathcal{L}^{\theta} f(\theta, x) = \sum_{i=1}^{d(\theta)} \frac{\partial f(\theta, x)}{\partial x^i} F^i(\theta, x) + \frac{1}{2} \sum_{i,j=1}^{d(\theta)} \frac{\partial^2 f(\theta, x)}{\partial x^i \partial x^j} G^i(\theta, x) G^j(\theta, x)^T, \tag{4.9}$$

$$\mathcal{J} f(\theta, x) = \sum_{\eta \neq \theta} \lambda_{\theta\eta} \left[\int_U f(\eta, x + C(\theta, \eta, x, u)) \cdot \mu(du) - f(\theta, x) \right]. \tag{4.10}$$

Proof: Following [29], $\{\theta_t\}$ can be written as solution of an SDE driven by $N(N-1)$ independent Poisson point processes $dp_{\theta\eta,t}(du)$, $\eta \neq \theta$, with rates $\lambda_{\theta\eta}$:

$$d\theta_t = \sum_{\substack{\theta,\eta \in \Theta \\ \eta \neq \theta}} \left[1\{\theta_{t-} = \theta\}(\eta - \theta) dp_{\theta\eta,t} \right], \tag{4.11a}$$

where $t- = \lim_{\Delta \downarrow 0}(t - \Delta)$, and the compensated processes $\{dp_{\theta\eta,t} - \lambda_{\theta\eta} dt\}$ are F-martingales. Moreover, (4.3) can be rewritten as:

$$dX_t = F(\theta_t, X_t)dt + G(\theta_t, X_t)dB_t + \sum_{\substack{\theta,\eta \in \Theta \\ \eta \neq \theta}} \left[1\{\theta_{t-} = \theta\} C(\theta, \eta, X_{t-}, u_{N_t}) dp_{\theta\eta,t} \right]. \tag{4.11b}$$

By extending each Poisson point process $dp_{\theta\eta,t}$ to a Poisson random measure $dp_{\theta\eta,t}(du)$ on Borel measurable mark space U, with mean measure $\lambda_{\theta\eta}\mu(du)dt$, (4.11a,b) can be written as:

$$d\theta_t = \sum_{\substack{\eta \in \Theta \\ \eta \neq \theta_{t-}}} \left[(\eta - \theta_{t-}) dp_{\theta_{t-}\eta,t}(U) \right], \tag{4.12a}$$

$$dX_t = F(\theta_t, X_t)dt + G(\theta_t, X_t)dB_t + \sum_{\substack{\eta \in \Theta \\ \eta \neq \theta_{t-}}} \left[C(\theta_{t-}, \eta, X_{t-}, u) dp_{\theta_{t-}\eta,t}(du) \right]. \tag{4.12b}$$

Following [30, pp. 169, 170], for SDE (4.12a, b) we obtain the infinitesimal generator Eqs. (4.8)–(4.10).

Q.E.D.

Proposition 4.2 Let assumptions A1–A4 be satisfied. Let $p_{\xi_t}(\xi)$ denote the probability mass-density of Feller process $\xi_t = (\theta_t, \xi_t)$, then its unconditional evolution satisfies:

$$\frac{\partial}{\partial t} p_{\xi_t}(\xi) = \mathcal{A}^* p_{\xi_t}(\xi), \tag{4.13}$$

where the formal adjoint $\mathcal{A}^*$ of $\mathcal{A}$ satisfies for twice-continuously differential and bounded probability mass-density $\rho(\theta, x)$:

$$\mathcal{A}^* \rho(\theta, x) = \mathcal{L}^{\theta *} \rho(\theta, x) + \mathcal{J}^* \rho(\theta, x), \tag{4.14}$$

with $\mathcal{L}^{\theta *}$ the mode-conditional Fokker-Planck operator, and $\mathcal{J}^*$ the Kolmogorov operator:

$$\begin{aligned} \mathcal{L}^{\theta *} \rho(\theta, x) = &- \sum_{i=1}^{d(\theta)} \frac{\partial}{\partial x^i} \left[F^i(\theta, x)\, \rho(\theta, x) \right] \\ &+ \frac{1}{2} \sum_{i,j=1}^{d(\theta)} \frac{\partial^2}{\partial x^i \partial x^j} \left[G^i(\theta, x) G^j(\theta, x)^T \rho(\theta, x) \right], \end{aligned} \tag{4.15}$$

$$\mathcal{J}^*\rho(\theta, x) = \sum_{\eta\neq\theta}\left[\lambda_{\eta\theta}\int_{\mathbf{R}^{d(\eta)}} p_{X_\tau|\theta_{\tau-},X_{\tau-},\theta_\tau}(x|\eta, y, \theta)\rho(\eta, y)dy - \lambda_{\theta\eta}\rho(\theta, x)\right], \tag{4.16}$$

where $p_{X_\tau|\theta_{\tau-},X_{\tau-},\theta_\tau}(x|\eta, y, \theta)$ is the conditional density of X_τ, given $\theta_{\tau-} = \eta, X_{\tau-} = y$, and $\theta_\tau = \theta$, for any stopping time τ at which $dN_\tau = 1$.

Proof: Fokker-Planck operator $\mathcal{L}^{\theta*}$ is well known as the formal adjoint of $\mathcal{L}^\theta$. It remains to be shown that the formal adjoint of operator $\mathcal{J}$ satisfies (4.16).

$$\begin{aligned}
\int_\Xi \rho(\xi)\mathcal{J}f(\xi)d\xi &= \sum_{\theta\in\Theta}\int_{\mathbf{R}^{d(\theta)}} \rho(\theta, x)\sum_{\eta\neq\theta}\lambda_{\theta\eta}\left[\int_U f(\eta, x + C(\theta, \eta, x, u))\cdot\mu(du) - f(\theta, x)\right]dx \\
&= \sum_{\theta\in\Theta}\int_{\mathbf{R}^{d(\theta)}} \rho(\theta, x)\sum_{\eta\neq\theta}\left[\int_U \lambda_{\theta\eta}f(\eta, x + C(\theta, \eta, x, u))\cdot\mu(du)\right]dx \\
&\quad - \sum_{\theta\in\Theta}\int_{\mathbf{R}^{d(\theta)}} \rho(\theta, x)\sum_{\eta\neq\theta}\lambda_{\theta\eta}f(\theta, x)dx \\
&= \sum_{\theta\in\Theta}\int_{\mathbf{R}^{d(\theta)}} \rho(\theta, x)\sum_{\eta\neq\theta}\left[\int_{\mathbf{R}^{d(\eta)}} \lambda_{\theta\eta}f(\eta, y)p_{X_\tau|\theta_{\tau-},X_{\tau-},\theta_\tau}(y|\theta, x, \eta)dy\right]dx \\
&\quad - \sum_{\theta\in\Theta}\int_{\mathbf{R}^{d(\theta)}} \rho(\theta, x)\sum_{\eta\neq\theta}\lambda_{\theta\eta}f(\theta, x)dx \\
&= \sum_{\eta\in\Theta}\left[\int_{\mathbf{R}^{d(\eta)}} f(\eta, y)\sum_{\theta\neq\eta}\int_{\mathbf{R}^{d(\theta)}} \rho(\theta, x)\lambda_{\theta\eta}p_{X_\tau|\theta_{\tau-},X_{\tau-},\theta_\tau}(y|\theta, x, \eta)dxdy\right] \\
&\quad - \sum_{\theta\in\Theta}\int_{\mathbf{R}^{d(\theta)}} \rho(\theta, x)\sum_{\eta\neq\theta}\lambda_{\theta\eta}f(\theta, x)dx \\
&= \sum_{\theta\in\Theta}\left[\int_{\mathbf{R}^{d(\theta)}} f(\theta, x)\sum_{\eta\neq\theta}\int_{\mathbf{R}^{d(\eta)}} \rho(\eta, y)\lambda_{\eta\theta}p_{X_\tau|\theta_{\tau-},X_{\tau-},\theta_\tau}(x|\eta, y, \theta)dydx\right] \\
&\quad - \sum_{\theta\in\Theta}\int_{\mathbf{R}^{d(\theta)}} f(\theta, x)\rho(\theta, x)\sum_{\eta\neq\theta}\lambda_{\theta\eta}dx \\
&= \sum_{\theta\in\Theta}\left[\int_{\mathbf{R}^{d(\theta)}} f(\theta, x)\left[\sum_{\eta\neq\theta}\int_{\mathbf{R}^{d(\eta)}} \rho(\eta, y)\lambda_{\eta\theta}p_{X_\tau|\theta_{\tau-},X_{\tau-},\theta_\tau}(x|\eta, y, \theta)dy - \rho(\theta, x)\sum_{\eta\neq\theta}\lambda_{\theta\eta}\right]dx\right] \\
&= \sum_{\theta\in\Theta}\left[\int_{\mathbf{R}^{d(\theta)}} f(\theta, x)\sum_{\eta\neq\theta}\left[\int_{\mathbf{R}^{d(\eta)}} \lambda_{\eta\theta}p_{X_\tau|\theta_{\tau-},X_{\tau-},\theta_\tau}(x|\eta, y, \theta)\rho(\eta, y)dy - \lambda_{\theta\eta}\rho(\theta, x)\right]dx\right] \\
&= \sum_{\theta\in\Theta}\left[\int_{\mathbf{R}^{d(\theta)}} f(\theta, x)\mathcal{J}^*\rho(\theta, x)dx\right].
\end{aligned}$$

Hence, Eq. (4.7) is satisfied for $\mathcal{J}^*$ given by Eq. (4.16).

Q.E.D.

Remark 4.1 Characterization of $p_{X_\tau|\theta_{\tau-},X_{\tau-},\theta_\tau}(x|\eta, y, \theta)$ in (4.16) yields for $\Gamma \in \beta(\mathbf{R}^{d(\theta)})$:

$$\int_{\Gamma} p_{X_\tau|\theta_{\tau-},X_{\tau-},\theta_\tau}(x|\eta, y, \theta)dx = \int_{U} 1\{[y + C(\eta, \theta, y, u)] \in \Gamma\}\mu(du). \tag{4.17}$$

4.3 Conditional Probability Mass-density Given Continuous-time Observations

At moment t, the sigma-algebra of continuous-time observations is $\mathcal{Y}_t = \sigma\{Y_s, s \in [0, t]\}$, with

$$dY_t = H(\xi_t)dt + dW_t, \tag{4.18}$$

where $\{Y_t\}$ is the $\mathbf{R}^m$-valued, $m \leq n$, observation process, $\{\xi_t\} = \{\theta_t, x_t\}$ is the stochastic process of (4.1)–(4.3), and $\{W_t\}$ is an m-dimensional standard Brownian motion that is independent of $\{B_t\}$, $\{\theta_t\}$, and ξ_0. Subsection 4.3.1 gives the nonlinear filter equations from literature. Subsection 4.3.2 uses these equations for the characterization of the joint conditional probability mass-density of a hybrid system with HJs. Subsection 4.3.3 characterizes the evolution of the mode- and observation-conditional density $p_{x_t|\theta_t,\mathcal{Y}_t}(x|\theta)$.

4.3.1 Nonlinear Filtering Equation

To develop nonlinear filter equation for $f(\xi_t)$ given $\mathcal{Y}_t$, we assume that $H(.)$ satisfies the following condition:

A5. $H(\theta, .)$ is measurable and satisfies a quadratic growth condition for each $\theta \in \Theta$.

Proposition 4.3 Let A1–A5 be satisfied, let $\{\xi_t\}$ be the solution of (4.1)–(4.3), and let $H(\xi)$ in (4.18) satisfy a quadratic growth condition. Then, for any bounded $f(\xi) \in D(\mathcal{A})$, the estimator $\widehat{f}(\xi_t) = E\{f(\xi_t)|\mathcal{Y}_t\}$ satisfies:

$$\widehat{f}(\xi_t) = \widehat{f}(\xi_0) + \int_0^t E\{\mathcal{A}f(\xi_{s-})|\mathcal{Y}_{s-}\}ds$$
$$+ \int_0^t \left[E\{f(\xi_{s-})H(\xi_{s-})|\mathcal{Y}_{s-}\} - \widehat{f}(\xi_{s-}) \cdot \widehat{H}(\xi_{s-})\right]^T dv_s, \tag{4.19}$$

where $\mathcal{A}$ is the infinitesimal generator of Proposition 4.1, and:

$$\widehat{H}(\xi_t) = E\{H(\xi_t)|\mathcal{Y}_t\}, \tag{4.20}$$

$$dv_t = dY_t - \widehat{H}(\xi_t)dt. \tag{4.21}$$

Proof: In the proof of Proposition 4.1 it has been shown that (4.12a, b) has the same solution as (4.1)–(4.3). For observations in (4.18) of Feller solution of (4.12a, b), nonlinear filter Eq. (4.19) has been derived by [22–24], under specific conditions posed on $H(.)$. The proof by [23, Theorem 4.5] used the more general condition A5.

Q.E.D.

4.3.2 Evolution of Joint Conditional Probability Mass-density

Now we characterize the evolution of the joint conditional probability mass-density $\hat{p}_{\xi_t}(\xi) \triangleq p_{\xi_t|\mathcal{Y}_t}(\xi)$ by making use of Proposition 4.2 for the formal adjoint $\mathcal{A}^*$ of $\mathcal{A}$.

Theorem 4.1 Let A1–A5 be satisfied, let $\{\xi_t\}$ be the solution of (4.1)–(4.3). Then the conditional probability mass-density $\hat{p}_{\xi_t}(\xi)$ satisfies:

$$\hat{p}_{\xi_t}(\xi) = \hat{p}_{\xi_0}(\xi) + \int_0^t \mathcal{A}^* \hat{p}_{\xi_{s-}}(\xi) ds + \int_0^t \hat{p}_{\xi_{s-}}(\xi)[H(\xi) - \hat{H}(\xi_{s-})]^T dv_s, \tag{4.22}$$

with $\hat{H}(\xi_t)$ and dv_t satisfying (4.20) and (4.21).

Proof: Replacing all expectations in (4.19) by integrations over $\hat{p}_{\xi_t}(\xi)$ yields:

$$\int_\Xi \hat{p}_{\xi_s}(\xi) f(\xi) d\xi = \int_\Xi \hat{p}_{\xi_0}(\xi) f(\xi) d\xi + \int_0^t \int_\Xi \hat{p}_{\xi_{s-}}(\xi) \mathcal{A} f(\xi) d\xi ds$$
$$+ \int_0^t \int_\Xi \hat{p}_{\xi_{s-}}(\xi)[f(\xi)H(\xi) - f(\xi)\hat{H}(\xi_{s-})]^T d\xi dv_s.$$

Since $\mathcal{A}^*$ in Proposition 4.2 is the formal adjoint of $\mathcal{A}$, for all $f(.) \in D(\mathcal{A})$:

$$\int_\Xi \hat{p}_{\xi_s}(\xi) \mathcal{A} f(\xi) d\xi = \int_\Xi f(\xi) \mathcal{A}^* \hat{p}_{\xi_s}(\xi) d\xi.$$

Substitution of the latter in the preceding equation yields:

$$\int_\Xi f(\xi) \hat{p}_{\xi_s}(\xi) d\xi = \int_\Xi f(\xi) \hat{p}_{\xi_0}(\xi) d\xi + \int_0^t \int_\Xi f(\xi) \mathcal{A}^* \hat{p}_{\xi_{s-}}(\xi) d\xi ds$$
$$+ \int_0^t \int_\Xi f(\xi) \hat{p}_{\xi_{s-}}(\xi)[H(\xi) - \hat{H}(\xi_{s-})]^T d\xi dv_s.$$

Since the latter holds true for all $f \in D(\mathcal{A})$, the weighted integration over $f(\xi)$ can be deleted in all four terms, which yields (4.22).

Q.E.D.

Corollary 4.1 The estimator $\hat{p}_{\theta_t}(\theta) \triangleq P\{\theta_t = \theta | \mathcal{Y}_t\}$ satisfies:

$$d\hat{p}_{\theta_t}(\theta) = \sum_{\eta \in \Theta} [\lambda_{\eta\theta} \hat{p}_{\theta_t}(\eta)] dt + \hat{p}_{\theta_t}(\theta)[\hat{H}(\theta, X_{t-}) - \hat{H}(\xi_{t-})]^T dv_t, \tag{4.23}$$

where: $\hat{H}(\theta, X_t) = E\{H(\theta_t, X_t) | \mathcal{Y}_t, \theta_t = \theta\}$. (4.24)

Proof: Since $\hat{p}_{\theta_t}(\theta) = \int_{\mathbb{R}^{d(\theta)}} \hat{p}_{\theta_t, x_t}(\theta, x) dx$, we get $d\hat{p}_{\theta_t}(\theta) = \int_{\mathbb{R}^{d(\theta)}} \hat{p}_{\theta_t, x_t}(\theta, x) dx$.

Substituting (4.22) and subsequent evaluation yields:

$$
\begin{aligned}
d\hat{p}_{\theta_t}(\theta) &= \int_{\mathbf{R}^{d(\theta)}} \left[\mathcal{A}^*\hat{p}_{\xi_{t-}}(\theta,x)dt + \hat{p}_{\xi_{t-}}(\theta,x)\left[H(\theta,x)-\hat{H}(\xi_{t-})\right]^T dv_t\right]dx \\
&= \int_{\mathbf{R}^{d(\theta)}} \left[\mathcal{J}^*\hat{p}_{\xi_{t-}}(\theta,x)dt + \hat{p}_{\theta_t}(\theta)\hat{p}_{X_{t-}|\theta_t}(x|\theta)\left[H(\theta,x)-\hat{H}(\xi_{t-})\right]^T dv_t\right]dx \\
&= \int_{\mathbf{R}^{d(\theta)}} \left[\sum_{\eta\neq\theta}\left[\lambda_{\eta\theta}\int_{\mathbf{R}^{d(\eta)}} p_{X_\tau|\theta_{\tau-},X_{\tau-},\theta_\tau}(x|\eta,y,\theta)\hat{p}_{\xi_{t-}}(\eta,y)dy - \lambda_{\theta\eta}\hat{p}_{\xi_{t-}}(\theta,x)\right]dt\right]dx \\
&\quad + \hat{p}_{\theta_t}(\theta)\int_{\mathbf{R}^{d(\theta)}} \left[\hat{p}_{X_{t-}|\theta_t}(x|\theta)\left[H(\theta,x)-\hat{H}(\xi_{t-})\right]^T dv_t\right]dx \\
&= \sum_{\eta\neq\theta}\left[\lambda_{\eta\theta}\int_{\mathbf{R}^{d(\eta)}} \hat{p}_{\xi_{t-}}(\eta,y)dy - \lambda_{\theta\eta}\hat{p}_{\theta_{t-}}(\theta)\right]dt + \hat{p}_{\theta_t}(\theta)\left[\hat{H}(\theta,x_t)-\hat{H}(\xi_{t-})\right]^T dv_t \\
&= \sum_{\eta\neq\theta}\left[\lambda_{\eta\theta}\hat{p}_{\theta_{t-}}(\eta) - \lambda_{\theta\eta}\hat{p}_{\theta_{t-}}(\theta)\right]dt + \hat{p}_{\theta_t}(\theta)\left[\hat{H}(\theta,x_t)-\hat{H}(\xi_{t-})\right]^T dv_t.
\end{aligned}
$$

Substituting $\sum_{\eta\neq\theta}\lambda_{\theta\eta} = -\lambda_{\theta\theta}$ yields (4.23).

Q.E.D.

4.3.3 Evolution of Mode- and Observation-conditional Density

This subsection characterizes the evolution of the mode- and observation-conditional density $\hat{p}_{X_t|\theta_t}(x|\theta) \triangleq p_{X_t|\theta_t,Y_t}(x|\theta)$ under the condition $\hat{p}_{\theta_s}(\theta) > 0$.

Theorem 4.2 Let A1–A5 be satisfied, let $\{\xi_t\}$ be the solution of (4.1)–(4.3), and let $\hat{p}_{X_t|\theta_t}(x|\theta)$ denote the conditional density of X_t, given $(\theta_t = \theta, \mathcal{Y}_t)$. If $\hat{p}_{\theta_s}(\theta) > 0$ for all $s \leq t$, then $d\hat{p}_{X_t|\theta_t}(x|\theta)$ satisfies:

$$
\begin{aligned}
d\hat{p}_{X_t|\theta_t}(x|\theta) &= \mathcal{L}^{\theta *}\hat{p}_{X_t|\theta_t}(x|\theta)dt + \widehat{M}_t^{\theta}\hat{p}_{X_t|\theta_t}(x|\theta)dt \\
&\quad + \hat{p}_{X_t|\theta_t}(x|\theta)\left[H(\theta,x)-\hat{H}(\theta,X_{t-})\right]^T dv_t(\theta),
\end{aligned}
\tag{4.25}
$$

where $\widehat{M}_t^{\theta}$ is the interaction operator:

$$
\widehat{M}_t^{\theta}\hat{p}_{X_t|\theta_t}(x|\theta) = \sum_{\eta\neq\theta}\lambda_{\eta\theta}\frac{\hat{p}_{\theta_t}(\eta)}{\hat{p}_{\theta_t}(\theta)}\left[\int_{\mathbf{R}^{d(\eta)}} p_{X_\tau|\theta_{\tau-},X_{\tau-},\theta_\tau}(x|\eta,y,\theta)\hat{p}_{X_t|\theta_t}(y|\eta)dy - \hat{p}_{X_t|\theta_t}(x|\theta)\right].
\tag{4.26}
$$

Proof: Applying Ito's differentiation rule to $\widehat{p}_{X_t|\theta_t}(x|\theta) = \widehat{p}_{X_t,\theta_t}(x,\theta)/\widehat{p}_{\theta_t}(\theta)$ yields:

$$
\begin{aligned}
d\widehat{p}_{X_t|\theta_t}(x|\theta) &= \frac{\partial \widehat{p}_{X_t|\theta_t}(x|\theta)}{\partial \widehat{p}_{\theta_t}(\theta)} d\widehat{p}_{\theta_t}(\theta) + \frac{\partial \widehat{p}_{X_t|\theta_t}(x|\theta)}{\partial \widehat{p}_{X_t,\theta_t}(x,\theta)} d\widehat{p}_{X_t,\theta_t}(x,\theta) \\
&+ \frac{1}{2}\frac{\partial^2 \widehat{p}_{X_t|\theta_t}(x|\theta)}{\partial^2 \widehat{p}_{X_t,\theta_t}(x,\theta)} d < m_J^c, m_J^c >_t + \frac{1}{2}\frac{\partial^2 \widehat{p}_{X_t|\theta_t}(x|\theta)}{\partial^2 \widehat{p}_{\theta_t}(\theta)} d < m_\theta^c, m_\theta^c >_t \\
&+ \frac{1}{2}\frac{\partial^2 \widehat{p}_{X_t|\theta_t}(x|\theta)}{\partial \widehat{p}_{\theta_t}(\theta)\partial \widehat{p}_{X_t,\theta_t}(x,\theta)} d < m_\theta^c, m_J^c >_t + \frac{1}{2}\frac{\partial^2 \widehat{p}_{X_t|\theta_t}(x|\theta)}{\partial \widehat{p}_{X_t,\theta_t}(x,\theta)\partial \widehat{p}_{\theta_t}(\theta)} d < m_J^c, m_\theta^c >_t.
\end{aligned}
$$

Using $\widehat{p}_{X_t|\theta_t}(x|\theta) = \dfrac{\widehat{p}_{X_t,\theta_t}(x,\theta)}{\widehat{p}_{\theta_t}(\theta)}$, yields:

$$
\frac{\partial \widehat{p}_{X_t|\theta_t}(x|\theta)}{\partial \widehat{p}_{\theta_t}(\theta)} = -\frac{\widehat{p}_{X_t,\theta_t}(x,\theta)}{\widehat{p}_{\theta_t}(\theta)^2}, \frac{\partial \widehat{p}_{X_t|\theta_t}(x|\theta)}{\partial \widehat{p}_{X_t,\theta_t}(x,\theta)} = \frac{1}{\widehat{p}_{\theta_t}(\theta)}, \frac{\partial^2 \widehat{p}_{X_t|\theta_t}(x|\theta)}{\partial^2 \widehat{p}_{X_t,\theta_t}(x,\theta)} = 0,
$$

$$
\frac{\partial^2 \widehat{p}_{X_t|\theta_t}(x|\theta)}{\partial^2 \widehat{p}_{\theta_t}(\theta)} = 2\frac{\widehat{p}_{X_t,\theta_t}(x,\theta)}{\widehat{p}_{\theta_t}(\theta)^3}, \frac{\partial^2 \widehat{p}_{X_t|\theta_t}(x|\theta)}{\partial \widehat{p}_{\theta_t}(\theta)\partial \widehat{p}_{X_t,\theta_t}(x,\theta)} = \frac{\partial^2 \widehat{p}_{X_t|\theta_t}(x|\theta)}{\partial \widehat{p}_{X_t,\theta_t}(x,\theta)\partial \widehat{p}_{\theta_t}(\theta)} = -\frac{1}{\widehat{p}_{\theta_t}(\theta)^2}.
$$

Substitution in $d\widehat{p}_{X_t|\theta_t}(x|\theta)$ yields:

$$
\begin{aligned}
d\widehat{p}_{X_t|\theta_t}(x|\theta) &= \frac{1}{\widehat{p}_{\theta_t}(\theta)} d\widehat{p}_{X_t,\theta_t}(x,\theta) - \frac{\widehat{p}_{X_t,\theta_t}(x,\theta)}{\widehat{p}_{\theta_t}(\theta)^2} d\widehat{p}_{\theta_t}(\theta) \\
&+ \frac{\widehat{p}_{X_t,\theta_t}(x,\theta)}{\widehat{p}_{\theta_t}(\theta)^3} d < m_\theta^c, m_\theta^c >_t - \frac{1}{2}\frac{1}{\widehat{p}_{\theta_t}(\theta)^2} d < m_\theta^c, m_J^c >_t \\
&- \frac{1}{2}\frac{1}{\widehat{p}_{\theta_t}(\theta)^2} d < m_J^c, m_\theta^c >_t \\
&= \frac{1}{\widehat{p}_{\theta_t}(\theta)} d\widehat{p}_{X_t,\theta_t}(x,\theta) - \widehat{p}_{X_t|\theta_t}(x|\theta)\frac{d\widehat{p}_{\theta_t}(\theta)}{\widehat{p}_{\theta_t}(\theta)} + < \text{terms} >,
\end{aligned} \tag{4.27}
$$

with: $< \text{terms} > \triangleq \widehat{p}_{X_t|\theta_t}(x|\theta)\dfrac{d < m_\theta^c, m_\theta^c >_t}{\widehat{p}_{\theta_t}(\theta)^2} - \dfrac{1}{2}\dfrac{d < m_\theta^c, m_J^c >_t}{\widehat{p}_{\theta_t}(\theta)^2} - \dfrac{1}{2}\dfrac{d < m_J^c, m_\theta^c >_t}{\widehat{p}_{\theta_t}(\theta)^2}$.

Substitution of (4.22) in (4.27) yields:

$$
\begin{aligned}
d\widehat{p}_{X_t|\theta_t}(x|\theta) &= \frac{\mathcal{A}^* \widehat{p}_{\xi_t}(\theta,x)}{\widehat{p}_{\theta_t}(\theta)} dt + \widehat{p}_{X_t|\theta_t}(x|\theta)[H(\theta,x) - \widehat{H}(\xi_{t-})]^T dv_t \\
&- \widehat{p}_{X_t|\theta_t}(x|\theta)\frac{d\widehat{p}_{\theta_t}(\theta)}{\widehat{p}_{\theta_t}(\theta)} + < \text{terms} >.
\end{aligned}
$$

Substitution of (4.14), (4.16) and (4.23), and subsequent evaluation yields:

$$
\begin{aligned}
d\widehat{p}_{X_t|\theta_t}(x|\theta) &= \mathcal{L}^{\theta*}\widehat{p}_{X_t|\theta_t}(x|\theta)dt - \frac{\widehat{p}_{X_t|\theta_t}(x|\theta)}{\widehat{p}_{\theta_t}(\theta)}\sum_{\eta\in\Theta}[\lambda_{\eta\theta}\widehat{p}_{\theta_t}(\eta)]dt + <\text{terms}> \\
&+ \frac{1}{\widehat{p}_{\theta_t}(\theta)}\sum_{\eta\neq\theta}\left[\lambda_{\eta\theta}\int_{\mathbf{R}^{d(\eta)}} p_{X_\tau|\theta_{\tau-},X_{\tau-},\theta_\tau}(x|\eta,y,\theta)\widehat{p}_{X_t,\theta_t}(y,\eta)dy - \lambda_{\theta\eta}\widehat{p}_{X_t,\theta_t}(x,\theta)\right]dt \\
&- \widehat{p}_{X_t|\theta_t}(x|\theta)[\widehat{H}(\theta,X_{t-}) - \widehat{H}(\xi_{t-})]^T dv_t + \widehat{p}_{X_t|\theta_t}(x|\theta)[H(\theta,x) - \widehat{H}(\xi_{t-})]^T dv_t \\
&= \mathcal{L}^{\theta*}\widehat{p}_{X_t|\theta_t}(x|\theta)dt + \sum_{\eta\neq\theta}\lambda_{\eta\theta}\frac{\widehat{p}_{\theta_t}(\eta)}{\widehat{p}_{\theta_t}(\theta)}\left[\int_{\mathbf{R}^{d(\eta)}} p_{X_\tau|\theta_{\tau-},X_{\tau-},\theta_\tau}(x|\eta,y,\theta)\widehat{p}_{X_t|\theta_t}(y|\eta)dy\right]dt \\
&+ \lambda_{\theta\theta}\widehat{p}_{X_t|\theta_t}(x|\theta)dt - \widehat{p}_{X_t|\theta_t}(x|\theta)\sum_{\eta\neq\theta}\left[\lambda_{\eta\theta}\frac{\widehat{p}_{\theta_t}(\eta)}{\widehat{p}_{\theta_t}(\theta)}\right]dt - \lambda_{\theta\theta}\widehat{p}_{X_t|\theta_t}(x|\theta)dt \\
&+ \widehat{p}_{X_t|\theta_t}(x|\theta)[H(\theta,x) - \widehat{H}(\theta,X_{t-})]^T dv_t + <\text{terms}> \\
&= \mathcal{L}^{\theta*}\widehat{p}_{X_t|\theta_t}(x|\theta)dt + \sum_{\eta\neq\theta}\lambda_{\eta\theta}\frac{\widehat{p}_{\theta_t}(\eta)}{\widehat{p}_{\theta_t}(\theta)}\left[\int_{\mathbf{R}^{d(\eta)}} p_{X_\tau|\theta_{\tau-},X_{\tau-},\theta_\tau}(x|\eta,y,\theta)\widehat{p}_{X_t|\theta_t}(y|\eta)dy\right. \\
&\left. - \widehat{p}_{X_t|\theta_t}(x|\theta)\right]dt + \widehat{p}_{X_t|\theta_t}(x|\theta)[H(\theta,x) - \widehat{H}(\theta,X_{t-})]^T dv_t + <\text{terms}> \\
&= \mathcal{L}^{\theta*}\widehat{p}_{X_t|\theta_t}(x|\theta)dt + \widehat{M}_t^\theta\widehat{p}_{X_t|\theta_t}(x|\theta)dt \\
&+ \widehat{p}_{X_t|\theta_t}(x|\theta)[H(\theta,x) - \widehat{H}(\theta,X_{t-})]^T dv_t + <\text{terms}>.
\end{aligned}
\tag{4.28}
$$

From (4.22) and (4.23), we know:

$$dm^c_{\theta,t} = \widehat{p}_{\theta_t}(\theta)[\widehat{H}(\theta,X_{t-}) - \widehat{H}(\xi_{t-})]^T dv_t, \tag{4.29}$$

$$dm^c_{J,t} = \widehat{p}_{\xi_t}(\xi)[H(\xi) - \widehat{H}(\xi_{t-})]^T dv_t. \tag{4.30}$$

Hence: $d<m^c_\theta,m^c_\theta>_t = \widehat{p}_{\theta_t}(\theta)^2[\widehat{H}(\theta,X_{t-}) - \widehat{H}(\xi_{t-})]^T[\widehat{H}(\theta,X_{t-}) - \widehat{H}(\xi_{t-})]dt,$

$$d<m^c_\theta,m^c_J>_t = \widehat{p}_{\theta_t}(\theta)\widehat{p}_{\xi_t}(\theta,X)[\widehat{H}(\theta,X_{t-}) - \widehat{H}(\xi_{t-})]^T[H(\theta,X) - \widehat{H}(\xi_{t-})]dt,$$

$$d<m^c_J,m^c_\theta>_t = \widehat{p}_{\theta_t}(\theta)\widehat{p}_{\xi_t}(\theta,X)[H(\theta,X) - \widehat{H}(\xi_{t-})]^T[\widehat{H}(\theta,X_{t-}) - \widehat{H}(\xi_{t-})]dt.$$

Further evaluation yields:

$$\widehat{p}_{X_t|\theta_t}(x|\theta)\frac{d<m^c_\theta,m^c_\theta>_t}{\widehat{p}_{\theta_t}(\theta)^2} = \widehat{p}_{X_t|\theta_t}(x|\theta)[\widehat{H}(\theta,X_{t-}) - \widehat{H}(\xi_{t-})]^T[\widehat{H}(\theta,X_{t-}) - \widehat{H}(\xi_{t-})]dt, \tag{4.31}$$

$$\frac{d<m^c_\theta,m^c_J>_t}{\widehat{p}_{\theta_t}(\theta)^2} = \widehat{p}_{X_t|\theta_t}(x|\theta)[\widehat{H}(\theta,X_{t-}) - \widehat{H}(\xi_{t-})]^T[H(\theta,x) - \widehat{H}(\xi_{t-})]dt, \tag{4.32}$$

$$\frac{d < m_J^c, m_\theta^c >_t}{\hat{p}_{\theta_t}(\theta)^2} = \hat{p}_{X_t|\theta_t}(x|\theta)[H(\theta, x) - \hat{H}(\xi_{t-})]^T [\hat{H}(\theta, X_{t-}) - \hat{H}(\xi_{t-})]dt. \tag{4.33}$$

Substitution of (4.31)–(4.33) in "< terms >" of Eq. (4.27) yields:

$$\begin{aligned} < \text{terms} > &= \hat{p}_{X_t|\theta_t}(x|\theta)\frac{d < m_\theta^c, m_\theta^c >_t}{\hat{p}_{\theta_t}(\theta)^2} - \frac{1}{2}\frac{d < m_\theta^c, m_J^c >_t}{\hat{p}_{\theta_t}(\theta)^2} - \frac{1}{2}\frac{d < m_J^c, m_\theta^c >_t}{\hat{p}_{\theta_t}(\theta)^2} \\ &= \hat{p}_{X_t|\theta_t}(x|\theta)[\hat{H}(\theta, X_{t-}) - \hat{H}(\xi_{t-})]^T [\hat{H}(\theta, X_{t-}) - \hat{H}(\xi_{t-})]dt \\ &\quad + \hat{p}_{X_t|\theta_t}(x|\theta)[H(\theta, x) - \hat{H}(\xi_{t-})]^T [\hat{H}(\theta, X_{t-}) - \hat{H}(\xi_{t-})]dt \\ &= \hat{p}_{X_t|\theta_t}(x|\theta)[H(\theta, x) - \hat{H}(\theta, X_{t-})]^T [\hat{H}(\theta, X_{t-}) - \hat{H}(\xi_{t-})]dt. \end{aligned} \tag{4.34}$$

Substituting (4.34) in (4.28), and subsequent evaluation yields:

$$\begin{aligned} d\hat{p}_{X_t|\theta_t}(x|\theta) &= \mathcal{L}^{\theta *}\hat{p}_{X_t|\theta_t}(x|\theta)dt + \hat{M}_t^\theta \hat{p}_{X_t|\theta_t}(x|\theta)dt \\ &\quad + \hat{p}_{X_t|\theta_t}(x|\theta)[H(\theta, x) - \hat{H}(\theta, X_{t-})]^T [dY_t - \hat{H}(\theta, X_t)]dt. \end{aligned}$$

Substituting: $d\nu_t(\theta) = dY_t - \hat{H}(\theta, X_t)dt$ completes the proof.

Q.E.D.

Remark 4.2 The appearance of the interaction operator $\hat{M}_t^\theta$ stands out in the sense that this is a stochastic operator involving the quotient $\dfrac{\hat{p}_{\theta_t}(\eta)}{\hat{p}_{\theta_t}(\theta)}$ within the summation over η.

4.4 IMM Estimation of MJLS with HJs

This section develops the exact IMM estimator in continuous time for a MJLS version of (4.1)–(4.3) and (4.18), satisfying:

$$F(\theta, x) = F_\theta x, \tag{4.35}$$

$$G(\theta, x) = G_\theta, \tag{4.36}$$

$$C(\theta, \eta, x, u) = C_{\theta\eta}^x x + C_{\theta\eta}^u u, \quad \text{with:} \quad C_{\theta\theta}^x = C_{\theta\theta}^u = 0, \tag{4.37}$$

$$H(\theta, x) = H_\theta x. \tag{4.38}$$

where F_θ, G_θ, $C_{\theta\eta}^x$, $C_{\theta\eta}^u$, and H_θ have finite-valued components.

Subsection 4.4.1 derives the exact IMM estimator for continuous-time MJLS. Subsection 4.4.2 extends this to discrete-time observations. Subsection 4.4.3 develops a numerical integration method. Subsection 4.4.4 specifies the HJ-IMM estimator.

4.4.1 Exact IMM Estimation Given Continuous-time Observations

This subsection considers Section 4.3 case of receiving continuous-time observations (4.18). To obtain the exact IMM estimator, we characterization the evolution of mode-conditional mean $\widehat{X}_t(\theta)$ and covariance $\widehat{R}_t(\theta)$, which are defined as:

$$\widehat{X}_t(\theta) = E\{X_t \,|\, \mathcal{Y}_t, \theta_t = \theta\}, \tag{4.39}$$

$$\widehat{R}_t(\theta) = E\left\{[X_t - \widehat{X}_t(\theta)][X_t - \widehat{X}_t(\theta)]^T \,|\, \mathcal{Y}_t, \theta_t = \theta\right\}. \tag{4.40}$$

Theorem 4.3 For an MJLS with HJs satisfying (4.1–4.3), (4.18), and (4.35–4.38), $\widehat{X}_t(\theta)$ and $\widehat{R}_t(\theta)$ of (4.39) and (4.40) are solutions of:

$$d\widehat{X}_t(\theta) = F_\theta \widehat{X}_t(\theta)dt + \sum_{\eta\neq\theta}\left[\lambda_{\eta\theta}\frac{\widehat{p}_{\theta_t}(\eta)}{\widehat{p}_{\theta_t}(\theta)}[\widehat{X}_t^\theta(\eta) - \widehat{X}_t(\theta)]\right]dt + \widehat{R}_t(\theta){H_\theta}^T d\nu_t(\theta), \tag{4.41}$$

$$\begin{aligned} d\widehat{R}_t(\theta) &= [F_\theta \widehat{R}_t(\theta) + \widehat{R}_t(\theta)F_\theta^T + G_\theta G_\theta^T]dt + \widehat{Z}_t(\theta)H_\theta^T d\nu_t(\theta) \\ &\quad + \sum_{\eta\neq\theta}\left[\lambda_{\eta\theta}\frac{\widehat{p}_{\theta_t}(\eta)}{\widehat{p}_{\theta_t}(\theta)}\left[\widehat{R}_t^\theta(\eta) - \widehat{R}_t(\theta) + [\widehat{X}_t^\theta(\eta) - \widehat{X}_t(\theta)][\widehat{X}_t^\theta(\eta) - \widehat{X}_t(\theta)]^T\right]\right]dt, \end{aligned} \tag{4.42}$$

with:

$$\widehat{X}_t^\theta(\eta) = \left[I + C_{\eta\theta}^x\right]\widehat{X}_t(\eta) + C_{\eta\theta}^u E\{u\}, \tag{4.43}$$

$$\widehat{R}_t^\theta(\eta) = \left[I + C_{\eta\theta}^x\right]\widehat{R}_t(\eta)\left[I + C_{\eta\theta}^x\right]^T + C_{\eta\theta}^u Cov\{u\}{C_{\eta\theta}^u}^T, \tag{4.44}$$

$$d\nu_t(\theta) = dY_t - H_\theta \widehat{X}_t(\theta), \tag{4.45}$$

$$\widehat{Z}_t(\theta) = E\left\{[X_t - \widehat{X}_t(\theta)][X_t - \widehat{X}_t(\theta)]^T \times [X_t - \widehat{X}_t(\theta)]^T \,|\, \mathcal{Y}_t, \theta_t = \theta\right\}, \tag{4.46}$$

where the tensor product $\times$ makes clear that $\widehat{Z}_t(\theta)$ is a three-dimensional array.

Proof: By the definitions of $\widehat{X}_t(\theta)$ and $\widehat{R}_t(\theta)$ we have:

$$\widehat{X}_t(\theta) = \int_{\mathbf{R}^{d(\theta)}} x\widehat{p}_{X_t|\theta_t}(x|\theta)dx,$$

$$\widehat{R}_t(\theta) = \int_{\mathbf{R}^{d(\theta)}} [x - \widehat{X}_t(\theta)][x - \widehat{X}_t(\theta)]^T \widehat{p}_{X_t|\theta_t}(x|\theta)dx.$$

Hence:

$$d\widehat{X}_t(\theta) = \int_{\mathbf{R}^{d(\theta)}} x d\widehat{p}_{X_t|\theta_t}(x|\theta)dx,$$

$$d\widehat{R}_t(\theta) = \int_{\mathbf{R}^{d(\theta)}} [x - \widehat{X}_t(\theta)][x - \widehat{X}_t(\theta)]^T d\widehat{p}_{X_t|\theta_t}(x|\theta)dx.$$

Substitution of the $d\hat{p}_{X_t|\theta_t}(x|\theta)$ characterization from Theorem 4.2 yields:

$$d\hat{X}_t(\theta) = \int_{\mathbf{R}^{d(\theta)}} x\mathcal{L}^{\theta*}\hat{p}_{X_t|\theta_t}(x|\theta)dtdx + \int_{\mathbf{R}^{d(\theta)}} x\hat{p}_{X_t|\theta_t}(x|\theta)[H(\theta,x) - \hat{H}(\theta,X_{t-})]^T d\nu_t(\theta)dx$$

$$+ \int_{\mathbf{R}^{d(\theta)}} x\sum_{\eta\neq\theta}\lambda_{\eta\theta}\frac{\hat{p}_{\theta_t}(\eta)}{\hat{p}_{\theta_t}(\theta)}\left[\int_{\mathbf{R}^{d(\eta)}} p_{X_\tau|\theta_{\tau-},X_{\tau-},\theta_\tau}(x|\eta,y,\theta)\hat{p}_{X_t|\theta_t}(y|\eta)dy - \hat{p}_{X_t|\theta_t}(x|\theta)\right]dtdx,$$

$$d\hat{R}_t(\theta) = \int_{\mathbf{R}^{d(\theta)}} C_t(\theta,x)\mathcal{L}^{\theta*}\hat{p}_{X_t|\theta_t}(x|\theta)dtdx$$

$$+ \int_{\mathbf{R}^{d(\theta)}} C_t(\theta,x)\sum_{\eta\neq\theta}\lambda_{\eta\theta}\frac{\hat{p}_{\theta_t}(\eta)}{\hat{p}_{\theta_t}(\theta)}\left[\int_{\mathbf{R}^{d(\eta)}} p_{X_\tau|\theta_{\tau-},X_{\tau-},\theta_\tau}(x|\eta,y,\theta)\hat{p}_{X_t|\theta_t}(y|\eta)dy\right.$$

$$\left. - \hat{p}_{X_t|\theta_t}(x|\theta)\right]dtdx + \int_{\mathbf{R}^{d(\theta)}} C_t(\theta,x)\hat{p}_{X_t|\theta_t}(x|\theta)$$

$$[H(\theta,x) - \hat{H}(\theta,X_{t-})]^T d\nu_t(\theta)dx,$$

with $C_t(\theta,x) \triangleq [x - \hat{X}_t(\theta)][x - \hat{X}_t(\theta)]^T$.

Evaluation for the jump linear system yields:

$$d\hat{X}_t(\theta) = F_\theta\hat{X}_t(\theta)dt + \sum_{\eta\neq\theta}\lambda_{\eta\theta}\frac{\hat{p}_{\theta_t}(\eta)}{\hat{p}_{\theta_t}(\theta)}[\hat{X}_t^\theta(\eta) - \hat{X}_t(\theta)]dt$$

$$+ \int_{\mathbf{R}^{d(\theta)}} x\hat{p}_{X_t|\theta_t}(x|\theta)[x - \hat{X}_t(\theta)]^T H_\theta^T d\nu_t(\theta)dx,$$

$$d\hat{R}_t(\theta) = [F_\theta\hat{R}_t(\theta) + \hat{R}_t(\theta)F_\theta{}^T + G_\theta G_\theta{}^T]dt$$

$$+ \sum_{\eta\neq\theta}\lambda_{\eta\theta}\frac{\hat{p}_{\theta_t}(\eta)}{\hat{p}_{\theta_t}(\theta)}\left[\hat{R}_t^\theta(\eta) - \hat{R}_t(\theta) + [\hat{X}_t^\theta(\eta) - \hat{X}_t(\theta)][\hat{X}_t^\theta(\eta) - \hat{X}_t(\theta)]^T\right]dt$$

$$+ \int_{\mathbf{R}^{d(\theta)}} [x - \hat{X}_t(\theta)][x - \hat{X}_t(\theta)]^T\hat{p}_{X_t|\theta_t}(x|\theta)[x - \hat{X}_t(\theta)]^T H_\theta^T d\nu_t(\theta)dx.$$

Q.E.D.

Remark 4.3 Exact IMM Eqs. (4.41, 4.42) have originally been developed in [9]. Equations (4.41) and (4.42) show the interaction/mixing weights $\lambda_{\eta\theta}\frac{\hat{p}_{\theta_t}(\eta)}{\hat{p}_{\theta_t}(\theta)}$. As is shown in [7, 8], for an MJLS without HJs the same interaction/mixing weights apply, though $\hat{X}_t^\theta(\eta)$ in Eq. (4.41) simplifies to $\hat{X}_t(\eta)$, and $\hat{R}_t^\theta(\eta)$ in Eq. (4.42) simplifies to $\hat{R}_t(\eta)$.

Remark 4.4 Due to the innovation term $\hat{Z}_t(\theta)H_\theta^T d\nu_t(\theta)$ in Eq. (4.42), the exact IMM Eqs. (4.41)–(4.46) do not form a finite-dimensional filter. The commonly adopted approximation is to assume $\hat{Z}_t(\theta) = 0$. The structure of Eq. (4.42) shows that this approximation does not influence the effect of HJs.

4.4.2 Exact IMM Estimation Given Discrete-time Observations

In practice, the observation process of a continuous-time stochastic process is in discrete-time. Let the MJLS process $\{\xi_t\} = \{\theta_t, x_t\}$ be observed at equidistant time moments $t_k = k\Delta, k = 1, 2, ...,$ by:

$$Y_k = H_{\theta_{t_k}} X_{t_k} + J_{\theta_{t_k}} V_k, \tag{4.47}$$

where observation matrix H_θ and the noise matrix J_θ are now functions of the regime θ_{t_k} at observation moment t_k, and the observation noise $\{V_k\}$ is a sequence of independent m-dimensional standard Gaussian variables. Hence the increasing sigma algebra of observations is now $\mathcal{Y}_t = \sigma\{Y_k, t_k \in [0, t]\}$.

In between subsequent observation moments t_{k-1} and t_k, the predicted conditional mass-density $\bar{p}_{\xi_{t_k}}(\theta, x)$ evolves according to Eq. (4.22) in case there are no observations, i.e.,

$$\bar{p}_{\xi_{t_k}}(\theta, x) = \hat{p}_{\xi_{t_{k-1}}}(\theta, x) + \int_{t_{k-1}}^{t_k} \mathcal{A}^* \bar{p}_{\xi_s}(\theta, x) ds. \tag{4.48}$$

At observation moment $t = t_k$, $\hat{p}_{\xi_t}(\xi)$ satisfies Bayes equation:

$$\hat{p}_{\xi_{t_k}}(\theta, x) = \frac{p_{Y_k|\xi_{t_k}}(Y_k|\theta, x)\bar{p}_{\xi_{t_k}}(\theta, x)}{p_{Y_k}(Y_k)}, \tag{4.49}$$

with normalization factor $p_{Y_k}(Y_k)$, and conditional Gaussian factor:

$$p_{Y_k|\xi_{t_k}}(Y_k|\theta, x) = \mathrm{N}\{Y_k; H_\theta x, J_\theta J_\theta{}^T\}. \tag{4.50}$$

For the linear situation of (4.35)–(4.37), the above implies that Eqs. (4.22) and (4.41)–(4.44), with $d\nu_t = d\nu_t(\theta) = 0$, apply for the prediction of $\bar{p}_{\theta_t}(\theta)$, $\bar{X}_t(\theta)$ and $\bar{R}_t(\theta)$ on the interval $(t_{k-1}, t_k]$, i.e.,

$$d\bar{p}_{\theta_t}(\theta) = \sum_{\eta\in\Theta} [\lambda_{\eta\theta} \bar{p}_{\theta_t}(\eta)] dt, \tag{4.51}$$

$$d\bar{X}_t(\theta) = F_\theta \bar{X}_t(\theta) dt + \sum_{\eta\neq\theta} \left[\lambda_{\eta\theta} \frac{\bar{p}_{\theta_t}(\eta)}{\bar{p}_{\theta_t}(\theta)} [\bar{X}_t^\theta(\eta) - \bar{X}_t(\theta)] \right] dt, \tag{4.52}$$

$$\begin{aligned} d\bar{R}_t(\theta) &= [F_\theta \bar{R}_t(\theta) + \bar{R}_t(\theta) F_\theta^T + G_\theta G_\theta^T] dt \\ &+ \sum_{\eta\neq\theta} \left[\lambda_{\eta\theta} \frac{\bar{p}_{\theta_t}(\eta)}{\bar{p}_{\theta_t}(\theta)} \left[\bar{R}_t^\theta(\eta) - \bar{R}_t(\theta) + [\bar{X}_t^\theta(\eta) - \bar{X}_t(\theta)][\bar{X}_t^\theta(\eta) - \bar{X}_t(\theta)]^T \right] \right] dt, \end{aligned} \tag{4.53}$$

$$\text{with: } \bar{X}_t^\theta(\eta) = \left[I + C_{\eta\theta}^x \right] \bar{X}_t(\eta) + C_{\theta\eta}^u E\{u\}, \tag{4.54}$$

$$\bar{R}_t^\theta(\eta) = \left[I + C_{\eta\theta}^x \right] \bar{R}_t(\eta) \left[I + C_{\eta\theta}^x \right]^T + C_{\eta\theta}^u Cov\{u\} C_{\eta\theta}^{u\ T}. \tag{4.55}$$

The latter also implies $\bar{X}_t^\theta(\theta) = \bar{X}_t(\theta)$ and $\bar{R}_t^\theta(\theta) = \bar{R}_t(\theta)$, i.e., nothing happens if the mode does not switch.

If, at observation moment t_k, the predicted conditional density $\bar{p}_{x_t|\theta_t}(x|\theta)$ is assumed to have a Gaussian shape, then the observation update equations satisfy those of the discrete-time IMM estimator:

$$\hat{X}_{t_k}(\theta) = \bar{X}_{t_k}(\theta) + K_k(\theta)\upsilon_k(\theta), \tag{4.56a}$$

$$\hat{R}_{t_k}(\theta) = \bar{R}_{t_k}(\theta) - K_k(\theta)\bar{R}_{t_k}(\theta){H_\theta}^T, \tag{4.56b}$$

$$\hat{p}_{\theta_{t_k}}(\theta) = \Lambda_k(\theta)\bar{p}_{\theta_{t_k}}(\theta)/c_{norm}^p, \tag{4.57}$$

with c_{norm}^p the normalization constant, and:

$$\upsilon_k(\theta) = Y_k - H_\theta \bar{X}_{t_k}(\theta), \tag{4.58}$$

$$U_k(\theta) = H_\theta \bar{R}_{t_k}(\theta){H_\theta}^T + J_\theta {J_\theta}^T, \tag{4.59}$$

$$K_k(\theta) = \bar{R}_{t_k}(\theta){H_\theta}^T U_k(\theta)^{-1}, \tag{4.60}$$

$$\Lambda_k(\theta) = \|U_k(\theta)\|^{-\frac{1}{2}} \exp\left\{-\frac{1}{2}\upsilon_k(\theta)^T U_k(\theta)^{-1}\upsilon_k(\theta)\right\}. \tag{4.61}$$

4.4.3 Numerical Integration

The quotient $\dfrac{\bar{p}_{\theta_t}(\eta)}{\bar{p}_{\theta_t}(\theta)}$ in Eqs. (4.52, 4.53) complicates a straightforward numerical integration on the interval $(t_k, t_{k+1}]$. A preparatory step in simplifying this, is to approximate the continuous time Markov chain by a discrete time Markov chain that makes jumps right after each measurement moment only, i.e., for $t \in (t_k, t_{k+1}]$:

$$\bar{p}_{\theta_t}(\theta) = \sum_{\eta\in\Theta}\left[\Pi_{\eta\theta}\hat{p}_{\theta_{t_k}}(\eta)\right], \tag{4.62}$$

with the transition probability matrix $\Pi_{\eta\theta}$ satisfying:

$$\Pi_{\eta\theta} = P\{\theta_{t_{k+1}} = \theta | \theta_{t_k} = \eta\} = \begin{cases} 1 - \exp\{-\lambda_{\eta\theta}\Delta\}, & \text{if } \eta \neq \theta \\ 1 - \sum_{\theta'\neq\theta}[1 - \exp\{-\lambda_{\theta\theta'}\Delta\}], & \text{if } \eta = \theta \end{cases}. \tag{4.63}$$

The implicit condition is that time interval Δ should be small enough to be covered by a single integration time step. Otherwise, the numerical integration should be conducted over a sequence of smaller sub-intervals. We also assume that dN_t in (4.2) may have discontinuities right after each measurement moment only. Hence, Eq. (4.3) yields for $t \in (t_k, t_{k+1}]$:

$$X_{t_k+} = X_{t_k} + C(\theta_{t_k}, \theta_{t_k+}, X_{t_k}, u_{t_k+})(N_{t_k+} - N_{t_k}), \tag{4.64a}$$

$$X_t = X_{t_k+} + \int_{t_k}^{t} F(\theta_s, X_s)ds + \int_{t_k}^{t} G(\theta_s, X_s)dB_s. \tag{4.64b}$$

Under linearity assumptions (4.35–4.37), Eq. (4.64b) implies for $t \in (t_k, t_{k+1}]$:

$$\bar{X}_t(\theta) = \widehat{X}^M_{t_k+}(\theta) + \int_{t_k}^{t} F_\theta \bar{X}_s(\theta)ds, \tag{4.65}$$

$$\bar{R}_t(\theta) = \widehat{R}^M_{t_k+}(\theta) + \int_{t_k}^{t} [F_\theta \bar{R}_t(\theta) + \bar{R}_t(\theta)F_\theta^T + G_\theta G_\theta^T]ds, \tag{4.66}$$

where: $\widehat{X}^M_{t_k+}(\theta) \triangleq E\{X_{t_k+}|\theta_{t_k+} = \theta, \mathcal{Y}_{t_k}\}$, (4.67)

$$\widehat{R}^M_{t_k+}(\theta) \triangleq E\left\{\left[X_{t_k+} - \widehat{X}^M_{t_k+}(\theta)\right]\left[X_{t_k+} - \widehat{X}^M_{t_k+}(\theta)\right]^T |\theta_{t_k+} = \theta, \mathcal{Y}_{t_k}\right\}. \tag{4.68}$$

Theorem 4.4 HJ Interaction/Mixing The mode prediction $\bar{p}_{\theta_{t_{k+1}}}(\theta)$, the mixed mean $\widehat{X}^M_{t_k+}(\theta)$ and the mixed covariance $\widehat{R}^M_{t_k+}(\theta)$, defined by (4.62) and (4.67, 4.68), satisfy:

$$\bar{p}_{\theta_{t_{k+1}}}(\theta) = \sum_{\eta\in\Theta}\left[\Pi_{\eta\theta}\widehat{p}_{\theta_{t_k}}(\eta)\right], \tag{4.69}$$

$$\widehat{X}^M_{t_k+}(\theta) = \sum_{\eta\in\Theta}\left[\Psi_{k,\eta\theta}\widehat{X}^\theta_{t_k+}(\eta)\right], \tag{4.70}$$

$$\widehat{R}^M_{t_k+}(\theta) = \sum_{\eta\in\Theta}\left[\Psi_{k,\eta\theta}\left[\widehat{R}^\theta_{t_k+}(\eta) + \left[\widehat{X}^\theta_{t_k+}(\eta) - \widehat{X}^M_{t_k+}(\theta)\right]\left[\widehat{X}^\theta_{t_k+}(\eta) - \widehat{X}^M_{t_k+}(\theta)\right]^T\right]\right], \tag{4.71}$$

with:

$$\Psi_{k,\eta\theta} \triangleq \Pi_{\eta\theta}\frac{\widehat{p}_{\theta_{t_k}}(\eta)}{\bar{p}_{\theta_{t_{k+1}}}(\theta)}, \quad \text{if } \bar{p}_{\theta_{t_{k+1}}}(\theta) > 0, \tag{4.72}$$

$$\widehat{X}^\theta_{t_k+}(\eta) = \left[I + C^x_{\eta\theta}\right]\widehat{X}_{t_k}(\eta) + C^u_{\eta\theta}E\{u\}, \tag{4.73}$$

$$\widehat{R}^\theta_{t_k+}(\eta) = \left[I + C^x_{\eta\theta}\right]\widehat{R}_{t_k}(\eta)\left[I + C^x_{\eta\theta}\right]^T + C^u_{\eta\theta}Cov\{u\}C^{u\ T}_{\eta\theta}. \tag{4.74}$$

Proof: Equation (4.69) follows straightforwardly from (4.62).

Equations for the mixed $\widehat{X}^M_{t_k+}(\theta)$ and $\widehat{R}^M_{t_k+}(\theta)$ are obtained as follows:

$$\begin{aligned}\widehat{X}^M_{t_k+}(\theta) &= E\{X_{t_k+}|\theta_{t_k+} = \theta, \mathcal{Y}_{t_k}\}\\ &= \sum_{\eta\in\Theta}\left[P\{\theta_{t_k} = \eta|\theta_{t_k+} = \theta, \mathcal{Y}_{t_k}\}E\{X_{t_k+}|\theta_{t_k+} = \theta, \theta_{t_k} = \eta, \mathcal{Y}_{t_k}\}\right]\\ &= \sum_{\eta\in\Theta}\left[P\{\theta_{t_k} = \eta|\theta_{t_{k+1}} = \theta, \mathcal{Y}_{t_k}\}\widehat{X}^\theta_{t_k+}(\eta)\right],\end{aligned}$$

$$\begin{aligned}\widehat{R}^M_{t_k+}(\theta) &= E\left\{\left[X_{t_k+} - \widehat{X}^M_{t_k+}(\theta)\right]\left[X_{t_k+} - \widehat{X}^M_{t_k+}(\theta)\right]^T|\theta_{t_k+} = \theta, \mathcal{Y}_{t_k}\right\}\\ &= \sum_{\eta\in\Theta}\left[P\{\theta_{t_k} = \eta|\theta_{t_k+} = \theta, \mathcal{Y}_{t_k}\}E\left\{\left[X_{t_k+} - \widehat{X}^M_{t_k+}(\theta)\right]\left[X_{t_k+} - \widehat{X}^M_{t_k+}(\theta)\right]^T|\theta_{t_k+} = \theta, \theta_{t_k} = \eta, \mathcal{Y}_{t_k}\right\}\right]\\ &= \sum_{\eta\in\Theta}\left[P\{\theta_{t_k} = \eta|\theta_{t_{k+1}} = \theta, \mathcal{Y}_{t_k}\}\left[\widehat{R}^\theta_{t_k+}(\eta) + \left[\widehat{X}^\theta_{t_k+}(\eta) - \widehat{X}^M_{t_k+}(\theta)\right]\left[\widehat{X}^\theta_{t_k+}(\eta) - \widehat{X}^M_{t_k+}(\theta)\right]^T\right]\right].\end{aligned}$$

Evaluation of the weights yields:

$$P\{\theta_{t_k} = \eta | \theta_{t_k+} = \theta, \mathcal{Y}_{t_k}\} = \frac{P\{\theta_{t_k} = \eta, \theta_{t_{k+1}} = \theta | \mathcal{Y}_{t_k}\}}{P\{\theta_{t_{k+1}} = \theta | \mathcal{Y}_{t_k}\}}$$

$$= \frac{\Pi_{\eta\theta} P\{\theta_{t_k} = \eta | \mathcal{Y}_{t_k}\}}{P\{\theta_{t_{k+1}} = \theta | \mathcal{Y}_{t_k}\}} = \frac{\Pi_{\eta\theta} \hat{p}_{\theta_{t_k}}(\eta)}{\overline{p}_{\theta_{t_{k+1}}}(\theta)} = \Psi_{k,\eta\theta}.$$

Substituting the latter in the equations for $\hat{X}^M_{t_k+}(\theta)$ and $\hat{R}^M_{t_k+}(\theta)$ yields (4.70) and (4.71).

Q.E.D.

Hence, numerical integration of (4.70) and (4.71) is accomplished in two fractional stages. The first stage conducts numerical evaluation of the summations according to Theorem 4.4. The second fractional step conducts first order Euler integration of the integral terms in (4.65, 4.66).

4.4.4 HJ-IMM Estimator

Algorithm 4.1 specifies Steps I–VI of the HJ-IMM estimator. Figure 4.1 presents the functional blocks of the HJ-IMM estimator, their inputs, outputs, and corresponding steps. Steps I–VI have been obtained as follows:

- Step I covers Eqs. (4.69, 4.72).
- Step II covers the HJ interaction/mixing Eqs. (4.70, 4.71) and (4.73, 4.74).
- Step III are discrete-time Kalman prediction equations. First order Euler integration of (4.65) over Δ yields equations for the predicted mean and the corresponding covariance.

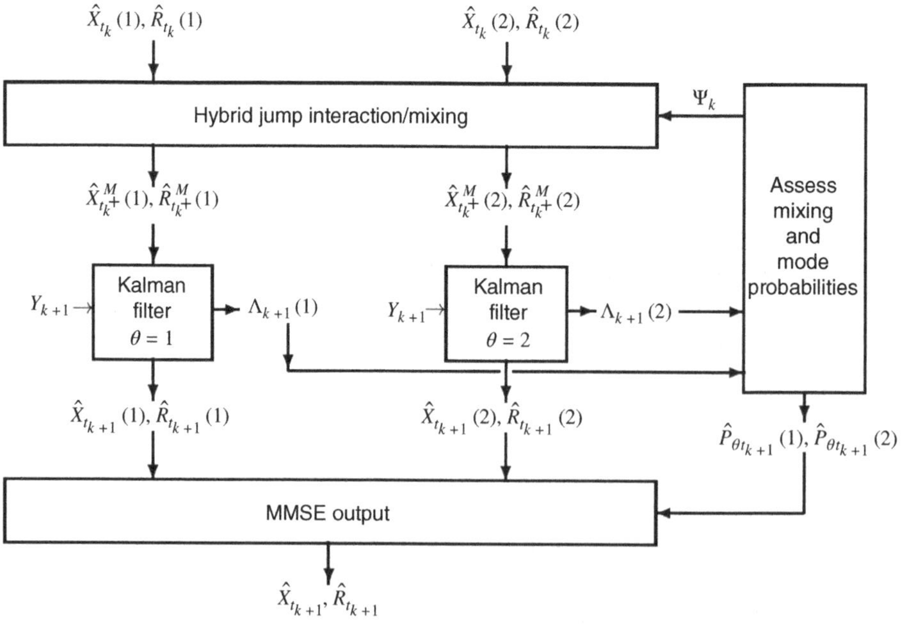

Figure 4.1 Functionalities of the HJ-IMM estimator for $\Theta = \{1, 2\}$. Comparison to the well-known IMM estimator, e.g., [31, pages 453–457], shows that HJs influence the interaction/mixing functionality only.

- Step IV covers the measurement update Eqs. (4.56a, b), (4.58–4.60), and (4.61) for $\Lambda_{k+1}(\theta)$.
- Step V covers the mode probability update Eq. (4.57).
- Step VI produces the MMSE output, by weighted averaging over the mode-specific estimates.

Of these steps, only Step II differs from those of the well know IMM estimator.

Algorithm 4.1: HJ-IMM estimator

Initialization. Set $k := 0$ and start with initial values for the mode probabilities $\hat{p}_{\theta_0}(\theta)$, and the mode conditional first and second moments $\hat{X}_0(\theta)$ and $\hat{R}_0(\theta)$.

HJ-IMM estimation cycle

I. Assess Mixing probabilities; for $\theta \in \Theta$:

$$\bar{p}_{\theta_{t_{k+1}}}(\theta) = \sum\nolimits_{\eta\in\Theta}\left[\Pi_{\eta\theta}\hat{p}_{\theta_{t_k}}(\eta)\right],$$

$$\Psi_{k,\eta\theta} = \Pi_{\eta\theta}\hat{p}_{\theta_{t_k}}(\eta)/\bar{p}_{\theta_{t_{k+1}}}(\theta).$$

II. State mixing; for $\theta \in \Theta$:

$$\hat{X}^M_{t_k+}(\theta) = \sum_{\eta\in\Theta}\left[\Psi_{k,\eta\theta}\hat{X}^{\theta}_{t_k+}(\eta)\right],$$

$$\hat{R}^M_{t_k+}(\theta) = \sum_{\eta\in\Theta}\left[\Psi_{k,\eta\theta}\left[\hat{R}^{\theta}_{t_k+}(\eta) + \left[\hat{X}^{\theta}_{t_k+}(\eta) - \hat{X}^M_{t_k+}(\theta)\right]\left[\hat{X}^{\theta}_{t_k+}(\eta) - \hat{X}^M_{t_k+}(\theta)\right]^T\right]\right],$$

where for $\eta, \theta \in \Theta$:

$$\hat{X}^{\theta}_{t_k}(\eta) = \left[I + C^x_{\eta\theta}\right]\hat{X}_{t_k}(\eta) + C^u_{\eta\theta}E\{u\},$$

$$\hat{R}^{\theta}_{t_k}(\eta) = \left[I + C^x_{\eta\theta}\right]\hat{R}_{t_k}(\eta)\left[I + C^x_{\eta\theta}\right]^T + C^u_{\eta\theta}Cov\{u\}{C^u_{\eta\theta}}^T.$$

III. Kalman filter prediction; for $\theta \in \Theta$:

$$\bar{X}_{t_{k+1}}(\theta) = [I + F_\theta\Delta]\hat{X}^M_{t_k+},$$

$$\bar{R}_{t_{k+1}}(\theta) = [I + F_\theta\Delta]\hat{R}^M_{t_k+}(\theta)[I + F^T_\theta\Delta] + G_\theta {G_\theta}^T\Delta.$$

IV. Kalman filter state update; for $\theta \in \Theta$:

$$\nu_{k+1}(\theta) = Y_{k+1} - H_\theta\bar{X}_{t_{k+1}}(\theta),$$

$$U_{k+1}(\theta) = H_\theta\bar{R}_{t_{k+1}}(\theta){H_\theta}^T + J_\theta {J_\theta}^T,$$

$$K_{k+1}(\theta) = \bar{R}_{t_{k+1}}(\theta){H_\theta}^T U_{k+1}(\theta)^{-1},$$

$$\hat{X}_{t_{k+1}}(\theta) = \bar{X}_{t_{k+1}}(\theta) + K_{k+1}(\theta)\nu_{k+1}(\theta),$$

$$\hat{R}_{t_{k+1}}(\theta) = \bar{R}_{t_{k+1}}(\theta) - K_{k+1}(\theta)H_\theta\bar{R}_{t_{k+1}}(\theta),$$

$$\Lambda_{k+1}(\theta) = \|U_{k+1}(\theta)\|^{-1/2}\exp\left\{-1/2\,\nu_{k+1}(\theta)^T U_{k+1}(\theta)^{-1}\nu_{k+1}(\theta)\right\}.$$

V. Mode probabilities update; for $\theta \in \Theta$:

$$\hat{p}_{\theta_{t_{k+1}}}(\theta) = \Lambda_{k+1}(\theta)\bar{p}_{\theta_{t_{k+1}}}(\theta)/c_{k+1},$$

with c_{k+1} the normalization constant.

VI. MMSE output:

$$\hat{X}_{t_{k+1}} = \sum_{\theta\in\Theta}\left[\hat{p}_{\theta_{t_{k+1}}}(\theta)\hat{X}_{t_{k+1}}(\theta)\right],$$

$$\hat{R}_{t_{k+1}} = \sum_{\theta\in\Theta}\left[\hat{p}_{\theta_{t_{k+1}}}(\theta)\left[\hat{R}_{t_{k+1}}(\theta) + [\hat{X}_{t_{k+1}}(\theta) - \hat{X}_{t_{k+1}}][\hat{X}_{t_{k+1}}(\theta) - \hat{X}_{t_{k+1}}]^T\right]\right].$$

Stop if $t_{k+1} \geq T$; otherwise set $k := k + 1$ and repeat the above filter cycle.

Remark 4.5 During HJ-IMM estimation, an output $\hat{X}_{t_{k+1}}(\theta)$ of Bayes state updating may yield an estimated value that falls outside the range of allowed values under regime θ. This we refer to as a "derailed regime estimate." In such case, the "derailed regime estimate" can simply be removed by enforcing $\hat{p}_{\theta_{t_{k+1}}}(\theta) := 0$, and to subsequently apply a renormalization, i.e., $\hat{p}_{\theta_{t_{k+1}}}(\theta) := \hat{p}_{\theta_{t_{k+1}}}(\theta)/\sum_{\eta\in\Theta}\hat{p}_{\theta_{t_{k+1}}}(\eta)$. Removal of a "derailed regime estimate" can best be done in between Steps V and VI in Algorithm 4.1.

Remark 4.6 In [13, 14], filtering of a pure discrete-time MJLS with HJs has been studied. The discrete-time generalized IMM estimation equations obtained are the same as those in Algorithm 4.1, with the exception that prediction Step III is an exact equation instead of the Euler integrated version.

4.5 Particle Filtering of Markov Switching Diffusion with HJs

In case of nonlinear (F, G, C, H), the prediction and Bayesian update equations of the exact IMM estimation equations can be implemented through a PF.

Subsection 4.5.1 describes the discrete time observations of (4.1)–(4.3). Subsection 4.5.2 develops the HJ-SIR PF. Subsection 4.5.3 develops the HJ-IMM PF.

4.5.1 Mode- and Observation-conditional Density Given Discrete-time Observations

We consider equidistant discrete-time observations Y_k, $t_k = k\Delta$, $k = 1, 2, .., $ i.e.,

$$Y_k = H(X_{t_k}, \theta_{t_k}) + J(X_{t_k}, \theta_{t_k})V_k, \tag{4.75}$$

of the Markov switching diffusion $\{\theta_t, X_t\}$, satisfying (4.1)–(4.3) of Section 4.2. The observation noise $\{V_k\}$ is a sequence of independent m-dimensional standard Gaussian variables, which are independent of $\{\theta_t, X_t\}$. The increasing sigma algebra of observations is $\mathcal{Y}_t = \sigma\{Y_k, t_k \leq t\}$.

In between subsequent observation moments t_k and t_{k+1}, the predicted conditional probability mass-density $\bar{p}_{\xi_t}(\theta, x) \triangleq p_{\xi_t|\mathcal{Y}_{t_k}}(\theta, x)$, $t > t_k$, evolves according to Eq. (4.22) for $dv_t = 0$, i.e.,

$$\bar{p}_{\xi_t}(\theta, x) = \hat{p}_{\xi_{t_k}}(\theta, x) + \int_{t_k}^{t} \mathcal{A}^*\bar{p}_{\xi_s}(\theta, x)ds, \tag{4.76}$$

with $\mathcal{A}^* \bar{p}_{\xi_s}(\theta, x) = \mathcal{L}^{\theta*} \bar{p}_{\xi_s}(\theta, x) + \mathcal{J}^* \bar{p}_{\xi_s}(\theta, x)$. By adopting the discrete-time Markov chain approximation of Subsection 4.3, the Kolmogorov part of Eq. (4.76) satisfies:

$$\begin{aligned}
\hat{p}_{\xi_{t_k+}}(\theta, x) &= \sum_{\eta\in\Theta} \left[\Pi_{\eta\theta} \int_{d(\eta)} p_{X_{t_k+}|\theta_{t_k}, X_{t_k}, \theta_{t_k+}}(x|\eta, y, \theta) \hat{p}_{\xi_{t_k}}(\eta, y) dy \right] \\
&= \sum_{\eta\in\Theta} \left[\Pi_{\eta\theta} \hat{p}_{\theta_{t_k}}(\eta) \int_{d(\eta)} p_{X_{t_k+}|\theta_{t_k}, X_{t_k}, \theta_{t_k+}}(x|\eta, y, \theta) \hat{p}_{X_{t_k}|\theta_{t_k}}(y|\eta) dy \right] \\
&= \sum_{\eta\in\Theta} \left[\Pi_{\eta\theta} \hat{p}_{\theta_{t_k}}(\eta) \hat{p}_{X_{t_k+}|\theta_{t_k}, \theta_{t_k+}}(x|\eta, \theta) \right].
\end{aligned} \tag{4.77}$$

By starting from $\hat{p}_{\xi_{t_k+}}(\theta, x)$, the diffusion part of Eq. (4.76) remains, i.e.,

$$\bar{p}_{\xi_t}(\theta, x) = \hat{p}_{\xi_{t_k+}}(\theta, x) + \int_{t_k}^{t} \mathcal{L}^{\theta*} \bar{p}_{\xi_s}(\theta, x) ds. \tag{4.78}$$

At observation moment $t = t_k$, $\hat{p}_{\xi_t}(\theta, x)$ satisfies Bayes equation:

$$\hat{p}_{\xi_{t_k}}(\theta, x) = \frac{p_{Y_k|\xi_{t_k}}(Y_k|\theta, x) \bar{p}_{\xi_{t_k}}(\theta, x)}{p_{Y_k}(Y_k)}, \tag{4.79}$$

with normalization factor $p_{Y_k}(Y_k)$, and conditional Gaussian density:

$$p_{Y_k|\xi_{t_k}}(Y_k|\theta, x) = \mathrm{N}\left\{Y_k; H(\theta, x), J(\theta, x)J(\theta, x)^T\right\}.$$

Hence, the joint conditional probability mass-density update satisfies:

$$\hat{p}_{\xi_{t_k}}(\theta, x) = \frac{1}{C_k} \bar{p}_{\xi_{t_k}}(\theta, x) \cdot \mathrm{N}\left\{Y_k; H(\theta, x), J(\theta, x)J(\theta, x)^T\right\}, \tag{4.80}$$

where C_k is the normalization constant.

HJ PFs use sets of particles to numerically evaluate (4.77), (4.78), and (4.80) in a sequential way.

4.5.2 HJ-SIR Particle Filter

The joint conditional probability mass-density $\hat{p}_{\xi_{t_{k-1}}}(\theta, x)$ is approximated by empirical density:

$$\tilde{p}_{X_{t_k}, \theta_{t_k}|Y_k}(X, \theta) = \sum_{j=1}^{N_P} w_{t_k}^j 1\{\theta = \theta_{t_k}^j\} \delta\left(X - X_{t_k}^j\right), \tag{4.81}$$

with $1\{\theta = \eta\} = 1$, if $\theta = \eta$; else $1\{\theta = \eta\} = 0$. In (4.81), $\tilde{p}_{X_{t_k}, \theta_{t_k}|Y_k}(X, \theta)$ is spanned by the set of N_p hybrid-state particles $\left\{w_{t_k}^j \in [0, 1], \theta_{t_k}^j \in \Theta, X_{t_k}^j \in \mathbf{R}^n; j = 1, \ldots, N_p\right\}$. Each particle consists of weight $w_{t_k}^j$, mode $\theta_{t_k}^j$, and Euclidean-valued state $X_{t_k}^j$. Algorithm 4.2 specifies the HJ-SIR PF, which involves five steps during each filter cycle. Step 1 evaluates (4.77) by simulating each particle according to Eqs. (4.62, 4.63) and the jump term in Eq. (4.64a). Step 2 evaluates (4.78) by simulating each particle according to a first order Euler integration of the drift and Brownian motion terms in

Eq. (4.64b). Step 3 evaluates (4.80) by applying a Bayesian update to the weight of each particle for discrete-time measurement Y_{k+1}. Next, Step 4 produces the estimation empirical density. Finally, Step 5 applies a multinomial resampling to the set of particles. The latter avoids impoverishment of the set of particles, in the sense that too many particles might end up with a very low weight.

Algorithm 4.2: HJ-SIR Particle Filter

Initialization: At t_0, for $j = 1, .., N_p$ (with N_p the number of particles) set $w^j_{t_0} = 1/N_p$ and $k = 0$, and simulate $\left(X^j_{t_0}, \theta^j_{t_0}\right)$ from $p_{X_{t_0},\theta_{t_0}}(.,\theta) = p_{X_{t_0}|\theta_{t_0}}(.|\theta)p_{\theta_{t_0}}(\theta)$.

Particles at moment t_k: $\left\{w^j_{t_k} \in [0,1], \theta^j_{t_k} \in \Theta, X^j_{t_k} \in \mathbf{R}^n; j = 1, \dots, N_p\right\}$.

HJ-SIR PF cycle:

- Step 1: Generate HJs for $j = 1, \dots, N_p$:
  Simulate: $\theta^j_{t_{k+1}} \sim p_{\theta_{t_k+1}|\theta_{t_k}}\left(\theta|\theta^j_{t_k}\right) = \Pi_{\theta^j_{t_k}\theta}$,

  $$u^j_{t_k} \sim \mu(du),$$

  $$\mathrm{X}^j_{t_k} := X^j_{t_k} + C\left(\theta^j_{t_k}, \theta^j_{t_{k+1}}, X^j_{t_k}, u^j_{t_k}\right),$$

  where $\mu(du)$ is the measure on mark space U.
- Step 2: Evolution for $j = 1, \dots, N_p$:
  Simulate: $B^j_k \sim N(0, I)$,

  $$\bar{X}^j_{t_{k+1}} = \mathrm{X}^j_{t_k} + F\left(\theta^j_{t_{k+1}}, \mathrm{X}^j_{t_k}\right)\Delta + G\left(\theta^j_{t_{k+1}}, \mathrm{X}^j_{t_k}\right)B^j_k\sqrt{\Delta}.$$

- Step 3: Bayes update correction for $j = 1, .., N_p$:

  $$\bar{w}^j_{t_{k+1}} = w^j_{t_k} \cdot \mathrm{N}\left\{Y_{k+1}; H\left(\theta^j_{t_{k+1}}, \bar{X}^j_{t_{k+1}}\right), J\left(\theta^j_{t_{k+1}}, \bar{X}^j_{t_{k+1}}\right)J\left(\theta^j_{t_{k+1}}, \bar{X}^j_{t_{k+1}}\right)^T\right\}/c_{k+1},$$

  with c_{k+1} such that $\sum_{j=1}^{N_P} \bar{w}^j_{t_{k+1}} = 1$.
- Step 4: Estimated empirical density:

  $$\tilde{p}_{X_{t_{k+1}},\theta_{t_{k+1}}|Y_{k+1}}(X,\theta) = \sum_{j=1}^{N_p}\left[w^j_{t_{k+1}} 1\left\{\theta = \theta^j_{t_{k+1}}\right\}\delta\left(X - X^j_{t_{k+1}}\right)\right].$$

- Step 5: Multinomial resampling for $j = 1, .., N_p$:

  $$w^j_{t_{k+1}} = 1/N_p,$$

  $$\left(X^j_{t_{k+1}}, \theta^j_{t+1}\right) \sim \tilde{p}_{X_{t_{k+1}},\theta_{t_{k+1}}|Y_{k+1}}(X,\theta).$$

Stop if $t_{k+1} \geq T$; otherwise set $k := k + 1$ and repeat the filter cycle above.

A weakness of a SIR PF is that for a low probability mode, the particles for this mode tend to disappear during resampling Step 5. To remedy this weakness for the situation without HJs, [32–34] have used Rao-Blackwellization to develop the IMM PF. In the next subsection, this approach is followed to obtain the HJ-IMM PF.

4.5.3 HJ-IMM Particle Filter

To improve the healthiness of particle weights, we apply a Rao-Blackwellization approach that uses analytical equations for the evolution of $\hat{p}_{\theta_{t_k}}(\theta)$, and particles for the numerical evolution of $\hat{p}_{X_{t_k}|\theta_{t_k}}(X|\theta)$. For each $\theta \in \Theta$ there is a set of particles $\left\{w_{t_k}^{\theta j} \in [0,1], X_{t_k}^{\theta j} \in \mathbf{R}^n; j = 1, \ldots, K\right\}$ which spans the empirical density approximation of $\hat{p}_{X_{t_k}|\theta_{t_k}}(X|\theta)$:

$$\hat{p}_{X_{t_k}|\theta_{t_k}}(X|\theta) \cong \sum_{j=1}^{K} w_{t_k}^{\theta,j} \delta\left(X - X_{t_k}^{\theta,j}\right). \tag{4.82}$$

The resulting HJ-IMM particle filter is specified in Algorithm 4.3.

- Step 1 in Algorithm 4.3 implements Eq. (4.62) for $\bar{\gamma}_{t_{k+1}}(\theta) \triangleq \bar{p}_{\theta_{t_{k+1}}|Y_{t_k}}(\theta)$.
- Step 2: To characterize the evolution of the Euclidean-valued particles, we divide the left and right terms in Eq. (4.77) by $\bar{p}_{\theta_{t_k+}}(\theta) = \bar{p}_{\theta_{t_{k+1}}}(\theta)$, which yields for $\bar{p}_{\theta_{t_{k+1}}}(\theta) > 0$:

$$\begin{aligned}\hat{p}_{X_{t_k+}|\theta_{t_k+}}(x|\theta) &= \bar{p}_{\theta_{t_{k+1}}}(\theta)^{-1} \sum_{\eta\in\Theta}\left[\Pi_{\eta\theta}\hat{p}_{\theta_{t_k}}(\eta)\hat{p}_{X_{t_k+}|\theta_{t_k},\theta_{t_k+}}(x|\eta,\theta)\right] \\ &= \sum_{\eta\in\Theta}\left[\Psi_{k,\eta\theta}\hat{p}_{X_{t_k+}|\theta_{t_k},\theta_{t_k+}}(x|\eta,\theta)\right],\end{aligned} \tag{4.83}$$

  with mixing weights $\Psi_{k,\eta\theta}$ satisfying (4.72). The particle version $\tilde{p}_{X_{t_k+}|\theta_{t_k},\theta_{t_k+}}(x|\eta,\theta)$ of $\hat{p}_{X_{t_k+}|\theta_{t_k},\theta_{t_k+}}(x|\eta,\theta)$ is obtained by simulating each Euclidean state particle according to the jump term in (4.64a). Together with (4.83), this yields Step 2.
- Step 3: Since Step 2 increases the number of particles by a factor $|\Theta|$, Step 3 applies multinomial resampling to bring the number of modes back to K particles per mode.
- Step 4 conducts Euler integration of the drift and diffusion terms in Eq. (4.64b).
- Step 5 is a Bayesian update of all particle weights from discrete-time measurement Y_{k+1}.
- Step 6 provides the estimated outputs.

Algorithm 4.3: HJ-IMM Particle Filter

Initialization: Set $k = 0$, $\hat{\gamma}_0(\theta) = p_{\theta_{t_0}}(\theta)$, and $w_{t_0}^{\theta,j} = 1/K$ for $\theta \in \Theta$, $j = 1, .., K = N_P/|\Theta|$, and simulate $X_{t_0}^{\theta,j}$ from $p_{X_{t_0}|\theta_{t_0},Y_{t_0}}(.|\theta)$.

Particles at moment t_k, for each $\theta \in \Theta$: $\left\{w_{t_k}^{\theta,j} \in [0,1],\ X_{t_k}^{\theta,j} \in \mathbf{R}^n;\ j = 1, \ldots, K\right\}$, spanning the empirical density: $\tilde{p}_{X_{t_k}|\theta_{t_k},Y_k}(X|\theta) = \sum_{j=1}^{K} w_{t_k}^{\theta,j}\delta\left(X - X_{t_k}^{\theta,j}\right)$.

HJ-IMM PF cycle:

- Step 1: Assess Mixing probabilities: For $\theta \in \Theta$:

  $\bar{\gamma}_{t_{k+1}}(\theta) = \sum_{\eta\in\Theta}\left[\Pi_{\eta\theta}\hat{\gamma}_{t_k}(\eta)\right],$

$$\Psi_{k,\eta\theta} = \Pi_{\eta\theta}\hat{\gamma}_{t_k}(\eta)/\bar{\gamma}_{t_{k+1}}(\theta), \quad \text{if } \bar{\gamma}_{t_{k+1}}(\theta) > 0.$$

- Step 2: Simulate mode switching effect on particles.
  For each $(\eta, \theta) \in \Theta^2$, and $j \in [1, K]$:

$$u_{t_k}^{\eta\theta,j} \sim \mu(du),$$

$$X_{t_k}^{\eta\theta,j} = X_{t_k}^{\eta,j} + C\left(\eta, \theta, X_{t_k}^{\eta,j}, u_{t_k}^{\eta\theta,j}\right),$$

$$\omega_{t_k}^{\eta\theta,j} = \begin{cases} \Psi_{k,\eta\theta} w_{t_k}^{\eta,j}, & \text{if } \bar{\gamma}_{t_{k+1}}(\theta) > 0 \\ 1/K, & \text{if } \bar{\gamma}_{t_{k+1}}(\theta) = 0 \end{cases},$$

with $\mu(du)$ the measure on mark space U.
  For each $\theta \in \Theta$, the resulting set of particles is

$$\left\{\omega_{t_k}^{\eta\theta,j}, X_{t_k}^{\eta\theta,j}, \eta \in \Theta, j \in [1, K]\right\}.$$

- Step 3: Multinomial resampling:

$$\bar{w}_{t_{k+1}}^{\theta,j} = 1/K$$

$$\bar{X}_{t_k}^{\theta,j} \sim \sum_{\eta\in\Theta} \sum_{j'=1}^{K} \omega_{t_k}^{\eta\theta,j'} \delta\left(X - X_{t_k}^{\eta\theta,j'}\right)$$

  For each $\theta \in \Theta$, the resulting set of particles is:

$$\left\{\bar{w}_{t_{k+1}}^{\theta,j}, \bar{X}_{t_k}^{\theta,j}, j \in [1, K]\right\}$$

- Step 4: Prediction for $j \in [1, K], \theta \in \Theta$:
  Simulate: $B_{t_k}^{\theta,j} \sim N(0, I)$,

$$X_{t_{k+1}}^{\theta,j} = \bar{X}_{t_k}^{\theta,j} + F\left(\theta, \bar{X}_{t_k}^{\theta,j}\right)\Delta + G\left(\theta, \bar{X}_{t_k}^{\theta,j}\right)B_{t_k}^{\theta,j}\sqrt{\Delta}.$$

- Step 5: Bayesian update correction: For $j \in [1, K], \theta \in \Theta$:

$$w_{t_{k+1}}^{\theta,j} = \bar{\gamma}_{t_{k+1}}(\theta) \cdot \mathrm{N}\left\{Y_{k+1}; H\left(\theta, X_{t_{k+1}}^{\theta,j}\right), J(\theta)J(\theta)^T\right\}/c_{k+1},$$

with c_{k+1} such that $\sum_{j=1}^{K} \sum_{\theta\in\Theta} w_{t_{k+1}}^{\theta,j} = 1$,

$$\hat{\gamma}_{t_{k+1}}(\theta) = \sum_{j=1}^{K} w_{t_{k+1}}^{\theta,j}.$$

- Step 6: Estimation output:
  Mode probabilities

$$\hat{p}_{\theta_{t_{k+1}}|Y_{k+1}}(\theta) = \hat{\gamma}_{t_{k+1}}(\theta).$$

  Mode conditional means and covariances:

$$\hat{X}_{t_{k+1}}(\theta) = \sum_{j=1}^{K}\left[w_{t_{k+1}}^{\theta,j} X_{t_{k+1}}^{\theta,j}\right],$$

$$\widehat{R}_{t_{k+1}}(\theta) = \sum_{j=1}^{K}\left[w_{t_{k+1}}^{\theta,j}\left[X_{t_{k+1}}^{\theta,j} - \widehat{X}_{t_{k+1}}(\theta)\right]\left[X_{t_{k+1}}^{\theta,j} - \widehat{X}_{t_{k+1}}(\theta)\right]^T\right].$$

MMSE output:

$$\widehat{X}_{t_{k+1}} = \sum_{\theta\in\Theta}[\widehat{\gamma}_{t_{k+1}}(\theta)\widehat{X}_{t_{k+1}}(\theta)],$$

$$\widehat{R}_{t_{k+1}} = \sum_{\theta\in\Theta}\left[\widehat{\gamma}_{t_{k+1}}(\theta)\left[\widehat{R}_{t_{k+1}}(\theta) + [\widehat{X}_{t_{k+1}}(\theta) - \widehat{X}_{t_{k+1}}][\widehat{X}_{t_{k+1}}(\theta) - \widehat{X}_{t_{k+1}}]^T\right]\right].$$

Stop if $t_{k+1} \geq T$; otherwise set $k := k + 1$ and repeat the filter cycle above.

Table 4.1 Comparison of HJ-SIR PF versus HJ-IMM PF.

Aspect	**HJ-SIR PF**	**HJ-IMM PF**
Particle has θ_t sample	Yes	No
Number of particles	Fixed number	Fixed number per mode
Mode Switching	Simulation	Analytical
Interaction/Mixing	No	Yes
State Prediction	Simulation	Simulation
Bayes update	Each simulated particle	Each simulated particle
Resampling after	Measurement update	State Mixing

Table 4.1 provides an overview of similarities and differences between HJ SIR PF in Algorithm 4.2 and HJ-IMM PF in Algorithm 4.3. The main improvements of the HJ-IMM PF relative to the HJ-SIR PF are:

- Evaluate mode probabilities instead of particle weights for each mode sample;
- Maintain a fixed number of particles per mode;
- Makes use of IMM's Interaction/Mixing;
- Resampling after interaction/mixing rather than after measurement update.

4.6 Conclusions

This chapter has presented an up-to-date derivation of IMM estimation for Markov switching diffusions with HJs, i.e., jumps in the Euclidean-valued X_t that occur simultaneously with θ_t switching. Section 4.2 has characterized the hybrid stochastic process $\{\theta_t, X_t\}$ as a Feller Markov process. Section 4.3 has addressed the nonlinear estimation of $\{\theta_t, X_t\}$ from continuous-time observations, using nonlinear filtering results for jump-diffusions [23]. This leads to a characterization of the mode- and observation-conditional density $\widehat{p}_{X_t|\theta_t}(x|\theta)$, which reveals the Interaction/mixing that is so typical for IMM estimation. Section 4.4 has used the latter for the development of IMM estimation equations for a MJLS with HJs. First, it was shown that in case of continuous-time observations, the exact estimator is an IMM estimator. Subsequently, the HJ-IMM estimator has been developed for the case of discrete-time observations.

Section 4.5 has developed two PFs given discrete-time observations of a Markov switching diffusion with HJs: HJ-SIR PF, and HJ-IMM PF. A key difference between these two PFs is that the latter makes use of IMM's Interaction/mixing.

An interesting direction for follow-on research is to extend the IMM feedback particle filtering [35] to the MJLS with HJs. Another interesting direction is to extend HJ-IMM estimation to the larger class of continuous-time Stochastic Hybrid Systems considered by Liu and Hwang, 2011 [24].

Funding: This research did not receive any specific grant from funding agencies in the public, commercial, or not-for-profit sectors.

References

1 X. R. Li and Y. Bar-Shalom. Multiple model estimation with variable structure. *IEEE Transactions on Automatic Control*, 41:478–493, 1996.
2 X. R. Li, Z. Zhao, and X. B. Li. General model-set design methods for multiple model approach. *IEEE Transactions Automatic Control*, 50:1260–1276, 2005.
3 J. Liu and R. Li. Hierarchical adaptive interacting multiple model algorithm. *IET Control Theory Applications*, 2:479–487, 2008.
4 Y. Tian, H. Jiang, Q. X. Ding, and G. W. Liang. Turn rate estimation based adaptive IMM algorithm for maneuvering target tracking. *Advanced Materials Research*, 383:5609–5614, 2012.
5 J. Lan, X. R. Li, and C. Mu. Best model augmentation for variable structure multiple-model estimation. *IEEE Transactions on Aerospace and Electronic Systems*, 59:4414–4433, 2011.
6 G. Li, S. Zhang, Y. Han, W. Sheng, and T. Kirubarajan. Reweighting interacting multiple-model algorithm to overcome model competition for target tracking in the hybrid system. *IEEE Sensors Journal*, 24:12689–12704, 2024.
7 H. A. P. Blom. The continuous time roots of the interacting multiple model filter. In *Proceedings on the 51st IEEE Conference on Decision and Control*. Maui, Hawaii, USA, 2012.
8 H. A. P. Blom. The continuous time roots of the Interacting Multiple Model filter. In M. Mallick, V. Krishnamurthy, and B. N. Vo, editors, *Integrated Tracking, Classification and Sensor Management*. Wiley-IEEE, Chapter 4:127–161, 2013.
9 H. A. P. Blom. Markov jump-diffusion models and decision-making-free filtering. In A. Bensoussan and J. L. Lions, editors, *Proceedings of the 6th International Conference Analysis and Optimization of Systems*, Springer, 1984, 568–580, 1984.
10 H. A. P. Blom. An efficient filter for abruptly changing systems. *Proceedings of the 23rd IEEE Conference on Decision and Control*. 1984: 656–658, 1984.
11 H. A. P. Blom and Y. Bar-Shalom. The interacting multiple model algorithm for systems with Markovian switching coefficients. *IEEE Transactions Automatic Control*, 33:780–783, 1988.
12 I. I. Gihman and A. V. Skorohod. *Stochastic Differential Equations*. Springer, Berlin, 1972.
13 H. A. P. Blom. An efficient decision-making-free filter for processes with abrupt changes. In *Proceedings of the 7th IFAC Symposium Identification and System Parameter Estimation*, pages 631–636, 1985.
14 H. A. P. Blom. Overlooked potential of systems with Markovian coefficients. In *Proceedings of the 25th IEEE Conference on Decision and Control*, pages 1758–1764, 1986.
15 H. A. P. Blom. Continuous-discrete filtering for systems with Markovian switching coefficients and simultaneous jumps. In *Proceedings of the 21st Annual Asilomar Conference On Signals, Systems and Computers*. Pacific Grove, USA, pages 261–265, 1987.
16 H. A. P. Blom. *Bayesian Estimation for Decision-directed Stochastic Control*, PhD thesis, Delft University of Technology, 1990.

17 H. A. P. Blom, R. A. Hogendoorn, and F. van Schaik. Bayesian multi-sensor tracking for advanced air traffic control systems. In A. Benoit, editors, *Aircraft Trajectories, Computation, Prediction, Control, Volume 2*, AGARDograph 301, AGARD, 34/1–8, 1990.

18 H. A. P. Blom, R. A. Hogendoorn, and B. A. van Doorn. Design of a multisensor tracking system for advanced air traffic control. In Y. Bar-Shalom, editors, *Multitarget-multisensor Tracking: Applications and Advances*, II. Artech House, 31–63, 1992.

19 E. A. Bloem, H. A. P. Blom and F. J. Van Schaik. Advanced data fusion for airport surveillance. In *Proceedings of the JISSA2001 International Conference on Airport Surveillance Sensors*, Paris, pages 106–111, 2001.

20 EUROCONTROL (2017). ARTAS Service Movement Surveillance, artas-articlepaper-update-08032017.pdf (eurocontrol.int).

21 H. Khaledian, H. A. P. Blom, J. Vila-Valls, and X. Prats. Interacting Multiple Model joint aircraft guidance mode and control set-point estimation. *Journal of Guidance, Control, and Dynamics*, accepted for publication on 8th February 2026.

22 D. R. Poklukar. Nonlinear filtering for jump-diffusions. *Journal of Computational and Applied Mathematics*, 197:558–567, 2006.

23 S. Popa and S. S. Sritharan. Nonlinear filtering of Ito-Levy stochastic differential equations with continuous observations. *Communications on Stochastic Analysis*, 3:313–330, 2009.

24 W. Liu and I. Hwang. On hybrid state estimation for stochastic hybrid systems. *IEEE Transactions on Automatic Control*, 59:2615–2628, 2014.

25 J. Lygeros and M. Prandini. Stochastic hybrid systems: a powerful framework for complex, large scale applications. *European Journal of Control*, 6:583–594, 2010.

26 J. Bect. A unifying formulation of the Fokker-Planck-Kolmogorov equation for general stochastic hybrid systems. *Nonlinear Analysis: Hybrid Systems*, 4:357–370, 2010.

27 J. Krystul, A. Bagchi, and H. A. P. Blom. On strong Markov property of solutions to stochastic differential equations on hybrid state spaces. Report 28th February 2011, Twente University, 2011. https://repository.tudelft.nl/islandora/object/uuid%3A948fc4dd-66f5-42bc-82e6-8770d371add3.

28 H. A. P. Blom. Feller property of regime-switching jump diffusion processes with hybrid jumps. *Stochastic Analysis and Applications*, 42:516–532, 2024.

29 R. W. Brockett and G. L. Blankenship. A representation theorem for linear differential equations with Markovian coefficients. In *Proceedings of the Allerton Conference on Circuits Systems Theory*, pages 671–679, 1977.

30 A. V. Skorohod. *Random Processes with Independent Increments*. Kluwer Academic Publishers, Dordrecht, 1991.

31 Y. Bar-Shalom, X. R. Li, and T. Kirubarajan. *Estimation with Applications to Tracking and Navigation*. Wiley, 2001.

32 H. Driessen and Y. Boers. An efficient particle filter for jump Markov nonlinear systems. In *Proceedings of the IEE Colloquium on Target Tracking*, 2004.

33 H. A. P. Blom and E. A. Bloem. Particle filtering for stochastic hybrid systems. In *Proceedings of the IEEE Conference on the Decision and Control*, Bahamas, 2004.

34 H. A. P. Blom, and E. A. Bloem. Exact Bayesian and particle filtering of stochastic hybrid systems. *IEEE Transactions on Aerospace and Electronic Systems*, 43:55–70, 2007.

35 T. Yang, H. A. P. Blom, and P. G. Mehta. Interacting multiple model-feedback particle filter for stochastic hybrid systems. In *Proceedings of the 52nd IEEE CDC*, Florence, Italy, pages 7065–7070, 2013.

5

Joint Target Tracking and Intent Inference Using a Destination-constrained Model

Linfeng Xu[1], Peijie Yang[1], and Mahendra Mallick[2]*

[1] School of Automation, Northwestern Polytechnical University, Xi'an, China
[2] Independent Consultant, Cambria, CA, USA

*Corresponding author: xulinf@gmail.com

5.1 Introduction

Goal-oriented behavior and destination-driven motion are ubiquitous in a wide range of human activities and physical processes. For instance, in air traffic control near airports, all arriving flights are directed toward runways; in maritime surveillance, divergent ship trajectories ultimately converge to ports; and in homeland defense, most incoming threats are aimed at specific targets such as ships, airports, or other military facilities. These goals or destinations reflect behavioral intents and provide predictive information about future actions. Therefore, they should be effectively utilized in various problems, such as acquiring accurate system models and state estimates [1–10]. Conversely, due to the coupling between an object's intent and its motion, we can infer the object's intent by observing its motion trajectory. In this chapter, we explore and utilize complete or incomplete intent information, such as destinations and future actions, to construct a destination-constrained (DC) model, and then address target tracking and intent inference problems using the DC model.

The investigation on DC estimation was pioneered by Castanon et al. [11], who proposed a forward-backward estimate fusion (FBEF) method for maritime surveillance based on the vessel's position measurements and possible destinations. This method obtains the DC state estimate by combining two separate estimates at a time of interest: one from forward estimation and the other from backward prediction that starts from the DC final state. Assuming both the initial state and the final state are random with given boundary conditions, Levy et al. [12] approached the destination-directed process as a Gaussian reciprocal process whose covariance matrix over the given interval has a cyclic block tridiagonal inverse. Despite its more potent descriptive capability than the Markov process, the reciprocal process demands greater complexity in its manipulations due to its dynamic-noise correlation and nearest-neighbor structure. Consequently, considerable effort has been devoted to exploring reciprocal processes over the last three decades and has yielded promising results. For example, White and Carravetta [13, 14] investigated modeling, estimation, and smoothing of reciprocal processes comprehensively. Fanaswala and Krishnamurthy [15, 16] adopted a reciprocal model to detect anomalous trajectory patterns in moving target surveillance by using knowledge of the intended destination. Recently, Rezaie and Li [17–19] recognized that the conditionally Markov (CM) process is a more natural, powerful, and systematic generalization of the Markov process than the reciprocal process. They studied in-depth the modeling by the CM sequence and presented characterizations of the general Gaussian CM sequence. It is worth noting that the proposed Gaussian CM process can be represented by a sum of a Gaussian Markov sequence and an uncorrelated Gaussian vector, making recursive estimation of the state based on the Gaussian CM model straightforward. Within a Bayesian framework, Ahmad et al. [20–22] proposed a bridge distribution (BD) model for object trajectories terminating at a destination, and further, Liang et al. [23, 24] proposed a pseudo-observation-based formulation of the BD models. Additionally, Zhou et al. [25, 26] treated the destination constraint as additional measurements instead of constraining the (unconstrained) dynamics and proposed a pseudo measurement (PM)-based DC estimation method to incorporate the destination constraint into the state estimation.

The above FBEF, reciprocal process-based, CM-based, and PM-based estimation methods all require knowing the arrival time of the destination [15]. However, this requirement is not practical for many DC problems since the arrival time is often unknown in advance. This is particularly true for adversarial tracking, where a defender may know an attacker's intended destination but not its arrival time in advance. To address this issue and incorporate the spatial information of the destination into modeling and estimation, two crucial challenges must be met, which are the topics of this chapter. The first involves modeling of the spatial-temporal relationship between the state at any time and at the destination, as the destination constraint actually affects the entire state evolution due to the interdependence of the state at different times. However, exactly establishing this relationship is difficult from a discrete-time state equation as in [9] because the actual arrival time is random and unlikely to be an integer multiple of the sampling interval. The second challenge pertains to estimation for DC processes, including estimation of the DC state and arrival time, where the complexity stems from the coupling of the arrival time and the state at each time.

Here we are concerned with the DC dynamic system with uncertain arrival time. To our knowledge, related studies are rare. Ahmad et al. [21] and Liang et al. [23, 24] devised various BD models within a Bayesian framework and employed them to refine state estimates, predict future trajectories, and estimate arrival time. This chapter presents a novel methodology for modeling and state estimation of DC systems with uncertain arrival time, based on which we address target tracking and intent inference problem.

The primary contributions of this chapter are threefold. The first is to propose a DC dynamic model from partially compatible dynamics (referred to as relaxed dynamics) by incorporating the destination constraint. Differing from the Bayesian perspective in [21, 23, 24], which uses the distribution of the state transition, our modeling is optimization-based and converts the relaxed dynamics to the DC model using only the first two moments of the state transition—it starts with a relaxed dynamics, represented by a state equation that is often readily available in practice, and then optimally fuses the constraint information using the nearest neighbor criterion. The resultant DC dynamic model is applicable and practical beyond those proposed in [9] and [27], and it allows the arrival time to be any future time, be it known or uncertain, discrete or continuous. Additionally, the superior properties of the model are presented, and they are very beneficial for recursive estimation. Second, we establish a connection between our DC model and a type of BD model proposed in [21, 23, 24]: Despite their different modeling prerequisites, derivations, and resultant model forms, for the linear Gaussian system with a linear destination constraint, the transition density of our DC model coincides with that of the pseudo-observation-based BD model. We also present the respective applicable scenarios for these two distinct modeling methods. The third contribution is to address estimation in practical cases, where the arrival time is partially known or random. Two effective DC estimation algorithms are developed by incorporating the spatial information of the destination constraints into state transition. In particular, a computationally efficient sampling-based estimation algorithm is developed to estimate the state and the arrival time simultaneously. This algorithm distinguishes from those proposed in [21, 23, 24] in the treatment of the integral: It employs Monte Carlo integration, whereas the latter utilizes quadrature integration. Finally, we apply the proposed modeling and DC state estimation to tackle the problem of adversarial tracking and specifically demonstrate their effectiveness and superiority for tracking guided targets.

5.2 Problem Formulation

Consider a dynamic system whose state vector $x(t) \in \mathbb{R}^n$ satisfies the destination constraint:

$$x(t_D) \in \mathcal{D}, \tag{5.1}$$

where $x(t_D)$ represents the state vector at the destination, t_D is the arrival time at the destination, and $\mathcal{D}$ is a known (constraint) set of states based on physical laws, mathematical properties, and/or other a priori information. Unlike our previous work [9] and [18], the arrival time t_D in this chapter is an arbitrary future time, not necessarily multiple of the sampling interval, and can be random and take on any discrete or continuous value. The arrival time t_D is in general assumed to be a continuous random variable with a known prior distribution. We focus on a linear destination constraint

$$\mathcal{D} = \{x(t_D): \quad Ax(t_D) = b\}, \tag{5.2}$$

where $A \in \mathbb{R}^{m \times n}$ (with $m \leq n$) and $b \in \mathbb{R}^m$ are known, and A is of full row rank.

A congruous DC system model satisfies the constraint; that is, all its possible state sequence realizations adhere to the constraint. However, as pointed out in [28], formulating such a model directly is often challenging, but a partially compatible dynamic model, for which some but not all trajectories adhere to the constraint, is often available or can be easily obtained. To come up with a congruous DC model, a partially compatible dynamic model can be conformed to the destination constraint (5.2) using the modeling framework presented in [29]. Note that the destination constraint (5.2) not only imposes on the final state but also may affect the entire state evolution due to the interdependence of the state at different times. Describing the state interdependence is thus essential, and stochastic differential equations (SDEs) are commonly used for this purpose, which will be elaborated in Section 5.3. Compared with that obtained from the difference equations in [9], the SDEs are more general, as they specify interdependence of the state at any time, not just at sampling time instants. Then, the following questions arise naturally: how does the destination constraint affect the state at any time, and how to incorporate the effects in the DC dynamic modeling?

Aside from modeling DC dynamics, this chapter also aims at optimal DC state estimation. Suppose that the state vectors are linearly observed by

$$z_k = H_k x_k + v_k, k = 1, 2, \dots, \tag{5.3}$$

where $z_k = z(t_k)$ and $x_k = x(t_k)$ represent the measurement and the state at time t_k, respectively. The measurement matrix $H_k \in \mathbb{R}^{n_z \times n}$ is known, and additive white noise $\langle v_k \rangle$ is assumed Gaussian with zero mean and covariance R_k. The problem of DC state estimation is to estimate the system state based on the SDE, the intended destination, and the noisy measurement sequence $\langle z_k \rangle$.

One way to handle the destination constraint is to treat the constraint as a perfect measurement; that is, augment the original measurement by a PM without error to account for the effect of the constraint, and then apply traditional state estimation (e.g., Kalman filtering). However, as pointed out in [9], the rationality of doing so is questionable because the constraint is a priori and should be incorporated into each predicted trajectory of the state apart from being used to correct the state estimate as an additional measurement. In view of this, we propose imposing the destination constraint on the state transition by using the congruous DC dynamic model. Even if t_D is random or unknown a priori, incorporating the spatial information of the destination constraint into the state transition can improve estimation.

5.3 Modeling of DC Dynamics

Without loss of generality, we choose the following SDE as partially compatible dynamics, which specifies the interdependence of the state at different times.

$$\dot{x}^o(t) = Bx^o(t) + \Gamma\tilde{w}(t), \tag{5.4}$$

where $\tilde{w}(t)$ is assumed to be zero-mean white noise with covariance $\boldsymbol{q}\delta(t - t')$ and is independent of the initial state $x^o(t_0)$. Here, $\boldsymbol{q}$ is the power spectral density matrix, and $\delta(\cdot)$ is the Dirac delta function. B and Γ are known matrices of appropriate dimensions.

We refer to the model (5.4) as the relaxed (i.e., unconstrained) dynamics to distinguish it from the constrained dynamic model. To avoid notational confusion, in the sequel, denote by x^o and x the relaxed state and the DC state, respectively. Next, we construct a congruous DC dynamic model by imposing the destination constraint (5.1) on the relaxed dynamics (5.4).

The solution to (5.4) is

$$x^o(t) = \Phi(t, t_0)x^o(t_0) + \int_{t_0}^{t} \Phi(t, \tau)\Gamma(\tau)\tilde{w}(\tau)\mathrm{d}\tau, \tag{5.5}$$

where the state transition matrix $\Phi(t, t_0) = e^{B(t-t_0)}$ transits the state from the initial time t_0 to any time t [30]. Equation (5.5) can be written for the discrete time as

$$x_k^o = \Phi_{k-1}x_{k-1}^o + w_{k-1}, \tag{5.6}$$

where $x_k^o \triangleq x^o(t_k)$, $\Phi_{k-1} \triangleq \Phi(t_k, t_{k-1})$, and

$$w_{k-1} \triangleq w(t_k, t_{k-1}) = \int_{t_{k-1}}^{t_k} e^{B(t_k-\tau)}\Gamma\tilde{w}(\tau)\mathrm{d}\tau. \tag{5.7}$$

Clearly, the discrete-time process noise w_{k-1} is zero-mean and white. Further, define an integrated noise as

$$w_j^l \triangleq w(t_l, t_j) = \int_{t_j}^{t_l} e^{B(t_l-\tau)}\Gamma\tilde{w}(\tau)\mathrm{d}\tau,\ j < l, \tag{5.8}$$

and let $w_{k-1}^D \triangleq w(t_D, t_{k-1})$. Then, the covariances and the cross-covariance of w_{k-1} and w_{k-1}^D are

$$Q_{k-1} = \int_{t_{k-1}}^{t_k} e^{B(t_k-\tau)}\Gamma\boldsymbol{q}\left[e^{B(t_k-\tau)}\Gamma\right]^\mathsf{T}\mathrm{d}\tau, \tag{5.9}$$

$$Q_{k-1}^D = \int_{t_{k-1}}^{t_D} e^{B(t_D-\tau)}\Gamma\boldsymbol{q}\left[e^{B(t_D-\tau)}\Gamma\right]^\mathsf{T}\mathrm{d}\tau, \tag{5.10}$$

$$Q_{k-1}^{k,D} \triangleq E\left[w_{k-1}\left(w_{k-1}^D\right)^\mathsf{T}\right] = Q_{k-1}\left(\Phi_k^D\right)^\mathsf{T}, \tag{5.11}$$

where $\Phi_k^D = \Phi(t_D, t_k)$. Additionally, they have the following relationship:

$$Q_{k-1}^D = \Phi_k^D Q_{k-1} (\Phi_k^D)^\top + Q_k^D. \tag{5.12}$$

5.3.1 Congruous DC Dynamics Modeling

Our DC dynamics modeling starts with a relaxed dynamics, which is fundamental for constructing the constrained dynamic model and affects the quality of the constructed model to some extent. Although many options for the relaxed dynamics are available in practice, specific guidelines can be offered. Typically, it should be among the best models possible to describe the actual system and usually be somewhat conservative to ensure the robustness and stability of the estimators using the resultant model [29].

On the other hand, the destination constraint (5.1) may affect the entire state evolution, which will be more evident in the DC model to be constructed. To accomplish this, we first augment the state to rewrite the destination constraint (5.1) as an equality constraint on the augmented state, aiming to figure out the components of the relaxed state at any time that are affected by the destination constraint. Subsequently, we adopt a DC modeling strategy akin to that presented in [9], which leads to an equivalent model of the relaxed dynamics and then imposes the destination constraint on the equivalent model. However, here we provide an optimization-based DC modeling methodology, with its specific process detailed as follows.

First, augment the state x_k by the final state $x_D = x(t_D)$ to yield $\mathbf{x}_k = [x_k^\top, x_D^\top]^\top$. Then, the destination constraint (5.2) can be reformulated equivalently as

$$\mathcal{D} = \{\mathbf{x}_k : \ \mathbf{A}\mathbf{x}_k = b\}, \tag{5.13}$$

where

$$\mathbf{A} = [\mathbf{0} \quad A], \operatorname{rank}(\mathbf{A}) = m.$$

The destination constraint (5.1) on the final state x_D is thus converted to a linear equality constraint (LEC) (5.13) on the augmented state vector $\mathbf{x}_k$.

Similarly, define the augmented relaxed state vector $x_k^o = [(x_k^o)^\top, (x_D^o)^\top]^\top$, where $x_D^o = x^o(t_D)$. Again, from the solution (5.5), x_D^o depends on the state at time t_{k-1} through

$$x_D^o = \Phi_{k-1}^D x_{k-1}^o + w_{k-1}^D. \tag{5.14}$$

Combining (5.14) with (5.6) yields

$$\mathbf{x}_k^o = \Psi_{k-1} x_{k-1}^o + \mathbf{w}_{k-1}, \tag{5.15}$$

where $\Psi_{k-1} = [\Phi_{k-1}^\top, (\Phi_{k-1}^D)^\top]^\top$. The augmented process noise $\mathbf{w}_{k-1} = [w_{k-1}^\top, (w_{k-1}^D)^\top]^\top$ has zero mean and covariance

$$\mathbf{Q}_{k-1} = \operatorname{cov}(\mathbf{w}_{k-1}) = \begin{bmatrix} Q_{k-1} & Q_{k-1}^{k,D} \\ (Q_{k-1}^{k,D})^\top & Q_{k-1}^D \end{bmatrix}, \tag{5.16}$$

whose entries are specified in (5.9)–(5.11).

Clearly, constraint (5.13) and relaxed dynamics (5.15) are two complementary parts of our prior knowledge of the state evolution, and thus a congruous DC model can be established by fusing them. A straightforward approach for modeling the constrained state $\mathbf{x}_k$ is to choose the point in the constraint subset (5.13) that is nearest to $\mathbf{x}_k{}^o$. To find a reasonable distance metric between $\mathbf{x}_k$ and $\mathbf{x}_k^o$, we turn to the modeling criterion proposed in [29] for constrained state evolution described as follows:

$$\mathbf{x}_k = \begin{cases} \arg\min_{\mathbf{x}_0 \in \mathcal{D}} \left\|\mathbf{x}_0 - \mathbf{x}_0^o\right\|_W^2, k = 0 \\ \arg\min_{\mathbf{x}_k \in \mathcal{D}} \|\mathbf{x}_k - (\Psi_{k-1} x_{k-1} + \mathbf{w}_{k-1})\|_W^2, \\ \qquad k = 1, 2, \ldots \end{cases} \tag{5.17}$$

where $\|x\|_W = \sqrt{x^\top W^{-1} x}$ is the W-norm and weight W can be an arbitrary symmetric and positive definite matrix. Note that the state x_{k-1}^o is replaced by x_{k-1} in (5.17) since the state is also constrained at each time prior to k.

Let $\mathbf{x}_{k|k-1}^o = \Psi_{k-1} x_{k-1} + \mathbf{w}_{k-1}$. Solving (5.17) with equality constraint (5.13) using the method of Lagrange multipliers yields

$$\mathbf{x}_k = \mathbf{P}^W \mathbf{x}_{k|k-1}^o + (I - \mathbf{P}^W)\mathbf{A}^\top(\mathbf{A}\mathbf{A}^\top)^{-1} b, k = 0, 1, \ldots, \tag{5.18}$$

where

$$\mathbf{P}^W = I - W\mathbf{A}^\top(\mathbf{A}W\mathbf{A}^\top)^{-1}\mathbf{A} \tag{5.19}$$

is a projector onto the null space null($\mathbf{A}$). Specifically, let $\mathbf{x}_{k|k-1}^o = \mathbf{x}_0^o$ for $k = 0$. Weight W in (5.19) is a user-defined weight matrix, and a different W usually leads to a different DC model. Similar to classic weighted least squares, there are several common choices for W in practice. That said, for some systems, selecting a suitable W can improve the modeling to some extent.

For relaxed dynamics (5.15) with additive noise, as suggested in [29], a good choice is $W = \mathbf{Q}_{k-1}$. Consequently, (5.18) becomes

$$\mathbf{x}_k = \mathbf{P}_{k-1}(\Psi_{k-1} x_{k-1} + \mathbf{w}_{k-1}) + (I - \mathbf{P}_{k-1})\mathbf{A}^\top(\mathbf{A}\mathbf{A}^\top)^{-1} b, \tag{5.20}$$

where

$$\mathbf{P}_{k-1} = I - \mathbf{Q}_{k-1}\mathbf{A}^\top(\mathbf{A}\mathbf{Q}_{k-1}\mathbf{A}^\top)^{-1}\mathbf{A}. \tag{5.21}$$

With this projector $\mathbf{P}_k$, the constrained model (5.20) has the smallest modeling error in the family of the constrained models (5.18), as $\text{cov}(\mathbf{P}_k\mathbf{w}_k) \leq \text{cov}(\mathbf{P}^W\mathbf{w}_k)$ holds for any positive definite W. (A proof of this inequality can be found in [31].) Additional advantages of the projector $\mathbf{P}_k$ for the resultant DC model will be presented in Section 5.3.3.

Thus far the constrained model (5.20), which describes the state evolution from x_{k-1} to the augmented $\mathbf{x}_k$, has been constructed by fusing the relaxed dynamics (5.14) and the LEC (5.13) through the nearest neighbor criterion (5.17). Lastly, we formulate the evolution from x_{k-1} to x_k using (5.20).

From (5.16) and letting

$$U_D = A^\top(AQ_{k-1}^D A^\top)^{-1} A, \tag{5.22}$$

the projector $\mathbf{P}_{k-1}$ is given by

$$\mathbf{P}_{k-1} = I - \begin{bmatrix} \mathbf{0} & Q_{k-1}^{k,D} U_D \\ \mathbf{0} & Q_{k-1}^D U_D \end{bmatrix} = \begin{bmatrix} I & -Q_{k-1}^{k,D} U_D \\ \mathbf{0} & I - Q_{k-1}^D U_D \end{bmatrix}. \tag{5.23}$$

Substituting (5.23) into (5.20) yields

$$\begin{bmatrix} x_k \\ x_D \end{bmatrix} = \mathbf{P}_{k-1}\left(\begin{bmatrix} \Phi_{k-1} \\ \Phi^D_{k-1} \end{bmatrix} x_{k-1} + \begin{bmatrix} w_{k-1} \\ w^D_{k-1} \end{bmatrix}\right) + (I - \mathbf{P}_{k-1})\begin{bmatrix} \mathbf{0} \\ \Lambda b \end{bmatrix}, \tag{5.24}$$

where $\Lambda = A^{\top}(AA^{\top})^{-1}$. It is evident that the first row of (5.24) corresponds to the state evolution from x_{k-1} to x_k, which conforms to the destination constraint (5.2). By simplifying the form, a congruous DC dynamic model is

$$x_k = F^D_{k-1} x_{k-1} + G^D_{k-1} b + \epsilon^D_{k-1}, \tag{5.25}$$

where the coefficient matrices are

$$F^D_{k-1} = \Phi_{k-1} - Q^{k,D}_{k-1} U_D \Phi^D_{k-1}, \tag{5.26}$$

$$G^D_{k-1} = Q^{k,D}_{k-1} U_D \Lambda, \tag{5.27}$$

and the integrated noise is

$$\epsilon^D_{k-1} = w_{k-1} - Q^{k,D}_{k-1} U_D w^D_{k-1}. \tag{5.28}$$

Figure 5.1 presents a graphical interpretation of the proposed modeling. The resultant model, along with its parameters, is summarized by (5.25)–(5.28). This model has several advantages:

1. It is more general than that in [9] since the arrival time t_D here as a model parameter can take an arbitrary, continuous value.
2. It fuses two complementary pieces of prior information about the DC state evolution—destination constraint (5.1) and relaxed dynamics (5.4). This leads to a model structure with a clear physical interpretation. Superior properties of the model will be elaborated further in Section 5.3.3.
3. Our modeling starts from a relaxed dynamics, which is readily available in practice, and a variety of existing relaxed dynamics can be utilized (see [32] for a survey).

This model has numerous promising applications, including estimation, control, dynamics analysis, and system identification. As a representative example, its capability for adversarial tracking will be demonstrated in Section 5.6.

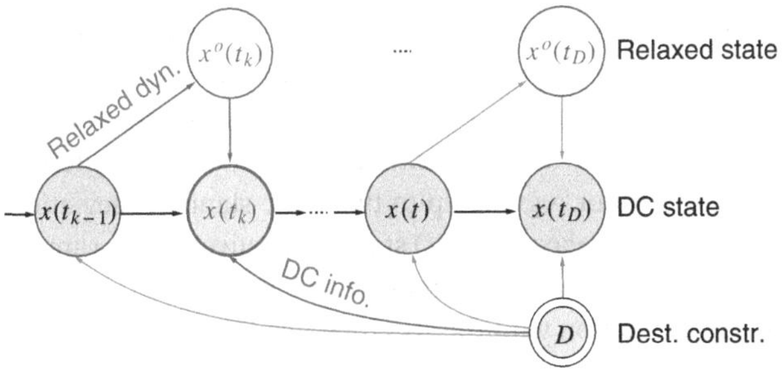

Figure 5.1 Graphical interpretation of the modeling of DC state evolution.

5.3.2 Illustrative Examples of DC Modeling

Consider a scenario in which an object moves in the 2D plane from the origin (0,0) km to the destination (10,10) km. Suppose that the object starts with the initial velocity (200,0) m/s and finally arrives at the destination with a heading angle θ (measured counterclockwise from due east) at time t_D. Distinct from its deterministic and exactly known value in [9], the arrival time t_D here is random and uniformly distributed over the interval [54, 60] s. In this example, assume a popular nearly constant velocity (NCV) model [30] serves as the relaxed model in the form of (5.4) with

$$\begin{aligned} B(t) &= I_2 \otimes \begin{bmatrix} 0 & 1 \\ 0 & 0 \end{bmatrix}, \ \Gamma(t) = I_2 \otimes \begin{bmatrix} 0 \\ 1 \end{bmatrix}, \\ \operatorname{cov}(\tilde{w}(t), \tilde{w}(t')) &= qI_2\delta(t - t') \end{aligned} \tag{5.29}$$

where $\otimes$ stands for the Kronecker product. The state $x^o = [\mathrm{x}, \dot{\mathrm{x}}, \mathrm{y}, \dot{\mathrm{y}}]^\mathrm{T}$ consists of position (x, y) and velocity $(\dot{\mathrm{x}}, \dot{\mathrm{y}})$. Let the intensity q be g^2 with the gravitational acceleration $\mathrm{g} = 9.8\ \mathrm{m/s}^2$. Discretize the continuous motion with sampling interval T. Define $\Delta_k^D = t_D - kT$. Then the system parameters are

$$\Phi_{k-1} = I_2 \otimes \begin{bmatrix} 1 & T \\ 0 & 1 \end{bmatrix}, \tag{5.30}$$

$$\Phi_k^D = I_2 \otimes \begin{bmatrix} 1 & \Delta_k^D \\ 0 & 1 \end{bmatrix}, \tag{5.31}$$

$$Q_{k-1} = qI_2 \otimes \begin{bmatrix} T^3/3 & T^2/2 \\ T^2/2 & T \end{bmatrix}, \tag{5.32}$$

$$Q_{k-1}^D = qI_2 \otimes \begin{bmatrix} \left(\Delta_{k-1}^D\right)^3/3 & (\Delta_{k-1}^D)^2/2 \\ \left(\Delta_{k-1}^D\right)^2/2 & \Delta_{k-1}^D \end{bmatrix}. \tag{5.33}$$

The object's state at the destination amounts to the destination constraint (5.2) with

$$A = \begin{bmatrix} 1 & 0 & 0 & 0 \\ 0 & 0 & 1 & 0 \\ 0 & 1 & 0 & -\cot(\theta) \end{bmatrix}, \quad b = \begin{bmatrix} 10\ \mathrm{km} \\ 10\ \mathrm{km} \\ 0\ \mathrm{m/s} \end{bmatrix}. \tag{5.34}$$

Following the proposed modeling and substituting (5.30)–(5.34) into (5.25), a DC model for the DC motion is obtained. Figure 5.2 illustrates the DC models with heading $\theta = 0°$ and $\theta = 90°$, respectively, at random arrival time. We assess these models by visualizing their sample trajectories from spatial-temporal perspectives. The solid lines are the position means of the two different dynamic models, and the two shaded areas stand for regions up to twice the standard deviation of the position components of the DC state and the relaxed (NCV) state, respectively. Five sample trajectories generated from each model are depicted as dashed lines, showing that the DC model ensures every sample trajectory reaches the destination with the specified heading while the NCV model does not. In addition, from the vertical axis, we can see that the destination arrival time of

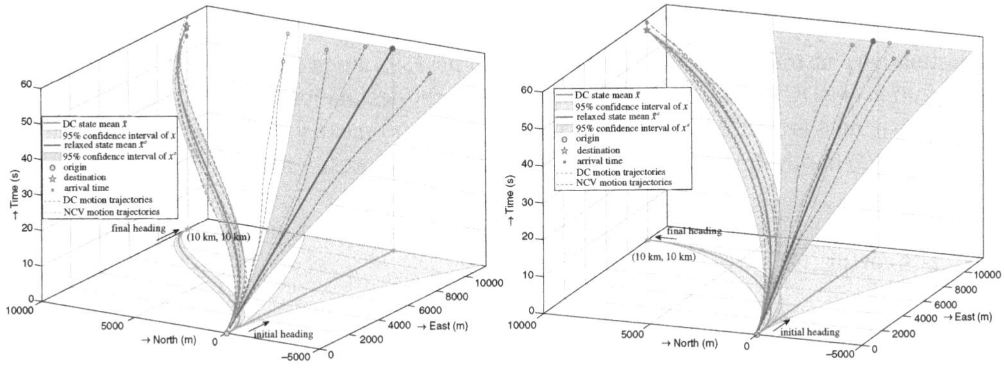

(a) (b)

Figure 5.2 Trajectories generated from the DC model and the NCV model. (a) Arriving at position (10 km, 10 km) with heading $\theta = 0°$. (b) Arriving at position (10 km, 10 km) with heading $\theta = 90°$.

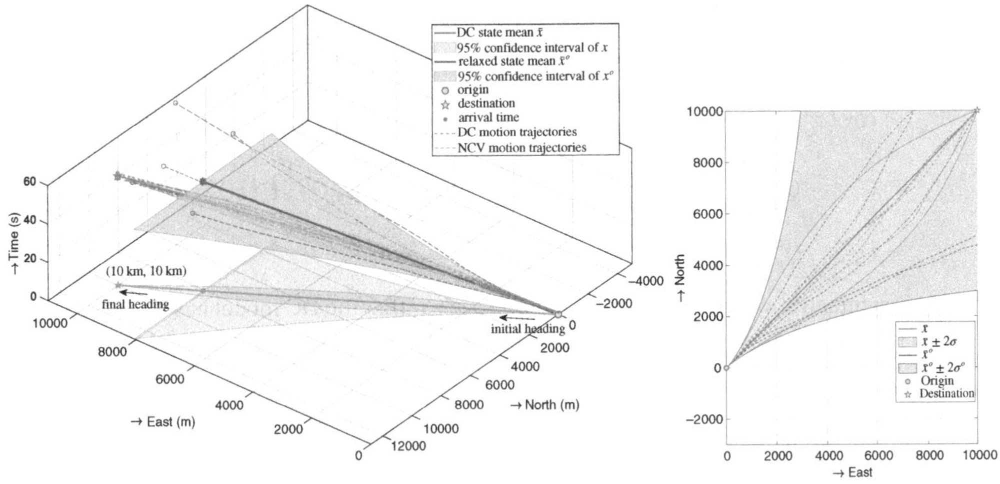

(a) (b)

Figure 5.3 Trajectories generated from the DC model and the NCV model when the initial heading is directed toward the destination. (a) Position versus time and (b) positions in 2D coordinates.

each sample trajectory of the DC model can be a continuous random variable—a more practical situation in application. Again, although the state components of the NCV model are uncoupled across different coordinates, the counterparts of the DC model become coupled across the spatial coordinates due to the arrival time synchronization of the motions at the specified location.

Further, we consider another motion where the initial heading is directed towards the destination. Such a motion is naturally described by the NCV model. However, the trajectory profile of the NCV model, shown by the shadow in Figure 5.3, diverges over time. Almost no sample NCV trajectories, depicted by the solid lines, reach the destination. In contrast, the resulting trajectory profile using our DC model, depicted by the shadow, shrinks towards the destination over time. The sample trajectories, represented by the dashed lines, converge to the destination.

5.3.3 Properties of the DC Model

The destination constraint (5.1) and the relaxed dynamics (5.4) are two complementary parts of the state evolution of a DC dynamic system. After imposing the destination constraint, the uncertainty in the DC dynamics is reduced, as expected. Given t_D, the zero-mean white noise ϵ_k^D of (5.28) in the DC model (5.25) has the following properties relative to the process noise w_k in (5.6).

Property 5.1 The DC model (5.25) has smaller process noise than the relaxed dynamics (5.6): $\mathrm{cov}(\epsilon_{k-1}^D) \le \mathrm{cov}(w_{k-1})$.

Proof: From (5.22), we have

$$\begin{aligned}
&U_D Q_{k-1}^D U_D^\top \\
&= A^\top (A Q_{k-1}^D A^\top)^{-1} A Q_{k-1}^D A^\top (A Q_{k-1}{}^D A^\top)^{-\top} A \\
&= A^\top (A Q_{k-1}^D A^\top)^{-1} A = U_D.
\end{aligned}$$

Then, the covariance of the integrated noise ϵ_{k-1}^D is

$$\begin{aligned}
\Xi_{k-1}^D &\triangleq \mathrm{cov}\left(\epsilon_{k-1}^D\right) = \mathrm{cov}\left(w_{k-1} - Q_{k-1}^{k,D} U_D w_{k-1}^D\right) \\
&= Q_{k-1} - Q_{k-1}^{k,D} U_D^\top \left(Q_{k-1}^{k,D}\right)^\top - Q_{k-1}^{k,D} U_D \left(Q_{k-1}^{k,D}\right)^\top + Q_{k-1}^{k,D} U_D Q_{k-1}^D U_D^\top \left(Q_{k-1}^{k,D}\right)^\top, \\
&= Q_{k-1} - Q_{k-1}^{k,D} U_D (Q_{k-1}^{k,D})^\top
\end{aligned} \tag{5.35}$$

where U_D is symmetric and positive semidefinite. It is clear that the DC model has smaller process noise than the relaxed dynamics:

$$\Xi_{k-1}^D \le Q_{k-1}.$$

In particular, if the arrival time is a given multiple of the sampling interval, i.e., $t_D = NT$, then the DC model (5.25) reduces to the model proposed in [9]. In this situation, the process noise at time t_{N-1} is $\epsilon_{N-1}^D = \left(I - Q_{N-1} A^\top (A Q_{N-1} A^\top)^{-1} A\right) w_{N-1}$, and it lies in the null space of A. Consequently,

$$\begin{aligned}
\Xi_{N-1}^D &\triangleq \mathrm{cov}\left(\epsilon_{N-1}^D\right) \\
&= Q_{N-1} - Q_{N-1} A^\top (A Q_{N-1} A^\top)^{-1} A Q_{N-1}
\end{aligned}$$

is singular and $\mathrm{rank}(\Xi_{N-1}^D) \le n - m$.

Property 5.2 The zero-mean integrated process noise ϵ_j^D term is uncorrelated with the (past and current) state x_k, i.e.,

$$E\left[x_k \left(\epsilon_j^D\right)^\top\right] = 0, \quad \text{for } k \le j.$$

Proof: From the definition (5.8) and by the assumed independence of the process noise and the initial state, we have

$$E\left[x_k \left(w_j^l\right)^\top\right] = 0, \quad \text{for } k \le j < l. \tag{5.36}$$

Therefore, by (5.28), we have

$$E\left[x_k\left(\epsilon_j^D\right)^{\mathsf{T}}\right] \overset{(5.28)}{=} E\left[x_k\left(w_j - Q_j^{j+1,D}U_D w_j^D\right)^{\mathsf{T}}\right]$$
$$= E\left[x_k w_j^{\mathsf{T}}\right] - E\left[x_k\left(w_j^D\right)^{\mathsf{T}}\right]U_D\left(Q_j^{j+1,D}\right)^{\mathsf{T}} \overset{(5.36)}{=} 0.$$

Property 5.3 The integrated process noise ϵ_k^D is white, i.e.,

$$\mathrm{cov}(\epsilon_j^D, \epsilon_k^D) = 0, \ \text{for } j \neq k, t_j, t_k < t_D.$$

Proof: See Appendix A.

Consider further uncertain arrival time t_D with a density $p(t_D)$. Then, the integrated process noise averaged over t_D is $\epsilon_k = E[\epsilon_k^D] = \int \epsilon_k^D p(t_D)\mathrm{d}t_D$. From Properties 2 and 3 and applying the law of total expectation [30], we can conclude that ϵ_k inherits the same properties, i.e.,

$$\mathrm{cov}(\epsilon_j, \epsilon_k) = 0, \ \text{for } j \neq k,$$

$$E\left[x_0\epsilon_k^{\mathsf{T}}\right] = 0.$$

Their derivations are direct and thus omitted.

The above properties of the process noise ϵ_k^D and its average counterpart ϵ_k are critical for recursive estimation. For instance, their whiteness and uncorrelatedness with the initial state perfectly fit the optimality condition of recursive estimators (e.g., the Kalman filter [KF]).

5.3.4 Connection with Pseudo-observation-based BD Model

BD models, originally developed in a Bayesian framework by Ahmad et al. [20, 21], ensure that an object eventually reaches a specified destination by capturing the underlying long-term dependencies in the DC trajectories. Recently, Liang et al. [23, 24] proposed a pseudo-observation-based formulation of the BD models. The destination information is integrated into (posterior) state transition in the pseudo-observation-based BD model. In the following, we briefly review the pseudo-observation-based BD model and establish a connection with our DC model.

In the Bayesian formulation, the state Eq. (5.6) under a linear Gaussian assumption corresponds to the transition density

$$p(x_k|x_{k-1}) = \mathcal{N}(x_k; \Phi_{k-1}x_{k-1}, Q_{k-1}). \tag{5.37}$$

Interpreting the destination constraint (5.2) as a pseudo-observation $z(t_D) = Ax(t_D) = b$ implies the likelihood function

$$p(z(t_D) = b|x(t_D)) = \mathcal{N}(Ax(t_D); b, 0). \tag{5.38}$$

Then, applying Bayes' rule, as shown by Eq. (14) in [23], the destination-conditioned state transition density can be expressed as

$$\begin{aligned} p(x_k|x_{k-1}, \mathcal{D}) &\propto p(z(t_D) = b|x_k)p(x_k|x_{k-1}) \\ &\propto \mathcal{N}(b; A\Phi_k^D x_k, AQ_k^D A^{\mathsf{T}})\mathcal{N}(x_k; \Phi_{k-1}x_{k-1}, Q_{k-1}), \\ &= \mathcal{N}(x_k; \mathbf{F}_{D,k-1}x_{k-1} + \mathbf{M}_{D,k-1}, \mathbf{Q}_{D,k-1}) \end{aligned} \tag{5.39}$$

where "∝" stands for "is proportional to," and

$$\mathbf{F}_{D,k-1} = \mathbf{Q}_{D,k-1}Q_{k-1}^{-1}\Phi_{k-1}, \tag{5.40}$$

$$\mathbf{M}_{D,k-1} = \mathbf{Q}_{D,k-1}\mathbf{L}b, \tag{5.41}$$

$$\mathbf{Q}_{D,k-1} = \left(Q_{k-1}^{-1} + \mathbf{L}A\Phi_k^D\right)^{-1}, \tag{5.42}$$

$$\mathbf{L} = \left(A\Phi_k^D\right)^\top \left(AQ_k^D A^\top\right)^{-1}.$$

The pseudo-observation-based BD model described above is obtained from a Bayesian perspective, while our DC dynamic model (5.25) is constructed from the relaxed dynamics through solving an optimization problem based only on the first two moments of the state transition, without needing the entire distributional information of the state transition. Despite their different modeling prerequisites, derivations, and resultant model forms, these two models share the following connection in certain special cases.

Proposition 5.1 For the linear Gaussian dynamic system with a linear destination constraint (5.2), the transition density (5.39) of the pseudo-observation-based BD model coincides with that of the DC model (5.25). That is, albeit in different forms, these two models are equivalent as models for the linear Gaussian DC system.

Proof: See Appendix B.

It is worth noting that the models produced by the above two modeling approaches generally differ. Each has its own applicable situations. Specifically, the pseudo-observation-based BD model is preferred when the state transition distribution and the likelihood function are both available and computationally tractable. In cases where the transition distribution and the likelihood function are unavailable, hard, or complicated to determine/evaluate, however, our optimization-based modeling approach is a more suitable alternative.

5.4 State Transition with Uncertain Arrival Time

In Section 5.3, we have presented a method for congruous DC modeling by constraining the relaxed dynamics (5.4) to the destination. The DC state estimation thereby becomes the state estimation using the constructed DC model and the measurements. However, as mentioned earlier, the arrival time t_D is often uncertain in real application. Due to the Markovian property of the state evolution, we consider one cycle of DC state transition with uncertain arrival time next.

The DC model (5.25) implies that the mean and covariance of x_k conditioned on x_{k-1} and t_D are

$$E[x_k|x_{k-1}, t_D] = F_{k-1}^D x_{k-1} + G_{k-1}^D b$$

$$\text{cov}(x_k|x_{k-1}, t_D) = \Xi_{k-1}^D.$$

Considering the randomness of the arrival time t_D and assuming it follows a density function $p(t_D)$, the conditional mean of x_k can be obtained by averaging over t_D using the law of total expectation

$$\begin{aligned} E[x_k|x_{k-1}] &= E[E[x_k|x_{k-1}, t_D]|x_{k-1}] \\ &= \int (F_{k-1}^D x_{k-1} + G_{k-1}^D b) p(t_D|x_{k-1}) \mathrm{d}t_D \\ &= \bar{F}_{k-1}x_{k-1} + \bar{G}_{k-1}b. \end{aligned} \tag{5.43}$$

where

$$\bar{F}_{k-1} = \int F_{k-1}^{D} p(t_D|x_{k-1}) \mathrm{d}t_D$$
$$\bar{G}_{k-1} = \int G_{k-1}^{D} p(t_D|x_{k-1}) \mathrm{d}t_D.$$

Accordingly, the conditional covariance of x_k that averages over t_D is

$$\operatorname{cov}(x_k|x_{k-1}) = \bar{\Xi}_{k-1} + \tilde{\bar{\Xi}}_{k-1}(x_{k-1}), \tag{5.44}$$

where

$$\begin{aligned} \bar{\Xi}_{k-1} &= E\left[\operatorname{cov}(x_k|x_{k-1}, t_D)|x_{k-1}\right] \\ &= \int \Xi_{k-1}^{D} p(t_D|x_{k-1}) \mathrm{d}t_D, \end{aligned} \tag{5.45}$$

$$\begin{aligned} \tilde{\bar{\Xi}}_{k-1}(x_{k-1}) &= \operatorname{cov}(E[x_k|x_{k-1}, t_D]|x_{k-1}) \\ &= E\Big[E[x_k|x_{k-1}, t_D](*)^{\mathsf{T}}|x_{k-1}\Big] - E[x_k|x_{k-1}](*)^{\mathsf{T}} \\ &= \int [(F_{k-1}^{D} x_{k-1} + G_{k-1}^{D} b)(*)^{\mathsf{T}}] p(t_D|x_{k-1}) \mathrm{d}t_D - E[x_k|x_{k-1}](*)^{\mathsf{T}}. \end{aligned} \tag{5.46}$$

Here, $(*)$ represents the term right before it.

Moreover, the above conditional means and conditional covariances might be analytically intractable and can be approximated numerically. Here, we opt for the Monte Carlo (MC) integration:

$$\begin{aligned} E[g(t_D, x_{k-1})|x_{k-1}] &= \int g(t_D, x_{k-1}) p(t_D|x_{k-1}) \mathrm{d}t_D \\ &\approx \frac{1}{N_s} \sum_{i=1}^{N_s} g(t_D^{(i)}, x_{k-1}) = \hat{g}(x_{k-1}) \end{aligned},$$

where $g(\cdot)$ is a given function (e.g., F_{k-1}^{D}, G_{k-1}^{D}, and Ξ_{k-1}^{D}), and a set of N_s sample points $\{t_D^{(i)}\}_{i=1}^{N_s}$ is drawn from the density $p(t_D|x_{k-1})$. The MC integration is preferred here because of its relative ease of implementation and acceptable computational costs. To simplify the above integrations, we approximate the conditional density $p(t_D|x_{k-1})$ by the prior density $p(t_D)$. Since the arrival time t_D is the *final* time, the condition $t_k \leq t_D$ should be maintained throughout sampling. Denote by $\mathcal{T}$ the allowable set of t_D determined in advance. Then, all sample points drawn satisfy

$$t_D^{(i)} \in \mathcal{T} \cap \{t_D : t_D \geq t_k\}, \quad i = 1, \ldots, N_s. \tag{5.47}$$

Thus far, a DC state transition with uncertain arrival time has been developed, imposing the destination constraint as spatial information on the state evolution. It can have many applications, such as motion analysis and trajectory prediction (TP) of guided targets, to be demonstrated in Section 5.6.

5.5 Estimation for DC Systems

The state transition of the DC system with uncertain arrival time is characterized by the conditional expectation (5.43) and covariance (5.44), which can be used to estimate the DC state recursively based on the noisy measurements. Let $Z^k = \langle z_\kappa \rangle_{\kappa \leq k}$ denote the measurement sequence up to time k. Under the jointly Gaussian assumption of the state x and the measurement z, and given the state estimate $\hat{x}_{k-1|k-1} = E[x_{k-1}|Z^{k-1}]$ and its covariance $P_{k-1|k-1} = \text{cov}(x_{k-1}|Z^{k-1})$ at time $k-1$, the predicted DC state and its covariance at time k are derived by

$$\begin{aligned} \hat{x}_{k|k-1} &= E\left[x_k|Z^{k-1}\right] = E[E[x_k|x_{k-1}]|Z^{k-1}] \\ &= \bar{F}_{k-1}\hat{x}_{k-1|k-1} + \bar{G}_{k-1}b, \end{aligned} \tag{5.48}$$

$$\begin{aligned} P_{k|k-1} &= \text{cov}(x_k|Z^{k-1}) \\ &= E\left[\text{cov}(x_k|x_{k-1})|Z^{k-1}\right] + \text{cov}(E[x_k|x_{k-1}]|Z^{k-1}) \\ &= \bar{\Xi}_{k-1} + E\left[\tilde{\Xi}_{k-1}(x_{k-1})|Z^{k-1}\right] + \bar{F}_{k-1}P_{k-1|k-1}\bar{F}_{k-1}^{\mathsf{T}}. \end{aligned} \tag{5.49}$$

The second term $E[\tilde{\Xi}_{k-1}(x_{k-1})|Z^{k-1}]$ above can be approximated by $\tilde{\Xi}_{k-1}(\hat{x}_{k-1|k-1})$, or more accurately using the sigma-point integration [33]. Subsequently, apply the KF or its variant to update the DC state x_k based on the current measurement, as outlined in steps 16–20 of Algorithm 5.1 in [9]. This chapter refers to such estimation for DC systems as the DC estimator (DCE) by averaging over t_D (DCEA).

In some situations, however, the arrival time t_D is a critical quantity for decision-making that a user desires to know and needs to be estimated. In the rest of this section, we present a sampling-based strategy to estimate the arrival time. Related work on arrival time estimation is rare, and [21, 23, 24] are the only publications we know that reported relevant work. They employed quadrature schemes to approximate the integrals, while we use sampling-based estimation that offers an effective paradigm for recursively updating the model parameter's distribution as measurements come in. Again, t_D and x_k are inherently interdependent and therefore should be estimated jointly. As a result, the state estimation accuracy can be improved using the posterior distribution of t_D compared with using its prior distribution. Next, we elaborate on the simultaneous estimation of the state and the arrival time (SESAT) for DC systems.

5.5.1 Estimation of DC State and Arrival Time

Recall that F_{k-1}^D and G_{k-1}^D along with ϵ_{k-1}^D in the DC model (5.25) all depend on t_D. Then we can rewrite the DC model as

$$x_k = F_{k-1}(t_D)x_{k-1} + G_{k-1}(t_D)b + \epsilon_{k-1}(t_D). \tag{5.50}$$

This model differs from (5.25) in that the arrival time t_D is treated as a random variable to be estimated. Taking into account their mutual dependence, the objective is to estimate x_k and t_D jointly based on the available measurements that obey measurement model (5.3). The joint estimation obtains the posterior density $p(x_k, t_D|Z^k)$ recursively. Generally, the recursive estimation consists of two stages: prediction and update. A number of sequential ways for joint state and parameter estimation are summarized in [34]. A representative is to construct a dynamics for the state x_k augmented by t_D according to (5.50) and then do nonlinear filtering. Here, we present a recursive joint estimation of x_k and t_D.

Suppose that $p(x_{k-1}, t_D|Z^{k-1})$ is given. The prediction uses model (5.50) to obtain the prediction density $p(x_k, t_D|Z^{k-1})$ at time t_k via the Chapman–Kolmogorov equation

$$\begin{aligned} & p(x_k, t_D|Z^{k-1}) \\ & = p(t_D|Z^{k-1})p(x_k|t_D, Z^{k-1}) \\ & = p(t_D|Z^{k-1}) \int p(x_k|x_{k-1}, t_D)p(x_{k-1}|t_D, Z^{k-1})\mathrm{d}x_{k-1} \end{aligned} \tag{5.51}$$

where the transition density $p(x_k|x_{k-1}, t_D)$ is defined by both the DC model (5.50) conditioned on t_D and the distribution of ϵ^D_{k-1}.

When a measurement z_k is received at time t_k, the density is updated using z_k via Bayes' rule

$$p(x_k, t_D|Z^k) = \alpha_k p(z_k|x_k)p(x_k, t_D|Z^{k-1}), \tag{5.52}$$

$$p(t_D|Z^k) = \beta_k p(z_k|t_D, Z^{k-1})p(t_D|Z^{k-1}), \tag{5.53}$$

where α_k and β_k are normalizing constants depending on neither x_k nor t_D. The likelihood function $p(z_k|x_k)$ can be computed by the measurement model (5.3) and the distribution of v_k. The likelihood function $L^D_k \triangleq p(z_k|t_D, Z^{k-1})$ is computed using the estimate $\hat{x}^D_{k-1|k-1}$ and covariance $P^D_{k-1|k-1}$ corresponding to t_D at time t_{k-1} as

$$L^D_k = p(z_k|t_D, \hat{x}^D_{k-1|k-1}, P^D_{k-1|k-1}). \tag{5.54}$$

Given the prior initial density $p(x_0, t_D)$, the recursion (5.51)–(5.53) forms the basis for the Bayesian estimation. In principle, knowledge of the posterior density $p(x_k, t_D|Z^k)$ enables us to calculate optimal estimates of the state and the arrival time under any optimality criterion. Assuming the goal is to obtain the minimum mean squares error (MMSE) estimate of the state, i.e., its conditional mean, we adopt the following two optimality criteria for the arrival time [30, 35]. The first is the maximum a posteriori (MAP) estimate, i.e.,

$$\hat{t}^{\text{MAP}}_{D|k} = \arg\max_{t_D} p(t_D|Z^k). \tag{5.55}$$

Following this, the state estimate is

$$\hat{x}^{\text{MMSE-MAP}}_{k|k} = E\left[x_k|\hat{t}^{\text{MAP}}_{D|k}, Z^k\right]. \tag{5.56}$$

Instead of such a hard decision, a second scheme is averaging over all the possibilities of the arrival time, i.e.,

$$\hat{t}^{\text{MMSE}}_{D|k} = E[t_D|Z^k] = \int t_D p(t_D|Z^k)\mathrm{d}t_D, \tag{5.57}$$

and then the state estimate is

$$\begin{aligned} \hat{x}^{\text{MMSE}}_{k|k} & = E[E[x_k|t_D, Z^k]|Z^k] \\ & = \iint x_k p(x_k|t_D, Z^k)p(t_D|Z^k)\mathrm{d}x_k\mathrm{d}t_D. \end{aligned} \tag{5.58}$$

Moreover, the associated covariances can also be obtained from the posterior densities. As the second scheme is more robust due to its soft decision, its corresponding algorithm will be developed subsequently.

5.5.2 Efficient Sampling-based DC Estimation

Although the recursion (5.51)–(5.53) can estimate both the state and the arrival time based on measurements, the integrals therein are analytically intractable in most cases due to the nonlinearity of model (5.50) with respect to random variable t_D. To tackle this problem, we again resort to the MC integration.

Meanwhile, from the analysis in Section 5.3.3, it is found that the proposed DC model (5.25) given arrival time t_D is linear Gaussian and, together with the measurement model (5.3), has representation

$$p(x_k|x_{k-1}, t_D) = \mathcal{N}(x_k; F^D_{k-1}x_{k-1} + G^D_{k-1}b, \Xi^D_{k-1}), \tag{5.59}$$

$$p(z_k|x_k) = \mathcal{N}(z_k; H_k x_k, R_k). \tag{5.60}$$

Clearly, model (5.50) is of *mixed linear/nonlinear* structure, which allows us to apply the KF to obtain analytically the posterior density of the state vector and to use the MC methods (e.g., the marginalized particle filter (PF) in [36]) for the arrival time t_D, which has a time-invariant distribution. This strategy is computationally more efficient because the marginalized PF concerning only the scalar-valued t_D needs much fewer particles than the standard PF for (x_k, t_D). Such an idea is certainly not new and has been well investigated in the literature (see, e.g., [36–40]). By comparison, the problem of jointly estimating x_k and t_D under consideration is analogous to the conventional ones except for the condition $t_D \geq t_k$ (as in Section 5.4). To highlight it, our primary focus will be on obtaining the posterior density $p(t_D|Z^k)$ through Monte Carlo approximation.

Assume that the density $p(t_D|Z^{k-1})$ is approximated by a set of N_s weighted particles $\{W^{(i)}_{k-1}, t^{(i)}_{D,k-1}\}_{i=1}^{N_s}$. To prevent PF degeneracy during static parameter estimation, a pragmatic approach is to introduce artificial dynamics for the parameter [39]. Among various artificial dynamics presented in [34], we opt to model t_D as a quantity with slight variations, leading the particles to propagate in the following manner:

$$t^{(i)}_{D,k} = t^{(i)}_{D,k-1} + \varepsilon_{k-1}, \tag{5.61}$$

where ε_k is artificially assigned zero-mean Gaussian white noise with a tiny variance. The constraint $t_D \geq t_k$ is accounted for by assigning zero weight to the particles that violate it. Then, the standard resampling procedure is implemented to duplicate particles with large weights and eliminate those with small weights, resulting in fresh particles with a weight of $1/N_s$ and their respective indices. A pseudocode of the constrained resampling procedure is given by Algorithm 5.1.

Next, rearrange $\{\hat{x}^{(i)}_{k-1|k-1}, P^{(i)}_{k-1|k-1}\}_{i=1}^{N_s}$ to match $\{t^{(i)}_{D,k}\}_{i=1}^{N_s}$ according to the new indices. By Kalman filtering under the linear Gaussian assumption, we have the likelihood function and the posterior density as

$$\begin{aligned} L^{(i)}_{D,k} &= p(z_k|Z^{k-1}, t^{(i)}_{D,k}) \\ &= \mathcal{N}(z_k; H\hat{x}^{(i)}_{k|k-1}, HP^{(i)}_{k|k-1}H^\top + R), \end{aligned}$$

$$p(x^{(i)}_k|Z^k, t^{(i)}_{D,k}) = \mathcal{N}(x^{(i)}_k; \hat{x}^{(i)}_{k|k}, P^{(i)}_{k|k})$$

where $(\hat{x}^{(i)}_{k|k-1}, P^{(i)}_{k|k-1})$ and $(\hat{x}_{k|k}{}^{(i)}, P^{(i)}_{k|k})$, respectively, represent the predicted state and the updated state as well as their error covariances for any given $t^{(i)}_{D,k}$, and their relationship is presented in

Algorithm 5.1: Constrained Resampling

// function $[\{W^{(j)}, t_D^{(j)}\}_{j=1}^{N_s}$, index] = CReS($\{W^{(i)}, t_D^{(i)}\}_{i=1}^{N_s}, t_k$)

Input: Weighted particle set $\{W^{(i)}, t_D^{(i)}\}_{i=1}^{N_s}$, time t_k

Output: Resampled particle set $\{W^{(j)}, t_D^{(j)}\}_{j=1}^{N_s}$, index

for $i = 1, \ldots, N_s$ **do in parallel**
    if $t_D^{(i)} < t_k$ **then**
        Set the weight zero: $W^{(i)} = 0$
    end
end
Normalize weights: $\{\tilde{W}^{(i)}\} = \{W^{(i)}\}/(\sum_{i=1}^{N_s} W^{(i)})$
Build cumulative weight sum (CWS) sequence: $\{\tilde{W}_c^{(i)}\} = \texttt{cumsum}(\{\tilde{W}^{(i)}\})$
Start at bottom of CWS: $k = 1$
Draw starting point: $u_1 \sim \mathcal{U}(0, 1/N_s)$
Original particle index set: index $= \{1, \ldots, N_s\}$
for $j = 1, \ldots, N_s$ **do in parallel**
    Move along CWS: $u_j = u_1 + (j-1)/N_s$
    while $u_j > \tilde{W}_c^{(k)}$ **do**
        $k = k + 1$
    end
    Create new index: index$(j) = k$
    Rearrange particles: $t_D^{(j)} = t_D^{(k)}$
    Reset weights: $W^{(j)} = 1/N_s$
end
return

Algorithm 5.2. Accordingly, the weight of each fresh particle $t_{D,k}^{(i)}$ at time t_k is updated by [34, 41]

$$W_k^{(i)} = \frac{L_{D,k}^{(i)}}{\sum_{j=1}^{N_s} L_{D,k}^{(j)}}.$$

Consequently, the arrival time is estimated as

$$\hat{t}_{D,k} = \sum_{i=1}^{N_s} W_k^{(i)} t_{D,k}^{(i)}. \tag{5.62}$$

Using (5.58), we have the MMSE estimate of x_k and its error covariance as

$$\hat{x}_{k|k} = \sum_{i=1}^{N_s} W_k^{(i)} \hat{x}_{k|k}^{(i)}, \tag{5.63}$$

$$P_{k|k} = \sum_{i=1}^{N_s} W_k^{(i)} [P_{k|k}^{(i)} + (\hat{x}_{k|k}^{(i)} - \hat{x}_{k|k})(\hat{x}_{k|k}^{(i)} - \hat{x}_{k|k})^\top]. \tag{5.64}$$

Note that the arrival time estimate $\hat{t}_{D,k}$ and the state estimate $\hat{x}_{k|k}$ along with their covariances are outputs at time k, but they are not necessarily utilized in the next recursion of the algorithm. To facilitate adoption, a pseudocode of the SESAT is provided in Algorithm 5.2.

Algorithm 5.2: Simultaneous Estimation of State and Arrival Time

---

Input: $\{A, b\}, \{B(t), \Gamma(t), q(t)\}, \{H_k, R_k\}, N_s$, and $\langle z_k \rangle$
Output: $\{\hat{x}_{k|k}, P_{k|k}\}$ and $\{\hat{t}_{D|k}, \sigma^2_{D|k}\}$

`// 1. Filter Initialization:`

Draw sample points $t^{(i)}_{D,0}$ with weights $W^{(i)}_0 = 1/N_s$ from prior distribution $p(t_D)$, and set initial state as $(\hat{x}^{(i)}_{0|0}, P^{(i)}_{0|0}) = (\hat{x}_{0|0}, P_{0|0})$ for $i = 1, \dots, N_s$

`// 2. DC estimation by Marginalized PF:`

Compute parameter: $\Lambda \leftarrow A^\top (AA^\top)^{-1}$

for $k = 1, \dots, N$ **do**

    `// 2.1 Constrained resampling:`

    Roughen particles: $t^{(i)}_{D,k} = t^{(i)}_{D,k-1} + \varepsilon^{(i)}_{D,k-1}$

    Multiply/suppress particles $t^{(i)}_{D,k}$ with weights $W^{(i)}_{k-1}$ by running `CReS` (Algorithm 5.1)

    Rearrange $\{\hat{x}^{(i)}_{k-1|k-1}, P^{(i)}_{k-1|k-1}\}_{i=1}^{N_s}$ corresponding to new index output by `CReS`

    `// 2.2 Marginalized particle filtering:`

    for $i = 1, \dots, N_s$ **do in parallel**

        Compute $\Phi^{D,(i)}_k$ and $\Phi^{D,(i)}_{k-1}$ using (5.31) for $t^{(i)}_{D,k}$ with initial times t_k and t_{k-1}, respectively

        Compute $Q^{k,D,(i)}_{k-1}$, $Q^{D,(i)}_{k-1}$, and $U^{(i)}_D$ by (5.10), (5.11), and (5.22), respectively

        Compute $F^{D,(i)}_{k-1}$, $G^{D,(i)}_{k-1}$, and $\Xi^{D,(i)}_{k-1}$ using (5.26), (5.27), and (5.35), respectively

        $\hat{x}^{(i)}_{k|k-1} \leftarrow F^{D,(i)}_{k-1} \hat{x}^{(i)}_{k-1|k-1} + G^{D,(i)}_{k-1} b$

        $P^{(i)}_{k|k-1} \leftarrow F^{D,(i)}_{k-1} P^{(i)}_{k-1|k-1} (F^{D,(i)}_{k-1})^\top + \Xi^{D,(i)}_{k-1}$

        $\nu^{(i)}_k \leftarrow z_k - H_k \hat{x}^{(i)}_{k|k-1}$

        $S^{(i)}_k \leftarrow H_k P^{(i)}_{k|k-1} H_k^\top + R_k$

        $K^{(i)}_k \leftarrow P^{(i)}_{k|k-1} H_k^\top (S^{(i)}_k)^{-1}$

        $\hat{x}^{(i)}_{k|k} \leftarrow \hat{x}^{(i)}_{k|k-1} + K^{(i)}_k {\nu_k}^{(i)}$

        $P^{(i)}_{k|k} \leftarrow P^{(i)}_{k|k-1} - K^{(i)}_k S^{(i)}_k ({K_k}^{(i)})^\top$

        $L^{(i)}_k \leftarrow \mathcal{N}\left(\nu^{(i)}_k; 0, S^{(i)}_k\right)$

    end

    Update weight: $W^{(i)}_k \leftarrow L^{(i)}_k / (\sum_{j=1}^{N_s} L^{(j)}_k)$

    `// 2.3 MMSE estimation:`

    Obtain arrival time estimate $\hat{t}_{D,k}$ using (5.62)

    Obtain state estimate $\hat{x}_{k|k}$ and its error covariance $P_{k|k}$ via (5.63) and (5.64), respectively

end

The next section will showcase the efficacy of the algorithm through simulations of tracking a guided target.

5.6 Illustrative Examples and Discussions

5.6.1 Tracking a Target with a Destination Constraint

Success of target tracking depends largely upon the quality of motion models, especially in the case of limited measurements with large errors. We continue with the example given in Section 5.3.2 to evaluate the performance of the proposed DCEs. As stated before, an object moves in the 2D plane from the origin $(0, 0)$ m with the initial velocity $(200, 0)$ m/s and finally arrives at the destination $(10, 10)$ km with a heading angle $\theta = 90°$. The moving object is monitored by a sensor, which provides position measurements at a sampling interval of $T = 1$ s. The measurement model is given by (5.3) with

$$H_k = \begin{bmatrix} 1 & 0 & 0 & 0 \\ 0 & 0 & 1 & 0 \end{bmatrix}.$$

The measurement noise $\upsilon \sim \mathcal{N}(0, \sigma^2 I_2)$ is white with standard deviation $\sigma = 120$ m and uncorrelated with the state.

In this simulation, we opt for the commonly used NCV model with parameters (5.29) as the relaxed dynamics based on the object's motion pattern and set the process noise intensity as $q = g^2$. Then, by the modeling of Section 5.3, a DC model conditioned on t_D is constructed as in (5.25). The true arrival time is $t_D = 55.7$ s.

In practice, arrival time t_D is uncertain and here it is assumed to have a uniform prior over the interval $[54, 60]$ s. We implement the DCEA and SESAT estimators, both using $N_s = 500$ sample points. To show the effects of the destination constraint on estimation performance, the KF without considering the destination constraint is compared. In addition, the DCE with known t_D is also implemented. To be fair, they all use the same initialization and are based on the identical NCV model with the same preset intensity q. The performance comparison of various tracking algorithms is based on Monte Carlo simulations. To help indicate the best possible performance in the scenario, we give the posterior Cramér–Rao lower bound (PCRLB) for the case that the true t_D is known [42, 43], which is a lower bound on mean squares error for filtering.

The performance comparison over 300 Monte Carlo runs is shown in Figure 5.4, and we can make the following observations:

1. Overall, the DCEs significantly outperform the unconstrained KF. Although the proposed DCEA and the SESAT estimators have worse performance than the DCE with exactly known arrival time due to their lack of knowledge about the arrival time, this still verifies that incorporation of the destination information enhances the estimation accuracy of the DC state.
2. The SESAT estimator outperforms the DCEA estimator as it updates the posterior density of t_D at each time using the data stream, resulting in a more accurate estimate of t_D. However, its computational cost increases substantially due to the online updating of the particle weights.
3. For the case of known t_D, the DCE using the constructed model (5.25) achieves the best performance. Its root mean square (RMS) error is very close to the square root of the PCRLB.

In addition, the SESAT estimator provides the empirical posterior density of the arrival time, as illustrated in Figure 5.5, which converges to the truth and its variance shrinks gradually.

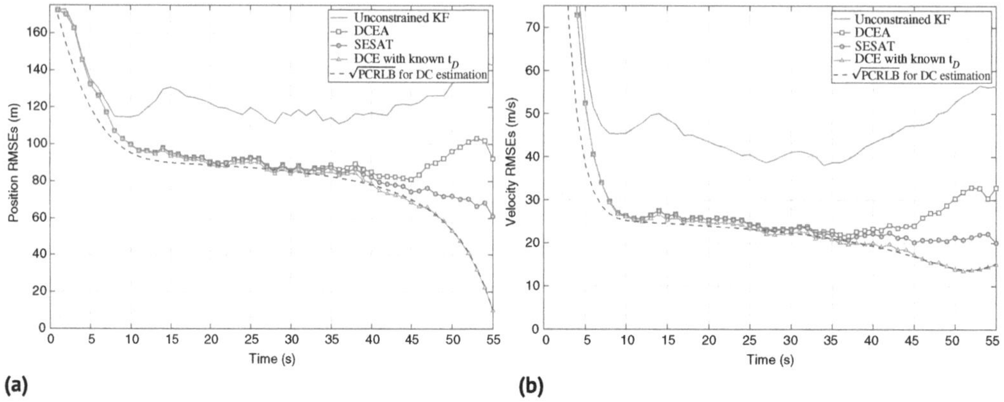

Figure 5.4 Performance comparison of four filters for tracking a target with destination constraints. (a) RMS position errors and (b) RMS velocity errors.

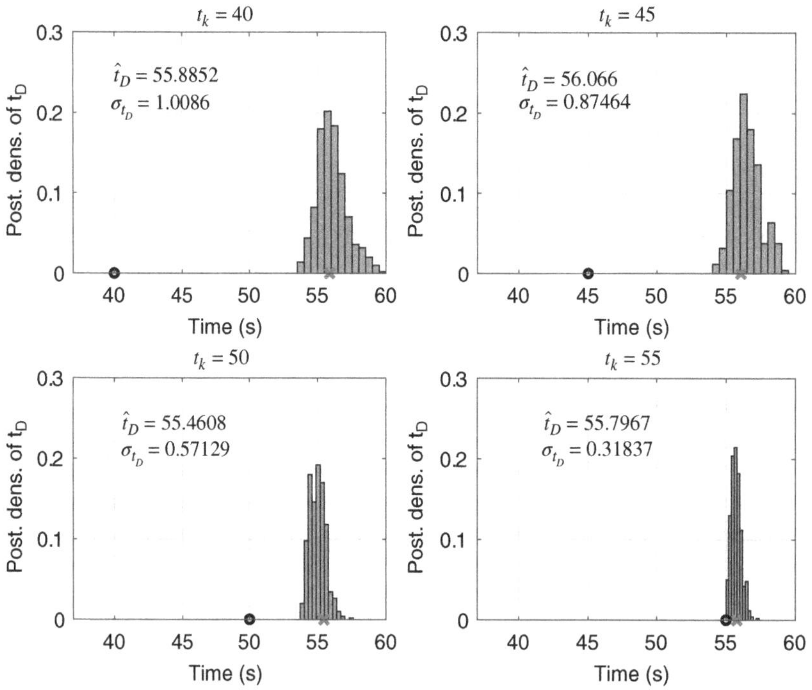

Figure 5.5 Sequential estimation of arrival time t_D (true $t_D = 55.70$ s). "×" indicates the estimate of t_D and "o" indicates the current time.

The SESAT estimator is superior in estimation accuracy and capability but at the expense of more computation, which largely depends on the number of sample points. To assess their computational demands, we implemented the DCEA, SESAT, DCE with known t_D, and the unconstrained KF in the MATLAB (R2021a) environment on a computer equipped with a 2.6 GHz CPU (the 12th Gen Intel Core i7-1255U processor). Table 5.1 presents the average run time of one cycle of these algorithms. We can see that their run time is much shorter than the sampling interval, which ensures their application in real time. Specifically, other than the unconstrained KF, the DCE with known arrival time has the least online computation. In cases where estimating the arrival time is unnecessary, the DCEA estimator is a good option due to its good balance between computation and accuracy.

5.6.2 Tracking a Guided Target with Uncertain Destinations

Our proposed DC dynamics modeling, along with estimation of the DC state and arrival time, promises to be highly effective in tracking guided targets, especially in their terminal flight phase.

As depicted in Figure 5.6, a target starts at the origin with an initial speed of $200\sqrt{2}$ m/s and a heading angle of 135 degrees, intending to reach one of three possible destinations D_1, D_2, and D_3, located at (12, 4) km, (10, 10) km, and (4, 12) km, respectively. To reach a destination, various guidance laws, including the (pure) proportional navigation (PN), have been developed via optimal control theory and other methods. The PN law directs the target's commanded acceleration perpendicular to its heading, with a magnitude in proportion to the product of the target's velocity and the line-of-sight angular rate [44]. In this simulation, a proportional factor of 6 is applied to the PN law, resulting in a successful trajectory leading the target to the intended location D_2 in 56.61 seconds. The trackers do not know any of the concrete guidance law, proportional factor, actual destination D_2, and the exact arrival time t_D, especially when dealing with noncooperative targets.

Table 5.1 Average run time of different algorithms.

KF (unconstrained)	DCEA (uncertain t_D)	SESAT (uncertain t_D)	DCE (known t_D)
6.32×10^{-5} s	4.38×10^{-3} s	1.40×10^{-2} s	10^{-4} s

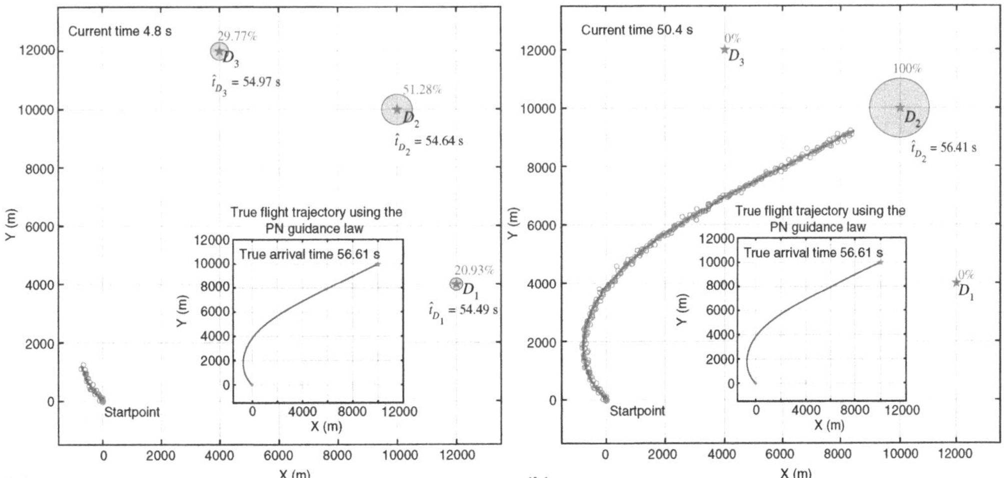

(a) **(b)**

Figure 5.6 Destination prediction and DC motion tracking for PN-guided flight: stars "★" indicate possible destinations, circles "o" represent measurement sequence, points "·" form filtered trajectory, shadowed circles of varying sizes correspond to probabilities of possible destinations, and curve in subplot illustrates true trajectory using PN guidance law. (a) At time $t = 4.8$ s. (b) At time $t = 50.4$ s.

Define the target state vector as $x = [\mathrm{x}, \dot{\mathrm{x}}, \mathrm{y}, \dot{\mathrm{y}}]^{\mathrm{T}}$ with position (x, y) and velocity ($\dot{\mathrm{x}}$, $\dot{\mathrm{y}}$). The three possible destinations are

$$\mathcal{D}_r = \{x(t_D) \colon Ax(t_D) = b_r\}, \quad r = 1, 2, 3 \tag{5.65}$$

with parameters

$$A = \begin{bmatrix} 1 & 0 & 0 & 0 \\ 0 & 0 & 1 & 0 \end{bmatrix}, \quad b_r = \begin{cases} [12, 4]^{\mathrm{T}} \text{ km}, & r = 1 \\ [10, 10]^{\mathrm{T}} \text{ km}, & r = 2 \\ [4, 12]^{\mathrm{T}} \text{ km}, & r = 3 \end{cases}. \tag{5.66}$$

Assuming perfect detection by a radar with zero false alarm rate and unity detection probability, the target's position is observed with a sampling interval of $T = 200$ ms using the same measurement model as in Section 5.6.1, with standard deviation $\sigma = 60$ m for measurement noise in each coordinate.

As mentioned above, the tracker does not know the specific guidance law but knows that the target is bound to navigate towards one of the three points. To identify D_r and estimate x and t_D, we choose the NCV model as the relaxed dynamics and construct three DC models following the procedures in Section 5.3, with three models each for destination constraint (5.65). These DC models with uncertain destinations can be treated as a hybrid system, in which the hybrid state includes the continuous-valued (x, t_D) and a discrete set for the destinations. For estimation of a hybrid system, the multiple model (MM) approach is cost-effective. Denote by M_r the DC model following which the target gets to destination D_r. Assuming equal a priori probability of the target reaching any of the three possible points, we initialize model probabilities $\Pr(M_r) = 1/3$, $r = 1, 2, 3$. From the law of total expectation, the estimate of the state based on the measurements up to time k is

$$\hat{x}_{k|k} = \sum_{r=1}^{3} E\left[E[x_k | t_D, M_r, Z^k] | M_r, Z^k\right] \Pr(M_r | Z^k). \tag{5.67}$$

The posterior probability of getting to destination D_r is

$$\begin{aligned}\Pr(M_r|Z^k) &\propto p(z_k|M_r, Z^{k-1}) \Pr(M_r|Z^{k-1}) \\ &= p(z_k|\hat{x}^r_{k-1|k-1}, P^r_{k-1|k-1}, M_r) \Pr(M_r|Z^{k-1}).\end{aligned}$$

That is, a bank of SESAT estimators runs in parallel at each time to estimate the state and the arrival time under model M_r, and then the overall state estimate $\hat{x}_{k|k}$ is a combination of these state estimates.

We assess the effectiveness of the SESAT-based MM estimator by visualizing the filtered trajectory of the PN-guided target. At 4.8 s and 50.4 s in the initial and ending phases, as shown in Figure 5.6, we observe that the filtered trajectory closely approximates the true flight trajectory shown in the subplot and is significantly smoother than the sequence of the measurements. Figure 5.7 shows the posterior model probabilities corresponding to different destinations, with destination D_2 correctly identified within 10 seconds. As time goes on, the estimate of t_D gradually converges to the true value of 56.61 s.

The standard KF with the NCV model is also implemented for comparison. Figure 5.8 shows RMS position errors of both estimators over 200 Monte Carlo runs. It can be observed that the KF with the NCV model performs well overall because the NCV model is a good approximation

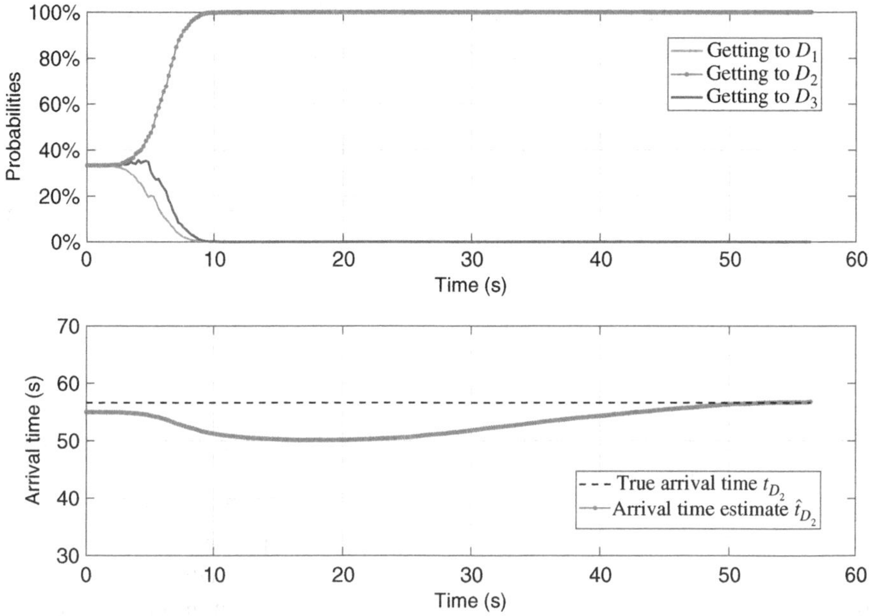

Figure 5.7 Top figure: probabilities of getting to different destinations. Bottom figure: sequential estimation of arrival time t_{D_2}.

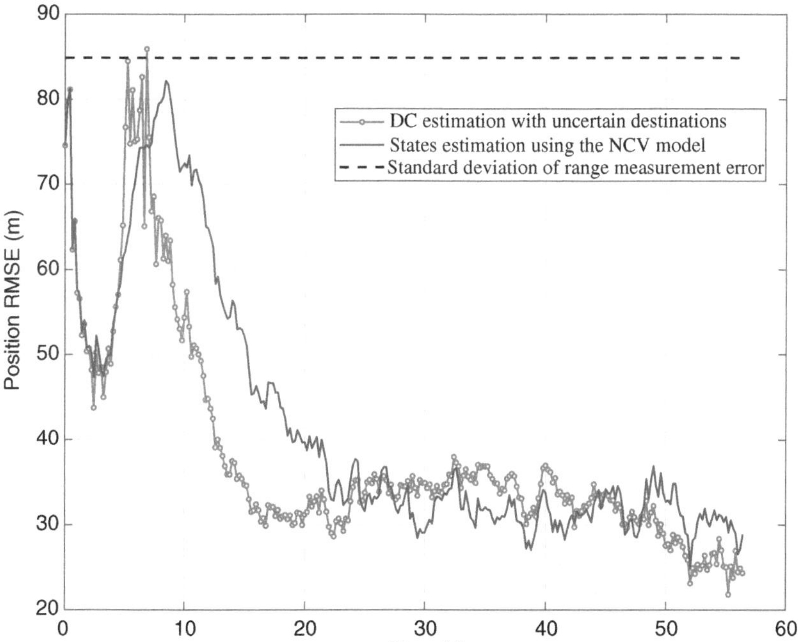

Figure 5.8 Estimation accuracy comparison between trackers using DC model and using NCV model.

of any trajectory for small sampling intervals. Nevertheless, the SESAT-based MM algorithm still outperforms the KF in state estimation accuracy because the intent information is incorporated. Additionally, the SESAT-based MM algorithm successfully identifies the destination and accurately estimates the arrival time, while the KF is unable to do these.

In addition, we use the DC model for TP of the PN-guided target. Figure 5.9 presents the logarithm of the average Euclidean error (AEE) [45] of the position prediction obtained at 11 s and 19.2 s based on the DC model with uncertain arrival time and the NCV model, respectively. We can clearly see that the former has much better prediction than the latter, thereby demonstrating the superiority of the DC model in TP for destination-directed motions.

5.6.3 Model Validation on Real Flight Data

Real data of aircraft flights in a terminal airspace surrounding an airport are used to validate the effectiveness of the proposed DC dynamic model. As reported in the Boeing statistical summary, 49% of all air-related fatal accidents worldwide occur in the final approach phase and the landing phase of the flight, even though these two phases constitute only 4% of the flight time [46].

Here, we take the Flight MU2111 from Xi'an to Beijing as an example and set up models to describe the plane's approach and landing at the Capital International Airport (PEK), Beijing. Once entering the approach phase, the plane first takes a U-turn maneuver to align its heading to the runway and then descends on the glide path till it lands on the runway. According to its flight schedule, the straight line extending from the runway and the center of the touchdown zone on the runway are determined as the waypoint constraint and the destination constraint in the two phases, respectively. Next, following the DC modeling proposed in the paper, dynamic models for landing at PEK

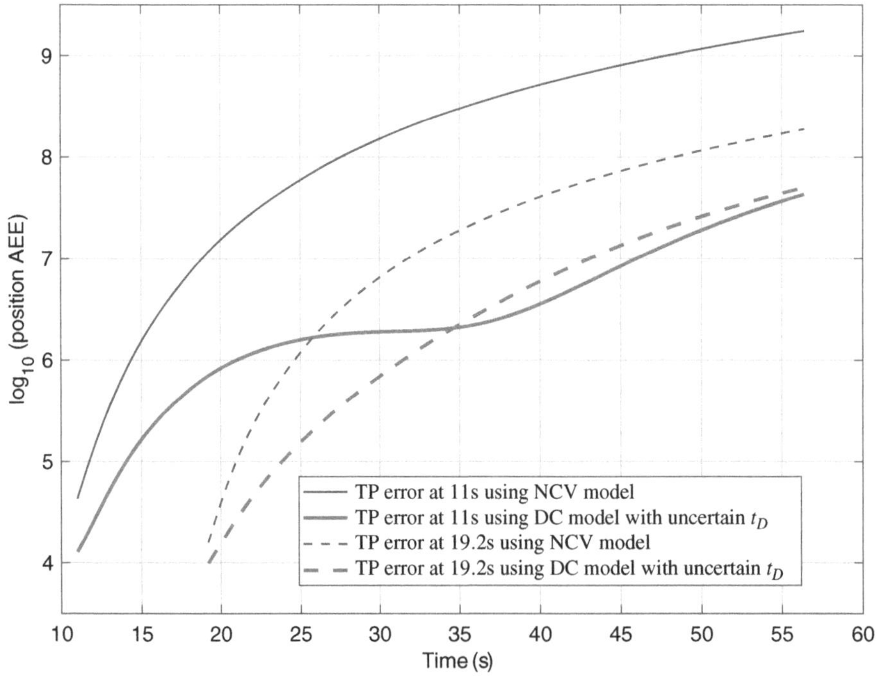

Figure 5.9 A comparison of position prediction errors between DC model and NCV model at two distinct time instants.

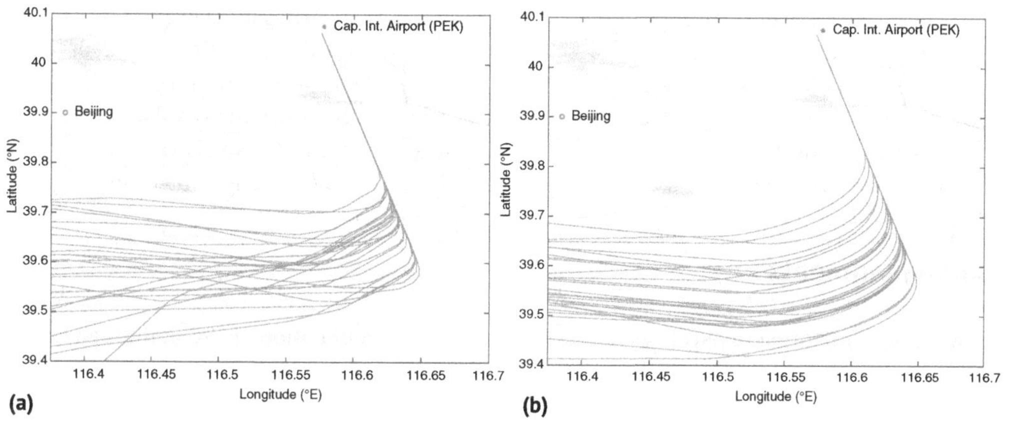

Figure 5.10 Model validation by comparing the generated flight tracks with the real flight tracks. (a) Bird's-eye view of generated landing flight tracks. (b) Bird's-eye view of real landing flight tracks.

are constructed. We independently sample trajectories from the models, and the bird's-eye view of the landing flight tracks is shown in Figure 5.10a. The vertical profile of the tracks is omitted for the sake of space.

For comparison, we collected flight data of MU2111 landing at PEK in the entire month of May 2019. The data were acquired from the ADS-B receiver, and each record contains timestamped entries of plane ID, position (latitude/longitude), pressure altitude, and lateral/vertical velocity. Figure 5.10b shows the bird's-eye view of the real flight tracks of the plane. It can be seen from the figures that the generated landing flight tracks match well with the real ones, which testifies to the high fidelity of the constructed models. Some nuances caused by atmospheric turbulence or pilots' specific operation are not taken into account in the modeling procedures. One point worth stressing is that our models are devised in advance merely using the scheduled waypoint/destination information along with the inherent flight behavior, rather than learning from the historical flight data as presented in [47, 48]. This means our modeling method is more applicable to the situations where historical data are lacking.

5.7 Conclusions

This chapter delved into the problem of modeling and estimation for DC systems. By fusing the destination constraint and the relaxed dynamics using the nearest neighbor criterion, a congruous DC model has been constructed. It is more applicable than that proposed in [9], as it allows the arrival time to be at any future continuous time. Superior properties of the proposed DC model have been presented. In addition, it has been shown that such a DC model is equivalent to the pseudo-observation-based BD model for the linear Gaussian system with a linear destination constraint. For realistic cases where the arrival time is either unknown or partially known, we have developed two DC estimation algorithms (i.e., the DCEA and the SESAT estimators) to incorporate the spatial information of the destination constraints while accounting for the arrival time uncertainty. Specifically, the SESAT estimator can simultaneously estimate the state and the arrival time if desired, while the DCEA estimator has an advantage in computational efficiency. Finally, we have demonstrated the effectiveness of the proposed estimation algorithms via examples of tracking a destination-guided target. The proposed DC model and the DC state estimation algorithms have promising applications in adversarial tracking, intelligent transportation, and other domains.

Our proposed DCEs not only solve the estimation problem with uncertain arrival times but can also be extended to practical cases where the destination is uncertain. We can also predict the destination location by augmenting it into the state vectors and estimate the augmented state.

Acknowledgments

This work was supported in part by National Natural Science Foundation of China through Grant 62273283, in part by Shanghai Aerospace Science and Technology Innovation Foundation through Grant SAST2025-009, and in part by Aeronautical Science Foundation of China through Grant 2024Z074053005.

Appendix

Proof of Property 3

From (5.22), it is clear that U_D depends on t_{k-1}. To avoid confusion, in the following proof, we use $U_{D|k-1}$ instead of U_D in definition (5.28) of ϵ_{k-1}^D.

Without loss of generality, let $j > k, \forall j, k$. Since

$$E[w_j(w_k^D)^\top] = Q_j\Phi^\top(t_D, t_{j+1}) = Q_j\left(\Phi_{j+1}^D\right)^\top$$

$$E[w_j^D(w_k^D)^\top] = \int_{t_j}^{t_D} \Phi(t_D, \tau)\Gamma(\tau)q[\Phi(t_D, \tau)\Gamma(\tau)]^\top \mathrm{d}\tau = Q_j^D,$$

we have

$$\begin{aligned}
&\mathrm{cov}(\epsilon_j^D, \epsilon_k^D) \\
&= E\left[\left(w_j - Q_j^{j+1,D}U_{D|j}w_j^D\right)\left(w_k - Q_k^{k+1,D}U_{D|k}w_k^D\right)^\top\right] \\
&= -E\left[w_j(w_k^D)^\top\right]\left(Q_k^{k+1,D}U_{D|k}\right)^\top + Q_j^{j+1,D}U_{D|j}E\left[w_j^D(w_k^D)^\top\right]\left(Q_k^{k+1,D}U_{D|k}\right)^\top \\
&= -Q_j\left(\Phi_{j+1}^D\right)^\top(Q_k^{k+1,D}U_{D|k})^\top + Q_j^{j+1,D}U_{D|j}Q_j^D(Q_k^{k+1,D}U_{D|k})^\top \\
&= \left[Q_j^{j+1,D}U_{D|j}Q_j^D - Q_j\left(\Phi_{j+1}^D\right)^\top\right]U_{D|k}^\top Q_k^{k+1,D} \\
&= \left[Q_j^{j+1,D}U_{D|j}Q_j^D - Q_j\left(\Phi_{j+1}^D\right)^\top\right]A^\top(AQ_{k-1}^D A^\top)^{-\top}AQ_k^{k+1,D}.
\end{aligned}$$

Note that

$$\begin{aligned}
&\left[Q_j^{j+1,D}U_{D|j}Q_j^D - Q_j\left(\Phi_{j+1}^D\right)^\top\right]A^\top \\
&\overset{(5.9)}{=} Q_j\left(\Phi_{j+1}^D\right)^\top A^\top\left(AQ_j^D A^\top\right)^{-1}AQ_j^D A^\top - Q_j\left(\Phi_{j+1}^D\right)^\top A^\top \\
&= 0.
\end{aligned}$$

Therefore, we have $\mathrm{cov}(\epsilon_j^D, \epsilon_k^D) = 0$.

Proof of Proposition 1

Using the matrix inversion lemma [30], the expression (5.42) can be computed by

$$
\begin{aligned}
\mathbf{Q}_{D,k-1} &= \left(Q_{k-1}^{-1} + \left(A\Phi_k^D\right)^{\top}\left(AQ_k^D A^{\top}\right)^{-1} A\Phi_k^D\right)^{-1} \\
&= Q_{k-1} - Q_{k-1}\left(A\Phi_k^D\right)^{\top} \\
&\quad \times (AQ_k^D A^{\top} + A\Phi_k^D Q_{k-1}\left(A\Phi_k^D\right)^{\top})^{-1} A\Phi_k^D Q_{k-1} \\
&\overset{(5.12)}{=} Q_{k-1} - Q_{k-1}(\Phi_k^D)^{\top} A^{\top}(AQ_{k-1}^D A^{\top})^{-1} A\Phi_k^D Q_{k-1} \\
&= Q_{k-1} - Q_{k-1}^{k,D} U_D (Q_{k-1}^{k,D})^{\top} \overset{(5.35)}{=} \operatorname{cov}(\epsilon_{k-1}^D).
\end{aligned} \tag{5.68}
$$

From (5.40), we find

$$
\begin{aligned}
\mathbf{F}_{D,k-1} &= \left(Q_{k-1} - Q_{k-1}^{k,D} U_D (Q_{k-1}^{k,D})^{\top}\right) Q_{k-1}^{-1} \Phi_{k-1} \\
&\overset{(5.11)}{=} \Phi_{k-1} - Q_{k-1}^{k,D} U_D (Q_{k-1}\left(\Phi_k^D\right)^{\top})^{\top} Q_{k-1}^{-1} \Phi_{k-1} \\
&= \Phi_{k-1} - Q_{k-1}^{k,D} U_D \Phi_{k-1}^D \overset{(5.26)}{=} F_{k-1}^D.
\end{aligned} \tag{5.69}
$$

From (5.41), we calculate the constant matrix by

$$
\begin{aligned}
\mathbf{M}_{D,k-1} &= [Q_{k-1} - Q_{k-1}^{k,D} U_D (Q_{k-1}^{k,D})^{\top}](A\Phi_k^D)^{\top}(AQ_k^D A^{\top})^{-1} b \\
&= Q_{k-1}(A\Phi_k^D)^{\top}(AQ_k^D A^{\top})^{-1} A\Lambda b - Q_{k-1}^{k,D} U_D (Q_{k-1}^{k,D})^{\top}(A\Phi_k^D)^{\top}(AQ_k^D A^{\top})^{-1} A\Lambda b \\
&\overset{(5.11)(5.22)}{=} Q_{k-1}^{k,D} A^{\top}[(AQ_k^D A^{\top})^{-1} - \tilde{\mathbf{Q}}] A\Lambda b
\end{aligned} \tag{5.70}
$$

where

$$
\begin{aligned}
\tilde{\mathbf{Q}} &\overset{(5.22)}{=} \left(AQ_{k-1}^D A^{\top}\right)^{-1} A(Q_{k-1}^{k,D})^{\top}\left(A\Phi_k^D\right)^{\top}\left(AQ_k^D A^{\top}\right)^{-1} \\
&= \left(AQ_{k-1}^D A^{\top}\right)^{-1} A\Phi_k^D Q_{k-1}\left(A\Phi_k^D\right)^{\top}\left(AQ_k^D A^{\top}\right)^{-1} \\
&\overset{(5.12)}{=} \left(AQ_{k-1}^D A^{\top}\right)^{-1}\left[A(Q_{k-1}^D - Q_k^D)A^{\top}\right]\left(AQ_k^D A^{\top}\right)^{-1} \\
&= \left(AQ_k^D A^{\top}\right)^{-1} - \left(AQ_{k-1}^D A^{\top}\right)^{-1}.
\end{aligned} \tag{5.71}
$$

Finally, substituting (5.71) into (5.70) yields

$$
\begin{aligned}
\mathbf{M}_{D,k-1} &= Q_{k-1}^{k,D} A^{\top}[(AQ_k^D A^{\top})^{-1} - \tilde{\mathbf{Q}}] A\Lambda b \\
&= Q_{k-1}^{k,D} A^{\top}\left(AQ_{k-1}^D A^{\top}\right)^{-1} A\Lambda b \\
&= Q_{k-1}^{k,D} U_D \Lambda b \overset{(5.27)}{=} G_{k-1}^D b.
\end{aligned} \tag{5.72}
$$

Since Eqs. (5.68), (5.69), and (5.72) hold, we conclude that the transition densities of the proposed DC model and the pseudo-observation-based BD model are mathematically identical.

References

1 B. I. Ahmad, S. J. Godsill, V. Krishnamurthy, P. Willett, and M. Rangaswamy. Foreword to the special section on meta-level and adversarial tracking. *IEEE Transactions on Aerospace and Electronic Systems*, 57(4):1994–1995, 2021.

2 L. Badriasl, S. Arulampalam, J. Hoek, and A. Finn. Bayesian WIV estimators for 3-D bearings-only TMA with speed constraints. *IEEE Transactions on Signal Processing*, 67(13):3576–3591, 2019.

3 A. Elnakeeb and U. Mitra. Line constrained estimation with applications to target tracking: Exploiting sparsity and low-rank. *IEEE Transactions on Signal Processing*, 66(24):6488–6502, 2018.

4 K. Li and G. Zhou. State estimation with a destination constraint imposed by proportional navigation guidance law. *IEEE Transactions on Aerospace and Electronic Systems*, 58(1):58–73, 2022.

5 K. Li and G. Zhou. Tracking filter with implicit constraints. In M. Mallick, B.-N. Vo, and R. Tharmarasa, editors, *Advanced Techniques in Tracking and Sensor Management: Theory and Applications*, chapter 6. Wiley/IEEE, 2025.

6 K. Li, T. Kirubarajan, and G. Zhou. State estimation with implicit constraints of circular trajectory using pseudomeasurements. *IEEE Transactions on Aerospace and Electronic Systems*, 56(6): 4406–4425, 2020.

7 B. O. S. Teixeira, J. Chandrasekar, H. J. Palanthandalam-Madapusi, L. A. B. Torres, L. A. Aguirre, and D. S. Bernstein. Gain-constrained Kalman filtering for linear and nonlinear systems. *IEEE Transactions* on *Signal Processing*, 56(9):4113–4123, 2008.

8 B. O. S. Teixeira, J. Chandrasekar, L. A. B. Torres, L. A. Aguirre, and D. S. Bernstein. State estimation for linear and non-linear equality-constrained systems. *International Journal of Control*, 82(5):918–936, 2009.

9 L. Xu, X. R. Li, Y. Liang, and Z. Duan. Modeling and state estimation of linear destination-constrained dynamic systems. *IEEE Transactions on Signal Processing*, 70:2374–2387, 2022.

10 L. Xu, X. R. Li, M. Mallick, and Z. Duan. Modeling and state estimation of destination-constrained dynamic systems. Part II: Uncertain arrival time. *IEEE Transactions on Signal Processing*, 73:633–648, 2025.

11 D. A. Castanon, B. C. Levy, and A. S. Willsky. Algorithms for the incorporation of predictive information in surveillance theory. *International Journal of Systems Science*, 16(3):367–382, 1985.

12 B. C. Levy, R. Frezza, and A. J. Krener. Modeling and estimation of discrete-time Gaussian reciprocal processes. *IEEE Transactions on Automatic Control*, 35(9):1013–1023, 1990.

13 F. Carravetta and L. B. White. Modelling and estimation for finite state reciprocal processes. *IEEE Transactions on Automatic Control*, 57(9): 2190–2202, 2012.

14 L. B. White and F. Carravetta. Optimal smoothing for finite state hidden reciprocal processes. *IEEE Transactions on Automatic Control*, 56(9):2156–2161, 2011.

15 M. Fanaswala and V. Krishnamurthy. Detection of anomalous trajectory patterns in target tracking via stochastic context-free grammars and reciprocal process models. *IEEE Journal of Selected Topics in Signal Processing*, 7(1):76–90, 2013.

16 M. Fanaswala and V. Krishnamurthy. Syntactic models for trajectory constrained track-before-detect. *IEEE Transactions on Signal Processing*, 62(23):6130–6142, 2014.

17 R. Rezaie and X. R. Li. Gaussian conditionally Markov sequences: Dynamic models and representations of reciprocal and other classes. *IEEE Transactions* on *Signal Processing*, 68:155–169, 2020.

18 R. Rezaie and X. R. Li. Gaussian conditionally Markov sequences: Singular/nonsingular. *IEEE Transactions on Automatic Control*, 65(5):2286–2293, 2020.

19 R. Rezaie and X. R. Li. Mathematical modeling and optimal inference of guided Markov-like trajectory. In *IEEE International Conference on Multisensor Fusion and Integration*, Karlsruhe, Germany, 2020.

20 B. I. Ahmad, J. K. Murphy, S. J. Godsill, P. M. Langdon, and R. Hardy. Intelligent interactive displays in vehicles with intent prediction: A Bayesian framework. *IEEE Signal Processing Magazine*, 34(2):82–94, 2017.

21 B. I. Ahmad, J. K. Murphy, P. M. Langdon, and S. J. Godsill. Bayesian intent prediction in object tracking using bridging distributions. *IEEE Transactions on Cybernetics*, 48(1):215–227, 2018.

22 R. Gan, B. I. Ahmad, and S. J. Godsill. Lévy state-space models for tracking and intent prediction of highly maneuverable objects. *IEEE Transactions on Aerospace and Electronic Systems*, 57(4):2021–2038, 2021.

23 R. Gan, J. Liang, B. I. Ahmad, and S. J. Godsill. Modeling intent and destination prediction within a Bayesian framework: Predictive touch as a usecase. *Data-Centric Engineering*, 1:e12, 2020.

24 J. Liang, B. I. Ahmad, R. Gan, P. Langdon, R. Hardy, and S. J. Godsill. On destination prediction based on Markov bridging distributions. *IEEE Signal Processing Letters*, 26(11):1663–1667, 2019.

25 G. Zhou, K. Li, X. Chen, L. Wu, and T. Kirubarajan. State estimation with a destination constraint using pseudo-measurements. *Signal Processing*, 145:155–166, 2018. ISSN 0165-1684.

26 G. Zhou, K. Li, and T. Kirubarajan. Constrained state estimation using noisy destination information. *Signal Processing*, 166:107226, 2020. ISSN 0165-1684.

27 L. Xu, Z. Hou, M. Mallick, and Y. Fang. Intention-aware motion modeling using GP priors with conditional kernels. In *25th International Conference on Information Fusion (FUSION)*, pages 1–7, 2022.

28 X. R. Li. Compatibility and modeling of constrained dynamic systems. In *19th International Conference on Information Fusion*, pages 240–247, 2016.

29 L. Xu, X. R. Li, Y. Liang, and Z. Duan. Constrained dynamic systems: Generalized modeling and state estimation. *IEEE Transactions on Aerospace and Electronic Systems,* 53(5):2594–2609, 2017.

30 Y. Bar-Shalom, X. R. Li, and T. Kirubarajan. *Estimation with Applications to Tracking and Navigation: Theory Algorithms and Software*. Wiley, New York, 2001.

31 L. Xu, X. R. Li, Z. Duan, and J. Lan. Modeling and state estimation for dynamic systems with linear equality constraints. *IEEE Transactions on Signal Processing*, 61(11):2927–2939, 2013.

32 X. R. Li and V. P. Jilkov. Survey of maneuvering target tracking. Part I: Dynamic models. *IEEE Transactions on Aerospace and Electronic Systems,* 39(4):1333–1364, 2003.

33 S. Julier, J. Uhlmann, and H. F. Durrant-Whyte. A new method for the nonlinear transformation of means and covariances in filters and estimators. *IEEE Transactions on Automatic Control*, 45(3):477–482, 2000. ISSN 0018-9286.

34 J. V. Candy. *Bayesian Signal Processing: Classical, Modern, and Particle Filtering Methods*. Wiley-IEEE Press, 2nd edition, 2016.

35 Y. Bar-Shalom, P. K. Willett, and X. Tian. *Tracking and Data Fusion: A Handbook of Algorithms*. YBS Publishing, 2011.

36 T. Schon, F. Gustafsson, and P.-J. Nordlund. Marginalized particle filters for mixed linear/nonlinear state-space models. *IEEE Transactions* on *Signal Processing,* 53(7):2279–2289, 2005.

37 D. Crisan and J. Míguez. Nested particle filters for online parameter estimation in discrete-time state-space Markov models. *Bernoulli*, 24(4A):3039–3086, 2018.

38 A. Doucet, N.J. Gordon, and V. Krishnamurthy. Particle filters for state estimation of jump Markov linear systems. *IEEE Transactions on Signal Processing*, 49(3):613–624, 2001.

39 N. Kantas, A. Doucet, S. S. Singh, J. Maciejowski, and N. Chopin. On particle methods for parameter estimation in state-space models. *Statistical Science*, 30(3):328–351, 2015.

40 L. Martino, J. Read, V. Elvira, and F. Louzada. Cooperative parallel particle filters for on-line model selection and applications to urban mobility. *Digital Signal Processing*, 60: 172–185, 2017.

41 B. Ristic, S. Arulampalam, and N. Gordon. *Beyond the Kalman Filter: Particle Filters for Tracking Applications*. Artech house, 2004.

42 M. Šimandl, J. Královec, and P. Tichavský. Filtering, predictive, and smoothing Cramér-Rao bounds for discrete-time nonlinear dynamic systems. *Automatica*, 37(11):1703–1716, 2001.

43 P. Tichavsky, C. H. Muravchik, and A. Nehorai. Posterior Cramer-Rao bounds for discrete-time nonlinear filtering. *IEEE Transactions on Signal Processing*, 46(5):1386–1396, 1998.

44 N. A. Shneydor. *Missile Guidance and Pursuit: Kinematics, Dynamics and Control*. Elsevier, 1998.

45 X. R. Li and Z. Zhao. Evaluation of estimation algorithms. Part I: Incomprehensive measures of performance. *IEEE Transactions on Aerospace and Electronic Systems*, 42(4):1340–1358, 2006.

46 B. Pasztor. *Statistical Summary of Commercial Jet Airplane Accidents Worldwide Operations (1959–2018)*. The Boeing Company, Chicago, IL, USA, 2019.

47 S. T. Barratt, M. J. Kochenderfer, and S. P. Boyd. Learning probabilistic trajectory models of aircraft in terminal airspace from position data. *IEEE Transactions on Intelligent Transportation Systems*, 20(9):3536–3545, 2019. ISSN 1558-0016.

48 Z. Mahboubi and M. J. Kochenderfer. Learning traffic patterns at small airports from flight tracks. *IEEE Transactions on Intelligent Transportation Systems*, 18(4):917–926, 2017.

6

Tracking Filter with Implicit Constraints

*Keyi Li and Gongjian Zhou**

School of Electronics and Information Engineering, Harbin Institute of Technology, Harbin, China

6.1 Introduction

To improve tracking accuracy, it is essential to fully utilize all the available prior information about the target motion. In many practical target tracking applications, the true trajectory of targets is significantly constrained by environmental limitations and inherent motion characteristics, rather than being unconstrained motion in free space [1]. For instance, in vehicle tracking, impassable terrain (including forest barriers, rivers/lakes), roadside buildings, and road boundaries all affect vehicle motion [2]. Roadmaps contain information on position and speed constraints [3, 4]. In outer space, spacecrafts like space stations and satellites follow predefined orbits (circular, elliptical, etc.) determined by orbital mechanics and mission requirements [5]. Aircraft flights are restricted by established routes [6] and may follow predetermined trajectories for speed and altitude reduction before landing [7]. The maritime navigation is constrained by shipping lanes and reefs [8]. In all these scenarios, target states conform to specific constraints that may be equality/inequality, linear/nonlinear, or time-varying/invariant. Despite their diversity, these constraints share a critical commonality: they contain valuable prior information about target states that can significantly enhance tracking accuracy when properly utilized.

Conventional tracking methods face three main limitations in addressing constrained tracking problems:

- Inability to accurately describe constraints, causing motion model mismatch;
- Failure to incorporate prior knowledge of constraints, leading to constraint information waste;
- Estimate to be unconstrained, resulting in estimation performance degradation.

To address the above issues, it is necessary to develop constrained target tracking methods. Current applications already demonstrate the advantage of incorporating constraints, such as airborne/spaceborne radars using road constraints for ground target tracking [9] and information-assisted ship monitoring in ports/rivers [10]. The constrained target tracking began attracting the attention of researchers in the 1990s and evolved into a key topic by the early twenty-first century. With sustained exploration in this field, it has emerged as a cutting-edge direction in target tracking research [11]. Existing advancements can be systematically categorized into two technical frameworks.

The first kind of method focuses on modifying motion models by incorporating constraint information to address constrained tracking problems using existing filtering frameworks. This can be further divided into the fusion of traditional motion models and constraints [12–14] or direct derivation of constrained dynamic models [15–20]. For example, techniques such as projection operators and Gram-Schmidt decomposition have been used to fuse unconstrained dynamic models with linear or quadratic equality constraints, generating constrained dynamic models [12–14]. For a detailed survey of constrained dynamic modeling, please refer to Chapter 5 [21]. Motion modeling based on one-dimensional (1-D) constrained coordinates [20, 22] shows significant potential due to its physical interpretability. However, the resulting 1-D state estimates require conversion to Cartesian coordinates through static transformation models. Under complex spatial constraints, such transformations may not have analytical solutions, which is a critical limitation hindering broader application. The second kind of approach introduces constraints into tracking systems by improving filtering algorithms [2, 23–25]. This strategy modifies conventional filters with constraint knowledge, offering greater design flexibility and implementation ease, thus gathering wide attention [26]. Although constrained dynamic models are not explicitly adopted, the incorporation of constraints into state estimation systems still achieves performance improvements. Commonly used equality-constrained filtering methods include the pseudo measurement method [23, 27] and the estimate projection method [1, 2, 24]. While these methods deliver enhanced estimation accuracy in constrained tracking scenarios, their performance is limited by the mismatched dynamic models, which inherently prevent the filtering methods from attaining theoretical optimality.

Most existing equality-constrained state estimation methods assume complete and accurate prior knowledge of target trajectory information corresponding to the constraints. These methods become inapplicable when such assumptions fail, particularly when complete trajectory information is unavailable. In practical applications, complete prior knowledge of target trajectories is usually not available, and the accuracy and reliability of information cannot always be guaranteed. Under these conditions, constraints cannot be abstracted into specific curves for explicit modeling, making conventional constrained tracking methods ineffective. Unlike explicit constraints, such constraints can only be described through implicit mathematical expressions, hence they are termed implicit constraints [28]. Proper extraction and utilization of implicit constraints can enhance tracking precision. Studies in [28–32] have conducted systematic research on implicit constraint modeling and state estimation problems, effectively leveraging additional target state information contained in various implicit constraints, including

- Destination constraints
- Trajectory shape constraints
- Guidance law constraints

The chapter is organized as follows. The modeling and state estimation with destination constraints are presented in Section 6.2. Section 6.3 gives the constraint models and corresponding filtering methods for tracking with trajectory shape constraints. The guidance law constraint models for several representative proportional navigation laws are presented in Section 6.4, along with the constrained filters. Conclusions are summarized in Section 6.5.

6.2 Tracking Filter with Destination Constraints

In certain target tracking scenarios, the destination information of a tracked target is known a priori or measurable. For example, to protect radars from the threat of anti-radiation missiles, interception or decoying tactics are usually applied, which requires the acquisition of the accurate position of anti-radiation missiles by using tracking systems [33, 34]. In such cases, the prior information

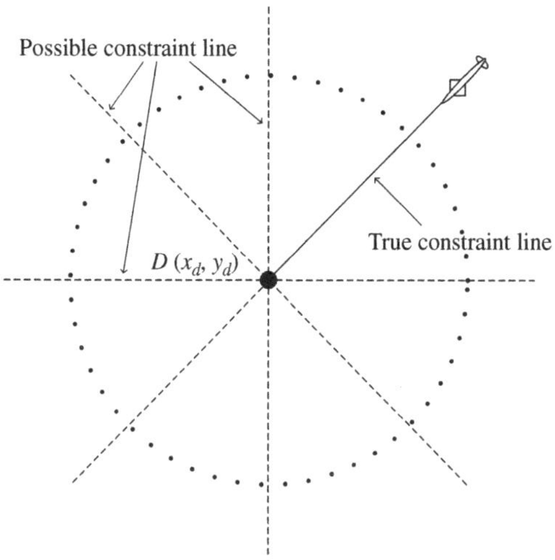

Figure 6.1 Geometry of possible and true constraint lines pointing to the destination.

about the positions of radars being protected is available in advance, while the trajectory shape can be inferred based on the motion characteristics of anti-radiation missiles guided by passive radar seekers. In the terminal phase of the missile flight, when the heading of the missile is aligned with the line-of-sight direction, the trajectory can typically be assumed as a straight line. Under these conditions, the missile state obeys an implicit constraint, referred to as the destination constraint, which contains destination and trajectory shape information. The effective utilization of destination constraints can improve tracking performance. In this subsection, the destination constraint models for straight trajectories and corresponding tracking methods are introduced.

A typical destination constrained target tracking scenario with a straight trajectory is illustrated in Figure 6.1. Here, D represents the destination, and the target moves along a straight trajectory toward the destination from an unknown initial position. When the straight trajectory is exactly known, it can be described as an explicit constraint and addressed using existing explicit constraint tracking methods [26, 35]. However, in destination constrained tracking scenarios, only one point on the trajectory, i.e., the destination, is known a priori. The direction from which the target may emerge cannot be predetermined. In this case, all straight trajectories passing through the known destination can be potential constraint lines, as indicated by the dashed lines in Figure 6.1. To effectively leverage constraint information for tracking performance enhancement, studies in [32, 36] have conducted systematic research on destination constraint modeling and state estimation.

6.2.1 Modeling of Destination Constraints

1. Modeling with straight trajectory

Generally, if two points (x_0, y_0) and (x_d, y_d) on the straight line are known, the position and velocity constraints can be expressed by a generic form of linear equality constraint as

$$\begin{cases} y_k - ax_k = b \\ \dot{y}_k - a\dot{x}_k = 0 \end{cases}, \tag{6.1}$$

where $a = \frac{y_0 - y_d}{x_0 - x_d}$, $b = y_n - \frac{y_0 - y_d}{x_0 - x_d} x_d$ stand for the slope and intercept of the straight line, respectively.

In the modeling of destination constraints, only one point on the trajectory, i.e., the destination (x_d, y_d), is known a priori. Therefore, the generic form cannot be employed to describe the destination constraint. One possible strategy is to find a substitution of (x_0, y_0) and modify the generic form. Here, the measurement point (x_k^m, y_k^m), which is noisy but also contains position information of the target, at the current time step is used to replace (x_0, y_0). The line connecting (x_d, y_d) and (x_k^m, y_k^m) is treated as an approximate constraint line, whose slope can be easily calculated and expressed as the slope of the true constraint line plus an error term, given by

$$\frac{y_k^m - y_d}{x_k^m - x_d} = \frac{y_k - y_d}{x_k - x_d} + e_k^a, \tag{6.2}$$

$$\frac{y_k^m - y_d}{x_k^m - x_d} = \frac{\dot{y}_k}{\dot{x}_k} + e_k^a, \tag{6.3}$$

where e_k^a is the error term arising from the measurement noise.

It should be noticed that (6.2) describes the relationship between the slopes of the approximate constraint line and the true constraint line, which contains no extra prior information about target position. While (6.3) expresses the constraint on the velocity of target, which contains extra information from destination, it is an effective destination constraint model. However, due to the existence of the uncertain term e_k^a, (6.3) is an approximate model that suffers from approximation errors. The effect of approximation errors becomes pronounced when the target is close to the destination, which significantly affects the effectiveness of the model.

Another strategy to model the destination constraint is to directly describe the constraint relationship using state components. Given the destination (x_d, y_d) and the straight trajectory shape, the relationship between the position and velocity components that are subject to a destination constraint can be written as

$$\frac{y_k - y_d}{x_k - x_d} = \frac{\dot{y}_k}{\dot{x}_k}, \tag{6.4}$$

which can be derived from (6.2) and (6.3). For a target traveling along a straight trajectory, the slope of the true constraint line (i.e., the left side of the equation) should always be equal to the tangent of target heading (i.e., the right side of the equation). Compared with the approximate model in (6.3), the formulation in (6.4) includes position components and removes the uncertain term, resulting in an accurate destination constraint model.

Besides the above two models, a more direct strategy to model the destination constraint is to augment the unknown parameters in (6.1) into the state vector and describe the constraint relationship based on the augmented state vector. In this case, the unknown slope a and intercept b are augmented into the state vector and are being estimated together with the original state components. Using the augmented state components to replace the true parameters, the destination constraint model can be formulated as

$$\begin{cases} y_k - a_k x_k = b_k \\ \dot{y}_k - a_k \dot{x}_k = 0 \end{cases}. \tag{6.5}$$

It can be seen that the model established based on the state augmentation is in a simple form, and the position constraint and velocity constraint are expressed separately, which is more intuitive and easy to understand. The destination information is not explicitly involved but used to calculate the initial estimates of augmented components. Furthermore, an obvious advantage of this model is that the state augmentation enables the estimation of extra trajectory parameters, which can be used as prior information in the subsequent tracking when the target is subject to the same constraint to enhance estimation performance.

In practical applications, the destination position is measured by some imperfect sensors, such as radar, satellite, or camera, which inevitably introduce noises into the destination information [32]. Assuming that the destination noise is zero-mean Gaussian distributed white noise, the noisy destination position is subject to the following distribution:

$$x_d^m \sim \mathcal{N}(x_d, \sigma_x^2), \qquad y_d^m \sim \mathcal{N}(y_d, \sigma_y^2), \tag{6.6}$$

where σ_x^2 and σ_y^2 are the known noise covariances.

The destination noises may lead to constraint errors when the noisy destination information is directly incorporated into the destination constraint model, resulting in deterioration in estimation performance. To describe the constraint relationship accurately, the destination position (x_k^d, y_k^d) is augmented into the state vector, and the constraint model is formulated using the state components instead of the noisy information, given by

$$\frac{y_k - y_k^d}{x_k - x_k^d} = \frac{\dot{y}_k}{\dot{x}_k}. \tag{6.7}$$

2. Modeling of destination constraint with circular trajectory

For a target traveling along a circular trajectory with center (x_c, y_c) and radius r, the explicit constraint imposed on the position components can be written as

$$(x_k - x_c)^2 + (y_k - y_c)^2 = r^2. \tag{6.8}$$

Taking the time derivative of (6.8), the velocity constraint can be obtained as

$$(x_k - x_c)\dot{x}_k + (y_k - y_c)\dot{y}_k = 0, \tag{6.9}$$

which indicates that the velocity vector is always vertical to the radius vector.

For the destination constraint with the circular trajectory, the center and the radius are usually not available. More equations are needed to formulate the constraint. By taking advantage of the destination position (x_d, y_d), which is on the circular trajectory and therefore subjected to (6.8), a new equation can be written as

$$(x_d - x_c)^2 + (y_d - y_c)^2 = r^2, \tag{6.10}$$

which shows the deterministic relationship between the destination, the center, and the radius.

The turn rate is also important to describe the target motion on the circular trajectory, which can be expressed by

$$\omega_k = \pm\sqrt{\frac{\dot{x}_k^2 + \dot{y}_k^2}{r^2}}, \tag{6.11}$$

and can also be written in a unique form

$$r = \sqrt{\frac{\dot{x}_k^2 + \dot{y}_k^2}{\omega_k^2}}. \tag{6.12}$$

Using the relationships in (6.8), (6.9), (6.10), and (6.12), the unknown center and the radius can be eliminated, and the direct form of destination constraint model can be derived as

$$\left(x_d - x_k + \frac{\dot{y}_k}{\omega_k}\right)^2 + \left(y_d - y_k - \frac{\dot{x}_k}{\omega_k}\right)^2 = \frac{\dot{x}_k^2 + \dot{y}_k^2}{\omega_k^2}, \tag{6.13}$$

which extracts all the valuable information expressed by (6.8), (6.9), (6.10), and (6.12).

Other cases, where extra trajectory information other than the destination and the trajectory shape is available, are also taken into account. When the true radius r_t is available, a variant formulation can be written as

$$\begin{cases} \left(x_d - x_k + \dfrac{\dot{y}_k}{\omega_k}\right)^2 + \left(y_d - y_k - \dfrac{\dot{x}_k}{\omega_k}\right)^2 = \dfrac{\dot{x}_k^2 + \dot{y}_k^2}{\omega_k^2} \\ (\dot{x}_k^2 + \dot{y}_k^2) - r_t^2\omega_k^2 = 0 \end{cases}, \tag{6.14}$$

which is actually a combination of (6.12) and (6.13) using r_t.

When the center of the circular trajectory is known a priori, the radius can also be determined using the destination, which converts the estimation problem to a conventional one with explicit constraints presented in (6.8) and (6.9).

In the case when the destination information is noisy, the unknown true destination is augmented into the state vector. The augmented state vector is represented as

$$\mathbf{x}_k^d = \begin{bmatrix} x_k & \dot{x}_k & y_k & \dot{y}_k & \omega_k & x_{d,k} & y_{d,k} \end{bmatrix}'. \tag{6.15}$$

According to the augmented state vector, the constraint relationship in (6.13) is modified to

$$\left(x_{d,k} - x_k + \frac{\dot{y}_k}{\omega_k}\right)^2 + \left(y_{d,k} - y_k - \frac{\dot{x}_k}{\omega_k}\right)^2 = \frac{\dot{x}_k^2 + \dot{y}_k^2}{\omega_k^2}. \tag{6.16}$$

When the true radius is known a priori, (6.14) is also changed to

$$\begin{cases} \left(x_{d,k} - x_k + \dfrac{\dot{y}_k}{\omega_k}\right)^2 + \left(y_{d,k} - y_k - \dfrac{\dot{x}_k}{\omega_k}\right)^2 = \dfrac{\dot{x}_k^2 + \dot{y}_k^2}{\omega_k^2} \\ (\dot{x}_k^2 + \dot{y}_k^2) - r_t^2\omega_k^2 = 0 \end{cases}. \tag{6.17}$$

Since the destination information is noisy, the radius cannot be determined even if the true center (x_c^t, y_c^t) is known. A direct form model is provided to utilize the center information when available, which is given by

$$\begin{cases} \left(x_{d,k} - x_k + \dfrac{\dot{y}_k}{\omega_k}\right)^2 + \left(y_{d,k} - y_k - \dfrac{\dot{x}_k}{\omega_k}\right)^2 = \dfrac{\dot{x}_k^2 + \dot{y}_k^2}{\omega_k^2} \\ x_c^t = x_k - \dfrac{\dot{y}_k}{\omega_k},\ y_c^t = y_k + \dfrac{\dot{x}_k}{\omega_k} \end{cases}. \tag{6.18}$$

Besides the direct form, the destination constraint model can also be derived using the parameter augmentation. The unknown center and radius of the circular trajectory as well as the destination position (when the prior information is noisy) can be augmented into the state vector as $(x_{c,k}, y_{c,k})$, r_k, and $(x_{d,k}, y_{d,k})$, resulting in state-augmentation form constraint models in different cases:

a. without any other information about the trajectory

$$\begin{cases} (x_k - x_{c,k})^2 + (y_k - y_{c,k})^2 = r_k^2 \\ (x_k - x_{c,k})\dot{x}_k + (y_k - y_{c,k})\dot{y}_k = 0 \\ (x_{d,k} - x_{c,k})^2 + (y_{d,k} - y_{c,k})^2 = r_k^2 \\ r_k = \sqrt{(\dot{x}_k^2 + \dot{y}_k^2)/\omega_k^2} \end{cases}, \tag{6.19}$$

b. with the radius information

$$\begin{cases} (x_k - x_{c,k})^2 + (y_k - y_{c,k})^2 = r_t^2 \\ (x_k - x_{c,k})\dot{x}_k + (y_k - y_{c,k})\dot{y}_k = 0 \\ (x_{d,k} - x_{c,k})^2 + (y_{d,k} - y_{c,k})^2 = r_t^2 \\ (\dot{x}_k^2 + \dot{y}_k^2) - r_t^2\omega_k^2 = 0 \end{cases}, \tag{6.20}$$

c. with the center information

$$\begin{cases} (x_k - x_c^t)^2 + (y_k - y_c^t)^2 = r_k^2 \\ (x_k - x_c^t)\dot{x}_k + (y_k - y_c^t)\dot{y}_k = 0 \\ (x_{d,k} - x_c^t)^2 + (y_{d,k} - y_c^t)^2 = r_k^2 \\ r_k = \sqrt{(\dot{x}_k^2 + \dot{y}_k^2)/\omega_k^2} \end{cases}, \tag{6.21}$$

6.2.2 Filtering Algorithms

1. State equation

The original state equation is given by

$$\mathbf{x}_{k+1} = \Phi_k \mathbf{x}_k + \Gamma_k \mathbf{v}_k, \tag{6.22}$$

where $\mathbf{x}_k$ denotes the state vector, the prior distribution of $\mathbf{x}_k$ is assumed to be Gaussian as $\mathbf{x}_0 \sim \mathcal{N}(\bar{\mathbf{x}}_0, \mathbf{P}_0)$, Φ_k stands for the state transition matrix, Γ_k represents the noise gain matrix, $\mathbf{v}_k$ is the process noise vector, the process noise is assumed to be zero mean white Gaussian distributed with covariance matrix given by

$$\mathbf{Q}_k = \Gamma_k \text{cov}(\mathbf{v}_k)(\Gamma_k)'. \tag{6.23}$$

In the state-augmentation form models, to formulate the constraint relationships, the target state vector $\mathbf{x}_k$ is augmented with constraint parameters, including the slope a_k and intercept b_k for the case with straight trajectory, center (x_k^c, y_k^c) and the radius r_k for the case with circular trajectory, as well as the destination (x_k^d, y_k^d) for the case with noisy destination information. The augmented state vector is given by

$$\mathbf{x}_k^{ad} = \begin{bmatrix} \mathbf{x}_k & \mathbf{c}_k & \mathbf{d}_k \end{bmatrix}', \tag{6.24}$$

where $\mathbf{c}_k$ contains components of constraint parameters, $\mathbf{d}_k$ contains components of destination position.

In this section, the target is assumed to be traveling along a deterministic straight/circular trajectory towards a fixed destination, which is common in some practical applications, such as guided target tracking. Due to the lack of the complete information about the trajectory, the constraint parameters are unknown but constant during the whole filtering procedure, thus are immune to the process noise. The corresponding augmented state equation can be written as

$$\mathbf{x}_{k+1}^{ad} = \Phi_k^{ad}\mathbf{x}_k^{ad} + \Gamma_k^{ad}\mathbf{v}_k, \tag{6.25}$$

where

$$\Phi_k^{ad} = \begin{bmatrix} \Phi_k & \mathbf{0}_{n\times(l+m)} \\ \mathbf{0}_{(l+m)\times n} & \mathbf{I}_{l+m} \end{bmatrix}, \quad \Gamma_k^{ad} = \begin{bmatrix} \Gamma_k \\ \mathbf{0}_{(l+m)\times m} \end{bmatrix}. \tag{6.26}$$

$0_{l_1\times l_2}$ is the zero matrix with dimension of $l_1 \times l_2$, n is the dimension of $\mathbf{x}_k$, l is the dimension of $\mathbf{c}_k$, m is the dimension of $\mathbf{d}_k$.

The augmentation state equation enables the simultaneous estimation of the target states and the constraint parameters, which can be used as prior information when tracking other targets traveling along the same trajectory.

2. Measurement equation

The radar measurements include range component r_k^m and azimuth component θ_k^m. Corresponding measurement noises are denoted by w_k^r and w_k^θ, which are assumed to be mutually independent zero-mean white Gaussian noises with variances σ_r^2 and σ_θ^2. The radar measurement equation is given by

$$\mathbf{z}_k = \begin{bmatrix} r_k^m \\ \theta_k^m \end{bmatrix} = \mathbf{h}_k(\mathbf{x}_k^{ad}) + \mathbf{w}_k = \begin{bmatrix} \sqrt{x_k^2 + y_k^2} \\ \tan^{-1}(y_k, x_k) \end{bmatrix} + \begin{bmatrix} w_k^r \\ w_k^\theta \end{bmatrix}. \tag{6.27}$$

The covariance matrix of $\mathbf{w}_k$ can be written as

$$\mathbf{R}_k = \begin{bmatrix} \sigma_r^2 & 0 \\ 0 & \sigma_\theta^2 \end{bmatrix}. \tag{6.28}$$

Pseudo measurements are constructed based on the constraint models in Subsection 6.2.1 to incorporate the constraint into the tracking process. For the destination constraint with exact destination information, by rearranging the terms in (6.14), the pseudo measurement for the direct form can be written as

$$\lambda_k = \left(x_d - x_k + \frac{\dot{y}_k}{\omega_k}\right)^2 + \left(y_d - y_k - \frac{\dot{x}_k}{\omega_k}\right)^2 - \frac{\dot{x}_k^2 + \dot{y}_k^2}{\omega_k^2} = 0. \tag{6.29}$$

While for the state-augmentation form model in (6.19), the corresponding pseudo measurements are given by

$$\begin{cases} \eta_k^1 = (x_k - x_{c,k})^2 + (y_k - y_{c,k})^2 - r_k^2 = 0 \\ \eta_k^2 = (x_k - x_{c,k})\dot{x}_k + (y_k - y_{c,k})\dot{y}_k = 0 \\ \eta_k^3 = (x_{d,k} - x_{c,k})^2 + (y_{d,k} - y_{c,k})^2 - r_k^2 = 0 \\ \eta_k^4 = \sqrt{(\dot{x}_k^2 + \dot{y}_k^2)/\omega_k^2} - r_k = 0 \end{cases}. \tag{6.30}$$

For other cases with extra prior information, pseudo measurements can be constructed in the same manner. The formulations are not provided here due to space limitation.

The pseudo measurements are used to update the state estimates in the filter procedure. Here, the original radar measurements and the pseudo measurements are processed sequentially. The reason will be explained later. For the pseudo measurements in (6.30), the corresponding measurement equation is

$$\mathbf{z}_k^p = \begin{bmatrix} \eta_k^1 \\ \eta_k^2 \\ \eta_k^3 \\ \eta_k^4 \end{bmatrix} = \mathbf{h}_k^p(\mathbf{x}_k^{ad}) = \begin{bmatrix} (x_k - x_{c,k})^2 + (y_k - y_{c,k})^2 - r_k^2 \\ (x_k - x_{c,k})\dot{x}_k + (y_k - y_{c,k})\dot{y}_k \\ (x_{d,k} - x_{c,k})^2 + (y_{d,k} - y_{c,k})^2 - r_k^2 \\ \sqrt{(\dot{x}_k^2 + \dot{y}_k^2)/\omega_k^2} - r_k \end{bmatrix}. \tag{6.31}$$

The corresponding measurement covariance matrix is $\mathbf{0}_{4\times 4}$ since the pseudo measurements are noiseless.

3. Constrained filtering

Since both the radar measurements and pseudo measurements are nonlinear functions of the Cartesian state, nonlinear filtering methods are required to conduct the constrained state estimation. Here, the unscented Kalman filter (UKF) [32, 37], is employed, which demonstrates better accuracy and stability compared to the extended Kalman filter (EKF) [38] in highly nonlinear problems, while maintains a lower computational load than the particle filter (PF) [39] under equivalent precision. Furthermore, the sequential processing [40] is adopted to handle nonlinear measurements in the filtering process. This approach effectively improves estimation accuracy while reducing computational complexity by processing nonlinear measurements sequentially in ascending order of nonlinearity. Compared with radar measurements, the pseudo measurements exhibit higher nonlinearity, thus are processed after radar measurements. The first-stage state update using radar measurements provides more accurate predicted values for the second-stage state update using pseudo measurements, thereby enhancing overall estimation precision. The decomposition of measurement vectors and covariance matrices avoids complex high-dimensional matrix operations, further reducing computational complexity.

The filtering procedure is given as follows:

a. Filter initialization

The initialization of the augmented state vector can be divided into several parts. First, the two-point differencing method [38] is employed to initialize original state estimate including the position and velocity components at the second time step using converted measurements [41, 42] at the first two time steps. Then, the constraint parameter estimates can be initialized using the initial

---

Algorithm 6.1: A recursion of the UDC-ASF using radar and pseudo measurements

---

Require: $\hat{\mathbf{x}}^{ad}_{k-1|k-1}$, $\mathbf{P}^{ad}_{k-1|k-1}$, $\mathbf{z}_k$, $\mathbf{R}_k$

Ensure: $\hat{\mathbf{x}}^{ad}_{k|k}$, $\mathbf{P}^{ad}_{k|k}$.

1: State Prediction

$$\hat{\mathbf{x}}^{ad}_{k-1|k-1,i} = \{\hat{\mathbf{x}}^{ad}_{k-1|k-1}, \hat{\mathbf{x}}^{ad}_{k-1|k-1} \pm (\sqrt{(n+\lambda)\mathbf{P}^{ad}_{k-1|k-1}})_j\}$$

$$\hat{\mathbf{x}}^{u}_{k|k-1,i} = \mathbf{f}^{ad}_k(\hat{\mathbf{x}}^{ad}_{k-1|k-1,i}),\ \hat{\mathbf{x}}^{u}_{k|k-1} = \sum_{i=0}^{2n} W^m_i \hat{\mathbf{x}}^{u}_{k|k-1,i}$$

$$\mathbf{P}^{u}_{k|k-1} = \mathbf{Q}^{ad}_{k-1} + \sum_{i=0}^{2n} W^c_i (\hat{\mathbf{x}}^{u}_{k|k-1,i} - \hat{\mathbf{x}}^{u}_{k|k-1})(\hat{\mathbf{x}}^{u}_{k|k-1,i} - \hat{\mathbf{x}}^{u}_{k|k-1})'$$

2: State Update Using Radar Measurement

(1) Measurement Prediction

$$\hat{\xi}^{u}_{k|k-1,i} = \left\{\hat{\mathbf{x}}^{u}_{k|k-1}, \hat{\mathbf{x}}^{u}_{k|k-1} \pm (\sqrt{(n+\lambda)\mathbf{P}^{u}_{k|k-1}})_j\right\}$$

$$\hat{\mathbf{z}}_{k|k-1,i} = \mathbf{h}_k(\hat{\xi}^{u}_{k|k-1,i}),\ \hat{\mathbf{z}}_{k|k-1} = \sum_{i=0}^{2n} W^m_i \hat{\mathbf{z}}_{k|k-1,i}$$

$$\mathbf{P}_{k,zz} = \mathbf{R}_k + \sum_{i=0}^{2n} W^c_i (\hat{\mathbf{z}}_{k|k-1,i} - \hat{\mathbf{z}}_{k|k-1})(\hat{\mathbf{z}}_{k|k-1,i} - \hat{\mathbf{z}}_{k|k-1})'$$

(2) Filter Gain Calculation

$$\mathbf{P}^{u}_{k,xz} = \sum_{i=0}^{2n} W^c_i (\hat{\mathbf{x}}^{u}_{k|k-1,i} - \hat{\mathbf{x}}^{u}_{k|k-1})(\hat{\mathbf{z}}_{k|k-1,i} - \hat{\mathbf{z}}_{k|k-1})',\ \mathbf{K}^u_k = \mathbf{P}^{u}_{k,xz}\mathbf{P}^{-1}_{k,zz}$$

(3) State Update

$$\hat{\mathbf{x}}^{u}_{k|k} = \hat{\mathbf{x}}^{u}_{k|k-1} + \mathbf{K}^u_k(\mathbf{z}_k - \hat{\mathbf{z}}_{k|k-1}),\ \mathbf{P}^{u}_{k|k} = \mathbf{P}^{u}_{k|k-1} - \mathbf{K}^u_k \mathbf{P}_{k,zz}(\mathbf{K}^u_k)'$$

3: State Update Using Pseudo Measurement

(1) Measurement Prediction

$$\hat{\mathbf{x}}^{u}_{k|k,i} = \left\{\hat{\mathbf{x}}^{u}_{k|k}, \hat{\mathbf{x}}^{u}_{k|k} \pm (\sqrt{(n+\lambda)\mathbf{P}^{u}_{k|k}})_j\right\},\ \hat{\mathbf{z}}^{p}_{k|k,i} = \mathbf{h}^p_k(\hat{\mathbf{x}}^{u}_{k|k,i})$$

$$\hat{\mathbf{z}}^{p}_{k|k} = \sum_{i=0}^{2n} W^m_i \hat{\mathbf{z}}^{p}_{k|k,i},\ \mathbf{P}^{p}_{k,zz} = \sum_{i=0}^{2n} W^c_i (\hat{\mathbf{z}}^{p}_{k|k,i} - \hat{\mathbf{z}}^{p}_{k|k})(\hat{\mathbf{z}}^{p}_{k|k,i} - \hat{\mathbf{z}}^{p}_{k|k})'$$

(2) Filter Gain Calculation

$$\mathbf{P}^{p}_{k,xz} = \sum_{i=0}^{2n} W^c_i (\hat{\mathbf{x}}^{u}_{k|k,i} - \hat{\mathbf{x}}^{u}_{k|k})(\hat{\mathbf{z}}^{p}_{k|k,i} - \hat{\mathbf{z}}^{p}_{k|k})',\ \mathbf{K}_k = \mathbf{P}^{p}_{k,xz}(\mathbf{P}^{p}_{k,zz})^{-1}$$

(3) State Update

$$\hat{\mathbf{x}}^{ad}_{k|k} = \hat{\mathbf{x}}^{u}_{k|k} + \mathbf{K}_k(\mathbf{z}^p_k - \hat{\mathbf{z}}^{p}_{k|k}),\ \mathbf{P}^{ad}_{k|k} = \mathbf{P}^{u}_{k|k} + (\hat{\mathbf{x}}_{k|k} - \hat{\mathbf{x}}^{u}_{k|k})(\hat{\mathbf{x}}_{k|k} - \hat{\mathbf{x}}^{u}_{k|k})'$$

---

estimate of original state and the unscented transformation (UT) [37]. The initialization of destination position components are directly accomplished using the noisy prior information.

b. Filter procedure

According to the augmented state equation in (6.25), the radar measurement equation and the pseudo measurement equation in (6.31), a recursion of the uncertain destination constrained augmented state filter, denoted by UDC-ASF, is shown in Algorithm 6.1.

6.2.3 Simulation Results

Simulation results are provided to show performance of the constraint models and filters introduced in this section. The simulation scenario is similar to that in [32] where the destination information is contaminated by noises. The algorithms include: the unconstrained UKF, two destination constraint Kalman filters (DCKF) that directly utilize noisy destination information, the least square destination constraint Kalman filter (LSDCKF) in [43], and two UDC-ASFs with destination position or slope and intercept augmented. The root mean square errors (RMSE) of position and velocity estimates are shown in Figure 6.2. The posterior Cramér–Rao lower bound (PCRLB) is provided as the theoretical benchmark [32].

It can be seen that all filtering methods demonstrate effectiveness in reducing measurement errors. The UDC-ASFs significantly outperform the unconstrained UKF approach. This validates the capability of constrained filtering methods to extract and utilize additional constraint information. In contrast, the DCF-Noiseless and LSDCKF methods exhibit larger position estimation errors after 30 seconds, primarily due to their direct utilization of noisy destination information, which leads to severe performance degradation. The key advantage of UDC-ASFs lies in its robust performance under noisy destination information conditions. Furthermore, both UDC-ASFs exhibit nearly identical estimation performance that is close to the PCRLB.

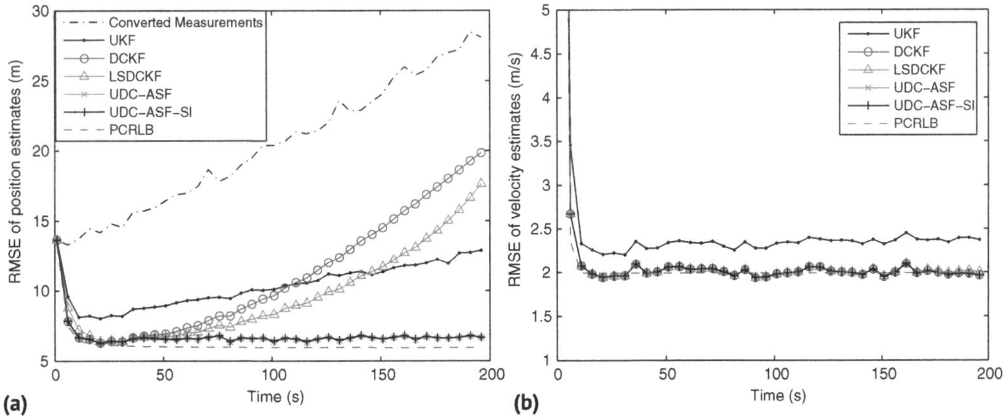

Figure 6.2 RMSE comparison of filters in the case with noisy destination information. (a) RMSE of position estimates and (b) RMSE of velocity estimates.

6.3 Tracking Filter with Trajectory Shape Constraints

In this section, the modeling and estimation with a more general kind of implicit constraints are introduced. Different from the destination constraints involving specific destination information introduced in the previous section, this kind of constraints only contain information about the shape of target trajectories, named trajectory shape constraints. Corresponding scenarios can be found in many practical tracking applications. For instance, in ground target tracking systems, one may know the shape of a road from an offline hard-copy map, but no more information is available. Moreover, in civilian vessel or aircraft tracking systems, the targets may travel along straight trajectories at a constant heading and speed during most of the trip, out of consideration for reducing cost. Due to the lack of specific information about the trajectory, the modeling and estimation with trajectory shape constraints are more complex and challenging. Nevertheless, trajectory shape constraints also contain additional information about target states beyond system dynamic models and measurements. Extracting and effectively utilizing this information can contribute to tracking performance improvement.

The geometry of multilane road and vehicle trajectory is illustrated in Figure 6.3. In this tracking scenario, a vehicle located at position V travels along a multilane straight highway, where the solid line C represents the center line of the road and the red dashed line T denotes the true trajectory of vehicle. Conventional explicit constraint tracking methods for road-constrained target tracking typically assume that the complete prior knowledge of road position and orientation is available. However, in multilane scenarios as illustrated in Figure 6.3, the specific lane occupied by the target is generally unknown. A common practice is to approximate the true constraint line using the road center line, yet deviations between the center line and the actual constraint line may induce significant performance degradation. Actually, the target state obeys a trajectory shape constraint under such conditions, where only the shape of the trajectory is known a priori, while both trajectory position and orientation remain uncertain. Studies in [28, 29, 31] provide several approaches for extracting and utilizing trajectory shape constraints to enhance tracking performance, including effective constraint modeling and state estimation methodologies.

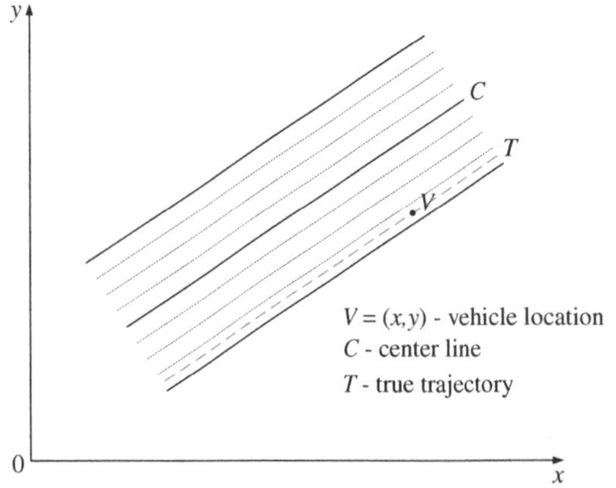

Figure 6.3 Geometry of multilane road and vehicle trajectory.

6.3.1 Modeling of Trajectory Shape Constraints

1. Modeling with straight trajectory

When the target trajectory is a straight line, the constraint imposed on the target states can be divided into two aspects: (1) all the points that represent the target position at different time steps must be on the same straight line; (2) the direction of the target velocity at different time steps must be the same as that of the straight line. Given a straight line trajectory, these two conditions are always satisfied by the states. Conversely, given these conditions, the trajectory shape is deterministic and the target states obey the trajectory shape constraint. Thus, these two conditions are equivalent to imposing the trajectory shape constraint and the constraint relationships can be formulated accordingly.

First, condition 1 can be transformed into an equivalent one that constrains any three successive points to the same line. Then, we can use the position components of the states at three successive time steps k, $k-1$ and $k-2$ to describe the position constraint as

$$\frac{y_k - y_{k-1}}{x_k - x_{k-1}} = \frac{y_{k-1} - y_{k-2}}{x_{k-1} - x_{k-2}}. \tag{6.32}$$

Then, according to condition 2, the corresponding velocity constraint, which constrains the direction of velocity at each point to the same as that determined by (6.32), is expressed as

$$\frac{y_k - y_{k-1}}{x_k - x_{k-1}} = \frac{\dot{y}_k}{\dot{x}_k}, \frac{y_k - y_{k-1}}{x_k - x_{k-1}} = \frac{\dot{y}_{k-1}}{\dot{x}_{k-1}}, \frac{y_k - y_{k-1}}{x_k - x_{k-1}} = \frac{\dot{y}_{k-2}}{\dot{x}_{k-2}}. \tag{6.33}$$

Equations (6.32) and (6.33) completely describe the constraint relationships among the state components, which contain extra information that can improve the estimation accuracy.

In (6.32), the constraint relationship between position components is described by using only the position components of the states. In this case, no less than three successive points are necessary. But if both the position and velocity constraints are taken into account, it is possible to simplify the formulation by describing the position constraint implicitly using the hybrid relationship among the state components of position and velocity. A simplified formulation, which uses two successive points at time step k and $k-1$ to describe the constraint relationship, is given by

$$\frac{y_k - y_{k-1}}{x_k - x_{k-1}} = \frac{\dot{y}_k}{\dot{x}_k} = \frac{\dot{y}_{k-1}}{\dot{x}_{k-1}}, \tag{6.34}$$

where the constraint relationship between the velocity components is expressed explicitly while that between the position components implicitly. The state-augmentation form constraint model can be formulated as

$$\begin{cases} (y_k - y_{k-1})\dot{x}_k - (x_k - x_{k-1})\dot{y}_k = 0 \\ \dot{y}_k \dot{x}_{k-1} - \dot{y}_{k-1}\dot{x}_k = 0 \end{cases}. \tag{6.35}$$

Similar to destination constraints, the modeling of trajectory shape constraints can also be achieved through parameter augmentation. For targets with straight trajectories, the constraint parameters are augmented into the state vector for joint estimation. By replacing the true trajectory parameters with these augmented components, a trajectory shape constraint model can be established as

$$\begin{cases} y_k - a_k x_k = b_k \\ \dot{y}_k - a_k \dot{x}_k = 0 \end{cases}, \tag{6.36}$$

which is the same with the destination constraint model in (6.5). However, due to the absence of destination information, different methods are required for initializing the parameter components during filtering. Furthermore, this model exhibits a critical limitation: when the true trajectory of target is perpendicular to the x-axis, the slope becomes infinite, and the y-axis intercept ceases to exist. If this is the case, the mathematical expressions lose physical meaning, leading to model failure and filter divergence. To address this issue, [29] conducted further research on parameter-augmented trajectory shape constraint models and proposed a generalized model expressed as

$$\begin{cases} c_1 x_k + c_2 y_k + c_3 = 0 \\ c_1 \dot{x}_k + c_2 \dot{y}_k = 0 \end{cases}, \tag{6.37}$$

where c_1, c_2, and c_3 are parameters related to the position of heading of the straight trajectory. Different from the model in (6.36), c_1, c_2, and c_3 always exist and are of finite values. When the trajectory of target is perpendicular to the x-axis, c_2 is zero-value while the values of c_1 and c_3 depend on the position of trajectory. Generally, the model established based on slope and y-intercept can be converted to the generalized model by using the following formulations:

$$\begin{cases} c_1 \neq 0 \\ c_2 = -c_1 \cdot a \\ c_3 = -c_1 \cdot b \end{cases}, \tag{6.38}$$

where c_1 can be arbitrary nonzero real value, the other two can be calculated according to c_1, slope and y-intercept.

Since the true values of c_1, c_2, and c_3 are unknown, they are augmented into the vector and being estimated together with the state, resulting in the parameter-augmentation form model as

$$\begin{cases} c_{k,1} x_k + c_{k,2} y_k + c_{k,3} = 0 \\ c_{k,1} \dot{x}_k + c_{k,2} \dot{y}_k = 0 \end{cases}. \tag{6.39}$$

2. Modeling with circular trajectory

For the case with circular trajectory, the connection between the destination constraint and the trajectory shape constraint is closer. Among the four destination constraint formulations in (6.19), three equations only contain the trajectory shape information, thus can direct compose a trajectory shape constraint model. Notably, since the number of equations is reduced, the direct form model cannot be obtained for the trajectory shape constraint.

Augmenting unknown center position or radius into the state vector, the parameter-augmentation form trajectory shape constraint model corresponding to different prior information are given by

a. without any other information about the trajectory

$$\begin{cases} (x_k - x_c)^2 + (y_k - y_c)^2 = r^2 \\ (x_k - x_c)\dot{x}_k + (y_k - y_c)\dot{y}_k = 0 \\ r = \sqrt{(\dot{x}_k^2 + \dot{y}_k^2)/\omega_k^2} \end{cases}, \tag{6.40}$$

b. with the radius information

$$\begin{cases} (x_k - x_c)^2 + (y_k - y_c)^2 = r_t^2 \\ (x_k - x_c)\dot{x}_k + (y_k - y_c)\dot{y}_k = 0, \\ (\dot{x}_k^2 + \dot{y}_k^2) - r_t^2\omega_k^2 = 0 \end{cases} \tag{6.41}$$

c. with the center information

$$\begin{cases} (x_k - x_c^t)^2 + (y_k - y_c^t)^2 = r^2 \\ (x_k - x_c^t)\dot{x}_k + (y_k - y_c^t)\dot{y}_k = 0. \\ r = \sqrt{(\dot{x}_k^2 + \dot{y}_k^2)/\omega_k^2} \end{cases} \tag{6.42}$$

The relationship between the destination constraint and the trajectory shape constraint can be interpreted from two different perspectives. On one hand, the destination constraint can be regarded as a special case of the trajectory shape constraint with the additional information about a single point on the trajectory. On the other hand, the trajectory shape constraint can be considered as a destination constraint with infinite uncertainty on the destination.

6.3.2 Filtering Algorithms

1. State equation

In the state augmentation constraint model, the state at $k-1$ time step is involved to describe the trajectory shape constraint relationship. To utilize the constraint model, the state at previous time step is augmented into the state vector forming the augmentation state vector as

$$\mathbf{x}_k^l = \begin{bmatrix} \mathbf{x}_k' & \mathbf{x}_{k-1}' \end{bmatrix}' = \begin{bmatrix} x_k & y_k & \dot{x}_k & \dot{y}_k & x_{k-1} & y_{k-1} & \dot{x}_{k-1} & \dot{y}_{k-1} \end{bmatrix}'. \tag{6.43}$$

The corresponding augmented state equation is given by

$$\mathbf{x}_{k+1}^l = \Phi_k^l \mathbf{x}_k^l + \Gamma_k^l \mathbf{v}_k = \begin{bmatrix} \Phi_k & 0_{n\times n} \\ \mathbf{I}_{n\times n} & 0_{n\times n} \end{bmatrix} \begin{bmatrix} \mathbf{x}_k \\ \mathbf{x}_{k-1} \end{bmatrix} + \begin{bmatrix} \Gamma_k \\ 0_{n\times m} \end{bmatrix} \mathbf{v}_k. \tag{6.44}$$

In the parameter augmentation constraint model, parameters are used to describe the constraint relationship between the state components, yet their true values are unknown. To utilize the constraint model to incorporate trajectory shape constraint into the tracking system, the parameter vector $\mathbf{c}_k$ is augmented into the state vector, resulting in an augmented state vector as

$$\mathbf{x}_k^a = \begin{bmatrix} x_k & y_k & \dot{x}_k & \dot{y}_k & \mathbf{c}_k \end{bmatrix}'. \tag{6.45}$$

The augmented state equation can be obtained as

$$\mathbf{x}_{k+1}^a = \Phi_k^a \mathbf{x}_k^a + \Gamma_k^a \mathbf{v}_k = \begin{bmatrix} \Phi_k & 0_{n\times l} \\ 0_{l\times n} & \mathbf{I}_{l\times l} \end{bmatrix} \begin{bmatrix} \mathbf{x}_k \\ \mathbf{c}_k \end{bmatrix} + \begin{bmatrix} \Gamma_k \\ 0_{l\times m} \end{bmatrix} \mathbf{v}_k. \tag{6.46}$$

2. Measurement equation

The original measurements reported by the radar include range component r_k^m and azimuth component θ_k^m as presented in (6.27). Constructing pseudo measurements according to the state-augmentation form model in (6.35) and augmenting them into the measurement equation, the augmented measurement equation can be written as

$$\mathbf{z}_k^l = \begin{bmatrix} r_k^m \\ \theta_k^m \\ \eta_k^{(1)} \\ \eta_k^{(2)} \end{bmatrix} = \mathbf{h}_k^l(\mathbf{x}_k^l) + \mathbf{w}_k^l = \begin{bmatrix} \sqrt{x_k^2 + y_k^2} \\ \tan^{-1}(y_k, x_k) \\ (y_k - y_{k-1})\dot{x}_k - (x_k - x_{k-1})\dot{y}_k \\ \dot{y}_k\dot{x}_{k-1} - \dot{y}_{k-1}\dot{x}_k \end{bmatrix} + \begin{bmatrix} w_k^r \\ w_k^\theta \\ 0 \\ 0 \end{bmatrix}, \tag{6.47}$$

where $\mathbf{z}_k^l$ is the state-augmented measurement vector and $\mathbf{w}_k^l$ is the state-augmented measurement noise vector. The pseudo measurements are noiseless. The corresponding measurement noise covariance matrix is given by

$$\mathbf{R}_k^l = \begin{bmatrix} R_{k,rr} & R_{k,r\theta} & R_{k,r\eta_1} & R_{k,r\eta_2} \\ R_{k,r\theta} & R_{k,\theta\theta} & R_{k,\theta\eta_1} & R_{k,\theta\eta_2} \\ R_{k,r\eta_1} & R_{k,\theta\eta_1} & R_{k,\eta_1\eta_1} & R_{k,\eta_1\eta_2} \\ R_{k,r\eta_2} & R_{k,\theta\eta_2} & R_{k,\eta_1\eta_2} & R_{k,\eta_2\eta_2} \end{bmatrix} = \begin{bmatrix} \sigma_r^2 & 0 & 0 & 0 \\ 0 & \sigma_\theta^2 & 0 & 0 \\ 0 & 0 & 0 & 0 \\ 0 & 0 & 0 & 0 \end{bmatrix}. \tag{6.48}$$

For the parameter-augmentation form models in (6.36), two pseudo measurements can be constructed as

$$\begin{cases} \xi_k^{(1)} = y_k - a_k x_k - b_k \\ \xi_k^{(2)} = \dot{y}_k - a_k \dot{x}_k \end{cases}, \tag{6.49}$$

where ξ_k and η_k are pseudo measurements with zero-value.

Based on the pseudo measurements, the heading constrained filtering problem is converted into a conventional filtering problem using the augmented measurement equation given by

$$\mathbf{z}_k^a = \begin{bmatrix} r_k^m \\ \theta_k^m \\ \xi_k^{(1)} \\ \xi_k^{(2)} \end{bmatrix} = \mathbf{h}_k^a(\mathbf{x}_k^a) + \mathbf{w}_k^a = \begin{bmatrix} \sqrt{x_k^2 + y_k^2} \\ \tan^{-1}(y_k, x_k) \\ y_k - a_k x_k - b_k \\ \dot{y}_k - a_k \dot{x}_k \end{bmatrix} + \begin{bmatrix} w_k^r \\ w_k^\theta \\ 0 \\ 0 \end{bmatrix}, \tag{6.50}$$

where $\mathbf{z}_k^a$ is the parameter-augmented measurement vector and $\mathbf{w}_k^a$ is the parameter-augmented measurement noise vector. The last two elements in $\mathbf{w}_k^a$ are noiseless, thus the covariance matrix of $\mathbf{w}_k^a$ is the same as $\mathbf{R}_k^l$ in (6.48).

3. Constrained filtering

In the case with trajectory shape constraints, the nonlinear filters are also required. For the case using the parameter-augmentation form models, the filter procedure is similar to that in Algorithm 6.1. For the state-augmentation form trajectory shape constraint model, the augmentation of previous states into the state vector leads to a significant increase in dimension. For high-dimensional nonlinear filtering problems, the cubature Kalman filter (CKF) [44] is a preferred option with enhanced numerical stability. Similar to the UKF, the CKF employs deterministic sampling strategies to estimate the mean and covariance of random variables. The spherical-radial cubature rule is employed in the CKF to select cubature points. Compared to the sigma points used in the UT, this sampling strategy ensures all cubature points have positive weights, thereby guaranteeing the positive definiteness of the posterior covariance matrix. Owing to this property, the CKF demonstrates superior numerical stability over the UKF when handling high-dimensional nonlinear filtering problems. Moreover, the CKF achieves comparable estimation accuracy to UKF. Consequently, the CKF is adopted in the filtering with the state-augmentation form constraint model. Taking the trajectory shape constraint augmented state filter, denoted as TSC-ASF, as an example, the filtering procedure is given as follows:

a. Filter initialization
The initial estimate and covariance at the current time step is carried out using the two-point differencing method [38]. The initial estimates and covariances at past time steps are all set to zero due to the lack of prior information.

b. Filter procedure
A recursion of the TSC-ASF is shown in Algorithm 6.2. The augmented state estimation simultaneously produces constrained filtering results and smoothing results, both derived within the same Bayesian framework. As previously discussed, the state-augmentation form constraint model is derived from the augmented state vector. Therefore, it is impossible to obtain constrained filtering results independently without computing the smoothing outcomes. In applications with less strict real-time requirements, this method can also be used as a trajectory shape constraint smoother (TSCS), with the smoothing results serving as the final output.

6.3.3 Simulation Results

Simulation results are provided to show performance of the constraint models and filters introduced in this section. The simulation scenario is similar to that in [29]. The algorithms include: the unconstrained converted measurement Kalman filter (CMKF) [41], the trajectory shape constraint Kalman filters (TSCKF) that using the state-augmentation form constraint model, the trajectory shape constrained generalized augmented state filter (TSC-GASF), and the trajectory shape constrained augmented state filter (TSC-ASF). The RMSEs of position and velocity estimates are shown in Figure 6.4.

It can be seen the trajectory shape constrained filtering methods outperform the unconstrained CMKF approach. This performance improvement validates the effectiveness of constrained filtering methods in leveraging trajectory shape information to reduce estimation errors. Furthermore, all three trajectory shape constrained filtering methods exhibit comparable performance approaching the PCRLB. Among these methods, the TSC-GASF estimates one more parameter compared to the TSC-ASF under equivalent information conditions, resulting in marginally inferior performance.

6.4 Tracking Filter with Guidance Law Constraints

Beyond the previously discussed implicit constraints, which correspond to fixed trajectory shapes, there is another kind of implicit constraints with uncertain trajectory shapes in practical target tracking applications. For instance, the tactical missile usually employs guidance laws during the terminal phase for guidance to move toward its target, where the trajectory shape depends on both

Algorithm 6.2: A recursion of the TSC-ASF using radar and pseudo measurements

Require: $\hat{\mathbf{x}}^l_{k-1|k-1}$, $\mathbf{P}^l_{k-1|k-1}$, $\mathbf{z}^l_k$, $\mathbf{R}^l_k$

Ensure: $\hat{\mathbf{x}}^l_{k|k}$, $\mathbf{P}^l_{k|k}$.

1: State Prediction

$$\hat{\mathbf{x}}^l_{k|k-1} = \Phi^l_{k-1}\hat{\mathbf{x}}^l_{k-1|k-1}$$

$$\mathbf{P}^l_{k|k-1} = \Phi^l_{k-1}\mathbf{P}^l_{k-1|k-1}(\Phi^l_{k-1})' + \mathbf{Q}^l_{k-1}$$

2: Measurement Prediction

$$\mathbf{P}^l_{k|k-1} = \mathbf{D}^l_{k|k-1}(\mathbf{D}^l_{k|k-1})'$$

$$\hat{\mathbf{x}}_{k|k-1,i} = \mathbf{D}^l_{k|k-1}\gamma_i + \hat{\mathbf{x}}^l_{k|k-1}$$

where $\gamma_i = \sqrt{n_l}\mathbf{e}_i, i = 1, 2, \cdots, 2n_l$, n_l is the dimension of the augmented state vector $\mathbf{x}^l_k$, $\mathbf{e}_i$ represents the ith column of the matrix $[\mathbf{I}^{n_l \times n_l}, -\mathbf{I}^{n_l \times n_l}]$.

$$\hat{\mathbf{z}}^l_{k|k-1,i} = \mathbf{h}^l_k(\hat{\mathbf{x}}^l_{k|k-1,i})$$

$$\bar{\mathbf{z}}^l_{k|k-1} = \frac{1}{2n_l}\sum_{i=1}^{2n_l}\hat{\mathbf{z}}^l_{k|k-1,i}$$

$$\mathbf{P}_{k,xz} = \frac{1}{2n_l}\sum_{i=1}^{2n_l}(\hat{\mathbf{x}}^l_{k|k-1,i})(\hat{\mathbf{z}}^l_{k|k-1,i})' - \hat{\mathbf{x}}^l_{k|k-1}(\hat{\mathbf{z}}^l_{k|k-1})'$$

3: Filter Gain Calculation

$$\mathbf{P}_{k,zz} = \mathbf{R}^l_k + \frac{1}{2n_l}\sum_{i=1}^{2n_l}(\hat{\mathbf{z}}^l_{k|k-1,i})(\hat{\mathbf{z}}^l_{k|k-1,i})' - \hat{\mathbf{z}}^l_{k|k-1}(\hat{\mathbf{z}}^l_{k|k-1})'$$

$$\mathbf{K}_k = \mathbf{P}_{k,xz}\mathbf{P}^{-1}_{k,zz}$$

4: State Update

$$\hat{\mathbf{x}}^l_{k|k} = \hat{\mathbf{x}}^l_{k|k-1} + \mathbf{K}_k(\mathbf{z}^l_k - \hat{\mathbf{z}}^l_{k|k-1})$$

$$\mathbf{P}^l_{k|k} = \mathbf{P}^l_{k|k-1} - \mathbf{K}_k\mathbf{P}_{k,zz}\mathbf{K}'_k$$

the guidance law and the target's position. It implies that the missile state is subject to an implicit constraint imposed by the guidance law with deterministic destinations and uncertain trajectory shapes, referred to as a guidance constraint. As discussed before, the implicit constraints contain valuable information about target states. When guidance law parameters and destination information are known, systematically extracting and integrating such constraint knowledge into tracking systems can enhance performance.

In [30], the constrained tracking problem focusing specifically on proportional navigation guidance constraints in three-dimensional space is investigated. The study develops constraint models to accurately describe guidance constraints with non-fixed trajectory shapes under proportional navigation, along with corresponding tracking methodologies.

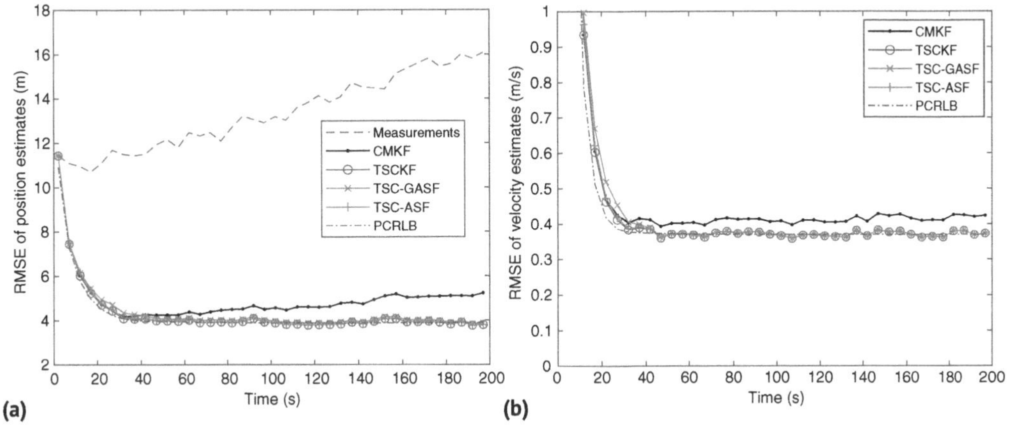

Figure 6.4 RMSE comparison of filters in the case with straight shape trajectory. (a) RMSE of position estimates and (b) RMSE of velocity estimates.

6.4.1 Modeling of Guidance Law Constraints

1. Model with Pure Proportional Navigation (PPN)

Figure 6.5 shows the geometry of a missile M flying with velocity v_M towards a stationary target T located at (x_d, y_d, z_d). According to the definition of PPN [45], the following equations can be written in the relative coordinate as

$$\dot{r} = -v_M \cos(\theta - \phi), \tag{6.51}$$

$$r\dot{\theta} = v_M \sin(\theta - \phi), \tag{6.52}$$

$$a_M = v_M\dot{\phi} = N v_M \dot{\theta}, \tag{6.53}$$

where r represents the range between the missile and target, θ stands for the LOS angle, ϕ denotes the heading angle, $\dot{\phi}$, $\dot{\theta}$ and $\dot{r}$ are corresponding rates, a_M is the commanded acceleration, N denotes the navigation constant.

According to (6.53), the following relationship can be obtained

$$\dot{\phi} = N\dot{\theta}, \tag{6.54}$$

which corresponds to the original definition of the law relating the rotating rate of the missile to the LOS angular rate through a proportional relationship with the navigation constant. Since the relationship between the two rates is linear, it can be directly extended to obtain the relationship between the changes of angles by integrating both sides of (6.54) over time, resulting in a new formulation given by

$$\Delta\phi = N\Delta\theta. \tag{6.55}$$

For missile states at arbitrary two time steps, the above constraint is satisfied, which is equivalent to the formulation given by

$$\phi_{k-1} - \phi_k = N(\theta_{k-1} - \theta_k). \tag{6.56}$$

Let $\mathbf{r}_k = [x_d - x_k, y_d - y_k, z_d - z_k]$ be the vector of LOS from M to T at time step k, the change of LOS angle over one sampling interval can be written as

$$\theta_k - \theta_{k-1} = \arccos\left(\frac{\mathbf{r}_{k-1} \cdot \mathbf{r}_k'}{\| \mathbf{r}_{k-1} \| \cdot \| \mathbf{r}_k \|}\right) \tag{6.57}$$

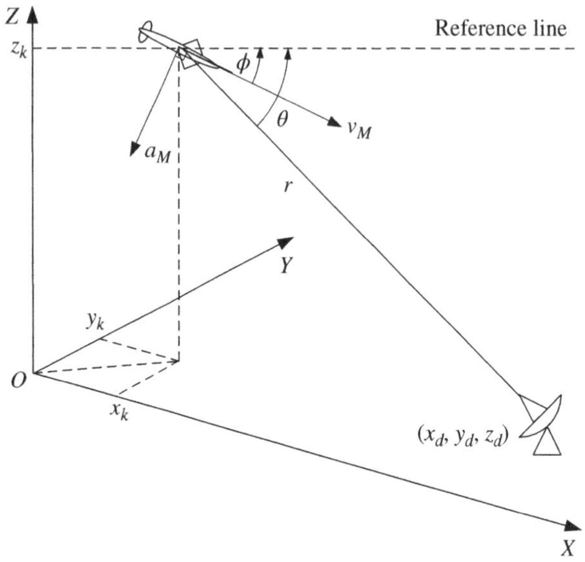

Figure 6.5 Geometry of PPN.

Let $\mathbf{v}_k^m = [\dot{x}_k, \dot{y}_k, \dot{z}_k]$ be the velocity vector of missile at time step k, the change of velocity direction over one sampling interval can be calculated in the same way and written as

$$\phi_k - \phi_{k-1} = \arccos\left(\frac{\mathbf{v}_{k-1}^m \cdot (\mathbf{v}_k^m)'}{\| \mathbf{v}_{k-1}^m \| \cdot \| \mathbf{v}_k^m \|}\right) \tag{6.58}$$

Substitute (6.57) and (6.58) into (6.56), the constraint model with PPN in 3-D space is given by

$$\arccos\left(\frac{\mathbf{v}_{k-1}^m \cdot (\mathbf{v}_k^m)'}{\| \mathbf{v}_{k-1}^m \| \cdot \| \mathbf{v}_k^m \|}\right) = N \arccos\left(\frac{\mathbf{r}_{k-1} \cdot \mathbf{r}_k'}{\| \mathbf{r}_{k-1} \| \cdot \| \mathbf{r}_k \|}\right) \tag{6.59}$$

It can be seen that the destination position (x_d, y_d, z_d), the navigation constant N and the state components, including position (x_k, y_k, z_k) and velocity components $(\dot{x}_k, \dot{y}_k, \dot{z}_k)$ at two adjacent time steps, are used to describe the constraint relationship.

2. Model with True Proportional Navigation (TPN)

In TPN, the direction of the commanded accelerations is normal to the LOS [46]. The magnitude of the acceleration is proportional to the product of the closing velocity and the LOS angular rate. Compared with the PPN, the TPN is less "natural" since forward acceleration/deceleration is required. Meanwhile, the TPN is more tractable in mathematics since it is LOS-referenced. The geometry of TPN in 3-D space is shown in Figure 6.6. According to the definition and geometry, TPN equations can be written as follows:

$$\dot{r} = -v_M \cos(\theta - \phi), \tag{6.60}$$

$$r\dot{\theta} = v_M \sin(\theta - \phi), \tag{6.61}$$

$$a_M = -N\dot{r}\dot{\theta}. \tag{6.62}$$

Substitute (6.60) into (6.62), the commanded acceleration is calculated as

$$a_M = N v_M \cos(\theta - \phi)\dot{\theta}. \tag{6.63}$$

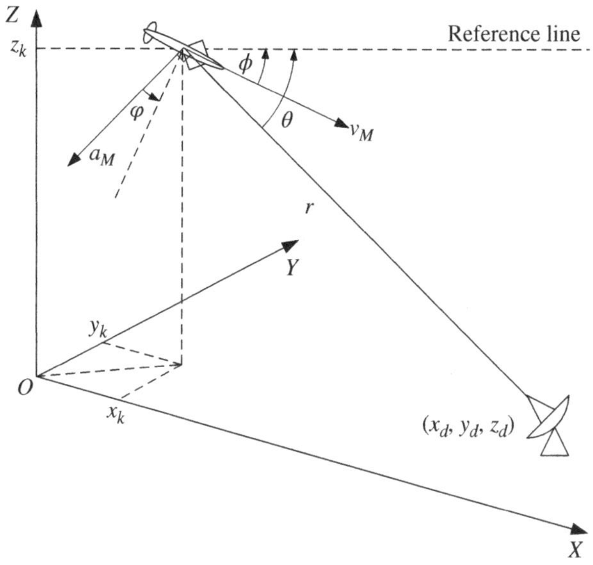

Figure 6.6 Geometry of TPN.

The component of the commanded acceleration in the normal direction of velocity is

$$a_{\perp} = a_M \cos(\varphi) = v_M \dot{\phi}, \tag{6.64}$$

where $\varphi = \theta - \phi$ is the angle between the acceleration direction and the normal direction of the LOS, which is equal to the angle between the heading and the direction of LOS.

Substitute (6.63) into (6.64), the relationship between the turning rate of missile and the LOS angular rate is given by

$$N \cos^2(\varphi)\dot{\theta} = \dot{\phi}. \tag{6.65}$$

The variation of angle φ is usually slight over one small sampling interval. Therefore, φ can be approximately treated as a constant. Integrating the two sides in (6.65) over one sampling interval, the constraint relationship can be expressed as

$$\Delta\phi = N \cos^2(\varphi)\Delta\theta. \tag{6.66}$$

Furthermore, the constraint relationship between states at two successive time steps can be described as

$$\phi_{k-1} - \phi_k = N \cos^2(\varphi_{k-1})(\theta_{k-1} - \theta_k), \tag{6.67}$$

where the value of φ over the sampling interval is approximated by φ_{k-1}.

The variations of ϕ and θ from time step $k-1$ to k are the same as those in (6.57) and (6.58). The approximate value φ_{k-1} can be calculated as

$$\varphi_{k-1} = \theta_{k-1} - \phi_{k-1} = \arccos\left(\frac{\mathbf{r}_{k-1} \cdot (\mathbf{v}_{k-1}^m)'}{\| \mathbf{r}_{k-1} \| \cdot \| \mathbf{v}_{k-1}^m \|} \right) \tag{6.68}$$

Substitute (6.57), (6.58) and (6.68) into (6.67), the constraint model for TPN is given by

$$\arccos\left(\frac{\mathbf{v}_{k-1}^m \cdot (\mathbf{v}_k^m)'}{\|\mathbf{v}_{k-1}^m\| \cdot \|\mathbf{v}_k^m\|}\right) = N\left(\frac{\mathbf{r}_{k-1} \cdot (\mathbf{v}_{k-1}^m)'}{\|\mathbf{r}_{k-1}\| \cdot \|\mathbf{v}_{k-1}^m\|}\right)^2 \arccos\left(\frac{\mathbf{r}_{k-1} \cdot \mathbf{r}_k'}{\|\mathbf{r}_{k-1}\| \cdot \|\mathbf{r}_k\|}\right) \tag{6.69}$$

3. Model with Generalized Proportional Navigation (GPN)
In the GPN guidance law, the direction of commanded accelerations is not necessarily normal to the LOS or the velocity direction [47]. There is a fixed bias angle η between the acceleration direction and the normal direction of the LOS. The acceleration magnitude is the same as that in the TPN. When the closing velocity is small, it is proven that the capture region of GPN is larger than that of TPN and the interception time is shorter. The geometry of GPN in 3-D space is shown in Figure 6.7. According to the definition, GPN equations can be written as follows:

$$\dot{r} = -v_M \cos(\varphi - \eta), \tag{6.70}$$

$$r\dot{\theta} = v_M \sin(\varphi - \eta), \tag{6.71}$$

$$a_M = -N\dot{r}\dot{\theta}. \tag{6.72}$$

Substitute (6.70) into (6.72), the commanded acceleration is calculated as

$$a_M = Nv_M \cos(\varphi - \eta)\dot{\theta}. \tag{6.73}$$

The component of commanded acceleration in the normal direction of velocity is given by

$$a_\perp = a_M \cos(\varphi - \eta) = v_M\dot{\phi}. \tag{6.74}$$

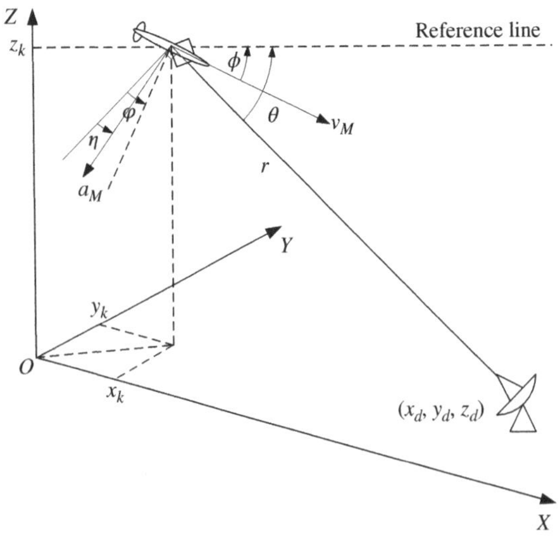

Figure 6.7 Geometry of GPN.

Substitute (6.73) into (6.74), the relationship between the turning rate of missile and the LOS angular rate is given by

$$N\cos^2(\varphi-\eta)\dot{\theta}=\dot{\phi}. \tag{6.75}$$

Treating φ approximately as a constant and integrating the two sides in (6.75) over time, the constraint relationship can be expressed as

$$\Delta\phi=N\cos^2(\varphi-\eta)\Delta\theta. \tag{6.76}$$

The constraint relationship between states at two successive time steps is given by

$$\phi_{k-1}-\phi_k=N\cos^2(\theta_{k-1}-\phi_{k-1}-\eta)(\theta_{k-1}-\theta_k). \tag{6.77}$$

Substituting (6.57), (6.58), (6.68) into (6.77), the constraint model is obtained as

$$\begin{aligned}\arccos\left(\frac{\mathbf{v}_{k-1}^m\cdot(\mathbf{v}_k^m)'}{\|\mathbf{v}_{k-1}^m\|\cdot\|\mathbf{v}_k^m\|}\right)&=N\cos^2\left(\arccos\left(\frac{\mathbf{r}_{k-1}\cdot(\mathbf{v}_{k-1}^m)'}{\|\mathbf{r}_{k-1}\|\cdot\|\mathbf{v}_{k-1}^m\|}\right)-\eta\right)\\&\quad\times\arccos\left(\frac{\mathbf{r}_{k-1}\cdot\mathbf{r}_k'}{\|\mathbf{r}_{k-1}\|\cdot\|\mathbf{r}_k\|}\right)\end{aligned} \tag{6.78}$$

In practical applications, the bias angle η is usually not available. To utilize the constraint model, the unknown η is augmented into the state vector as a state component η_k, resulting in a realistic constraint model given by

$$\begin{aligned}\arccos\left(\frac{\mathbf{v}_{k-1}^m\cdot(\mathbf{v}_k^m)'}{\|\mathbf{v}_{k-1}^m\|\cdot\|\mathbf{v}_k^m\|}\right)&=N\cos^2\left(\arccos\left(\frac{\mathbf{r}_{k-1}\cdot(\mathbf{v}_{k-1}^m)'}{\|\mathbf{r}_{k-1}\|\cdot\|\mathbf{v}_{k-1}^m\|}\right)-\eta_k\right)\\&\quad\times\arccos\left(\frac{\mathbf{r}_{k-1}\cdot\mathbf{r}_k'}{\|\mathbf{r}_{k-1}\|\cdot\|\mathbf{r}_k\|}\right)\end{aligned} \tag{6.79}$$

When the destination information is noisy, the distributions of the position components is assumed to be

$$x_d^m\sim\mathcal{N}(x_d,\sigma_x^2),\qquad y_d^m\sim\mathcal{N}(y_d,\sigma_y^2),\qquad z_d^m\sim\mathcal{N}(z_d,\sigma_z^2), \tag{6.80}$$

where (x_d^m,y_d^m,z_d^m) is the destination position with noises, σ_x^2, σ_y^2, and σ_z^2 represent the known destination noise variances.

To avoid the effect of destination noises and navigation constant bias on the constraint modeling accuracy, these parameters are augmented into the state vector to be estimated as state components. The true destination position (x_d,y_d,z_d) and navigation constant N are replaced by $(x_{d,k},y_{d,k},z_{d,k})$ and n_k in the constraint models.

Taking the guidance constraint model in (6.81) as an example, the corresponding parameter-augmentation form model is

$$\arccos\left(\frac{\mathbf{v}_{k-1}^m\cdot(\mathbf{v}_k^m)'}{\|\mathbf{v}_{k-1}^m\|\cdot\|\mathbf{v}_k^m\|}\right)=n_k\arccos\left(\frac{\bar{\mathbf{r}}_{k-1}\cdot\bar{\mathbf{r}}_k'}{\|\bar{\mathbf{r}}_{k-1}\|\cdot\|\bar{\mathbf{r}}_k\|}\right) \tag{6.81}$$

where $\bar{\mathbf{r}}_k=[x_{d,k}-x_k,y_{d,k}-y_k,z_{d,k}-z_k]$, $\bar{\mathbf{r}}_{k-1}=[x_{d,k}-x_{k-1},y_{d,k}-y_{k-1},z_{d,k}-z_{k-1}]$.

6.4.2 Filtering Algorithms

1. State equation

In the guidance constraints, the previous state, destination position (when the prior information is noisy), navigation constant and biased angle (when the prior information is not available) are augmented into the state vector, resulting in a series of augmented state vector:

$$\mathbf{x}_k^{\text{an}} = \begin{bmatrix} \mathbf{x}_k' & \mathbf{x}_{k-1}' & n_k \end{bmatrix}', \tag{6.82}$$

$$\mathbf{x}_k^{\text{ad}} = \begin{bmatrix} \mathbf{x}_k' & \mathbf{x}_{k-1}' & x_{d,k} & y_{d,k} & z_{d,k} \end{bmatrix}', \tag{6.83}$$

$$\mathbf{x}_k^{\text{adn}} = \begin{bmatrix} \mathbf{x}_k' & \mathbf{x}_{k-1}' & x_{d,k} & y_{d,k} & z_{d,k} & n_k \end{bmatrix}', \tag{6.84}$$

$$\mathbf{x}_k^{\text{adne}} = \begin{bmatrix} \mathbf{x}_k' & \mathbf{x}_{k-1}' & x_{d,k} & y_{d,k} & z_{d,k} & n_k & \eta_k \end{bmatrix}'. \tag{6.85}$$

Here, the destination position and navigation constant are assumed to be fixed. Taking $\mathbf{x}_k^{\text{adn}}$ as an example, the corresponding state equation can be written as

$$\mathbf{x}_{k+1}^{\text{adn}} = \mathbf{f}_k^{\text{adn}}(\mathbf{x}_k^{\text{adn}}) + \Gamma_k^{\text{adn}}\mathbf{v}_k, \tag{6.86}$$

where

$$\mathbf{f}_k^{\text{adn}}(\mathbf{x}_k^{\text{adn}}) = \begin{bmatrix} \mathbf{f}_k(\mathbf{x}_k) \\ \mathbf{x}_k \\ x_{d,k} \\ y_{d,k} \\ z_{d,k} \\ n_k \end{bmatrix}, \quad \Gamma_k^{\text{adn}} = \begin{bmatrix} \Gamma_k \\ \mathbf{0}_{(n+4)\times 3} \end{bmatrix}. \tag{6.87}$$

The state equation corresponding to $\mathbf{x}_{k+1}^{\text{an}}$, $\mathbf{x}_{k+1}^{\text{ad}}$, and $\mathbf{x}_{k+1}^{\text{adne}}$ can be formulated in the same manner, where the dimensions of matrices match the dimensions of state vector.

2. Measurement equation

Pseudo measurements are constructed to incorporate the guidance constraint information into the filtering procedure. When the destination information is noisy and the navigation constant is unknown, the constraint model with PPN is expressed in (6.88). The corresponding pseudo measurement is given by

$$\mu_k = n_k \arccos\left(\frac{\bar{\mathbf{r}}_{k-1} \cdot \bar{\mathbf{r}}_k'}{\| \bar{\mathbf{r}}_{k-1} \| \cdot \| \bar{\mathbf{r}}_k \|} \right) - \arccos\left(\frac{\mathbf{v}_{k-1}^m \cdot (\mathbf{v}_k^m)'}{\| \mathbf{v}_{k-1}^m \| \cdot \| \mathbf{v}_k^m \|} \right) = 0 \tag{6.88}$$

While the pseudo measurement corresponding to the TPN guidance constraint model is formulated as

$$\rho_k = n_k \left(\frac{\bar{\mathbf{r}}_{k-1} \cdot (\mathbf{v}_{k-1}^m)'}{\| \bar{\mathbf{r}}_{k-1} \| \cdot \| \mathbf{v}_{k-1}^m \|} \right)^2 \arccos\left(\frac{\bar{\mathbf{r}}_{k-1} \cdot \bar{\mathbf{r}}_k'}{\| \bar{\mathbf{r}}_{k-1} \| \cdot \| \bar{\mathbf{r}}_k \|} \right) - \arccos\left(\frac{\mathbf{v}_{k-1}^m \cdot (\mathbf{v}_k^m)'}{\| \mathbf{v}_{k-1}^m \| \cdot \| \mathbf{v}_k^m \|} \right) = 0 \tag{6.89}$$

Under the same condition, the pseudo measurement corresponding to the GPN guidance constraint model is

$$\begin{aligned}\epsilon_k &= n_k \cos^2\left(\arccos\left(\frac{\bar{\mathbf{r}}_{k-1} \cdot (\mathbf{v}_{k-1}^m)'}{\|\bar{\mathbf{r}}_{k-1}\| \cdot \|\mathbf{v}_{k-1}^m\|}\right) - \eta_{k-1}\right) \\ &\times \arccos\left(\frac{\bar{\mathbf{r}}_{k-1} \cdot \bar{\mathbf{r}}_k'}{\|\bar{\mathbf{r}}_{k-1}\| \cdot \|\bar{\mathbf{r}}_k\|}\right) - \arccos\left(\frac{\mathbf{v}_{k-1}^m \cdot (\mathbf{v}_k^m)'}{\|\mathbf{v}_{k-1}^m\| \cdot \|\mathbf{v}_k^m\|}\right) = 0\end{aligned} \tag{6.90}$$

The 3-D radar reports range measurement r_k^m, azimuth measurement a_k^m, and elevation measurement e_k^m. The measurement equation is given by

$$\mathbf{z}_k = \begin{bmatrix} r_k^m \\ a_k^m \\ e_k^m \end{bmatrix} = \mathbf{h}_k(\mathbf{x}_k) + \mathbf{w}_k = \begin{bmatrix} \sqrt{x_k^2 + y_k^2 + z_k^2} \\ \tan^{-1}(y_k, x_k) \\ \tan^{-1}(z_k, \sqrt{x_k^2 + y_k^2}) \end{bmatrix} + \begin{bmatrix} w_k^r \\ w_k^a \\ w_k^e \end{bmatrix}, \tag{6.91}$$

where w_k^r, w_k^a, and w_k^e are measurement noises corresponding to range, azimuth, and elevation measurements, assuming to be mutually independent zero-mean white Gaussian noises.

The measurement equation corresponding to the pseudo measurement in (6.88) is

$$\begin{aligned}\mathbf{z}_k^{\text{PPN}} &= \mu_k = \mathbf{h}_k^{\text{PPN}}(\mathbf{x}_k^{\text{adn}}) \\ &= n_k \arccos\left(\frac{\bar{\mathbf{r}}_{k-1} \cdot \bar{\mathbf{r}}_k'}{\|\bar{\mathbf{r}}_{k-1}\| \cdot \|\bar{\mathbf{r}}_k\|}\right) - \arccos\left(\frac{\mathbf{v}_{k-1}^m \cdot (\mathbf{v}_k^m)'}{\|\mathbf{v}_{k-1}^m\| \cdot \|\mathbf{v}_k^m\|}\right)\end{aligned} \tag{6.92}$$

3. Constrained filtering

The nearly constant acceleration (NCA) model is adopted to approximately describe the target motion. The radar measurements and pseudo measurements are processed sequentially in the filtering procedure to enhance estimation accuracy and reduce computational load. The CMKF is employed to handle the radar measurements while the CKF is employed to deal with nonlinearity relationship between the augmented state vector and pseudo measurements.

a. Filter initialization

The filter initialization is divided into several parts. First, the TPD method [38] is used to initialize the original state using converted measurements [41, 42] at the first two time steps. Then, the single-point initialization method [48] is used to initialize the acceleration and unknown navigation constant and biased angle components. The state components at the previous time step are initialized as zero values with zero covariances. Finally, the destination position components are initialized using the noisy prior information.

b. Filter procedure

During the filtering procedure, the 3-D radar measurements in (6.91) are converted into the Cartesian coordinates using the modified unbiased converted measurement method [41]. The converted measurement equation is given by

$$\mathbf{z}_k^c = \begin{bmatrix} x_k^m \\ y_k^m \\ z_k^m \end{bmatrix} = \mathbf{H}_k^c \mathbf{x}_k^{\text{adn}} + \mathbf{w}_k^c, \tag{6.93}$$

Algorithm 6.3: A recursion of the GC-ASF using radar and pseudo measurements

Require: $\hat{\mathbf{x}}^{adn}_{k-1|k-1}$, $\mathbf{P}^{adn}_{k-1|k-1}$, $\mathbf{z}^c_k$, $\mathbf{R}^c_k$

Ensure: $\hat{\mathbf{x}}^{adn}_{k|k}$, $\mathbf{P}^{adn}_{k|k}$.

1: State Prediction

$$\hat{\mathbf{x}}^u_{k|k-1} = \Phi^{adn}_{k-1}\hat{\mathbf{x}}^{adn}_{k-1|k-1}$$

$$\mathbf{P}^u_{k|k-1} = \Phi^{adn}_{k-1}\mathbf{P}^{adn}_{k-1|k-1}(\Phi^{adn}_{k-1})' + \mathbf{Q}^{adn}_{k-1}$$

2: State Update Using Radar Measurement

(1) Measurement Prediction

$$\hat{\mathbf{z}}^c_{k|k-1} = \mathbf{H}^c_k\hat{\mathbf{x}}^u_{k|k-1}$$

$$\mathbf{S}_k = \mathbf{H}^c_k\mathbf{P}^u_{k|k-1}(\mathbf{H}^c_k)' + \mathbf{R}^c_k$$

(2) Filter Gain Calculation

$$\mathbf{K}^u_k = \mathbf{P}^u_k(\mathbf{H}^c_k)'\mathbf{S}^{-1}_k$$

(3) State Update

$$\hat{\mathbf{x}}^u_{k|k} = \hat{\mathbf{x}}^u_{k|k-1} + \mathbf{K}^u_k(\mathbf{z}^c_k - \hat{\mathbf{z}}^c_{k|k-1})$$

$$\mathbf{P}^u_{k|k} = \mathbf{P}^u_{k|k-1} - \mathbf{K}^u_k\mathbf{S}_k(\mathbf{K}^u_k)'$$

3: State Update Using Pseudo Measurement

(1) Measurement Prediction

$$\mathbf{P}^u_{k|k} = \mathbf{D}_{k|k}(\mathbf{D}_{k|k})', \hat{\mathbf{x}}^u_{k|k,i} = \mathbf{D}_{k|k}\gamma_i + \hat{\mathbf{x}}^u_{k|k}$$

$$\hat{\mathbf{z}}^p_{k|k,i} = \mathbf{h}^{\text{PPN}}_k(\hat{\mathbf{x}}^u_{k|k,i}), \hat{\mathbf{x}}^u_{k|k,i} = \mathbf{D}_{k|k}\gamma_i + \hat{\mathbf{x}}^u_{k|k}$$

$$\mathbf{P}^p_{k,zz} = \frac{1}{2n_1}\sum_{i=1}^{2n_1}\hat{\mathbf{z}}^p_{k|k,i}(\hat{\mathbf{z}}^p_{k|k,i})' - \hat{\mathbf{z}}^p_{k|k}(\hat{\mathbf{z}}^p_{k|k})'$$

(2) Filter Gain Calculation

$$\mathbf{P}^p_{k,xz} = \frac{1}{2n_1}\sum_{i=1}^{2n_1}\hat{\mathbf{x}}^u_{k|k,i}(\hat{\mathbf{z}}^p_{k|k,i})' - \hat{\mathbf{x}}^u_{k|k}(\hat{\mathbf{z}}^p_{k|k})',$$

$$\mathbf{K}_k = \mathbf{P}^p_{k,xz}(\mathbf{P}^p_{k,zz})^{-1}.$$

(3) State Update

$$\hat{\mathbf{x}}^{\text{adn}}_{k|k} = \hat{\mathbf{x}}^u_{k|k} + \mathbf{K}_k(\mathbf{z}^{\text{PPN}}_k - \hat{\mathbf{z}}^p_{k|k}),$$

$$\mathbf{P}^{\text{adn}}_{k|k} = \mathbf{P}^u_{k|k} - \mathbf{K}_k\mathbf{P}^p_{k,zz}\mathbf{K}'_k.$$

where

$$\mathbf{H}_k^c = \begin{bmatrix} 1 & 0 & 0 & 0 & 0 & 0 & 0 & 0 & 0 & \mathbf{0}_{1\times(n+4)} \\ 0 & 0 & 0 & 1 & 0 & 0 & 0 & 0 & 0 & \mathbf{0}_{1\times(n+4)} \\ 0 & 0 & 0 & 0 & 0 & 0 & 1 & 0 & 0 & \mathbf{0}_{1\times(n+4)} \end{bmatrix}. \tag{6.94}$$

x_k^m, y_k^m, and z_k^m are the converted position measurements; $\mathbf{w}_k^c$ is the converted measurement noise vector.

Based on the augmented state equation in (6.86), the converted measurement equation in (6.93), and the pseudo measurement equation in (6.92), a recursion of the GC-ASF is shown in Algorithm 6.3, where the generating of sampling points in the CKF is in the same manner as that in Algorithm 6.2.

6.4.3 Simulation Results

Simulation results are provided to show performance of the constraint models and filters introduced in this section. The simulation scenario is similar to that in [30]. The algorithms include the unconstrained CMKF, the IMM method using NCV and NCA models, and GC-ASF. The RMSEs of position and velocity estimates are shown in Figure 6.8.

It can be seen that constrained estimation algorithms outperform unconstrained approaches, i.e., the CMKF and IMM methods, due to the incorporation of guidance constraints. The significant performance improvement validates the effectiveness of constrained estimation methods in extracting and utilizing the guidance and noisy destination information to mitigate the effect of destination noise and enhance estimation accuracy. Furthermore, the RMSE of constrained estimation methods approaches the PCRLB, demonstrating their superior performance. The fluctuation observed in the velocity error curves may have arisen from the dynamic model mismatch and the strong nonlinearity in pseudo measurement formulations.

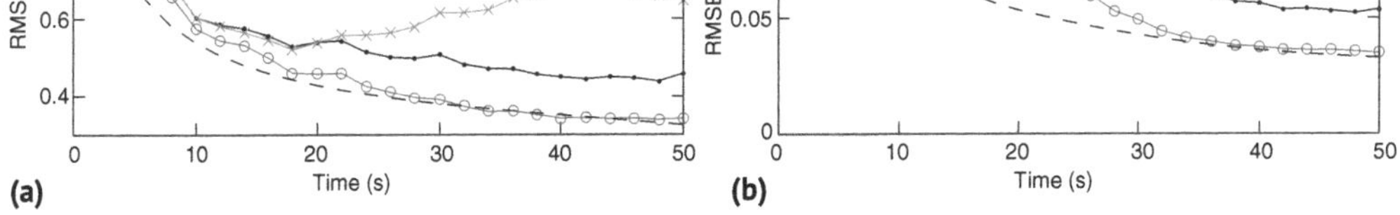

Figure 6.8 RMSE comparison of filters in the case with PPN guidance law. (a) RMSE of position estimates and (b) RMSE of velocity estimates.

6.5 Conclusions

This chapter explores the modeling of implicit constraints and the incorporation of constraints into target tracking frameworks to enhance estimation accuracy. The models of three classes of implicit constraints, including the destination constraints, the trajectory shape constraints, and the guidance constraints, are derived based on partial information about target trajectories. These implicit constraints all contain extra prior information about the target state and thus are valuable for state estimation. The investigation of such constraints enables broader utilization of constraint information to enhance estimation performance, demonstrating significant application potential.

From the introduction of implicit constraint models, it is revealed that the state augmentation is an effective way to describe constraint relationships. By augmenting previous states into the state vector, the constraint information implicitly contained in the state evolution imposed by the fixed trajectory shape or guidance law can be expressed. By augmenting the trajectory parameters into the state vector, the implicit constraint relationships can be described using augmented state components, which significantly enhances the robustness of constraint models in conditions with unreliable or biased prior information. In this case, the parameters are estimated along with the state vector, which can be used as prior information in the subsequent tracking with the same constraint.

For the incorporation of constraints, the pseudo measurement method is employed to conduct the constrained filter in this chapter. The equality constraint models can be easily transformed to pseudo measurements and augmented into the measurement vector, transforming the constrained filtering problem to an unconstrained one. It appears from the results that incorporating the pseudo-measurements into the UKF/CKF is an effective way to utilize the constraints, leading to superior performance over the unconstrained estimation methods. The developed algorithms for the straight/circular trajectory or PN guidance law have the potential to be extended to address constrained estimation problems with more complex trajectory shapes or advanced guidance laws with analytical expressions.

The future work will consider the modeling of the variation of trajectory shape, such as the stage transformation and maneuvering in ballistic target tracking. The state augmentation method will be employed to describe the shape variation. The association of implicit constraints with targets in multiple target tracking is also an open problem that deserves investigation. New principles and metrics will be defined to determine the association relationships.

References

1 S. J. Julier and J. J. LaViola. On kalman filtering with nonlinear equality constraints. *IEEE Transactions on Signal Processing*, 55(6):2774–2784, 2007.

2 C. Yang and E. Blasch. Kalman filtering with nonlinear state constraints. *IEEE Transactions on Aerospace and Electronic Systems*, 45(1):70–84, 2009.

3 J. Lopez-Araquistain, A. J. Jarama, J. A. Besada, G. de Miguel, and J. R. Casar. A new approach to map-assisted Bayesian tracking filtering. *Information Fusion*, 45:79–95,2019

4 D. Song, R. Tharmarasa, T. Kirubarajan, and X. N. Fernando. Multi-vehicle tracking with road maps and car-following models. *IEEE Transactions on Intelligent Transportation Systems*, 19(5):1375–1386, 2018.

5 B. O. S. Teixeira, M. A. Santillo, R. S. Erwin, and D. S. Bernstein. Spacecraft tracking using sampled-data kalman filters. *IEEE Control Systems Magazine*, 28(4):78–94, 2008.

6 M. Rotea and C. Lana. State estimation with probability constraints. *International Journal of Control*, 81(6):920–930, 2008.

7 L. Xu, Y. Liang, Z. Duan, and G. Zhou. Route-based dynamics modeling and tracking with application to air traffic surveillance. *IEEE Transactions on Intelligent Transportation Systems*, 21(1):1–13, 2020.

8 G. Battistello and M. Ulmke. Exploitation of a-priori information for tracking maritime intermittent data sources. In *14th International Conference on Information Fusion*, pages 1–8, 2011.

9 T. Kirubarajan, Y. Bar-Shalom, K. R. Pattipati, and I. Kadar. Ground target tracking with variable structure IMM estimator. *IEEE Transactions on Aerospace and Electronic Systems*, 36(1):26–46, 2000.

10 E. Sviestins and V. Pirard. Constraint aware dynamics in target tracking. In *Proceedings of the 16th International Conference on Information Fusion*, pages 572–579, 2013.

11 B. I. Ahmad, S. J. Godsill, V. Krishnamurthy P. Willett, and M. Rangaswamy. Foreword to the special section on meta-level and adversarial tracking. *IEEE Transactions on Aerospace and Electronic Systems*, 57(4):1994–1995, 2021.

12 Z. Duan and X. R. Li. Analysis, design and estimation of linear equality constrained dynamic systems. *IEEE Transactions on Aerospace and Electronic Systems*, 51(4):2723–2746, 2015.

13 S. Ko and R. R. Bitmead. State estimation for linear systems with state equality constraints. *Automatica*, 43(8):1363–1368, 2007.

14 L. Xu, X. R. Li, Y. Liang, and Z. Duan. Constrained dynamic systems: Generalized modeling and state estimation. *IEEE Transactions on Aerospace and Electronic Systems*, 53(5): 2594–2609,2017

15 R. Gan, B. I. Ahmad, and S. J. Godsill. Lévy state-space models for tracking and intent prediction of highly maneuverable objects. *IEEE Transactions on Aerospace and Electronic Systems*, 57(4):2021–2038, 2021.

16 R. Rezaie, X. R. Li, and V. P. Jilkov. Conditionally Markov modeling and optimal estimation for trajectory with waypoints and destination. *IEEE Transactions on Aerospace and Electronic Systems*, 57(4):2006–2020, 2021.

17 L. Xu, Z. Hou, M. Mallick, and Y. Fang. Intention-aware motion modeling using GP priors with conditional kernels. In *Proceedings of the 25th International Conference on Information Fusion*, pages 1–7, 2022.

18 L. Xu, X. R. Li, Y. Liang, and Z. Duan. Modeling and state estimation of linear destination-constrained dynamic systems. *IEEE Transactions on Signal Processing*, 61(11): 2927–2939,2022

19 L. Xu, X. R. Li, M. Mallick, and Z. Duan. Modeling and state estimation of destination-constrained dynamic systems part II: uncertain arrival time. *IEEE Transactions on Signal Processing*, 73:633–648, 2025.

20 C. Yang, M. Bakich, and E. Blasch. Nonlinear constrained tracking of targets on roads. In *Proceedings of the 8th International Conference on Information Fusion*, 2005.

21 L. Xu, P. Yang, and M. Mallick. Joint target tracking and intent inference using a destination-constrained model. In M. Mallick, B. -N. Vo, and R. Tharmarasa, editors, *Advanced Techniques in Tracking and Sensor Management: Theory and Applications*, Wileys/IEEE, 5, 2025.

22 M. Ulmke and W. Koch. Road-map assisted ground moving target tracking. *IEEE Transactions on Aerospace and Electronic Systems*, 42(4):2594–2609, 2006.

23 Z. Duan and X. R. Li. The role of pseudo measurements in equality-constrained state estimation. *IEEE Transactions on Aerospace and Electronic Systems*, 49(3):1654–1666, 2013.

24 H. Mohammadi, H. Yao, G. Khademi, T. T. Nguyen, D. Simon, and H. Richter, Extended Kalman filtering for state estimation of a hill muscle model. *IET Control Theory and Applications*, 12(3):384–394, 2018.

25 G. Wang, Z. Zhu, C. Yang, L. Ma, W. Dai, and X. Chen. Distributed multi-kernel maximum correntropy state-constrained Kalman filter under deception attacks. *IEEE Transactions on Network Science and Engineering*, 12(1):533–546, 2025.

26 D. Simon. Kalman filtering with state constraints: A survey of linear and nonlinear algorithms. *IET Control Theory and Applications*, 4(8):1303–1318, 2010.

27 L. S. Wang, Y. T. Chiang, and F. R. Chang. Filtering method for nonlinear systems with constraints. *IEE Proceedings – Control Theory and Applications*, 149(6):525–531, 2002.

28 K. Li, T. Kirubarajan, and G. Zhou. State estimation with implicit constraints of circular trajectory using pseudomeasurements. *IEEE Transactions on Aerospace and Electronic Systems*, 56(6):4406–4425, 2020.

29 K. Li and G. Zhou. Robust state estimation with trajectory shape constraints. *IET Radar, Sonar and Navigation*, 14(7):1046–1054, 2020.

30 K. Li and G. Zhou. State estimation with a destination constraint imposed by proportional navigation guidance law. *IEEE Transactions on Aerospace and Electronic Systems*, 58(1):58–73, 2021.

31 G. Zhou, K. Li, T. Kirubarajan, and L. Xu. State estimation with trajectory shape constraints using pseudomeasurements. *IEEE Transactions on Aerospace and Electronic Systems*, 55(5):2395–2407, 2019.

32 G. Zhou, K. Li, and T. Kirubarajan. Constrained state estimation using noisy destination information. *Signal Processing*, 166:107226, 2020.

33 A. K. Rutkowski, M. Czyzewski, A. Witczak, and A. Kawalec. Protection of radar against anti-radiation missile using single electromagnetic decoy. In *Proceedings of the 16th International Radar Symposium*, pages 973–978, 2015.

34 W. Zhou, J. Luo, Y. Jia, and H. Wang. Performance evaluation of radar and decoy system counteracting antiradiation missile. *IEEE Transactions on Aerospace and Electronic Systems*, 47(3):2026–2036, 2011.

35 L. Xu, X. R. Li, Z. Duan, and J. Lan. Modeling and state estimation for dynamic systems with linear equality constraints. *IEEE Transactions on Signal Processing*, 61(11):2927–2939, 2013.

36 G. Zhou, K. Li, X. Chen, L. Wu, and T. Kirubarajan. State estimation with a destination constraint using pseudo-measurements. *Signal Processing*, 145:155–166, 2018.

37 S. J. Julier. The scaled unscented transformation. In *Proceedings of the American Control Conference*, pages 4555–4559, 1999.

38 Y. Bar-Shalom, X. R. Li, and T. Kirubarajan. *Estimation with Applications to Tracking and Navigation: Theory, Algorithms, and Software*. Wiley, New York, 2001.

39 M. S. Arulampalam, S. Maskell, N. Gordon, and T. Clapp, A tutorial on particle filters for online nonlinear/non-gaussian Bayesian tracking. *IEEE Transactions on Signal Processing*, 50(2):174–188, 2002.

40 M. Lei and C. Han. Sequential nonlinear tracking using UKF and raw range-rate measurements. *IEEE Transactions on Aerospace and Electronic Systems*, 43(1):239–250, 2007.

41 Z. Duan, C. Han, and X. R. Li. Comments on 'unbiased converted measurements for tracking'. *IEEE Transactions on Aerospace and Electronic Systems*, 40(4):1374–1376, 2004.

42 M. Longbin, S. Xiaoquan, Z. Yiyu, S. Z. Kang, and Y. Bar-Shalom, Unbiased converted measurements for tracking. *IEEE Transactions on Aerospace and Electronic Systems*, 34(3):1023–1027, 1998.

43 Y. Huang, X. Wang, Y. Guo, and W. An. State estimation with incomplete linear constraint. In *Proceedings of the 20th International Conference on Information Fusion*, pages 1–8, 2017.

44 I. Arasaratnam and S. Haykin. Cubature kalman filters. *IEEE Transactions on Automatic Control*, 54(6):1254–1269, 2009.

45 K. Becker. Closed-form solution of pure proportional navigation. *IEEE Transactions on Aerospace and Electronic Systems*, 26(3):526–533, 1990.

46 K. Li, W. Su, and L. Chen. Performance analysis of realistic true proportional navigation against maneuvering targets using Lyapunov-like approach. *Aerospace Science and Technology*, 69:333–341, 2017.

47 D. Ghose. On the generalization of true proportional navigation. *IEEE Transactions on Aerospace and Electronic Systems*, 30(2):545–555, 1994.

48 M. Mallick and B. F. La Scala. Comparison of single-point and two-point difference track initiation algorithms using position measurements. *Acta Automatica Sinica*, 34(3):258–265, 2008.

7

Particle Filter Convergence: Various Notes

Yvo Boers[1] and Pranab K. Mandal[2]*

[1] *Department of Applied Mathematics, University of Twente, The Netherlands*
[2] *Department of Mathematics and Computer Science, Eindhoven University of Technology, The Netherlands*

7.1 Introduction

Particle filters (PFs) have been developed, studied, and used by both theoreticians and practitioners for quite some time now. Especially since the seminal PF paper [1] from 1993.

PFs have been successfully applied in different fields, topics, and products. These include robotics, target tracking, (radar) track-before-detect, localization and navigation as well as microscopic bio-imaging, in particular, automated cell counting and tracking. See, for example, [2–4].

The celebrated Kalman filter (KF), see [5] and [6], works under the (implicit) assumptions that the system under consideration is linear and all noises are additive and Gaussian. Although various extensions, such as the extended Kalman filter (EKF), the unscented Kalman filter (UKF), and the cubature Kalman filter (CKF), see [7–9], have been developed to handle some nonlinearities, all these Kalman-based filtering methods still assume that the a posteriori density is Gaussian, and thus unimodal. These methods track the unique single mode of the density through a time-varying mean and covariance. In situations where the true a posteriori density resembles a Gaussian, KF type of filters can be sufficient, appropriate and maybe even all that is needed and therefore preferred. However, when a system or problem formulation leads to a non-Gaussian and/or multimodal posterior, these techniques will fail to correctly capture the true density through the estimated mean, the so-called minimum mean squared error (MMSE) estimator, and the corresponding covariance estimates. See [10] for several examples where multimodality appears naturally in the a posteriori density and where KF type of filters are insufficient to capture these densities.

One such situation is tracking a target that moves under (hard) constraints. We provide an example below.

Consider the situation in [11, Chapter 6]. A ship is to be tracked. The ship is moving through the exit of a harbor into a channel, where immediately after the exit the channel splits into two branches. Also, at the onset of the split of the waterway, sensor measurements are missing. See Figure 7.1. Naturally, the movement of the ship is restricted to water only. This has been modeled by posing hard inequality constraints on the position/state of the system. On the other hand, given the (lack of enough) measurements, the a posteriori density of the position of the ship while passing through the split will be bimodal since (initially, after the split) there will be uncertainty

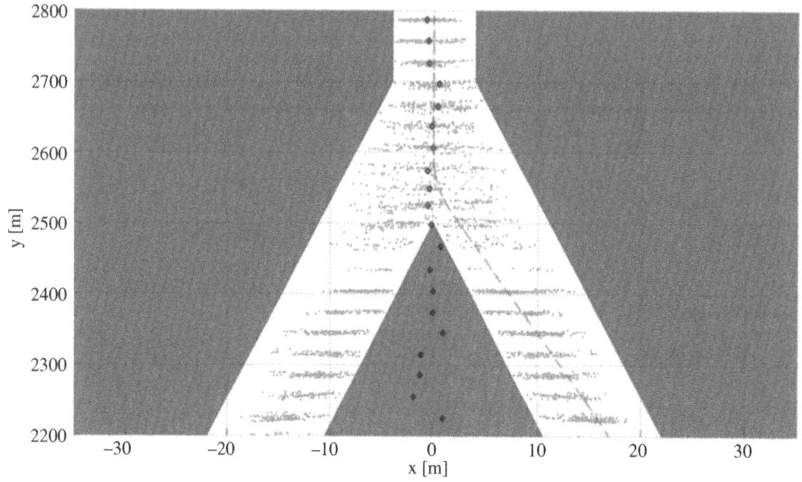

Figure 7.1 Tracking of a ship exiting an harbor (top-side) and going downward into one of the channels: the ground truth (dashed line), PF outputs representing the approximate posterior distribution (smaller dots/particles within the channels), and the (PF-based empirical) mean (larger dots). *Source*: Reproduced from [11] / with permission of Centre for Telematics and Information Technology (CTIT).

about which channel branch it went into. We point out that the PF captures this bimodality quite well, as the particles and thus its probability mass is "distributed" between the two branches of the channel. As a consequence, the (PF-based) MMSE estimator will estimate the location of the ship somewhere in between the channel branches, on land which is a physical impossibility. See the larger dots in Figure 7.1. As also can be seen in Figure 7.1, the PF, given by the smaller dots/particles within the channels, provides an approximate description of the true a posteriori density. Note that this probability density function (PDF) describes the information, using all knowledge available, i.e., both the measurements *and* the constraints. See, e.g., [12] and [13] for a treatment on this matter as well as a proof thereof. In [12], it has been formally shown (proved) that there are two inherent correct and equivalent ways of applying hard inequality constraints to the dynamical system. This can be done in the prediction step or in the update step. It has been proved that the two modeling/implementation choices lead to identical a posteriori pdfs and thus provide the exact same filtering solutions.

An important lesson to be learned from this example is that even if the filter is (approximately) exact, the MMSE output (based on this optimal/exact filter) can still be inappropriate. Thus, a "wrong" outcome here is not due to a flaw in the PF, but it is due to naively using the MMSE estimate as an output, where it is not appropriate. We shall discuss this issue further in this chapter through other examples. See, e.g., Section 7.3.

The Gaussian sum filter (GSF), see [14] and [15], can handle some multi-modalities by approximating the true posterior as a weighted sum of Gaussians. Subsequently, the Gaussian components are estimated through EKF-type of filter equations. In essence, the GSF (and also the EKF, UKF, and CKF) first approximate the true posterior by a (sum of) Gaussian distributions/densities and then provide the estimates of the mean and variances of these Gaussian distributions/densities, calculated as exact solutions of other intermediary equations.

PFs, on the other hand, can accommodate, at least theoretically, any nonlinearity, non-Gaussianity or multimodality without resorting to a Gaussian approximation of the true posterior. A PF provides the posterior directly, by (approximately) providing/calculating the posterior in the form of a set of representative (weighted) samples. In order to generate these samples, called particles, it uses the sequential Monte Carlo technique. These particles are then used, for example in target tracking, to capture and represent the tracks. See [10] for an example of terrain-based navigation and [2] for an example of roadmap aided car positioning.

The straightforward and (mathematically/theoretically) sound interpretation of the PF has contributed to its popularity. Equally important has been the easier/cheaper access to larger computational power in recent times. This has also propelled the popularity of PFs. The PF has become an important "standard" method (complementary to the Kalman-based filters) for nonlinear filtering and applications like target tracking, see also [16] and [17].

It must be stated that even if a PF is used, or for that matter an exact filter, a good choice for obtaining/selecting an appropriate/meaningful point estimate of the state (e.g., object position in the case of target tracking) from the (approximate) posterior is not obvious at all. And the plain vanilla mean, the MMSE estimator oftentimes is not a good choice at all, as we will see in the proceedings.

In the ship example in Figure 7.1, an MMSE estimator (the most popular estimator) obtained from the PF output will show the track to be on land and is therefore not appropriate for this specific application. A maximum a posteriori (MAP) estimate may be a better choice in this case.

A (theoretical) illustration of a bimodal (a posteriori) pdf and the corresponding MMSE and MAP estimators is shown in Figure 7.2. It is obvious that the MMSE estimator is in a low probability area, whereas the MAP is in a high probability area. An algorithm to calculate the PF-based MAP has been provided in [18]. In Section 7.3, we illustrate this issue further by means of a realistic example of labeled target tracking.

The PF represents the posterior density/distribution through a finite set of particles. An *important* question is whether this representation (asymptotically) converges to the true, yet unknown, a posteriori density/distribution, as the number of particles increases. And if it does, in which sense this convergence holds. These questions have been studied and partially answered by various authors in different publications. See, e.g., [19–27].

In this chapter, we will review and discuss convergence results from [24–26]. In addition, we will derive a (new) corollary of some key theorems in [25]. We will also discuss the implications, interpretations and limitations of the different convergence results.

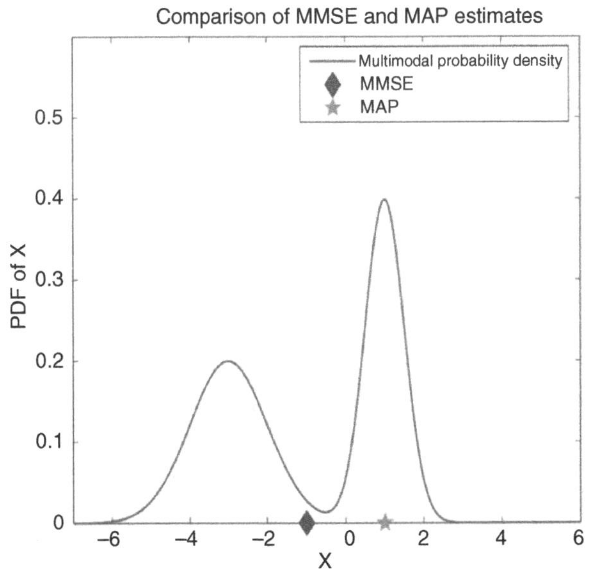

Figure 7.2 MMSE and MAP estimates.

We assert that the convergence results provide more understanding of the theory as well as practical handles for engineers designing, e.g., target tracking algorithms based on PFs, as well as other applications.

Moreover, the presented convergence results have also been instrumental in deriving some new (theoretical) results, see [28], where a PF has been used to calculate/approximate the likelihood ratio (LR), used for a Neyman–Pearson optimal detector, see [29]. It has been proved in [28] that the PF-based LR converges (weakly) to the true, yet unknown, LR. See also Example 7.4 in Section 7.7.

When it comes to convergence, in the context of randomness, the law of large numbers (LLN) comes first to mind, and also, is the easiest to interpret. See, e.g., [30] and [31]. It is about the convergence of the sample average to the population average as the sample size increases. We will show that the convergence results under consideration all relate to certain specific types of LLNs for the PF samples (i.e., the particles). While a specific version of the strong law of large numbers (SLLN) for PFs can be derived straightforwardly from the main result of [24], such an SLLN type result is not immediate from [25]. We will derive a specific version of the weak law of large numbers (WLLN) in this case. See Section 7.5.

The results in [24] and [25] require, as with many stochastic convergence results, that the so-called test functions are continuous and bounded for the convergence of empirical means (based on the PF samples). More precisely, if one is interested in the convergence of the average of the $\phi(x^i)$'s to $E(\phi(x))$, where the x^i's are the samples (particles), the function $\phi(\cdot)$ needs to be bounded. It is worthwhile to note that this boundedness requirement prevents one to formally state the convergence of the PF-based empirical estimate, $\bar{x}_t$, to the true mean of the state, $E(x_t)$. This is because the corresponding function $\phi(x) = x$, the identity function, is unbounded and therefore, some of the convergence results do not apply. We also refer to the Remarks and the Discussion Section in [24]. We note that the authors of [24], in their paper, in the Discussion Section, note that it is disappointing that for convergence results one seems to have to assume boundedness of the test function. They also noted that the research on PF convergence probably would lead to more and different results in the future. As we know now, it indeed did. We also note that the excellent seminal PF convergence paper [24] is 23 years old by now.

The results in [26] provide some relaxation of this boundedness restriction in certain situations. We will come back to this in Section 7.6.

Practitioners/Engineers have already for a long time been "ignoring" this boundedness requirement, knowingly or not. We will show and explain why and how this "practitioners approach" can be justified also theoretically by now. As will be seen, the theoretical results in [26] allow for a specific set of unbounded test functions for which a formal proof of convergence of the PF-based empirical mean to the true mean exists.

The remainder of this chapter is organized as follows. We first provide, in Section 7.2, the general setup of the system, an introduction of the basic terminologies, that we will use throughout the chapter and describe the general PF Algorithm. In Section 7.3, we look into another example elaborating on the potential drawbacks of using the MMSE estimator, as well as some other issues related to the PF. Thereafter, we (re)state the main convergence results from [24] and [25] in Sections 7.4 and 7.5. We describe various interpretations and implications of these results for the (practical) use of PFs. In Section 7.5, we also prove explicitly a (new) corollary to the main results in [25].

In Section 7.6 we formulate and discuss the convergence results from [26]. These results allow to calculate the PF-based empirical means for specific classes of *unbounded* test functions and still have a convergence guarantee of some kind. This is in contrast to results from [24] and [25], but also many other works, where the convergence results are restricted to bounded test functions. We explain why this relaxation to the unbounded test function is especially important for practitioners.

We will prove that systems with additive Gaussian measurement noise, even when the system is nonlinear, satisfy the criteria for which unbounded test functions can be admissible. We provide several explicit examples of these cases in Section 7.7. Finally, we end the chapter with some conclusions.

7.2 Preliminaries

Let us consider the following general nonlinear system. For $t \in \mathbb{N}^+$:

$$x_t \backsim p(x_t \mid x_{t-1}), \quad \leftarrow \text{dynamic model} \tag{7.1}$$

$$y_t \backsim p(y_t \mid x_t), \quad \leftarrow \text{measurement model} \tag{7.2}$$

$$x_0 \backsim p(x_0), \quad \leftarrow \text{initial density} \tag{7.3}$$

where the state $x_t \in \mathcal{S} \subset \mathbb{R}^n$, and the measurement $y_t \in \mathbb{R}^m$. We shall denote by $y_{1,\dots,t}$, or equivalently, $y_{1:t}$, the set of measurements from time 1 to time t. In filtering literature, these are often referred to as the measurement history.

The quantity of interest is the conditional distribution or density of the state at time t, x_t, conditioned on the measurement history up to and including that time. This filtering solution is denoted either by the a posteriori distribution:

$$P(x_t \mid y_{1,\dots,t}), \tag{7.4}$$

or, by the a posteriori density (if it exists):

$$p(x_t \mid y_{1,\dots,t}). \tag{7.5}$$

We do stress the fact that, although expressions (7.4) and/or (7.5) may seem simple, explicit formulas only exist for a very limited number of system classes. The most well-known class is the class of Linear Gaussian Systems. For this class, the a posteriori quantities are also Gaussian with the mean and covariance given by the popular KF equations, see, e.g., [5] and [7]. The PF provides these quantities through a set of weighted particles, representing a sample drawn from the a posteriori distribution. Below, we provide the most commonly used PF, the so-called Sampling Importance Resampling PF (SIR-PF). See, e.g., [7]. There exist other versions of the PF. One alternative, that is also employed in [25], is the so-called Accept–Reject PF (AR-PF). We refer to [7] and [32] for the details and other alternatives.

Algorithm 7.1: ***(SIR-PF)***:

---

1. initialization: *at* $t = 0$ *draw* N *particles* $x_0^i \backsim p(x_0)$ *and set* $w_0^i := \frac{1}{N}$.
2. *if* $t \leq T$, $\forall i \in \{1, 2, \dots, N\}$ *draw particles* $x_t^i \backsim p(x_t \mid x_{t-1}^i)$
3. assign particle weights : $w_t^i := w_{t-1}^i p(y_t \mid x_t^i)$
4. normalize weights : $w_t^i := \dfrac{w_t^i}{\sum_{j=1}^N w_t^j}$
5. *output* : $\{x_t^i, w_t^i\}_{i=1,\dots,N}$
6. *if required, resample with replacement* N *times to arrive at a new set* $\{x_t^i, w_t^i\}_{i=1,\dots,N}$ *with* $w_t^i := \frac{1}{N}$
7. *set* $t := t + 1$
8. *go to 2.*

In Algorithm 7.1, the number of particles, N, is a design parameter to be selected by the user and T is the length of the time interval being considered. We would also like to point out that even though the resampling step (in Step 6 above) is an integral part for the algorithm to work and in particular, to avoid degeneracy of the particles, it nevertheless introduces an extra variability in the particles. This has been (rigorously) proved, see [33]. As a result, intermediate relevant estimates at time t are better derived from the particles in Step 5 of the algorithm rather than using the resampled particles from Step 6. This is also the reason that the SIR-PF Algorithm, specified above, outputs the set of weighted samples/particles from Step 5 (before the resampling at Step 6).

As mentioned before, the samples (particles) of the PF can be regarded as if they had been drawn (approximately) according to the true, yet unknown, a posteriori density/distribution. See, also, the seminal PF paper [1]. Thus, although we do not have an explicit expression for (7.4) or (7.5), the PF provides samples drawn (approximately) according to the *unknown* expressions (7.4) or (7.5). The convergence results of interest, in this chapter, assert that as the sample size N grows, this approximation becomes better and better. Figure 7.3 depicts the visualization of a true a posteriori distribution and its PF representation. We emphasize that this is merely a visualization for illustrative purposes only and does not represent an example of the model of some real physical system. If one is interested in such an example/system, we refer to the ship tracking example at the beginning of this section. In particular, see Figure 7.1.

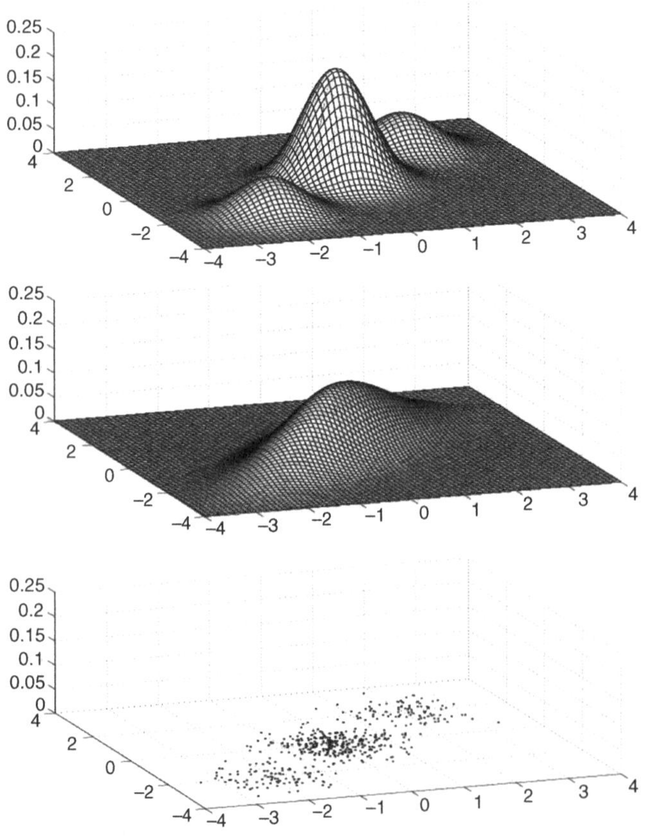

Figure 7.3 Top: true a posteriori density, middle: KF generated density, and bottom: PF samples.

Popular estimators such as the MMSE estimator or the MAP estimator can be calculated based on the PF output. See [18] for the PF-based MAP estimator. As for the MMSE estimator of the state, it is theoretically given by (see, e.g., [5]) the conditional mean:

$$\hat{x}_t^{MMSE} = E_{p(x_t|y_{1,\dots,t})}(x_t). \tag{7.6}$$

Given the PF output at time t, $\{(x_t^i, w_t^i), i = 1, 2, \dots, N\}$, the PF-based MMSE estimator is calculated as:

$$\hat{x}_t^N := \sum_{i=1}^{N} w_t^i x_t^i. \tag{7.7}$$

There is more to say about the convergence, as N grows, of the quantity in (7.7) to the one in (7.6). This is discussed in more detail in Sections 7.4 and 7.6.

First, let us consider a generalized expression, for an arbitrary function ϕ:

$$\hat{\phi}^N(x_t) := \sum_{i=1}^{N} w_t^i \phi(x_t^i). \tag{7.8}$$

In fact, this is an estimator for $E(\phi(x_t)|y_1, \dots, y_t)$ based on the PF output. Notice that when ϕ is the identity function, i.e., $\phi(x) = x$, then this reduces to the PF-based MMSE estimator given by (7.7).

All the convergence results in the following sections will involve the convergence of the (PF-based) empirical expectation of ϕ, given by Eq. (7.8), to the true a posteriori expectation, $E(\phi(x_t)|y_1, \dots, y_t)$. In order to state these results properly, we introduce the following necessary definitions and notations.

Definition 7.1 We denote by $C_b(\mathcal{S})$, the set of continuous and bounded (test) functions defined on the state space, $\mathcal{S}$

Remark 7.1 We note that typically $\mathcal{S} \subset \mathbb{R}^n$.

Definition 7.2 *(Weak convergence)* For a probability distribution $P : \mathcal{S} \to \mathbb{R}^+$ and a sequence of probability distributions $P^N, N \in \mathbb{N}^+$, we say that P^N converges weakly or in distribution to P as $N \to \infty$ if $\forall \phi \in C_b(\mathcal{S})$:

$$\lim_{N\to\infty} \int_{\mathcal{S}} \phi(x) dP^N(x) = \int_{\mathcal{S}} \phi(x) dP(x). \tag{7.9}$$

Remark 7.2 Convergence, as introduced in Definition 7.2, is oftentimes denoted as: $P^N \Rightarrow P$. In probability literature, it is also often referred to as: Convergence in distribution.

We also take a moment to introduce some additional notations to be used in Section 7.6 of this chapter. These notations are from [24] and [26] and follow their exact nomenclature. Although this means introducing two different notations for the same quantity, it helps in comprehending the notations in the key convergence results.

Let $\rho(y_t \mid x_t)$ be the alternate notation for the likelihood density of the system Eq. (7.2). Similarly, let the transition kernel $K(\cdot|\cdot)$ denote the transition density of the system Eq. (7.1). Then the predictive distribution, $\pi_{t|t-1}$, of the state, x_t, at time t using the measurements up to time $(t - 1)$ and the filter distribution, $\pi_{t|t}$, of x_t using all the measurements up to time t, are governed by (see, e.g., [24, Section II] or [26, Section II]) the following *prediction* and *update* equations:

$$\pi_{t|t-1}(dx_t) = \int_{\mathcal{S}} \pi_{t-1|t-1}(dx_{t-1})K(dx_t \mid x_{t-1}), \qquad \text{[Prediction]} \tag{7.10}$$

$$\pi_{t|t}(dx_t) = \frac{\rho(y_t \mid x_t)\pi_{t|t-1}(dx_t)}{\int_{\mathcal{S}} \rho(y_t \mid \tilde{x}_t)\pi_{t|t-1}(d\tilde{x}_t)}. \qquad \text{[Update]} \tag{7.11}$$

Moreover, given a measure ν, a function ϕ and a transition kernel K. We now write/define:

$$(\nu, \phi) := \int_{\mathcal{S}} \phi(x)\nu(dx), \quad \text{and} \quad K\phi(x) := \int K(dz \mid x)\phi(z).$$

Using this notation, the filtering Eqs. (7.10) and (7.11) can be reformulated even more compactly as:

$$(\pi_{t|t-1}\,,\ \phi) = (\pi_{t-1|t-1}\,,\ K\phi), \tag{7.12}$$

$$(\pi_{t|t}\,,\ \phi) = \frac{(\pi_{t|t-1}\,,\ \phi\rho)}{(\pi_{t|t-1}\,,\ \rho)}. \tag{7.13}$$

Before we present the various convergence results for PFs, we consider another example where multimodality appears inherently and how PFs can be useful in this situation.

7.3 Multimodality and the PF

In Section 7.1, we have seen a specific example of tracking a single target (a ship), where the ship moves under certain physical constraints. This leads to multimodality in the posterior. In this section, we investigate another example, this time without any constraint in the movements but in the presence of multiple targets. We analyze and show that in this situation as well, multimodality surfaces inherently. We discuss how PFs can help to improve the tracking.

Suppose two targets are being tracked, using a joint state multi-target filter, where the targets start well separated; subsequently, move in close proximity for a while and then separate again. Such examples have been discussed and analyzed in [18, 34–37]. It has been shown that in this case, a(n) (E)KF-based joint state tracker suffers from so-called track coalescence. The result is that the (labeled) track estimates will end up "on top" of each other, whereas in reality, the true targets are well separated. The main reason is that even after the targets separate, the (E)KF-based tracker continues to provide the MMSE estimator despite the fact that the true a posteriori density is bimodal. See Figure 7.4 for such an illustration, also called mixed labeling. A mathematical illustration/proof of this phenomenon has been provided in [34, Section 5]. We will elaborate on this a little bit more here as it helps understanding this specific example, and also, to illustrate how other PF-based estimators such as MAP can potentially help in this situation.

Let us denote the positions of the two targets as the joint state vector: $(x^{t1}, x^{t2})^T$, where the locations x^{t1} and x^{t1} themselves are vector-valued. Refer, again, to Figure 7.4. The squares represent the first target (the $t1$-part of the joint state vector) and the circles represent the second target ($t2$-part of the joint state). As can be seen, initially (i.e., on the left side of the plot) the labeling is unambiguous. All targets estimated at the top are circles, indicating that the $t2$-labeled partial particles/hypotheses correspond to the same top target and to the x^{t2} part of the joint state. When the targets move into close proximity, ambiguity starts to develop. This ambiguity remains long after the

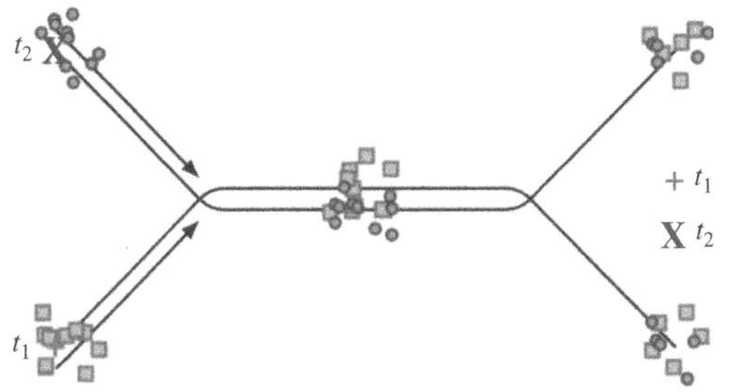

Figure 7.4 Illustration of mixed labeling in a joint two target particle filter. The targets move from the left to the right. The MMSE tracks for the first and the second target are indicated by the squares and circles, respectively.

targets separate again. Some joint particles/hypotheses have the original target $t2$, correctly labeled as $t2$ (circle), whereas some joint particles/hypotheses now have the original $t2$ target labeled as $t1$ (square), and the same for the target 1. Subsequently, a mixed labeling situation appears; see [38]. We can identify this situation with the two modes of a bimodal bivariate density, each mode corresponding to a specific permutation of the two (true) single target states: $(x^{t1}, x^{t2})^T$ or $(x^{t2}, x^{t1})^T$. See also [34]. Generally, when there are M targets, there will be $M!$ modes in the (exact) posterior, where each permutation of the different partial states will correspond to one of the modes.

In this example, see Figure 7.4, taking an MMSE estimate of the joint state vector will result in a joint state estimate, where the target state estimates end up being located somewhere in the middle of the true positions of the two targets. Clearly, such an MMSE estimate should be avoided in a tracking system where situational awareness is desired, such as in air traffic control. See also, [35, 36, 39] and [40].

However, a (PF-based) MAP estimator will avoid the track coalescence phenomenon altogether, as it will pick one point (estimate) from the multivariate state-distribution, presumably near the highest peak of the bimodal distribution.

The complete tracking of the targets based on a PF-based MMSE and PF-based MAP estimators is shown in Figures 7.5 and 7.6. Note that Figure 7.6 shows that two targets cross their paths often (after separation). This is due to the fact that even when a MAP estimator is used, the uncertainty of the labeling, i.e., the mixed labeling problem, will exist inherently. As a result, the MAP estimate at one time-moment may "jump" from one set of labels/mode to the other at the very next time-moment. This is also partially due to the use of a crude PF as well as a crude MAP estimator. Other sophisticated point-estimators such as MMOSPA may mitigate this problem, see [37, 39] and [40]. Although very interesting and practically very useful, especially in combination with joint state multitarget filters, we will not elaborate more on this estimator here in this chapter, as it is not our main focus.

Another note of interest is that even though multimodalities appear in both the ship example in the introduction and the labeled tracking of this section, there are some fundamental differences as well. In the single-target ship example, multimodality appeared purely because of the uncertainty in the dynamics, modeled through the process noise (including the restriction in the movements) and the measurement noise. In the example of this section, multimodality appears when one wants to assign an identity or label to the targets. This problem has been treated in detail in the Random Finite Set framework in [36]. Once again, we leave it to the reader to refer to that paper for more detail.

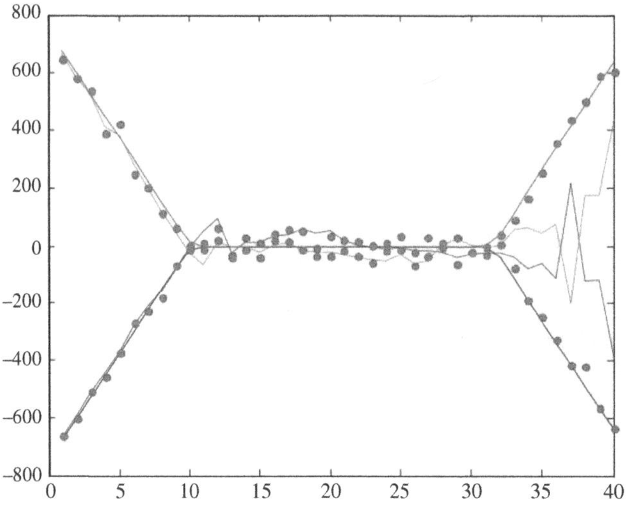

Figure 7.5 Joint state estimation of two targets with PF and MMSE estimate—(smooth) solid lines are the true trajectories, dots are the measurements, and the wiggly lines are the two estimated tracks. The horizontal axis is the time step(scan), the vertical axis is the object position [m].

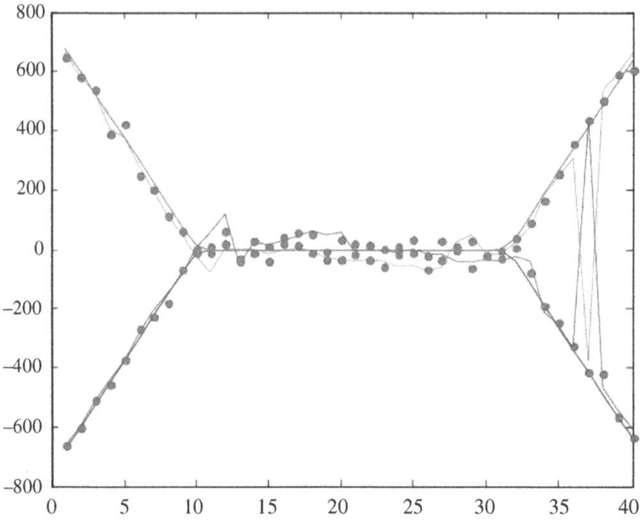

Figure 7.6 Joint state estimation of two targets with PF and MAP estimate—(smooth) solid lines are the true trajectories, dots are the measurements, and the wiggly lines are the two estimated tracks. The horizontal axis is the time step(scan), the vertical axis is the object position [m].

In any case, it is clear that when multimodality is present in the posterior, an alternate estimator such as MAP will do a better job as the estimate of locations than the MMSE estimator. However, to have a good point estimate, one needs the approximation of the a posteriori distribution to be close to the true a posteriori distribution. One way to achieve this is to use more particles/samples to represent the distribution, provided, of course, certain convergence results hold. The latter is exactly the topic of this chapter, and we will present and discuss such results in the following sections.

7.4 Convergence of the PF-based Distribution

In this section, we present a convergence result from [24] and discuss some of its implications. The type of convergence considered is weak convergence, or equivalently: convergence in distribution; see Definition 7.2. First, let us define the empirical distribution corresponding to a PF.

Definition 7.3 ***(Empirical Distribution)*** The empirical distribution corresponding to a weighted N-sample $\{x^i, w^i\}_{i=1,\dots,N}$, is defined as

$$\hat{P}^N(x) := \sum_{i=1}^{N} w^i 1_{\{x^i \leq x\}}. \tag{7.14}$$

Additionally, in the next theorem, the following property will be imposed on the transition density of the system.

Property 7.1 ***(Feller)*** A transition density/kernel $p(\cdot \mid \cdot)$ is said to be *Feller* if the following holds:

$$\forall \phi \in C_b(\mathcal{S}): \quad \psi(\cdot) := \int_{\mathcal{S}} p(\cdot \mid x_{t-1})\phi(x_{t-1})dx_{t-1} \in C_b(\mathcal{S}). \tag{7.15}$$

We are now ready to formulate a main theorem from [24] on PF convergence.

Theorem 7.1 ***(Crisan and Doucet, [24, Theorem 1])*** Consider the dynamical system governed by Eqs. (7.1)–(7.3). Assume further that the likelihood and the transition density satisfy the following conditions.

- *Likelihood condition:*
  The likelihood $p(y_t \mid x_t)$ is strictly greater than zero, continuous and bounded as a function of x_t.
- *Transition density condition:*
  The transition density $p(x_t \mid x_{t-1})$ is Feller.
- Let $\{x_t^i, w_t^i\}_{i=1,\dots,N}$ be the weighted particles from a SIR-PF, as given by Algorithm 7.1.

Then for any $\phi \in C_b(\mathcal{S})$:

$$\lim_{N\to\infty} \sum_{i=1}^{N} w_t^i \phi(x_t^i) = \int_{\mathcal{S}} \phi(x_t)dP(x_t \mid y_{1,\dots,t}) \qquad \text{almost surely.} \tag{7.16}$$

In the following, we shall discuss some implications of Theorem 7.1 through a series of remarks.

Remark 7.3 A comment is due for the phrase "almost surely" in the theorem above. Usually, in the context of PFs, there are two types of randomness. One type that originates from the system itself such as from the measurement noise (for example, from the imperfections in the measurement sensors) and/or the process noise related to the hidden/unknown state, e.g., caused by the uncertainty of the future location of a moving object. This "system" randomness is, in a sense, of truly physical origin; it accounts for the measurement errors or for unknown disturbances on the dynamics of the system. The other type of randomness enters the PF output through the Monte Carlo (re)generation of the particles $\{x_t^i\}_{i=1,2,\dots,N}$ from the PF Algorithm. For example, every "draw" of Steps 1 and 2 of Algorithm 7.1 introduces randomness in the output. So does the resampling in Step 6, as it can be seen as drawing from a discrete probability distribution. The "almost sure" in the theorem refers to the latter randomness in the PF Algorithm and not to the former "system" randomness of physical origin.

One way to look at/interpret the phrase "almost sure" is as follows. Consider the event that a realization $\{x_t^i\}_{i=1,2,\dots,N}$ is generated by the PF Algorithm for which the limiting property (16) does *not* hold. Then the probability of the union of all such events is zero.

Remark 7.4 The quantity on the LHS of Eq. (7.16) is the expectation of $\phi(x_t)$ with respect to the empirical (posterior) distribution corresponding to the PF output

$$\{x_t^i, w_t^i\}_{i=1,2,\dots,N}.$$

The LHS of Eq. (7.16) equals:

$$\int_{\mathcal{S}} \phi(x_t) d\hat{P}^N(x_t). \tag{7.17}$$

Thus, Theorem 7.1 states that the PF-based empirical mean of $\phi(x_t)$ is asymptotically equal to the true mean, $E_{p(x_t|y_{1,\dots,t})}[\phi(x_t)]$, provided, of course, the function ϕ is continuous and bounded. In other words, the empirical distribution $\hat{P}^N$ converges weakly to the true a posteriori distribution of x_t conditioned on all measurements $y_{1:t}$.

Remark 7.5 We stress that Theorem 7.1 does not hold for the empirical mean of the state itself: $E_{p(x_t|y_{1,\dots,t})}[x_t]$. This is because, in this case, the corresponding test function is the identity function: $\phi(x) = x$, and the identity function is unbounded. Since $\phi \notin C_b(\mathcal{S})$, the theorem does not apply. This has already been pointed out by the authors of [24].

Consequently, the standard LLN type result, i.e., empirical mean of x_t converging to the true mean, cannot be concluded directly from the theorem. However, in [26], it has been shown that a similar convergence result holds for a specific class of unbounded functions ϕ and in particular $\phi(x) = x$ can be in this class as we will see in Sections 7.6 and 7.7.

Remark 7.6 Another point to note is that the LLNs in standard Statistics or Probability literature assume that the samples are independent and identically distributed (IID), see, e.g., [41]. However, the samples (particles) from a SIR-PF are not IID. Still the theorem provides a specific type of the LLN for PFs, see Eq. (7.16). Specific here means that $\phi \in C_b(\mathcal{S})$.

Remark 7.7 Results like the one formulated in Theorem 7.1 have been instrumental in deriving practical and even theoretical results in different fields. For example, in the field of signal processing, the authors of [28] use Theorem 7.1 extensively to derive and prove new results for a PF-based (Neyman–Pearson) optimal detector. In particular, they show that the PF-based LR converges to the true, yet unknown, LR almost surely. See also Example 7.4 in Section 7.7.

7.5 L_1-Convergence of the PF-based a Posteriori Density

In this section, we state and discuss the convergence results from [25]. We shall show that these results also lead to an LLN-type result for PFs. To start with it is assumed the true a posteriori distribution has a density. On the other hand, the empirical distribution $\hat{P}^N(x)$ (see Definition 7.3) does not admit any density. Consequently, the author in [25] defines the posterior density at time t using the particles from time $(t-1)$ as follows:

$$\hat{p}^N(x_t) := p(y_t \mid x_t)\frac{1}{N}\sum_{j=1}^{N} p(x_t \mid x_{t-1}^j), \tag{7.18}$$

where the particles x_{t-1}^j are available from the PF at time step $(t-1)$. We refer to [25] for more detail. In [25], the probability density in (7.18) is studied. Its distance from the true, but unknown, a posteriori density is considered for both the SIR-PF and an alternative PF, called the AR-PF. We will not go into the details of the AR-PF in here. We only mention that the particles generated by AR-PF are IID, whereas the SIR-PF generated samples are not. See also Remark 7.6. For the detailed AR-PF Algorithm we refer the reader to [7, pages 29, 30].

For completeness, we also give the definition of the L_1-norm of a function. This norm will be used to express the distance between two densities in the convergence results presented below.

Definition 7.4 For a function f defined on $\mathcal{S}$, the L_1-norm is defined as

$$\| f \|_1 := \int_{\mathcal{S}} |f(s)| ds,$$

provided the integral on the RHS exist.

In particular, the convergence results in [25] are related to the distance between the PF-based density, given by (7.18), and the true, yet unknown, a posteriori density

$$\| \hat{p}^N(\cdot) - p(\cdot \mid y_{1,\dots,t}) \|_1, \tag{7.19}$$

for $N \to \infty$.

Below we state two theorems from [25]. We emphasize that the first theorem is specifically for the AR-PF case, where the samples (particles) are IID. See comments at the start of Section 4.1.1 in [25]. The second theorem covers the SIR-PF case, where the particles are not IID, therefore the second theorem has more/extra conditions compared to the first one below.

Theorem 7.2 *(Künsch AR-PF, [25, Theorem 1])* Suppose that the transition density $p(x_t \mid x_{t-1})$ is continuous w.r.t the L_1-norm in the following sense:

$$\| p(\cdot \mid x_{t-1}) - p(\cdot \mid \tilde{x}_{t-1}) \|_1 \to 0 \qquad \text{for} \quad \tilde{x}_{t-1} \to x_{t-1}, \tag{7.20}$$

and for all t and measurements y_t, the likelihood $p(y_t \mid x_t)$ is, as a function of x_t strictly positive and bounded. Let $\hat{p}^N(\cdot)$ be the posterior density based on the AR-PF generated samples. Then $\forall t \in \mathbb{N}^+$ and $y_{1,\dots,t}$:

$$\lim_{N\to\infty} \| \hat{p}^N(\cdot) - p(\cdot \mid y_{1,\dots,t}) \|_1 = 0 \qquad \text{in probability.} \tag{7.21}$$

Remark 7.8 The convergence "in probability" in Theorem 7.2 is, as in the case of Theorem 7.1, with respect to the randomness that arises during the Monte Carlo drawings of particles in the PF Algorithm. See, also, Remark 7.3 in Section 7.4.

The following result states that under some additional assumptions, the difference in L_1-norm between the PF-based density and the true posterior density approaches zero exponentially fast, even when the PF does not generate IID samples, such as the SIR-PF.

Theorem 7.3 *(Künsch, SIR-PF, [25, Theorem 2])* Suppose that the transition density $p(x_t \mid x_{t-1})$ is continuous w.r.t the L_1-norm in the sense of (7.20). Suppose further that there exist a pdf $h(\cdot)$ and finite constants c_p, C_p with $0 < c_p < C_p$, such that for all $x, \tilde{x}$, the transition density satisfies

$$c_p h(x) \leq p(x \mid \tilde{x}) \leq C_p h(x).$$

In addition, assume that the likelihood function $p(y \mid x)$ satisfies

$$C_l := \sup_{x,\tilde{x},y} \frac{p(y \mid x)}{p(y \mid \tilde{x})} < \infty.$$

Let $\hat{p}^N(\cdot)$ be the (SIR) PF-based posterior density. Then for any $\epsilon > 0$ there exist positive constants c_1, c_2 such that for all $t \in \mathbb{N}^+$, $y_{1,\dots,t}$ and N:

$$Prob(\| \hat{p}^N(\cdot) - p(\cdot \mid y_{1,\dots,t}) \|_1 > \epsilon) \leq c_1 \exp(-c_2 N). \tag{7.22}$$

Remark 7.9 Since for any given $\epsilon > 0$, the RHS of (7.22) converges to zero as $N \to \infty$, (7.21) of Theorem 7.2 follows directly from (7.22) and the definition of convergence in probability.

Remark 7.10 We note that even though not stated in the form of a convergence of the empirical mean of some test functions ϕ, both theorems above do imply an LLN-type of result, albeit in the weak sense. This is formally stated in the following corollary.

Corollary 7.1 *(Corollary to Theorems 7.2 and 7.3)* Suppose that the assumptions of either Theorem 7.2 or Theorem 7.3 hold. Then for all $\phi \in C_b(\mathcal{S})$:

$$\lim_{N\to\infty} \int_{\mathcal{S}} \phi(x_t)\hat{p}^N(x_t)dx_t = \int_{\mathcal{S}} \phi(x_t)p(x_t \mid y_{1,\dots,t})dx_t \qquad \textit{in probability}. \tag{7.23}$$

Proof: First note that, through Remark 7.9, the assumptions of either Theorem 7.2 or Theorem 7.3 imply (7.21). Next, let us consider the absolute value of the difference between the PF-based mean of ϕ and the true mean of ϕ:

$$\begin{aligned}
&\left| \int_{\mathcal{S}} \phi(x_t)\hat{p}^N(x_t)dx_t - \int_{\mathcal{S}} \phi(x_t)p(x_t \mid y_{1,\dots,t})dx_t \right| \\
&= \left| \int_{\mathcal{S}} \phi(x_t)[\hat{p}^N(x_t) - p(x_t \mid y_{1,\dots,t})]dx_t \right| \\
&\leq \int_{\mathcal{S}} | \phi(x_t) | \left| \hat{p}^N(x_t) - p(x_t \mid y_{1,\dots,t}) \right| dx_t \\
&\leq \sup_{x_t \in \mathcal{S}} \{| \phi(x_t) |\} \cdot \int_{\mathcal{S}} \left| \hat{p}^N(x_t) - p(x_t \mid y_{1,\dots,t}) \right| dx_t \\
&= \sup_{x_t \in \mathcal{S}} \{| \phi(x_t) |\} \cdot \| \hat{p}^N(x_t) - p(x_t \mid y_{1,\dots,t}) \|_1 \\
&\leq U \cdot \| \hat{p}^N(x_t) - p(x_t \mid y_{1,\dots,t}) \|_1, \quad \text{for some positive constant } U \text{, such } U \text{ exists, as } \phi \in C_b(\mathcal{S}).
\end{aligned}$$

Since, by assumption, we have: $\| \hat{p}^N(x_t) - p(x_t \mid y_{1,\dots,t}) \|_1 \to 0$ in probability, the result, Eq. (7.23), follows.

Remark 7.11 The statement in (7.23) is about *convergence of the empirical mean* of ϕ to the true a posteriori expectation of ϕ. Therefore, it is an LLN-type of result. However, the convergence holds "only" *in probability* and not *almost surely* as in Section 7.4. In this sense, this is a weak LLN (WLLN) type of result and not an SLLN type of result, see Section 7.4.

We emphasize that the difference between the two types of LLNs is mostly of a theoretical nature and not of a practical one.

The concern of Remark 7.5 still remains. In particular, the results in [24] and in [25] allow "only" continuous bounded test functions. As a consequence, the convergence of the empirical mean of the state x_t cannot yet be concluded on the basis of these results. As stated in the Discussion Section of their paper [24, Section VII], the authors note that: "... from a practitioners viewpoint, it seems highly unsatisfactory to have to assume that $\phi(x)$ is bounded to obtain some convergence results ..." This issue can be partially solved with the help of the convergence results from [26]. We will discuss these in the next section. However, as we shall see, allowing for unbounded test functions comes at the cost of some necessary additional assumptions on the systems.

7.6 PF Convergence for Unbounded Test Functions

In the previous two sections we have presented convergence results that hold for continuous bounded test functions ϕ. This leaves us still in an unsatisfactory situation, namely that there is no theoretical guarantee for the convergence of the simple MMSE estimate of the state to its true value, as the latter corresponds to the *unbounded* function $\phi(x) = x$. See Remark 7.5 in Section 7.4. Practitioners have been ignoring this boundedness knowingly, or possibly, in anticipation of the existence of such a result.

In this section, we present and discuss convergence results, that at least partially underpin/justify this "practitioners approach" theoretically. We shall show that the additional assumptions on the system that are required for the convergence result are satisfied for an important class of systems. This class includes systems with additive Gaussian measurement noise.

In [26] a slightly modified PF is used. In this algorithm, a few new design parameters are introduced, namely, the weights α_j^i's satisfying $\alpha_j^i \geq 0$, $\sum_{i=1}^N \alpha_j^i = 1$ and $\sum_{j=1}^N \alpha_j^i = 1$, and threshold parameters γ_t which are to be chosen by experience.

We point out that although we are still considering, as in the previous sections, the system governed by Eqs. (7.1)–(7.3), we will use an alternative notation. In particular, we adopt the notations from [26]; see also Section 7.2 for details. The transition density/kernel is denoted by $K(x_t|x_{t-1})$, the likelihood function by $\rho(y_t|x_t)$ and the true posterior distribution by $\pi_{t|t}$.

Algorithm 7.2: *(PF-MOD)*:

---

1. Initialization: *at* $t = 0$ draw N particles $x_0^i \backsim p(x_0)$ and set $w_0^i := \frac{1}{N}$.
2. If $t < T, \forall i \in \{1, 2, \dots, N\}$ draw particles :

   $\tilde{x}_t^i \backsim \sum_{j=1}^N \alpha_j^i K(dx_t \mid x_{t-1}^j)$
3. If $\frac{1}{N} \sum_{i=1}^N \rho(y_t \mid \tilde{x}_t^i) \geq \gamma_t$ rename $x_t^i := \tilde{x}_t^i$, proceed to 4; otherwise, go to 2
4. Assign particle weights : $w_t^i := \rho(y_t \mid x_t^i)$
5. Normalize weights : $w_t^i := \dfrac{w_t^i}{\sum_{j=1}^N w_t^j}$
6. Output : $\{x_t^i, w_t^i\}_{i=1,\dots,N}$

7. Resample with replacement N times from $\{x_t^i, w_t^i\}_{i=1,\ldots,N}$ to arrive at: $\{x_t^i, w_t^i\}_{i=1,\ldots,N}$ with $w_t^i := \frac{1}{N}$

8. *Set* $t := t + 1$

9. Go to 2.

For the convergence result of [26], the following Assumptions will be imposed on the system and the test function:

A1: For any given $s = 1, \ldots, t$ and $y_{1:s}$, the design parameter γ_s of Algorithm 7.2 satisfies:

$$0 < \gamma_s < (\pi_{s|s-1}, \rho). \tag{7.24}$$

A2: For given $y_{1:s}$, $s = 1, \ldots, t$, the likelihood and transition density functions $\rho(y_s \mid x_s)$ and $K(x_s \mid x_{s-1})$ are bounded.

A3: For some constant $p > 1$, the test function $\phi(\cdot)$ satisfies

$$\sup_{x_s} \mid \phi(x_s) \mid^p \rho(y_s \mid x_s) < C(y_{1:s}), \tag{7.25}$$

for some constant $C(y_{1:s})$, $s = 1, \ldots, t$.

Following [26], we denote the set of functions ϕ satisfying (7.25) as $L_t^p(\rho)$.

A few observations on the assumptions, especially A1 and A3 are due.

Remark 7.12

(a) Assumption A1 guarantees that there is no division by zero in the calculation of the filtering solution, see Eq. (7.13) in Section 7.2.

(b) We emphasize that Assumption A1 is not directly verifiable, because $\pi_{s|s-1}$ cannot be expressed directly in terms of the system quantities. It is very intricately related to the system through the recursive relations (7.12) and (7.13) of Section 7.2. Usually, one has to choose $\{\gamma_t\}$ by experience/trial-and-error so that this condition is satisfied. See [24, comments immediately after equation (10)].

Note that the Assumption in [24] to prevent division by zero is in terms of the likelihood ρ, namely, the (strict) positivity of the latter; see Theorem 7.1 and [24, equation (22)]. This makes it readily verifiable, but stronger than what is imposed in Assumption A1.

(c) Assumption A1 also prevents Algorithm 7.2 from running into an infinite loop between step 2 and step 3. This has been explicitly proved in [26, Proposition 3.1].

Remark 7.13 The restriction on the test functions imposed by Assumption A3 is in terms of its growth in comparison to that of the likelihood function. As we shall see later in this section, this is much weaker than the condition of boundedness. In particular, we shall show in Theorem 7.5 that certain unbounded test functions in combination with additive (Gaussian) measurement noise models satisfy A3.

We now state a main convergence theorem from [26].

Theorem 7.4 *(Theorem 3.2, [26])* Consider a system for which the Assumptions A1 and A2 hold for some $\gamma_s > 0$, $s = 1, 2, \ldots, t$. Let $\pi_{t|t}^N$ be the empirical posterior distribution based on the PF output, given by Algorithm 7.2. Then for any ϕ satisfying A3 for some $p \geq 2$, [i.e., for any $\phi \in L_t^p(\rho)$ with $p \geq 2$]:

$$\lim_{N\to\infty} (\pi_{t|t}^N, \phi) = (\pi_{t|t}, \phi) \qquad \text{almost surely.} \tag{7.26}$$

Remark 7.14 It is fairly straightforward to see that Theorem 7.4 implies an SLLN for the PF-based empirical mean for test functions from the class $L_t^p(\rho)$, which may include unbounded functions. In this sense, this can be seen as an extension of Theorem 7.1. Also, here it is an SLLN, because of the "almost surely" addition.

We now show that for systems with additive Gaussian measurement noise, Assumption A3 will be satisfied. These types of likelihoods arise in many (physical) systems.

Theorem 7.5 Suppose that the likelihood ρ of a dynamical system is given by

$$\rho(y \mid x) = \frac{1}{\sqrt{(2\pi)^m \, |\Sigma|}} \exp\left\{-\frac{1}{2}(y - h(x))^T \Sigma^{-1}(y - h(x))\right\}, \tag{7.27}$$

where $h(x)$ is of polynomial order. Then assumption A3 is satisfied for any $\phi(x)$ of polynomial order.

Proof: Let us first consider the case when the state and measurements are one-dimensional, i.e., $x, y \in \mathbb{R}$. Also, suppose that $h(x) = a_0 + a_1x + \cdots + a_kx^k$ and $\phi(x) = b_0 + b_1x + \cdots + b_lx^l$, where $k, l \geq 1$. Then the likelihood function ρ will simplify to

$$\rho(y \mid x) = \frac{1}{\sqrt{2\pi\sigma}} \exp\left\{-\frac{1}{2\sigma^2}\left(y - a_0 - a_1x - \cdots - a_kx^k\right)^2\right\}. \tag{7.28}$$

Since both $\phi(\cdot)$ and $\rho(y|\cdot)$ are continuous and we are interested in the bounds of the $|\phi(x)|^p\rho(y \mid x)$, we have to ensure that the latter is bounded as $x \to \pm\infty$. To this end, note that the tail behavior of the expression $|\phi(x)|^p\rho(y \mid x)$ will be determined by the function of the form $x^c \exp(-dx^{2k})$, where c, d are real numbers. Since

$$x^c \exp(-dx^{2k}) \to 0 \qquad \text{as } x \to \pm\infty, \tag{7.29}$$

we conclude that (7.25) is satisfied.

Remark 7.15 The case of a higher dimensional system can be proven similarly using the one-dimensional limit of (7.29).

Remark 7.16 An immediate consequence of Theorem 7.5 is that in case of additive Gaussian measurement noise, the identity function $\phi(x) = x$ will satisfy Assumption A3. More generally, for every $q = 1, 2, \ldots$, the test function $\phi(x) = x^q$ will satisfy Assumption A3. Moreover, convergence of the qth empirical moment, based on the PF, to the true qth moment is guaranteed.

As an elaboration and a visualization of the argument used in the proof of Theorem 7.5 to justify Assumption A3, Figure 7.7 shows two plots of $|\phi(x)|^4\rho(y \mid x)$, where $\phi(x) = x$ and ρ is the standard Gaussian likelihood, i.e., $\sigma = 1$ and $h(x) = x$, for two different values of y: $y = 0$ and $y = 1$. We can see from the plots that the functions are bounded with the values decreasing to zero as $x \to \pm\infty$. The maximum values are different for different y's and they occur at different values of x's.

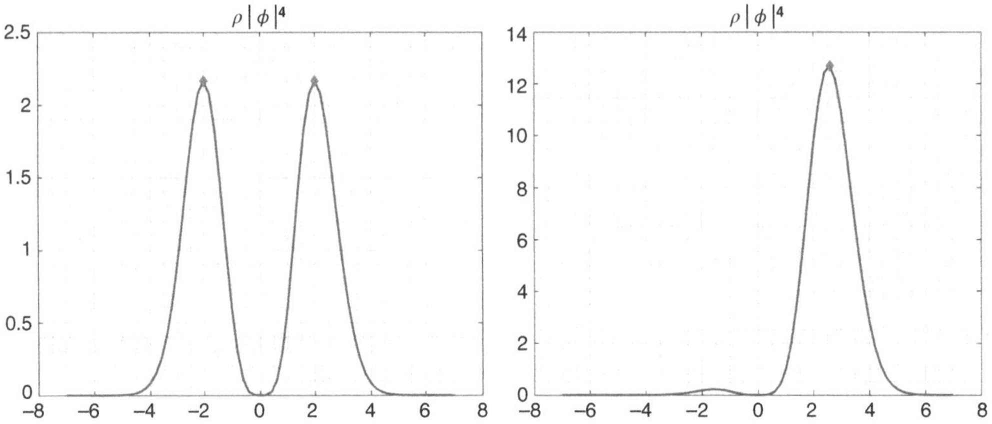

Figure 7.7 plot of x (horizontal axis) vs. $|\phi(x)|^p \rho(y \mid x)$ (vertical axis) for $\phi(x) = x$, $p = 4$, ρ the standard Gaussian likelihood and $y = 0$ (left) and $y = 1$ (right).

Remark 7.17 We emphasize here that the vertical and horizontal axes of the plots in Figure 7.7 have no labels and units, as they are not related to some real physical system. Here they are used only for the purpose of illustrating some properties of Theorem 7.5.

Theorems 7.4 and 7.5 together guarantee convergence of the PF-based empirical mean for a large class of possibly unbounded test functions, including the often-used identity function in obtaining the PF-based MMSE estimator given by (7.7). Therefore, even though in practice a PF-based empirical mean of the system-state was and is frequently used without having convergence guarantees, we can now do so with full confidence, thanks to results of this section.

Finally, we want to note that the PF used in [26], which is provided by Algorithm 7.2, is a modified version of the SIR-PF of Algorithm 7.1. Both these algorithms use the predictive density as the so-called *importance function* (see, e.g., [32]). In [27], some of these convergence results are extended to PFs that incorporate more general importance functions.

In Section 7.7, we provide several examples and applications of the discussed convergence results.

7.7 Examples

In this section, we provide several examples of systems that have guaranteed convergence for a PF-based empirical estimate of an unbounded test function to the true expected value of that test function.

Remark 7.18 In line with Remark 7.16, we will show some examples illustrating the above-defined convergence results. Most of these applications are scalar time series, used/presented in PF literature. Note that the one-dimensional states and measurements are not derived from specific physical models. Therefore, for most examples, no labels/units have been provided.

To ensure the convergence of empirical estimates (of the unbounded test functions), we shall apply Theorem 7.4. Consequently, we need only to verify the Assumptions A1, A2, and A3. However, as mentioned in Remark 7.12(b) Assumption A1 is not readily verifiable. One actually has to build the full filter and "experimentally" choose suitable $\{\gamma_t\}$ for Algorithm 7.2. If this can be done,

A1 is satisfied. This is indeed the case for the models/examples given in this section. Furthermore, the measurement models in these examples all have additive Gaussian noise and dependencies of the measurement y on the state x are of polynomial type. As a result, from Theorem 7.5, Assumption A3 will also be satisfied. Thus, it will remain to check Assumption A2 only.

The first example is the "famous" PF benchmark one for a nonlinear system, amongst others used in [1] and also used, e.g., in [32].

Example 7.1 Suppose the system is given by:

$$x_t = \frac{x_{t-1}}{2} + \frac{25x_{t-1}}{1+x_{t-1}^2} + 8\cos(1.2t) + w_{t-1}, \tag{7.30}$$

$$y_t = \frac{x_t^2}{20} + e_t, \tag{7.31}$$

where w_t and e_t are the process and measurement noises with variances Q_t and R_t, respectively. Both noises are assumed to be Gaussian in this example. However, due to nonlinearities in the system, the a posteriori density of this system is highly non-Gaussian. See [1] for more background/detail.

This system, classically described by a dynamic equation and a measurement equation, see, e.g., [17], is oftentimes denoted in probability literature and/or works on PF as

$$x_t \backsim \mathcal{N}(x_t; f_t(x_{t-1}, t), Q_{t-1}) \tag{7.32}$$

$$y_t \backsim \mathcal{N}(y_t; \frac{x_t^2}{20}, R_t), \tag{7.33}$$

And

$$f_t(x_{t-1}, t) = \frac{x_{t-1}}{2} + \frac{25x_{t-1}}{1+x_{t-1}^2} + 8\cos(1.2t). \tag{7.34}$$

To verify assumption A2 for this system note first that the transition kernel is given by

$$K(x_t \mid x_{t-1}) = \frac{1}{\sqrt{2\pi Q_{t-1}}} \exp\left\{-\frac{1}{2Q_{t-1}}\left(x_t - \frac{x_{t-1}}{2} - \frac{25x_{t-1}}{1+x_{t-1}^2} - 8\cos(1.2t)\right)^2\right\}.$$

Next we observe that as a function of (x_t, x_{t-1}), this is continuous with the property that as any of x_t and x_{t-1} approaches $\pm\infty$, the function converges to zero. Furthermore, for this system, the likelihood as a function of x_t is bounded, see also Eq. (7.33). This implies that Assumption A2 is met for this system.

Remark 7.19 We want to point out that strictly speaking, the system considered in the above example does not fall under the purview of the systems mentioned in Section 7.1. This is because the transition kernel/density in this example violates Eq. (7.1) in that it depends explicitly on time t through the presence of the term "cos(1.2 t)" in (7.34).

However, all the results discussed in this chapter, except for the convergence of the SIR-PF-based density given in Theorem 7.3, will hold even when the transition density/kernel and the likelihood are of the form $p_t(x_t \mid x_{t-1})$ and $p_t(y_t \mid x_t)$, respectively. In particular, the convergence of the PF-based MMSE, which is guaranteed by Theorem 7.4, will still hold for this example.

The next example is from [10, Section 5.1]. The state transition density for this system is bimodal by nature. As a result, the posterior is expected to be bimodal as well.

Example 7.2 Suppose the system is governed by

$$\text{[state dynamics]} \qquad x_t = 0.4x_{t-1} + w_{t-1}, \tag{7.35}$$

$$\text{[measurement]} \qquad y_t = x_t + e_t, \tag{7.36}$$

where the initial state $x_0 \backsim \mathcal{N}(0, 0.1)$, measurement noise $e_t \backsim \mathcal{N}(0, 2^2)$ and the process noise w_t is distributed as $\mathcal{N}(-1, 0.1^2)$ with probability $1/2$ and as $\mathcal{N}(1, 0.1^2)$ with probability $1/2$.

To verify Assumption A2 for this system we note that the transition kernel is given by

$$K(x_t \mid x_{t-1}) = \frac{1}{2}\frac{1}{\sqrt{2\pi}\cdot 0.1}\left[\exp\left\{-\frac{(x_t - 0.4\,x_{t-1} + 1)^2}{2\cdot 0.01}\right\} + \exp\left\{-\frac{(x_t - 0.4\,x_{t-1} - 1)^2}{2\cdot 0.01}\right\}\right].$$

Note that as a function of (x_t, x_{t-1}) the kernel is continuous and converges to zero as any of x_t or x_{t-1} approaches $\pm\infty$, we conclude, as in Example 7.1, that the kernel satisfies A2.

Our next model is a simple, one-dimensional linear system with additive Gaussian process and measurement noise. Consequently, the KF will provide the true quantities—the posterior density, the posterior mean, etc. Recall that for a non-linear and/or non-Gaussian model, the true posterior density does not have a closed form solution, nor is it easy to obtain. As a result, it is difficult to visualize or verify numerically the convergence discussed in the previous sections. That is also the main reason for considering this simple example. For this example, we can compare the numerical results with the true quantities. In what follows, we shall consider this model and present some numerical results confirming the (theoretical) convergence of different quantities to the corresponding true values.

Example 7.3 Suppose the system is governed by:

$$x_t = x_{t-1} + w_{t-1}, \qquad \text{and} \qquad y_t = x_t + e_t, \tag{7.37}$$

where $w_{t-1} \backsim \mathcal{N}(0, \sigma_w^2)$, $e_t \backsim \mathcal{N}(0, \sigma_e^2)$. Following the same arguments as in Examples 7.1 and 7.2, it is easy to verify that Assumption A2 is satisfied.

Figures 7.8 and 7.9 show the PF-based empirical state estimates (i.e., the PF-based MMSE estimates) and the true posterior expectations of the state obtained using the KF. These plots are from [28, Section IV]. The parameters used for these simulations are $\sigma_w = \sigma_e = 0.5$. We can see that PF-based state estimates using $N = 5,000$ particles (Figure 7.9) are much closer to the corresponding KF estimates than they are when using $N = 100$ particles (Figure 7.8). This is in line with what has been established, theoretically, in Remark 7.16 in Section 7.6, namely, as $N \to \infty$, the PF-based empirical mean converges to the true, unknown posterior mean. Additional extensive Monte Carlo simulations can be found in [28, Section IV].

The next example, although related to the convergence of PF-based quantities, is not about the convergence of empirical means for some test function. This example can be considered as an application of the convergence results to obtain other new and useful results. In many real life applications, one must choose between two diverging hypotheses, on the basis of observed data. In these scenarios related to optimal detection problems, the so-called LRs play an important role in taking the decisions. The PF-based track-before-detect (PFTBD) method for radar is one such an area, where the PF-based LR has been successfully applied. See, for example, [11] and [42]. Another such

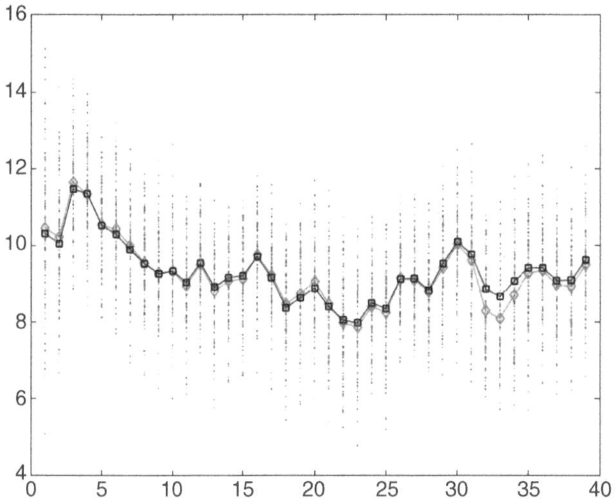

Figure 7.8 Kalman filter output and particle filter (state) output for 100 particles. The single dots are the particles, squares are the particle cloud means, and the diamonds are the Kalman filter state estimates. The horizontal axis represents the time index.

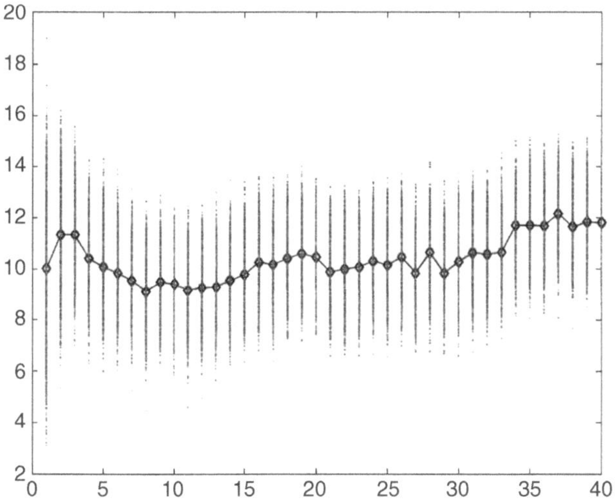

Figure 7.9 KF output and particle filter (state) output for 5,000 particles. The single dots are the particles, squares are the particle cloud means, and the diamonds are the Kalman filter state estimates. The horizontal axis represents the time index.

application is in medical image processing, see [4]. Naturally, from the "practitioners" point of view, it would be good to have a theoretical justification that the PF-based LR will be close to the true, unknown, LR at least when the number of particles grows large. Both theoretical and numerical results confirm this. The (theoretical) proof can be found in [28]. We note that the idea of PF TBD has been presented/published first at the 2001 American Control Conference, independently by [43] and [44]. Since, the idea has been widely applied. Also, see [42] for a good overview, as well as for an explanation of its (performance) benefits over classical detect-then-track approaches and comparison to other TBD methods.

Example 7.4 Following up on Remark 7.7, in Section 7.4, we consider the main result from [28]. Using a slightly extended version of Theorem 7.1, it has been proven that the PF based LR converges to the true, but unknown LR, almost surely. For the detailed theoretical exposition we refer the reader to [28]. Below, we reproduce some simulation results from [28].

The model considered for the numerical results is again the same one as in Example 7.3. The one dimensional linear Gaussian "toy" system model given in (7.37). In this case, the true LR has a closed form expressions involving the likelihoods $p(y_t \mid y_{1,\ldots,t-1})$. The latter can in turn be calculated/evaluated exactly by using the expressions for the innovations from a running KF, see [5]. The parameter settings are the same as in [28, Section IV].

Figures 7.10 and 7.11 show the PF-based likelihoods $p(y_t \mid y_{1,\ldots,t-1})$ (indicated by the diamonds) when using $N = 100$ and $N = 5,000$ particles, respectively. The figures also show the corresponding ground truth quantities (indicated by the squares) as obtained from the KF.

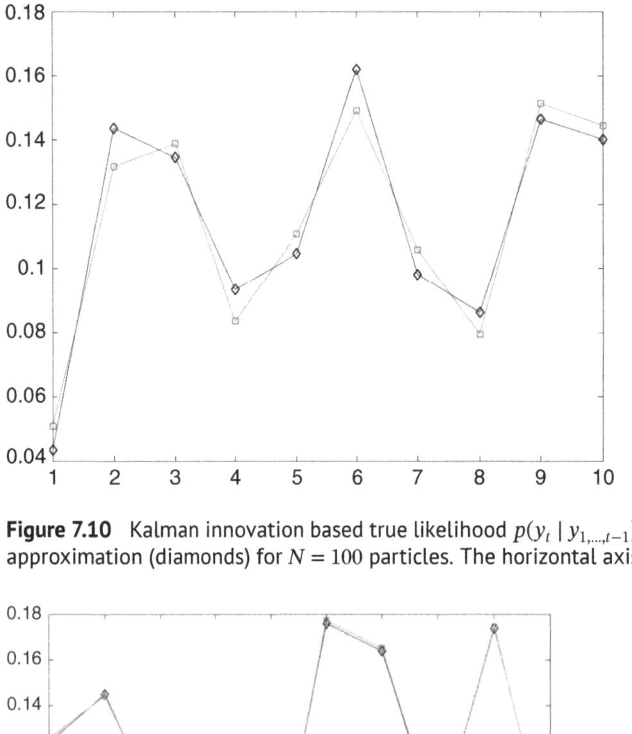

Figure 7.10 Kalman innovation based true likelihood $p(y_t \mid y_{1,\ldots,t-1})$, (squares) and particle-based approximation (diamonds) for $N = 100$ particles. The horizontal axis represents the time index.

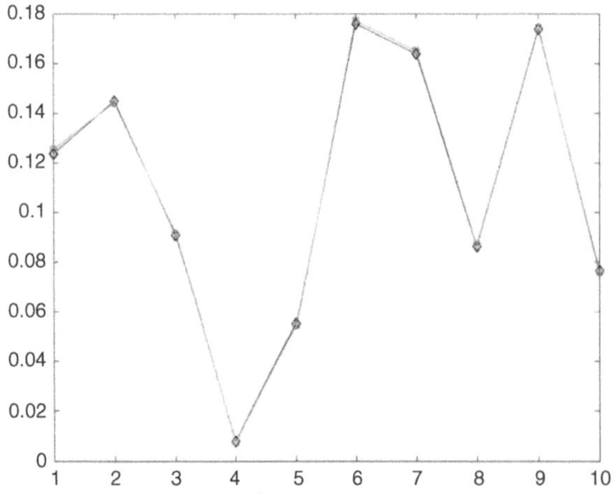

Figure 7.11 True, Kalman innovation based, likelihood $p(y_t \mid y_{1,\ldots,t-1})$(squares) and particle-based approximation (diamonds) for $N = 5,000$ particles. The horizontal axis represents the time index.

The figures are identical to the ones from [28, Section IV]. We can see, again, that PF-based quantities when using $N = 5,000$ particles (Figure 7.11) are much closer to the corresponding true values than they are when using $N = 100$ particles (Figure 7.10). This being all in line with (theoretical) PF convergence results, treated in this chapter.

7.8 Conclusions

In this chapter, we have taken a look at PF convergence. We have reviewed different types of PF convergence results from various publications. We have discussed the extent of their applicability. For example, whether a result can be used to conclude the convergence of the PF-based MMSE estimate to the true expected state, and more generally, which results can handle unbounded test functions and under what conditions. We have shown via a new theorem that the required conditions are satisfied for an important and large class of systems and thereby confirming the convergence of the PF-based MMSE for this class. We have also provided some examples of such systems.

Also, through the examples in the introduction, Section 7.1 as well as in Section 7.3, we have illustrated that care must be taken w.r.t. the choice of (the right) filter as well as w.r.t. choosing an appropriate point estimator of the state, for the system under consideration.

Acknowledgments

- The first author thanks Professor Hans Zwart for hosting him as a guest researcher in the group he chairs: Mathematics of Systems Theory (MAST), part of the Department of Applied Mathematics at the University of Twente. Also all other MAST group members are thanked as well for providing a nice and stimulating work and research environment.
- The first author thanks the following (former) PhD students in the MAST group for their help/support with mathematical details and Latex issues: Philip Preußler, Dr. Alexander Wierzba and Jens de Vries.
- Dr. Ir. Gjerrit Meinsma is greatly acknowledged for several discussions on mathematical/ modeling issues.
- The authors thank Dr. Melanie Bocquel for granting them the use of the constrained system PF example in Section 7.1.
- The authors thank Ms. Maryann Bjorklund MBA for proofreading and correcting the document for proper English. Any errors left, language-wise or otherwise, are the sole responsibility of the authors.

References

1 N. J. Gordon, D. J. Salmond, and A. F. M. Smith. Novel approach to nonlinear/non-Gaussian Bayesian state estimation. *IEE Proceedings F Radar Signal Process. UK*, 140(2):107, 1993.

2 F. Gustafsson. Particle filter theory and practice with positioning applications. *IEEE Aerospace and Electronic Systems Magazine*, 25(7):53–82, 2010.

3 F. Gustafsson, F. Gunnarsson, Niclas Bergman, U. Forssell, J. Jansson, R. Karlsson, and P.-J. Nordlund. Particle filters for positioning, navigation, and tracking. *IEEE Transactions on Signal Processing*, 50(2):425–437, 2002.

4 I. Smal, E. Meijering, K. Draegestein, N. Galjart, I. Grigoriev, A. Akhmanova, M. E. van Royen, A. B. Houtsmuller, and W. Niessen. Multiple object tracking in molecular bioimaging by Rao-Blackwellized marginal particle filtering. *Medical Image Analysis*, 12(6):764–777, 2008. Special issue on information processing in medical imaging.

5 B. D. O. Anderson and J. B. Moore. *Optimal Filtering*. Prentice-Hall, 1979.

6 R. E. Kalman. A new approach to linear filtering and prediction problems. *Journal of Basic Engineering*, 82(1):35, 1960.

7 Z. Chen. Bayesian filtering: from Kalman filters to particle filters, and beyond. *Statistics*, 182, (1):1–69, 2003.

8 S. J. Julier and J. K. Uhlmann. Unscented filtering and nonlinear estimation. *Proceedings of the IEEE*, 92(3):401–422, 2004.

9 I. Arasaratnam and S. Haykin. Cubature Kalman filters. *IEEE Transactions on Automatic Control*, 54(6):1254–1269, 2009.

10 G. Hendeby, R. Karlsson, F. Gustafsson, and N. Gordon. Performance issues in non-Gaussian filtering problems. In *Proceedings of the IEEE Nonlinear Statistical Signal Processing Workshop*, pages 65–68, 2006.

11 M. A. E. Bocquel. *Random Finite Sets in Multi-target Tracking Efficient Sequential MCMC implementation*, PhD thesis, University of Twente, Enschede, The Netherlands, 2013. CTIT Ph.D. Thesis Series No. 13-270, ISBN 978-90-365-0578-9.

12 F. Papi, M. Podt, Y. Boers, G. Battistello, and M. Ulmke. On constraints exploitation for particle filtering based target tracking. In *15th International Conference on Information Fusion*, pages 455–462, 2012.

13 F. Papi, M. A. E. Bocquel, M. Podt, and Y. Boers. Fixed-lag smoothing for Bayes optimal knowledge exploitation in target tracking. *IEEE Transactions on Signal Processing*, 62(12): 3143–3152, 2014.

14 D. Alspach and H. Sorenson. Nonlinear Bayesian estimation using Gaussian sum approximations. *IEEE Transactions on Automatic Control*, 17(4):439–448, 1972.

15 J. H. Kotecha and P. M. Djuric. Gaussian sum particle filtering. *IEEE Transactions on Signal Processing*, 51(10):2602–2612, 2003.

16 Y. Bar-Shalom, X.R. Li, and T. Kirubarajan. *Estimation with Applications to Tracking and Navigation*. Wiley, 2001.

17 B-N. Vo, M. Mallick, Y. Bar-Shalom, S. Coraluppi, R. Osborne III, R. Mahler, and B-T Vo. Multitarget tracking. In *Wiley Encyclopedia of Electrical and Electronics Engineering*, 1–25. John Wiley & Sons, Inc., 2015.

18 J. N. Driessen and Y. Boers. MAP estimation in particle filter tracking. In *Proceedings of the IET Seminar on Target Tracking and Data Fusion: Algorithms and Applications*, pages 39–39, 2008.

19 P. Del Moral. Measure-valued processes and interacting particle systems. Application to nonlinear filtering problems. *The Annals of Applied Probability*, 8(2):438–495, 1998.

20 N. Whiteley. Stability properties of some particle filters. *The Annals of Applied Probability*, 23(6):2500–2537, 2013.

21 P. Del Moral and M. Ledoux. Convergence of empirical processes for interacting particle systems with applications to non-linear filtering. *Journal of Theoretical Probability*, 13:225–257, 2000.

22 R. Douc, E. Moulines, and J. Olsson. Long term stability of sequential Monte Carlo under verifiable conditions. *The Annals of Applied Probability*, 24(5):1767–1802, 2014.

23 D. Kelly and A. Stuart. Ergodicity and accuracy of optimal particle filters for Bayesian data assimilation. *Chinese Annals of Mathematics, Series B*, 40:811–842, 2019.

24 D. Crisan and A. Doucet. A survey of convergence results on particle filtering methods for practitioners. *IEEE Transactions on Signal Processing*, 50(3):736–746, 2002.

25 H. R. Künsch. Recursive Monte Carlo filters: algorithms and theoretical analysis. *Annals of Statistics*, 33, (5):1983–2021, 2006.

26 X. Hu, T. Schön, and L. Ljung. A general convergence result for particle filtering. *IEEE Transactions on Signal Processing*, 59:3424–3429, 2011.

27 I. S. Mbalawata and S. Sarkka. Moment conditions for convergence of particle filters with unbounded importance weights. *Signal Processing*, 118:133–138, 2016.

28 Y. Boers and P. K. Mandal. Optimal particle-filter-based detector. *IEEE Signal Processing Letters*, 26(3):435–439, 2019.

29 H. V. Poor. *An Introduction to Signal Detection and Estimation*. Springer -Verlag, Berlin, Heidelberg, 1988.

30 P. Billingsley. *Probability and Measure*. A Wiley-Interscience publication. Wiley, New York, 1995.

31 E. Seneta. A Tricentenary history of the Law of Large Numbers. *Bernoulli*, 19(4):1088–1121, 2013.

32 M. S. Arulampalam, S. Maskell, N. Gordon, and T. Clapp. A tutorial on particle filters for online nonlinear/non-Gaussian Bayesian tracking. *IEEE Transactions on Signal Processing*, 50(2):174–188, 2002.

33 J. Carpenter, P. Clifford, and P. Fearnhead. An improved particle filter for non-linear problems. *IEE Proceedings Radar, Sonar Navigation*, 146(9):2000.

34 H. A. P. Blom and E. A. Bloem. Permutation invariance in Bayesian estimation of two targets that maneuver in and out formation flight. In *Proceedings of the 12th International Conference on Information Fusion*, Seattle, WA, pages 1296–1303, 2009.

35 H. A. P. Blom, E. A. Bloem, Y. Boers, and J. N. Driessen. Tracking closely spaced targets: Bayes outperformed by an approximation? In *Proceedings of the 11th International Conference on Information Fusion*, Cologne, Germany, pages 1–8, 2008.

36 E. H. Aoki, P. K. Mandal., L. Svensson, Y. Boers, and A. Bagchi. Labeling uncertainty in multitarget tracking. *IEEE Transactions on Aerospace and Electronic Systems*, 52(3):1006–1020, 2016.

37 R. Georgescu, P. Willett, L. Svensson, and M. Morelande. Two linear complexity particle filters capable of maintaining target label probabilities for targets in close proximity. In *Proceedings of the 15th International Conference on Information Fusion*, pages 2370–2377, 2012.

38 Y. Boers, E. Sviestins, and J. N. Driessen. Mixed labelling in multitarget particle filtering. *IEEE Transactions on Aerospace and Electronic Systems*, 46(2):792–802, 2010.

39 M. Guerriero, L. Svensson, D. Svensson, and P. Willett. Shooting two birds with two bullets: how to find minimum mean OSPA estimates. In *Proceedings of the 13th International Conference on Information Fusion*, Edinburgh, UK, pages 1–8, 2010.

40 M. Baum, P. Willett, and U. D. Hanebeck. Calculating some exact MMOSPA estimates for particle distributions. In *Proceedings of the 15th International Conference on Information Fusion*, pages 847–853, 2012.

41 A. van der Vaart and J. Wellner. *Weak Convergence and Empirical Processes*. Springer, New York, 1st edition, 1996.

42 S. J. Davey, M. G. Rutten, and N. J. Gordon. Track-before-detect techniques. In M. Mallick, V. Krishnamurthy, and B. N. Vo, editors, *Integrated Tracking, Classification and Sensor Management. Theory and Applications*, 311–344. John Wiley & Sons, Inc., Hoboken, NJ, 2013.

43 D. J. Salmond and H. Birch. A particle filter for track-before-detect. In *Proceedings of the American Control Conference*, pages 3755–3760, 2001.

44 Y. Boers and J. N. Driessen. Particle filter based detection for tracking. In *Proceedings of the American Control Conference*, pages 4393–4397, 2001.

8

Posterior Cramér–Rao Bounds in Cluttered Environments with Measurement Origin and Accuracy Uncertainties

Marcel Hernandez[1] *and Alfonso Farina*[2]

[1] *Department of Electrical Engineering and Electronics, University of Liverpool, Liverpool, UK*
[2] *Independent Researcher, Rome, Italy*

8.1 Introduction

The posterior Cramér–Rao lower bound (PCRLB) provides a bound on the mean squared error (MSE) accuracy of any unbiased estimator on an unknown parameter vector [1]. In the context of target tracking, the PCRLB is a powerful tool for quantifying the optimal achievable accuracy of target state estimation. This provides a mechanism for establishing the maximum degree to which suboptimal filtering algorithms could potentially be improved (e.g., [2]). Furthermore, the PCRLB enables system designers to identify whether the required tracking accuracy is theoretically achievable by a prototype sensor system. Moreover, the PCRLB has been used as a predictive performance metric in sensor management applications (e.g., see [3, 4]).

There has been an explosion of interest in the PCRLB in the last 25 years as a result of the establishment of an efficient recursive formula for the bound for nonlinear filtering [5]. This has led to the extensive development of PCRLBs for cluttered environments in which there can be missed detections (i.e., $P_d < 1$) and spurious false alarms (i.e., $P_{fa} > 0$); see [6] and references therein.

This chapter is concerned with calculating the PCRLB for tracking a noncooperative target in a cluttered environment in which there can be measurement origin or measurement accuracy uncertainties. To this end, Section 8.2 presents preliminaries (i.e., the efficient recursion of [5]). Following this, Section 8.3 reviews seminal approaches for the calculation of the PCRLB in the presence of measurement origin uncertainty. These approaches exploit information reduction factors (IRFs) to scale the contribution of each measurement according to the magnitude of the measurement origin uncertainty. The IRF was first identified in cluttered environments in [7], with the IRF PCRLB established in [8, 9]. Later, conditional approaches were used to determine less optimistic performance bounds [10–12].

Section 8.4 considers the calculation of the PCRLB when there are accuracy uncertainties associated with target generated measurements. Such uncertainties (referred to herein as "biases") can be the result of, e.g., intermittent sensor faults (e.g., [13]), or spoofing by an adversary to manipulate the sensor signal (e.g., [14]). In such cases, inspired by the conditional approaches developed in cluttered environments, a bound is determined conditional on the bias sequence of the measurements, with an unconditional bound calculated as a weighted average of the conditional bounds. As a baseline, a naive bound (analogous to the IRF PCRLB [8, 9]) is determined that does not directly take into account the bias sequence, but instead uses the expected information content of each measurement to account for potential measurement bias. The naive approach is computationally simple, but provides an optimistic bound because it fails to take into account the potential sequences of measurement biases, which can have a critical impact on performance if they occur shortly after a target has been detected (and has not yet been accurately localized) or if biased measurements are generated for extended periods (thereby continually degrading estimation performance).

In Section 8.5, we review recent developments in the calculation of the PCRLB when target generated measurements are intermittently degraded by autocorrelated specular multipath [15, 16] as a result of signal reflection at a surface boundary. In order to maintain the Markov property necessary to implement the efficient recursion of [5], the target state is augmented to include the biases created by the presence of multipath. The PCRLB is then calculated by conditioning on multipath sequence realizations and an unconditional bound is again calculated as a weighted average of the conditional bounds.

Simulations demonstrate the approaches, considering the following scenarios: (1) estimating the impact point of a ballistic missile in a cluttered environment (scenario #1; in Section 8.6); (2) tracking a ground-based vehicle in the presence of radar spoofing (scenario #2; in Section 8.7); (3) tracking a low flying airborne target in the presence of autocorrelated specular multipath (scenario #3; in Section 8.8). The simulation sections use a tutorial-style to fully explain the implementation of each approach. For scenario #1, the CRLBs are compared with the performance of a maximum likelihood estimator (MLE) [17]. For scenarios #2 and #3, the PCRLBs are compared to the performance of an interacting multiple model (IMM) filter [18] that uses two models to account for the presence or absence of biased/multipath measurements. As we shall see, in all three cases, the performance of the estimator is in close agreement with the bound.

8.2 Posterior Cramér–Rao Lower Bound

8.2.1 Definition

The PCRLB [1] for the estimation of the target state X_k at sampling time k is defined to be the inverse J_k^{-1} of the Fisher information matrix (FIM) J_k, i.e.,:

$$C_k \triangleq \mathbb{E}_{X_{0:k}, Z_{1:k}}[(X_k - \hat{X}_k)(X_k - \hat{X}_k)'] \geq J_k^{-1}, \tag{8.1}$$

where $\hat{X}_k$ denotes any unbiased estimator of X_k based on the sequence of measurements $Z_{1:k}$ up to sampling time k; and $X_{0:k}$ denotes the entire target state evolution up to sampling time k. The inequality in (8.1) denotes that the difference is a positive semi-definite matrix. The notation "$\mathbb{E}_{\phi}$" denotes mathematical expectation with respect to ϕ.

8.2.2 Computationally Efficient Formulation

Consider the discrete-time, nonlinear filtering problem with additive Gaussian process and measurement noise. The target state evolution is given by:

$$X_{k+1} = f_k(X_k) + \omega_k, \tag{8.2}$$

where $X_k \in \mathbb{R}^n$ is the target state at sampling time k, n is the dimensionality of the target state, $h_k(\cdot)$ is a potentially nonlinear function which models the state evolution in time, and $\omega_k \sim \mathcal{N}(0, \Sigma_k)$.

Measurements are generated at the sampling times, with the measurement sequence denoted by $Z_k, k = 1, 2, \ldots$.

Exploiting the Markov property and a block diagonal decomposition of the matrix inverse, the FIM, J_{k+1} at time $k + 1$ is given by the following recursion [5]:

$$J_{k+1} = D_k^{33} - (D_k^{12})'\left[J_k + D_k^{11}\right]^{-1} D_k^{12} + J_Z(k+1), \tag{8.3}$$

where:

$$D_k^{11} = \mathbb{E}_{X_k, X_{k+1}}\left[[\nabla_{X_k} \ln\ p(X_{k+1}|X_k)][\nabla_{X_k} \ln\ p(X_{k+1}|X_k)]'\right], \tag{8.4}$$

$$D_k^{12} = \mathbb{E}_{X_k, X_{k+1}}\left[[\nabla_{X_k} \ln\ p(X_{k+1}|X_k)][\nabla_{X_{k+1}} \ln\ p(X_{k+1}|X_k)]'\right], \tag{8.5}$$

$$D_k^{33} = \mathbb{E}_{X_k, X_{k+1}}\left[[\nabla_{X_{k+1}} \ln\ p(X_{k+1}|X_k)][\nabla_{X_{k+1}} \ln\ p(X_{k+1}|X_k)]'\right], \tag{8.6}$$

$$J_Z(k+1) = \mathbb{E}_{X_{k+1}, Z_{k+1}}\left[[\nabla_{X_{k+1}} \ln\ p(Z_{k+1}|X_{k+1})][\nabla_{X_{k+1}} \ln\ p(Z_{k+1}|X_{k+1})]'\right]. \tag{8.7}$$

and ∇_ζ is a first-order partial derivative operator with respect to the parameter vector ζ.

If the target dynamics are linear, with additive Gaussian noise, we can write:

$$X_{k+1} = A_k X_k + \omega_k. \tag{8.8}$$

It is then straightforward to show that (e.g., [9]):

$$J_{k+1} = \left(\Sigma_k + A_k J_k^{-1} A_k'\right)^{-1} + J_Z(k+1). \tag{8.9}$$

Equation (8.9) also holds if there is no process noise (i.e., $\Sigma_k = 0$; [19]). The recursive formula is initialized with $J_0 = C_0^{-1}$ where C_0 is the initial target state uncertainty covariance. C_0 is typically given either: (1) a nominal large value (reflecting a high degree of uncertainty in the initial target state), with the bound typically decreasing rapidly as the sampling time increases; (2) a value that reflects the target state uncertainty subsequent to track initialization (e.g., via the two-point method).

8.3 Posterior Cramér–Rao Lower Bounds with Measurement Origin Uncertainty

8.3.1 Measurement Model

We consider the problem of tracking a single target in a cluttered environment, in which there can be both missed detections and false alarms. Purely for notational simplicity, the theory is presented for a system composed of a single sensor. The extension to multisensor systems is straightforward provided that the sensors have independent measurement processes (conditional on the target state).

We use a common measurement model [8, 9, 11, 20], for which there can be a maximum of one target generated measurement per sampling time, which occurs with constant probability P_d. Each target generated measurement, $Z_k^{TG} \in \mathbb{R}^d$ is a nonlinear function of the target state, corrupted by white Gaussian noise, i.e.,:

$$Z_k^{TG} = h_k(X_k) + \epsilon_k \qquad \text{with } \epsilon \sim \mathcal{N}(0, R_k). \tag{8.10}$$

At each sampling time, the number of false alarm measurements has a Poisson distribution with mean λV, where V denotes the volume of the surveillance region A, and λ is the density of false alarms per unit volume. Each false alarm measurement is generated at random within the surveillance region.

In the PCRLB recursion (8.3), only the $J_Z(.)$ matrix is dependent on the measurement model, with the D-matrices dependent solely on the target dynamical model. In the following sections, several different formulations of the PCRLB are presented in cluttered environments. These formulations differ in the way in which they condition on the measurement sequence.

8.3.2 Information Reduction Factor Approach

8.3.2.1 General Approach

The IRF PCRLB is given by J_k^{-1}, with the sequence of FIMs given by the recursion (8.3), or by the recursion (8.9) if the target dynamics are linear. The measurement contribution is calculated as follows [8, 9, 20]:

$$J_Z(k) = E_{X_k, Z_k}\left[[\nabla_{X_k} \ln\ p(Z_k|X_k)]\ [\nabla_{X_k} \ln\ p(Z_k|X_k)]'\right], \tag{8.11}$$

$$= \sum_{m_k=1}^{\infty} p(m_k) J_Z(k : m_k), \tag{8.12}$$

where m_k is the total number of measurements (each of dimensionality d) generated in a *gated* observation region[1] at sampling time k; and $p(m_k)$ is the probability that there are m_k measurements in the gated observation region; $J_Z(k : m_k)$ is the measurement contribution given that there are m_k measurements at that time. This is given as follows:

$$J_Z(k : m_k) = E_{X_k, Z_k}\left[[\nabla_{X_k} \ln\ p(Z_k|X_k, m_k)]\ [\nabla_{X_k} \ln\ p(Z_k|X_k, m_k)]'\right]. \tag{8.13}$$

The measurement likelihood, $p(Z_k|X_k, m_k)$ is a mixture distribution, i.e., a mixture of the PDFs of target generated measurements and false alarms. This makes $J_Z(k : m_k)$ difficult to calculate. However, conditional on several unrestrictive assumptions, it can be shown that [9, 11]:

$$J_Z(k : m_k) = q_k(m_k) E_{X_k}\left[H_k' R_k^{-1} H_k\right] \text{ for } m_k > 0, \tag{8.14}$$

where $q_k(m_k) \in [0, 1]$ is an IRF conditional on there being m_k measurements at sampling time k [11]; and $H_k = \nabla_{X_k}\left[h_k(X_k)'\right]'$ is the Jacobian of the measurement vector (of target generated measurements). Clearly, $J_Z(k : 0) = 0$, i.e., there is no contribution if there are no measurements.

It follows from (8.12) and (8.14) that:

$$J_Z(k) = q_k E_{X_k}\left[H_k' R_k^{-1} H_k\right], \tag{8.15}$$

where:

$$q_k = \sum_{m_k=1}^{\infty} p(m_k) q_k(m_k). \tag{8.16}$$

The parameter $q_k \in [0, 1]$ is the overall IRF [9]. The IRFs q_k and $q_k(m_k)$ ($m_k > 0$) are typically independent of the target state, and quantify the effect of missed detections and false alarms. If there is no clutter and there are no missed detections, then $m_k = 1$ and $q_k(m_k) = q_k = 1$. In all other cases,

---

1 This region is a d-dimensional hyper-cube centered on the ground-truth measurement, with edge lengths $2g\sigma_1, \ldots, 2g\sigma_d$; where σ_i is the error standard deviation of the i-th component of each measurement; and g is a constant (typically $g = 4$).

$q_k(m_k)$ and q_k are less than unity; and the IRF reduces the measurement contribution (compared to the no clutter case), and therefore increases the PCRLB. If there are multiple sensors, with independent measurement processes, Eq. (8.15) is replaced by the sum total of the contributions of each individual sensor.

In Section 8.3.2.2, Eq. (8.18) provides a general expression for the IRF $q_k(m_k)$ (taken from [11]) and Eq. (8.22) then provides a numerical approximation of Eq. (8.18). Of particular note, if the measurement, clutter and detection models are time invariant then so too is the IRF $q_k(m_k)$, which need then only be calculated once (for each sensor). This calculation can be performed offline, enabling the PCRLB to be calculated efficiently, and implemented in real-time applications, e.g., [3].

An extension of the IRF approach also exists for the problem of tracking multiple targets in cluttered environments [21]. The sequence of manipulations is considerably more complex in that case, and the measurement origin uncertainty can be shown to manifest itself as an information reduction matrix.

8.3.2.2 Calculation of the IRFs

General Expression

The IRFs are given as follows (e.g., see Eq. (54) in [11]):

$$q_k(0) = 0, \tag{8.17}$$

$$q_k(m) = \frac{\epsilon(m)^2 |R_k|^{(m-2)/2}}{m V_g^{2m-2} (2\pi)^d} \left\{ \int_{Z_{md}=-g}^{g} \cdots \int_{Z_{m1}=-g}^{g} \cdots \int_{Z_{1d}=-g}^{g} \cdots \int_{Z_{11}=-g}^{g} \frac{Z_{11}^2 \exp\left\{-\sum_{i=1}^{d} Z_{1i}^2\right\}}{\left[\frac{(1-\epsilon(m))}{V_g^m} + \frac{\epsilon(m)}{m V_g^{m-1} \sqrt{(2\pi)^d |R_k|}} \sum_{r=1}^{m} \exp\left\{-\frac{1}{2} \sum_{j=1}^{d} Z_{rj}^2\right\}\right]} dZ_{11} \ldots dZ_{md} \right\}, \tag{8.18}$$

for $m > 0$, where:

- V_g is the volume of the gated observation region, given by:

$$V_g = (2g)^d \prod_{i=1}^{d} \sigma_i, \tag{8.19}$$

- $\epsilon(m_k)$ is the probability that one of the m_k measurements is target generated, given as follows:

$$\epsilon(m_k) = \frac{P_d^g}{p(m_k)} \frac{(\lambda V_g)^{m_k-1} \exp(-\lambda V_g)}{(m_k - 1)!}, \tag{8.20}$$

- $p(m_k)$ is the probability of m_k measurements in the gated observation region, given as follows:

$$p(m_k) = (1 - P_d^g) \frac{(\lambda V_g)^{m_k} \exp(-\lambda V_g)}{m_k!} + P_d^g \frac{(\lambda V_g)^{m_k-1} \exp(-\lambda V_g)}{(m_k - 1)!}, \tag{8.21}$$

- P_d^g is the probability of obtaining a target generated measurement in the gated observation region. Clearly, provided g is not small (typically $g = 4$) then $P_d^g \approx P_d$.

It is again noted that if the measurement, clutter and detection models are time invariant then so too are the IRFs, $q_k(m_k)$, $m_k > 0$, which then need only be calculated once.

Numerical Approximation

A Monte Carlo approximation of (8.18) is given as follows (see Eq. (58) in [11]):

$$q_k(m_k) \approx \frac{(2g)^{dm_k}\epsilon(m_k)^2|R_k|^{(m_k-2)/2}}{m_k V_g^{2m_k-2}(2\pi)^d} \times \frac{1}{N_p}\sum_{l=1}^{N_p} \frac{U_{1l}[l]^2 \exp\left\{-\sum_{i=1}^{d} U_{1i}[l]^2\right\}}{\left[\frac{(1-\epsilon(m_k))}{V_g^{m_k}} + \frac{\epsilon(m_k)}{m_k V_g^{m_k-1}\sqrt{(2\pi)^d|R_k|}}\sum_{r=1}^{m_k}\exp\left\{-\frac{1}{2}\sum_{j=1}^{d} U_{rj}[l]^2\right\}\right]}, \tag{8.22}$$

where $U_{ij}[l]$, $i = 1, \ldots, m_k$; $j = 1, \ldots, d$; $l = 1, \ldots, N_p$ are independent and identically distributed random variables drawn from a Uniform distribution on $[-g, g]$, i.e., each $U_{ij}[l] \sim U[-g, g]$.

8.3.3 Measurement Sequence Conditioning Approach

8.3.3.1 General Approach

Let $m_{1:k}$ denote the sequence comprising of the number of measurements available at sampling times 1 through k, i.e., $m_{1:k} \triangleq \{m_1, \ldots, m_k\}$. The measurement sequence conditioning (MSC) approach [11] creates a MSE estimator performance bound as follows:

$$E\left[(\hat{X}_k - X_k)(\hat{X}_k - X_k)'\right] = E_{m_{1:k}}\left[E\left[(\hat{X}_k - X_k)(\hat{X}_k - X_k)'\middle|m_{1:k}\right]\right] \tag{8.23}$$

$$\geq E_{m_{1:k}}\left[J_k^{-1}(m_{1:k})\right], \tag{8.24}$$

$$= \sum_{m_{1:k}}\left[Pr(m_{1:k}) \times J_k^{-1}(m_{1:k})\right], \tag{8.25}$$

$$\triangleq \text{MSC PCRLB}, \tag{8.26}$$

where $Pr(m_{1:k})$ is the probability of the sequence of measurements $m_{1:k}$ occurring, i.e.,:

$$Pr(m_{1:k}) = \prod_{i=1}^{k} p(m_i), \tag{8.27}$$

with $p(m_i)$ again given by (8.21). $J_k(m_{1:k})$ is the FIM conditional on the measurement sequence $m_{1:k}$.

Conditioning on $m_{1:k}$, and repeating the derivation of [5], the recursive formula for the conditional FIM is given by:

$$J_k(m_{1:k}) = D_{k-1}^{33} - (D_{k-1}^{12})'\left[J_{k-1}(m_{1:k-1}) + D_{k-1}^{11}\right]^{-1} D_{k-1}^{12} + J_Z(k : m_k), \tag{8.28}$$

$J_Z(k : m_k)$ is again the measurement contribution at sampling time k given that there are m_k measurements at that time (see Eqs. (8.13) and (8.14)). The initial FIM is given by $J_0 = C_0^{-1}$ irrespective of the sequence $m_{1:k}$.

If there are multiple sensors, with independent measurement processes, each measurement realization is given by a tuple of the number of measurements generated by each of the sensors. The Riccati recursion (8.28) is again applied, with the measurement contribution on each iteration (8.14) replaced by the sum total of the contributions of each individual sensor.

If the target dynamics are linear, then (8.28) can again be simplified as follows:

$$J_k(m_{1:k}) = \left(\Sigma_{k-1} + A_{k-1}J_{k-1}(m_{1:k-1})^{-1}A_{k-1}'\right)^{-1} + J_Z(k : m_k). \tag{8.29}$$

Finally, (8.25) is approximated by sampling measurement sequence (rather than enumerating the expectation for every possible measurement sequence), i.e.,:

$$\text{MSC PCRLB} \approx \frac{1}{n_m}\sum_{j=1}^{n_m} J_k^{-1}(m_{1:k}^j), \tag{8.30}$$

with $m_{1:k}^j \sim Pr(m_{1:k})$ for $j = 1, \dots, n_m$.

The MSC approach generalizes the enumeration approach of [10] to the case in which there are both false alarms and missed detections. To see this, if there are no false alarms (i.e., $\lambda = 0$), then $m_k = 0$ or 1 and it can easily be shown that $q_k(1) = 1$. Hence the approach of conditioning on the measurement sequence reduces to conditioning on the detection sequence, as in [10].

8.3.4 Measurement Existence Sequence Conditioning Approach

The measurement existence sequence conditioning (MESC) approach [12] is similar in principle to the MSC approach [11], but rather than condition on the number of measurements generated at each sampling time, the MESC approach conditions on the *existence* of at least one measurement. Therefore, instead of conditioning on $m_{1:k}$, the MESC approach conditions on $e_{1:k} \triangleq \{e_1, \dots, e_k\}$ where $e_k = 1$ if $m_k \geq 1$ and $e_k = 0$ otherwise. The performance bound is then given as follows:

$$\text{MESC PCRLB} \triangleq \sum_{e_{1:k}} \left[Pr(e_{1:k}) \times J_k^{-1}(e_{1:k})\right]. \tag{8.31}$$

Repeating the derivation of the recursive formula (8.3), conditional this time on $e_{1:k}$, it can easily be shown that:

$$J_k(e_{1:k}) = D_{k-1}^{33} - (D_{k-1}^{12})'\left[J_{k-1}(e_{1:k-1}) + D_{k-1}^{11}\right]^{-1} D_{k-1}^{12} + J_Z(k : e_k). \tag{8.32}$$

Again, the measurement contribution is zero if there are no measurements, i.e., $J_Z(k : e_k = 0) = 0$. In the case $e_k = 1$, it can then be shown that [12]:

$$J_Z(k : e_k = 1) = \frac{q_k}{Pr(e_k = 1)} E_{X_k}\left[H_k' R_k^{-1} H_k\right], \tag{8.33}$$

where $Pr(e_k = 1) = [1 - p(0)]$, with $p(0)$ given by (8.21). The IRF, q_k is again as given in Eq. (8.16). Finally, if the target dynamics are linear, Eq. (8.32) can again be simplified as follows:

$$J_k(e_{1:k}) = \left(\Sigma_{k-1} + A_{k-1} J_{k-1}(e_{1:k-1})^{-1} A_{k-1}'\right)^{-1} + J_Z(k : e_k). \tag{8.34}$$

The extension to multisensor systems is again straightforward but is omitted for brevity.

8.3.5 Relationships Between the Performance Bounds

The following relationships, proven in [12] hold between the three performance bounds:

$$\text{IRF PCRLB} \leq \text{MESC PCRLB} \leq \text{MSC PCRLB} \quad \text{for all sampling times, } k. \tag{8.35}$$

It is noted that IRF PCRLB ≤ MSC PCRLB was first proven in [22] for the case with no false alarms, and later proven in [11] in the general case. It is also noted that the inequalities (8.35) hold in general, and are not reliant on the measurement model prescribed in Section 8.3.1. Indeed, the proofs of the inequalities (8.35) are reliant only on convexity of the matrix inversion operation and Jensen's inequality (e.g., see Appendix A in [22]).

Three special cases are worthy of mention. These are as follows:

- *Case 1*: There are no false alarms (i.e., $\lambda = 0$). In this case, $m_k = e_k$ $(= 0$ or $1)$ for all k and the MSC and MESC approaches give identical bounds, equal to the enumeration bound of [10].
- *Case 2*: There are no missed detections (i.e., $P_d = 1$). In this case $e_k = 1$ for all k, $Pr(e_k = 1) = 1$ and the IRF and MESC approaches give identical bounds.
- *Case 3*: There are no false alarms and no missed detections. In this case, $m_k = e_k = 1$ for all k; and $q_k = q_k(1) = 1$. Clearly, in this case, all three performance bounds are identical, with each giving the PCRLB in a clutter-free environment.

8.3.6 Computational Considerations

The IRF PCRLB is computationally the simplest of the three performance bounds, because it only uses one Riccati recursion. The approach does not consider the impact of the actual sequence of measurements (it simply uses the expected information content of a measurement at each sampling time, averaged over the number of potential measurements). However, the actual sequence of measurements can have a critical impact on tracking performance, particularly after track formation, in which target state uncertainty can be high. At such times, the IRF PCRLB can be optimistic compared to the MSC and MESC PCRLBs [11, 12].

The MESC approach has the advantage that it can be enumerated exactly, because there are precisely 2^k existence sequences $e_{1:k}$ at each sampling time k. By comparison, the MSC approach has an infinite number of potential measurement sequences. However, many of these sequences have an extremely low probability of occurrence and can be instantly discounted. Although the MSC approach produces the least optimistic performance bound, simulations have yielded insignificant differences between the MESC and MSC bounds [12]. Furthermore, as the number of sampling times increases, all three bounds converge to the same steady-state value [11, 12]. Hence, in assessing the long-term performance of an estimator, the computationally simple IRF PCRLB is often sufficient.

8.4 Posterior Cramér–Rao Lower Bounds with Intermittently Inflated Measurement Errors

8.4.1 Measurement Model

Nonlinear measurements $Z_k \in \mathbb{R}^d$ of the target state $X_k \in \mathbb{R}^n$ are available at discrete sampling times $k = 1, 2, \ldots$, with each measurement corrupted by additive white Gaussian noise $\epsilon_k \sim \mathcal{N}(0, R_k)$, as in (8.10). At each sampling time k, the measurement error covariance R_k can take one of two values:

1. $R_k = R_k^0$ if the sensor is operating normally.
2. $R_k = R_k^1$ (with $R_k^1 \gg R_k^0$) if the sensor is operating in an abnormal/faulty manner.

Inflation of the measurement error covariance can occur for a number of reasons, including, e.g., faulty operation of the sensor (e.g., [13]), spoofing by an adversary to manipulate the sensor signal (e.g., [14] and references therein), or signal multipath (e.g., [16, 23]).[2] The inflation of the measurement error covariance is assumed to occur intermittently, not persistently.

2 It is noted that the analysis of autocorrelated multipath effects is considered in scenario #3 presented in Section 8.8.

The indicator variable b_k (termed "bias") is used to denote the presence (denoted $b_k = 1$) or absence (denoted $b_k = 0$) of the inflated covariance at each sampling time k. The transitions between the two error models are governed by a Markov chain with known transition probabilities, $p_{ij} \triangleq Pr(b_{k+1} = j|b_k = i)$.

It is assumed that $P_d \approx 1.0$ and $P_{fa} \approx 0.0$. This allows the degradation in tracking performance as a result of the inflation in the measurement error covariance to be analyzed in isolation. It is noted that the approaches developed in this section can be combined with the PCRLB approaches described in Section 8.3 if there is also measurement origin uncertainty resulting in missed detections and false alarms.

8.4.2 Posterior Cramér–Rao Lower Bounds

8.4.2.1 Naive Approach

The target dynamics are independent of the measurement model. Hence, the FIM, J_{k+1} at time $k+1$ is again given by the recursion (8.3), i.e.,:

$$J_{k+1} = D_k^{33} - (D_k^{12})' \left[J_k + D_k^{11}\right]^{-1} D_k^{12} + J_Z(k+1). \tag{8.36}$$

Similar to (8.12) we can write:

$$J_Z(k) = \sum_{b_k=0}^{1} p(b_k) J_Z(k : b_k), \tag{8.37}$$

with (similar to (8.13)):

$$J_Z(k : b_k) = E_{X_k, Z_k}\left[\left[\nabla_{X_k} \ln\ p(Z_k|X_k, b_k)\right] \left[\nabla_{X_k} \ln\ p(Z_k|X_k, b_k)\right]'\right]. \tag{8.38}$$

It then follows from (8.14) that:

$$J_Z(k : b_k) = E_{X_k}\left[H_k'(R_k^{b_k})^{-1} H_k\right] \text{for } b_k \in \{0, 1\}, \tag{8.39}$$

where to remind the reader, $H_k = \nabla_{X_k}\left[h_k(X_k)'\right]'$.

Calculating the PCRLB via Eqs. (8.36), (8.38), and (8.39) is referred to as the naive error covariance averaging (NECA) PCRLB approach.

8.4.2.2 Conditioning on the Measurement Sequence

The primary drawback with the naive approach introduced in Section 8.4.2.1 is that it does not take into account the actual sequence of inflated measurement errors, which can have a critical impact on performance, particularly initially when the target state may be highly uncertain.

Motivated by this, and analogous to the MSC approach, the measurement error covariance sequence conditioning (MECSC) approach is developed. Define $b_{1:k} \triangleq \{b_1, \dots, b_k\}$. The MECSC PCRLB is then given as follows:

$$E\left[(\hat{X}_k - X_k)(\hat{X}_k - X_k)'\right] = E_{b_{1:k}}\left[E\left[(\hat{X}_k - X_k)(\hat{X}_k - X_k)' \middle| b_{1:k}\right]\right] \tag{8.40}$$

$$\geq E_{b_{1:k}}\left[J_k^{-1}(b_{1:k})\right], \tag{8.41}$$

$$= \sum_{b_{1:k}} \left[Pr(b_{1:k}) \times J_k^{-1}(b_{1:k})\right], \tag{8.42}$$

$$\triangleq \text{MECSC PCRLB}, \tag{8.43}$$

where $Pr(b_{1:k})$ is the probability of the sequence $b_{1:k}$ and $J_k(b_{1:k})$ is the FIM conditional on the measurement sequence $b_{1:k}$.

Conditioning on $b_{1:k}$, and repeating the derivation of [5], the recursive formula for the conditional FIM is given by:

$$J_k(b_{1:k}) = D_{k-1}^{33} - (D_{k-1}^{12})' \left[J_{k-1}(b_{1:k-1}) + D_{k-1}^{11}\right]^{-1} D_{k-1}^{12} + J_Z(k : b_k), \tag{8.44}$$

with $J_Z(k : b_k)$ given by Eq. (8.39).

If the target dynamics are linear, then Eq. (8.44) can be simplified as follows:

$$J_k(b_{1:k}) = \left(\Sigma_{k-1} + A_{k-1} J_{k-1}(b_{1:k-1})^{-1} A'_{k-1}\right)^{-1} + J_Z(k : b_k). \tag{8.45}$$

The initial FIM is given by $J_0 = C_0^{-1}$ irrespective of the sequence $b_{1:k}$.

Rather than enumerate (8.43) for all possible sequences $b_{1:k}$, we use the following approximation:

$$\text{MECSC PCRLB} \approx \frac{1}{n_b} \sum_{j=1}^{n_b} J_k^{-1}(b_{1:k}^j), \tag{8.46}$$

where $b_{1:k}^j \sim Pr(b_{1:k})$, $j = 1, \dots, n_b$ are samples of the bias sequence up to sampling time k.

8.4.2.3 Relationship Between the Two Bounds

It is straightforward to show that:

$$\text{NECA PCRLB} \leq \text{MECSC PCRLB} \qquad \text{for all sampling times, } k. \tag{8.47}$$

The proof exploits convexity of the matrix inversion operation, Jensen's inequality and the fact that by definition $J_Z(k) = \mathbb{E}_{b_k}[J_Z(k, b_k)]$. The proof is almost identical to those presented in [11, 22] and is omitted for brevity.

Hence, the naive approach determines a PCRLB that is more optimistic than the MECSC approach. However, it should be noted that the naive approach is computational much simpler than the MECSC approach, because it does not need to sample sequences $b_{1:k}$, and may be a suitable approximation in some applications (e.g., in providing a predictive performance metric for online sensor management in time critical applications, such as those presented in [4]). It should also be noted that the MECSC approach implicitly knows b_k. Therefore, it is likely to offer an optimistic bound in scenarios in which it is challenging to accurately infer b_k.[3]

8.5 Posterior Cramér–Rao Bound with Autocorrelated Multipath Measurements

8.5.1 Multipath Measurement Model

Again, nonlinear measurements $Z_k \in \mathbb{R}^d$ of the target state $X_k \in \mathbb{R}^n$ are available at discrete sampling times k (= 1, 2, ...), with each measurement corrupted by additive white Gaussian noise $\epsilon_k \sim \mathcal{N}(0, R_k)$ (again, see Eq. (8.10)). The multipath measurement model (see [15, 16]) is a generalization of the model introduced in [23], and is given as follows.

---

3 e.g., in tracking a target that intermittently performs maneuvres. In this case, it may be difficult to distinguish between the following hypotheses: (1) highly inaccurate measurements of a non-maneuvring target, (2) accurate measurements of a maneuvering target.

- A multipath effect r_k may be present, or absent, at each sampling time k. The transitions between the two measurement models: "multipath effect absent" (denoted by $r_k = 0$) and "multipath effect present" (denoted by $r_k = 1$) are governed by a Markov chain with known transition probabilities, $p_{ij} \triangleq Pr(r_{k+1} = j | r_k = i)$.
- If a multipath effect is present, each measurement is further corrupted by an additional measurement noise term $w_M(k) \in \mathbb{R}^{d_M}$, which is an autocorrelated sequence.[4]

The autocorrelated sequence of multipath effects is given as follows:

$$w_M(k+1) = r_{k+1}\left[\alpha_k w_M(k) + v_M(k)\right], \tag{8.48}$$

where α_k is the autocorrelation between successive multipath induced biases; and $v_M(k) \sim \mathcal{N}(0, \Sigma_M)$. It is assumed that ϵ_k and $v_M(k)$ are independent of the target state. It is noted that if $\alpha_k = 0$, the measurement model reduced to the model in Section 8.4.1, with Σ_M giving the increase in the measurement error covariance when $r_k = b_k = 1$.

Each measurement can be written as follows:

$$Z_k = h_k(X_k) + \epsilon_k + I'_{d_M \times d} w_M(k), \tag{8.49}$$

where $h_k(X_k) \triangleq (h_k^1(X_k) \dots h_k^d(X_k))'$ is the ground-truth, error-free measurement. $I_{d_M \times d}$ is a $(d_M \times d)$ matrix with $\left[I_{d_M \times d}\right]_{ij} = 1$ only if the i-th bias term affects the j-th measurement component, and $\left[I_{d_M \times d}\right]_{ij} = 0$ otherwise. It is noted that if $d_M = d$, then $I_{d_M \times d} = I_d$, where I_d denotes the d-dimensional identity matrix.

Further assumptions—We assume that the values of α_k and Σ_M are known (and constant throughout). In cases for which α_k and Σ_M are unknown, they may be approximated using suitably large values that allow the estimation algorithm to manage the tradeoff between exploitation (of the measurements) and robustness (i.e., non-divergent behavior). It is noted that in the most general case, α_k will be dependent on the target state X_k, and Σ_M will be dependent on both α_k and X_k. This significantly complicates the estimation problem and is left for future work, as noted in [16].

8.5.2 Posterior Cramér–Rao Lower Bound

8.5.2.1 Augmentation of the Target State

Let $\bar{X}_k = (X_k' \; w_M(k)')'$ denote the augmented target state, that includes the multipath measurement bias, at sampling time k. Let the target state evolution be given by Eq. (8.8),[5] with the bias evolution given by Eq. (8.48). The evolution of the augmented target state can then be written as follows:

$$\bar{X}_{k+1} = \bar{A}_k(r_{k+1})\bar{X}_k + \bar{v}_k(r_{k+1}), \tag{8.50}$$

where $\bar{v}_k(r_{k+1}) \sim \mathcal{N}(0, \bar{\Sigma}_k(r_{k+1}))$, and:

$$\bar{A}_k(r_{k+1}) = \begin{pmatrix} A_k & 0_{d \times d_M} \\ 0_{d_M \times d} & r_{k+1}\alpha_k I_{d_M} \end{pmatrix}, \tag{8.51}$$

---

4 It is noted that multipath may not affect every component of each measurement. Hence, the number of nonzero multipath induced noise terms may be less than the dimensionality of each measurement, i.e., $d_M \leq d$.

5 It is noted that if the target state evolution is nonlinear, the approach developed herein can still be applied, but the conditional Riccati equation then follows from Eq. (8.3).

$$\bar{\Sigma}_k(r_{k+1}) = \begin{pmatrix} \Sigma_k & 0_{d \times d_M} \\ 0_{d_M \times d} & r_{k+1}\Sigma_M \end{pmatrix}, \tag{8.52}$$

where $0_{d_M \times d}$ denotes a $d_M \times d$ matrix of zeros.

This augmentation was introduced in [15, 16]. It is noted that augmenting the target state to include additional variables has previously been performed in [24] in order to calculate PCRLB-like measures for tracking a maneuvering target whose motion is governed by a quantized acceleration model.

8.5.2.2 Conditioning on the Multipath Sequence

Let $r_{1:k} \triangleq \{r_1, \dots, r_k\}$ denote a sequence of multipath effects (present/absent) up to sampling time k. We then have:

$$C(\hat{X}_k; r_{1:k}) \triangleq \mathbb{E}_{\bar{X}_{0:k}, Z_{1:k} | r_{1:k}}\left[(\bar{X}_k - \hat{X}_k)(\bar{X}_k - \hat{X}_k)' \middle| r_{1:k}\right] \geq \bar{J}_k(r_{1:k})^{-1}, \tag{8.53}$$

where $\hat{X}_k$ is any unbiased estimator of $\bar{X}_k$ conditional on the sequence of multipath effects $r_{1:k}$ and $\bar{J}_k(r_{1:k})$ denotes the FIM at sampling time k conditional on $r_{1:k}$.
A recursive formula for $\bar{J}_k(r_{1:k})$ conditional on $r_{1:k}$ is then, according to (8.9), given as follows [5][6]:

$$\bar{J}_k(r_{1:k}) = \left(\bar{\Sigma}_{k-1}(r_k) + \bar{A}_{k-1}(r_k)\bar{J}_{k-1}(r_{1:k-1})^{-1}\bar{A}_{k-1}(r_k)'\right)^{-1} + \bar{J}_Z(k, r_k), \tag{8.54}$$

where $\bar{J}_Z(k, r_k)$ denotes the conditional measurement contribution:

$$\bar{J}_Z(k, r_k) = \mathbb{E}_{\bar{X}_k, Z_k | r_k}\left[[\nabla_{\bar{X}_k} \ln p(Z_k|\bar{X}_k, r_k)][\nabla_{\bar{X}_k} \ln p(Z_k|\bar{X}_k, r_k)]'\right]. \tag{8.55}$$

It is shown in [15] that:

$$\bar{J}_Z(k, r_k) = \begin{pmatrix} B_{11} & r_k B_{12} \\ r_k B_{12}' & r_k^2 B_{22} \end{pmatrix}, \tag{8.56}$$

with the constituent $(n \times n)$, $(n \times d_M)$ and $(d_M \times d_M)$ submatrices are given as follows:

$$B_{11} = \mathbb{E}_{X_k}\left[H_k' R_k^{-1} H_k\right], \tag{8.57}$$

$$B_{12} = \mathbb{E}_{X_k}\left[H_k' R_k^{-1} I_{d_M \times d}'\right], \tag{8.58}$$

$$B_{22} = I_{d_M \times d} R_k^{-1} I_{d_M \times d}', \tag{8.59}$$

where $H_k = \nabla_{X_k}\left[h_k(X_k)'\right]'$ is again the Jacobian of a target generated measurement. It is noted that the conditional measurement contribution $\bar{J}_Z(k, r_k)$ is independent of the multipath noise covariance Σ_M, which is modeled as additional process noise.

---

6 It is noted that without the augmentation, conditional on the target state X_k, each measurement Z_k would depend on the entire multipath sequence $r_{1:k}$ (because of the autocorrelation across sampling times). This would preclude exploitation of the Markov property, as in [5], and the conditional FIM $\bar{J}_k(r_{1:k})$ would have to be calculated from the full $(kn \times kn)$ FIM for the estimation of the entire target state evolution up to sampling time k.

Without loss of generality, it is assumed that the multipath effect does not occur initially, because of the difficulties this would subsequently cause for track initialization. Therefore, each conditional FIM is initialized as follows:

$$\bar{J}_0 = \text{diag}(\Delta_1, \dots, \Delta_n, 1/\delta, \dots, 1/\delta), \tag{8.60}$$

where δ again takes a nominal small value in order to ensure that all matrix inversions remain non-singular. The parameters Δ_i, $i = 1, \dots, n$ take nominal small values to signify that there is no prior knowledge regarding the initial target state.

As non-multipath measurements become increasingly accurate (i.e., $R_k \to 0$), both the lower-right block of $\bar{J}_Z(k, r_k)$ and the lower-right block of the conditional FIM $\bar{J}_k(r_{1:k})$ increase in value. Consequently, the lower-right block of the conditional PCRLB decreases in value. Qualitatively, this shows that as the (non-multipath) measurement accuracy improves, the ability to estimate the multipath biases also improves. A plausible explanation for this is that when the measurement errors are lower, the multipath effect is more easily observed/identified and therefore more easily mitigated/estimated.

8.5.2.3 Unconditional Posterior Cramér–Rao Lower Bound

Let us make the following definitions:

- $\check{\mathbf{X}}_k$ is the set of all unbiased estimators $\check{X}_k$ of the augmented target state $\bar{X}_k$ at sampling time k in the case that the sequence $r_{1:k}$ is <u>not known</u>.
- $\widehat{\mathbf{X}}_k$ is the set of all unbiased estimators $\hat{X}_k$ of the augmented target state $\bar{X}_k$ at sampling time k in the case that the sequence $r_{1:k}$ <u>is known</u>.

Clearly $\check{\mathbf{X}}_k \subseteq \widehat{\mathbf{X}}_k$, because the set $\widehat{\mathbf{X}}_k$ will also contain estimators that do not exploit the fact that the sequence $r_{1:k}$ is known, and these estimators constitute the set $\check{\mathbf{X}}_k$.

It then follows that:

$$C(\check{X}_k; r_{1:k}) \geq \min_{\hat{X}_k \in \widehat{\mathbf{X}}_k} C(\hat{X}_k; r_{1:k}), \tag{8.61}$$

$$\geq \bar{J}_k(r_{1:k})^{-1} \quad \text{(from (8.53))} \qquad \text{for } \check{X}_k \in \check{\mathbf{X}}_k. \tag{8.62}$$

Taking the expectations of both sides of Eq. (8.62) with respect to $r_{1:k}$, it follows that:

$$\mathbb{E}_{\bar{X}_{0:k}, Z_{1:k}}\left[\left(\bar{X}_k - \check{X}_k\right)\left(\bar{X}_k - \check{X}_k\right)'\right] \geq P_k \qquad \text{for } \check{X}_k \in \check{\mathbf{X}}_k, \tag{8.63}$$

where:

$$P_k = \mathbb{E}_{r_{1:k}}\left[\bar{J}_k(r_{1:k})^{-1}\right]. \tag{8.64}$$

It is clear from inequality (8.63) that P_k is a PCRLB at sampling time k for any unconditional, unbiased estimator of the augmented target state $\bar{X}_k$.

The unconditional PCRLB (8.64) is approximated as follows:

$$P_k \approx \frac{1}{N_m} \sum_{j=1}^{N_m} \left[\bar{J}_k(r^j_{1:k})^{-1}\right], \tag{8.65}$$

where $r^j_{1:k}$, $j = 1, \dots, N_m$ are sampled realizations of the multipath sequence up to time k, and N_m is the number of multipath sequences sampled.

8.6 Simulation Scenario #1–Impact Point of a Ballistic Missile

8.6.1 Scenario Overview

This scenario is concerned with estimating the impact point of a ballistic missile (see also [11, 17]). It is noted that the CRLB is calculated, not the PCRLB, because, as we shall see, the target motion is deterministic with no random perturbations. We assume that the time of launch of the missile is known.

In [17], a heuristic CRLB and a MLE were developed for this scenario, and assumed that a maximum of one false alarm was generated per time step, with probability $P_{FA}^{g} \in \{0.0, 0.01, 0.05\}$. In [11], for comparison, the IRF CRLB and MSC CRLB were calculated for this scenario, also assuming a maximum of one false alarm measurement per sampling time and using approximations valid when the false alarm rate is low (see Section VIII. in [11]). Herein, approximations are not used, and instead the average number of false alarms in the gated observation region per sampling time is set equal to P_{FA}^{g}, i.e., $\lambda V_g = P_{FA}^{g}$. The IRF and MSC CRLBs are then calculated via the methodology in Sections 8.3.2 and 8.3.3, respectively.[7]

It is noted that the IRF and MSC approaches allow more than one false alarm to be generated per sampling time, something the heuristic approach of [17] is unable to deal with. Hence, as well as comparing the IRF and MSC CRLBs to the heuristic CRLB and MLE introduced in [17], we also calculate the IRF and MSC CRLBs for scenarios with higher false alarm rates, and demonstrate the degradation in optimal estimation performance in high-clutter environments.

8.6.2 Target Motion

The state of a ballistic missile is denoted by $X_k = (x_k\ \dot{x}_k\ y_k\ \dot{y}_k\ z_k\ \dot{z}_k)'$ where the notation is standard. The ballistic missile is subjected to an initial instantaneous thrust, and there is no drag.[8] Hence the velocity components $\dot{x}_k$ and $\dot{y}_k$ are constant. The component $\dot{z}_k$ is subjected to gravitational pull, with the acceleration due to gravity, $a_g = 9.81$ m/s^2.

Hence, the deterministic target motion can be written as:

$$X_{k+1} = AX_k + Y, \tag{8.66}$$

where:

$$A_k = I_3 \otimes \begin{pmatrix} 1 & t \\ 0 & 1 \end{pmatrix} \qquad Y = \left(0,\ 0,\ 0,\ 0,\ -\frac{1}{2}a_g t^2,\ -a_g t\right)'. \tag{8.67}$$

$\otimes$ denotes the Kronecker product, and t is the sampling interval. The elevation of the missile as a function of time is shown in Figure 8.1a, with the missile trajectory shown in Figure 8.2a,b.

8.6.3 Measurement Model

An active radar generates measurements of azimuth (θ), elevation (ϕ), and range (ρ). Target generated measurements are then given as follows:

$$Z_k = h_k(X_k) + \varepsilon_k \qquad \varepsilon_k \sim N(0, R_k), \tag{8.68}$$

7 The MESC CRLB is indistinguishable from the MSC CRLB for all parameter values considered, and is not shown.
8 To remind the reader, Chapter 2 considered the tracking a ballistic missile (during the reentry phase), and rigorously calculated the process noise covariance when drag was included in the model (see also [25]).

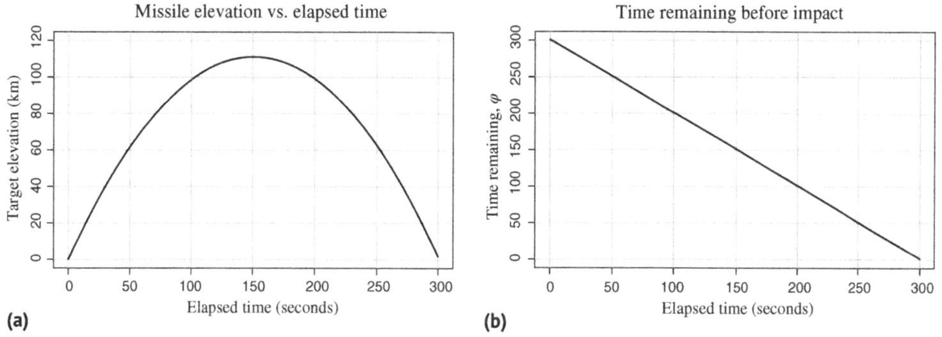

Figure 8.1 (a) The target elevation and (b) the time remaining before impact φ (in seconds), each plotted against elapsed time.

where:

$$h_k(X_k) \triangleq \begin{pmatrix} \theta_k \\ \phi_k \\ \rho_k \end{pmatrix} = \begin{pmatrix} \tan^{-1}(y_k - y_k^S, x_k - x_k^S) \\ \tan^{-1}(z_k - z_k^S, r_{xy}) \\ \sqrt{r_{xy}^2 + (z_k - z_k^S)^2} \end{pmatrix}, \tag{8.69}$$

where r_{xy} denotes the flat-Earth distance between the radar and the missile; $x^S(k) \triangleq (x_k^S, y_k^S, z_k^S)$ denotes the Cartesian coordinates of the radar at sampling time k, and $\tan^{-1}$ is the two-argument inverse tangent function (e.g., see [26] and references therein). The measurement error standard deviations of the azimuth, elevation, and range measurements are given by σ_θ, σ_ϕ and σ_ρ respectively, and $R = \text{diag}(\sigma_\theta^2, \sigma_\phi^2, \sigma_\rho^2)$.

At each sampling time, the probability of detecting the target is denoted by P_D. In calculating the CRLBs, the number of false alarm measurements has a Poisson distribution with mean λV. Each false alarm measurement is generated at a random location in the surveillance region.

8.6.4 Summary of Parameter Settings

The parameter values used in the simulations are summarized in Table 8.1.

8.6.5 Calculation of the Cramér–Rao Lower Bounds

8.6.5.1 Cramér–Rao Bounds at Each Sampling Time

Under the IRF approach, the CRLB (J_k^{-1}) is given by Eq. (8.9) with $A_k = A$ and $\Sigma_k = 0$ for all sampling times k. When using the MSC approach in which we condition on the measurement sequence, the CRLB is given by Eq. (8.25). The recursive formula that specifies each $J_k(m_{1:k})$ is then given by Eq. (8.29), again with $A_k = A$ and $\Sigma_k = 0$ for all sampling times k.

We note that in each formulation, the CRLB is independent of the gravitational pull (given by Y in (8.66) and (8.67)). To see why this is so, referring to the FIM recursion (8.3) and its constituent matrices (8.4)–(8.6), it can be seen that the contribution of the uncertain target dynamics is given by the derivatives of the log-likelihood: $\ln p(X_{k+1}|X_k)$ with respect to X_k and X_{k+1}. Clearly, these derivatives do not include the constant term Y.

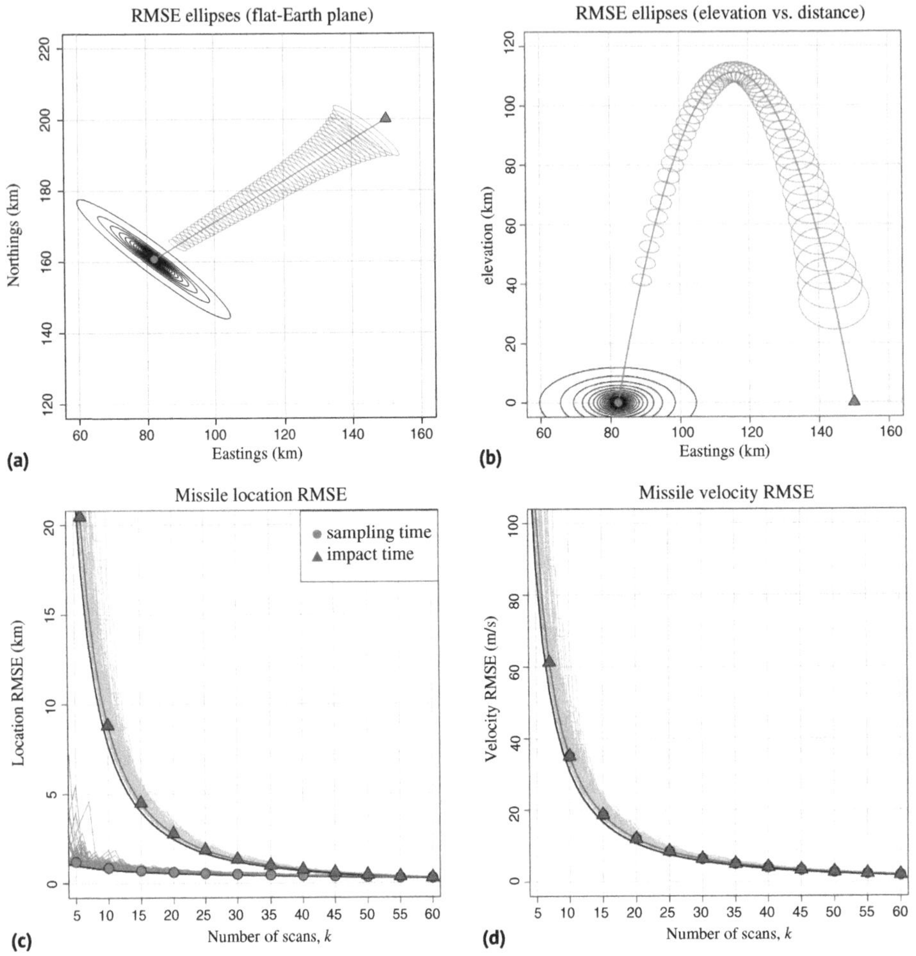

Figure 8.2 CRLB values calculated using the MSC approach, for a scenario in which $P_D = 0.9$ and $P^g_{FA} = 0.05$. (a,b) Ballistic missile location uncertainty at each sampling time (10× RMSE ellipses; in gray) and the associated impact point location uncertainties (1× RMSE ellipses; in black). Each ellipses is centered on the target coordinates at each sampling time. The solid line shows the missile trajectory, with the triangle marking the launch location and the circle showing the impact location. (c) Ballistic missile location RMSE at each sampling time and at the impact time. (d) Ballistic missile velocity RMSE; note that the motion of the missile is deterministic, therefore the velocity RMSE values have the same value at the sampling time and the impact time. In (c) and (d) the gray lines show the conditional RMSE values for 1,000 measurement sequence realizations.

Using the IRF approach, the measurement contribution $J_Z(k)$ is given by Eq. (8.15). Using the MSC approach, the measurement contribution $J_Z(k : m_k)$ is given by Eq. (8.14). We note that the expectations "$\mathbb{E}_{X_k}$" in Eqs. (8.14) and (8.15) are redundant here because the target path X_k is deterministic. Therefore, we have:

$$J_Z(k) = q_k H_k(X_k^\star)' R_k^{-1} H_k(X_k^\star), \tag{8.70}$$

$$J_Z(k : m_k) = q_k(m_k) H_k(X_k^\star)' R_k^{-1} H_k(X_k^\star), \tag{8.71}$$

where $X_k^\star$ denotes the ballistic missile location at sampling time k.[9]

The 6×3 Jacobian matrix $H_k(X_k)' = \nabla_{X_k}\left[h_k(X_k)'\right]$ is given as follows:

$$H_k(X_k)' = \begin{pmatrix} \frac{-(y_k - y_k^S)}{r_{xy}^2} & \frac{-(x_k - x_k^S)(z_k - z_k^S)}{\rho_k^2 r_{xy}} & \frac{(x_k - x_k^S)}{\rho_k} \\ 0 & 0 & 0 \\ \frac{(x_k - x_k^S)}{r_{xy}^2} & \frac{-(y_k - y_k^S)(z_k - z_k^S)}{\rho_k^2 r_{xy}} & \frac{(y_k - y_k^S)}{\rho_k} \\ 0 & 0 & 0 \\ 0 & \frac{r_{xy}}{\rho_k^2} & \frac{(z_k - z_k^S)}{\rho_k} \\ 0 & 0 & 0 \end{pmatrix}. \tag{8.72}$$

For the MSC approach, each IRF, $q_k(m_k)$, is calculated via (8.22). For the IRF approach, the overall IRF q_k is given by (8.16).

Each bound is initialized with $J_0 = C_0^{-1}$ where the initial target state covariance, C_0 is given a very large value, which is indicative of only minimal prior knowledge. The initial uncertainties in the x- and y- Cartesian coordinate directions each have a standard deviation of a massive 100 km. The initial uncertainty in the z-coordinate direction is lower (i.e., with a standard deviation 100 meters) because it is known that the ballistic missile will be launched from the ground. The uncertainties in the initial velocity components of the missile are also large, a further indication of the lack of prior knowledge (again, see Table 8.1 for a summary of parameter values used in the simulations).

Table 8.1 Parameter settings used in the simulations (unless stated otherwise).

Parameter	Value
Sensor location, (x^S, y^S, z^S)	(0 km, 0 km, 0 km)
Initial target location, (x_0, y_0, z_0),	(150 km, 200 km, 10 m)
Initial target velocity, $(\dot{x}_0, \dot{y}_0, \dot{z}_0)$	(−225.6 m/s, −130.2 m/s, 1477.2 m/s)
Initial target location covariance, C_0^l	diag(10,000 km^2, 10,000 km^2, 10 km^2)
Initial target velocity covariance, C_0^v	diag(25,000 m^2/s^2, 25,000 m^2/s^2, 25,000 m^2/s^2)
Measurement error standard deviation, $(\sigma_\theta, \sigma_\phi, \sigma_\rho)$	(0.3^o, 0.3^o, 100 m)
Size of measurement gate, g	4
Volume of the gated observation region, V_g	$8g^3\sigma_\theta\sigma_\phi\sigma_\rho$
False alarm probability (in gated region), P_{FA}^g	0.0, 0.01, 0.05, 0.5
Average false alarms (in gated region), λV_g	0.0, 0.01, 0.05, 0.5
Probability of detection, P_D	0.3, 0.5, 0.7, 0.9, 1.0
Maximum number of sensor scans	60
Sampling interval, t	5 seconds

9 In applications in which the target path is non-deterministic, (8.14) and (8.15) can be calculated using target state evolutions sampled from the prior target state distribution, e.g., see scenarios #2 and #3 in Sections 8.7 and 8.8, respectively.

8.6.5.2 Cramér–Rao Bounds for Impact Point Estimation

The error bound for the impact point $X_{t(I)}$ is now calculated (see also [17]), where excusing a slight abuse of notation, $t(I)$ denotes the elapsed time at which the ballistic missile strikes the surface of the Earth. This error bound is based on measurements generated at sampling times $k = 1, \dots, K$.

To recap, in Section 8.6.5.1, CRLBs were determined for the target state X_K at each sampling time K, based on the measurements available at times $1, \dots, K$. These CRLBs for X_K can then be extrapolated forwards to determine error bounds for $X_{t(I)}$, enabling direct comparison of the performance of the IRF and MSC CRLB approaches with the heuristic CRLB and MLE developed in [17].

It follows from (8.66) and (8.67) that:

$$X_{t(I)} = \hat{A}^{\varphi} X_K + \hat{Y}^{\varphi}, \tag{8.73}$$

where $\varphi = (t(I) - Kt)$ is the time remaining between the K-th measurement and the impact time. In Figure 8.1b, the time remaining before impact is plotted against the elapsed time for the focal scenario.

The matrices $\hat{A}^{\varphi}$ and $\hat{Y}^{\varphi}$ are then given by:

$$\hat{A}^{\varphi} = I_3 \otimes \begin{pmatrix} 1 & \varphi \\ 0 & 1 \end{pmatrix}, \qquad \hat{Y}^{\varphi} = \left(0,\ 0,\ 0,\ 0,\ -\frac{1}{2}a_g\varphi^2,\ -a_g\varphi\right)'. \tag{8.74}$$

Finally, in the absence of measurements in the time interval $[Kt, t(I)]$, and with $\Sigma_k = 0$, it clearly follows from Eq. (8.9) that the CRLB for the estimation of the impact point is given by:

$$\text{CRLB}_{t(I)} = \hat{A}^{\varphi} \times \text{CRLB}_K \times \left[\hat{A}^{\varphi}\right]', \tag{8.75}$$

where CRLB_K denotes the CRLB value (i.e., IRF CRLB or MSC CRLB) at sampling time K.

8.6.6 Simulation Results

8.6.6.1 High Signal-to-noise Ratio Scenarios

This section considers high signal-to-noise ratio (SNR) scenarios, characterized by a high probability of detection and a low incidence of false alarms. In Table 8.2, the IRFs are shown for the four combinations of (P_D, P_{FA}^g) considered.

It can be seen that as the probability of detection decreases and/or the false alarm probability increases, the probability of a target generated measurement (ϵ) decreases, and the IRFs $q_k(1), q_k(2)$ and q_k each decrease in value. This signifies a reduction in the information content of the available measurements.

Table 8.2 IRFs for different values combinations P_D and P_{FA}^g. Each $q_k(.)$ is approximated via (8.22) using $N_p = 10^5$ sample points. Also shown are the probabilities $\epsilon(1)$ and $\epsilon(2)$ that a target measurement is generated, and the probabilities $p(1)$ and $p(2)$ that there are one and two measurements respectively.

P_D	P_{FA}^g	$p(1)$	$p(2)$	$\epsilon(1)$	$\epsilon(2)$	$q_k(1)$	$q_k(2)$	q_k
1.0	0.0	1.0	0.0	1.0	N/A	1.0	N/A	1.0
0.9	0.0	0.9	0.0	1.0	N/A	1.0	N/A	0.9
0.9	0.01	0.89	0.009	0.99	0.999	0.97	0.88	0.87
0.9	0.05	0.86	0.04	0.95	0.997	0.89	0.87	0.80

In Figure 8.2, the MSC CRLB root mean squared error (RMSE) values are shown for the missile location both in-flight (i.e., at each sampling time) and upon impact, each as a function of the number of scans. The MSC CRLB is approximated via (8.30) using $n_m = 10^4$ measurement sequences. Also shown in Figure 8.2c, d are 1,000 of the measurement sequence realizations used in calculating the MSC CRLB. The IRF CRLB is more optimistic than the MSC PCRLB and is not shown. The MESC CRLB is indistinguishable from the MSC CRLB, as previously observed in [12]. It can be seen from Figure 8.2a,c that the CRLB RMSE is greater for the impact location than for the missile location in-flight. This is because of uncertainty in the missile velocity at each sampling time, which creates additional uncertainty with regard to the motion in the period between the sampling time and missile impact. Conversely, the velocity RMSE upon impact is the same as the velocity RMSE at the sampling time. This is because the missile trajectory is deterministic. Unsurprisingly, the RMSEs decrease as the number of scans increases, indicating that estimating both the missile location in-flight and the impact point are made easier by exploiting a greater number of measurements.

In Figure 8.3, we consider the estimation of the impact point with just $K \in \{6, 7, 8\}$ sensor scans. Shown are the impact location RMSE values provided by the IRF CRLB, MSC CRLB, and a heuristic CRLB specifically constructed for this scenario [17]. Also shown is the performance of the MLE implemented in [17]. It is observed that the MSC CRLB compares favorably with the heuristic bound and the MLE performance [17], with the MSC location RMSE always within 10% of the MLE RMSE, and differences frequently much smaller. Hence, the MSC approach provides a tight performance bound in these scenarios. Moreover, in 2 of the 4 scenarios considered (i.e., shown

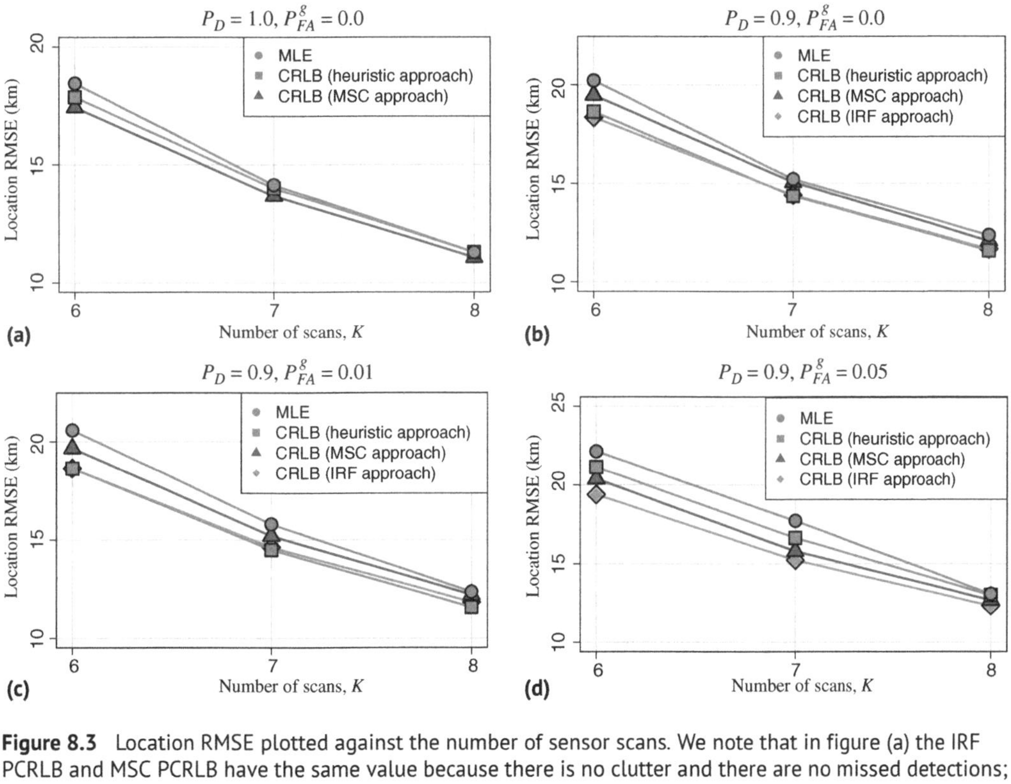

Figure 8.3 Location RMSE plotted against the number of sensor scans. We note that in figure (a) the IRF PCRLB and MSC PCRLB have the same value because there is no clutter and there are no missed detections; (b) $P_D = 0.9,\ P_{FA} = 0.0$; (c) $P_D = 0.9,\ P_{FA} = 0.01$; (d) $P_D = 0.9,\ P_{FA} = 0.05$.

in Figure 8.3b,c) the MSC CRLB is a better predictor of estimator performance than the heuristic bound introduced in [17], despite the fact that the bound [17] was specifically constructed for these high SNR scenarios. Similar findings were previously reported in [11].

Figure 8.4 shows the rate of convergence of the MSC CRLB as a function of the number of measurement sequence realizations n_m for $K \in \{6, 8\}$. In each case it can be see that the CRLB estimate becomes stable after sampling around 3,000 measurement sequences, with the estimate then consistently within ±0.25% of the estimated value calculated using $n_m = 10^4$ measurement sequence realizations.

8.6.6.2 Lower Signal-to-noise Ratio Scenarios

This section considers lower SNR scenarios, with $P_D \in \{0.3, 0.5, 0, 7, 0.9\}$ and an average of $\lambda V_g = 0.5$ false alarms per sampling time. It is noted that these scenarios are unsuitable for the heuristic approach of [17] that is limited to a maximum of one false measurement per time step.

Figure 8.5a shows the IRFs, $q_k(m_k)$ for $m_k = 1, \ldots, 15$ for these lower SNR scenarios. Each IRF is again calculated using (8.22) with $N_p = 10^5$ samples. Figure 8.5b shows the probability that a target measurement is generated, for each value of m_k and each probability of detection P_D. Interestingly, when the probability of detection is low (i.e., $P_D \in \{0.3, 0.5\}$) the IRF $q_k(m_k)$ initially increases as m_k increases from 1 to 2 (and from 2 to 3 for $P_D = 0.3$). This is a result of the sharp increases in the probability that a target measurement is generated in these cases (shown in Figure 8.5b).

Figure 8.6 shows the MSC PCRLB RMSE values at the time of impact, as a function of the number of scans, for the highest and lowest SNR cases (i.e., $P_d \in \{0.3, 0.9\}$). The bound is again approximated via (8.30) using $n_m = 10^4$ measurement sequences. Also shown are 1,000 of the measurement sequence realizations used in calculating each PCRLB. Figure 8.6 enables one to quantify the magnitude of the performance degradation if the probability of detection is low/reduced, and it can be seen that, unsurprisingly, RMSE values are significantly greater than in the higher SNR case considered in Figure 8.2.

Figure 8.7 shows the MSC PCRLB RMSE values at the time of impact for an "enhanced" system with either two or four sensors. In these cases, it is assumed that the measurement processes at each sensor are independent, which allows the MSC to be applied by expressing the measurement realizations as tuples of the number of measurements at each of the sensors. The Riccati recursion (8.28) is then applied as before, with the measurement contribution on each iteration now

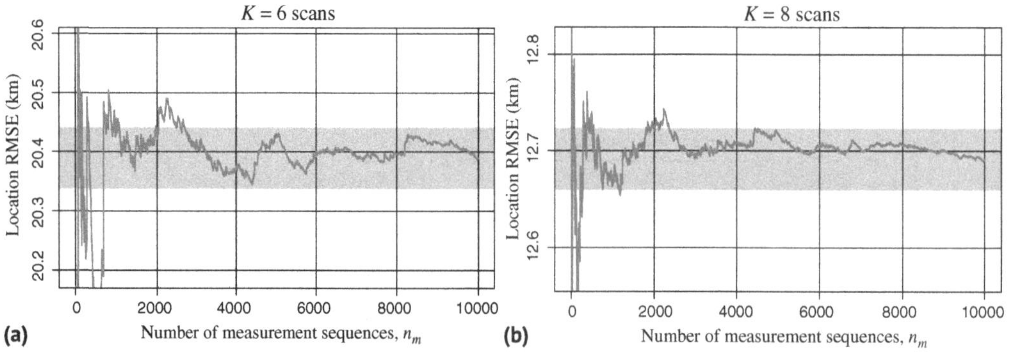

Figure 8.4 Convergence of the MSC CRLB as the number of measurement sequences n_m increases, for $K = 6$ and $K = 8$ scans. In each case $P_D = 0.9$ and $P_{FA}^g = 0.05$, and the gray region shows ±0.25% of the value of the location RMSE calculated using $n_m = 10^4$ measurement sequences.

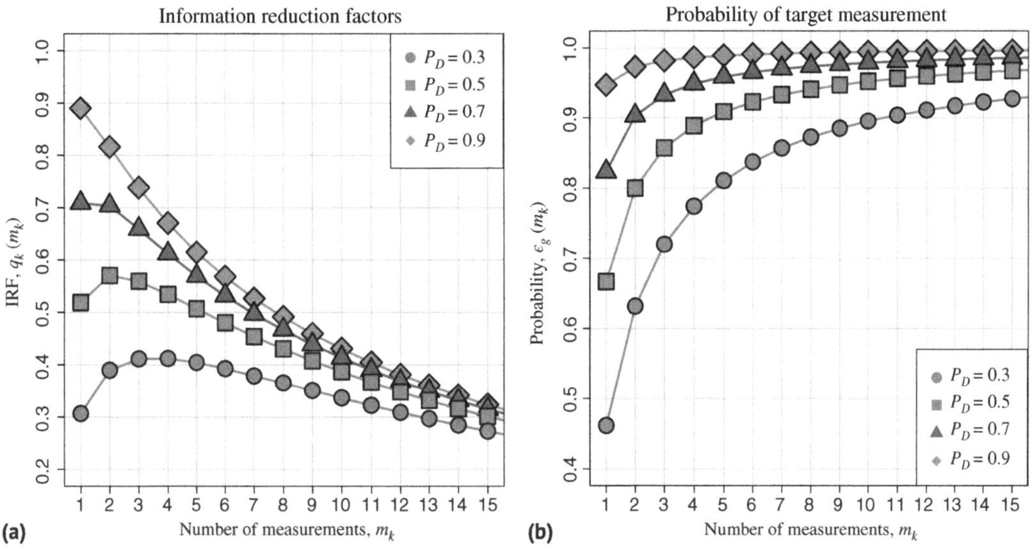

Figure 8.5 (a) Information reduction factors ($q_k(m_k)$) and (b) probability that a measurement is generated by the target ($\epsilon_g(m_k)$), each plotted against the number of measurements (m_k).

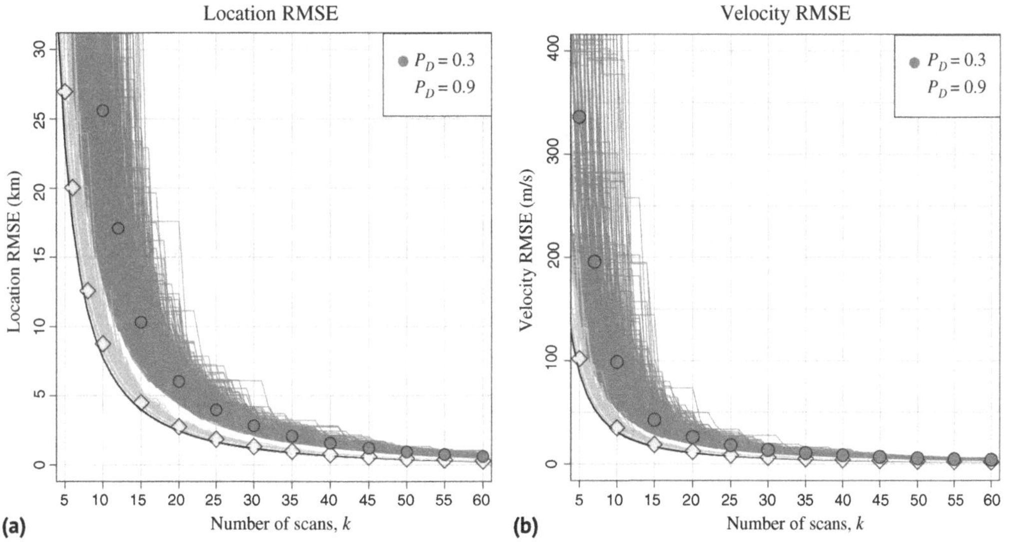

Figure 8.6 Impact location and velocity RMSE, given by the MSC CRLB, each plotted against the number of sensor scans, for $P_D \in \{0.3, 0.9\}$ and $\lambda V_g = 0.5$. The gray lines show the conditional RMSE values for 1,000 measurement sequence realizations, for the two values of P_D (in light and dark gray respectively). The black lines show the corresponding RMSE values for the case $P_D = 1.0$, $\lambda V_g = 0.0$.

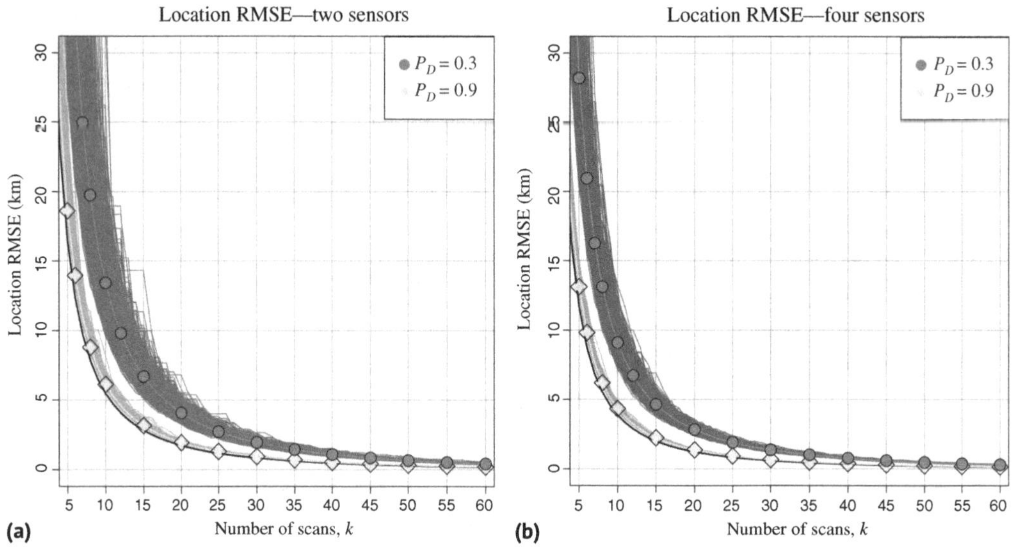

Figure 8.7 Improvement in impact location estimation accuracy as a result of using additional sensors. (a) Two sensors are used, located at coordinates (0, 0, 0) and (100, 0, 0) respectively; (b) four sensors are used, located at coordinates (0, 0, 0), (100, 0, 0), (200, 0, 0), (300, 0, 0) respectively. The gray lines show the conditional RMSE values for 1,000 measurement sequence realizations, for the two values of P_D (in light and dark gray respectively). All sensors have the same P_D value, and in each case $\lambda V_g = 0.5$.

given by the sum total of the contributions of each individual sensor. Such an analysis allows a system designer to identify the number of sensors (and their configuration) required for the system to potentially achieve a desired level of performance (e.g., the estimation accuracy required for weapons targeting).[10]

8.7 Simulation Scenario #2—Tracking a Ground-based Vehicle in the Presence of Radar Spoofing

8.7.1 Scenario Specification

An airborne UAV tracks a ground-based vehicle that is moving in a horizontal plane with nearly constant velocity (NCV) [27] (see Figure 8.8a). The target state is given by $X_k \triangleq (x_k\ \dot{x}_k\ y_k\ \dot{y}_k\ z_k\ \dot{z}_k)'$ where notation is again standard. Hence, the target state evolution is given by (8.8), with:

$$A_k = I_3 \otimes \begin{pmatrix} 1 & t \\ 0 & 1 \end{pmatrix}, \tag{8.76}$$

$$\Sigma_k = q_1 \begin{pmatrix} 1 & 0 & 0 \\ 0 & 1 & 0 \\ 0 & 0 & 0 \end{pmatrix} \otimes \begin{pmatrix} t^3/3 & t^2/2 \\ t^2/2 & t \end{pmatrix}, \tag{8.77}$$

10 Such an analysis may also take into account the possible suboptimality of the estimation algorithm utilized by the system.

where $\otimes$ again denotes the Kronecker product, t is the sampling interval, and q_1 is the power spectral density of the corresponding continuous process noise in the x- and y- coordinate directions. It is noted that the bottom left 2×2 block of A_k implies that the target has CV motion in the z-coordinate direction. This is true, albeit the z-coordinate velocity component is zero, with no process noise.

The UAV uses an active radar system that generates measurements of azimuth, elevation and range. The ground-truth measurements are as given in (8.69). The radar is subjected to intermittent "spoofing" by an adversary. The effect of the spoofing is to inflate the measurement error covariance. If spoofing occurs at sampling time k, then $b_k = 1$. For brevity, we refer to this effect as measurement "bias." The measurement error standard deviations of the azimuth, elevation, and range measurements under normal operation ($b = 0$) and when subject to spoofing/bias ($b = 1$) are given by σ_θ^b, σ_ϕ^b and σ_ρ^b respectively. Consequently, $R_k^b = \text{diag}((\sigma_\theta^b)^2, (\sigma_\phi^b)^2, (\sigma_\rho^b)^2)$ for $b \in \{0, 1\}$. The transitions between the two measurement error models are governed by a Markov chain with time-invariant, known transition probabilities, $p_{ij} \triangleq Pr(b_{k+1} = j | b_k = i)$.

The parameter settings in the simulations are summarized in Table 8.3.

In Figure 8.8a, the sensor trajectory is shown, together with the 500 ($\triangleq n_t$) randomly sampled target trajectories. These trajectories are used in estimating the measurement contribution $J_Z(k : b_k)$ in the PCRLBs. In Figure 8.8b, the range error of exemplar measurements is shown (by circles) along with the average measurement error (bold line; averaged over 5,000 measurement sequence realizations)[11]; for an exemplar measurement bias sequence (the bias occurs at the sampling times shaded in gray). The azimuth and elevation measurement errors are not shown, but follow a similar pattern.

8.7.2 Interacting Multiple Model Filter Implementation

An IMM filter [18] is implemented to estimate the target state. The IMM uses two models, representing the two possible measurement error covariances: "low covariance" (denoted by $b_k = 0$) and "high covariance" (denoted by $b_k = 1$) and runs two EKFs in parallel, one for each model. Each EKF uses the spherical polar coordinate (SPC) measurements, with the system model as given in (8.76) and (8.77), and the measurement model as given in Section 8.4.1.

Table 8.3 Parameter settings used in the simulations.

Parameter	Value
Initial UAV/sensor location, (x_0^S, y_0^S, z_0^S)	(1 km, 2 km, 2 km)
UAV/sensor trajectory	Due North at 55 m/s (≈124 mph)
Initial target location, (x_0, y_0, z_0)	(3 km, 3 km, 0 km)
Initial target trajectory	Due North at 25 m/s (≈56 mph)
Power spectral density, q_1 (of NCV x/y motion)	3 m^2/s^3
Sampling interval, t	1 second (i.e., 1 Hz)
Number of sampling times	100
Measurement errors (accurate), $(\sigma_\theta^0, \sigma_\phi^0, \sigma_\rho^0)$	$(0.5^o, 0.5^o, 10\text{ m})$
Measurement errors (inaccurate), $(\sigma_\theta^1, \sigma_\phi^1, \sigma_\rho^1)$	$(0.5\sqrt{5}^o, 0.5\sqrt{5}^o, 10\sqrt{5}\text{ m})$
Measurement model transition probability, p_{ij}	0.1 ($i \neq j$)

11 The range measurement error (in meters) has a half-normal distribution with mean $\sigma_\rho^b\sqrt{2/\pi} = 7.98$ (when $b = 0$) or 17.84 (when $b = 1$).

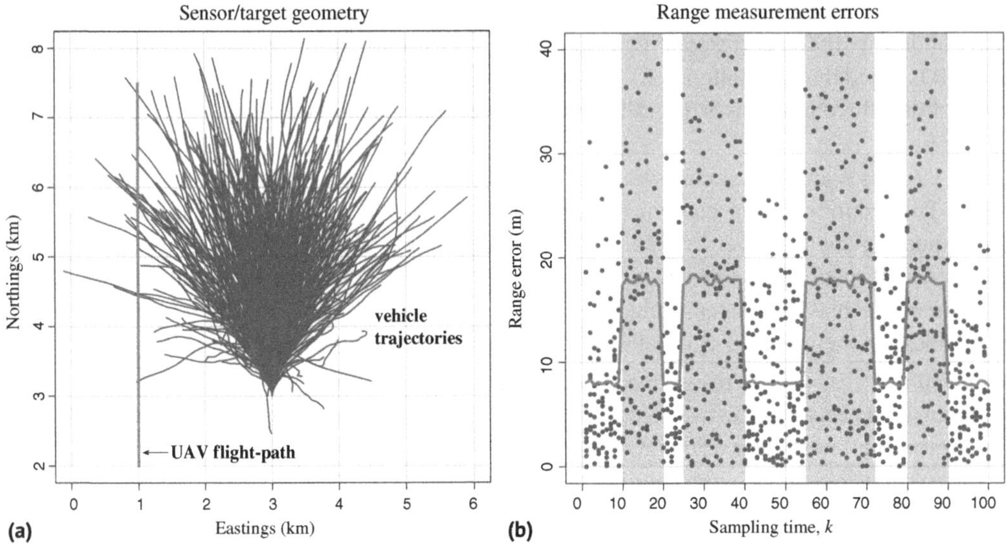

Figure 8.8 (a) Sensor trajectory and $n_t = 500$ candidate target trajectories. (b) Range error of exemplar measurements (circles) and average measurement error (bold line; averaged over 5,000 measurement sequences); for an exemplar measurement bias sequence (the bias occurs at the sampling times shaded in gray).

The IMM performs "mixing" and "merging" steps in order to provide an augmented target state estimate that is a weighted mixture of Gaussian probability density functions. For brevity, the IMM-EKF equations are omitted, and the reader is referred to, e.g., [27]. The implementation of the IMM filter has the following features (e.g., again see [27] for details of the various steps).

1. Nearest neighbor data association is performed (with a gate of $g = 4$ standard deviations), and each track state estimate is updated via its own interacting multiple model extended Kalman filter (IMM-EKF).[12]
2. Track initialization uses "two-out-of-two" logic.
3. Each initial target state estimate uses two-point initialization to set the target velocity, and the most recent measurement to set the target location.
4. Each initial target state estimate covariance is calculated using the unbiased SPC to Cartesian coordinate conversion measurement error covariance given in [28].
5. A track is deleted if no measurement is associated for 5 successive scans.

In the simulations herein, false tracks are rare, and are most prevalent in the period immediately following the inflation of the measurement error covariance. Figure 8.9 shows simulation results of the performance of the IMM-EKF tracker for an exemplar bias sequence realization ($b_{1:k}$), and demonstrates the degradation in tracking performance caused by intermittent inflation of the measurement error covariance.

8.7.3 Calculation of the Posterior Cramér–Rao Lower Bounds

The NECA PCRLB is calculated via Eqs. (8.36), (8.38), and (8.39). This bound is shown in Figure 8.10b,d. In the calculation of the MECSC PCRLB, each conditional PCRLB is given by the

12 It is noted that the algorithm allows multiple tracks to be generated even though the focal scenario considers just a single target.

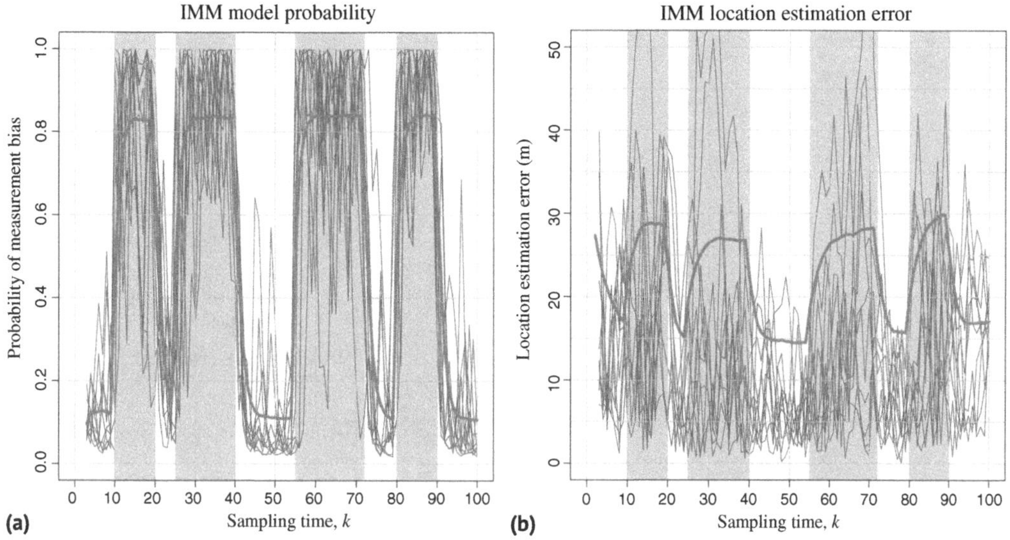

Figure 8.9 (a) IMM-EKF tracker estimates of the probability of measurement bias for an exemplar measurement bias sequence. (b) IMM tracker location estimation error for the exemplar bias sequence. The bias occurs at the sampling times shaded in gray; and the gray lines show the IMM estimates for 10 runs of the tracker. The bold lines show the average probability of measurement bias (in (a)), and the location RMSE (in (b)) respectively, each averaged over 5,000 simulations.

inverse of the conditional FIM (8.45). Figure 8.10a,c show the conditional PCRLB for one bias sequence realization, with Figure 8.10b,d showing the conditional PCRLB for 50 bias sequence realizations (gray lines). The overall MECSC PCRLB in Figure 8.10b,d is then calculated via (8.46), averaged over $n_b = 2{,}000$ bias sequence realizations.

For both the NECA PCRLB and MECSC PCRLB approaches, the measurement contribution $J_Z(k : b_k)$ is given by Eq. (8.39), which is calculated using n_t (= 500) samples $X^j_{1:k}$, $j = 1, \dots, n_t$, of the target trajectory at sampling times 1 through k (shown in Figure 8.8a).

Hence, the measurement contribution is given as follows:

$$J_Z(k : b_k) \approx \frac{1}{n_t} \sum_{j=1}^{n_t} \left[H_k(X^j_k)'(R^{b_k}_k)^{-1} H_k(X^j_k) \right] \quad \text{for } b_k \in \{0, 1\}. \tag{8.78}$$

The Jacobian of the measurement vector (i.e., $H_k(X^j_k)$) is given by (8.72).

It is noted that the measurement contribution for the estimation of a deterministic ballistic missile trajectory exploited the ground-truth target trajectory (see Eqs. (8.70) and (8.71)). However, in this second scenario, the target motion is non-deterministic (i.e., NCV motion) and the measurement contribution is calculated in (8.78) by averaging over a sufficiently large number of potential target trajectories.

8.7.4 Performance Metrics

8.7.4.1 Estimation Accuracy

The location and velocity RMSE values of the IMM-EKF are compared to the NECA and MECSC PCRLBs. In this scenario, the target is ground-based, hence the elevation and vertical velocity

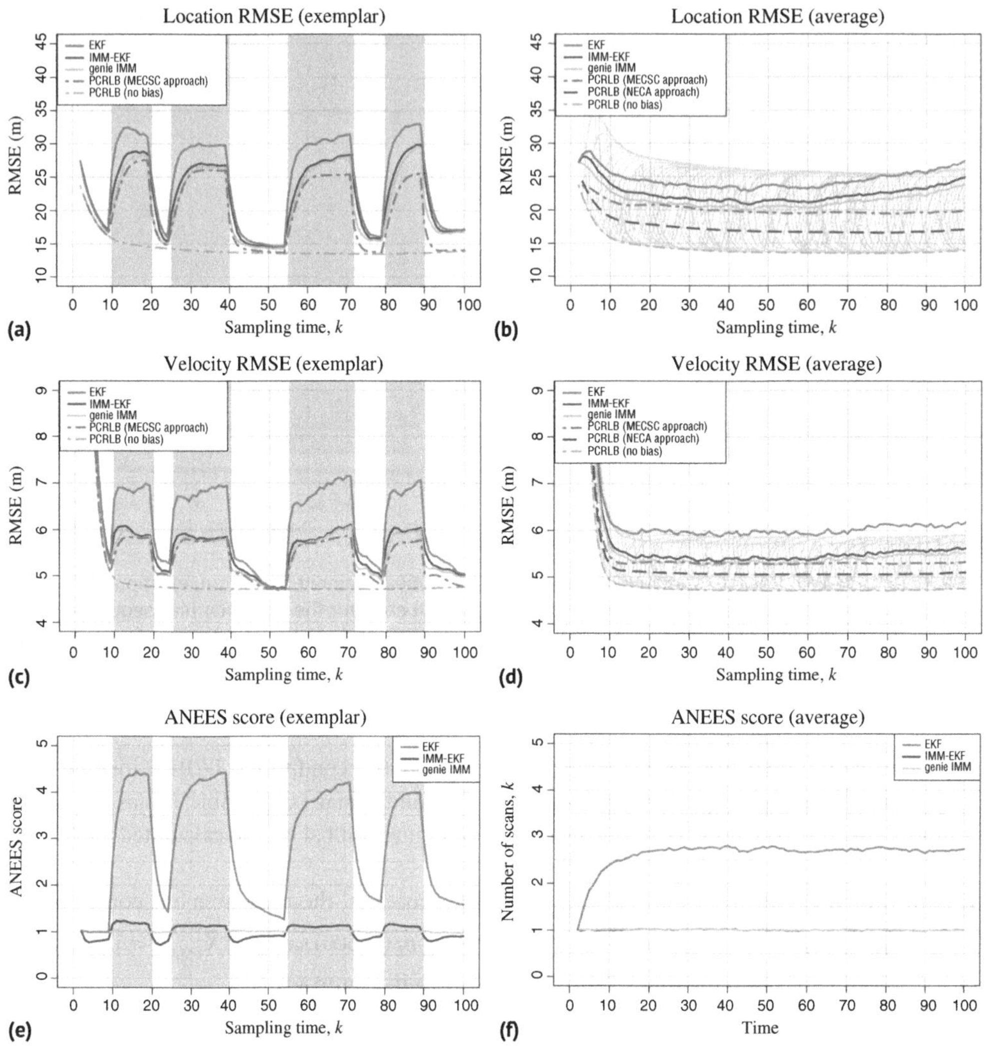

Figure 8.10 RMSE values plotted against sampling time. *Left-hand column:* values calculated for an exemplar bias sequence, with the sampling times at which measurement bias occurs shaded in gray, averaged over 5,000 simulations of the target state evolution; the RMSEs of ascending magnitude are (1) no bias PCRLB, (2) MECSC PCRLB, (3) genie-IMM tracker, (4) IMM-EKF tracker, (5) EKF tracker. *Right-hand column:* values averaged over 5,000 measurement bias sequence realizations, with a randomly sampled target trajectory used in each realization; the RMSEs of ascending magnitude are (1) no bias PCRLB, (2) NECA PCRLB, (3) MECSC PCRLB, (4) genie-IMM tracker, (5) IMM-EKF tracker, (6) EKF tracker. (a) location RMSE (exemplar), (b) location RMSE (average), (c) velocity RMSE (exemplar), (d) velocity RMSE (average), (e): ANEES score (exemplar), (f): ANEES score (average).

components are not used in each calculation. Therefore, for each measurement bias sequence $b_{1:k}$, the conditional PCRLB location and velocity RMSEs are calculated as follows:

$$\text{location RMSE} = \left\{ \left[J_k(b_{1:k})^{-1} \right]_{1,1} + \left[J_k(b_{1:k})^{-1} \right]_{3,3} \right\}^{1/2} \tag{8.79}$$

$$\text{velocity RMSE} = \left\{ \left[J_k(b_{1:k})^{-1} \right]_{2,2} + \left[J_k(b_{1:k})^{-1} \right]_{4,4} \right\}^{1/2} \tag{8.80}$$

where $[J_k(b_{1:k})^{-1}]_{i,i}$ denotes the i-th diagonal entry of $J_k(b_{1:k})^{-1}$ (calculated via (8.44) or (8.45)). The overall long-run (unconditional) MECSC PCRLB RMSEs are calculated by replacing the $J_k(b_{1:k})^{-1}$ entries in (8.79) and (8.80) with the corresponding entries from the approximation (8.46). The naive PCRLB RMSEs are correspondingly calculated via the relevant entries from (8.36).

The corresponding tracker location and velocity RMSEs are calculated in a similar fashion. If multiple tracks exist at any given sampling time (which is a rare occurrence due to the lack of clutter), the RMSE values are calculated for the track that is closest to the target in terms of the Mahalanobis distance.

8.7.4.2 Credibility Measure

To assess whether the tracker is being truthful, i.e., whether the tracker overall covariance accurately represents the estimation error, or at the very least is not optimistic, the average normalized estimation error (ANEES) score [29] is calculated for the target location estimation. The ANEES score at sampling time k is given as follows:

$$\text{ANEES}(k) = \frac{1}{3N_s}\sum_{i=1}^{N_s}\left\{(X_i^l(k) - \hat{X}_i^l(k|k))' P_i^l(k|k)^{-1}(X_i^l(k) - \hat{X}_i^l(k|k))\right\}, \tag{8.81}$$

where $X_i^l(k) \in \mathbb{R}^3$ is the ground-truth target location at sampling time k, for the i-th scenario (i.e., target ground-truth trajectory and bias sequence realization); $\hat{X}_i^l(k|k)$ is the corresponding tracker location estimate; and $P_i^l(k|k)$ is the corresponding sub-matrix consisting of the tracker covariance terms for the target location. The parameter N_s is the number of scenarios considered.

An ANEES score value close to unity indicates that the value lies within a chi-squared distributed confidence interval (and is therefore credible). ANEES scores lower than unity indicate that the covariance estimate is pessimistic (i.e., greater than it needs to be), which can be appealing as in such cases the covariance based confidence region is pessimistically large, and therefore more likely (all else being equal) to contain the target (i.e., provides a "truthful" estimate of a region in which the target is likely to reside).

8.7.5 Simulation Results

Simulation results are shown in Figure 8.10. These results are representative of a number of sensor/target geometries (some of which are shown in Figure 8.8a. The performance of the IMM-EKF is compared to that of an EKF that does not account for covariance inflation, and a "genie" IMM-EKF that is provided with the correct measurement covariance at each sampling time. The suboptimality of the EKF demonstrates the difficulties created when inflated measurement errors are not accounted for. The "genie" IMM provides a bound on the performance of the IMM-EKF methodology. Figure 8.10 also shows the PCRLB in the absence of measurement bias, to indicate the optimal estimation accuracy if bias effects can be mitigated.

Figure 8.10a,c show the direct impact inflated measurement errors can have on target state estimation accuracy, with the target location and velocity RMSEs of the trackers (EKF, IMM-EKF and genie IMM) and the conditional PCRLB increasing rapidly (i.e., giving the sharp peaks) subsequent to the measurement error inflation occurring. Furthermore, it can be seen from Figure 8.10a,c that:

1. The IMM-EKF performs well in comparison with both the PCRLB and the "genie" IMM. Indeed, the RMSE performance of the IMM-EKF is typically within a few percent of the performance of the "genie" IMM with both algorithms typically having RMSEs within 20% of the PCRLB.
2. The EKF performs poorly, because it is unable to adapt to the inflation of the measurement errors.

In Figure 8.10b,d overall location and velocity RMSEs are shown, calculated using $n_b = 5{,}000$ bias sequence realizations, with a randomly sampled target trajectory used in each realization. These figures demonstrate that overall, the RMSE performance of the IMM-EKF is again within a few percent of the performance of the "genie" IMM with both algorithms again typically within 25% of the MECSC PCRLB. The largest performance differences occur towards the end of the scenarios, and are the result of a small number of unfavorable geometries in which the sensor/target separation becomes large. As noted in Section 8.4.2.3, the naive approach generates a more optimistic PCRLB that offers only limited value in assessing the performance of tracking algorithms, but might provide a reasonable approximation of the PCRLB in time critical applications (e.g., when using the PCRLB as a predictive metric in sensor management applications [4]).

In the scenarios considered, the IMM-EKF suffered very few lost tracks. It is also noted that the IMM will typically take one to five scans to accurately infer a switch in the measurement error model (see Figure 8.9a). Therefore, performance is likely to degrade if the model switches rapidly between scans. Performance is also likely to degrade if the target has the capability to perform maneuvers. At such times, it is difficult to ascertain whether unexpectedly large measurement perturbations are the result of spoofing, or the result of the target performing a maneuver. In such cases, an alternative approach, such as a multiple hypothesis tracker, is likely to prove more effective.

In Figure 8.10e and f the tracker ANEES scores are calculated, for the exemplar bias sequence (in (e)); and averaged across the 5,000 bias sequence realizations (in (f)). Given that the IMM-EKF requires several time steps to react to a switch in measurement model, and even then has a typical residual model error probability of 0.1-0.2 (again, see Figure 8.9a), at sampling times for which spoofing occurs the IMM-EKF underestimates the error covariance of each measurement, and becomes overly confident/optimistic. This results in a ANEES score greater than unity, and (slightly) overconfident estimation. Conversely, when spoofing bias does not occur, the IMM-EKF is (slightly) pessimistic, and has an ANEES score lower than unity. When the ANEES score is averaged across 5,000 bias sequence realizations, these effects average out, and the ANEES score is close to unity (see Figure 8.10f). By comparison, the genie-IMM uses the correct measurement error covariance, and so always generates consistent estimates with an ANEES score close to unity. The EKF is unable to adapt to the inflated measurement error covariance, which results in hugely overconfident estimates with consistently high ANEES scores at sampling times on which spoofing occurs.

8.8 Simulation Scenario #3—Tracking a Low-flying Airborne Target in the Presence of Specular Multipath

8.8.1 Scenario Specification

A airborne radar tracks a low-altitude (i.e., 100 meter altitude) airborne target that is moving in a horizontal plane with NCV motion (see Figure 8.11a). In Figure 8.11a, the sensor trajectory is shown, together with 500 ($\triangleq n_t$) randomly sampled target trajectories. These trajectories are used in estimating the measurement contribution $\bar{J}_Z(k : r_k)$ in the PCRLBs via (8.57)–(8.59). The low altitude of the target results in each measurement potentially being corrupted by multipath due to specular reflection of the radar signal at the (ground) surface boundary (see Figure 8.11b).

It is assumed $P_d \approx 1.0$ and $P_{fa} \approx 0.0$. A potential avenue for future work is to address the more general case in which $P_d < 1.0$ and $P_{fa} > 0$, by conditioning on the number of measurements generated at each sampling time, using the MSC approach presented in Section 8.3.3, as noted in [11].

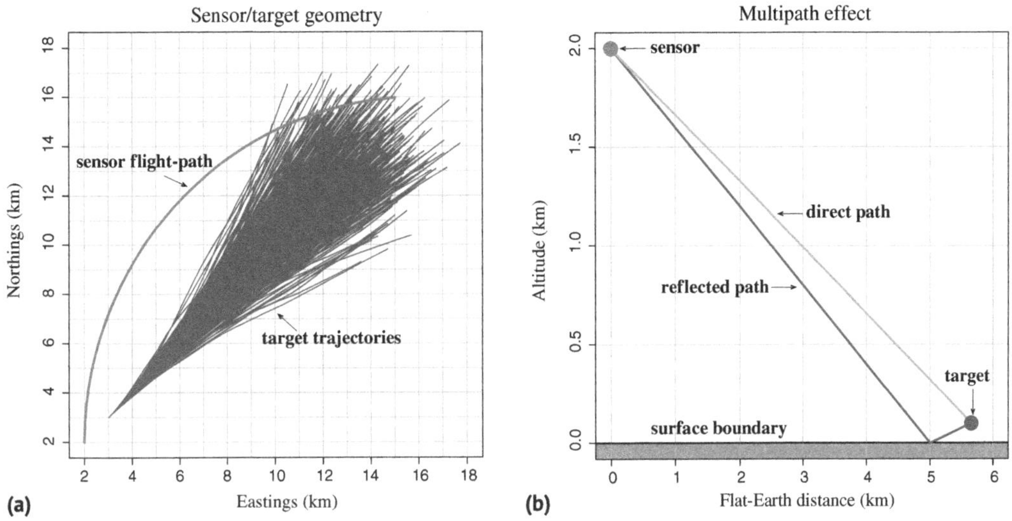

Figure 8.11 (a) Sensor trajectory and 500 potential target trajectories. (b) Multipath effect.

However, the intention here is to demonstrate the direct impact of specular multipath alone, without including further complicating factors that will degrade performance further.

Measurements of azimuth, elevation and range are generated by the radar. When multipath is not present (denoted $r_k = 0$) the error covariance of each measurement is $R_k = \text{diag}(\sigma_\theta^2, \sigma_\phi^2, \sigma_\rho^2)$ where notation is standard. The measurements of range and elevation are intermittently subject to multipath (denoted $r_k = 1$), which is autocorrelated across sampling times (as in (8.48)), with time-invariant autocorrelation $\alpha_k = \alpha$. Reflections are assumed to occur in the vertical plane containing the sensor and target. Therefore, the azimuth measurements are not subject to multipath. Hence, the multipath induced bias term is given by $w_M = (w_\phi \; w_r)'$, and $\Sigma_M = \text{diag}((\sigma_\phi^M)^2, (\sigma_\rho^M)^2)$. The transitions in and out of the multipath model are governed by a Markov chain with time invariant, known transition probabilities, $p_{ij} \triangleq Pr(r_{k+1} = j | r_k = i)$.

The parameter settings used in the simulations are summarized in Table 8.4.

In Figure 8.12, the range and elevation errors of exemplar measurements are shown (by circles) along with the average measurement errors (bold lines; averaged over 5,000 measurement sequence realizations), for an exemplar multipath sequence.

If multipath effects occur during the track initialization phase, it is likely that extremely inaccurate estimates of the target state would be generated, and as a result, the track would be quickly deleted. Successful tracking is therefore dependent on multipath being absent initially. Consequently, without loss of generality, the multipath effect is not present at the beginning of each simulation.

It is straightforward to show that (from (8.48)) the absolute range and elevation measurement errors each have a half-normal distribution with mean $\sigma\sqrt{2/\pi}$, where:

$$\sigma = \left(\sigma_\xi^2 + (\sigma_\xi^M)^2 \sum_{k=1}^{\Delta_t} \alpha^{2(k-1)} \right)^{1/2}, \qquad \text{for } \xi = \rho, \phi, \tag{8.82}$$

Table 8.4 Parameter settings used in the simulations.

Parameter	Value
Initial sensor location, (x_0^S, y_0^S, z_0^S)	(2 km, 2 km, 2 km)
Sensor trajectory	parabolic trajectory at 210 m/s
Initial target location, (x_0, y_0, z_0),	(3 km, 3 km, 0.1 km)
Initial target trajectory	North East at 150 m/s
Power spectral density, q_1 (of NCV x/y motion)	$6\ \text{m}^2/\text{s}^3$
Sampling interval, t	1 second (i.e., 1 Hz)
Number of sampling times	100
Measurement errors, $(\sigma_\theta, \sigma_\phi, \sigma_\rho)$	$(0.5^o, 0.5^o, 10\ \text{m})$
Multipath bias autocorrelation, α	0.5
Elevation multipath bias standard deviation, σ_ϕ^M	1°
Range multipath bias standard deviation, σ_ρ^M	20 m
Multipath model transition probability, p_{ij}	0.1 $(i \neq j)$

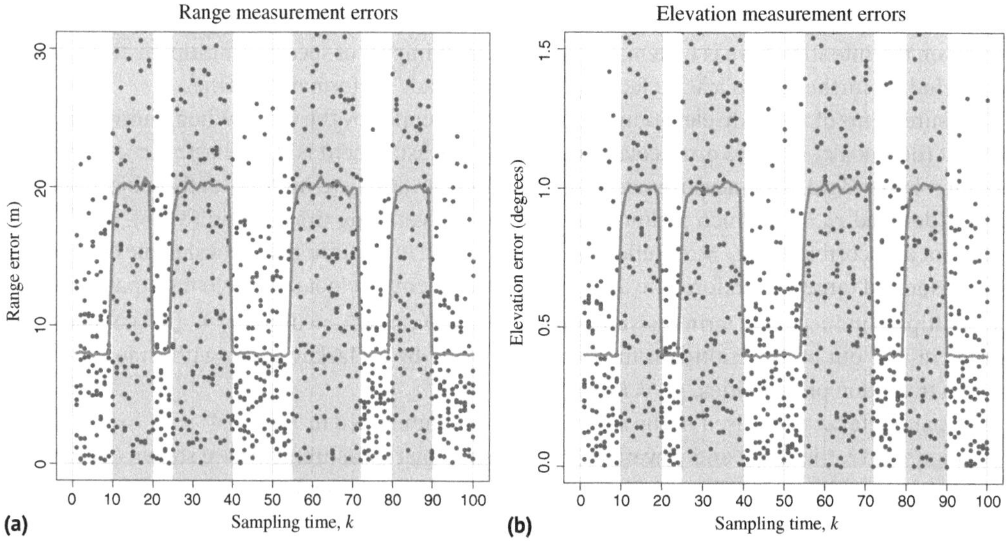

Figure 8.12 Exemplar range and elevation measurement errors (absolute values; circles) and the average range and elevation measurement errors (bold lines; averaged over 5,000 measurement sequence realizations); for a single exemplar multipath sequence. Multipath occurs at sampling times shaded in gray.

and Δ_t is the number of successive sampling times on which the multipath effect has occurred. It is noted that $\sigma \to \left(\sigma_\xi^2 + (\sigma_\xi^M)^2/(1-\alpha^2)\right)^{1/2}$ as $t \to \infty$. Hence, for the range measurements, the average absolute error has a maximum value of around 20 meters, and for the elevation measurements the average absolute error has a maximum value of around 1 degree. The simulation results shown in Figure 8.12 are consistent with these analytical results.

8.8.2 Interacting Multiple Model Filter Implementation

An IMM-EKF algorithm [18] is again implemented, to estimate the augmented target state given by $\bar{X}_k = (X_k' \; w_M(k)')'$. The IMM-EKF uses two models, representing the hypotheses "multipath effect not present" (denoted by $r_k = 0$) and "multipath effect present" (denoted by $r_k = 1$) and runs two EKFs in parallel, one for each hypothesis. The augmented system model is as given in (8.50)–(8.52), for $r_k = 0, 1$. The initial multipath bias estimates are set to zero, with the initial multipath bias covariance terms given nominal large values.

The remaining steps in the implementation of the IMM-EKF algorithm are as described in Section 8.7.2. As in the case of scenario #2, false tracks being rarely generated, and only ever occur in the period immediately following a switch to the multipath model. Figure 8.13 shows simulation results of the performance of the IMM-EKF tracker for an exemplar bias sequence realization, and demonstrates the degradation in tracking performance when the multipath effect occurs.

It is noted that in [23], an IMM-KF implementation also augmented the target state to include specular multipath bias, for a simpler scenario concerned only with the estimation of target altitude.

8.8.3 Calculation of the Posterior Cramér–Rao Lower Bounds

In the calculation of the PCRLB, each conditional PCRLB is given by the inverse of the conditional FIM (8.54). Figure 8.14 shows the conditional PCRLB for an exemplar multipath sequence realization. Subsequently, Figure 8.15 shows the conditional PCRLB for 50 multipath sequence realizations (gray lines) and the overall PCRLB calculated via (8.65), averaged over $N_m = 2{,}000$ multipath sequence realizations. It is noted that the augmentation of the target state to include the multipath

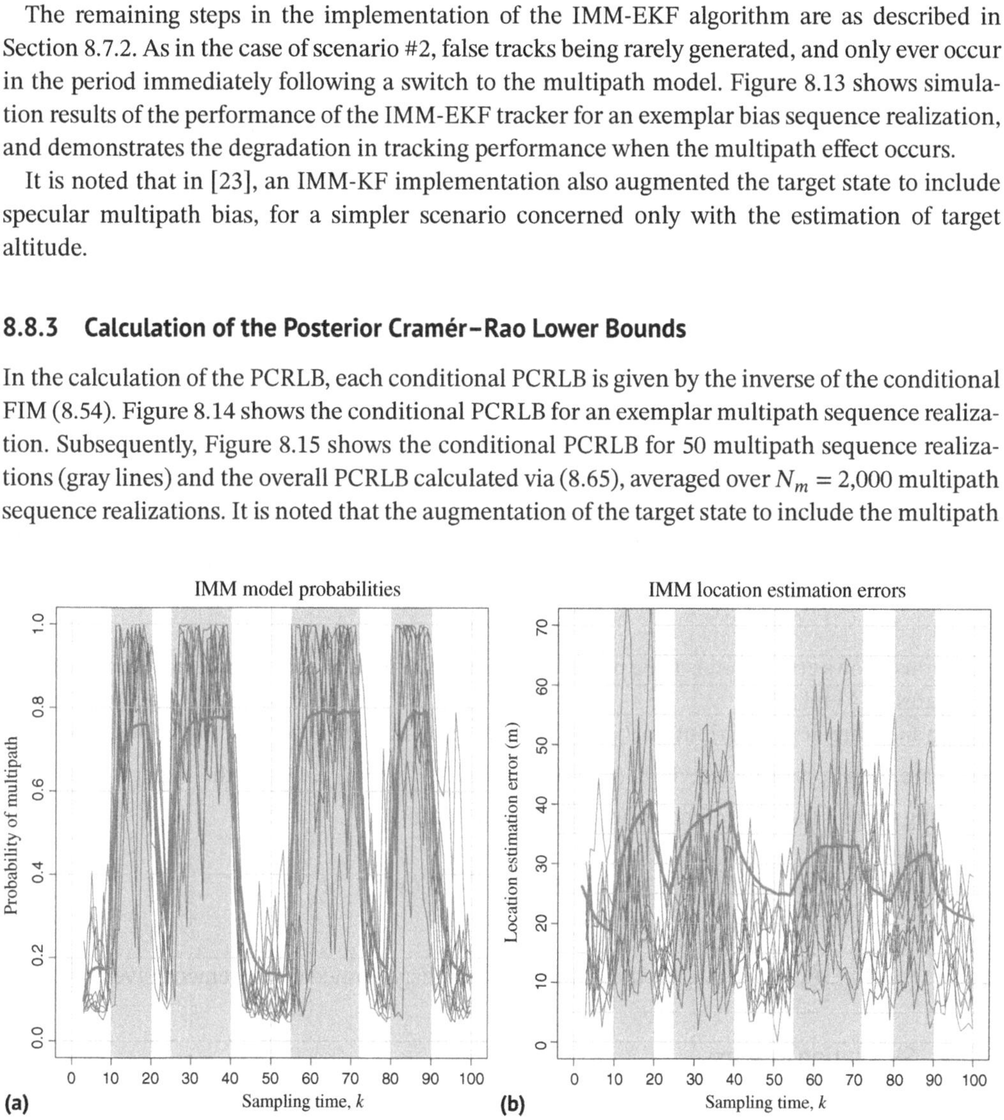

Figure 8.13 (a) IMM-EKF tracker estimates of the probability of multipath for an exemplar multipath sequence. (b) IMM tracker location estimation error for the exemplar multipath sequence. Multipath occurs at the sampling times shaded in gray; and the gray lines show the estimates for 10 runs of the tracker. The bold lines show the average probability of multipath (in (a)), and the location RMSE (in (b)) respectively, each averaged over 5,000 simulations.

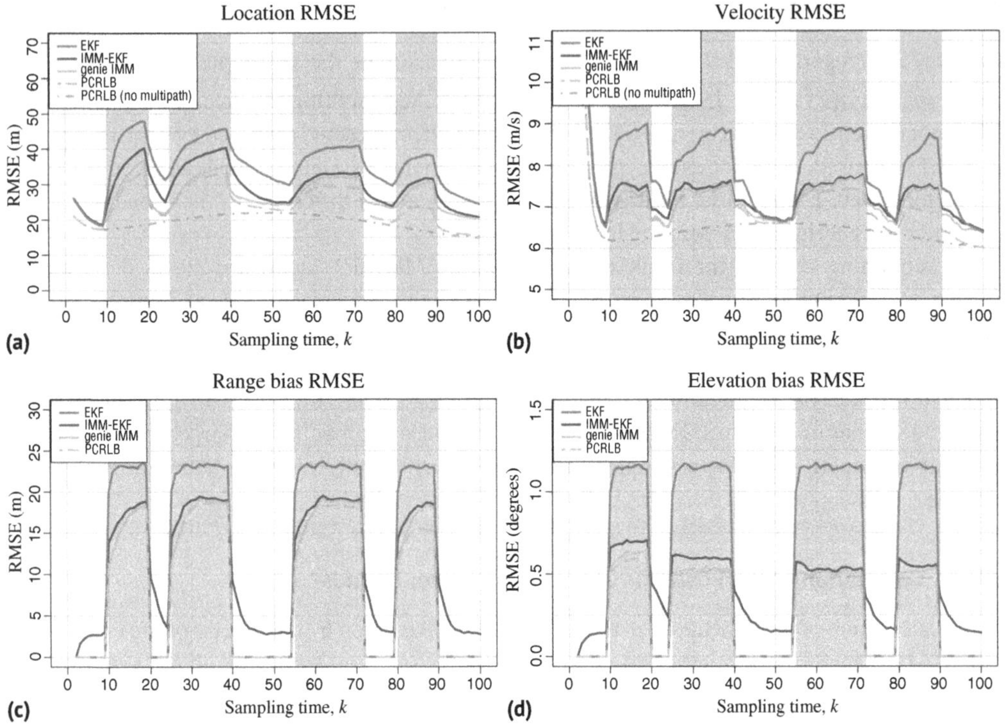

Figure 8.14 RMSE values plotted against sampling time for an exemplar multipath sequence, with the sampling times at which multipath occurs shaded in gray, averaged over 5,000 simulations (i.e., target state realizations). The RMSEs of ascending magnitude are (1) PCRLB in the absence of multipath ((a) and (b) only), (2) PCRLB, (3) genie-IMM tracker, (4) IMM-EKF tracker, (5) EKF tracker.

bias terms allows error bounds to be provided for the estimation of these bias terms, as also shown in Figures 8.14 and 8.15.

Each measurement contribution $\bar{J}_Z(k : r_k)$ is given by Eqs. (8.56)–(8.59). Equations (8.57) and (8.58) are then approximated via n_t (= 500) samples $X^j_{1:k}$, $j = 1, \dots, n_t$, of the target trajectory at sampling times 1 through k (shown in Figure 8.11a), i.e.,:

$$B_{11} \approx \frac{1}{n_t}\sum_{j=1}^{n_t} H_k(X_k^j)' R_k^{-1} H_k(X_k^j) \qquad B_{12} \approx \frac{1}{n_t}\sum_{j=1}^{n_t} H_k(X_k^j)' R_k^{-1} I'_{d_M \times d}, \tag{8.83}$$

where $H_k = \nabla_{X_k}\left[h_k(X_k)'\right]'$ is again the Jacobian of a target generated measurement, given by (8.72).

8.8.4 Simulation Results

Simulation results are presented in Figures 8.14 and 8.15. These results are again representative of a number of sensor/target geometries (again, see Figure 8.11a for 500 of the sensor/target geometries considered).

The performance of the IMM-EKF is again compared to that of an EKF that does not account for multipath, and a “genie” IMM-EKF that is provided with the correct multipath model at each sampling time. Figures 8.14 and 8.15 also show the PCRLB in the absence of multipath effects, to indicate the optimal estimation accuracy if multipath can be mitigated. Location and velocity

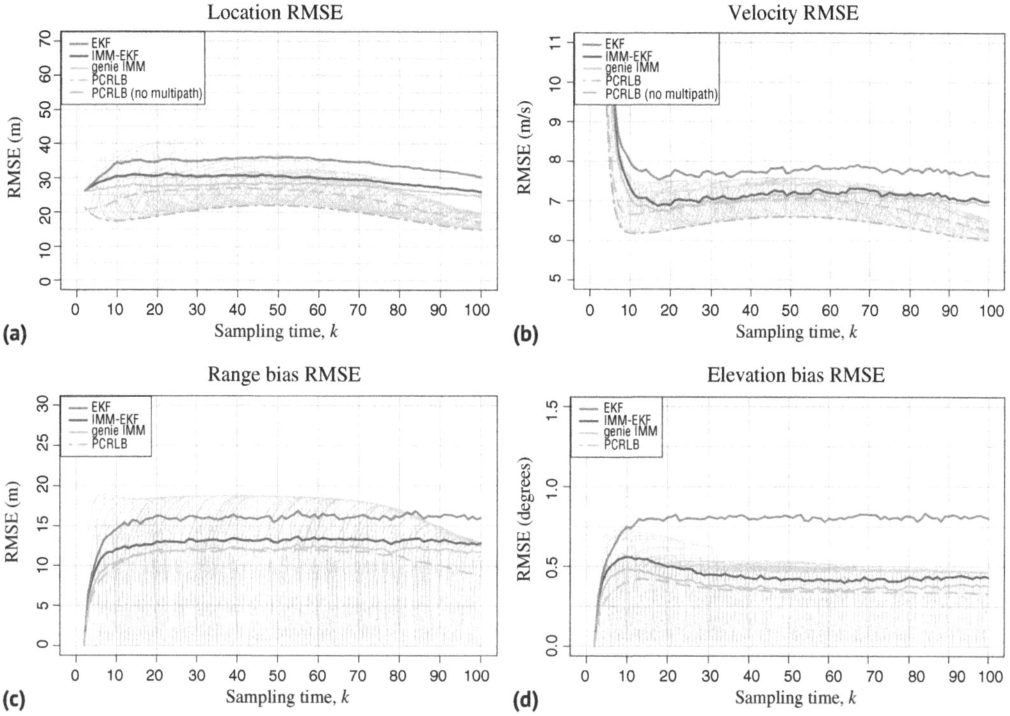

Figure 8.15 RMSE values plotted against sampling time averaged over 5,000 realizations of the target state and multipath sequence. The RMSEs of ascending magnitude are (1) PCRLB in the absence of multipath ((a) and (b) only), (2) PCRLB, (3) genie-IMM tracker, (4) IMM-EKF tracker, (5) EKF tracker.

RMSE are based on 3D Euclidean distance. The ANEES scores are similar to those for scenario #2 and are omitted for brevity.

Figure 8.14 demonstrates the direct impact of multipath on estimation accuracy, with the target state and multipath bias RMSEs of the trackers (EKF, IMM-EKF, and genie IMM) and the PCRLB increasing significantly subsequent to multipath occurring. Furthermore, it can be seen that:

1. As in scenario #2, the IMM-EKF performs well in comparison with both the PCRLB and the "genie" IMM, with the target state estimation performance of the IMM-EKF typically within a few percent of the performance of the "genie" IMM and within 20% of the PCRLB;
2. Again, the EKF performs poorly, because it is unable to adapt to the inflation of the measurement errors.
3. The IMM-EKF performs extremely well in estimating the range and elevation multipath biases. Indeed, the bias RMSEs of the IMM-EKF are only marginally greater than the RMSE of the PCRLB.[13]

In Figure 8.15, overall estimation RMSEs are shown. The PCRLB is based on $N_m = 2{,}000$ multipath sequence realizations, and the tracker RMSEs are calculated via 5,000 simulations, with each simulation using a randomly sampled target trajectory and a randomly sampled multipath sequence realization. These figures demonstrate that overall, the RMSE performance of the IMM-EKF is

13 It is noted that estimating the biases is easier than estimating the target state, because the bias values are reset to zero when multipath is absent.

again within a few percent of the performance of the "genie" IMM with the target state estimation typically within 25% of the PCRLB. The range and elevation multipath bias estimates are also near optimal and only marginally greater than the PCRLB. However, it is again noted that the performance of the IMM-EKF is likely to degrade compared to the PCRLB if (1) the multipath model can switch rapidly between scans, (2) the target has the ability to perform significant maneuvers (rather than just have NCV motion).

8.9 Conclusions

This chapter has been concerned with the calculation of the PCRLB for tracking a noncooperative target in environments in which there can be measurement origin or measurement accuracy uncertainties. Measurement origin uncertainty typically occurs in cluttered environments, resulting in missed detections and false alarms. Measurement accuracy uncertainty can be the result of sensor faults, spoofing by an adversary to manipulate the sensor signal, or multipath resulting from signal reflection at a surface boundary (etc.).

Seminal PCRLB approaches in cluttered environments [8, 9, 11] have been presented, along with a recent approach developed for scenarios with autocorrelated specular multipath [15, 16] that augments the state with the measurement bias and determine the bound via sampling multipath sequence realizations. Inspired by these approaches, herein two approaches have been developed for calculating the PCRLB for scenarios with intermittent sensor faults/spoofing. The first (naive) approach is analogous to the IRF approach [8, 9] in cluttered environments, and simply uses the expected information content of each measurement within a standard Riccati recursion. The second approach (measurement error sequence conditioning) is analogous to the MSC approach [11] and samples realizations of the measurement sequence. It is computationally more expensive, but provides a less optimistic lower bound.

The approaches were demonstrated in the following scenarios: (1) estimating the impact point of a ballistic missile in a cluttered environment; (2) tracking a ground-based vehicle in the presence of radar spoofing; (3) tracking a low flying airborne target in the presence of multipath. In each case, the PCRLB was shown to offer an accurate bound on estimator performance, with estimator RMSEs typically within 20% of the bound. It is noted that performance differences can be greater in low SNR environments, for which even the MSC approach [11] can be optimistic. Also, in scenarios for which there is measurement accuracy uncertainty, approaches that condition on the measurement error sequence can be optimistic if it is challenging to accurately identify the times at which the measurement error is inflated, e.g., when tracking a maneuvering target. Further work is required to develop consistently tight performance bounds in these scenarios.

References

1 H. L. Van Trees, K. L. Bell, and Z. Tian. *Detection Estimation and Modulation Theory, Part I: Detection, Estimation, and Filtering Theory*.John Wiley & Sons, New York, USA, 2nd edition, 2013.

2 N. Bergman. Posterior Cramér-Rao bounds for sequential estimation. In A. Doucet, N. de Freitas, and N. Gordon, editors, *Sequential Monte Carlo Methods in Practice*, Springer, 321–338, 2001.

3 M. L. Hernandez, T. Kirubarajan, and Y. Bar-Shalom. Multisensor resource deployment using posterior Cramér-Rao bounds. *IEEE Transactions on Aerospace and Electronic Systems*, 40(2):399–416, 2004.

4 M. L. Hernandez. Performance bounds for target tracking: computationally efficient formulations and associated applications. In M. Mallick, V. Krishnamurthy, and B.-N. Vo, editors, *Integrated Tracking, Classification, and Sensor Management: Theory and Applications*. John Wiley and Sons, Ltd. / IEEE Press, New York, USA, 2013.

5 P. Tichavský, C. H. Muravchik, and A. Nehorai. Posterior Cramér-Rao bounds for discrete-time nonlinear filtering. *IEEE Transactions on Signal Processing*, 46(5):1386–1396, 1998.

6 M. Hernandez, A. Farina, and B. Ristic. Posterior Cramér-Rao bounds for target tracking. In R. Klemm, U. Nickel, C. Gierull, P. Lombardo, H. Griffiths, and W. Koch, editors, *Novel Radar Techniques and Applications*. SciTech Publishing, 263–302, 2017.

7 R. Niu, P. Willett, and Y. Bar-Shalom. Matrix CRLB scaling due to measurements of uncertain origin. *IEEE Transactions on Signal Processing*, 497:1325–1335, 2001.

8 X. Zhang and P. K. Willett, Cramer-Rao bounds for discrete time linear filtering with measurement origin uncertainty. In *Proceedings of the Workshop on Estimation, Tracking and Fusion: A Tribute to Yaakov Bar-Shalom,* Monterey, California, USA, pages 546–561, 2001.

9 M. L. Hernandez, A. D. Marrs, N. J. Gordon, S. R. Maskell, and C. M. Reed, Cramér-Rao bounds for non-linear filtering with measurement origin uncertainty. In *Proceedings of the Fifth International Conference on Information Fusion,* vol. 1, Annapolis, Maryland, USA, pages 18–25, 2002.

10 A. Farina, B. Ristic, and L. Timmoneri. Cramer-Rao bound for nonlinear filtering with Pd<1 and its application to target tracking. *IEEE Transactions on Signal Processing*, 50(8):1916–1924, 2002.

11 M. L. Hernandez, A. Farina, and B. Ristic. PCRLB for tracking in cluttered environments: Measurement sequence conditioning approach. *IEEE Transactions on Aerospace and Electronic Systems*, 42(2):680–704, 2006.

12 H. Meng, M. L. Hernandez, Y. Liu, and X. Wang. Computationally efficient PCRLB for tracking in cluttered environments: measurement existence conditioning approach. *IET Signal Processing*, 3(2):133–149, 2009.

13 I. Rapoport and Y. Oshman. A Cramér-Rao-type estimation lower bound for systems with measurement faults. *IEEE Transactions on Automatic Control*, 50(9):1234–1245, 2005.

14 A. J. Kerns, D. P. Shepard, J. A. Bhatti, and T. E. Humphreys. Unmanned aircraft capture and control via GPS spoofing. *Journal of Field Robotics*, 31(4):617–636, 2014.

15 M. L. Hernandez and A. Farina, Posterior Cramér-Rao bound for target tracking in the presence of multipath. In *2018 21st International Conference on Information Fusion (FUSION),* IEEE, pages 151–158, 2018.

16 M. Hernandez and A. Farina. PCRB and IMM for target tracking in the presence of specular multipath. *IEEE Transactions on Aerospace and Electronic Systems*, 56(3):2437–2449, 2020.

17 A. Farina, A. Di Lallo, L. Timmoneri, T. Volpi, and B. Ristic. CRLB and ML for parametric estimate: new results. *Signal Processing*, 86(4):804–813, 2006.

18 H. A. P. Blom and Y. Bar-Shalom. The interacting multiple model algorithm for systems with Markovian switching coefficients. *IEEE Transactions on Automatic Control*, 33(8):780–783, 1988.

19 J. Taylor. The Cramer-Rao estimation error lower bound computation for deterministic nonlinear systems. *IEEE Transactions on Automatic Control*, 24(2):343–344, 2003.

20 X. Zhang, P. Willett, and Y. Bar-Shalom. Dynamic Cramer-Rao bound for target tracking in clutter. *IEEE Transactions on Aerospace and Electronic Systems*, 41(4):1154–1167, 2005.

21 R. Tharmarasa, T. Kirubarajan, M. L. Hernandez, and A. Sinha. PCRLB-based multisensor array management for multitarget tracking. *IEEE Transactions on Aerospace and Electronic Systems*, 43(2):539–555, 2007.

22 M. L. Hernandez, B. Ristic, A. Farina, and L. Timmoneri. A comparison of two Cramér-Rao bounds for nonlinear filtering with $P_d < 1$. *IEEE Transactions on Signal Processing*, 52(9):2361–2369, 2004.

23 Y. Bar-Shalom, A. Kumar, W. D. Blair, and G. W. Groves. Tracking low elevation targets in the presence of multipath propagation. *IEEE Transactions on Aerospace and Electronic Systems*, 30(3):973–979, 1994.

24 M. L. Hernandez, B. Ristic, A. Farina, T. Sathyan, and T. Kirubarajan. Performance measure for Markovian switching systems using best-fitting Gaussian distributions. *IEEE Transactions on Aerospace and Electronic Systems*, 44(2):724–747, 2008.

25 M. Mallick, X. Tian, and L. Xu. Reentry vehicle filtering using a passive angle-only sensor. *IEEE Transactions on Aerospace and Electronic Systems*, 60(6):8539–8554, 2024.

26 M. Mallick. A note on bearing measurement model. *ResearchGate*, 2018.

27 Y. Bar-Shalom, X. R. Li, and T. Kirubarajan. *Estimation with Applications to Tracking and Navigation: Theory Algorithms and Software*. John Wiley & Sons, 2001.

28 Z. Duan, C. Han, and X. R. Li. Comments on "unbiased converted measurements for tracking". *IEEE Transactions on Aerospace and Electronic Systems*, 40(4):1374, 2004.

29 Y. Bar-Shalom and K. Birmiwal. Consistency and robustness of PDAF for target tracking in cluttered environments. *Automatica*, 19(4):431–437, 1983.

9

Tensor Decomposition in Point-mass Filters

Ondřej Straka, Jakub Matoušek, Ivo Punčochář, and Jindřich Duník

Department of Cybernetics, University of West Bohemia, Pilsen, Czech Republic

9.1 Introduction

This chapter deals with the state estimation of nonlinear stochastic dynamic systems. State estimation has been a subject of considerable research interest for the last decades. It plays an important role in signal processing, fault detection, and optimal control in areas such as navigation, tracking, image and speech processing, weather forecasting, biology, and economics.

The emphasis is laid on the Bayesian point-mass (or grid-based) filters, which are particularly suitable for accurate and consistent state estimation of highly nonlinear or non-Gaussian models. The point-mass filters (PMFs) solve the Bayesian recursive relations (BRRs) for calculation of the conditional probability density function (PDF) using deterministic integration rules evaluated in a grid and stored in the form of *tensors*. As a consequence, the PMF suffers from the curse of dimensionality.

Therefore, in the literature, a variety of approaches for reducing the computational and memory complexity have been proposed. The approaches range from relatively basic sparse grid methods through the model-specific methods to the most recent tensor decomposition methods. Compared to other approaches, the tensor decomposition methods alleviate both the computational complexity and memory requirements and allow the use of the PMFs for higher-dimensional models. The tensor decomposition has been, however, independently utilized in various steps of the estimator and also in model design.

The goal of the chapter is to provide a unifying framework for tensor decomposition-based PMF design. The proposed PMF algorithms use tensor decompositions to approximate the dynamic model and to represent the conditional PDF of the state. In the first case, the decomposition divides the transient density into two cores representing the current and future states. In the second case, the tensor train (TT) decomposition is used to approximate the PDF. The PMFs utilizing the decompositions are evaluated using a tracking scenario.

9.2 Model, Bayesian Estimation, and Numerical Solution

Consider a discrete-time stochastic system described by a nonlinear state-space model with additive noises.

$$\mathbf{x}_{k+1} = \mathbf{f}_k(\mathbf{x}_k, \mathbf{u}_k) + \mathbf{w}_k, \tag{9.1a}$$

$$\mathbf{z}_k = \mathbf{h}_k(\mathbf{x}_k) + \mathbf{v}_k, \tag{9.1b}$$

where $\mathbf{x}_k \in \mathbb{R}^{n_x}$, $\mathbf{u}_k \in \mathbb{R}^{n_u}$ and $\mathbf{z}_k \in \mathbb{R}^{n_z}$ represent the hidden state of the system, known input, and the available measurement at time instant $k = 0, 1, \ldots, T$, respectively. The functions $\mathbf{f}_k : \mathbb{R}^{n_x \times n_u} \mapsto \mathbb{R}^{n_x}$ and $\mathbf{h}_k : \mathbb{R}^{n_x} \mapsto \mathbb{R}^{n_z}$ are assumed known. The state noise $\mathbf{w}_k \in \mathbb{R}^{n_x}$ and measurement noise $\mathbf{v}_k \in \mathbb{R}^{n_z}$ are described by known PDFs $p_{\mathbf{w}_k}$ and $p_{\mathbf{v}_k}$. The initial state $\mathbf{x}_0$ is given by known PDF $p_{\mathbf{x}_0}$. Both noises are assumed to be white, mutually independent, and independent of the initial state. Since the input $\mathbf{u}_k$ is assumed to be known, its effect can be reflected through the time-varying function $\mathbf{f}_k$. Hence, it will be omitted in the following for convenience. The model (9.1) can be, then, expressed using transition PDF (9.2a) and measurement PDF (9.2b)

$$p(\mathbf{x}_{k+1}|\mathbf{x}_k) = p_{\mathbf{w}_k}(\mathbf{x}_{k+1} - \mathbf{f}_k(\mathbf{x}_k)), \tag{9.2a}$$

$$p(\mathbf{z}_k|\mathbf{x}_k) = p_{\mathbf{v}_k}(\mathbf{z}_k - \mathbf{h}_k(\mathbf{x}_k)). \tag{9.2b}$$

9.2.1 Bayesian State Estimation

The goal of state estimation is to infer the posterior PDF $p(\mathbf{x}_z|\mathbf{z}^k)$ of the state $\mathbf{x}_k$ given all the measurements available up to time k denoted as $\mathbf{z}^k := [\mathbf{z}_1^\top, \mathbf{z}_2^\top, \cdots, \mathbf{z}_k^\top]^\top$. The general solution to the problem is provided by the BRRs consisting of the Bayes' rule (9.3a) and the Chapman–Kolmogorov equation (CKE) (9.3b)

$$p(\mathbf{x}_k|\mathbf{z}^k) = \frac{p(\mathbf{z}_k|\mathbf{x}_k)p(\mathbf{x}_k|\mathbf{z}^{k-1})}{p(\mathbf{z}_k|\mathbf{z}^{k-1})}, \tag{9.3a}$$

$$p(\mathbf{x}_k|\mathbf{z}^{k-1}) = \int p(\mathbf{x}_k|\mathbf{x}_{k-1})p(\mathbf{x}_{k-1}|\mathbf{z}^{k-1})\mathrm{d}\mathbf{x}_{k-1}, \tag{9.3b}$$

where $p(\mathbf{z}_k|\mathbf{z}^{k-1})$ is the evidence given by $p(\mathbf{z}_k|\mathbf{z}^{k-1}) = \int p(\mathbf{z}_k|\mathbf{x}_k)p(\mathbf{x}_k|\mathbf{z}^{k-1})\mathrm{d}\mathbf{x}_k$. The initial condition for the BRRs is $p(\mathbf{x}_0|\mathbf{z}^{-1}) = p_{\mathbf{x}_0}(\mathbf{x}_0)$. The calculation of the BRRs thus involves alternating the filtering step (9.3a) calculating the filtering PDF $p(\mathbf{x}_k|\mathbf{z}^k)$ and the prediction step (9.3b) yielding the predictive PDF $p(\mathbf{x}_k|\mathbf{z}^{k-1})$.

Unfortunately, the closed-form solution to the BRRs is possible for a limited set of models only, such as a linear Gaussian (or Gaussian sum) model. In other cases, an approximate solution has to be used [1]. In the literature, many approximate (local and global) solutions have been proposed, offering a trade-off between estimation performance and computational complexity. Selection of the most suitable estimation algorithm should reflect not only the available computational power but also the model properties. For highly nonlinear models and/or multi-modal PDFs, a global PMF, which is based on a grid-based numerical solution to the BRRs, is a good choice due to its robustness and versatility [2–5].

9.2.2 Numerical Solution to BRRs and PMF Design

The grid-based solution to the BRRs starts with an approximation $\hat{p}(\mathbf{x}_0|\mathbf{z}^{-1}; \Xi_0)$ of an initial conditional PDF $p(\mathbf{x}_0|\mathbf{z}^{-1})$ in the form of a point-mass density (PMD), which is interpreted here as a *piece-wise constant* approximation of a PDF [6]. The PMD is given by a set of N grid points $\Xi_0 = \{\xi_0^{(i)}\}_{i=1}^{N}$, $\xi_0^{(i)} \in \mathbb{R}^{n_x}$, approximating and discretizing the continuous state-space, as

$$\hat{p}(\mathbf{x}_0|\mathbf{z}^{-1};\Xi_0) := \sum_{i=1}^{N} \mathcal{P}_{0|-1}(i)\mathbf{1}_{\mathcal{D}_0^{(i)}}(\mathbf{x}_0), \tag{9.4}$$

where $N = \prod_{i=1}^{n_x} N_i$ is a total number of points, N_i is a user-defined number[1] of grid points in ith dimension of the state-space, $\mathcal{P}_{0|-1}(i) \propto p(\xi_0^{(i)}|\mathbf{z}^{-1})$ is a normalized[2] value of the PDF $p(\mathbf{x}_0|\mathbf{z}^{-1})$ evaluated at the ith grid point $\xi_0^{(i)}$, further referred to as *weight*, and $\mathbf{1}_{\mathcal{D}_0^{(i)}}(\mathbf{x}_0)$ is the indicator function that equals to 1 if $\mathbf{x}_0 \in \mathcal{D}_0^{(i)}$ and otherwise is zero. That is, the $\hat{p}(\mathbf{x}_0|\mathbf{z}^{-1};\Xi_0)$ is constant in the neighborhood $\mathcal{D}_0^{(i)}$.

In this chapter, we assume an equidistant grid, i.e., the grid where each neighborhood $\mathcal{D}_0^{(i)}$ is an n_x-dimensional hyper-rectangle centered on its grid point $\xi_0^{(i)}$ and has sides of lengths specified as elements of vector $\mathbf{\Delta}_0 \in \mathbb{R}^{n_x}$. The volume of each neighborhood $\mathcal{D}_0^{(i)}$ is the same $\delta_0 = \prod_{j=1}^{n_x} \mathbf{\Delta}_0(j), \forall i$, as illustrated in Figure 9.1. The notation $\mathbf{a}(j)$ and $\mathbf{A}(i,j)$ denotes the jth element of the vector $\mathbf{a}$ and the element on the ith row and jth column of the matrix $\mathbf{A}$, respectively. The analogous notation of an element is also adopted for tensors. Note that, following this notation, the weights $\mathcal{P}_{0|-1}$ are treated as a vector, whereas the grid points Ξ_0 are treated as a set, which simplifies the further derivations.

Having the initial PMD (9.4), the grid-based numerical solution to the Bayes' rule (9.3a) leads to the *filtering* PMD $\hat{p}(\mathbf{x}_k|\mathbf{z}^k;\Xi_k) = \sum_i \mathcal{P}_{k|k}(i)\mathbf{1}_{\mathcal{D}_k^{(i)}}(\mathbf{x}_k)$ with the weight

$$\mathcal{P}_{k|k}(i) \propto \mathcal{P}_{k|k-1}(i)p\left(\mathbf{z}_k|\mathbf{x}_k = \xi_k^{(i)}\right), \tag{9.5}$$

where $p(\mathbf{z}_k|\mathbf{x}_k = \xi_k^{(i)})$ is the measurement PDF (9.2b) evaluated at the ith grid point. Thus, the grid Ξ_k remains untouched in the filtering step; only the weights are updated w.r.t the measurement (9.1b).

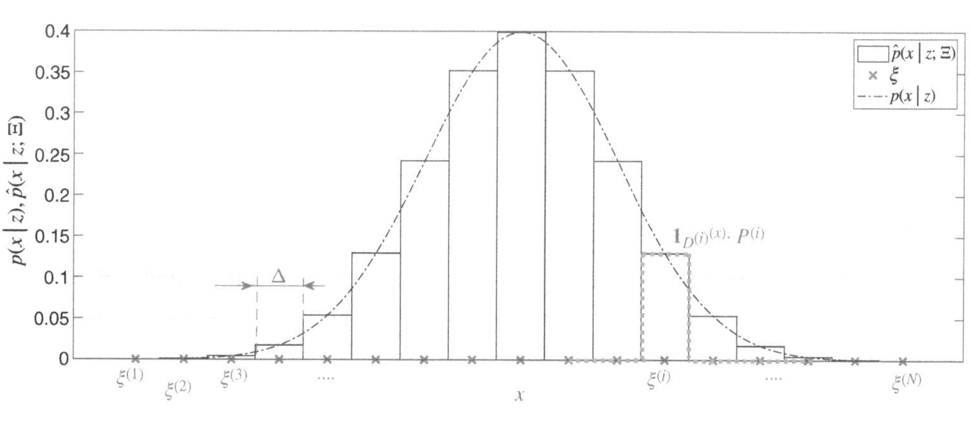

Figure 9.1 Point-mass density approximation illustration.

1 N_i is usually time independent to achieve constant computational complexity.
2 Due to point-wise approximation of the PDF, the PMD should be normalized to ensure that $\int \hat{p}(\mathbf{x}_0|\mathbf{z}^{-1};\Xi_0)d\mathbf{x}_0 = 1$.

Following the Eulerian[3] perspective to the PMF design, the prediction step starts with *defining* a new grid $\Xi_{k+1} = \{\xi_{k+1}^{(i)}\}_{i=1}^{N}, \xi_{k+1}^{(i)} \in \mathbb{R}^{n_x}$, by the user, such that it sufficiently covers[4] the expected support of the PDF $p(\mathbf{x}_{k+1}|\mathbf{z}_k)$. Then, the *predictive* PMD weights are

$$\mathcal{P}_{k+1|k}(j) = \sum_{i=1}^{N} p\left(\xi_{k+1}^{(j)}|\mathbf{x}_k = \xi_k^{(i)}\right)\mathcal{P}_{k|k}(i), \tag{9.6}$$

where $p\left(\xi_{k+1}^{(j)}|\mathbf{x}_k = \xi_k^{(i)}\right)$ is the evaluation of the transition PDF (9.2a) at the new grid-point $\xi_{k+1}^{(j)}$. The predictive weights $\mathcal{P}_{k+1|k}(j)$ inform on how probable it is that the true state $\mathbf{x}_{k+1}$ is in the vicinity of $\xi_{k+1}^{(j)}$ given the filtering PMD (giving information on the probable location of the past state $\mathbf{x}_k$) and the dynamics (9.1a).

The point-wise relations (9.5), (9.6) can be written in a convenient vector/matrix form as

$$\mathcal{P}_{k|k} \propto \mathbf{L}_k \odot \mathcal{P}_{k|k-1}, \tag{9.7a}$$

$$\mathcal{P}_{k+1|k} = \mathbf{T}_k \mathcal{P}_{k|k}, \tag{9.7b}$$

where $\mathbf{L}_k \in \mathbb{R}^N$ is the likelihood vector and $\mathbf{T}_k \in \mathbb{R}^{N\times N}$ is the transition probability matrix (TPM) with elements defined as

$$\mathbf{L}_k(i) = p\left(\mathbf{z}_k|\mathbf{x}_k = \xi_k^{(i)}\right), \tag{9.8a}$$

$$\mathbf{T}_k(j,i) = p\left(\xi_{k+1}^{(j)}|\mathbf{x}_k = \xi_k^{(i)}\right), \tag{9.8b}$$

respectively. The notation $\odot$ stands for the Hadamard (point-wise) product.

The PMF recursively calculates the estimates in the form of the PMDs. The conditional mean $\hat{\mathbf{x}}_{k|\ell} = \mathsf{E}[\mathbf{x}_k|\mathbf{z}^\ell]$ and covariance matrix $\hat{\mathbf{P}}_{k|\ell} = \mathsf{E}[\mathbf{x}_k|\mathbf{z}^\ell]$ can readily be computed if required [9]

$$\hat{\mathbf{x}}_{k|\ell} \approx \delta_k \sum_{i=1}^{N} P_{k|\ell}(i)\xi_k^{(i)}, \tag{9.9a}$$

$$\mathbf{P}_{k|\ell} \approx \delta_k \sum_{i=1}^{N} P_{k|\ell}(i)\left(\xi_k^{(i)} - \hat{\mathbf{x}}_{k|\ell}\right)\left(\xi_k^{(i)} - \hat{\mathbf{x}}_{k|\ell}\right)^T. \tag{9.9b}$$

9.2.3 PMF Algorithm Summary

A standard form of the PMF algorithm can be summarized by the following steps.

Algorithm 9.1: Point-mass Filter

---

1. *Initialization*: Define numbers of grid points $N_i, \forall i$, set $k = 0$, construct the initial grid of points Ξ_0, and on the basis of the initial PDF $p(\mathbf{x}_0|\mathbf{z}^{-1})$ calculate the initial PMD vector $\mathcal{P}_{0|-1}$ evaluated at Ξ_0.

3 The name comes from Eulerian and Lagrangian specification of the flow field. Whereas in the Lagrangian definition, the observer is moving according to the flow, in the Eulerian definition the observer observes a given space [7].

4 The support well-covering the predictive PDF can be based on predictive mean and covariance matrix calculated by the extended or unscented Kalman filter or on transformed grid Ξ_k via the system dynamics [8].

2. *Measurement update*: Using the actual measurement $\mathbf{z}_k$ and measurement equation (9.1b), construct the likelihood vector $\mathbf{L}_k$ using (9.8a) and calculate the filtering PMD $\mathcal{P}_{k|k}$ vector using (9.7a).
3. *Time update*:
   - Based on the filtering moments $\hat{\mathbf{x}}_{k|k}$ and $\mathbf{P}_{k|k}$ (9.9) and state equation (9.1a), calculate the predictive moments and determine the grid Ξ_{k+1}.
   - Using the grids Ξ_k, Ξ_{k+1} and state equation (9.1a), construct the TPM $\mathbf{T}_k$ (9.8b) and calculate the predictive PMD vector $\mathcal{P}_{k|k}$ (9.7b).
4. Set $k = k + 1$ and go to Step 2.

9.3 Overcoming PMF Complexity: Sparse, Smart, and Compressed

The PMF is an accurate and robust estimation algorithm, especially due to the sophisticated design of the grid [10, 11]. The grid covers not only parts of the state-space covering high-density regions of the conditional PMF but also the tails with "low probability" of occurrence of $\mathbf{x}_k$. As such, the PMF is an algorithm resilient w.r.t. the measurement outliers. On the other hand, the grid design is the root cause of the curse of dimensionality in both *memory* and *computational* requirements of the PMF.

9.3.1 PMF Complexity

Assuming an equidistant grid that can be efficiently stored using a reference grid point along with number of grid points per dimension $N_i, \forall i$, and the neighborhood $\boldsymbol{\Delta}_k$, the *memory complexity* arises from the requirement to store the weights $\mathcal{P}_k(j)$ of a PMD for $j = 1, \ldots, N$. For example, considering equal number of grid points along dimensions, i.e., $N_1 = N_2 = \cdots = N_{n_x}$ and a rather modest number $N_i = 10^2$ points per dimension, almost 1GB storage is needed for one PMD with $N = N_i^4 = 10^8$ weights in double precision in a rather low-dimensional state-space with $n_x = 4$. For a $n_x = 6$ dimensional state-space, we need a few TBs to store the PMD with $N = N_i^6 = 10^{12}$. The memory requirements thus grow exponentially with the state dimension n_x, i.e., the complexity is $\mathcal{O}(N_i^{n_x})$.

The PMF *computational complexity* is determined by the complexity of evaluation of the transition and measurement PDFs (9.2). While the measurement PDF in (9.2b) is evaluated at all N grid points Ξ_k, the transition PDF in (9.2a) must be evaluated for all combinations of the grid points Ξ_k and Ξ_{k+1}. It indicates that the computational requirements of the standard Eulerian PMF scale with the square of the number of grid points N, meaning the complexity is $\mathcal{O}(N^2)$. Due to its complexity, the application of the standard PMF is limited. For example, the application in terrain-aided navigation is restricted to modeling only a two-dimensional state vector representing the horizontal vehicle's position, and there is no room for additional error states [10].

9.3.2 Approaches to Overcome PMF Complexity

The memory and computational requirements are the main limitations of PMF when applied to models with $n_x \geq 4$. As a result, various approaches have been developed to reduce the complexity through either specific grid designs or the *smart* use of the specific properties of the state-space model. The former approaches include *sparse* and multi-grid designs, which allow the state-space to be covered by a reduced number of grid points with a little impact on the PMD approximation error [5, 8, 12]. The latter approaches include:

- Rao-Blackwellization, also known as marginalization, which utilizes the conditional linearity of models and splits the model into a nonlinear sub-model, which is estimated using an expensive PMF, and a linear sub-model estimated by cheap Kalman filters (KFs) [9, 13].
- Thrifty convolution, which evaluates the transition PDF (9.2a) for the grid points in Ξ_k and in Ξ_{k+1} that are statistically close to each other given the properties of the transition PDF $p(\mathbf{x}_{k+1}|\mathbf{x}_k)$ [5].
- Copula decomposition, which decomposes the joint conditional density into the marginal densities (allowing efficient prediction) and an easy-to-calculate copula density function. As a consequence, the computational complexity grows only linearly with the state dimension, which is a great simplification compared to the exponential growth of the standard convolution in other PMF methods [14].

All of these approaches help reduce both computational and memory requirements. However, apart from the Rao-Blackwellization and the copula decomposition, they do not overcome the curse of dimensionality, meaning that the computational complexity remains high. The Rao-Blackwellization reduces the model dimensionality, but its usage is limited to models in a specific conditionally linear form. The copula decomposition propagates the marginal densities only, and thus it provides an approximate solution, ignoring the dependence among the state variables in the prediction step.

Another smart approach to reduce the PMF computational complexity is the Lagrangian approach. It splits the prediction solution using the CKE (9.3b) into two parts; advection solved by a carefully designed grid movement through the function $\mathbf{f}_k$ and diffusion reflecting the noise properties using a convolution efficiently solved by the Fourier transform [15]. Consequently, the Lagrangian approach reduces the computational complexity to $\mathcal{O}(N\log(N))$, but the memory requirements are the same. The Lagrangian approach is applicable to linear dynamics only.

9.3.3 Complexity Reduction by Tensor Decomposition

Recently, several approaches to reducing the PMF complexity have emerged utilizing tensor decompositions [15–18]. These approaches interpret densities evaluated over a grid using a tensor format and represent the tensor in an approximate decomposed form.[5] In this chapter, we introduce two tensor decomposition-based concepts of PMF design, namely

- Functional tensor decomposition of the dynamic model, i.e., the transition PDF [18],
- TT decomposition of all conditional PMDs [15].

Both concepts are presented using a unified notation, and their properties are analyzed in detail. The discussed codes are supplemented with MATLAB® implementation.

5 This form can be seen as a generalization of the matrix singular value decomposition to higher dimensions.

9.4 Tensor Decomposition of the Dynamic Model

To attain a model structure that facilitates the computational complexity reduction, the approach approximates the transition PDF by a functional decomposition[6] as

$$p(\mathbf{x}_{k+1}|\mathbf{x}_k) \approx \hat{p}(\mathbf{x}_{k+1}|\mathbf{x}_k) = \sum_{r=1}^{R} \mathcal{F}^r(\mathbf{x}_{k+1})\bar{\mathcal{G}}^r(\mathbf{x}_k), \tag{9.10}$$

where $\mathcal{F}^r : \mathbb{R}^{n_x} \mapsto \mathbb{R}^+$ and $\bar{\mathcal{G}}^r : \mathbb{R}^{n_x} \mapsto \mathbb{R}^+$, $r = 1, \dots, R$ are suitable non-negative functions, designed in advance, and R is the order of the approximation called a *rank*. Given the decomposition (9.10), the CKE (9.3b) can be written as

$$p(\mathbf{x}_{k+1}|\mathbf{z}^k) \approx \sum_{r=1}^{R} \mathcal{F}^r(\mathbf{x}_{k+1}) \int \bar{\mathcal{G}}^r(\mathbf{x}_k)p(\mathbf{x}_k|\mathbf{z}^k)d\mathbf{x}_k. \tag{9.11}$$

The simplification of using (9.11) in (9.6) is that R integrals in (9.3b) can be computed at the grid points $\xi_k^{(i)} \in \Xi_k$ only, without the need to consider all combinations of grid points $\xi_{k+1}^{(j)} \in \Xi_{k+1}$ and grid points $\xi_k^{(i)} \in \Xi_k$.

For the point-mass approximation

$$\hat{p}(\mathbf{x}_k|\mathbf{z}^k; \Xi_k) := \sum_{i=1}^{N} \mathcal{P}_{k|k}(i)\mathbf{1}_{\mathcal{D}_k^{(i)}}(\mathbf{x}_k), \tag{9.12}$$

of $p(\mathbf{x}_k|\mathbf{z}^k)$ the integral in (9.11) is[7]

$$\int \bar{\mathcal{G}}^r(\mathbf{x}_k)p(\mathbf{x}_k|\mathbf{z}^k)d\mathbf{x}_k \approx \sum_{i=1}^{N} \bar{\mathcal{G}}^r(\xi_k^{(i)})\mathcal{P}_{k|k}(i)\delta_k =: \mathcal{I}_k^r. \tag{9.13}$$

The values of the predictive PDF (9.11) are then given as

$$\mathcal{P}_{k+1|k}(j) = \sum_{r=1}^{R} \mathcal{F}^r\left(\xi_{k+1}^{(j)}\right) \sum_{i=1}^{N} \bar{\mathcal{G}}^r\left(\xi_k^{(i)}\right)\mathcal{P}_{k|k}(i)\delta_k = \sum_{r=1}^{R} \mathcal{F}^r\left(\xi_{k+1}^{(j)}\right)\mathcal{I}_k^r. \tag{9.14}$$

The effect of transition PDF decomposition on the evaluation of the predictive PDF in the CKE is illustrated in Figure 9.2.

The model (9.1a) with additive noise is advantageous for the decomposition as the transition PDF depends only on the difference $\mathbf{x}_{k+1} - \mathbf{f}_k(\mathbf{x}_k)$ as

$$p(\mathbf{x}_{k+1}|\mathbf{x}_k) = p_{\mathbf{w}_k}(\mathbf{x}_{k+1} - \mathbf{f}_k), \tag{9.15}$$

where the notation $\mathbf{f}_k := \mathbf{f}_k(\mathbf{x}_k)$ is introduced for convenience. Expressing the transition PDF (9.15) in terms of $\mathbf{x}_{k+1}$ and $\mathbf{f}_k$ is substantially simpler[8] than expressing the PDF $p(\mathbf{x}_{k+1}|\mathbf{x}_k) = p_{\mathbf{w}_k}(\mathbf{x}_{k+1} - \mathbf{f}_k)$ in terms of $\mathbf{x}_{k+1}$ and $\mathbf{x}_k$, especially for highly nonlinear $\mathbf{f}_k$. Thus, the approximate transition PDF will be further expressed in terms of $\mathbf{x}_{k+1}$ and $\mathbf{f}_k$ as follows

$$\hat{p}(\mathbf{x}_{k+1}|\mathbf{x}_k) = \hat{p}(\mathbf{x}_{k+1}|\mathbf{f}_k) = \sum_{r=1}^{R} \mathcal{F}^r(\mathbf{x}_{k+1})\mathcal{G}^r(\mathbf{f}_k), \tag{9.16}$$

6 Relation (9.10) [18] can be seen as a particular case of the functional TT decomposition [19] where the functions $\mathcal{F}^r$ and $\mathcal{G}^r$ are not decomposed further to functions of single elements of $\mathbf{x}_{k+1}$ and $\mathbf{x}_k$.

7 For convenience, the constant number N of grid points of Ξ_{k+1} and Ξ_k is considered.

8 Note that $\mathbf{f}_k$ is a random variable obtained by transforming the random variable $\mathbf{x}_k$ through $\mathbf{f}_k(\cdot)$.

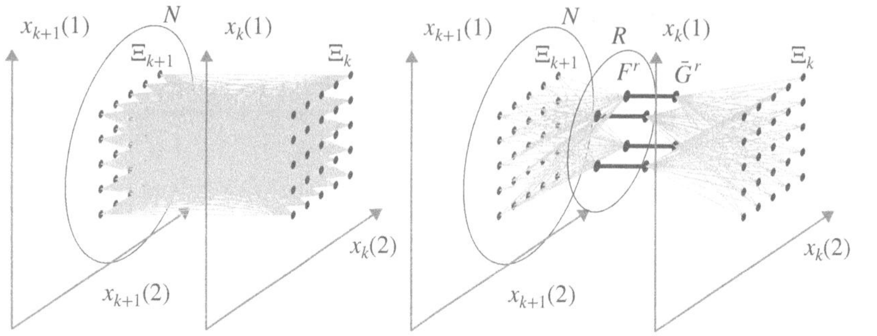

Evaluation of $p(\mathrm{x}_{k+1}|\mathrm{x}_k)$ at Ξ_{k+1} and Ξ_k without transition density decomposition ($n_x = 2$)

(a)

Evaluation of $p(\mathrm{x}_{k+1}|\mathrm{x}_k)$ at Ξ_{k+1} and Ξ_k with transition PDF decomposition ($n_x = 2$)

(b)

Figure 9.2 Illustration of evaluating the predictive PDF without and with transition density decomposition.

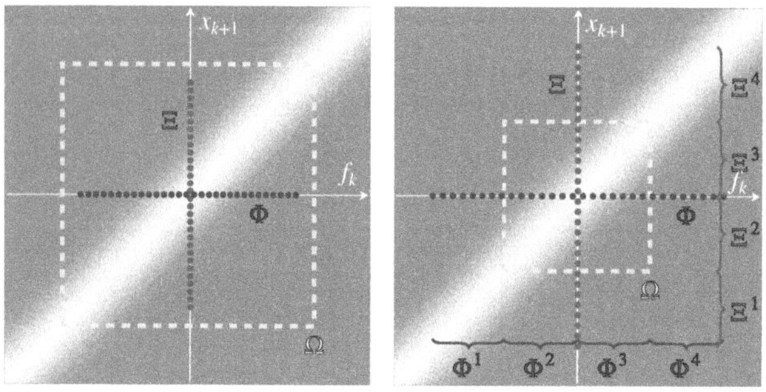

(a) Single grids Ξ and Φ **(b)** Split grids Ξ and Φ

Figure 9.3 Illustration of the Gaussian transition PDF $p(x_{k+1}|f_k) = p_{w_k}(x_{k+1} - f_k)$ (shaded background)), the approximation region Ω (dashed), and the grids Ξ (dotted) and Φ (dotted) for x_{k+1} and f_k, respectively.

where $\bar{\mathcal{G}}^r(\mathbf{x}_k) = \mathcal{G}^r(\mathbf{f}_k(\mathbf{x}_k))$.

The grid Ξ_k for $\mathbf{x}_k$ can then be used to obtain the grid $\boldsymbol{\Phi}_k := \{\boldsymbol{\phi}_k^{(i)}\}_{i=1}^N$ for $\mathbf{f}_k$. Remark that the grid $\boldsymbol{\Phi}_k$ serves as a foundation for designing the grid Ξ_{k+1}. The approximation (9.16) is calculated in an approximation region $\Omega_k \subset \mathbb{R}^{2n_x}$ for $\mathbf{x}_{k+1}$ and $\mathbf{f}_k$. For $n_x = 1$ and standard Gaussian PDF p_{w_k}, the transition PDF $p(x_{k+1}|x_k) = p_{w_k}(x_{k+1} - f_k)$, the grids Ξ_{k+1}, $\boldsymbol{\Phi}_k$ and the approximation region Ω_k are illustrated in Figure 9.3a.

The approximation region should cover the space where the PDF $p_{\mathbf{w}_k}$ is non-negligible. Its size affects the rank R of the approximation. Enlarging the region while keeping the approximation error fixed leads to an increase in the rank, which should be avoided, especially for higher dimensions n_x. When the PDF $p_{\mathbf{w}_k}$ is time-invariant, the region can be *fixed*, $\Omega_k = \Omega$, $\forall k$. Hence, if either grid points of Ξ_{k+1} or grid points of $\boldsymbol{\Phi}_k$ do not fit into the fixed region Ω, they can be conveniently shifted by some $\boldsymbol{\zeta} \in \mathbb{R}^{n_x}$ to the region as

$$p_{\mathbf{w}_k}(\boldsymbol{\xi}_{k+1}^{(j)} - \boldsymbol{\phi}_k^{(i)}) = p_{\mathbf{w}_k}\left((\boldsymbol{\xi}_{k+1}^{(j)} - \boldsymbol{\zeta}) - (\boldsymbol{\phi}_k^{(i)} - \boldsymbol{\zeta})\right).$$

Shifting each pair of the grid points $\boldsymbol{\xi}_{k+1}^{(j)}$ and $\boldsymbol{\phi}_k^{(i)}$, $i, j = 1, \dots, N$ individually by a different value $\boldsymbol{\zeta}^{i,j}$ for each pair $\boldsymbol{\xi}_{k+1}^{(j)}$ and $\boldsymbol{\phi}_k^{(i)}$ is highly inefficient. This is because there are N^2 different pairs, and the number of shifts would also be N^2. The decomposition aims to avoid such combinations when calculating the predictive PDF values (9.6). Hence, the strategy is to shift the entire grids Ξ_{k+1} and $\mathbf{\Phi}_k$ by a single value ζ calculated from averages of all points in both grids

$$\zeta = \frac{1}{2N}\left(\sum_{i=1}^{N} \phi_k^{(i)} + \sum_{j=1}^{N} \xi_{k+1}^{(i)}\right). \tag{9.17}$$

Shifting both grids using this value positions their centers as close as possible to the origin, where the center of the approximation region Ω is located, see Figure 9.3a for illustration.

9.4.1 Two-level Convolution

In some cases, the shifted grids may cover an area larger than Ω. Expanding Ω results in a higher rank R, which subsequently increases computational costs, requiring $\mathcal{O}((2N+1)R)$ evaluations of the functions $\mathcal{G}^r$ and $\mathcal{F}^r$. The idea of [20] is to split the grids Ξ and $\mathbf{\Phi}$ into subgrids and treat each subgrid individually. For convenience, the time index subscripts will be dropped in the rest of the subsection, e.g., $\Xi := \Xi_{k+1}$. In particular, when $n_x = 1$, the grid Ξ is split into N_Ξ subgrids $\{\Xi^j\}_{j=1}^{N_\Xi}$ such that

$$\xi' < \xi'', \forall \xi' \in \Xi^i, \ \forall \xi'' \in \Xi^j, \ i < j$$

and by analogy, the grid $\mathbf{\Phi}$ is split into N_Φ subgrids $\{\mathbf{\Phi}^j\}_{j=1}^{N_\Phi}$. Each subgrid should fit into the approximation region Ω. The splitting is illustrated in Figure 9.3b for $n_x = 1$. The description of the splitting for $n_x > 1$ will follow later.

The value of predictive PDF (9.14) at a point $\xi \in \Xi^j$ can then be written as

$$\mathcal{P}_{k+1|k}(\xi) = \sum_{r=1}^{R} \mathcal{F}^r(\xi) \sum_{i=1}^{N_\Phi} \sum_{\phi \in \Phi^i} \mathcal{G}^r(\phi)\mathcal{P}_{k|k}(\phi) = \sum_{i=1}^{N_\Phi}\left(\sum_{r=1}^{R} \mathcal{F}^r(\xi) \sum_{\phi \in \Phi^i} \mathcal{G}^r(\phi)\mathcal{P}_{k|k}(\phi)\right), \tag{9.18}$$

where $\mathcal{P}_{k+1|k}(\xi)$ and $\mathcal{P}_{k|k}(\phi)$ denote the weights corresponding to ξ and ϕ, respectively. Note that the term $\sum_{i=1}^{N_\Phi} \sum_{\phi \in \Phi^i} \mathcal{G}^r(\phi)$ (summing over subgrids Φ^i and the points in each) is equivalent to the term $\sum_{i=1}^{N} \mathcal{G}^r(\phi^{(i)})$ (summing over all points in Φ). Then, for each combination of the subgrids Ξ^j and $\mathbf{\Phi}^j$, $j = 1, \dots, N_\Xi$, $i = 1, \dots, N_\Phi$ a single shift vector $\zeta^{i,j}$ is calculated similarly to that in (9.17).

The computation of the predictive PDF value (9.18) can be viewed as proceeding through *two levels* of convolution.

- The *upper-level convolution* combines each subgrid $\mathbf{\Phi}^i$, $i = 1, \dots, N_\Phi$ with all subgrids Ξ^j, $j = 1, \dots, N_\Xi$ by computing the values $\zeta^{i,j}$ shifting these subgrids.
- The *lower-level convolution* employs the functional decomposition (the parenthesis in (9.18)) by processing the gridpoints of Ξ^j and the gridpoints of Φ^i:

$$\mathcal{P}_{k+1|k}(\xi) = \sum_{i=1}^{N_\Phi}\left(\sum_{r=1}^{R} \mathcal{F}^r\left(\xi - \zeta^{i,j}\right) \sum_{\phi \in \Phi^i} \mathcal{G}^r\left(\phi - \zeta^{i,j}\right)\mathcal{P}_{k|k}(\phi)\right), \tag{9.19}$$

where $\xi \in \Xi^j$.

As the construction of Ξ_{k+1} proceeds from $\mathbf{\Phi}_k$ [20] as $\mathbf{x}_{k+1} = \mathbf{f}_k + \mathbf{w}_k$, the grids usually have significant overlap. This can increase efficiency in the predictive PDF calculation by assessing the closeness of individual subgrids. Here, the closeness relates to the covariance matrix of $\mathbf{w}_k$ and associated confidence sets. If the norm $\| \xi - \phi \|_{(\Sigma_k^{\mathbf{w}})^{-1}}$ with $\Sigma_k^{\mathbf{w}} = \text{var}[\mathbf{w}_k]$ is large $\forall \xi \in \Xi^j$ and $\forall \phi \in \mathbf{\Phi}^i$, then the value of the transition PDF (9.2a) is close to zero and this combination of subgrids can be omitted from the calculation. Thus, some combinations of subgrids Ξ^j, $j = 1, \ldots, N_\Xi$ and $\mathbf{\Phi}^i$, $i = 1, \ldots, N_\Phi$ can be dropped in (9.19). This approach is similar to the thrifty convolution proposed in [8] with the difference of proceeding in the upper-level convolution for the whole subgrids.

As stated, the subgrids of Ξ and $\mathbf{\Phi}$ must be designed to fit into the approximation region Ω after the shifting. This ensures that processing the subgrids Ξ^j and $\mathbf{\Phi}^i$ through the lower-level convolution, which involves the functional decomposition, does not result in a zero prediction PDF due to the limited size of the approximation region. Reducing the size of the subgrids increases their number N_Ξ and N_Φ, respectively. This results in more shifts and greater computational complexity. As $N_\Xi \to N$ and $N_\Phi \to N$, the complexity of this two-level convolution approaches that of the full convolution.

In terms of the number of function evaluations, the full calculation requires $\mathcal{O}(N^2)$ evaluations of the transition PDF. When using the decomposition of the transition PDF $\mathcal{O}((2N+1)R)$ evaluations of the functions $\mathcal{F}^r$ and $\mathcal{G}^r$ are needed. Thus, the savings due to the decomposition can be expected for $N \gg R$. Assuming subgrids $\{\Xi^j\}_{j=1}^{N_\Xi}$ having an equal number of grid points N/N_Ξ and subgrids $\{\mathbf{\Phi}^j\}_{j=1}^{N_\Phi}$ have an equal number of grid points N/N_Φ, then the proposed two-level convolution requires

$$\mathcal{O}\left(N_\Xi N_\Phi \left(\frac{N}{N_\Xi} R + \frac{N}{N_\Phi} R + R\right)\right) = \mathcal{O}((N_\Phi N + N_\Xi N + 1)R) \tag{9.20}$$

operations. The complexity depends on the number of subgrids N_Ξ and N_Φ. Note that when the subgrids have only a single grid point (i.e., $N_\Xi = N, N_\Phi = N$), then the number of function evaluations is $\mathcal{O}((2N^2+1)R)$, which means higher computational complexity than the full convolution. If, on the other hand, there are only single subgrids (i.e., $N_\Xi = 1, N_\Phi = 1$), then the number of function evaluations is $\mathcal{O}((2N+1)R)$, which means the same computational complexity as the convolution without the grid splitting. The complexity described in (9.20) of the proposed two-level convolution may increase with the same rank R, the two-level convolution allows the use of a small approximation region Ω, and a smaller rank. As a result, this reduces complexity.

Now, the subgrid construction will be generalized to higher dimensions. The usual approach to design the grid[9] Ξ is to

1. specify a grid for each axis $\Psi^{(i)} = \{\psi^{(i)}(j_i)\}$, $j_i = 1, \ldots N_{\Psi(i)}$ $i = 1, \ldots n_x$ (further called *axis grid*) and
2. combine the points of the axis grids as

$$\xi^{(\ell)} = [\psi^{(1)}(j_1), \psi^{(2)}(j_2), \ldots, \psi^{(n_x)}(j_{n_x})]^\mathrm{T}, \ j_i = 1, \ldots N_{\Psi(i)}, \ i = 1, \ldots n_x.$$

The number of points of Ξ is then $N = \prod_{i=1}^{n_x} N_{\Psi(i)}$.

The direct construction of the subgrids Ξ_{k+1}^j, $j = 1, \ldots N_\Xi$ facilitates streamlining the creation of a new grid Ξ_{k+1} and its splitting. First, each axis grid $\Psi^{(i)}$ is split into $M(i)$ subgrids. Combining these

9 The time index is again dropped here for convenience.

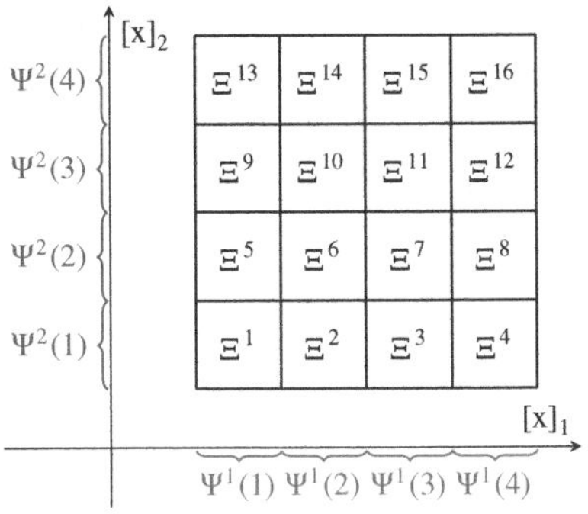

Figure 9.4 Illustration of construction of subgrids Ξ^{ℓ} for $n_x = 2$ from axis subgrids $\Psi^1(i)$ and $\Psi^2(j)$.

subgrids, one for each dimension, will result in a subgrid Ξ^{ℓ}. The total number of these subgrids is given by $N_{\Xi} = \prod_{i=1}^{n_x} M(i)$. The procedure is illustrated in Figure 9.4 for $n_x = 2$.

9.4.2 Algorithm of PMF with Two-level convolution

The algorithm of the PMF with the two-level convolution and full combination of the subgrids can be stated as follows in Algorithm 9.2.

Algorithm 9.2: PMF with Tensor Decomposition of Dynamics and Two-level Convolution

---

1. *Initialization*: (same as in Algorithm 9.1)
2. *Measurement update*: (same as in Algorithm 9.1)
3. *Grid construction:* Construct the new grid Ξ_{k+1} and split it into $N_{\Xi_{k+1}}$ subgrids Ξ^j_{k+1}, $j = 1, \dots, N_{\Xi_{k+1}}$.
4. *Time update*:
   - **(a)** Propagate the grid points $\xi_k^{(i)}$ of Ξ_k through the dynamics function $\mathbf{f}_k(\xi_k^{(i)})$ to obtain the grid points $\phi_k^{(i)}$ of $\mathbf{\Phi}_k$.
   - **(b)** Split $\mathbf{\Phi}_k$ into $N_{\mathbf{\Phi}_k}$ subgrids $\mathbf{\Phi}_0^i$, $i = 1, \dots, N_{\mathbf{\Phi}_k}$.
   - **(c)** For each pair of subgrids $\mathbf{\Phi}_k^i$ and Ξ^j_{k+1}, $i = 1, \dots, N_{\mathbf{\Phi}_k}$ and $j = 1, \dots, N_{\Xi_k}$ calculate the shift value $\zeta^{i,j}$.
   - **(d)** Compute the weights predictive PMD (9.6) at grid points of Ξ_{k+1} according to (9.19).
5. Set $k = k + 1$ and go to Step 2.

Note that using the thrifty combination of the subgrids would affect step (4d).

9.5 Tensor-train Decomposition of Point-mass Densities

In the previous approach, the computational complexity of the PMF prediction step was solved at the level of the model, where the transition PDF (9.10) was decomposed. In this approach, the transition PDF and the whole state space model (9.1) remain *unchanged*, but all the PMD-related quantities are represented in the form of *full* and latter *decomposed* tensors [21]. In particular, the TPM $\mathbf{T}_k$, the likelihood $\mathbf{L}_k$, and the PMD weights $\mathcal{P}_{k|k}$, $\mathcal{P}_{k+1|k}$ in (9.7a), (9.7b) represented as tensors reads

$$\widetilde{\mathbf{T}}_k \in \mathbb{R}^{N_1 \times N_2 \dots \times N_{n_x} \times N_1 \times N_2 \dots \times N_{n_x}}, \tag{9.21a}$$

$$\widetilde{\mathbf{L}}_k \in \mathbb{R}^{N_1 \times N_2 \dots \times N_{n_x}}, \tag{9.21b}$$

$$\widetilde{\mathcal{P}}_{k|k} \in \mathbb{R}^{N_1 \times N_2 \dots \times N_{n_x}}, \tag{9.21c}$$

$$\widetilde{\mathcal{P}}_{k+1|k} \in \mathbb{R}^{N_1 \times N_2 \dots \times N_{n_x}}. \tag{9.21d}$$

These tensors, further referred to as *uncertainty quantification tensors*, take advantage of the multilinear indexing. That means the relation between the linear index j of the PMD weight vector $\mathcal{P}_{k|k}(j)$ and the index of the tensorized PMD element

$$\widetilde{\mathcal{P}}_{k|k}(i_1, i_2, \dots, i_{n_x}), \tag{9.22}$$

in the row-major style is

$$\begin{aligned} j &= i_1 \cdot (N_2 \cdot N_3 \cdot \dots \cdot N_{n_x}) + i_2 \cdot (N_3 \cdot \dots \cdot N_{n_x}) + \dots i_{n_x-1} \cdot N_{n_x} + i_{n_x} \\ &= \sum_{\ell=1}^{n_x} i_\ell \left(\prod_{m=\ell+1}^{n_x} N_m \right). \end{aligned} \tag{9.23}$$

The vectorized and tensorized representations of a PMD is illustrated in Figure 9.5 for $n_x = \{2, 3\}$. Note that a detailed description and illustration of these tensor-based expressions can be found, e.g., in [21], and in the attached codes.

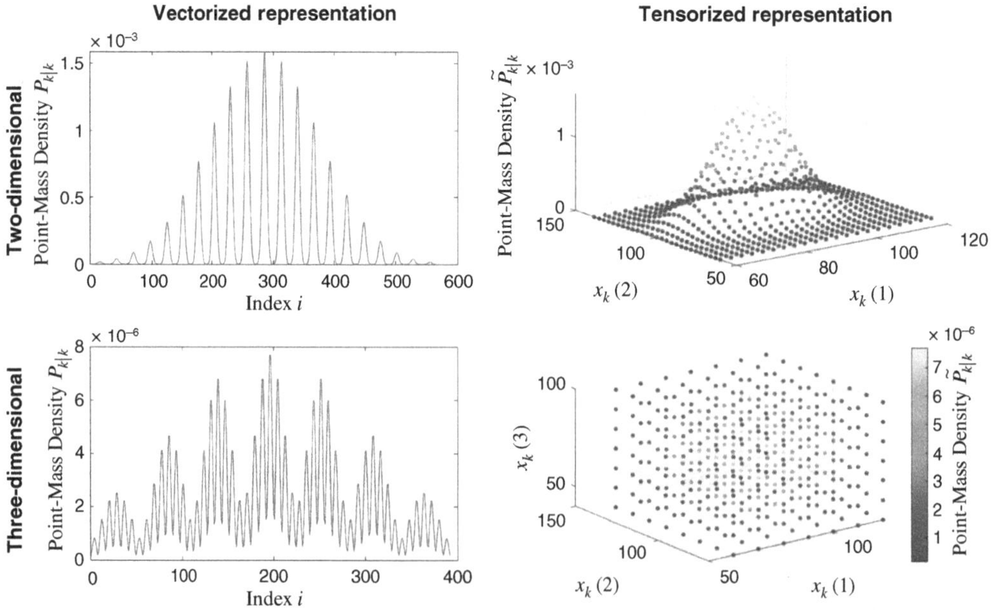

Figure 9.5 Possible representations of a PMD for two- and three-dimensional state.

9.5.1 Full Tensor Representation in PMF Design

Considering the *full* tensor representation of all related quantities (9.21), the *measurement update* (i.e., the filtering) (9.7a) in tensor form can be simply performed as

$$\widetilde{\mathcal{P}}_{k|k} = \widetilde{\mathbf{L}}_k \odot \widetilde{\mathcal{P}}_{k|k-1}. \tag{9.24}$$

The *time update* (or prediction) (9.7b) in the tensor form can be efficiently expressed using the Einstein summation notation as [21]

$$\widetilde{\mathcal{P}}_{k+1|k} = \sum_{i_1, i_2, \dots, i_{n_x}} \widetilde{\mathbf{T}}_k(i_1 i_2 \dots i_{2n_x}) \widetilde{\mathcal{P}}_{k|k}(i_{n_x+1} i_{n_x+2} \dots i_{2n_x}). \tag{9.25}$$

9.5.2 TT Approximation

Although the tensor representation allows efficient handling of the PDFs, it does not eliminate the curse of dimensionality. The tensor representation is "just" another representation of vectors and matrices storing all the probability-related quantities such that PMDs, likelihood, and transition probabilities.

To make a first step towards breaking the curse of dimensionality of tensorized PMF, we can approximate the tensorized quantities by the TT decomposition [22, 23]. An approximation of a tensor in $\mathbb{R}^{40\times30\times10}$ using TT decomposition with three cores (or "carriages") connected via the ranks $R_1 = 4$, $R_2 = 3$ is illustrated in Figure 9.6. We can see that the TT decomposition allows us to reduce the number of full tensor elements from $N = 40 \times 30 \times 10 = 12,000$ to $N_{TT} = 550$ elements at the cost of losing some information about the stored PMD.

In general, any d-dimensional tensor $\widetilde{\mathcal{A}} \in \mathbb{R}^{N_1 \times N_2 \times \dots \times N_d}$ can be decomposed using the TT into the product of the input matrix G_1, $(d-2)$ three-dimensional tensors $\tilde{G}_\ell$, $\ell = 2, 3, \dots, (d-1)$, and output matrix G_d, defined as

$$A(i_1, i_2, \dots, i_d) \approx \sum_{r_0 \dots r_d}^{\mathcal{R}_1 \cdots \mathcal{R}_d} G_1(r_0, i_1, r_1)\widetilde{G}_2(r_1, i_2, r_2) \dots \widetilde{G}_{d-1}(r_{d-2}, i_{d-1}, r_{d-1}) G_d(r_{d-1}, i_d, r_d). \tag{9.26}$$

where R_ℓ, $\forall \ell$, stands for the decomposition rank. The first and last ranks are $R_0 = R_d = 1$, and the remaining can be seen as user-defined parameters. The lower rank leads to lower dimensional decomposition with lower requirements on available storage, however, with higher loss of the information about the PMD. However, as the PMD is representing a PDF, which is rather a smooth (and possibly also close to symmetric) function, the decomposition is typically very accurate even for low ranks R_i. Note that the PMD of a Gaussian PDF can be well approximated by rank one due to its symmetry.

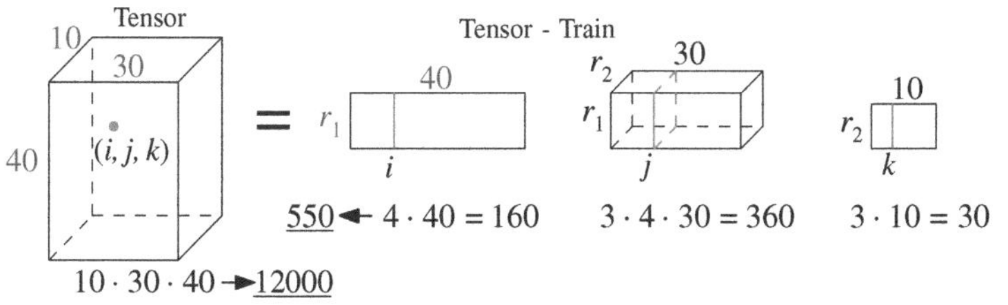

Figure 9.6 Tensor approximation by tensor-train decomposition with ranks $R_1 = 4$, $R_2 = 3$.

9.5.3 Tools for TT Decomposition

Design of the PMF based on TT decomposition, further denoted tensor train point-mass filter (TT-PMF), is based on a simple idea: *to decompose all probability-related quantities into the low-rank TTs and perform the measurement and time updates directly in the decomposed tensor format.*

To implement the TT-PMF efficiently, we can use a set of tools operating on tensors. Namely, we adopt the following functions from the MATLAB® TT Toolbox [24]:

- TT (greedy-cross[10]) approximation, routine `greedy2_cross`,
- TTs dot product, routine `dot`,
- TTs (core-wise) product, routine `times`,
- TT rank reduction with required precision, routine `round`,
- TT division by a number, routine `mrdivide`,
- Summation of elements of a tensor calculated from a TT, routine `sum`,
- TT second power, routine `power`.

Besides these toolbox functions, we also need the routine `EinsteinSum` for Einstein summation of two tensors over a subset in dimensions, which was described and implemented in [21].

9.5.4 TT-PMF Design

Having all the tools for tensor decomposition and manipulation, we can start with the development of the TT-PMF. The first step is the *initialization*. The weights $\mathcal{P}^{(i)}_{0|-1}$ of the $\hat{p}(\mathbf{x}_0|\mathbf{z}^{-1};\Xi_0)$ (9.4) calculated at grid Ξ_0 are transformed into the tensor format $\widetilde{\mathcal{P}}^{(i)}_{0|-1}$ (9.21d). Then, the initial tensor $\widetilde{\mathcal{P}}^{(i)}_{0|-1}$ is decomposed using the routine `greedy2_cross` into TT format (9.26), further denoted as $\breve{\mathcal{P}}^{(i)}_{0|-1}$. The TT ranks are selected automatically by the routine to meet the user-defined approximation accuracy ε.

In the *measurement update* we construct the likelihood tensor $\widetilde{\mathbf{L}}_k$ (9.21b) using the actual measurement $\mathbf{z}_k$ as in (9.8a) and (9.7a), and decompose it using `greedy2_cross` into TT format $\breve{\mathbf{L}}_k$. Then, the filtering TT is calculated as

$$\breve{\mathcal{P}}_{k|k} = \texttt{mrdivide}\left(\texttt{times}\left(\breve{\mathbf{L}}_k, \breve{\mathcal{P}}_{k|k-1}\right), c\right), \tag{9.27}$$

where $c = \delta_k \texttt{sum}(\texttt{times}(\breve{\mathbf{L}}_k, \breve{\mathcal{P}}_{k|k-1}))$ is a normalization constant.

Note that the TT multiplication `times` in (9.27) is realized using the Kronecker product of particular cores of the factors $\breve{\mathbf{L}}_k$ and $\breve{\mathcal{P}}_{k|k-1}$ [23]. As a consequence, the ranks of resulting cores of $\breve{\mathcal{P}}_{k|k}$ are higher than those of the factors. In particular, if a core of both $\breve{\mathbf{L}}_k$ and $\breve{\mathcal{P}}_{k|k-1}$ has the rank R_ℓ, then the respective core of $\breve{\mathcal{P}}_{k|k}$ has rank R_ℓ^2. Without treating this rank growth, the complexity of the TT-PMF would grow exponentially. To keep computational complexity constant and acceptable, we have to, therefore, reduce the rank of the filtering TT (9.27) using the routine `round`. The routine `round` reduces the rank of the TT to meet the required approximation accuracy ε.

The *time update* is preceded by creation of the "new" grid Ξ_{k+1}. The *new grid construction* is based on the moment-based approach according to the following steps

---

10 We prefer *black-box* TT decomposition approximation [25], where the only requirement for the decomposition is the ability to evaluate the tensor at an arbitrary position with evaluation of as low a number of entries as possible. These algorithms are usually based on a matrix approximate (i.e., suboptimal) decompositions. Out of these sub-optimal cross interpolation algorithms, we adopted the *greedy restricted cross interpolation algorithm* for TT, where the cross selection is performed by random selection of the rows and columns [26].

- From filtering PMD in the TT format $\check{\mathcal{P}}_{k|k}$ (9.27) into the tensor or vector format, calculate filtering mean $\hat{\mathbf{x}}_{k|k}$ and diagonal of the covariance matrix $\mathbf{P}_{k|k}$ of the filtering PMD according to

$$\hat{\mathbf{x}}_{k|k}(i) = \mathtt{dot}(\check{\mathcal{P}}_{k|k}, \check{\Xi}_{i,k})\delta_k, \tag{9.28a}$$

$$\mathbf{P}_{k|k}(i,i) = \mathtt{dot}(\check{\mathcal{P}}_{k|k}, \mathtt{power}(\check{\Xi}_{i,k}))\delta_k - (\hat{\mathbf{x}}_{k|k}(i))^2, \tag{9.28b}$$

where $i = 1, \dots, n_x$ and $\check{\Xi}_{i,k}$ represents the vector $[\xi_k^{(1)}(i), \xi_k^{(2)}(i), \dots, \xi_k^{(N)}(N)]$ of ith elements of all grid points expressed in the TT format calculated by the routine `greedy2_cross`.
- Use a prediction step of the KF or a local filter (such as the extended or unscented KF) to compute the predictive state moments $\hat{\mathbf{x}}_{k+1|k}$ and $\mathbf{P}_{k+1|k}$, which determines the region of the state-space where the predictive PMD is expected to lie [27].
- Based on the predictive moments, construct the new grid Ξ_{k+1} well-covering the support of the predictive PMD $\hat{p}(\mathbf{x}_{k+1}|\mathbf{z}^k; \Xi_{k+1})$.

Having the grids Ξ_k and Ξ_{k+1}, we can calculate TPM $\mathbf{T}_k$ as in (9.8b) and (9.7b), and decompose it using `greedy2_cross` into TT format $\check{\mathbf{T}}_k$. The predictive TT is, then, calculated as

$$\check{\mathcal{P}}_{k+1|k} = \mathtt{EinsteinSum}(\check{\mathbf{T}}_k, \check{\mathcal{P}}_{k|k}, [i_1, i_2, \dots, i_{n_x}]). \tag{9.29}$$

Similarly to the measurement update, the predictive TT shall be normalized using the routine `sum` and its rank reduced using the routine `round`.

9.5.5 TT-PMF Algorithm Summary and Notes

9.5.5.1 Algorithm Summary

The algorithm of the TT-PMF can be summarized as follows in Algorithm 9.3.

Algorithm 9.3: Point-mass Filter in TT Format

---

1. *Initialization*: Set required accuracy of the TT approximation ε, set $k = 0$, construct the initial grid of points Ξ_k, and define the initial PMD $\check{\mathcal{P}}_{k|k-1}$ in TT format.
2. *Measurement update*: Using the actual measurement $\mathbf{z}_k$ and measurement equation (9.1b), construct the likelihood tensor $\widetilde{\mathbf{L}}_k$ w.r.t. (9.8a) and (9.22), and its TT approximation $\check{\mathbf{L}}_k$. Calculate the filtering TT using (9.27) and reduce its rank.
3. *Time update*:
   - Based on the filtering moments $\hat{\mathbf{x}}_{k|k}$ and $\mathbf{P}_{k|k}$ (9.28) and state equation (9.1a), calculate the predictive moments and determine the grid Ξ_{k+1}.
   - Using the grids Ξ_k, Ξ_{k+1} and state equation (9.1a), construct the TPM $\widetilde{\mathbf{T}}_k$ w.r.t. (9.8b) and (9.22), and its TT approximation $\check{\mathbf{T}}_k$. Calculate the predictive TT according to (9.29).
4. Set $k = k + 1$ and go to step 2.

9.5.5.2 Computational and Memory Complexity Reduction

As can be seen in Figure 9.7, the TT formulation of the PMF allows significant reduction of both the computational and memory requirements. The *left* plot of the figure shows memory requirements to store one TPM and one PMD with $N_i = 101, \forall i$ in the tensor format required for the PMF and in TT format required for the TT-PMF. The curves for the standard PMF end at dimension $n_x = 4$

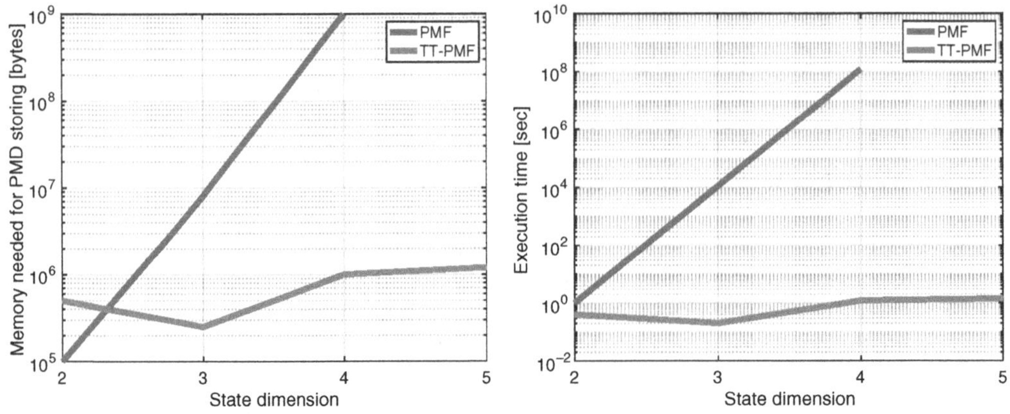

Figure 9.7 PMF and TT-PMF complexity as a function of state dimension n_x. Left plot: memory required for storing a PMD and a TPM in tensor and TT formats. Right plot: computational time required for one measurement and time update.

as there was not enough memory to save all the data needed for higher-dimensional states, but they can be easily extrapolated. It can be seen that the required memory by the proposed TT-PMF does not vary much with the dimension. The *right* plot shows the time needed for execution of one filtering and prediction step of the PMF and TT-PMF depending on n_x. We can see that PMF complexity grows exponentially, whereas TT-PMF complexity grows at a much slower rate.

However, it should be noted that these results depend on the model, initial condition, selected approximation accuracy ε (and thus TT ranks), and realization of the *random* greedy restricted cross TT approximation algorithm. The "unpredictable" complexity of the greedy cross algorithm is demonstrated by a drop in TT-PMF-related curves in $n_x = 3$.

We should also mention that similarly to the PMF, the most demanding step of the TT-PMF is the calculation of the prediction step, namely the calculation of the TPM in tensor format $\widetilde{\mathbf{T}}_k$ (9.7b). However, if the grid remains constant (e.g., when tracking an object using a static camera), then the TPM and its TT representation are also constant and can be decomposed offline. Similarly, the tensor of grid points in measurement space needed to compute likelihoods can be precomputed offline, as demonstrated in [28] for the canonical polyadic decomposition.

9.5.5.3 Alternative Tensor Decompositions

The TT decomposition is one of the options for storing a tensor in an efficient decomposed form. Other decompositions are, for example, CANDECOMP, Tucker, or hierarchical Tucker decomposition [29]. Each of these decompositions has some advantages but also limitations w.r.t. TT [21], which *currently* makes them slightly less suitable candidates for tensor decomposition in PMF design.

9.6 Numerical Illustration

The performance of the PMF algorithms described in this chapter is compacted using a numerical tracking example. The description of symbols used in the illustration and their units are summarized in Table 9.1. Consider an object moving in a two-dimensional plane with its position at time t_k denoted by $[\mathrm{x}_k, \mathrm{y}_k]$. The object's position is observed by a radar sensor providing range ρ_k and

Table 9.1 Table of symbols and units.

Symbol	Description	Units
x_k	Horizontal position in north direction	[m]
y_k	Horizontal position in east direction	[m]
γ_k	Bearing measurement	[deg]
ρ_k	Range measurement	[m]
$\bar{\upsilon}^\gamma_k$	Bearing measurement bias	[deg]
T_s	Sampling period	[s]
β	Time constant	[s^{-1}]

bearing γ_k. Both range and bearing measurements are subject to white noise with known variances. In addition, the bearing measurement is affected by an unknown bias denoted by $\bar{\upsilon}^b_k$. Therefore, the state $\mathbf{x}_k$ of the model is defined as $\mathbf{x}_k = [x_k, y_k, \bar{\upsilon}^\gamma_k]^\top$. Its dynamic equation is of the form:

$$\underbrace{\begin{bmatrix} x_{k+1} \\ y_{k+1} \\ \bar{\upsilon}^\gamma_{k+1} \end{bmatrix}}_{\mathbf{x}_{k+1}} = \underbrace{\begin{bmatrix} 1.1 & 0 & 0 \\ 0 & 1.1 & 0 \\ 0 & 0 & \exp(-1/\beta)T_s \end{bmatrix}}_{\mathbf{F}} \underbrace{\begin{bmatrix} x_k \\ y_k \\ \bar{\upsilon}^\gamma_k \end{bmatrix}}_{\mathbf{x}_k} + \mathbf{w}_k, \tag{9.30}$$

where the process noise $\mathbf{w}$ is white zero-mean Gaussian with a known covariance matrix $\mathbf{Q} = 10^{-1}\begin{bmatrix} 5 & 0 & 0 \\ 0 & 5 & 0 \\ 0 & 0 & 0.5 \end{bmatrix}$. The parameter β is set to $\beta = 100$, and the sampling period is $T_s = 1$. The measurement equation is then given by

$$\underbrace{\begin{bmatrix} \rho_k \\ \gamma_k \end{bmatrix}}_{\mathbf{z}_k} = \underbrace{\begin{bmatrix} \sqrt{x_k^2 + y_k^2} \\ atan2(y_k, x_k) + \bar{\upsilon}^\gamma_k \end{bmatrix}}_{\mathbf{h}(\mathbf{x}_k)} + \underbrace{\begin{bmatrix} \upsilon^\rho_k \\ \upsilon^\gamma_k \end{bmatrix}}_{\mathbf{v}_k} \tag{9.31}$$

with $\mathbf{v}_k$ being a white zero-mean Gaussian measurement noise with covariance matrix $\mathbf{R} = \begin{bmatrix} (0.1)^2 & 0 \\ 0 & \left(0.01\frac{\pi}{180}\right)^2 \end{bmatrix}$. The process measurement noises are mutually independent and independent of the initial condition $\mathbf{x}_0$, which is Gaussian with mean $[10,\ 10,\ 1]^\top$ and covariance matrix $\text{cov}[\mathbf{x}_0] = 0.1 \cdot \mathbf{I}_3$.

The following algorithms were compared:

- Bootstrap particle filter (PF) (BPF) with 12 000 particles as a baseline,
- PMF with functional tensor decomposition of dynamic model (PMF-FTD),
- tensor train point-mass filter (TT-PMF),
- Fourier transform-based PMF [6] (FT-PMF).

All PMF algorithms considered a grid with 21 points per axis, leading to 21^3 grid points. The system was simulated for $k = 1, \dots, 20$ in $N_{MC} = 100$ Monte Carlo (MC) simulations, which were used to calculate the root mean squared error

$$\text{RMSE}_k = \sqrt{\frac{1}{N_{MC}} \sum_{\ell=1}^{N_{MC}} (\mathbf{x}_{k,\ell} - \hat{\mathbf{x}}_{k,\ell})^2}, \tag{9.32}$$

where $\mathbf{x}_{k,\ell}$ is the true state in ℓth MC simulation and $\hat{\mathbf{x}}_{k,\ell}$ is its estimate obtained as the conditional mean $\mathsf{E}[\mathbf{x}_k|\mathbf{z}^k]$ computed from the conditional posterior PDF $p(\mathbf{x}_k|\mathbf{z}^k)$. The conditional covariances $\text{cov}[\mathbf{x}_k|\mathbf{z}^k]$ obtained from $p(\mathbf{x}_k|\mathbf{z}_k)$ were used to assess the consistency of the estimates.

The time averages of RMSE, time averages of square roots of average conditional covariances (average standard deviation)

$$\text{ASTD}(i) = \frac{1}{20} \sum_{k=1}^{20} \text{ASTD}_k(i), \quad \text{ASTD}_k(i) = \sqrt{\frac{1}{N_{MC}} \sum_{\ell=1}^{N_{MC}} \text{cov}_\ell[\mathbf{x}_k(i)|\mathbf{z}^k]}, \tag{9.33}$$

and average computational times of single runs of the algorithms are given in Figure 9.8 and Table 9.2.

From the results, it follows that the algorithms achieve estimates of similar quality, with the best RMSE achieved by the BPF and FT-PMF. The estimated errors of all algorithms are consistent, as the covariances reported by the filters are similar to the RMSEs. The BPF and FT-PMF algorithms also have the lowest computational costs. It must be noted, however, that utilization of FT-PMF

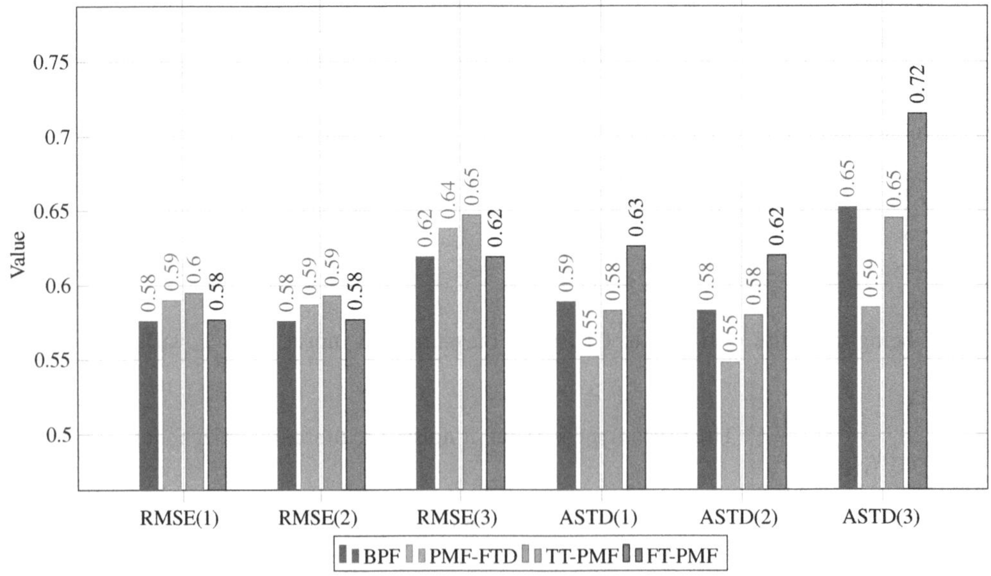

Figure 9.8 Comparison of RMSE and ASTD.

Table 9.2 Comparison of computational time.

Filter	BPF	PMF-FTD	TT-PMF	FT-PMF
Time	0.0022	0.4630	0.3743	0.0024

assumes linear dynamics with an invertible dynamics matrix [6]. The estimates produced by the BPF are random in the sense that for a fixed set of measurements, different estimates are produced with each run of the BPF, which prevents the PFs from being used in applications requiring some certification.

From the table, it is evident that the algorithms yield estimates of comparable quality, with the best RMSE achieved by the BPF and FT-PMF methods. The reported error estimates are consistent across all algorithms, as the covariances provided by the filters closely align with the observed RMSE values. Notably, the BPF and FT-PMF also demonstrate the lowest computational costs. However, it is important to note that the FT-PMF relies on the assumption of linear dynamics with an invertible dynamics matrix [6]. Additionally, the estimates produced by the BPF are inherently stochastic—given a fixed set of measurements, each run of the BPF can yield different results. This randomness limits the applicability of PFsin scenarios where certified or deterministic outputs are required. Consequently, the PMF algorithms proposed in this chapter offer a practical and scalable alternative, especially for applications characterized by higher-dimensional state spaces.

9.7 Conclusions

This chapter has presented a unified framework for the design of Bayesian PMFs based on tensor decompositions, aimed at addressing the curse of dimensionality in nonlinear stochastic state estimation. By leveraging tensor decomposition techniques, the proposed methods significantly reduce computational and memory demands while maintaining estimation accuracy.

Overall, the proposed tensor-based PMFs offer a viable and scalable alternative to conventional filtering approaches, particularly in applications involving complex, nonlinear, or high-dimensional systems. The general framework introduced here lays the groundwork for further research and development of efficient and flexible filtering algorithms leveraging modern numerical techniques such as tensor decompositions.

Acknowledgments

O. Straka and I. Punčochář have been supported by the Ministry of Education, Youth and Sports of the Czech Republic under the project ROBOPROX - Robotics and Advanced Industrial Production CZ.02.01.01/00/22_008/0004590. J. Duník and J. Matoušek have been supported by the Czech Science Foundation (GACR) under grant GA 25-16919J.

References

1 B. D. O. Anderson and J. B. Moore. *Optimal Filtering*. Prentice Hall, New Jersey, 1979.
2 R. S. Bucy and K. D. Senne. Digital synthesis of non-linear filters. *Automatica*, 7:287–298, 1971.
3 S. C. Kramer and H. W. Sorenson. Recursive Bayesian estimation using piece-wise constant approximations. *Automatica*, 24(6):789–801, 1988.
4 H. W. Sorenson. *Recursive Estimation for Nonlinear Dynamic Systems*, pages 127–165. Marcel Dekker, New York, 1988.
5 M. Šimandl, J. Královec, and T. Söderström. Anticipative grid design in point-mass approach to nonlinear state estimation. *IEEE Transactions on Automatic Control*, 47(4): 699–702, 2002.

6 J. Matoušek, J. Duník, and M. Brandner. Design of efficient point-mass filter with illustration in terrain aided navigation. In *26th International Conference on Information Fusion (FUSION)*, Charleston, USA, 2023.

7 J. Maljaars, *When Euler meets Lagrange - Particle-Mesh Modeling of Advection Dominated Flows.* PhD thesis, Technische Universiteit Delft, Nederland, 2019.

8 M. Šimandl, J. Královec, and T. Söderström. Advanced point-mass method for nonlinear state estimation. *Automatica*, 42(7):1133–1145, 2006.

9 J. Duník, M. Soták, M. Veselý, O. Straka, and W. J. Hawkinson. Design of Rao-Blackwellised point-mass filter with application in terrain aided navigation. *IEEE Transactions on Aerospace and Electronic Systems*, 55(1):251–272, 2019.

10 K. B. Anonsen and O. Hallingstad. Terrain aided underwater navigation using point mass and particle filters. In *2006 IEEE/ION Position, Location, And Navigation Symposium*, pages 1027–1035, 2006. https://doi.org/10.1109/PLANS.2006.1650705.

11 U. Orguner, P. Skoglar, D. Törnqvist, and F. Gustafsson, Combined point-mass and particle filter for target tracking. 2010.

12 C. Kalender and A. Schottl. Sparse grid-based nonlinear filtering. *IEEE Transactions on Aerospace and Electronic Systems*, 49(4):2386–2396, 2013.

13 V. Šmídl and M. Gašperin. Rao-Blackwellized point mass filter for reliable state estimation. In *Proceedings of the 16th International Conference on Information Fusion*, pages 312–318, 2013.

14 J. Duník, O. Straka, J. Matoušek, and E. Blasch. Copula-based convolution for fast point-mass prediction. *Signal Processing*, 192:108367, 2022.

15 J. Matoušek, J. Duník, and O. Straka. Lagrangian grid-based filters with application to terrain-aided navigation. *IEEE Signal Processing Magazine*, 42(2):98–104, 2019.

16 B. Demissie, M. A. Khan, and F. Govaers. Nonlinear filter design using Fokker-Planck propagator in Kronecker tensor format. In *2016 19th International Conference on Information Fusion (FUSION)*, pages 1–8, 2016.

17 F. Govaers, B. Demissie, M. A. Khan, M. Ulmke, and W. Koch. Tensor decomposition-based multitarget tracking in cluttered environments. *Journal of Advances in Information Fusion*, 14(1):86–97, 2019.

18 P. Tichavský, O. Straka, and J. Duník. Point-mass filter with decomposition of transient density. In *Proceedings of the 2022 IEEE International Conference on Acoustics, Speech and Signal Processing (ICASSP)*, pages 5752–5756, 2022.

19 A. A. Gorodetsky, S. Karaman, and Y. M. Marzouk. A continuous analogue of the tensor-train decomposition. arXiv:1510.09088 [math.NA], 2018.

20 O. Straka, J. Duník, and P. Tichavský. Point-mass filter with functional decomposition of transient density and two-level convolution. *IFAC-PapersOnLine*, 56(2):6934–6939, 2023. https://doi.org/10.1016/j.ifacol.2023.10.509. ISSN 2405-8963. 22nd IFAC World Congress.

21 J. Matoušek, M. Brandner, J. Duník, and I. Punčochář. Tensor train discrete grid-based filters: Breaking the curse of dimensionality. *IFAC-PapersOnLine*, 58(15):19–24, 2024. ISSN 2405-8963. https://doi.org/10.1016/j.ifacol.2024.08.498. 20th IFAC Symposium on System Identification SYSID 2024.

22 I. V. Oseledets. Tensor-train decomposition. *SIAM Journal on Scientific Computing*, 33(5):2295–2317, 2011.

23 J. Ajgl and O. Straka. Approximate fusion of probability density functions using Gaussian copulas. In *2023 26th International Conference on Information Fusion (FUSION)*, pages 1–7, 2023. https://doi.org/10.23919/FUSION52260.2023.10224201.

24 I. V. Oseledets. Tensor train decompostion toolbox. https://github.com/oseledets/TT-Toolbox, 2023.

25 L. Grasedyck, D. Kressner, and C. Tobler. A literature survey of low-rank tensor approximation techniques. *GAMM-Mitteilungen*, 36(1):53–78, 2013.

26 D. V. Savostyanov. Quasioptimality of maximum-volume cross interpolation of tensors. *Linear Algebra and its Applications*, 458:217–244, 2014. ISSN 0024-3795.

27 M. Šimandl and J. Královec. Multigrid design in point-mass approach to nonlinear state estimation. In *System Identification*,Elsevier, Oxford, 1693–1698, 2004.

28 F. Govaers. On canonical polyadic decomposition of non-linear gaussian likelihood functions. 2018.

29 I. Oseledets and E. Tyrtyshnikov. TT-cross approximation for multidimensional arrays. *Linear Algebra and its Applications*, 432(1):70–88, 2010. ISSN 0024-3795.

Part II

Multitarget Tracking

10

Bayesian Multitarget Tracking via Labeled Random Finite Set

B.-N. Vo, B.-T. Vo, T.T.D. Nguyen, C. Shim, and H.V. Nguyen

School of Electrical Engineering, Computing and Mathematical Sciences, Curtin University, WA, Australia

10.1 Introduction

From its origin in the aerospace industry, multitarget tracking (MTT) has grown to encompass a wide range of applications, among which are surveillance, oceanography, cell biology, computer vision, and autonomous vehicles/drones [1–4]. The primary objective of MTT is to determine the trajectories of an unknown and dynamically changing number of targets. Challenges such as false alarms, missed detections, and data association uncertainty make MTT significantly more difficult than tracking a single target [1–3]. Numerous MTT techniques have been developed, primarily stemming from three core approaches: multiple hypothesis tracking, joint probabilistic data association (JPDA), and random finite set (RFS), see, e.g., the texts [2], [3], and also [5] for a brief overview.

In MTT, the number of targets and their states are unknown and vary randomly with time, as illustrated in Figure 10.1. At any given moment, the underlying system state, called the *multitarget state*, is the *set of existing target states*, and the underlying system trajectory, called multitarget trajectory, is the set of target trajectories. The multitarget state generates at each sampling instant, an observation set, wherein some observations may be false alarms, not linked to any target, and not all of the targets generate observations. Adding to the challenges is the *data association uncertainty*, meaning it is not known which observation corresponds to which target. To capture real-world phenomena, the multitarget trajectory must have provisions for trajectory crossings due to multiple targets sharing the same attributes [6, 7], fragmentation due to targets exiting and reentering the state space [8–10], and ancestries for spawning targets [11–13]. Note that tracking is not limited to physical targets, in which case targets may merge and share attribute states like in cloud or storms tracking. Depending on the application, targets may also share physical states, as seen in collisions or 3D merging scenarios. The phenomenon of targets exiting and reentering the state space is especially prevalent in computer vision applications.

Fundamentally, a multitarget state space model (SSM) should mimic the classical SSM, where the state history corresponds to the trajectory. However, unlike the traditional SSM, the multitarget state history does not necessarily represent the multitarget trajectory, see Figure 10.2. Fortunately, this conceptual challenge is solved by simply augmenting distinct labels or provisional identities to each target state [2, 14], as illustrated in Figure 10.2, and invoking the multitarget estimation tools

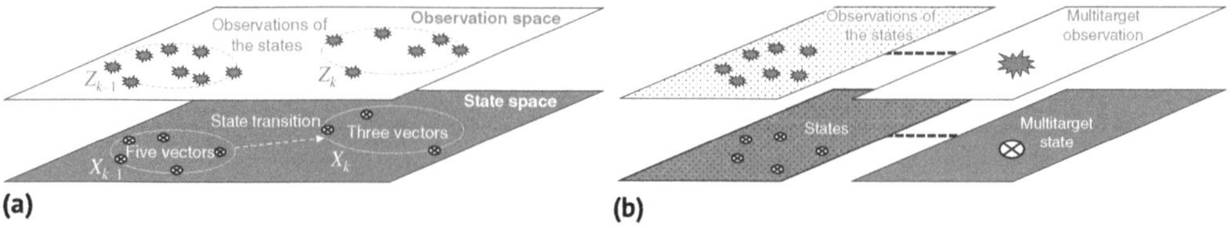

Figure 10.1 Multitarget SSM. The number of targets and observations vary with time. Existing targets may not be detected, new targets may appear, false observations may occur, and it is not known which observation originated from which target: (a) multitarget transition and observation; (b) Reconceptualizing the state as a finite set.

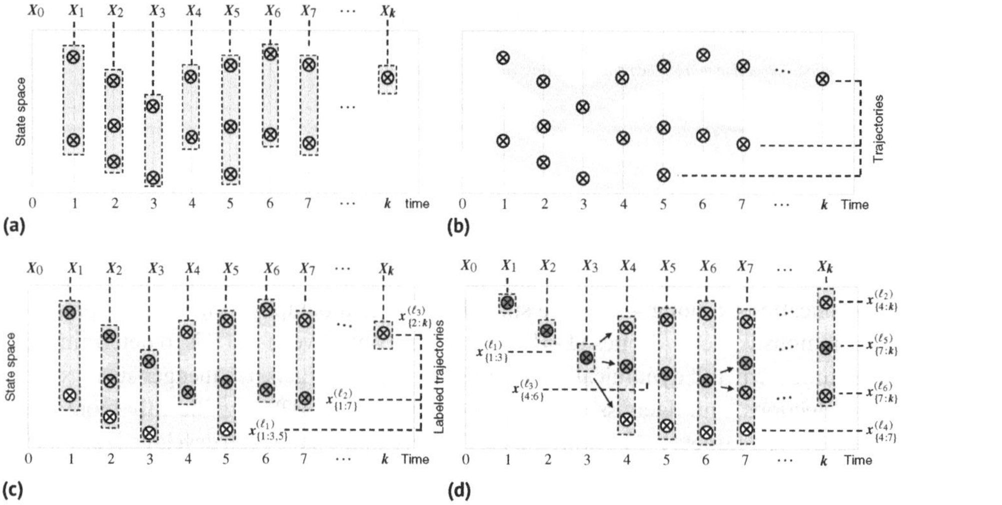

Figure 10.2 States and trajectory of a 1D multitarget system (note $X_0 = \{\}$ in this example). Targets may enter, spawn, exit, or reenter the state space. Without labels the multitarget state history $X_{0:k}$ does not represent (nor contains sufficient information to construct) the multitarget trajectory. (a) Multitarget state history $X_{0:k} = X_0, \ldots, X_k$; (b) multitarget trajectory; (c) labeled multitarget states and trajectory; and (d) spawning with the labeled representation.

from the RFS framework. Apart from a solid mathematical foundation, this so-called labeled RFS (LRFS) approach:

- Enables principled MTT with bounded complexity per time instant via filtering or smoothing within limited moving windows [15], without post-processing heuristics;
- Provides a statistical framework for characterization of uncertainty for multitarget trajectories, offering new conceptual tools;
- Seamlessly handles track crossings, fragmentations, and lineages, as well as diverse observation types;
- Admits the concept of joint existence probability for the elements of a multitarget state, and hence, a meaningful notion of the most probable multitarget state(s), fundamental to multitarget SSM;
- Enables principled approximations of multitarget densities as well as facilitating efficient solutions, scalable in the numbers of targets, detections, sensors, and scans;
- Reduces the transition density from a combinatorial sum to a single term and the measurement likelihood function to a sum of symmetric terms, under standard multitarget model assumptions.

This chapter discusses key advancements in the LRFS approach [15, 16] to MTT over the past decade, which culminated in a comprehensive suite of tools, expanded applications, and increased commercial interest. The capability of this approach to mitigate crucial computational challenges was demonstrated in an experiment that successfully tracked over a million targets simultaneously, even under demanding conditions involving about a quarter of a million false detections per scan and moderate detection probability [17]. We aim to highlight key developments to give readers an intuitive understanding and a nuanced appreciation of this approach to MTT. For a concise and comprehensive examination of the LRFS framework, we refer the reader to [18].

10.2 Bayesian MTT

We start this section by revising the Bayesian approach to tracking a single target. We then formalize the multitarget state/trajectory representation, which leads to the Bayesian formulation of MTT.

10.2.1 Single-target Tracking

In single-target tracking, the target state x_k, at time k, is a vector in the state space $\mathbb{X}$, which evolves from its previous value x_{k-1} according to the Markov *transition density* $f_k(x_k|x_{k-1})$. Further, it generates an observation z_k in an observation space $\mathbb{Z}$, according to the measurement *likelihood function* $g_k(z_k|x_k)$.

In the Bayesian paradigm, all information about the trajectory up to time k is encapsulated in the *posterior* $p_{0:k}(x_{0:k}) \triangleq p_{0:k}(x_{0:k}|z_{1:k})$. Here, we omit the dependence on $z_{1:k}$ for convenience. The posterior can be propagated forward in time via the Bayes *posterior recursion*:

$$p_{0:k}(x_{0:k}) \propto g_k(z_k|x_k) f_k(x_k|x_{k-1}) p_{0:k-1}(x_{0:k-1}). \tag{10.1}$$

The dimensionality of the posterior grows with time, leading to a growth in computational and memory requirements at each time step. As a result, the above recursion is unsuitable for real-time operations, which demand a computational complexity per time step that does not grow with time [19, pp. 53, 54]. A more suitable alternative is the *filtering* density $p_k(x_k) \triangleq \int p_{0:k}(x_{0:k}) dx_{0:k-1}$, which encapsulates all information about the current state. It can be propagated by the Bayes *filtering recursion*:

$$p_k(x_k) \propto g_k(z_k|x_k) \int f_k(x_k|y) p_{k-1}(y) dy, \tag{10.2}$$

and has a fixed computational complexity per time step [19]. The conditional mean/mode of the filtering/posterior density can be used as the state/trajectory estimate. Moreover, a time sequence of individual state estimates can be used as the trajectory estimate.

10.2.2 Multitarget State and Trajectory

Unlike single-target tracking, where the state vector of the only target is considered, MTT involves a finite set of state vectors corresponding to multiple targets. When more than one target occupies the same physical state (see Figure 10.3), due to mergings/collisions, and occlusions, especially in a finite state space, their trajectories may overlap or cross [6, 7]. Additionally, trajectories may be fragmented due to targets disappearing and reappearing in the state space (particularly if it is bounded), creating the reappearance and reidentification challenges in computer vision [8–10].

Figure 10.3 Occlusion in pixel space. The two targets merge and share the same state (pixel coordinate) in the last frame.

Most importantly, similar to single-target tracking, we need the multitarget trajectory to be determined from the multitarget state history to ensure that MTT algorithms with fixed complexity per time step can be devised. To address these, we use the following multitarget state representation.

We define a *trajectory* in a state space $\mathbb{X}$ and a discrete-time window $\mathbb{W}$ as a mapping $\tau : \mathbb{W} \to \mathbb{X}$ to cover *fragmented trajectories* arising from targets physically exiting and reentering the state space [17], as shown in Figure 10.2b. The set of instances when the target exists is given by the *domain* $\mathcal{D}(\tau) \subseteq \mathbb{W}$ of τ.

In MTT, the state of a target is represented by an *attribute* $x \in \mathbb{X}$ and a discrete *label* ℓ [2]. The attribute can include kinematics, such as position, velocity, acceleration, or visual features like shape, size, color, texture, or other appearance-based traits. Attributes might also include thermal signatures, material properties, or other specific features unique to the target or the application domain. The label is a provisional identity from the label space $\mathbb{L}$, e.g., $\ell = (s, \iota)$, where s is the time of birth and ι is an index to distinguish targets born at the same time [15, 20]. We define the *state* of a target by $\boldsymbol{x} = (x, \ell) \in \mathbb{X} \times \mathbb{L}$. This allows a trajectory $\tau_\ell : i \mapsto x_i, i \in \mathcal{D}(\tau_\ell)$ to be represented by a sequence $[(x_s, \ell), \dots, (x_t, \ell)]$ of time-stamped states with a common label.

A multitarget state $\boldsymbol{X}$ is a finite subset of the product space $\mathcal{X} = \mathbb{X} \times \mathbb{L}$, where each element of $\boldsymbol{X}$ has a *distinct label*, ensuring no two elements share the same label [14, pp. 135, 196–197], [21, pp. 506], [15, 20]. The labeled multitarget representation enables a sequence $\boldsymbol{X}_{j:k}$ of multitarget states to fully define the multitarget trajectory on the interval $\{j:k\}$ analogous to the single-target case, by grouping the states based on their labels (see Figure 10.2c). Formally, the trajectory of target ℓ that exists on $T(\ell) \subseteq \{j:k\}$ is the time-stamped state sequence $\boldsymbol{x}^{(\ell)}_{T(\ell)} = [(x_i, \ell) \in \boldsymbol{X}_i]_{i \in T(\ell)}$, which defines the mapping $\tau_\ell : i \mapsto x_i, i \in T(\ell)$. Let $\mathcal{L}:(x, \ell) \mapsto \ell$ denote the *label projection*. Then $\boldsymbol{X}_{j:k}$ is equivalently represented as the SoTs of the (targets with) labels $\ell \in \cup_{i=j}^{k} \mathcal{L}(\boldsymbol{X}_i)$, i.e.,

$$\boldsymbol{X}_{j:k} \equiv \left\{\boldsymbol{x}^{(\ell)}_{T(\ell)} : \ell \in \cup_{i=j}^{k} \mathcal{L}(\boldsymbol{X}_i)\right\}. \tag{10.3}$$

This allows MTT (filtering or smoothing on moving windows, see Figure 10.4a) with computational complexity per time step that does not grow with time, and modeling of fragmented trajectories and crossing trajectories. Note that certain applications involve target states that naturally contain distinct labels, e.g., in [22] each target has a unique radio-frequency identification (RFID) from a finite set of RFIDs.

Remark 10.1 Labeling is not only sufficient but necessary for consistent multitarget state and trajectory representation. If we remove the labels from a multitarget trajectory, a member trajectory of length L is characterized by its starting time t and a sequence $x_{1:L}$ of consecutive attributes, i.e.,

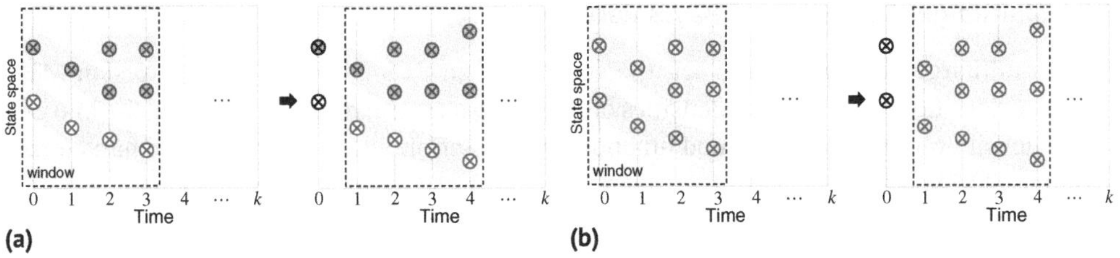

Figure 10.4 MTT with moving windows. Trajectories cannot be estimated without labels, because there is no means to link the trajectories between the windows. (a) Labeled and (b) unlabeled.

$(t, x_{1:L})$. In addition to the inability to represent the multitarget trajectory as a sequence of states, this unlabeled multitarget trajectory representation, also known as (unlabeled) set of trajectory (SoT) [23], is fundamentally inconsistent with the multitarget state, as illustrated below.

Example 10.1 Consider the following set of three distinct elements

$$\left\{ \begin{bmatrix} 1 \\ a \end{bmatrix}, \begin{bmatrix} 1 \\ a \\ b \end{bmatrix}, \begin{bmatrix} 1 \\ a \\ b \\ c \end{bmatrix} \right\} \subseteq \mathbb{T},$$

where $\mathbb{T}$ denotes the space of unlabeled trajectories. In this unlabeled SoT, the 1st trajectory begins at time 1 and has state a at time 1; the 2nd begins at time 1 and has states a and b at times 1 and 2; and the 3rd begins at time 1 with states a, b and c at times 1, 2, and 3. Note that all three targets exist at time 1, two exist at time 2, and one exists at time 3. However, given that a set cannot contain duplicate elements, the multitarget states at times 1, 2, and 3 are the sets $\{a\}$, $\{b\}$, and $\{c\}$ of unlabeled states, which implies that only one target exists at each time, contradicting the information in the corresponding SoT.

Remark 10.2 Apart from the mathematical inconsistency, the (unlabeled) SoT representation is impractical for large windows, even in the simplest case of a single target [19], which necessitates using shorter moving windows and linking trajectories between the windows together. However, without the labels, they cannot be consistently linked (unless ad-hoc post-processing is applied), as illustrated in Figure 10.4b. Indeed, certain SoT-based algorithms even resort to labeling the unlabeled trajectories with unique labels (just like the well-established labeled representation) but refer to these labels as "auxiliary variables," see for example [24]. Another drawback of the SoT representation is that it cannot accommodate fragmented trajectories since each trajectory is a consecutive sequence of states, thereby limiting their use in scenarios where targets can move in and out of the state space.

Beyond providing a mathematically consistent and versatile multitarget state/trajectory representation, the labeled formulation enables the modeling and estimation of target ancestry and lineage [11, 13]. Tracking the sequence of ancestors—i.e., a target's lineage—is crucial in various applications, including cell biology [25–27] and space situational awareness [28]. Labels inherently provide a mechanism for encoding lineage information into the states of individual targets, which can then be inferred from data [13]. Labeling is also essential in real MTT systems, for display and communication purposes as well as numerical tractability [2]. Displaying sequences of (unlabeled) states is not very informative/useful to the operators. Hence, labels are encoded in the display via colors or markers for different trajectories.

10.2.3 Multitarget Densities and Bayes Recursions

Since the multitarget state is a finite set, the concept of an RFS is needed for a Bayesian MTT formulation. Specifically, probability densities of RFSs play a crucial role in prescribing the Markov transition density, observation likelihood function, and posterior/filtering density that characterize the multitarget SSM.

The probability density function of an RFS is well-defined on the space $\mathcal{F}(\mathcal{X})$ of finite subsets of $\mathcal{X}$ [29–31], with respect to (w.r.t.) the distribution of a unit Poisson RFS–the cardinality is Poisson with unit rate and, conditional on the cardinality, the elements are uniformly identically and independently distributed (i.i.d.) [32]. This probability density is equivalent to the finite set statistic (FISST) belief density [3, 18, 32] and together they are referred to as multitarget densities. A multitarget trajectory density is simply the joint probability density of consecutive multitarget states, like the single-target case.

It is imperative to note that multitarget densities (and multitarget trajectory densities) are zero at invalid/disallowed multitarget states (and multitarget trajectories), i.e., those with non-distinct labels [15, 16, 18], see also Section 10.3. Note also that a multitarget density is a set-function—that is, a function of sets—and is therefore invariant to the ordering of its constituent elements. In practice, multitarget densities such as the posterior and filtering densities above are invariably made up of large sums, and require truncation for tractability. However, truncation does not necessarily preserve symmetry, which can invalidate the truncated sum as a set-function, see also [18]. Careful consideration is needed to ensure the truncated expression remains a valid set-function.

Analogous to Bayesian single-target tracking, all information on the multitarget trajectory is captured in the multitarget posterior density $\pi_{0:k}(\boldsymbol{X}_{0:k}) \triangleq \pi_{0:k}(\boldsymbol{X}_{0:k}|Z_{1:k})$ (we omitted the dependence on observation history $Z_{1:k}$ for convenience). This posterior can be propagated forward in time by the Bayes multitarget posterior recursion [33],

$$\pi_{0:k}(\boldsymbol{X}_{0:k}) \propto \boldsymbol{g}_k(Z_k|\boldsymbol{X}_k)\boldsymbol{f}_k(\boldsymbol{X}_k|\boldsymbol{X}_{k-1})\pi_{0:k-1}(\boldsymbol{X}_{0:k-1}), \tag{10.4}$$

where $\boldsymbol{f}_k(\cdot|\cdot)$ denotes the multitarget transition density that describes the targets' motion, appearance/disappearance, and $\boldsymbol{g}_k(\cdot|\cdot)$ denotes the multitarget observation likelihood that describes the underlying detections/misdetections, clutter, and data association uncertainty. Similar to its single-target counterpart, the computational complexity per iterate grows with time. A more tractable alternative is the multitarget filtering density, $\pi_k(\boldsymbol{X}_k) \triangleq \int \pi_{0:k}(\boldsymbol{X}_{0:k})\delta\boldsymbol{X}_{0:k-1}$, where the integral is a set integral defined in [3], which can be propagated by the Bayes multitarget filtering recursion [15, 20].

$$\pi_k(\boldsymbol{X}_k) \propto \boldsymbol{g}_k(Z_k|\boldsymbol{X}_k)\int \boldsymbol{f}_k(\boldsymbol{X}_k|\boldsymbol{Y})\pi_{k-1}(\boldsymbol{Y})\delta\boldsymbol{Y}, \tag{10.5}$$

with a time-independent per-iterate computational complexity.

Remark 10.3 Removing labels introduces theoretical problems in multitarget trajectory densities. If target crossings are allowed, then the unit Poisson RFS and multitarget trajectory densities can be defined on the space $\mathcal{F}(\mathbb{T})$ of finite SoTs (provided their distributions are absolutely continuous w.r.t. the unit Poisson RFS distribution). However, this SoT representation violates the set representation of the multitarget state (see Example 10.1), because multitarget states may contain duplicate elements. Consequently, the multitarget filtering density does not exist [29–31]. Conversely, if target crossings are disallowed, the space of non-crossing SoTs is not $\mathcal{F}(\mathbb{T})$, and a Poisson RFS on this space does not exist, as the elements of any non-crossing SoT are inherently dependent. Ultimately, this dependency prevents the formulation of a multitarget trajectory density, which requires not only the existence of the Poisson RFS of trajectories but also absolute continuity w.r.t. this distribution [29–31]. Note that if we were to have multitarget trajectory densities on $\mathcal{F}(\mathbb{T})$ (assuming they exist w.r.t. the Poisson RFS distribution), then these densities must be zero at disallowed SoTs, i.e., those with track crossings. However, this is not the case as shown in Example 10.2 (in the next subsection).

Remark 10.4 Although the inconsistencies of the SoT approach are now well-recognized [34], no satisfactory solutions/remedies have been developed. Instead, the prevailing assumption is that any SoT involving target crossings "cannot occur," even if such configurations have positive likelihoods (see Section III.C, [34]). The justification offered—that these SoTs represent zero-probability events—is misguided because it conflates two distinct concepts: the probability density of a random variable and the probability of its realization. If this line of reasoning were valid, by the same argument, no SoT could ever occur. For a continuous random variable, it is imperative to distinguish between realizations with zero probability density and those with zero probability. The fact that $Pr(X = x) = 0$ does not mean the realization x can be ignored, e.g., in a zero-mean Gaussian random variable, $Pr(X = 0) = 0$, yet 0 remains the most likely/probable realization. Similarly, an SoT with a target crossing can be more probable than a valid SoT (see Example 10.2), and cannot be dismissed simply because it has zero probability. Further, from a practical perspective, target crossings are common, particularly in 2D spaces where targets frequently experience mutual occlusions.

10.3 LRFS Tracking Filters and Smoothers

LRFS offers a suite of models and mathematical tools that enable tractable Bayesian estimation of multitarget states and trajectories. In Subsection 10.3.1, we introduce the labeled multi-Bernoulli (LMB), generalized LMB (GLMB), and multiscan GLMB models, which are fundamental to tractable MTT. The standard multitarget SSMs are then presented in Subsection 10.3.2. In Subsection 10.3.3, we introduce the GLMB filtering recursions–exact solutions to the Bayes filter, while in Subsection 10.3.4 we discuss the GLMB posterior recursion, i.e., the GLMB smoother. In Subsection 10.3.5, we demonstrate how otherwise intractable recursions with non-standard models can be efficiently approximated using LMB or GLMB. In Subsection 10.3.6, we discuss different types of estimators for LRFS, while in Subsection 10.3.7 we provide efficient numerical solutions for the discussed filters and smoother. Finally, in Subsection 10.3.8, we conclude this section with the control formulation or sensor scheduling for MTT.

10.3.1 GLMB Models

We start by defining the basic building blocks of the LMB and GLMB density: the *multitarget exponential* $h^X \triangleq \prod_{x\in X} h(x)$, with the convention $h^{\emptyset} = 1$; and the *multiscan multitarget exponential* as the product of the values of h at every trajectory in $\boldsymbol{X}_{j:k}$ [33], i.e.,

$$h^{\boldsymbol{X}_{j:k}} \triangleq h^{\{\boldsymbol{x}^{(\ell)}_{T(\ell)}:\ell\in\mathcal{L}(\boldsymbol{X}_{j:k})\}} = \prod\nolimits_{\ell\in\mathcal{L}(\boldsymbol{X}_{j:k})} h(\boldsymbol{x}^{(\ell)}_{T(\ell)}), \tag{10.6}$$

for any real function h that operates on trajectories and $\mathcal{L}(\boldsymbol{X}_{j:k}) = (\mathcal{L}(\boldsymbol{X}_j), ..., \mathcal{L}(\boldsymbol{X}_k))$.[1] For convenience, if $\boldsymbol{x}^{(\ell)}_{T(\ell)} = [(x_i, \ell)]_{i\in T(\ell)}$, we write $h(\boldsymbol{x}^{(\ell)}_{T(\ell)})$ as $h(x_{T(\ell)}; \ell)$, or $h(x_{s(\ell):t(\ell)}; \ell)$ if the trajectory is unfragmented, i.e., $T(\ell) = \{s(\ell) : t(\ell)\}$, where $s(\ell) \triangleq \min T(\ell)$, $t(\ell) \triangleq \max T(\ell)$.

1 $\ell \in \mathcal{L}(\boldsymbol{X}_{j:k})$ means that ℓ belongs to any of the sets $\mathcal{L}(\boldsymbol{X}_j), ..., \mathcal{L}(\boldsymbol{X}_k)$.

The LMB and GLMB multitarget densities serve as fundamental building blocks for tractable MTT. An LMB multitarget density can be written as [15]:

$$\pi(\boldsymbol{X}) = \Delta(\boldsymbol{X})\left[\prod_{i\in\mathbb{L}}(1-r^{(i)})\prod_{\ell\in\mathcal{L}(\boldsymbol{X})}\frac{\mathbf{1}_{\mathbb{L}}(\ell)r^{(\ell)}}{1-r^{(\ell)}}\right]p^{\boldsymbol{X}}, \tag{10.7}$$

where $\Delta(\boldsymbol{X})$ is the distinct label indicator, $p(\cdot,\ell)$ is a probability density on $\mathbb{X}$, and $\mathbf{1}_X(x)$ is the indicator function: $\mathbf{1}_X(x)=1$ if $x\in X$; $\mathbf{1}_X(x)=0$ otherwise.[2] In the above expression, $r^{(\ell)}$ is referred to as the existence probability of target (with label) ℓ. For convenience, an LMB can be written in terms of its parameters as $\{r^{(\ell)}, p(\cdot,\ell) : \ell\in\mathbb{L}\}$.

The GLMB is a more versatile class of LRFS, whose multitarget density is given by the following mixture [15]:

$$\pi(\boldsymbol{X}) = \Delta(\boldsymbol{X})\sum_{\xi\in\Xi} w^{(\xi)}(\mathcal{L}(\boldsymbol{X}))[p^{(\xi)}]^{\boldsymbol{X}}, \tag{10.8}$$

where Ξ is an arbitrary discrete space, $p^{(\xi)}(\cdot,\ell)$ is a probability density on $\mathbb{X}$, and $w^{(\xi)}\geq 0$ are the weights that satisfy $\sum_{\xi\in\Xi}\sum_{L\subseteq\mathbb{L}} w^{(\xi)}(L)=1$. Intuitively, each pair (ξ, L) of "event" ξ and label set L can be interpreted as a hypothesis whose probability is $w^{(\xi)}(L)$, and $p^{(\xi)}(\cdot,\ell)$ is the attribute probability density of target ℓ in this hypothesis. For implementation, we use the δ-form of the GLMB density

$$\pi(\boldsymbol{X}) = \Delta(\boldsymbol{X})\sum_{(\xi,I)\in\Xi\times\mathcal{F}(\mathbb{L})} w^{(\xi)}(I)\delta_I[\mathcal{L}(\boldsymbol{X})][p^{(\xi)}]^{\boldsymbol{X}}, \tag{10.9}$$

where $\delta_X[Y]$ is the general Kronecker delta function: $\delta_X[Y]=1$ if $X=Y$; $\delta_X[Y]=0$ otherwise. Similar to an LMB, a GLMB density is completely characterized by its parameters; hence, we use the notation

$$\pi = \{(w^{(\xi)}(I), p^{(\xi)}) : (\xi, I)\in\Xi\times\mathcal{F}(\mathbb{L})\}. \tag{10.10}$$

Note that each term of the GLMB (10.8) covers a wide class of LRFSs, including the LMB, the labeled i.i.d. clusters (and labeled Poisson), and their disjoint union. The unlabeled version of the GLMB is not an RFS in general but is a class of non-simple point processes, including the Poisson multi-Bernoulli mixture [35].

The GLMB multitarget density only captures the information of the multitarget state at the current time. The multiscan GLMB [33] offers a tractable means to capture information on a sequence of multitarget states. A multiscan GLMB density over the interval $\{j:k\}$ is a JoM density on the space $\bowtie_{i=j}^{k}\mathcal{F}(\mathbb{X}\times\mathbb{L}_i)$ (i.e., the Cartesian product of $\mathcal{F}(\mathbb{X}\times\mathbb{L}_i)$ for i from j to k), given by

$$\pi(\boldsymbol{X}_{j:k}) = \Delta(\boldsymbol{X}_{j:k})\sum_{\xi\in\Xi} w^{(\xi)}(\mathcal{L}(\boldsymbol{X}_{j:k}))[p^{(\xi)}]^{\boldsymbol{X}_{j:k}}, \tag{10.11}$$

where $\Delta(\boldsymbol{X}_{j:k})\triangleq\prod_{i=j}^{k}\Delta(\boldsymbol{X}_i)$ is the multiscan distinct label indicator, $w^{(\xi)}(I_{j:k})\geq 0$ are weights such that $\sum_{\xi\in\Xi}\sum_{I_{j:k}\in\bowtie_{i=j}^{k}\mathcal{F}(\mathbb{L}_i)} w^{(\xi)}(I_{j:k})=1$, and $p^{(\xi)}(x_{s(\ell):t(\ell)};\ell)$ is a joint probability density of the attribute sequence $x_{s(\ell):t(\ell)}$, for each $\ell\in I_{j:k}$. $s(\ell)$ is the birth time and $t(\ell)$ is the termination

2 The original expression of an LMB involves a mapping $\alpha:\mathbb{L}\to\Psi$ between an index set Ψ and the label set $\mathbb{L}$. Here we assume an identity mapping α to simplify the notation.

time of target ℓ. Note that $s(\ell)$ and $t(\ell)$ implicitly depend on $(\xi, I_{j:k})$. Similar to the GLMB, the multiscan GLMB can also be written in the δ-form, i.e.,

$$\pi(\boldsymbol{X}_{j:k}) = \Delta(\boldsymbol{X}_{j:k}) \sum_{\xi, I_{j:k}} w^{(\xi, I_{j:k})} \delta_{I_{j:k}}[\mathcal{L}(\boldsymbol{X}_{j:k})][p^{(\xi)}]^{\boldsymbol{X}_{j:k}}. \tag{10.12}$$

Remark 10.5 Note that the distinct label indicator ensures disallowed multitarget states/trajectories have zero likelihood of occurring. In contrast, in the SoT formulation, disallowed SoTs do not necessarily have zero likelihood, and may, in fact, be more probable than a valid SoT. Indeed, the following example shows an invalid SoT that is the most "probable," and is actually defined to be an estimate for RFS of trajectories [36].

Example 10.2 Recall that an unlabeled trajectory $(t, x_{1:L})$ has length L, starting time t and consecutive attributes $x_{1:L}$. According to [36], a Poisson RFS of trajectories on $\mathcal{F}(\mathbb{T})$ has density

$$v(\{T_1, \ldots, T_n\}) = e^{-\lambda_v} \lambda_v^n \prod_{j=1}^{n} \breve{v}(T_j),$$

where $\int \breve{v}(T)dT = 1$ and $\lambda_v \geq 0$. Consider the specific case:

$$\lambda_v = 2, \quad \breve{v}(T) = 0 \text{ for } T \neq (2, x_1) \text{ or } T \neq (1, x_{1:2}),$$

$$\breve{v}((2, x_1)) = \mathcal{N}(x_1; 0, 1), \quad \breve{v}((1, x_{1:2})) = \mathcal{N}(x_{1:2}; [0, 0]^T, I), I = \text{diag}([1, 1]^T),$$

where $\mathcal{N}(\cdot; \mu, P)$ denotes a Gaussian probability density with mean μ and covariance P. While the authors assert that any SoT involving crossings is impossible, the Poisson RFS formulation in [36] does not enforce this restriction. In fact, under the specified formulation the trajectory probability hypothesis density (PHD) estimator [36] identifies the most probable SoT at time 2 as $\{(2, [0]), \left(1, [0, 0]^T\right)\}$, which contains a crossing. If crossings were truly prohibited, such configurations should either be excluded from the domain or assigned zero density. Instead, the model assigns positive density and even ranks a crossing SoT as the most probable, directly contradicting the assumption that crossings cannot occur.

10.3.2 Standard Multitarget SSMs

10.3.2.1 Standard Multitarget Dynamic Model

Similar to classical SSMs, the multitarget state $\boldsymbol{X}_{k-1}$ at time $k-1$, evolves to $\boldsymbol{X}_k$ at the current time k, according to the multitarget Markov transition density $\boldsymbol{f}_k(\boldsymbol{X}_k|\boldsymbol{X}_{k-1})$. For brevity, we use the subscript "-" to indicate variables from the previous time step, and omit the subscript for the current time. The transition density characterizes the transition of the target attributes as well as target appearances and disappearances. Note that the label of a target is unchanged over the course of its existence.

The standard multitarget dynamic model described in the following can cater for a wide range of MTT applications. In this model, at each time, a set of newborn targets with labels from the label space $\mathbb{B}$ appears in the state space, and is modeled as a GLMB, described by the multitarget birth density $\boldsymbol{f}_B$. In practice the LMB birth model with parameters $\{P_B(\ell), p(\cdot, \ell)\}_{\ell \in \mathbb{B}}$ is widely used ($P_B(\ell)$ is referred to as the birth probability of label ℓ). In addition to newborn targets, there are also targets that survived from the previous time. A target $\boldsymbol{x}_- = (x_-, \ell) \in \boldsymbol{X}_-$ at time $k-1$ could survive to the current time step with probability $P_S(\boldsymbol{x}_-)$, or cease to exist with probability $1 - P_S(\boldsymbol{x}_-)$. If the target survives, its attribute is distributed according to the transition density $f_S(x|x_-, \ell)$, independent of the other targets. Assuming the newborn and surviving targets are mutually independent (they are also disjoint because their label spaces are disjoint, i.e., $\mathbb{L}_- \cap \mathbb{B} = \emptyset$), the transition density of the multitarget state is given by

$$\boldsymbol{f}(\boldsymbol{X}|\boldsymbol{X}_-) = \boldsymbol{f}_B(\boldsymbol{X} \cap (\mathbb{X} \times \mathbb{B}))\boldsymbol{f}_S(\boldsymbol{X} \cap (\mathbb{X} \times \mathbb{L}_-)|\boldsymbol{X}_-), \tag{10.13}$$

where $\boldsymbol{f}_S(\boldsymbol{X}|\boldsymbol{X}_-) = \Delta(\boldsymbol{X})[1_{\mathcal{L}(\boldsymbol{X}_-)}]^{\mathcal{L}(\boldsymbol{X})}[\boldsymbol{f}_S(\boldsymbol{X};\cdot)]^{\boldsymbol{X}_-}$, and

$$\boldsymbol{f}_S(\boldsymbol{X};\zeta,\ell) = \begin{cases} P_S(\zeta,\ell)f_S(x|\zeta,\ell), & \text{if } (x,\ell) \in \boldsymbol{X} \\ 1 - P_S(\zeta,\ell), & \text{if } \ell \notin \mathcal{L}(\boldsymbol{X}) \end{cases}. \tag{10.14}$$

Note the remarkably simple form of the multitarget transition density, but when the labels are removed, it becomes an intractable combinatorial sum of products [18].

10.3.2.2 Standard Multitarget Observation Model

At each time step, the multitarget state $\boldsymbol{X}$ generates an observation Z–a set of measurements—according to the multitarget measurement likelihood function $\boldsymbol{g}(Z|\boldsymbol{X})$. The multitarget observation model, described by $\boldsymbol{g}$, governs the detection process, miss-detection, false alarms, and data association uncertainty.

In the standard multitarget observation model, each target $\boldsymbol{x}$ could be detected and generates a single measurement with probability $P_D(\boldsymbol{x})$, or miss-detected with probability $1 - P_D(\boldsymbol{x})$. If a target is detected, the likelihood of it generating a measurement z is given by the single-target likelihood function $g(z|\boldsymbol{x})$. Assuming each target generates a measurement independent of others, the set of measurements generated from $\boldsymbol{X}$ is an LMB with parameters $\{P_D(\boldsymbol{x}), g(\cdot|\boldsymbol{x}) : \boldsymbol{x} \in \boldsymbol{X}\}$. In addition to target-generated measurements, the observation could include clutter measurements that are not associated with any targets. This set is modeled by a Poisson RFS with intensity function κ.

For a set of measurements $Z = \{z_1, ..., z_m\}$ and a multitarget state $\boldsymbol{X}$, we define an *association map* $\theta : \mathbb{L} \to \{0 : m\}$ that associates the labels of $\boldsymbol{X}$ to the measurement indices or miss-detection, such that θ is one-to-one on the set of labels that are mapped to positive values. If $\theta(\ell) = 0$ the target (with label) ℓ is miss-detected. If $\theta(\ell) > 0$, the target ℓ is associated with the measurement $z_{\theta(\ell)}$. The multitarget measurement likelihood function is written as

$$\boldsymbol{g}(Z|\boldsymbol{X}) \propto \sum_{\theta \in \Theta(\mathcal{L}(\boldsymbol{X}))} \left[\Psi_Z^{(\theta \circ \mathcal{L}(\cdot))}(\cdot)\right]^{\boldsymbol{X}}, \tag{10.15}$$

where $\Theta(L) \subseteq \Theta$ is the set of association maps with domain L, and

$$\Psi_Z^{(j)}(\boldsymbol{x}) = \begin{cases} \dfrac{P_D(\boldsymbol{x})g(z_j|\boldsymbol{x})}{\kappa(z_j)}, & \text{if } j \in \{1:|Z|\} \\ 1 - P_D(\boldsymbol{x}), & \text{if } j = 0 \end{cases}. \tag{10.16}$$

In the above expression, $\Psi_Z^{(\theta \circ \mathcal{L})}$ is referred to as the signal-to-noise function.

The standard multitarget observation model can also be extended to multisensor. The observation in this case is $Z = (Z^{(1)}, ..., Z^{(V)})$, where $Z^{(v)}$ is the set of measurements from sensor v. If $Z^{(1)}, ..., Z^{(V)}$ are independent conditioned on $\boldsymbol{X}$, the multisensor multitarget likelihood function is the product

$$\boldsymbol{g}(Z^{(1)}, \dots, Z^{(V)}|\boldsymbol{X}) \propto \prod_{v=1}^{V} \boldsymbol{g}^{(v)}(Z^{(v)}|\boldsymbol{X}). \tag{10.17}$$

Let us abbreviate $\theta \triangleq (\theta^{(1)}, \dots, \theta^{(V)})$ and

$$\Psi^{(j^{(1)}, \dots, j^{(V)})}_{(Z^{(1)}, \dots, Z^{(V)})}(\boldsymbol{x}) \triangleq \prod_{v=1}^{V} \Psi^{(v,j^{(v)})}_{Z^{(v)}}(\boldsymbol{x}), \tag{10.18}$$

where $\Psi^{(v,j^{(v)})}_{Z^{(v)}}$ is the SNR function of the vth sensor. Then, the multisensor likelihood function can be written in the same form as the single-sensor version (10.15).

Since each term in (10.15) exhibits symmetry, truncation can be performed without concern for the symmetry of the resulting sum. However, this does not hold true when the labels are removed, see also [18].

10.3.3 GLMB Filtering Recursions

Under the standard multitarget SSM discussed above, the GLMB recursion is an exact closed-form solution to Bayes MTT filter. We first present the closure of the GLMB density under the Chapman–Kolmogorov prediction and Bayes update. The prediction and update can be combined into a single recursion. We also present LMB filtering recursion as an approximation, and the multiscan GLMB recursion as an extension of the GLMB filter to multiple scans.

For convenience, we define the label-marginal of a function $f : \mathcal{F}(\mathbb{X} \times \mathbb{L}) \to \mathbb{R}$ as

$$\left\langle f \right\rangle(\{\ell_1, \dots, \ell_n\}) \triangleq \int f(\{(x_1, \ell_1), \dots, (x_n, \ell_n)\}) dx_{1:n}. \tag{10.19}$$

10.3.3.1 Chapman–Kolmogorov Prediction

Starting with a GLMB density at time $k-1$, with parameters

$$\pi_- = \{(w_-^{(\xi)}(I_-), p_-^{(\xi)}) : (\xi, I_-) \in \Xi \times \mathcal{F}(\mathbb{L}_-)\}, \tag{10.20}$$

and applying the Chapman–Kolmogorov prediction $\boldsymbol{\pi}(\boldsymbol{X}) = \int \boldsymbol{f}(\boldsymbol{X}|\boldsymbol{X}_-)\boldsymbol{\pi}_-(\boldsymbol{X}_-)\delta\boldsymbol{X}_-$ yields the following GLMB predicted density [15]

$$\pi = \{(w^{(\xi)}(I), p^{(\xi)}) : (\xi, I) \in \Xi \times \mathcal{F}(\mathbb{L})\}, \tag{10.21}$$

where

$$w^{(\xi)}(I) = w_B(I \cap \mathbb{B})[\left\langle P_S\, p_-^{(\xi)} \right\rangle]^{I \cap \mathbb{L}_-} \sum_{I_- \supseteq I \cap \mathbb{L}_-} [1 - \left\langle P_S\, p_-^{(\xi)} \right\rangle]^{I_- - (I \cap \mathbb{L}_-)} w_-^{(\xi)}(I_-), \tag{10.22}$$

$$p^{(\xi)}(x, \ell) = \mathbf{1}_{\mathbb{B}}(\ell) p_B(x, \ell) + \mathbf{1}_{\mathbb{L}_-}(\ell) \frac{\left\langle f_S(x|\cdot) P_S(\cdot) p_-^{(\xi)}(\cdot) \right\rangle(\ell)}{\left\langle P_S\, p_-^{(\xi)} \right\rangle(\ell)}. \tag{10.23}$$

10.3.3.2 Bayes Update

Given the predicted GLMB density with parameters given in (10.21) and a set of measurements Z, applying Bayes rule, $\boldsymbol{\pi}(\boldsymbol{X}|Z) \propto \boldsymbol{g}(Z|\boldsymbol{X})\boldsymbol{\pi}(\boldsymbol{X})$, yields the following updated GLMB density [15]

$$\boldsymbol{\pi}(\cdot|Z) \propto \{(w_Z^{(\xi,\theta)}(I), p_Z^{(\xi,\theta)}) : (\xi, \theta, I) \in \Xi \times \Theta \times \mathcal{F}(\mathbb{L})\}, \tag{10.24}$$

where

$$w_Z^{(\xi,\theta)}(I) = w^{(\xi)}(I)\mathbf{1}_{\Theta(I)}(\theta)\prod_{\ell\in I}\left\langle p^{(\xi)}\Psi_Z^{(\theta(\ell))}\right\rangle(\ell), \tag{10.25}$$

$$p_Z^{(\xi,\theta)}(x,\ell) = \frac{p^{(\xi)}(x,\ell)\Psi_Z^{(\theta(\ell))}(x,\ell)}{\left\langle p^{(\xi)}\Psi_Z^{(\theta(\ell))}\right\rangle(\ell)}. \tag{10.26}$$

10.3.3.3 Joint Prediction and Update Recursion

The prediction and update can be combined into a single recursion that solves the multitarget filtering recursion (10.5). Given the previous GLMB filtering density (10.20) and the new set Z of measurements, the current filtering GLMB density is given by [37]

$$\pi = \{(w_Z^{(\xi,\theta)}(I), p_Z^{(\xi,\theta)}) : (\xi,\theta,I)\in\Xi\times\Theta\times\mathcal{F}(\mathbb{L})\}, \tag{10.27}$$

where, using $B(I) = I\cap\mathbb{B}$ to denote the birth labels of I, $S(I) = I\cap\mathbb{L}_-$ to denote the surviving labels of I, and $D(I_-) = I_- - (I\cap\mathbb{L}_-)$ to denote labels of I_- that died,

$$w_Z^{(\xi,\theta)}(I) \propto \mathbf{1}_{\Theta(I)}(\theta)\left[\bar{\Psi}_Z^{(\xi,\theta(\cdot))}(\cdot)\right]^I w_B(B(I))[\langle P_S\, p_-^{(\xi)}\rangle]^{S(I)}\sum\nolimits_{I_-\supseteq I\cap\mathbb{L}_-}[1-\langle P_S\, p_-^{(\xi)}\rangle]^{D(I_-)}w_-^{(\xi)}(I_-), \tag{10.28}$$

$$p_Z^{(\xi,\theta)}(x,\ell) \propto \Psi_Z^{(\theta(\ell))}(x,\ell)\left[\mathbf{1}_{\mathbb{B}}(\ell)p_B(x,\ell) + \mathbf{1}_{\mathbb{L}_-}(\ell)\frac{\langle f_S(x|\cdot)P_S(\cdot)p_-^{(\xi)}(\cdot)\rangle(\ell)}{\langle P_S\, p_-^{(\xi)}\rangle(\ell)}\right], \tag{10.29}$$

$$\bar{\Psi}_Z^{(\xi,\theta(\ell))}(\ell) = \int \Psi_Z^{(\theta(\ell))}(x,\ell)\left[\mathbf{1}_{\mathbb{B}}(\ell)p_B(x,\ell) + \mathbf{1}_{\mathbb{L}_-}(\ell)\frac{\langle f_S(x|\cdot)P_S(\cdot)p_-^{(\xi)}(\cdot)\rangle(\ell)}{\langle P_S\, p_-^{(\xi)}\rangle(\ell)}\right]dx. \tag{10.30}$$

While there is no analytic advantage, the joint prediction and update facilitates more efficient implementations [37].

10.3.3.4 LMB Filtering Recursion

The LMB filtering recursion approximates the GLMB filtering density by an LMB with matching first moment [38]. Given an LMB filtering density with parameters $\pi_- = \{(r_-(\ell), p_-(\cdot,\ell)) : \ell\in\mathbb{L}_-\}$ from the previous time step, applying the Chapman–Kolmogorov prediction yields the LMB prediction density

$$\pi = \{(r(\ell), p(\cdot,\ell)) : \ell\in\mathbb{L}\}, \tag{10.31}$$

where

$$r(\ell) = r_-(\ell)\langle P_S p_-\rangle(\ell), \tag{10.32}$$

$$p(x,\ell) = \frac{\langle f_S(x|\cdot)P_S(\cdot)p_-(\cdot)\rangle(\ell)}{\langle P_S\, p_-\rangle(\ell)}. \tag{10.33}$$

Converting π to a one-term GLMB and applying the GLMB update (10.24) yields the GLMB filtering density $\{(w_Z^{(\theta)}(I), p_Z^{(\theta)}) : (\theta, I)\in\Theta\times\mathcal{F}(\mathbb{L})\}$ (note that LMBs are not closed under the Bayes update), which can then be approximated by an LMB with matching first moment

$$\pi(\cdot|Z) = \{(r_Z(\ell), p_Z(\cdot,\ell)) : \ell\in\mathbb{L}\}, \tag{10.34}$$

where

$$r_Z(\ell) = \sum_{\theta\in\Theta} \sum_{I\subseteq\mathbb{L}} \mathbf{1}_I(\ell) w_Z^{(\theta)}(I), \tag{10.35}$$

$$p_Z(y,\ell) \propto \sum_{\theta\in\Theta} p_Z^{(\theta)}(y,\ell) \sum_{I\subseteq\mathbb{L}} \mathbf{1}_I(\ell) w_Z^{(\theta)}(I). \tag{10.36}$$

Due to its simplicity, the LMB filter has been used in many applications.

10.3.4 GLMB Posterior Recursion

The multiscan GLMB recursion is an analytic solution to the multitarget posterior recursion (10.4) [33]. Starting with a multiscan GLMB, via the standard models, a multitarget trajectory density at any time is a multiscan GLMB. Similar to a GLMB, the multiscan GLMB at time $k-1$ can be written in terms of its parameters as

$$\pi_{0:k-1} = \left\{ (w_-^{(\xi)}(I_{0:k-1}), p_-^{(\xi)}) : (\xi, I_{0:k-1}) \in \Xi \times \bigtimes_{i=0}^{k-1} \mathcal{F}(\mathbb{L}_i) \right\}. \tag{10.37}$$

At time k, the updated multiscan GLMB is [33]

$$\pi_{0:k}(\cdot|Z) \propto \left\{ (w_Z^{(\xi,\theta)}(I_{0:k}), p_Z^{(\xi,\theta)}) : (\xi, \theta, I_{0:k}) \in \Xi \times \Theta \times \bigtimes_{i=0}^{k} \mathcal{F}(\mathbb{L}_i) \right\}. \tag{10.38}$$

If we denote $B(I_k) = I_k \cap \mathbb{B}_k$, $S(I_k) = I_k \cap \mathbb{L}_{k-1}$, and $D(I_{k-1}) = I_{k-1} - (I_k \cap \mathbb{L}_{k-1})$, then

$$w_Z^{(\xi,\theta)}(I_{0:k}) = \mathbf{1}_{\Theta(I_k)}(\theta) \left[\bar{\Psi}_Z^{(\xi,\theta(\cdot))}(\cdot) \right]^{I_k} w_B(B(I_k)) \left[\bar{P}_S^{(\xi)} \right]^{S(I_k)} \left[1 - \bar{P}_S^{(\xi)} \right]^{D(I_{k-1})} w_-^{(\xi)}(I_{0:k-1}), \tag{10.39}$$

$$\bar{\Psi}_Z^{(\xi,\theta(\ell))}(\ell) = \begin{cases} \displaystyle\int \Psi_Z^{(\theta(\ell))}(x_k,\ell) f_S(x_k|x_{k-1},\ell) P_S(x_{k-1},\ell) \\ \quad p_-^{(\xi)}(x_{s(\ell):k-1},\ell) dx_{s(\ell):k}, & s(\ell) < t(\ell) = k, \\ \displaystyle\int \Psi_Z^{(\theta(\ell))}(x_k,\ell) p_B(x_k,\ell) P_B(\ell) dx_k, & s(\ell) = t(\ell) = k \end{cases} \tag{10.40}$$

$$\bar{P}_S^{(\xi)}(\ell) = \int P_S(x_{k-1},\ell) p_-^{(\xi)}(x_{s(\ell):k-1},\ell) dx_{s(\ell):k-1}, \tag{10.41}$$

$$p_Z^{(\xi,\theta)}(x_{s(\ell):t(\ell)},\ell) \propto \begin{cases} p_-^{(\xi)}(x_{s(\ell):t(\ell)},\ell), & t(\ell) < k-1 \\ (1 - P_S(x_{t(\ell)},\ell)) p_-^{(\xi)}(x_{s(\ell):t(\ell)},\ell), & t(\ell) = k-1 \\ \Psi_Z^{(\theta(\ell))}(x_k,\ell) f_S(x_k|x_{k-1},\ell) P_S(x_{k-1},\ell) \\ \quad p_-^{(\xi)}(x_{s(\ell):k-1},\ell), & s(\ell) < t(\ell) = k \\ \Psi_Z^{(\theta(\ell))}(x_k,\ell) p_B(x_k,\ell) P_B(\ell), & s(\ell) = t(\ell) = k \end{cases}. \tag{10.42}$$

Note that $\bar{\Psi}_Z^{(\xi,\theta(\ell))}(\ell)$ above is the same as (10.30) for the GLMB filtering density, and $\bar{P}_S^{(\xi)}(\ell)$ is the same as $\left\langle P_S\, p_-^{(\xi)} \right\rangle(\ell)$ in the GLMB prediction density.

As discussed in [18], the multiscan GLMB posterior recursion presented above is algebraically simpler than the GLMB filtering recursion, as it does not require marginalizing out past states. Specifically, when updating the multitarget trajectory density at each time step, the current weights are obtained by simply multiplying the previous weights by the weight increments. If a target is newly born, its posterior is computed as the product of the birth existence probability, the birth attribute density, and the sensor SNR. If a target has survived from the previous time step, its posterior is updated by multiplying it with the survival probability, the transition density, and the sensor SNR. Conversely, if a target dies at the current time step, its posterior density is scaled by the probability of death. However, if a target had already died before the current time step, its posterior density remains unchanged.

10.3.5 Approximations with LMB, GLMB, and Multiscan GLMB

In practice, the number of components in a GLMB can become intractably large. To manage this complexity, insignificant components can be removed while preserving its validity as a set function. Furthermore, the truncation error can be quantified in a similar manner as in a Fourier series approximation as follows. The L_1-error between two unnormalized GLMBs $\pi^{(\mathbb{H})} = \{(w^{(\xi)}(L), p^{(\xi)}) : (\xi, L) \in \mathbb{H}\}$ and $\pi^{(\mathbb{T})} = \{(w^{(\xi)}(L), p^{(\xi)}) : (\xi, L) \in \mathbb{T} \subset \mathbb{H}\}$ is [16]

$$||\pi^{(\mathbb{H})} - \pi^{(\mathbb{T})}||_1 = \sum_{(\xi,L)\in\mathbb{H}-\mathbb{T}} w^{(\xi)}(L), \tag{10.43}$$

where $\| \boldsymbol{f} \|_1 \triangleq \int |\boldsymbol{f}(\boldsymbol{X})| \delta \boldsymbol{X}$ is the L_1-norm of some function $\boldsymbol{f}$ on $\mathcal{F}(\mathbb{X} \times \mathbb{L})$. The error bound for their normalized versions is

$$\left\| \frac{\pi^{(\mathbb{H})}}{||\pi^{(\mathbb{H})}||_1} - \frac{\pi^{(\mathbb{T})}}{||\pi^{(\mathbb{T})}||_1} \right\|_1 \leq 2 \frac{||\pi^{(\mathbb{H})}||_1 - ||\pi^{(\mathbb{T})}||_1}{||\pi^{(\mathbb{H})}||_1}. \tag{10.44}$$

Examining the error expressions reveals that deleting components with the smallest weights minimizes the L_1-error between the original and the truncated GLMBs. This result also holds for multiscan GLMB [33].

While standard multitarget dynamic and observation models can accommodate most scenarios in MTT, certain applications require more sophisticated models. Examples include cell division [13], lineage estimation [11], target occlusion or interaction [10, 39, 40], tracking with merged measurements [41], and super-positional or image measurements [42] (see Section 10.5 for more details). These advanced models often lead to intractable multitarget densities. Therefore, it is crucial to approximate these densities using tractable LRFSs, such as the LMB and GLMB, including the multiscan extension.

An arbitrary labeled multitarget density can be approximated by an LMB through first-moment matching, enabling the closed-form recursion for the LMB filter in the previous section. However, the cardinality distribution of an LMB density lacks the flexibility to match all multitarget densities, and it cannot capture dependencies between target attributes. On the other hand, the cardinality distribution of a GLMB has a more flexible form, capable of matching the cardinality distributions of a wider class of multitarget densities. Moreover, the structure of the GLMB also allows the approximation of attribute dependencies among jointly existing targets. As a result, the GLMB often provides an effective balance between versatility and tractability in many applications.

The GLMB (10.9) can be approximated by a marginalized-GLMB (M-GLMB) [43], which is a GLMB with the discrete state ξ marginalized out. This M-GLMB is designed to match both the first moment and the cardinality of the GLMB while also minimizing its Kullback–Leibler divergence (KLD), similar to how the product of marginals minimizes the KLD from the joint density [44]. As shown in [43], given an arbitrary multitarget density, the parameters of an M-GLMB can be selected to match both the first moment and the cardinality distribution of the original density while minimizing its KLD. However, there are applications where the multitarget density takes on the generic multimodal form

$$\pi(\boldsymbol{X}) = \Delta(\boldsymbol{X}) \sum_{\xi \in \Xi} w^{(\xi)}(\mathcal{L}(\boldsymbol{X}))p^{(\xi)}(\boldsymbol{X}), \tag{10.45}$$

where each ξ corresponds to a mode, $\sum_{\xi \in \Xi} \sum_{L \subseteq \mathbb{L}} w^{(\xi)}(L) = 1$, and $\langle p^{(\xi)} \rangle (I) = 1$. The M-GLMB density cannot capture these modes (and the information therein) since the modes are marginalized out. In this case, one can approximate the multimodal density by a GLMB that matches its first moment and cardinality distribution [45]. Extensions of these results to the multiscan case can be found in [46].

10.3.6 LRFS Estimators

The traditional marginal multitarget (MaM) estimate, defined as the most probable multitarget state conditioned on the most probable cardinality, is an optimal technique for extracting a multitarget estimate from a given multitarget density [21]. LRFS offers an alternative to the MaM estimator, known as the label-MaM estimator, which is defined as the most probable multitarget state given the most probable label set [18]. Given a multitarget density π, the probability that only the labels in the set $L \subseteq \mathbb{L}$ exist (and the rest do not exist), i.e., the joint existence probability of L, is given by

$$w(L) \triangleq \langle \pi \rangle (L). \tag{10.46}$$

Given the most probable label set L^*, i.e., L^* maximizes (10.46), the estimated attributes of the label-MaM estimator are given by the mode of the corresponding label-conditioned joint attribute density. If there is more than one such label set with the highest probability, we choose the one with the most probable cardinality. If the most probable label set is still not unique, we choose the one that has the highest label-conditioned joint attribute density [18].

10.3.6.1 GLMB Estimators

For a GLMB of the form (10.8), the joint existence probability of a label set L is $w(L) = \sum_{\xi \in \Xi} w^{(\xi)}(L)$, and for an LMB of the (10.7), $w(L) = [1 - r^{(\cdot)}]^{\mathbb{L}-L}[r^{(\cdot)}]^L$. However, given the most probable label set L^*, determining the mode of the label-conditioned joint attribute density may be intractable. A suboptimal alternative is to find the mean or mode of $p(\cdot, \ell)$ for each $\ell \in L^*$ where [18]

$$p(x, \ell) = \frac{\sum_{\xi \in \Xi} p^{(\xi)}(x, \ell) \sum_{L \subseteq \mathbb{L}} \mathbf{1}_L(\ell) w^{(\xi)}(L)}{\sum_{\xi \in \Xi} \sum_{L \subseteq \mathbb{L}} \mathbf{1}_L(\ell) w^{(\xi)}(L)}. \tag{10.47}$$

Apart from label-MaM estimator, another intuitive suboptimal estimator for GLMB based on MaM estimator is the following estimator. Condition on the most probable cardinality n^*, where the cardinality distribution is given by

$$\rho(n) = \sum_{L \subseteq \mathbb{L}} \sum_{\xi \in \Xi} \delta_n[|L|] w^{(\xi)}(L), \tag{10.48}$$

we determine the most probable hypothesis (ξ^*, I^*) with $|I^*| = n^*$. The attribute of each $\ell \in I^*$ is the mean or mode of the attribute density $p^{(\xi^*)}(\cdot, \ell)$ [15].

10.3.6.2 PHD Estimators

For an arbitrary LRFS density, its PHD $\boldsymbol{v}$ can be used to construct a tractable suboptimal MaM or joint multitarget (JoM) estimator [18]. The existence probability and the attribute density of a target with label ℓ can be expressed in terms of the PHD, respectively, as [18]

$$r(\ell) = \int \boldsymbol{v}(x, \ell)dx \quad \text{and} \quad p(\cdot, \ell) = \boldsymbol{v}(\cdot, \ell)/r(\ell). \tag{10.49}$$

A suboptimal JoM estimate includes the set of labels whose existence probabilities exceed a specified threshold, along with their corresponding estimated attributes. A suboptimal MaM estimate consists of the n^* most probable labels, where n^* is the most probable cardinality (computed either from the cardinality distribution if available, or from the multi-Bernoulli cardinality approximation using the existence probabilities). Each selected label is paired with its corresponding estimated attributes, which are derived from the means or modes of the respective attribute densities.

10.3.6.3 Multiscan GLMB Estimators

The label-MaM estimator can be extended to the multiscan GLMB (10.11) as follows. First, determine the most probable label set sequence $I^*_{j:k}$ by maximizing $\sum_\xi w^{(\xi)}(I_{j:k})$, either with or without the constraint that the number of trajectories is equal to the most probable trajectory cardinality n^*. Second, compute the estimated attribute of the trajectory with label ℓ as the mean or mode of

$$p(\cdot; \ell) \propto \sum_\xi w^{(\xi)}(I^*_{j:k})p^{(\xi)}(\cdot; \ell). \tag{10.50}$$

Alternatively, condition on the most probable label set L^* that maximizes the joint existence probability

$$\sum_{\xi, I_{j:k}} \delta_{\cup_{i=j}^k I_i}[L]w^{(\xi)}(I_{j:k}), \tag{10.51}$$

we seek the most probable length m^* for each $\ell \in L^*$ by maximizing

$$\sum_{\xi, I_{j:k}} \mathbf{1}_{\cup_{i=j}^k I_i}(\ell)\delta_{t(\ell)-s(\ell)+1}[n]w^{(\xi)}(I_{j:k}). \tag{10.52}$$

The estimated attribute of the trajectory ℓ is taken as the mean or mode of

$$p_{m^*}(\cdot; \ell) \propto \sum_{\xi, I_{j:k}} \mathbf{1}_{\bigcup_{i=j}^{k} I_i}(\ell)w^{(\xi)}(I_{j:k})\delta_{t(\ell)\text{-}s(\ell)+1}[m^*]\, p^{(\xi)}(\cdot; \ell). \tag{10.53}$$

Note that L^* can also be the most probable label set conditioned on n^*. A simpler estimator for multiscan GLMB density is to determine $(\xi^*, I^*_{j:k})$, the most probable hypothesis, either with or without the most probable cardinality n^*. The estimated attribute for each trajectory $\ell \in I^*_{j:k}$ is taken to be the mean or mode of $p^{(\xi^*)}(\cdot; \ell)$.

10.3.7 LRFS Tracker Implementations

Since the number of components in the GLMB filtering and posterior densities grows superexponentially with time, mixture truncation is necessary in numerical implementation. Note that each term or component of a GLMB is symmetric, and hence truncation does not suffer from the non-symmetric problem discussed in Subsection 10.2.3. Truncation by keeping the components with highest weights minimizes L_1 approximation error [33].

10.3.7.1 Implementation of GLMB Filters

Consider a component (ξ_{k-1}, I_{k-1}) of the GLMB filtering density at time $k-1$ which generates a set $\{(\xi_{k-1}, I_{k-1}, \theta_k, I_k) : (\theta_k, I_k) \in \Theta_k \times \mathcal{F}(\mathbb{L}_k)\}$ of children components at time k. The objective of truncation is to select a subset of $(\theta_k, I_k) \in \Theta_k \times \mathcal{F}(\mathbb{L}_k)$ such that the filtering weights (10.28), which can be written as

$$w_{Z_k}^{(\xi_k,\theta_k)}(I_k) \propto w_{Z_k}^{(\xi_{k-1},I_{k-1},\theta_k,I_k)} w_{k-1}^{(\xi_{k-1})}(I_{k-1}), \tag{10.54}$$

are significant [37]. Note that the weight increment $w_{Z_k}^{(\xi_{k-1},I_{k-1},\theta_k,I_k)}$ is the only term influenced by choosing $(\theta_k, I_k) \in \Theta_k \times \mathcal{F}(\mathbb{L}_k)$. Each child component can equivalently be represented by an *extended association* $\gamma_k : I_{k-1} \uplus \mathbb{B}_k \rightarrow \{-1 : |Z_k|\}$, defined by

$$\gamma_k(\ell) \triangleq \begin{cases} \theta_k(\ell), & \text{if } \ell \in \mathcal{D}(\theta_k) \\ -1, & \text{if } \ell \in (I_{k-1} \uplus \mathbb{B}_k) - \mathcal{D}(\theta_k) \end{cases}, \tag{10.55}$$

which inherits from θ_k the *positive 1-1* property, i.e., each target generates at most one measurement and each measurement is associated with at most one target. Let $Z_k = \{z_{1:M_k}\}$, enumerate $I_{k-1} \uplus \mathbb{B}_k = \{\ell_{1:P_k}\}$, and define

$$\eta_k^{(i)}(j) \triangleq \begin{cases} 1 - \left\langle P_S p_{k-1}^{(\xi)} \right\rangle(\ell_i), & \ell_i \in I_{k-1}, j < 0 \\ 1 - P_B(\ell_i), & \ell_i \in \mathbb{B}_k, j < 0 \\ \left[\mathbf{1}_{\mathbb{B}_k}(\ell_i) P_B(\ell_i) + \mathbf{1}_{I_{k-1}}(\ell_i) \left\langle P_S p_{k-1}^{(\xi)} \right\rangle(\ell_i)\right] \overline{\Psi}_{Z_k}^{(\xi,j)}(\ell_i), & \ell_i \in I_{k-1} \uplus \mathbb{B}_k, j \geq 0 \end{cases}, \tag{10.56}$$

where $\bar{\Psi}_{Z_k}^{(\xi,j)}$ is given by (10.30). Then, the weight $w_{Z_k}^{(\xi_{k-1},I_{k-1},\theta_k,I_k)}$ increment in (10.54) can equivalently be written as [37]

$$\omega(\gamma_k) \triangleq \mathbf{1}_{\Gamma_k}(\gamma_k) \prod_{i=1}^{P} \eta_k^{(i)}(\gamma_k(\ell_i)). \tag{10.57}$$

For $i \in \{1 : P_k\}$ and $j \in \{-1 : M_k\}$, each $\eta_k^{(i)}(j)$ is precomputed and stored in the $P_k \times (M_k + 2)$ *association score matrix*, see Figure 10.5. The problem of GLMB truncation then reduces to finding a set of γ_k with highest weights $\omega(\gamma_k)$ which can be cast as a *ranked assignment problem* [37].

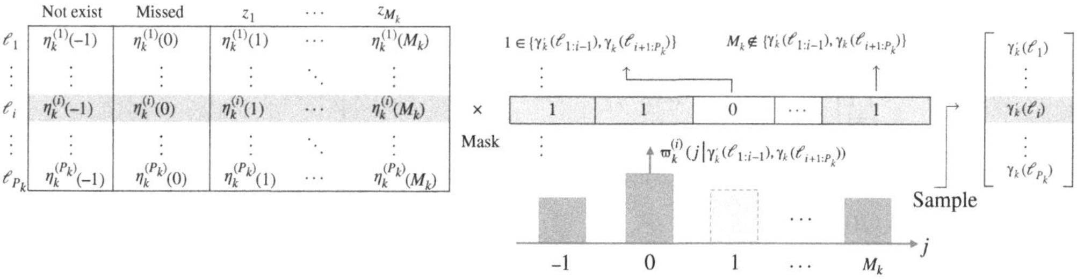

Figure 10.5 For GLMB truncation, the ith unnormalized conditional $\varpi_k^{(i)}$ at time k is obtained by the element-wise product of the ith row of the association score matrix and a binary mask vector determined by $\gamma'_k(\ell_{1:i-1}) = (\gamma'_k(\ell_1), \ldots, \gamma'_k(\ell_{i-1}))$ and $\gamma_k(\ell_{i+1:P_k}) = (\gamma_k(\ell_{i+1:}), \ldots, \gamma_k(\ell_{P_k}))$. *Source*: [18] / IEEE / CC BY 4.0.

Murty's algorithm [47] can solve the ranked assignment problem with $\mathcal{O}\left(K(M_k + 2P_k)^4\right)$ complexity, or $\mathcal{O}\left(K(M_k + 2P_k)^3\right)$ complexity with more efficient algorithms [48, 49]. Interested readers are referred to [16, 37].

Alternatively, Gibbs sampling (GS) [50, 51] can be employed as a computationally efficient Markov chain Monte Carlo (MCMC) technique to generate significant GLMB components in [37]. Specifically, a set of γ_k is sampled from some probability distribution ϖ_k on $\{-1\colon M_k\}^{P_k}$ with significant $\omega(\gamma_k)$. The distribution ϖ_k is defined over the set of positive 1-1 P_k-tuples, where sampling is performed by sequentially selecting measurements for targets while restricting (masking) measurements already assigned to other targets in previous iterations, see Figure 10.5 and Algorithm 10.1. This particular GS implementation of the GLMB filter requires $\mathcal{O}(TP_k^2M_k)$ complexity, where T is the number of GS iterations, and the complexity was further reduced to $\mathcal{O}(T(P_k + M_k))$ using tempered GS in [52].

Algorithm 10.1: Systematic-scan GS [37]

Input : $T, P_k, M_k, \gamma_k^{(0)}, \eta_k=[\eta_k^{(i)}]_{i=1}^{P_k}$
Output: $\gamma_k^{(1)}, \gamma_k^{(2)}, \dots, \gamma_k^{(T)}$

```
foreach t = 1 : T do
    foreach i = 1 : P_k do
        foreach j = 1 : M_k do // O(M_k P_k) complexity [37] or O(M_k) complexity [52]
            η_k^(i)(j) ← η_k^(i)(j)(1 − 1_{γ_k^(t)(ℓ_{1:i-1}), γ_k^(t-1)(ℓ_{i+1:P_k})}(j));
        γ_k^(t)(ℓ_i) ~ Categorical([−1:M_k], η_k^(i));
    γ_k^(t) ← [γ_k^(t)(ℓ_i)]_{i=1}^{P_k};
```

10.3.7.2 Implementation of Multiscan GLMB Filters

The key difference between single-scan and multiscan GLMB truncation is that the latter can correct past associations but the former cannot. In implementation, this entails solving a multidimensional ranked assignment problem, which is NP-hard for more than two dimensions. Consider a time window $\{0\colon n\}$, and time instants $k \in \{0\colon n\}$. The extended association $\gamma_k\colon I_{k-1} \uplus \mathbb{B}_k \to \{-1\colon |Z_k|\}$ in (10.55) represents (θ_k, I_k) with the domain $\mathcal{D}(\theta_k)$ given by the live labels of γ_k, defined as $\mathcal{L}(\gamma_k) \triangleq \{\ell \in I_{k-1} \uplus \mathbb{B}_k : \gamma_k(\ell) \geq 0\}$. For each $k \in \{0\colon n\}$, we can represent γ_k as a P_k-tuple in $\{-1\colon M_k\}^{P_k}$ by enumerating $\mathcal{L}(\gamma_{k-1}) \uplus \mathbb{B}_k = \{\ell_{1:P_k}\}$ and $Z_k = \{z_{1:M_k}\}$. From an initial multiscan GLMB $\pi_0 = \{(\omega_0(\gamma_0), p_0^{(\gamma_0)})\colon \gamma_0 \in \Gamma_0\}$, the propagation to $w_{Z_n}^{(\xi_n,\theta_n)}(I_{0:n})$ is given by [33]

$$\omega_{0:n}(\gamma_{0:n}) \triangleq \prod_{k=1}^{n}\left[\mathbf{1}_{\Gamma_k}(\gamma_k)\mathbf{1}_{\mathcal{F}(\mathcal{L}(\gamma_{k-1})\uplus\mathbb{B}_k)}(\mathcal{L}(\gamma_k)) \prod_{i=1}^{P_k} \eta_k^{(i)}(\gamma_k(\ell_i))\right]\omega_0(\gamma_0), \tag{10.58}$$

where $\eta_k^{(i)}(\gamma_k(\ell_i))$ has the same expression as (10.56) with $\bar{\Psi}_{Z_k}^{(\xi,j)}$ defined in (10.40) and $\left\langle P_S p_{k-1}^{(\xi)}\right\rangle$ replaced by $\bar{P}_S^{(\xi)}$ in (10.41).

Analogous to the GS approach for the single-scan truncation, implementation requires sampling significant association histories from some discrete probability distribution $\varpi = \omega_{0:n}$ on $\bigtimes_{k=0}^{n}\{-1\colon M_k\}^{P_k}$, which ensures only valid association histories are chosen and those with high weights are more likely. The two steps in multiscan GLMB implementation are (1) sequential factor

Algorithm 10.2: Sequential Factor Sampling [33]

Input : $R, n, G_0^{(0)} = (\gamma_0^{(0)}, w_0^{(0)}, \tau_0^{(0)})$ in [33]
Output: $G_{0:n}^{(R)}$

$r \leftarrow 1$;
foreach $k = 1 : n$ **do**
  $P_k \leftarrow |\mathbb{B}_k \uplus \mathcal{L}(\gamma_{k-1}^{(r-1)})|$; $M_k \leftarrow |Z_k|$; $\gamma_k^{(r-1)} \leftarrow$ zeros(P_k, 1);
  foreach $r = 1 : R$ **do**
    $\gamma_k^{(r)} \leftarrow [\ ]$;
    foreach $i = 1 : P_k$ **do**
      foreach $j = -1 : M_k$ **do**
        Compute $\varpi_k^{(i)}(j|\gamma_k^{(r)}(\ell_{1:i-1}), \gamma_k^{(r-1)}(\ell_{i+1:P_k}), \gamma_{0:k-1}^{(r-1)})$; // via (56) in [33]
      $\gamma_k^{(r)}(\ell_i) \sim$ Categorical($[-1{:}M_k]$, $\varpi_k^{(i)}$);
    $\gamma_{0:k}^{(r)} \leftarrow [\gamma_{0:k-1}^{(r-1)}, \gamma_k^{(r)}]$;
    Compute $w_{0:k}, \tau_{0:k}$ from $\gamma_{0:k}^{(r)}$; // via (49) and (51) in [33]
    $G_{0:k}^{(r)} \leftarrow (\gamma_{0:k}^{(r)}, w_{0:k}^{(r)}, \tau_{0:k}^{(r)})$;

sampling (multiscan initialization) and (2) systematic-scan GS (multiscan correction). The first stage uses the decomposition

$$\varpi(\gamma_{0:n}) = \prod_{k=1}^{n} \varpi_k(\gamma_k|\gamma_{0:k-1})\varpi_0(\gamma_0), \tag{10.59}$$

where

$$\varpi_k(\gamma_k|\gamma_{0:k-1}) \propto \mathbf{1}_{\Gamma_k}(\gamma_k)\mathbf{1}_{\mathcal{F}(\mathcal{L}(\gamma_{k-1})\uplus\mathbb{B}_k)}(\mathcal{L}(\gamma_k)) \prod_{i=1}^{P_k} \eta_k^{(i)}(\gamma_k(\ell_i)),$$

and simply setting the factors $\varpi_0 = \omega_0$. A computationally inexpensive method to generate valid association histories is to sequentially sample each association $\gamma_k : \mathcal{L}(\gamma_{k-1}) \uplus \mathbb{B}_k \rightarrow \{-1 : |Z_k|\}$ from the factor $\varpi_k(\cdot|\gamma_{0:k-1})$ sequentially from $k = 1 : n$, yielding $\varpi = \omega_{0:n}$. Notice that sampling from

$$\varpi_k(\gamma_k|\gamma_{0:k-1}) \propto \mathbf{1}_{\Gamma_k}(\gamma_k) \prod_{i=1}^{P_k} \eta_k^{(i)}(\gamma_k(\ell_i)), \tag{10.60}$$

can be accomplished by the single-scan GS [33]. The initialization routine is described in Algorithm 10.2.

The second stage also uses a systematic-scan GS-based procedure to correct the initialization $\gamma_{0:n}$ generated from the first stage. Specifically, systematic-scan GS constructs a Markov chain by iteratively sampling the coordinates $\gamma'_k(\ell_i)$ of $\gamma'_{1:n}$ for $k = 1{:}n$, and $i = 1{:}P_k$ from the conditional distributions

$$\varpi_k^{(i)}(j|\overbrace{\gamma'_{0:k-1}}^{\text{past}}, \overbrace{\underbrace{\gamma'_k(\ell_{1:i-1})}_{\text{processed}}, \underbrace{\gamma_k(\ell_{i+1:P_k})}_{\text{unprocessed}}}^{\text{current}}, \overbrace{\gamma_{k+1:n}}^{\text{future}}) \propto \varpi(\gamma'_{0:k-1}, \gamma'_k(\ell_{1:i-1}), j, \gamma_k(\ell_{i+1:P_k}), \gamma_{k+1:n}), \tag{10.61}$$

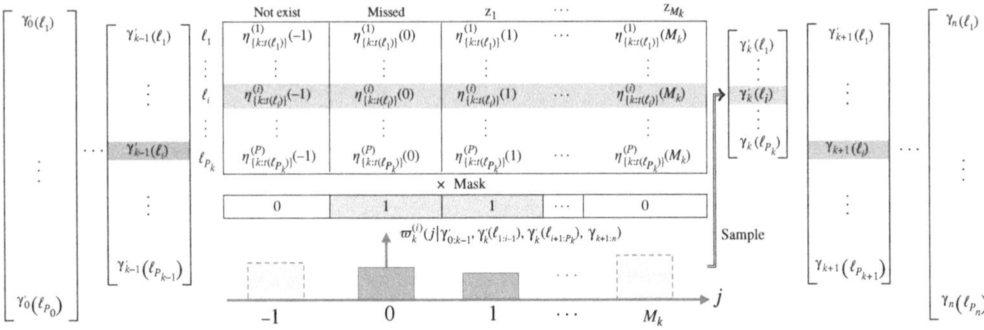

Figure 10.6 For multiscan GLMB truncation, the unnormalized conditional $\varpi_k^{(i)}$ for the ith track at time k is derived from the ith row of the kth association score matrix. *Source*: [18] / IEEE / CC BY 4.0. It is imposed that a label surviving at time $k+1$ must be a live track at time k, i.e., $\gamma_k'(\ell_i) > -1$, consistent with the unfragmented trajectory assumption of the standard multitarget dynamic model.

where the conditional distribution $\varpi_k^{(i)}$ is derived from the multiscan association score matrix for time k as illustrated in Figure 10.6. Consistent with the standard measurement model and as per (10.56), $\eta_k^{(i)}(j_k)$ can be written as $\eta_k^{(i)}(j_{s(\ell_i):k})$, which explicitly includes the dependence of ℓ_i associated with measurements up to k. Then, the score for the (i, j)th multiscan association is given by

$$\eta^{(i)}_{\{k:t(\ell_i)\}}(j) \triangleq \prod_{m=k}^{t(\ell_i)} \eta_m^{(i)}(\gamma'_{0:k-1}(\ell_i), j, \gamma_{k+1:m}(\ell_i)), \tag{10.62}$$

where $t(\ell_i)$ is the latest time ℓ_i exists on $\{0:n\}$. The unnormalized conditional $\varpi_k^{(i)}$ is generated from the ith row of the multiscan association score matrix at time k with a masking operation: zeroing the entry of every $j > 0$ that has been associated with a label other than ℓ_i at time k (positive 1-1 constraint), and any $j < 0$, if ℓ_i is a live label at $k+1$ (live time continuity). Further details for the multiscan GLMB recursion based on systematic-scan GS is given in Algorithm 10.3. A multisensor version of the multiscan GLMB truncation technique in [33] was developed in [53], which incurs $\mathcal{O}(LTVP^2M)$ complexity per time step with a moving window of a fixed size of L, or $\mathcal{O}(LTV(P+M))$ time using tempered GS [52], where T is total iterations of GS, V is total sensors, P is the maximum number of hypothesized trajectories, and M is the maximum number of measurements across all sensors.

The following example demonstrates the capabilities of the GLMB MTT approach. In particular, it highlights the ability of the GLMB filters to deal with challenging signal scenarios with very low detection probability.

Example 10.3 In this example, we apply the GLMB filter to track 12 targets in 3D space with a 3D positional sensor that has a detection probability of 0.5 and a clutter rate of 10. The scenario spans 100 seconds over a $2 \times 2 \times 2$ km region, with targets appearing randomly from four birth regions. Figure 10.7 shows the ground truth and estimated trajectories from the GLMB filters, along with the four birth locations.

Evaluating MTT performance requires meaningful metrics to quantify estimation accuracy. The optimal sub-pattern assignment (OSPA) metric assesses multitarget state errors [54], and its extension accommodates multitarget trajectory errors [17]. These metrics are widely used in the MTT literature because they provide a principled way to measure both localization and tracking errors

Algorithm 10.3: Multiscan GS [33]

Input : $T, n, G_{0:n}^{(0)} = (\gamma_{0:n}^{(0)}, w_{0:n}^{(0)}, \tau_{0:n}^{(0)})$ in [33]
Output: $G_{0:n}^{(1)}, G_{0:n}^{(2)}, \dots, G_{0:n}^{(T)}$

foreach $t = 1 : T$ **do**
  foreach $k = 1 : n$ **do**
    $P_k \leftarrow |\mathbb{B}_k \uplus \mathcal{L}(\gamma_{k-1}^{(t-1)})|$; $M_k \leftarrow |Z_k|$; $\gamma_k^{(t)} \leftarrow \gamma_k^{(t-1)}$;
    foreach $i = 1 : P_k$ **do**
      foreach $j = -1 : M_k$ **do**
        Compute $\varpi_k^{(i)}(j|\gamma_{0:k-1}^{(t)}, \gamma_k^{(t)}(\ell_{1:i-1}), \gamma_k^{(t-1)}(\ell_{i+1:P_k}), \gamma_{k+1:n}^{(t-1)})$;   `// via (47) in [33]`
      $\gamma_k^{(t)}(\ell_i) \sim$ Categorical($[-1{:}M_k]$, $\varpi_k^{(i)}$);
    $\gamma_{0:k}^{(t)} \leftarrow [\gamma_{0:k-1}^{(t-1)}, \gamma_k^{(t)}]$;
    Compute $w_{0:k}, \tau_{0:k}$ from $\gamma_{0:k}^{(t)}$;   `// via (49) and (51) in [33]`
  $G_{0:n}^{(t)} \leftarrow (\gamma_{0:k}^{(t)}, w_{0:k}^{(t)}, \tau_{0:k}^{(t)})$;

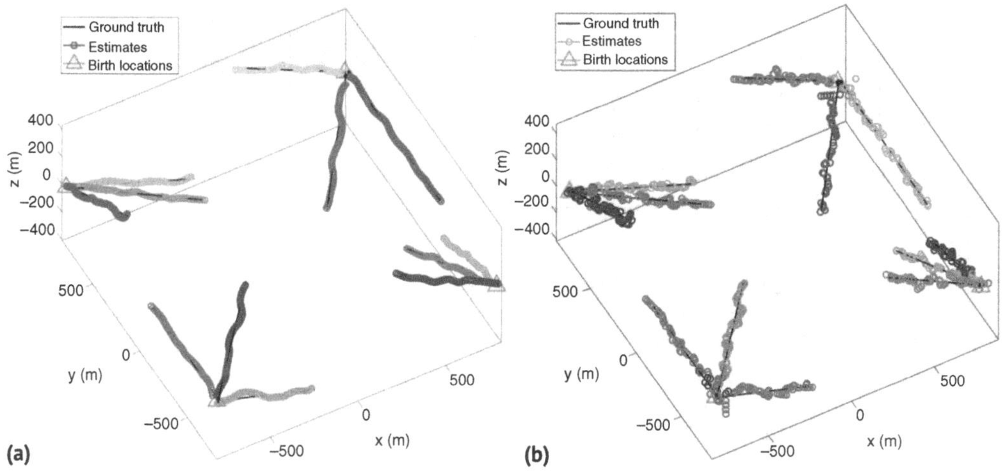

Figure 10.7 Ground truth and estimated trajectories from the 3D tracking scenario with a low detection probability sensor. (a) GLMB smoother and (b) GLMB filter.

that capture track fragmentations and track switching. While these metrics are not discussed in detail here, we refer the reader to [55] for a comprehensive study.

Figure 10.8 shows the estimated cardinality and the OSPA/OSPA$^{(2)}$ errors (with a cutoff of 100 m and order of 1). The plots indicate slight delays in target initialization as well as early track termination, which can be attributed to the sensor's severely low detection probability. Nevertheless, the multiscan GLMB filter tracks all targets with a high level of accuracy. This demonstrates the applicability of the tracking techniques introduced in this chapter in severe tracking scenarios. Further, to showcase the capability of the multiscan GLMB in expressing the posterior statistics, we also plot the posterior distributions of the length, and of the numbers of births/deaths at each time step in Figure 10.9.

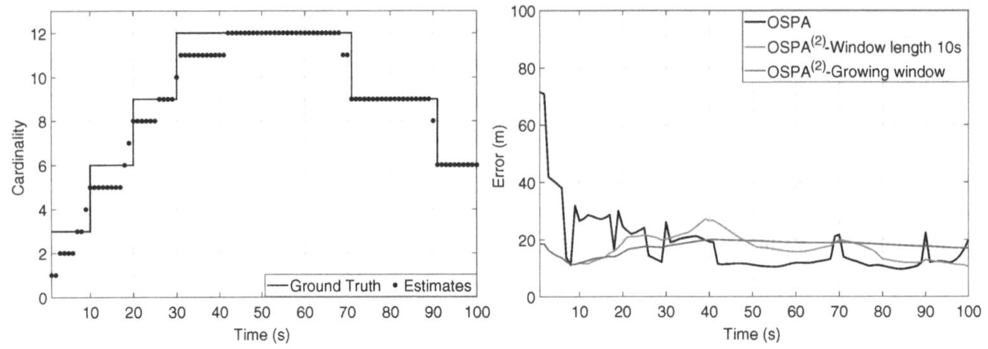

Figure 10.8 Estimated cardinality (left) and OSPA, OSPA$^{(2)}$ errors (right) from GLMB smoother.

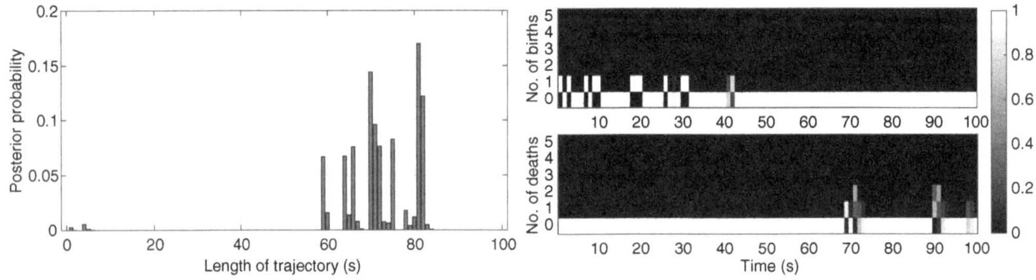

Figure 10.9 Left: posterior distribution of trajectory lengths. Right: posterior distributions of numbers of births (top) and deaths (bottom) at each time step.

10.3.8 Sensor Control and Management for MTT

Sensor control is a critical aspect of modern surveillance systems, aimed at optimizing sensor actions—such as movement, orientation, and operational modes—to enhance overall system performance while adhering to constraints like physical limitations, energy consumption, and operational policies. The challenge lies in computing optimal control actions to improve searching and tracking of multiple targets in complex environments, typically achieved by maximizing a reward function or minimizing a cost function that quantifies uncertainty or estimation error. As a sequential decision-making process, sensor control integrates new measurements to refine subsequent estimations, subject to uncertainty in both state and measurement [56].

Model predictive control (MPC) is an effective strategy for stochastic control and is extensively adopted in practice [57]. In comparison to meta-heuristic or bio-inspired techniques—such as genetic algorithms and particle swarm optimization [58]—which are often computationally prohibitive for real-time deployment in dynamic environments like MTT, MPC offers significant computational advantages. Notably, the MPC problem can be formulated as a partially observable Markov decision process (POMDP), a formulation that has garnered substantial interest as a real-time planning framework [59].

Computing optimal actions within the POMDP framework necessitates a representation that can capture the multitarget information state effectively. Among several MTT approaches, the LRFS framework uniquely provides a rigorous multitarget filtering density encompassing all relevant information about the trajectories of targets and inherently accounts for their temporal evolution. Conceptually, the POMDP paradigm, when integrated with LRFS SSM, enables the natural generalization of sensor management strategies from single-target to multitarget scenarios [60].

Sensor control and management usually rely on information divergence to formulate the reward functions. The idea of using Cauchy–Schwarz divergence (CS-D) to control the sensors in multitarget localization was introduced in [61], where the CS-D between Poisson densities was derived. In the LRFS literature, the analytical form of CS-D between GLMBs was used for sensor control problem in [62]. A sensor management framework for the GLMB filter based on CS-D was proposed in [60]. Recently, a solution to the passive sensor management problem was also proposed for the GLMB filter in [63].

Nonetheless, computing the CS-D between the GLMBs is expensive. In [64], Gostar et al. proposed an alternative based on the LMB filter. In particular, they approximated the LMBs by Poisson densities which the CS-D between them could be efficiently computed [61]. When the objective of the control action is to maximize the accuracy of the estimates, the authors of [65] proposed measuring the dispersion of the labeled multitarget density about its statistical mean using the OSPA distance. Similarly, techniques for multisensor control/selection that optimize a metric-based objective function were also proposed in [66, 67].

In [68], Rényi divergence (Rényi-D) was used as the reward function for multisensor tasking. However, since jointly computing the Rényi-D over all control actions for multiple sensors is expensive, the authors proposed an approximation that sums over the divergences computed independently at each sensor node. In [69], Zhang et al. suggested using Rényi-D or statistical distance as the reward function to control a sensor with the GLMB filter. In multisensor control, the authors considered the action at each sensor independently. Alternatively, in [70], a multisensor control solution for the LMB filter based on generalized covariance intersection (GCI) fusion and a task-driven reward function was proposed. Panicker et al. [71] also proposed a multisensor control method based on GCI fusion with the objective to minimize the ratio of non-existence over existence probability for a subset of targets of interest in the LMB. A solution to this problem with CS-D as the reward function is also given in [72]. In [73], a multi-objective (i.e., tracking and discovery) reward function to control limited field-of-view (FoV) sensors based on mutual information and differential entropy was developed. The authors also proved that this function is a monotone submodular set function, which allows the optimal control action to be efficiently obtained via greedy algorithms.

The POMDP framework is described by the following tuple $\langle \mathcal{F}(\mathcal{X}), \mathcal{F}(\mathbb{Z}), \mathcal{A}, \boldsymbol{f}, \boldsymbol{g}, \boldsymbol{\rho}, H \rangle$, where $\mathcal{F}(\mathbb{Z})$ denotes multitarget observation space; $\mathcal{A}$ is the set of admissible control actions (e.g., for a single agent, a discrete control action set might be $\mathcal{A} = \{\uparrow, \rightarrow, \downarrow, \leftarrow\}$). The functions $\boldsymbol{f}_k(\boldsymbol{X}_k|\boldsymbol{X}_{k-1}, a_{k-1})$ and $\boldsymbol{g}_k(Z_k|\boldsymbol{X}_k, a_{k-1})$ specify the multitarget transition density and measurement likelihood, given action a_{k-1} has been taken. The immediate reward $\boldsymbol{\rho}_k(Z_k, \boldsymbol{X}_k, a_{k-1})$ quantifies the reward of observing Z_k from $\boldsymbol{X}_k$, under action a_{k-1}, and the scalar H defines a look ahead planning horizon. The POMDP formulation readily extends to multiple sensors (or, equivalently, multiagent in sensor control), where the action space becomes $\mathcal{A} = \mathcal{A}^{(1)} \times \ldots \mathcal{A}^{(V)}$ with V denoting the number of agents. The objective is to determine a sequence of control actions $a_{k,[H]} = (a_k, \ldots, a_{k+H-1})$ that maximizes a reward (value) function $V(a_{k,[H]})$, constructed from the evolving information states (i.e., multitarget density $\boldsymbol{\pi}_{i,a_{i-1}}(\cdot|Z_i)$, which now depends on the action taken) and an immediate reward $\boldsymbol{\rho}_i$ over a planning horizon H as follows:

$$V(a_{k,[H]}) = \sum_{i=k+1}^{k+H} \iint \boldsymbol{\rho}_i(Z_i, \boldsymbol{X}_i, a_{i-1})\, \boldsymbol{\pi}_{i,a_{i-1}}(\boldsymbol{X}_i|Z_i)\, g_i(Z_i|\boldsymbol{X}_i, a_{i-1})\, \delta\boldsymbol{X}_i\, \delta Z_i. \tag{10.63}$$

In general, the value function defined in (10.63) has no closed-form solution and is computationally intractable. In practice, a popular approximation strategy is to set $a_j = a$ for $j = k : k + H - 1$, and identify the action a that optimizes the resulting value function, and apply this action over $M < H$ time steps before replanning at time $M + 1$. Another widely adopted approximation strategy involves the use of the predicted ideal measurement set (PIMS) [21]. Starting from the information state $\pi_{k,a_{k-1}}(\cdot|Z_k)$ at time k, one first computes the multitarget prediction density

$$\hat{\pi}_{a,j}(\boldsymbol{X}_j) = \int \prod_{i=k+1}^{j} \boldsymbol{f}_i(\boldsymbol{X}_i|\boldsymbol{X}_{i-1}, a)\pi_{k,a_{k-1}}(\boldsymbol{X}_k)\, \delta\boldsymbol{X}_{k:j-1}, \tag{10.64}$$

where the predicted multitarget state $\widehat{\boldsymbol{X}}_{a,j}$ is obtained via some multitarget estimator. The PIMS $\widehat{Z}_{a,j}$ corresponds to the noise-free and clutter-free observation of $\widehat{\boldsymbol{X}}_{a,j}$ with known target-to-measurement associations, after executing action a from time steps k to $k + j - 1$ [3]. The PIMS-based value function is given by

$$V(a) = \sum_{j=k+1}^{k+H} \int \rho_j\,(\widehat{Z}_{a,j}, \boldsymbol{X}_j, a)T(\pi_{j,a}(\boldsymbol{X}_j|\widehat{Z}_{a,j}))\delta\boldsymbol{X}_j, \tag{10.65}$$

where T denotes some transformation of the information state, such as a distribution constructed from its moments.

Remark 10.6 For tractability, the ideal measurements are used in place of the best available informative measurements. In practice, real sensor observations (subject to noise, false alarms, misdetections, and ambiguous origins) can be considered to be more informative if they are "closer" to the PIMS. Therefore, optimizing the PIMS-based value function naturally rewards actions that led to the most informative measurements. Consequently, excluding noise and false-alarms/misdetections from the ideal measurements does not imply that the PIMS approach disregards the inherent sensor limitations. Rather, the reason is to encourage actions that might generate (real) measurements with reduced noise and fewer false-alarms/misdetections.

Example 10.4 To illustrate the above multitarget control formulation, consider the task of searching and tracking a time-varying number of targets using multiple agents, each equipped with limited FoV sensors. Note that the searching and tracking objectives are competing with each other. To address this, we use a multi-objective value function that combines tracking and discovery objectives [73].

Tracking value function: We maximize the mutual information $I(\boldsymbol{X}; Z) = h(\boldsymbol{X}) - h(\boldsymbol{X}|Z)$ between the multi-object state $\boldsymbol{X}$ and the measurement Z, where $h(\cdot)$ denotes the differential entropy, to reduce state uncertainty. Since $h(\boldsymbol{X})$ is independent of control actions, maximizing $I(\boldsymbol{X}; Z)$ is equivalent to minimizing the conditional differential entropy $h(\boldsymbol{X}|Z)$. To ensure computational tractability, we adopt the PIMS approach [3] with the immediate reward (10.65). The resulting tracking value function is the cumulative differential entropy over the horizon:

$$\mathcal{J}_1(a) = -\sum_{j=k+1}^{k+H} h(\boldsymbol{X}_j|\widehat{Z}_{a,j}). \tag{10.66}$$

Discovery value function: We employ an entropy-based discovery value function using a dynamic occupancy grid to model undetected objects outside the sensor FoV. The search area $G \subset \mathbb{R}^{d_x}$ is partitioned into disjoint cells $\{\varkappa^{(i)}\}_{i=1}^{N}$. The *occupancy* of cell $\varkappa^{(i)}$ at time k is a Bernoulli random variable $O^{(i)}$. Let $Y_{a\_}^{(i)}$ denote a binary observation indicating if measurements originated from $\varkappa^{(i)}$. The discovery value function $\mathcal{J}_2$ is defined as the cumulative entropy of the occupancy grid:

$$\mathcal{J}_2(a) = -\sum_{j=k+1}^{k+H} \sum_{i=1}^{N} h(O_j^{(i)}|Y_{a,j}^{(i)}). \tag{10.67}$$

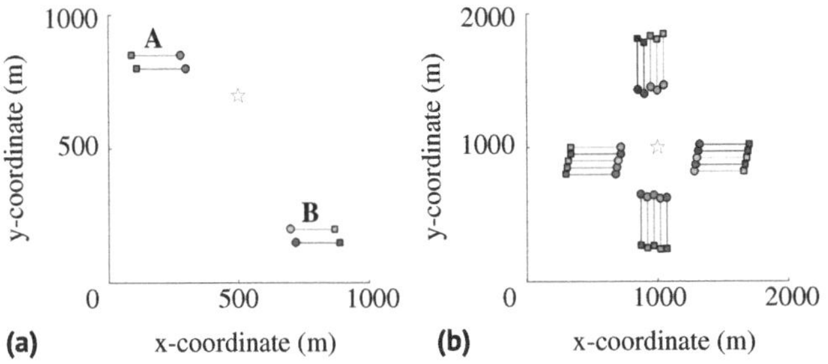

Figure 10.10 Experimental settings. (a) **Scenario 1 (Opposite)**: Two groups moving rapidly in opposing directions. (b) **Scenario 2 (Explosion)**: Four groups of 20 fast objects dispersing radially in opposite directions. Red ★ denotes agent start positions; ○/□ denote object Start/Stop positions.

Multi-objective sensor control: We balance the competing objectives via the vector $\mathcal{J}(a) = [\mathcal{J}_1(a), \mathcal{J}_2(a)]^\top$. Utilizing the global criterion method (GCM) to identify a Pareto-optimal solution, the combined value function is:

$$\mathcal{J}_{mo}(a) = \sum_{i=1}^{2} \frac{\mathcal{J}_i(a) - \min_{a\in\mathcal{A}} \mathcal{J}_i(a)}{\max_{a\in\mathcal{A}} \mathcal{J}_i(a) - \min_{a\in\mathcal{A}} \mathcal{J}_i(a)}. \tag{10.68}$$

While maximizing (10.68) is generally NP-hard, if $\mathcal{J}_{mo}$ is monotone submodular, a greedy algorithm provides a solution with guaranteed optimality bounds and quadratic complexity w.r.t. the number of sensors.

Experimental settings: A team of agents (e.g., drones) surveils a 1000×1000 m area for Scenario 1 and a 2000×2000 m area for Scenario 2, as depicted in Figure 10.10. The drone altitudes are stratified (with a minimum altitude at 30 m and 5 m separation) to prevent collisions. The objects follow a 2D constant velocity model $x_{k+1} = Fx_k + q_k$. Here, $F = [1, T; 0, 1] \otimes I_2$ with $T = 1$ s being the measurement sampling interval and I_2 denoting a 2×2 identity matrix. Each agent carries a sensor that has a limited range $r_D = 200$ m and is subject to Poisson clutter ($\lambda = 15$) with detection probability $P_D = 0.9$ and position-dependent noise. The occupancy grid size is 100×100.

Evaluation: We evaluate the proposed $\mathcal{J}_{mo}$ against single-objective baselines ($\mathcal{J}_1$ for tracking, $\mathcal{J}_2$ for discovery) and a theoretical upper bound ($\mathcal{J}_{mo}^*$ with known data association). Tracking performance is measured using the OSPA and OSPA$^{(2)}$ metrics ($c = 50$ m, $p = 1$) and averaged over 100 Monte-Carlo runs.

Figure 10.11 visually demonstrates the effectiveness of the multi-objective multisensor control algorithm in Scenario 1: $\mathcal{J}_{mo}$ successfully discovers the out-of-range group B while maintaining tracks on group A, whereas $\mathcal{J}_1$ ignores the unobserved region most of the time and $\mathcal{J}_2$ only performs extensive exploration instead of tracking multiple objects. Furthermore, Figure 10.12 demonstrates scalability in Scenario 2. As the number of sensors increases ($V = 2 \to 10$), $\mathcal{J}_{mo}$ retains its advantage. Notably, the performance of the discovery-only $\mathcal{J}_2$ approaches $\mathcal{J}_{mo}$ at higher sensor counts, as the increased sensor density renders the search space effectively fully covered.

10.4 MTT with Non-standard Models

The standard dynamic model described in Subsection 10.3.2 can accommodate maneuvering targets by augmenting the target state with a motion mode [74–76]. Similarly, the standard multitarget observation model in Subsection 10.3.2 seamlessly handles glint targets that generate heavy-tailed

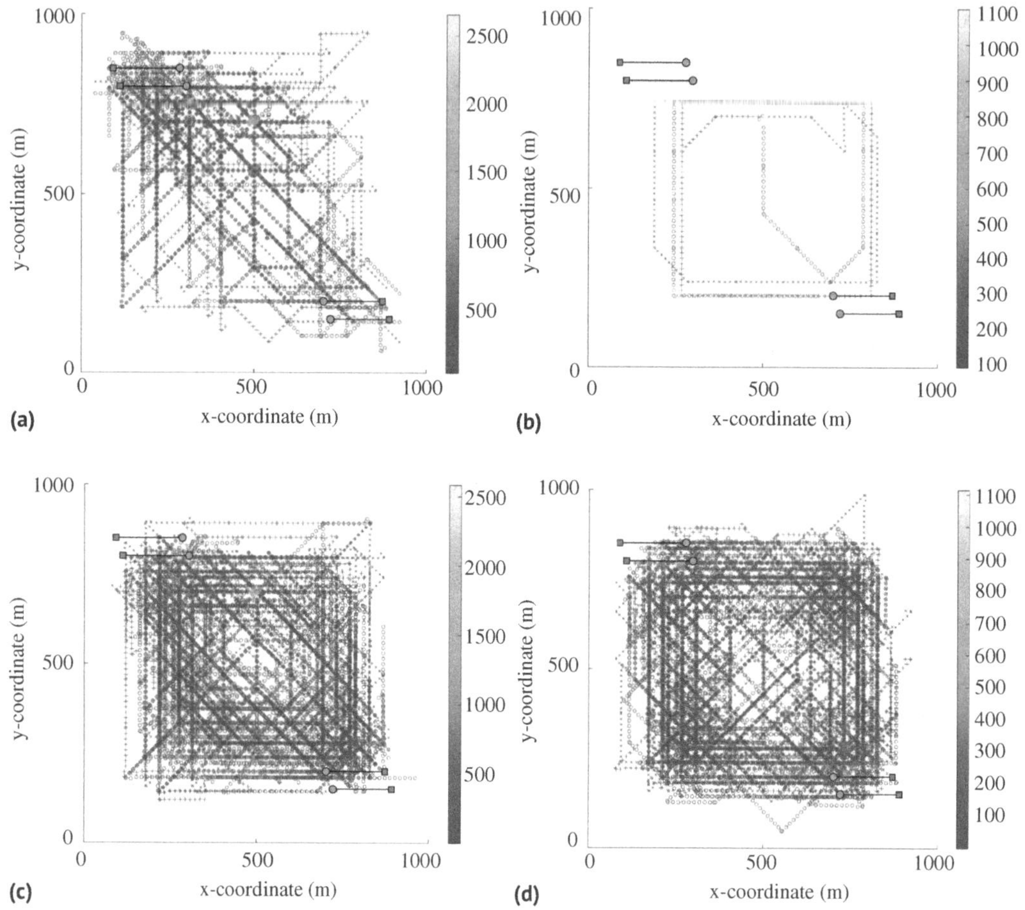

Figure 10.11 **Scenario 1 (Opposite)** sensor trajectories ($V = 3$, $r_D = 200$ m): (a) $\mathcal{J}_1$ fails to discover group B most of the time; (b) $\mathcal{J}_2$ explores extensively; (c) $\mathcal{J}_{mo}$ tracks both groups while discovering B; (d) $\mathcal{J}_{mo}^*$ (upper bound).

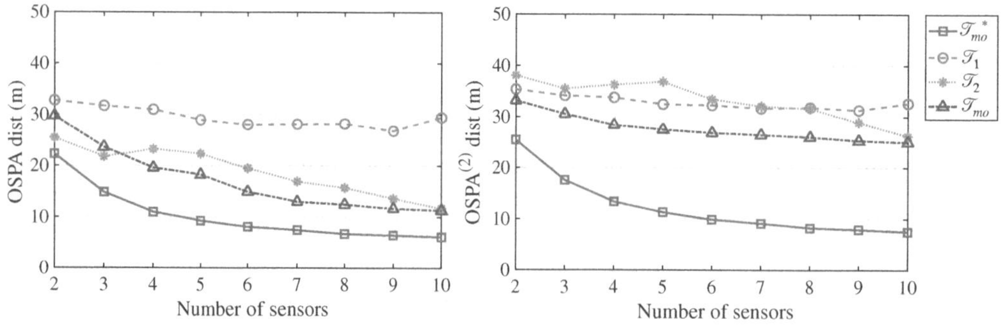

Figure 10.12 **Scenario 2 (Explosion)**: OSPA Dist and OSPA$^{(2)}$ Dist vs. the number of sensors ($V \in [2, 10]$). $\mathcal{J}_{mo}$ consistently outperforms single-objective baselines.

observation noise by employing the Student's t-distribution instead of the conventional Gaussian distribution [77]. However, several practical scenarios remain challenging, where these standard dynamic and measurement models are insufficient, as outlined below.

10.4.1 Interacting Multitarget Models

When the targets interact with each other, the multitarget transition density for the set of surviving targets cannot be written in a separable form $[f_S(X;\cdot)]^{X_-}$, but a more general expression $\tilde{f}_S(X|X_-)$ since the transition of a target depends on other targets in the multitarget state. Under this general dynamic model, closed-form filtering solution for the LMB/GLMB filter is not available. It was proposed in [39, 78], that the approximate single-target transition density only depends on the estimated states of the other targets, which allows the multitarget transition density to have a separable form. Groups of targets were jointly tracked using graph theory [79] or clustering techniques [80] by also approximating the joint-target transition density by a separable one. Alternatively, one can apply the general dynamic model to compute the multitarget density. Then, statistical approximation can be applied at some stage to achieve closed-form filtering solution.

10.4.2 Spawning Models

Beyond surviving and newly born targets, a spawning model is needed to account for the possibility of multiple targets being spawned from the same parent. At time k, a target spawned from a parent with label ℓ is assigned a unique label $\varsigma = (\ell, k, i)$ where i is an index to differentiate between multiple spawned targets by the same parent at time k. Hence, the spawn labels include two parts: an ancestral component (i.e., the parent's label) and a non-ancestral component to uniquely identify multiple spawned targets originating simultaneously from the same parent. Therefore, given the label space at the previous time is $\mathbb{L}_{k-1}$, the label space $\mathbb{T}_k$ for targets spawned at the current time k is defined as $\mathbb{T}_k = \mathbb{L}_{k-1} \times \{k\} \times \mathbb{N}$. For brevity, the time k is omitted and time $k-1$ is denoted by the subscript "-" when no ambiguity arises.

Suppose each $\boldsymbol{x}_- \in \boldsymbol{X}_-$ spawns new targets independently, and the set of spawn targets from parent $\mathbf{x}_-$ can be modeled by an LMB with parameters $\{(P_T(\boldsymbol{x}_-;\varsigma), f_T(\cdot|\boldsymbol{x}_-;\varsigma)) : \varsigma \in \mathbb{T}(\mathcal{L}(\boldsymbol{x}_-))\}$, where $\mathbb{T}(\ell) \triangleq \{(\ell, k)\} \times \{1 : M_\ell\} \subset \mathbb{T}$ and M_ℓ is the maximum number of spawned targets from parent ℓ. Then, the multitarget density of the spawned target set is given by [11]

$$\boldsymbol{f}_T(\boldsymbol{V}|\boldsymbol{X}_-) = \Delta(\boldsymbol{V})\mathbf{1}_{\mathbb{T}(\mathcal{L}(\boldsymbol{X}_-))}(\mathcal{L}(\boldsymbol{V}))[\Phi_T(\boldsymbol{V}|\cdot)]^{\boldsymbol{X}_-}, \tag{10.69}$$

where

$$\Phi_T(\boldsymbol{V}|\boldsymbol{x}_-) = [\Phi_T(\boldsymbol{V} \cap (\mathbb{X} \times \mathbb{T}(\mathcal{L}(\boldsymbol{x}_-))|\boldsymbol{x}_-;\cdot)]^{\mathbb{T}(\mathcal{L}(\boldsymbol{x}_-))}, \tag{10.70}$$

$$\Phi_T(\boldsymbol{V}|\boldsymbol{x}_-;\varsigma) = \sum_{(x,\ell)\in \boldsymbol{V}} \delta_\varsigma(\ell) P_T(\boldsymbol{x}_-;\varsigma) f_T(x|\boldsymbol{x}_-;\varsigma) + [1 - \mathbf{1}_{\mathcal{L}(\boldsymbol{V})}(\varsigma)](1 - P_T(\boldsymbol{x}_-;\varsigma)). \tag{10.71}$$

Since the sets of newborn, surviving and spawned targets are mutually disjoint (because their label spaces $\mathbb{B}$, $\mathbb{L}_-$, and $\mathbb{T}$ are disjoint), the multitarget state transition is given by

$$\boldsymbol{f}(\boldsymbol{X}|\boldsymbol{X}_-) = \boldsymbol{f}_B(\boldsymbol{X} \cap (\mathbb{X} \times \mathbb{B}))\boldsymbol{f}_S(\boldsymbol{X} \cap (\mathbb{X} \times \mathbb{L}_-)|\boldsymbol{X}_-)\boldsymbol{f}_T(\boldsymbol{X} \cap (\mathbb{X} \times \mathbb{T})|\boldsymbol{X}_-). \tag{10.72}$$

Bryant et al. [11] proposed the aforementioned target spawning model and approximating the multitarget filtering density (also captures the parent-spawn relations) with a GLMB.

Similarly, Nguyen et al. [13] introduced a mitosis model for biological cells in which a target ℓ may die, survive while retaining the same label ℓ, or evolve into a set of multiple targets

$L = \{\ell_1, \ldots, \ell_M\}$ with the parent label ℓ disappearing post-spawning (i.e., $\ell \notin L$). They derived the exact propagation of the multitarget filtering density and developed corresponding tracking solutions that approximate the prediction or filtering density with a GLMB. The following example illustrates MTT with spawning.

Example 10.5 Consider a cell biology experiment where the cells could migrate in a specific direction and mitosis is likely to occur in a prescribed region where a certain type of chemical is present. The goal is to infer the migration and mitosis behavior of the cell ensembles from a sequence of charge-coupled device (CCD) microscope images—an example is shown in Figure 10.13. A spot detector is used to generate the input to the MTT filter.

The GLMB-based cell tracking filter with the mitosis model proposed in [13] is used to compute the multitarget density for each frame. From this density, we generate a position heatmap, representing the accumulated intensity of cell positions over time (i.e., sum of cell position intensity over time), along with an accumulated velocity heatmap. These two heatmaps are displayed in Figure 10.14 and reveal a higher concentration of cells in the bottom half of the image, with cells tending to move directly downward.

To analyze where mitosis tends to occur, we compute a mitosis location heatmap (sum of cell intensity at the time of mitosis), which is shown in Figure 10.15 alongside the probability of a cell undergoing mitosis after entering the tracking region. The heatmap indicates that mitosis tends to occur when cells are in the middle of the region, while the right plot in Figure 10.15 shows that cells typically enter mitosis after appearing in the tracking scene for between 20 and 40 frames.

Figure 10.13 An example of the CCD microscopy image.

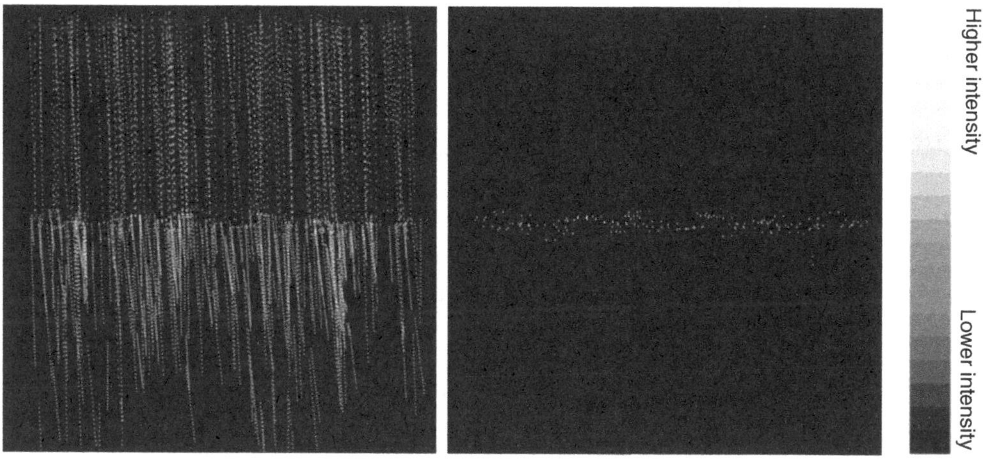

Figure 10.14 Heatmaps of cell position (left) and mitosis location (right) based on cell intensity, computed from the GLMB density.

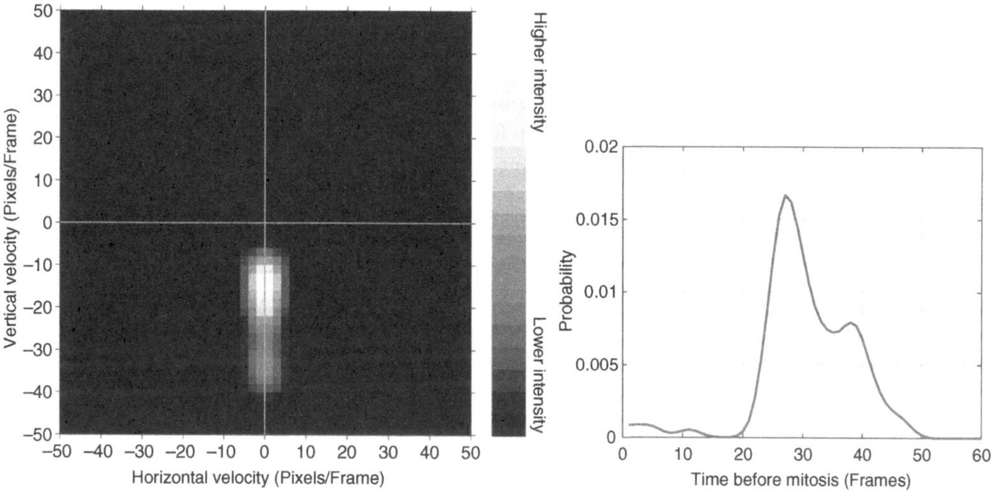

Figure 10.15 Heatmap of cell velocity based on cell intensity (left), and probability of time to mitosis after entering the region (right), computed from the GLMB density.

10.4.3 Extended Targets Measurement Models

An extended target observation model is a non-standard observation model, where each target can produce *multiple* measurements. This scenario challenges several conventional MTT algorithms since it violates the standard MTT assumption that each target can produce at most one measurement per scan. In particular, an extended target $\boldsymbol{x}$ could be detected with probability $P_D(\boldsymbol{x})$ and generates a set of observations W with observation likelihood $\tilde{g}(W|\boldsymbol{x})$ or misdetected with probability $1 - P_D(\boldsymbol{x})$. Similar to the standard observation, the false-alarm observations are modeled by a Poisson RFS with intensity function κ.

Let $\mathcal{P}_n(Z)$ be the set of all partitions of the measurement set Z into n disjoint subsets, and let $\mathcal{U}(Z) = \{W_1, \dots, W_n\} \in \mathcal{P}_n(Z)$ be a particular partition of Z. Given multitarget state $\boldsymbol{X}$ and measurement set Z, there exists a positive $1-1$ association map $\theta : \mathcal{L}(\boldsymbol{X}) \rightarrow \{0, 1, \dots, |\mathcal{U}(Z)|\}$ such that

$\theta(\ell) = \theta(\ell') > 0 \Rightarrow \ell = \ell'$. Denoting the set of all association mappings for the partition $\mathcal{U}(Z)$ as $\Theta_{\mathcal{U}(Z)}$, the multitarget likelihood function is [81]

$$g(Z|X) \propto \sum_{n=1}^{|X|+1} \sum_{\substack{\mathcal{U}(Z) \in \mathcal{P}_n(Z) \\ \theta \in \Theta_{\mathcal{U}(Z)}(\mathcal{L}(X))}} \left[\Psi_{\mathcal{U}(Z)}^{(\theta \circ \mathcal{L}(\cdot))}(\cdot)\right]^X \text{ where } \Psi_{\mathcal{U}(Z)}^{j}(x) = \begin{cases} \dfrac{P_D(x)\tilde{g}(W_j|x)}{\kappa^{W_j}}, & j = \{1 : |\mathcal{U}(Z)|\} \\ 1 - P_D(x), & j = 0 \end{cases}. \tag{10.73}$$

In [81], Gaussian (for target centroid), gamma (for measurement rate), and inverse Wishart (for target extent) (GGIW) distributions are used to model single extended target states. Further, the GLMB filter admits closed-form filtering solution for the type of likelihood function in (10.4.3) (also for the LMB filter with the GLMB to LMB conversion at the update step). In [82], B-spline functions were used to model the shape of the extended targets.

10.4.4 Merged Measurement Models

Another non-standard multitarget observation model is one for merged measurements, wherein multiple targets share a single measurement. In particular, a group of targets Y could be detected with probability $P_D(Y)$ and generate a common measurement z with observation likelihood $\ddot{g}(z|Y)$ or misdetected with probability $1 - P_D(Y)$. Similar to the standard observation model, the false-alarm observations are modeled by a Poisson RFS with intensity function κ. When merging happens, standard MTT algorithms often exhibit degraded performance due to premature track termination.

Given a multitarget state X, let $\mathcal{P}(X)$ be the set of all partitions of X and $\mathcal{U}(X) = \{Y_1, \ldots, Y_{|\mathcal{U}(X)|}\} \in \mathcal{P}(X)$ be a particular partition. We define a positive 1-1 association map $\theta : \mathcal{U}(\mathcal{L}_X) \to \{0, 1, \ldots, |Z|\}$ as a mapping such that $\theta(I) = \theta(L)$ implies $I = L$, where Z denotes the measurement set, and denote the set of all such association maps by $\Theta(\mathcal{U}(\mathcal{L}_X))$. Then, the merged measurement likelihood function is given by [41]

$$g(Z|X) \propto \sum_{\substack{\mathcal{U}(X) \in \mathcal{P}(X) \\ \theta \in \Theta(\mathcal{U}(\mathcal{L}_X))}} \left[\Psi_Z^{(\theta)}(\cdot)\right]^{\mathcal{U}(X)} \text{where } \Psi_Z^{(\theta)}(Y) = \begin{cases} \dfrac{P_D(Y)\ddot{g}(z_{\theta(\mathcal{L}_Y)}|Y)}{\kappa(z_{\theta(\mathcal{L}_Y)})}, & \theta(\mathcal{L}_Y) > 0 \\ 1 - P_D(Y), & \theta(\mathcal{L}_Y) = 0 \end{cases}. \tag{10.74}$$

Note that the multitarget filtering density under the merged measurement model is no longer a GLMB. In [41], the authors proposed approximating this filtering density with a GLMB. They also relaxed the one-to-one condition of the track to measurement assignment to efficiently select high weight components.

10.4.5 Measurement Models for Track-before-detect

A track-before-detect (TBD) MTT algorithm operates directly on pre-detection sensor observations rather than processing discrete detections. TBD is often required in low signal-to-noise scenarios, but is computationally demanding. The RFS framework naturally supports TBD and, more importantly, seamless fusion of TBD measurements with traditional detections, whereas other popular

MTT frameworks, such as MHT and JPDA, are inherently restricted to detection-based measurements. Indeed, even with a single sensor, RFS methods permit algorithms that switch between TBD and detection-based tracking, depending on the quality of the measurements [83].

Separable likelihoods: Given a multitarget state $\boldsymbol{X}$ and a raw measurement z. If the multitarget prior density is a GLMB and the targets are well-separated such that the multitarget likelihood can be expressed as $g(z|\boldsymbol{X}) \propto \gamma_z^{\boldsymbol{X}} = \prod_{\boldsymbol{x}\in\boldsymbol{X}} \gamma_z(\boldsymbol{x})$, then the multitarget posterior GLMB density is [45]

$$\pi(\boldsymbol{X}|z) \propto \Delta(\boldsymbol{X}) \sum_{c\in\mathbb{C}} w_z^{(c)}(\mathcal{L}(\boldsymbol{X}))\left[p_z^{(c)}(\cdot \mid z)\right]^{\boldsymbol{X}}, \tag{10.75}$$

where $w_z^{(c)}(L) = [\eta_z]^L w^{(c)}(L)$; $p_z^{(c)}(x, \ell \mid z) = p^{(c)}(x, \ell)\gamma_z(x, \ell)/\eta_z(\ell)$; $\eta_z(\ell) = \langle p^{(c)}(\cdot, \ell), \gamma_z(\cdot, \ell)\rangle$. In general, the multitarget likelihood is usually not separable; however, assuming separability can serve as a reasonable approximation when the targets are well-separated in the measurement space and their observations do not overlap.

General likelihood models: Several attempts have been made to develop efficient implementations of the multitarget Bayes filter with the general TBD observation model. In [42], a particle filter for superpositional sensors, whose proposal density was either the LMB or GLMB density computed based on the superpositional cardinalized probability hypothesis density (CPHD) filter was developed. Papi et al. [45] proposed approximating the filtering density after the update step with a GLMB. A similar solution based on the LMB approximation was also proposed in [84]. Mahler [85] also proposed a more efficient solution for the LMB filter using heuristic approximation at the price of lower performance. Given a generic observation model, Li et al. [86] derived an exact solution to propagate the multitarget filtering density, and approximated the density by an LMB which can be divided into (statistically independent) groups for parallel computation.

10.4.6 Occlusion Measurement Models

To explicitly capture the spatial relationships amongst targets, one can incorporate occlusion effects that model how targets obscure each other. Unlike approaches that handle occlusions through external modules, RFS-based MTT algorithms can directly incorporate occlusion model via multitarget-dependent detection probabilities, which consider the statistical dependence among targets commonly encountered in practice. Specifically, the occlusion model accommodates both partial and full occlusions, determined by the overlapping area between measurement regions—such as bounding 2D polygons—of the occluded and occluding targets. The greater the overlap, the lower the detection probability of the occluded target. Importantly, a target can only be occluded by those positioned in front of it (i.e., closer to the sensor).

Formally, let $\mathrm{Fr}(\boldsymbol{x};\boldsymbol{X})$ denote the subset of targets in the multitarget state $\boldsymbol{X}$ that are in front of the target with state $\boldsymbol{x}$. Let $\phi(\boldsymbol{x})$ be the 2D polygon bounding the target $\boldsymbol{x}$ in the measurement plane, and let $\mathrm{Area}(S)$ indicate the area of polygon S. The occlusion score is defined as [10]

$$O(\boldsymbol{x};\boldsymbol{X}) = \frac{1}{\mathrm{Area}(\phi(\boldsymbol{x}))} \times \mathrm{Area}\Big(\phi(\boldsymbol{x}) \cap \big(\cup_{\boldsymbol{y}\in\mathrm{Fr}(\boldsymbol{y};\,\boldsymbol{X})}\, \phi(\boldsymbol{y})\big)\Big). \tag{10.76}$$

Since the occluded target is less likely to be detected, the detection probability of a target with state $\boldsymbol{x}$ is bounded within the interval $(\epsilon, 1-\epsilon)$ and is given by

$$P_D(\boldsymbol{x};\boldsymbol{X}) = \max(\epsilon, 1-\epsilon - O(\boldsymbol{x};\boldsymbol{X})) \text{ where } \epsilon \in (0, 1/2). \tag{10.77}$$

Given the above occlusion model, each target $\boldsymbol{x} \in \boldsymbol{X}$ can be detected with probability $P_D(\boldsymbol{x};\boldsymbol{X})$ and generates a measurement z with observation likelihood $g(z|\boldsymbol{x})$ or misdetected with probability 1 –

$P_D(\boldsymbol{x};\boldsymbol{X})$. Suppose that $Z = \{z_1, \dots, z_{|Z|}\}$ is the measurement set generated from $\boldsymbol{X}$, the multitarget likelihood function with occlusion model is given by [10]

$$g(Z|\boldsymbol{X}) \propto \sum_{\theta \in \Theta(\mathcal{L}(\boldsymbol{X}))} \left[\Psi_{Z,\boldsymbol{X}}^{(\theta \circ \mathcal{L}(\cdot))}(\cdot)\right]^{\boldsymbol{X}} \text{ where } \quad \Psi_{Z,\boldsymbol{X}}^{(j)}(\boldsymbol{x}) = \begin{cases} \dfrac{P_D(\boldsymbol{x};\boldsymbol{X})g(z_j|\boldsymbol{x})}{\kappa(z_j)}, & j = \{1 : |Z|\} \\ 1 - P_D(\boldsymbol{x};\boldsymbol{X}), & j = 0 \end{cases}. \tag{10.78}$$

Ong et al. [40] proposed an occlusion model in which the detection probability has the form $P_D(\boldsymbol{x}|\boldsymbol{X} - \{\boldsymbol{x}\})$, a formulation that is primarily accurate for full occlusions. In contrast, the occlusion model in (10.77) by [10] supports both partial and full occlusions.

For closed-form GLMB filtering using multitarget likelihood function in (10.78), the detection probability is often approximated as a constant computed from the (statistical) mean/mode of the predicted states [40]. Another method to resolve occlusion is to store the truncated tracks then recall them later by matching the target appearance and motion feature [87]. This technique induces the dependence among the targets due to the recall process. Nevertheless, the authors proposed using only the multitarget estimates for matching to neglect the statistical dependence.

10.5 Applications of LRFS MTT

LRFS finds its applications in a range of real-world MTT problems. In this section, we discuss the applications of LRFS in distributed MTT and a variety of other practical domains including computer vision, cell biology, space situational awareness, simultaneous localization and mapping (SLAM), autonomous driving, and audio processing.

10.5.1 Distributed MTT

The idea of using GCI fusion (i.e., Kullback–Leibler averaging) for multitarget densities was popularized by Mahler in [88]. This GCI fusion rule indeed minimizes the KL-D between the fused density and the local ones, and it obeys the principle of minimum discrimination information (PMDI). The notion of GCI fusion was later extended to LRFS in [43]. Specifically, the fused density is the solution of the optimization problem

$$\bar{\pi}_{GCI} = \arg\min_{\pi} \sum_{s \in S} \omega^{(s)} D_{KL}(\pi || \pi^{(s)}), \tag{10.79}$$

where $\omega^{(s)}$ is user-defined fusion weight, and S is a discrete index set, recall that $D_{KL}(\cdot,\cdot)$ is the KL-D between two densities. The derivations in [43] show that if the local densities are either M-GLMBs or LMBs, the GCI-fused density also inherits the same form. Further, assume the local densities are LMBs, Wang et al. [89] adaptively compute the weights $\omega^{(s)}$ for each component of the LMBs using the information gain at the local update step. In [90], an event-trigger strategy was also proposed to reduce the communication burden.

In decentralized MTT, the filtering densities are computed independently, and the inferred labels can differ between the nodes. It was noted by Li et al. [91] that the GCI rule is highly sensitive to the label mismatch/inconsistency since fusion needs to be done label-wise. This phenomenon diminishes the accuracy of the fused density even though the local densities could be accurate. A heuristic remedy is to first compute the unlabeled multitarget densities from the local (labeled) ones using an unlabeled multi-Bernoulli approximation, then fuse them together via GCI, and augment the labels to yield a global GLMB density [91]. Alternatively, one can derive a label inconsistency indicator and minimize such indicator to obtain optimal label matches for label-wise fusion. Li et al. [92] proposed using the minimum KL-D between the unlabeled counterparts of the local labeled densities to quantify the label inconsistency. They showed that if the local densities are LMBs, label matching can be cast as a linear assignment problem, hence allowing efficient computation. Alternatively, in the scenarios where sensors have different FoVs, Shen et al. [93] maintained a list of label space matching and extended the label space of each node to include labels of targets that cannot be observed by the local sensor.

Remark 10.7 Contrary to the belief that label mismatch results from nodes having different label spaces, we emphasize that the label space is inherently tied to the multitarget state space and remains independent of the sensors or nodes. Like the attribute space, the label space is common across all nodes and sensors, and observing the multitarget state from different nodes does not alter it. Instead, label mismatch stems from the node's poor observability of the labels, leading to varying label estimates across different nodes.

Based on the PMDI, the minimum information loss (MIL) fusion rule was introduced by Gao et al. in [94]. Using this rule, the fused density is given by

$$\bar{\pi}_{MIL} = \arg\min_{\pi} \sum_{s\in S} \omega^{(s)} D_{KL}(\pi^{(s)}||\pi). \tag{10.80}$$

Note that the MIL fusion rule differs from the GCI in the exchange of the arguments in the KL-D. MIL fusion preserves both common and exclusive information of the local densities, unlike the GCI rule that hinders the exclusive information and reinforces the common one. From an information theoretic point of view, the GCI rule reduces the uncertainty of the common information among the local densities (i.e., true tracks) while increasing the uncertainty of the exclusive information (i.e., false tracks at different nodes). Conversely, the MIL rule preserves the information on both the true and false tracks. Such behavior is due to the fact that the MIL-fused density is the arithmetic mean of the local densities rather than the geometric mean as for the GCI rule. Li et al. [95] showed that both GCI and MIL fusion methods are fundamentally the Fréchet mean of local densities.

Based on the fact that all local densities have the same label space, and imposing certain constraints on the optimization problem in (10.80), Gao et al. [94] derived the MIL fusion formulas for the M-GLMBs and LMBs such that the fused density is of the same family as the local ones. To address label mismatch, the authors resorted to the label-matching strategy in [92]. Further, since components of the LMB density with the lower number of existing components will always be matched, they proposed measuring the Jensen–Shannon divergence between the state distributions of the components to eliminate the unlikely matches. When sensors have different FoVs, Gao et al. [96] proposed decomposing the local densities into sub-densities such that the sub-densities have the minimum KL-D to the original ones. MIL fusion rule was then used to fuse all sub-densities together, and the labels matched as per [92]. In [97], both GCI and MIL fusion rules were applied to track targets that can be detected in different observation modes. The GCI fusion was used to fuse the densities of confirmed targets while MIL was used for the unconfirmed ones. In [98], Li et al. also combined the GCI and MIL fusion rules to track targets of different classes and modes using distributed LMB filters. Interested readers can refer to [99] for a comprehensive analysis of multisensor fusion within the RFS framework.

Fusion rule based on CS-D can also preserve the information difference among the local densities. In centralized sensor networks, Gostar et al. [100] demonstrated the effectiveness of such method in the case where sensors have different FoVs. They approximated the local densities by Poisson densities that match the first moments, so that CS-D can be computed efficiently as per

[61]. Further, the authors proposed a fusion solution that fuses the LMB components corresponding to a particular label from different nodes into a single component. A method based on Bayes parallel combination rule to compute the global posterior density using synchronous centralized sensors was proposed for GLMB filter in [101].

Noting that multitarget density fusion is computationally demanding, not suitable for most real-life networks with limited computation and communication resources, Nguyen et al. [102] and Chen et al. [103] proposed alternatives that fuse only the estimates from the local nodes. Since the complexity is much lower than fusing the densities, the fusion of estimates can be efficiently extended to multiple scans, which significantly improves the tracking performance. Further, since only the track estimates are fused, the fusion algorithm is agnostic to specific tracking techniques used at the local nodes. Recently, Nguyen et al. [104] introduced the concept of a mean of multi-target trajectories using the Fréchet mean, and developed efficient techniques for multitarget trajectory consensus from distributed nodes.

10.5.2 Other Applications

Computer vision has advanced along with technologies in target detection/tracking. LRFS is an emerging approach to visual tracking. A computationally tractable filter was developed by exploiting the separable form of the multitarget likelihood function, and was subsequently applied to industrial environments as a comprehensive solution [105]. The authors also devised an online GLMB filtering algorithm for occlusion handling in visual tracking [87]. The proposed birth process model covers all regions by considering targets that can occlude each other, and it allows target re-detection after occlusion. Another GLMB visual tracker was proposed in [106] to resolve occluded and miss-detected targets. In particular, the number of ID switching is reduced and missed IDs are recovered by the proposed modules using spatial and feature information. In 3D visual MTT, the state-of-the-art GLMB method in [40] is able to track 3D targets from multi-view 2D detections. Building on this, an efficient algorithm that seamlessly performs automatic target initialization/termination method, occlusion handling, and re-ID algorithm using deep target appearance features was developed in [107].

Cell biology is another prominent area requiring visual tracking, where well-established methods have been developed within the LRFS framework. The LRFS-based tracking approach offers principled solutions to modeling and estimation of cell counts, survival rates, birth rates, position intensity, and migration patterns—key parameters in cell biology [108]. To enhance cell tracking performance, a measurement-driven birth model was incorporated into the GLMB filter in [109] further extended the model to account for elliptic cell shapes. Using the target spawning model of [11], a GLMB approximation was proposed for tracking biological cells and inferring their lineage, under unknown clutter rate and detection profile [12]. This line of research has been further advanced with a new spawning model that better captures cell mitosis in Nguyen et al. [13].

Space situational awareness is a well-established application where the LRFS framework proves particularly effective, especially for tracking space debris, which naturally requires target labeling. In [110], a distributed approach to space debris tracking using a consensus LMB filter was studied. The devised filter is designed to address the challenges of tracking space targets in a distributed setting, using the consensus LRFS filtering approach in [43] to fuse information from different sensors. This line of research was further refined and extended in subsequent works by Wei and Nener [111]. Additional recent contributions include [112, 113], and the Poisson labeled multi-Bernoulli (PLMB) filter [114], using the probabilistic admissible region method and partially uniform birth model.

SLAM is a long-standing and fundamental problem in robotics, where the goal is to estimate a robot's trajectory using sensor measurements while simultaneously constructing a map of the unknown environment. Among the various approaches developed to address this challenge, methods based on the LRFS framework have also emerged. In [115], the authors applied a Rao-Blackwellized implementation of the LMB filter to SLAM, offering an alternative that avoids the approximations required in PHD SLAM. Another LRFS-based SLAM method called δ-GLMB-SLAM2.0 was proposed in [116]. This method combines an optimal kernel-based Rao-Blackwellized particle filter for robot trajectory estimation with an efficient δ-GLMB filter for landmark mapping. In [117] a two-stage algorithm for SLAM was proposed using the LRFS framework. In the first stage, path planning is performed, followed by the second stage where an LMB filter is used for estimating the state of the sensors and feature points. To enhance obstacle avoidance, the sensor path is further refined using B-spline curve smoothing.

Autonomous vehicles is a promising application area for MTT, with numerous studies leveraging the LRFS framework—most commonly through the LMB [22, 39, 73, 118–120]. In [118] a variational radar model was introduced specifically for automotive applications, utilizing a variational Gaussian mixture and integrating it into an extended target LMB filter. Sensor fusion challenges in vehicle tracking have also been explored, see e.g., [121, 122]. In the context of unmanned aerial vehicles (UAVs), an online path planning method using the jump Markov system (JMS) TBD-LMB filter was developed in [22] to address joint detection and tracking of multiple targets. The proposed JMS TBD-LMB filter addresses the problem of joint detection and tracking of multiple targets. This research was later extended to multi-objective planning framework based on a multisensor GLMB filter, treating multiple UAVs as mobile sensors [73].

Audio processing involves the analysis and manipulation of audio signals, where the LRFS framework can provide versatile tools for areas such as source localization, tracking, and separation. For example, in [123], multi-acoustic array localization and tracking using the GLMB filter was proposed using GS for measurement association, and incorporates both time-difference-of-arrival (TDOA) and angle-of-arrival (AOA) measurements to track multiple acoustic sources. In [124], a multisource tracking filter was developed based on the multisensor GLMB tracker. The preprocessing pipeline includes short-time Fourier transform, steered-response power phase transform, and stochastic region contraction. For source separation, the method employs a generalized sidelobe canceler along with time-frequency masking. This work was further extended into an online audio-visual speech separation solution that integrates image data from multiple cameras [125]. The inclusion of visual measurements enhances the GLMB tracker's ability to estimate source trajectories and maintain speaker identification.

10.6 Conclusions

This chapter presents the key concepts and advantages of the LRFS framework for MTT. This framework is built on a rigorous mathematical foundation and offers a structured approach to tracking multiple targets over time. One of its most significant strengths is its ability to naturally handle complex tracking scenarios, including trajectory crossings, track fragmentation, and target spawning—situations that are common in real-world applications but challenging for traditional methods. Moreover, LRFS supports the fusion of heterogeneous observations collected at different times or from different sensors in a principled manner. This flexibility is crucial in modern tracking systems, where data may arrive asynchronously or from multiple modalities (e.g., radar, camera, lidar). A major advantage of LRFS is its capability for accurate multitarget trajectory estimation with bounded computational complexity per time step. This efficiency eliminates the need for heuristic post-processing steps—such as track stitching or identity management—that are often required in unlabeled RFS approaches. Another salient feature of the LRFS framework is the introduction of the joint existence probability for multitarget labels, which admits a meaningful notion of the most probable multitarget states fundamental for multitarget estimation. As a result, LRFS provides a more elegant and scalable solution to MTT. The fundamental importance of labels is emphasized by highlighting serious technical problems when labels are removed, such as the inability to consistently link tracks between windows for MTT with bounded computational complexity per time step, the incompatibility with fragmented tracks and target crossings, and theoretical problems in defining multitarget trajectory densities.

References

1 Y. Bar-Shalom, P. Willett, and X. Tian. *Tracking and Data Fusion: A Handbook of Algorithms*. YBS Publishing, 2011.

2 S. S. Blackman and R. Popoli. *Design and Analysis of Modern Tracking Systems*. Artech House, 1999.

3 R. Mahler. *Advances in Statistical Multisource-multitarget Information Fusion*. Artech House, 2014.

4 M. Mallick, V. Krishnamurthy, and B.-N. Vo. *Integrated Tracking, Classification, and Sensor Management: Theory and Applications*. John Wiley Sons, 2012.

5 B.-N. Vo, M. Mallick, Y. Bar-Shalom, S. Coraluppi, R. Osborne, R. Mahler, and B.-T. Vo. Multitarget tracking, In J. G. Webster, editors, *Wiley Encyclopedia of Electrical and Electronics Engineering*, 2015.

6 C. B. Storlie, T. C. Lee, J. Hannig, and D. Nychka. Tracking of multiple merging and splitting targets: a statistical perspective. *Statistica Sinica*, 19:1–31, 2009.

7 A. Andriyenko, S. Roth, and K. Schindler. An analytical formulation of global occlusion reasoning for multi-target tracking. In *2011 IEEE International Conference on Computer Vision Workshops (ICCV Workshops)*, Barcelona, Spain, pages 1839–1846, 2011.

8 Y. Zhang, C. Wang, X. Wang, W. Zeng, and W. Liu. FairMOT: on the fairness of detection and re-identification in multiple object tracking. *International Journal of Computer Vision*, 129:3069–3087, 2021.

9 C. Liang, Z. Zhang, X. Zhou, B. Li, S. Zhu, and W. Hu. Rethinking the competition between detection and reid in multiobject tracking. *IEEE Transactions on Image Processing*, 31:3182–3196, 2022

10 L. Van Ma, T. T. D. Nguyen, B.-N. Vo, H. Jang, and M. Jeon. Track initialization and re-identification for 3D multi-view multi-object tracking. *Information Fusion*, 111:102496, 2024.

11 D. S. Bryant, B.-T. Vo, B.-N. Vo, and B. A. Jones. A generalized labeled multi-Bernoulli filter with object spawning. *IEEE Transactions on Signal Processing*, 66(23):6177–6189, 2018.

12 T. T. D. Nguyen, C. Shim, and W. Kim. Biological cell tracking and lineage inference via random finite sets. In *IEEE 18th International Symposium on Biomedical Imaging (ISBI)*, Nice, France, pages 339–343, 2021.

13 T. T. D. Nguyen, B.-N. Vo, B.-T. Vo, D. Y. Kim, and Y. S. Choi. Tracking cells and their lineages via labeled random finite sets. *IEEE Transactions on Signal Processing*, 69:5611–5626, 2021.

14 I. Goodman, R. Mahler, and H. Nguyen. *Mathematics of Data Fusion*. Springer Netherlands, 2010.

15 B.-T. Vo and B.-N. Vo. Labeled random finite sets and multi-object conjugate priors. *IEEE Transactions on Signal Processing*, 61(13):3460–3475, 2013.

16 B.-N. Vo, B.-T. Vo, and D. Phung. Labeled random finite sets and the Bayes multi-target tracking filter. *IEEE Transactions on Signal Processing*, 62(24):6554–6567, 2014.

17 M. Beard, B. T. Vo, and B.-N. Vo. A solution for large-scale multi-object tracking. *IEEE Transactions on Signal Processing*, 68:2754–2769, 2020.

18 B.-N. Vo, B.-T. Vo, T. T. D. Nguyen, and C. Shim. An overview of multi-object estimation via labeled random finite set. *IEEE Transactions on Signal Processing*, 72:4888–4917, 2024.

19 S. Särkkä. *Bayesian Filtering and Smoothing*. Cambridge University Press, 2013.

20 B.-T. Vo and B.-N. Vo. A random finite set conjugate prior and application to multi-target tracking. In *International Conference on Intelligent Sensors, Sensor Networks and Information Processing*, pages 431–436, 2011.

21 R. Mahler. *Statistical Multisource-multitarget Information Fusion*. Artech House, 2007.

22 H. V. Nguyen, H. Rezatofighi, B.-N. Vo, and D. C. Ranasinghe. Online UAV path planning for joint detection and tracking of multiple radio-tagged objects. *IEEE Transactions on Signal Processing*, 67(20):5365–5379, 2019.

23 L. Svensson and M. Morelande. Target tracking based on estimation of sets of trajectories. In *17th International Conference on Information Fusion (FUSION)*, pages 1–8, 2014.

24 Á. F. García-Fernández, L. Svensson, J. L. Williams, Y. Xia, and K. Granström. Trajectory Poisson multi-Bernoulli filters. *IEEE Transactions on Signal Processing*, 68:4933–4945, 2020.

25 K. Li, M. Chen, and T. Kanade. Cell population tracking and lineage construction with spatiotemporal context. In *10th International Conference on Medical Image Computing and Computer-assisted Intervention - MICCAI*, pages 295–302, 2007.

26 E. Meijering, O. Dzyubachyk, and I. Smal. Methods for cell and particle tracking. *Methods in Enzymology.*, 504:183–200, 2012.

27 M. A. Lodato, M. B. Woodworth, S. Lee, G. D. Evrony, B. K. Mehta, A. Karger, S. Lee, T. W. Chittenden, A. M. D'Gama, X. Cai, L. J. Luquette, E. Lee, P. J. Park, and C. A. Walsh. Somatic mutation in single human neurons tracks developmental and transcriptional history. *Science*, 350(6256):94–98, 2015.

28 Paul W. Schumacher Jr. Tracking satellite break-ups. In *Sixth US/Russian Space Surveillance Workshop Conference Proceedings*, pages 174–182, 1996.

29 J. MØller and R. P. Waagepetersen. Modern statistics for spatial point processes. *Scandinavian Journal of Statistics*, 34(4):643–684, 2007.

30 D. Daley and D. Vere-Jones. *An Introduction to the Theory of Point Processes*. Springer-Verlag, 1988.

31 S. Chiu, D. Stoyan, W. Kendall, and J. Mecke. *Stochastic Geometry and Its Applications*. John Wiley Sons, 2013.

32 B.-N. Vo, S. Singh, and A. Doucet. Sequential Monte Carlo methods for multitarget filtering with random finite sets. *IEEE Transactions on Aerospace and Electronic Systems*, 41(4):1224–1245, 2005.

33 B.-N. Vo and B.-T. Vo. A multi-scan labeled random finite set model for multi-object state estimation. *IEEE Transactions on Signal Processing*, 67(19):4948–4963, 2019.

34 K. Granström, L. Svensson, Y. Xia, J. Williams, and Á. F. García-Fernández. Poisson multi-Bernoulli mixtures for sets of trajectories. *IEEE Transactions on Aerospace and Electronic Systems*, 61(2):5178-5194, 2025

35 A. F. García-Fernández, J. L. Williams, K. Granstrőm, and L. Svensson. Poisson multi-Bernoulli mixture filter: direct derivation and implementation. *IEEE Transactions on Aerospace and Electronic Systems*, 54(4):1883–1901, 2018.

36 Á. F. García-Fernández and L. Svensson. Trajectory PHD and CPHD filters. *IEEE Transactions on Signal Processing*, 67(22):5702–5714, 2019.
37 B.-N. Vo, B.-T. Vo, and H. G. Hoang. An efficient implementation of the generalized labeled multi-Bernoulli filter. *IEEE Transactions on Signal Processing*, 65(8):1975–1987, 2017.
38 S. Reuter, B.-T. Vo, B.-N. Vo, and K. Dietmayer. The labeled multi-Bernoulli filter. *IEEE Transactions on Signal Processing*, 62(12):3246–3260, 2014.
39 N. Ishtiaq, A. K. Gostar, A. Bab-Hadiashar, and R. Hoseinnezhad. Interaction-aware labeled multi-Bernoulli filter. *IEEE Transactions on Intelligent Transportation Systems*, 24(11):11668–11681, 2023.
40 J. Ong, B.-T. Vo, B.-N. Vo, D. Y. Kim, and S. Nordholm. A Bayesian filter for multi-view 3D multi-object tracking with occlusion handling. *IEEE Transactions on Pattern Analysis and Machine Intelligence*, 44(5):2246-2263, 2022.
41 M. Beard, B.-T. Vo, and B.-N. Vo. Bayesian multi-target tracking with merged measurements using labelled random finite sets. *IEEE Transactions on Signal Processing*, 63(6):1433–1447, 2015.
42 F. Papi and D. Y. Kim. A particle multi-target tracker for superpositional measurements using labeled random finite sets. *IEEE Transactions on Signal Processing*, 63(16):4348–4358, 2015.
43 C. Fantacci, B.-N. Vo, B.-T. Vo, G. Battistelli, and L. Chisci. Robust fusion for multisensor multiobject tracking. *IEEE Signal Processing Letters*, 25(5):640–644, 2018.
44 J.-F. Cardoso. Dependence, correlation and Gaussianity in independent component analysis. *Journal of Machine Learning Research*, 4: 1177–1203, 2003.
45 F. Papi, B.-N. Vo, B.-T. Vo, C. Fantacci, and M. Beard. Generalized labeled multi-Bernoulli approximation of multi-object densities. *IEEE Transactions on Signal Processing*, 63(20):5487–5497, 2015.
46 T. H. T. Nguyen, B.-N. Vo, and B.-T. Vo. Tractable approximation of labeled multi-object posterior densities. *arXiv preprint arXiv:2509.18780*, 2025.
47 K. G. Murty. An algorithm for ranking all the assignments in order of increasing cost. *Operations Research*, 16(3):682–687, 1968.
48 M. L. Miller, H. S. Stone, and I. J. Cox. Optimizing Murty's ranked assignment method. *IEEE Transactions on Aerospace and Electronic Systems*, 33(3):851–862, 1997.
49 C. R. Pedersen, L. R. Nielsen, and K. A. Andersen. An algorithm for ranking assignments using reoptimization. *Computers & Operations Research*, 35(11):3714–3726, 2008.
50 G. Casella and E. I. George. Explaining the Gibbs sampler. *American Statistician*, 46(3):167–174, 1992.
51 S. Geman and D. Geman. Stochastic relaxation, Gibbs distributions, and the Bayesian restoration of images. *IEEE Transactions on Pattern Analysis and Machine Intelligence*, 6:721–741, 1984.
52 C. Shim, B.-T. Vo, B.-N. Vo, J. Ong, and D. Moratuwage. Linear complexity Gibbs sampling for generalized labeled multi-Bernoulli filtering. *IEEE Transactions on Signal Processing*, 71:1981–1994, 2023.
53 D. Moratuwage, B.-N. Vo, B.-T. Vo, and C. Shim. Multi-scan multi-sensor multi-object state estimation. *IEEE Transactions on Signal Processing*, 70:5429–5442, 2022.
54 D. Schuhmacher, B.-T. Vo, and B.-N. Vo. A consistent metric for performance evaluation of multi-object filters. *IEEE Transactions on Signal Processing*, 56(8):3447–3457, 2008.
55 T. T. D. Nguyen, H. Rezatofighi, B.-N. Vo, B.-T. Vo, S. Savarese, and I. Reid. How trustworthy are performance evaluations for basic vision tasks? *IEEE Transactions on Pattern Analysis and Machine Intelligence*, 45(7):8538–8552, 2023.
56 D. A. Castañón and L. Carin. Stochastic control theory for sensor management," In A. O. Hero, D. A. Castañón, D. Cochran, and K. Kastella, editors, *Foundations and Applications of Sensor Management*, 7-32, Springer, 2008.

57 C. E. Garcia, D. M. Prett, and M. Morari. Model predictive control: Theory and practice–a survey. *Automatica*, 25(3):335–348, 1989.

58 V. Roberge, M. Tarbouchi, and G. Labonté. Comparison of parallel genetic algorithm and particle swarm optimization for real-time UAV path planning. *IEEE Transactions on Industrial Informatics*, 9(1):132–141, 2012.

59 D. Hsu, W. S. Lee, and N. Rong. A point-based POMDP planner for target tracking. In *IEEE International Conference on Robotics and Automation,* pages 2644–2650, 2008.

60 R. Mahler. A GLMB filter for unified multitarget multisensor management. In *Signal Processing, Sensor/Information Fusion, and Target Recognition*, 1101, 8123–134, 2019.

61 H. G. Hoang, B.-N. Vo, B.-T. Vo, and R. Mahler. The Cauchy-Schwarz divergence for Poisson point processes. *IEEE Transactions on Information Theory*, 61(8):4475–4485, 2015.

62 M. Beard, B.-T. Vo, B.-N. Vo, and S. Arulampalam, Sensor control for multi-target tracking using Cauchy-Schwarz divergence. In *2015 18th International Conference on Information Fusion (Fusion),* pages 937–944, 2015.

63 Y. Zhu, S. Liang, M. Gong, and J. Yan. Decomposed POMDP optimization-based sensor management for multi-target tracking in passive multi-sensor systems. *IEEE Sensors Journal*, 22(4):3565–3578, 2022.

64 A. K. Gostar, R. Hoseinnezhad, T. Rathnayake, X. Wang, and A. Bab-Hadiashar. Constrained sensor control for labeled multi-Bernoulli filter using Cauchy-Schwarz divergence. *IEEE Signal Processing Letters*, 24(9):1313–1317, 2017.

65 A. K. Gostar, R. Hoseinnezhad, A. Bab-Hadiashar, and W. Liu. Sensor-management for multitarget filters via minimization of posterior dispersion. *IEEE Transactions on Aerospace and Electronic Systems*, 53(6):2877–2884, 2017

66 F. Lian, L. Hou, J. Liu, and C. Han. Constrained multi-sensor control using a multi-target MSE bound and a δ-GLMB filter. *Sensors*, 18(7):2308, 2018.

67 F. Lian, L. Hou, B. Wei, and C. Han. Sensor selection for decentralized large-scale multi-target tracking network. *Sensors*, 18(12):4115, 2018.

68 H. Cai, S. Gehly, Y. Yang, R. Hoseinnezhad, R. Norman, and K. Zhang. Multisensor tasking using analytical Rényi divergence in labeled multi-Bernoulli filtering. Journal of Guidance, Control, and Dynamics, 42(9):2078–2085, 2019.

69 G. Zhang, F. Lian, C. Han, H. Chen, and N. Fu. Two novel sensor control schemes for multi-target tracking via delta generalised labelled multi-Bernoulli filtering. *IET Signal Processing*, 12(9):1131–1139, 2018.

70 X. Wang, R. Hoseinnezhad, A. K. Gostar, T. Rathnayake, B. Xu, and A. Bab-Hadiashar. Multi-sensor control for multi-object Bayes filters. *Signal Processing*, 142:260–270, 2018.

71 S. Panicker, A. K. Gostar, A. Bab-Hadiashar, and R. Hoseinnezhad. Tracking of targets of interest using labeled multi-Bernoulli filter with multi-sensor control. *Signal Processing*, 171:107451, 2020.

72 N. Ishtiaq, S. Panicker, A. K. Gostar, A. Bab-Hadiashar, and R. Hoseinnezhad. Selective sensor control via Cauchy Schwarz divergence. In *Smart Trends in Computing and Communications*, 113–124, 2021.

73 H. V. Nguyen, B.-N. Vo, B.-T. Vo, H. Rezatofighi, and D. C. Ranasinghe. Multi-objective multi-agent planning for discovering and tracking multiple mobile objects. *IEEE Transactions on Signal Processing*, 72:3669–3685, 2024.

74 Y. Punchihewa, B.-N. Vo, and B.-T. Vo. A generalized labeled multi-Bernoulli filter for maneuvering targets. In *19th International Conference on Information Fusion (FUSION),* pages 980–986, 2016.

75 W. Yi, M. Jiang, and R. Hoseinnezhad. The multiple model Vo-Vo filter. *IEEE Transactions on Aerospace and Electronic Systems*, 53(2):1045–1054, 2017

76 C. Cao and Y. Zhao. A multiple-model generalized labeled multi-Bernoulli filter based on blocked Gibbs sampling for tracking maneuvering targets. *Signal Processing*, 186:108119, 2021.

77 P. Dong, Z. Jing, H. Leung, K. Shen, and M. Li. The labeled multi-Bernoulli filter for multitarget tracking with glint noise. *IEEE Transactions on Aerospace and Electronic Systems*, 55(5): 2253–2268, 2018.

78 A. K. Gostar, T. Rathnayake, C. Fu, A. Bab-Hadiashar, G. Battistelli, L. Chisci, and R. Hoseinnezhad. Interactive multiple-target tracking via labeled multi-Bernoulli filters. In *2019 International Conference on Control, Automation and Information Sciences*, pages 1–6, 2019.

79 W. Liu and Y. Chi. Resolvable group state estimation with maneuver based on labeled RFS and graph theory. *Sensors*, 19(6):1307, 2019.

80 X. Xue, S. Huang, J. Xie, J. Ma, and N. Li. Resolvable cluster target tracking based on the DBSCAN clustering algorithm and labeled RFS. *IEEE Access*, 9:43364–43377, 2021.

81 M. Beard, S. Reuter, K. Granström, B.-T. Vo, B.-N. Vo, and A. Scheel. Multiple extended target tracking with labeled random finite sets. *IEEE Transactions on Signal Processing*, 64(7):1638–1653, 2015.

82 A. Daniyan, S. Lambotharan, A. Deligiannis, Y. Gong, and W.-H. Chen. Bayesian multiple extended target tracking using labeled random finite sets and splines. *IEEE Transactions on Signal Processing*, 66(22):6076–6091, 2018.

83 D. Y. Kim, B.-N. Vo, B.-T. Vo, and M. Jeon. A labeled random finite set online multi-object tracker for video data. *Pattern Recognition*, 90:377–389, 2019.

84 A. F. García-Fernández. Track-before-detect labeled multi-Bernoulli particle filter with label switching. *IEEE Transactions on Aerospace and Electronic Systems*, 52(5):2123–2138, 2016.

85 R. Mahler. A fast labeled multi-Bernoulli filter for superpositional sensors. In *Signal Processing, Sensor/Information Fusion, and Target Recognition*, 1064, 6, pages 113–124, 2018.

86 S. Li, W. Yi, R. Hoseinnezhad, B. Wang, and L. Kong. Multiobject tracking for generic observation model using labeled random finite sets. *IEEE Transactions on Signal Processing*, 66(2):368–383, 2017.

87 T. Rathnayake, A. K. Gostar, R. Hoseinnezhad, R. Tennakoon, and A. Bab-Hadiashar. On-line visual tracking with occlusion handling. *Sensors*, 20(3):929, 2020.

88 R. Mahler. Optimal/robust distributed data fusion: A unified approach. In *Signal Processing, Sensor/Information Fusion, and Target Recognition*, 4052: 128–138, 2000

89 X. Wang, A. K. Gostar, T. Rathnayake, B. Xu, A. Bab-Hadiashar, and R. Hoseinnezhad. Centralized multiple-view sensor fusion using labeled multi-Bernoulli filters. *Signal Processing*, 150:75–84, 2018.

90 K. Shen, C. Zhang, P. Dong, Z. Jing, and H. Leung. Consensus-based labeled multi-Bernoulli filter with event-triggered communication. *IEEE Transactions on Signal Processing*, 70:1185–1196, 2022.

91 S. Li, W. Yi, R. Hoseinnezhad, G. Battistelli, B. Wang, and L. Kong. Robust distributed fusion with labeled random finite sets. *IEEE Transactions on Signal Processing*, 66(2):278–293, 2017.

92 S. Li, G. Battistelli, L. Chisci, W. Yi, B. Wang, and L. Kong. Computationally efficient multi-agent multi-object tracking with labeled random finite sets. *IEEE Transactions on Signal Processing*, 67(1):260–275, 2018.

93 K. Shen, P. Dong, Z. Jing, and H. Leung. Consensus-based labeled multi-Bernoulli filter for multitarget tracking in distributed sensor network. *IEEE Transactions on Cybernetics*, 52(12):12722–12733, 2021.

94 L. Gao, G. Battistelli, and L. Chisci. Fusion of labeled RFS densities with minimum information loss. *IEEE Transactions on Signal Processing*, 68:5855–5868, 2020.

95 T. Li, X. Wang, Y. Liang, and Q. Pan. On arithmetic average fusion and its application for distributed multi-Bernoulli multitarget tracking. *IEEE Transactions on Signal Processing*, 68:2883–2896, 2020.

96 L. Gao, G. Battistelli, and L. Chisci. Fusion of labeled RFS densities with different fields of view. *IEEE Transactions on Aerospace and Electronic Systems*, 58(6):5908–5924, 2022.

97 L. Gao, G. Battistelli, L. Chisci, and A. Farina. Fusion-based multidetection multitarget tracking with random finite sets. *IEEE Transactions on Aerospace and Electronic Systems*, 57(4):2438–2458, 2021.

98 G. Li, G. Battistelli, L. Chisci, L. Gao, and P. Wei. Distributed joint detection, tracking, and classification via labeled multi-Bernoulli filtering. *IEEE Transactions on Cybernetics*, 54(3):1429–1441, 2024.

99 T. Li, K. Da, H. Fan, and B. Yu. Multisensor random finite set information fusion: Advances, challenges, and opportunities. In *Secure and Digitalized Future Mobility*, 33–64, CRC Press, 2022.

100 A. K. Gostar, T. Rathnayake, R. Tennakoon, A. Bab-Hadiashar, G. Battistelli, L. Chisci, and R. Hoseinnezhad. Cooperative sensor fusion in centralized sensor networks using Cauchy-Schwarz divergence. *Signal Processing*, 167:107278, 2020.

101 M. Herrmann, C. Hermann, and M. Buchholz. Distributed implementation of the centralized generalized labeled multi-Bernoulli filter. *IEEE Transactions on Signal Processing*, 69:5159–5174, 2021.

102 H. V. Nguyen, H. Rezatofighi, B.-N. Vo, and D. C. Ranasinghe. Distributed multi-object tracking under limited field of view sensors. *IEEE Transactions on Signal Processing*, 69:5329–5344, 2021.

103 F. Chen, H. Van Nguyen, A. S. Leong, S. Panicker, R. Baker, and D. C. Ranasinghe. Distributed multi-object tracking under limited field of view heterogeneous sensors with density clustering. *Signal Processing*, 228:109703, 2025.

104 T. T. D. Nguyen, B. T. Vo, B.-N. Vo, H. Van Nguyen, and C. Shim. The mean of multi-object trajectories. *IEEE Transactions on Signal Processing*, 74: 531–544, 2026.

105 T. Rathnayake, R. Tennakoon, A. K. Gostar, A. Bab-Hadiashar, and R. Hoseinnezhad. Information fusion for industrial mobile platform safety via track-before-detect labeled multi-Bernoulli filter. *Sensors*, 19(9):2016, 2019.

106 M. Abbaspour and M. A. Masnadi-Shirazi. Online multi-object tracking with δ-GLMB filter based on occlusion and identity switch handling. *Image Vision Computing*, 127:104553, 2022.

107 L. Van Ma, T. T. D. Nguyen, C. Shim, D. Y. Kim, N. Ha, and M. Jeon. Visual multi-object tracking with re-identification and occlusion handling using labeled random finite sets. *Pattern Recognition*, 156:110785, 2024.

108 W. J. Hadden, J. L. Young, A. W. Holle, M. L. McFetridge, D. Y. Kim, P. Wijesinghe, H. Taylor-Weiner, J. H. Wen, A. R. Lee, K. BiebackB. N. Vo, D. D. Sampson, B. F. Kennedy, J. P. Spatz, A. J. Engler, and Y. S. Choi. Stem cell migration and mechanotransduction on linear stiffness gradient hydrogels. *Proceedings of the National Academy of Sciences*, 114(22):5647–5652, 2017.

109 M. I. Hossain, A. K. Gostar, A. Bab-Hadiashar, and R. Hoseinnezhad, Visual mitosis detection and cell tracking using labeled multi-Bernoulli filter. In *International Conference* on *Information Fusion,* pages 2472–2476, 2018.

110 B. Wei and B. Nener. Consensus labeled multi-Bernoulli filtering for distributed space debris tracking. In *2017 International Conference on Control, Automation and Information Sciences (ICCAIS),* pages 203–208, 2017.

111 B. Wei and B. Nener. Distributed space debris tracking with consensus labeled random finite set filtering. *Sensors*, 18(9):3005, 2018.

112 D. Moratuwage, M. Adams, and L. Cament. Space object tracking using a jump Markov system based -GLMB filter for space situational awareness. In *Advanced Maui Optical and Space Surveillance Technologies Conference*, pages 32, 2019.

113 J. A. Gaebler and P. Axelrad. Identity management of clustered satellites with a generalized labeled multi-Bernoulli filter. *Journal of Guidance, Control, and Dynamics*, 43(11):2046–2057, 2020.

114 L. Cament, M. Adams, and P. Barrios. Space debris tracking with the Poisson labeled multi-Bernoulli filter. *Sensors*, 21(11):3684, 2021.

115 H. Deusch, S. Reuter, and K. Dietmayer. The labeled multi-Bernoulli SLAM filter. *IEEE Signal Processing Letters*, 22(10):1561–1565, 2015.

116 D. Moratuwage, M. Adams, and F. Inostroza. δ-generalized labeled multi-Bernoulli simultaneous localization and mapping with an optimal kernel-based particle filtering approach. *Sensors*, 19(10):2290, 2019.

117 J. Yang and W. Liu. A two-stage feature point detection and marking approach based on the labeled multi-Bernoulli filter. *Sensors*, 22(14):5083, 2022.

118 A. Scheel and K. Dietmayer. Tracking multiple vehicles using a variational radar model. *IEEE Transactions on Intelligent Transportation Systems*, 20(10):3721–3736, 2019.

119 W. J. Park and C. G. Park. Road constrained labeled multi Bernoulli filter based on PDF truncation for multi-target tracking. *IFAC-PapersOnLine*, 53(2):15739–15744, 2020.

120 K. Dai, Y. Wang, S. Wang, and C. Yin. Inter-target occlusion handling in multi-vehicle corner tracking based on labeled multi-Bernoulli filter with detection probability optimization model. In *International Conference on Intelligent Transportation Engineering*, pages 178–189, 2022.

121 A. K. Gostar, T. Rathnayake, R. Tennakoon, A. Bab-Hadiashar, G. Battistelli, L. Chisci, and R. Hoseinnezhad. Centralized cooperative sensor fusion for dynamic sensor network with limited field-of-view via labeled multi-Bernoulli filter. *IEEE Transactions on Signal Processing*, 69:878–891, 2020.

122 J. Klupacs, A. K. Gostar, A. Bab-Hadiashar, J. Palmer, and R. Hoseinnezhad, Distributed complementary fusion for connected vehicles. In *International Conference on Control, Automation and Information Sciences*, pages 316–321, 2022.

123 Z. Tian, W. Liu, and X. Ru, Multi-acoustic array localization and tracking method based on Gibbs-GLMB. In *International Conference on Control, Automation and Information Sciences*, pages 1–6, 2019.

124 J. Ong, B. T. Vo, and S. Nordholm. Blind separation for multiple moving sources with labeled random finite sets. *IEEE/ACM Transactions on Audio, Speech, and Language Processing*, 29:2137–2151, 2021.

125 J. Ong, B. T. Vo, S. Nordholm, B.-N. Vo, D. Moratuwage, and C. Shim. Audio-visual based online multi-source separation. *IEEE/ACM Transactions on Audio, Speech and Language Processing*, 30:1219–1234, 2022.

11

Bayesian Track-before-detect for Maritime Radar

Du Yong Kim[1], Branko Ristic[1], and Luke Rosenberg[2]

[1]*School of Engineering, RMIT University, Melbourne, Victoria, Australia*
[2]*Advanced Systems & Technologies, Lockheed Martin Australia, Adelaide, Australia*

11.1 Introduction

Airborne platforms for maritime surveillance traditionally fly at low altitudes and use non-coherent, single-channel, high-resolution radar systems for detection and tracking of surface targets. Radar signal processing in the described setup is challenging for two reasons. First, the sea clutter is affected by spatiotemporal correlation, depending on environmental conditions (waves, sea swell, wind speed) and radar operation (range resolution, grazing angle) [1]. Second, the amplitude distribution of sea clutter is non-Gaussian, due to characteristically long tails resulting from the occasional strong (target-like) radar returns.

Furthermore, there has been an increasing operational requirement to conduct maritime surveillance using unmanned aerial vehicles. In order to enable longer endurance flights, the unmanned aerial platforms are required to operate at higher altitudes. This further exacerbates detection and tracking of weak maritime targets because the higher altitudes increase the grazing angles, causing even stronger levels of sea clutter.

In order to deal effectively with the difficulties described above, it has become important to investigate the track-before-detect (TBD) method for detection and tracking of weak (dim) surface targets. Unlike the conventional tracking algorithms, which input the point radar measurements (PRMs) produced by peak extraction and threshold detection applied to consecutive radar frames, TBD techniques [2] use the entire radar frames as input measurements. By exploiting the full information contained in radar data, TBD has been shown in many applications to improve the tracking performance of weak (dim) targets. Early studies on TBD mainly focused on detection of small moving objects using optical and infrared imaging sensors [3, 4]. Pioneering works on TBD in the context of radar were reported in [5–7]. In general, there are two main TBD techniques in the radar context: dynamic programming-based TBD [7, 8], and the recursive Bayesian TBD [9, 10, chapter 11]. Dynamic programming is a fairly cumbersome framework for on-line implementation of TBD, mainly due to the impracticality of discretizing the state space in realistic problems and a need for high-dimensional maximization. The recursive Bayesian TBD, on the other hand, can be implemented in a straightforward manner using the particle filter and will be the focus of this chapter. The foundation block is the optimal Bayesian TBD filter for a single-target, known as the *Bernoulli TBD filter*.

The chapter represents a compilation of authors' research, published in several papers [11–16]. Section 11.2 presents an overview of maritime radar data collected by an airborne platform, and introduces the mathematical model of sea clutter distribution. The theoretical formulation of the Bernoulli TBD filter is presented in Section 11.3 and includes in Section 11.3.5 a comparison with the standard point-measurement based Bayesian trackers. A multi-target Bernoulli TBD tracker for interacting targets is proposed in Section 11.4. Exploitation and the benefits of Doppler information in the slow scan mode are studied in Section 11.5. A Bernoulli TBD filter for a multichannel maritime radar is considered in Section 11.6. Finally, the conclusions are drawn in Section 11.7.

11.2 Maritime Radar Data

11.2.1 Overview

Suppose the platform is traveling with velocity $\mathbf{v}_\mathrm{p}$ and a scanning radar is pointing at an angle θ_{plat} relative to the platform motion [17], as illustrated in Figure 11.1a. The angle between the radar look direction and the wind is denoted by ψ. As the radar scans with a rate ω_0 [rad/s], there are $N_\mathrm{p} = 2\pi\, f_\mathrm{r}/\omega_0$ pulses in a given scan, where f_r is the pulse repetition frequency (PRF) of the radar. Also, if the two-way azimuth 3 dB beamwidth is $\phi_{3\mathrm{dB}}$, then a point target will be in the beam for $T_\tau = \phi_{3\mathrm{dB}}/\omega_0$ seconds (dwell time) or $N_\tau = \mathrm{floor}(T_\tau \cdot f_\mathrm{r})$ pulses.

Throughout the chapter we consider two scanning modes. Most of the chapter will deal with the fast scanning mode, where the dwell time T_τ is short, which corresponds to the typical non-coherent high-resolution operation. In this case, only the range-azimuth data are available for detection and tracking. For example, consider a fast scanning mode with $\omega_0 = 100$ rpm, $f_\mathrm{r} = 1500$ Hz and other simulation parameters given in Table 11.1. With this set of parameters, the scan time is 0.6 s, with $N_\tau = 7$ pulses on the target. For this work, we consider the magnitude of the received radar data. A typical range-azimuth radar image, created using the high fidelity radar

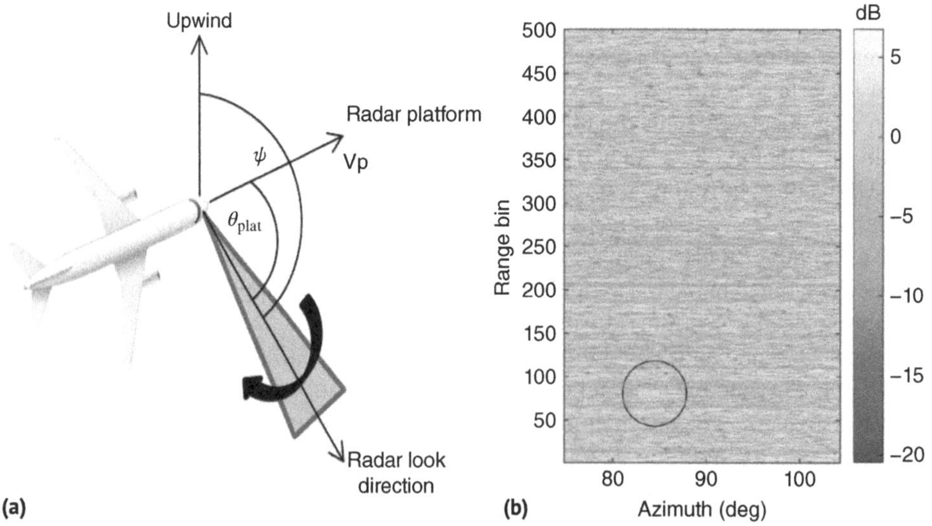

Figure 11.1 Scanning airborne radar: (a) Operational scenario and (b) Typical range-azimuth radar image, obtained in the fast scanning mode ($N_\mathrm{r} = 500$ range bins, $N_\mathrm{a} = 75$ azimuth bins). Target with SIR = 4 dB indicated by the circle. *Source*: [34] / IEEE / CC BY 4.0.

Table 11.1 Parameters of the scanning radar.

Parameter	Value
Platform speed ($\lvert \mathbf{v}_\mathrm{p} \rvert$)	100 m/s
Platform heading ($\angle \mathbf{v}_\mathrm{p}$)	0° (North direction)
Wind and swell direction (ψ)	270°
Azimuth look direction (θ_plat)	90°
Grazing angle	30°
Slant range offset to sea (km)	20 km
Sea state	3
Radar scan rate (ω_0)	100 rpm = 10.47 rad/s
Azimuth 3 dB beamwidth (ϕ_3dB)	3°
Radar carrier frequency (f_0)	10 GHz
Bandwidth (B)	100 MHz
Pulse repetition frequency (f_r)	1500 Hz
Number of range bins (N_r)	500
Scanning region	30°
Clutter-to-noise ratio	20 dB

simulator described in [17], is shown in Figure 11.1b. There are $N_\mathrm{r} = 500$ range bins and $N_\mathrm{a} = 75$ azimuth beams, and it corresponds to a side looking region in the upwind direction, centered at the azimuth look direction of 90° and spread over the scanning region of 30°. An outbound target is present on the sea surface, with a signal-to-interference ratio (SIR) of 4 dB, located at 85° azimuth and range bin 80 (indicated by a circle). As the radar scans, each pulse corresponds to a different azimuth angle with resolution in degrees of $\omega_0/f_r \cdot (\frac{180}{\pi}) = 0.4°$. The range cell size is 1.5 m and is determined by the range resolution, $c/(2B)$, where c is the speed of light and B is the radar bandwidth. Spatial correlation of the mean clutter level in Figure 11.1b is due to the fluctuations in the swell.

In the slow scanning mode, the Doppler information can be obtained by taking the fast Fourier transform (FFT) of a block of pulses centered on the current azimuth location. The slow scanning mode and the exploitation of Doppler in TBD will be discussed in Section 11.5.

11.2.2 Clutter Magnitude Modeling

Consider a range-azimuth image (map), such as the one in Figure 11.1b. In the absence of any targets, the measured magnitude z in any cell of the range-azimuth image is due to background clutter-plus-noise (jointly referred to as the *interference*). The magnitude distribution of the interference is formulated using the compound Gaussian model [1], which comprises a speckle component for modeling the small ripples on top of larger waves. These larger waves are then modeled by a slowly varying texture which can be assumed constant over a typical radar dwell period.

A common model for the probability density function (PDF) of the *sea clutter* magnitude is the K-distribution [1]

$$K(z; \nu_0, b) = \frac{4z^{\nu_0}}{\sqrt{b}^{\nu_0+1}\,\Gamma(\nu_0)} \mathcal{K}_{\nu_0-1}\left(\frac{2z}{\sqrt{b}}\right), \tag{11.1}$$

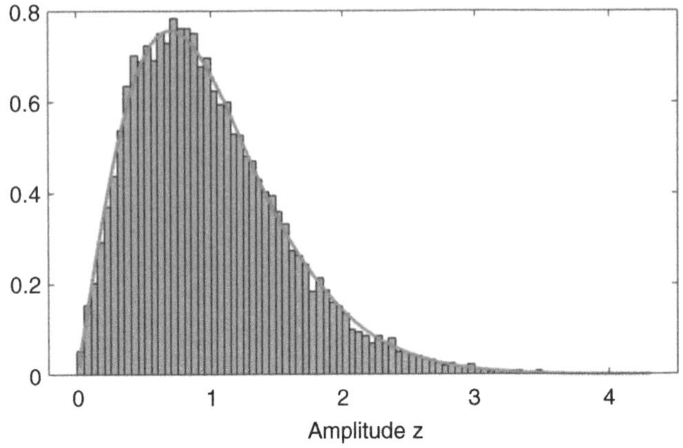

Figure 11.2 Histogram of interference magnitude samples and a K-distribution fit.

where $\Gamma(\cdot)$ denotes the gamma function, $\nu_0 > 0$ and $b > 0$ are the shape and scale parameters, respectively, and $\kappa_n(\cdot)$ is the modified Bessel function of the second kind. Its mean value is given by

$$\mu = \mathbb{E}[z] = \sqrt{b}\,\frac{\Gamma(1.5)\,\Gamma(\nu_0 + 0.5)}{\Gamma(\nu_0)}. \tag{11.2}$$

For $\nu_0 \to \infty$, the K-distribution approaches the Rayleigh distribution.

The PDF of the interference magnitude can also be represented by K-distribution, where the shape parameter ν_0 is replaced by an *effective* shape ν. By matching the moments of the PDF with and without thermal noise [1], one can obtain the following relationship: $\nu = \nu_0(1 + 1/C)^2$, where C is the clutter to noise ratio (CNR). For the remainder of the chapter, we will use ν as the shape parameter without explicitly defining the CNR.

Figure 11.2 shows the histogram of $N_r N_a = 50,000$ magnitude samples from a normalized range-azimuth image[1] created using the high fidelity simulator [17], in the absence of a target. The K-distribution fit, indicated with a solid line, is remarkably accurate. Estimation of the parameters of K-distribution will be discussed in Section 11.3.3. For the distribution fit in Figure 11.2, the calculated estimates were $\hat{\nu} = 4.37$ and $\hat{b} = 0.31$.

11.3 Bernoulli TBD Filter for Maritime Radar

The Bernoulli filter is the optimal recursive Bayesian estimator for an appearing / disappearing target [18]. It is based on the Bernoulli random finite set (RFS) model to represent the target state at time k. This state is fully defined by two quantities. The first is the posterior probability of target presence $q_{k|k} = P\{\epsilon_k = 1|\mathbf{z}_{1:k}\}$, where $\epsilon \in \{0, 1\}$ is a binary random variable which equals 1 if the target is present (otherwise it is 0) and $\mathbf{z}_{1:k} = \mathbf{z}_1, \mathbf{z}_2, \dots, \mathbf{z}_k$ is a sequence of measurements collected at discrete times $1, 2, \dots, k$. In the TBD context, a measurement $\mathbf{z}_k$ is a radar magnitude map, such as the range-azimuth image in Figure 11.1b. The second quantity is the posterior spatial probability distribution $s_{k|k}(\mathbf{x}) = p_k(\mathbf{x}|\mathbf{z}_{1:k})$, where $\mathbf{x} \in \mathcal{X}$ is the target state vector, and $\mathcal{X}$ is the state space.

1 By normalization, we mean that the magnitude values of the range-azimuth map are divided by their sample mean.

The transitions of the *target presence* variable ϵ_k are typically modeled by the first-order two-state Markov chain. Its transitional probability matrix (TPM) is given by:

$$\mathbf{\Pi} = \begin{bmatrix} 1 - p_\text{b} & p_\text{b} \\ 1 - p_\text{s} & p_\text{s} \end{bmatrix}, \tag{11.3}$$

where $p_\text{b} = P\{\epsilon_{k+1} = 1 | \epsilon_k = 0\}$ and $p_\text{s} = P\{\epsilon_{k+1} = 1 | \epsilon_k = 1\}$ are referred to as the probability of *target birth* and the probability of *target survival*, respectively.

If at time $k-1$, both $q_{k-1|k-1}$ and $s_{k-1|k-1}(\mathbf{x})$ are available, then the recursive computation of $q_{k|k}$ and $s_{k|k}(\mathbf{x})$ at time k is carried out in two steps: prediction and update.

11.3.1 Theoretical Formulation

The prediction equations of the Bernoulli filter are as follows [18]:

$$q_{k|k-1} = p_b \left(1 - q_{k-1|k-1}\right) + p_s \, q_{k-1|k-1}, \tag{11.4}$$

$$s_{k|k-1}(\mathbf{x}) = \frac{p_b \left(1 - q_{k-1|k-1}\right) \int \varphi_{k|k-1}(\mathbf{x}|\mathbf{x}') \, b_{k-1|k-1}(\mathbf{x}') d\mathbf{x}'}{q_{k|k-1}} + \frac{p_s \, q_{k-1|k-1} \int \varphi_{k|k-1}(\mathbf{x}|\mathbf{x}') \, s_{k-1|k-1}(\mathbf{x}') d\mathbf{x}'}{q_{k|k-1}}, \tag{11.5}$$

where two statistical models feature: $\varphi_{k|k-1}(\mathbf{x}|\mathbf{x}')$ is the state transition density (a target motion model), and $b_{k-1|k-1}(\mathbf{x})$ is the target birth density (a model of the state of the target appearing at $k-1$).

After receiving measurement $\mathbf{z}_k$, the two quantities are updated as follows [18]:

$$q_{k|k} = \frac{q_{k|k-1} \int \ell_k(\mathbf{z}_k|\mathbf{x}) \, s_{k|k-1}(\mathbf{x}) \, d\mathbf{x}}{1 - q_{k|k-1} + q_{k|k-1} \int \ell_k(\mathbf{z}_k|\mathbf{x}) \, s_{k|k-1}(\mathbf{x}) \, d\mathbf{x}}, \tag{11.6}$$

$$s_{k|k}(\mathbf{x}) = \frac{\ell_k(\mathbf{z}_k|\mathbf{x}) \, s_{k|k-1}(\mathbf{x})}{\int \ell_k(\mathbf{z}_k|\mathbf{x}) \, s_{k|k-1}(\mathbf{x}) \, d\mathbf{x}}, \tag{11.7}$$

where $\ell_k(\mathbf{z}_k|\mathbf{x}_k)$ is the likelihood ratio and integration is over the state space $\mathcal{X}$. Assuming the statistical independence of the cells of $\mathbf{z}_k$, the likelihood ratio can be expressed as:

$$\ell_k(\mathbf{z}_k|\mathbf{x}_k) = \prod_{m=1}^{N_\text{a}} \prod_{n=1}^{N_\text{r}} \frac{p_1^{(m,n)}\left(z_{m,n}|\mathbf{x}_k\right)}{p_0^{(m,n)}\left(z_{m,n}\right)}, \tag{11.8}$$

where $z_{m,n}$ is the measured magnitude in the (m, n)th cell of the radar image $\mathbf{z}_k$, while $p_1^{(m,n)}(\cdot|\mathbf{x}_k)$ and $p_0^{(m,n)}(\cdot)$ represent the PDFs of the measured magnitude in the (m, n)th cell of $\mathbf{z}_k$ for the case when a target in state $\mathbf{x}_k$ is present and the case of interference-only (target absent), respectively. According to Section 11.2.2, the interference-only PDF is the K-distribution with parameters ν and b, i.e. $p_0^{(m,n)}(z) = K(z; \nu, b)$. The target-plus-interference PDF $p_1^{(m,n)}(z|\mathbf{x}_k)$ will be discussed in the next section.

The likelihood ratio (11.8) plays the crucial role in the Bernoulli filter for TBD. Note that in practice, the parameters ν and b are unknown and need to be estimated from the data. The Bernoulli filter reports a track only if $q_{k|k} > \xi$, where ξ is the reporting threshold.

11.3.2 Mathematical Models

The observed region on the sea surface is arc-shaped, and we adopt the state vector in polar coordinates. The state vector at discrete-time k is specified as:

$$\mathbf{x}_k = [x_k \;\; \dot{x}_k \;\; y_k \;\; \dot{y}_k \;\; \chi_k]^{\mathsf{T}}, \tag{11.9}$$

where x_k, $\dot{x}_k$, y_k and $\dot{y}_k$ are target azimuth, azimuth-rate, range and range-rate, respectively. The unit of both x_k and y_k is the *cell*. Given the cell size in azimuth and range, one could easily obtain the range, range-rate, azimuth and azimuth-rate values in their corresponding physical units. The last component of the state vector, χ_k, is a target feature referred to as its mean amplitude: a measure of how much on average it reflects the radar signals.

Because of the large distance to the observed region on the sea surface, the nearly constant velocity target motion model is approximated with a nearly constant range-rate and a nearly constant azimuth-rate model. Hence the transition density, which features in (11.5), is approximated with

$$\varphi_{k|k-1}(\mathbf{x}_k|\mathbf{x}_{k-1}) \approx \mathcal{N}(\mathbf{x}_k; \mathbf{F}\mathbf{x}_{k-1}, \mathbf{Q}), \tag{11.10}$$

where $\mathcal{N}(\mathbf{x}; \boldsymbol{\mu}, \boldsymbol{\Sigma})$ denotes the Gaussian PDF, evaluated at $\mathbf{x}$, with mean $\boldsymbol{\mu}$ and covariance matrix $\boldsymbol{\Sigma}$. Furthermore, $\mathbf{F}$ and $\mathbf{Q}$ represent the transition matrix and the process noise covariance matrix, respectively, given by [19]:

$$\mathbf{F} = \begin{bmatrix} \boldsymbol{\Phi} & 0 & 0 \\ 0 & \boldsymbol{\Phi} & 0 \\ 0 & 0 & 1 \end{bmatrix}, \qquad \boldsymbol{\Phi} = \begin{bmatrix} 1 & T \\ 0 & 1 \end{bmatrix}, \tag{11.11}$$

$$\mathbf{Q} = \begin{bmatrix} \boldsymbol{\Psi}_x & 0 & 0 \\ 0 & \boldsymbol{\Psi}_y & 0 \\ 0 & 0 & q_\chi T \end{bmatrix}, \qquad \boldsymbol{\Psi}_u = q_{\mathrm{u}} \begin{bmatrix} T^3/3 & T^2/2 \\ T^2/2 & T \end{bmatrix}, \tag{11.12}$$

where T is the sampling interval, q_{u} and q_χ denote the process noise intensity in the spatial (range for $u = r$ or azimuth for $u = a$) domain and the target amplitude domain.

The birth density $b_{k-1|k-1}(\mathbf{x})$, which features in (11.5), is typically measurement driven, that is constructed from $\mathbf{z}_{k-1}$. Its practical implementation is discussed in Section 11.3.3.

Next we introduce the model of likelihood function $p_1^{(m,n)}(z|\mathbf{x}_k)$, which features in (11.8). Let the contribution of a target in state $\mathbf{x}$ to the measured magnitude z of the (m, n)th cell of $\mathbf{z}_k$ be denoted $h^{(m,n)}(\mathbf{x})$. Typically, this contribution is modeled by a Gaussian point spread function (PSF). For a 2D range-azimuth map, it is given by

$$h^{(m,n)}(\mathbf{x}) = \frac{\chi}{2\pi\sigma_x\sigma_y} \exp\left[-\frac{(m-x)^2}{2\sigma_x^2} - \frac{(n-y)^2}{2\sigma_y^2} \right], \tag{11.13}$$

where $\sigma_x \ll N_{\mathrm{a}}$ and $\sigma_y \ll N_{\mathrm{r}}$ determine the spread in azimuth and range, respectively. For the case of a Swerling 1 target in K-distributed interference, $p_1^{(m,n)}(z|\mathbf{x}_k) = L\left(z; h^{(m,n)}(\mathbf{x}), \nu, b\right)$ is referred to as the L-distribution, derived in [20, Appendix]. Unfortunately, this distribution can be expressed only in an integral form as [14, 20]:

$$L(z; h, \nu, b) = \frac{2z}{b^\nu \Gamma(\nu)} \int_0^\infty \frac{\eta^{\nu-1}}{\eta + h} \exp\left[-\frac{z^2}{\eta + h} - \frac{\eta}{b} \right] d\eta. \tag{11.14}$$

By setting $h = 0$ in (11.14), the L-distribution reduces to the K-distribution.

11.3.3 Practical Considerations

11.3.3.1 Particle Filter Implementation

Note that there is no closed-form analytic solution for the prediction and update of the spatial density $s_{k|k}(\mathbf{x})$, see (11.5) and (11.7), and for the update of the probability of target presence, see (11.6). All three equations involve multidimensional integration and need to be computed numerically. For this purpose, we adopt the sequential Monte Carlo method [10] and implement the Bernoulli TBD filter as a particle filter. The spatial posterior density of a target $s_{k-1|k-1}(\cdot)$ is thus approximated by a weighted random sample from this density $\{(w^{(i)}_{k-1|k-1}, \mathbf{x}^{(i)}_{k-1|k-1})\}_{1\leq i\leq N_{k-1}}$, where i is the sample (particle) index, and N_{k-1} is the number of particles at $k-1$. By adopting the sequential importance resampling paradigm [10], the predicted density $s_{k|k-1}(\cdot)$ is approximated by $\{(w^{(i)}_{k|k-1}, \mathbf{x}^{(i)}_{k|k-1})\}_{1\leq i\leq N_{k-1}}$, where $w^{(i)}_{k|k-1} = w^{(i)}_{k-1|k-1}$ and $\mathbf{x}^{(i)}_{k|k-1} \sim \varphi_{k|k-1}(\cdot|\mathbf{x}^{(i)}_{k-1|k-1})$, for $i = 1, \dots, N_{k-1}$. The predicted probability of target presence $q_{k|k-1}$ is trivially computed from (11.4) and the weighs are updated according to (11.7) using the likelihood ratio: $\tilde{w}^{(i)}_{k|k} = \ell_k\left(\mathbf{z}_k|\mathbf{x}^{(i)}_{k|k-1}\right)w^{(i)}_{k|k-1}$, $i = 1, \dots, N_{k-1}$. The integral $I = \int \ell_k(\mathbf{z}_k|\mathbf{x})\, s_{k|k-1}(\mathbf{x})\, d\mathbf{x}$, which features in (11.6) and (11.7) is approximated by the sum $\sum_{i=1}^{N_{k-1}} \ell_k(\mathbf{z}_k|\mathbf{x}^{(i)}_{k|k-1})\, w^{(i)}_{k|k-1}$. Full details of TBD particle filter implementation can be found in [18, Sec. VIII.A].

11.3.3.2 Numerical Approximation of the Likelihood Ratio

Computation of the integral which features in (11.14) is carried out using the trapezoidal numerical integration. This has to be done for every particle in order to compute its updated importance weight.

11.3.3.3 Estimation of ν and b Parameters

Parameters of the K-distributed interference are unknown in practice and depend on the radar data collection geometry and environmental conditions. The fundamental limitation is that the Cramér–Rao lower bound for estimation of ν is proportional to ν^2/S [21], where S is the sample size. This implies that it is more difficult to estimate ν when it has a large value (i.e. when the K-distribution approaches a Rayleigh) and for small sample sizes, S. We adopted the moment based method [21] to estimate ν and b. Let $z_1, \cdots, z_S$ be a sample from the K-distribution. Then

$$\hat{\nu} = \frac{1}{4}\left[\ln\left(\frac{\pi\,\hat{m}_2}{4\,\hat{m}_1^2}\right)\right]^{-1}, \text{ where } \hat{m}_1 = \frac{1}{S}\sum_{s=1}^{S} z_s, \quad \hat{m}_2 = \frac{1}{S}\sum_{s=1}^{S} z_s^2. \tag{11.15}$$

The scale estimate determined by $\hat{b} = \hat{m}_2/\hat{\nu}$.

11.3.3.4 Measurement-driven Birth Density

The N_b particles from the birth density $b_{k|k}(\mathbf{x})$ are formed using the radar map $\mathbf{z}_k$. First, the cells of $\mathbf{z}_k$ whose magnitudes are above a certain threshold $\hbar > 0$ are found. Birth particles in the (x, y) subspace are drawn at random from these cells. They indicate the region of the range-azimuth space where a target can appear at time k. We can use a constant false alarm rate (CFAR) algorithm to estimate the mean interference level $\boldsymbol{\mu}$, and then set the threshold $\hbar = \zeta\boldsymbol{\mu}$, with ζ chosen so that a target above a certain SIR value is detected and tracked. Because $\mathbf{z}_k$ does not contain information about target velocity, the birth-particle components $(\dot{x}, \dot{y})$ are a random draw from $[-\upsilon_{\max}, \upsilon_{\max}]$, where $\upsilon_{\max}$ is the maximum expected target speed. Finally, the target intensity of a birth particle χ

is drawn from a gamma distribution with parameters (α_b, θ_b) chosen so that the target SIR values of interest are within the support of this distribution. The weights of birth particles are uniform, and equal to $1/N_b$.

11.3.4 Numerical Results

We consider a generic scanning radar which reports range-azimuth maps of size 64×64, with the sampling interval $T = 1$s. The clutter is modeled with unity mean, while two values of the shape parameter were considered: $\nu = 2$ (spiky clutter), and $\nu = 8$ (moderately spiky clutter). A sequence of 40 range-azimuth maps is generated for every run of the algorithm, and the target is present throughout the observation period. Its initial position and velocity are given by (18.2, 7.2) and (0.45, 0.25), respectively. The SIR is defined (in dB) as SIR=$10 \log_{10} \chi/\mu$, where μ, given by (11.2), is set to 1, while χ is varied so that we consider targets with SIR values of 3, 5 and 8 dB. The point-spread function parameters were set to $\sigma_x = \sigma_y = 1$.

The following parameters were used in the TBD algorithm (given in arbitrary units): process noise $q_u = 0.002$ and $q_\chi = 0$; $p_b = 0.01$, $p_s = 0.98$; the number of particles that is used for the representation of spatial density $s_{k|k}(\mathbf{x})$ was fixed to $N_k = 5000$; number of birth particles $N_b = 5000$, maximum target speed $v_{max} = 0.6$, birth threshold parameter $\zeta = 1.9$ and the reporting threshold $\xi = 0.7$.

The joint detection and tracking performance is measured using the mean optimal sub-pattern assignment (OSPA) metric [22]. This metric penalizes both the error in the cardinality estimate (number of targets, i.e. detection) and the error in the target position, but does not consider the target velocity and the target amplitude estimates in the performance assessment. The OSPA metric parameters are as follows: order $p = 2$ and cut-off $c = 30$. Figure 11.3 displays the mean OSPA error curves, obtained by averaging over 100 Monte Carlo runs. In addition, we have also measured the probability of false track initiation, in the case of target being absent. This probability, at each time step, equals 0.009 and 0.001 for $\nu = 2$ and $\nu = 8$, respectively.

The OSPA errors are initially high, because of a delay in establishing the track. Overall, as expected, the OSPA errors reduce quicker and achieve a lower steady-state level when the target SIR is higher. Comparing the OSPA results in Figure 11.3 for $\nu = 2$ and $\nu = 8$, we observe that the difference is small. However, for $\nu = 2$, the probability of false track initiation is nine times higher, reflecting the difficulties encountered when clutter is spiky.

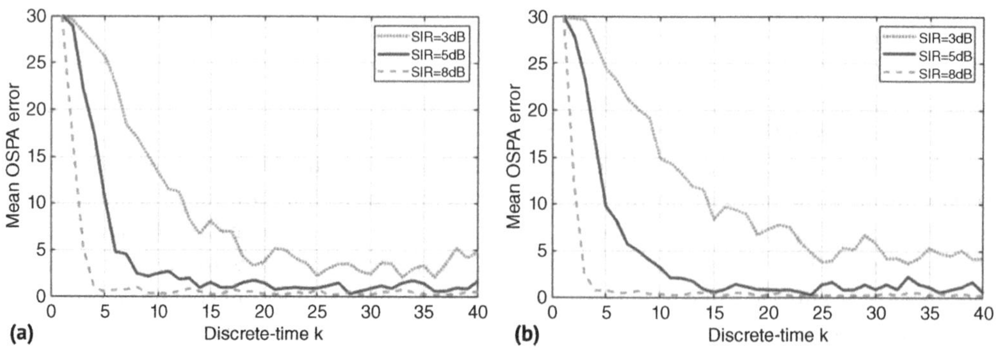

Figure 11.3 Mean OSPA error curves: (a) $\nu = 2$; (b) $\nu = 8$.

11.3.5 Comparison with Point-measurement Tracking

Standard tracking algorithms process the point radar measurements (PRMs), obtained from a detector applied to the measured range-azimuth maps. Using PRMs for tracking offers a computational advantage, but we also expect some performance degradation compared to TBD, because by extracting the PRMs we potentially lose useful information. For the sake of comparison against the Bernoulli TBD filter, we will consider two tracking algorithms using PRMs. The first is the Bernoulli filter which is using the positional PRMs *together with* their magnitude values. This is a hybrid Bernoulli filter, in the sense that, similarly to the TBD, it takes advantage of the magnitude information, while operating in the framework of conventional Bayesian tracking using point measurements. The hybrid Bernoulli filter is formulated via a Rao-Blackwell decomposition: a particle filter estimates the target amplitude, while the target position and velocity are estimated using the Bernoulli Gaussian sum filter (GSF), conditioned on the amplitude. This algorithm is described in [23]. The second algorithm for positional (range-azimuth) PRMs (without using amplitude information) is the straightforward Bernoulli GSF, described in [18, Sec.VIII.B].

We consider the same simulation setup as described in Section 11.3.4. The CFAR threshold for obtaining the PRMs was found experimentally as a tradeoff between the computational load and tracking accuracy, and results in the probability of false alarm of $P_{\mathrm{FA}}^{\tau} \approx 0.07$. The OSPA error curves of the three contesting algorithms are shown in Figure 11.4 for SIR = 3 dB, 5 dB and 8 dB, respectively. The colors used in Figure 11.4 are as follows: the Bernoulli TBD—solid curve, the hybrid Bernoulli filter—dashed curve, and the Bernoulli GSF—dotted curve. Observe that the OSPA error performance of the hybrid Bernoulli filter is always between the two extremes: it is worse than that of the Bernoulli TBD, but better than the Bernoulli GSF. It can also be noted that in terms of its error performance, the hybrid filter is approaching that of the Bernoulli TBD at higher SIR values and higher values of the shape parameter ν.

Finally, a note on the average execution time of the three Bernoulli filters: the hybrid Bernoulli filter is 4 times slower than the Bernoulli GSF, and 11 times faster than the Bernoulli TBD filter.

11.4 A Multitarget Bernoulli TBD Tracker

For well separated targets (i.e. with non-overlapping regions of target influence), then the multitarget Bernoulli TBD filter is relatively straightforward to design: it can be formulated as a collection of independent Bernoulli TBD filters with unique labels. However, when targets are closely spaced (interacting), the problem is more difficult as the measured intensity maps represent a superposition of individual target contributions, immersed in the background interference. The first tractable implementation for RFS-based multi-object filtering using superpositional measurements was proposed in [24] assuming Gaussian sensor noise. Pappi and Kim [25] have also developed an integrated track management by using labeled RFS for the super-positional measurement model, where an efficient importance density is designed for the multitarget particle filter.

This section presents a simple, yet efficient, approximation for handling interacting targets in the context of the Bayesian TBD. The main idea is to update each track, by treating all nearby (interacting) targets as contributors to the background interference. A similar idea was used in [26], in the context of point measurements.

11.4.1 Modeling: Multitarget State

Let $\rho_k \in \{0, 1, 2 \ldots, \}$ be the number of targets that are present in the surveillance volume at time k. Both the number of targets ρ_k and their individual states $\mathbf{x}_k^{(r)} \in \mathcal{X}$, $r = 1, \ldots, \rho_k$, are considered

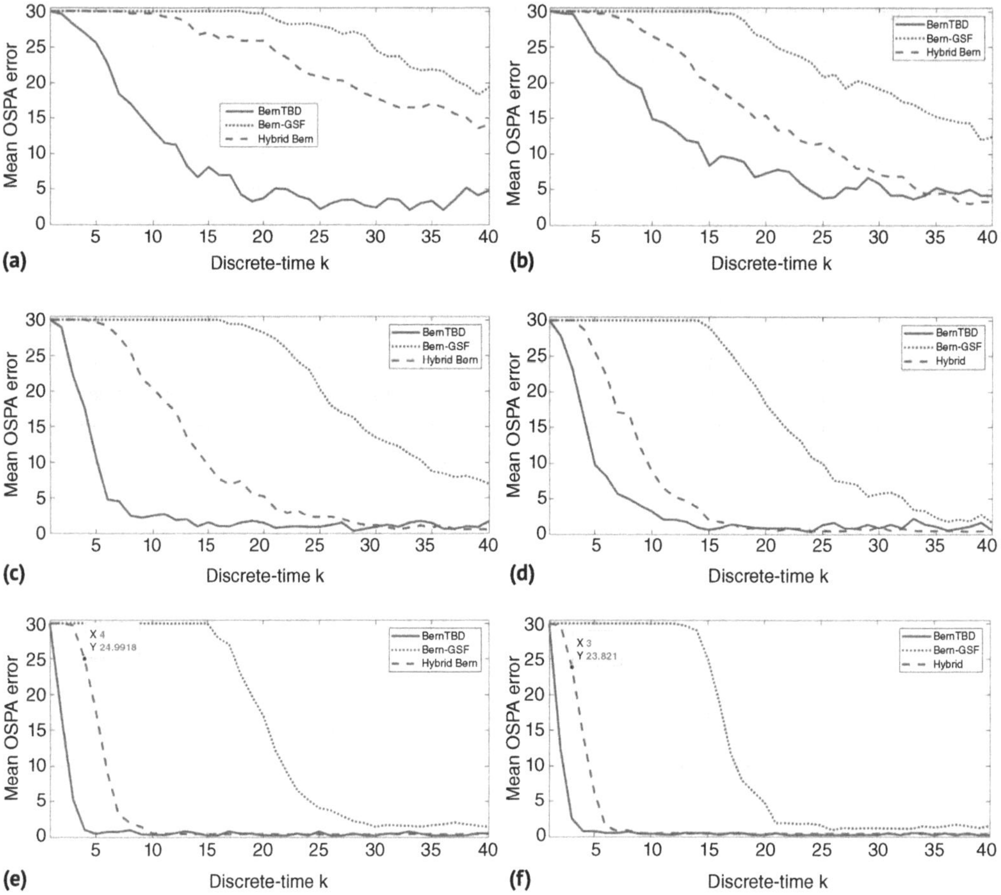

Figure 11.4 Mean OSPA error of three tracking algorithms (target-plus-interference case); (a) SIR = 3 dB, $\nu = 2$; (b) SIR = 3 dB, $\nu = 8$; (c) SIR = 5 dB, $\nu = 2$; (d) SIR = 5 dB, $\nu = 8$; (e) SIR = 8 dB, $\nu = 2$; (f) SIR = 8 dB, $\nu = 8$.

as random and time-varying. The multitarget state at discrete-time k can conveniently be modelled as a RFS on $\mathcal{X}$ [27]. We continue using the Bernoulli RFS model to represent a single target. If the individual targets are modeled as Bernoulli RFSs, denoted as $\mathbf{X}_k^{(r)}$, where $i = 1, \dots, \rho_k$, then the multi-Bernoulli RFS is defined as $\mathbf{X}_k = \bigcup_{r=1}^{\rho_k} \mathbf{X}_k^{(r)}$. Each independent Bernoulli RFS is characterized by the probability of target presence $q_k^{(r)}$ and the spatial PDF $s_k^{(r)}(\cdot)$ over $\mathcal{X}$. Hence the multi-Bernoulli RFS is completely specified by the set of pairs $\{(q_k^{(r)}, s_k^{(r)})\}_{1 \leq r \leq \rho_k}$. The multitarget state at time k will be represented by the multi-Bernoulli RFS, $\mathbf{X}_k$.

We adopt the same mathematical models of target state evolution and radar measurements, specified in Section 11.3.2. According to (11.9), the rth target state is then

$$\mathbf{x}_k^{(r)} = \left[x_k^{(r)} \ \dot{x}_k^{(r)} \ y_k^{(r)} \ \dot{y}_k^{(r)} \ \chi_k^{(r)} \right]^\intercal, \quad r = 1, \dots, \rho_k, \tag{11.16}$$

where $(x_k^{(r)}, y_k^{(r)})$ are the Cartesian coordinates of the rth target position, $(\dot{x}_k^{(r)}, \dot{y}_k^{(r)})$ is its velocity vector and $\chi_k^{(r)}$ is the mean amplitude of its radar return.

The contribution of a target in state $\mathbf{x} \in \mathbf{X}_k$ to the range-azimuth radar map $\mathbf{z}_k$ is given by the PSF, adopted as in (11.13). The contribution of a multitarget state $\mathbf{X}_k$ to the measured magnitude in the (m, n)th cell at time k is represented by a superposition of individual target contributions [24, 25, 28]:

$$h^{(m,n)}(\mathbf{X}_k) = \sum_{\mathbf{x} \in \mathbf{X}_k} h^{(m,n)}(\mathbf{x}). \tag{11.17}$$

The PDF of measured magnitude z in cell (m, n) is given by (11.14), with h given by (11.17):

$$p_1^{(m,n)}(z|\mathbf{X}_k) = L\big(z; h^{(m,n)}(\mathbf{X}_k), \nu, b\big). \tag{11.18}$$

We will assume that the targets move independently from each other. The main difficulty in dealing with multitarget TBD tracking is in the update stage, due to the superposition (11.17).

11.4.2 A Formulation of the Multitarget Bernoulli TBD Tracker

Suppose at time $k-1$ there are ρ_{k-1} tracks (i.e. Bernoulli components in $\mathbf{X}_k$). Track r at time k, where $r = 1, \dots, \rho_{k-1}$, is specified by $\mathcal{T}_{k-1}^{(r)} = \left(\mathcal{L}^{(r)}, q_{k-1|k-1}^{(r)}, s_{k-1|k-1}^{(r)}\right)$, where $\mathcal{L}^{(r)}$ is a unique track label (invariant in time), $q_{k-1|k-1}^{(r)}$ is the posterior probability of the rth target being present and $s_{k-1|k-1}^{(r)}(\cdot)$ is its posterior PDF over the state space $\mathcal{X}$. The set of all tracks at $k-1$ is then specified as

$$\mathcal{T}_{k-1} = \left\{\left(\mathcal{L}^{(r)}, q_{k-1|k-1}^{(r)}, s_{k-1|k-1}^{(r)}\right)\right\}_{1 \le r \le \rho_{k-1}}. \tag{11.19}$$

At the next time step k, when a new measured range-azimuth map $\mathbf{z}_k$ becomes available, the proposed multitarget Bernoulli TBD (MTB-TBD) tracker is executed using the processing steps listed in Algorithm 11.1.

Algorithm 11.1: Processing steps of the MTB-TBD tracker at time k

1: **Input:** $\mathcal{T}_{k-1}, \mathbf{z}_k$
2: $\mathcal{T}_{k|k-1}$ = Predict_Tracks ($\mathcal{T}_{k-1}$)
3: $[\mathcal{T}_{k|k-1}, \mathcal{C}_k]$ = Clustering ($\mathcal{T}_{k|k-1}$)
4: $[\mathcal{T}_{k|k}, \mathbf{z}_k^*]$ = Update_Tracks ($\mathcal{T}_{k|k-1}, \mathcal{C}_k, \mathbf{z}_k$)
5: $\widetilde{\mathcal{T}}_k$ = Track_Management ($\mathcal{T}_{k|k}$)
6: Track_Reporting ($\widetilde{\mathcal{T}}_k$)
7: $\mathcal{V}_k$ = Init_NewTracks ($\mathbf{z}_k^*$)
8: $\mathcal{T}_k$ = Track_Union ($\widetilde{\mathcal{T}}_k, \mathcal{V}_k$)
9: **Output:** $\mathcal{T}_k$

Initially, at $k = 1$, $\mathcal{T}_{k-1} = \emptyset$.

In line 2 of Algorithm 11.1, each track $r = 1, \dots, \rho_{k-1}$ is predicted as follows [27, p. 662]:

$$q_{k|k-1}^{(r)} = p_{\text{S}} \cdot q_{k-1|k-1}^{(r)}, \tag{11.20}$$

$$s_{k|k-1}^{(r)}(\mathbf{x}) = \int \varphi_{k|k-1}(\mathbf{x}|\mathbf{x}')\, s_{k-1|k-1}^{(r)}(\mathbf{x}')\, d\mathbf{x}', \tag{11.21}$$

while the track label does not change. Note that this is a simplified version of (11.4) and (11.5), with p_{b} set to zero (target birth is treated differently, see line 7).

If the rth track is well separated from all other tracks in $\mathcal{T}_{k|k-1}$, the computation of the likelihood function for this track can be approximated by (11.8) and we can directly apply the single-target formulae (11.6) and (11.7). The notion of "well-separated tracks" follows from the clustering of tracks. We adopt the following heuristic for track clustering: targets in state $\mathbf{x}_k^{(r)}$ and $\mathbf{x}_k^{(\ell)}$ at time k are well separated if

$$|x_k^{(r)} - x_k^{(\ell)}| > 6\sigma_x \text{ or } |y_k^{(r)} - y_k^{(\ell)}| > 6\sigma_y, \tag{11.22}$$

where σ_x and σ_y are the PSF parameters, see (11.13). If this is true for every pair of targets (r, ℓ), such that $r \neq \ell$ and $r, \ell \in \{1, \dots, \rho_k\}$, then all targets are well separated. Condition (11.22) practically means that for every cell of the measured range-azimuth map $\mathbf{z}_k$, at most one target contributes to the cell magnitude.

In summary, for a track r which does not interact with any other tracks in $\mathcal{T}_{k|k-1}$, the likelihood ratio to be used in the track update is approximated with

$$\ell_r(\mathbf{z}_k|\mathbf{x}) = \prod_{m\in\mathcal{S}_r(x)} \prod_{n\in\mathcal{S}_r(y)} \frac{L\left(z^{(m,n)}; h^{(m,n)}(\mathbf{x}), \nu, b\right)}{K(z^{(m,n)}; \nu, b)}, \tag{11.23}$$

where $\mathcal{S}_r(x)$ and $\mathcal{S}_r(y)$ are the sets of cell indices, covering the support of the PDF $s_{k|k-1}^{(r)}$, in the x and y directions, with the spread parameters σ_x and σ_y. The track update equations for this track (used in line 4 of Algorithm 11.1) are as follows [28]:

$$q_{k|k}^{(r)} = \frac{q_{k|k-1}^{(r)} \int \ell_r(\mathbf{z}_k|\mathbf{x})\, s_{k|k-1}^{(r)}(\mathbf{x})\, d\mathbf{x}}{1 - q_{k|k-1}^{(r)} + q_{k|k-1}^{(r)} \int \ell_r(\mathbf{z}_k|\mathbf{x})\, s_{k|k-1}^{(r)}(\mathbf{x})\, d\mathbf{x}}, \tag{11.24}$$

$$s_{k|k}^{(r)}(\mathbf{x}) = \frac{\ell_r(\mathbf{z}_k|\mathbf{x})\, s_{k|k-1}^{(r)}(\mathbf{x})}{\int \ell_r(\mathbf{z}_k|\mathbf{x})\, s_{k|k-1}^{(r)}(\mathbf{x})\, d\mathbf{x}}. \tag{11.25}$$

If there are interacting (closely spaced) targets among $\mathcal{T}_{k|k-1}$, forming a cluster of tracks, then the targets in the cluster need to be updated jointly. A track $\mathcal{T}_k^{(i)}$ belongs to cluster $\mathcal{C}_k^*$ at time k if there is at least one other track $\mathcal{T}_k^{(j)}$ in $\mathcal{C}_k^*$ such that the pair (i, j) does not satisfy condition (11.22). Note that a cluster at time k can contain 1, 2, and up to $|\mathcal{T}_k|$ tracks. When all target tracks are well separated, the number of clusters is the same as the number of tracks (i.e. each cluster contains only one track).

Suppose there are n_{cl} targets in a cluster of tracks $\mathcal{C}_k = \{\mathcal{T}_k^{(i)}\}_{1\leq i\leq n_{cl}}$ and they need to be updated using the range-azimuth map $\mathbf{z}_k$. Consider a cell of map $\mathbf{z}_k$ which contains a non-negligible magnitude contribution from the targets of this cluster. The individual contribution of targets in states $\mathbf{x}_{k,1}, \mathbf{x}_{k,2}, \dots \mathbf{x}_{k,n_{cl}}$ are denoted $h(\mathbf{x}_{k,1}), h(\mathbf{x}_{k,2}), \dots h(\mathbf{x}_{k,n_{cl}})$, respectively, in accordance with (11.13). The multitarget likelihood ratio of a magnitude measurement for the cluster in this cell, can be expressed as:

$$\ell(z|\mathbf{X}_{k,cl}) = \frac{L\left(z; \sum_{r=1}^{n_{cl}} h(\mathbf{x}_{k,r}), \nu, b\right)}{K(z; \nu, b)}. \tag{11.26}$$

Consider the case with $n_{cl} = 2$ targets in the cluster, i.e. $\mathbf{X}_{k,cl} = \{\mathbf{x}_{k,1}, \mathbf{x}_{k,2}\}$. In updating the states of these two targets, ideally, we would like to know $\ell(z|\mathbf{x}_{k,1})$ and $\ell(z|\mathbf{x}_{k,2})$, because then we could apply the update equations given by (11.6) and (11.7). The problem, however, is that we do not measure magnitudes $z|\mathbf{x}_1$ and $z|\mathbf{x}_2$, but only their combined effect $z|\mathbf{x}_1, \mathbf{x}_2$.

In order to update a track in a cluster, we need a single-object likelihood ratio as expressed by (11.23). The multi-object likelihood ratio of (11.26) for $n_{cl} = 2$ is

$$\ell(z|\mathbf{X}_{k,cl=2}) = \frac{L(z; h(\mathbf{x}_{k,1}) + h(\mathbf{x}_{k,2}), \nu, b)}{K(z; \nu, b)}, \tag{11.27}$$

while the likelihood ratio for a single target $\mathbf{x}_{k,1}$ is

$$\ell(z|\mathbf{x}_{k,1}) = \frac{L(z;\, h(\mathbf{x}_{k,1}),\, \nu,\, b)}{K(z;\, \nu,\, b)}. \tag{11.28}$$

We need $\ell(z|\mathbf{x}_{k,1})$ to update the Bernoulli filter for track 1, but only $\ell(z|\mathbf{X}_{k,cl=2})$ given by (11.27) for $n_{cl} = 2$ targets, is available.

The proposed solution is to treat the other targets in the cluster of tracks as the contributors to the increased magnitude level of the interference in the cell. Thus we use the L-distribution in the denominator of the likelihood function. For example, in the case of two targets, this means (we drop index k on target state vectors for clarity):

$$\ell(z|\mathbf{x}_1) \approx \frac{L(z;\, h(\mathbf{x}_1),\, \nu,\, b)}{L(z;\, h(\mathbf{x}_2),\, \nu,\, b)}, \tag{11.29}$$

and for a cluster with $n_{cl} > 1$ tracks, the likelihood ratio for track r can be generalized for a single cell of $\mathbf{z}_k$ as

$$\ell(z|\mathbf{x}_r) \approx \frac{L(z;\, h(\mathbf{x}_r),\, \nu,\, b)}{L\left(z;\, \sum_{\ell=1,\ell\neq r}^{n_{cl}} h(\mathbf{x}_\ell),\, \nu,\, b\right)}. \tag{11.30}$$

The likelihood function of the entire map $\mathbf{z}_k$ for rth track is then

$$\ell_r(\mathbf{z}_k|\mathbf{x}_r) = \prod_{m\in\mathcal{S}_r(x)} \prod_{n\in\mathcal{S}_r(y)} \frac{L\left(z^{(m,n)};\, h^{(m,n)}(\mathbf{x}_r),\, \nu,\, b\right)}{L\left(z^{(m,n)};\, \sum_{\ell=1,\ell\neq r}^{n_{cl}} h^{(m,n)}(\mathbf{x}_\ell),\, \nu,\, b\right)}, \tag{11.31}$$

This likelihood is to be used in (11.24) and (11.25). The proposed method is general, in the sense that it applies to any $p_1^{(m,n)}(z|\mathbf{x})$ and $p_0^{(m,n)}(z)$. Furthermore, its computational complexity grows linearly with the number of tracks.

11.4.3 Practical Considerations

We use the Monte Carlo method to represent the spatial density of every track in $\mathcal{T}_k$. Thus, the spatial density of track $r = 1, \ldots, \rho_k$, that is $s_{k|k}^{(r)}(\mathbf{x})$, is approximated with the set of particles $\{w_{k|k}^{(r,i)}, \mathbf{x}_{k|k}^{(r,i)}\}_{1\leq i\leq N_k^r}$. The propagation of weighted particle sets follows the same steps as described for the single target case in Section 11.3.3, except that the prediction and update equations are now given in Section 11.4.2. The computation of (11.24) is then approximated with

$$q_{k|k}^{(r)} \approx \frac{q_{k|k-1}^{(r)} \sum_{i=1}^{N_{k-1}^r} \ell_r(\mathbf{z}_k|\mathbf{x}_{k|k-1}^{(r,i)})\, w_{k|k-1}^{(r,i)}}{1 - q_{k|k-1}^{(r)} + q_{k|k-1}^{(r)} \sum_{i=1}^{N_{k-1}^r} \ell_r(\mathbf{z}_k|\mathbf{x}_{k|k-1}^{(r,i)})\, w_{k|k-1}^{(r,i)}}. \tag{11.32}$$

Clustering of tracks at each time k is carried out using the density-based spatial clustering algorithm referred to as DBSCAN [29], that is widely used in machine learning applications. Its main advantage is robustness against noise and its flexibility with an unknown number of clusters.

Observe from Algorithm 11.1, that the output of the track update step (line 4) is not only the set of updated tracks $\mathcal{T}_{k|k}$, but also a modified range-azimuth map $\mathbf{z}_k^*$. This map is subsequently used to initialize new tracks in line 7 of Algorithm 11.1. The map $\mathbf{z}_k^*$ is initially set to $\mathbf{z}_k$. Then, during the track update step, those cells of $\mathbf{z}_k$ that have been used in the computation of the likelihoods $\ell_r\left(\mathbf{z}_k|\mathbf{x}_{k|k-1}^{(r,i)}\right)$, for $r = 1, \ldots, |\mathcal{T}_{k|k-1}|$ and $i = 1, \ldots, N_{k-1}^r$, are set to zero in $\mathbf{z}_k^*$ (and therefore not used in the initiation of new tracks). Clearly, if $\mathcal{T}_{k-1} = \emptyset$, $\mathbf{z}_k^*$ remains equal to $\mathbf{z}_k$.

The track management step in line 5 of Algorithm 11.1 is explained next. Each new track (initialized in line 7 of Algorithm 11.1) is given a new label and the status of a *tentative* track. If, after the track update step (line 4), a track is characterized with $q_{k|k}^{(r)} \geq \xi_c$, then its status is changed to *confirmed*. On the other hand, the status of the rth track with $q_{k|k}^{(r)} < \xi_d$, is changed to *deleted*, where ξ_c and ξ_d are track confirmation and termination thresholds, respectively. The set of tracks $\widetilde{\mathcal{T}}_k$, obtained after the track management step, contains only tentative and confirmed tracks. Among them, only the confirmed tracks are reported (see line 6 of Algorithm 11.1) and used in the performance evaluation.

New tracks are initialized in line 7 of Algorithm 11.1 using a modified map $\mathbf{z}_k^*$ as follows. First, the cells of $\mathbf{z}_k^*$ with magnitude higher than a certain threshold $\hbar_k > 0$ are found. They indicate the regions of the range-azimuth space where new targets can potentially appear. A typical map $\mathbf{z}_k^*$ and the corresponding binary map $\mathbf{b}_k$, obtained after thresholding $\mathbf{z}_k^*$, is shown in Figure 11.5. The applied threshold is $\hbar = \zeta\hat{\mu}$, where $\hat{\mu}$ is the sample mean of $\mathbf{z}_k^*$ and $\zeta = 1.9$ is chosen so that targets with SIR greater than 2.8 dB can be detected and tracked.

For each connected component of the binary map $\mathbf{b}_k$, we initialize a new track. For example, the binary map in Figure 11.5b initializes 236 tentative tracks. In the (x, y) subspace of the state space, the particles are drawn with equal probability from the connected component of the binary map $\mathbf{b}_k$. However, since $\mathbf{z}_k$ does not contain information about target velocity, a uniform distribution is adopted for the initial $\dot{x}$ and $\dot{y}$. Its support is from $-v_{\max}$ to $+v_{\max}$, where $v_{\max}$ is the maximum expected target speed. In the subspace of the mean target amplitude χ, the birth density is a gamma distribution with parameters (α_b, θ_b), chosen so that it covers the SIR values of interest. The weights of new target particles are uniform and the probability of a new target presence is set to p_b, see (11.3).

Finally, according to lines 8 and 9 of Algorithm 11.1, the output of the MTB-TBD tracker is a union of surviving and initialized tracks with unique labels.

11.4.4 Numerical Results

Three testing scenarios in the context of maritime radar are considered: (1) a single target scenario; (2) three well-separated targets; (3) two crossing targets (thus interacting when they are crossing). The purpose of the first scenario is to compare the OSPA error performance of the MTB-TBD tracker against the optimal single-target Bernoulli TBD filter, introduced in Section 11.3. The second scenario shows the performance of the MTB-TBD when targets are not interacting. The third scenario demonstrates the performance for interacting targets.

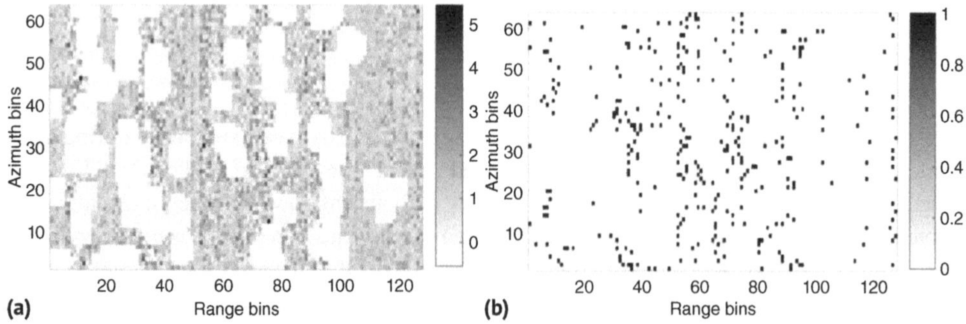

Figure 11.5 New track initiation: (a) map $\mathbf{z}_k^*$; (b) binary map $\mathbf{b}_k$ obtained after thresholding $\mathbf{z}_k^*$: *Source*: [13] / with permission of IEEE.

11.4.4.1 Single-target Scenario

The target initial state vector (11.16) is $\mathbf{x}_0 = [14, \ \ 0.45, \ \ 7.2, \ \ 0.25, \ \ \chi]^\mathsf{T}$, where the mean amplitude χ is varied as a function of the specified SIR and the shape parameter is set to $\nu = 2$. The mean amplitude χ is constant during the length of the scenario and hence the SIR is also constant. Figure 11.6 compares the mean OSPA errors of the Bernoulli TBD algorithm (Section 11.3) and the MTB-TBD tracker, discussed in this section, for (a) SIR = 5 dB and (b) SIR = 8 dB. As expected, the (single-target optimal) Bernoulli TBD performs better. However, the mean OSPA error of the MTB-TBD tracker is only slightly higher. This degradation in the error performance of the MTB-TBD is due to the occasional false-track declarations, which is to be expected from a multitarget tracking algorithm.

11.4.4.2 Three Well-separated Targets

The testing scenario for well separated (non-interacting) targets is shown Figure 11.7 and includes three targets moving with nearly constant velocity. Their mean amplitudes χ_k are equal and constant over time, determined from the specified SIR. Figure 11.7 also displays both the ground truth

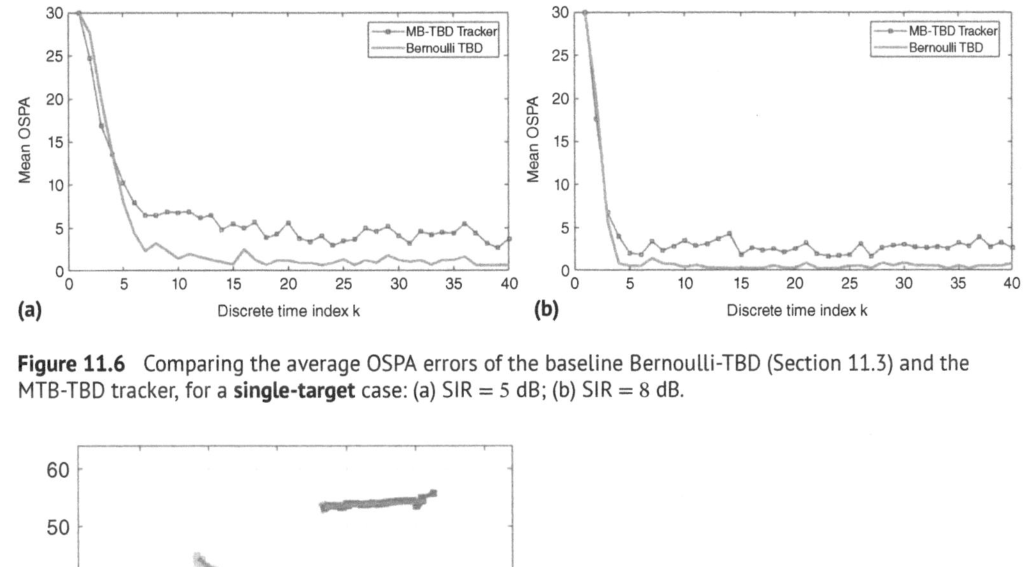

Figure 11.6 Comparing the average OSPA errors of the baseline Bernoulli-TBD (Section 11.3) and the MTB-TBD tracker, for a **single-target** case: (a) SIR = 5 dB; (b) SIR = 8 dB.

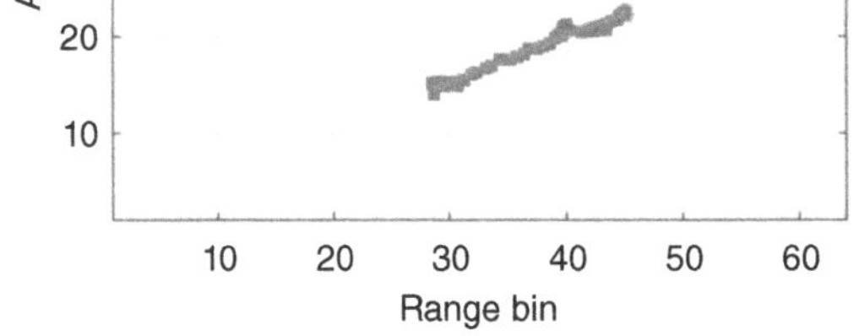

Figure 11.7 A top-down view of the testing scenario with three well-separated targets, moving along trajectories indicated by gray solid lines. The estimated tracks from a single run of the MTB-TBD algorithm are also shown.

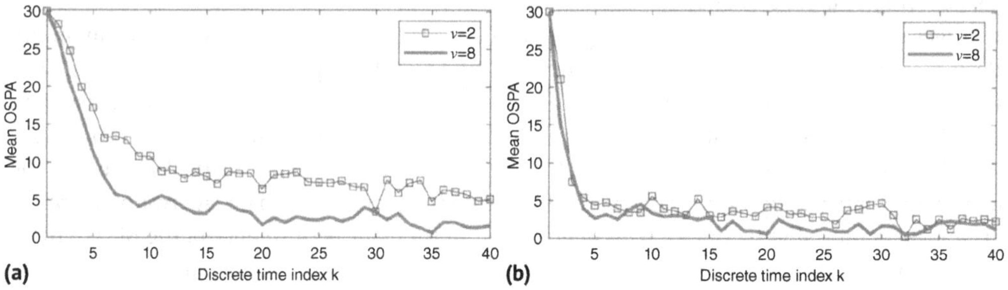

Figure 11.8 Average OSPA errors of the MTB-TBD tracker for the scenario with three targets, shown in Figure 11.7: (a) SIR = 5 dB; (b) SIR = 8 dB.

tracks (solid gray lines), and the three estimated tracks, produced by a single run of the MTB-TBD tracker. The parameters used in these runs are SIR = 5 dB and $\nu = 2$.

Figure 11.8 shows the mean OSPA error of the MTB-TBD versus time for the scenario shown in Figure 11.7 and two values of the shape parameter: $\nu = 2$ and $\nu = 8$. The results are again presented for two cases: (a) SIR = 5 dB and (b) SIR = 8 dB. In the absence of any targets, the MTB-TBD filter rarely declares false tracks. When the targets are present, see Figure 11.8a, b, we observe that initially, the OSPA errors are high due to a small delay in establishing and confirming the tracks. Note that when the SIR = 5 dB, a better performance is achieved for higher values of ν (i.e. for less spiky clutter). Finally, observe that for SIR = 8 dB in Figure 11.8b, the filter performance improves and becomes similar for both $\nu = 2$ and $\nu = 8$. This means that for high SIR targets, the spikiness of clutter has a less detrimental effect.

11.4.4.3 Interacting Targets

Next, we compare the performance of the MTB-TBD tracker with and without the modification of the likelihood for interacting targets (described at the end of Section 11.4.2). The testing scenario includes two crossing targets moving with nearly constant velocities. Their mean amplitudes, χ_k are constant and equal, determined from the specified SIR. Figure 11.9 displays the top-down view of the true tracks in this scenario (solid gray lines) as well as the estimated tracks produced by a single run of the two versions of the MTB-TBD tracker. Observe from Figure 11.9 that the MTB-TBD which assumes that targets are always well separated (non-interacting), loses one track when the two targets come close to each other. On the other hand, the MTB-TBD for interacting targets (with the modification of the likelihood functions) successfully tracks both targets.

The resulting mean OSPA errors, obtained using the testing scenario in Figure 11.9, are shown Figure 11.10 for MTB-TBD with and without the presented modification for interacting targets. The parameters used in Monte Carlo runs include SIR = 5 dB, with clutter shape values, $\nu = 2$ and 8. One can observe from 11.10 that in the first half of the scenario, before the crossing, the targets are well separated and the error performance is similar in both versions of the MTB-TBD. However, as the targets come close to each other (especially after the crossing), the MTB-TBD with the modification is clearly superior, because the version without the described modification is unable to maintain both tracks.

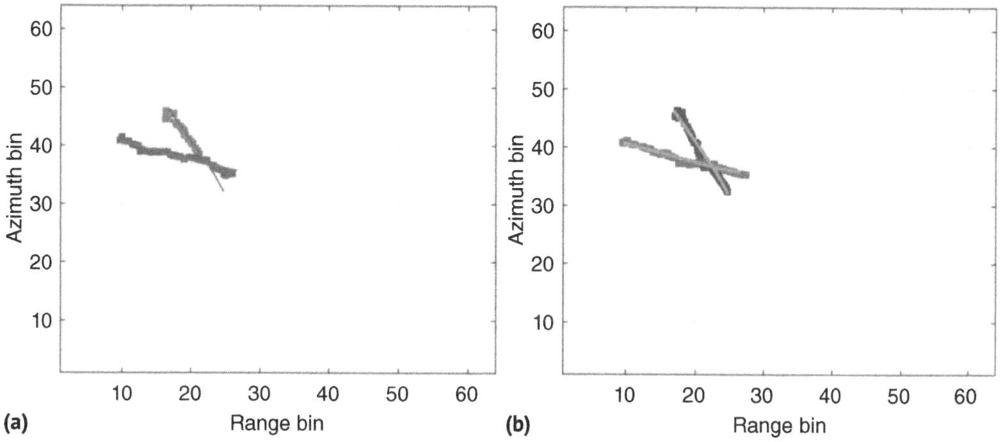

Figure 11.9 A top-down view of the testing scenario involving two interacting targets: ground truth indicated by gray solid lines; (a) Estimated tracks using MTB-TBD **without** the modification of the likelihood for interacting targets; (b) Estimated tracks **with** the modification of the likelihood for interacting targets.

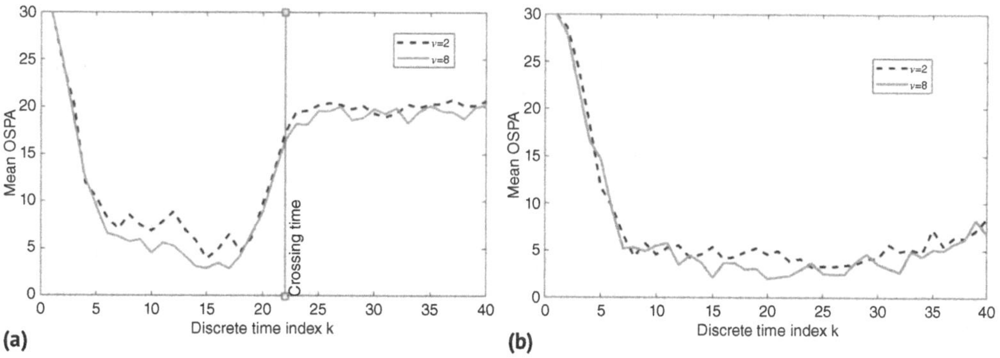

Figure 11.10 Mean OSPA error of the MB-TBD at SIR = 5 dB: (a) **without** and (b) **with** the modification for interacting targets.

11.5 Exploiting Doppler in the Bernoulli TBD

11.5.1 Doppler Information

In the presentation so far, we have focused on the fast radar scanning mode. Doppler information can be exploited only in a slow radar scanning mode, where there are sufficient returns from a target to use coherent integration. Consider a radar scanning rate of $\omega_0 = 20$ rpm ≈ 2.1 rad/s, with all other parameters given in Table 11.1. Radar data from a single scan comprises slow-time and range, with the slow time dimension representing the radar return from different azimuth locations. To include Doppler, a sliding window of N_D pulses is Fourier transformed and used to construct a three-dimensional complex-valued "frame" with dimensions of range, azimuth and Doppler frequency. The radar revisits each azimuth look direction periodically, and hence the fourth dimension is the scan number. With this set of parameters, the radar scan time (an interval of time between two successive scans) is 3 s, with $N_\mathrm{p} = 4500$ pulses. Figure 11.11a shows the radar looking to a side region in the upwind direction, centered at the azimuth look direction of 90° and spread

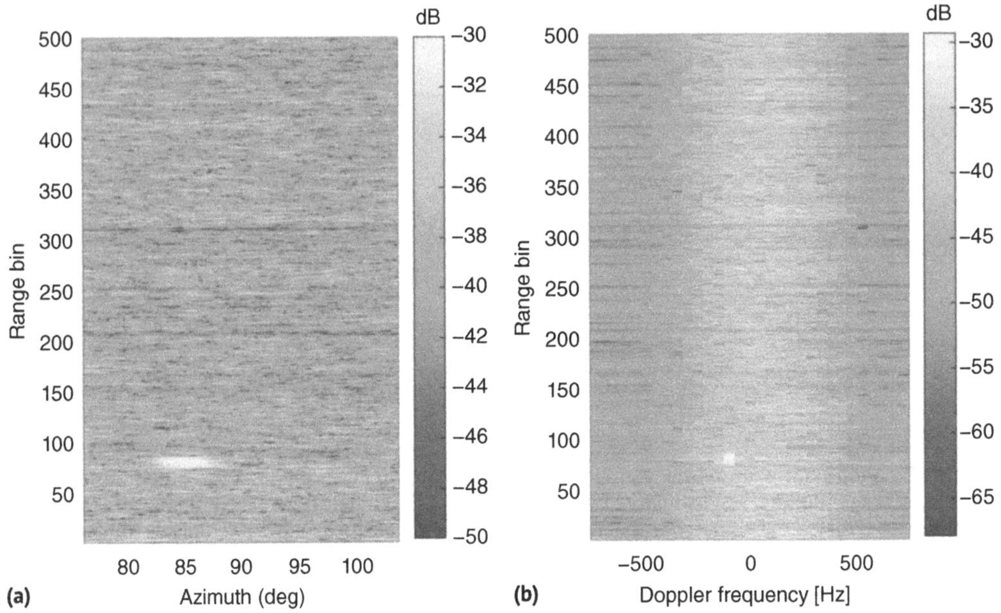

Figure 11.11 Illustration of radar data obtained in the **slow** scanning mode. An SIR = 4 dB target is in range bin 80 with azimuth 85° and a Doppler frequency of −150 Hz: (a) range-azimuth image (modulus) at Doppler bin corresponding to the frequency of −150 Hz; (b) range-Doppler image centered at the azimuth bin corresponding to 85°. *Source*: [34] / IEEE / CC BY 4.0.

over the scanning region of 30°. In this scene, an outbound target is present with an SIR of 4 dB, moving with a radial velocity of 2.2 m/s (corresponding to a Doppler frequency of −150 Hz) and is located at 85° azimuth and range bin 80. The range-azimuth image in Figure 11.11a represents the magnitude of the received radar signal in the range-azimuth subspace, taken in the Doppler cell containing the target Doppler of −150 Hz. The target appears spread along the azimuth direction, as it is illuminated by $N_\tau = 37$ pulses. The coherent processing exploits this effect by integrating with a window of $N_\mathrm{D} = 32$ pulses. Figure 11.11b illustrates the range-Doppler image (modulus of radar data) with the sliding window centered at the azimuth bin corresponding to 85°. Note that x-axis in Figure 11.11b spans from $-f_\mathrm{r}/2$ to $+f_\mathrm{r}/2$, where the PRF is $f_\mathrm{r} = 1500$ Hz.

Now compare Figure 11.11 with Figure 11.1b, which shows the range-azimuth map obtained using the same parameters and the same target, but in the fast scanning mode. Even visually, we can observe that the target is more pronounced in the slow scanning mode. In addition, the slow scanning mode provides a measurement of the target range-rate (via Doppler). The fast scanning mode, on the other hand, results in a higher measurement update rate (in this case, five times higher).

Clutter magnitude modeling follows from Section 11.2.2, albeit with one difference: the parameters of the K-distribution (the shape ν and scale b), which are homogeneous across the range and azimuth, vary with the Doppler frequency [1, 30, 31]. Figure 11.12 displays the shape and scale, estimated from five scans of radar data (in the absence of a target and using the slow scanning mode with $\omega_0 = 20$ rpm, as a function of Doppler frequency. The shape parameter has a peak around −300 Hz where thermal noise has the greatest contribution. In this example, there is aliasing present due to the choice of PRF and the broad Doppler spectrum which reduces the shape parameter outside of the clutter region. Parameter estimation was carried out using the moment based

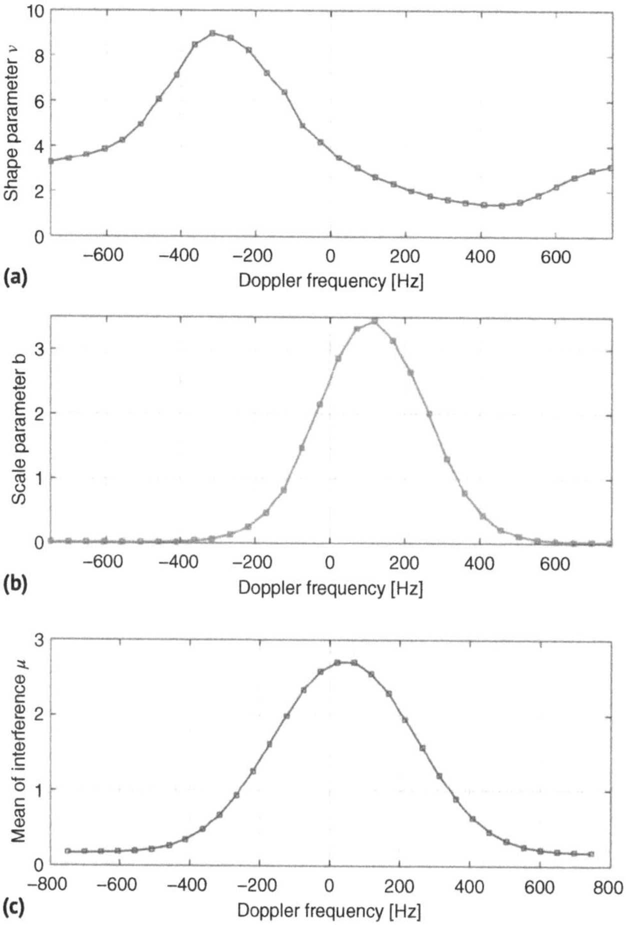

Figure 11.12 Estimated values of (a) shape $\hat{\nu}_d$ and (b) scale $\hat{b}_d$ and (c) the mean of the interference as a function of Doppler frequency. Parameter estimation was carried out using five scans of radar data in the absence of a target in the slow scanning mode ($\omega_0 = 20$ rpm). The remaining parameters given in Table 11.1.

method [21], explained in Section 11.3.3. However, this time, the samples[2] are collected for each Doppler cell $d = 1, \dots, N_D$, and therefore the parameters estimates $\hat{\nu}_d$ and $\hat{b}_d$ have a subscript d. Figure 11.12 also displays the resulting estimate of the interference mean, as a function of Doppler frequency, computed using (11.2). Note that the model parameters represent an underlying swell in the scene, and hence the interference mean is slightly shifted toward the positive Doppler frequency.

11.5.2 Bernoulli TBD Using Azimuth-range-Doppler Data

The theoretical formulation of the Bernoulli TBD filter follows from Section 11.3.1. The only difference is that now the input radar frame at each sampling time is an azimuth-range-Doppler (ARD) magnitude map $\mathbf{z}_k$. The PDF of a measured value $z \geq 0$ in the (m, n, d)th cell of $\mathbf{z}_k$, for $m = 1, \dots, N_r$,

2 The number of samples used for the calculation of estimates $\hat{\nu}_d$ and $\hat{b}_d$ for Doppler indices $d = 1, \dots, N_D$ in Figure 11.12 is $5\, N_a\, N_r = 187,500$.

$n = 1, \dots, N_a$ and $d = 1, \dots, N_D$, is again presented for two cases: the interference only case and the target-plus-interference case. The measured magnitude z, due to interference only, is modeled with the K-distribution,

$$p_0^{(m,n,d)}(z) = K(z; \nu_d, b_d). \tag{11.33}$$

The contribution of a target in the state $\mathbf{x} = [x, \dot{x}, y, \dot{y}, \chi]^\intercal$ to the measured value z in the (m, n, d)th cell, is expressed by a Gaussian PSF:

$$h^{(m,n,d)}(\mathbf{x}) = \frac{\chi}{(2\pi)^{3/2}\sigma_x\sigma_y\sigma_D} \exp\left[-\frac{(m-x)^2}{2\sigma_x^2} - \frac{(n-y)^2}{2\sigma_y^2} - \frac{(d-f_d)^2}{2\sigma_D^2}\right], \tag{11.34}$$

where the Doppler frequency f_d is linearly related to the range-rate, that is $f_d = -2\dot{y}f_0/B$, where f_0 is the carrier frequency and B is the radar bandwidth (see Table 11.1). The spread of the Gaussian PSF models the target amplitude contribution to its neighboring cells in azimuth, range, and Doppler. This spread is particularly large in azimuth, as we discussed in relation to Figure 11.11. The PDF of measured magnitude z in the (m, n, d)th cell of $\mathbf{z}_k$, for the case of a Swerling 1 target in K-distributed interference, is the L-distribution [14],

$$p_1^{(m,n,d)}(z|\mathbf{x}) = L\left(z; h^{(m,n,d)}(\mathbf{x}), \nu_d, b_d\right). \tag{11.35}$$

The prediction equations of the Bernoulli TBD are given by (11.4) and (11.5). Update equations are given by (11.6) and (11.7), with the likelihood ratio expressed as:

$$\ell(\mathbf{z}_k|\mathbf{x}) = \prod_{m=1}^{N_a}\prod_{n=1}^{N_r}\prod_{d=1}^{N_D} \frac{p_1^{(m,n,d)}(z|\mathbf{x})}{p_0^{(m,n,d)}(z)}. \tag{11.36}$$

Practical implementation follows the ideas presented in Section 11.3.3, except the measurement $\mathbf{z}_k$ is now three-dimensional. The spatial distribution $s_{k|k}(\cdot)$ is approximated by a set of weighted particles (samples) $\{w_{k|k}^{(i)}, \mathbf{x}_{k|k}^{(i)}\}_{1\le i\le N_k}$, where $\mathbf{x}_{k|k}^{(i)}$ is the ith particle and $w_{k|k}^{(i)}$ is its normalized weight. The likelihood ratio (11.36) needs to be computed for every particle. First, the index triple (m_i, n_i, d_i), corresponding to the particle $\mathbf{x}_{k|k}^{(i)}$, $i = 1, \dots, N_k$ is determined. Then, in order to speed up the computation of (11.8), each product in (11.36) is computed for a subset of indices in the vicinity of (m_i, n_i, d_i), that is for $m = m_i - U_m, \dots, m_i + U_m$, $n = n_i - U_n, \dots, n_i + U_n$ and $d = d_i - U_D, \dots, d_i + U_D$. The values of U_m, U_n and U_D are adopted to correspond to the spread factors of the Gaussian PSF, i.e. σ_x, σ_y and σ_D in (11.34), respectively.

The birth density $b_{k-1|k-1}(\mathbf{x})$ is measurement driven [32] and is approximated with a weighted random sample using $\mathbf{z}_{k-1}$ as follows. In the $(x, y, \dot{y})$ subspace, the samples are drawn to correspond to the cells of $\mathbf{z}_{k-1}$ which have a magnitude that is above a Doppler frequency dependent threshold $\hbar_d = \zeta\mu_d$, where threshold multiplier $\zeta > 1$, for all range-azimuth indices. In order to illustrate this, in Figure 11.13 we show a slice of Figure 11.11b at range cell 80 (line with circles); this is the Doppler spectrum at a range-azimuth cell where the 4 dB target is present at Doppler frequency 150 Hz. Importantly, Figure 11.13 illustrates the threshold $\hbar_d$ for $d = 1, \dots, N_D$ (line with squares), obtained by multiplying the mean μ_d, shown in Figure 11.12c, with the coefficient ζ set to 5.5. In order to approximate $b_{k-1|k-1}(\mathbf{x})$, in every cell (m, n, d) of $\mathbf{z}_k$, for which the measured magnitude $z^{(m,n,d)} > \hbar$, we draw random samples in the x, y and $\dot{y}$ subspace of the state space, corresponding to (m, n, d) and taking into account the resolution of the radar cell. In the subspace of $\dot{x}$ (i.e. the speed in the azimuth direction), a uniform distribution is adopted with an adequate interval (corresponding to the maximum expected speed of the target). Finally, in the subspace of the mean target intensity χ, the birth density is a gamma distribution with parameters (α_b, θ_b), where α_b is proportional to the estimated mean μ_d (see Figure 11.12c), while θ_b is chosen so that it covers the SIR values of interest.

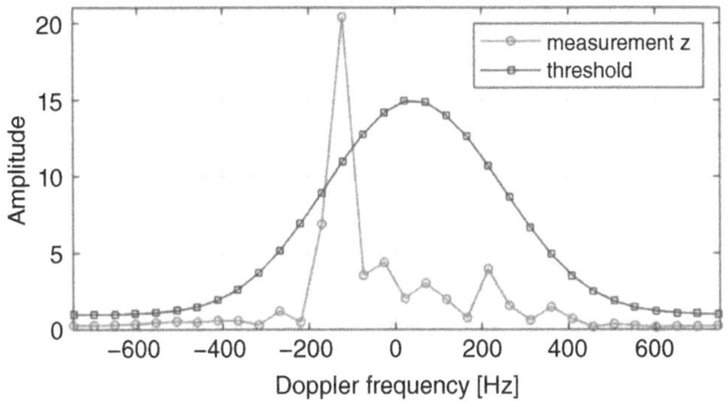

Figure 11.13 Illustration of the birth density design: line with circles is the Doppler spectrum at the range-azimuth cell where the target is present; the line with squares is the threshold $\hbar$.

11.5.3 Numerical Results

We compare the error performance of two Bernoulli TBD filters. The first uses radar data collected in the slow scan mode, and exploiting the Doppler information. We refer to this case as Bern-TBD-Slow. The second Bernoulli filter, uses radar data (range-azimuth maps) collected in the fast scan mode and was described in Section 11.3. This case is referred to as Bern-TBD-Fast. Note that in the fast scanning mode, the radar measurements and the update steps are more frequent.

The radar parameters used in simulations are the same as in Table 11.1, with $N_a = 375$ and two values of ω_0. The target is outbound, moving from the range bin 80 with a radial velocity of 2.2 m/s, but stays in the azimuth bin corresponding to 85°. The scenario lasts 90 seconds. In the slow scanning mode with $\omega_0 = 20$ rpm, the scan time is $T = 60/\omega_0 = 3$ s resulting in 30 radar measurements $\mathbf{z}_k$ available for processing during the observation period. In the fast scanning mode with $\omega_0 = 100$ rpm, the scan time is $T = 60/\omega_0 = 0.6$ s, and hence 150 measurements are available during the observation period of 90 seconds.

The parameters used in both Bernoulli TBD filters include: process noise parameters $q_u = 0.1$, $q_\chi = 1.5$; dynamic model parameters $p_b = 0.01$, $p_s = 0.99$; the number of particles fixed at 7500; the Gaussian PSF parameters: $\sigma_x = 2$ for Bern-TBD-Fast and $\sigma_x = 6$ for Bern-TBD-Slow, $\sigma_y = 1.2$ (for both) and $\sigma_D = 0.7$ (for Bern-TBD-Slow). Accordingly we adopted $U_m = 4$, $U_n = 1$ and $U_d = 1$. The birth density parameters α_b and θ_b, as well as the reporting threshold ξ, were tuned so that both Bernoulli TBD filters achieve a similar false track declaration rate. For the Bern-TBD-Slow, the parameter α_b depends on Doppler, and is set to $\alpha_b(d) = 9\ \mu_d$, with $\theta_b = 3$ and $\xi = 0.85$.

The detection and tracking performance is measured using the mean OSPA error [22], obtained by averaging over 100 Monte Carlo runs, using parameters $p = 2$ and $c = 100$ cells for the positional errors and $c = 10$ cells/s for the velocity error. The results are shown in Figure 11.14. In the absence of a target, both Bernoulli TBDs initialize a false track with a similar rate of 0.002 per second. The mean OSPA error for this case is shown in Figure 11.14a. The mean OSPA error curves in the presence of a target with SIRs equal to 0, 2, 4 and 6 dB are shown in Figure 11.14b–f. In all cases, the initial mean OSPA error equals $c = 100$, because the Bernoulli TBD filter requires at least one scan of radar data to establish a track. For the case of a 0 dB, the Bern-TBD-Slow is capable of establishing a track on the target with a delay of about 40 s, see panel (b). Once the track has been established, the OSPA error remains somewhat elevated (at the level of about 10-15), because the posterior probability of track presence $q_{k|k}$ occasionally falls below the reporting threshold ξ. Note that the Bern-TBD-Fast is unable to reliably detect and track the target at SIR = 0 dB. The situation

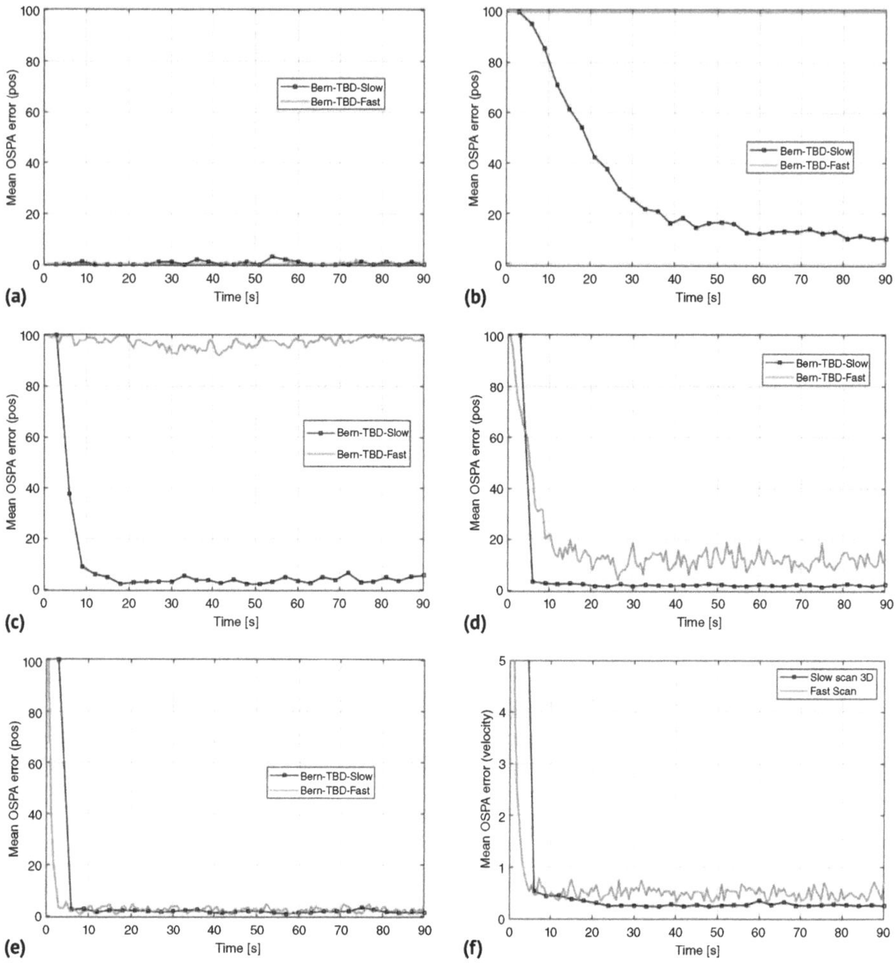

Figure 11.14 Mean OSPA errors for Bernoulli TBD, comparing slow vs fast scanning modes. Positional OSPA error in panels (a)–(e), velocity OSPA error in panel (f): (a) no target present; (b) SIR = 0 dB; (c) SIR = 2 dB; (d) SIR = 4 dB; (e) SIR = 6 dB; (f) SIR = 6 dB.

is somewhat similar for the case of a target at SIR = 2 dB, see panel (c), although here the positional OSPA error of the Bern-TBD-Slow drops to very low levels in less than 10 s, and remains low throughout the observation period. For the case of a target at SIR = 4 dB, see panel (d), Bern-TBD-Slow needs about 6 s to establish a track on the target and subsequently maintain the track with a minimal error. At SIR = 4 dB even the performance of the Bern-TBD-Fast improves considerably: because of the higher update rate (i.e. shorter scan time), the track is established almost as quickly as in the Bern-TBD-Slow, although the steady-state mean OSPA error remains about two to three times higher. For a target at SIR = 6 dB (or higher), the Bern-TBD-Fast performs better than Bern-TBD-Slow: their steady-state OSPA errors are the same. However, due to the higher update rate, Bern-TBD-Fast establishes a track quicker (3 s instead of 6 s). Finally, panel (f) displays the mean OSPA curves for velocity error in the presence of a target characterized with SIR = 6 dB. We can observe from panel (f) that the velocity errors are smaller for Bern-TBD-Slow, because it exploits the Doppler information. Note, however, that the primary purpose of coherent radar processing and exploitation of Doppler in the slow scanning mode is not to improve the estimation error in velocity, but to detect and track targets at lower SIR.

11.6 Bernoulli TBD for an Airborne Multichannel Radar

In this section we explore a further enhancement in detection and tracking of low SIR targets, by exploiting both coherent processing and the spatial degrees of freedom available in multichannel radar systems [33, 34]. Two-stage radar data processing is adopted. The first stage is space-time adaptive processing (STAP) of multichannel radar data, based on the normalized adaptive matched filter (NAMF) [35]. Target tracking using the STAP output is then performed in the second stage, using one of two alternate schemes.

The first scheme applies a detection threshold to the output of STAP to obtain point measurements. The threshold is set to a relatively low value to enable detection of targets with low SIRs. This choice introduces additional challenges because the false alarm rate is high and there are multiple detections per target.[3] In target tracking literature, this problem is referred to as *extended target tracking* [36–38]. Grouping of detections is carried out using a clustering algorithm, thereby generating an equivalent single point detection per target. Target tracking can then be performed using standard techniques for point measurements, such as the Bernoulli GSF (ignoring the magnitude of the point measurements).

The second tracking scheme is based on the TBD concept, applied to the raw (unthresholded) STAP output, reminiscent of [39–41]. For this purpose we again adopt the Bernoulli TBD filter, described in Section 11.3. In this case, however, the distributions $p_1^{(m,n)}(z|\mathbf{x})$ and $p_0^{(m,n)}(z)$ will be different.

11.6.1 Radar Data

The radar data is again generated using a high fidelity X-band multichannel simulation (see Sections 11.2.1 and 11.5.1) with the bi-modal evolving Doppler spectrum model, and K-distribution for the inference fluctuation and spatial correlation modeled by a Gaussian function [17].

Table 11.2 provides the key radar and environmental parameters, which are different or additional compared to those given in Table 11.1. The airborne radar comprises a linear array with $N_c = 4$ antennas and an inter-element spacing of $d = c/(2\,f_0) = 0.015$ m. The radar is horizontally polarized, with a PRF of 2000 Hz and scans at a rate of $\omega_0 = 20$ rpm. An azimuth region of 20° has been simulated in the upwind direction with $\psi = 0°$ and a 30° grazing angle in sea state 3. The clutter-to-noise ratio is 20 dB and the K-distribution shape parameter is set to $\nu = 5$. This region is centered at an angle of $\theta_{\text{plat}} = 90°$ and offers a side-looking geometry for the detection analysis.

The return from each radar scan is collected in the slow-time domain and corresponds to different azimuth angles. In addition to the interference, a point target signature has been modeled along both range and Doppler. Figure 11.15a illustrates the magnitude map of a typical range-azimuth (slow-time) return from channel 1, with a target at range bin 250 and azimuth bin 185. The number of range bins (cells) is 500, while the number of azimuth bins (cells) is 332. The target SIR in Figure 11.15a is 0 dB and is illuminated by $N_\tau = 37$ pulses in each scan. Figure 11.15b shows the corresponding angle-Doppler optimal power spectrum (at one range-azimuth bin), which is dominated by the sea clutter. This representation is optimal in the sense that the target SIR is maximum [42]. The goal of STAP filtering, which will be described in the next section, is to use

3 As the radar scans the sea surface, the target is illuminated over several azimuth angles, resulting in multiple detections.

Table 11.2 Radar and environmental parameters (the rest as in Table 11.1).

Parameter	Value
Bandwidth (B)	200 MHz
Pulse repetition frequency (f_r)	2000 Hz
Azimuth 3 dB beamwidth (ϕ_{3dB})	4°
Scan rate (ω_0)	20 rpm
Number of spatial channels (N_c)	4
Polarization	Horizontal
Inter-element spacing	0.015 m
Target range	20 km
Wind direction (ψ)	0°
K-distribution shape parameter (ν)	5

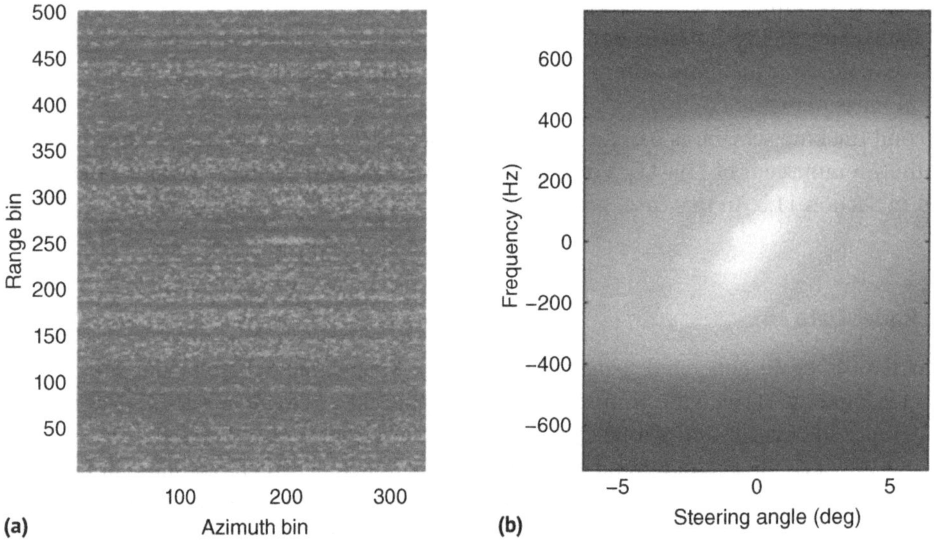

Figure 11.15 Typical radar data: (a) range-azimuth magnitude map (target with SIR = 0 dB is in azimuth bin 185 and range bin 250); (b) optimal angle-Doppler power spectrum. *Source*: [14] / with permission of IEEE.

this angle-Doppler representation to filter out the sea clutter and enable detection of slow-moving targets.

11.6.2 Joint Space and Time Processing

For each pulse, there are N_r fast-time samples (corresponding to range-gates). The coherent processing interval (CPI) is formed by taking a window of N_D pulses at each azimuth angle. Angle-Doppler beamforming is carried out using a *data cube* [43] centered at a particular range-azimuth cell, referred to as the cell under test (CUT). The data cube dimension is $N_a \times N_D \times N_r$. The received signal for a single range gate is an $N_a \times N_D$ slice of the data cube, expressed as

$$\mathbf{Z} = \alpha\mathbf{S} + \mathbf{C} + \mathbf{N}, \tag{11.37}$$

where α is the complex target amplitude, $\mathbf{S}$ is the target space-time steering vector, $\mathbf{C}$ is the clutter return and $\mathbf{N}$ is the thermal noise. The space-time steering vector can be written in terms of the spatial steering angle, θ and the Doppler frequency, f_d, as:

$$\mathbf{S}(\theta, f_\mathrm{d}) = \mathbf{s}_\mathrm{sp}(\theta)\, \mathbf{s}_\mathrm{t}^\mathrm{T}(f_\mathrm{d}), \tag{11.38}$$

where $\mathbf{s}_\mathrm{sp}(\theta)$ and $\mathbf{s}_\mathrm{t}(f_\mathrm{d})$ are the spatial and temporal steering vectors defined as:

$$\mathbf{s}_\mathrm{sp}(\theta) = \begin{bmatrix} 1 & e^{j2\pi f_\mathrm{s}} & \dots & e^{j2\pi(N_\mathrm{a}-1)f_\mathrm{s}} \end{bmatrix}^\mathrm{T}, \tag{11.39}$$

$$\mathbf{s}_\mathrm{t}(f_\mathrm{d}) = \begin{bmatrix} 1 & e^{j2\pi f_\mathrm{d}} & \dots & e^{j2\pi(N_\mathrm{D}-1)f_\mathrm{d}} \end{bmatrix}^\mathrm{T}. \tag{11.40}$$

The spatial frequency f_s, which features in (11.39), and the Doppler frequency f_d are given by

$$f_\mathrm{s} = \frac{d}{\lambda_\mathrm{c}} \sin\theta, \qquad f_\mathrm{d} = \frac{2v_\mathrm{r}}{c}, \tag{11.41}$$

where λ_c is the radar wavelength, v_r is the target radial velocity and c is the speed of light. Note that this array only steers in azimuth and not elevation.

To improve the detection of slow-moving targets, space-time processing can be used to suppress the clutter spectrum, shown in Figure 11.15b. Let us introduce the following notation: $\mathbf{z} = \mathrm{vec}(\mathbf{Z})$, where $\mathrm{vec}(\cdot)$ simply stacks the columns of matrix $\mathbf{Z}$ introduced in (11.37). This equation can then be expressed in a vector form as:

$$\mathbf{z} = \alpha\mathbf{s} + \mathbf{c} + \mathbf{n}, \tag{11.42}$$

where $\mathbf{s} = \mathrm{vec}(\mathbf{S})$, $\mathbf{c} = \mathrm{vec}(\mathbf{C})$ and $\mathbf{n} = \mathrm{vec}(\mathbf{N})$.

The target detection problem is generally formulated as a hypothesis test. The likelihood ratio test statistic is defined as

$$\ell(\mathbf{z}) = \frac{p(\mathbf{z}|H_1)}{p(\mathbf{z}|H_0)}, \tag{11.43}$$

where $p(\mathbf{z}|H_1)$ and $p(\mathbf{z}|H_0)$ are the likelihood functions under hypotheses H_1 (target is present) and H_0 (target is not present), respectively. Under the assumption of Gaussian and homogeneous clutter, this likelihood ratio can be approximated by detectors that have the constant false alarm rate property. One such detector that is suitable for the maritime domain, is the NAMF [44]. Its test statistics are given by,

$$y_\mathrm{NAMF}(\theta, f_\mathrm{d}) = \frac{|\mathbf{s}^H\, \mathbf{\Sigma}^{-1}\, \mathbf{z}|^2}{(\mathbf{s}^H\, \mathbf{\Sigma}^{-1}\mathbf{s})(\mathbf{z}^H\, \mathbf{\Sigma}^{-1}\, \mathbf{z})}, \tag{11.44}$$

where superscript H denotes Hermitian transpose and $\mathbf{\Sigma}$ is the interference covariance matrix. In practice, this covariance matrix is unknown and must be estimated from the data by forming a set of training data comprising G range gates surrounding the CUT. If the CUT range bin has index r, the training data vector is:

$$\mathbf{Z}_r = \begin{bmatrix} \mathbf{z}_{r-G/2}, \dots, \mathbf{z}_{r-1}, \mathbf{z}_{r+1}, \dots, \mathbf{z}_{r+G/2} \end{bmatrix}. \tag{11.45}$$

The sample covariance matrix (SCM) estimate is then given by, $\widehat{\mathbf{\Sigma}} = \frac{1}{G}\mathbf{Z}_r\, \mathbf{Z}_r^H$.

To implement the STAP output in a scanning radar, the value of test statistic $y_\mathrm{NAMF}(\theta, f_\mathrm{d})$ is determined over the Doppler frequencies and steering angles, followed by its maximization.

For illustration, the STAP output corresponding to the range-azimuth magnitude map shown in Figure 11.15a, is displayed in Figure 11.16a. This result is obtained using radar data from $N_c = 4$ channels with the number of pulses in the CPI set to $N_D = 16$.

The SCM, $\hat{\boldsymbol{\Sigma}}$ is computed using $G = 2N_D N_c = 128$ range cells. To speed up the computational time, the range of steering angles is reduced to $[-3°, -2°, \dots, 2°, 3°]$ and the Doppler frequencies to $[-250, -200, \dots, 200, 250]$ Hz, which correspond to radial velocities between −4 to 4 m/s. The blank areas along the edges in Figure 11.16a are due to insufficient samples required to estimate the SCM. Comparing Figure 11.15a, which was obtained from a single channel, and Figure 11.16a, one can observe that the target is more pronounced in the latter case. Application of a detection threshold to the STAP output in Figure 11.16a, with the probability of false alarm set to $P_{FA}^{\tau} = 10^{-3}$, results in point detections shown in Figure 11.16b. Note that the target in range bin 250 is detected in $N_\tau = 37$ azimuth cells and there is a relatively large number of false detections in the scan (i.e. 166).

Next we consider tracking methods, to be applied to the STAP output (illustrated in Figure 11.16) over a temporal sequence of radar scans.

11.6.3 Tracking Using Point Measurements

The first tracking approach is based on standard point measurements (without using the magnitude information), and is obtained by thresholding the STAP output illustrated in Figure 11.16b. Tracking is carried out using the Bernoulli GSF [18, Sec.VIII.B], with the state vector $\mathbf{u}_k = [x_k \;\; \dot{x}_k \;\; y_k \;\; \dot{y}_k]^\top$. The set of point measurements at time k is $\mathbf{W}_k = \{\mathbf{p}_{k,1}, \mathbf{p}_{k,2}, \dots, \mathbf{p}_{k,m_k}\}$, where $\mathbf{p} \in \mathbf{W}_k$ represents the range-azimuth polar coordinates, $\mathbf{p} = [X, \;\; Y]^\top$. The distribution of false detections is assumed to be uniform with $c_p(\mathbf{p}) = (N_a(N_r - G))^{-1}$, and the count of false detections is modeled by a Poisson distribution with the mean λ. The probability of detection is a fixed value P_D^{τ}.

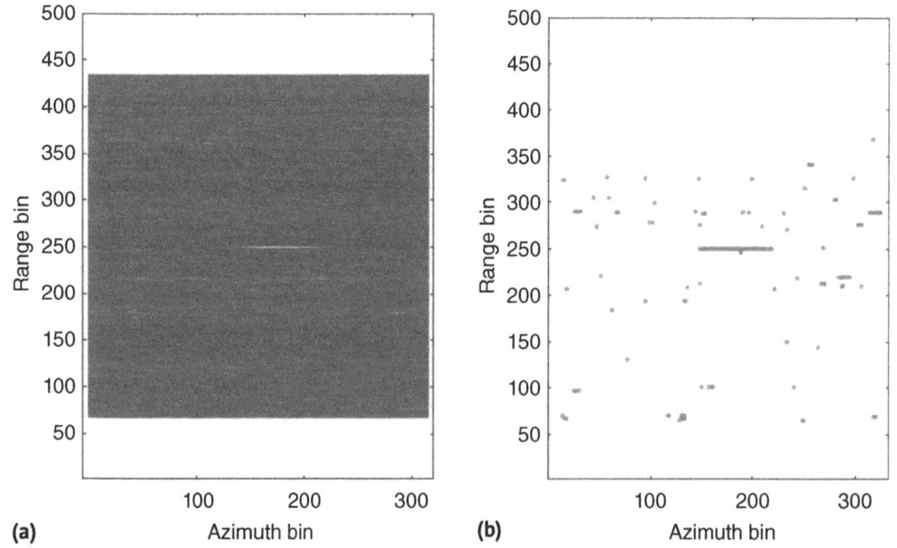

Figure 11.16 (a) STAP generated range-azimuth map (SIR= 0 dB is in azimuth bin 185 and range bin 250); (b) detections obtained by thresholding the output STAP range-azimuth map with $P_{FA}^{\tau} = 10^{-3}$.

We saw earlier that the target can be detected multiple times (in N_τ azimuth cells). Therefore, before we apply the Bernoulli GSF, it is necessary to ensure that at most a single detection is reported for a target. For this purpose, we use a clustering algorithm, with the objective to recognize and group the point measurements into clusters. Each declared cluster is then transformed into an "equivalent" point measurement and processed using the Bernoulli GSF, in order to establish and maintaining a track on the target.

Clustering of detections is carried out using the DBSCAN algorithm [29]. Its parameters are: (1) the minimum number of points for each cluster (denoted κ_{db}), and (2) the maximum distance between two points in the same cluster (denoted ε_{db}). The parameter κ_{db} should be selected as a trade-off between the ability to form a cluster on a target even when the SIR is low and the desire for a small number of falsely declared clusters. For the selection of the maximum distance ε_{db}, we adopt the method in [45] to calculate the average distance between each point and its κ_{db} nearest neighbors. Once these two parameters are determined, it is necessary to tune the value of P^τ_{FA} (to be discussed in Section 11.6.5.1). Figure 11.17 shows the result of DBSCAN clustering using parameters $\kappa_{db} = 15$ and $\varepsilon_{db} = 10$ for a target with SIR $= -2$ dB and different values of the probability of false alarm: (a) $P^\tau_{\mathrm{FA}} = 10^{-4}$; (b) $P^\tau_{\mathrm{FA}} = 0.5 \cdot 10^{-3}$, (c) $P^\tau_{\mathrm{FA}} = 10^{-3}$. The dots indicate detections that are declared by DBSCAN as being outliers.

Let us analyze the results shown in Figure 11.17. For the value of $P^\tau_{\mathrm{FA}} = 10^{-4}$, see panel (a), the detection threshold is too high and no cluster has been found (i.e. all detections are declared to be outliers). For $P^\tau_{\mathrm{FA}} = 0.5 \cdot 10^{-3}$, see panel (b), DBSCAN correctly declares one cluster of detections (shown as stars) at the true target position. Finally, with $P^\tau_{\mathrm{FA}} = 10^{-3}$, the number of detections are the highest and DBSCAN declares, according to panel (c), the existence of two clusters (stars and crosses).

Each cluster declared by the DBSCAN algorithm is converted to a point measurement to be subsequently processed by the Bernoulli GSF. The centroid of the cluster determines the coordinates of the point measurement, while the corresponding measurement covariance matrix is estimated from the size of the cluster. If we denote the standard deviation of a measurement error in azimuth and range by σ_x and σ_y, respectively, then the measurement covariance matrix is a diagonal matrix $\mathbf{R} = \mathrm{diag}(\sigma_x^2 \;\; \sigma_y^2)$. For example, if the cluster consists of four detections at range bins $\{45, 44, 44, 43\}$, then we calculate their corrected sample standard deviation as $\sigma_y = 0.82$. In the pathological case,

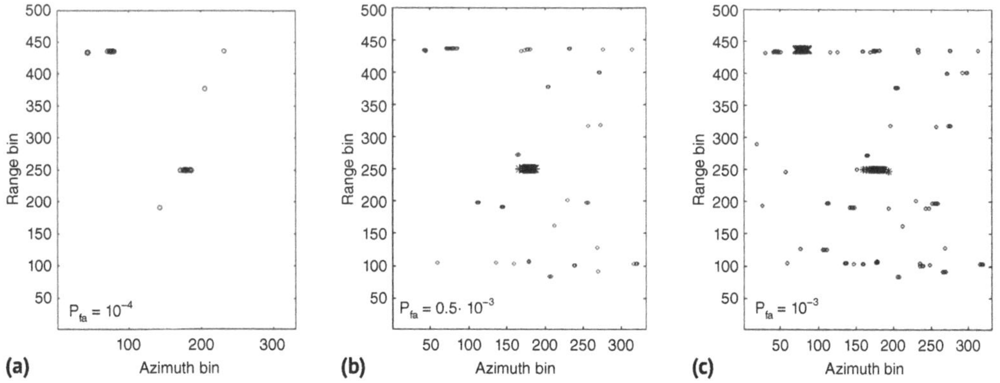

Figure 11.17 Clustering by DBSCAN algorithm for: (a) $P^\tau_{\mathrm{FA}} = 10^{-4}$; (b) $P^\tau_{\mathrm{FA}} = 0.5 \cdot 10^{-3}$, (c) $P^\tau_{\mathrm{FA}} = 10^{-3}$. Target with SIR $= -2$ dB is in azimuth bin 185 and range bin 250. Dots indicate detections, declared to be outliers by DBSCAN. Number of clusters declared: (a) none, (b) one, (c) two.

where all detections appear in the same range bin, we adopt $\sigma_y = 0.7$. Similar reasoning applies to the azimuth domain. Occasional falsely declared clusters are dealt with the Bernoulli GSF, via false alarm modeling parameters $c_p(\mathbf{p})$ and λ.

11.6.4 TBD Applied to the STAP Output

Tracking using TBD is now carried out using the sequence of range-azimuth maps produced by STAP, similar to the one illustrated in Figure 11.16a. For this purpose, we adopt the Bernoulli TBD filter described in Section 11.3. The state vector is given by (11.9), the transition density by (11.10), the target presence TPM by (11.3). The measurement $\mathbf{z}_k$ is the STAP output, and what remains to be done is determine the analytical expressions for PDFs $p_0(\cdot)$ and $p_1(\cdot|\mathbf{x})$. The point spread function is given by (11.13).

The PDF p_0 is postulated by fitting a distribution to a large sample of magnitude values from a STAP generated range-azimuth map $\mathbf{z}_k$, in the absence of a target. While a few well known PDFs provide a good fit to the data, we have adopted the inverse Gaussian (a.k.a. Wald) distribution for p_0. This is a convenient choice because it provides a simple extension to $p_1(\cdot|\mathbf{x})$ and will lead to an analytic expression for the likelihood ratio $p_1(z|\mathbf{x})/p_0(z)$, which features in (11.8). The inverse Gaussian distribution $\mathcal{I}(z; \mu_w, \lambda_w)$ has two parameters, the mean $\mu_w > 0$ and the shape $\lambda_w > 0$, while its support is $(0, \infty)$. The magnitude distribution $p_0(z)$ is specified as:

$$p_0(z) = \mathcal{I}(z; \mu_0, \lambda_0) = \sqrt{\frac{\lambda_0}{2\pi z^3}} \exp\left[-\frac{\lambda_0 (z-\mu_0)^2}{2\mu_0^2 z}\right], \quad (z > 0). \tag{11.46}$$

Figure 11.18 illustrates a normalized histogram of 22, 800 magnitude values from a STAP generated range-azimuth map $\mathbf{z}_k$ in the absence of a target, together with an inverse Gaussian model fit (solid line), whose estimated parameters are $\hat{\mu}_0 = 7.96$ and $\hat{\lambda}_0 = 45.49$. The mean estimate $\hat{\mu}_0$ is calculated as a sample mean, while $\hat{\lambda}_0$ is estimated using the maximum likelihood method.

The distribution $p_1^{(m,n)}(\cdot|\mathbf{x})$ is adopted as the inverse Gaussian distribution (11.46), but shifted by the contribution of a target $h^{(m,n)}$, specified by (11.13), that is:

$$p_1^{(m,n)}(z|\mathbf{x}_k) = \mathcal{I}(z + h^{(m,n)}; \mu_0, \lambda_0).$$

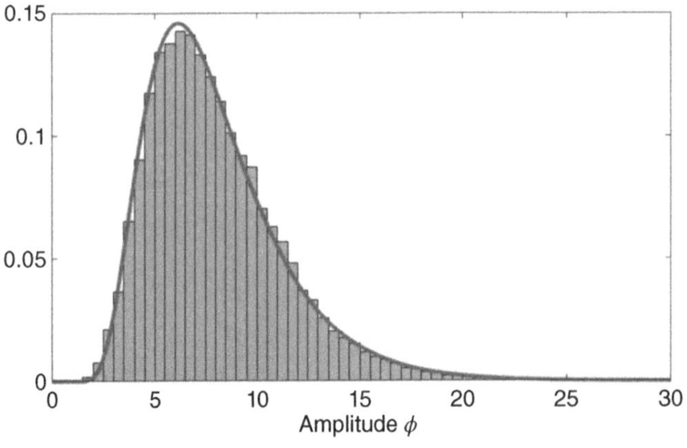

Figure 11.18 A normalized histogram of the samples from STAP generated range-azimuth map $\mathbf{z}_k$ in the absence of a target. A fitting inverse Gaussian PDF indicates a solid line.

We also approximate this distribution with an inverse Gaussian, whose parameters are μ_1 and λ_1, i.e.

$$p_1^{(m,n)}(z|\mathbf{x}) \approx \mathcal{I}(z; \mu_1, \lambda_1), \tag{11.47}$$

where the mean is $\mu_1 = \mu_0 + h^{(m,n)}$, while the shape parameter λ_1 can be approximated using the property that the variance remains unchanged, after shifting the distribution. In the case of the inverse Gaussian distribution, this property can be expressed as:

$$\frac{\mu_0^3}{\lambda_0} = \frac{\mu_1^3}{\lambda_1}, \text{ which leads to: } \lambda_1 = \frac{\left(\mu_0 + h^{(m,n)}\right)^3}{\mu_0^3} \lambda_0. \tag{11.48}$$

Now it is straightforward to derive the analytic expression for the likelihood ratio in the (m, n)th cell, which features in (11.8), as:

$$\frac{p_1^{(m,n)}(z|\mathbf{x})}{p_0(z)} \approx \left(\frac{\mu_1}{\mu_0}\right)^{3/2} \exp\left\{-\frac{\lambda_1(z-\mu_1)^2}{2\mu_1^2 z} + \frac{\lambda_0(z-\mu_0)^2}{2\mu_0^2 z}\right\}. \tag{11.49}$$

When the target is far from the (m, n)th cell, then $h^{(m,n)} \approx 0$ and the ratio in (11.49) approximately equals 1.

The practical implementation of the Bernoulli TBD filter follows the description in Section 11.3.3.

11.6.5 Numerical Results

11.6.5.1 The Experimental Setup and Parameters

The error performance is measured using the mean OSPA, obtained by averaging over 50 Monte Carlo runs. Table 11.2 shows the radar parameters, with the total observation period consisting of 15 scans. Since the scan rate is 20 rpm, the interval between the scans is 3 s. An outbound Swerling 0 target is present throughout the observation period with its starting position at range bin 125 and azimuth bin 185. It moves in range with a speed of 3 m/s, but remains in the same azimuth bin.

During the calibration phase of the algorithms, one scan of radar data *without a target present* is made available for estimation of the SCM, $\widehat{\boldsymbol{\Sigma}}$ (required for NAMF), to determine the detection threshold prior to DBSCAN (given P_{FA}^{τ}) and estimate the parameters μ_0 and λ_0 (required for the Bernoulli TBD filter). The parameters used for the estimation of the SCM have been specified in Section 11.6.2.

The following parameters were used in algorithms described in Sec.11.4.2 and 11.6.4: $p_s = 0.99$, $p_b = 0.01$, $q_0 = 0.01$ and $\xi = 0.75$. The specific parameters of Algorithm 11.1 include the probability of detection, $P_{\text{D}}^{\tau} = 0.95$ and the mean count of false point measurements per scan $\lambda = 1.5$. The spatial distribution of false point measurements c_p is uniform over the range-azimuth map and a Gaussian mixture component is eliminated if its weight is below 0.001.

In a typical detection scheme, the probability of false alarm P_{FA}^{τ} is fixed and subsequently the detection threshold is determined. However, for this algorithm, P_{FA}^{τ} is tuned to allow improved tracking performance, for details see [13]. The selected values is $P_{\text{FA}}^{\tau} = 0.7 \cdot 10^{-3}$, resulting in a detection threshold of 24.2.

The specific parameters of Algorithm 11.2 include the number of particles, $N_k = 10{,}000$ and the value of the threshold $\hbar$, for the creation of the birth density particles. This threshold is found experimentally as $\hbar \approx 24.7$, for details see [13]. Other parameters of Algorithm 11.2 include the parameters of the PSF, which are set to $\sigma_x = 8$ azimuth cells and $\sigma_y = 1$ range cell. Finally, the amplitude of birth particles is sampled from a gamma distribution, that covers the range of values that target amplitude χ can take; the shape parameter is set to 30, while the scale parameter is 0.7.

11.6.5.2 Mean OSPA Curves

The first result shows the proposed tracking algorithms evaluated in the absence of a target. The resulting mean OSPA errors (not shown) are practically equal to zero throughout the observation period, for both Algorithms 11.1 and 11.2. This indicates a very small probability of false track declarations by either algorithm.

Numerical results for a target characterized by SIR values of −4, −3, −2 and −1 dB are shown in Figure 11.19. Panel (a) displays the result for a target with an SIR value of −4 dB. Observe that both algorithms perform poorly in this case, because both mean OSPA error curves remains high throughout the observation interval. Algorithm 11.2, nevertheless, performs slightly better because its mean OSPA error is reducing with time: after 14 scans it drops to 50, meaning that with a probability of 0.5 it can detect and track the target by that time. Panel (b) shows the result for an SIR value of −3 dB. Algorithm 11.1 is still unable to detect and establish the existence of the target. Algorithm 11.2 in this case reliably detects the target and establishes a track on it, albeit with a delay of about nine scans. Panel (c) shows the result for a target with an SIR = −2 dB. Observe that the OSPA error of Algorithm 11.1 drops after several scans, because the target is occasionally detected and tracked properly. However, due to the occasional breaks in the track (caused by missed target detections), the OSPA error of Algorithm 11.1 remains relatively high. Algorithm 11.2, on the other

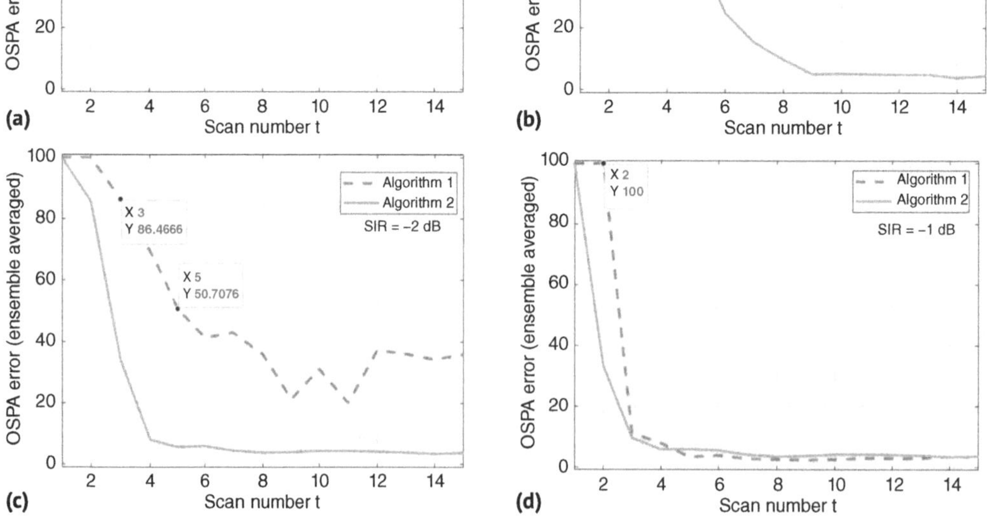

Figure 11.19 Mean OSPA error curves for multichannel radar data: (a) SIR = −4 dB; (b) SIR = −3 dB; (c) SIR = −2 dB; (d) SIR = −1 dB.

hand, detects and tracks the target reliably with its OSPA error reaching the steady state minimum after four scans. This steady state value is above zero, mainly for two reasons: (1) the uncertainty caused by the spread of the target in azimuth and (2) the level of process noise in the particle filter. Finally, panel (d) demonstrates good performance for both algorithms when the target SIR is −1 dB. Both algorithms need three to four scans to establish a good quality track on the target. The tracking error of Algorithm 11.1 (dashed line) is somewhat smaller than the tracking error of Algorithm 11.2 (solid line), because the Gaussian sum filter better handles the uncertainty caused by the spread of the target in azimuth.

This section demonstrated that both proposed tracking algorithms achieve good performance, even at negative SIR values. Algorithm 11.2 (based on the TBD scheme), performs better: it is able to establish a reliable track even with an SIR of −3 dB. Algorithm 11.1 is less accurate at very low SIR values because of thresholding. Regarding the computational time, Algorithm 11.2 is about three times slower to run than Algorithm 11.1.

11.7 Conclusions

The chapter presented a compilation of recent work on the topic of track-before-detect for maritime radar. The fundamental building block was the Bernoulli TBD filter, as the optimal sequential Bayesian estimator for detection and tracking using a temporal sequence of radar intensity maps. Various extensions of this fundamental algorithm were presented in the chapter, backed up by error performance evaluations. Numerical results demonstrated that TBD approaches provide better error performance than the conventional tracking methods for low signal-to-interference targets. Their main drawback is the higher computational cost and a requirement for accurate statistical models of radar magnitude maps, both in the absence and in the presence of a target. Future work will focus toward a practical real-time implementation of the Bernoulli TBD filter.

References

1 K. D. Ward, R. J. A. Tough, and S. Watts. *Sea Clutter: Scattering, the K-Distribution and Radar Performance*. The Institute of Engineering Technology, 2nd edition, 2013.

2 S. Blackman and R. Popoli. *Design and Analysis of Modern Tracking Systems*. Artech House, 1999.

3 Y. Barniv. Dynamic programming solution for detecting dim moving targets. *IEEE Transactions on Aerospace and Electronic Systems*, 21(1):144–156, 1985.

4 I. S. Reed, R. M. Gagliardi, and L. B. Stotts. A recursive moving-target-indication algorithm for optical image sequences. *IEEE Transactions on Aerospace and Electronic Systems*, 26(3):434–440, 1990.

5 W. R. Wallace. The use of track-before-detect in pulse-doppler radar. In *2002 International Radar Conference*, pages 315–319, 2002.

6 J. D. R. Kramer and W. S. Reid. Track-before-detect processing for an airborne type radar. In *IEEE International Radar Conference*, pages 422–427, 1990.

7 S. Buzzi, M. Lops, and L. Venturino. Track-before-detect procedures for early detection of moving target from airborne radars. *IEEE Transactions on Aerospace and Electronic systems*, 41(3):937–954, 2005.

8 W. Yi, H. Jiang, T. Kirubarajan, L. Kong, and X. Yang. Track-before-detect strategies for radar detection in G0-distributed clutter. *IEEE Transactions on Aerospace and Electronic Systems*, 53(5):2516–2533, 2017.

9 D. J. Salmond and H. Birch. A particle filter for track-before-detect. In *Proceedings of the American Control Conference*, volume 5, pages 3755–3760, 2001.

10 B. Ristic, S. Arulampalam, and N. Gordon. *Beyond the Kalman Filter: Particle Filters for Tracking Applications*. Artech House, 2004.

11 D. Y. Kim, B. Ristic, X. Wang, L. Rosenberg, J. Williams, and S. Davey. A comparative study of track-before-detect algorithms in radar sea clutter. In *Proceedings of the International Radar Conference*, Toulon, France, 2019.

12 D. Y. Kim, B. Ristic, R. Guan, and L. Rosenberg. A Bernoulli track-before-detect filter for interacting targets in maritime radar. *IEEE Transactions on Aerospace and Electronic Systems*, 57(3): 1981–1991, 2021.

13 D. Y. Kim, L. Rosenberg, B. Ristic, and R. P. Guan. Track-before-detect using an airborne multichannel radar in the maritime domain. *IEEE Transactions on Aerospace and Electronic Systems*, 59(3):2221–2231, 2023.

14 B. Ristic, L. Rosenberg, D. Y. Kim, X. Wang, and J. Williams. A Bernoulli track-before-detect filter for maritime radar. *IET Radar Sonar Navigation*, 14(3):356–363, 2020.

15 B. Ristic, R. Guan, D. Y. Kim, and L. Rosenberg. Bernoulli track-before-detect smoothing for maritime radar. *IET Radar, Sonar & Navigation*, 16(6):953–960, 2022.

16 B. Ristic, D. Y. Kim, L. Rosenberg, and R. Guan. Exploiting Doppler in Bernoulli track-before-detect for a scanning maritime radar. *IEEE Transactions on Aerospace and Electronic Systems*, 58(1):720–728, 2022.

17 L. Rosenberg, S. Watts, and S. Bocquet. Scanning radar simulation in the maritime environment. In *Proceedings of the IEEE Radar Conference*, Florence, Italy, 2020.

18 B. Ristic, B.-T. Vo, B.-N. Vo, and A. Farina. A tutorial on Bernoulli filters: theory, implementation and applications. *IEEE Transactions on Signal Processing*, 61(13):3406–3430, 2013.

19 Y. Bar-Shalom, X. R. Li, and T. Kirubarajan. *Estimation with Applications to Tracking and Navigation*. John Wiley & Sons, 2001.

20 J. Sun, C. Liu, Q. Li, and X. Chen. Labelled multi-Bernoulli filter with amplitude information for tracking marine weak targets. *IET Radar, Sonar & Navigation*, 13(6):983–991, 2019.

21 D. A. Abraham and A. P. Lyons. Novel physical interpretations of K-distributed reverberation. *IEEE Journal of Oceanic Engineering*, 27(4):800–813, 2002.

22 D. Schuhmacher, B.-T. Vo, and B.-N. Vo. A consistent metric for performance evaluation of multi-object filters. *IEEE Transactions on Signal Processing*, 56(8):3447–3457, 2008.

23 B. Ristic, L. Rosenberg, D. Y. Kim, and R. Guan. Bernoulli filter for tracking maritime targets using point measurements with amplitude. *Signal Processing*, 181:107919, 2021.

24 S. Nannuru, M. Coates, and R. Mahler. Computationally-tractable approximate PHD and CPHD filters for superpositional sensors. *IEE Journal of Selected Topics in Signal Processing*, 7(3): 410–420, 2013.

25 F. Papi and D. Y. Kim. A particle multi-target tracker for superpositional measurements using labeled random finite sets. *IEEE Transactions on Signal Processing*, 63(16):4348–4358, 2015.

26 D. Mušicki and B. La Scala. Multi-target tracking in clutter without measurement assignment. *IEEE Transactions on Aerospace and Electronic Systems*, 44(2): 877–896, 2008.

27 R. Mahler. *Statistical Multisource Multitarget Information Fusion*. Artech House, 2007.

28 B.-N. Vo, B.-T. Vo, N.-T. Pham, and D. Suter. Joint detection and estimation of multiple objects from image observations. *IEEE Transactions on Signal Processing*, 58(10):5129–5141, 2010.

29 M. Ester, H.-P. Kriegel, J. Sander, and X. Xu. A density-based algorithm for discovering clusters in large spatial databases with noise. In *Proceedings of the 2nd International Conference Knowledge Discovery and Data Mining*, pages 226–231, 1996.

30 S. Watts. Modeling and simulation of coherent sea clutter. *IEEE Transactions on Aerospace and Electronic Systems*, 48(4):3303–3317, 2012.

31 S. Watts, L. Rosenberg, S. Bocquet, and M. Ritchie. Doppler spectra of medium grazing angle sea clutter; part 2: model assessment and simulation. *IET Radar, Sonar & Navigation*, 10(1):32–42, 2016.

32 B. Ristic, D. Clark, B-N Vo, and B-T Vo. Adaptive target birth intensity for PHD and CPHD filters. *IEEE Transactions on Aerospace and Electronic Systems*, 48(2):1656–1668, 2012.

33 M. K. McDonald and D. Cerutti-Maori. Coherent radar processing in sea clutter environments, part 1: modelling and partially adaptive STAP performance. *IEEE Transactions on Aerospace and Electronic Systems*, 52(4):1797–1817, 2016.

34 L. Rosenberg and S. Watts. Coherent detection in medium grazing angle sea-clutter. *IET Radar Sonar and Navigation*, 11(9):1340–1348, 2017.

35 S. Kraut and L. L. Scharf. The CFAR adaptive subspace detector is a scale-invariant GLRT. *IEEE Transactions on Signal Processing*, 47(9):2538–2541, 1999.

36 B. Ristic and J. Sherrah. Bernoulli filter for joint detection and tracking of an extended object in clutter. *IET Radar, Sonar & Navigation*, 7(1):26–35, 2013.

37 M. Beard, S. Reuter, K. Granström, B.-T. Vo, B.-N. Vo, and A. Scheel. Multiple extended target tracking with labeled random finite sets. *IEEE Transactions on Signal Processing*, 64(7): 1638–1653, 2015.

38 K. Granstrom, M. Baum, and S. Reuter. Extended object tracking: introduction, overview and applications. *arXiv preprint arXiv:1604.00970*, 2016.

39 D. Orlando, L. Venturino, M. Lops, and G. Ricci. Track-before-detect strategies for STAP radars. *IEEE Transactions on Signal Processing*, 58(2):933–938, 2010.

40 D. Orlando, G. Ricci, and Y. Bar-Shalom. Track-before-detect algorithms for targets with kinematic constraints. *IEEE Transactions on Aerospace and Electronic Systems*, 47(3):1837–1849, 2011.

41 N. Fiscante, P. Addabbo, C. Clemente, F. Biondi, G. Giunta, and D. Orlando. A track-before-detect strategy based on sparse data processing for air surveillance radar applications. *Remote Sensing*, 13(4):662, 2021.

42 J. R. Guerci. *Space-time Adaptive Processing for Radar*. Artech House, 2nd edition, 2015.

43 W. L. Melvin. A STAP overview. *IEEE Aerospace and Electronic Systems Magazine*, 19(1):19–35, 2004.

44 E. Conte, M. Lops, and G. Ricci. Asymptotically optimum radar detection in compound-Gaussian clutter. *IEEE Transactions on Aerospace and Electronic Systems*, 31(2):617–625, 1995.

45 N. Rahmah and I. S. Sitanggang. Determination of optimal epsilon (eps) value on DBSCAN algorithm to clustering data on peatland hotspots in sumatra. In *IOP Conference Series: Earth and Environmental Science*, 31: 012012, 2016.

12

Moving Target Tracking Using ViSAR Imagery

Xiaoqing Tian[1], Jing Liu[2], and Mahendra Mallick[3]*

[1] School of Electrical and Control Engineering, Xi'an University of Science and Technology, Xi'an, China
[2] School of Automation Science and Engineering, Xi'an Jiaotong University, Xi'an, China
[3] Independent Consultant, Cambria, CA, USA
*Corresponding author: elelj20080730@xjtu.edu.cn

12.1 Introduction

Synthetic aperture radar (SAR) as a high resolution sensor is of great importance in surveillance, since it is capable of operating at long distances, in all-weather, and in day and night conditions. The SAR system has been used widely in the battlefields to identify and track different kinds of targets in complex environments. However, the conventional SAR system has limitations in tracking moving targets due to the short dwell time of the standard linear SAR operation mode. Recently, many researchers have focused on developing the video SAR (ViSAR) [1], which collects data on targets of interest while the aircraft either flies by or circles around the region of interest. Compared with the conventional SAR, the synthetic aperture time of ViSAR is greatly reduced. As a result, a single flight consists of multiple synthetic apertures and the ViSAR can form sequential frame images at a high frame rate. Therefore, ViSAR allows for effective detection and tracking of moving targets over short time periods.

Among the research works on ViSAR videos, moving target detection and tracking are of great interest to researchers. The moving targets in ViSAR videos mainly include vehicles and helicopters, whose locations can be effectively inferred from their shadows [2]. Moving-target tracking using ViSAR images is a typical dim target tracking problem. In the context of this chapter, "dim targets" correspond to shadows of moving vehicles [3] cast onto the stationary background scene. The "target shadows" appear to move in the same way as the vehicles, whereas the actual signals scattered from the moving vehicles move in an entirely different manner. Specifically, in the context of the ViSAR imagery, the actual signals can be viewed essentially as an unwanted source of bright moving visual "clutter." The reason that the moving target shadows are found in the correct location is that the shadows correspond to an absence of radar signals from a stationary part of the scene. The reason that the signals scattered from a target fall in the wrong location is due to the fact that the signals scattered from the target are Doppler shifted [4] (due to the cross-track velocities of targets). This results in large azimuth shifts in the corresponding focused contributions to the radar image data.

Compared with conventional object tracking in computer vision, moving target tracking using ViSAR videos is more challenging due to the following factors: (1) Generally, the background is complex, and the dimensions of the vehicle shadows are small. Moreover, the intensities of pixels centered in vehicle shadows are lower than those centered at noise and clutter, resulting in a low signal-to-clutter-plus-noise ratio; (2) Due to the speckle noise in ViSAR images, the intricate textures of vehicle shadows are masked by and even confused with those of man-made structures; (3) Due to the motion of the platform and vehicles, the scales, deformations and orientations of the vehicle shadows are time-varying.

This chapter is based on our previous work on simultaneous detection and tracking [5] and real-time tracking [6] of moving target shadows in ViSAR imagery. Simultaneous detection and tracking of moving target shadows are based on the DP-TBD technique [5]. The second part of the chapter addresses the correlation filtering-based real-time tracking method for moving target shadows in ViSAR imagery [6].

This chapter is based on the following key contributions of our work:

1. At present, publications on ViSAR focus either on the detailed introduction of basic ViSAR mode [1] or on subsequent applications of generated video frames [7–9], e.g., autofocus, coherent change detection, and moving target shadow detection. In this chapter, we focus on joint detection and tracking of multiple dim targets in ViSAR imagery. Specifically, we adopt the track-before-detect (TBD) technique to simultaneously detect targets and estimate target tracks. Secondly, we propose a novel algorithm, the expanding and shrinking-strategy-based particle filter/dynamic-programming-based TBD (ES-TBD) algorithm, to track a fixed number of slowly maneuvering targets. In the ES-TBD algorithm, an expanding and shrinking strategy is adopted by combining the particle filter (PF) and the dynamic-programming (DP) algorithms. The ES-TBD algorithm improves the merit integration efficiency of the targets and obtains more accurate searching ranges. Thus, the ES-TBD algorithm is more robust to track slowly maneuvering targets.
2. TBD-based algorithms process a batch of foreground images generated from original video data by a series of operations, e.g., registration [10], background differencing, constant false-alarm rate (CFAR) detection [11], among others. These operations increase the computational cost and decrease the computation speed. To reduce the complexity and increase the speed of the algorithm, tracking directly using the original video data is desired. When the correlation filtering (CF)-based algorithm is applied to ViSAR videos, the multipeak problem may degrade the performance of the CF. We propose a novel appearance-distance information-assisted probabilistic data association (AD-PDA) algorithm that addresses the multipeak problem. Secondly, by combining the AD-PDA and background-aware correlation filter (BACF) based with the target dynamic model, we propose a novel AD-PDA-BACF algorithm that can accurately estimate the states of targets using original ViSAR images. In contrast to the traditional image difference-based algorithms [9], neither image registration nor background extraction operations are required in the AD-PDA-BACF algorithm. This largely reduces the algorithmic complexity and accelerates the tracking process.

Remaining sections are organized as follows. The tracking algorithms of the ViSAR is presented in Section 12.2 and track filtering models and algorithms are described in Section 12.3. Sections 12.4 and 12.5 discuss the ES-TBD and AD-PDA-BACF algorithms, respectively. We present experimental results in Section 12.6 and conclusions are drawn in Section 12.7.

12.2 Tracking Algorithms

The modeling-based algorithms for ViSAR target tracking can be divided into three categories: (1) detect-before-track (DBT)-based algorithms; (2) TBD-based algorithms; and (3) CF-based algorithms.

12.2.1 Tracking with Thresholded Measurements

The DBT technique is popular in the field of radar target tracking. In 2009, the Air Force Research Laboratory of the USA released the Gotcha data set [12] to support the development of ViSAR ground moving target indication (GMTI) algorithms. Many researchers have attempted to develop GMTI algorithms for ViSAR, which can obtain a moving target's track for a long time [13–19]. The detect-before-track (DBT) technique first utilizes the GMTI technique to detect the moving targets, and then uses traditional tracking algorithms to estimate the target tracks, using the thresholded measurements. Examples of such algorithms include multiple hypotheses tracking (MHT) [20, 21] and joint probabilistic data association filter (JPDAF) [22]. Although DBT-based algorithms have good performance in tracking high signal-to-noise ratio (SNR) targets, they perform poorly in tracking very low SNR targets due to the thresholding procedures [23].

We estimate the states of a target using the probabilistic data association (PDA) algorithm [22] due to its satisfactory tracking performance for a single-target in heavy clutter. The PDA algorithm makes the following assumptions: At time k, the number of validated measurements falling in a validated region, m_k, is not less than 1 and the number of validated measurements originating from the target is no more than 1. We denote the set of m_k validated measurements at time k as

$$\mathbf{Z}_k := \{\mathbf{z}_{k,1}, \mathbf{z}_{k,2}, \dots, \mathbf{z}_{k,m_k}\}, \tag{12.1}$$

where $\mathbf{z}_{k,j}$ denotes the jth validated measurement vector. The set of measurements up to and including time k is denoted by

$$\mathbf{Z}_{1:k} := \{\mathbf{Z}_1, \mathbf{Z}_2, \dots, \mathbf{Z}_k\}. \tag{12.2}$$

The PDA algorithm considers a set of exhaustive events $\{\epsilon_{k,0}, \epsilon_{k,1}, \dots, \epsilon_{k,j}, \dots, \epsilon_{k,m_k}\}$, which are expressed as

$$\epsilon_{k,j} = \begin{cases} \mathbf{z}_{k,j} \text{ is target-originated}, & j = 1, 2, \dots, m_k \\ \text{none of the measurements is target-originated}, & j = 0 \end{cases}. \tag{12.3}$$

According to the total probability theorem [24], the state update equation of PDA can be calculated as

$$\hat{\mathbf{x}}_{k|k} = E[\mathbf{x}_k \mid \mathbf{Z}_{1:k}] = \sum_{j=0}^{m_k} E[\mathbf{x}_k \mid \epsilon_{k,j}, \mathbf{Z}_{1:k}]P\{\epsilon_{k,j} \mid \mathbf{Z}_{1:k}\} = \sum_{j=0}^{m_k} \hat{\mathbf{x}}_{k|k,j}\theta_{k,j}. \tag{12.4}$$

Here, $\mathbf{x}_k$ denotes the target state vector at time k. For the events $\epsilon_{k,j}$, $j = 1, 2, \dots, m_k$, $\hat{\mathbf{x}}_{k|k,j}$ is the updated state conditioned on $\epsilon_{k,j}$, and the association probability $\theta_{k,j} \triangleq P\{\epsilon_{k,j} \mid \mathbf{Z}_{1:k}\}$ is the conditional probability of $\epsilon_{k,j}$. For the event $\epsilon_{k,0}$, the PDA algorithm calculates the predicted state as the updated state, i.e., $\hat{\mathbf{x}}_{k|k,0} = \hat{\mathbf{x}}_{k|k-1}$, and $\theta_{k,0}$ is defined similarly to $\theta_{k,j}$ ($j = 1, 2, \dots, m_k$). In addition, the association probabilities satisfy $\sum_{j=0}^{m_k} \theta_{k,j} = 1$. We refer the reader to [22] for details on the PDA algorithm.

12.2.2 Tracking with Non-thresholded Measurements

The TBD technique [25, 26] has been proposed and widely applied in detecting and tracking targets with low SNR. The TBD technique estimates the multitarget states based on a number of consecutive frames. The dynamic programming-based TBD (DP-TBD) algorithms approximate the joint posterior probability density function (pdf) of states by discretizing the state space, searching the paths, and integrating merits of all the paths recursively [27]. In the conventional DP-TBD algorithm, state transitions of a target are limited. Moreover, the velocities are quantified, so that the conventional DP-TBD algorithm can track non-maneuvering or slowly maneuvering targets. In order to improve the accuracy of tracking maneuvering targets by the DP-TBD algorithm, researchers have proposed several methods [27–32]. Compared to the DP-TBD algorithm, the particle filtering-based TBD (PF-TBD) algorithm has received great attention due to its capability of detecting and tracking a time-varying number of dim targets in a high cluttered environment. The PF-TBD algorithm detects and tracks targets by estimating a multimodal posterior pdf of states and existence probability [33–35]. However, if the number of targets is large and no prior information regarding a newborn target is available, then the computational complexity of PF-TBD algorithm increases dramatically.

The DP-TBD technique has been applied to the target tracking in ViSAR images in recent years. To the best of our knowledge, the DP-TBD technique was first applied to the problem of detecting and tracking dim targets with slow maneuvers in ViSAR imagery in our paper [5]. Qin et al. [36] proposed a joint processing strategy-based DP-TBD (JP-DP-TBD) algorithm to track high-maneuvering targets using both location and radial velocity information in the video SAR image and range-Doppler (RD) spectrum, respectively. This algorithm first conducts a broad preliminary search for the possible states of maneuvering targets by increasing the number of target transition states. It then screens the effective transition states of the targets based on the joint intensity values in both domains. Finally, it accumulates the normalized joint intensity as the value function to enhance the algorithm's detection and tracking performance for highly maneuverable targets and its ability to suppress false alarms.

12.2.3 Tracking with Appearance Information

The TBD technique has produced promising results on the simultaneous detection and tracking of low SNR targets in ViSAR data [5, 36]. TBD-based algorithms process a batch of foreground images generated from original video data by a series of operations, e.g., registration [10], background differencing, constant false-alarm rate (CFAR) detection [11], among others. These operations increase the computational cost and decrease the computation speed. To reduce the complexity and increase the speed of the algorithm, tracking directly using the original video data is desired.

Correlation filtering (CF)-based algorithms are widely used for visual object tracking, whose central strategy is learning a prominent target template [37–45]. Recently, a number of CF-based algorithms have been developed for tracking targets in ViSAR data [46–48]. To correct the model drift of correlation filters, Zhao et al. [46] fused the shadow tracking results obtained by the spatial-temporal regularized BACF (ST-BACF) with the detection results obtained by the multi-scale patch-based contrast measure (MPCM) method. This fusion approach accurately locates the shadow targets. The effectiveness of the algorithm was verified on a public dataset released by Sandia National Laboratories (SNL) [49] and a private dataset released by Institute 23, China Aerospace Science and Industry Corporation (CASIC). Zhong et al. [47] proposed a joint kernelized correlation filters (JKCF) framework for moving target tracking. In this framework, the shadow features of the target in consecutive SAR images and the energy features in the RD domain are first extracted.

Then, the target is tracked jointly in the two domains based on JKCF to obtain the preliminary tracking positions in both the image domain and the RD domain. To address the issues of tracking drift and failure, an interactive processing technique is employed to correct the preliminary tracking results in the two domains, thereby improving the accuracy of target localization. Liu et al. [48] combined correlation filtering techniques with the interacting multiple model probabilistic data association filter (IMM-PDAF) algorithm. They assumed that the target motion model could be a stopping model, a constant acceleration model, or a nearly constant velocity (NCV) model. By integrating the target's appearance and motion features, they achieved robust tracking for weak targets in video SAR image sequences and validated the method on a public dataset released by SNL.

12.3 Track Filtering

12.3.1 Dynamic Models

We assume that M targets move independently in the XY-plane with the NCV model [24, 50]. The state vector of the nth target at discrete time k, is defined by

$$\mathbf{x}_k^n := [x_k^n \quad \dot{x}_k^n \quad y_k^n \quad \dot{y}_k^n]', \quad n = 1, 2, ..., M, \tag{12.5}$$

where (x^n, y^n) and $(\dot{x}^n, \dot{y}^n)$ are the Cartesian position coordinates and corresponding velocity components, respectively. Moreover, $k = 1, ..., K$, where K is the number of consecutive frames in ViSAR imagery. We use the direct discrete kinematic model [24] and assume a constant measurement time interval T. The dynamic model of the target is given by [24]

$$\mathbf{x}_k^n = \mathbf{F}\mathbf{x}_{k-1}^n + \mathbf{G}\mathbf{w}_{k-1}^n, \tag{12.6}$$

where $\mathbf{F}$, $\mathbf{G}$, and $\mathbf{w}_{k-1}^n$ are the state transition matrix, process noise gain, and zero-mean white Gaussian process noise with covariance matrix $\mathbf{Q}$, respectively. The matrices $\mathbf{F}$ and $\mathbf{G}$ are then given by

$$\mathbf{F} = \mathbf{I}_2 \otimes \begin{bmatrix} 1 & T \\ 0 & 1 \end{bmatrix}, \tag{12.7}$$

$$\mathbf{G} = \mathbf{I}_2 \otimes \begin{bmatrix} T^2/2 \\ T \end{bmatrix}, \tag{12.8}$$

where $\otimes$ denotes the Kronecker product [51]. The process noise forms a white noise sequence, and $\mathbf{Q}$ [24] is given by

$$\mathbf{Q} = \mathbf{G}\,\mathrm{diag}(\sigma_x^2, \sigma_y^2)\,\mathbf{G}', \tag{12.9}$$

where σ_x^2 and σ_y^2 are the variances of the acceleration process noise along the X and Y directions respectively.

In this chapter, we use $\mathbf{X}^n$ to denote the states of the nth target from frame 1 to K, namely $\mathbf{X}^n := \left(\mathbf{x}_1^n, \mathbf{x}_2^n, \cdots, \mathbf{x}_K^n\right)$. We use $\mathbf{X}_k$ to denote the states of all M targets in the kth frame, namely $\mathbf{X}_k := ((\mathbf{x}_k^1)', (\mathbf{x}_k^2)', \cdots, (\mathbf{x}_k^M)')'$. In addition, we use $\mathbf{X}$ to denote the states of all M targets from frame 1 to K, $\mathbf{X} := ((\mathbf{X}^1)', (\mathbf{X}^2)', \cdots, (\mathbf{X}^M)')' = (\mathbf{X}_1, \mathbf{X}_2, \cdots, \mathbf{X}_K)$.

12.3.2 Measurement Models

12.3.2.1 Image Registration Model

In the ViSAR system, the sub-aperture images are formed in a local output coordinate system, and rotate with the rotation of the local output coordinate system. Thus, the sub-aperture images need to be calibrated to the ground output coordinate system through image registration. We apply a joint registration and reconstruction method [52] to the ViSAR sub-aperture images first. The method employs the framework of common and innovative visuals (CIV). In the CIV framework, we consider a consecutive set of original ViSAR sub-aperture images $\{\boldsymbol{f}_k\}_{k=1}^{K}$, where $\boldsymbol{f}_k$ denotes the kth original image. Each original image $\boldsymbol{f}_k$ is decomposed into two components: the common frame $\boldsymbol{c}$ and innovative frame $\boldsymbol{i}_k$. The common frame $\boldsymbol{c}$ is composed of the background that is common (not changing) among the input original images, accounting for the temporal correlation. The innovative frame $\boldsymbol{i}_k$ comprises the moving targets, clutter, speckle noise, object deformation and variations in illumination, which is specific for each sub-aperture image.

The relationship between the original image $\boldsymbol{f}_k$, and the common and innovative frames can be modeled by

$$\boldsymbol{f}_k = E(\boldsymbol{c}, \boldsymbol{\Theta}_k) + \boldsymbol{i}_k, \quad k = 1, 2, \ldots, K, \tag{12.10}$$

where each image $\boldsymbol{f}_k$ is synthesized by applying an alignment function E on the common frame $\boldsymbol{c}$ with a set of alignment parameters $\boldsymbol{\Theta}_k$, and then adding the respective innovative frame $\boldsymbol{i}_k$. The registration shifts are represented by the alignment function E, which could be either a linear or a nonlinear image registration algorithm and is capable of nonrigid transformations.

For simplicity of implementation, we adopt an affine transform [52] for the registration algorithm E. Let $\boldsymbol{c}' = E(\boldsymbol{c}, \boldsymbol{\Theta}_k)$ be the aligned common frame with alignment parameters $\boldsymbol{\Theta}_k = (a_k, b_k, \alpha_k, \theta_k)$, such that $\boldsymbol{c}'$ can be obtained from $\boldsymbol{c}$, using circular shifting by a_k and b_k pixels in the X and Y directions, scaling by α_k, and finally rotating by an angle θ_k. More specifically, a pixel in (c_1, c_2) of $\boldsymbol{c}$ is mapped into (c_1', c_2') of $\boldsymbol{c}'$ by

$$\begin{bmatrix} c_1' \\ c_2' \end{bmatrix} = \alpha_k \begin{bmatrix} \cos\theta_k & -\sin\theta_k \\ \sin\theta_k & \cos\theta_k \end{bmatrix} \begin{bmatrix} c_1 \\ c_2 \end{bmatrix} + \begin{bmatrix} a_k \\ b_k \end{bmatrix}. \tag{12.11}$$

Considering that images and frames are denoted in the vectorized form (e.g. raster scan), the affine transform E is a linear function of the common frame $\boldsymbol{c}$. Thus E can be represented by a matrix operation, i.e., there exists a matrix $\boldsymbol{A}(\boldsymbol{\Theta}_k)$ such that $E(\boldsymbol{c}, \boldsymbol{\Theta}_k) = \boldsymbol{A}(\boldsymbol{\Theta}_k)\boldsymbol{c}$. Hence, the original ViSAR image $\boldsymbol{f}_k$ can be represented by

$$\boldsymbol{f}_k = \boldsymbol{A}(\boldsymbol{\Theta}_k)\boldsymbol{c} + \boldsymbol{i}_k, \quad k = 1, 2, \ldots, K. \tag{12.12}$$

The key to morph all the original images $\{\boldsymbol{f}_k\}_{k=1}^{K}$ in the same geometry is to find the corresponding alignment parameters $\{\boldsymbol{\Theta}_k\}_{k=1}^{K}$. The registration and reconstruction for the common and innovative frames can be represented in a unified framework, as

$$\begin{bmatrix} \boldsymbol{f}_1 \\ \boldsymbol{f}_2 \\ \vdots \\ \boldsymbol{f}_K \end{bmatrix} = \begin{bmatrix} \boldsymbol{A}(\boldsymbol{\Theta}_1) & \boldsymbol{I} & \boldsymbol{0} & \ldots & \boldsymbol{0} \\ \boldsymbol{A}(\boldsymbol{\Theta}_2) & \boldsymbol{0} & \boldsymbol{I} & \ldots & \boldsymbol{0} \\ \vdots & \vdots & \vdots & \ddots & \vdots \\ \boldsymbol{A}(\boldsymbol{\Theta}_K) & \boldsymbol{0} & \boldsymbol{0} & \ldots & \boldsymbol{I} \end{bmatrix} \begin{bmatrix} \boldsymbol{c} \\ \boldsymbol{i}_1 \\ \vdots \\ \boldsymbol{i}_K \end{bmatrix}. \tag{12.13}$$

This system of equations can be solved by the existing scheme proposed in [52], which jointly obtains $\{\boldsymbol{\Theta}_k\}_{k=1}^{K}$ and reconstructs $\boldsymbol{c}$ and $\{\boldsymbol{i}_k\}_{k=1}^{K}$. Finally, using the obtained alignment parameters $\{\boldsymbol{\Theta}_k\}_{k=1}^{K}$, all the original images can be morphed in the same geometry by

$$\boldsymbol{y}_k = E^{-1}(\boldsymbol{f}_k, \boldsymbol{\Theta}_k), \quad k = 1, 2, \ldots, K, \tag{12.14}$$

where E^{-1} stands for the inverse affine transform of E and $\boldsymbol{y}_k$ is the registered ViSAR image.

To further reduce noise and clutter, we employ a two-parameter CFAR detector [53]. Finally, we perform a morphological opening process [54] to eliminate false alarms and obtain multiple candidate targets in each frame. The obtained candidate targets include moving targets and clutter.

We discretize the observation region into an $N_x \times N_y$ grid of cells, where N_x and N_y are the numbers of cells along the X and Y directions respectively. We use $\mathbf{y}_k$ to denote the kth image frame, which is an $N_x \times N_y$ array. For the ViSAR images, only intensity information is available. Let $\mathbf{y}_k(l, m) \in \mathcal{R}$ denote the amplitude of reflected signals recorded in the (l, m)th resolution cell of $\mathbf{y}_k$, where $l = 1, 2, \ldots, N_x$, $m = 1, 2, \ldots, N_y$. We use $\mathbf{Y}$ to denote the frame sequence from 1 to K, $\mathbf{Y} := (\mathbf{y}_1, \mathbf{y}_2, \cdots, \mathbf{y}_K)$.

12.3.2.2 Measurement Model

The measurement model is described by [24]

$$\mathbf{z}_k = \mathbf{H}\mathbf{x}_k + \mathbf{n}_k, \tag{12.15}$$

where $\mathbf{z}_k := \left[z_{x,k}, z_{y,k}\right]^{\mathrm{T}}$ denotes the position measurement at time k and $\mathbf{H}$ and $\mathbf{n}_k$ denote the constant measurement matrix and the zero-mean white Gaussian measurement noise with covariance $\mathbf{R}$, respectively. The matrices $\mathbf{H}$ and $\mathbf{R}$ are defined as

$$\mathbf{H} = \begin{bmatrix} 1 & 0 & 0 & 0 \\ 0 & 0 & 1 & 0 \end{bmatrix}, \tag{12.16}$$

$$\mathbf{R} = \sigma_z^2 \mathbf{I}_2, \tag{12.17}$$

where σ_z^2 is the variance of the measurement noise.

12.3.3 Joint Detection and Tracking Using Track-before-detect

The DP-TBD algorithm [50] estimates the states of multiple targets by maximizing the joint posterior pdf, $p(\mathbf{X}|\mathbf{Y})$, as

$$\hat{\mathbf{X}} = \arg \max_{\mathbf{X} \in \overline{\mathcal{R}^s}^{MK}} p(\mathbf{X}|\mathbf{Y}), \text{ s.t. } p(\hat{\mathbf{X}}|\mathbf{Y}) > V_T, \tag{12.18}$$

where $\hat{\mathbf{X}}$ denotes the estimated states of multiple targets, V_T is a threshold, and $\overline{\mathcal{R}^s}^{MK}$ represents the discretized state space of $\mathbf{X}$.

At discrete time k, the observation region R_k can be partitioned into $M + 1$ mutually disjoint partitions, defined as $\Omega_k^0, \Omega_k^1, \ldots, \Omega_k^M$, such that

$$R_k = \bigcup_{n=0}^{M} \Omega_k^n \text{ and } \Omega_k^n \cap \Omega_k^m = \emptyset \text{ for } m \neq n, \tag{12.19}$$

where Ω_k^0 denotes the sub-region where no target is located, and Ω_k^n $(n = 1, 2, \ldots, M)$ denotes a rectangular sub-region centered at the position of the nth target. Let $\mathbf{y}_k(\Omega_k^n)$ denote the measurements in Ω_k^n, which consists of measurements for the nth target and noise. For this kind of scenario, $p(\mathbf{X}|\mathbf{Y})$ can be approximately calculated as

$$p(\mathbf{X}|\mathbf{Y}) \tilde{\propto} \prod_{n=1}^{M} \pi(\mathbf{X}^n|\mathbf{Y}(\Omega^n)), \tag{12.20}$$

where $\tilde{\propto}$ means "is approximately proportional to," Ω^n denotes the sub-regions centered at the positions of the nth target, and $\pi(\mathbf{X}^n|\mathbf{Y}(\Omega^n))$ denotes the posterior pdf of the nth target state. According to (12.20), we can estimate the multitarget states individually using a single-target DP-TBD (ST-DP-TBD) algorithm [55], which implements merit integration over all K consecutive frames. According to Bayes' rule [24], $\pi(\mathbf{X}^n|\mathbf{Y}(\Omega^n))$ can be expressed as

$$\pi(\mathbf{X}^n|\mathbf{Y}(\Omega^n)) \propto \prod_{k=1}^{K} \pi(\mathbf{x}_k^n|\mathbf{y}_k(\Omega_k^n)) \propto \prod_{k=1}^{K} \pi(\mathbf{y}_k(\Omega_k^n)|\mathbf{x}_k^n)p(\mathbf{x}_k^n|\mathbf{x}_{k-1}^n). \tag{12.21}$$

Consequently, the merit integration in the ST-DP-TBD algorithm is decomposed into K subprocesses, and is implemented recursively. The ST-DP-TBD algorithm decides the existence of a target by thresholding the integrated merit functions with a constant value V_T. If the integrated merit value exceeds V_T, then the ST-DP-TBD algorithm declares the target's presence; otherwise we say that no target is present in the scene [56]. The detection and final state estimate of a target can be formulated as,

$$\hat{\mathbf{x}}_K^n = \arg \max_{\mathbf{x}_K^n \in \mathcal{R}^s} I(\mathbf{x}_K^n|\mathbf{y}_K(\Omega_K^n)), \text{ s.t. } I(\hat{\mathbf{x}}_K^n|\mathbf{y}_K(\Omega_K^n)) > V_T, \tag{12.22}$$

where $I(\mathbf{x}_K^n|\mathbf{y}_K(\Omega_K^n))$ denotes the merit function of $\mathbf{x}_K^n$. The estimated states of a target, $\{\hat{\mathbf{x}}_k^n\}_{k=1}^{K-1}$, are obtained by exploiting a retracing function $\Psi_k(\mathbf{x}_k^n)$.

The detailed procedures of the conventional DP-TBD algorithm [57] are described in Algorithm 12.1. Here the amplitude of reflected signals is chosen as a merit function since only intensity information is available in the ViSAR images. For simplicity, we use $I(\mathbf{x}_k^n)$ to replace $I(\mathbf{x}_k^n|\mathbf{y}_k(\Omega_k^n))$. In Algorithm 12.1, $\mathcal{R}^s$ represents the discretized state space, and N_C denotes the number of candidate targets (including true targets and clutter) in the first frame. Let $\tau(\mathbf{x}_k^n)$ denote a collection of discrete target states at discrete time $k-1$, from which a transition to $\mathbf{x}_k^n$ is possible. Generally, we choose the threshold V_T according to a minimum Bayesian risk criterion or some false-alarm specifications [58]. Due to the merit function expansion, there remains inappropriate tracks, e.g., similar tracks. Thus, a sorting and merging process [50] is performed to remove the inappropriate tracks.

In Algorithm 12.1, the position components x and y are quantified with the range cell in the discretized state space, such that $(x, y) \in \{(x, y) : x \in [N_x], y \in [N_y]\}$. The notation $[N]$ denotes the set $\{1, 2, \ldots, N\}$. The velocity components $\dot{x}$ and $\dot{y}$ are also quantified with the range cell [59]. If the maximum velocity is $v_{\max}$, the ranges of $\dot{x}$ and $\dot{y}$ can be $\dot{x}, \dot{y} \in [-v_{\max}, v_{\max}]$. In addition, for each state $\mathbf{x}_{k-1} = [x, \dot{x}, y, \dot{y}]_{k-1}^{'}$, N_s valid transition states $\mathbf{x}_k^i$ $(i = 1, 2, \ldots, N_s)$ are chosen at discrete time k. For the simplest scenario of tracking a slowly maneuvering target, we set $N_s = 4$, $T = 1$, and draw quantified acceleration process noise $\delta\dot{x}, \delta\dot{y} \in \{0, 1\}$. The valid transition states are $\mathbf{x}_k^i \in \{[x + (\dot{x} + \delta\dot{x}), \dot{x} + \delta\dot{x}, y + (\dot{y} + \delta\dot{y}), \dot{y} + \delta\dot{y}]_k^{'} : \delta\dot{x}, \delta\dot{y} \in \{0, 1\}\}$.

In practice, it is hard to obtain accurate Ω_k in (12.21), thus Ω_k is replaced by its estimate $\hat{\Omega}_k$. Usually the likelihood $\pi(\mathbf{y}_k(\Omega_k)|\mathbf{x}_k)$ is approximated by $\pi(\mathbf{y}_k(\hat{\Omega}_k)|\mathbf{x}_k)$ in the discrete space. The conventional DP-TBD algorithm is suitable for tracking non-maneuvering targets since $\hat{\Omega}_k$ consists of N_s rectangular sub-regions centered at the position of $\mathbf{x}_k^i$ $(i = 1, 2, \ldots, N_s)$ at discrete time k. Thus $\hat{\Omega}_k$ is determined by the $\delta\dot{x}$, $\delta\dot{y}$, and N_s. If N_s, $\delta\dot{x}$, and $\delta\dot{y}$ are small, $\hat{\Omega}_k$ covers the true positions of non-maneuvering targets. However, the conventional DP-TBD algorithm has limitation in maneuvering target tracking since $\hat{\Omega}_k$ may not cover the true positions based on small N_s, $\delta\dot{x}$, and $\delta\dot{y}$. We increase N_s, $\delta\dot{x}$, and $\delta\dot{y}$ to larger values, which results in an expanding $\hat{\Omega}_k$. The expanding $\hat{\Omega}_k$ will include more false paths, and the merit integration in $\hat{\Omega}_k$ will degrade in performance.

Moreover, the quantified acceleration process noise components $\delta\dot{x}$ and $\delta\dot{y}$ bring estimation errors in the tracking process. Thus, an improved DP-TBD algorithm is required to solve the problem of detecting and tracking maneuvering targets.

Algorithm 12.1: Conventional DP-TBD Algorithm

Inputs: The prior distribution $p(\mathbf{x}_1^n)$ and the measurements $\mathbf{Y}$.
Outputs: Estimate of the number of targets M, and estimated states $\hat{\mathbf{X}}$.

for $n = 1$ to N_C **do**
    Initialization: at discrete time $k = 1$, $I(\mathbf{x}_1^n) = \mathbf{y}_1(l, m)$, $\Psi_k(\mathbf{x}_1^n) = 0$.
    Recursion:
    for $k = 2$ to K **do**
        For all $\mathbf{x}_k^n \in \overline{\mathcal{R}^s}$,

$$I(\mathbf{x}_k^n) = \max_{\mathbf{x}_{k-1}^n \in \tau(\mathbf{x}_k^n)} [I(\mathbf{x}_{k-1}^n)] + \mathbf{y}_k(l, m),$$

$$\Psi_k(\mathbf{x}_k^n) = \arg \max_{\mathbf{x}_{k-1}^n \in \tau(\mathbf{x}_k^n)} [I(\mathbf{x}_{k-1}^n)].$$

    end for
    Termination: For a threshold V_T find $\hat{\mathbf{x}}_K^n = \{\mathbf{x}_K^n : I(\mathbf{x}_K^n) > V_T\}$.
    Backtracking:
    for $k = K - 1$ to 1 **do**
        $\hat{\mathbf{x}}_k^n = \Psi_{k+1}(\mathbf{x}_{k+1}^n)$.
    end for
    Obtain the estimated states of the nth target $\hat{\mathbf{X}}^n := \left(\hat{\mathbf{x}}_1^n, \hat{\mathbf{x}}_2^n, \ldots, \hat{\mathbf{x}}_K^n\right)$.
end for
Perform sorting and merging to detect the targets and estimate corresponding tracks $\hat{\mathbf{X}}$.

12.3.4 Real-time Tracking Using CF

Among these CF-based algorithms, the BACF algorithm [44] has demonstrated accurate and robust tracking results. Thus, the BACF is a good choice for tracking targets using ViSAR videos.

We track targets using the BACF tracker [44] due to its high accuracy and robustness against the challenges of deformation, scaling, and background clutter. The BACF tracker aims to separate each target from the surrounding background by learning/updating a multichannel convolution filter using a number of training samples. Specifically, the training samples consist of all positive and negative feature samples extracted from the image.

Generally, there are two stages for the application of the BACF, i.e., the training and tracking. In the training process, the filter $\mathbf{h}$ is trained by minimizing the following objective function:

$$E(\mathbf{h}) = \frac{1}{2} \sum_{c=1}^{C} ||\mathbf{y}(c) - \sum_{\ell=1}^{L} \mathbf{h}_\ell^{\mathrm{T}} \mathbf{P} \mathbf{u}_\ell[\Delta\tau_c]||_2^2 + \frac{\gamma}{2} \sum_{\ell=1}^{L} ||\mathbf{h}_\ell||^2, \tag{12.23}$$

where C is the length of the feature sample $\mathbf{u}$, L is the number of feature channels, and $\mathbf{u}_\ell \in \mathbb{R}^C$ and $\mathbf{h}_\ell \in \mathbb{R}^U$ denote the ℓth channel of the feature sample $\mathbf{u}$ and filter $\mathbf{h}$, respectively. The binary matrix $\mathbf{P} \in \mathbb{R}^{U \times C}$ helps choose U elements in the middle of $\mathbf{u}_l$, where $C \gg U$. $\mathbf{y} \in \mathbb{R}^C$ is the desired correlation response, whose cth element is denoted by $\mathbf{y}(c)$. The symbol $\mathbf{u}_\ell[\Delta\tau_c]$ denotes the c-step discrete circularly shifted sample of $\mathbf{u}_\ell$. The variable γ is a regularization parameter that controls the overall complexity of the objective function. The superscript T denotes the transpose of a complex vector or matrix, and $|| \cdot ||_2^2$ denotes the L2-norm of a variable.

In the tracking process, the correlation response of the filter $\mathbf{h}$ on a test feature sample $\boldsymbol{\nu}$, $S_{\mathbf{h}}(\boldsymbol{\nu})$, is calculated as

$$S_{\mathbf{h}}(\boldsymbol{\nu}) = \sum_{\ell=1}^{L} \mathbf{h}_\ell^{\mathrm{T}} \mathbf{P} \boldsymbol{\nu}_\ell, \tag{12.24}$$

where $\boldsymbol{\nu}_\ell \in \mathbb{R}^C$ denotes the ℓth channel of $\boldsymbol{\nu}$.

Furthermore, using the convolution property of discrete Fourier transform (DFT), the classification scores at all spatial locations in a frame are calculated as

$$\mathbf{S}_{\mathbf{h}}(\boldsymbol{\nu}) = \mathcal{F}^{-1}\left[\sum_{\ell=1}^{L} \bar{\boldsymbol{\nu}}_\ell \cdot \left(\sqrt{C}\mathbf{D}\mathbf{P}^{\mathrm{T}}\mathbf{h}_\ell\right)\right], \tag{12.25}$$

where $\bar{\boldsymbol{\nu}}_\ell$ denotes the DFT of $\boldsymbol{\nu}_\ell$, $\cdot$ denotes pointwise multiplication, $\mathbf{D}$ denotes the $C \times C$ DFT matrix, and $\mathcal{F}^{-1}$ denotes the inverse DFT. For details of the BACF tracker, please see [44].

However, when applied to ViSAR videos, there exist a number of issues that degrade the performance of the BACF. In Figure 12.1, target detections in original frames 7 and 53 using the BACF algorithm are shown in the first column. Correspondingly, the response maps are shown in the second column. In the original images, the bounding boxes indicate the true and estimated positions, respectively. The acronym GT is an abbreviation for *ground truth*. In the response maps, the arrows indicate the correlation outputs of the true and estimated targets, respectively. It's observed that there are probably multiple peaks in the response map, and the highest peak does not correspond to the true target. We define this as the multipeak problem in this study (as shown in Figure 12.1c, d). The multipeak problem is encountered frequently with CF-based algorithms, especially when

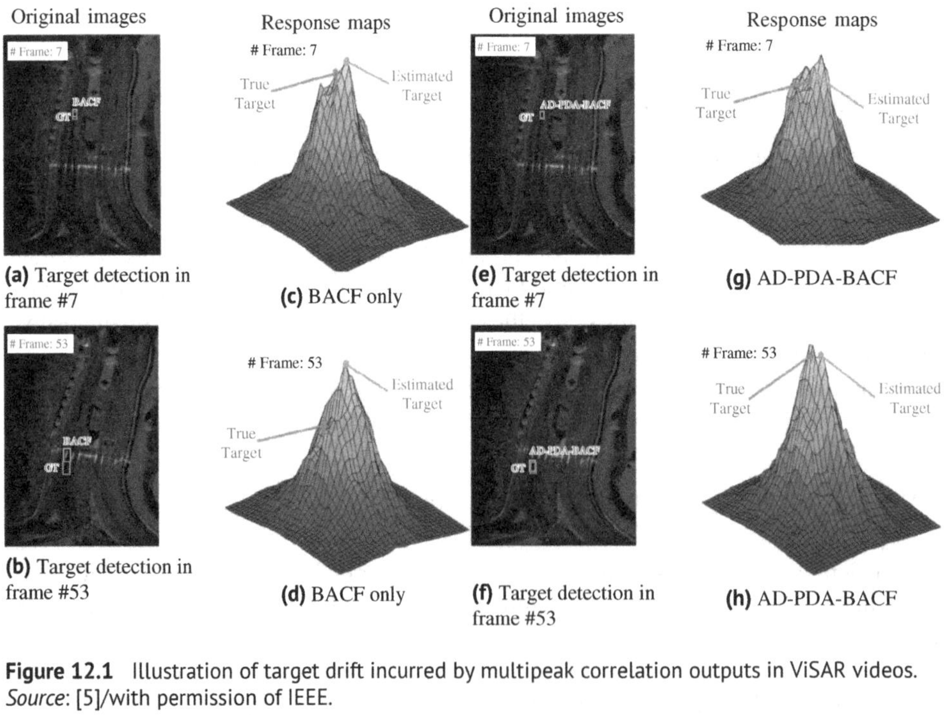

(a) Target detection in frame #7 **(c)** BACF only **(e)** Target detection in frame #7 **(g)** AD-PDA-BACF

(b) Target detection in frame #53 **(d)** BACF only **(f)** Target detection in frame #53 **(h)** AD-PDA-BACF

Figure 12.1 Illustration of target drift incurred by multipeak correlation outputs in ViSAR videos. *Source*: [5]/with permission of IEEE.

tracking dim and small targets in complex environments with many kinds of interference, such as strong clutter, nearby similar objects, or both. In addition, the estimated target position becomes farther away from the true target position over time (from frames 7–53, as shown in Figure 12.1a, b). This is because incorrect detections generate template model-based errors, which lead to tracker drift during long-term tracking. Thus, it is highly desirable to address the multipeak problem to update the target template model with high confidence and few errors.

The CF-based methods mentioned in Section 12.2.3 are not designed to address multipeak problem. In this chapter, we propose a novel AD-PDA algorithm that is able to take advantage of the appearance and distance information and localizes targets more accurately despite the multipeak response. The proposed algorithm can be combined with any CF-based algorithm when the validated measurements contain at most one target-originated measurement.

12.4 TBD Algorithms

We propose a novel ES-TBD algorithm to tackle the multiple maneuvering target tracking problem under low SNR conditions. We assume that the number of targets is fixed during the tracking process. The key idea of the proposed ES-TBD algorithm is the expanding and shrinking strategy, which is introduced in Section 12.4.1. The details of the proposed algorithm are given in Section 12.4.2.

12.4.1 Expanding and Shrinking Strategy

The goal of this section is to present the expanding and shrinking strategy, which is adopted to approximate Ω_k^n in the ES-TBD algorithm.

First, at discrete time $k-1$, we perform the expanding operation. We assign an independent PF [60] to each candidate target. Each PF consists of N particles $\{\chi_{k-1}^{n,i}, i = 1, 2, \ldots, N\}$. The particles in a PF are sampled from a prior distribution of the state $\mathbf{x}_{k-1}^n$. Let Ω_{k-1}^n denote a rectangular sub-region centered at the position of $\mathbf{x}_{k-1}^n$. The expanded sub-region of the nth target, $\Omega_{k-1}^{n,e}$, is defined as the union of N rectangular sub-regions,

$$\Omega_{k-1}^{n,e} = \bigcup \Omega_{k-1}^{n,i}, i = 1, 2, \ldots, N, \tag{12.26}$$

where $\Omega_{k-1}^{n,i}$ is a rectangular sub-region centered at the position of $\chi_{k-1}^{n,i}$. Compared to Ω_{k-1}^n, we can see that $\Omega_{k-1}^{n,e}$ is expanded. Next, at discrete time k, we perform the state evolution based on $\Omega_{k-1}^{n,e}$. All PFs (assigned to true targets or clutter) evolve according to the dynamic models of targets (12.6). The ith predicted particle in the nth PF is

$$\chi_{k|k-1}^{n,i} = \mathbf{F}\chi_{k-1}^{n,i} + \mathbf{G}\mathbf{w}_{k-1}^{n,i}. \tag{12.27}$$

Each predicted particle in a PF represents a possible transition state of the target at k. Since $\mathbf{w}_{k-1}^{n,i}$ is not quantified with the range cell, we can select the acceleration process noise components $\delta\dot{x}$ and $\delta\dot{y}$ according to the target dynamic model. Thus, at k, the transition states $\{\chi_{k|k-1}^{n,i}, i = 1, 2, \ldots, N\}$ are more accurate than those in the conventional DP-TBD algorithm. The predicted expanded sub-region $\Omega_{k|k-1}^{n,e}$ is defined as the union of N rectangular sub-regions, with each centered at the position of $\chi_{k|k-1}^{n,i}$.

Secondly, we perform the shrinking operation to reduce computational burden during the merit integration. The posterior probability density of each predicted particle, $\pi(\chi_{k|k-1}^{n,i}|\mathbf{y}_k(\Omega_{k|k-1}^{n,e}))$, is calculated in the Bayesian framework. A resampling step [60] is performed to concentrate on particles with large weights, generating particles $\{\chi_k^{n,i*}, i = 1, 2, ..., N\}$ by

$$\pi\left(\chi_k^n|\mathbf{y}_k\left(\Omega_{k|k-1}^{n,e}\right)\right) \approx \sum_{i=1}^{N} \pi\left(\chi_{k|k-1}^{n,i}|\mathbf{y}_k\left(\Omega_{k|k-1}^{n,e}\right)\right) \times \delta\left(\chi_k^n - \chi_{k|k-1}^{n,i}\right), \tag{12.28}$$

so that

$$\mathrm{P_r}\left(\chi_k^{n,i*} = \chi_{k|k-1}^{n,i}\right) = \pi\left(\chi_{k|k-1}^{n,i}|\mathbf{y}_k\left(\Omega_{k|k-1}^{n,e}\right)\right). \tag{12.29}$$

The resampled particles in a PF represent the valid transition states of a target at k. Let $\Omega_{k|k-1}^{n,s}$ denote the sub-region consisting of N rectangular sub-regions. Each rectangular sub-region centers on the position of $\chi_k^{n,i*}$. The resampled particles contain many repeated points. The number of different resampled particles, denoted as N_s, is smaller than N. Thus $\Omega_{k|k-1}^{n,s}$ shrinks. In other words, $\Omega_{k|k-1}^{n,s}$ represents the predicted shrunken sub-region. Meanwhile, the number of possible transition states in $\Omega_{k|k-1}^{n,s}$ is reduced. The resampling step ensures that the possible transition states with higher merit values are selected. Thus after the shrinking operation, the searching ranges in $\Omega_{k|k-1}^{n,s}$ can cover the true paths. We denote the N_s different resampled particles as $\chi_k^{n,j}$, $j = 1, 2, ..., N_s$. At discrete time k, the N_s particles are considered as the valid transition states of a target.

Finally, we obtain the approximated region $\hat{\Omega}_k^n$, which is defined as the union of N_s rectangular sub-regions. Each rectangular sub-region centers on the position of $\chi_k^{n,j}$. Obviously, $\hat{\Omega}_k^n \subset \Omega_{k|k-1}^{n,s}$. Based on $\hat{\Omega}_k^n$, the ES-TBD algorithm can perform merit integration efficiently.

Figure 12.2 shows the comparison of the transition states between the conventional DP-TBD (Figure 12.2a) and the proposed ES-TBD (Figure 12.2b) algorithms. The parameters are set as $\mathbf{x}_{k-1} = [4, 3, 6, 3]'$, and $N_s = 4$. The true cell occupied by a target is marked with a star. The solid rectangles represent the valid transition states. The dashed rectangles represent the expanded searching ranges in the proposed ES-TBD algorithm. The states of targets are propagated according to the dynamic model for both the conventional DP-TBD and the proposed ES-TBD algorithms. In the conventional DP-TBD algorithm, the quantified acceleration process noise components are chosen as $\delta\dot{x}, \delta\dot{y} \in [0, 1]$. We can see that the valid transition states cannot cover the true state of the targets at discrete time $k + 1$. In the proposed ES-TBD algorithm, the acceleration process noise is selected according to the dynamic model. The valid transition states are obtained using the expanding and shrinking strategy. Compared with the conventional DP-TBD algorithm

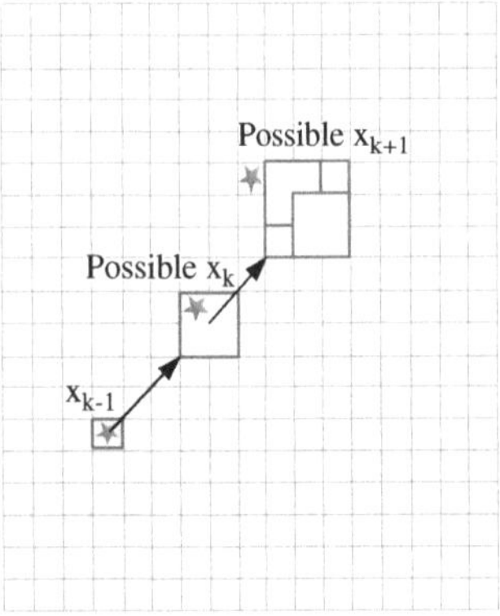

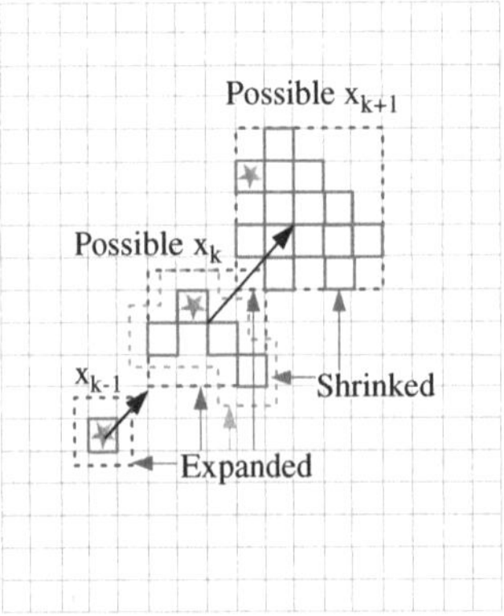

Figure 12.2 Comparison of the transition states between the (a) conventional DP-TBD and the (b) proposed ES-TBD algorithms.

(Figure 12.2a), the searching procedure of the proposed ES-TBD algorithm (Figure 12.2b) is performed in an expanded range. Therefore, the ES-TBD is more robust to detect and track targets with maneuvers. A number of valid transition states with high posterior probability densities in the predicted expanded range are selected to form a predicted shrunken range. The valid transition states are selected from the possible transition states and the merit integration is performed in a shrunken range, resulting in a more efficient DP integration.

Remark 12.1: The proposed ES-TBD algorithm adopts the resampling step, which is a fundamental component of the standard PF and does not affect the convergence property of a PF [61]. The impact of the resampling step is that the particles with small weights are eliminated and those particles with large weights are duplicated many times. Thus the distribution area of particles shrinks, concentrating on the particles with large weights. The readers can refer to [60] for more details of the resampling step of PF.

12.4.2 Details of the Proposed ES-TBD Algorithm

Next we describe five main steps of the proposed algorithm.

(1) Track Initiation: At discrete time $k = 1$, we assume that there are N_C candidate targets (including true targets and clutter) in the first frame, which are labeled sequentially. Let $\mathbf{x}_1^n$ ($n = 1, 2, ..., N_C$) denote the initial state of a candidate target. First, the merit function of each candidate target is initialized as in the conventional DP-TBD algorithm. Secondly, each candidate target is assigned with an independent PF, which consists of N particles. Thus, there are totally N_C PFs. The ith particle in the nth PF is denoted as $\chi_1^{n,i}$, where $i = 1, 2, ..., N$. We assume that the particles in each PF are drawn from a prior distribution, for example, $\chi_1^{n,i} \sim \mathcal{N}(\mathbf{x}_1^n, \mathbf{P}_1^n)$, where $\mathbf{P}_1^n$ denotes the covariance matrix of the nth target. The initial weights of the particles in each PF $w_1^{n,i}$ are set to $\frac{1}{N}$, $i = 1, 2, ..., N$.

(2) Prediction of PFs: At discrete time $k > 1$, all the C PFs evolve from $k - 1$ to k according to the NCV model (12.6). In each PF, we obtain the predicted particles $\{\chi_{k|k-1}^{n,i}, i = 1, 2, ..., N\}$, which are considered as the possible transition states of the target at k.

(3) Weight Calculation in Each PF: At discrete time $k > 1$, we calculate the non-normalized weights of the predicted particles $\{\chi_{k|k-1}^{n,i}, i = 1, 2, ..., N\}$ in each PF as

$$\hat{w}_k^{n,i} = w_{k-1}^{n,i} * p\left(\mathbf{y}_k(l, m)|\chi_{k|k-1}^{n,i}\right), \tag{12.30}$$

where $w_{k-1}^{n,i}$ denotes the weight of the ith particle in the nth PF at $k - 1$, and the likelihood [62] is calculated as

$$p(\mathbf{y}_k(l, m)|\chi_{k|k-1}^{n,i}) = \exp([-I_n(I_n - 2\mathbf{y}_k(l, m))]/2\sigma^2). \tag{12.31}$$

In the above equation σ^2 is the variance of the measurement noise, and I_n denotes the intensity of the nth target. The normalized weights of predicted particles in each PF are

$$w_k^{n,i} = \hat{w}_k^{n,i} / \sum_{j=1}^{N} \hat{w}_k^{n,j}. \tag{12.32}$$

(4) Resampling: The discrete distribution in each PF $\{w_k^{n,i}, i = 1, 2, ..., N\}$ is resampled to obtain the resampled particles $\{\chi_k^{n,i*}, i = 1, 2, ..., N\}$ of the target.

(5) DP Integration: Among the resampled particles, N_s different particles $\{\chi_k^{n,j}, j = 1, 2, ..., N_s\}$ are considered as valid transition states. Merit integration is performed as the Recursion step in the conventional DP-TBD algorithm (Algorithm 12.1).

After implementing the expanding and shrinking strategy for all the candidate targets, the ES-TBD algorithm detects a number of targets. Algorithm 12.2 presents the detailed procedures of the proposed ES-TBD algorithm.

Algorithm 12.2: ES-TBD Algorithm

Inputs: The prior distribution $p(\mathbf{x}_1^n)$ and the measurements $\mathbf{Y}$.
Outputs: Estimate of the number of targets M, and estimated states $\hat{\mathbf{X}}$.

for $n = 1$ to N_C **do**
  Initialization of the Merit Function of State $\mathbf{x}_1^n$: at discrete time $k = 1$, $I(\mathbf{x}_1^n) = \mathbf{y}_1(l, m)$.
  Initialization of Independent PF: at discrete time $k = 1$, draw particles $\chi_1^{n,i} \sim p(\mathbf{x}_1^n)$ and set the weight of each particle $w_1^{n,i}$ to $\frac{1}{N}$, $i = 1, 2, \ldots, N$.
  for $k = 2$ to K **do**
    Initialization of Independent PF: at discrete time $k > 1$, draw particles $\chi_{k-1}^{n,i} \sim p(\mathbf{x}_{k-1}^n)$ and set the weight of each particle $w_{k-1}^{n,i}$ to $\frac{1}{N}$, $i = 1, 2, \ldots, N$.
    Prediction of Independent PF: at discrete time $k > 1$, evolve the nth PF from $k-1$ to k according to the NCV model (12.6), so that the particles $\{\chi_{k-1}^{n,i}, i = 1, 2, \ldots, N\}$ are propagated to the predicted particles $\{\chi_{k|k-1}^{n,i}, i = 1, 2, \ldots, N\}$.
    Weight Calculation in Independent PF: calculate the normalized weight of each predicted particle in the nth PF based on (12.32).
    Resampling: resample the predicted particles in the nth PF to generate $\{\chi_k^{n,i*}, i = 1, 2, \ldots, N\}$, and set the weight of each resampled particle $w_k^{n,i}$ to $\frac{1}{N}$.
    DP Integration: Among the resampled particles, N_s different particles $\{\chi_k^{n,j}, j = 1, 2, \ldots, N_s\}$ in the nth PF are considered as the valid transition states of the target. Perform the merit integration as the Recursion step in Algorithm 12.1, and obtain the merit function $I(\mathbf{x}_k^n)$.
  end for
end for
Perform sorting and merging to detect the targets and estimate corresponding tracks $\hat{\mathbf{X}}$.

12.5 CF Algorithms

We first propose a novel AD-PDA algorithm to address the multipeak problem using multiple peaks of correlation outputs instead of the highest peak. Furthermore, we propose a novel AD-PDA-BACF algorithm for ViSAR shadow tracking by combining the AD-PDA and BACF algorithms.

The main procedures of the proposed AD-PDA-BACF algorithm consist of (1) the selection of multiple peak locations; (2) state estimation using AD-PDA; and (3) the combination of AD-PDA and BACF. The above main procedures are introduced in Sections 12.5.1–12.5.3.

12.5.1 Selection of Multiple Peak Locations

In CF-based algorithms, the location of a target is estimated based on the highest peak of the response map. However, the detection results relying on the highest peak may be erroneous due to similar objects, strong clutter, or both. This leads to erroneous localization of the target. Furthermore, erroneous localization results contaminate the target template due to negative training samples. As shown in Figure 12.1a–d, the peaks in the response map corresponding to the samples centered at similar objects or strong clutter, may approach or even surpass the peak corresponding to the sample centered at the true target. Thus, we consider selecting multiple peaks based on the correlation outputs of the BACF algorithm.

First, we calculate the classification scores at all spatial locations of a test sample $\boldsymbol{\nu}$ in the current input frame at time k. The classification scores $\mathbf{S}_{\mathbf{h}}(\boldsymbol{\nu})$ are calculated according to (12.25).

Second, we calculate the detection scores at all pixel locations in the test sample $\boldsymbol{\nu}$. Using the trigonometric polynomial interpolation strategy [63], the detection score $\mathbf{S}_D$ at the continuous location $(x_k, y_k) \in [0, M) \times [0, N)$ in $\boldsymbol{\nu}$, $\mathbf{S}_D(x_k, y_k)$, is interpolated as

$$\mathbf{S}_D(x_k, y_k) = \frac{1}{MN} \sum_{m=0}^{M-1} \sum_{n=0}^{N-1} \bar{\mathbf{S}}(m, n) e^{i2\pi(\frac{m}{M}x_k + \frac{n}{N}y_k)}, \tag{12.33}$$

where $\bar{\mathbf{S}}$ represents the DFT of $\mathbf{S}_{\mathbf{h}}(\boldsymbol{\nu})$ and the superscript $i = \sqrt{-1}$.

Then, the detection scores are normalized to vary within [0, 1] using the Min-Max method as

$$\mathbf{S}_N := (\mathbf{S}_D - \mathbf{S}_{D,\min})/(\mathbf{S}_{D,\max} - \mathbf{S}_{D,\min}), \tag{12.34}$$

where $\mathbf{S}_{D,\min}$ and $\mathbf{S}_{D,\max}$ represent the minimum and maximum detection scores, respectively.

Due to the nonnegativity of the detection scores, $\mathbf{S}_{D,\min} = 0$. To obtain $\mathbf{S}_{D,\max}$, we first calculate the pixel location (x_k^*, y_k^*) with the maximum score, among all pixel locations in the test sample $\boldsymbol{\nu}$ by solving the following objective function:

$$(x_k^*, y_k^*) = \arg \max_{(x_k, y_k)} \mathbf{S}_D(x_k, y_k). \tag{12.35}$$

The spatial location $(x_k^{(0)}, y_k^{(0)})$ with the maximal score is used as the initial estimate. Then, we iteratively maximize (12.33) using Newton's method, starting at $(x_k^{(0)}, y_k^{(0)})$. The gradient and Hessian in each iteration are computed by analytically differentiating (12.33). The solution to the above function (12.35) converges after a few iterations. Based on the obtained solution (x_k^*, y_k^*), $\mathbf{S}_{D,\max}$ is then calculated as $\mathbf{S}_{D,\max} = \mathbf{S}_D(x_k^*, y_k^*)$.

Finally, we select multiple peak locations based on the normalized detection scores. The peak locations with the first m_k maximum detection scores are selected as the current validated measurements. In detail, given the normalized detection response map $\mathbf{S}_N(\boldsymbol{\nu})$, we calculate the multipeaks by

$$P_N(\boldsymbol{\nu}) = \mathbf{S}_N(\boldsymbol{\nu}) \circ \mathbf{B}. \tag{12.36}$$

In (12.36), $\mathbf{B}$ is a binary matrix with the same size as $\mathbf{S}_N$ that identifies the locations of local maxima in $\mathbf{S}_N$ and $\circ$ represents the elementwise multiplication operator. The elements at the locations of local maxima in $\mathbf{B}$ are set to 1, while others are set to 0. Thus, all nonzero elements in $P_N(\boldsymbol{\nu})$ indicate multiple peaks in the normalized response map of $\boldsymbol{\nu}$.

Using $P_N(\boldsymbol{\nu})$, we obtain a set of pixel locations corresponding to the first m_k maximum detection scores, i.e., $(\tilde{x}_{k,j}, \tilde{y}_{k,j})$, $j = 1, 2, \ldots, m_k$, which are defined as

$$\begin{aligned} \{(\tilde{x}_{k,j}, \tilde{y}_{k,j})\}_{j=1}^{m_k} \;:=\; & \{(\tilde{x}_{k,j}, \tilde{y}_{k,j}) | (\tilde{x}_{k,j}, \tilde{y}_{k,j}) \in [0, M) \times [0, N), \\ & \mathbf{S}_N(\tilde{x}_{k,j}, \tilde{y}_{k,j}) \in P_N(\boldsymbol{\nu}), j = 1, 2, \ldots, m_k\}, \end{aligned} \tag{12.37}$$

where $\mathbf{S}_N(\tilde{x}_{k,j}, \tilde{y}_{k,j})$, $j = 1, 2, \ldots, m_k$ satisfies

$$\begin{aligned} &\mathbf{S}_N(\tilde{x}_{k,1}, \tilde{y}_{k,1}) \geq \mathbf{S}_N(\tilde{x}_{k,2}, \tilde{y}_{k,2}) \geq \ldots \geq \mathbf{S}_N(\tilde{x}_{k,j-1}, \tilde{y}_{k,j-1}) \\ &\geq \mathbf{S}_N(\tilde{x}_{k,j}, \tilde{y}_{k,j}) \geq \ldots \geq \mathbf{S}_N(\tilde{x}_{k,m_k}, \tilde{y}_{k,m_k}). \end{aligned} \tag{12.38}$$

Based on the selected grid locations from (12.38), we obtain the validated measurements at time k, i.e., $\mathbf{z}_{k,j} = [\tilde{x}_{k,j}, \tilde{y}_{k,j}]^{\mathrm{T}}$, $j = 1, 2, \ldots, m_k$.

Remark 12.2: The parameter m_k is used to determine the area of the validated region in the AD-PDA algorithm, which consists of the measurements with large detection scores. In the AD-PDA algorithm, the probability mass function of the number of false measurements in the validated region is modeled as a Poisson distribution with spatial density λ_c. A smaller value of m_k implies that fewer false measurements will fall into the validated region. Consequently, the false measurements that are far away from the predicted location are not considered in the AD-PDA algorithm [22]. In contrast, a larger value of m_k implies that more false measurements will fall into the validated region. To balance the area of the validated region and the tracking accuracy, m_k is fixed manually to $m_k = 30$ in this study.

Remark 12.3: It should be emphasized that, the multipeak problem is different from the high false alarm rate defined in [11]. The difference between the multipeak problem and high false alarm rate are summarized as: (1) The multipeak problem occurs when CF-based algorithms track dim and small targets in complex environments with many kinds of interference, such as strong clutter, nearby similar objects, or both. On the contrary, in practice, CFAR-based algorithms demand that the false alarm rate should be independent of the distribution of the background interference [64]. (2) For the CF-based algorithms, multiple peaks in the response maps are determined by features (Felzenszwalb's histogram of oriented gradients (FHOG) and intensity) of the test samples and learned target templates. For the CFAR-based algorithms, high false alarm rate is obtained when the detection thresholds are low. (3) In the proposed AD-PDA-BACF algorithm, multiple peaks are considered in the PDA framework to estimate the state of the true target. In the CFAR-based algorithms, the false alarms generated by strong clutter/noise will be suppressed in the subsequent procedures.

12.5.2 State Estimation Using AD-PDA

The traditional PDA algorithm considers all of the validated measurements in the validated region. In PDA, the incorrect (nontarget-originated) measurements are assumed to be uniformly distributed in the validated region, while the correct (target-originated) measurement is assumed to be Gaussian distributed with the mean as the predicted measurement. The above assumptions imply that the distance between the correct and the predicted measurements is more likely smaller than that between the incorrect and predicted measurements. Thus, this important prior is introduced as a distance weight to calculate the association probability. Moreover, the detection scores are also introduced as an appearance weight to calculate the association probability. Finally, we propose a novel AD-PDA algorithm that uses a new expression of appearance-distance weight to associate the validated measurements with the true target. The main procedures of the AD-PDA algorithm are shown in Figure 12.3.

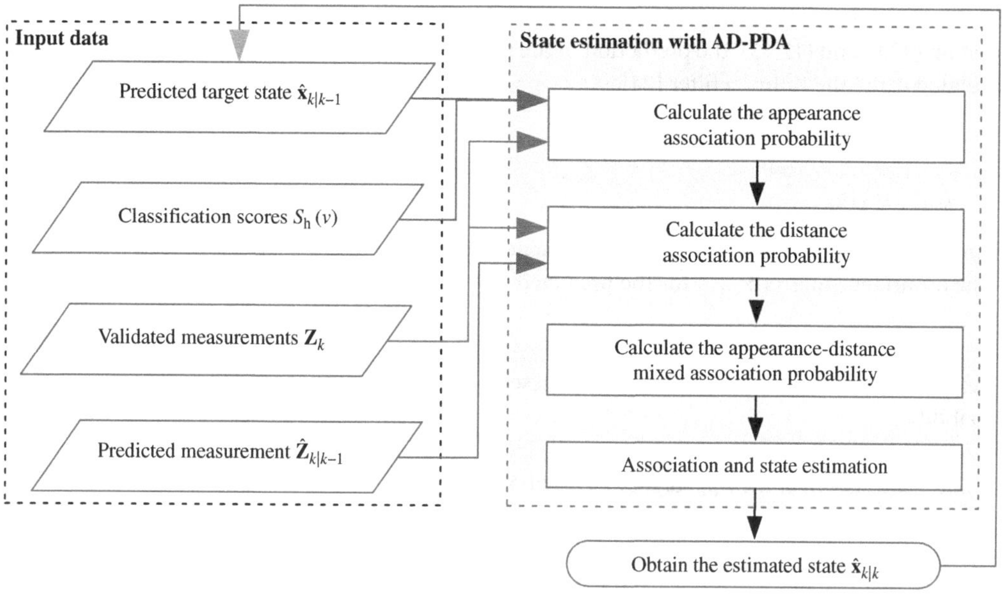

Figure 12.3 Illustration of state estimation with AD-PDA.

12.5.2.1 General Framework for Estimating the Target State

The target state $\hat{\mathbf{x}}_{k|k}$ is estimated based on both the visual similarity densities and the position distribution of the validated measurements [65],

$$\hat{\mathbf{x}}_{k|k} = E\{\mathbf{x}_k \mid \mathbf{V}_k(\mathbf{Z}), \mathbf{Z}_{1:k}\}, \tag{12.39}$$

where the validated measurement set $\mathbf{Z}_k = \{\mathbf{z}_{k,j}\}_{j=1}^{m_k}$ is returned by the BACF detection process and filtered through the validated measurement selection process (Subsection 12.5.1). The variable $\mathbf{V}_k(\mathbf{Z})$ denotes the observed visual similarities for $\mathbf{Z}_k$, i.e., $\mathbf{V}_k(\mathbf{Z}) := \{\mathbf{V}_k(\mathbf{z}_j)\}_{j=1}^{m_k} = \{\mathbf{V}(\mathbf{z}_{k,1}), \mathbf{V}(\mathbf{z}_{k,2}), \dots, \mathbf{V}(\mathbf{z}_{k,m_k})\}$. According to the total probability theorem [24] for the set of events $\{\epsilon_{k,j}\}_{j=0}^{m_k}$, (12.39) can be rewritten as

$$\hat{\mathbf{x}}_{k|k} = \sum_{j=0}^{m_k} E\{\mathbf{x}_k \mid \epsilon_{k,j}, \mathbf{V}_k(\mathbf{Z}), \mathbf{Z}_{1:k}\} P\{\epsilon_{k,j} \mid \mathbf{V}_k(\mathbf{Z}), \mathbf{Z}_{1:k}\} = \sum_{j=0}^{m_k} \hat{\mathbf{x}}_{k|k,j} \beta_{k,j}, \tag{12.40}$$

where $\hat{\mathbf{x}}_{k|k,j} = E\{\mathbf{x}_k \mid \epsilon_{k,j}, \mathbf{V}_k(\mathbf{Z}), \mathbf{Z}_{1:k}\}$ denotes the updated state conditioned on event $\epsilon_{k,j}$ and $\beta_{k,j} \triangleq P\{\epsilon_{k,j} \mid \mathbf{V}_k(\mathbf{Z}), \mathbf{Z}_{1:k}\}$ is the mixed appearance-distance association probability. Applying Bayes' formula [24], we obtain the following expression:

$$\beta_{k,j} = \frac{P\{\mathbf{V}_k(\mathbf{Z}) \mid \epsilon_{k,j}, \mathbf{Z}_{1:k}\} P\{\epsilon_{k,j} \mid \mathbf{Z}_{1:k}\}}{P\{\mathbf{V}_k(\mathbf{Z}) \mid \mathbf{Z}_{1:k}\}} \propto P\{\mathbf{V}_k(\mathbf{Z}) \mid \epsilon_{k,j}, \mathbf{Z}_{1:k}\} P\{\epsilon_{k,j} \mid \mathbf{Z}_{1:k}\}, \quad j = 0, 1, \dots, m_k. \tag{12.41}$$

The first term after the proportionality symbol in (24), $P\{\mathbf{V}_k(\mathbf{Z}) \mid \epsilon_{k,j}, \mathbf{Z}_{1:k}\}$, represents the joint probability density function (pdf) of the visual similarities conditioned on the event $\epsilon_{k,j}$, which is defined as the appearance association probability. The second term, $P\{\epsilon_{k,j} \mid \mathbf{Z}_{1:k}\}$, corresponds to the conditional probability of the event $\epsilon_{k,j}$ given the validated measurements, which is defined as the distance association probability.

12.5.2.2 Prediction of State and Measurement Vectors

Based on (12.6) and (12.15), the prediction of the state and measurement vectors from $k-1$ to k is calculated using the Kalman filter [24]

$$\hat{\mathbf{x}}_{k|k-1} = \mathbf{F}\hat{\mathbf{x}}_{k-1|k-1}, \tag{12.42}$$

$$\hat{\mathbf{z}}_{k|k-1} = \mathbf{H}\hat{\mathbf{x}}_{k|k-1}, \tag{12.43}$$

where $\hat{\mathbf{x}}_{k|k-1}$ and $\hat{\mathbf{z}}_{k|k-1}$ denote the predicted state and measurement vectors, respectively, at time k. The covariance matrix $\mathbf{P}_{k|k-1}$ for the predicted state is calculated as

$$\mathbf{P}_{k|k-1} = \mathbf{F}\mathbf{P}_{k-1|k-1}\mathbf{F}^{\mathrm{T}} + \mathbf{Q}. \tag{12.44}$$

Next, we introduce the calculation processes of the appearance and distance association probabilities.

12.5.2.3 Calculation of the Appearance Association Probability

The observed visual similarities show the relationships between the test samples and the learned template. Given the learned template, the visual similarities depend mainly on the test samples, which are real patches extracted from each frame in the BACF. Considering the independence between the test samples, we assume that the observed visual similarities are conditionally independent. Under the assumption of conditional independence, $P\{\mathbf{V}_k(\mathbf{Z}) \mid \epsilon_{k,j}, \mathbf{Z}_{1:k}\}$ is decomposed as follows:

$$\begin{aligned} P\{\mathbf{V}_k(\mathbf{Z}) \mid \epsilon_{k,j}, \mathbf{Z}_{1:k}\} &= P\{\mathbf{V}(\mathbf{z}_{k,j}) \mid \mathbf{z}_{k,j}, \epsilon_{k,j}, \mathbf{Z}_{1:k-1}\} \prod_{r\neq j, r=1}^{m_k} P\{\mathbf{V}(\mathbf{z}_{k,r}) \mid \mathbf{z}_{k,r}, \epsilon_{k,j}, \mathbf{Z}_{1:k-1}\} \\ &= \frac{I_s(\mathbf{z}_{k,j})}{I_{\bar{s}}(\mathbf{z}_{k,j})} \prod_{r=1}^{m_k} I_{\bar{s}}(\mathbf{z}_{k,r}),\ j = 1, 2, \dots, m_k. \end{aligned} \tag{12.45}$$

Here, $I_s(\mathbf{z}_{k,j}) = P\{\mathbf{V}(\mathbf{z}_{k,j}) \mid \mathbf{z}_{k,j}, \epsilon_{k,j}, \mathbf{Z}_{1:k-1}\}$ is the pdf of the visual similarity for the validated measurement $\mathbf{z}_{k,j}$, which is defined as the sample feature density, and $I_{\bar{s}}(\mathbf{z}_{k,r}) = P\{\mathbf{V}(\mathbf{z}_{k,r}) \mid \mathbf{z}_{k,r}, \epsilon_{k,j}, \mathbf{Z}_{1:k-1}\}$ is the pdf of the visual similarity for measurement $\mathbf{z}_{k,r}$ when it is assumed to be false. According to the total probability theorem, $I_{\bar{s}}(\mathbf{z}_{k,r}) = 1 - I_s(\mathbf{z}_{k,r})$.

The BACF algorithm learns the visual appearance of the target as a template, and the normalized detection scores imply the visual similarity between the test samples and the learned template. Specifically, the normalized detection score $\mathbf{S}_N(x_k, y_k)$ evaluates the similarity between the feature vector at pixel location (x_k, y_k) of the test sample and the target reference feature vector. Thus, the visual similarity of a sample $\mathbf{V}(x_k, y_k)$ located at pixel (x_k, y_k), can be calculated based on the normalized detection score $\mathbf{S}_N(x_k, y_k)$ as

$$\mathbf{V}(x_k, y_k) = -\log \mathbf{S}_N(x_k, y_k). \tag{12.46}$$

Inspired by [66], the visual similarity is mapped to the probability using an inverse Gaussian distribution as

$$P_{\mathbf{V}}(\mathbf{V}) = \sqrt{\frac{\lambda}{2\pi\mathbf{V}^3}}\exp\left\{\frac{-\lambda(\mathbf{V}-\mu)^2}{2\mu^2\mathbf{V}}\right\}, \tag{12.47}$$

where $\lambda > 0$ and $\mu > 0$ are designed parameters.

Based on (12.46) and (12.47), the normalized detection score of a feature vector $\mathbf{S}_N(\mathbf{z}_{k,j})$ located at a validated measurement $\mathbf{z}_{k,j}$, can be interpreted as the probability $P_{\mathbf{V}}[\mathbf{V}(\mathbf{z}_{k,j})]$ that the validated measurement visually corresponds to the target. The larger the normalized detection score is, the higher the probability of similarity for the measurement, and vice versa. Thus, we choose $P_{\mathbf{V}}[\mathbf{V}(\mathbf{z}_{k,j})]$ as the sample feature density of the validated measurement $\mathbf{z}_{k,j}$, i.e.,

$$I_s(\mathbf{z}_{k,j}) = P_{\mathbf{V}}[\mathbf{V}(\mathbf{z}_{k,j})]. \tag{12.48}$$

Remark 12.4: The distribution of visual similarities is approximated by a parametric cumulative distribution function – an inverse Gaussian distribution in this work. The distribution is inspired by an idea from [66]. In detail, for design reasons, a small similarity value implies a larger visual similarity. To enforce this property, the negative logarithm is adopted in this work. Thus, the negative logarithm of the normalized detection scores (in (12.46)) is calculated as the visual similarity. Then, the visual similarity is mapped to a likelihood. Since a CF-based algorithm uses the feature vectors, the likelihood can be determined based on the visual similarity between the feature vector of a detection and the visual state of the target. Specifically, for a target, we first calculate the visual similarities of the candidates and sort them in ascending order. Then, we generate an empirical distribution function using the visual similarities on the X-axis, which is then approximated by a parametric cumulative distribution function. According to [66], the best fit for the likelihood of the visual similarity obtained using a CF-based algorithm is an inverse Gaussian distribution (12.47).

12.5.2.4 Calculation of the Distance Association Probability

The calculation of $P\{\epsilon_{k,j} \mid \mathbf{Z}_{1:k}\}$ can be referred to as PDA estimation [22]. For the Poisson clutter model, $P\{\epsilon_{k,j} \mid \mathbf{Z}_{1:k}\}$ is denoted by $\theta_{k,j}$ and calculated as

$$\theta_{k,j} = \begin{cases} e_j, & j = 1, 2, \dots, m_k \\ b, & j = 0 \end{cases}, \tag{12.49}$$

where

$$e_j := \mathcal{N}[\boldsymbol{\nu}_{k,j}; 0, \mathbf{J}_k] = \exp\left(-\frac{1}{2}\boldsymbol{\nu}_{k,j}^{\mathrm{T}}\mathbf{J}_k^{-1}\boldsymbol{\nu}_{k,j}\right), \tag{12.50}$$

$$b := \lambda_c |2\pi \mathbf{J}_k|^{\frac{1}{2}}(1 - P_D P_G)/P_D. \tag{12.51}$$

In (12.50) and (12.51), J_k denotes the innovation covariance [22, 24], $\mathbf{J}_k = \mathbf{H}_k \mathbf{P}_{k|k-1}\mathbf{H}_k^{\mathrm{T}} + \mathbf{R}$; $\boldsymbol{\nu}_{k,j} \triangleq \mathbf{z}_{k,j} - \hat{\mathbf{z}}_{k|k-1}$ is the innovation; λ_c denotes the clutter density; P_D is the target detection probability; and P_G is the probability that the target measurement falls into the validated region.

Remark 12.5: By establishing the validated region centered at the predicted measurement $\hat{\mathbf{z}}_{k|k-1}$ at time k, the validated measurements $\mathbf{z}_{k,j}$, $j = 1, 2, \dots, m_k$, are associated probabilistically to the target track with (12.49). The shorter the distance between $\hat{\mathbf{z}}_{k|k-1}$ and $\mathbf{z}_{k,j}$ is, the smaller the innovation is, thereby resulting in a higher distance association probability.

12.5.2.5 Calculation of the Mixed Appearance-distance Association Probability

For a validated measurement $\mathbf{z}_{k,j}$ conditioned on the event $\epsilon_{k,j}$, $j = 1, 2, \dots, m_k$, the mixed appearance-distance association probability, $\beta_{k,j}$, is obtained by substituting (12.45) and (12.50) into (12.41):

$$\beta_{k,j} = \frac{I_s(\mathbf{z}_{k,j})}{I_{\bar{s}}(\mathbf{z}_{k,j})} \prod_{r=1}^{m_k} I_{\bar{s}}(\mathbf{z}_{k,r}) \times \exp\left(-\frac{1}{2}\boldsymbol{\nu}_{k,j}^{\mathrm{T}}\mathbf{J}_k^{-1}\boldsymbol{\nu}_{k,j}\right). \tag{12.52}$$

Under the condition of event $\epsilon_{k,0}$, obviously, there is no observed visual similarity since no measurements are taken. The mixed appearance-distance association probability reduces to the following:

$$\beta_{k,0} = \lambda_c |2\pi \mathbf{J}_k|^{\frac{1}{2}} (1 - P_D P_G)/P_D. \tag{12.53}$$

Finally, the mixed appearance-distance association probabilities are normalized to satisfy $\sum_{j=0}^{m_k} \beta_{k,j} = 1$ as

$$\beta_{k,j} = \beta_{k,j} / \sum_{r=0}^{m_k} \beta_{k,r}, \; j = 0, 1, \dots, m_k. \tag{12.54}$$

12.5.2.6 Association and State Estimation

Using the mixed association probabilities, we estimate the target position $\hat{\rho}_k$ according to (12.40) as

$$\hat{\rho}_{k|k} := \left[\hat{x}_{k|k}, \hat{y}_{k|k}\right]^{\mathrm{T}} = \hat{\mathbf{z}}_{k|k-1} + \sum_{j=1}^{m_k} \beta_{k,j} \mathbf{z}_{k,j}. \tag{12.55}$$

Consequently, the target velocity $\hat{v}_k$ is estimated as

$$\hat{v}_{k|k} := \left[\hat{v}_{x,k|k}, \hat{v}_{y,k|k}\right]^{\mathrm{T}} = \hat{\rho}_{k|k} - \hat{\rho}_{k-1|k-1}. \tag{12.56}$$

12.5.3 Combination of AD-PDA and BACF

Finally, by combining AD-PDA and BACF, we propose a novel AD-PDA-BACF algorithm that can accurately estimate the states of targets by integrating the motion and appearance characteristics of the targets. The architecture of the proposed AD-PDA-BACF algorithm is shown in Figure 12.4, which consists of three modules, i.e., a position initialization module, feature representation module and AD-PDA-BACF module.

12.5.3.1 Position Initialization Module

In the tracking framework, the BACF algorithm needs to initialize the filter $\mathbf{h}_1$ and the target template model $\hat{\mathbf{u}}_1^{\text{model}}$, which inherently depends on the target positions in the first frame. Thus, the

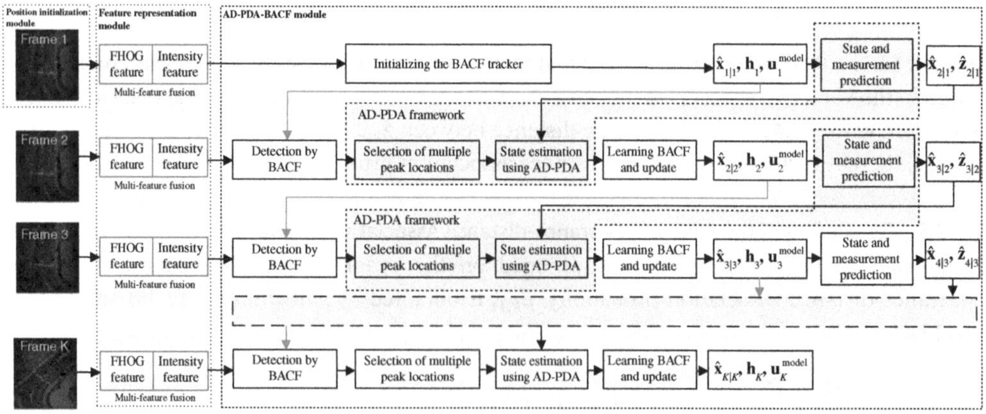

Figure 12.4 Flowchart of the proposed AD-PDA-BACF algorithm. *Source*: [6]/with permission of IEEE.

initial positions of target shadows are manually set by defining a bounding box at the beginning of the BACF in this study. Notably, an external detector is typically used to initialize the target positions. For example, background subtraction followed by CFAR detection [11, 47] is used to automatically obtain reliable initial target positions in ViSAR.

12.5.3.2 Feature Representation Module

Since different features can provide complementary supporting information, in the proposed AD-PDA-BACF algorithm, multifeature fusion is performed to enhance the ability to represent the target and improve the tracking success rate and precision. Considering that the target shadow intensity, shape and gradation features may vary with the target type and motion state during the tracking process, two typical features, i.e., the intensity and 31-dimensional FHOG features [67], are fused for ViSAR videos. The intensity is a typical single-channel feature that describes the global pixel information of the target. The FHOG feature represents the local edge information of the target. Thus, a multichannel fusion feature is obtained by cascading the intensity and FHOG features in the dimension of the channel. Assuming that the total number of channels of the multichannel feature vector is L, the feature representation of a sample is denoted as $\mathbf{u} = [\mathbf{u}_{\text{intensity}}, \mathbf{u}_{\text{FHOG}}] = [\mathbf{u}_1, \mathbf{u}_2, \ldots, \mathbf{u}_L]$.

12.5.3.3 AD-PDA-BACF Module

In AD-PDA-BACF, we first select multiple validated measurements based on the correlation outputs of the BACF algorithm. Second, the predicted state and measurement are obtained based on the target dynamic model and measurement model, respectively. Third, the mixed association probabilities are calculated, based on which the target states are then estimated, in the proposed AD-PDA framework. Finally, based on the estimated state at time k, the feature of the target, $\mathbf{u}_k$, is calculated, and the target template model $\mathbf{u}_k^{\text{model}}$ is updated with the online adaptation rate η as, $\mathbf{u}_k^{\text{model}} = (1-\eta)\mathbf{u}_{k-1}^{\text{model}} + \eta\mathbf{u}_k$. Furthermore, we adopt the alternating direction method of multipliers (ADMM) algorithm [68] to calculate the filter $\mathbf{h}_k$ of the BACF [44]. The AD-PDA-BACF algorithm is presented in detail as Algorithm 12.3.

12.6 Examples of TBD and CF Tracking Methods

12.6.1 Example 1: TBD Tracking for Multiple Targets

12.6.1.1 Datasets and Compared Algorithms

We conduct some experiments to test the performance of the proposed algorithms, using real data, "highway", provided by SNL [49]. The data is taken by a ViSAR system, which shows the shadows of a number of vehicles moving in various directions. We take sub-aperture images every five frames from the data to generate image samples for the experiment. We consider a scenario of detecting and tracking a fixed number of targets. We use Ti to denote the ith target. The first 12 image samples are used for the experiment of detecting and tracking a fixed number of targets. The initial sub-aperture images at discrete times $k = 1$, and $k = 12$ are shown in the first column of Figure 12.5. We recommend that the readers download and view the video data prior to analyzing Figure 12.5, from the website in [49]. Six moving targets (vehicles) appear during the observation period. True positions of six targets T1–T6 are marked in colored rectangles in the first column corresponding to the initial sub-aperture images. We can see that T1–T5 exist during the entire observation period, moving closely toward the south. The along-track velocities of T1–T5 are within the range of [3, 41] m/s. Moreover, targets T1–T2 turn to right and move with slow maneuvers. In addition, the target T6

Algorithm 12.3: AD-PDA-BACF Algorithm

Inputs: The initial state of the target $\mathbf{x}_1$, and the ViSAR image sequences $\{\boldsymbol{f}_1, \boldsymbol{f}_2, \dots, \boldsymbol{f}_K\}$.
Outputs: Estimated states of the target $\{\hat{\mathbf{x}}_{2|2}, \hat{\mathbf{x}}_{3|3}, \dots, \hat{\mathbf{x}}_{K|K}\}$.
1: **Initialization:** At time $k = 1$, we train a BACF and obtain the filter $\mathbf{h}_1$ based on a set of training samples $\{\mathbf{u}_c\}_{c=1}^{C}$ in $\boldsymbol{f}_1$ according to (12.23).
2: **for** $k = 2$ to K **do**
3: For a test sample $\boldsymbol{\nu} \in \mathbb{R}^C$ in $\boldsymbol{f}_k$, compute its classification scores $\mathbf{S}_{\mathbf{h}}(\boldsymbol{\nu})$ and detection scores $\mathbf{S}_D$ according to (12.25) and (12.33), respectively;
4: Select multiple peak locations $\{\mathbf{z}_{k,j}\}_{j=1}^{m_k}$ in the correlation outputs using (12.38);
5: Obtain the predicted state $\hat{\mathbf{x}}_{k|k-1}$ according to the dynamic model (12.42);
6: Obtain the predicted measurement $\hat{\mathbf{z}}_{k|k-1}$ according to the measurement model (12.43);
7: Estimate the position $\left[\hat{x}_{k|k}, \hat{y}_{k|k}\right]^{\mathrm{T}}$ and velocity $\left[\hat{v}_{x,k|k}, \hat{v}_{y,k|k}\right]^{\mathrm{T}}$ with (12.55) and (12.56) to obtain $\hat{\mathbf{x}}_{k|k} = [\hat{x}_{k|k}, \hat{v}_{x,k|k}, \hat{y}_{k|k}, \hat{v}_{y,k|k}]^{\mathrm{T}}$ using AD-PDA;
8: Extract the sample $\mathbf{u}$ of the target in $\boldsymbol{f}_k$ from the image patch centered at $\hat{\mathbf{x}}_{k|k}$, and update the target template model $\mathbf{u}_k^{\text{model}}$ based on $\mathbf{u}$;
9: Calculate the filter $\mathbf{h}_k$ in the BACF by solving (12.23) using the ADMM algorithm.
10: **end for**

exists during the time interval [1, 8], moving toward the north. In the experiment, since the TBD based methods integrate the merit function of targets over 12 frames, the merit value of T6 is lower than the threshold V_T. Thus, the TBD-based methods cannot detect T6.

12.6.1.2 Datasets and Compared Algorithms

First, we perform an initial detection procedure. We register the 12 images using a joint registration and reconstruction method, and further process using a two-parameter CFAR detector. After the CFAR detection, we obtain a number of candidate targets. The initial detection results are shown in the second column of Figure 12.5. We can see that the five moving targets are still submerged in the noisy background. Secondly, based on the initial detection results, we use the proposed ES-TBD algorithm to detect and track the moving targets. Moreover, for comparison, we perform the conventional DP-TBD, two-stage detection architecture based DP-TBD (Two-Stage-TBD), and time-series-predicting strategy based DP-TBD (TSP-TBD) algorithms to detect and track the moving targets. Next we describe the experimental parameters of the four algorithms.

12.6.1.3 Parameter Settings

In the proposed ES-TBD algorithm, the particles in each PF are sampled from a prior Gaussian distribution $p(\mathbf{x}_1^n)$, where the initial standard deviation of states is $[0.2\ \text{m}, 0.1\ \text{m/s}, 0.2\ \text{m}, 0.1\ \text{m/s}]'$. The number of particles in each PF is $N = 100$. The standard deviations of the acceleration process noise along the X and Y axes are $\sigma_x = 0.85\ \text{m/s}^2$ and $\sigma_y = 0.85\ \text{m/s}^2$. In the conventional DP-TBD algorithm, the number of valid transition state is $N_s = 9$. The parameters of the NCV model are same with those in the proposed ES-TBD algorithm. In the Two-Stage-TBD algorithm, the maximum number of consecutive misses and the minimum number of plots equal 0, and 8, respectively. In the TSP-TBD algorithm, the weight is 0.1 for the time series prediction model.

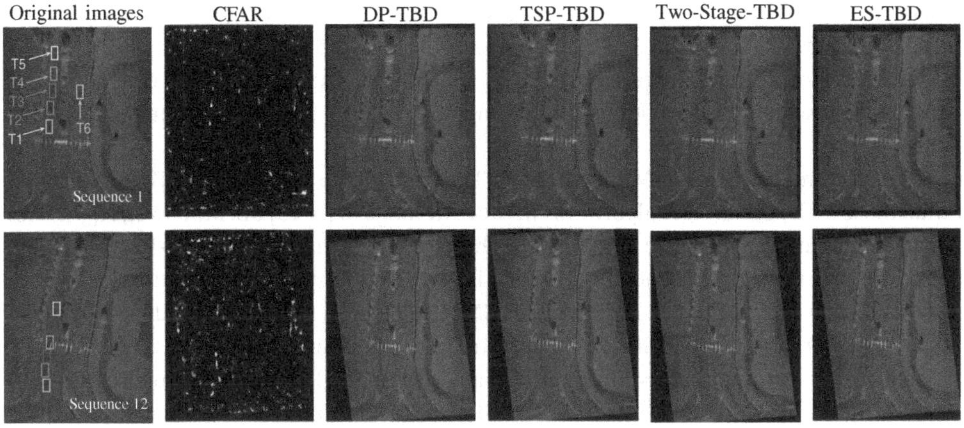

Figure 12.5 Comparison of the detection results using the conventional DP-TBD, TSP-TBD, Two-Stage-TBD, and proposed ES-TBD algorithms. *Source*: [5]/with permission of IEEE.

Table 12.1 Performance metrics of four algorithms.

Algorithm	N_{vd}	N_{fd}	**RMSE (cells)**
ES-TBD (proposed)	5	0	0.71
Conventional DP-TBD	0	1.13	N/A
TSP-TBD	3	0.375	135.6
Two-Stage-TBD	3	0.50	6.51

12.6.1.4 Evaluation Metrics

The detection and tracking results are obtained from one trial. We use the following metrics to evaluate the performance of the three algorithms: (1) The number of declared valid detections N_{vd}, which denotes the number of detected true targets [50]. (2) The number of declared false detections N_{fd}, which denotes the number of detected false targets [50]. (3) Tracking error, which is represented by the RMSE of positions of detected targets [22].

12.6.1.5 Comparison Results on Highway

The detection results using the conventional DP-TBD, TSP-TBD, Two-Stage-TBD, and proposed ES-TBD algorithms are shown in the third, fourth, fifth, and last columns of Figure 12.5, respectively. The estimated positions of the true and false targets are marked in colored rectangles and crosses, respectively. We can see that all the algorithms declare T1–T5, and among them the error between the true and estimated positions of the proposed ES-TBD algorithm is the lowest. The proposed ES-TBD algorithm estimates the positions of T1–T5 accurately at each time. Since ES-TBD adopts the expanding and shrinking strategy, the searching ranges cover the true positions of the slowly maneuvering targets consistently. The proposed ES-TBD algorithm declares no false targets, whereas the other three algorithms declare a number of false targets.

The four algorithms are compared based on the above performance metrics in Table 12.1, which shows that the proposed ES-TBD algorithm can track the slowly maneuvering targets more accurately.

The performance metrics of four algorithms versus SNR are plotted in Figure 12.6. Figure 12.6a shows the curves of N_{vd} versus SNR. The proposed ES-TBD algorithm can detect all the targets under a lower SNR condition. The reason is that the ES-TBD algorithm adopts the expanding and shrinking strategy, which ensures that the searching ranges cover the true positions of targets at each time.

Figure 12.6b shows the curves of N_{fd} versus SNR. The curve of the proposed ES-TBD algorithm is lower than those of the other three algorithms in the SNR range of [2, 13] dB. Moreover, $N_{fd} = 0$ for ES-TBD algorithm in the SNR range of [7, 13] dB. This is due to the fact that the ES-TBD algorithm evaluates the searching ranges at each time, which is beneficial in suppressing clutter.

Figure 12.6c shows the curves of RMSE versus SNR. The RMSE of the proposed ES-TBD algorithm is much lower than that of TSP-TBD algorithm in the SNR range of [7, 13] dB. This is due to the missed detection of targets at $k = 1$ in TSP-TBD algorithm. The RMSE of the proposed ES-TBD algorithm is slightly lower than that of Two-Stage-TBD algorithm in the SNR range of [5, 13] dB. Since the ES-TBD algorithm performs the state transition using PFs, it can obtain more accurate state estimates.

12.6.2 Example 2: Real-time CF Tracking for a Single Target

12.6.2.1 Datasets and Compared Algorithms

In this section, we conduct a number of experiments to test the performance of the proposed algorithm and 13 state-of-the-art algorithms, using the public ViSAR video datasets, Victr [69], provided by SNL [49].

Inspired by the multitarget dynamic programming based TBD tracking framework [50], we use multiple parallel AD-PDA-BACF trackers to track multiple targets. An AD-PDA-BACF tracker is established for each target, and the six trackers work independently.

Moreover, we compare the results of our tracker with those of 13 state-of-the-art algorithms: such as the spatially regularized discriminative CF (SRDCF) [43], scale-adaptive with multiple features tracker (SAMF) [40], spatial–temporal regularized CF (STRCF) [70], sum of template and pixel-wise learners (STAPLE) [71], discriminative scale space tracking (DSST) [39], reliable patch trackers (RPT) [72], BACF [44], spatial–temporal regularized BACF (ST-BACF) [46], multipeak detection strategy-based BACF (MPD-BACF) [73], joint kernelized correlation filter (JKCF) [47], faster regions with convolutional neural networks and a bidirectional long short-term memory network (Faster RCNN+Bi-LSTM) [74], and ES-TBD [5]. Among the compared algorithms, there is one TBD-based algorithm, i.e., ES-TBD, and two deep learning (DL)-based algorithms, i.e., Faster RCNN+Bi-LSTM and Faster RCNN+multi-object tracking (MOT). The remaining algorithms,

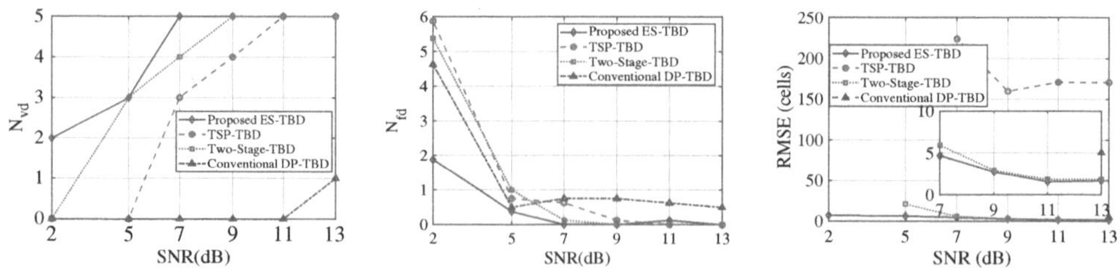

(a) (b) (c)

Figure 12.6 The performance metrics versus SNR: (a) N_{vd} versus SNR, (b) N_{fd} versus SNR, (c) RMSE versus SNR.

i.e., SRDCF, SAMF, STRCF, STAPLE, DSST, RPT, BACF, ST-BACF, MPD-BACF, and JKCF, are CF-based algorithms. Specifically, the MPD-BACF exploits a multimodal detection technique to prevent model drift under multipeak conditions.

12.6.2.2 Parameter Settings

The proposed algorithm and the state-of-the-art algorithms are implemented in MATLAB R2019b on a computer with an i7-7700HQ (2.81 GHz) CPU. The proposed AD-PDA-BACF algorithm uses handcrafted features, i.e., intensity and FHOG, to represent each target. The parameter settings in the AD-PDA-BACF are shown in Table 12.2. For the DL-based algorithms (Faster RCNN+Bi-LSTM and Faster RCNN+MOT), we conduct off-line retraining to obtain the Faster RCNN models for Highway and Victr, respectively, using the images from the two datasets.

12.6.2.3 Evaluation Metrics

Two metrics, i.e., success rate and precision, are used to quantitatively evaluate the tracking performance of all algorithms. In addition, the FPS is used to evaluate the speed of the algorithms.

(1) For an estimated bounding box B_k and the corresponding ground-truth bounding box B_g, the overlap ratio is defined as

$$\text{OP}_k = \frac{\text{area}(B_k \cap B_g)}{\text{area}(B_k \cup B_g)}, \tag{12.57}$$

where $\cap$ and $\cup$ represent the intersection and union, respectively, of two regions, and area$(\cdot)$ denotes the number of pixels in the specified region. Tracking is regarded as successful for a given frame if the overlap ratio is larger than a given threshold. The success rate is defined as the ratio of

Table 12.2 Parameter settings in AD-PDA-BACF.

Module		Parameter	Value
Position initialization module		Target size $w \times h$ (pixels)	Ground-truth size
Feature representation module		Cell size of FHOG	2×2
AD-PDA-BACF module	BACF	Regularization factor γ	0.05
		Number of scales	5
		Scale step	1.05
		Bandwidth of ideal correlation output	$\sqrt{wh}/16$
		Number of iterations	2
		Maximum penalty factor	10^4
		Online adaption rate η	10^{-4}
	AD-PDA	Sampling period T (s)	1
		Probability of detection P_D	1
		Gating probability P_G	0.99
		Standard deviation of the measurement noise (pixels) σ_z	2
		Clutter density λ_c (pixel^{-2})	10^{-4}
		Number of validated measurements m_k	30

the number of successful frames to the total number of frames. For better visualization, we plot the success rates with respect to the overlap ratio thresholds varying in [0, 1]. The area under the curve (AUC) of the success plot is calculated, which equals the average overlap ratio across the image sequences [75]. For simplicity, we refer to the AUC value as the success rate AUC score in the next subsections.

(2) The center location error (CLE) is defined as the Euclidean distance between the center of the estimated bounding box and that of the ground-truth bounding box. For the ViSAR data, the ground-truth center locations are labeled manually for performance evaluation. The precision metric is the ratio of the number of frames whose CLE is within a given threshold to the total number of frames. Conventionally, the CLE threshold is set to 20 pixels [75]. For better visualization, we plot the precision with respect to the CLE threshold varying in [0, 50]. The distance precision score is also presented, which is calculated similarly to the success rate AUC score.

(3) The FPS is defined as the number of frames that a tracker can process per second.

A smaller average CLE for tracking accuracy and a larger FPS value for tracking speed are desired.

12.6.2.4 Comparison Results on Victr

Figure 12.7a shows the precision plots on the Victr dataset. Among the compared algorithms, the Faster RCNN+Bi-LSTM, AD-PDA-BACF, and JKCF are the top three trackers, with high precision curves and large distance precision scores. Figure 12.7b shows the success plots for the Victr dataset. Among the compared algorithms, the Faster RCNN+Bi-LSTM, AD-PDA-BACF, and ES-TBD are the top three trackers in terms of success plot curves, with corresponding success rate AUC scores of 55.79, 31.33, and 21.66%, respectively.

The overall performance, including FPS values and average CLEs, are shown in Table 12.3. In terms of average CLE, the Faster RCNN+Bi-LSTM, AD-PDA-BACF, and JKCF are the top three trackers. Correspondingly, their average CLEs are 4.60, 12.65, and 16.06, respectively. It is observed that the AD-PDA-BACF tracker achieves the best results in terms of the average CLE among the model-based algorithms.

In terms of FPS, the three fastest algorithms are the DSST (FPS: 147.71), SAMF (FPS: 139.03) and JKCF (FPS: 43.05). However, the DSST and SAMF have very small success rates and very large average CLEs. The slowest algorithm is the Faster RCNN+Bi-LSTM (FPS: 1.48). The speed of the proposed AD-PDA-BACF tracker (FPS: 29.98) meets the demands of real-time tracking at the 20 Hz-level ViSAR imaging frame rate.

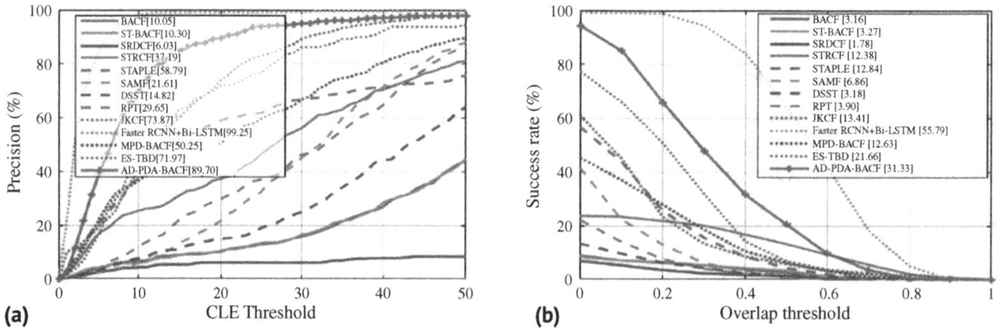

Figure 12.7 Quantitative performance evaluation of AD-PDA-BACF and several state-of-the-art trackers on the Victr dataset. (a) Precision plots and (b) success plots.

Table 12.3 Comparison of FPS values and average CLEs of the proposed AD-PDA-BACF and several state-of-the-art trackers on Victr.

	FPS	Avg. CLE (in pixel)
BACF	33.24	59.97
ST-BACF	31.33	59.61
SRDCF	27.49	168.57
STRCF	22.97	31.37
STAPLE	34.11	35.46
SAMF	139.03	31.07
DSST	147.71	44.92
RPT	13.22	33.16
JKCF	43.05	16.06
Faster RCNN+Bi-LSTM	1.48	4.60
MPD-BACF	18.96	25.16
ES-TBD	2.93	17.92
Proposed AD-PDA-BACF	29.98	12.65

12.7 Conclusions

Video synthetic aperture radar (ViSAR) is a new radar imaging technology developed to detect and track ground moving targets. ViSAR is capable of real-time and persistent monitoring of the area of interest in all-weather. Detecting and tracking dim, small, maneuvering and time-varying number of targets accurately in a low signal-to-noise ratio environment and complex dynamic background using ViSAR pose a great challenge.

We develop two moving target tracking algorithms using ViSAR imagery: (1) joint detection and tracking of dim and small targets based on pre-detected image sequence; and (2) real-time tracking of targets based on the original image sequence. For the benefit of the reader, we present a survey of the current state-of-the-art research in ViSAR target tracking so that the chapter is self-contained. We provide extensive references on various track-before-detect (TBD) and correlation filter (CF) based algorithms.

Acknowledgments

This work was supported in part by National Natural Science Foundations of China under Grant 62173267, and Grant 61573276.

References

1 A. Damini, B. Balaji, C. Parry, and V. Mantle. A videoSAR mode for the X-band wideband experimental airborne radar. In E. G. Zelnio and F. D. Garber, editors, *Algorithms for Synthetic Aperture Radar Imagery XVII*, volume 7699, page 76990E. SPIE, 2010. https://doi.org/10.1117/12.855376.

2 A. M. Raynal, D. L. Bickel, and A. W. Doerry. Stationary and moving target shadow characteristics in synthetic aperture radar. In K. I. Ranney and A. Doerry, editors, *Radar Sensor Technology XVIII*, volume 9077, pages 413–427. International Society for Optics and Photonics, SPIE, 2014. https://doi.org/10.1117/12.2049729.

3 H. Xu, Z. Yang, G. Chen, G. Liao, and M. Tian. A ground moving target detection approach based on shadow feature with multichannel high-resolution synthetic aperture radar. *IEEE Geoscience and Remote Sensing Letters*, 13(10): 1572–1576, 2016. https://doi.org/10.1109/LGRS.2016.2597159.

4 H. Wang, Z. Chen, and S. Zheng. Preliminary research of low-RCS moving target detection based on Ka-band video SAR. *IEEE Geoscience and Remote Sensing Letters*, 14(6):811–815, 2017. https://doi.org/10.1109/LGRS.2017.2679755.

5 X. Tian, J. Liu, M. Mallick, and K. Huang. Simultaneous detection and tracking of moving-target shadows in ViSAR imagery. *IEEE Transactions on Geoscience and Remote Sensing*, 59(2): 1182–1199, 2021. https://doi.org/10.1109/TGRS.2020.2998782.

6 X. Tian, J. Liu, M. Mallick, J. Wang, X. Ji, L. Su, and J. Si. A novel AD-PDA-BACF algorithm for real-time moving target shadow tracking using visar imagery. *IEEE Transactions on Geoscience and Remote Sensing*, 61: 1–17, 2023.

7 R. Linnehan, J. Miller, and A. Asadi. Map-drift autofocus and scene stabilization for video-SAR. In *2018 IEEE Radar Conference*, pages 1401–1405, 2018. https://doi.org/10.1109/RADAR.2018.8378769.

8 A. Damini, V. Mantle, and G. Davidson. A new approach to coherent change detection in videoSAR imagery using stack averaged coherence. In *2013 IEEE Radar Conference*, pages 1–5, 2013. https://doi.org/10.1109/RADAR.2013.6586152.

9 Z. Liu, D. An, and X. Huang. Moving target shadow detection and global background reconstruction for videoSAR based on single-frame imagery. *IEEE Access*, 7:42418–42425, 2019. https://doi.org/10.1109/ACCESS.2019.2907146.

10 X. Huang, Z. Xu, and J. Ding. Video SAR image despeckling by unsupervised learning. *IEEE Transactions on Geoscience and Remote Sensing*, 59(12):10151–10160, 2021. https://doi.org/10.1109/TGRS.2020.3045284.

11 C. Zhong, J. Ding, and Y. Zhang. Joint tracking of moving target in single-channel video SAR. *IEEE Transactions on Geoscience and Remote Sensing*, 60:1–18, 2022. https://doi.org/10.1109/TGRS.2021.3115491.

12 S. M. Scarborough, C. H. Casteel, Jr., L. Gorham, M. J. Minardi, U. K. Majumder, M. G. Judge, E. Zelnio, M. Bryant, H. Nichols, and D. Page. A challenge problem for SAR-based GMTI in urban environments. In E. G. Zelnio and F. D. Garber, editors, *Algorithms for Synthetic Aperture Radar Imagery XVI*, volume 7337, page 73370G. SPIE, 2009. https://doi.org/10.1117/12.823461.

13 S. Suchandt, H. Runge, H. Breit, U. Steinbrecher, A. Kotenkov, and U. Balss. Automatic extraction of traffic flows using TerraSAR-X along-track interferometry. *IEEE Transactions on Geoscience and Remote Sensing*, 48(2):807–819, 2010. https://doi.org/10.1109/TGRS.2009.2037919.

14 D. Henke, C. Magnard, M. Frioud, D. Small, E. Meier, and M. E. Schaepman. Moving-target tracking in single-channel wide-beam SAR. *IEEE Transactions on Geoscience and Remote Sensing*, 50(11): 4735–4747, 2012. https://doi.org/10.1109/TGRS.2012.2191561.

15 D. Henke, E. Mendez Dominguez, D. Small, M. E. Schaepman, and E. Meier. Moving target tracking in single- and multichannel SAR. *IEEE Transactions on Geoscience and Remote Sensing*, 53(6): 3146–3159, 2015.https://doi.org/10.1109/TGRS.2014.2369060.

16 D. Henke, E. Mendez Dominguez, D. Small, M. E. Schaepman, and E. Meier. Moving target tracking in SAR data using combined exo- and endo-clutter processing. *IEEE Transactions on Geoscience and Remote Sensing*, 56(1): 251–263, 2018. https://doi.org/10.1109/TGRS.2017.2746262.

17 Y. Zhang, S. Yang, H. Li, and Z. Xu. Shadow tracking of moving target based on CNN for video SAR system. In *IGARSS 2018 – 2018 IEEE International Geoscience and Remote Sensing Symposium*, pages 4399–4402, 2018. https://doi.org/10.1109/IGARSS.2018.8518431.

18 C. K. Kim, M. Y. Park, G. H. Shin, and S. O. Park. An improved technique for single-channel video-SAR based on fractional Fourier transform. *IEEE Transactions on Aerospace and Electronic Systems*, 58(5):4044–4052, 2022. https://doi.org/10.1109/TAES.2022.3157657.

19 R. K. Hersey and E. Culpepper. Radar processing architecture for simultaneous SAR, GMTI, ATR, and tracking. In *2016 IEEE Radar Conference*, pages 1–5, 2016. https://doi.org/10.1109/RADAR.2016.7485076.

20 M. Mallick, S. Coraluppi, and C. Carthel. Multitarget tracking using multiple hypothesis tracking. In M. Mallick, V. Krishnamurthy, and B.-N. Vo, editors, *Integrated Tracking, Classification, and Sensor Management: Theory and Applications, Chapter 5*. Wiley and IEEE, 2012. ISBN 9781118450550.

21 T. Kurien. Issues in the design of practical multitarget tracking algorithms. In Y. Bar-Shalom, editor, *Multitarget-multisensor Tracking : Advanced Applications*, Chapter 3, pages 43–83. Artech House Publishers, 1990.

22 Y. Bar-Shalom, P. Willett, and X. Tian. *Multitarget-multisensor Tracking: Aplications and Advances*. YBS Publishing, 2011.

23 H. Askar, X. Li, and Z. Li. Performance analysis of dim moving point target detection algorithms. In *IEEE 2002 International Conference on Communications, Circuits and Systems and West Sino Expositions*, volume 1, pages 605–609 vol.1, 2002. https://doi.org/10.1109/ICCCAS.2002.1180692.

24 Y. Bar-Shalom, X.-R. Li, and T. Kirubarajan. *Estimation with Applications to Tracking and Navigation*. John Wiley and Sons, New York, 2001.

25 L. Ubeda-Medina, F. García-Fernández, and J. Grajal. Adaptive auxiliary particle filter for Track-Before-Detect with multiple targets. *IEEE Transactions on Aerospace and Electronic Systems*, 53(5): 2317–2330, 2017. https://doi.org/10.1109/TAES.2017.2691958.

26 D. Yong Kim, B. Ristic, and L. Rosenberg. Bayesian track-before-detect for maritime radar. In M. Mallick, B.-N. Vo, and R. Tharmarasa, editors, *Advanced Techniques in Tracking and Sensor Management: Theory and Applications*, Chapter 9. Wiley/IEEE, 2025.

27 W. Yi, Z. Fang, W. Li, R. Hoseinnezhad, and L. Kong. Multi-frame Track-Before-Detect algorithm for maneuvering target tracking. *IEEE Transactions on Vehicular Technology*, 69: 4104–4118, 2020. https://doi.org/10.1109/TVT.2020.2976095.

28 E. Grossi, M. Lops, and L. Venturino. A novel dynamic programming algorithm for Track-Before-Detect in radar systems. *IEEE Transactions on Signal Processing*, 61(10): 2608–2619, 2013. https://doi.org/10.1109/TSP.2013.2251338.

29 S. Chen, F. Luo, L. Zhang, C. Hu, and S. Chen. Maneuvering target Track-Before-Detect method using dynamic programming. *Journal of Xidian University (Natural Science)*, 44(5): 39–45, 2017.

30 X. Li, S. Wang, and D. Zheng. A DP-TBD algorithm with adaptive state transition set for maneuvering targets. In *CIE International Conference on Radar*, 2017.

31 D. Zheng, S. Wang, and Q. Meng. Dynamic programming Track-Before-Detect algorithm for radar target detection based on polynomial time series prediction. *IET Radar, Sonar Navigation*, 10(8): 1327–1336, 2016. https://doi.org/10.1049/iet-rsn.2015.0332.

32 J. Wang, W. Yi, T. Kirubarajan, and L. Kong. An efficient recursive multiframe Track-Before-Detect algorithm. *IEEE Transactions on Aerospace and Electronic Systems*, 54(1): 190–204, 2018. https://doi.org/10.1109/TAES.2017.2741898.

33 M. G. Rutten, N. J. Gordon, and M. Simon. Particle-based Track-Before-Detect in Rayleigh noise. In O. E. Drummond, editor, *Signal and Data Processing of Small Targets 2004*, volume 5428, pages 509–519. SPIE, 2004.

34 L. Frederic. Recursive Bayesian filtering for multitarget Track-Before-Detect in passive radars. *IEEE Transactions on Aerospace & Electronic Systems*, 48(3): 2458–2480, 2012.

35 S. P. Ebenezer and A. Papandreou-Suppappola. Generalized recursive Track-Before-Detect with proposal partitioning for tracking varying number of multiple targets in low SNR. *IEEE Transactions on Signal Processing*, 64(11): 2819–2834, 2016.

36 S. Qin, J. Ding, L. Wen, and M. Jiang. Joint track-before-detect algorithm for high-maneuvering target indication in video SAR. *IEEE Journal of Selected Topics in Applied Earth Observations and Remote Sensing*, 14: 8236–8248, 2021. https://doi.org/10.1109/JSTARS.2021.3104603.

37 D. S. Bolme, J. Ross Beveridge, B. A. Draper, and Y. M. Lui. Visual object tracking using adaptive correlation filters. In *2010 IEEE Computer Society Conference on Computer Vision and Pattern Recognition*, pages 2544–2550, 2010. https://doi.org/10.1109/CVPR.2010.5539960.

38 J. F. Henriques, P. Caseiro, R. Martins, and J. Batista. Exploiting the circulant structure of tracking-by-detection with kernels. In A. Fitzgibbon, S. Lazebnik, P. Perona, Y. Sato, and C. Schmid, editors, *Computer Vision – ECCV 2012*, Springer Berlin Heidelberg, Berlin, Heidelberg, 2012.

39 M. Danelljan, G. Häger, F. S. Khan, and M. Felsberg. Accurate scale estimation for robust visual tracking. In *Proceedings of the British Machine Vision Conference*. BMVA Press, 2014.

40 Y. Li and J. Zhu. A scale adaptive kernel correlation filter tracker with feature integration. In L. Agapito C. Bronstein, M. M. Rother, editors, *Computer Vision – ECCV 2014 Workshops*, Springer International Publishing, Cham, 2015. ISBN 978-3-319-16181-5.

41 H. K. Galoogahi, T. Sim, and S. Lucey. Correlation filters with limited boundaries. In *2015 IEEE Conference on Computer Vision and Pattern Recognition (CVPR)*, pages 4630–4638, 2015. https://doi.org/10.1109/CVPR.2015.7299094.

42 J. F. Henriques, R. Caseiro, P. Martins, and J. Batista. High-speed tracking with kernelized correlation filters. *IEEE Transactions on Pattern Analysis and Machine Intelligence*, 37(3):583–596, 2015. https://doi.org/10.1109/TPAMI.2014.2345390.

43 M. Danelljan, G. Häger, F. S. Khan, and M. Felsberg. Learning spatially regularized correlation filters for visual tracking. In *2015 IEEE International Conference on Computer Vision (ICCV)*, pages 4310–4318, 2015. https://doi.org/10.1109/ICCV.2015.490.

44 H. K. Galoogahi, A. Fagg, and S. Lucey. Learning background-aware correlation filters for visual tracking. In *2017 IEEE International Conference on Computer Vision (ICCV)*, pages 1144–1152, 2017. https://doi.org/10.1109/ICCV.2017.129.

45 F. Li, C. Tian, W. Zuo, L. Zhang, and M.-H. Yang. Learning spatial-temporal regularized correlation filters for visual tracking. In *2018 IEEE/CVF Conference on Computer Vision and Pattern Recognition*, pages 4904–4913, 2018. https://doi.org/10.1109/CVPR.2018.00515.

46 B. Zhao, Y. Han, H. Wang, L. Tang, X. Liu, and T. Wang. Robust shadow tracking for video SAR. *IEEE Geoscience and Remote Sensing Letters*, 18(5): 821–825, 2021. https://doi.org/10.1109/LGRS.2020.2988165.

47 C. Zhong, J. Ding, and Y. Zhang. Video SAR moving target tracking using joint kernelized correlation filter. *IEEE Journal of Selected Topics in Applied Earth Observations and Remote Sensing*, 15: 1481–1493, 2022. https://doi.org/10.1109/JSTARS.2022.3146035.

48 Z. Liu, Y. Fang, L. Zhang, J. Li, and J. Hu. A robust videosar single target tracker by integrating correlation filter and imm-pdaf. *IEEE Geoscience and Remote Sensing Letters*, 19:1–5, 2022.

49 National Technology and LLC. Engineering Solutions of Sandia. Eubank gate and traffic videosar. http://www.sandia.gov/radar/video.

50 W. Yi, M. R. Morelande, L. Kong, and J. Yang. An efficient multi-frame track-before-detect algorithm for multi-target tracking. *IEEE Journal of Selected Topics in Signal Processing*, 7(3): 421–434, 2013. https://doi.org/10.1109/JSTSP.2013.2256415.

51 R. A. Horn and C. R.Johnson. *Topics in Matrix Analysis.* Cambridge University Press, 1991.

52 A. A. Moghadam, M. Kumar, and H. Radha. Common and innovative visuals: a sparsity modeling framework for video. *IEEE Transactions on Image Processing*, 23(9): 4055–4069, 2014. https://doi.org/10.1109/TIP.2014.2321476.

53 L. M. Novak, M. C. Burl, and W. W. Irving. Optimal polarimetric processing for enhanced target detection. *IEEE Transactions on Aerospace and Electronic Systems*, 29(1): 234–244, 1993. https://doi.org/10.1109/7.249129.

54 R. C. Gonzalez and R. E. Woods. *Digital Image Processing*. Pearson education, Inc, 2008.

55 J. Arnold, S. W. Shaw, and H. Pasternack. Efficient target tracking using dynamic programming. *IEEE Transactions on Aerospace and Electronic Systems*, 29(1): 44–56, 1993.

56 S. Buzzi, M. Lops, and L. Venturino. Track-Before-Detect procedures for early detection of moving target from airborne radars. *IEEE Transactions on Aerospace and Electronic Systems*, 41(3): 937–954, 2005. https://doi.org/10.1109/TAES.2005.1541440.

57 L. A. Johnston and V. Krishnamurthy. Performance analysis of a dynamic programming track before detect algorithm. *IEEE Transactions on Aerospace and Electronic Systems*, 38(1): 228–242, 2002. https://doi.org/10.1109/7.993242.

58 Y. Wei, L. Kong, J. Yang, and X. Deng. A tracking approach based on dynamic programming track-before-detect. In *2009 IEEE Radar Conference*, Pasadena, CA, pages 1–4, 2009.

59 S. M. Tonissen and R. J. Evans. Peformance of dynamic programming techniques for Track-Before-Detect. *IEEE Transactions on Aerospace and Electronic Systems*, 32(4): 1440–1451, 1996. https://doi.org/10.1109/7.543865.

60 M. S. Arulampalam, S. Maskell, N. Gordon, and T. Clapp. A tutorial on particle filters for online nonlinear/non-Gaussian Bayesian tracking. *IEEE Transactions on Signal Processing*, 50(2): 174–188, 2002. https://doi.org/10.1109/78.978374.

61 D. Crisan and A. Doucet. A survey of convergence results on particle filtering methods for practitioners. *IEEE Transactions on Signal Processing*, 50(3): 736–746, 2002. https://doi.org/10.1109/78.984773.

62 D. J. Salmond and H. Birch. A particle filter for Track-Before-Detect. In *Proceedings of the 2001 American Control Conference. (Cat. No.01CH37148)*, volume 5, pages 3755–3760, 2001.

63 M. Danelljan, A. Robinson, F. S. Khan, and M. Felsberg. Beyond correlation filters: learning continuous convolution operators for visual tracking. In B. Leibe, J. Matas, N. Sebe, and M. Welling, editors, *Computer Vision – ECCV 2016*, pages 472–488, Cham, 2016. Springer International Publishing.

64 H. L. Van Trees, K. L. Bell, and Z. Tian. *Detection, Estimation, and Modulation Theory.* John Wiley & Sons, Inc, 1968

65 C.-M. Huang, D. Liu, and L.-C. Fu. Visual tracking in cluttered environments using the visual probabilistic data association filter. *IEEE Transactions on Robotics*, 22(6):1292–1297, 2006. https://doi.org/10.1109/TRO.2006.882960.

66 T. Kronauer, F. Pfaff, B. Noack, W. Tiant, G. Maier, and U. D. Hanebeck. Feature-aided multitarget tracking for optical belt sorters. In *2019 22th International Conference on Information Fusion (FUSION)*, pages 1–8, 2019.

67 P. F. Felzenszwalb, R. B. Girshick, D. McAllester, and D. Ramanan. Object detection with discriminatively trained part-based models. *IEEE Transactions on Pattern Analysis and Machine Intelligence*, 32(9): 1627–1645, 2010. https://doi.org/10.1109/TPAMI.2009.167.

68 S. Boyd, N. Parikh, E. Chu, B. Peleato, and J. Eckstein. *Distributed Optimization and Statistical Learning via the Alternating Direction Method of Multipliers.* Now Foundations and Trends, 2011.

69 National Technology and Engineering Solutions of Sandia, LLC. Velocity independent continuous tracking radar mode. http://www.sandia.gov/radar/assets/videos/victr.mp4.

70 S. Feng, K. Hu, E. Fan, L. Zhao, and C. Wu. Kalman filter for spatial-temporal regularized correlation filters. *IEEE Transactions on Image Processing*, 30: 3263–3278, 2021. https://doi.org/10.1109/TIP.2021.3060164.

71 L. Bertinetto, J. Valmadre, S. Golodetz, O. Miksik, and P. H. S. Torr. Staple: Complementary learners for real-time tracking. In *2016 IEEE Conference on Computer Vision and Pattern Recognition (CVPR)*, pages 1401–1409, 2016. https://doi.org/10.1109/CVPR.2016.156.

72 Y. Li, J. Zhu, and S. C. H. Hoi. Reliable patch trackers: robust visual tracking by exploiting reliable patches. In *2015 IEEE Conference on Computer Vision and Pattern Recognition (CVPR)*, pages 353–361, 2015. https://doi.org/10.1109/CVPR.2015.7298632.

73 M. Wang, Y. Liu, and Z. Huang. Large margin object tracking with circulant feature maps. In *2017 IEEE Conference on Computer Vision and Pattern Recognition (CVPR)*, pages 4800–4808, 2017. https://doi.org/10.1109/CVPR.2017.510.

74 J. Ding, L. Wen, C. Zhong, and O. Loffeld. Video SAR moving target indication using deep neural network. *IEEE Transactions on Geoscience and Remote Sensing*, 58(10): 7194–7204, 2020. https://doi.org/10.1109/TGRS.2020.2980419.

75 Y. Wu, J. Lim, and M. Yang. Object tracking benchmark. *IEEE Transactions on Pattern Analysis and Machine Intelligence*, 37(9): 1834–1848, 2015. https://doi.org/10.1109/TPAMI.2014.2388226.

13

Space Object Tracking

Brandon A. Jones and Benjamin L. Reifler

Department of Aerospace Engineering and Engineering Mechanics University of Texas at Austin, Austin, TX, USA

13.1 Introduction

Space object tracking refers to the practice of collecting and processing observations of anthropogenic space objects (ASOs) for situational awareness. Space object tracking shares similarities with the broader multi-target tracking community, but nuances of the problem prevent the simple transition of algorithms from the fusion community to the space domain. The dominant physics of satellite motion is well known, but operator decisions, such as maneuvers, cause a spacecraft to deviate from predicted trajectories. As in other domains, data association, the identification of the ASO that generated a measurement, is a critical component of the tracking problem. However, data sparsity combined with unpredictable operator decisions can challenge the most robust and efficient state-of-the-art tracking algorithms. This chapter will introduce the reader to the basic principles of space object tracking and the ongoing and future challenges in this discipline.

Space situational awareness (SSA) is the detection, tracking, and characterization of ASOs. It is an interdisciplinary field that combines astrodynamics, space weather, satellite systems, astronomy, uncertainty quantification, remote sensing, and information fusion. A history of space surveillance in the United States (up to its publication) may be found in [1]. The ability to track an ASO refers to producing an estimated position and velocity state with an accuracy sufficient to predict its location for follow-up measurements. As of March, 2025, the United States Space Force (USSF) has cataloged and tracks 29,568 ASOs[1] as small as 10 cm in low Earth orbit (LEO) via radar observations and approximately 1 m in geosynchronous Earth orbit (GEO) with optical data. Unfortunately, impacts with objects as small as 1 mm can result in the end of a spacecraft's operations. Hence, there is a critical gap between capabilities and the space community's needs for space object tracking.

Future SSA systems must deal with increasing measurement sparsity. Sensors regularly provide measurements of spacecraft, but as the ASO catalog grows and the network workload increases, the time between measurements of individual satellites will increase. SSA also has military implications that place geopolitical limits on the locations of ground-based sensors. This gap in sensor coverage yields regions of space that are observed less frequently than others. These factors combined create

---

1 *Orbital Debris Quarterly News*, Vol. 29, No. 2, May 2025, https://orbitaldebris.jsc.nasa.gov/quarterly-news/pdfs/ODQNv29i2.pdf.

issues of measurement density in both time and space. To maintain SSA despite this sparsity issue, space object trackers (SOTs) will require improvements in information extraction.

The goal of this chapter is to introduce space object tracking to a larger community, and provide a primer for those first working on the problem. To this end, we will introduce the foundational principles of satellite motion, the topic of orbit determination (OD) and its relationship with track initiation and information fusion, and some recently developed methods to address problems in space object tracking. We note that the chapter is not comprehensive, and there are some topics where we refer the reader to a more detailed treatment of the material. This chapter assumes the reader is familiar with the fundamentals of information fusion and multi-target tracking, i.e., the topics covered in other chapters of this book. In Section 13.2, we provide a basic introduction to modeling the dynamics of space objects to help put future sections in the proper context. Section 13.3 provides a brief introduction to the types of observations typically used for ASO tracking. Section 13.4 introduces the reader to the initial OD problem, i.e., track initiation, and recently developed methods for sparse OD. With the pieces defined, Section 13.5 describes the current state of the practice and newly developed methods for multi-target tracking of space objects. Section 13.6 illustrates one possible combination of the capabilities and algorithms from the previous sections to create an SOT, and then presents an example when tracking over 16,300 ASOs in LEO. Finally, Section 13.7 summarizes and highlights new and ongoing challenges in SSA.

13.2 Modeling and Dynamics of Space Objects

Dynamic model fidelity and the parameterization of the translational state greatly influence computational efficiency and performance of an SOT. Dynamic signatures vary with orbit altitude, yielding varying time scales for required measurement frequency to maintain custody of ASOs. A spacecraft in LEO (<1000 km in altitude) will have an orbit period of approximately 100 minutes, while communication satellites in GEO have a 24-hour period. Observation frequencies are determined by the relative dynamics of ground-/space-based sensors and an ASO, with possibly days between observations of a single object. The potential long gaps in data combined with the nonlinear dynamic environment make the accuracy of the SOT's time update a primary driver of performance. In this section, we highlight some of the dominant forces and considerations in orbit propagation. This topic is covered in more detail in other references, inter alia, [2, 3]. The interested reader is referred to those and related references for further details.

To create a common notation for a prediction model over all sections of this chapter, let a solution of the orbit propagation problem be denoted:

$$\boldsymbol{x}_k = \boldsymbol{f}(\boldsymbol{x}_{k-1}, t_{k-1}, \boldsymbol{v}_{k-1}; \boldsymbol{p}), \tag{13.1}$$

where $\boldsymbol{x}_k \in \mathbb{R}^{n_x}$ is the state at time t_k (see Section 13.2.3), $\boldsymbol{v}_{k-1}$ is a process noise vector (see Section 13.2.5), and $\boldsymbol{p}$ are constant parameters, such as gravitational parameters, that influence the orbit. Multiple choices for $\boldsymbol{f}$ exist based on, for example, model truncation, simplifying assumptions, and process noise modeling. We denote the engineer's chosen solver as $\boldsymbol{f}$. Evaluation of $\boldsymbol{f}$ is usually performed via a numeric method, but (semi-)analytic methods exist under specific simplifying assumptions (e.g., see [2] and the SGP4 propagator in Section 13.5.1). The remainder of Section 13.2 is based on propagation via a differential equation and the physics-based models used in its definition.

13.2.1 Central and N-body Dynamics

The second-order, ordinary differential equation describing the motion of an object in orbit about a central body may be expressed as:

$$a = \ddot{r} = -\frac{\mu}{\| r\|^3} r + a_{3B} + a_{\text{Asphere}} + a_{\text{drag}} + a_{\text{SRP}} + \cdots + B(t)v(t), \tag{13.2}$$

where $\mu = GM$ is the gravitational parameter, G is Newton's gravitational constant, M is the mass of the central body, r is the position of the object relative to the center of mass of the body, a_* are perturbing accelerations including, inter alia, other celestial bodies, drag, etc., $B(t) \in \mathbb{R}^{3\times3}$ is a known matrix, and $v(t) \in \mathbb{R}^3$ is random process noise with $B(t)v(t)$ modeled as a perturbing acceleration. When $v(t)$ is included, (13.2) is a stochastic differential equation. These terms, including statistics on v, are defined below. For simplicity in presentation, we will assume the primary body is the Earth and Figure 13.1 illustrates the relative magnitude of these perturbations as a function of altitude above the Earth's surface.

The first term in (13.2) is dubbed the *two-body term*, and only in the case where $a_* = 0$ is there an analytic solution to this problem. In real-world cases, this does not occur, and we require semi-analytic or numeric methods to produce an approximate solution to the differential equation. The most common approach uses the ubiquitous Runge-Kutta integrator to numerically propagate an orbit, but alternatives such as Gauss-Jackson methods often produce more computationally efficient solutions [3]. Evaluation of such an integrator with input state $x_{k-1} = \left[r_{k-1}^T \quad \dot{r}_{k-1}^T\right]^T$ and a realization of v_{k-1} is one example of f in (13.1).

For orbits near the Earth, models of the conservative forces acting on the satellite are well understood and accurately modeled at the fidelity required for an SOT. The parameter $\mu = 398600.4415 \pm 0.0008$ km^3/s^2 (e.g., [2]), and its precision indicates a strong confidence in the two-body term. Other celestial bodies, such as the Sun and Moon, also exert an appreciable gravitational force on Earth-orbiting spacecraft. Let r_3 and r_E be the inertial-frame locations of a third body and the Earth, respectively, and μ_3 is the gravitational parameter for that third body. The gravitational acceleration of the spacecraft due to the third body is

$$a_{3B} = \mu_3\left(\frac{r_{s3}}{\| r_{s3}\|^3} - \frac{r_{3E}}{\| r_{3E}\|^3}\right), \tag{13.3}$$

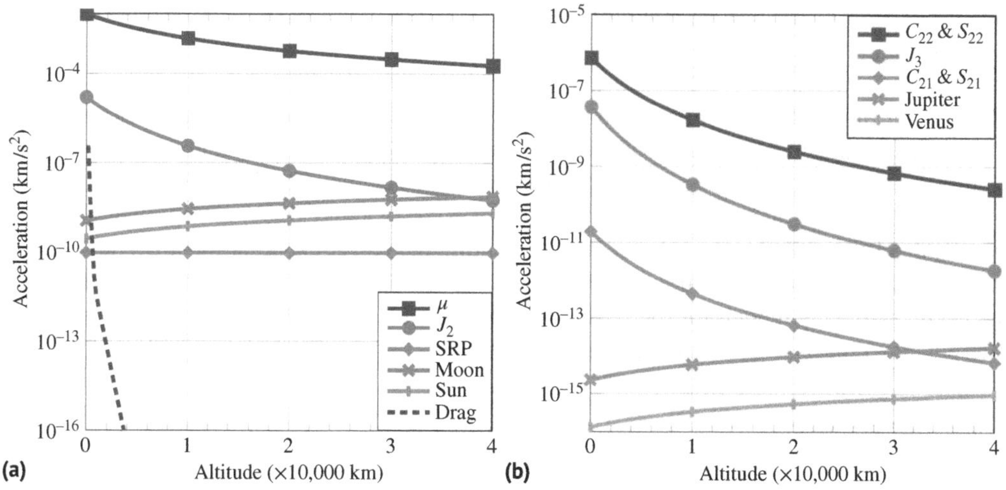

Figure 13.1 Comparison of magnitudes of different accelerations affecting an object in Earth orbit.

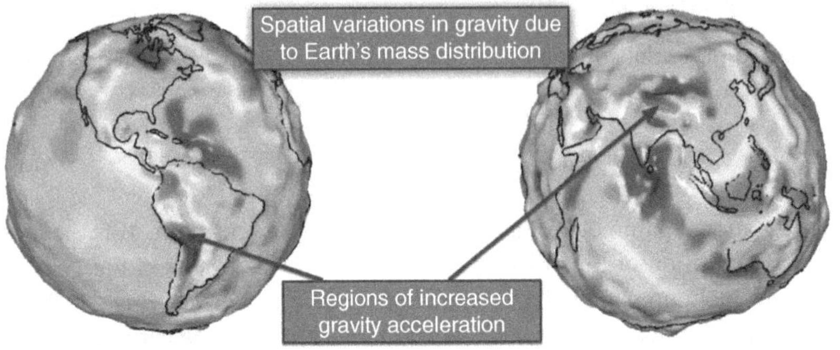

Figure 13.2 Illustration of gravity anomalies for the Earth. Image courtesy of NASA/JPL/University of Texas Center for Space Research.

where

$$\boldsymbol{r}_{s3} = \boldsymbol{r} - \boldsymbol{r}_3, \tag{13.4}$$

$$\boldsymbol{r}_{3E} = \boldsymbol{r}_3 - \boldsymbol{r}_E. \tag{13.5}$$

Locations of the celestial bodies are commonly computed using the SPICE library,[2] which uses precise solutions produced by the Jet Propulsion Laboratory [4].

Derivation of the two-body term assumes a point mass for the central body, so higher-fidelity models include a perturbing acceleration $\boldsymbol{a}_{\text{Asphere}}$ due to the nonuniform mass distribution within a circumscribing sphere of radius R. These variations to the gravitational potential U are most commonly modeled using the spherical harmonics model:

$$U(r, \phi, \lambda) = \frac{\mu}{r} \sum_{n=2}^{n_{\max}} \sum_{m=0}^{n} \left(\frac{R}{r}\right)^n P_{nm}[\sin(\phi)](C_{nm}\cos(m\lambda) + S_{nm}\sin(m\lambda)), \tag{13.6}$$

where r, ϕ, and λ are the radius, geocentric latitude, and longitude of the spacecraft, respectively, n and m are the degree and order of the expansion, $n_{\max}$ is the maximum degree of the model, P_{nm} are the associated Legendre functions, and C_{nm} and S_{nm} are the Stokes coefficients that quantify the perturbed gravity potential for a given n and m. The $J_n = -C_{n0}$ coefficients are referred to as *zonal* terms and determine the magnitude of potential variations symmetric about the body's pole (since the $m = 0$ terms in (13.6) are independent of λ). The perturbing acceleration is

$$\boldsymbol{a}_{\text{Asphere}} = \nabla U \tag{13.7}$$

evaluated at the satellite's location relative to the Earth, where ∇ denotes the gradient operator. Full details on the derivation and software implementation of this model can be found in [2, 6]. Figure 13.2 illustrates the gravity anomalies for the Earth, i.e., the effect of the Stokes coefficients on U near the surface of the Earth. Several sets of Stokes coefficients exist for orbit propagation with the GGM05 [7] and EGM2008 [8] models representing the current state-of-the-art. The required model maximum degree $n_{\max}$ varies with accuracy requirements, orbit distance, and compute costs, but even low-fidelity models for the Earth must include oblateness (the $C_{2,0}$ term). Alternatively, interpolation models (e.g., [9, 10]) may be used for faster computation.[3]

2 https://naif.jpl.nasa.gov
3 Fortran software for the model of [9] may be found at https://sites.utexas.edu/russell/publications/code/fetch/

13.2.2 Nonconservative Forces

Nonconservative forces acting on an ASO are the largest source of modeling error in orbit propagation. These are dependent on the ASO's size, shape, orientation, material properties, solar weather, modes of satellite operation, and other phenomena that can be difficult to predict and estimate. SOT algorithms typically use simplified models, e.g., assume a spherical spacecraft, that provide a means for low- to medium-fidelity modeling of these forces.

For objects in LEO, the residual atmosphere induces a drag force on the satellite that depends on an ASO's attitude, shape, and surface properties. The *residual atmosphere* refers to the low-density portions of the Earth's atmosphere that extend up to 100s of kilometers above the planet's surface. Due to a lack of observability with traditional measurements, SOTs provide limited information on these spacecraft properties except under simplifying assumptions. Instead, the community relies on a model that assumes a constant area-to-mass ratio A/m and drag coefficient C_D; the so-called *cannonball model* for drag

$$\boldsymbol{a}_{\text{drag}} = -\frac{1}{2}\frac{C_D A}{m}\rho(\boldsymbol{r}, t) \parallel \boldsymbol{v}_{\text{rel}} \parallel \boldsymbol{v}_{\text{rel}}, \tag{13.8}$$

where $\rho(\cdot, \cdot)$ is the atmospheric density as a function of time and position, and $\boldsymbol{v}_{\text{rel}}$ is the velocity of the spacecraft relative to the atmosphere. For spheres, C_D is typically in the range 2.0 to 2.1, while A/m can vary significantly with the object. Typical values are in the range 10^{-2}–10^{-3} m^2/kg but can be as high as 100 m^2/kg [11]. Due to the linear relationship in the model, C_D, A, and m cannot be uniquely estimated. Instead, the ballistic coefficient $B = \frac{m}{C_D A}$ is sometimes included in the estimated state vector $\boldsymbol{x}$ in the OD process (see Section 13.4). When attempting to characterize an ASO, estimating B can only help to constrain the mass and size of a satellite from the tracking data. Multiple models of the Earth's atmospheric density ρ exist with varying accuracy and computational complexity (e.g., [12–14]).

Accounting for space weather's effects on the atmosphere is difficult when predicting $\boldsymbol{a}_{\text{drag}}$. Density models used in operations ingest one or more *space weather indices* that quantify the effects of the space environment on the Earth's atmosphere. For example, the $F_{10.7}$ index is a measure of the solar flux from the Sun at a wavelength of 10.7 cm, and is one of several values used in density models to account for the effects of space environment on the atmosphere. Larger $F_{10.7}$ values indicate an increased atmospheric density at higher altitudes, i.e., a larger drag perturbation. Space weather indices can be difficult to predict with large daily variations occurring with solar storms. Due to space weather prediction errors and the simplifying assumptions with the cannonball model, atmospheric drag is the primary source of propagation error for SOTs.

In higher-altitude orbits, photons from the Sun induce the largest nonconservative force. This solar radiation pressure (SRP) depends on space weather, satellite shape in the direction of the source of illumination, and surface properties. The cannonball model for SRP is

$$\boldsymbol{a}_{\text{SRP}} = -p_{\text{SRP}}\frac{C_R A}{m}\frac{\boldsymbol{r}_{s\odot}}{\parallel \boldsymbol{r}_{s\odot} \parallel}, \tag{13.9}$$

where p_{SRP} is the solar pressure per unit area, C_R is a coefficient of reflectivity, and $\boldsymbol{r}_{s\odot}$ is the position of the spacecraft relative to the Sun. While $p_{\text{SRP}} = 4.57 \times 10^{-6}$ N/m^2 is often used [2], time-dependent models account for variations such as the change in the distance between the Earth and Sun. Note that in addition to assuming a spherical body, (13.9) assumes a Lambertian reflectance model for how the spacecraft surface absorbs and reflects the incoming photons. Other, more complex, reflectance models can instead be used under additional surface property assumptions [15].

13.2.3 State Representations

The translational state $\boldsymbol{x}$ of the satellite can be parameterized in many ways. Cartesian coordinates are common, but Keplerian orbital elements or equinoctial elements use coordinates rooted in the geometry of the problem of orbit propagation. They provide an intuitive understanding of the orbit (size, shape, etc.) and a parameter set where a subset of elements vary slowly in time. This aids in visualization of the orbits and uncertainty propagation (see Section 13.4.3), but at the potential cost of propagator complexity or increased nonlinearity in data fusion. The selection of the orbit coordinates/parameters defines the state $\boldsymbol{x}$ in (13.1), and it can vary between SOTs. Note that the survey paper [16] lists 22 different parameter sets.

Figure 13.3 illustrates the Keplerian orbital elements, which parameterize the orbit's size as the semimajor axis of the orbit ellipse a, its shape by the eccentricity of the ellipse $e = \| \boldsymbol{e} \|$, the location of the spacecraft in the orbit via its true anomaly ν, and a 3-1-3 Euler sequence (Ω, i, and ω) that describes the orientation of the orbit relative to a planet-centered inertial frame. These elements allow for quick computation of the orbit's distance of closest approach ($a(1-e)$), the distance of furthest approach ($a(1+e)$), and other values that provide an intuitive understanding of the orbit geometry. In the case of the two-body problem, all parameters are constant except ν. For near-Earth orbits with perturbations, the other elements vary slowly over time. However, they do contain singularities, e.g., ω is undefined for a circular orbit when $\boldsymbol{e} = \boldsymbol{0}$. A common alternative, the modified equinoctial elements, are less intuitive but only have one singularity at a point in the parameter space that corresponds to an orbit that requires too much fuel to be practical. Hence, in operational and research settings for ASO tracking, equinoctial elements are commonly used (e.g., [5, 17]).

High-fidelity propagation of equinoctial elements often uses the Cartesian-based model of (13.2). To leverage these models in, for example, a sigma-point or ensemble filter, points in equinoctial elements are converted to Cartesian coordinates and propagated to the desired time. The propagated point is then converted from Cartesian to equinoctial, and the predicted density (or particle ensemble) is represented in the element set. This allows the SOT to leverage the accuracy of propagators defined in Cartesian coordinates with a set of slowly varying orbit elements. We expand on the benefits of elements for probability density function (PDF) propagation in Section 13.4.3.

13.2.4 Coordinate and Time Systems

Software implementation of astrodynamics algorithms requires accurate bookkeeping of time and coordinate systems. The nuanced differences between similar coordinate systems such as the "J2000" and "International Celestial Reference Frame" (see Figure 13.4) can initially be difficult to

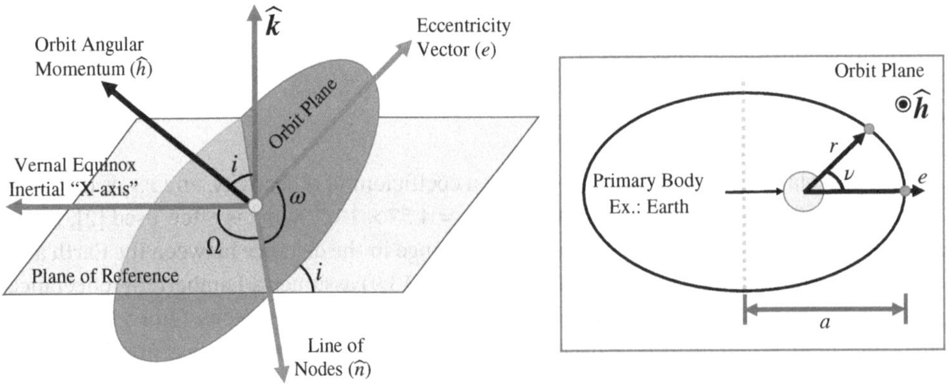

Figure 13.3 Keplerian Orbital Elements—Euler angles on the left and orbit size/shape parameters on right.

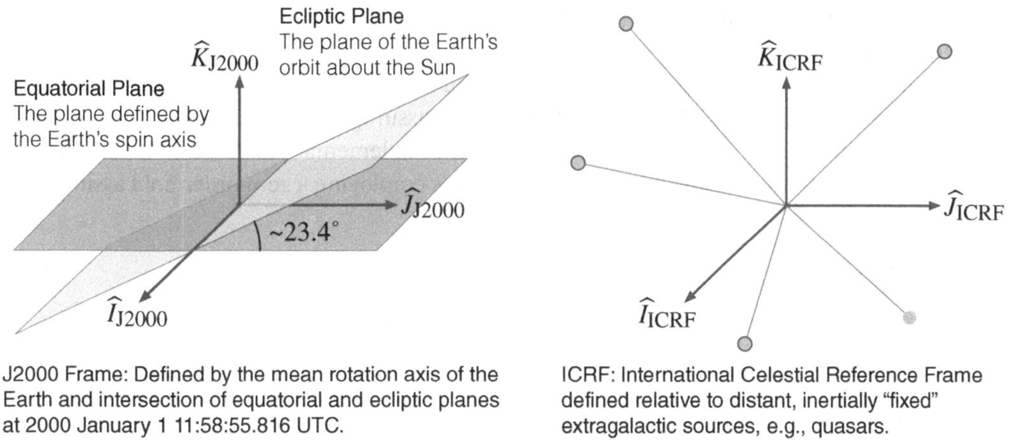

Figure 13.4 Two examples of Earth centered inertial references frames that agree to within 0.1 arcseconds [18]: J2000 (left) and ICRF (right).

navigate. Truncating the small variations in the transformation between an inertial and an Earth-fixed frame, i.e., a frame that rotates with the Earth's surface, can yield errors up to 18 km for the inertial location of a ground station [19]. A common mistake made by new astrodynamicists is misusing (or mislabeling) data with the wrong coordinate system, creating issues with seemingly simple comparisons of propagator results or in data fusion.

Even time has different definitions that must be accounted for in astrodynamics. On the surface of the Earth, the International System of Units (SI) defines a second by International Atomic Time (TAI). Civil time, i.e., the time one usually uses to schedule their day, is based on Coordinated Universal Time (UTC), which includes a time-varying number of *leap seconds* relative to TAI. Due to relativistic effects, the independent variable of motion for celestial bodies is based on *dynamical time*, which does not use the SI second. When querying the SPICE library for a planet's position to evaluate (13.3), software must convert from TAI (or UTC) to an appropriate dynamical time. Differences between these time systems can be as large as ~85 seconds (see figure 3.24 of [2]).

In summary, time and coordinate systems are a common source of error for those first entering the field of astrodynamics. A proper treatment of the subject could be its own chapter, and we only allude to some of the challenges. We refer the reader to [2] and [20] for further details, and recommend that they not underestimate its importance.

13.2.5 Process Noise Modeling

The estimator employed, the truncation of the modeled forces, and the average propagation time determine the process noise statistics in an SOT. Not all SOTs use process noise. When including process noise in an SOT, the engineer must derive or define a mathematical formulation that is consistent with the modeling assumptions, test the propagator with appropriate force-model truncation, and revisit the assumptions based on system validation results. The function $\boldsymbol{f}$ in (13.1) assumes the engineer has an implementation of the chosen model given a realization $\boldsymbol{v}_{k-1}$.

Methods based on differential correction (see Section 13.4) often include no process noise in the formulation and mitigate propagation errors through observation batch sizes. More observations over a longer time period provide more information for the estimator, but the longer propagation time accumulates propagated state error that biases the solution. The operator balances these two properties through the batch size and then chains batches together as necessary to process all available measurements.

When modeled in, for example, an extended Kalman filter (EKF), the process noise model used in the engineer's chosen $\boldsymbol{f}$ (13.1) varies with the model assumptions. As presented in (13.2), $\boldsymbol{v}(t)$ is a continuous-time process, but that is not practical in implementation. Process noise models in an SOT are often assumed to be white, Gaussian models employing a zero-order hold assumption when propagating from t_{k-1} to t_k, i.e.,

$$\boldsymbol{v}_{k-1} \sim \mathcal{N}(\boldsymbol{0}, \boldsymbol{Q}_{k-1}), \tag{13.10}$$

where $\mathcal{N}$ denotes the Gaussian distribution with a given mean and covariance:

$$\mathbb{E}\left[\boldsymbol{v}_i \boldsymbol{v}_j^{\mathrm{T}}\right] = \boldsymbol{Q}_i \delta_{ij}, \tag{13.11}$$

and δ_{ij} is the Kronecker delta. The mapping of random inputs $\boldsymbol{v}_{k-1}$ to a perturbing acceleration or the use of $\boldsymbol{Q}_{k-1}$ in the filter time update can take multiple forms. The form in (13.2) is one choice, but *it is not always the most appropriate for a given application.* Many in the astrodynamics community refer to two primary categories of process noise models: state noise compensation (SNC) and dynamic model compensation (DMC). Some adaptive forms of these methods work well in traditional OD and navigation (e.g., [21]), but one should acknowledge and mitigate potential performance limitations with sparse data.

SNC adds a term to the discrete-time covariance time update to compensate for $\boldsymbol{v}_{k-1}$ defined as in (13.2). While the name SNC is used by the astrodynamics community, the technique is common in the broader estimation literature. The method propagates the covariance matrix via, for example,

$$\begin{aligned}\overline{\boldsymbol{P}}_{k|k-1} = {} & \boldsymbol{F}(t_k, t_{k-1})\boldsymbol{P}_{k-1|k-1}\boldsymbol{F}^T(t_k, t_{k-1}) \\ & +\boldsymbol{\Gamma}(t_k, t_{k-1})\boldsymbol{Q}_{k-1}\boldsymbol{\Gamma}^T(t_k, t_{k-1}),\end{aligned} \tag{13.12}$$

where $\boldsymbol{F}$ is the state transition matrix produced via numeric integration (see Section 4.2.1 of [22]), $\boldsymbol{\Gamma}(t_k, t_{k-1})$ is a process noise transition matrix, and $\boldsymbol{P}_{k-1|k-1}$ is the posterior covariance at time t_{k-1} based on data up to that time. Note that, in an SOT, $\boldsymbol{\Gamma}(t_k, t_{k-1})$ should be numerically integrated instead of using an analytic function of the time between measurements. In the context of (13.2) with $\boldsymbol{x}$ defined by the Cartesian position and velocity states, this differential equation is

$$\dot{\boldsymbol{\Gamma}}(t, t_{k-1}) = \boldsymbol{A}(t)\,\boldsymbol{\Gamma}(t, t_{k-1}) + \begin{bmatrix} \boldsymbol{0}_{3\times3} \\ \boldsymbol{B}(t) \end{bmatrix}, \tag{13.13}$$

where

$$\boldsymbol{A}(t) = \begin{bmatrix} \boldsymbol{0}_{3\times3} & \mathbb{I}_{3\times3} \\ \dfrac{\partial \ddot{\boldsymbol{r}}}{\partial \boldsymbol{r}} & \dfrac{\partial \ddot{\boldsymbol{r}}}{\partial \dot{\boldsymbol{r}}} \end{bmatrix}_{\boldsymbol{r}=\boldsymbol{r}(t),\dot{\boldsymbol{r}}=\dot{\boldsymbol{r}}(t)} \tag{13.14}$$

is the Jacobian matrix of a first-order, coupled linear system based on (13.2). This numerical integration would be done simultaneously with the deterministic terms in (13.2). More details on this process can be found in Chapters 4 and 6 of [22]. When process noise is not of the form in (13.2) or a given estimator requires a different approach, an engineer will require an alternative solution.

DMC instead adds elements to $\boldsymbol{x}$ corresponding to an additional perturbing acceleration that is modeled as a Gauss-Markov process with a random noise in its differential equation. The perturbing acceleration is then estimated in the filter as a solve-for parameter. Details on DMC with a first-order Gauss-Markov process can be found in [22], and higher-order methods may be leveraged as appropriate (e.g., [23]).

The value of $\boldsymbol{Q}_k$ depends on the assumptions in the force model and the times between the measurements. Hence, we cannot provide universally applicable values. As illustrated in Figure 13.1, truncation of, for example, atmospheric drag at low altitudes creates an acceleration error on the order of 10^{-7} km/s^2. Orbit motion about the primary body causes gravity perturbations $\boldsymbol{a}_{\text{Asphere}}$ to be periodic with a near-zero average. Hence, their effects require a smaller $\boldsymbol{Q}_k$ than Figure 13.1 implies when one or more revolutions occur between observations. This culminates in matrices $\boldsymbol{Q}_k$ customized to the scenario/system being considered. For example, in a multi-target scenario for ASO tracking, [24] and [25] used $\boldsymbol{Q}_{k-1} = \sigma^2 \mathbb{I}_{3\times3}$ with σ values of 8×10^{-9} km/s^2 and 10^{-12} km/s^2, respectively. Using this matrix with (13.10), realizations of $\boldsymbol{v}_{k-1}$ are produced (as sigma points [24] or random samples [25]) and used in the direct evaluation of $\boldsymbol{f}$ as a solution to (13.2) with $\boldsymbol{B}(t) = \mathbb{I}_{3\times3}$. The former included more force-model truncation than the latter, resulting in a larger process noise statistic. Operationally, $\boldsymbol{Q}_k$ is adjusted to yield approximately Gaussian, zero mean prediction residuals in the estimator, while balancing this precision with the propagator's computation time in the SOT.

Finally, for active ASOs, modeling unknown maneuvers as process noise creates challenges for SOTs. In this context, a maneuver refers to the use of a propulsion system to change the velocity or attitude of an ASO. Maneuvers usually occur during observation gaps and the resulting propagation errors can create data association issues. Except when an operator provides information on maneuvers or they take place at regular and known intervals, they occur at any time and with any thrust vector. Mono- and bi-propellant propulsion systems provide short-duration maneuvers (seconds to minutes) with magnitudes in the range 10^{-1}–10^{4} N (e.g., see tables 18-5 and 18-6 of [26]). Newer satellites leverage fuel-efficient electric propulsion systems that use long duration maneuvers (minutes to hours) with magnitudes closer to 10^{-3}–10^{0} N (tables 18-12 and 18-13 of [26]). The wide range of available propulsion systems and variations in how they operate create systematic errors in propagator force models that cannot easily be modeled as process noise. Methods of robust OD can help mitigate this, but this is an ongoing area of research with sparse measurements (e.g., [27, 28]). Estimating the thrust vector using optimal control theory or thrust Fourier coefficients provide candidate approaches to improve robustness and characterize possible maneuvers, but such algorithms assume a known data association (e.g., [29, 30]).

13.2.6 Open-source Orbit Propagation Software

The astrodynamics community has produced multiple open-source software options. The utility of these models vary, and each requires at least some basic knowledge of orbit propagation and careful review of documentation to use the tools correctly. Some candidate tools include:

- **Orekit**[4] is an open-source software package (Apache License V2.0) for space dynamics modeling written in Java with a Python wrapper available. The community of users includes academics and commercial companies, and multiple efforts have validated the software through comparison with other tools. It is often used in academic SSA research. Some open-source toolboxes available online use this tool as the core propagator, e.g., `orbdetpy`.[5]
- **General Mission Analysis Tool (GMAT)**[6] is a NASA-developed toolbox (Apache License V2.0) primarily used for mission analysis and design, but it includes validated high-fidelity propagators written in C/C++ that can be incorporated into an SOT toolbox. Developers support interfaces to Python, MATLAB, and Java. The software includes extensive tutorials and a community wiki to assist new users.

4 https://www.orekit.org
5 https://github.com/ut-astria/orbdetpy
6 https://software.nasa.gov/software/GSC-17177-1

- **Basilisk (BSK)**[7] (ISC License) is designed for spacecraft simulations, including Monte-Carlo analysis, in support of space missions at the University of Colorado Boulder and the Laboratory for Atmospheric and Space Physics. Written in C/C++, it includes Python interfaces for scripting and integration with other tools. It includes software for both orbit and attitude propagation.
- **TU Delft Astrodynamics Toolbox (Tudat)**[8] is also written in C/C++ with a Python interface under the BSD 3-Clause license. While open to the public, this software is developed for research at TU Delft with a focus on enabling work at that institution.

The reader should note that these tools provide advanced capabilities and augment, not replace, the practitioner's knowledge of astrodynamics. They will provide the user with validated software tools capable of incorporating complex dynamic models that can take years to implement, debug, and validate.

13.3 Observing Space Objects

SOTs use data from a combination of Earth- and space-based sensors for sensor fusion. These measurements come in several modalities, but are most commonly radar or electro-optical. Both modalities exist in a passive or active setting, where the former listens for natural or human-made signals and the latter emits energy that is reflected and received by a detector. This section provides a brief overview of common observations used in an SOT. The reader should note that sensors are an ongoing area of research with new capabilities in development to address new challenges in ASO tracking. This includes, inter alia, spectroscopy to fingerprint ASOs, event-based cameras for higher dynamic range, and hypertemporal imaging.

For this presentation, we will express the measurement model as:

$$\boldsymbol{z}_k = \boldsymbol{h}(\boldsymbol{x}_k; \boldsymbol{o}) + \boldsymbol{w}_k, \tag{13.15}$$

where $\boldsymbol{h}$ is the measurement function, $\boldsymbol{o}$ is a vector of known model parameters (e.g., sensor location), and the errors are white with $\boldsymbol{w}_k \sim \mathcal{N}(\boldsymbol{0}, \boldsymbol{R}_k)$. The accuracy of the noise assumptions depends on the sensor and how it is used.

13.3.1 Existing Networks

Sensor networks for ASO tracking predate the launch of Sputnik [31], and currently include (1) exquisite sensors designed for tracking and (2) units repurposed to fulfill tracking needs. Early studies debated the use of radar versus optical sensors for tracking [31], but both are now used in tandem based on their relative strengths. In the United States, what was initially a collection of ballistic-missile warning sensors evolved into the USSF's Space Surveillance Network (SSN). Today, that network includes ground- and space-based sensors. Using the SSN, the USSF maintains its Space Object Catalog, which is used to provide collision warnings to the broader space community. Russia, the European Union, and China have similar networks for SSA.

Several commercial networks provide data and processed products for government and commercial customers. ExoAnalytic Solutions[9] and Slingshot Aerospace each operate a network of ground-based optical sensors distributed over the globe. LeoLabs uses a network of phased-array radars to

7 https://avslab.github.io/basilisk/
8 https://docs.tudat.space/en/latest/
9 ExoAnalytic Solutions was acquired by Anduril Industries in March, 2026.

track the LEO population and provide collision warning services to commercial customers. These and other entities sell observations, orbit solutions, collision warning data, and other services as part of their suite of products.

Academic networks extend across international boundaries with the goal of conducting research and training students in space operations, orbital debris, and SSA. The International Scientific Optical Network (ISON) network [32] includes electro-optical sensors from approximately 10 countries. Managed by the Keldysh Institute of Applied Mathematics, this loose collective of sensors collaborates on small- and large-scale surveys of the GEO belt and other regions of space. The Falcon Telescope Network [33] is managed by the United States Air Force Academy with sensors in the United States, Australia, and Germany. Multiple isolated sensors gather data for smaller-scale studies such as those at the VaDeR Laboratory at the University of Colorado Boulder or the Purdue Optical Ground Station (POGS).

13.3.2 Radar Observations

A *radar observation* refers to radio waves detected by a sensor that provides information on the translational state of an ASO. This modality is most commonly associated with active sensing of a known signal combined with time-of-flight principles to measure distance traveled, i.e., transmitting a signal from an emitter, receiving the reflected or relayed signal at a known location, and the measured time delay. The known signal speed allows for approximating the distance traveled. While it shares some similarities with electro-optical measurements, satellite laser ranging is often considered as a radar-like observation. Due to path loss, radar observatories focus on spacecraft in LEO, but some high-power, precise systems can reach high-Earth orbits [34]. Passive sensing uses time difference of arrival (TDOA) and frequency difference of arrival (FDOA) measurements, which require two or more sensors at different locations receiving and comparing the same signal. These active measurements allow for tracking operational spacecraft even in very high altitude orbits. The random measurement errors $\boldsymbol{w}_k$ depend on the sensor. Typical noise values used in the literature are 10–100 m in range, 0.01–0.2° for direction of the signal, and 0.1–3 m/s in range-rate (e.g., see [25, 27]).

Radar observation vectors can include a combination of angles, range, and Doppler observables. Ground-based tracking of ASOs requires careful consideration of light-time delays and media effects in measurement modeling. Satellites may be 100s to 1,000s of kilometers away from the sensor, so the time of signal propagation must be accounted for in the modeling. Depending on the frequency of the signal, different atmospheric and space weather effects must also be modeled. For one-way range observations [22]

$$z_k = \| \boldsymbol{r}_{\text{rec}}(t_k) - \boldsymbol{r}_{\text{tran}}(t_k - \tau) \| + \epsilon_{\text{trop}} + \epsilon_{\text{iono}} + w_k, \tag{13.16}$$

where $\boldsymbol{r}_{\text{tran}}$ is the position of the signal's origin, $\boldsymbol{r}_{\text{rec}}$ is the receiver's location, τ is the signal propagation time, ϵ_{trop} is the bias due to the troposphere path delay, and ϵ_{iono} is the bias due to the ionosphere path delay. The satellite is either the transmitter or the receiver depending on the scenario. We use the prior information on the satellite and fixed-point iteration to approximate τ within the measurement model [3]. Troposphere and ionosphere delays are based on frequency-appropriate refraction models (e.g., see Chapter 3 of [22]). Troposphere biases vary with elevation, and range from 2 m at ~90° elevation to tens of meters at less than 10° above the horizon [3]. Ionosphere biases depend on frequency of the signal. For the L-band (most satellite navigation signals), these errors typically fall in the range 2–30 m [3, 35], but can escalate with solar weather events that increase the total electron count in the ionosphere. Other model biases, such as multipath, may also need to be modeled to achieve system accuracy requirements.

Two-way ranging is then the sum of the uplink and downlink range [22]:

$$z_k = \| \boldsymbol{r}_{\text{rec}}(t_k) - \boldsymbol{r}(t_k - \tau_d) \| + \| \boldsymbol{r}(t_k - \tau_d) - \boldsymbol{r}_{\text{tran}}(t_k - \tau_u - \tau_d) \| \\ + \delta_{\text{trop}} + \delta_{\text{iono}} + w_k, \tag{13.17}$$

where τ_d and τ_u are the signal propagation times for the downlink and uplink paths, respectively, and δ_* are the accumulated atmospheric effects on the signal (up and down). Note that this model assumes no delay when the signal arrives at the satellite. Two fixed-point iterations are required to approximate τ_u and then τ_d. Ideal measurements, which are used for algorithm development before testing with real data, assume that all τ_* and media delays are zero.

13.3.3 Electro-optical Observations

Electro-optical observations use principles of astrometry with electromagnetic signals typically in the visual spectrum. Most often passive, a properly designed telescope will receive sunlight reflected by an ASO. Its location relative to the stars in the image provides a measurement of the ASO's location relative to the sensor and in an inertial frame.[10] For example, the same satellite seen by one telescope at the equator and a second at the south pole will observe different locations relative to the stars, but both unit vectors (from the sensor to the object) will be in coordinates of the same inertial frame. According to [2], optical sensors in the SSN achieve noise errors of ~10 arcsec in each direction, but the author notes that these values are estimates and may no longer be correct. ExoAnalytic produces optical measurements accurate to 0.25 arcsec,[11] and Slingshot Aerospace advertises a similar sub-arcsec accuracy.[12]

Longer exposure times (up to tens of seconds) are usually required to detect ASOs. When satellites move quickly relative to the sensor, the movement of the photons' impact location on the sensor over the exposure time can produce a streak in the image instead of a point-source measurement. Hence, fast-moving LEO objects can be observed, but care must be taken in the sensor instrumentation or in the data processing to handle these streaks. It is easier and more common to use optical sensors for slow-moving objects further from the sensor where radar observations begin to fail.

Optical observations provide angles-only measurements of an ASO, usually topocentric right ascension $\alpha \in [-\pi, \pi)$ rad and declination $\delta \in [-\pi/2, \pi/2]$ rad. A time history of these single-point observations may be preprocessed before OD to produce angles and angle-rates [37]. Similarly, a streak in the image can provide information on angle rates under simplifying assumptions. Some ongoing work attempts to use light curves, i.e., time variations in brightness, for attitude and ASO size and shape estimation, but we will not discuss that here for brevity. Like radar measurements, the Earth's atmosphere affects the measurement quality. Some of these can be accounted for in the sensor, e.g., adaptive optics to correct for aberration and reduce $\boldsymbol{R}_k$. Assuming media effects are addressed in the sensor design and post-processing, optical observations are modeled via

$$\boldsymbol{z}_k = \begin{bmatrix} \alpha_k \\ \delta_k \end{bmatrix} + \boldsymbol{w}_k = \begin{bmatrix} \tan^{-1}\left(\dfrac{\rho_J}{\rho_I}\right) \\ \sin^{-1}\left(\dfrac{\rho_K}{\| \boldsymbol{\rho} \|}\right) \end{bmatrix} + \boldsymbol{w}_k, \tag{13.18}$$

10 The measurements may need to be converted to the desired inertial frame using, for example, Section 7.2.2 of [36]. See discussion in Section 13.2.4.

11 https://exoanalytic.com/space-domain-awareness/ [Accessed June 6, 2025]

12 https://www.slingshot.space/solutions/data-insights [Accessed June 6, 2025]

where

$$\rho = \begin{bmatrix} \rho_I & \rho_J & \rho_K \end{bmatrix}^\mathsf{T} = \boldsymbol{r}(t_k - \tau) - \boldsymbol{r}_{\text{rec}}(t_k). \tag{13.19}$$

Like in the radar case, ideal measurements assume $\tau = 0$. Note that software implementation should use the `atan2()` function to resolve the angle ambiguity in the right ascension value.

13.4 Orbit Determination

In the astronautics community, OD refers to the estimation of a spacecraft orbit using observations. Its scope includes both deterministic and statistical approaches to nonlinear estimation. This problem dates back to Laplace's studies of celestial bodies and is a fundamental component of an SOT, satellite navigation, and scientific studies of exoplanets. Interested readers are referred to [38] for a summary of OD up to its publication. Several excellent books on the subject provide more details (e.g., [2, 3, 22, 39]). This section summarizes the elements of OD required for ASO tracking.

13.4.1 Initial Orbit Determination

Initial orbit determination (IOD) algorithms produce an estimate of an object's orbit from a small number of measurements with no prior information. This domain-specific form of track initiation is nontrivial to execute in practice, and is complicated by nonlinear propagation models and data sparsity. Traditional methods use deterministic algorithms to produce a full position/velocity state to seed a nonlinear maximum likelihood estimator. The output of this estimator then serves as a prior in an SOT. Newer methods combine measurements and operator-defined constraints to define a probabilistic prior for follow-up estimation.

Deterministic solutions to IOD assume (1) perfect observations and (2) two-body dynamics ($\boldsymbol{a}_* = \mathbf{0}$) to produce a position and velocity at a prescribed time. Table 13.1 summarizes representative algorithms grouped by observation type. This list only includes the commonly referenced methods and is not exhaustive. More details on these algorithms may be found in [2, 36]. The shared two-body assumption creates issues in solution accuracy and convergence. While measurements closely spaced in time mitigate the accumulated effects of perturbations, they introduce a new challenge in IOD: the *too short arc* problem.

Table 13.1 Summary of common traditional IOD algorithms.

Measurement	Algorithm	Notes
Position (radar)	Gibbs	Vectors $\boldsymbol{r}_i$ ordered in time and coplanar [2, 36]
	Herrick-Gibbs	Like Gibbs, but over short time period [2, 36]
	Lambert's problem	Only requires two vectors, but an ambiguity in number of orbit revolutions between them [2]
Angles-only (optical)	Gauss's	Best for closely spaced observations, but multiple solutions due to range ambiguity [2, 36]
	Laplace's	Requires initial guess at range, which yields issues with robustness and singularities [36]
	Double-*r* iteration	Better performance than Gauss for observations far apart, but possible convergence issues [36]
	Gooding's	Combines double-*r* iteration and Lambert's problem to improve robustness [40]
	Olbers's	Assumes parabolic motion [41]

The *too short arc* problem refers to the lack of convergence in an orbit solution from a small number of observations over a small arc of an orbit. Conceptually, noisy observations over a short arc of a spacecraft orbit provide insufficient information to establish custody. To illustrate its challenges, let us consider Gibbs method for IOD. Assume that we have three measurements of position $\boldsymbol{r}_j$, $j = 1, 2, 3$ derived from radar measurements (two angles and range) at distinct times. Gibbs method uses these position estimates to solve for $\boldsymbol{v}_2$ under an assumption of two-body dynamics. An intermediate step in the algorithm computes

$$\boldsymbol{D} = \boldsymbol{r}_1 \times \boldsymbol{r}_2 + \boldsymbol{r}_2 \times \boldsymbol{r}_3 + \boldsymbol{r}_3 \times \boldsymbol{r}_1, \tag{13.20}$$

and the final solution for $\boldsymbol{v}_2 \propto \|\boldsymbol{D}\|^{-1/2}$. As the angles between vectors $\boldsymbol{r}_j$ approaches 0, then $\|\boldsymbol{D}\| \to 0$ and $\boldsymbol{v}_2$ is sensitive to small changes in $\boldsymbol{r}_j$. This yields a sensitivity to measurement errors that ultimately corrupts the solution when provided observations of a short arc of the spacecraft orbit, thereby making the estimated trajectory insufficient for follow-up tracking. This is demonstrated in Figure 13.5. Since LEO spacecraft have velocities on the order of 7.5 km/s, the velocity estimates at small apparent viewing angles reach 10–100% error. Figure 13.5 implies that observing a spacecraft as it passes over the full sky with larger angle separation mitigates this issue, but operational and geometric constraints make this impractical for some cases. In practice, an IOD solution often requires measurements from multiple arcs to produce a reliable solution, making measurement-to-measurement association a critical component of an SOT. A track of measurements not associated with another arc or an object in the catalog becomes an uncorrelated track (UCT).

To help mitigate the issue of observation association between tracks, recent methods of single-observation track initiation use an *admissible region* to describe the space of feasible states and produce a prior for follow-up tracking. These admissible region approaches were first introduced

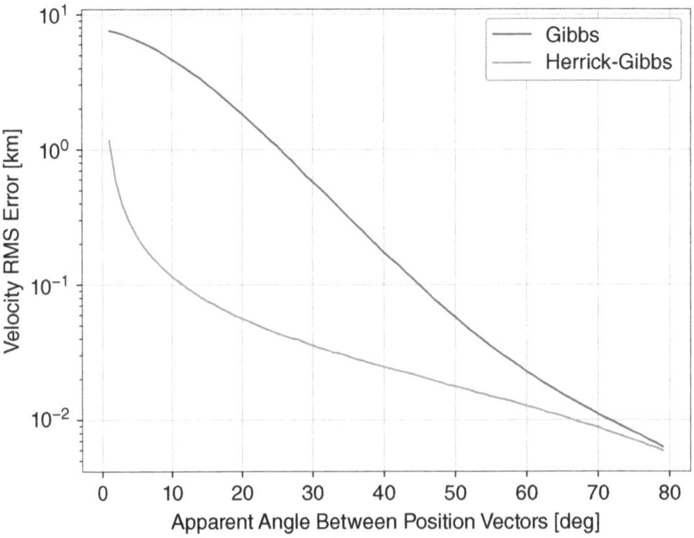

Figure 13.5 Effect of apparent angle between noisy radar observations on $\boldsymbol{v}_2$ estimated using the Gibbs and Herrick-Gibbs methods for a 400 km altitude satellite passing over a ground station. Standard deviations of the observation components are 0.15° in angles and 100 m in range. We use 10,000 Monte Carlo trials to produce the root mean square (RMS) error.

for asteroid tracking [42] and are used operationally for such purposes [43]. Using the formulation of [44], the state is partitioned into

$$\boldsymbol{x} = \begin{bmatrix} \boldsymbol{\theta}^\top & \boldsymbol{\phi}^\top \end{bmatrix}^\top, \tag{13.21}$$

where

$$\boldsymbol{\theta} = \boldsymbol{h}^{-1}(\boldsymbol{z}_k; \boldsymbol{o}), \tag{13.22}$$

i.e., $\boldsymbol{\theta}$ is an element of the subspace of $\mathbb{R}^{n_x}$ that uniquely determines the measurement $\boldsymbol{z}_k$. For example, if $\boldsymbol{z}_k$ is an optical measurement, then $\boldsymbol{\theta}$ is the unit vector from the sensor to the object and $\boldsymbol{\phi}$ encodes information on distance and velocity. The admissible region is then

$$\mathcal{R} = \mathcal{R}_1 \cap \mathcal{R}_2 \cap \cdots \cap \mathcal{R}_C, \tag{13.23}$$

where

$$\mathcal{R}_i = \{\boldsymbol{\phi} \mid c_i(\boldsymbol{\theta}, \boldsymbol{\phi}; \boldsymbol{o}) \leq 0\}, \tag{13.24}$$

and c_i are physics-based constraints on the unobservable space. These constraints may be based on orbit properties, sensor detection limitations, or other domain-specific quantities. The admissible region is then the union over all points $\boldsymbol{\phi}$ that satisfy these constraints. An example based on the semimajor axis and eccentricity constraints is provided in Figure 13.6. With these admissible regions defined, they may be used to instantiate a prior PDF:

$$p(\boldsymbol{x}) = p(\boldsymbol{\theta}, \boldsymbol{\phi}) \equiv \mathcal{N}(\boldsymbol{\theta}; \boldsymbol{z}_k, \boldsymbol{R}_k) p(\boldsymbol{\phi}; \mathcal{R}), \tag{13.25}$$

where $p(\boldsymbol{\phi}; \mathcal{R}) = 0$ if $\boldsymbol{\phi} \notin \mathcal{R}$, and $\mathcal{N}$ with three arguments denotes evaluation of the Gaussian PDF. In the SOT, this PDF is parameterized using samples [42] or kernel functions [45]. For an example use of the admissible region for multi-target tracking, see [46, 47].

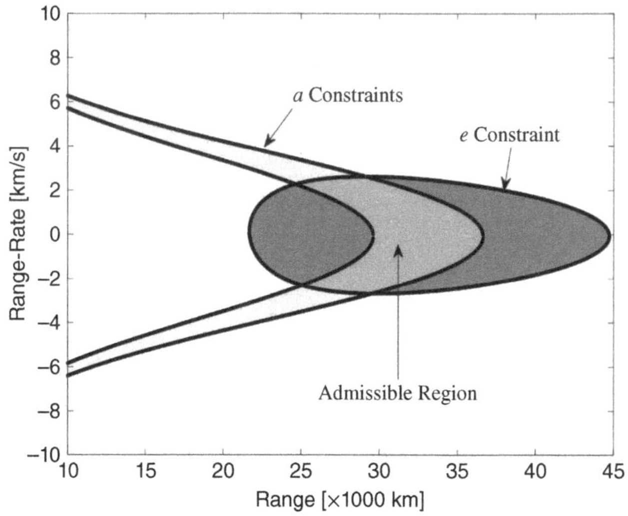

Figure 13.6 Range/range-rate admissible region illustration with an optical measurement (right ascension, declination, and their rates) and operator-defined semimajor axis and eccentricity constraints.

13.4.2 OD with Sparse Observations

The traditional approach to OD solves the nonlinear least-squares problem

$$\begin{aligned}\hat{\boldsymbol{x}}_0 = \arg \underset{\boldsymbol{x}_0}{max}\Big\{&(\boldsymbol{x}_0 - \overline{\boldsymbol{x}}_0)^{\mathsf{T}} \overline{\boldsymbol{P}}_0^{-1} (\boldsymbol{x}_0 - \overline{\boldsymbol{x}}_0) \\ &+ \sum_{k=1}^{K} (\boldsymbol{z}_k - \boldsymbol{h}(\boldsymbol{x}_k))^{\mathsf{T}} \boldsymbol{R}_k^{-1} (\boldsymbol{z}_k - \boldsymbol{h}(\boldsymbol{x}_k))\Big\},\end{aligned} \tag{13.26}$$

with prior information $\boldsymbol{x}_0 \sim \mathcal{N}\left(\overline{\boldsymbol{x}}_0, \overline{\boldsymbol{P}}_0\right)$ at time t_0, and K measurements $\boldsymbol{z}_k$ at times t_k. This prior can be the output of the maximum-likelihood IOD process or a previous least-squares solution. Practitioners usually solve (13.26) with Gauss–Newton iteration and linearization of the propagation function $\boldsymbol{f}$ and the measurement function $\boldsymbol{h}$ about the reference solution $\overline{\boldsymbol{x}}_0$ (e.g., see Ch. 4 of [22]). In the astrodynamics community, this is referred to as *batch least squares* or *differential correction*. In operations, measurement sparsity may prevent a differential corrector from converging. A key element that distinguishes space object tracking from other single- and multiple-target tracking problems is the emphasis on accurately propagating a state PDF to mitigate these sparsity issues. To this end, novel methods rooted in non-Gaussian estimation and/or unique parameterizations of the state produce algorithms that minimize information loss in prediction and increase information extraction in the update step. This section focuses on these newer methods.

Recent ASO tracking research shifts the OD paradigm from a batch approach, rooted in Gaussian distributions, to recursive estimation with mixture methods, e.g., Gaussian mixtures. In this context, let the prior PDF that has been predicted from time t_{k-1} to t_k be [48]

$$p_{k|k-1}(\boldsymbol{x}_k) = \sum_{n=1}^{N} w_{k|k-1}^{(n)} \mathcal{N}\left(\boldsymbol{x}_k; \boldsymbol{x}_{k|k-1}^{(n)}, \boldsymbol{P}_{k|k-1}^{(n)}\right), \tag{13.27}$$

where $0 \leq w_{k|k-1}^{(n)} \leq 1$ are the predicted weights, $\sum_{n=1}^{N} w_{k|k-1}^{(n)} = 1$, and $\boldsymbol{x}_{k|k-1}^{(n)}$ and $\boldsymbol{P}_{k|k-1}^{(n)}$ are the nth mixand's mean vector and covariance matrix, respectively. For a linear measurement update and additive, Gaussian measurement noise, the Gaussian mixture distribution is a conjugate prior for the measurement likelihood function [48]. Here, we present how a nonlinear measurement update may be used to produce an approximate updated state PDF. For additive Gaussian measurement errors but a nonlinear model, the posterior distribution is of the form:

$$p_{k|k}(\boldsymbol{x}_k) = \sum_{n=1}^{N} w_{k|k}^{(n)} \mathcal{N}\left(\boldsymbol{x}_k; \boldsymbol{x}_{k|k}^{(n)}, \boldsymbol{P}_{k|k}^{(n)}\right), \tag{13.28}$$

where

$$\boldsymbol{x}_{k|k}^{(n)} = \boldsymbol{x}_{k|k-1}^{(n)} + \boldsymbol{K}_k^{(n)}\left(\boldsymbol{z}_k - \overline{\boldsymbol{z}}_k^{(n)}\right), \tag{13.29}$$

$$\boldsymbol{P}_{k|k}^{(n)} = \boldsymbol{P}_{k|k-1}^{(n)} - \boldsymbol{K}_k^{(n)} \boldsymbol{P}_{zz}^{(n)} \left(\boldsymbol{K}_k^{(n)}\right)^{\mathsf{T}}, \tag{13.30}$$

$$\boldsymbol{K}_k^{(n)} = \boldsymbol{P}_{xz}^{(n)} \left(\boldsymbol{P}_{zz}^{(n)}\right)^{-1}, \tag{13.31}$$

and $\overline{\boldsymbol{z}}_k$, $\boldsymbol{P}_{zz}^{(n)}$ and $\boldsymbol{P}_{xz}^{(n)}$ are the expected measurement, innovations covariance, and cross-covariance matrices, respectively. The updated weights are:

$$w_{k|k}^{(n)} = \frac{w_{k|k-1}^{(n)} q^{(n)}(\boldsymbol{z}_k)}{\sum_{j=1}^{N} w_{k|k-1}^{(j)} q^{(j)}(\boldsymbol{z}_k)}, \tag{13.32}$$

$$q^{(n)}(\boldsymbol{z}_k) = \mathcal{N}\left(\boldsymbol{z}_k; \overline{\boldsymbol{z}}_k^{(n)}, \boldsymbol{P}_{zz}^{(n)}\right). \tag{13.33}$$

When using an unscented Kalman filter (UKF) or cubature Kalman filter (CKF) update, then $\bar{z}_k$, $P_{zz}^{(n)}$, and $P_{xz}^{(n)}$ are computed using sigma or cubature points, respectively. If using an EKF update, then

$$\bar{z}_k^{(n)} = h\left(x_{k|k-1}^{(n)}; o\right), \tag{13.34}$$

$$P_{zz}^{(n)} = H_k^{(n)} P_{k|k-1}^{(n)} \left(H_k^{(n)}\right)^{\mathsf{T}} + R_k, \tag{13.35}$$

$$P_{xz}^{(n)} = P_{k|k-1}^{(n)} \left(H_k^{(n)}\right)^{\mathsf{T}} \left(P_{zz}^{(n)}\right)^{-1}, \tag{13.36}$$

where

$$H_k^{(n)} \equiv \left[\frac{\partial h(x; o)}{\partial x}\right]_{x = x_{k|k-1}^{(n)}} \tag{13.37}$$

is the measurement Jacobian matrix evaluated at the mean vector for the corresponding mixand. The mixture filter reduces error with nonlinear models since each mixand is a local solution over a smaller subdomain of the state space.

New methods of OD for an SOT place a strong emphasis on improving accuracy of the time update step, i.e., the prediction of a PDF to produce the prior in (13.28). The AEGIS approach [49] autonomously increases the number of mixture components during the prediction step to mitigate the effects of nonlinear dynamics. Information loss under the assumption of a local Gaussian distribution is approximated by comparing a partial second-order propagated covariance (via the unscented transformation) to a first-order solution (linearization). When the difference in differential entropy of the two solutions exceeds a threshold, i.e., too much information is lost, the mixand is split into smaller components to maintain accuracy. The resulting mixture undergoes a traditional measurement update and mixand refinement through common merge, prune, and truncate operations. This mixture splitting method continues until the time of a measurement. The AEGIS algorithm creates an adaptive approach to increase N to meet accuracy requirements based on the physics of the problem, but can yield a rapid growth in N for long propagation times or eccentric orbits.

Alternatively, new methods based on regularized particle filters demonstrate improved robustness to information sparsity. These approaches combine the strengths of the particle filter in the prediction step with a Gaussian mixture measurement update. Let $x_{k-1|k-1}^{(n)} \sim p_{k-1|k-1}, n = 1, \ldots, N$ be samples from the posterior at t_{k-1}. We propagate each particle to the next time using a selected model f. The predicted samples $x_{k|k-1}^{(n)}$ are then combined in a chosen algorithm to form a Gaussian mixture. The ensemble Gaussian mixture filter (EnGMF) [17] produces the prior mixture density:

$$p_{k|k-1}(x_k) = \frac{1}{N} \sum_{n=1}^{N} \mathcal{N}\left(x_k; x_{k|k-1}^{(n)}, \beta P'_{k|k-1}\right) \tag{13.38}$$

where $P'_{k|k-1}$ is the empirical covariance computed using the propagated samples, and

$$\beta = \left(\frac{4}{N(n_x + 2)}\right)^{2/(n_x+4)}, \tag{13.39}$$

which is Silverman's rule of thumb and is used in common kernel density estimation techniques. The UKF, CKF, or EKF mixture update is then used to get $p_{k|k}$, and new samples are generated for the next prediction step. The particle Gaussian mixture filter (PGMF) follows a similar process, but uses k-means clustering to produce $p_{k|k-1}$ [50]. Both approaches demonstrate increased robustness to data sparsity with manageable increases in runtime. The EnGMF has been shown to maintain a statistically consistent, or slightly pessimistic, solution with up to 20 orbits between measurement arcs [17]. It is also embedded in the multi-target tracker used in the tracking example provided in Section 13.6.1.

New methods of OD extend beyond Gaussian mixture methods. Examples include, inter alia, higher-order Taylor series expansions about reference solutions [51], polynomial functions of the system's random variables [52], particle flow [53], cubature methods [54], and Koopman operator theory [55]. The problem of sparse OD is an active area of research to address rapid population growth.

13.4.3 Uncertainty Representation, Propagation, and Measurement Updates

Parameterization of the orbit translational state, e.g., Cartesian or orbital elements, influences the tractability and accuracy of the OD process. With potentially long gaps between measurements, accurate PDF propagation is critical to produce a statistically consistent, recursive Bayesian estimator. As described in Section 13.2.3, elements rooted in the geometry of the orbit propagation problem yield slowly varying elements of the state. This produces a state representation that is nearly linear in time, which simplifies the propagation problem in weakly perturbed cases. Said differently, when the state is described using slow-varying parameters and is initially Gaussian, the information loss when assuming a Gaussian propagated PDF is reduced and possibly negligible. For example, Figure 13.7 illustrates propagated random samples from a Gaussian prior PDF for near-circular and highly elliptical orbits. Propagation in orbital elements remains Gaussian after 10 days for the near-circular case. However, they still deviate from a Gaussian distribution in a highly elliptical orbit over the same duration. Both cases fail to remain Gaussian for the Cartesian case. The Gauss von Mises filter was designed for sparse OD with these slow-varying parameters,

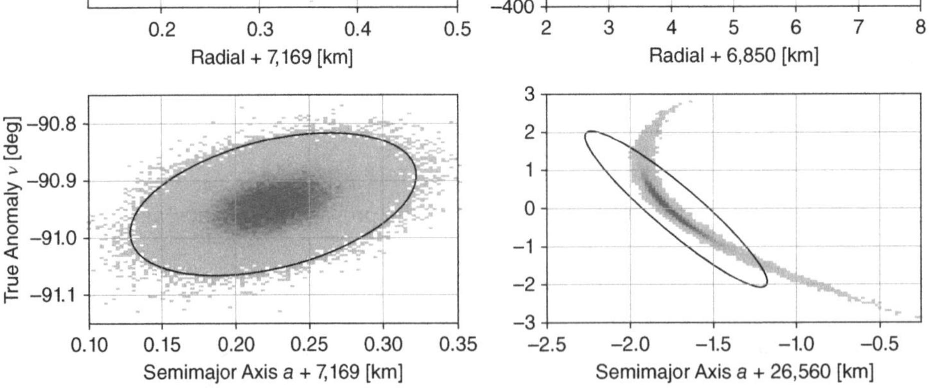

Figure 13.7 Histogram of propagated samples in Cartesian coordinates (top) and orbital elements (bottom). Images on the left are for a Sun-synchronous orbit (near-circular) and the right are for a Molniya orbit (highly eccentric).

and one of its key innovations is leveraging a state PDF consistent with the definition of angle quantities on the unit circle [56].

Changing the parameterization of the orbit requires adjusting the measurement update step in an OD filter. To illustrate this point, consider Doppler measurements of a spacecraft, which provide information on the velocity along the line-of-sight direction. This is naturally a function of the Cartesian position. However, expressing this measurement as a function of the angles in the Keplerian orbital elements introduces additional nonlinear terms that reduce information extracted from the measurement in a linear update. The EnGMF and PGMF mitigate this through the mixture's more compact representation on a per-component basis.

13.5 Multi-target Tracking

In 2012, the National Research Council (NRC) produced an independent assessment of what was then the United States Air Force Space Command's astrodynamics standards and algorithms, i.e., the software tools and processes used at the time for SSA [57]. That report noted that, at the time, data association leveraged nonstatistical, gating-based methods. The performance of this was sufficient to meet the demands of SSA up to that time, but the report recommended that the larger astrodynamics and SSA communities leverage advanced data association methods such as those in multiple hypothesis tracking. Around the same time, two SOTs were being developed leveraging established multi-target tracking methods. The Air Force Research Laboratory led an effort to develop a tracker based on joint probabilistic data association (JPDA) [58] that incorporates admissible regions for spacecraft IOD. That method is described below. Additionally, the space object tracking division of the Numerica Corporation (later acquired by Slingshot Aerospace) developed an SOT based on multi-frame assignment [59], which they dubbed the Multiple Frame Assignment Space Tracker (MFAST). Initial developments based on their approach may be found in [60, 61]. Since details on their approach are limited in the public domain, we do not expand on MFAST in this chapter. Finally, given the timing of the identified need for advanced methods and the state-of-the-art tracking algorithms being developed in the information fusion community, most academic research into rigorous SOTs uses random finite sets (RFSs) and relatively new methods. We briefly summarize the RFS-based methods as well.

13.5.1 Operational Approaches

While the precise methodology used by the USSF to generate their widely-used orbital element sets has not been published, it is known to involve fitting arcs of measurements using a high-fidelity dynamics model, then generating pseudo-observations based on the fit orbit, which are then fit using the Simplified General Perturbations 4 (SGP4) model [62]. In general, SSA data providers first collect sensor observations, whether they are radar returns, optical streaks, or short arcs of dense optical measurements. When observations are sufficiently dense, the sensor can generally associate short arcs of measurements into tracklets. Then, these tracklets are compared to the catalog of known space objects to determine whether or not they are likely to correspond to a tracked object. When a set of measurements cannot be associated to a tracked object, it is designated a UCT.

A UCT is typically the result of a spacecraft maneuver or the appearance of a new ASO. The term *UCT processing* refers to various methods that are used to resolve UCTs, whether by determining that the UCT is a known object whose orbit has changed or by acquiring custody of the new object via OD. These techniques often involve associating UCTs to one another (a form of measurement-to-measurement association) to construct longer arcs of data for better OD (i.e., to mitigate the too short arc problem). When such an orbit estimate converges but the identity of the object is still unknown, the USSF will sometimes provide an orbital element set for the object in the *analyst objects* category in the space object catalog. Confirming the identity of a new ASO requires additional information from other sources and is beyond the scope of this chapter. In the event of a maneuver or breakup of a tracked object, the process may involve identifying missed detections of cataloged objects and finding corresponding UCTs.

13.5.2 Constrained Admissible Region Multiple Hypothesis Filter (CAR-MHF)

In the early 2010s, the Air Force Research Laboratory developed an ASO tracker based on JPDA: the constrained admissible region multiple hypothesis filter (CAR-MHF) [47]. This algorithm for UCT processing uses (1) admissible regions for track initialization, (2) a variation on JPDA they dubbed multiple hypothesis JPDA (MH-JPDA), and (3) the UKF for subsequent hypothesis prediction and update. The filter is typically combined with a fixed-interval smoother as a refinement step. Note that, in the context of CAR-MHF, *multiple hypothesis* refers to considering a variety of initial orbital state PDFs for each object. The tracker does not maintain knowledge of data association hypotheses over the set of objects being tracked, as it would in the context of multiple hypothesis multi-target tracking. The initialization and evolution of these state hypotheses are illustrated in Figure 13.8. The following description of CAR-MHF is adapted from [47]. Note that, for the sake of a simpler notation in this section, we do not include subscripts denoting time. This presentation of the CAR-MHF focuses on fusion with measurements at a single time, and can be extended to multiple times by repeating the process after propagation. We also note that, while literature outside of the space object tracking community has established that multiple hypothesis methods outperform JPDA in dense and high-clutter environments (e.g., [63]), we include the CAR-MHF method here because (1) it has been increasingly adopted in the government and by industry, and (2) it demonstrates the integration of a key astrodynamics tool (admissible regions) with an established tracking algorithm.

Given a UCT, admissible region constraints are computed as described in Section 13.4.1. The precise set of constraints that are used depend on the type of measurement data and the specific use case, so we refer the reader to [42, 45, 64, 65] for mathematical definitions of some common admissible region constraints. A grid of points is sampled over the admissible region. Each of these points is concatenated with the measurement to form the mean estimate of the object's state for a single hypothesis. Each hypothesis is also assigned a covariance based on the measurement noise and grid resolution. This yields a prior density of the form (13.25) where the marginal PDF of ϕ

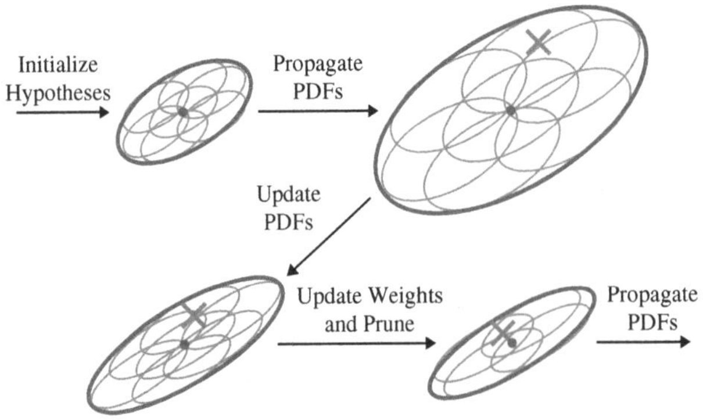

Figure 13.8 Illustration of multiple hypothesis filter. Dark lines indicate the overall PDF estimate, light gray indicates the individual hypothesis PDFs, and the X indicates the measurement. Adapted from [47].

is a Gaussian mixture and each mixand is a hypothesis. The resulting PDF is transformed via the unscented transform to the Cartesian state space. The overall state estimate for object r is given by the minimum mean square error estimate:

$$\hat{\boldsymbol{x}}_r = \sum_{j=1}^{n_r} \hat{\mu}_{r,j} \hat{\boldsymbol{x}}_{r,j}, \tag{13.40}$$

where n_r is the number of hypotheses for object r, $\hat{\mu}_{r,j}$ is the weight of the jth hypothesis subject to

$$\sum_{j=1}^{n_r} \hat{\mu}_{r,j} = 1, \tag{13.41}$$

and $\hat{\boldsymbol{x}}_{r,j}$ is the mean of the corresponding PDF. Each hypothesis starts with equal weight.

When new UCT measurements are processed, each hypothesis is predicted via the UKF prediction step, with the chosen propagation model $\boldsymbol{f}$, to the measurement time. Hypothesis weights do not change during prediction. As is common in multi-target tracking, the first step of data association is to rule out unlikely association possibilities by comparing measurements to a *gate* based on each object's hypothesis PDFs. This is done by transforming each PDF into the measurement space and computing the Mahalanobis distance between it and each measurement:

$$d = \sqrt{(\boldsymbol{z}_m - \bar{\boldsymbol{z}}_r)^{\mathsf{T}} \boldsymbol{P}_{zz,r}^{-1} (\boldsymbol{z}_m - \bar{\boldsymbol{z}}_r)}, \tag{13.42}$$

where $\boldsymbol{z}_m$ is the mth measurement at t_k, $\bar{\boldsymbol{z}}_r$ is the expected measurement of object r, and $\boldsymbol{P}_{zz,r}$ is its innovation covariance. If a measurement falls within a user-defined distance of any one of an object's hypothesis PDFs, then it is considered a possible association for MH-JPDA.

Given the set of possible measurement–object associations, the likelihood that hypothesis j of object r generated measurement m is

$$\psi_{r,m,j} = \bar{\mu}_{r,j} \times \begin{cases} P_{D,r} \kappa^{-1} \mathcal{N}(\boldsymbol{z}_m; \bar{\boldsymbol{z}}_{r,j}, \boldsymbol{P}_{zz,r,j}), & m > 0, \\ 1 - P_{D,r}, & m = 0, \end{cases} \tag{13.43}$$

with $m = 0$ indicating the event that object r was not detected, $\bar{\mu}_{r,j}$ is the prior weight for hypothesis (r, j), $\bar{\boldsymbol{z}}_{r,j}$ is the expected measurement of object r under hypothesis j and $\boldsymbol{P}_{zz,r,j}$ is its covariance, $P_{D,r}$ is the probability of detecting object r, and κ is the sensor's clutter intensity. The updated hypothesis weights are

$$\hat{\mu}_{r,j} = \sum_{m=1}^{M} \frac{\beta_{m,r}}{\psi_{r,m}} \psi_{r,m,j}, \tag{13.44}$$

where M is the number of possible measurements,

$$\beta_{m,r} = \sum_{\Theta:\theta_{r,m} \in \Theta} \left(\prod_{s=1}^{S} \psi_{s,m} \right), \tag{13.45}$$

$\{\Theta : \theta_{r,m} \in \Theta\}$ is the set of possible *joint events* Θ in which object r is associated with measurement m (or not detected, if $m = 0$), S is the number of objects, and

$$\psi_{r,m} = \sum_{j=1}^{n_r} \psi_{r,m,j}. \tag{13.46}$$

A joint event is a combination of possible association events for each object. For example, if the filter is tracking two objects, a joint event may be that Object 1 is associated to Measurement 1 and Object 2 is not detected. At this point, hypotheses with updated weights below some user-defined threshold are pruned and the weights of the remaining hypotheses are normalized to ensure that (13.41) holds.

The probability that hypothesis j of object r generated measurement m is

$$\beta_{m,r,j} = \frac{\beta_{m,r}\psi_{r,m,j}}{\psi_{r,m}\,\hat{\mu}_{r,j}}. \tag{13.47}$$

We can now compute the weighted innovation for hypothesis (r, j):

$$\boldsymbol{\nu}_{r,j} = \sum_{m=1}^{M} \beta_{m,r,j}\left(\boldsymbol{z}_m - \bar{\boldsymbol{z}}_{r,j}\right). \tag{13.48}$$

Finally, we update the mean and covariance of this hypothesis:

$$\hat{\boldsymbol{x}}_{r,j} = \bar{\boldsymbol{x}}_{r,j} + \boldsymbol{K}_{r,j}\boldsymbol{\nu}_{r,j}, \tag{13.49}$$

$$\hat{\boldsymbol{P}}_{r,j} = \bar{\boldsymbol{P}}_{r,j} - \left(\sum_{m=1}^{M} \beta_{m,r,j}\right)\boldsymbol{K}_{r,j}\boldsymbol{P}_{zz,r,j}\boldsymbol{K}_{r,j}^{\mathsf{T}} + \tilde{\boldsymbol{P}}, \tag{13.50}$$

where $\boldsymbol{K}_{r,j} = \boldsymbol{P}_{xz,r,j}\boldsymbol{P}_{zz,r,j}^{-1}$ is the UKF gain and

$$\tilde{\boldsymbol{P}} = \boldsymbol{K}_{r,j}\left(\left(\sum_{m=1}^{M} \beta_{m,r,j}\boldsymbol{\nu}_{m,r,j}\boldsymbol{\nu}_{m,r,j}^{\mathsf{T}}\right) - \left(\boldsymbol{\nu}_{r,j}\boldsymbol{\nu}_{r,j}^{\mathsf{T}}\right)\right)\boldsymbol{K}_{r,j}^{\mathsf{T}}. \tag{13.51}$$

The admissible region-based IOD used in CAR-MHF will generally result in high initial state uncertainty for newly detected objects, which decreases over subsequent measurement updates. After applying the prediction and update procedure described above to a sequence of measurements, if the estimate has converged sufficiently to prevent false associations, the estimated state trajectory can be refined via a backward smoothing pass. Reference [47] elaborates on when and how to use a Rauch–Tung–Striebel (RTS) UKF smoother for this purpose [66].

13.5.3 RFS Methods

As described previously, most recent research in space object tracking focuses on RFSs and finite set statistics (FISST)-based methods [67]. An RFS is a random variable over the set of finite subsets of a space, and can be thought of as a set of random vectors whose cardinality is itself a random variable. More formally, let the function $\mathcal{F}$ denote the set of all finite subsets of a space. Given a space $\mathbb{X}$, e.g., $\mathbb{X} = \mathbb{R}^{n_x}$, then an RFS is a random variable on $\mathcal{F}(\mathbb{X})$. An RFS can represent the set of states of objects that exist (the *multi-target state*), the set of observations collected by a sensor at a particular time, and other useful quantities for multi-target tracking.

Both the probability hypothesis density (PHD) [68] and cardinalized probability hypothesis density (CPHD) filters [69] have been leveraged to produce new SOTs. These RFS-based approaches provide a tractable approach to estimate the density of objects in the surveillance region, but at the cost of losing the identity of the targets. A two-tier augmented CPHD filter [70] tracks unconfirmed objects using a PHD filter and confirmed objects using a CPHD filter with a mechanism to promote objects from one to the other. This method allows unconfirmed objects to be accounted for in sensor tasking without potentially degrading the multi-target estimate due to false alarms. A naive combination of the admissible region with the PHD or CPHD filter yields excessive clutter tracks, and [46] presented a rigorous approach to mitigate this issue for ASO tracking. Motivated by the difficulty in determining when an ASO is detectable, a modified PHD filter has been developed that accounts for uncertainty in the probability of detection function itself [71]. The key disadvantage of these approaches is the lack of a rigorous identity for each extracted object in the multi-target state.

Instead, newer methods use labeled RFSs that augment each state vector with a discrete label [72]. In the SSA context, the ability to assign a distinct label to each object in the multi-target state estimate is critical to maintain the space object catalog, identify the actions of actors operating in the domain, and conduct collision risk assessment. It also allows objects to be assigned different properties, such as maneuvering capabilities and shape models, based on their identities. Given the state space $\mathbb{X}$ and label space $\mathbb{L}$, a labeled RFS is a random variable on $\mathcal{F}(\mathbb{X} \times \mathbb{L})$. A labeled RFS realization with n elements has the form:

$$\boldsymbol{X} = \{(\boldsymbol{x}_1, \ell_1), \dots , (\boldsymbol{x}_n, \ell_n)\} \subset \mathbb{X} \times \mathbb{L}, \tag{13.52}$$

where $\ell_j \in \mathbb{L}$ is the label for the object with state $\boldsymbol{x}_j \in \mathbb{X}$. The use of labeled RFSs is described in Chapter 8 of this book [73], and we refer the reader there for details on the formulation and implementation of these filters. The recent survey paper [74] also provides details on labeled RFS-based approaches. Additionally, Chapter 16 of this book looks at using the Poisson labeled multi-Bernoulli filter for SSA [75].

Newer methods of ASO tracking address issues of combinatorial growth in hypotheses in the generalized labeled multi-Bernoulli (GLMB) [72, 76, 77] and similar filters. These issues are caused by the large number of objects that need to be tracked and by field of view (FOV) and detection limitations due to sensor geometry and varying illumination conditions. The field-of-view-partitioned GLMB filter (FP-GLMBF) [78] partitions the label space $\mathbb{L}$ based on sensors' FOVs. This technique was developed for large-scale ASO tracking problems (e.g., [25]), and is also used in the tracking example in Section 13.6.1. The randomized finite set statistics (R-FISST) filter uses unlabeled tracks to estimate the distribution of highly probable data association hypotheses via smart sampling Markov-chain Monte Carlo [79]. Though it is not RFS-based, hypothesized and independent stochastic populations (HISP) separates tracks for objects that can be distinctly identified from the overall distribution of indistinguishable objects to improve scalability [80]. This is conceptually similar to the two-tier augmented CPHD filter. Other relevant work, based on the theory of outer probability measures, has resulted in the development of uncertain finite sets and the possibilistic Bernoulli filter [81]. That work further developed into the possibilistic GLMB filter [82]. These possibilistic filters can be used when birth, death, dynamics, detection, and measurement models are not precisely known.

13.6 Space Object Tracking

The tools described in Sections 13.2–13.5 combine to form the principal elements of an SOT. Figure 13.9 illustrates one possible configuration of these tools. The sensor network generates heterogeneous observations of ASOs, which are passed to the SOT for centralized fusion. An ASO catalog stores all knowledge of the space object population, including the translational-state PDF for all hypothesized tracks, track labels, measurement association history, maneuver capabilities, etc. The multi-target tracker updates the catalog using the observations while accounting for data association ambiguity, clutter, and detection probability models. ASOs rarely "die" in the multi-target tracking sense, so assuming a survival probability slightly less than unity, e.g., $P_S = 1 - 10^{-6}$,

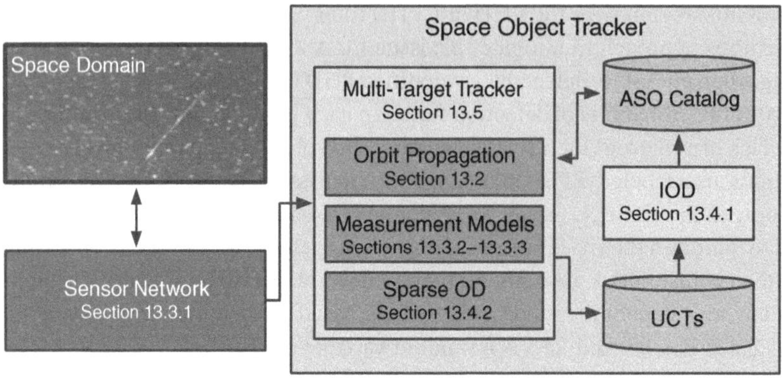

Figure 13.9 Illustration of a sample space object tracker.

is valid. Within the tracker, a sparse OD algorithm updates the single-target PDF for each track that has been associated with an observation. Use of any of the described sparse OD algorithms is analogous to choosing between a particle filter, EKF, UKF, CKF, or Gaussian mixture prediction and update approach in existing multi-target tracking algorithms, such as the GLMB filter. The tracker and OD algorithm internally use the orbit propagator ($\boldsymbol{f}$) to predict the state PDF, e.g., to propagate a particle ensemble for the EnGMF. The predicted state PDF, the measurement models, and the sparse OD algorithm produce updated tracks within the multi-target state hypotheses. The probability of each hypothesis is updated in a manner consistent with its formulation, e.g., the hypothesis weights $\hat{\mu}_{r,j}$ in CAR-MHF or the component weights in the GLMB filter. Any measurements not associated with an ASO are deemed a UCT and stored. As UCTs are added to the database, an appropriate IOD algorithm attempts to correlate measurement tracks with each other and produce an initial prior density. When the deterministic approach is successful, the output position/velocity state is used as an initial solution for a nonlinear maximum likelihood estimator (e.g., least squares). This produces the probabilistic prior required for follow-up tracking. Alternatively, a single tracklet may instead be used to instantiate a prior based on the admissible region methods. Any hypothesized new objects are then added to the ASO catalog for future data association and tracking. The following subsection illustrates this process (without IOD or object birth) for a large population of objects in LEO.

13.6.1 Tracking Example

This section provides example results of a large-scale space object tracking simulation using the FP-GLMBF. These results were originally presented in [25]. The population of objects to be tracked consists of:

- a constellation of 1,000 satellites whose orbits all have semimajor axis 7528.1363 km and inclination 53°, and
- 15,302 real space objects in LEO taken from the public USSF catalog.

Figure 13.10 illustrates the distribution of the objects in two of the Keplerian orbital elements: semimajor axis a and inclination i. Like the true ASO population, many objects are in similar orbits based on mission needs and orbit properties. Simulated measurements of azimuth, elevation, range, and range-rate are produced every minute by a network of ten sensors based on the SSN. The simulated scenario covers 72 hours of operation. Each sensor has a different field of regard and

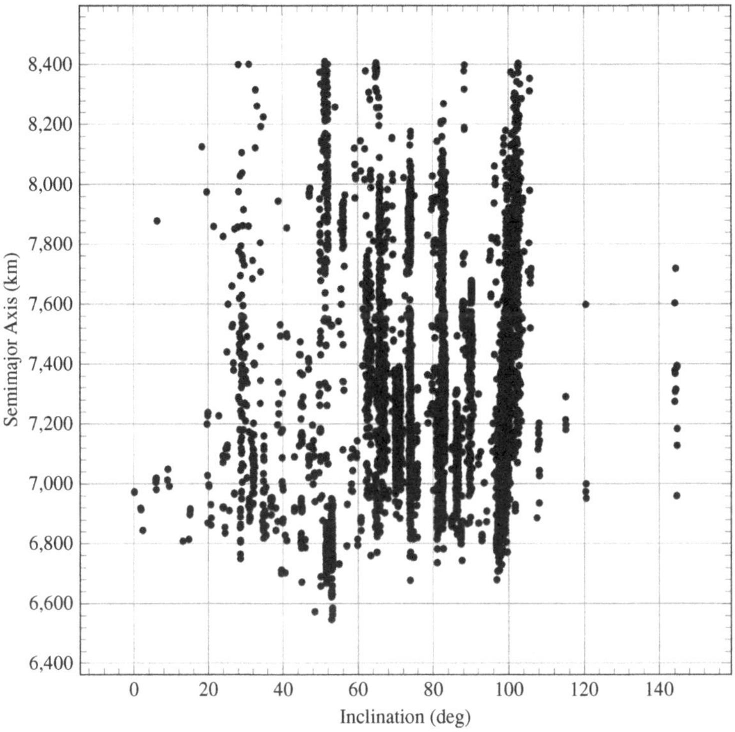

Figure 13.10 Semimajor axis and inclination for the population of objects in the tracking example. Originally published in [25]. *Source*: [25] / with permission of the International Society of Information Fusion.

measurement noise covariance. All sensors have constant probability of detection 0.98 for objects in their FOV and clutter returns are not generated. The sensors are tasked using a greedy algorithm based on the time since each object was last in the FOV of any sensor. The FP-GLMBF in this example features the EnGMF as its single-target filter, augmented with multi-fidelity orbit uncertainty propagation [83] and a method for adjusting the number of EnGMF samples N over time. We refer the reader to [25] for additional details regarding the scenario design and filter implementation.

The simulation is run on a server with four 2.4 GHz CPUs that can support a total of 192 concurrent threads. For one simulation run, the total filtering time (i.e., not including propagating objects' true orbits, tasking sensors, generating measurements, or computing tracking error) is 10 hours. Tracking error is calculated using the optimal subpattern assignment (OSPA) metric with a cutoff of 15 km and order 2 [84].

Figure 13.11 shows the tracking error over one run of the simulation. Overall error and localization (position) errors increase slightly over the first 6 hours because the procedure used to initialize the objects' state PDFs assumes that all objects are observed simultaneously. Due to FOV limitations and sensor tasking, only an average of 11 objects are observed at each timestep. Overall tracking error reaches its minimum value approximately two days into the simulation, after which it oscillates but does not increase significantly. Because the multi-target state is initialized with the correct cardinality and the cardinality is not expected to change over the course of the simulation (no birth or death), cardinality error is zero for most of the simulation. The one exception occurs between approximately 34 and 52 hours, where one object disappears from the multi-target state estimate. However, the fact that the object's estimate later improves shows that the filter has not actually lost custody of it.

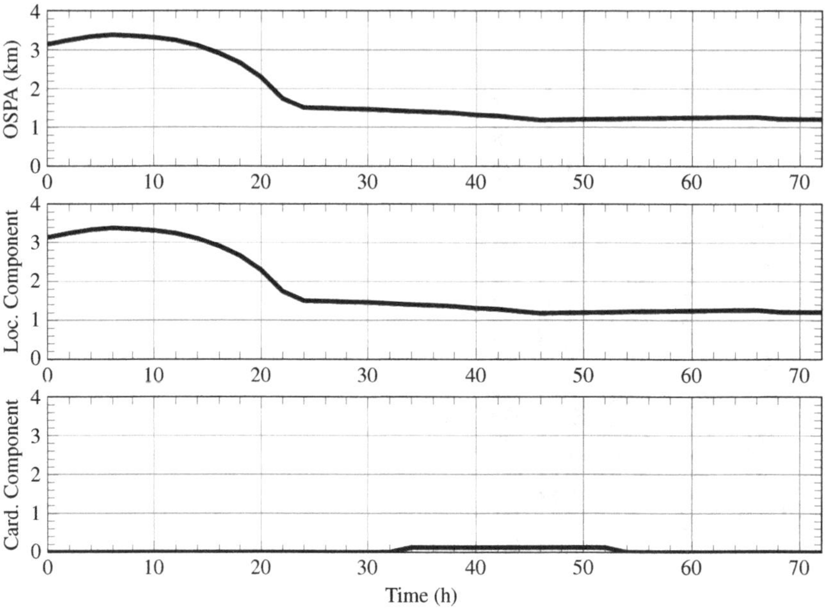

Figure 13.11 Tracking error over time. Top: Overall OSPA error. Middle: Localization error. Bottom: Cardinality error. Originally published in [25]. *Source* : [25] / with permission of the International Society of Information Fusion.

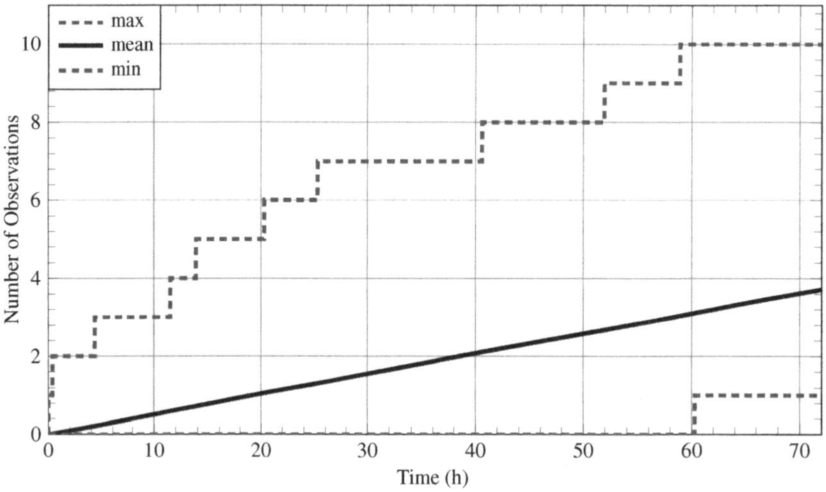

Figure 13.12 Observation rate statistics over time for the example target tracking scenario. Originally published in [25]. *Source* : [25] / with permission of the International Society of Information Fusion.

Figure 13.12 provides statistics on the number of observations per ASO in the simulation. On average, each object generates one observation each day over the 72-hour simulation, illustrating the sparsity of the space object tracking problem. One object is not observed until 60 hours into the simulation. Each of these ASOs completes approximately 14 revolutions per day, i.e., the SOT must maintain custody of the population with multiple revolutions between measurements.

This case demonstrates that a combination of tools designed for accurate uncertainty propagation, sparse OD, and a multi-target tracker designed for space object tracking can maintain custody of more than 16,000 space objects in LEO. This is accomplished without high-performance computing resources. Ongoing research in the astrodynamics community continues to enhance this and similar SOTs for challenge problems, including satellite fragmentation events (explosions similar to spawning in the larger fusion literature) and maneuvering target tracking.

13.7 Conclusions

This chapter provided an overview of the current challenges in space object tracking. Measurement sparsity requires accurate prediction of the translational state PDF for data association and non-Gaussian methods of Bayesian fusion. The nonlinear dynamics coupled with the long propagation times require accurate orbit propagators and tools for uncertainty quantification. Expressing the state using orbital elements can reduce the complexity of the prediction step, but introduces nonlinearity in the measurement update. Non-Gaussian methods for the measurement update improve accuracy and robustness to measurement sparsity, but increase runtime and algorithm complexity.

The advent of large constellations poses a new challenge for SSA over the coming decade. Thanks to rideshare opportunities and reusable launch vehicles, the cost of launching spacecraft into LEO has greatly decreased and the space industry is taking advantage of this change. In early 2019, approximately 19,400 satellites were tracked by the USSF. In December 2024, that number increased to 29,228; this is a 50% increase in almost 5 years. The largest source of that growth is the SpaceX Starlink constellation, which has over 7,000 satellites operating around 500–600 km above the Earth's surface (as of May 2025). SpaceX aspires to increase that to 34,400. Amazon Leo launched its first satellite in 2025 with plans to constitute a 3,236 satellite constellation. The Spacesail Constellation (otherwise known as Thousand Sails) plans to expand to 15,000 satellites. Current SOTs have adjusted to these changes by cooperating with satellite operators, but this will likely become difficult over time. The SSA community will have to adapt to this unprecedented growth in the ASO catalog.

Another new challenge for SSA is tracking objects in cislunar space. All but a handful of ASOs are currently within 40,000 km of the Earth, but humanity is extending its reach to the Moon ($\sim$380,000 km away) with plans for a permanent presence. This will create satellite traffic between the Earth and the Moon in the coming decade, and SOTs will need to expand their capabilities to compensate. In several regimes of interest, the dynamics are chaotic, which will stress methods of PDF propagation. The timescales of motion at further distances from the Earth will create long periods where satellites cannot be observed by electro-optical sensors due to persistent light from other objects (Sun and Moon). Research in cislunar SSA is accelerating rapidly to address these and other problems for future SSA.

References

1 P. W. Schumacher, Jr., F. R. Hoots, and K. T. Alfriend. *History of U.S. Space Surveillance and Satellite Cataloging: A Long and Winding Road*. American Institute of Aeronautics and Astronautics, New York, NY, 2025.

2 D. A. Vallado and W. D. McClain. *Fundamentals of Astrodynamics and Applications*. Microcosm Press and Springer, Hawthorne, CA and New York, NY, 3rd edition, 2007.

3 O. Montenbruck and E. Gill. *Satellite Orbits: Models, Methods, Applications*. Springer, Berlin Heidelberg New York, 2005.

4 R. S. Park, W. M. Folkner, J. G. Williams, and D. H. Boggs. The JPL planetary and lunar ephemerides DE440 and DE441. *The Astronomical Journal*, 161:(3)1–15, 2021. https://doi.org/10.3847/1538-3881/abd414.

5 W. Park and K. T. Alfriend. Dynamic model fidelity effects on covariance based track association. *Journal of Guidance, Control, and Dynamics*, 47(7):1298–1308, 2024. https://doi.org/10.2514/1.G007871.

6 B. A. Jones. *Efficient Models for the Evaluation and Estimation of the Gravity Field*. Ph.D. thesis, University of Colorado at Boulder, Boulder, CO, 2010.

7 J. C. Ries, S. V. Bettadpur, R. J. Eanes, K. Zhigui, U. Ko, C. McCullough, N. Peter, N. Pei, S. Poole, T. Richter, H. Save, and B. D. Tapley. Technical Report CSR-16-02, Center for Space Research, The University of Texas at Austin, Austin, Texas, 2016.

8 N. K. Pavlis, S. A. Holmes, S. C. Kenyon, and J. K. Factor. The development and evaluation of the Earth Gravitational Model 2008 (EGM2008). *Journal of Geophysical Research: Solid Earth*, 117(B4):B04406, 2012. https://doi.org/10.1029/2011JB008916.

9 N. Arora and R. P. Russell. Efficient interpolation of high-fidelity geopotentials. *Journal of Guidance, Control, and Dynamics*, 39(1):128–143, 2016. https://doi.org/10.2514/1.G001291.

10 B. A. Jones, G. H. Born, and G. Beylkin. Comparisons of the cubed-sphere gravity model with the spherical harmonics. *Journal of Guidance, Control, and Dynamics*, 33(2):415–425, 2010. https://doi.org/10.2514/1.45336.

11 T. Schildknecht, R. Musci, M. Ploner, G. Beutler, W. Flury, J. Kuusela, J. de Leon Cruz, and L. de Fatima Dominguez Palmero. Optical observations of space debris in GEO and in highly-eccentric orbits. *Advances in Space Research*, 34(5):901–911, 2004. https://doi.org/10.1016/j.asr.2003.01.009.

12 L. G. Jacchia. New static models of the thermosphere and exosphere with empirical temperature profiles. Special Report 313, Smithsonian Astrophysical Observatory, 1970.

13 J. M. Picone, A. E. Hedin, and D. P. Drob. NRLMSISE-00 empirical model of the atmosphere: Statistical comparisons and scientific issues. *Journal of Geophysical Research*, 107(A12):1–16, 2002. https://doi.org/10.1029/2002JA009430.

14 B. Bowman, W. Kent Tobiska, F. A. Marcos, and C. Valladares. The JB2006 empirical thermospheric density model. *Journal of Atmospheric and Solar-Terrestrial Physics*, 70:774–793, 2008. https://doi.org/10.1016/j.jastp.2007.10.002.

15 C. J. Wetterer, R. Linares, J. L. Crassidis, T. M. Kelecy, M. K. Ziebart, M. K. Jah, and P. J. Cefola. Refining space object radiation pressure modeling with bidirectional reflectance distribution functions. *Journal of Guidance, Control, and Dynamics*, 37(1):185–196, 2014. https://doi.org/10.2514/1.60577.

16 G. R. Hintz. Survey of orbit element sets. *Journal of Guidance, Control, and Dynamics*, 31(3):785–790, 2008. https://doi.org/10.2514/1.32237.

17 S. Yun, R. Zanetti, and B. A. Jones. Kernel-based ensemble Gaussian mixture filtering for orbit determination with sparse data. *Advances in Space Research*, 69(12):4179–4197, 2022. https://doi.org/10.1016/j.asr.2022.03.041.

18 G. H. Kaplan. *The IAU Resolutions on Astronomical Reference Systems, Time Scales, and Earth Rotation Models: Explanation and Implementation. Circular 179,* United States Naval Observatory, Washington, D.C., 2005.

19 B. K. Bradley, D. A. Vallado, A. Sibois, and P. Axelrad. Earth orientation parameter considerations for precise spacecraft operations. In *AAS/AIAA Astrodynamics Specialist Conference,* Girdwood, AK, 2011.

20 G. Petit and B. Luzum. IERS Conventions (2010). IERS Technical Note 36, International Earth Rotation and Reference Systems Service (IERS), Frankfurt am Main, Germany, 2010.

21 N. Stacey and S. D'Amico. Adaptive and dynamically constrained process noise estimation for orbit determination. *IEEE Transactions on Aerospace and Electronic Systems*, 57(5):2920–2937, 2021. https://doi.org/10.1109/TAES.2021.3074205.

22 B. D. Tapley, B. E. Schutz, and G. H. Born. *Statistical Orbit Determination.* Elsevier Academic Press, Burlington, MA, 2004.

23 J. M. Leonard, F. G. Nievinski, and G. H. Born. Gravity error compensation using second-order Gauss–Markov processes. *Journal of Spacecraft and Rockets*, 50(1):217–229, 2013.

24 N. Ravago and B. A. Jones. Risk-aware sensor scheduling and tracking of large constellations. *Advances in Space Research*, 68(6):2530–2550, 2021. https://doi.org/10.1016/j.asr.2021.04.042.

25 B. L. Reifler, A. A. Popov, B. A. Jones, and R. Zanetti. Large-scale space object tracking in a proliferated leo scenario. In *2023 26th International Conference on Information Fusion (FUSION),* Charlotte, NC, pages 1–8, 2023.

26 J. R. Wertz, D. F. Everett, and J. J. Puschell, editors. *Space Mission Engineering: The New SMAD.* Microcosm Press, Torrance, CA, 2011.

27 E. M. Zucchelli and B. A. Jones. A Bayesian approach to low-thrust maneuvering spacecraft tracking. *Journal of Guidance, Control, and Dynamics*, 47(8):1586–1601, 2024. https://doi.org/10.2514/1.G007849.

28 G. M. Goff, J. T. Black, and J. A. Beck. Orbit estimation of a continuously thrusting spacecraft using variable dimension filters. *Journal of Guidance, Control, and Dynamics*, 38(12):2407–2420, 2015. https://doi.org/10.2514/1.G001091.

29 D. P. Lubey and D. J. Scheeres. Identifying and estimating mismodeled dynamics via optimal control policies and distance metrics. *Journal of Guidance, Control, and Dynamics*, 37(5): 1512–1523, 2014. https://doi.org/10.2514/1.G000369.

30 H. C. Ko and D. J. Scheeres. Maneuver detection with event representation using thrust Fourier coefficients. *Journal of Guidance, Control, and Dynamics*, 39(5):1080–1091, 2018/10/17 2016. https://doi.org/10.2514/1.G001463.

31 W. R. Corliss. Historics of the space tracking and data acquisition network (STADAN), the manned space flight network (MSFN), and the NASA communications network (NASCOM). Technical Report CR-140390, NASA, 1974.

32 I. Molotov, V. Agapov, V. Titenko, Z. Khutorovsky, Yu. Burtsev, I. Guseva, V. Rumyantsev, M. Ibrahimov, G. Kornienko, A. Erofeeva, V. Biryukov, V. Vlasjuk, R. Kiladze, R. Zalles, P. Sukhov, R. Inasaridze, G. Abdullaeva, V. Rychalsky, V. Kouprianov, O. Rusakov, E. Litvinenko, and E. Filippov. International scientific optical network for space debris research. *Advances in Space Research*, 41(7):1022–1028, 2008. https://doi.org/10.1016/j.asr.2007.04.048.

33 K. C. Gresham, C. Palma, D. E. Polsgrove, F. K. Chun, D. J. Della-Rose, and R. D. Tippets. Education and outreach using the Falcon telescope network. *Acta Astronautica*, 129:130–134, 2016. https://doi.org/10.1016/j.actaastro.2016.09.006.

34 J. A. Haimerl and G. P. Fonder. Space fense system overview. In *Proceedings of the 2015 Advanced Maui Optical and Space Surveillance Technologies Conference,* Hawaii, 2015.

35 P. Misra and P. Enge. *Global Positioning System: Signals, Measurements, and Performance.* Ganga-Jamuna Press, Lincoln, MA, 2001.

36 P. R. Escobal. *Methods of Orbit Determination.* John Wiley & Sons, Inc., New York, 1965.

37 J. M. Maruskin, D. J. Scheeres, and K. T. Alfriend. Correlation of optical observations of objects in earth orbit. *Journal of Guidance, Control, and Dynamics*, 32(1):194–209, 2009. https://doi.org/10.2514/1.36398.

38 G. H. Born and B. A. Jones. Satellite orbit determination. In *Encyclopedia of Aerospace Engineering*, Chapter 255, John Wiley and Sons, Ltd., volume 5, 2011.

39 J. L. Crassidis and J. L. Junkins. *Optimal Estimation of Dynamic Systems*. Applied Mathematics and Nonlinear Science. Chapman and Hall/CRC, 2004.

40 R. H. Gooding. A new procedure for the solution of the classical problem of minimal orbit determination from three lines of sight. *Celestial Mechanics and Dynamical Astronomy*, 66(4):387–423, 1996. https://doi.org/10.1007/BF00049379.

41 A. E. Roy. *Orbital Motion*. Taylor and Francis Group, LLC, New York, NY, USA, 4th edition, 2005.

42 A. Milani, G. F. Gronchi, M. D. Vitturi, and Z. Knežević. Orbit determination with very short arcs. i admissible regions. *Celestial Mechanics and Dynamical Astronomy*, 90(1):57–85, 2004. https://doi.org/10.1007/s10569-004-6593-5.

43 D. Farnocchia, S. R. Chesley, A. Milani, G. F. Gronchi, and P. W. Chodas. Orbits, long-term predictions, and impact monitoring. *Asteroids IV*. The University of Arizona Press, Tucson, AZ, 2015.

44 J. L. Worthy and M. J. Holzinger. Incorporating uncertainty in admissible regions for uncorrelated detections. *Journal of Guidance, Control, and Dynamics*, 38(9):1673–1689, 2015. https://doi.org/10.2514/1.G000890.

45 K. J. DeMars and M. K. Jah. Probabilistic initial orbit determination using Gaussian mixture models. *Journal of Guidance, Control, and Dynamics*, 36(5):1324–1335, 2013. https://doi.org/10.2514/1.59844.

46 B. A. Jones. CPHD filter birth modeling using the probabilistic admissible region. *IEEE Transactions on Aerospace and Electronic Systems*, 54(3):1456–1469, 2018. https://doi.org/10.1109/TAES.2018.2793378.

47 J. Stauch, T. Bessell, M. Rutten, J. Baldwin, M. Jah, and K. Hill. Joint probabilistic data association and smoothing applied to multiple space object tracking. *Journal of Guidance, Control, and Dynamics*, 41(1):19–33, 2017. https://doi.org/10.2514/1.G002230.

48 H. W. Sorenson and D. L. Alspach. Recursive Bayesian estimation using Gaussian sums. *Automatica*, 7(4):465–479, 1971. https://doi.org/10.1016/0005-1098(71)90097-5.

49 K. J. DeMars, R. H. Bishop, and M. K. Jah. Entropy-based approach for uncertainty propagation of nonlinear dynamical systems. *Journal of Guidance, Control, and Dynamics*, 36(4):1047–1057, 2013. https://doi.org/10.2514/1.58987.

50 D. Raihan and S. Chakravorty. Particle Gaussian mixture filters-I. *Automatica*, 98:331–340, 2018. https://doi.org/10.1016/j.automatica.2018.07.023.

51 M. Majji, J. L. Junkins, and J. D. Turner. A high order method for estimation of dynamic systems. *Journal of the Astronautical Sciences*, 56(3):401–440, 2008. https://doi.org/10.1007/BF03256560.

52 S. Servadio, R. Zanetti, and B. A. Jones. Nonlinear filtering with a polynomial series of Gaussian random variables. *IEEE Transactions on Aerospace and Electronic Systems*, 57(1):647–658, 2021. https://doi.org/10.1109/TAES.2020.3028487.

53 K. C. Ward and K. J. DeMars. Information-based particle flow with convergence control. *IEEE Transactions on Aerospace and Electronic Systems*, 58(2):1377–1390, 2022. https://doi.org/10.1109/TAES.2021.3123296.

54 N. Adurthi, P. Singla, and T. Singh. Conjugate Unscented Transformation: Applications to Estimation and Control. *Journal of Dynamic Systems, Measurement, and Control*, 140(3):030907, 2017. https://doi.org/10.1115/1.4037783.

55 S. Servadio, G. Lavezzi, C. Hofmann, D. Wu, and R. Linares. Analytical uncertainty propagation and maximum a posteriori filtering with the Koopman operator. *IEEE Transactions on Aerospace and Electronic Systems*, 61(5):11512–11527, 2025. https://doi.org/10.1109/TAES.2025.3566685.

56 J. T. Horwood and A. B. Poore. Gauss von Mises distribution for improved uncertainty realism in space situational awareness. *SIAM/ASA Journal on Uncertainty Quantification*, 2(1):276–304, 2014. https://doi.org/10.1137/130917296.

57 Committee for the Assessment of the U.S. Air Force's Astrodynamic Standards; Aeronautics and Space Engineering Board; Division on Engineering and Physical Sciences; National Research Council. *Continuing Kepler's Quest: Assessing Air Force Space Command's Astrodynamics Standards*. The National Academies Press, 2012.

58 T. E. Fortmann, Y. Bar-Shalom, and M. Scheffe. Sonar tracking of multiple targets using joint probabilistic data association. *IEEE Journal of Oceanic Engineering*, 8(3):173–184, 1983. https://doi.org/10.1109/JOE.1983.1145560.

59 A. B. Poore and N. Rijavec. A Lagrangian relaxation algorithm for multidimensional assignment problems arising from multitarget tracking. *SIAM Journal on Optimization*, 3(3):544–563, 1993. https://doi.org/10.1137/0803027.

60 A. B. Poore and J. T. Horwood. Multiple hypothesis correlation for space situational awareness. Final Technical Report AFRL-OSR-VA-TR-2012-0284, Numerica Corporation, Loveland, CO, 2011.

61 N. Singh, J. T. Horwood, J. M. Aristoff, A. B. Poore, C. Sheaff, and M. K. Jah. Multiple hypothesis tracking (MHT) for space surveillance: Results and simulation studies. In *Proceedings of the 2013 Advanced Maui Optical and Space Surveillance Technologies Conference,* Wailea, Maui, Hawaii, 2013.

62 M. D. Hejduk, S. J. Casali, D. A. Cappellucci, N. L. Ericson, and D. E. Snow. A catalogue-wide implementation of general perturbations orbit determination extrapolated from higher order orbital theory solutions. In *Proceedings of the 2013 AAS/AIAA Space Flight Mechanics Meeting,* pages 1–14, 2013. Paper no. AAS 13-240.

63 S. Blackman and R. Popoli. *Design and Analysis of Modern Tracking Systems*. Artech House, Norwood, MA, 1999.

64 G. Tommei, A. Milani, and A. Rossi. Orbit determination of space debris: admissible regions. *Celestial Mechanics and Dynamical Astronomy*, 97(4):289–304, 2007. https://doi.org/10.1007/s10569-007-9065-x.

65 D. Farnocchia, G. Tommei, A. Milani, and A. Rossi. Innovative methods of correlation and orbit determination for space debris. *Celestial Mechanics and Dynamical Astronomy*, 107(1–2):169–185, 2010. https://doi.org/10.1007/s10569-010-9274-6.

66 S. Särkkä. Unscented Rauch–Tung–Striebel smoother. *IEEE Transactions on Automatic Control*, 53(3):845–849, 2008. https://doi.org/10.1109/TAC.2008.919531.

67 R. P. S. Mahler. *Statistical Multisource-multitarget Information Fusion*. Artech House, Boston, Massachusetts, 2007.

68 R. P. S. Mahler. Multitarget bayes filtering via first-order multitarget moments. *IEEE Transactions on Aerospace and Electronic Systems,* 39(4):1152–1178, 2003. https://doi.org/10.1109/TAES.2003.1261119.

69 R. P. S. Mahler. PHD filters of higher order in target number. *IEEE Transactions on Aerospace and Electronic Systems*, 43(4):1523–1543, 2007. https://doi.org/10.1109/TAES.2007.4441756.

70 S. Gehly, B. A. Jones, and P. Axelrad. Search-detect-track sensor allocation for geosynchronous space objects. *IEEE Transactions on Aerospace and Electronic Systems*, 54(6):2788–2808, 2018. https://doi.org/10.1109/TAES.2018.2830578.

71 R. R. Sanaga and C. Frueh. Probability hypothesis density filter with uncertainty in the probability of detection. *Advances in Space Research*, 67:1437–1453, 2020. https://doi.org/10.1016/j.asr.2020.11.035.

72 B.-T. Vo and B.-N. Vo. Labeled random finite sets and multi-object conjugate priors. *IEEE Transactions on Signal Processing*, 61(13):3460–3475, 2013. https://doi.org/10.1109/TSP.2013.2259822.

73 B.-N. Vo, B.-T. Vo, T. T. D. Nguyen, C. Shim, and H. V. Nguyen. Bayesian multi-target tracking via labeled random finite sets. In M. Mallick, B.-N. Vo, and R. Tharmarasa, editors, *Advanced Techniques in Tracking and Sensor Management: Theory and Applications*, Chapter 8. Wiley/IEEE, 2025.

74 B.-N. Vo, B.-T. Vo, T. T. D. Nguyen, and C. Shim. An overview of multi-object estimation via labeled random finite set. *IEEE Transactions on Signal Processing*, 72:4888–4917, 2024. https://doi.org/10.1109/TSP.2024.3472068.

75 M. Adams, L. Cament, A. Pak, and J. Correa. Space surveillance via poisson labeled multi-bernoulli tracking. In M. Mallick, B.-N. Vo, and R. Tharmarasa, editors, *Advanced Techniques in Tracking and Sensor Management: Theory and Applications*, Wiley/IEEE, 2025

76 B.-N. Vo, B.-T. Vo, and D. Phung. Labeled random finite sets and the bayes multi-target tracking filter. *IEEE Transactions on Signal Processing*, 62(24):6554–6567, 2014. https://doi.org/10.1109/TSP.2014.2364014.

77 B.-N. Vo, B.-T. Vo, and H. G. Hoang. An efficient implementation of the generalized labeled multi-bernoulli filter. *IEEE Transactions on Signal Processing*, 65(8):1975–1987, 2017. https://doi.org/10.1109/TSP.2016.2641392.

78 B. L. Reifler and B. A. Jones. Improving labeled random finite-set-based filtering with limited fields of view. *IEEE Transactions on Aerospace and Electronic Systems*, 60(6):9275–9290, 2024. https://doi.org/10.1109/TAES.2024.3439264.

79 W. R. Faber, U. R Mishra, S. Chakravorty, and Islam I. Hussein. Application of a randomized-finite set statistics technique (R-FISST) to space situational awareness. *Journal of the Astronautical Sciences*, 69:1149–1178, 2022. https://doi.org/10.1007/s40295-022-00331-1.

80 E. Delande, J. Houssineau, J. Franco, C. Frueh, D. Clark, and M. Jah. A new multi-target tracking algorithm for a large number of orbiting objects. *Advances in Space Research*, 64:645–667, 2019. https://doi.org/10.1016/j.asr.2019.04.012.

81 B. Ristic, J. Houssineau, and S. Arulampalam. Target tracking in the framework of possibility theory: The possibilistic Bernoulli filter. *Information Fusion*, 62:81–88, 2020. https://doi.org/10.1016/j.inffus.2020.04.008.

82 H. Cai, J. Houssineau, B. A. Jones, M. Jah, and J. Xhang. Possibility generalized labeled multi-Bernoulli filter for multi-target tracking under epistemic uncertainty. *IEEE Transactions on Aerospace and Electronic Systems*, 59(2):1312–1326, 2023. https://doi.org/10.1109/TAES.2022.3200022.

83 B. A. Jones and R. Weisman. Multi-fidelity orbit uncertainty propagation. *Acta Astronautica*, 155:406–417, 2019. https://doi.org/10.1016/j.actaastro.2018.10.023.

84 D. Schuhmacher, B.-T. Vo, and B.-N. Vo. A consistent metric for performance evaluation of multi-object filters. *IEEE Transactions on Signal Processing*, 56(8):3447–3457, 2008. https://doi.org/10.1109/TSP.2008.920469.

14

Generalized Bernoulli Filters for Challenging Sensing Conditions

Ronald Mahler

Independent Researcher, Eagan MN, USA

14.1 Introduction

Suppose that a single appearing or reappearing target is observed by (as is presumed by most target tracking algorithms) a detection-type sensor. That is, at any given time, t_k a detection process is applied to a sensor signature, resulting in a finite set $Z_k \subseteq \mathbb{Z}$ of point measurements ("detections"). Also assume the "small target" model: a single point target can generate at most a single point detection, and any target detection can be generated by only a single point target [1, p. 433].

Given such a model, the Bernoulli filter is a Bayes-optimal approach for tracking the target if, at any given time t_k, the target detection profile $p_{D,k}(x)$ is known; and the clutter process is independent of the target measurement-generation process and has known distribution $\kappa_k(Z)$.

The single-target recursive Bayes (STRB) filter—see Section 14.2.1—is the usual theoretical basis for optimal single-sensor, single-target tracking. The Bernoulli filter generalizes it to considerably more challenging sensing conditions. Specifically, the STRB filter is the special case of the Bernoulli filter when, for any $k \geq 1$: (1) the clutter distribution is trivial, $\kappa_k(\emptyset) = 1$ (i.e., no clutter or false alarms); (2) the probability of detection is $p_{D,k}(x) = 1$ (i.e., the target is perfectly detected); and (3) the probability of target survival from time t_{k-1} to time t_k is $p_{S,k|k-1}(x) = 1$ (i.e., the target cannot disappear).

An important point is that the Bernoulli filter requires no heuristic improvisations. It is simply the special case of the (basic) multi-target recursive Bayes (MTRB) filter, Eqs. (14.8–14.10), given that (1) the sensor is detection-type, (2) target-generated and clutter-generated measurements are independent, and (3) target number cannot exceed 1.

The purpose of this paper is to summarize and clarify, in a single publication, three recent Bayes-optimal generalizations of the Bernoulli filter itself to still more challenging sensing conditions:

1. The pairwise-Markov Bernoulli (PMB) filter: Extension to non-Markovian targets, unknown motion noise, and unknown (including "colored") measurement noise [2].
2. The dyadic filter: Extension to at most two closely-spaced targets [3].
3. The set-valued STRB filter: Extension to ambiguously known measurements and/or ambiguously known target states [4].

These are also exact closed-form solutions of the MTRB filter (in the sense of [5]).

Because of page limitations, it is not feasible to include summaries of topics such as random finite sets (RFSs) and labeled RFSs (LRFSs). Readers are instead directed to the publications [1, 6, 7].

Organization of the paper: Mathematical background (Section 14.2); Bernoulli filter (Section 14.3); PMB filter (Section 14.4); dyadic filter (Section 14.5); set-valued Bernoulli filters (Section 14.6); mathematical proofs (Section 14.7); and Conclusions (Section 14.8).

14.2 Mathematical Background

Organization of the section: the STRB filter (Section 14.2.1); the MTRB filter (Section 14.2.2); and random closed sets (RCSs) (Section 14.2.3).

14.2.1 The Single-target Recursive Bayes (STRB) Filter

The single-target state space is $\mathbb{X}_0$ and the single-target measurement space is $\mathbb{Z}$, both of which are assumed to be Euclidean.

In the pure single-target case, there is no clutter and the target is perfectly detected and always present. At time t_k the target state is an unknown random variable (RV) $\mathbf{X}_{k|k} \in \mathbb{X}_0$ with probability distribution $f_{k|k}(x|z_{1:k}) = f_{\mathbf{X}_{k|k}}(x)$ where $z_{1:k} : z_1, ..., z_k \in \mathbb{Z}$ are the measurements collected from the target at times $t_1, ..., t_k$. In this case, target tracking is accomplished using the STRB filter:[1]

$$f_{k|k-1}(x|z_{1:k-1}) = \int f_{k|k-1}(x|x_{k-1}) f_{k-1|k-1}(x_{k-1}|z_{1:k-1}) dx_{k-1} \tag{14.1}$$

$$f_{k|k}(x|z_{1:k}) = \frac{f_k(z_k|x) f_{k|k-1}(x|z_{1:k-1})}{f_k(z_k|z_{1:k-1})} \tag{14.2}$$

$$f_k(z_k|z_{1:k-1}) = \int f_k(z_k|x) f_{k|k-1}(x|z_{1:k-1}) dx \tag{14.3}$$

with $k \geq 1$ and initial distribution $f_{0|0}(x)$ and $z_{1:0} \overset{\text{def.}}{=} z_1$.

In Section 14.6, the following additive Markov state-transition and sensor measurement models will be assumed:

$$\mathbf{X}_{k|k-1} = \varphi_{k|k-1}(x) + \mathbf{V}_{k|k-1} \tag{14.4}$$

$$\mathbf{Z}_k = \eta_k(x) + \mathbf{W}_k, \tag{14.5}$$

in which case the corresponding Markov and measurement densities have the form:

$$M_x(x_{k-1}) = f_{k|k-1}(x|x_{k-1}) = f_{\mathbf{V}_{k|k-1}}(x - \varphi_{k|k-1}(x_{k-1})) \tag{14.6}$$

$$L_z(x) = f_k(z|x) = f_{\mathbf{W}_k}(z - \eta_k(x)), \tag{14.7}$$

where $\mathbf{V}_{k|k-1}, \mathbf{W}_k$ are, respectively, zero-mean measurement and motion ("plant") noise RVs; $\varphi_{k|k-1}(x_{k-1})$ is a state-transition function; and $\eta_k(x)$ a measurement function.

Other than in Section 14.6, $f_{k|k-1}(x|x_{k-1})$ and $f_k(z|x)$ are arbitrary.

1 Eqs. (14.1-14.3) define the basic STRB filter. The general single-target Markov and measurement densities have the form $f_{k|k-1}(x|x_{k-1}, z_{1:k-1})$ and $f_k(z|x, z_{1:k-1})$ respectively—see [1, Eqs. (3.54, 3.56)].

14.2.2 The MTRB Filter

The MTRB filter is the obvious multi-target generalization of the STRB filter:[2]

$$f_{k|k-1}(X|Z_{1:k-1}) = \int f_{k|k-1}(X|X_{k-1})\, f_{k-1|k-1}(X_{k-1}|Z_{1:k-1})\delta X_{k-1} \tag{14.8}$$

$$f_{k|k}(X|Z_{1;k}) = \frac{f_k(Z_k|X)\, f_{k|k-1}(X|Z_{1:k-1})}{f_k(Z_k|Z_{1:k-1})} \tag{14.9}$$

$$f_k(Z_k|Z_{1:k-1}) = \int f_k(Z_k|X)\, f_{k|k-1}(X|Z_{1:k-1})\delta X \tag{14.10}$$

with initial distribution $f_{0|0}(X)$ and $Z_{1:0} \overset{\text{def.}}{=} Z_1$. Here, (1) $f_{k|k}(X|Z_{1;k})$ is the probability (density) that the finite set $X \subseteq \mathbb{X}_0$ of targets is present at time t_k if the sequence $Z_{1;k} : Z_1, ..., Z_k$ of measurement-sets has been collected; (2) $f_k(Z|X)$ is the probability (density) that the measurement-set Z will collected at time t_k if X is present at time t_k; and (3) $f_{k|k-1}(X|X_{k-1})$ is the probability (density) that the state-set X_{k-1} at time t_{k-1} will evolve to the state-set X at time t_k.

In Eqs. (14.8, 14.10) the indicated integrals are L/RFS set integrals. In the LRFS case, this is defined as:

$$\int f(X)\delta X = \sum_{n\geq 0} \frac{1}{n!} \sum_{(l_1,...,l_n)\in \mathbb{L}^n} \int f(\{(x_1, l_1), ..., (x_n, l_n)\})dx_1 \cdots dx_n \tag{14.11}$$

where by convention $f(\{(x_1, l_1), ..., (x_n, l_n)\}) = f(\emptyset)$ if $n = 0$.

Ordinarily, $f_{k|k-1}(X|X_{k-1})$ is assumed to be the multi-target Markov density for the (labeled) "standard multi-target motion model" [7, pp. 480–481]. This model presumes that target motions are independent and that the target-appearance process is LMB (labeled multi-Bernoulli). However, the Bernoulli, PMB, and dyadic filters reduce to this model only when there are no target appearances. This is because the number of targets cannot exceed 1 or 2, and thus target appearance must be appropriately counterbalanced by target disappearance—which in turn forces target motions to be correlated.

Likewise, $f_k(Z|X) = L_Z(X)$ will be the multi-target likelihood function of the (labeled) "standard multi-target measurement model" [7, pp. 474–476]—but with general (rather than Poisson) clutter. In this case, $f_k(Z|\emptyset) = \kappa_k(Z)$.

14.2.3 Random Closed Sets (RCSs)

The mathematical foundation of the L/RFS approach is RCS theory [1, 8], [9, Appx. F]. Given a space $\mathbb{Y}$, an RCS $\Psi \subseteq \mathbb{Y}$ is an RV whose realizations are closed subsets of $\mathbb{Y}$. RCSs subsume RFSs since finite subsets $Y \subseteq \mathbb{Y}$ are closed. In particular, if $\mathbf{Y} \in \mathbb{Y}$ is an RV then $\{\mathbf{Y}\} \subseteq \mathbb{Y}$ is an RCS.

A major reason why RCSs are useful is that they provide Bayesian formulations of various expert-systems concepts—see Section 14.6.2 and [1]. They are also important for the statistical modeling of multi-target systems [1] and, in particular, for measurement modeling for the Bernoulli filter—see Remark 14.2. Specifically, such modeling requires the RCS-based *belief measure* (a.k.a. belief-mass function) of an RFS $\Xi \subseteq \mathbb{Y}$ [1, Eq. (F.13)]:

$$\beta_\Xi(T) = \Pr(\Xi \subseteq T), \tag{14.12}$$

which is defined on all closed subsets $T \subseteq \mathbb{Y}$ and which completely characterizes the statistics of Ξ.

2 Eqs. (14.8–14.10) define the basic MTRB filter. The general multi-target Markov and measurement densities are $f_{k|k-1}(X|X_{k-1}, Z_{1:k-1})$ and $f_k(Z|X, Z_{1:k-1})$—again see [1, Eqs. (3.54, 3.56)].

A concept that will be central in Section 14.6 is Goodman's *one-point covering function* (OPCF). It is, roughly speaking, a fuzzy-set-valued average value of an RCS [1, Eq. (4.20)], [9, p. 178]:

$$\mu_\Psi(x) \stackrel{\text{def.}}{=} \Pr(x \in \Psi) = \mathrm{E}[\mathbf{1}_\Psi(x)] \tag{14.13}$$

where the second equation states that $\mu_\Psi(x)$ is the expected value of the random integer $\mathbf{1}_\Psi(x) \in \{0, 1\}$ where $\mathbf{1}_S(x)$ denotes the set-indicator function of $S \subseteq \mathbb{X}_0$.

Suppose that Ψ_1, Ψ_2 are statistically independent RCSs. Then, as follows from [9, p. 12, Prop. 1.19],

$$\mu_{\Psi_1 \cap \Psi_2}(x) = \mu_{\Psi_1}(x)\, \mu_{\Psi_2}(x). \tag{14.14}$$

Suppose that Ψ is an RFS: $|\Gamma| < \infty$. Then $\mu_\Psi(x) = 0$ identically since "$x \in \Psi$" is a zero-probability event. This situation will not occur because the realizations of true RCSs will be assumed to have a continuously infinite number of elements.

14.3 The Bernoulli Filter

As previously noted, the Bernoulli filter is a special case of the MTRB filter when target number does not exceed 1. At any measurement-collection time t_k it allows (1) targets to appear and disappear and (2) the detection profiles $p_{D,k}(x)$ and clutter distributions $\kappa_k(Z)$ to be arbitrary.

The Bernoulli filter was independently and contemporaneously devised in 2007 by B.-T. Vo [10] and Mahler [1, Sec. 14.7].[3] A tutorial introduction can be found in the paper [11] by Ristic, Vo, Vo, and Farina.

As noted in [1, p. 515], the Bernoulli filter generalized the 1994 integrated probabilistic data association (IPDA) filter [12] of Musicki, Evans and Stankovic. The IPDA filter was originally derived using bottom-up methods. In 2002, Challa and Vo [13] showed that it can be derived using top-down RFS methods.

In either version, the IPDA filter assumes that, at any measurement-collection time t_k, (1) the target can disappear but not reappear; (2) $p_{D,k}(x) = p_{D,k}$ is constant; (3) the clutter process is Poisson (i.e., $\kappa_k(Z) = e^{-\lambda_k} \prod_{z \in Z} \kappa_k(z)$ where $\kappa_k(z)$ is the clutter intensity function and $\lambda_k = \int \kappa_k(z)dz$ is the clutter rate); and (4) clutter is spatially uniform[4] (i.e., $\kappa_k(z) = \kappa_k$ is constant).

It has been claimed that the Bernoulli filter was actually invented in 2005 in [14]. That paper described a track-before-detect (TBD) filter, assuming a specific radar-signature model. However, the Bernoulli filter presumes a detection-type sensor—i.e., it is *detect-before-track*. It therefore cannot be a TBD filter and hence the algorithm in [14] cannot be the Bernoulli filter. Also, while [14] does employ an "existence variable to determine the presence of a target in the data," this was true of the IPDA filter in 1994 and 2002.

There are four mathematically equivalent versions of the Bernoulli filter, as follows.

14.3.1 Bernoulli Filter: Original Version

Because the Bernoulli filter is a special case of the MTRB filter when $|X| \leq 1$, it must be the case that either $X = \emptyset$ or $X = \{x\}$ for some $x \in \mathbb{X}_0$. Equivalently, $f_{k|k-1}(X|Z_{1:k-1}) = f_{k|k}(X|Z_{1:k}) = 0$ if $|X| \geq 2$. From the unlabeled version of the set integral, Eq. (14.11), this implies that:

3 In [1, Sec. 14.7] it was called the "JoTT filter."
4 This implicitly presumes that $\mathbb{Z}$ is bounded.

$$1 = \int f_{k|k}(X|Z_{1:k})\delta X = f_{k|k}(\emptyset|Z_{1:k}) + \int f_{k|k}(\{x\}|Z_{1:k})dx \tag{14.15}$$

$$= 1 - q_{k|k} + q_{k|k} \int s_{k|k}(x)dx \tag{14.16}$$

where

$$q_{k|k} \overset{\text{def.}}{=} 1 - f_{k|k}(\emptyset|Z_{1:k}) = \int f_{k|k}(\{x\}|Z_{1:k})dx \tag{14.17}$$

is the probability that the target exists at time t_k. If it does, its spatial distribution is:

$$s_{k|k}(x) \overset{\text{def.}}{=} \frac{f_{k|k}(\{x\}|Z_{1:k})}{q_{k|k}}. \tag{14.18}$$

Thus, the MTRB filter propagates *Bernoulli RFSs* $\Xi \subseteq \mathfrak{X}_0$, with distributions:

$$f_\Xi(X) = \begin{cases} 1-q & \text{if} \quad X = \emptyset \\ q\, s(x) & \text{if} \quad X = \{x\}. \\ 0 & \text{if} \quad |X| \geq 2 \end{cases} \tag{14.19}$$

Note that this Bernoulli structure—and the target-existence variable q in particular—are not heuristic improvisations. Rather, *given that* $|X| \leq 1$, *the MTRB filter dictates them.*

To emphasize that the original Bernoulli filter is not heuristic, it is worth revisiting its filtering equations, Eqs. (14.26, 14.28, 14.33, 14.34), with greater exactitude.

14.3.1.1 Original Bernoulli Filter: Time-update

Consider the MTRB filter time-update equation, Eq. (14.8), and define:

$$p_S(x_{k-1}) \overset{\text{abbr.}}{=} p_{S,k|k-1}(x_{k-1}) = 1 - f_{k|k-1}(\emptyset|\{x_{k-1}\}) = \int f_{k|k-1}(\{x\}|\{x_{k-1}\})dx. \tag{14.20}$$

It is the probability that a target with state x_{k-1} at time t_{k-1} will survive into time t_k. If so, the probability (density) that it will have state x is given by the Markov density:

$$f_{k|k-1}(x|x_{k-1}) \overset{\text{def.}}{=} \frac{f_{k|k-1}(\{x\}|\{x_{k-1}\})}{p_{S,k|k-1}(x_{k-1})}. \tag{14.21}$$

Similarly

$$p_{B,k|k-1} \overset{\text{def.}}{=} 1 - f_{k|k-1}(\emptyset|\emptyset) = \int f_{k|k-1}(\{x\}|\emptyset)dx \tag{14.22}$$

is the probability that, if the target was not present at time t_{k-1}, then it will appear at time t_k. If so, its spatial density is:

$$s_{B,k|k-1}(x) \overset{\text{def.}}{=} \frac{f_{k|k-1}(\{x\}|\emptyset)}{p_{B,k|k-1}}. \tag{14.23}$$

Given this, some algebra shows that when $X = \{x\}$, Eq. (14.8) reduces to:

$$q_{k|k-1}\, s_{k|k-1}(x) = (1 - q_{k-1|k-1})\, B_{k|k-1}(x) + q_{k-1|k-1}\, s_{k-1|k-1}[p_S M_x] \tag{14.24}$$

where

$$B_{k|k-1}(x) \overset{\text{def.}}{=} p_{B,k|k-1}\, s_{B,k|k-1}(x) \tag{14.25}$$

is the target-appearance (birth) density and $M_x(x_{k-1})$ was defined in Eq. (14.6) and $s_{k-1|k-1}[h] \overset{\text{def.}}{=} \int h(x)\, s_{k-1|k-1}(x)dx$. Integrating both sides of Eq. (14.24) we get:

$$q_{k|k-1} = p_{B,k|k-1}(1 - q_{k-1|k-1}) + q_{k-1|k-1}\, s_{k-1|k-1}[p_S] \tag{14.26}$$

$$\leq \max\{p_{B,k|k-1},\ s_{k-1|k-1}[p_S]\} \leq 1. \tag{14.27}$$

Substituting this into Eq. (14.24), we get the following formula for $s_{k|k-1}(x)$:

$$s_{k|k-1}(x) = \frac{(1 - q_{k-1|k-1})\, B_{k|k-1}(x) + q_{k-1|k-1}\, s_{k-1|k-1}[p_S M_x]}{p_{B,k|k-1}(1 - q_{k-1|k-1}) + q_{k-1|k-1}\, s_{k-1|k-1}[p_S]}. \tag{14.28}$$

Remark 14.1: In Eq. (14.26), if $p_{B,k|k-1} = p_{S,k|k-1}(u) = p_0$ then $q_{k|k-1} = p_0$ regardless of the value of $q_{k-1|k-1}$. Other poor choices for $p_{B,k|k-1}$ and $p_{S,k|k-1}(u)$ can result in up or down biases in the target-number estimates. A possible choice is $p_{B,k|k-1} = \varepsilon \cong 0$ and $p_{S,k|k-1}(u) = 1 - p_{B,k|k-1} = 1 - \varepsilon$. In this case, target appearance and disappearance are exactly synchronized and, if $\varepsilon \ll q_{k-1|k-1}$, $q_{k|k-1} \cong q_{k-1|k-1}$.

14.3.1.2 Original Bernoulli Filter: Measurement-update

Now consider the MTRB filter measurement-update, Eq. (14.9). When $|X| \leq 1$ the measurement density for the standard multi-target measurement model can be shown to reduce to $f_k(Z|\emptyset) = \kappa_k(Z)$ and [1, Eqs. (14.181,14.182)]:

$$f_k(Z|\{x\}) \overset{\text{abbr.}}{=} L_Z(x) = \kappa_k(Z)\, \hat{L}_Z(x) \tag{14.29}$$

$$\hat{L}_Z(x) = 1 - p_{D,k}(x) + p_{D,k}(x) \sum_{z \in Z} \frac{L_z(x)\, \kappa_k(Z - \{z\})}{\kappa_k(Z)} \tag{14.30}$$

where $L_z(x)$ was defined in Eq. (14.7).

It follows from Eqs. (14.29,14.30,14.10) that:

$$f_k(Z_k|Z_{1:k-1}) = \kappa_k(Z_k)\left(1 - q_{k|k-1} + q_{k|k-1} \int \hat{L}_{Z_k}(x)\, s_{k|k-1}(x)dx\right) \tag{14.31}$$

and so since $q_{k|k}\, s_{k|k}(x) = f_{k|k}(\{x\}|Z_{1:k})$, from Eq. (14.9) we get:

$$q_{k|k}\, s_{k|k}(x) = \frac{\hat{L}_{Z_k}(x)\, q_{k|k-1}\, s_{k|k-1}(x)}{1 - q_{k|k-1} + q_{k|k-1}\, s_{k|k-1}[\hat{L}_{Z_k}]} \tag{14.32}$$

where $s_{k|k-1}[h] \overset{\text{def.}}{=} \int h(x)\, s_{k|k-1}(x)dx$. Integrating both sides of this yields:

$$q_{k|k} = \frac{q_{k|k-1}\, s_{k|k-1}[\hat{L}_{Z_k}]}{1 - q_{k|k-1} + q_{k|k-1}\, s_{k|k-1}[\hat{L}_{Z_k}]} \leq 1. \tag{14.33}$$

Substituting this into Eq. (14.32) we get:

$$s_{k|k}(x) = \frac{\hat{L}_{Z_k}(x)\, s_{k|k-1}(x)}{s_{k|k-1}[\hat{L}_{Z_k}]}. \tag{14.34}$$

Remark 14.2: Equations (14.29,14.30) are not heuristic improvisations. Rather, they follow directly from the RFS measurement model $\Sigma_k(x) = \Upsilon_k(x) \cup C_k$ for a detection-type sensor, where (1) C_k is the clutter RFS, (2) $\Upsilon_k(x)$ with $|\Upsilon_k(x)| \leq 1$ is the RFS of the measurements generated by target x, and (3) $\Upsilon_k(x), C_k$ are assumed independent. Given this, Eqs. (14.29,14.30) are derived from the belief measure $\beta_k(T|x) = \Pr(\Sigma_k(x) \subseteq T)$ of $\Sigma_k(x)$ using the RFS set derivative [1, Eq. (9.15)]:

$$f_k(Z|\{x\}) = \left[\frac{\delta\beta_k}{\delta Z}(T|x)\right]_{T=\emptyset} = \frac{\delta\beta_k}{\delta Z}(\emptyset|x). \tag{14.35}$$

14.3.1.3 Original Bernoulli Filter: State Estimation

The target is Bayes-optimally estimated to exist if its existence probability exceeds its nonexistence probability: $q_{k|k} > 1 - q_{k|k}$—i.e., if $q_{k|k} > 1/2$; and in which case the estimated target state is:

$$\hat{x}_{k|k} = \arg\sup_{x\in\mathbb{X}_0} s_{k|k}(x). \tag{14.36}$$

14.3.2 Bernoulli Filter: PHD Version

This appeared in [2, Eqs. (109, 110)]. Since $\int f_{k|k}(\{x\}|Z_{1:k})dx = q_{k|k}$, $q_{k|k}$ is actually redundant as a filtering variable. Thus define:

$$D_{k|k}(x|Z_{1:k}) \overset{\text{def.}}{=} q_{k|k}\, s_{k|k}(x). \tag{14.37}$$

This is a probability hypothesis density (PHD) [1, Sec. 16.2] with the additional property that $\int D_{k|k}(x|Z_{1:k})dx \le 1$. It is easily seen that Eqs. (14.24, 14.32) can be rewritten as the time- and measurement-update equations of a PHD filter:

$$D_{k|k-1}(x) = \left(1 - D_{k-1|k-1}[1]\right) B_{k|k-1}(x) + D_{k-1|k-1}[p_S M_x] \tag{14.38}$$

$$D_{k|k}(x) = \frac{\hat{L}_{Z_k}(x)\, D_{k|k-1}(x)}{1 - D_{k|k-1}[1] + D_{k|k-1}[\hat{L}_{Z_k}]} \tag{14.39}$$

where $D_{k-1|k-1}[1] = q_{k-1|k-1}$ and $D_{k|k-1}[1] = q_{k|k-1}$; and where $M_x(x_{k-1})$ was defined in Eq. (14.6); and where:

$$B_{k|k-1}(x) \overset{\text{def.}}{=} p_{B,k|k-1} s_{B,k|k-1}(x) \tag{14.40}$$

is the target-appearance PHD, where $p_{B,k|k-1} \le 1$ is the target-appearance probability. Note that $D_{k-1|k-1}[p_S]/D_{k-1|k-1}[1] \le 1$ so that $\int D_{k|k-1}(x)dx \le 1$.

Remark 14.3: In regard to Eq. (14.38), note that the predicted PHD $D_{k|k-1}(x)$ is a weighted average of the target-appearance PHD $B_{k|k-1}(x)$ and the target-persistence PHD $D_{k-1|k-1}[p_S M_x]/D_{k-1|k-1}[1]$, with weights $q_{k-1|k-1}$ and $1 - q_{k-1|k-1}$. If the target-existence probability $q_{k-1|k-1}$ is large then it weights $D_{k-1|k-1}[p_S M_x]/D_{k-1|k-1}[1]$ more heavily than $B_{k|k-1}(x)$ and vice-versa. It is important to note that this averaging mechanism is not a heuristic improvisation. Rather, it is dictated by $f_{k|k-1}(X|X_{k-1})$ given that $|X|, |X_{k-1}| \le 1$.

14.3.3 Bernoulli Filter: Single-step Version

This appeared in [2, Eqs. (8,111)] as a precondition for the derivation of the PMB filter. It results from substituting Eq. (14.38) into Eq. (14.39). First define:

$$\hat{R}_Z(u) \overset{\text{def.}}{=} 1 - p_{S,k|k-1}(u) + p_{S,k|k-1}(u) \int \hat{L}_Z(x)\, M_x(u)dx. \tag{14.41}$$

Then the update for the PHD is [2, Eq. (111)]:

$$D_{k|k}(x) = \frac{\hat{L}_{Z_k}(x)\, B_{k|k-1}(x)\,(1 - D_{k-1|k-1}[1]) + \hat{L}_{Z_k}(x)\, D_{k-1|k-1}[p_S M_x]}{(1 - B_{k|k-1}[1] + B_{k|k-1}[\hat{L}_{Z_k}])(1 - D_{k-1|k-1}[1]) + D_{k-1|k-1}[\hat{R}_{Z_k}]}. \tag{14.42}$$

14.3.4 Bernoulli Filter: LRFS Version

This version was introduced in [3] as part of the derivation of the dyadic filter (Section 14.5) but does not differ substantively from the original Bernoulli filter.

Specifically, if ϖ is the single target's label, then substituting (x, ϖ) for x in the Bernoulli filter equations results in the labeled Bernoulli filter. The main difference is that the definition of a labeled Bernoulli distribution differs slightly from that for a Bernoulli distribution, Eq. (14.19), as follows [3, Eq. (1)]:

$$f_{\varpi}(X) = \begin{cases} 1 - q_{\varpi} & \text{if} \quad X = \emptyset \\ \delta_{l,\varpi}\, q_{\varpi}\, s_{\varpi}(x) & \text{if} \quad X = \{(x, l)\} \\ 0 & \text{if} \quad |X| \geq 2 \end{cases} \tag{14.43}$$

where $\int s_{\varpi}(x)dx = 1$. This is necessary to ensure that $\int f_{\varpi}(X)\delta X = 1$.

14.3.5 Bernoulli Filter: Multiple Correlated Sensors

These were introduced in [15, Eqs. (44–46)].

14.4 PMB Filter

Recall that a hidden Markov model (HMM) has the following form. At time t_k, the target's state x_k depends only on its previous state x_{k-1}, as governed by the Markov density $f_{k|k-1}(x|x_{k-1})$; and the current measurement z_k depends only on the current state x, as characterized by the measurement density $f_k(z|x)$.

Pairwise-Markov models (PMMs) are a generalization of HMMs originated by Pieczynski [16, 17]. The target and its measurements are regarded as a joint time-evolving system with a random joint state $(\mathbf{X}_k, \mathbf{Y}_k)$ for $k = 0, 1, \ldots$ for $\mathbf{X}_k \in \mathbb{X}_0$ and $\mathbf{Y}_k \in \mathbb{Z}$. The joint density $f_{\mathbf{X}_k,\mathbf{Y}_k}(x, y)$ is an unknown that must be estimated along with the unknown random state variables $\mathbf{X}_k$ and $\mathbf{Y}_k$.

This section describes a PMM generalization of the Bernoulli filter. It is organized as follows: PMMs (Section 14.4.1); PMM Bernoulli filter (Section 14.4.2); and PMB filters for multiple correlated sensors (Section 14.4.3).

14.4.1 Pairwise-Markov Models

The dynamics of a PMM are specified by a Bayesian Markov transition function written in the following factorized form:

$$f_{k|k-1}(x, y|x_{k-1}, y_{k-1}) = f_{k|k-1}(x|x_{k-1}, y_{k-1})\, f_k(y|x, x_{k-1}, y_{k-1}). \tag{14.44}$$

A PMM can be recast as an HMM by introducing the measurement density $f_k(z|x_k, y_k) = \delta_{y_k}(z)$ where $\delta_y(z)$ denotes the Dirac delta concentrated at y [15, p. 168232, col. 2].

Remark 14.4: The notations z_k and y_k differ in that y_k is an unknown realization $\mathbf{Y}_k = y_k$ of the random state variable $\mathbf{Y}_k$ whereas z_k is a known realization $\mathbf{Y}_k = z_k$ collected by the sensor [15], Remark 2]. In the PMM literature, the notation z_k is typically suppressed and only y_k is used. In what follows, both notations will be used as appropriate since confusion can result otherwise.

A PMM reduces to an HMM if, for $k \geq 1$, [15, Eq. (35)]:

$$f_{k|k-1}(x|x_{k-1}, y_{k-1}) = f_{k|k-1}(x|x_{k-1}) \tag{14.45}$$

$$f_k(y|x, x_{k-1}, y_{k-1}) = f_k(y|x). \tag{14.46}$$

The Bayesian PMM update equation is, for $k \geq 1$ [15, Eq. (32)]:

$$f_k(x, y|z_{1:k-1}) \tag{14.47}$$

$$= \frac{\int f_{k|k-1}(x, y|x_{k-1}, z_{k-1})\, f_{k-1}(x_{k-1}, z_{k-1}|z_{1:k-2}) dx_{k-1}}{f_{k-1}(z_{k-1}|z_{1:k-2})} \tag{14.48}$$

$$f_{k-1}(z_{k-1}|z_{1:k-2}) = \int f_{k-1}(x_{k-1}, z_{k-1}|z_{1:k-2}) dx_{k-1}.$$

The initial PMM distribution can be chosen as $f_0(x, y) = f_0(y|x)\, f_0(x)$ where $f_0(x)$ is an initial target distribution and $f_0(y|x)$ is an initial measurement density. Also, $z_{1:0} \overset{\text{def.}}{=} z_1$.

The measurement density at time t_k can be estimated as [15, Eq. (33)]:

$$f_k(y|x, z_{1:k-1}) = \frac{f_k(x, y|z_{1:k-1})}{f_k(x|z_{1:k-1})} = \frac{f_k(x, y|z_{1:k-1})}{\int f_k(x, y|z_{1:k-1}) dy}. \tag{14.49}$$

Note that this has the form of a general Bayes measurement density in the sense of [1, Eq. (3.56)]. The Markov density at time t_k can be similarly estimated as [15, Eq. (34)].[5]

The posterior distribution $f_k(x|z_{1:k})$ of a single target with state x and PMM dynamics can be propagated using the Bayesian recursion [18, Eq. (12)], [15, Eq. (37)]:

$$f_k(x|z_{1:k}) = \frac{\int f_{k|k-1}(x, z_k|x_{k-1}, z_{k-1}) f_{k-1}(x_{k-1}|z_{1:k-1}) dx_{k-1}}{\int f_{k|k-1}(z_k|x_{k-1}, z_{k-1})\, f_{k-1}(x_{k-1}|z_{1:k-1}) dx_{k-1}} \tag{14.50}$$

$$f_{k|k-1}(z_k|x_{k-1}, z_{k-1}) = \int f_{k|k-1}(x_k, z_k|x_{k-1}, z_{k-1})\, dx_k. \tag{14.51}$$

PMMs significantly generalize HMMs in three respects:

1. The current state x can depend on the previous measurement z_{k-1} and not just on the previous state x_{k-1}, as described by $f_{k|k-1}(x|x_{k-1}, y_{k-1})$. That is, the target can be non-Markovian. Nevertheless, this is still consistent with HMM Bayes filtering theory in the sense that $f_{k|k-1}(x|x_{k-1}, z_{k-1})$ is a special case of the general Bayesian state-transition function $f_{k|k-1}(x|x_{k-1}, z_{1:k-1})$ [1, Eq. (3.54)].
2. The current measurement y can depend not just on the current state x but also the previous state x_{k-1} and previous measurement z_{k-1}, as described by $f_k(y|x, x_{k-1}, y_{k-1})$. That is, measurement noise can be colored and correlated with motion noise. This generalizes HMM Bayes filtering theory in the sense that $f_k(y|x, x_{k-1}, z_{k-1})$ is not a special case of the general Bayes measurement density $f_k(y|x, z_{1:k-1})$ [1, Eq. (3.56)].
3. The Markov density $f_{k|k-1}(x|x_{k-1})$ and measurement-noise density $f_k(y|x)$ are recursively estimated rather than assumed a priori.

5 The proof requires the identity $f_{kk-1}(x_k, y_k|x_{k-1}, y_{k-1}) = f_{kk-1}(x_k, y_k|x_{k-1}, y_{k-1}, z_{1:k-2})$.

Remark 14.5: Pieczynski and Desbouvries have described a Kalman filter-based single-target PMM tracker in [19]. Petetin and Pieczynski have described a PMM version of the PHD filter and applied it to multi-target tracking in [18]. Derode and Pieczynski have applied PMM techniques to signal and image segmentation in [20].

14.4.2 The PMB Filter

In [15], a single-step Bernoulli filter with PMM statistics was derived [15, Eq. (10)]:

$$D_{k|k}(x) = \frac{\hat{L}_{Z_k}(x)\, B_{k|k-1}(x)\,(1 - D_{k-1|k-1}[1]) + D_{k-1k-1}[p_S \hat{R}_{Z_k,Z_{k-1},x}]}{\begin{pmatrix} (1 - B_{k|k-1}[1] + B_{k|k-1}[\hat{L}_{Z_k}])(1 - D_{k-1|k-1}[1]) \\ +D_{k-1k-1}[1 - p_S + p_S \hat{R}_{Z_k,Z_{k-1}}] \end{pmatrix}}. \tag{14.52}$$

Here, (1) $p_D(x) \overset{\text{abbr.}}{=} p_{D,k}(x)$ and $p_S(x_{k-1}) \overset{\text{abbr.}}{=} p_{S,k|k-1}(x_{k-1})$ are as in Eqs. (14.30,14.20); (2) $B_{k|k-1}(x)$ is the target-appearance (birth) PHD at time t_k; (3) $M_x(x_{k-1})$ was defined in Eq. (14.6); and where:

1. $\hat{L}_{Z_k}(x)$ [15, Eq. (2)]:

$$\hat{L}_{Z_k}(x) = 1 - p_{D,k}(x) + p_{D,k}(x) \sum_{z \in Z_k} \frac{L_z(x)\, \kappa_k(Z_k - \{z\})}{\kappa_k(Z_k)} \tag{14.53}$$

   where the summation vanishes if $Z_k = \emptyset$.
2. $\hat{R}_{Z_k,Z_{k-1},x}(x_{k-1})$: If $Z_{k-1} = \emptyset$ then [15, Eq. (4)]

$$\hat{R}_{Z_k,\emptyset,x}(x_{k-1}) = \hat{L}_{Z_k}(x)\, M_x(x_{k-1}) \tag{14.54}$$

   whereas if $Z_{k-1} \neq \emptyset$ and

$$f_{k|k-1}(x|x_{k-1}, z_{k-1}) = \int f_{k|k-1}(x, y|x_{k-1}, z_{k-1})dy \tag{14.55}$$

   then [15, Eq. (3)]

$$\begin{aligned} &\hat{R}_{Z_k,Z_{k-1},x}(x_{k-1}) \\ &= \frac{1 - p_{D,k}(x)}{|Z_{k-1}|} \sum_{z_{k-1} \in Z_{k-1}} f_{k|k-1}(x|x_{k-1}, z_{k-1}) \\ &\quad + \frac{p_{D,k}(x)}{|Z_{k-1}|} \sum_{z_{k-1} \in Z_{k-1}} \sum_{z \in Z_k} \frac{f_{k|k-1}(x, z|x_{k-1}, z_{k-1})\, \kappa_k(Z_k - \{z\})}{\kappa_k(Z_k)} \end{aligned} \tag{14.56}$$

   where the second summation vanishes if $Z_k = \emptyset$.
3. $\hat{R}_{Z_k,Z_{k-1}}$: [15, Eq. (8)]: If $Z_{k-1} = \emptyset$ then [15, Eq. (8)][6]

$$\hat{R}_{Z_k,\emptyset}(x_{k-1}) = \int \hat{L}_{Z_k}(x)\, M_x(x_{k-1})dx \tag{14.57}$$

6 *Errata*: There was a typo in the denominator of the fraction in the second summation in [15, Eq. (3)]. It has been corrected in the current Eq. (14.56).

whereas if $Z_{k-1} \neq \emptyset$, [15, Eq. (7)]

$$\hat{R}_{Z_k,Z_{k-1}}(x_{k-1}) = \int \hat{R}_{Z_k,Z_{k-1},x}(x_{k-1})dx \tag{14.58}$$

$$= \frac{1}{|Z_{k-1}|} \sum_{z_{k-1} \in Z_{k-1}} \int (1 - p_{D,k}(x)) f_{k|k-1}(x|x_{k-1}, z_{k-1})dx \tag{14.59}$$

$$+ \frac{1}{|Z_{k-1}|} \sum_{z_{k-1} \in Z_{k-1}} \sum_{z \in Z_k} \frac{\kappa_k(Z_k - \{z\}) \int p_{D,k}(x) f_{k|k-1}(x, z|x_{k-1}, z_{k-1})dx}{\kappa_k(Z_k)}$$

where the second summation vanishes if $Z_k = \emptyset$.

The computational order of the PMB filter at time t_k is $|Z_{k-1}|\ |Z_k|$ [15, Eq. (67)].

Remark 14.6: If the PMM is actually an HMM then Eq. (14.52) reduces to the single-step Bernoulli filter, Eq. (14.42)—see [15, Eqs. (12, 200)].

14.4.3 Multisensor PMB Filter

The multisensor PMM Bernoulli filter for correlated sensors was introduced in [15, Eq. (54)]. There it was also shown that it reduces to the corresponding multisensor Bernoulli filter if the PMM is actually an HMM [15, Eqs. (55–66)].

14.5 The Dyadic Filter

As previously noted, the Bernoulli filter is the special case of the MTRB filter, Eqs. (14.8–14.10), assuming that $|X| \leq 1$ and that the single-target detection profile $p_{D,k}(x)$ and the distribution $\kappa_k(Z)$ of the clutter process are known and that the clutter and target-measurement processes are independent. Also, the structure of the Bernoulli filter is not heuristic or otherwise ad hoc, but dictated by these assumptions.

It is natural to ask: If we assume that $|X| \leq 2$ along with assumptions similar to those made for the Bernoulli filter, what filtering structure will be dictated? The answer to this question is the subject of this section: the *dyadic filter* [3].

Organization of the section: dyadic RFSs (Section 14.5.1); dyadic filter time-update (Section 14.5.2); dyadic filter measurement-update (Section 14.5.3); and state estimation for the dyadic filter (Section 14.5.4).

14.5.1 Dyadic LRFSs

Just as the assumption $|X| \leq 1$ leads to Bernoulli RFSs, the assumption $|X| \leq 2$ leads to *dyadic LRFSs*. Given a single target with label ϖ, a Bernoulli LRFS Ξ_ϖ has realizations of the form $\Xi_\varpi = \emptyset$ or $\Xi_\varpi = \{(x, \varpi)\}$. Likewise, given two targets with labels $\varpi \neq \varpi'$, a dyadic LRFS $\Xi_{\varpi,\varpi'}$ has realizations of the form $\Xi_{\varpi,\varpi'} = \emptyset$, $\Xi_{\varpi,\varpi'} = \{(x, \varpi)\}$, $\Xi_{\varpi,\varpi'} = \{(x, \varpi')\}$, or $\Xi_{\varpi,\varpi'} = \{(x_1, \varpi), (x_2, \varpi')\}$.

Given this, the distribution of $\Xi_{\varpi,\varpi'}$ has the form [3, Eqs. (2-5, 36)]:[7]

$$f_{\varpi,\varpi'}(X) = \begin{cases} 1 - \tilde{q}_{\varpi,\varpi'} - \tilde{q}_{\varpi',\varpi} - q_{\varpi,\varpi'} & \text{if } X = \emptyset \\ \begin{cases} \delta_{l,\varpi}\, \tilde{D}_{\varpi,\varpi'}(x) \\ +\delta_{l,\varpi'}\, \tilde{D}_{\varpi',\varpi}(x) \end{cases} & \text{if } X = \{(x,l)\} \\ \begin{cases} \delta_{l_1,\varpi}\, \delta_{l_2,\varpi'} \\ \cdot D_{\varpi,\varpi'}(x_1,x_2) \\ +\delta_{l_1,\varpi'}\, \delta_{l_2,\varpi} \\ \cdot D_{\varpi,\varpi'}(x_2,x_1) \end{cases} & \text{if } \begin{cases} X = \{(x_1,l_1),(x_2,l_2)\}, \\ l_1 \neq l_2 \end{cases} \\ 0 & \text{if } |X| \geq 3 \end{cases} \tag{14.60}$$

where [3, Eqs. (6–9)]

$$\tilde{D}_{\varpi,\varpi'}(x) = \tilde{q}_{\varpi,\varpi'}\, \tilde{s}_{\varpi,\varpi'}(x) \tag{14.61}$$

$$\tilde{D}_{\varpi',\varpi}(x) = \tilde{q}_{\varpi',\varpi}\, \tilde{s}_{\varpi',\varpi}(x) \tag{14.62}$$

$$D_{\varpi,\varpi'}(x_1,x_2) = \tilde{q}_{\varpi,\varpi'}\, s_{\varpi,\varpi'}(x_1,x_2) \tag{14.63}$$

and where

$$\tilde{q}_{\varpi,\varpi'}, \tilde{q}_{\varpi',\varpi}, q_{\varpi,\varpi'}, \geq 0 \tag{14.64}$$

$$\tilde{q}_{\varpi,\varpi'} + \tilde{q}_{\varpi',\varpi} + q_{\varpi,\varpi'} \leq 1 \tag{14.65}$$

$$q_{\varpi,\varpi'} = q_{\varpi',\varpi} \tag{14.66}$$

$$\int \tilde{s}_{\varpi,\varpi'}(x)dx = 1 \tag{14.67}$$

$$\int \tilde{s}_{\varpi',\varpi}(x)dx = 1 \tag{14.68}$$

$$s_{\varpi,\varpi'}(x_1,x_2) = s_{\varpi',\varpi}(x_2,x_1) \tag{14.69}$$

$$\int s_{\varpi,\varpi'}(x_1,x_2)dx_1dx_2 = 1. \tag{14.70}$$

Here $\tilde{q}_{\varpi,\varpi'}, \tilde{q}_{\varpi',\varpi}, q_{\varpi,\varpi'}$ are, respectively, the probabilities that (1) ϖ but not ϖ' exists and, if so, its spatial distribution is $\tilde{s}_{\varpi,\varpi'}(x)$; (2) ϖ' but not ϖ exists and, if so, its spatial distribution is $\tilde{s}_{\varpi,'\varpi}(x)$; and (3) ϖ, ϖ' both exist and their joint distribution is $s_{\varpi,\varpi'}(x_1,x_2)$.

Remark 14.7: As a simple example, assume that ϖ, ϖ' are (1) independent and (2) labeled-Bernoulli as defined in Eq. (14.43). Then [3, Eqs. (42–44)]

$$\tilde{q}_{\varpi,\varpi'} = q_{\varpi}(1-q_{\varpi'}), \tilde{q}_{\varpi',\varpi} = (1-q_{\varpi})q_{\varpi'}, q_{\varpi,\varpi'} = q_{\varpi}q_{\varpi'} \tag{14.71}$$

$$\tilde{s}_{\varpi,\varpi'}(x) = s_{\varpi}(x), \tilde{s}_{\varpi',\varpi}(x) = s_{\varpi'}(x), s_{\varpi,\varpi'}(x_1,x_2) = s_{\varpi}(x_1)\, s_{\varpi'}(x_2). \tag{14.72}$$

---

7 In Eq. (14.60), the formula for $f_{\varpi,\varpi'}(X)$ when $X = \{(x_1,l_1),(x_2,l_2)\}$ is equivalent to [3, Eq. (4)].

It follows that $f_{\varpi,\varpi'}(X)$ is completely determined by three ordinary density functions defined on $\mathfrak{X}_0$ [3, Eqs. (10–12)]:

$$\tilde{D}_{\varpi,\varpi'}(x) = f_{\varpi,\varpi'}(\{(x,\varpi)\}) = \tilde{q}_{\varpi,\varpi'}\,\tilde{s}_{\varpi,\varpi'}(x) \tag{14.73}$$

$$\tilde{D}_{\varpi',\varpi}(x) = f_{\varpi,\varpi'}(\{(x,\varpi')\}) = \tilde{q}_{\varpi',\varpi}\,\tilde{s}_{\varpi',\varpi}(x) \tag{14.74}$$

$$D_{\varpi,\varpi'}(x_1,x_2) = f_{\varpi,\varpi'}(\{(x_1,\varpi),(x_2,\varpi')\}) = q_{\varpi,\varpi'}\,s_{\varpi,\varpi'}(x_1,x_2). \tag{14.75}$$

Equation (14.73) describes the probability of existence and the spatial distribution of ϖ if it exists. Equation (14.74) describes the probability of existence and the spatial distribution of ϖ' if it exists. Equation (14.75) describes the correlation between ϖ,ϖ' if they both exist. The time-update and measurement-update for $f_{\varpi,\varpi'}(X)$ are equivalent to, respectively, time-updates and measurement-updates for these three densities.

In what follows the following abbreviations will be employed:

$$f^{k|k-1}_{\varpi,\varpi'}(X) \overset{\text{abbr.}}{=} f^{k|k-1}_{\varpi,\varpi'}(X|Z_{1:k-1}), \quad f^{k|k}_{\varpi,\varpi'}(X) \overset{\text{abbr.}}{=} f^{k|k}_{\varpi,\varpi'}(X|Z_{1:k}) \tag{14.76}$$

and

$$\tilde{D}^{k|k-1}_{\varpi,\varpi'}(x) = f^{k|k-1}_{\varpi,\varpi'}(\{(x,\varpi)\}) \tag{14.77}$$

$$\tilde{D}^{k|k-1}_{\varpi',\varpi}(x) = f^{k|k-1}_{\varpi,\varpi'}(\{(x,\varpi')\}) \tag{14.78}$$

$$D^{k|k-1}_{\varpi,\varpi'}(x_1,x_2) = f^{k|k-1}_{\varpi,\varpi'}(\{(x_1,\varpi),(x_1,\varpi')\}) \tag{14.79}$$

and similarly for $\tilde{D}^{k|k}_{\varpi,\varpi'}(x)$, $\tilde{D}^{k|k}_{\varpi',\varpi}(x)$, and $D^{k|k}_{\varpi,\varpi'}(x_1,x_2)$.

14.5.2 Dyadic Filter Time-update

The time-update for the Bernoulli filter was completely dictated by the structure of the MTRB filter's Markov density $f_{k|k-1}(X|X_{k-1})$ when $|X|,|X_{k-1}| \le 1$. This turned out to be a weighted average of the target-disappearance and target-persistence processes, with the weights being the target-existence probability $q_{k-1|k-1}$ and its conjugate $1-q_{k-1|k-1}$.

The dyadic filter time-update presented in [3] was partially heuristic. This section presents a new and fully rigorous derivation that is completely dictated by (1) the structure of $f_{k|k-1}(X|X_{k-1})$ when $|X|,|X_{k-1}| \le 2$; and (2) the (labeled) standard multi-target motion model in the special case when there are no target births.

In [3, Eqs. (50, 52, 54)] the dyadic filter time-update was derived from the standard motion model when there are no target births. This derivation is theoretically correct. The new general formulas, which follow in Eqs. (14.96–14.98), must reduce to that result when there are no target births.

Begin with the following definitions and assumptions, all based on $f_{k|k-1}(X|X_{k-1})$:

First, define the birth densities:

$$\tilde{B}^{k|k-1}_{\varpi,\varpi'}(x) = f_{k|k-1}(\{(x,\varpi)\}|\emptyset) \tag{14.80}$$

$$\tilde{B}^{k|k-1}_{\varpi',\varpi}(x) = f_{k|k-1}(\{(x,\varpi')\}|\emptyset) \tag{14.81}$$

$$B^{k|k-1}_{\varpi,\varpi'}(x_1,x_2) = f_{k|k-1}(\{(x_1,\varpi),(x_2,\varpi')\}|\emptyset) \tag{14.82}$$

or, equivalently,

$$f_{k|k-1}(\{(x,l)\}|\emptyset) = \delta_{l,\varpi}\,\tilde{B}^{k|k-1}_{\varpi,\varpi'}(x) + \delta_{l,\varpi'}\,\tilde{B}^{k|k-1}_{\varpi',\varpi}(x) \tag{14.83}$$

$$f_{k|k-1}(\{(x_1, l_1), (x_2, l_2)\}|\emptyset) = \delta_{l_1,\varpi}\, \delta_{l_2,\varpi'}\, B_{\varpi,\varpi'}(x_1, x_2) \\ + \delta_{l_1,\varpi'}\, \delta_{l_2,\varpi}\, B_{\varpi,\varpi'}(x_2, x_1). \tag{14.84}$$

Second, define the target birth probabilities

$$\tilde{b}^{k|k-1}_{\varpi,\varpi'} = \int \tilde{B}^{k|k-1}_{\varpi,\varpi'}(x)dx \tag{14.85}$$

$$\tilde{b}^{k|k-1}_{\varpi',\varpi} = \int \tilde{B}^{k|k-1}_{\varpi',\varpi}(x)dx \tag{14.86}$$

$$b^{k|k-1}_{\varpi,\varpi'} = \int B^{k|k-1}_{\varpi,\varpi'}(x_1, x_2)dx_1dx_2, \tag{14.87}$$

in which case the LRFS identity $\int f_{k|k-1}(X|\emptyset)\delta X = 1$ tells us that

$$\tilde{b}^{k|k-1}_{\varpi,\varpi'} + \tilde{b}^{k|k-1}_{\varpi',\varpi} + b^{k|k-1}_{\varpi,\varpi'} \leq 1. \tag{14.88}$$

Third, to be consistent with the no-birth equations [3, Eqs. (50, 52, 54)], the following must be true:

$$f_{k|k-1}(\{(x, \varpi)\}|\{(u, l)\}) = \delta_{l,\varpi}\, p^{\varpi}_{S,k|k-1}(u)\, f^{k|k-1}_{\varpi}(x|u) \tag{14.89}$$

where the probability of target survival $p^{\varpi}_S(u) \overset{\text{abbr.}}{=} p^{\varpi}_{S,k|k-1}(u)$ and the Markov density $f^{k|k-1}_{\varpi}(x|u)$ are defined as:

$$p^{\varpi}_{S,k|k-1}(u) = \int f_{k|k-1}(\{(x, \varpi)\}|\{(u, \varpi)\})dx \leq 1 \tag{14.90}$$

$$M^{\varpi}_x(u) = f^{k|k-1}_{\varpi}(x|u) = \frac{f_{k|k-1}(\{(x, \varpi)\}|\{(u, \varpi)\})}{p^{\varpi}_{S,k|k-1}(u)} \tag{14.91}$$

and similarly for $M^{\varpi'}_x(u)$, $p^{\varpi'}_{S,k|k-1}(u)$, and $f^{k|k-1}_{\varpi'}(x|u)$.

Fourth, to be consistent with [3, Eqs. (50, 52, 54)] the following must be true:

$$f_{k|k-1}(\{(x, \varpi)\}|\{(u_1, l_1), (u_2, l_2)\}) \\ = \delta_{l_1,\varpi}\, p^{\varpi}_{S,kk-1}(u_1)\, f^{k|k-1}_{\varpi}(x|u_1)\,(1 - p^{\varpi}_S(u_2)) \\ + \delta_{l_2,\varpi}\,(1 - p^{\varpi}_{S,k|k-1}(u_1))\, p^{\varpi}_S(u_2)\, f^{k|k-1}_{\varpi}(x|u_2). \tag{14.92}$$

That is, only one of l_1 and l_2 can transition to ϖ and not the other. A similar formula applies to $f_{k|k-1}(\{(x, \varpi')\}|\{(u_1, l_1), (u_2, l_2)\})$.

Fifth, to be consistent with [3, Eqs. (50, 52, 54)] the following must be true:

$$f_{k|k-1}(\{(x_1, \varpi), (x_2, \varpi')\}|\{(u_1, \varpi), (u_2, \varpi')\}) \\ = p^{\varpi}_{S,k|k-1}(u_1)\, M^{\varpi}_{x_1}(u_1)\, p^{\varpi'}_{S,k|k-1}(u_2)\, M^{\varpi'}_{x_2}(u_2) \tag{14.93}$$

or, equivalently,

$$f_{k|k-1}(\{(x_1, \varpi), (x_2, \varpi')\}|\{(u_1, l_1), (u_2, l_2)\}) \\ = \delta_{l_1,\varpi}\delta_{l_2,\varpi'}\, p^{\varpi}_{S,k|k-1}(u_1)\, M^{\varpi}_{x_1}(u_1)\, p^{\varpi'}_{S,k|k-1}(u_2)\, M^{\varpi'}_{x_2}(u_2) \\ + \delta_{l_2,\varpi}\delta_{l_1,\varpi'}\, p^{\varpi}_{S,k|k-1}(u_2)\, M^{\varpi}_{x_2}(u_2)\, p^{\varpi'}_{S,k|k-1}(u_1)\, M^{\varpi'}_{x_1}(u_1). \tag{14.94}$$

That is, l_1 and l_2 transition independently to ϖ and ϖ', respectively.

Sixth, denote:

$$\bar{q}^{k-1|k-1}_{\varpi,\varpi'} = 1 - f^{k-1|k-1}_{\varpi,\varpi'}(\emptyset) = \tilde{q}^{k-1|k-1}_{\varpi,\varpi'} + \tilde{q}^{k-1|k-1}_{\varpi',\varpi} + q^{k-1|k-1}_{\varpi,\varpi'} \leq 1. \tag{14.95}$$

Abbreviate $p^{\varpi}_S(u) = p^{\varpi}_{S,k|k-1}(u)$ and $p^{\varpi'}_S(u) = p^{\varpi'}_{S,k|k-1}(u)$. Then, in Section 14.7, it is shown that:

1. Time-update for $\tilde{D}^{k|k-1}_{\varpi,\varpi'}(x)$:

$$\tilde{D}^{k|k-1}_{\varpi,\varpi'}(x) = \tilde{B}^{k|k-1}_{\varpi,\varpi'}(x)\,(1 - \bar{q}^{k-1|k-1}_{\varpi,\varpi'}) \tag{14.96}$$
$$+\tilde{D}^{k-1|k-1}_{\varpi,\varpi'}[p^{\varpi}_S M^{\varpi}_x] + D^{k-1|k-1}_{\varpi,\varpi'}[p^{\varpi}_S M^{\varpi}_x (1 - p^{\varpi'}_S)].$$

2. Time-update for $\tilde{D}^{k|k-1}_{\varpi',\varpi}(x)$:

$$\tilde{D}^{k|k-1}_{\varpi',\varpi}(x) = \tilde{B}^{k|k-1}_{\varpi',\varpi}(x)\,(1 - \bar{q}^{k-1|k-1}_{\varpi,\varpi'}) \tag{14.97}$$
$$+\tilde{D}^{k-1|k-1}_{\varpi',\varpi}[p^{\varpi'}_S M^{\varpi'}_x] + D^{k-1|k-1}_{\varpi,\varpi'}[(1 - p^{\varpi}_S) p^{\varpi'}_S M^{\varpi'}_x)].$$

3. Time-update for $D^{k|k-1}_{\varpi,\varpi'}(x_1, x_2)$:

$$D^{k|k-1}_{\varpi,\varpi'}(x_1, x_2) = B^{k|k-1}_{\varpi,\varpi'}(x_1, x_2)\,(1 - \bar{q}^{k-1|k-1}_{\varpi,\varpi'}) + D^{k-1|k-1}_{\varpi,\varpi'}[p^{\varpi}_S M^{\varpi}_{x_1} p^{\varpi'}_S M^{\varpi'}_{x_2}]. \tag{14.98}$$

4. Well-definedness: Let $\tilde{q}^{k|k-1}_{\varpi,\varpi'} = \int \tilde{D}^{k|k-1}_{\varpi,\varpi'}(x)dx$, $\tilde{q}^{k|k-1}_{\varpi',\varpi} = \int \tilde{D}^{k|k-1}_{\varpi',\varpi}(x)dx$, and $q^{k|k-1}_{\varpi,\varpi'} = \int D^{k|k-1}_{\varpi,\varpi'}(x_1, x_2)dx_1dx_2$. Then, as must be the case if $\tilde{q}^{k|k-1}_{\varpi,\varpi'}$, $\tilde{q}^{k|k-1}_{\varpi',\varpi}$, and $q^{k|k-1}_{\varpi,\varpi'}$ have been well-defined,

$$\tilde{q}^{k|k-1}_{\varpi,\varpi'} + \tilde{q}^{k|k-1}_{\varpi',\varpi} + q^{k|k-1}_{\varpi,\varpi'} \leq 1. \tag{14.99}$$

Note that, as required, Eqs. (14.96–14.98) reduce to [3, Eqs. (50, 52, 54)] when $\tilde{B}^{k|k-1}_{\varpi,\varpi'}(x)$, $\tilde{B}^{k|k-1}_{\varpi,\varpi'}(x)$, and $B^{k|k-1}_{\varpi',\varpi}(x_1, x_2)$ all vanish identically.

14.5.3 Dyadic Filter Measurement-update

This is given by [3, Eq. (68)]:

$$f^{k|k}_{\varpi,\varpi'}(X) = K^{-1}_{Z_{1:k}}\, \hat{L}_{Z_k}(X)\, f^{k|k-1}_{\varpi,\varpi'}(X) \text{ where} \tag{14.100}$$

$$K_{Z_{1:k}} = \int \hat{L}_{Z_k}(X)\, f^{k|k-1}_{\varpi,\varpi'}(X)\delta X \tag{14.101}$$

and thus equivalently by the following equations:

$$f^{k|k}_{\varpi,\varpi'}(\emptyset) = K^{-1}_{Z_{1:k}}\, \hat{L}_{Z_k}(\emptyset)\, f^{k|k-1}_{\varpi,\varpi'}(\emptyset) = K^{-1}_{Z_{1:k}}\, \hat{L}_{Z_k}(\emptyset)\,(1 - \bar{q}^{k|k-1}_{\varpi,\varpi'}) \tag{14.102}$$
$$\tilde{D}^{k|k}_{\varpi,\varpi'}(x) = K^{-1}_{Z_{1:k}}\, \hat{L}_{Z_k}(\{(x, \varpi)\})\, \tilde{D}^{k|k-1}_{\varpi,\varpi'}(x) \tag{14.103}$$
$$\tilde{D}^{k|k}_{\varpi',\varpi}(x) = K^{-1}_{Z_{1:k}}\, \hat{L}_{Z_k}(\{(x, \varpi')\})\, \tilde{D}^{k|k-1}_{\varpi',\varpi}(x) \tag{14.104}$$
$$\tilde{D}^{k|k}_{\varpi,\varpi'}(x_1, x_2) = K^{-1}_{Z_{1:k}}\, \hat{L}_{Z_k}(\{(x_1, \varpi), (x_2, \varpi')\})\, D^{k|k-1}_{\varpi,\varpi'}(x_1, x_2). \tag{14.105}$$

Note that $\hat{L}_{Z_k}(\emptyset) = 1$ since $L_{Z_k}(\emptyset) = \kappa_k(Z_k)$ and $\hat{L}_{Z_k}(\emptyset) = L_{Z_k}(\emptyset)/\kappa_k(Z_k)$. It follows from this and Eq. (14.11) that:

$$\begin{aligned} K_{Z_{1:k}} &= 1 - \bar{q}^{k|k-1}_{\varpi,\varpi'} \\ &+ \int \hat{L}_{Z_k}(\{(x,\varpi)\})\, \tilde{D}^{k|k-1}_{\varpi,\varpi'}(x)dx \\ &+ \int \hat{L}_{Z_k}(\{(x,\varpi')\})\, \tilde{D}^{k|k-1}_{\varpi',\varpi}(x)dx \\ &+ \int \hat{L}_{Z_k}(\{(x_1,\varpi),(x_2,\varpi')\})\, D^{k|k-1}_{\varpi,\varpi'}(x_1,x_2)dx_1dx_2 \end{aligned} \tag{14.106}$$

where:

1. *Case 1*: $X = \{(u,l)\}$ with $l \in \{\varpi, \varpi'\}$ [3, Eq. (69)]:

$$\hat{L}_{Z_k}(X) = 1 - p_{D,k}(u,l) + p_{D,k}(u,l) \sum_{z \in Z_k} \frac{L_z(u,l)\, \kappa_k(Z_k - \{z\})}{\kappa_k(Z_k)} \tag{14.107}$$

   and where the summation vanishes if $Z_k = \emptyset$.
2. *Case 2*: $X = \{(u_1,l_1),(u_2,l_2)\}$ with $\{l_1,l_2\} = \{\varpi,\varpi'\}$ [3, Eq. (70)]:

$$\begin{aligned} &\hat{L}_{Z_k}(X) \\ &= (1 - p_{D,k}(u_1,l_1))(1 - p_D(u_2,l_2)) \\ &\quad + p_{D,k}(u_1,l_1)(1 - p_{D,k}(u_2,l_2)) \sum_{z \in Z_k} \frac{L_z(u_1,l_1)\, \kappa_k(Z_k - \{z\})}{\kappa_k(Z_k)} \\ &\quad + (1 - p_{D,k}(u_1,l_1))\, p_{D,k}(u_2,l_2) \sum_{z \in Z_k} \frac{L_z(u_2,l_2)\, \kappa_k(Z_k - \{z\})}{\kappa_k(Z_k)} \\ &\quad + p_{D,k}(u_1,l_1)\, p_{D,k}(u_2,l_2) \sum_{z \neq z' \in Z_k} \frac{L_{\{z,z'\}}(u_1,l_1,u_2,l_2)\, \kappa_k(Z_k - \{z,z'\})}{\kappa_k(Z_k)} \end{aligned} \tag{14.108}$$

   where

$$L_{\{z,z'\}}(u_1,l_1,u_2,l_2) = L_z(u_1,l_1)\, L_{z'}(u_2,l_2) + L_{z'}(u_1,l_1)\, L_z(u_2,l_2) \tag{14.109}$$

   and where the third summation is taken over all $z, z' \in Z_k$ with $z \neq z'$; and where all of the summations vanish if $Z_k = \emptyset$.

The computational complexity of the dyadic filter is dominated by the third summation in Eq. (14.108), which has $|Z_k|\,(|Z_k| - 1)/2$ terms.

14.5.4 Dyadic Filter State Estimation

The JoM (Joint Multi-target) and MaM (Marginal Multi-target) multi-target state estimators [1, pp. 497–500] are provably Bayes-optimal. When restricted to the case $|X| \leq 2$ they reduce to [3, Eqs. (76–81)].[8]

8 The Bernoulli filter estimator of Section 14.3.1.3 is a MaM estimator restricted to the case $|X| \leq 1$.

14.6 Set-valued Bernoulli Filters

Conventional single-target tracking algorithms are recursive point estimators whose input data consists of point measurements. The STRB filter and its special case, the Kalman filter, are the most prominent examples.

Less familiar approaches employ recursive *set estimators* whose input data can be either points or sets. Early examples are set-valued Kalman filters that presume that the state variable is a closed ellipse or ellipsoid—which, in turn, is interpreted as a geometrical bound containing the actual point state. Instances include [21–24]. Especially notable is [21], in which measurement and motion noise are modeled using deterministic spherical constraints rather than probabilistic constraints $z \sim f_{\mathbf{W}_k}$ and $x \sim f_{\mathbf{V}_{k|k-1}}$.

More recently, a Bayes-optimal generalized Kalman filter (the Kalman evidential filter or KEF) was devised to track fuzzy Dempster-Shafer (FDS) targets using FDS measurements [1, pp. 199–309]. Another filter allows optimal tracking of fuzzy targets using fuzzy measurements when both measurement and motion noise are fuzzy (see Section 14.6.10).

In [4], it was demonstrated that these and other approaches can be subsumed as special cases of a general algorithm—the *recursive RCS propagator*—that allows direct propagation of single RCS targets using single RCS measurements.

The RCS propagator's time-update and measurement-update formulas are conceptually and algebraically simple—see Eqs. (14.125, 14.126). They can be further extended to incorporate RCS models of ambiguously known measurement noise and motion noise—see Eqs. (14.154, 14.155).

The primary purpose of this section is to summarize the RCS propagator, along with a few specific single-target trackers based on it.

Organization of the section: RCSs and RCS uncertainty models (Section 14.6.1); examples of RCSs (Section 14.6.2); GLFs (Section 14.6.3); Bayes filtering with RCS measurements (Section 14.6.4); the RCS propagator (Section 14.6.5); the OPCF propagator (Section 14.6.6); the RCS probability distribution function (PDF) filter (Section 14.6.7); the FDS recursive filter (Section 14.6.8); the RCS propagator for ambiguous motion and measurement noise (Section 14.6.9); and the fuzzy recursive propagator (Section 14.6.10).

14.6.1 RCSs and RCS Uncertainty Models

Theoretically grounded single-target tracking algorithms typically propagate probability distributions $f_{\mathbf{X}_{1|1}}(x|z_1)$,..., $f_{\mathbf{X}_{k|k}}(x|z_{1:k})$. The Kalman filter, which propagates state vectors and their covariance matrices using measurement vectors, is an exception. So is the RCS propagator, which propagates RCS states using RCS measurements:

RCS State: a nonempty RCS[9] $\Gamma \subseteq \mathbb{X}_0$. The actual random state $\mathbf{X} \in \mathbb{X}_0$ is assumed to be contained by Γ, in such a manner that Γ specifies (somewhat like a covariance) its allowable realizations. It is a bounding constraint $\Gamma \ni \mathbf{X}$.

RCS Measurement: a nonempty RCS $\Theta \subseteq \mathbb{Z}$. It is assumed to contain the actual random measurement $\mathbf{Z} \in \mathbb{Z}$ and also (again somewhat like a covariance) to constrain its allowable realizations: $\Theta \ni \mathbf{Z}$.

9 An RCS Γ is nonempty if none of its realizations are the empty set.

14.6.2 Examples of RCSs

As noted earlier, RCSs are useful in that they provide probabilistic—indeed, Bayesian—formulations of various expert-systems concepts. Here we address only the two most pertinent for what follows: fuzzy membership functions (FMFs) and fuzzy Dempster-Shafer basic mass assignments (FDSBMAs).

14.6.2.1 RCS Representation of FMFs

Let $\mathbb{Y}$ be a space and $h : \mathbb{Y} \to [0, 1]$ a fuzzy membership function (FMF). If $h(y) = 1$ for some $y \in \mathbb{Y}$ then h is said to be "normal" and is sometimes called a "possibility distribution." Let $A \in [0, 1]$ be a uniformly distributed RV. Then

$$\Sigma_A(h) = \{y \in \mathbb{Y} : A \leq h(y)\} \subseteq \mathbb{Y} \tag{14.110}$$

is the RCS representation of h [1, Eq. (4.21)], [7, Eq. (22.36)]. It is easily shown that the OPCF of $\Sigma_A(h)$ is (Eq. (14.13, [1, Eq. (4.22)], [7, Eq. (22.37)]),

$$\mu_{\Sigma_A(h)}(y) = h(y). \tag{14.111}$$

The Zadeh fuzzy logic is defined by $(h_1 \wedge h_2)(y) = \min\{h_1(t), h_2(y)\}$, $(h_1 \vee h_2)(y) = \max\{h_1(t), h_2(y)\}$, and $h^c(y) = 1 - h(y)$. The RCS representation is mostly consistent with it in the sense that $\Sigma_A(1) = \mathbb{Y}$, $\Sigma_A(0) = \emptyset$, and [1, p. 126]

$$\Sigma_A(h_1 \wedge h_2) = \Sigma_A(h_1) \cap \Sigma_A(h_2) \tag{14.112}$$

$$\Sigma_A(h_1 \vee h_2) = \Sigma_A(h_1) \cup \Sigma_A(h_2). \tag{14.113}$$

However, $\Sigma_A(h^c) \neq \Sigma_A(h)^c$.

Remark 14.8: The "generalized fuzzy sets" of Y. Li have RCS representations that are completely consistent with Zadeh logic—see [1, Sec. 4.4].

14.6.2.2 RCS Representation of FDSBMAs

An FDSBMA on $\mathbb{Y}$ is a function $m(h) \geq 0$ defined on the FMFs $h : \mathbb{Y} \to [0, 1]$ of $\mathbb{Y}$ such that $m(h) \neq 0$ for only a finite number of the h's (the "focal fuzzy subsets" of m) and such that $\sum_h m(h) = 1$. Let $h_1, \ldots, h_n$ be the focal fuzzy subsets and let $J \in \{1, \ldots, n\}$ be an RV with $\Pr(J = j) = m(h_j) = m_j$. If $h_1, \ldots, h_n$ are normal then h_J is a "random possibility distribution" (RPD); in which case FDSBMAs and RPDs differ only by a change of notation and terminology.

If A, J are independent then the RCS representation of m is [7, Eq. (22.51)]:

$$\Sigma_A(h_J) = \{y \in \mathbb{Y} : A \leq h_J(y)\}. \tag{14.114}$$

It is easily shown that

$$\mu_{\Sigma_A(h_J)}(y) = \sum_{j=1}^{n} m_j \, h_j(y). \tag{14.115}$$

FDSBMAs on $\mathbb{Z}$ can be used to mathematically represent natural-language statements such as [1, Eq. (3.34)], [7, Eq. (22.50)]:

"Gustav is probably near the tower,
but it could be the smokestack, (14.116)
it's so foggy I can't say for sure."

This statement consists of three hypotheses regarding Gustav's location. They can be mathematically represented as normal FMFs $g_1(z)$, $g_2(z)$, $g_3(z)$, with g_1 corresponding to the tower, g_2 corresponding to the smokestack, and $g_3(z) = 1$ (the "null hypothesis") representing complete uncertainty. The statement itself has the form $g_1(z)$ OR $g_2(z)$ OR $g_3(z)$. It thus can be represented as an FDSBMA with fuzzy focal sets $g_1(z)$, $g_2(z)$, $g_3(z)$ and subsequently as an RCS measurement. See [1, pp. 103–106] for more detail.

14.6.3 Generalized Likelihood Functions

Recall Eqs. (14.4–14.7). If Θ is an RCS measurement, its *generalized likelihood function* (GLF) is [1, Eq. (3.34)], [7, Eq. (22.88)]:

$$\rho_k(\Theta|x) = \Pr(\mathbf{Z}_k \in \Theta | \mathbf{X}_{k|k-1} = x) \tag{14.117}$$

$$= \Pr(\eta_k(x) \in \Theta - \mathbf{W}_k) \tag{14.118}$$

$$= \int \mu_\Theta(z)\, f_k(z|x) dz \tag{14.119}$$

where $\Theta - \mathbf{W}_k \overset{\text{abbr.}}{=} \{z - \mathbf{W}_k : z \in \Theta\}$ and where Eq. (14.119) is true in the case that Θ, $\mathbf{W}_k$ are independent.

Suppose that $f_{\mathbf{W}_k}(z) = \delta_0(z)$ where $\delta_0(z)$ is the Dirac delta function concentrated at $0 \in \mathbb{Z}$. Then the GLF reduces to

$$\rho_k(\Theta|x) = \mu_\Theta(\eta_k(x)). \tag{14.120}$$

See Eq. (14.137) for a concrete example.

The complete RCS measurement is actually $\Theta - \mathbf{W}_k$ and not just Θ. The RV $\mathbf{W}_k$ models its stochastic uncertainty. The RCS Θ, on the other hand, models its "epistemic uncertainty"—i.e., the uncertainty involved in constructing a mathematical model of the "nontraditional measurement" represented by Θ (e.g., quantized data, attributes, features, natural language statements, inference rules).

Thus $\Theta - z$ for $z \in \mathbb{Z}$ are the realizations of the RV $\Theta - \mathbf{W}_k$ and are therefore actually deterministic. In particular, the "noiseless" RCS Θ for $z = 0$ is deterministic. Since stochastic and epistemic uncertainty should be uncorrelated, Θ, $\mathbf{W}_k$ can be assumed to be independent. See [1, Sec. 3.3] for more detail.

14.6.4 Bayes Filtering With RCS Measurements

Let us be given a prior distribution $f_0(x)$, the measurement model $\mathbf{Z} = \eta(x) + \mathbf{W}$, and a noisy RCS measurement $\Theta - \mathbf{W}$. Then it can be shown [7, Sec. 22.3.4] that the probability density defined by $f_+(x) \propto \rho(\Theta|x)\, f_0(x)$ is actually the posterior probability density $f_+(x) = f(x|\mathbf{Z} \in \Theta) \overset{\text{abbr.}}{=} f(x|\Theta)$.

This fact leads to the following generalization of Eqs. (14.1–14.3)—a fully Bayesian and provably Bayes-optimal [7, pp. 779–782] STRB filter for RCS measurements [1, p. 116], [7, Eqs. (22.82, 22.83)]:[10]

$$f_{k|k-1}(x|\Theta_{1:k-1}) = \int f_{k|k-1}(x|u)\, f_{k-1|k-1}(u|\Theta_{1:k-1}) du \tag{14.121}$$

$$f_{k|k}(x|\Theta_{1:k}) \propto \rho_k(\Theta_k|x)\, f_{k|k-1}(x|\Theta_{1:k-1}) \tag{14.122}$$

10 This is the only filter in Section 14.6 that does not arise from the RCS propagator.

where $\Theta_{1:k} : \Theta_1, ..., \Theta_k$ is a sequence of RCS measurements; where $f_{k|k-1}(x|\Theta_{1:k-1}) = f_{\mathbf{X}_{k|k-1}}(x)$, $f_{k|k}(x|\Theta_{1:k}) = f_{\mathbf{X}_{k|k}}(x)$; and where $\mathbf{X}_{k|k-1}, \mathbf{X}_{k|k} \in \mathbb{X}_0$ denote the unknown random target states.

Using Eqs. (14.121, 14.122), single point targets can be tracked using "nontraditional measurements"—e.g., quantized data, attributes, features, natural language statements, and inference rules [1, Chs. 3–8, pp. 726–732]. The technique can be extended to the multi-target case [7, pp. 805–823], including the Bernoulli filter [7, Sec. 22.10.4].

Remark 14.9: The paper [25] reports a practical implementation of Eq. (14.122) when the (noiseless) RCS measurements Θ model natural-language statements (including fuzzy inference rules and sentences such as Eq. [14.116]).

Remark 14.10: If $L_z = \sup_{x \in \mathbb{X}_0} f_k(z|x) < \infty$ define [4, p. 68665, col. 2]

$$\rho_k(z|x) = f_k(z|x)/L_z \leq 1. \tag{14.123}$$

Given this, Eq. (14.122) can be applied to point measurements $z \in \mathbb{Z}$; in which case it reduces to the STRB filter measurement-update, Eq. (14.2). More generally, conventional measurement densities $f_k(z|x)$ can be regarded as limiting cases of GLFs [4, Eq. (9)].

14.6.5 The RCS Propagator

At time t_k the unknown state is an RCS $\Gamma_{k|k} \subseteq \mathbb{X}_0$ and the measurement collected from it is an RCS $\Theta_k \subseteq \mathbb{Z}$. The conditional RCS state, given the RCS measurement-sequence $\Theta_{1:k} : \Theta_1, ..., \Theta_k$ of collected RCS measurements, is denoted $(\Gamma_{k|k}|\Theta_{1:k})$. The recursive RCS propagator has the form

$$... \to (\Gamma_{k-1|k-1}|\Theta_{1:k-1}) \to (\Gamma_{k|k-1}|\Theta_{1:k-1}) \to (\Gamma_{k|k}|\Theta_{1:k}) \to ... \tag{14.124}$$

Here it is assumed that $\Gamma_{k|k-1} \neq \emptyset$ and $\Gamma_{k|k} \neq \emptyset$ for all k—i.e., that the target always exists (which, note, the STRB and Kalman filters also presume).

The *RCS propagator time-update and measurement-update equations* with initial RCS state $\Gamma_{0|0}$ are [4, Eqs. (30, 31)]:

$$\Gamma_{k|k-1} = \varphi_{k|k-1}(\Gamma_{k-1|k-1}) + \mathbf{V}_{k|k-1} \tag{14.125}$$

$$\Gamma_{k|k} = \Gamma_{k|k-1} \cap \eta_k^{-1}(\Theta_k - \mathbf{W}_k) \tag{14.126}$$

where

$$\varphi_{k|k-1}(\Gamma_{k-1|k-1}) + \mathbf{V}_{k|k-1} = \{\varphi_{k|k-1}(x) + \mathbf{V}_{k|k-1} : x \in \Gamma_{k-1|k-1}\} \tag{14.127}$$

$$\Theta_k - \mathbf{W}_k = \{z - \mathbf{W}_k : z \in \Theta_k\} \tag{14.128}$$

$$\eta_k^{-1}(\Theta_k - \mathbf{W}_k) = \{x \in \mathbb{X}_0 : \eta_k(x) \in \Theta_k - \mathbf{W}_k\}. \tag{14.129}$$

Eqs. (14.125, 14.126) have the following intuitive interpretations:

1. $\varphi_{k|k-1}(\Gamma_{k-1|k-1}) + \mathbf{V}_{k|k-1}$ is the RCS of all possible predicted states $\varphi_{k|k-1}(x)$ that originate from $x \in \Gamma_{k-1|k-1}$, all of which are identically randomly perturbed by $\mathbf{V}_{k|k-1}$.
2. $\Gamma_{k|k-1} \cap \eta_k^{-1}(\Theta_k - \mathbf{W}_k)$ is the RCS of all of the states $x \in \Gamma_{k|k}$ that are (1) predicted states $x \in \Gamma_{k|k-1}$ and (2) can also feasibly generate a measurement constrained by Θ—i.e., $\mathbf{Z}_k = \eta_k(x) + \mathbf{W}_k \in \Theta_k$.

Remark 14.11: Section V of [4] sketches approaches for approximately implementing the RCS propagator by assuming that RCSs are ellipsoidal.

Remark 14.12: The mathematical preconditions for the validity of Eqs. (14.125, 14.126) are listed in the first paragraph of Section III-A of [4]. If $\mathbb{X}_0$ is Euclidean then any $\varphi_{k|k-1} : \mathbb{X}_0 \to \mathbb{X}_0$ that is invertible and physically reasonable will be theoretically valid. Also, the assumption $\Gamma_{k|k} \neq \emptyset$ implies that the target is always detected (which the STRB and Kalman filters also presume).

14.6.6 The OPCF Propagator

It can be shown that the RCS propagator equations induce the following algebraically simple propagation equations for the corresponding OPCFs [4, Eqs. (32, 33)]:

$$\mu_{\Gamma_{k|k-1}}(x) = \int \mu_{\Gamma_{k-1|k-1}}(\varphi_{k|k-1}^{-1}(x-u))\, f_{\mathbf{V}_{k|k-1}}(u)du \tag{14.130}$$

$$\mu_{\Gamma_{k|k}}(x) = \rho_k(\Theta_k|x)\, \mu_{\Gamma_{k|k-1}}(x) \tag{14.131}$$

where Eq. (14.131) follows from Eq. (14.14). Note that if $f_{\mathbf{V}_{k|k-1}}(x) = \delta_0(x)$ then Eq. (14.130) reduces to the intuitively reasonable $\mu_{\Gamma_{k|k-1}}(x) = \mu_{\Gamma_{k-1|k-1}}(\varphi_{k|k-1}^{-1}(x))$.

14.6.7 The RCS PDF Propagator

Equations (14.130, 14.131) lead to a recursive Bayes filter for point targets and RCS measurements different than Eqs. (14.121, 14.122). Assume $\int \mu_{\Gamma_{k|k-1}}(x)dx < \infty$, $\int \mu_{\Gamma_{k|k}}(x)dx < \infty$ for all $k \geq 1$. Then $\Gamma_{k|k}$ induces a probability density function (PDF):

$$f_{k|k}(x|\Theta_{1:k}) = f_{\Gamma_{k|k}}(x) = \frac{\mu_{\Gamma_{k|k}}(x)}{\int \mu_{\Gamma_{k|k}}(x)dx} \tag{14.132}$$

and likewise for $f_{k|k-1}(x|\Theta_{1:k-1})$.

Equation (14.132) has the following interpretation. There are actually an infinite number of RVs $\mathbf{X}_{k|k} \in \mathbb{X}_0$ such that $\mathbf{X}_{k|k} \in \Gamma_{k|k}$ [9, p. 32]—and not just the particular $\hat{\mathbf{X}}_{k|k} \in \Gamma_{k|k}$ that the RCS state $\Gamma_{k|k}$ is assumed to imprecisely constrain. Because $\mu_{\Gamma_{k|k}}(x) = \Pr(x \in \Gamma_{k|k})$ for all $x \in \mathbb{X}_0$, the PDF $f_{\Gamma_{k|k}}(x)$ can be interpreted as an estimate of the PDF $f_{\hat{\mathbf{X}}_{k|k}}(x)$ of $\hat{\mathbf{X}}_{k|k}$.

It can be shown that the RCS PDF propagator has the following time-update and measurement-update equations [4, Eqs. (69, 66)]:

$$f_{k|k-1}(x|\Theta_{1:k-1}) \propto \int f_{k-1|k-1}(\varphi_{k|k-1}^{-1}(x-u)|\Theta_{1:k-1})\, f_{\mathbf{V}_{k|k-1}}(u)du \tag{14.133}$$

$$\propto \int f_{k|k-1}(x|u)\, f_{k-1|k-1}(u|\Theta_{1:k-1})\, |J_{\varphi_{k|k-1}}(u)|du \tag{14.134}$$

$$f_{k|k}(x|\Theta_{1:k}) \propto \rho_k(\Theta_k|x)\, f_{k|k-1}(x|\Theta_{1:k-1}) \tag{14.135}$$

where $|J_{\varphi_{k|k-1}}(u)|$ is the Jacobian determinant of $\varphi_{k|k-1}$.

If $\varphi_{k|k-1}(x) = F_{k|k-1}x$ is linear with nonsingular matrix $F_{k|k-1}$ then $|J_{\varphi_{k|k-1}}(u)| = \det F_{k|k-1}$ is constant and Eq. (14.134) reduces to Eq. (14.121):

$$f_{k|k-1}(x|\Theta_{1:k-1}) = \int f_{k|k-1}(x|u)\, f_{k-1|k-1}(u|\Theta_{1:k-1})du. \tag{14.136}$$

Remark 14.13: If Eq. (14.134) is replaced by Eq. (14.136) then it is the conventional prediction integral, Eq. (14.121). Eqs. (14.134, 14.136) are both theoretically valid but are semantically different. Equation (14.134) describes the evolution of $f_{\Gamma_{k|k-1}}(x)$ to $f_{\Gamma_{k|k}}(x)$ whereas Eq. (14.121) describes the evolution of $f_{\hat{\mathbf{X}}_{k|k-1}}(x)$ to $f_{\hat{\mathbf{X}}_{k|k}}(x)$. In particular: If $\Gamma_{k|k-1} = \{\hat{\mathbf{X}}_{k|k-1}\}$ and $\Gamma_{k|k} = \{\hat{\mathbf{X}}_{k|k}\}$ then $f_{\Gamma_{k|k-1}}(x)$, $f_{\Gamma_{k|k}}(x)$ do not even exist since $\mu_{\Gamma_{k|k-1}}(x) = \mu_{\Gamma_{k|k}}(x) = 0$ identically. For the same reason, Eqs. (14.135, 14.122) are semantically different even though algebraically identical.

14.6.8 The FDS Recursive Filter

Equation (14.116) is an example of a *noiseless FDS measurement*. In general, an FDS measurement is an FDSBMA $\upsilon : \mathbb{Z} \to [0,1]$ such that $\upsilon(g) \neq 0$ for only a finite number of the g's (the "focal fuzzy subsets" of m) and such that $\sum_g \upsilon(g) = 1$. Let $g_1, \ldots, g_m$ be the focal fuzzy subsets and let $J \in \{1, \ldots, m\}$ be an RV with $\Pr(J = j) = \upsilon(g_j) = \upsilon_j$. Then $\Theta = \Sigma_A(g_J)$ is the RCS measurement that represents the FDS measurement υ.

From Eq. (14.120) the GLF of Θ is [1, Eqs. (5.74–5.77)], [7, Eq. (22.77)]:

$$\rho_k(\Theta|x) = \sum_{j=1}^{m} \upsilon(g_j) \int g_j(z)\, f_k(z|x)dz \tag{14.137}$$

$$= \sum_{j=1}^{m} \upsilon(g_j)\, g_j(\eta_k(x)). \tag{14.138}$$

where Eq. (14.138) follows if Θ is noiseless—i.e., $f_{\mathbf{W}_k}(z) = \delta_0(z)$.

A *noiseless FDS state* is a function $w(h) \geq 0$ defined on FMFs $h : \mathbb{X}_0 \to [0,1]$ such that (1) $w(h) > 0$ for only a finite number of h; (2) $\sum_h w(h) = 1$; and (3) $|h| = \int h(x)dx < \infty$ if $w(h) > 0$. The intuitive meaning of $w(h)$ is this: it is a ranked list $h_1, \ldots, h_n$ of fuzzy constraints on the unknown point state x with ranked weights $w_i = w(h_i)$. The RCS model of w is as in Eq. (14.114): $\Gamma = \Sigma_A(h_J)$.

Given this and Eq. (14.132), $w(h)$ induces a PDF

$$f_w(x) = f_{\Sigma_A(h_J)}(x) = \frac{\sum_{i=1}^{n} w(h_i)\, h_i(x)}{\sum_{j=1}^{n} w(h_j)\, |h_j|}. \tag{14.139}$$

The *FDS filter* is a special case of the RCS PDF propagator of Section 14.6.7 when the RCS states and RCS measurements are, respectively, representations of noiseless FDS measurements and noiseless FDS states. It propagates the distributions of FDS states $w^{k|k} = \{(w_i^{k|k}, h_i^{k|k})\}_{i=1}^{n_{k|k}}$ (as defined in Eq. [14.139]):

$$\ldots \to f_{w^{k-1|k-1}}(x|\upsilon^{1:k-1}) \to f_{w^{k|k-1}}(x|\upsilon^{1:k-1}) \to f_{w^{k|k}}(x|\upsilon^{1:k}) \to \ldots \tag{14.140}$$

or, more succinctly,

$$\ldots \to (w^{k-1|k-1}|\upsilon^{1:k-1}) \to (w^{k|k-1}|\upsilon^{1:k-1}) \to (w^{k|k}|\upsilon^{1:k}) \to \ldots \tag{14.141}$$

FDS Filter Time-update: Let the FDS state at time t_{k-1} be $n_{k-1|k-1}$, $w_i^{k-1|k-1}$, and $h_i^{k-1|k-1}(x)$ where $n_{k-1|k-1}$ is the number of focal FMFs and $w_i^{k-1|k-1}$ are the weights of the focal FMFs $h_i^{k-1|k-1}$ for $i = 1, \ldots, n_{k-1|k-1}$. Then from Eq. (14.133), the time-update can be shown to have the simple form [4, Eqs. (81–83)]:[11]

11 *Errata*: [4, Eq. (83)] had a serious typo. It is corrected in the current Eq. (14.144).

$$n_{k|k-1} = n_{k-1|k-1} \tag{14.142}$$

$$w_i^{k|k-1} = w_i^{k-1|k-1} \tag{14.143}$$

$$h_i^{k|k-1}(x) = \int h_i^{k-1|k-1}(\varphi_{k|k-1}^{-1}(x-u))\, f_{\mathbf{V}_{k|k-1}}(u)du \leq 1. \tag{14.144}$$

Note that if $f_{\mathbf{V}_{k|k-1}}(x) = \delta_0(x)$ then $h_i^{k|k-1}(x) = h_i^{k-1|k-1}(\varphi_{k|k-1}^{-1}(x))$.

When applied to Eq. (14.139), the alternative time-update Eq. (14.121) leads to the original FDS filter time-update as presented in [1, pp. 726–728], [4, Eqs. (15–18)].

FDS Filter Measurement-update: Since Eqs. (14.135, 14.122) are algebraically identical, the FDS measurement-update is algebraically identical to that for the original FDS filter [4, Eqs. (19–20)].

Let the FDS state at time t_{k-1} be $n_{k|k-1}$, $w_i^{k|k-1}$, and $h_i^{k|k-1}(x)$ where $n_{k|k-1}$ is the number of focal FMFs and $w_i^{k|k-1}$ are the weights of the focal FMFs $h_i^{k|k-1}$ for $i = 1, \ldots, n_{k|k-1}$. Let the noiseless FDS measurement $\upsilon^k(g)$ be collected and let $g_1^k, \ldots, g_{m_k}^k$ be its focal FMFs with respective weights $\upsilon_1^k, \ldots, \upsilon_{m_k}^k$. Then [4, Eq. (74)]:

$$n_{k|k} = n_{k|k-1}\, m_k \tag{14.145}$$

$$w_{i,j}^{k|k} = w_i^{k|k-1}\, \upsilon_j^k \tag{14.146}$$

$$h_{i,j}^{k|k}(x) = h_i^{k|k-1}(x)\, g_j^k(\eta_k(x)). \tag{14.147}$$

These equations can be compactly rewritten as [4, Eq. (24)]

$$w^{k|k} = (\eta_k^{-1}\upsilon^k) * w^{k|k-1} \tag{14.148}$$

where $\eta_k^{-1}\upsilon^k$ is the FDS state defined by [4, Eq. (25)]

$$(\eta_k^{-1}\upsilon^k)(h) = \sum_{g:\eta_k \circ g = h} \upsilon^k(g), \; (\eta_k \circ g)(x) = g(\eta_k(x)) \tag{14.149}$$

and where the FDS version of Dempster's rule of combination is [4, Eq. (22)]

$$(w_1 * w_2)(h'') = \alpha_{w_1,w_2}^{-1} \sum_{h,h':h\cdot h'=h''} w_1(h)\, w_2(h') \tag{14.150}$$

$$(h \cdot h')(x) = h(x)\, h'(x), \tag{14.151}$$

assuming that the "FDS agreement" is nonzero [4, Eq. (23)]:

$$\alpha_{w_1,w_2} = \sum_{h,h':h\cdot h' \neq 0} w_1(h)\, w_2(h') \neq 0 \tag{14.152}$$

and where $h \cdot h' \neq 0$ means that $h(x)\, h'(x) \neq 0$ for some $x \in \mathbb{X}_0$.

The KEF [1, pp. 205–208] is an exact closed-form implementation of the time-update [1, pp. 726–728]) and Eqs. (14.145–14.147), when all focal state and measurement FMFs have the normalized-Gaussian form [4, Eq. (27)]:

$$\hat{N}_C(x) = \exp\left(-\frac{1}{2}x^T C^{-1} x\right). \tag{14.153}$$

Remark 14.14: Algebraically speaking, the KEF reduces to the Kalman filter if (1) the measurements are points, (2) the initial FDS state is unifocal (i.e., $n_{0|0} = 1$) and (3) the GLF is chosen as in Eq. (14.123) [4], p. 18666, col. 2], [1, p. 209]. However, the semantics are quite different. Whereas the state variable $\mathbf{X}_{k|k}$ of the Kalman filter is stochastic, the state variable $w^{k|k}(\cdot)$ of the KEF is deterministic.

14.6.9 RCS Propagator for Ambiguous Noise

Suppose that measurement and motion noises $\mathbf{V}_{k|k-1}, \mathbf{W}_k$ are ambiguous, in that $f_{\mathbf{V}_{k|k-1}}(x)$ and $f_{\mathbf{W}_k}(x)$ are imperfectly understood. This ambiguousness can be modeled by replacing the probabilistic constraints $x \sim f_{\mathbf{V}_{k|k-1}}$ and $z \sim f_{\mathbf{W}_k}$ with RCS constraints $x \in \Omega_{k|k-1} \subseteq \mathbb{X}_0$ and $z \in \Psi_k \subseteq \mathbb{Z}$.

Since $\mathbb{X}_0, \mathbb{Z}$ are Euclidean and if RCS realizations are closed and bounded, then the RCS propagator, Eqs. (14.125, 14.126), can be generalized to [4, Eqs. (92, 93)]

$$\Gamma_{k|k-1} = \varphi_{k|k-1}(\Gamma_{k-1|k-1}) + \Omega_{k|k-1} \tag{14.154}$$

$$\Gamma_{k|k} = \Gamma_{k|k-1} \cap \eta_k^{-1}(\Theta_k - \Psi_k) \tag{14.155}$$

where, for closed and bounded sets $S_1, S_2 \subseteq \mathbb{Y}$, Minkowski addition and subtraction are defined as [4, Eq. (89)]

$$S_1 + S_2 = \{y_1 + y_2 : y_1 \in S_1,\ y_2 \in S_2\} \tag{14.156}$$

$$S_1 - S_2 = \{y_1 - y_2 : y_1 \in S_1,\ y_2 \in S_2\}. \tag{14.157}$$

Since $\mathbb{Y}$ is Euclidean, $S_1 + S_2$ and $S_1 - S_2$ are closed and bounded if S_1, S_2 are closed and bounded.

Equations (14.154, 14.155) reduce to the basic RCS propagator, Eqs. (14.125, 14.126), when $\Omega_{k|k-1} = \{\mathbf{V}_{k|k-1}\}$ and $\Psi_k = \{\mathbf{W}_k\}$. Also, let $\Omega_{k|k-1} = V_{k|k-1}$ and $\Psi_k = W_k$ for fixed and bounded closed sets $V_{k|k-1} \subseteq \mathbb{X}_0$ and $W_k \subseteq \mathbb{Z}$. Then, as with the set-valued Kalman filter in [21], the stochastic constraints $x \sim f_{\mathbf{V}_{k|k-1}}$ and $z \sim f_{\mathbf{W}_k}$ are replaced by deterministic set-valued constraints $V_{k|k-1} \ni x$ and $W_k \ni z$.

14.6.10 Theoretically Rigorous Fuzzy Recursive Propagator

The GLF of Eq. (14.119) can be extended to RCS noise models Ψ_k [4, Eqs. (96–98)]:

$$\tilde{\rho}_k(\Theta|x) = \int \mu_{\Theta_k - \Psi_k}(z)\ f_k(z|x)dz. \tag{14.158}$$

Likewise for the OPCF and RCS PDF measurement-updates [4, Eqs. (99, 100)]:

$$\mu_{\Gamma_{k|k}}(x) = \tilde{\rho}_k(\Theta|x)\ \mu_{\Gamma_{k|k-1}}(x) \tag{14.159}$$

$$f_{\Gamma_{kk}}(x) \propto \tilde{\rho}_k(\Theta|x)\ f_{\Gamma_{k|k-1}}(x). \tag{14.160}$$

Unfortunately, however, no similar extensions seem to be possible in general for either the OPCF or PDF time-updates (see [4, Sec. IV-D]).

Nevertheless, it turns out that a generalization of the OPCF propagator is possible provided that the RCS measurements Θ, the RCS states Γ, and the RCS noise models Ω and Ψ are all fuzzy in the sense of Eq. (14.110):

$$\Theta_k = \Sigma_A(g_k) \tag{14.161}$$

$$\Gamma_{k|k} = \Sigma_A(h_{k|k}),\ \Gamma_{k|k-1} = \Sigma_A(h_{k|k-1}) \tag{14.162}$$

$$\Omega_{k|k-1} = \Sigma_A(\omega_{k|k-1}),\ \Psi_k = \Sigma_{A'}(\psi_k). \tag{14.163}$$

Here, $A, A' \in [0, 1]$ are uniformly distributed, independent RVs. Also, g_k is a fuzzy measurement; $h_{k|k}$ and $h_{k|k-1}$ are fuzzy states; $\omega_{k|k-1}$ is fuzzy motion noise; and ψ_k is fuzzy measurement noise.

Specifically, using Eqs. (14.112, 14.113) it can be shown that the OPCF propagator equations, Eqs. (14.130, 14.131), reduce to the following simple and algebraically exact closed-form FMF propagator equations [4, Eqs. (114–116)]:

$$h_{k|k-1}(x) = \sup_{u \in \mathbb{X}_0} \min\{h_{k-1|k-1}(\varphi^{-1}_{k|k-1}(x-u)),\ \omega_{k|k-1}(u)\} \tag{14.164}$$

$$h_{k|k}(x) = \tilde{\rho}_k(g_k|x)\, h_{k|k-1}(x) \tag{14.165}$$

$$\tilde{\rho}_k(g_k|x) = \sup_{w \in \mathbb{Z}} \min\{g_k(\eta_k(x)+w),\ \psi_k(w)\}. \tag{14.166}$$

This propagator is a theoretically rigorous fuzzy-set analog of the STRB filter, Eqs. (14.1–14.3).

14.7 Mathematical Derivations

14.7.1 Proof of Eq. (14.98)

Note that since $D^{k|k-1}_{\varpi,\varpi'}(x_1,x_2) = f^{k|k-1}_{\varpi,\varpi'}(\{(x_1,\varpi),(x_2,\varpi')\})$ we have from Eq. (14.11):

$$D^{k|k-1}_{\varpi,\varpi'}(x_1,x_2) = \int f_{k|k-1}(\{(x_1,\varpi),(x_2,\varpi')\}|U)\, f^{k-1|k-1}_{\varpi,\varpi'}(U)\delta U \tag{14.167}$$

$$= f_{k|k-1}(\{(x_1,\varpi),(x_2,\varpi')\}|\emptyset)\, f^{k-1|k-1}_{\varpi,\varpi'}(\emptyset) \tag{14.168}$$

$$+\sum_l \int f_{k|k-1}(\{(x_1,\varpi),(x_2,\varpi')\}|\{(u,l)\})\, f^{k-1|k-1}_{\varpi,\varpi'}(\{(u,l)\})du$$

$$+\frac{1}{2}\sum_{l_1\neq l_2} \int f_{k|k-1}(\{(x_1,\varpi),(x_2,\varpi')\}|\{(u_1,l_1),(u_2,l_2)\})$$

$$\cdot f^{k-1|k-1}_{\varpi,\varpi'}(\{(u_1,l_1),(u_2,l_2)\})du_1du_2$$

$$= B^{k|k-1}_{\varpi,\varpi'}(x_1,x_2)(1-\bar{q}_{k-1|k-1})$$

$$+\sum_l \int f_{k|k-1}(\{(x_1,\varpi),\{(x_2,\varpi')\}|\{(u,l)\})\, f^{k-1|k-1}_{\varpi,\varpi'}(\{(u,l)\})du \tag{14.169}$$

$$+\sum_{l_1\neq l_2} \int f_{k|k-1}(\{(x_1,\varpi),\{(x_2,\varpi')\}|\{(u_1,l_1),(u_2,l_2)\})$$

$$\cdot D^{k-1|k-1}_{l_1,l_2}(u_1,u_2)du_1du_2.$$

where

$$D^{k-1|k-1}_{l_1,l_2}(u_1,u_2) \overset{\text{abbr.}}{=} f^{k-1|k-1}_{\varpi,\varpi'}(\{(u_1,l_1),(u_2,l_2)\}). \tag{14.170}$$

The first summation must vanish identically. For, let $l=\varpi$. Then the left marginal

$$D_1(x_1,\varpi|u,\varpi) \overset{\text{abbr.}}{=} \int f_{k|k-1}(\{(x_1,\varpi),(x_2,\varpi')\}|\{(u,\varpi)\})dx_2 \tag{14.171}$$

has the form of a state transition of ϖ to itself. On the other hand, the right marginal

$$D_2(x_2,\varpi'|u,\varpi) \overset{\text{abbr.}}{=} \int f_{k|k-1}(\{(x_1,\varpi),(x_2,\varpi')\}|\{(u,\varpi)\})dx_1 \tag{14.172}$$

has the form of a state transition from ϖ to ϖ'. Since this is physically impossible, $D_2(x_2, \varpi'|u, \varpi) = 0$. The same reasoning applies when $l = \varpi'$. Thus

$$\int f_{k|k-1}(\{(x_1, \varpi), (x_2, \varpi')\}|\{(u, l)\})dx_1 dx_2 = 0 \tag{14.173}$$

from which follows

$$f_{k|k-1}(\{(x_1, \varpi), \{(x_2, \varpi')\}|\{(u, l)\}) = 0 \tag{14.174}$$

almost everywhere.

Now assume that there are no target births, in which case $B^{k|k-1}_{\varpi,\varpi'}(x_1, x_2) = 0$. Given this and Eq. (14.174), it follows that Eq. (14.167) cannot be consistent with the no-birth equations [3, Eqs. (50, 52, 54)] unless the second summation is given by Eq. (14.94):

$$\begin{aligned}&\sum_{l_1 \neq l_2} \int \delta_{l_1,\varpi}\delta_{l_2,\varpi'}\, p_S^{\varpi}(u_1)\, p_S^{\varpi'}(u_2)\, M_{x_1}^{\varpi}(u_1)\, M_{x_2}^{\varpi'}(u_2)\, D^{k-1|k-1}_{l_1,l_2}(u_1, u_2)du_1 du_2 \\ &= \int p_S^{\varpi}(u_1)\, p_S^{\varpi'}(u_2)\, M_{x_1}^{\varpi}(u_1)\, M_{x_2}^{\varpi'}(u_2)\, D^{k-1|k-1}_{\varpi,\varpi'}(u_1, u_2)du_1 du_2.\end{aligned} \tag{14.175}$$

This establishes the result.

14.7.2 Proof of Eq. (14.96)

Note that the proof that follows applies similarly to Eq. (14.97).

Begin by noting that since $\tilde{D}^{k|k-1}_{\varpi,\varpi'}(x) = f^{k|k-1}_{\varpi,\varpi'}(\{(x, \varpi)\})$ from Eq. (14.11) we have

$$\tilde{D}^{k|k-1}_{\varpi,\varpi'}(x) = \int f_{k|k-1}(\{(x, \varpi)\}|U)\, f^{k-1|k-1}_{\varpi,\varpi'}(U)\delta U \tag{14.176}$$

$$\begin{aligned}&= f_{k|k-1}(\{(x, \varpi)\}|\emptyset)\, f^{k-1|k-1}_{\varpi,\varpi'}(\emptyset) \\ &+ \sum_l \int f_{k|k-1}(\{(x, \varpi)\}|\{(u, l)\})\, \tilde{D}^{k-1|k-1}_l(u)du \\ &+ \sum_{l_1 \neq l_2} \int f_{k|k-1}(\{(x, \varpi)\}|\{(u_1, l_1), (u_2, l_2)\})\, D^{k-1|k-1}_{l_1,l_2}(u_1, u_2)du_1 du_2\end{aligned} \tag{14.177}$$

where $\tilde{D}^{k|k-1}_{\varpi,\varpi'}(x) = f^{k|k-1}_{\varpi,\varpi'}(\{(x, \varpi)\})$ and

$$\tilde{D}^{k-1|k-1}_l(u) \overset{\text{abbr.}}{=} f^{k-1|k-1}_{\varpi,\varpi'}(\{(u, l)\}) \tag{14.178}$$

$$D^{k-1|k-1}_{l_1,l_2}(u_1, u_2) \overset{\text{abbr.}}{=} f^{k-1|k-1}_{\varpi,\varpi'}(\{(u_1, l_1), (u_2, l_2)\}). \tag{14.179}$$

If this is to be consistent with the no-birth equations [3, Eqs. (50, 52, 54)], from Eqs. (14.89, 14.92, 14.95) it must be the case that:

$$\begin{aligned}\tilde{D}^{k|k-1}_{\varpi,\varpi'}(x) &= \tilde{B}^{k|k-1}_{\varpi,\varpi'}(x)(1 - \tilde{q}_{k-1|k-1}) \\ &+ \sum_l \int \delta_{l,\varpi}\, f^{k|k-1}_{\varpi}(x|u)\, D^{k-1|k-1}_l(u)du \\ &+ \sum_{l_1 \neq l_2} \int f_{k|k-1}(\{(x, \varpi)\}|\{(u_1, l_1), (u_2, l_2)\})\, D^{k-1|k-1}_{l_1,l_2}(u_1, u_2)du_1 du_2\,.\end{aligned} \tag{14.180}$$

Using Eqs. (14.66, 14.69) and the facts that $\{l_1, l_2\} = \{\varpi, \varpi'\}$ and

$$D^{k-1|k-1}_{l_1,l_2}(u_1, u_2) = D^{k-1|k-1}_{l_2,l_1}(u_2, u_1), \tag{14.181}$$

we get the desired result:

$$\begin{aligned}\tilde{D}^{k|k-1}_{\varpi,\varpi'}(x) = \tilde{B}^{k|k-1}_{\varpi,\varpi'}(x)(1 - \bar{q}_{k-1|k-1}) \\ + \sum_l \int \delta_{l,\varpi}\ p_S^{\varpi}(u)\, f_{\varpi}^{k|k-1}(x|u)\, D_l^{k-1|k-1}(u)du \\ + \sum_{l_1 \neq l_2} \int \delta_{l_1,\varpi}\ \delta_{l_2,\varpi'}\ p_S^{\varpi}(u_1)(1 - p_S^{\varpi}(u_2)) \\ \cdot f_{\varpi}^{k|k-1}(x|u_1)\, D^{k-1|k-1}_{l_1,l_2}(u_1, u_2)du_1 du_2\end{aligned} \tag{14.182}$$

$$\begin{aligned}= \tilde{B}^{k|k-1}_{\varpi,\varpi'}(x)(1 - \bar{q}_{k-1|k-1}) \\ + \int p_S^{\varpi}(u)\, f_{\varpi}^{k|k-1}(x|u)\, \tilde{D}^{k-1|k-1}_{\varpi,\varpi'}(u)du \\ + \int p_S^{\varpi}(u_1)(1 - p_S^{\varpi'}(u_2))\, f_{\varpi}^{k|k-1}(x|u_1)\, D^{k-1|k-1}_{\varpi,\varpi'}(u_1, u_2)du_1 du_2.\end{aligned} \tag{14.183}$$

14.7.3 Proof of Eq. (14.99)

Integrate Eqs. (14.96–14.98) to get:

$$\tilde{q}^{k|k-1}_{\varpi,\varpi'} = \tilde{b}^{k|k-1}_{\varpi,\varpi'}(1 - \bar{q}_{k-1|k-1}) + \tilde{D}^{k-1|k-1}_{\varpi,\varpi'}[p_S^{\varpi}] + D^{k-1|k-1}_{\varpi,\varpi'}[p_S^{\varpi}(1 - p_S^{\varpi'})] \tag{14.184}$$

$$\begin{aligned}\tilde{q}^{k|k-1}_{\varpi',\varpi} = \tilde{b}^{k|k-1}_{\varpi',\varpi}(1 - \bar{q}_{k-1|k-1}) \\ + \tilde{D}^{k-1|k-1}_{\varpi',\varpi}\left[p_S^{\varpi'}\right] + D^{k-1|k-1}_{\varpi,\varpi'}\left[\left(1 - p_S^{\varpi'}\right)p_S^{\varpi'}\right]\end{aligned} \tag{14.185}$$

$$q^{k|k-1}_{\varpi,\varpi'} = b^{k|k-1}_{\varpi,\varpi'}\left(1 - \overline{q}_{k-1|k-1}\right) + D^{k|k-1}_{\varpi,\varpi'}\left[p_S^{\varpi} p_S^{\varpi'}\right]. \tag{14.186}$$

It then follows that:

$$\begin{aligned}&\tilde{q}^{k|k-1}_{\varpi,\varpi'} + \tilde{q}^{k|k-1}_{\varpi',\varpi} + q^{k|k-1}_{\varpi',\varpi} \\ =\ &\left(\tilde{b}^{k|k-1}_{\varpi,\varpi'} + \tilde{b}^{k|k-1}_{\varpi',\varpi} + b^{k|k-1}_{\varpi,\varpi'}\right)\left(1 - \bar{q}^{k-1|k-1}_{\varpi,\varpi'}\right) \\ &+\tilde{D}^{k-1|k-1}_{\varpi,\varpi'}[p_S^{\varpi}] + \tilde{D}^{k-1|k-1}_{\varpi',\varpi}\left[p_S^{\varpi'}\right] \\ &+D^{k-1|k-1}_{\varpi,\varpi'}\left[p_S^{\varpi}\left(1 - p_S^{\varpi'}\right) + \left(1 - p_S^{\varpi'}\right)p_S^{\varpi'} + p_S^{\varpi} p_S^{\varpi'}\right]\end{aligned} \tag{14.187}$$

$$\begin{aligned}= \left(\tilde{b}^{k|k-1}_{\varpi,\varpi'} + \tilde{b}^{k|k-1}_{\varpi'\varpi} + b^{k|k-1}_{\varpi,\varpi'}\right)\left(1 - \bar{q}^{k-1|k-1}_{\varpi,\varpi'}\right) \\ +\tilde{D}^{k-1|k-1}_{\varpi,\varpi'}[p_S^{\varpi}] + \tilde{D}^{k-1|k-1}_{\varpi',\varpi}\left[p_S^{\varpi'}\right] \\ +D^{k-1|k-1}_{\varpi,\varpi'}\left[1 - (1 - p_S^{\varpi})\left(1 - p_S^{\varpi'}\right)\right]\end{aligned} \tag{14.188}$$

$$\leq 1 - \bar{q}^{k-1|k-1}_{\varpi,\varpi'} + \tilde{D}^{k-1|k-1}_{\varpi,\varpi'}[1] + \tilde{D}^{k-1|k-1}_{\varpi',\varpi}[1] + D^{k-1|k-1}_{\varpi,\varpi'}[1] \quad (14.189)$$

$$= 1 - \bar{q}^{k-1|k-1}_{\varpi,\varpi'} + \bar{q}^{k-1|k-1}_{\varpi,\varpi'} + \bar{q}^{k-1|k-1}_{\varpi',\varpi} + q^{k-1|k-1}_{\varpi,\varpi'} \quad (14.190)$$

$$= 1 - \bar{q}^{k-1|k-1}_{\varpi,\varpi'} + \bar{q}^{k-1|k-1}_{\varpi,\varpi'} = 1 \quad (14.191)$$

where Eq. (14.189) follows from Eq. (14.88) and Eq. (14.191) from Eq. (14.95).

14.8 Conclusions

This paper has summarized, for the first time in a single publication, three Bayes-optimal generalizations of the Bernoulli filter to increasingly challenging sensing conditions:

1. The PMB filter; which optimally extends the Bernoulli filter to non-Markovian targets, unknown motion noise, and unknown (including colored) measurement noise.
2. The dyadic filter; which optimally extends the Bernoulli filter from at most one target to at most two closely-spaced targets.
3. The set-valued STRB filter; which optimally subsumes under a single conceptually and algebraically simple algorithm—i.e., the recursive RCS propagator—a variety of single-target tracking approaches ranging from set-valued Kalman filters to filters that can process nonstandard measurements (i.e., quantized data, features, attributes, natural-language statements, and inference rules).

References

1 R. Mahler. *Statistical Multisource-multitarget Information Fusion*. Artech House, Norwood, MA, 2007.

2 R. Mahler. The pairwise-Markov Bernoulli filter. *IEEE Access*, 8:168229–168245, 2020. https://doi.org/10.1109/ACCESS.2020.3022752.

3 R. Mahler. Bayes-optimal tracking of two statistically correlated targets in general clutter. *IET Signal Processing*, 16:1034–1049,2022.https://doi.org/10.1049/sil2.12150.

4 R. Mahler. Bayes-optimal set-valued tracking of single point targets. *IEEE Access*, 9:68663–68674,2021.https://doi.org/10.1109/ACCESS.2021.3077458.

5 R. Mahler. Exact closed-form multitarget Bayes filters. *Sensors*, 19(12):2019. https://doi.org/10.3390/s19122818.

6 B.-N. Vo, B.-T. Vo, T. Nguyen, and C. Shim. An overview of multi-object estimation via labeled random finite set. *IEEE Transactions on Signal Processing*, 2024, https://www.ieeexplore.ieee.org/stamp/stamp.jsp?tp=&arnumber=10704579.

7 R. Mahler. *Advances in Statistical Multisource-multitarget Information Fusion*. Artech House, Norwood, MA, 2014.

8 G. Matheron. *Random Sets and Integral Geometry*. J. Wiley, New York, 1975.

9 I. Molchanov. *Theory of Random Sets*. Springer, New York, 2005.

10 B.-T. Vo. Random finite sets in multi-object filtering. Ph.D. Dissertation. School of Electrical, Electronic and Computer Engineering, The University of Western Australia, 254 pages, 2008.

11 B. Ristic, B.-T. Vo, B.-N. Vo, and A. Farina. A tutorial on Bernoulli filters: Theory, implementation, and applications. *IEEE Transactions on Signal Processing*, 61(13):3406–3430, 2012.

12 D. Musicki, R. Evans, and S. Stankovic. Integrated probabilistic data association. *IEEE Transactions on Automatic Control*, 39(6):1237–1241, 1994.

13 S. Challa and B.-N. Vo, Bayesian approaches to track existence—IPDA and random sets. In *Proceedings of the Fifth International Conference on Information Fusion,* Annapolis MD, 1228–1235, 2002.

14 M. Rutten, N. Gordon, and S. Maskell. Recursive track-before-detect with target amplitude fluctuations. *IEEE Processing: Radar, Sonar, and Navigation*, 152(5):345–352, 2005.

15 R. Mahler. Bernoulli filters for multiple correlated sensors. *IEEE Access*, 9:2310–2316, 2021. https://doi.org/10.1109/ACCESS.2020.3046631.

16 W. Pieczynski. Pairwise Markov chains. *IEEE Transactions on Pattern Analysis and Machine Intelligence*, 25(5):634–639, 2003.

17 W. Pieczynski, Pairwise Markov chains and Bayesian unsupervised fusion. In *Proceedings of the 3rd International Conference on Information Fusion,* Paris, France, 2000.

18 Y. Petetin and F. Desbouvries. Bayesian multi-object filtering for pairwise Markov chains. *IEEE Transactions on Signal Processing*, 61(18):4481–4490, 2013.

19 W. Pieczynski and F. Desbouvries. Kalman filtering using pairwise Markov models. In *Proceedings of the International Conference on Acoustics, Speech and Sensor Signal Processing (ICASSP2003)*, 6, pages 57–60, 2003.

20 S. Derrode and W. Pieczynski. Signal and image processing using pairwise Markov chains. *IEEE Transactions on Signal Processing*, 52(9):2477–2489, 2004.

21 G. Calafiore. Reliable localization using set-valued nonlinear filters. *IEEE Transactions on Signal Processing*, 35(2):189–197, 2005.

22 D. Shi, T. Chen, and L. Shi. On set-valued Kalman filtering and its application to event-based state estimation. *IEEE Transactions on Automatic Control*, 60(5):1275–1290, 2015.

23 A. Bishop and B. Ristic. Fusion of spatially referring natural language statements with random set theoretic likelihoods. *IEEE Transactions on Aerospace and Electronic Systems*, 49(2):932–944, 2013.

24 D. Morrell and W. Stirling, Set-valued filtering and smoothing. In *Proceedings of the 22nd Asilomar Conference Signals, Systems and Computers,* Pacific Grove, CA, 236–240, 1988.

25 D. Morrell and W. Stirling. Set-valued filtering and smoothing. *IEEE Transactions on Systems, Man, and Cybernetics: Systems,* 21(1):184–193, 1991.

15

Space Surveillance via Poisson Labeled Multi-Bernoulli Tracking

Martin Adams[1], Leonardo Cament[2], and Javier Correa[3]

[1,2] *Department of Electrical Engineering and Advanced Mining Technology Center Universidad de Chile, Santiago, Chile*
[3] *Galgo Chile, Santiago, Chile*

15.1 Introduction

The precise detection, continuous tracking, and systematic cataloging of orbiting resident space objects (RSOs) are critical components of effective space situational awareness (SSA). The escalating threat posed by the accumulation of defunct RSOs has intensified the urgency to enhance current methods for orbital state estimation. Notably, incidents such as the 2007 Chinese anti-satellite missile test and the accidental collision between Cosmos-2251 and Iridium-33 in 2009 are believed to have generated approximately 36% of all currently monitored objects in low Earth orbit (LEO). At present, the U.S. Space Surveillance Network (SSN) tracks over 21,000 space objects (SOs), primarily consisting of debris with effective diameters exceeding 10 cm.

A March 2016 report from Russia's Automated Space Danger Warning System [1] highlighted 50 instances in which RSOs encroached upon the international space station's (ISS's) 15 km safety perimeter, with 5 of these incursions approaching within 4 km. Additionally, 75 debris fragments resulting from the disintegration of the NOAA-16 satellite were identified during the same month. In contrast, only 26 objects deorbited, leading to a net growth in the orbital debris population. Consequently, there is an increasing demand for next-generation RSO tracking technologies that deliver greater accuracy and computational efficiency in object detection and orbit tracking.

Certain components of the forces influencing RSOs exhibit stochastic behavior, leading to gradual alterations in their orbital paths. As a result, recursive Bayesian estimation frameworks have been widely employed to support the detection, tracking, and continual state estimation of RSOs [2–4]. Under this paradigm, a probability density function (PDF) describing the multi-object state distribution for RSOs entering a sensor's field of view (FoV) is dynamically propagated using incoming observational data. These observations are then used to update the recursive Bayesian filter, allowing for refined predictions of RSO trajectories over time.

However, the effectiveness of this approach is hindered by several inherent challenges. Significant variability in orbital parameters, narrow sensor FoVs, and a typically sparse number of observations per RSO pass, all contribute to difficulties in reliably initiating new object tracks or updating existing ones. This is especially problematic when no prior information about the RSOs is available, leading to substantial uncertainty in data association.

These constraints give rise to several critical issues in the domain of orbital object tracking, namely:

- The inherently *nonlinear dynamics* of orbital motion models, coupled with the influence of external forces beyond simple gravitational attraction, such as solar radiation pressure and perturbations from third bodies, significantly increase the computational demands of state propagation algorithms.
- Customized *initial orbit determination (IOD)* techniques become essential when complete state information, specifically, both position and velocity, is not directly obtainable from sensor measurements.
- Limited observation duration and the narrow FoV of individual sensors result in RSOs remaining undetected for the majority of their orbital paths.
- While *measurement noise* and *environmental clutter* are common challenges in tracking applications, their impact is amplified in SO tracking due to the compounded difficulties posed by sparse observations and limited sensor coverage.

Accordingly, this chapter aims to present a multi-object tracking framework for space debris utilizing the Poisson labeled multi-Bernoulli (PLMB) filter. This approach advances target state estimation by propagating labeled multi-Bernoulli (LMB) densities over time and incorporating a Poisson intensity function to represent the likelihood of newly appearing targets. Given the inherent nonlinearity in both motion dynamics and measurement models, particle-based representations are employed to model the state distribution of each individual object.

The core components of the proposed tracking concept include:

- Development of a reference database constructed from publicly available two-line element (TLE) orbital datasets.
- Generation of synthetic observation data reflecting the characteristics and geographic positions of twelve observatories within the Falcon telescope network (FTN) [5].
- Implementation of a hybrid filtering approach: A particle-based distribution is used during the prediction phase, while a Gaussian distribution is applied for the update step. All state vectors are expressed in the Earth-centered inertial (ECI) coordinate system.
- Adoption of a dynamic motion model that incorporates external perturbative forces, including lunar and solar gravity as well as solar radiation pressure, acting on RSOs [6]. Additional implementation specifics are available in [3, 6–9]. This approach contrasts with prior models that considered only Earth's gravitational influence [10].

  It should be noted that, although for LEO RSOs atmospheric drag plays a significant role, it is not considered in this chapter. It was initially intended to include this important perturbative force within the dynamic motion model, based on the United States naval research laboratory mass spectrometer and incoherent scatter radar exosphere (2000) (NRLMSISE-00) empirical atmospheric model [11]. However, in its available form, the NRLMSISE-00 model fails to yield air density values at lower LEO altitudes. Since this chapter implements a particle-based filtering framework, it was determined that particles representing the RSO state with orbits within the lower regions of LEO, remained without assigned air density values and hence appropriate drag forces could not be calculated for them. Recent work which addresses a significant modification of the NRLMSISE-00 model to include drag force modeling can be found in [12].
- Application of a measurement model that emulates optical data acquisition by simulating the output of telescope-based imaging systems.

This chapter offers the following key contributions:

- Introduction of a partially uniform birth (PUB)-probabilistic admissible region (PAR)-PLMB filter, building upon the frameworks established in [13, 14]. This approach effectively models the initial uncertainty regions of RSOs within the PLMB filtering context. By narrowing the search space for candidate birth densities, it outperforms traditional Gaussian-based birth models in both efficiency and accuracy.
- Implementation of a PAR derived from a finely discretized uniform grid, which is then transformed into a sparse weighted particle distribution.

To provide context for the presented research, Section 15.2 offers a concise overview of track initiation strategies for RSOs and the broader domain of multi-target tracking (MTT). Section 15.3 introduces and explains the PAR and PUB models, referencing foundational work by Jones et al. [13]. In Section 15.4, the prediction and update mechanisms of the PLMB filter are reviewed, based on methods described in [14]. Section 15.5 presents a detailed dynamic model for RSOs, tailored to accommodate extended intervals between observational updates. Section 15.6 outlines an observation model based on telescopic imagery, which provides angular data from the initial and final points of streaks generated by moving RSOs. The method for extracting both individual and multi-target state estimates from the PLMB posterior, is detailed in Section 15.7 and key performance metrics for multi-target filtering are presented in Section 15.8. Finally, Sections 15.9 and 15.10 show the results and conclusions respectively.

15.2 A Brief Overview of SSA Research

An overview of the SSA, MTT problem and its distinct challenges can be found in [15, 16]. To address the specific challenges in initiating orbital tracks, the admissible region (AR) method was introduced [4]. This technique limits the possible orbits of resident RSOs by selecting a subset of valid range and range-rate combinations for optical sensors or RADEC rate pairs for radar systems [4]. This approach was later refined by adding specific orbital constraints, such as limits on the semimajor axis and eccentricity, leading to the constrained admissible region (CAR) method [4].

To make the CAR method more compatible with Bayesian tracking systems that use optical data, its rigid constraints were transformed into a probabilistic framework, known as the PAR approach [17].

Among the widely adopted solutions for MTT problems is multiple hypothesis tracking (MHT) [18, 19]. More recently, random finite set (RFS) algorithms based on finite set statistics (FISST) [20, 21] have gained traction in solving complex MTT tasks. These methods estimate both the number of targets and their states in a unified Bayesian framework, circumventing the dependency on separate, often fragile, data association techniques.

Several Bayesian tracking models for RSOs now utilize the CAR and PAR frameworks, including newer RFS-based methods [13, 22, 23]. For example, Jones and colleagues implemented the cardinalized probability hypothesis density (CPHD) filter for RSO tracking, integrating both PAR and PUB models. However, studies have demonstrated that the performance of probability hypothesis density (PHD) and CPHD filters is inferior to more advanced labeled RFS filters [14, 24, 25].

In 2015, the Poisson multi-Bernoulli (PMB) [26] filter was introduced, employing the loopy belief propagation (LBP) algorithm to perform measurement-to-state assignments. To yield a precisely defined PDF the use of state labels was introduced in the PLMB[1] filter [28], serving as an efficient approximation of the labeled Poisson multi-Bernoulli mixture (PMBM) filter. While the PLMB filter maintains a single LMB distribution, the PMBM filter propagates multiple multi-Bernoulli (MB) distributions.

According to [14], the LBP-based PLMB filter matches the tracking accuracy of the PMBM filter while delivering a higher computational efficiency. This performance-to-efficiency trade-off strongly supports the adoption of the PLMB filter in practical MTT applications.

This chapter therefore employs the PLMB filter [14], an RFS-based tracker, to process optical data from multiple spatially distributed observatories. Not only does the PLMB filter provide labeled tracking capabilities, but it also surpasses the performance of both PHD and CPHD filters [24]. Furthermore, it offers comparable accuracy to other leading labeled RFS filters including the GLMB and LMB filters [14], while uniquely benefiting from a Poisson birth model that enables a straightforward integration with the PUB framework [13].

15.3 Track Initialization for Multiple RSOs

The initialization of multiple RSOs involves two main elements: Employing a PUB during the prediction phase and using the PAR strategy to define spatial density parameters for individual RSOs during the update phase.

Drawing from the PUB-CPHD filter framework outlined in [13], this chapter introduces a PUB-PLMB filter integrated with the PAR technique. This combination is utilized to model the initial states derived from telescope-based measurements. The PAR method operates as an AR, assuming independence between newly detected and existing targets, as discussed in [13].

Unlike the PUB-CPHD approach, the newly proposed PUB-PLMB filter supports full MTT by assigning a distinct label to each target at every time instance, enabling precise target identification throughout the tracking process.

15.3.1 Multi-target Initiation with a PUB Model

The PUB intensity function defines a uniform birth density across the sensor's FoV. It is expressed as:

$$D_{\mathcal{B}}(\mathbf{x}) = \lambda_{\beta} \cdot \mathcal{U}(\theta(\mathbf{x}); \mathcal{B}) \cdot \sum_{i=1}^{J_B} w_b^{(i)} \mathcal{N}(\boldsymbol{\phi}(\mathbf{x}); \bar{\boldsymbol{\phi}}^{(i)}, \mathbf{P}_{\phi}^{(i)}). \tag{15.1}$$

$\mathbf{x}$ denotes the target's state in the ECI coordinate frame. The function $\theta(\mathbf{x})$ extracts the observable components of the target's state, specifically for telescope imaging scenarios, and is defined as $\theta(\mathbf{x}) = [\alpha, \beta, \dot{\alpha}, \dot{\beta}]^{\mathsf{T}}$.

Meanwhile, the unobservable components, represented by $\boldsymbol{\phi}(\mathbf{x}) = [s, \dot{s}]^{\mathsf{T}}$, include the distance s between the observer and the target, and the target's radial velocity $\dot{s}$. Together, the combined vector $[\theta(\mathbf{x})^{\mathsf{T}}, \boldsymbol{\phi}(\mathbf{x})^{\mathsf{T}}]^{\mathsf{T}}$ represents the full target state within the telescopic camera's coordinate frame.

The parameter λ_{β} denotes the Poisson rate for the expected number of targets. The term $\mathcal{U}(\theta(\mathbf{x}); \mathcal{B})$ models a uniform distribution for the observable state variables, bounded by $\mathcal{B}$, which corresponds to the sensor's FoV. The unobservable part is characterized by a Gaussian Mixture

1 Note that [27] suggests that the PLMB filter is a special case of the δ-generalized labeled multi-Bernoulli (GLMB) filter.

(GM) model with J_B components. Each component $\mathcal{N}(\boldsymbol{\phi}(\mathbf{x}); \bar{\boldsymbol{\phi}}^{(i)}, \mathbf{P}_{\phi}^{(i)})$ includes a weight $w_b^{(i)}$, a mean vector $\bar{\boldsymbol{\phi}}^{(i)}$, and a covariance matrix $\mathbf{P}_{\phi}^{(i)}$.

15.3.2 New Target Density Approximation via PARs

The PAR framework defines an orbital region that aligns with the assumption that newly detected targets are statistically independent from existing RSOs. These RSOs are constrained by orbital dynamics, specifically, limits on semimajor axis length and orbital eccentricity, within a predefined set of orbits. This study focuses on RSOs in LEO.

To model the distribution of new RSOs, the method involves multiplying the birth intensity function, $D_{\mathcal{B}}(\mathbf{x})$, by the measurement likelihood $l_z(\mathbf{z}|\mathbf{x})$. The resulting formulation, to be elaborated in Section 15.4, is:

$$D_{\mathcal{B}}(\mathbf{x})l_z(\mathbf{z}|\mathbf{x}) \approx \lambda_{\mathcal{B}} \cdot \mathcal{N}(\theta(\mathbf{x}); \mathbf{z}, \mathbf{R}) \cdot \sum_{i=1}^{J_B} w_b^{(i)} \mathcal{N}(\boldsymbol{\phi}(\mathbf{x}); \bar{\boldsymbol{\phi}}^{(i)}, \mathbf{P}_{\phi}^{(i)}), \tag{15.2}$$

Here, the Gaussian distribution $\mathcal{N}(\theta(\mathbf{x}); \mathbf{z}, \mathbf{R})$ captures the likelihood of observables $\theta(\mathbf{x})$ conditioned on the measurement $\mathbf{z}$, incorporating sensor uncertainty via the noise covariance matrix $\mathbf{R}$ (as described in [13], p. 1459).

To infer the unobservable components of the state vector, denoted by $\boldsymbol{\phi}(\mathbf{x}) = [s, \dot{s}]^{\mathrm{T}}$, based on constraints derived from sensor data, the PAR approach is employed. This technique, grounded in earlier work [2, 13, 29], leverages prior knowledge of orbital mechanics and sensor measurements.

Further information on the CAR and PAR methods, including key equations and their associated parameters, can be found in Appendix 15.A. Illustrative examples are provided in Figure 15.1a, b, which visualize constraint 1 (semimajor axis length) and constraint 2 (eccentricity), based on simulated data from two randomly selected satellites in the TLE dataset.

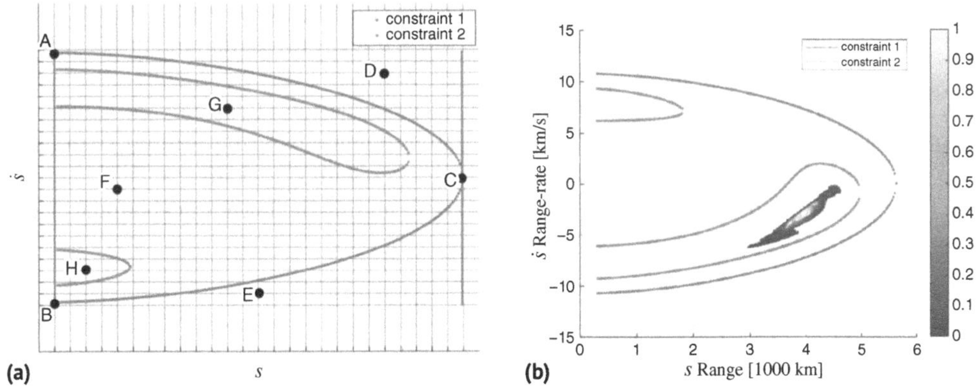

Figure 15.1 A graphical representation of PARs. In the right hand figure, the shaded region represents $p(a, e)$ as indicated by the vertical bar on the right of the graph and determined from (15.6), (15.7), and Appendix 15.C. The white region shows particles for which γ is lower than a minimum acceptable value. (a) Grid construction and constraints 1 and 2 $(s, \dot{s})$ for satnum = 42,583. (b) Weighted particles in $(s, \dot{s})$ space accordingto PAR for satnum = 337.

15.3.2.1 Constraint Definitions for the AR

The first constraint, associated with the semimajor axis of a hypothetical RSO orbit, is mathematically defined as:

$$c_1(\theta, \phi; \mathbf{q}) = \dot{s}^2 + w_1\dot{s} + F(s) + \frac{\mu}{a} = 0. \tag{15.3}$$

This equation characterizes the region of feasible orbits based on the semimajor axis a, with μ representing Earth's gravitational constant, valued at $3.986004418 \times 10^5 \left[\frac{\text{km}^3}{\text{s}^2}\right]$.

The second constraint addresses the orbital eccentricity and is given by:

$$c_2(\theta, \phi; \mathbf{q}) = \gamma_4\dot{s}^4 + \gamma_3\dot{s}^3 + \gamma_2\dot{s}^2 + \gamma_1\dot{s} + F(s)U(s) + \mu^2(1 - e^2) = 0. \tag{15.4}$$

Here, e is the orbital eccentricity. This constraint, as the first, involves the state variables s and $\dot{s}$, and incorporates a nonlinear relationship.

Both Eqs. 15.3 and 15.4 form a coupled, two-dimensional nonlinear system. The parameters w_1, $F(s)$, $U(s)$ and the coefficients γ_1 through γ_4 are detailed in Appendix 15.A. Complete derivations and mathematical justifications can be found in prior works [13, 17, 30], with solutions also provided in Appendix 15.A.

Figure 15.1b illustrates the constraint curves for a specific RSO, using simulated observational data for satellite number 337 from the TLE catalog. The outer and inner curves represent Constraints 1 and 2, respectively. The figure also displays the posterior distribution of the RSO, depicted through a particle-based model, where particle weights are shaded according to the vertical bar on the right of the graph. The state variables and weights of these particles are computed through the following steps:

1. Using a measurement $\mathbf{z}$, compute the parameters for both Constraints 1 and 2.
2. Segment the $(s, \dot{s})$ space into a grid, as illustrated in Figure 15.1a. This requires identifying the coordinates of boundary points A, B, and C:
   a. First, determine where the constraint curves intersect the edge of the defined region. This is based on limits of $a_{\max} = 43,000$ km and $e_{\max} = 0.4$, under the assumption that LEO RSOs must satisfy $a \leq a_{\max}$ and $e \leq e_{\max}$.
   b. Next, locate the rightmost point of Constraint 1, labeled as point C, by differentiating Constraint 1 with respect to $\dot{s}$, setting the derivative to zero, and solving the resulting sixth-order polynomial for s. This gives the coordinate $(s^{c1}_{\text{right}}, \dot{s}^{c1}_{\text{right}})$ with $\dot{s}^{c1}_{\text{right}} = -w_1/2$.
   c. Then, for a minimum value of $s_{\min} = 300$ km (representing the lower LEO boundary), calculate $\dot{s}^{c1}_{\text{upper}}$ and $\dot{s}^{c1}_{\text{lower}}$ as follows:

$$\dot{s}^{c1}_{\text{upper}}, \dot{s}^{c1}_{\text{lower}} = -\frac{w_1}{2} \pm \sqrt{\left(\frac{w_1}{2}\right)^2 - F(s) - \frac{\mu}{a_{\max}}} \tag{15.5}$$

3. For every point on the grid $(s, \dot{s})$, draw samples $\tilde{\mathbf{z}}$ from the Gaussian distribution $\mathcal{N}(\cdot; \mathbf{z}, \mathbf{R})$ and recalculate the parameters for the constraint equations.
4. At each grid point:
   a. Compute the semimajor axis a and eccentricity e by solving (15.3) and (15.4), giving the following equations:

$$a = -\frac{\mu}{\dot{s}^2 + w_1\dot{s} + F(s)} \tag{15.6}$$

$$e = \sqrt{1 + \frac{\gamma_4\dot{s}^4 + \gamma_3\dot{s}^3 + \gamma_2\dot{s}^2 + \gamma_1\dot{s} + F(s)U(s)}{\mu^2}}. \tag{15.7}$$

b. Identify valid LEO RSO points by verifying that both conditions $a \leq a_{\max}$ and $e \leq e_{\max}$ are met. Points such as D and E, which lie outside both constraint zones, are invalid. Point F, although within the Constraint 1 boundary, falls outside Constraint 2 and is also invalid. Only points such as G and H, which satisfy both constraints, are deemed valid.

c. Assign a weight of $\tilde{\gamma} = 0$ to invalid points. For valid points, compute the probability $\tilde{\gamma} = p(a, e)$, where $p(a, e)$ is the joint probability distribution derived from TLE-based statistics, as described in [13] and detailed further in Appendix 15.C.

5. Construct the birth particle distribution $\{\gamma^j, \mathbf{x}^j\}_1^{J_B}$:

$$\gamma^j = \frac{\tilde{\gamma}^j}{\sum_{j=1}^{J_B} \tilde{\gamma}_j}, \quad \mathbf{x}^j = f_{\text{t-radec}}^{\text{eci}}([s^j, \dot{s}^j, \tilde{\mathbf{z}}^j]), \tag{15.8}$$

where $f_{\text{t-radec}}^{\text{eci}}$ transforms coordinates from the topocentric right ascension and declination (RADEC) frame to the ECI frame.

15.4 Poisson Labeled Multi-Bernoulli Filter

This section introduces the PLMB filter, which serves as the foundation for the proposed multi-object tracking algorithm. The PLMB filter is designed to process an arbitrary number of potential new targets, while also enabling the tracking and identification of each target via a unique label. It integrates seamlessly with the PUB model, as previously demonstrated in conjunction with the CPHD filter in [13].

The MTT state is modeled using an LMB distribution, characterized by its probability generating functional (PGFl), denoted as $G_{\mathcal{X}}^{\text{lmb}}[h]$ [21, 31]:

$$G_{\mathcal{X}}^{\text{lmb}}[h] = \prod_{\ell \in L} (1 - r_\ell + r_\ell \langle f_\ell(\mathbf{x}), h(\mathbf{x}) \rangle), \tag{15.9}$$

where $\mathcal{X}$ represents the RFS of targets, f_ℓ is the PDF for a single target with label ℓ, r_ℓ is the existence probability for that target, and $h(\mathbf{x})$ is a bounded function over the target state space $(0 \leq h(\mathbf{x}) \leq 1)$. The notation $\langle \xi, \eta \rangle$ corresponds to the inner product, defined by the integral $\int \xi(\mathbf{x})\eta(\mathbf{x}), d\mathbf{x}$.

The birth model employs a Poisson process, defined by the intensity function:

$$D_{\mathcal{B}}(\mathbf{x}) = \lambda_B f_B(\mathbf{x}), \tag{15.10}$$

where λ_B is the expected number of target births, and $f_B(\mathbf{x})$ represents their spatial distribution.

The union of the Poisson birth model with the LMB distribution yields the overall PLMB PGFl:

$$G_{\mathcal{X}}^{\text{plmb}}[h] = G_{\mathcal{B}}^{\text{p}}[h] G_{\mathcal{Y}}^{\text{lmb}}[h], \tag{15.11}$$

where the Poisson PGFl component is given by:

$$G_{\mathcal{B}}^{\text{p}}[h] = e^{D_{\mathcal{B}}[h-1]}, \tag{15.12}$$

and $G_{\mathcal{Y}}^{\text{lmb}}[h]$ corresponds to the previously defined LMB PGFl. Equations (15.9) to (15.12), define the PGFl-based formulation of the general PLMB filter. Implementation details are presented in the subsequent subsections.

15.4.1 PLMB Multi-target Prediction

Given a prior LMB density characterized by the parameters (r_ℓ, f_ℓ) for each $\ell \in L$, where L denotes the set of all target labels, the predicted parameters (r'_ℓ, f'_ℓ) of the updated LMB density can be computed as follows:

$$r'_\ell = r_\ell \langle P_S, f_\ell \rangle, \quad f'_\ell(\mathbf{x}) = \frac{\langle P_S l_x(\mathbf{x}|\cdot), f_\ell \rangle}{\langle P_S, f_\ell \rangle}. \tag{15.13}$$

$P_S(\mathbf{x})$ denotes the survival probability, while $l_x(\mathbf{x}_k|\mathbf{x}_{k-1})$ represents the motion model. This prediction mechanism is both computationally efficient and aligns with the labeled probability hypothesis density (LPHD) filter prediction approach detailed in [24, 31].

15.4.2 PLMB Multi-target Update

The Bayesian update of the Poisson multi-Bernoulli PGFl defined in Eq. (15.11), results in a labeled multi-Bernoulli mixture (LMBM) PGFl, which is then approximated by an LMB PGFl. This updated LMBM PGFl comprises three main components: Newly appearing targets, successfully detected targets, and those that were missed, as detailed in [14]:

$$G^+_{\mathcal{X}|\mathcal{Z}}[h] = \prod_{\ell \in L} (1 - r^+_\ell + r^+_\ell \langle f^+_\ell, h \rangle). \tag{15.14}$$

Here, (r^+_ℓ, f^+_ℓ) denote the updated probability of existence and the corresponding spatial probability density for each target with label ℓ.

To compute the updated existence probabilities r^+_ℓ, an assignment problem must be solved using a cost matrix $\mathbf{W}$. This is typically achieved using LBP [26], though alternative approaches such as Murty's algorithm [32] and Gibbs sampling [33] are also effective.

The N by M cost matrix $\mathbf{W}$ contains the weights with elements $w_{n,m}$ associated with various pairings of targets ℓ_n, for $n = 1, \dots, N$, and observations $\mathbf{z}_m$, for $m = 1, \dots, M$. Targets that go undetected are represented by the indices $m = 0$, $n \geq 1$, while new targets not present in the prior are represented by $n = 0$, $m \geq 1$.

The parameters for misdetected targets are defined as follows:

$$w_{n,0} = 1 - r'_{\ell_n} + r'_{\ell_n} \left\langle (1 - P_D) f'_{\ell_n}, 1 \right\rangle, \tag{15.15}$$

$$r^+_{n,0} = \frac{r'_{\ell_n} \left\langle (1 - P_D) f'_{\ell_n}, 1 \right\rangle}{w_{n,0}}, \; f^+_{n,0}(\mathbf{x}) = \frac{(1 - P_D(\mathbf{x})) f'_{\ell_n}(\mathbf{x})}{\left\langle (1 - P_D) f'_{\ell_n}, 1 \right\rangle}. \tag{15.16}$$

The parameters for targets that have been detected are:

$$w_{n,m} = r'_{\ell_n} \langle P_D l_z(\mathbf{z}_m|\cdot) f'_{\ell_n}, 1 \rangle, \tag{15.17}$$

$$r^+_{n,m} = 1, \; f^+_{\ell_n,m}(\mathbf{x}) = \frac{P_D(\mathbf{x}) l_z(\mathbf{z}_m|\mathbf{x}) f'_{\ell_n}(\mathbf{x})}{\langle P_D l_z(\mathbf{z}_m|\cdot) f'_{\ell_n}, 1 \rangle}. \tag{15.18}$$

Newly born target labels are defined as a triplet (k, m, o), representing the current time k, measurement index m, and sensor index o, with label $\ell_{N+m} = (k, m, o)$. The corresponding parameters are:

$$w_{0,m} = D_{\mathcal{K}}(\mathbf{z}_m) + \langle D_{\mathcal{B}} l_z(\mathbf{z}_m|\cdot), 1 \rangle, \tag{15.19}$$

$$r_{0,m}^{+} = \frac{\langle D_{\mathcal{B}} l_z(\mathbf{z}_m|\cdot), 1\rangle}{w_{0,m}}, \; f_{0,m}^{+}(\mathbf{x}) = \frac{D_{\mathcal{B}}(\mathbf{x}) l_z(\mathbf{z}_m|\mathbf{x}, \ell_{N+m})}{\langle D_{\mathcal{B}} l_z(\mathbf{z}_m|\cdot), 1\rangle}. \tag{15.20}$$

Clutter is modeled using a uniform distribution $D_{\mathcal{K}}(\mathbf{z}) \approx \lambda_\kappa / V_\kappa$, where λ_κ represents the expected number of clutter measurements and V_κ the area of the sensor's FoV. The density $f_{0,m}^{+}(\mathbf{x})$ is computed using the PAR method described in Section 15.3.

The intensity of the PUB model, as outlined in Eq. (15.1), is divided into observed and unobserved components:

$$D_{\mathcal{B}}(\mathbf{x}) = \lambda_\beta \mathcal{U}(\theta(\mathbf{x})) f(\phi(\mathbf{x})), \tag{15.21}$$

yielding:

$$\begin{aligned} \langle D_{\mathcal{B}}(\mathbf{x}) l_z(\mathbf{z}_m|\mathbf{x}), 1\rangle &\approx \lambda_\beta \left\langle \frac{1}{V_\kappa} l_z(\mathbf{z}_m|\theta(\mathbf{x})), 1\right\rangle \left\langle \sum_{i=1}^{J_B} w_b^{(i)} \mathcal{N}(\phi(\mathbf{x}); \bar{\boldsymbol{\phi}}^{(i)}, \mathbf{P}_\phi^{(i)}), 1\right\rangle \\ &= \lambda_\beta \frac{1}{V_\kappa} \times 1 = \frac{\lambda_\beta}{V_\kappa}. \end{aligned} \tag{15.22}$$

Thus, the probability of existence for new targets is:

$$r_{0,m}^{+} = \frac{\lambda_\beta}{\lambda_\kappa + \lambda_\beta} \tag{15.23}$$

and the corresponding cost value is:

$$w_{0,m} = \frac{\lambda_\kappa + \lambda_\beta}{V_\kappa}. \tag{15.24}$$

The LBP algorithm computes assignment weights $p_{n,m}$ for each label and measurement pair. The detailed pseudo-code can be found in [26], page 20.

Using these weights, the posterior parameters for existing targets are:

$$r_{\ell_n}^{+} = \sum_{m=0}^{M} p_{n,m} r_{n,m}^{+}, \; f_{\ell_n}^{+}(\mathbf{x}) = \sum_{m=0}^{M} p_{n,m} f_{n,m}^{+}(\mathbf{x}), \tag{15.25}$$

and for new targets:

$$r_{\ell_{N+m}}^{+} = p_{0,m} r_{0,m}^{+}, \; f_{\ell_{N+m}}^{+}(\mathbf{x}) = f_{0,m}^{+}(\mathbf{x}). \tag{15.26}$$

These expressions define the parameters for the posterior multi-target distribution as formulated in Eq. (15.14).

15.5 RSO Motion Prediction Model

In the ECI system, the coordinate origin aligns with Earth's center of mass. The z-axis points along the Earth's rotational axis, while the y-axis lies in the equatorial plane and is orthogonal to the x-axis, which is oriented toward a fixed celestial reference. Unlike the Earth-centered, Earth-fixed (ECEF) frame, the ECI's frame does not rotate with the Earth and remains fixed relative to inertial space (see the Earth WGS84 ellipsoid model in Figure 15.2). This chapter adopts the true equator,

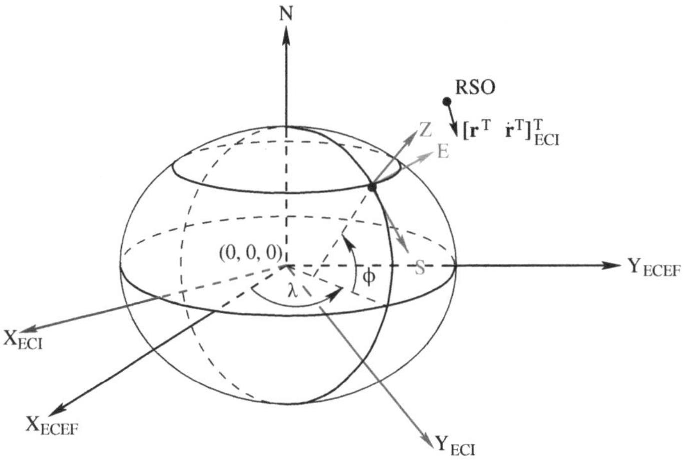

Figure 15.2 WGS84 Earth ellipsoid model with south-east-zenith (SEZ), ECEF, and ECI coordinate systems.

mean equinox (TEME) variant of the ECI frame. This is because, the TEME position-velocity coordinate system is adopted by libraries that convert TLEs into coordinates at specified times. Subsequently, the TEME coordinates are converted to an ECI J2000 coordinate system.

Additionally, the SEZ coordinate system serves as a localized Cartesian reference for the sensor, with axes directed toward the south, east, and zenith, respectively. Detailed coordinate transformations necessary for converting between these frames are provided in Appendix 15.B.

The RSO's motion model is assumed to be corrupted by an acceleration noise component. The unmodeled perturbation acceleration $\mathbf{a}_\epsilon$ used in this work, is based on the formulation in [23, 34]:

$$\mathbf{a}_\epsilon(\mathbf{r}(t), \dot{\mathbf{r}}(t), t, \omega(t_{k-1})) = f_{\text{ric}}^{\text{eci}(\mathbf{r}(t),\dot{\mathbf{r}}(t))}((t - t_{k-1})\omega(t_{k-1})) \quad \text{for} \quad t \in [t_{k-1},\ t_k]. \tag{15.27}$$

where $\mathbf{r}(t) = [x(t)\ \ y(t)\ \ z(t)]_{\text{ECI}}^{\text{T}}$, i.e., the RSO's x, y, z coordinates in the ECI frame and $\dot{\mathbf{r}}(t) = [\dot{x}(t)\ \ \dot{y}(t)\ \ \dot{z}(t)]_{\text{ECI}}^{\text{T}}$ is its velocity, also in the ECI frame. The random component within the acceleration $\mathbf{a}_\epsilon(\mathbf{r}(t), \dot{\mathbf{r}}(t), t, \omega(t_{k-1}))$ is modeled by ω, which represents zero-mean Gaussian noise applied to the second component in the satellite's radial-intrack-crosstrack (RIC) coordinate system at time t_{k-1}:

$$\omega(t_{k-1}) \sim \mathcal{N}(\mathbf{0}, \mathbf{Q}). \tag{15.28}$$

i.e., $\omega(t_{k-1})$ is sampled from a zero mean, multivariate Gaussian distribution with covariance matrix

$$\mathbf{Q} = \sigma_\omega^2 \mathbf{I}_3, \tag{15.29}$$

where σ_ω^2 is the variance parameter corresponding to the assumed piecewise constant noise term $\omega(t_{k-1})$ within the acceleration term given in (15.27). Note that $\omega(t_{k-1})$ corresponds to a piece-wise constant random variable in the rate of change of acceleration (jerk—having units m/s^3). Hence in the RHS of (15.27), it is multiplied by the time component $t - t_{k-1}$, meaning that for longer times $t - t_{k-1}$, it has a larger effect on $\mathbf{a}_\epsilon()$. Here, as in [34], it has been assumed that the along-track, cross-track, and radial perturbations are the same. Since, in reality, the along-track perturbation maybe more dominant than its cross-track and radial counterparts, it would be interesting to include relevant statistics in future work.

$f_{\text{ric}}^{\text{eci}}$ defines the transformation of a vector from the RIC frame into the ECI frame. Note that the RIC noise term ω is evaluated at time t_{k-1} and is assumed to remain constant from time t_{k-1} to t_k

(piece-wise constant) for integration purposes [23]. Refer to Appendix 15.B.4 for transformation details. Note that potentially more precise white noise or Brownian motion increment models are discussed in [35].

According to Newton's second law, valid in the ECI inertial frame, the total acceleration acting on the RSO is given by [23]:

$$\ddot{\mathbf{r}}(t) = -\frac{\mu}{\| \mathbf{r}(t) \|^3}\mathbf{r}(t) + \delta_p(\mathbf{r}(t), \dot{\mathbf{r}}(t)) + \mathbf{a}_\epsilon. \tag{15.30}$$

The term $\delta_p(\mathbf{r}(t), \dot{\mathbf{r}}(t))$ accounts for various perturbation forces from external influences. The vectors $\mathbf{r}(t)$ and $\dot{\mathbf{r}}(t)$ correspond to the satellite's position and velocity, forming the state vector $\mathbf{x}$, defined as $\mathbf{x} = [\mathbf{r}(t)^{\top}, \dot{\mathbf{r}}(t)^{\top}]^{\top}$. This six-dimensional state vector is expressed within the ECI coordinate frame.

The state $\mathbf{x}_k$ evolves in time according to:

$$\mathbf{x}_k = \begin{bmatrix} \mathbf{r}(t_{k-1}) \\ \dot{\mathbf{r}}(t_{k-1}) \end{bmatrix} + \int_{t_{k-1}}^{t_k} \begin{bmatrix} \dot{\mathbf{r}}(t) \\ \ddot{\mathbf{r}}(t) \end{bmatrix} dt, \tag{15.31}$$

where t_k denotes the timestamp corresponding to the current time k. Note that the acceleration term $\ddot{\mathbf{r}}(t)$ within the integral of (15.31) contains the random noise component $\mathbf{a}_\epsilon$ defined in (15.27) and (15.30).

Each particle in the system is propagated forward in time using the integration method outlined in [6], which employs the Shampine-Gordon integrator [36].[2]

The force model used in the propagator accounts for[3]:

- Earth's central gravitational pull,
- nonspherical Earth influences[4] including geopotential variations, solid and oceanic tides,
- gravitational effects from the Moon and Sun,
- solar radiation pressure.

No atmospheric drag was used in this work, due to the failure when obtaining the air density using the NRLMSISE-00 model, for particles representing low altitudes. In future works, drag acceleration will be incorporated in the motion model. Additional implementation specifics are available in [3, 6–9].

For a target represented by a particle set $\{\gamma_{k-1}, \hat{\mathbf{x}}_{k-1}\}_{j=1}^{J}$, where γ_k is the particle's weight at time k, the predicted particle distribution evolves as:

$$\gamma_{k|k-1} = \gamma_{k-1} \tag{15.32}$$

$$\hat{\mathbf{x}}_{k|k-1} = \hat{\mathbf{x}}_{k-1} + \int_{t_{k-1}}^{t_k} \begin{bmatrix} \dot{\mathbf{r}}(t) \\ \ddot{\mathbf{r}}(t) \end{bmatrix} dt. \tag{15.33}$$

2 Unlike the Runge-Kutta method, which discards previous data at each step, the Shampine-Gordon integrator is a multistep technique that leverages historical computations. This makes it a more computationally efficient choice. The C code for the Shampine-Gordon integrator and the RSO propagator utilized in this study is accessible at https://www.researchgate.net/publication/340793133_High_Precision_Orbit_Propagator_C_code, (retrieved on 10 June 2025). This integrator has also been successfully applied in advanced orbital models such as DROMO [37], SPOOK [38–40], and ZUNIEM [41], where its efficiency and accuracy have been well documented.

3 As stated in the introduction, due to the lack of air density information at LEO based on the NRLMSISE-00 model, the significant contribution of atmospheric drag is currently ignored in this chapter.

4 Based on the GGM03S spherical harmonic model with degree $n = 36$ and order $m = 36$.

15.6 RSO Measurement Model

In telescopic imagery, an RSO can be identified as a linear streak. This chapter characterizes RSO detection based on the angular coordinates $[\alpha_1, \beta_1, \alpha_2, \beta_2]$, which define the end points, measured in pixels, of the streak relative to the center of the image (refer to Figure 15.3).

The telescope's FoV spans angular dimensions $\alpha \in [-A/2, A/2]$ horizontally and $\beta \in [-B/2, B/2]$ vertically. Measurements are not instantaneous as they are taken over an exposure time interval Δt_{exp}. Accordingly, the recorded measurement vector becomes the initial pixel coordinate of the detected streak, $[\alpha_1, \beta_1]^{\mathsf{T}}$, augmented with the rate of change in pixel coordinates of the streak, calculated over the exposure time Δt_{exp}, $\left[(\alpha_2 - \alpha_1)/\Delta t_{\text{exp}},\ (\beta_2 - \beta_1)/\Delta t_{\text{exp}}\right]^{\mathsf{T}}$, i.e.,

$$[\alpha, \beta, \dot{\alpha}, \dot{\beta}]^{\mathsf{T}} = \left[\alpha_1, \beta_1, \frac{\alpha_2 - \alpha_1}{\Delta t_{\text{exp}}}, \frac{\beta_2 - \beta_1}{\Delta t_{\text{exp}}}\right]^{\mathsf{T}}. \tag{15.34}$$

The observation model, detailed in [42], is described by:

$$\mathbf{z} = \mathbf{H} f_{\text{eci}}^{\text{cam}}(\mathbf{x}) + w \tag{15.35}$$

where $f_{\text{eci}}^{\text{cam}}(\mathbf{x})$ projects the state vector into the camera's spherical coordinate system, and $\mathbf{H}$ is the observation matrix:

$$\mathbf{H} = \begin{bmatrix} 0 & 1 & 0 & 0 & 0 & 0 \\ 0 & 0 & 1 & 0 & 0 & 0 \\ 0 & 0 & 0 & 0 & 1 & 0 \\ 0 & 0 & 0 & 0 & 0 & 1 \end{bmatrix}. \tag{15.36}$$

The measurement model for a single target utilizes a Gaussian likelihood function:

$$l_z(\mathbf{z}|\mathbf{x}) = \mathcal{N}(\mathbf{z}; \mathbf{H} f_{\text{eci}}^{\text{cam}}(\mathbf{x}), \mathbf{R}), \tag{15.37}$$

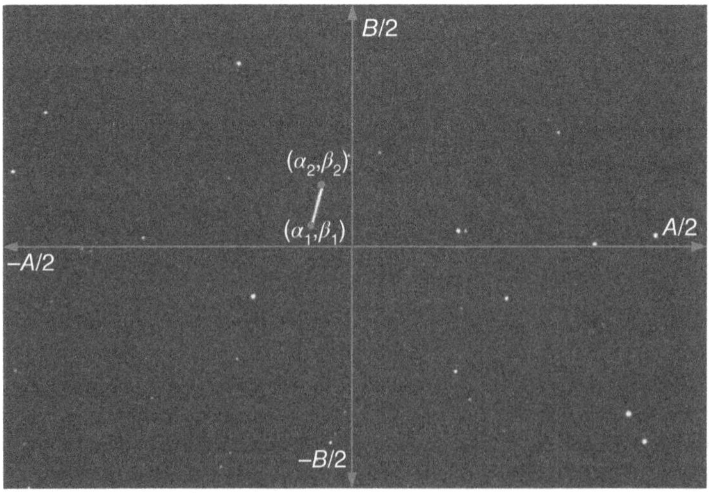

Figure 15.3 Image captured using sidereal tracking of NORAD satellite 25,853 by the Mitre Corporation. The lines with arrows represent the angular axes, while the ends of the line segment denote satellite pixel coordinates: (α_1, β_1) at time t_k and (α_2, β_2) at time $t_k + \Delta t_{\text{exp}}$. *Source* : With permission of The MITRE Corporation.

with $\mathbf{R}$ representing the covariance matrix of the measurement noise w in (15.35).

To define the complete covariance noise matrix $\mathbf{R}$, the matrix $\mathbf{R}_{1D}$ is defined as:

$$\mathbf{R}_{1D} = \sigma_r^2 \begin{bmatrix} 1 & -1/\Delta t_{\text{exp}} \\ -1/\Delta t_{\text{exp}} & 2/(\Delta t_{\text{exp}}^2) \end{bmatrix}, \tag{15.38}$$

where σ_r^2 is the sensor noise variance. The matrix $\mathbf{R}$ is then formed using the Kronecker product:

$$\mathbf{R} = \mathbf{R}_{1D} \otimes \mathbf{I}_2, \tag{15.39}$$

where $\mathbf{I}_2$ is the 2×2 identity matrix. The complete components of $\mathbf{R}$ are given in Table 15.2.

Each predicted measurement $\hat{\mathbf{z}}_k^j = \left[\alpha_k^j, \beta_k^j, \dot{\alpha}_k^j, \dot{\beta}_k^j\right]^{\mathsf{T}}$ can also be described in terms of the telescope's image frame as:

$$\left(\alpha_{1,k}^j, \beta_{1,k}^j, \alpha_{2,k}^j, \beta_{2,k}^j\right) = \left[\alpha_k^j, \beta_k^j, \alpha_k^j + \Delta t_{\text{exp}}\dot{\alpha}_k^j, \beta_k^j + \Delta t_{\text{exp}}\dot{\beta}_k^j\right]^{\mathsf{T}}. \tag{15.40}$$

The probability of detecting a particle with predicted state $\hat{\mathbf{x}}_{k|k-1}^j$ is defined by:

$$P_D(\hat{\mathbf{x}}_{k|k-1}^j) = \begin{cases} P_D, & \text{if } |\alpha_{1,k}^j|, |\alpha_{2,k}^j| \leq A/2 \ \& \ |\beta_{1,k}^j|, |\beta_{2,k}^j| \leq B/2 \\ 0, & \text{otherwise.} \end{cases} \tag{15.41}$$

Consequently, the average detection probability across all predicted particles $\{\gamma_{k|k-1}^j, \hat{\mathbf{x}}_{k|k-1}^j\}_{j=1}^J$ is computed as:

$$\overline{P}_D\left(\{\gamma_{k|k-1}^j, \hat{\mathbf{x}}_{k|k-1}^j\}_{j=1}^J\right) = \sum_{j=1}^J \gamma_{k|k-1}^j P_D\left(\hat{\mathbf{x}}_{k|k-1}^j\right). \tag{15.42}$$

The computation of single-target posterior densities follows the approach described in [23]:

1. Convert the particle states into topocentric camera coordinates. Adjust the weights based on the detection probability: $\left\{\gamma_{k|k-1}^j, \hat{\mathbf{x}}_{k|k-1}^j\right\}_{j=1}^J \rightarrow \left\{P_D\left(\hat{\mathbf{x}}_{k|k-1}^j\right)\gamma_{k|k-1}^j, f_{\text{eci}}^{\text{cam}}\left(\hat{\mathbf{x}}_{k|k-1}^j\right)\right\}_{j=1}^J$.
2. Approximate the result with a Gaussian characterized with mean $\boldsymbol{\mu}_{k|k-1}$ and covariance $\mathbf{P}_{k|k-1}$.
3. Apply a linear Kalman filter to update the estimate.
4. Draw samples from the updated Gaussian: $\left\{1/J, \mathbf{x}_{k,\text{cam}}^j\right\}_{j=1}^J$. Note that radial components $s, \dot{s}$ remain unchanged since they are unobservable.
5. Convert the sampled particles from camera coordinates back to the ECI frame: $\left\{1/J, f_{\text{cam}}^{\text{eci}}\left(\mathbf{x}_{k,\text{cam}}^j\right)\right\}_{j=1}^J$.

For targets that are misdetected, the particle state is unchanged: $\mathbf{x}_k^j = \hat{\mathbf{x}}_{k|k-1}^j$, and their updated weights are given by:

$$\gamma_k^j = \frac{\left[1 - P_D\left(\hat{\mathbf{x}}_{k|k-1}^j\right)\right]\gamma_{k|k-1}^j}{\sum_{j'=1}^J \left[1 - P_D\left(\hat{\mathbf{x}}_{k|k-1}^{j'}\right)\right]\gamma_{k|k-1}^{j'}}. \tag{15.43}$$

15.7 Multi-SO State Extraction

To estimate the target states, the particle distribution in the ECI coordinate system is transformed into a Gaussian distribution in the RADEC frame.

While a common technique for this is to average particles in Euclidean space, this approach is suboptimal for orbital trajectory calculations. Directly averaging particles converted to RADEC coordinates also poses challenges due to discontinuities at $\alpha_{\text{ra}} = 2\pi$ and $\beta_{\text{dec}} = \pi$.

To overcome these issues, methods from Riemannian geometry are used, as detailed in [43, 44]. Let $\{\gamma^j, \mathbf{x}^j\}_{j=1}^J$ denote the particle distribution for a specific target. The mean is computed in spherical RADEC coordinates, where $\mathbf{x} = [r, \alpha_{\text{ra}}, \beta_{\text{dec}}]^\top$. Each particle's position on the unit sphere is given by the unit vector $\mathbf{q}_j$:

$$\mathbf{q}_j = \left[\cos\left(\alpha_{\text{ra}}^j\right)\sin\left(\beta_{\text{dec}}^j\right)\sin\left(\alpha_{\text{ra}}^j\right)\sin\left(\beta_{\text{dec}}^j\right)\cos\left(\beta_{\text{dec}}^j\right)\right]^\top. \tag{15.44}$$

The logarithmic map relative to a reference unit vector $\mathbf{q}_0$ is defined as:

$$\boldsymbol{\theta} := \text{Log}_{\mathbf{q}_0}(\mathbf{q}) = \mathbf{u}\theta, \quad \mathbf{u} = \frac{\mathbf{q} - \mathbf{q}_0\cos(\theta)}{\sin(\theta)}, \quad \theta = \arccos(\mathbf{q}_0^\top\mathbf{q}), \tag{15.45}$$

and the corresponding exponential map is:

$$\mathbf{q} := \text{Exp}_{\mathbf{q}_0}(\boldsymbol{\theta}) = \Re(\mathbf{q}_0\cos(\theta) + \mathbf{u}\sin(\theta)), \quad \theta = \| \boldsymbol{\theta} \|, \quad \mathbf{u} = \frac{\boldsymbol{\theta}}{\theta}. \tag{15.46}$$

To compute the weighted average [44], the Gauss-Newton iterative method is applied. Initialization can be either $\bar{\mathbf{q}} \leftarrow \mathbf{q}_1$ or $\bar{\mathbf{q}} \leftarrow \mathbf{q}_{j^*}$ with $j^* = \arg_j \max \{\gamma^j\}^J$. The iterations proceed until the difference between successive estimates falls below a specified threshold:

$$\tau = \sum_{j=1}^{J} \gamma^j \text{Log}_{\bar{\mathbf{q}}}(\mathbf{q}_j), \quad \bar{\mathbf{q}} \leftarrow \text{Exp}_{\bar{\mathbf{q}}}(\tau). \tag{15.47}$$

For $\bar{\mathbf{q}} = [\bar{q}_x, \bar{q}_y, \bar{q}_z]^\top$, the conversion back to RADEC coordinates is carried out using:[5]

$$\bar{\alpha}_{ra} = \arctan\left(\frac{\bar{q}_y}{\bar{q}_x}\right), \tag{15.48}$$

$$\bar{\beta}_{ra} = \frac{\pi}{2} - \arccos(\bar{q}_z), \tag{15.49}$$

$$\bar{r} = \sum_{j=1}^{J} r^j \tag{15.50}$$

Finally, the result is transformed back into Cartesian ECI coordinates via:

$$\bar{\mathbf{x}} = f_{\text{radec}}^{\text{eci}}([\bar{r}, \bar{\alpha}_{\text{ra}}, \bar{\beta}_{\text{ra}}]^\top). \tag{15.51}$$

where the function $f_{\text{radec}}^{\text{eci}}$ is defined by (15.57) and (15.58) in Appendix 15.A.

15.7.1 Estimate-to-Ground Truth Distance Definitions

To assess the effectiveness of the PLMB filter, appropriate distance metrics are necessary. In this work, the metrics employed are optimal sub-pattern assignment (OSPA) and CLEAR MOT, both of which rely on the computation of individual error distances between estimated RSO states and their ground truth counterparts.

Given that this chapter focuses on estimating the orbital trajectories of RSOs, it is more appropriate to define error distances based on their respective orbital paths, rather than using direct

5 Note that in the software implementation, this would be coded as atan2($\bar{q}_y$, $\bar{q}_x$).

Euclidean separation. For instance, if the true location of an RSO is on one side of Earth and the estimate places it on the opposite side, the error should logically correspond to half the orbit's circumference.

When two objects are positioned at different radial distances from Earth (r_1 and r_2), the spatial error between the vectors $\mathbf{x}_1$ and $\mathbf{x}_2$ can be represented as the arc length of a spiral path. This is modeled by the equation:

$$r(\theta) = a + b\theta. \tag{15.52}$$

Setting $r(0) = a = \min(r_1, r_2)$, and defining the angle between the two position vectors as:

$$\theta_{12} = \arccos \frac{\mathbf{x}_1 \cdot \mathbf{x}_2}{\| \mathbf{x}_1 \| \| \mathbf{x}_2 \|}. \tag{15.53}$$

It is also required that $r(\theta_{12}) = \max(r_1, r_2)$, which yields the parameter $b = (\max(r_1, r_2) - \min(r_1, r_2))/\theta_{12}$.

The arc length, computed via line integral, becomes:

$$\begin{aligned} \int_0^{\theta_{12}} r(\theta) d\theta &= \int_0^{\theta_{12}} a + b\theta d\theta = a\theta + b\frac{\theta^2}{2}\Big|_0^{\theta_{12}} \\ &= \theta_{12} \min(r_1, r_2) + \frac{\max(r_1, r_2) - \min(r_1, r_2)}{\theta_{12}} \frac{\theta_{12}^2}{2} = \theta_{12}\frac{r_1 + r_2}{2}, \end{aligned} \tag{15.54}$$

Therefore, the final expression for the trajectory-based distance metric is:

$$d_{\circ}(\mathbf{x}_1, \mathbf{x}_2) = \theta_{12}\frac{r_1 + r_2}{2} = \arccos\left(\frac{\mathbf{x}_1}{\| \mathbf{x}_1 \|} \cdot \frac{\mathbf{x}_2}{\| \mathbf{x}_2 \|}\right) \frac{\| \mathbf{x}_1 \| + \| \mathbf{x}_2 \|}{2}. \tag{15.55}$$

This formulation provides the error distance between $\mathbf{x}_1$ and $\mathbf{x}_2$ and is applied in the multi-target RSO evaluation to quantify the accuracy of the PLMB filter.

15.8 Multi-SO Filter Performance Metrics

The performance of multi-RSO tracking is evaluated using the OSPA[45] and OSPA$^{(2)}$[46, 47] metrics, as well as the multi-object tracking precision (MOTP) and multi-object tracking accuracy (MOTA) CLEAR MOT metrics [48].

Both the OSPA and OSPA$^{(2)}$ metrics provide unified measures at each time that reflect both the localization accuracy and the set cardinality error between the estimated and ground truth targets. The key distinction is that OSPA$^{(2)}$ accounts for label consistency in track evaluation, whereas the original OSPA metric is label-agnostic.

The MOTP metric quantifies localization error only for correctly identified targets, while the MOTA score reflects overall tracking performance by incorporating missed detections, false positives, and label mismatches.

15.9 Results

15.9.1 Telescopic Measurement Database Construction

The dataset used in this work is constructed from simulated telescopic observations of LEO RSOs, derived from a TLE file dated 4 October 2019. The specific TLE file applied in the experiments spans trajectory data from 2 to 6 October 2019. This file includes information on over 5, 000 RSOs.

Table 15.1 Details of observatories within the FTN [5] with their simulated azimuthal and elevational pointing directions.

Telescope location (city, state)	Country	Longitude (East)	Latitude	Altitude [m]	Azimuth [°]	Elevation [°]
Woodland Park, CO	USA	255.01	39.01	2,790	179	5
Yoder, CO	USA	255.80	38.89	1,961	180	5
Grand Junction, CO	USA	251.76	39.96	1,380	176	5
Durango, CO	USA	252.13	37.27	1,880	173	5
Sterling, CO	USA	256.80	40.65	1,177	181	5
La Junta, CO	USA	256.46	37.97	1,221	183	5
State College, PA	USA	282.17	40.86	317	184	5
Vicuna	Chile	289.32	−29.99	1,139	4	5
Canberra	Australia	149.17	−35.29	600	260	5
Gingin	Australia	115.71	−31.36	18	230	40
Braunschweig	Germany	10.55	52.28	73	69	5
Cape Town	South Africa	18.46	−33.96	110	103	12

Simulated measurements are generated by projecting the motion paths of the RSOs onto the image planes of 12 telescopic cameras within the FTN network. Table 15.1 provides details on each telescope's name, geographic location, and respective simulated pointing directions.
In the table, the longitude, latitude, and altitude are expressed with respect to the WGS84 Earth ellipsoid model (see Figure 15.2).

For the experiment, a selected subset of nine RSOs was utilized. These particular RSOs were chosen based on their higher frequency of detections from the telescopic network, assuming a sampling rate of seven frames per second (fps).

Figure 15.4 presents the experimental outcomes at various measurement update instances: (a) and (b) correspond to $k = 1$, (c) depicts results at $k = 2$, and (d) shows the scenario at $k = 11$. The ground truth trajectories are depicted in as curved orbits (with only orbital sections shown in figs. (b), (c) and (d)). The telescope positions, marked by stars ($*$), shift over time due to the Earth's rotation. This variation is noticeable when comparing the different plots. Circles represent the true positions of the RSO, with each being labeled by the index m corresponding to the mth measurement update. Dots, which intersect the orbits, in the radial direction with respect to the active telescope illustrate the particle distribution for the RSO.

In Figure 15.4a,b, the particle cloud (dots) associated with a newly detected target displays a high standard deviation in the radial direction, denoted σ_r, relative to the telescope's location (highlighted by the radial solid line connecting the active telescope and the observed point on the orbital trajectory). The error between the ground truth and the mean particle position at this point is approximately 52 km.

Figure 15.4c captures the subsequent update, where a notable reduction in σ_r is observed. At time $k = 11$ in Figure 15.4d, the measurement for satnum = 337 results from a different telescope than in the previous graphs (b and c), as indicated by the updated radial line, from the different telescope (star). At this stage, the error has decreased substantially to 0.68 km, accompanied by a further reduction in σ_r.

15.9.2 Probabilistic AR Grid Size Comparisons

The computation of the particle distribution, comprising state vectors and their corresponding weights, can be efficiently parallelized. The PAR algorithm was implemented in C++ and executed on a system equipped with 16 processing threads.

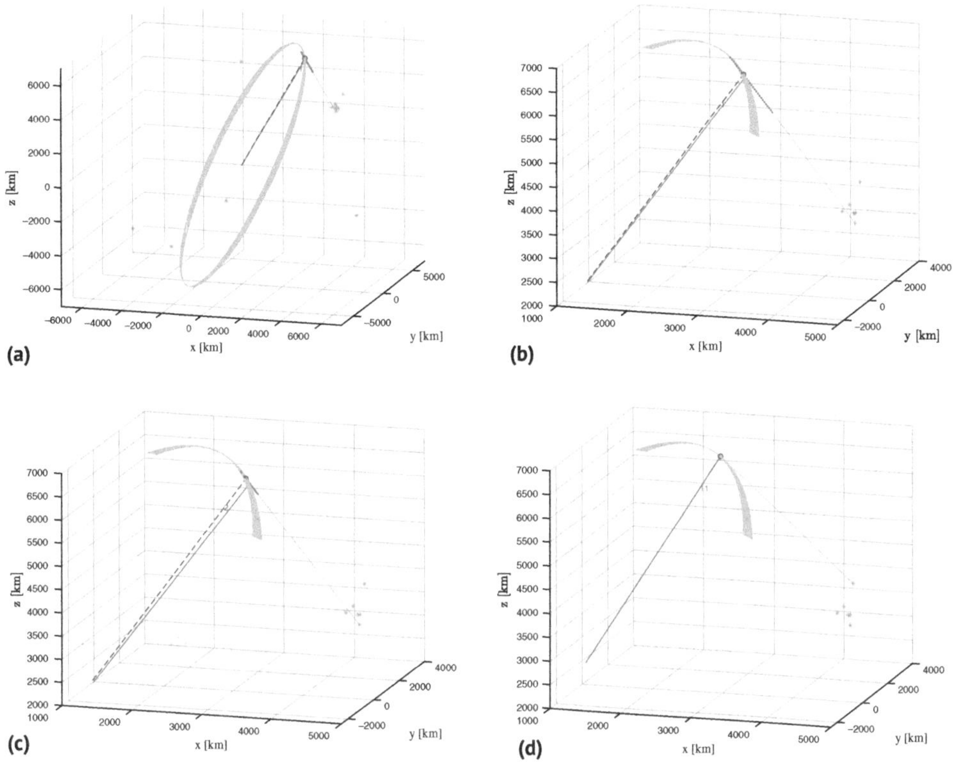

Figure 15.4 Graphs (a) to (d) display experimental RSO tracking results at three distinct times for the object with `satnum` = 337 from the TLE dataset. Graph (b) provides a magnified view of the RSO trajectory depicted in Graph (a). The ground truth orbital paths, spanning multiple revolutions over a 24-hour period, are illustrated in as curves. Telescope positions are marked with stars (∗). Circles indicate the ground truth positions of the RSO, with each one labeled by the index m, representing the mth measurement update. The dots, intersecting the orbital paths in the radial direction with respect to the active telescope correspond to the particle representation of the RSO. (a) k = 1, (b) k = 1, zoomed plot, (c) k = 2, zoomed plot, and (d) k = 11, zoomed plot.

Using a uniform grid over the $(s, \dot{s})$ space, the values of $p(a, e)$ were computed, leading to the generation of weighted particle states of the form $\{\gamma, [s, \alpha, \beta, \dot{s}, \dot{\alpha}, \dot{\beta}]^{\mathsf{T}}\}$. Due to the fact that a large portion of the weights γ are nearly zero, those associated particles are removed to reduce the computational burden and size of the particle set.

To evaluate computational performance and particle set density, three experiments were conducted using different grid resolutions for $(s, \dot{s})$:

- A 100×100 grid yielded an average of 136 ± 29 weighted particles and executed in 0.04 s. This corresponds to 1.36 ± 0.29 of the total grid elements.
- A 200×200 grid produced an average particle count of 465 ± 130, completing in 0.15 s. The resulting set represented 1.16 ± 0.33 of the grid.
- A 300×300 grid resulted in 964 ± 308 weighted particles, requiring 0.34 s to process. This accounted for 1.07 ± 0.34 of all grid points.

It is worth noting that, although the total number of particles generally scales with the grid resolution, the final number of weighted particles typically remains under 1.5% of the full grid cell number. Additionally, not all $(s, \dot{s})$ combinations within the grid need to be evaluated. Only those within the valid region defined by the PAR constraints, specifically, the overlapping area enclosed by the magenta and red boundaries in Figure 15.1a, such as grid cells containing points G and H, are necessary for computation.

15.9.3 GM Distribution Representation

During the update step, the predicted state, initially represented as a set of weighted particles, is transformed into a GM distribution. This transformation utilizes the fixed weighted-data EM (FWD-EM) method, as proposed in [49], and incorporates a modification that applies Fuzzy C-means clustering for initialization, following the recommendation in [50], which demonstrated improved performance.

The modified FWD-EM algorithm operates under the assumption of a fixed number of Gaussian components. In this chapter, GMs with component counts ranging from 1 to 10 were evaluated. Their alignment with the particle distribution was assessed using the Bayesian information criterion (BIC) metric.

For the single-component Gaussian case, the mean and standard deviation were computed directly from the weighted statistics of the particles. When more than one component was involved, the full FWD-EM algorithm was applied.

Interestingly, across all conversions tested, the BIC consistently favored the single-Gaussian model. Although this may appear counterintuitive, it is attributed to the fact that the GM modeling is performed in spherical rather than Cartesian coordinates. Consequently, all subsequent experiments employ a single-component Gaussian model.

15.9.4 Multi-SO Tracking

A total of ten Monte Carlo (MC) experiments were conducted using the PLMB filter, each incorporating randomly simulated observations based on the parameters listed in Table 15.2.

The tracking performance was assessed using both the OSPA and CLEAR MOT evaluation metrics.

The distances between the estimated and ground truth target positions were computed using Eq. (15.55). The MOTA score achieved was $96.34 \pm 1.09\%$, while the MOTP result was 3.14 ± 0.24 km. This implies that, once detected by a telescope, each target was accurately tracked more than 96% of the time, with positional accuracy ranging from approximately 2.9 to 3.4 km.

Figure 15.5 illustrates the average cardinality over time. It shows that the estimated number of targets was generally correct, with occasional discrepancies typically limited to a single target. These deviations are also reflected in the OSPA cardinality error plot in the third row of Figure 15.6.

Figure 15.6 presents the OSPA and OSPA$^{(2)}$ results. The top row displays total OSPA errors, while the second and third rows break these down into localization and cardinality components, respectively. The similarity between the first and second rows suggests that localization errors dominate, as cardinality discrepancies remained minimal throughout the tests.

Notably, OSPA errors are especially low between 7 and 11 hours. To understand this pattern, individual tracking results are shown in Figures 15.7 through 15.9.

Table 15.2 PLMB filter parameters.

Parameter	Symbol	Value
Process noise standard deviation[a]	σ_ω	10^{-7} ms^{-3}
Sensor noise standard deviation	σ	2 arcsec
Sample time	Δt	7 s
Exposure time	Δt_{exp}	4 s
Sensor noise covariance matrix	$\mathbf{R}$	$\sigma^2 \begin{bmatrix} 1 & 0 & -\Delta t_{\text{exp}}^{-1} & 0 \\ 0 & 1 & 0 & -\Delta t_{\text{exp}}^{-1} \\ -\Delta t_{\text{exp}}^{-1} & 0 & 2\Delta t_{\text{exp}}^{-2} & 0 \\ 0 & -\Delta t_{\text{exp}}^{-1} & 0 & 2\Delta t_{\text{exp}}^{-2} \end{bmatrix}$
Sensor FoV for α		$[-2.2, 2.2]^\circ$
Sensor FoV for β		$[-1.2, 1.2]^\circ$
Expected no. of new targets	λ_β	8.1×10^{-5}
Expected no. of clutter measurements	λ_κ	0.1
Probability of detection	P_D	0.8
Probability of survival	P_S	0.99999
Maximum number of particles	J	100
Area to mass ratio	AMR	0.01105
PAR grid size		300×300

a. Equation (15.27) shows that the acceleration noise is the multiplication of time difference $t - t_{k-1}$ by ω. Thus, the units of σ_ω are ms^{-3}.

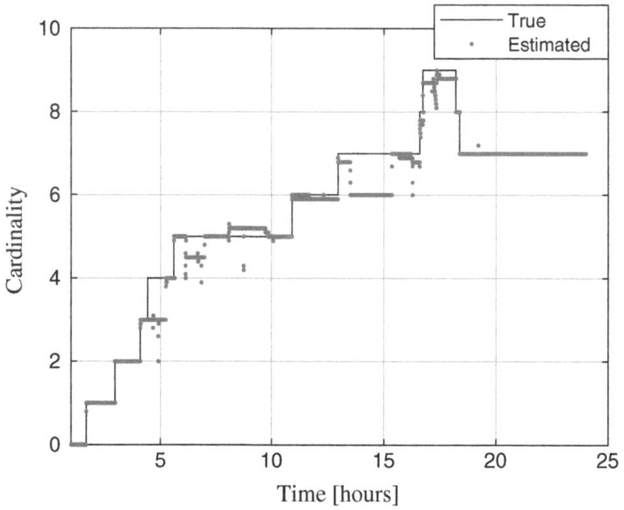

Figure 15.5 RSO average cardinality for the MC experiments at each time. The thin curve shows the ground-truth cardinality; the thick curve shows the estimated cardinality.

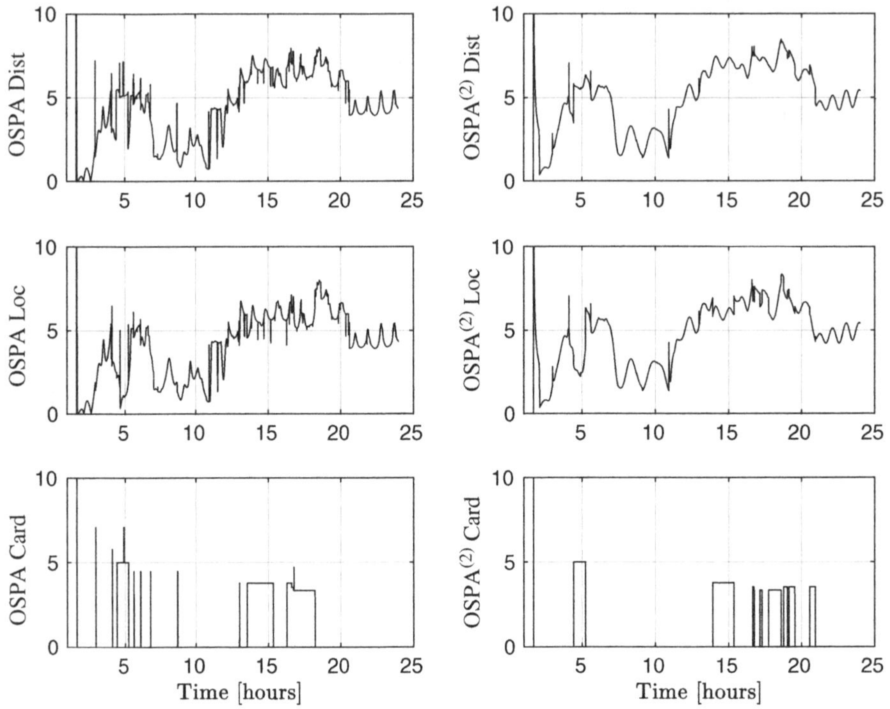

Figure 15.6 OSPA metric values (first column) and OSPA$^{(2)}$ values (second column). Top row: total OSPA errors. Middle row: localization error. Bottom row: cardinality error. Parameters: $c = 10$ km, $p = 2$, and OSPA$^{(2)}$ window size = 200 time steps (1 400 s).

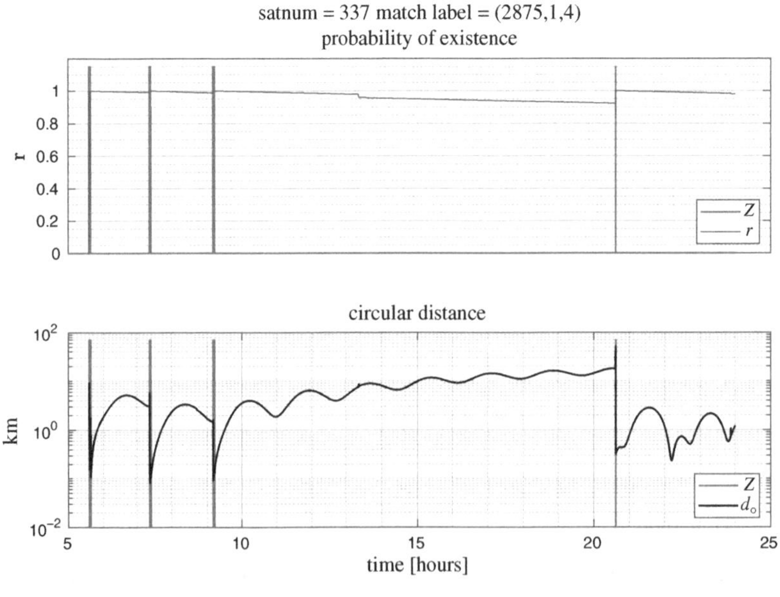

Figure 15.7 Trajectory results for satnum = 337.

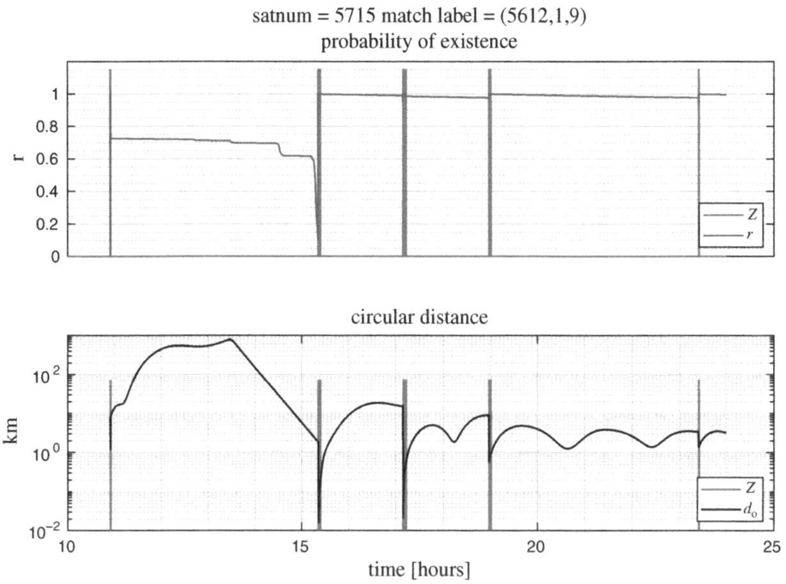

Figure 15.8 Trajectory results for satnum = 5715, first MC realization.

Each RSO begins tracking from the moment its first observation is recorded. The initial estimate, based on the PAR model, typically exhibits low uncertainty in angular coordinates but high uncertainty along the radial direction relative to the observing telescope, as illustrated in Figure 15.1b. This effect is apparent in Figure 15.8, where between 11 and 16 hours, the radial error approaches 100 km. However, as additional measurements are incorporated, the uncertainty decreases significantly, reducing the error to less than 1 km. Consequently, the radial distance between estimated and actual positions remains under 1 km for the majority of the observation period.

15.9.5 Individual SO Track Identification

In Figures 15.7–15.9, the top row of each graph displays the estimated probability of existence r for a given RSO (in continuous curves) alongside the observation windows (in thin vertical rectangles), which indicate time intervals when measurements were acquired. The second row illustrates the orbital distance between the estimate and ground truth, calculated using Eq. (15.55), which serves as the distance metric in both the OSPA and CLEAR MOT evaluations.

Figure 15.7 presents the tracking performance for the object with satnum 337. This target generated seven consecutive observations shortly after initialization via the PAR method, allowing for a rapid reduction in initial distance error. During the interval from 9 to 20 hours, no additional measurements were received, resulting in an increased estimation error. However, the track was effectively maintained and updated at 20.5 hours.

Figure 15.8 shows the tracking performance from a MC realization for satnum 5715, where only a single measurement was captured immediately following PAR initialization. In this case, the upper plot shows that the existence probability r remained around 0.7 until shortly after 15 hours, after which it decreased until another measurement was received, at which point r increased sharply to nearly unity. This increase corresponds with the estimated distribution intersecting the telescope's FoV, triggering a higher detection probability, as per Eqs. (15.41) and (15.42).

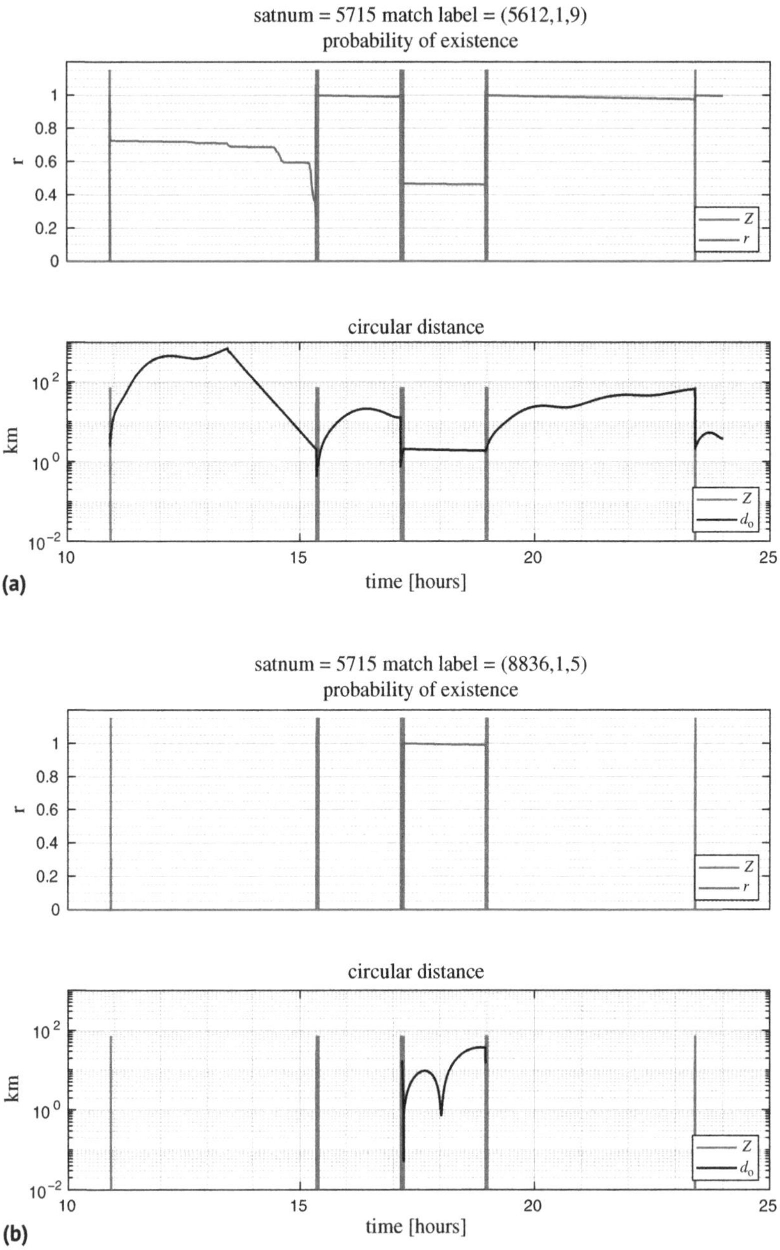

Figure 15.9 Trajectory results for satnum = 5715, second MC realization. (a) First hypothesis; (b) second hypothesis.

Figure 15.9 illustrates a less successful tracking scenario for the same satellite (satnum 5715) in a different MC trial. Here, multiple labels are mistakenly assigned to a single object, indicating a failure in track continuity. In Figure 15.9a, the track labeled $(k, m, o) = (5612, 1, 9)$ experiences a drop in r at approximately 17 hours before recovering at approximately 19 hours. Simultaneously, Figure 15.9b reveals that another track, labeled (8836, 1, 5), is assigned to the same object with a higher probability of existence during the same interval. This overlap indicates a mislabeling or fragmentation of the target's track.

To address such issues, several enhancements could be implemented. These include adopting a more robust state extractor that incorporates track history and hysteresis, fusing fragmented tracks, or employing trajectory sets as proposed in [51]. Additionally, refining the state representation by using a GM density, rather than a single Gaussian distribution, or employing a distribution that more accurately models the curved or "banana-shaped" uncertainty commonly associated with RSOs, may yield further improvements.

15.9.6 Computational Performance

To evaluate computational efficiency, the average execution time was measured over 10 simulation runs. On average, each cycle of the PUB-PAR-PLMB filter took approximately 2.5 s. This time is distributed across the filter's components as follows:

- **Prediction:** 79.94%
  - RSO particle propagation using the dynamic model (Section 15.5), $\mathcal{O}(N)$: 71.49%
  - Others: 0.14%
- **Update:** 19.95%
  - PAR algorithm (Section 15.3), $\mathcal{O}(M)$: 15.88%
  - LBP for measurement-to-track association (Section 15.4.2), $\mathcal{O}(NM)$: 0.02%
  - P_D **calculation (Section 15.6):** 0.15%
  - Coordinate transformation $f_{\text{eci}}^{\text{cam}}$ (Section 15.6): 0.30%
  - Others: 3.60%
- **State Extraction:** 0.02%
- **Others**[6] **:** 0.09%

It is evident that around 70% of the filter's runtime is dedicated to numerical integration within the prediction stage. This heavy computational load stems from the RSO propagation model itself and is not an inherent characteristic of the PUB-PAR-PLMB filtering approach.

While optimizing the propagator is outside the scope of this chapter, it is worth noting that computation time could be reduced by adopting alternative predictors, such as the unscented Kalman filter (UKF), allowing the particle approach used here to be replaced by sigma points. Further, a state distribution better suited to the non-Gaussian ("banana-shaped") profile of typical RSO distributions could be used [12].

15.10 Conclusions

This work serves as a proof of concept for integrating the PLMB filter with the PUB and PAR algorithms for multi-RSO tracking.

A simulated database was constructed using TLE data, where LEO satellites and debris were observed via simulated telescopes modeled on the FTN. Each satellite detection, derived from both ends of image streaks, was transformed into celestial coordinates. These coordinate pairs served as inputs to the PLMB multi-RSO tracking algorithm.

6 "Others" include data logging, system overhead, and auxiliary operations.

The experiment simulated observations from 12 telescopes and implemented a particle filter-based version of the PLMB filter. A multi-sensor tracking strategy was employed, wherein each sensor performed iterative multi-target updates using the LBP algorithm. Future comparisons with more advanced multi-sensor tracking methods, such as those involving Gibbs sampling as proposed in [25], could offer additional insights.

The use of PAR and PUB for modeling target birth within the PLMB framework demonstrated high accuracy and good computational efficiency. The PLMB filter produced promising tracking results, even during prolonged intervals without observations. Further improvements may be possible by adopting a different single-target state distribution that better captures the curved, "banana-shaped" characteristics of RSO distributions. Additional gains may be realized through enhanced state extraction techniques, trajectory-set modeling, or the replacement of the current propagator or integrator.

15.A PAR Constraint Equations Solution

The celestial two-body energy equation used to derive the constraint equations is:

$$\varepsilon = \frac{\| \dot{\mathbf{r}} \|^2}{2} - \frac{\mu}{\| \mathbf{r} \|}. \tag{15.56}$$

Here, ε denotes the specific orbital energy, μ is the gravitational parameter of the Earth, $\mathbf{r}$ is the position vector, and $\dot{\mathbf{r}}$ represents the velocity vector, all expressed in the ECI frame. The conceptual overview of these constraints is illustrated in Figure 15.10.

Measurements are assumed to be captured within a topocentric RADEC coordinate system, which is a spherical coordinate system aligned with the ECI frame. The transformation from measurements to state vectors is given by [13]:

$$\mathbf{r} = \mathbf{o} + s\mathbf{u}_s \tag{15.57}$$

$$\dot{\mathbf{r}} = \dot{\mathbf{o}} + \dot{s}\mathbf{u}_s + s(\dot{\alpha}\mathbf{u}_\alpha + \dot{\beta}\mathbf{u}_\beta) \tag{15.58}$$

where the directional unit vectors used in the coordinate transformation are defined as:

- $\mathbf{u}_s = [\cos(\alpha)\cos(\beta)\sin(\alpha)\cos(\beta)\sin(\beta)]^\mathrm{T}$
- $\mathbf{u}_\alpha = [-\sin(\alpha)\cos(\beta)\cos(\alpha)\cos(\beta)0]^\mathrm{T}$
- $\mathbf{u}_\beta = [-\cos(\alpha)\sin(\beta) - \sin(\alpha)\sin(\beta)\cos(\beta)]^\mathrm{T}$,

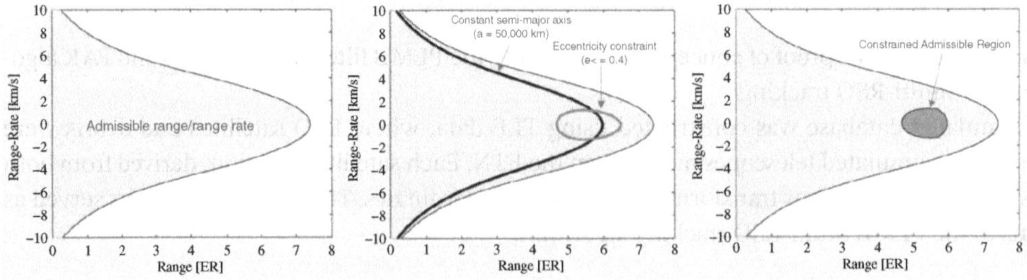

Figure 15.10 The CAR/PAR approach for initializing target tracks.

The position vector relative to the observer is given by $\mathbf{s} = s\mathbf{u}_s$, where $s = \| \mathbf{s} \|$ is the radial distance. The observer's position in the ECI frame is $\mathbf{o}$, and its velocity is computed as $\dot{\mathbf{o}} = \omega_\oplus \times \mathbf{o}$, where $\omega_\oplus$ is Earth's angular velocity vector in the ECI frame. Further details regarding this computation are provided in [52].

15.A.1 Semimajor Axis Constraint

The semimajor axis length constraint is provided by the energy:

$$\varepsilon = -\frac{\mu}{2a} \tag{15.59}$$

where substituting (15.57) and (15.58) into (15.56), for $a \leq a_{\max}$, results in:

$$c_1(\theta, \phi; \mathbf{q}) = \dot{s}^2 + w_1\dot{s} + F(s) + \frac{\mu}{a_{\max}} \leq 0 \tag{15.60}$$

where

$$F(s) = w_2 s^2 + w_3 s + w_4 - \frac{2\mu}{\sqrt{s^2 + w_5 s + w_0}} \tag{15.61}$$

with:

$$\begin{aligned} &w_0 = \| \mathbf{o} \|^2, \qquad w_1 = 2\dot{\mathbf{o}} \cdot \mathbf{u}_s, \qquad w_2 = \dot{\alpha}^2 \cos^2(\beta) + \dot{\beta}^2 \\ &w_3 = 2\dot{\alpha}(\dot{\mathbf{o}} \cdot \mathbf{u}_\alpha) + 2\dot{\beta}(\dot{\mathbf{o}} \cdot \mathbf{u}_\beta), \qquad w_4 = \| \dot{\mathbf{o}} \|^2, \qquad w_5 = 2\mathbf{o} \cdot \mathbf{u}_s. \end{aligned} \tag{15.62}$$

15.A.2 Eccentricity Constraint

The eccentricity constraint is given by the angular momentum $\mathbf{h} = \mathbf{r} \times \dot{\mathbf{r}}$, where the energy is:

$$\varepsilon = -\frac{\mu^2(1 - e^2)}{2 \| \mathbf{h} \|^2} \tag{15.63}$$

and substituting (15.57) and (15.58) into (15.56), for $e \leq e_{\max}$ results in:

$$c_2(\theta, \phi; \mathbf{q}) = \gamma_4\dot{s}^4 + \gamma_3\dot{s}^3 + \gamma_2\dot{s}^2 + \gamma_1\dot{s} + \gamma_0 \leq 0 \tag{15.64}$$

where

$$\begin{aligned} &\gamma_0 = F(s)U(s) + \mu^2(1 - e_{\max}^2), \qquad \gamma_1 = F(s)P(s) + w_1 U(s), \\ &\gamma_2 = U(s) + c_0 F(s) + w_1 P(s), \qquad \gamma_3 = P(s) + c_0 w_1, \qquad \gamma_4 = c_0, \end{aligned} \tag{15.65}$$

where the coefficients c_i are calculated from the angular momentum:

$$\mathbf{h} = \dot{s}\mathbf{h}_1 + s^2\mathbf{h}_2 + s\mathbf{h}_3 + \mathbf{h}_4 \tag{15.66}$$

and:

$$P(s) = c_1 s^2 + c_2 s + c_3 \tag{15.67}$$

$$U(s) = c_4 s^4 + c_5 s^3 + c_6 s^2 + c_7 s + c_8 \tag{15.68}$$

with:

$$c_0 = ||\mathbf{h}_1||^2, \quad c_1 = 2\mathbf{h}_1 \cdot \mathbf{h}_2, \quad c_2 = 2\mathbf{h}_1 \cdot \mathbf{h}_3, \quad c_3 = 2\mathbf{h}_1 \cdot \mathbf{h}_4, \quad c_4 = ||\mathbf{h}_2||^2,$$
$$c_5 = 2\mathbf{h}_2 \cdot \mathbf{h}_3, \quad c_6 = 2\mathbf{h}_2 \cdot \mathbf{h}_4 + ||\mathbf{h}_3||^2, \quad c_7 = 2\mathbf{h}_3 \cdot \mathbf{h}_4, \quad c_8 = ||\mathbf{h}_4||^2 \tag{15.69}$$

and

$$\mathbf{h}_1 = \mathbf{o} \times \mathbf{u}_s, \quad \mathbf{h}_2 = \mathbf{u}_s \times (\dot{\alpha}\mathbf{u}_\alpha + \dot{\beta}\mathbf{u}_\beta)$$
$$\mathbf{h}_3 = \mathbf{u}_s \times \dot{\mathbf{o}} + \mathbf{o} \times (\dot{\alpha}\mathbf{u}_\alpha + \dot{\beta}\mathbf{u}_\beta), \quad \mathbf{h}_4 = \mathbf{o} \times \dot{\mathbf{o}}. \tag{15.70}$$

15.B Coordinate Transforms

The variables, regarding Earth orbit variations, used in the coordinate transformations, on 4 October 2019 (the date of our experiment) were:

- $t_{\text{lod}} = --0.0003216\ s$
- $x_p = 0.195512'' = 9.478689242 \times 10^{-7}$ rad
- $y_p = 0.308733'' = 1.496779822 \times 10^{-6}$ rad
- UTC1-UTC: $dutc1 = -0.1536855$ s
- $ddpsi = --0.116622'' = --5.65399411183563 \times 10^{-7}$ rad
- $ddeps = --0.009989'' = --4.84280386060315 \times 10^{-8}$ rad
- TAI-UTC: $dat = 37$ s

15.B.1 Position and Velocity of the Observer

The Earth's angular velocity is expressed as $\omega_E = 7.29211514670698 \times 10^{-5} \times \left(1 - \frac{t_{\text{lod}}}{86400}\right)$, where t_{lod} is the excess length-of-day in seconds. Represented as a vector in the ECEF frame, it takes the form $\omega_E = [0, 0, \omega_E]^{\mathsf{T}}$.

To convert this to the ECI frame, the Earth's angular rotation vector becomes $\omega_\oplus = \mathbf{R}_{\text{pns}}\,\omega_E$, where $\mathbf{R}_{\text{pns}}$ is the time-dependent transformation matrix composed of precession, nutation, and sidereal rotation matrices: $\mathbf{R}_{\text{pns}} = \mathbf{R}_{\text{prec}}\mathbf{R}_{\text{nute}}\mathbf{R}_{\text{st}}$. These matrices are computed using the MATLAB functions *precess.m*, *nutation.m*, and *sidereal.m*, as described in Vallado [53]. Since $\mathbf{R}_{\text{pns}}$ varies with time, it must be recalculated at every time.

The observer's position in the ECI frame is determined by the equation $\mathbf{o} = \mathbf{R}_{\text{pns}}\mathbf{P}_M\mathbf{o}_{\text{ECEF}}$, where $\mathbf{P}_M$ represents the polar motion matrix (15.71) and $\mathbf{o}_{\text{ECEF}}$ is the position in the ECEF frame (15.72).

Finally, the observer's velocity in the ECI frame is the cross product: $\dot{\mathbf{o}} = \omega_\oplus \times \mathbf{o}$.

15.B.2 Polar Motion Matrix

The polar motion matrix is:

$$\mathbf{P}_M(x_p, y_p) = \begin{bmatrix} \cos x_p & 0 & -\sin x_p \\ \sin x_p \sin y_p & \cos y_p & \cos x_p \sin y_p \\ \sin x_p \cos y_p & -\sin y_p & \cos x_p \cos y_p \end{bmatrix} \approx \begin{bmatrix} 1 & 0 & -x_p \\ 0 & 1 & y_p \\ x_p & -y_p & 1 \end{bmatrix}. \tag{15.71}$$

15.B.3 Geodetic Coordinate to ECEF Frame

$\mathbf{o}_{\text{ECEF}} = f_{\text{geo}}^{\text{ecef}}(\mathbf{o}_{\text{geo}})$, where $\mathbf{o}_{\text{geo}}$ is the observer position in latitude ϕ, longitude λ (see Figure 15.2) and altitude h_{obs}.

$$\mathbf{o}_{ECEF} = \begin{bmatrix} (N + h_{\text{obs}}) \cdot \cos(\phi) \cdot \cos(\lambda) \\ (N + h_{\text{obs}}) \cdot \cos(\phi) \cdot \sin(\lambda) \\ ((1 - e_{\oplus}^2) \cdot N + h_{\text{obs}}) \cdot \sin(\phi) \end{bmatrix}, \tag{15.72}$$

where

- the radius of curvature at prime meridian $N = \dfrac{R_{\oplus}}{\sqrt{1 - e_{\oplus}^2 \sin^2(\phi)}}$,
- the equatorial Earth radius $R_{\oplus} = 6378.137$ km,
- the Earth is assumed to be an ellipsoid with eccentricity $e_{\oplus} = 8.1819190842622 \times 10^{-2}$.

15.B.4 RIC to ECI Frame Transformation

For a target $\mathbf{x} = [\mathbf{r}^{\top}, \dot{\mathbf{r}}^{\top}]^{\top}$, the mapping function from the RIC to the ECI frame is given by:

$$f_{\text{ric}}^{\text{eci}}(\mathbf{y}) = [\mathbf{uvw}]\mathbf{y}, \tag{15.73}$$

where

$$\mathbf{u} = \frac{\mathbf{r}}{\| \mathbf{r} \|}, \quad \mathbf{w} = \frac{\mathbf{r} \times \dot{\mathbf{r}}}{\| \mathbf{r} \times \dot{\mathbf{r}} \|}, \quad \mathbf{v} = \frac{\mathbf{w} \times \mathbf{u}}{\| \mathbf{w} \times \mathbf{u} \|}, \tag{15.74}$$

and $\mathbf{y}$ is sampled from a zero mean, multivariate Gaussian distribution with covariance matrix:

$$\mathbf{Q} = \sigma_{\omega}^2 \mathbf{I}_3, \tag{15.75}$$

where σ_{ω}^2 is the variance corresponding to the assumed piecewise constant noise term $\omega(t_{k-1})$ within the acceleration term given in (15.27).

15.C Semimajor Axis Length and Eccentricity GM Distribution

The generation of the $p(a, e)$ density follows the method outlined in [13]. This density, denoted as $p(\mathbf{q}_s) = p(a, e)$, is a GM approximation representing the distribution of semimajor axis a and eccentricity e, derived from the public TLE catalog dated 4 October 2019. From this catalog, 1, 118 samples were selected that satisfy the a and e constraints defined in [13], table I. The expectation maximization (EM) algorithm was then applied with a convergence threshold of 10^{-7} to fit a GM consisting of 30 Gaussian components.

Figure 15.11 visualizes the a–e dataset comprising over 10,000 SOs from the TLE file, along with the resulting GM probability density.

Although the TLE data directly specifies the orbital eccentricity, it does not include the semimajor axis length a. Instead, it provides the mean motion n, defined as the number of orbital revolutions per day. Applying Kepler's third law, $n^2 a^3 = \mu$, allows for the calculation of a from the provided n values.

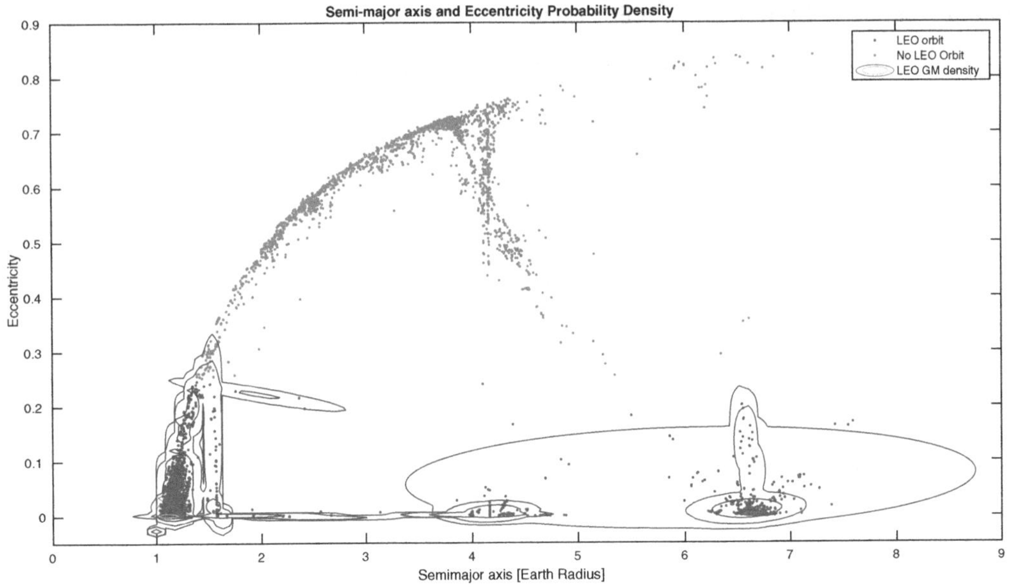

Figure 15.11 Semimajor axis (in Earth radii) and eccentricity of LEO objects' PDF. Points with eccentricity value greater than 0.25 represent *non-LEO targets*, while points with eccentricity value below 0.25 markers indicate *LEO targets* from 2 to 6 October 2019 TLE file. Contour lines represent the GM density estimate using 30 components, with a convergence tolerance of 10^{-7}, repeated over 10 trials.

Acknowledgments

This work was supported by the Southern Office of Aerospace Research and Development (SOARD) Project FA9550-23-1-0592, ANID (National Research Agency, Chile) PIA Project AFB180004 and ANID FONDECYT project 1231658.

References

1 Events in the near-Earth space (in Russian). Technical report, Russian Central Research Institute's of Machine Building, Automated Space Danger Warning System, 2016.

2 K. J. DeMars and M. K. Jah. Probabilistic initial orbit determination using gaussian mixture models. *Journal of Guidance, Control, and Dynamics*, 36(5):1324–1335, 2013.

3 E. D. Delande, C. Frueh, J. Houssineau, and D. E. Clark. Multi-object filtering for space situational awareness. In *AAS Space Flight Mechanics Meeting*, 2015.

4 A. Milani, G. F. Gronchi, M. D. M. Vitturi, and Z. Knežević. Orbit determination with very short arcs. I admissible regions. In *Celestial Mechanics and Dynamical Astronomy*, Universita di Pisa, Pisa, Italy, 59–87, 2004.

5 F. K. Chun, R. D. Tippets, D. M. Strong, D. J. Della-Rose, D. E. Polsgrove, K. C. Gresham, J. A. Reid, C. P. Christy, M. Korbitz, J. Gray, and S. Gartin. A new global array of optical telescopes: The Falcon telescope network. *Publications of the Astronomical Society of the Pacific*, 130(991):095003, 2018.

6 M. Mahooti and M. Sharifi. Variable and fixed step integrations for dynamic orbit determination of LEO satellites. In *The 9th Conference of Iranian Aerospace Society*, 2009.

7 M. Mallick, S. Rubin, and B.-N. Vo. An introduction to force and measurement modeling for space object tracking. In *Proceedings of the 16th International Conference on Information Fusion*, IEEE, pages 1013–1020, 2013.

8 G. Beutler, T. Schildknecht, U. Hugentobler, and W. Gurtner. Orbit determination in satellite geodesy. *Advances in Space Research*, 31(8):1853–1868, 2003.

9 T. D. Papanikolaou and D. Tsoulis. Assessment of numerical integration methods in the context of low earth orbits and inter-satellite observation analysis. *Acta Geodaetica et Geophysica*, 51(4):619–641, 2016.

10 A. Pak, J. Correa, and M. Adams. Robust joint target detection and tracking for space situational awareness. *Journal of Guidance, Control, and Dynamics*, 41(1):119–136, 2018.

11 J. Picone, A. E. Hedin, D. Drob, and A. Aikin. NRLMSISE-00 empirical model of the atmosphere: Statistical comparison and scientific issues. *Journal of Geophysical Research*, 107, 2002.

12 L. Cament, P. Barrios, and M. Adams. A Riemannian geometric approach to space debris orbit determination. In *IEEE International Conference on Control, Automation and Information Sciences (ICCAIS)*, Jeju Island, South Korea, 2025.

13 B. A. Jones. CPHD filter birth modeling using the probabilistic admissible region. *IEEE Transactions on Aerospace and Electronic Systems*, 54(3):1456–1469, 2018.

14 L. Cament, J. Correa, M. Adams, and C. Pérez. The histogram Poisson, labeled multi-Bernoulli multi-target tracking filter. *Signal Processing*, 176:107714, 2020.

15 J. A. Kennewell and B.-N. Vo. An overview of space situational awareness. In *Proceedings of the 16th International Conference on Information Fusion*, pages 1029–1036, 2013.

16 B. A. Jones, D. S. Bryant, B.-T. Vo, and B.-N. Vo. Challenges of multi-target tracking for space situational awareness. In *18th International Conference on Information Fusion*, pages 1278–1285, 2015.

17 I. Hussein, C. Roscoe, M. Wilkins, and P. Schumacher. Probabilistic Admissible Region for Short-Arc Angles-Only Observations. In *Proceedings of the Advanced Maui Optical and Space Surveillance Technologies Conference*, 2014.

18 Y. Bar-Shalom, P. Willet, and T. Xin. *Tracking and Data Fusion: A Handbook of Algorithms*. YBS Publishing, Bradford, U.K., 2011.

19 S. S. Blackman. Multiple hypothesis tracking for multiple target tracking. *IEEE Aerospace and Electronic Systems Magazine*, 19(1):5–18, 2004.

20 R. P. S. Mahler. *Statistical Multisource-multitarget Information Fusion*. Artech House Publishers, 2007.

21 R. P. S. Mahler. *Advances in Statistical Multisource-multitarget Information Fusion*. Artech House, Norwood, MA, USA, 2014. ISBN 9781608077984.

22 B. A. Jones, B.-T. Vo, and B.-N. Vo. Generalized Labeled Multi-Bernoulli Space-Object Tracking with Joint Prediction and Update. In *AIAA/AAS Astrodynamics Specialist Conference*, American Institute of Aeronautics and Astronautics. Long Beach, CA, USA, pages 1177–1194, 2016.

23 E Delande, J. Houssineau, J. Franco, C. Frueh, D. Clark, and M. Jah. A new multi-target tracking algorithm for a large number of orbiting objects. *Advances in Space Research*, 64(3):645–667, 2019.

24 Á. F. García-Fernández, J. L. Williams, K. Granstrom, and L. Svensson. Poisson multi-Bernoulli mixture filter: direct derivation and implementation. *IEEE Transactions on Aerospace and Electronic Systems*, 54(4):1883–1901, 2018.

25 B.-N. Vo, B.-T. Vo, and M. Beard. Multi-sensor multi-object tracking with the generalized labeled multi-Bernoulli filter. *IEEE Transactions on Signal Processing*, 67(23):5952–5967, 2019.

26 J. L. Williams. Marginal multi-Bernoulli filters - RFS derivation of MHT, JIPDA, and association-based member. *IEEE Transactions on Aerospace and Electronic Systems*, 51(3):1664–1687, 2015.

27 B.-T. Vo and B.-N.Vo. Labeled random finite sets and multi-object conjugate priors. *IEEE Transactions on Signal Processing*, 61(13):3460–3475, 2013.

28 B.-N. Vo, B.-T. Vo, T. T. D. Nguyen, and C. Shim. An overview of multi-object estimation via labeled random finite set. *IEEE Transactions on Signal Processing*, 72:4888–4917, 2024.

29 I. I. Hussein, C. W. T. Roscoe, M. P. Wilkins, and P. W. Schumacher Jr. Probabilistic admissibility in angles-only initial orbit determination. In *Proceedings of the 24th International Symposium on Space Flight Dynamics*, pages 5–9, 2014.

30 J. Franco, E. D. Delande, C. Frueh, J. Houssineau, and D. E. Clark. A spherical co-ordinate space parameterisation for orbit estimation. In *2016 IEEE Aerospace Conference,* IEEE, pages 1–12, 2016.

31 R. Mahler. On CPHD filters with track labeling. In *Signal Processing, Sensor/Information Fusion, and Target Recognition XXVI,* volume 10200, International Society for Optics and Photonics, page 102000E, 2017.

32 K. G. Murthy. An algorithm for ranking all the assignments in order of increasing costs. *Operations Research*, 16(3): 682–687, 1968.

33 S. Reuter, A. Danzer, M. Stuebler, A. Scheel, and K. Granström. A fast implementation of the Labeled Multi-Bernoulli filter using gibbs sampling. In *2017 IEEE Intelligent Vehicles Symposium (IV),* pages 765–772, 2017.

34 E. Delande, J. Houssineau, and M. Jah. Physics and human-based information fusion for improved resident space object tracking. *Advances in Space Research*, 62(7):1800–1812, 2018.

35 Y. Bar-Shalom, X.-R. Li, and T. Kirubarajan. *Estimation with Applications to Tracking and Navigation: Theory, Algorithms and Software*. John Wiley & Sons, Inc., New York, 2001.

36 M. M. Berry. *A Variable-step Double-integration Multi-step Integrator*. PhD thesis, Virginia Tech, 2004.

37 H. Urrutxua, M. Sanjurjo-Rivo, and J. Peláez. Dromo propagator revisited. *Celestial Mechanics and Dynamical Astronomy*, 124(1):1–31, 2016.

38 O. R. Fernandez, J. Utzmann, and U. Hugentobler, Spook-a comprehensive space surveillance and tracking analysis tool. *Acta Astronautica*, 158: 178–184, 2019.

39 F. Schiemenz, J. Utzmann, and H. Kayal. Propagation of grid-scale density model uncertainty to orbital uncertainties. *Advances in Space Research*, 65(1):407–418, 2020.

40 I. U. Dundar. Improvement of a space surveillance tracking analysis tool. Master's thesis, Lulea University of Technology, Department of Computer Science, Electrical and Space Engineering, Space Technology, 2018.

41 M. Möckel, J. Bennett, E. Stoll, and K. Zhang. High performance orbital propagation using a generic software architecture. In *Proceedings of the Advanced Maui Optical and Space Surveillance Technologies Conference,* pages 20–23, 2016.

42 J. S. McCabe, K. J. DeMars, and C. Frueh. Integrated detection and tracking for multiple space objects. In *AAS Space Flight Mechanics Meeting,* 2015.

43 J. Sola, J. Deray, and D. Atchuthan. A micro Lie theory for state estimation in robotics. *arXiv preprint arXiv:1812.01537,* 2018.

44 S. Calinon and N. Jaquier. Gaussians on Riemannian manifolds for robot learning and adaptive control. *arXiv preprint arXiv:1909.05946,* 2019.

45 D. Schuhmacher, B.-T. Vo, and B.-N. Vo. A consistent metric for performance evaluation of multi-object filters. *IEEE transactions on signal processing*, 56(8):3447–3457, 2008.

46 M. Beard, B. T. Vo, and B.-N. Vo. A solution for large-scale multi-object tracking. *IEEE Transactions on Signal Processing*, 68:2754–2769, 2020.

47 T. T. D. Nguyen, H. Rezatofighi, B.-N. Vo, B.-T. Vo, S. Savarese, and I. Reid. How trustworthy are performance evaluations for basic vision tasks? *IEEE Transactions on Pattern Analysis and Machine Intelligence*, 45(7):8538–8552, 2023.

48 K. Bernardin and R. Stiefelhagen. Evaluating multiple object tracking performance: the CLEAR MOT metrics. *EURASIP Journal on Image and Video Processing*, 2008:1–10, 2008.

49 I. D. Gebru, X. Alameda-Pineda, F. Forbes, and R. Horaud. Em algorithms for weighted-data clustering with application to audio-visual scene analysis. *IEEE Transactions on Pattern Analysis and Machine Intelligence*, 38(12):2402–2415, 2016.

50 S. Askari.Fuzzy C-means clustering algorithm for data with unequal cluster sizes and contaminated with noise and outliers: Review and development. *Expert Systems with Applications*, 165: 113856, 2021.

51 K. Granström, L. Svensson, Y. Xia, J. Williams, and Á. F. García-Femández. Poisson multi-Bernoulli mixture trackers: Continuity through random finite sets of trajectories. In *2018 21st International Conference on Information Fusion (FUSION)*, IEEE, pages 1–5, 2018.

52 L. Cament, M. Adams, and P. Barrios. Space debris tracking with the Poisson labeled multi-Bernoulli filter. *Sensors*, 21(11):3684, 2021. https://doi.org/10.3390/s21113684.ISSN 1424-8220.

53 D. A. Vallado and W. D. McClain. *Fundamentals of Astrodynamics and Applications*, volume The Space Technology Library. Microcosm Press, 4th edition, 2013.

16

Extended and Group Target Modeling and Estimation

Weifeng Liu[1]*, Yun Zhu*[2]*, and Xiaomeng Cao*[1]

[1] *Electrical and Control Engineering, Shaanxi University of Science and Technology, Xi'an, China*
[2] *School of Computer Science, Shaanxi Normal University, Xi'an, China*

16.1 Introduction

When radar is positioned at a considerable distance from a target, the target typically exhibits point source characteristics and is thus commonly modeled as a point target. Consequently, conventional radar target tracking methods [1–4] generally assume that a target produces at most one measurement. However, with the rapid advancement of radar technology and the continuous improvement of radar resolution, radars are now capable of receiving multiple echo signals (measurements) from a single target. These measurements not only capture the target's motion dynamics but also convey information about its structural features, such that the target exhibits extended target characteristics [5, 6]. For example, when detecting an aircraft, radar echoes may originate from multiple scattering points such as the aircraft nose, fuselage, wings, and tail, resulting in an extended target representation. Group targets are typically composed of multiple cooperative sub-targets [7]. If each sub-target is located within a different resolution cell of the sensor, the group target is referred to as a resolvable group target. Conversely, if all sub-targets fall within the same resolution cell, the group target is classified as an unresolvable one, while other cases falling in between are referred to as partially resolvable group targets [8]. Each sub-target within the group generates at most one measurement, and the group as a whole produces multiple measurements [9]. In general, extended targets and group targets share many similar characteristics in terms of dynamic modeling, state estimation, and shape estimation. First, both generate multiple measurements that reflect certain aspects of the target's shape or structure [10]. Second, under unresolvable conditions, they both exhibit similar extended target behaviors. The core objective of extended/group target estimation is to infer their shape, structure, overall state, and the states of individual sub-targets. During the motion of extended or group targets, the cooperative interactions among multiple groups can vary dynamically—for example, a group may split into multiple smaller groups, or multiple groups may merge into a single larger group. The study of group merging and splitting has become a significant research focus.

16.1.1 Extended and Group Target Estimation

The estimation of the states and cardinality of extended or group targets primarily refers to the estimation of their positions, velocities, trajectories, and the number of targets. In recent years, significant progress and breakthroughs have been achieved in this area, leading to the emergence of a wide range of methods. Based on their underlying theoretical frameworks, these methods can be broadly categorized into three groups: data association-based methods, random finite set (RFS)-based methods, and machine learning-based methods.

1. Data Association-based Methods: Most of these methods are developed from traditional multi-object tracking algorithms. By establishing the correspondence between targets and measurements, and integrating this with state estimation techniques, these methods enable the estimation of the states of extended or group targets. Classic algorithms include the nearest neighbor (NN) [11, 12], the global nearest neighbor (GNN) [13], the probabilistic data association (PDA) [14], the joint probabilistic data association (JPDA) [2, 15], the multiple hypothesis tracking (MHT) [3], the probabilistic multiple hypothesis tracking (PMHT) [16–19], and the n-dimensional assignment (n-dimension assignment) [19]. In 1999, D. J. Salmond et al. [20] first introduced the Bayesian framework into group target tracking and proposed a particle filter (PF)-based algorithm capable of jointly estimating the states and cardinality of group targets. To enhance the estimation accuracy of the original PF method, [21] further developed two improved approaches: the PF algorithm weighted by a generalized likelihood (GL) function, and the box particle filter (Box-PF). Both algorithms improve estimation accuracy and reduce peak estimation errors, albeit at the cost of increased computational time. In group target tracking methods based on data association, measurements must be associated with targets, but the computational complexity of this association process increases exponentially with the number of extended measurements. To mitigate this issue, heuristic clustering algorithms are typically employed to facilitate the association process.
2. RFS-based Methods: RFS theory [22] constitutes another major framework for addressing group target tracking problems. These approaches model multiple, extended, or group targets as a set-valued random variable, where both the number of targets and their states are random. A key advantage of this framework is to establish the probabilistic description model under multi-target scenarios, thereby avoiding explicit data association challenges between multiple targets and multiple measurements. The concept of random sets was first introduced by the French geologist Matheron [23] in the 1950s, who also proposed the notion of random set functions, laying an important theoretical foundation for subsequent research on RFS. Mahler [22, 24] established the theoretical foundation for the engineering application of RFSs and applied it to the field of target tracking. In 2009, Mahler proposed the extended target probability hypothesis density (ET-PHD) filter [25], based on a Poisson-distributed extended target model. To address the integral computation challenges of the ET-PHD filter, a Gaussian Mixture Probability Hypothesis Density filter for extended target tracking (ET-GM-PHD) was introduced in [26, 27]. Under the linear Gaussian assumptions, this filter effectively addresses the problem of multi-extended target tracking in cluttered environments. Under nonlinear and non-Gaussian conditions, Ref. [28] proposed a partially resolvable group target tracking algorithm based on the sequential Monte Carlo probability hypothesis density filter (SMC-PHDF). This method employs Gaussian mixture models (GMMs) to fit the resampled particle distributions from the SMC-PHDF, where the number of mixture components and their parameters correspond to the number and states of the groups, respectively. The algorithm is capable of efficiently tracking partially resolvable group targets under nonlinear and non-Gaussian scenarios, albeit at the cost of higher

computational complexity. Furthermore, to enhance the tracking performance of the traditional ET-GM-PHD method, Ref. [29] proposed the extended target cubature information Gaussian mixture PHD (ET-CIGM-PHD) filter. This filter utilizes cubature information filtering (CIF) and gating techniques to predict and update the Gaussian mixture components of the ET-GM-PHD method. Additionally, an adaptive birth intensity initialization method was introduced. The ET-CIGM-PHD filter is suitable for estimating both the number and states of multiple extended/group targets in nonlinear scenarios. Nevertheless, extended/group target tracking algorithms based on the PHD filter remain sensitive to variations in the number of targets and are highly susceptible to clutter and false alarms. To address these issues, literature [30] proposed the extended target cardinalized probability hypothesis density (ET-CPHD) filter. The PHD/CPHD-based group target tracking algorithms are first-order moment approximations of RFS theory and lack higher-order moment information. To address this limitation, Mahler proposed the multi-Bernoulli multi-target tracking algorithm in 2007 [24]. In 2013, B. Ristic et al. [31] introduced the extended target Bernoulli filter (ET-BF), which can effectively track a single extended target but is incapable of tracking multiple extended targets. To overcome this issue, References [32, 33] developed the measurement update equations for the cardinality balanced multi-Bernoulli (CBMeMBer) filter based on a Poisson extended target measurement model. They improved and extended the original CBMeMBer filter to the multi-extended target tracking scenario and proposed a multi-extended target tracking algorithm based on the CBMeMBer filter. Subsequently, further improvements were made by other researchers. For example, [34] addressed the limitation of the CBMeMBer algorithm assuming that the number of measurements generated by an extended target follows a Poisson distribution and proposed a CBMeMBer filter based on a binomial measurement model. In [35], to tackle the inaccuracy of measurement locations in extended target tracking, a box particle-based CBMeMBer filter was proposed. To enhance the tracking performance of CBMeMBer-based filters, in 2020, Granström K combined the Bernoulli and Poisson distributions and proposed the Poisson multi-Bernoulli mixture (PMBM) conjugate prior algorithm for multi-extended target tracking, which achieves higher estimation accuracy and lower errors compared to CBMeMBer-based filters [36].

The aforementioned RFS-based group target tracking algorithms focus on estimating the set of targets, where the elements within the set are unordered. Consequently, there is no direct correspondence between elements in the estimated sets at consecutive time steps, which leads to only obtaining the target states at each time instant without reconstructing the individual target trajectories. To address this limitation, Vo et al. introduced the concept of the labeled random finite set (Labeled-RFS, L-RFS) [37, 38], aiming to assign a unique label to each target, thereby enabling the extraction of complete target trajectories. They proposed the generalized labeled multi-Bernoulli (GLMB) filter. However, the GLMB filter incurs a high computational cost and poses certain challenges in practical implementation. To reduce the computational burden, Reuter et al. proposed the labeled multi-Bernoulli (LMB) filter [39], which allows for dynamic grouping and parallel computation, significantly reducing execution time. Nevertheless, since the GLMB/LMB algorithms were originally developed for point target tracking, their direct application to group target tracking is subject to certain limitations. After extensive research efforts, many scholars have successfully extended the GLMB/LMB frameworks to the domain of group target tracking.

In 2008, Koch et al. [5] modeled group targets using a Gaussian inverse-Wishart (GIW) distribution, where the kinematic state of the group target was modeled by a Gaussian distribution and the shape was modeled by an inverse-Wishart (IW) distribution. They also derived update equations for the group target's state and shape. However, this model faced challenges in accurately estimating the measurements of the group target. To address this issue, Granström et al. [40] extended the GIW model by introducing a Gamma distribution to model the number of measurements, thereby proposing the Gamma Gaussian Inverse-Wishart (GGIW) model. The GGIW model enables the simultaneous estimation of the kinematic state, shape, and measurement intensity of the group target. Beard et al. [41, 42] further advanced this framework by modeling the centroid state of the group target using a GLMB distribution, while maintaining the group shape modeling with the GGIW distribution. They proposed a GGIW-GLMB based group target tracking algorithm, which not only estimates the group target's state, cardinality, and motion trajectories but does so at the group level rather than at the level of individual members. To further describe the internal structure among group members, references [43–46] introduced graph theory to characterize the cooperative relationships within the group. They proposed a tracking algorithm for distinguishable group targets, enabling the estimation of the states, number of targets, and number of subgroups.

3. Machine Learning-based Methods: In recent years, with the rapid development of neural networks and deep learning, outstanding performance has been achieved across various fields, including target tracking. In 2013, Wang et al. [47] applied deep learning to single-target tracking and achieved promising results. In 2017, Milan et al. [48] employed two recurrent neural networks (RNNs) for multi-target tracking: one RNN was used to predict target states, while the other performed sample association. These two networks worked collaboratively to accomplish effective multi-target tracking. Bae et al. [49, 50] replaced traditional features such as histogram of oriented gradients (HOG) and local binary patterns (LBP) with deep features, achieving notable improvements. Although deep learning-based target tracking algorithms emerged relatively recently, a small amount of data may be one of the main problems limiting the development of radar target tracking.

16.1.2 Extended/Group Target Extension Shape Estimation

Another important direction in group target tracking is estimating the extended shape of group targets. Typically, when radar scans a group target, it produces multiple measurements including both targets and clutter. Group target shape estimation uses these measurements to estimate the extended shape of the group. Current estimation algorithms are mainly divided into three categories based on group shape modeling methods: random matrix model-based shape modeling methods, random hypersurface model-based shape modeling methods, and star-convex model-based shape modeling methods.

1. Random Matrix Model-based Shape Modeling Methods: The shape modeling approach based on the random matrices (RM) framework is widely adopted due to its simple filtering equations, computational efficiency, and strong robustness. In 2008, Koch et al. [5, 6] introduced the concept of indistinguishable targets and integrated it with random matrix theory, proposing an extended target tracking algorithm within the classical Bayesian framework. This method generalizes conventional single-point target tracking algorithms to accommodate single extended targets, allowing for the simultaneous estimation of both the centroid and the approximate elliptical shape of the target. However, its applicability is restricted to single extended targets in clutter-free environments. Subsequently, refinements proposed in [51, 52] further enhanced the model's accuracy and robustness. To address the challenge of estimating multiple extended targets in cluttered environments, [53] developed a multi-extended-target

shape modeling and tracking algorithm based on the RFS framework and Gibbs sampling, employing equivalent measurements to estimate the targets' elliptical shapes.

2. Random Hypersurface Model-based Shape Modeling Methods: The Random Hypersurface Model (RHM) constitutes an alternative approach for group shape modeling, offering higher estimation accuracy compared to random matrix-based methods. In 2009, Baum et al. [54, 55] introduced the concept of the extended target RHM, wherein each measurement source is assumed to be an element randomly generated from a hypersurface. The validity of this approach was demonstrated through the estimation of elliptical-shaped targets. However, it is not well suited for estimating group target shapes under nonlinear conditions. To address this limitation, an improved adaptive Extended Kalman Filter based on the RHM was proposed in [56]. Both filters exhibit strong performance in estimating group target shapes in nonlinear environments. Moreover, to further improve the efficiency of group shape estimation within nonlinear dynamic systems, a convolution particle filter (CPF) was developed in [57]. Nevertheless, it should be noted that these modeling methods are primarily restricted to the estimation of group targets with elliptical shapes.
3. Star-convex Model-based Shape Modeling Methods: Most group target shapes estimated using modeling approaches based on the random matrix model (RMM) and the RHM are elliptical, which limits their ability to accurately capture star-convex shaped group targets. To address this, a group shape modeling method based on a star-convex model was proposed in [58]. In [59], a star-convex random hypersurface model (SRHM) was introduced to model the spatial diffusion of measurements and estimate the group target shape. For the problem of star-convex group shape estimation under nonlinear system conditions, extensive research and analysis have been conducted in [60–63], leading to the development of appropriate estimation algorithms, such as the SRHM-based Rao-Blackwellised particle filter (RPF), GM-PHD filter, SMC-PHD filter, and the expectation-maximization (EM) algorithm.

16.2 Problem Description

16.2.1 Morphological Description for Extended Targets

Extended target tracking morphing method. In computer graphics, there is a simple and effective method for deforming shapes, which involves changing the positions of several control points on the shape. Using the moving least squares method, after moving several control points on a simple shape to new positions, a new and different shape will be generated. Therefore, when given an undeformed simple shape, the deformed complex shape can be fully characterized by these control points, thereby simplifying the description of the shape. Based on this, a morphing method for extended target tracking is proposed.

Consider the following system model:

$$x_{k+1} = f(x_k, w_k) \tag{16.1}$$

$$z_l^k = h\left(x_k, v_l^k\right), \quad l = 1, \dots, N \tag{16.2}$$

The equation above describes the target dynamic model and the sensor measurement model, where x_{k+1} and x_k are the state vector at time $k+1$ and k respectively, f is the state transition function. A total of N extended target measurements were obtained at time k, with z_l^k being one of them. In (17.2), $h(\cdot, \cdot)$ represents the measurement equation. w_k and v_l^k represent mutually independent process noise and lth measurement noise at time k, both with Gaussian distributions.

For the sake of notational brevity, the time k will be omitted in the subsequent analysis, as the description applies to an arbitrary time frame. Unlike point target tracking, in extended target tracking, the state vector $x = \left[(x^c)^{\mathrm{T}}, (x^s)^{\mathrm{T}}\right]^{\mathrm{T}}$ to be estimated consists of two parts, where $x^c = \left[p_x, \dot{p}_x, p_y, \dot{p}_y\right]^T$ represents the state of the target center, which includes the position $\left[p_x, p_y\right]^{\mathrm{T}}$ and velocity $\left[\dot{p}_x, \dot{p}_y\right]^{\mathrm{T}}$ of the target center. $(x^s)^T$ is a random vector that characterizes the target's shape. $Z = \{z_l\}_{l=1}^{n}$ represents the set of these n measurements. Since the measurement z_l^* comes from the noise-free measurement, z_l from the extended target, the measurement model can be further written as:

$$z_l = z_l^* + v_l \tag{16.3}$$

The actual target shape can be viewed as being obtained by deforming a simple, smooth figure. The simple, smooth figure is called the reference figure. Let n control point set be $p = \{p_1, p_2, p_3, \cdots, p_n\}$ on the boundary of the reference figure. By moving the control points p to new positions $q = \{q_1, q_2, \cdots, q_n\}$ and applying a deformation function, the original reference figure will be deformed to generate a new figure, as shown in Figure 16.1.

In this model, given the deformation function, the deformed figure is influenced by three factors: the reference figure, the control points p, and the new positions q of the control points after movement. In this model, the reference figure and control points are assumed to be known: the reference shape is an ellipse and the control points p lie on the boundary of this ellipse, with the n lines connecting the n control points and the center of the ellipse dividing 2π into n equal parts. Therefore, as long as the new positions q of the control points are obtained, the target shape at next time can be completely characterized. So, the parameter vector x^s of the shape can be written as:

$$x^s = \left[(p_1)^T, (p_2)^T, \cdots, (p_n)^T\right]^T \tag{16.4}$$

The real target shape can be regarded as obtained by deforming the reference shape, so we can derive that the noise-free measurement z_l^* on the extended target is obtained by deforming a point y_l on the reference figure. This point is called the data true source. Therefore, the measurement equation is:

$$z_l = z_l^* + v_l = g(y_l, p, q) + v_l \tag{16.5}$$

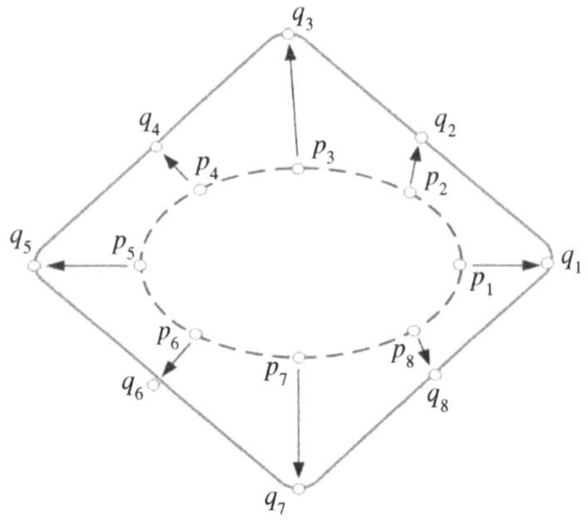

Figure 16.1 Morphological transformation.

Where $g(y_l, p, q)$ is a deformation function that maps the point y_l in the undeformed figure to the deformed figure. When control points p are mapped to q, the point y_l will be mapped to the noise-free measurement z_l^* through the deformation function, which can be obtained based on the moving least squares method.

When the data true source y_l is given, the deformation function $g(y_l, p, q)$ is linear with respect to the moved control points q. Based on this, the measurement model in the above equation is a linear equation with respect to the shape parameters FF to be estimated, and the measurement model will be further expressed as:

$$z_l = H_l x + \upsilon_l \tag{16.6}$$

Where z_l is the lth measurement, H_l is observation matrix, υ_l is the noise vector.

16.2.2 Group Targets

Based on the relationship between sensor detection group measurements and resolution units, group targets can be classified into resolvable group targets, partially resolvable group targets, and unresolvable group targets. Specifically:

1. Resolvable Group Targets:

At different positions along their trajectories, target measurements fall into different resolution cells. As illustrated in Figures 16.1 and 16.2, two targets (represented by a triangle and a circle) are moving horizontally to the right at constant velocities. The distances between the targets and the radar are continuously changing, and the geometric configuration formed by the two targets and the radar also varies over time. Consequently, the resolution probability between targets fluctuates accordingly.

To simplify the resolvability of group targets, the resolvable coefficient of radar group targets is introduced. Since the radar's angular resolution increases as the detection distance increases, when the distance between the target and the radar increases during movement, it will also lead to increased detection errors. Assuming this error follows a normal distribution $\mathcal{N}(\mu, \sigma^2)$, the radar's observation covariance is shown as follows:

$$R_k^{i,s} = \mathrm{cov}\left(\upsilon_k^{i,s}, \upsilon_k^{i,s}\right) = \left(I_k + D_k(i,s)L^{-1}\right)R_0^s \tag{16.7}$$

$$D_k(i,s) = \mathrm{cov}\left(x_{k,i}, x_{k,s}\right) \tag{16.8}$$

$$L = \mathrm{diag}\left(\left[L_x, L_y\right]\right) \tag{16.9}$$

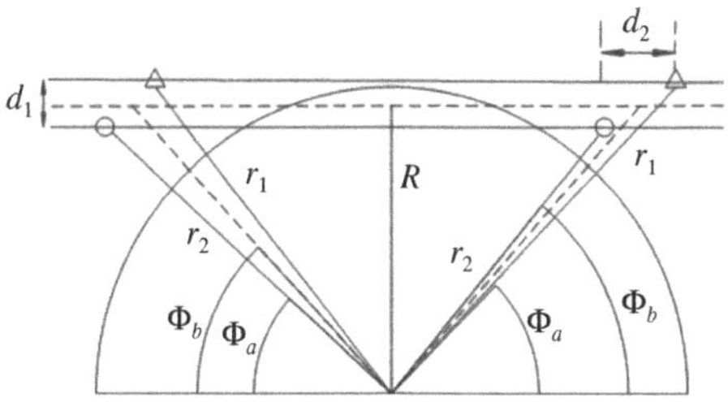

Figure 16.2 Scene illustration [64] / with permission of Springer Nature.

Where $v_k^{i,s}$ represents noise the measurement noise vector corresponding to the ith target detected by the sth radar at time step k; I_k represents the identity matrix at time k; L is a constant matrix; L_x and L_y are the scaling parameters for the x and y dimensions, respectively; $D_k(i, s)$ represents the state difference between target i and radar s; R_s^0 is the radar's constant matrix specific of the sth radar and $\|\cdot\|_2$ represents the 2-norm.

Through Eq. (16.7), the following covariance matrix can be obtained:

$$R = \begin{bmatrix} r_x^2 & 0 \\ 0 & r_y^2 \end{bmatrix} \tag{16.10}$$

Based on the elements of this covariance matrix, the noise standard deviation can be obtained as shown below:

$$\sigma = \sqrt{r_x^2 + r_y^2} \tag{16.11}$$

Assuming the noise follows a Gaussian distribution, as shown in Figure 16.3, using the 3σ criterion, suppose the resolvable distance between two targets is $r_{i,j}$, and it satisfies the following equation:

$$r_{i,j} = 3\sigma_i + 3\sigma_j \tag{16.12}$$

$$r_{i,j} = 3\left\{[tr(R_i)]^{\frac{1}{2}} + tr\left(R_j\right)^{\frac{1}{2}}\right\} \tag{16.13}$$

Where i and j represent target i and target j, respectively; $tr(\cdot)$ represents the trace operation of a matrix. Let the distance between the two targets be $d_{i,j} = ||x_i - x_j||_2$. For example, as shown in Figure 16.4, when the distance between targets A and B is greater than r, it means these two targets are resolvable.

The resolvability coefficient is defined through the ratio of the distance between targets $d_{i,j}$ and the resolvable distance $r_{i,j}$. Overall, the resolvability of group targets is defined as follows:

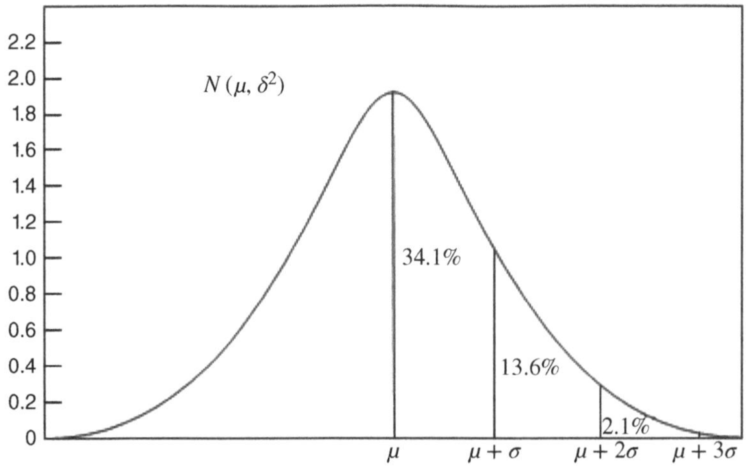

Figure 16.3 Schematic diagram of the Gaussian distribution.

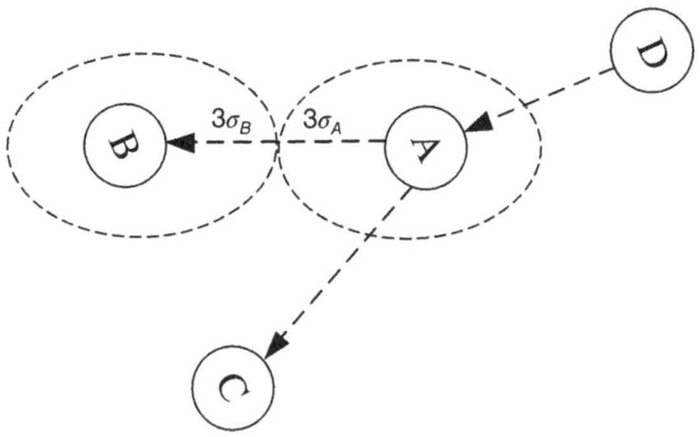

Figure 16.4 Resolvable group targets.

- Resolvable Group Targets: for any two targets i and $j(i \neq j)$ in a group target, if the formula (16.14) is satisfied, then the group target is regarded as a resolvable group.

$$\min_{i \neq j}\left(\frac{d_{i,j}}{r_{i,j}}\right) > 1 \tag{16.14}$$

where min(·) represents the minimum value for all i and j.

- Partially Resolvable Group Targets: for all targets i and $j(i \neq j)$ in a group target, if the formula (16.15) is satisfied, then the group target is called a partially resolvable group:

$$\min_{i \neq j}\left(\frac{d_{i,j}}{r_{i,j}}\right) \leq 1, \max_{i \neq j}\left(\frac{d_{i,j}}{r_{i,j}}\right) > 1 \tag{16.15}$$

where max(·) represents the maximum value for all i and $j(i \neq j)$.

- Unresolvable Group Targets: when the resolvability coefficient satisfies Eq. (16.16), the group target is called an unresolvable group target.

$$\max_{i \neq j}\left(\frac{d_{i,j}}{r_{i,j}}\right) \leq 1 \tag{16.16}$$

16.2.3 Graph Theory Model of Group Target Structure

The structure of group targets is closely related to graph theory, which provides a theoretical foundation for processing dependencies within group targets. At time k, a graph consists of some vertices and the connections between these vertices, which are called edges. Usually a graph is denoted as $G_k = (V_k, E_k)$. Where $V_k = \{v_{k,1}, \cdots, v_{k,n}\}$ represents the set of vertices, $E_k = V_k \times V_k$ represents the edges of the graph, representing the dependency relationship between two vertices. When viewing group targets from the perspective of graph theory, group members are equivalent to vertices in graph theory, and the cooperative relationships between group members are equivalent to edges in graph theory. According to whether the edges have direction, graphs can be divided into directed graphs and undirected graphs. The adjacency matrix of a directed graph is often asymmetric, while the adjacency matrix of an undirected graph is often symmetric.

The adjacency matrix describes whether there is a relationship between graph nodes. Its element $a(i, j)$ represents the connection status between the ith and jth target, In an unweighted graph, $a(i, j) = 1$ indicates a connection exists, where $a(i, j) = 0$ indicates no connection. The diagonal elements are set to 0, implying that a target does not have a dependency relationship with itself. For an undirected graph with symmetric relationships, the adjacency matrix A_d is given by:

$$A_d = \begin{bmatrix} 0 & a(1,2) & \cdots & a(1,n) \\ a(2,1) & 0 & \cdots & a(2,n) \\ \vdots & \vdots & \ddots & \vdots \\ a(n,1) & a(n,2) & \cdots & 0 \end{bmatrix} \tag{16.17}$$

If considering the parent-child relationship between nodes, a directed adjacency matrix can be used. When the ith target is the parent node of the jth target, $a(i,j) = 1$, otherwise $a(i,j) = 0$. For example: the group relationship consisting of seven targets is shown in Figure 16.5, the arrow points from the parent node to the child node, and the adjacency matrix corresponding to this group can be expressed as matrix (16.18):

$$A_d = \begin{bmatrix} 0 & 1 & 1 & 0 & 0 & 0 & 0 \\ 0 & 0 & 0 & 1 & 1 & 0 & 0 \\ 0 & 0 & 0 & 0 & 0 & 1 & 1 \\ 0 & 0 & 0 & 0 & 0 & 0 & 0 \\ 0 & 0 & 0 & 0 & 0 & 0 & 0 \\ 0 & 0 & 0 & 0 & 0 & 0 & 0 \\ 0 & 0 & 0 & 0 & 0 & 0 & 0 \end{bmatrix} \tag{16.18}$$

16.2.4 Mathematical Model of Resolvable Group Target Motion

The set of parent nodes of the ith node in a resolvable group target is $P(i)$, considering the following dynamic modeling:

$$x_{+,i} = \sum_{l \in P(i)} \omega(l,i)[F_l x_l + b(l,i)] + B_i w_i \tag{16.19}$$

$$z_i = H x_i + v_i \tag{16.20}$$

The weight coefficients must satisfy the following normalization constraint:

$$x_i \in X, \ \sum_{l \in P(i)} \omega(l,i) = 1, \omega(l,i) \in [0,1] \tag{16.21}$$

For simplicity, let $x_{+,i}$ represent the state of target i at the next moment step $k+1$, including the position and velocity; z_i represents the measurement information generated by target i at the current time step k; $l$represents the parent node of target i, $b(k,l)$ is the displacement vector

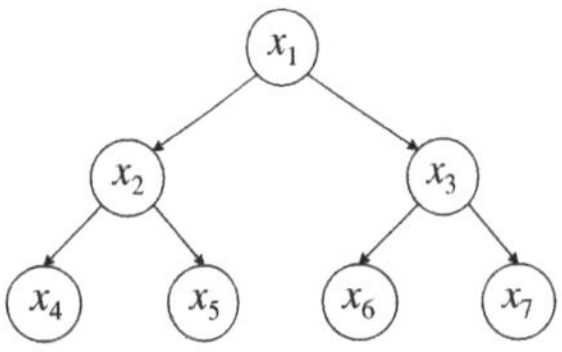

Figure 16.5 Schematic diagram of group target structure model.

between target i and its parent node l. $P(i)$ represents the set of all parent nodes of target i; $\omega(l, i)$ is the weight coefficient, representing the degree of dependence of the target on its parent node; F_l and H represent the state transition matrix and observation matrix respectively; B_i is the state noise coefficient matrix; w_i and v_i following a normal distribution, are the system and observation noise of target i respectively.

A target with no parent nodes is called a head node, and its motion is not affected by other targets. Therefore, in the head node motion model, the displacement vector $b(l, i)$ and x_i is the state of target i at the current moment, that is:

$$x_{+,i} = F_i x_i + B_i w_i \tag{16.22}$$

If a target has only one parent node, its linear motion model is given as follows:

$$x_{+,i} = F_l x_l + b(l, i) + B_i w_i \tag{16.23}$$

16.3 Extended/Group Target Tracking

16.3.1 GGIW Distribution Function

An extended/group target consists of multiple point targets, especially for unresolvable group targets, which present the same extended properties as extended targets, and their tracking methods adopt the same methods as extended targets [5, 6]. Assuming that the number of measurements generated by the cluster target at the current moment is n, that is $Z = \{z_i\}_{i=1}^{r}$, the attribute variables describing the cluster include three: the cluster target measurement number variable r, the cluster center point motion state x, and the group shape variable Σ. For this, it is necessary to establish the joint distribution function of these three types of variables $\xi = (r, x, \Sigma)$. The commonly used distribution function description model is the GGIW density function. Therefore, the extended variables of each extended target can be regarded as following a GGIW distribution.

1. The Number of Measurements of Group Targets r: the range of the gamma distribution is $(0, +\infty)$, that is, it only takes positive values, which is suitable for describing some nonnegative random variables, such as waiting time, life span, etc. The probability density function is given by:

$$G(r; \lambda, \alpha) = \frac{\lambda^{\alpha}}{\Gamma(\alpha)} r^{\alpha-1} e^{-\lambda r} \tag{16.24}$$

where $r > 0, \lambda > 0, \alpha > 0, \lambda$ is the rate parameter, α is the shape parameter, $\Gamma(\alpha)$ is the gamma function defined as:

$$\Gamma(\alpha) = \int_{0}^{\infty} t^{\alpha-1} e^{-t} dt \tag{16.25}$$

By changing the shape parameter α and the rate parameter λ, the gamma distribution can present various shapes, including unimodal, skewed, etc. When α is small, the distribution shows right skewness; as α increases, the distribution gradually becomes symmetric.

2. Extended/Group State x: this variable describes the state evolution of the group centroid, which follows μ Gaussian distribution function $\mathcal{N}(x; \mu, P)$ with mean parameter μ and a positive definite covariance matrix P.

3. Extended/Group Extension W: in Bayesian analysis, the inverse Wishart distribution is commonly employed to model the covariance matrix of a multivariate normal distribution. It updates the prior knowledge of the covariance matrix by incorporating observational data to obtain the posterior distribution. The distribution function is described as $\Sigma \sim W^{-1}(n, \Psi)$ below:

$$f(\Sigma) = \frac{|\Psi|^{\frac{n}{2}}}{2^{\frac{np}{2}}\Gamma_p\left(\frac{n}{2}\right)}|\Sigma|^{-\frac{n+p+1}{2}}\exp\left\{-\frac{1}{2}tr\left(\Psi\Sigma^{-1}\right)\right\} \tag{16.26}$$

Where, n denotes the degrees of freedom, Ψ is the scale matrix. The marginal distribution of the Inverse Wishart distribution remains in the form of an inverse Wishart distribution; that is, if $\Sigma \sim W^{-1}(n, \Psi)$, then the Σ submatrix also follows an inverse Wishart distribution.

Based on the above distribution, the distribution function of the extended/group shape variable under Hypothesis $\xi = (r, x, \Sigma)$ with parameter set $\theta = (\lambda, \alpha, \mu, P, n, \Psi)$ follows a GGIW distribution:

$$\begin{aligned} p(\xi) &= p(r, x, \Sigma) \\ &= p(r)p(x|\Sigma)p(\Sigma) \\ &= G(r; \lambda, \alpha)N(x; \mu, P)W^{-1}(\Sigma; n, \Psi) \\ &= \mathrm{GGIW}(\xi; \theta) \end{aligned} \tag{16.27}$$

16.3.2 Bayesian Estimation of Group Targets

Assume that the prior distribution of the group target is given by function $p(\xi_-) = p(r_-, x_-, \Sigma_-)$ and follows a GGIW distribution. Furthermore, assume that the group transition density, represented by function $f(\xi|\xi_-)$, satisfies the following conditions:

$$f(\xi|\xi_-) = f(r|r_-)f(x, \Sigma|x_-, \Sigma_-) \tag{16.28}$$

In the above expression, the number of measurements r depends only on the previous measurement number variable r_-, while the state variable x and the shape variable depend on the previous state variable and shape variable. Therefore, based on the Kolmogorov–Chapman equation, the predicted density is obtained as:

$$\begin{aligned} p_+(\xi) &= \int f(\xi|\xi_-)p(\xi_-)d\xi_- \\ &= \int f(r|r_-)p(r_-)dr_- \int f(x, \Sigma|x_-, \Sigma_-)p(x_-, \Sigma_-)dx_-d\Sigma_- \end{aligned} \tag{16.29}$$

Assuming that the prior distribution $p(\xi_-)$ follows a GGIW distribution, under certain assumptions, $p_+(\xi)$ also follows a GGIW distribution [27]:

$$\begin{aligned} p_+(\xi) &= \mathrm{GGIW}(\xi; \theta) \\ &= G(r; \lambda, \alpha)N(x; \mu, P)W^{-1}(n, \Sigma) \end{aligned} \tag{16.30}$$

The parameters for the prediction step are updated as follows:

$$\lambda = \lambda_-\left(1 - \frac{1}{\omega}\right) \tag{16.31}$$

$$\alpha = \alpha_-\left(1 - \frac{1}{\omega}\right) \tag{16.32}$$

$$\mu = F\mu_- \tag{16.33}$$

$$P = FP_-F^T + Q \tag{16.34}$$

$$n = e^{-T\tau}n_- \tag{16.35}$$

$$\Sigma = \frac{n - d - 1}{n_- - d - 1}\Sigma_- \tag{16.36}$$

Within the Bayesian framework, the posterior density of the group can be formulated as:

$$p(\xi_+|Z) = \frac{1}{C}g(Z|\xi)p_+(\xi) \tag{16.37}$$

Proposition 1 Assume that the measurements $z \in Z$ are mutually independent and follow a Gaussian distribution. The measurement likelihood function is given by [5].

$$g(Z|\xi) = \prod_{z \in Z} N(z; Hx, \Sigma) \propto N\left(z; Hx, \frac{\Sigma}{|Z|}\right)W^{-1}(Z; |Z| - 1, \Sigma) \tag{16.38}$$

Given the prior distribution $p(\xi)$ follows a GGIW distribution, the posterior distribution $p(\xi_+|Z)$ also follows a GGIW distribution [42]:

$$\begin{aligned} p(\xi_+|Z) &= \mathrm{GGIW}(\xi_+; \theta_+) \\ &= G(r_+; \lambda_+, \alpha_+)N(x_+; \mu_+, P_+)W^{-1}(n, \Sigma_+) \end{aligned} \tag{16.39}$$

The parameters for the update step are calculated as follows:

$$\lambda_+ = \lambda + |Z| \tag{16.40}$$

$$\alpha_+ = \alpha + 1 \tag{16.41}$$

$$\mu_+ = \mu + K(\bar{z} - H\mu) \tag{16.42}$$

$$P_+ = P - KSK^T \tag{16.43}$$

$$n_+ = n + |Z| \tag{16.44}$$

$$\Sigma_+ = \Sigma + S^{-1}(\bar{z} - H\mu)(\bar{z} - H\mu)^T + \sum_{z \in Z}(z - \bar{z})(z - \bar{z})^T \tag{16.45}$$

16.3.3 Learning-based Extended Target Tracking Algorithms

1. Finite Mixture Model for Extended Targets

At current time step, the extended target measurement set $z_i \in Z_e = \{Z_{e,1}, \cdots, Z_{e,n}\}$ has the following mixture distribution function:

$$f(z_i|\Theta) = \sum_{m=1}^{M} \omega_m f(z_i|\theta_m) \tag{16.46}$$

Here, M denotes the number of the extended target; ω_m denotes the mixture weight of the mth component of the extended target, and θ_m represents the parameter set of the mth component.

Assuming that the distribution of multiple extended targets consists of a uniform clutter distribution and several Gaussian distributions, where $Z = Z_e \cup Z_c$, Z_c denote the uniform clutter measurement sets, the mixture distribution of multiple extended targets can be expressed as follows:

$$f(z_i|\Theta) = \omega_1 \mathcal{U}(z_i|V) + \sum_{m=2}^{M} \omega_{\mathrm{m}} \mathcal{N}(z_i; u_{\mathrm{m}}, D_{\mathrm{m}}) \tag{16.47}$$

Here, m represents the mth measurement at the current time; $\theta_m = (u_m, D_m)$ represents the mean vector u_m and covariance matrix D_m of the mth Gaussian component, respectively. The extended target estimation can be expressed through the following Bayesian equation:

$$p(\Theta|Z) \propto g(Z|\Theta)p(\Theta) \tag{16.48}$$

Assuming that the measurements are independent of each other, the measurement likelihood function is expressed by the following formula:

$$g(Z|\Theta) = \prod_{i}^{n} f(z_i|\Theta) = \prod_{i}^{n} \left[\omega_1 \mathrm{U}(z_i) + \sum_{j=2}^{m} \omega_j N(z_i; u_j, D_j) \right] \tag{16.49}$$

Here, $\Theta = \{\omega_1, \omega_j, u_j, D_j\}_{j=2}^{m}$, the prior distribution of the extended target parameters is a joint distribution, making it difficult to directly estimate individual parameters. Therefore, a conditional distribution approach is adopted.

$$p(\Theta) = p(m)p(\omega|m)p(D|\omega, m)p(u|D, \omega, m) \tag{16.50}$$

Here, $\omega = \{\omega_1, \cdots \omega_m\}$, $u = \{u_2, \cdots u_m\}$, $D = \{D_2, \cdots D_m\}$ denote the set of mixture weights, the set of mean vectors, and the set of covariance matrices respectively.

The posterior distribution of the parameters is derived as follows:

$$p(\Theta|Z) = p(\omega|Z)p(D|\omega, Z)p(u|D, \omega, Z) \tag{16.51}$$

The covariance matrix follows an inverse Wishart distribution with degrees of freedom equal to β_j:

$$p\left[(D_j)^{-1} | \omega_j, m_j\right] = W(V_j, \beta_j) \tag{16.52}$$

2. Multiple Extended Target Association Diagram

In the GLMB filtering algorithm, the association graph plays a crucial role in the analysis of the RFS multi-target likelihood function. The association graph represents a mapping: $\upsilon : \mathbb{L} \rightarrow \{0, 1 \cdots, |Z|\}$, where $\upsilon(i) = \upsilon(i') > 0$, which implies $i = i'$. The association graph space is represented by $\mathbb{V}$, and $\mathbb{V}(I)$ denotes a subset I. In this context, the relationship between measurements and targets is described by the association graph. The measurement generated by target l is represented by $z_{\upsilon(l)} \in Z$, and targets that are not detected are marked with 0. Assuming that the detection of clutter and target measurements is independent, the multi-target likelihood function is as follows:

$$g(Z|X) = e^{-\langle k,1 \rangle} k^Z \sum_{\upsilon \in \mathbb{V}(L(X))} [\psi_Z(\cdot; \upsilon)]^X \tag{16.53}$$

where

$$\psi_Z(x, l; \upsilon) = \begin{cases} \dfrac{p_D(x, l) g(z_{\upsilon(l)}|x, l)}{k(z_{\upsilon(l)})}, & if \quad \upsilon(l) > 0 \\ 1 - p_D(x, l), & if \quad \upsilon(l) = 0 \end{cases} \tag{16.54}$$

It is assumed that the measurements of multiple extended targets are represented by $Z_e = \{Z_{e,1}, \cdots, Z_{e,n}\} \in \mathcal{F}(z)$, where $Z_{e,n}$ denotes the set of measurements generated by the nth extended target at the present time.

$$Z_{e,n} = \{z^e_{n,1}, \cdots, z^e_{n,m_n}\} \tag{16.55}$$

This paper establishes a measurement-limited mixture model and employs learning algorithms to derive the equivalent measurement set for extended targets:

$$\hat{Z} = \{\hat{z}_1, \cdots, \hat{z}_n\} \tag{16.56}$$

Here, $\hat{z}_n$ represents the equivalent measurement of the nth extended target at the current time.

The multi-extended target association graph is a mapping: $\upsilon_e : \mathbb{L} \to \{0, 1 \cdots, |\hat{Z}|\}$, where $\upsilon_e(i) = \upsilon_e(i') > 0$, indicating $i = i'$. The association graph space for multiple extended targets is represented by the set $\mathbb{V}_e$, with $\mathbb{V}_e(I)$ denoting the subset $\upsilon(i) = \upsilon(i') > 0$. The correspondence between extended target l and the individual measurements is described by the association graph space. The measurements generated by extended target l are represented by $z_{\upsilon_e(l)} \in Z$, and undetected extended targets are marked with 0. Similarly, assuming that clutter and target measurements are independent, the likelihood function for multiple extended targets is expressed by the Eq. (16.58):

$$g(\hat{Z}|X) = e^{-\langle k,1 \rangle} k^{\hat{Z}} \sum_{\upsilon_e \in \mathbb{V}_e(L(X))} [\psi_{\hat{Z}}(\cdot; \upsilon_e)]^X \tag{16.57}$$

where

$$\psi_{\hat{Z}_k}(x, l; \upsilon_e) = \begin{cases} \dfrac{p_D(x, l) g(z_{\upsilon_e(l)}|x, l)}{k(z_{\upsilon_e(l)})} & if \quad \upsilon_e(l) > 0 \\ 1 - p_D(x, l), & if \quad \upsilon_e(l) = 0 \end{cases} \tag{16.58}$$

The posterior probability representation of multiple extended targets is given as follows:

$$\pi(X|Z_e) = \frac{g(Z_e|X)\pi(X)}{\int g(Z_e|X)\pi(X)\delta X} \tag{16.59}$$

In the above equation, the set integral $\int$ in the denominator is defined by the following expression:

$$\int f(X)\delta X = \sum_{i=0}^{\infty} \frac{1}{i!} \sum_{(\ell_1,\cdots \ell_n)\in \mathbb{L}^i} \times \int_{\mathbb{X}^i} f(\{x_1, \ell_1\}, \cdots \{x_i, \ell_i\}) d(x_1, \cdots x_i) \tag{16.60}$$

3. Extended Target GLMB Filtering

The GLMB filtering algorithm is used to track extended targets. The GLMB filter mainly consists of two steps: the prediction step and the update step. Let the extended target state be denoted as (x, ℓ) and the probability of the target being detected be $p_D(x, \ell)$; then the probability of the target not being detected is $q_D(x, \ell) = 1 - p_D(x, \ell)$. The standard GLMB filtering algorithm is closed under Bayesian recursion, and its algorithm is defined as follows [65]:

$$\pi(\mathbf{X}) = \Delta(\mathbf{X}) \sum_{c\in C} \omega^{(c)}(\mathcal{L}(\mathbf{X}))\left[p^{(c)}\right]^{\mathbf{X}} \tag{16.61}$$

The computation of the above formula is relatively complicated. Vo reformulated Eq. (16.61) into the following form, which is referred to as δ-GLMB filtering:

$$\pi(\mathbf{X}) = \Delta(\mathbf{X}) \sum_{(I,\xi)\in F(\mathbb{L})\times\Xi} \omega^{(I,\xi)} \delta_I(\mathcal{L}(\mathbf{X}))\left[p^{(\xi)}\right]^{\mathbf{X}} \tag{16.62}$$

Assuming that the prior distribution of the target follows Distribution δ-GLMB, the prediction step for multiple extended targets is expressed as shown in Eq. (16.63):

$$\pi_+(\mathbf{X}_+) = \Delta(\mathbf{X}_+) \sum_{(I_+,\xi)\in F(L)\times\Xi} \omega_+{}^{(I_+,\xi)} \times \delta_{I_+}(L(X_+))\left[p_+^{(\xi)}\right]^{X_+} \tag{16.63}$$

When the GLMB prediction density for multiple extended targets satisfies Eq. (16.63), the update step follows Eq. (16.64). During the update step, it is necessary to partition and associate multiple measurements generated by the multiple extended targets. However, due to the large number of possible association events, prediction-based methods lack sufficient accuracy and suffer from high computational cost. Therefore, a Gibbs sampling iterative learning algorithm is adopted to implement the update step.

$$\pi(\mathbf{X}|Z) \approx \Delta(\mathbf{X}) \sum_{(I,\xi)\in F(\mathbb{L})\times\Xi} \sum_{\theta\in\Theta^{(M)}} \tilde{\omega}^{(I,\xi,\theta)} \times \delta_I(L(\mathbf{X}))\left[p^{(\xi,\theta)}\right]^{X} \tag{16.64}$$

Where $\theta : \mathbb{L} \to \{0, 1, \cdots, |Z|\}$, $\theta(i) = \theta(i') > 0$ indicates $i = i'$, and the corresponding parameters are expressed as follows:

$$w^{(I,\xi,\theta)}(Z) = \frac{\delta_{\theta^{-1}(\{0:|Z|\})}(I) w^{(I,\xi)}\left[\eta_Z^{(\xi,\theta)}\right]^I}{\sum_{(I,\xi)\in\mathcal{F}(\mathbb{L})\times\Xi} \sum_{\theta\in\Theta} \delta_{\theta^{-1}(\{0:|Z|\})}(I) w^{(I,\xi)}\left[\eta_Z^{(\xi,\theta)}\right]^I} \tag{16.65}$$

$$p^{(\xi,\theta)}(x, l|Z) = \frac{p^{(\xi)}(x, l)\psi_Z(x, l; \theta)}{\eta_Z^{(\xi,\theta)}(l)} \tag{16.66}$$

$$\eta_Z^{(\xi,\theta)}(l) = \left\langle p^{(\xi)}(\cdot, l)\psi_Z(\cdot, l; \theta)\right\rangle \tag{16.67}$$

$$\psi_Z(x, l; \theta) = \delta_0(\theta(l))q_D(x, l) + (1 - \delta_0(\theta(l)))\frac{p_D(x, l)g\left(z_{\theta(l)}|x, l\right)}{k\left(z_{\theta(l)}\right)} \tag{16.68}$$

Building upon the estimation of the target state, further learn the shape of the extended target.

4. Extended Target Shape Estimation

The Gibbs sampling algorithm is used to estimate the mean, covariance, and weights of the Gaussian distributions. Then, the Bayesian information criterion (BIC) is employed to assess the fitting accuracy of the Gaussian distributions. For modeling the shape of the extended target, an elliptical model is adopted, where the mean is used to replace the center measurement of the extended target, i.e., the center point of the ellipse. The shape of the extended target is iteratively learned. The overall process of the estimation algorithm is shown in Table 16.1.

Table 16.1 Complete procedure of extended target tracking estimation algorithms.

Algorithm 1: Extended target tracking and estimation

Input: *Initial target states, Prior parameters*
Output: *Estimated tracks and target shapes at each time step k*

1. **Initialize** Time Step $k = 1$
2. **Loop**
3. **Step** 1: **Prediction**
4. Perform GLMB Prediction Process
5. **Step** 2: **Partitioning and Optimization**
6. BIC Gibbs Sampling Algorithm and BIC Evaluation
7. **Step** 3: **Measurement Processing**
8. Acquiring extended target measurements and Computation of Equivalent Measurements to Replace Extended Measurements
9. **Step** 4: **Update**
10. Perform GLMB Update Process
11. **Output** Estimates for Current Step
12. $k \leftarrow k + 1$
13. **End Loop**

Gibbs Sampling is a Markov Chain Monte Carlo (MCMC) method used for sampling from a multivariate probability distribution. For a joint probability distribution with multiple variables, Gibbs Sampling generates samples by successively sampling from the conditional distribution of each variable at each step. That is, given the current values of the other variables, a new value is drawn from the conditional distribution of each variable. Gibbs sampling can obtain posterior distribution samples of parameters under the condition of known covariance data and prior distributions of parameters. Based on Gibbs sampling, the BIC is used to evaluate the goodness of fit between the extended target finite mixture model and the true data distribution. The BIC criterion is defined by the following formula:

$$\mathrm{BIC}(m, \Theta, Z) = -2 \log L(\Theta, m|Z) + M \ln(n) \tag{16.69}$$

Where M denotes the number of independent parameters, Θ represents the parameter set, and $\log L(\Theta, m|Z)$ refers to the log-likelihood function of the number of elements m.

$$M = 3m + 2 \tag{16.70}$$

$$L(\Theta, m|Z) = \prod_{i=1}^{n} f(z_i|\Theta, m) \tag{16.71}$$

When the BIC value reaches its minimum, priority should be given to estimating the number of extended target elements. The value of m is related to the parameter learning process. Once the learning process stabilizes, the learning process halts, and the estimation of the extended target parameters is derived from the following expression:

$$(\hat{m}, \hat{\Theta})s = \operatorname{argmin}_{(m,\Theta)} \mathrm{BIC}(m, \Theta) \tag{16.72}$$

Extended Target Mixture Model:

$$f(z|\hat{m}, \hat{\Theta}) \propto \omega_1 U(S) + \sum_{j=2}^{\hat{m}} \omega_j \mathcal{N}(z, \mu_j, D_j) \tag{16.73}$$

Through the above algorithm, the weights, means, covariances, and BIC values are obtained. The equivalent measurement $\hat{z}_j$ of the jth extended target is used to replace the measurement of the extended target. Taking the means and covariances as input variables, and using the mean as the center point, the shape of the extended target is approximated by an ellipse. The Gibbs parameter learning algorithm is then employed to estimate the shape of the extended target given in Table 16.2.

Table 16.2 Steps of the Gibbs sampling and BIC algorithm.

Algorithm 2: The Gibbs sampling and BIC

Gibbs sampling

Input: $\mu_j \sim U(S), D_j = S^{\frac{1}{z_{\dim} I}}/4$

Output: $\{\omega_j^{(q)}, \mu_j^{(q)}, D_j^{(q)}, \mathrm{BIC}(m), \hat{z}_{k,j}\}$

Initialize $q = 0,$

    for $q \leftarrow 0$ to $I, I = 20$ **do**

        $t = t + 1$

        $(\omega_1, \cdots, \omega_m) \sim D\left(a_1 + l_1^{(t-1)}, \cdots, a_m + l_m^{(t-1)}\right)$

        for $j \leftarrow 1$ to m ***do***

            $u_{i,j}^{(q)} = N\left(z_i; u_j^{(t-1)}, D_j^{(t-1)}\right)$

            $e_{i,j}^{(q)} = \omega_j u_{i,j}^{(q)} / \sum_j \omega_j u_{i,j}^{(q)}$

            $l_j^{(q)} = \sum_i e_{i,j}^{(q)}$

            $\xi_j^{(q)} = \sum_i e_{i,j}^{(q)} z_i / l_j^{(q)}$

            $K_j^{(q)^2} = \sum_i \left(z_i - \xi_j^{(q)}\right)\left(z_i - \xi_j^{(q)}\right)^T e_{i,j}^{(q)} / l_j^{(q)}$

            $D_j^{(q)^{-1}} \sim W\left(I + K_j^{(q)^2}/2, \beta_0 + l^q/2\right)$

            $u_j^{(q)} \sim N\left(\xi_j^{(q)}, D_j^{(q)}\right)$

        End

    End

BIC calculation

$\mathrm{BIC}(m, \Theta, Z) = -2 \log L(\Theta, m|Z) + M \ln(n)$

16.4 Conclusions

Extended and group targets are characterized by three random parameters: the number of measurements, the motion (centroid) state, and the shape, all of which are modeled using the Generalized Gamma Inverse Wishart (GGIW) distribution. Based on this, this section first discusses the resolvability properties of extended targets, the dynamic modeling of group targets, and the methods for state representation and estimation.

The main approach for extended target tracking is based on the GGIW distribution. Due to the presence of shape attributes in the extended target, the parameters describing the shape are modeled using the inverse Wishart distribution, which generates RM that are symmetric and positive definite. This implies that the extended target is approximated using a convex elliptical shape. For targets with more complex external shapes, improvements can be made by modifying the single inverse Wishart matrix.

References

1 Y. Bar-Shalom. Tracking methods in a multitarget environment. *IEEE Transactions on Automatic Control*, 23(4):618–626, 1978.

2 T. Fortmann, Y. Bar-Shalom, and M. Scheffe. Sonar tracking of multiple targets using joint probabilistic data association. *IEEE Journal of Oceanic Engineering*, 8(3):173–184, 1983.

3 D. Reid. An algorithm for tracking multiple targets. *IEEE Transactions on Automatic Control*, 24(6):843–854, 1979.

4 S. S. Blackman. Multiple hypothesis tracking for multiple target tracking. *IEEE Aerospace and Electronic Systems Magazine*, 19(1):5–18, 2004.

5 J. W. Koch. Bayesian approach to extended object and cluster tracking using random matrices. *IEEE Transactions on Aerospace and Electronic Systems*, 44(3):1042–1059, 2008.

6 M. Feldmann, D. Fränken, and W. Koch. Tracking of extended objects and group targets using random matrices. *IEEE Transactions on Signal Processing*, 59(4):1409–1420, 2011.

7 W. Liu, S. Zhu, C. Wen, and Y. Yu. Structure modeling and estimation of multiple resolvable group targets via graph theory and multi-bernoulli filter, Automatica, 89:274–289, 2018.

8 F. Lian, C.-Z. Han, W-F. Liu, and X-H. Yuan. Tracking partly resolvable group targets using SMC-PHDF. Acta Automatica Sinica, 36(5):731—741, 2010.

9 H. Chen, F-Q. Wei, and C-Z. Han. Sensor control based on multiple feature optimization in multiple extended targets tracking. *Journal of Electronics & Information Technology*, 45(1):191–199, 2023.

10 W-F. Liu, Z. Chai, and C-L. Wen. Multi-measurement target tracking by using random sampling approach. *Acta Automatica Sinica*, 39(2):168–178, 2013.

11 R. A. Singer and R. G. Sea. A new filter for optimal tracking in dense multitarget environments. In *Annual Allerton Conference on Circuit and System Theory, 9th*. Monticello. IEEE, pages 201–211, 1972.

12 R. A. Singer and J. J. Stein. An optimal tracking filter for processing sensor data of imprecisely determined origin in surveillance systems. In *1971 IEEE Conference on Decision and Control*. IEEE, Miami Beach, pages 171–175, 1971.

13 P. D. Konstantinova, A. Udvarev, and T. Semerdjiev. A study of a target tracking algorithm using global nearest neighbor approach. In *Proceedings of the 4th International Conference on Computer Systems and Technologies: E-Learning*. Association for Computing Machinery, Rousse, pages 290–295, 2003.

14 Y. Bar-Shalom and E. Tse. Tracking in a cluttered environment with probabilistic data association. *Automatica*, 11(5):451–460, 1975.

15 S. He, H. S. Shin, and A. Tsourdos. Information-theoretic joint probabilistic data association filter. *IEEE Transactions on Automatic Control*, 66(3):1262–1269, 2020.

16 R. L. Streit and T. E. Luginbuhl. Maximum likelihood method for probabilistic multihypothesis tracking. In *Signal and Data Processing of Small Targets 1994*. SPIE, Orlando, 2235, pages 394–405, 1994.

17 R. L. Streit and T. E. Luginbuhl. A probabilistic multi-hypothesis tracking algorithm without enumeration and pruning. In *Proceedings of the Sixth Joint Service Data Fusion Symposium*. IEEE, Miami. pages 1015–1024, 1993.

18 H. X. Gaetjens, S. J. Davey, S. Arulampalam, F. K. Fletcher, and C-C. Lim. Histogram-PMHT for fluctuating target models. *IET Radar, Sonar & Navigation*, 11(8):1292–1301, 2017.

19 K. R. Pattipati, S. Deb, Y. Bar-Shalom, and R. B. Washburn. A new relaxation algorithm and passive sensor data association. *IEEE Transactions on Automatic Control*, 37(2):198–213, 1992.

20 D. J. Salmond and N. J. Gordon. Group and extended object tracking. In *IEE Colloquium on Target Tracking: Algorithms and Applications*, pages 16–17, 1999.

21 Z-X. Li, J.-M. Liu, S. Li, D.-Y. Bai, and P. Ni. Group targets tracking algorithm based on box particle filter. *Acta Automatica Sinica*, 41(4):785—798, 2015.

22 R. Mahler. "Statistics 102" for multisource-multitarget detection and tracking. *IEEE Journal of Selected Topics in Signal Processing*, 7(3):376–389, 2013.

23 G. Matheron. Random sets and integral geometry. *Mining, Metallurgy & Exploration*, 4:96–102, 1987.

24 R. P. S. Mahler. *Statistical Multisource-multitarget Information Fusion*. Artech House, Norwood, MA, USA, 2007, 108–119.

25 R. Mahler. PHD filters for nonstandard targets, I: Extended targets. In *2009 12th International Conference on Information Fusion*. IEEE, Seattle, pages 915–921, 2009.

26 K. Granström, C. Lundquist, and U. Orguner. A gaussian mixture PHD filter for extended target tracking. In *2010 13th International Conference on Information Fusion*. IEEE, Edinburgh, pages 1–8, 2010.

27 K. Granström, C. Lundquist, O. Orguner. Extended target tracking using a Gaussian-mixture PHD filter. *IEEE Transactions on Aerospace and Electronic Systems*, 48(4):3268–3286, 2012.

28 K. Punithakumar, T. Kirubarajan, and A. Sinha. A sequential Monte Carlo probability hypothesis density algorithm for multitarget track-before-detect. In *Signal and Data Processing of Small Targets 2005*. SPIE, Philadelphia, pages 587–594, 2005.

29 Z. Liu, L. Ji, F. Yang, X. Qu, Z. Yang, and D. Qin. Cubature information Gaussian mixture probability hypothesis density approach for multi extended target tracking. *IEEE Access*, 7:103678–103692, 2019.

30 U. Orguner, C. Lundquist, and K. Granström. Extended target tracking with a cardinalized probability hypothesis density filter. In *14th International Conference on Information Fusion*. IEEE, Chicago, pages 1–8, 2011.

31 B. Ristic and J. Sherrah. Bernoulli filter for detection and tracking of an extended object in clutter. In *2013 IEEE Eighth International Conference on Intelligent Sensors, Sensor Networks and Information Processing*. IEEE, Melbourne, 306–311, 2013.

32 G. Zhang, F. Lian, and C. Han. CBMeMBer filters for nonstandard targets, I: extended targets. In *17th International Conference on Information Fusion*. IEEE, Salamanca, pages 1–6, 2014.

33 F. Lian, D-D. Ma, X-H. Yuan, W. Chen, and C-Z. Han. CBMeMBer filter for extended targets and its Gaussian mixture implementations. *Control and Decision*, 30(4):611–616, 2015.

34 W. Li, H. Gu, W. Su, J. Yang, and M. Xia. CBMeMBer filter for extended target tracking based on binomial measurement number model. In *2019 IEEE MTT-S International Microwave Biomedical Conference (IMBioC)*. IEEE, Nanjing, 1, pages 1–5, 2019.

35 S. Wei, X. Feng, and Y. Kong. CBMeMBer filter for extended object tracking using box particle. In *2016 First IEEE International Conference on Computer Communication and the Internet (ICCCI)*. IEEE, Wuhan, pages 454–458, 2016.

36 K. Granström, M. Fatemi, and L. Svensson. Poisson multi-Bernoulli mixture conjugate prior for multiple extended target filtering. *IEEE Transactions on Aerospace and Electronic Systems*, 56(1): 208–225, 2019.

37 B-T. Vo and B-N. Vo. A random finite set conjugate prior and application to multi-target tracking. In *2011 Seventh International Conference on Intelligent Sensors, Sensor Networks and Information Processing*. IEEE, Adelaide, pages 431–436, 2011.

38 B-N. Vo and B-T. Vo. Phung D. Labeled random finite sets and the Bayes multi-target tracking filter. *IEEE Transactions on Signal Processing*, 62(24): 6554–6567, 2014.

39 S. Reuter, B-T. Vo, B-N. Vo, and K. Dietmayer. The labeled multi-Bernoulli filter. *IEEE Transactions on Signal Processing*, 62(12): 3246–3260, 2014.

40 K. Granström and U. Orguner. Estimation and maintenance of measurement rates for multiple extended target tracking. In *2012 15th International Conference on Information Fusion*. IEEE. Singapore, pages 2170–2176, 2012.

41 M. Beard, S. Reuter, K. Granström, B.-T. Vo, B.-N. Vo, and A. Scheel. A generalised labelled multi-Bernoulli filter for extended multi-target tracking. In *2015 18th International Conference on Information Fusion (Fusion)*. IEEE, Washington, pages 991–998, 2015.

42 M. Beard, S. Reuter, K. Granström, B.-T. Vo, B.-N. Vo, and A. Scheel. Multiple extended target tracking with labeled random finite sets. *IEEE Transactions on Signal Processing*, 64(7): 1638–1653, 2015.

43 A. Gning, L. Mihaylova, S. Maskell, S. K. Pang, and S. Godsill Group object structure and state estimation with evolving networks and Monte Carlo methods. *IEEE Transactions on Signal Processing*, 59(4): 1383–1396, 2010.

44 S-J. Zhu, W-F. Liu, and H-L. Cui. Multiple resolvable groups tracking using the GLMB Filter. *Acta Automatica Sinica*, 43(12): 2178–2189, 2017.

45 W. Liu, Y. Chi, and G. Zhang. Multiple resolvable group estimation based on the GLMB filter with graph structure. In *2018 IEEE 8th Annual International Conference on CYBER Technology in Automation, Control, and Intelligent Systems (CYBER)*. IEEE, Tianjin, pages 960–964, 2018.

46 Y. Chi and W. Liu. Resolvable group state estimation with maneuver movement based on labeled RFS. In *2018 International Conference on Control, Automation and Information Sciences (ICCAIS)*. IEEE, Zhengzhou, pages 249–254, 2018.

47 N. Wang and D. Y. Yeung. Learning a deep compact image representation for visual tracking. *Advances in Neural Information Processing Systems*, 26, 2013.

48 A. Milan, S. H. Rezatofighi, and A. Dick. Online multi-target tracking using recurrent neural networks. In *Proceedings of the AAAI Conference on Artificial Intelligence*. AAAI, San Francisco, 31, pages 4225–4232, 2017.

49 S. H. Bae and K. J. Yoon. Confidence-based data association and discriminative deep appearance learning for robust online multi-object tracking. *IEEE Transactions on Pattern Analysis and Machine Intelligence*, 40(3): 595–610, 2017.

50 A. Sadeghian, A. Alahi, and S. Savarese. Tracking the untrackable: Learning to track multiple cues with long-term dependencies. In *Proceedings of the IEEE International Conference on Computer Vision*. IEEE, Himeji, pages 300–311, 2017.

51 M. Feldmann and D. Franken. Advances on tracking of extended objects and group targets using random matrices. In *2009 12th International Conference on Information Fusion*. IEEE, Seattle, pages 1029–1036, 2009.

52 J. Lan and X. R. Li. Tracking of extended object or target group using random matrix: New model and approach. *IEEE Transactions on Aerospace and Electronic Systems*, 52(6): 2973–2989, 2016.

53 Y-M. Chen, W-F. Liu, M-X. Kong, and G-L. Zhang. A modeling and tracking algorithm of finite mixture models for multiple extended target based on the GLMB Filter and gibbs sampler. *Acta Automatica Sinica*, 46(7): 1445–1456, 2020.

54 M. Baum and U. D. Hanebeck. Random hypersurface models for extended object tracking. In *2009 IEEE International Symposium on Signal Processing and Information Technology (ISSPIT)*. IEEE, Ajman, pages 178–183, 2009.

55 M. Baum, B. Noack, and U. D. Hanebeck. Extended object and group tracking with elliptic random hypersurface models. In *2010 13th International Conference on Information Fusion*. IEEE. Edinburgh, pages 1–8, 2010.

56 T. Ma, Q. Zhang, C. Chen, and S. Gao. Tracking of maneuvering star-convex extended target using modified adaptive extended kalman filter. *IEEE Access*, 8: 214030–214038, 2020.

57 D. Angelova, L. Mihaylova, N. Petrov, and A. Gning. A convolution particle filtering approach for tracking elliptical extended objects. In *Proceedings of the 16th International Conference on Information Fusion*. IEEE, Istanbul, pages 1542–1549, 2013.

58 M. Baum and U. D. Hanebeck. Shape tracking of extended objects and group targets with star-convex RHMs. In *14th International Conference on Information Fusion*. IEEE, Chicago, pages 1–8, 2011.

59 H. Chen, J-R. Du, and C-Z. Han. A multiple extended target multi-Bernouli Filter based on star-convex random hypersurface model. *Acta Automatica Sinica*, 46(5): 909–922, 2020.

60 E. Özkan, N. Wahlström, and S. J. Godsill. Rao-Blackwellised particle filter for star-convex extended target tracking models. In *19th International Conference on Information Fusion (FUSION)*. IEEE, Heidelberg, pages 1193–1199, 2016.

61 J. Steinbring, M. Baum, A. Zea, F. Faion, and U. D. Hanebeck. A closed-form likelihood for particle filters to track extended objects with star-convex RHMs. In *2015 IEEE International Conference on Multisensor Fusion and Integration for Intelligent Systems (MFI)*. IEEE, San Diego, pages 25–30, 2015.

62 L. Sun, J. Zhang, H. Yu, Z. Fu, and Z. He. Maneuvering extended object tracking with modified star-convex random hypersurface model based on minimum cosine distance. *Remote Sensing*, 14(17): 4376, 2022.

63 H. Kaulbersch, M. Baum, and P. Willett. EM approach for tracking star-convex extended objects. In *2017 20th International Conference on Information Fusion (Fusion)*. IEEE, Xi'an, pages 1–7, 2017.

64 W. Koch. *Tracking and Sensor Data Fusion*. John Wiley & Sons, Hoboken, NJ, USA, 2014.

65 B-T. Vo and B-N. Vo. Labeled random finite sets and multi-object conjugate priors. *IEEE Transactions on Signal Processing*, 61(13): 3460–3475, 2013.

17

Random Finite Sets Meet Simultaneous Localization and Mapping

Martin Adams[1], *Felipe Inostroza*[2], *and Keith Leung*[3]

[1] *Department of Electrical Engineering and Advanced Mining Technology Center Universidad de Chile, Santiago, Chile*
[2] *Advanced Mining Technology Center, Universidad de Chile, Santiago, Chile*
[3] *Applanix Corporation (Trimble), Richmond Hill, ON, Canada*

17.1 Introduction

Navigation, mapping, and tracking are fundamental state estimation challenges that play crucial roles in various domains, including mobile robotics, computer vision, and aerospace. Navigation involves determining one's position and orientation (pose), commonly known as the localization problem. Mapping focuses on identifying the positions of typically stationary objects of interest. When these two estimation tasks are integrated, the problem is referred to as simultaneous localization and mapping (SLAM) in robotics and structure from motion (SFM) in computer vision. If the map contains dynamic objects of interest, the problem is usually termed multi-target tracking.

The robotic SLAM problem is one of the fundamental problems in autonomous navigation and its automatic solution is crucial for almost any mobile robotic application. It has therefore been considered by some to be the "Holy Grail" of mobile robotics [1].

Probabilistic methods are commonly used to address these state estimation challenges. SLAM and multi-target tracking have historically been addressed via Bayesian estimation, as summarized in [2]. This chapter therefore begins with a brief survey of historical probabilistic SLAM solutions and an introduction to the core SLAM problem.

In recent years, nonlinear optimization (or batch estimation) methods that have traditionally been used in solving SFM and photogrammetry problems, have become more prominent in solving feature-based SLAM problems. Under this premise, remarkable demonstrations of large-scale feature-based SLAM have been made in which, given data association (DA), the SLAM problem can be solved by use of nonlinear least square solvers, often referred to as the SLAM *back-end*. These methods rely on external methods to solve both the DA and map management problems, which collectively comprise the SLAM *front-end*. Despite impressive random vector (RV)-SLAM demonstrations under favorable environmental conditions, convergence failures are common when these front-end heuristics fail, particularly when feature detection uncertainty increases.

Irrespective as to whether filtering or smoothing is adopted, the problems have traditionally been formulated using RVs to represent the state. More recently, the random finite set (RFS) formulation was introduced to the target tracking and, to a lesser extent, the feature-based SLAM community. It treats the map feature state as a finite-set-valued random variable, and hence the cardinality (size) of the set (number of map features) is also a random variable. The RFS estimation framework, calculates the multi-feature distribution and provides not only spatial estimates but also estimates of the number of expected map features. Fundamentally, it circumvents the necessity of often fragile external[1] DA decisions, thus eliminating the need for a SLAM front-end.

Therefore, the first major objective of this chapter is an evaluation of the strong connection between RFS-SLAM and the more conventional RV-SLAM frameworks. This analysis ultimately shows that the feature-based RFS-SLAM approach serves as a generalization of its traditional RV-SLAM predecessor. At the same time, fundamental questions are raised regarding the strict validity of how DA is addressed in state-of-the-art RV-SLAM, with RFS-SLAM concepts providing answers.

The remarkable performance of nonlinear least-squares, batch-based SLAM solvers on large data sets, combined with the strengths of RFS-based SLAM solutions in avoiding the reliance on potentially fragile SLAM front-ends, creates a strong motivation to develop batch RFS-based SLAM algorithms.

Consequently, the second major objective of this chapter is to develop a batch RFS-based SLAM concept. In particular, the RFS framework is integrated with graph optimization methods traditionally applied in RV-based batch SLAM solutions. This fusion results in the development of the batch, RFS-based, joint vector-set SLAM (JVS-SLAM) algorithm. This algorithm exploits the correlations among SLAM state components and capitalizes on the sparsity of the SLAM state through the use of the sparse information matrix.

It will be demonstrated that JVS-SLAM provides a tractable solution with high-accuracy SLAM results, validated with the challenging EuRoC data set. In this data set, a drone equipped with a stereo camera performs SLAM while undergoing rapid accelerations. The results illustrate that JVS-SLAM matches or exceeds the performance of leading-edge SLAM methods across varying levels of uncertainty, while eliminating the need for explicit DA.

17.2 A Brief History of SLAM

During the last decades, remarkable progress has been made in SLAM research and this section reviews some of the most important advances.

17.2.1 State-of-the-art RV-based SLAM Solutions

Over the past decades, multiple solutions to the SLAM problem, which model the SLAM state as a RV, have been proposed. Survey papers which describe these solutions, together with advantages and disadvantages are given in [2–5].

The following SLAM solutions are of direct relevance to the concepts introduced in this chapter.

Thrun and Montemerlo [6] proposed a strategy to solve the SLAM problem, referred to as GraphSLAM. This is also referred to in the literature as batch estimation or bundle adjustment in the field of computer vision. GraphSLAM consists of progressively building a graph of soft constraints (referred to in [6] as information constraints). Using this graph, the SLAM solution is obtained by minimizing a cost function, which considers all possible measurements to SLAM state association possibilities and odometry information over time, based on an initial anchoring constraint (assumed initial robot pose). The minimization problem can be solved in many ways, but in [6], it was solved via linearization and solving it in the information space. Then, not only the mode is recovered but also its associated covariance.

1 External to the Bayesian estimation process.

Kümmerle, Grisetti, Strasdat, Konolige, and Burgard [7] introduced a library, named g2o, to solve general least squares optimization problems such as the one introduced in [6]. In this library, both odometry measurements, as well as landmark measurements, relate different temporal SLAM states to each other. It is capable of using the special structure of the landmarks in SLAM, which do not have measurements that relate them to each other, to increase the efficiency of the solver.

Kaess, Ranganathan, and Dellaert [8] introduced incremental smoothing and mapping (iSAM), which solves SLAM using a Bayes tree which enables iSAM to solve the optimization formulation in an incremental fashion. The Bayes tree formulation is equivalent to the Cholesky square root of the Hessian matrix. This allows the solver to partially use the previous solution to incrementally solve the next optimization when adding new poses and measurements to the optimization problem. iSAM2 [9] is a follow-up to iSAM and it uses the Bayes tree representation while checking dynamically whether the linearization point of each measurement needs to be updated.

Mur-Artal, Montiel, and Tardós [10] proposed a complete monocular visual SLAM solution, using g2o [7] as a back-end and the oriented FAST rotated brief (ORB) keypoint detector and descriptor pair [11] as its front-end. In this system, ORB keypoints are extracted from each image and are matched using their descriptors. Importantly, this system uses three separate threads to accomplish real-time robust performance, *tracking*, *local mapping*, and *loop closing*. In the *tracking* thread, only the latest camera pose is estimated using the set of already estimated keypoints as a known map. The *tracking* thread determines whether or not the current frame should be a keyframe and therefore be optimized as a part of the trajectory, based on how many of the detections in the current frame are seen by keyframes already in the trajectory estimate. The *local mapping* thread inserts the keyframes given by the tracking thread into the g2o optimizer and performs the map management, i.e., creating new keypoints and removing existing keypoints from the map. The *loop closing* thread takes the latest keyframe from the *local mapping* thread and tries to detect loop closures between it and any keypose in the trajectory estimate. To achieve this, the ORB features are compared using bag of words approach.

In [12], the oriented FAST rotated brief SLAM (ORBSLAM) algorithm is extended to use stereo or RGB-D cameras and to be able to run the algorithm in localization-only mode with a fixed map. Finally, in [13], the algorithm is further extended to multisession SLAM in which the robot can start in the previously mapped area and is able to localize itself and continue the SLAM process.

17.2.2 State-of-the-art RFS-based SLAM Solutions

Mullane, Vo, and Adams [14] introduced the concept of RFSs to model the SLAM problem. By recognizing that the SLAM state is more naturally represented by a set, instead of a vector, they were able to include the DA and map management problems into the Bayesian estimation paradigm, thus eliminating the need for separate SLAM back and (heuristic-based) front-ends. This resulted in Rao-Blackwellized (RB)-probability hypothesis density (PHD)-SLAM and demonstrated significant improvements over its RV-based SLAM algorithm counterparts, under high false positive and low detection probabilities [15].

Leung, Inostroza, and Adams [16] showed that RB-PHD-SLAM is closely related to factored solution to SLAM (FastSLAM), and that RB-PHD-SLAM is equivalent to FastSLAM under ideal detection conditions, but offers significant generalizations and improvements under realistic, nonideal detection conditions. This work forms the basis of Section 17.4 in which fundamental similarities and differences between the RFS and RV SLAM concepts will be explained.

Deusch, Reuter, and Dietmayer [17] applied the labeled multi-Bernoulli (LMB) filter to SLAM and showed that it has better performance than PHD-SLAM in simulation [17].

Moratuwage, Adams, and Inostroza [18] applied the δ-generalized labeled multi-Bernoulli (δ-GLMB) filter to SLAM, also with a Rao-Blackwellized formulation.

Gao, Battistelli, and Chisci [19] improved RB-PHD-SLAM by using a proposal distribution that is closer to the posterior distribution on the latest robot pose, in a manner similar to the way that FastSLAM-2 improves FastSLAM.

To the best knowledge of the authors, Inostroza and Adams [20] introduced the first published batch-based maximum likelihood (ML)-RFS-SLAM solution, capable of employing well-established sparse graph optimization techniques, while circumventing the need for external DA and map-management heuristics. Referred to as JVS-SLAM, Section 17.5.1 builds on this concept, demonstrating highly performing visual SLAM results.

17.3 Bayesian-based SLAM Fundamentals

17.3.1 General Concepts and Definitions

SLAM is a state estimation problem, in which the best estimate of the robot trajectory and map feature positions is sought over time, using all sensor measurements. In general, the underlying stochastic system representing the robot's pose component of SLAM using nonlinear discrete-time equations, is represented as:

$$\mathbf{x}_k = \mathbf{g}(\mathbf{x}_{k-1}, \mathbf{u}_k, \delta_k) \tag{17.1}$$

$$\mathbf{z}_k^i = \mathbf{h}(\mathbf{x}_k, \mathbf{m}_k^j, \epsilon_k), \tag{17.2}$$

where:

- $\mathbf{x}_k$ represents the robot pose (spatial coordinates and orientation)[2] at discrete time k,
- $\mathbf{g}$ is the robot motion model,
- δ_k is the process noise at time k,
- $\mathbf{u}_k$ is the odometry input,[3]
- $\mathbf{z}_k^i$ is the ith measurement vector at time k,
- $\mathbf{h}$ is the sensor-specific measurement model relating measurement and state in the spatial domain,
- $\mathbf{m}_k^j$ is a RV representing the position of landmark j at time k,
- ϵ_k is the spatial measurement noise at time k.

The number of measurements/detections, n_k, produced by the measurement/detection process is a random variable. The RFS $\mathcal{Z}_k$ of measurement/detection vectors at time k is therefore given by:

$$\mathcal{Z}_k \equiv \{\mathbf{z}_k^1, \mathbf{z}_k^2, \ldots, \mathbf{z}_k^{n_k}\}. \tag{17.3}$$

Note that the model representing the collected measurements/detections, based on a sensor scan, should fundamentally be an RFS rather than an RV, since their number can, in general, change from scan to scan, and the order in which they occur has no significant meaning [15, 21]. Without loss

2 Note that, although in general, the SLAM state could include the robot's velocity and higher order components, in SLAM it is only its pose which is of interest.

3 Note that the odometry information is available at time k, corresponding to the relative motion from time $k-1$ to time k. For this reason, it cannot be considered as a measurement in the sense of (17.2), since it depends on both $\mathbf{x}_k$ *and* $\mathbf{x}_{k-1}$. In SLAM it is mathematically considered as a deterministic input, while the inherent corresponding uncertainty is modeled with δ_k.

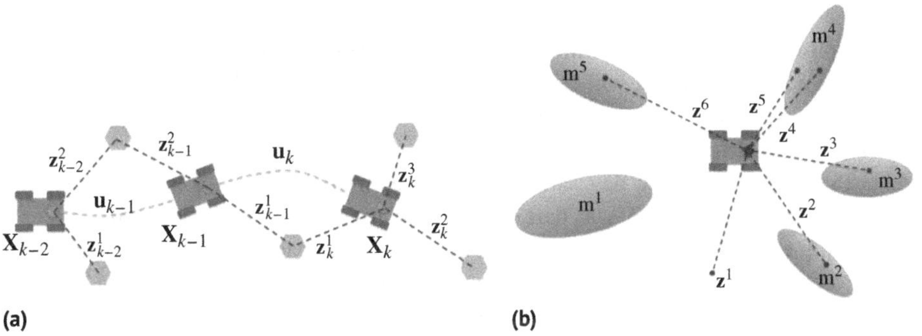

Figure 17.1 2D schematic representations of SLAM. (a) Map features with locations m_k^j are shown as gray hexagons. (b) DA in SLAM. In this example, $\theta_k^1 = 0, \theta_k^2 = 2, \ldots \theta_k^6 = 5$.

of generality, and for simplicity, a 2D schematic diagram depicting the evolution of the SLAM state and corresponding measurements is shown in Figure 17.1a. Note that map features with locations $\mathbf{m}_k^j$ are shown as gray hexagons.

17.3.2 RV-based SLAM

Traditional RV-based approaches to feature-based SLAM, concatenate random state vectors that model the robot and landmarks, in a single vector that is used for the estimation process. At discrete time k, the map component of this RV is:

$$\mathbf{M}_k \equiv \left[\mathbf{m}_k^{1\,\mathsf{T}}\ \mathbf{m}_k^{2\,\mathsf{T}}\ \cdots\ \mathbf{m}_k^{m\,\mathsf{T}}\right]^{\mathsf{T}}. \tag{17.4}$$

Let θ_k represent the correspondences, where $\theta_k^j = i$ indicates that measurement $\mathbf{z}_k^j$ from measurement RFS $\mathcal{Z}_k$ (17.3) is associated with map element $\mathbf{m}_k^i$ from the RV map (17.4). An example scenario is shown in Figure 17.1b. $\theta_k^j = 0$ if measurement $\mathbf{z}_k^j$ is not associated, as can be seen for example with measurement $\mathbf{z}_k^1$. In the RV formulation, the goal is to calculate the posterior that represents the estimated state of the vehicle $\mathbf{x}_{0:k}$, the state of all map elements that have been observed $\mathbf{M}_k$ and the unknown association history $\theta_{1:k}$:

$$p\left(\mathbf{x}_{0:k}, \mathbf{M}_k, \theta_{1:k} \middle| \mathcal{Z}_{1:k}, \mathbf{u}_{0:k}\right) \tag{17.5}$$

where the notation $k_1{:}k_2$ denotes the inclusive time interval from k_1 to k_2 for the referred quantity (e.g., $\mathbf{x}_{0:k} = \{\mathbf{x}_0, \mathbf{x}_1, \ldots, \mathbf{x}_k\}$). The estimates are made relative to the initial robot's pose, at each time step.

Equation (17.5) represents the posterior density associated with the multi-hypothesis correlation (MHC) framework for estimation. A nuanced consideration concerning (17.5) is highlighted in Mahler [22] (p. 340):

"It is not possible to estimate $\theta_{1:k}$ [the association hypothesis] as a state parameter without implicitly introducing knowledge of the observation process into the definition of a multi-target state."

The sequence $\theta_{1:k}$ inherently relies on a specific arrangement of the measurement or detection elements, represented as $\mathbf{z}_k^1, \mathbf{z}_k^2, \ldots, \mathbf{z}_k^{n_k}$ in (17.3). However, since the collected measurements or detections are unordered by nature, they are appropriately modeled as a set, as indicated in (17.3). Consequently, incorporating the association variable $\theta_{1:k}$ directly into the state representation would introduce a contradiction: The state variable would become dependent on the ordering of observations, despite the fact that the estimation process in Bayes' rule aims to be independent of such order. This conceptual conflict will be revisited in Section 17.4.1, where the measurement likelihood function based on RFS theory is explored.

In most RV-based SLAM frameworks, DA, the process of estimating or computing $\theta_{1:k}$, is typically carried out separately from the Bayesian update procedure that estimates the joint RV state $(\mathbf{x}_{0:k}, \mathbf{M}_k)$. The problem of treating $\theta_{1:k}$ as part of the state variable is generally ignored.

Before conducting estimation updates, carried out in the SLAM back-ends, to establish correspondences between map elements and measurements, dedicated DA algorithms (SLAM front-ends), such as those proposed in [23, 24], are often employed. These algorithms identify the associations independently of the state estimation process. Only after these association variables are determined is Bayes' theorem or ML estimation applied to update the joint distribution of the RV $(\mathbf{x}_{0:k}, \mathbf{M}_k)$. Consequently, the distribution that undergoes the actual update is[4]:

$$p(\mathbf{x}_{0:k}, \mathbf{M}_k | \mathcal{Z}_{1:k}, \mathbf{u}_{0:k}, \theta_{1:k}). \tag{17.6}$$

Within a probabilistic framework employing a filtering approach, this posterior is computed recursively through Bayesian estimation, conditioned on the assumed DA hypothesis $\theta_{1:k}$:

$$\begin{aligned} &p\left(\mathbf{x}_{0:k}, \mathbf{M}_k \middle| \mathcal{Z}_{1:k-1}, \mathbf{u}_{0:k}, \theta_{1:k-1}\right) \\ &= p\left(\mathbf{x}_k \middle| \mathbf{x}_{k-1}, \mathbf{u}_k\right) p\left(\mathbf{x}_{0:k-1}, \mathbf{M}_k \middle| \mathcal{Z}_{1:k-1}, \mathbf{u}_{0:k-1}, \theta_{1:k-1}\right) \end{aligned} \tag{17.7}$$

$$\begin{aligned} &p\left(\mathbf{x}_{0:k}, \mathbf{M}_k \middle| \mathcal{Z}_{1:k}, \mathbf{u}_{0:k}, \theta_{1:k}\right) \\ &= \frac{p\left(\mathcal{Z}_k \middle| \mathbf{x}_{0:k}, \mathbf{M}_k, \theta_k\right) p\left(\mathbf{x}_{0:k}, \mathbf{M}_k \middle| \mathcal{Z}_{1:k-1}, \mathbf{u}_{0:k}, \theta_{1:k-1}\right)}{p\left(\mathcal{Z}_k \middle| \mathcal{Z}_{1:k-1}, \mathbf{x}_{0:k}, \mathbf{u}_{0:k}\right)}. \end{aligned} \tag{17.8}$$

Alternatively, following the same reasoning concerning the DA variables $\theta_{1:k}$, SLAM is frequently addressed using a nonlinear least squares (batch) estimation method [6, 8]:

$$\begin{aligned} &\arg\max_{\mathbf{x}_{0:k}, \mathbf{M}_k} p\left(\mathbf{x}_{0:k}, \mathbf{M}_k \middle| \mathcal{Z}_{1:k}, \mathbf{u}_{0:k}, \theta_{1:k}\right) \\ &= \arg\max_{\mathbf{x}_{0:k}, \mathbf{M}_k} p(\mathbf{x}_0, \mathbf{M}_k) \prod_{l=1}^{k} p\left(\mathcal{Z}_l \middle| \mathbf{x}_l, \mathbf{M}_k, \theta_l\right) p\left(\mathbf{x}_l \middle| \mathbf{x}_{l-1}, \mathbf{u}_l\right). \end{aligned} \tag{17.9}$$

In this formulation, the factors within the product represent the likelihoods associated with exteroceptive measurements and the vehicle's motion model.

17.3.3 RFS-based SLAM

In the RFS approach, in contrast to (17.4), the observed landmarks up to and including time k, are mathematically defined and treated as a set

$$\mathcal{M}_k \equiv \{\mathbf{m}_k^1, \mathbf{m}_k^2, \ldots, \mathbf{m}_k^m\}, \tag{17.10}$$

4 It is important to note that (17.6) also presents a subtle problem, as the association variable $\theta_{1:k}$ implicitly assumes knowledge of the ordering of elements within $\mathbf{M}_k$, which, in a strict mathematical sense, contradicts the premise of state estimation [22, 25]. Despite this inconsistency, the joint density is presented here to illustrate the typical formulation of RV-based SLAM approaches.

where the number of landmarks, $|\mathcal{M}_k| = m$, is now a random variable. In general, the landmark from which a measurement is generated is unknown. Furthermore, there is a probability of detection, P_D, associated with every landmark, implying that it may be undetected with probability $1 - P_D$. Measurements may also be generated from sensor noise or objects of noninterest (referred to as false positives or clutter), with assumed known distributions. One could question why the use of RFSs to model the map is necessary, when traditionally the map is considered static and therefore the map size is fixed. However, in the SLAM definition, even though the map is often assumed static, it is unknown. In a manner similar to the way in which the uncertainty in the landmark positions should be represented via a distribution, even though they are assumed static, the uncertainty in the size of the map also needs to be modeled even though the map size does not change. RFS distributions jointly model the uncertainty both in the map size and spatial uncertainty in the map elements, i.e., the landmarks. In addition to this, map elements coming into the field of view (FoV) of the robot for the first time can be modeled as new map elements being born. With these definitions, a new, set-based, measurement model can be defined:

$$\mathcal{Z}_k \equiv \mathcal{H}(\mathbf{x}_k, \mathcal{M}_k, \epsilon_k) \cup \mathcal{E}_k, \tag{17.11}$$

where $\mathcal{E}_k$ represents the clutter and ϵ_k represents the spatial noise of the measurements.

Using the RFS map model (17.10), the required posterior can then be rewritten as[5]:

$$p\left(\mathbf{x}_{0:k}, \mathcal{M}_k \middle| \mathcal{Z}_{1:k}, \mathbf{u}_{0:k}\right), \tag{17.12}$$

which can also be solved by Bayesian estimation:

$$p\left(\mathbf{x}_{0:k}, \mathcal{M}_k \middle| \mathcal{Z}_{1:k-1}, \mathbf{u}_{0:k}\right) = p\left(\mathbf{x}_k \middle| \mathbf{x}_{k-1}, \mathbf{u}_k\right) p\left(\mathbf{x}_{0:k-1}, \mathcal{M}_k \middle| \mathcal{Z}_{1:k-1}, \mathbf{u}_{0:k-1}\right), \tag{17.13}$$

$$p\left(\mathbf{x}_{0:k}, \mathcal{M}_k \middle| \mathcal{Z}_{1:k}, \mathbf{u}_{0:k}\right) = \frac{p\left(\mathcal{Z}_k \middle| \mathbf{x}_{0:k}, \mathcal{M}_k\right) p\left(\mathbf{x}_{0:k}, \mathcal{M}_k \middle| \mathcal{Z}_{1:k-1}, \mathbf{u}_{0:k}\right)}{p\left(\mathcal{Z}_k \middle| \mathcal{Z}_{1:k-1}, \mathbf{x}_{0:k}, \mathbf{u}_{0:k}\right)}. \tag{17.14}$$

The RFS batch estimation framework can be represented as follows:

$$\underset{\mathbf{x}_{0:k}, \mathcal{M}_k}{\arg\max}\, p\left(\mathbf{x}_{0:k}, \mathcal{M}_k \middle| \mathcal{Z}_{1:k}, \mathbf{u}_{0:k}\right) = \underset{\mathbf{x}_{0:k}, \mathcal{M}_k}{\arg\max}\, p(\mathbf{x}_0, \mathcal{M}_k) \prod_{l=1}^{k} p\left(\mathcal{Z}_l \middle| \mathbf{x}_l, \mathcal{M}_k\right) p\left(\mathbf{x}_l \middle| \mathbf{x}_{l-1}, \mathbf{u}_l\right). \tag{17.15}$$

The batch optimization process is challenging due to the necessary joint consideration of continuous variables related to the vehicle and map state, and the discrete variable representing map cardinality. Nevertheless, solutions will be discussed in Section 17.5.

One advantage of RFS estimation is its ability to include detection statistics into the set measurement likelihood. The *probability of detection*, denoted as $P_D(\mathbf{x}_k, \mathbf{m}_k^i)$, represents the likelihood of obtaining a measurement based on a specific robot pose, $\mathbf{x}_k$, and the map element state, $\mathbf{m}_k^i$. Additionally, $p_\kappa(\mathcal{Z}_k)$ indicates the likelihood that a set of measurements corresponds to false positives.

17.4 Relationships Between RV and RFS SLAM

This section aims to demonstrate the close relationship between the RFS estimation concept and the RV framework, while highlighting key differences. It will also illustrate that by imposing certain

5 Note that (17.12) is a finite set statistics (FISST) density [22] which was shown in [26] to be equivalent to a standard probability density with respect to an unnormalized Poisson point process.

assumptions on detection and false positive statistics, as well as on the set cardinality within the RFS framework, we can establish equivalences with the RV framework for both stochastic filtering and batch estimation. In this context, the RFS model emerges as a robust generalization of its RV counterpart.

Beginning with the RFS framework, the Bayes filter equations presented in (17.13) and (17.14), as well as the batch estimation equation in (17.15), incorporate several RFS probability density functions (PDFs). In the Bayes filter prediction equation (17.13), the conditional PDF is expressed over the vehicle trajectory and the map:

$$p\left(\mathbf{x}_{0:k}, \mathcal{M}_k \middle| \mathcal{Z}_{1:k-1}, \mathbf{u}_{0:k}\right). \tag{17.16}$$

Both the filtering and batch estimation approaches share a common set measurement likelihood:

$$p\left(\mathcal{Z}_k \middle| \mathbf{x}_{0:k}, \mathcal{M}_k\right), \tag{17.17}$$

which appears in both (17.14) and (17.15). Note that (17.17) is the probability density of the measurement set, given the multi-target state. Importantly, RFS concepts capture measurement uncertainty in the measurement likelihood function in a rigorous probabilistic context, because the set integral of the measurement likelihood function, over all measurement sets, equals unity [22].

Additionally, the denominator of (17.14) includes a normalizing factor:

$$p\left(\mathcal{Z}_k \middle| \mathcal{Z}_{1:k-1}, \mathbf{x}_{0:k}, \mathbf{u}_{0:k}\right). \tag{17.18}$$

This normalizing factor is consistent across both the RFS and RV formulations, appearing in the denominators of both (17.8) and (17.14). However, the conditions under which the distributions in (17.16) and (17.17) can be considered equivalent within the RFS and RV frameworks are now examined.

Theorem 17.1 If the joint distribution for the vehicle and the RFS map assumes a deterministic cardinality for the RFS map component, and the elements of the map can only be arranged in a specific order within a RV, then the joint distribution becomes equivalent to that of the vehicle and the RV map.

Proof: Assuming that the joint distribution for the vehicle and RV map can be expressed as:

$$\begin{aligned} &p\left(\mathbf{x}_{0:k}, \mathbf{M}_k \middle| \mathcal{Z}_{1:k-1}, \mathbf{u}_{0:k}, \theta_{1:k-1}\right) \\ &= p\left(\mathbf{x}_{0:k}, \left(\mathbf{m}_k^1, \mathbf{m}_k^2, \dots, \mathbf{m}_k^m\right) \middle| \mathcal{Z}_{1:k-1}, \mathbf{u}_{0:k}, \theta_{1:k-1}\right), \end{aligned} \tag{17.19}$$

then using the finite set statistics (FISST) [25] framework, the distribution over the vehicle and the map can be written as[6]:

$$\begin{aligned} &p\left(\mathbf{x}_{0:k}, \mathcal{M}_k \middle| \mathcal{Z}_{1:k-1}, \mathbf{u}_{0:k}\right) \\ &= m_k!\; p_{|\mathcal{M}_k|}(m_k) p\left(\mathbf{x}_{0:k}, \left(\mathbf{m}_k^1, \dots, \mathbf{m}_k^{m_k}\right) \middle| \mathcal{Z}_{1:k-1}, \mathbf{u}_{0:k}\right). \end{aligned} \tag{17.20}$$

6 Noting that the density for the RFS map is a Janossy density.

Here, m_k is a random variable representing the cardinality of the RFS map. If m_k is deterministic and equal to m, then $p_{|\mathcal{M}_k|}(m) = 1$, simplifying (17.20) to:

$$m!p\left(\mathbf{x}_{0:k}, (\mathbf{m}_k^1, \dots, \mathbf{m}_k^m)\middle|\mathcal{Z}_{1:k-1}, \mathbf{u}_{0:k}\right). \tag{17.21}$$

The factorial term $m!$ accounts for all possible permutations of the elements in $\mathcal{M}_k$, with each permutation having the same likelihood, indicating symmetry. If only one specific ordering is permitted, the likelihoods of all other orderings become zero, leading to:

$$p\left(\mathbf{x}_{0:k}, (\mathbf{m}_k^1, \dots, \mathbf{m}_k^m)\middle|\mathcal{Z}_{1:k-1}, \mathbf{u}_{0:k}\right). \tag{17.22}$$

It is important to note that if (17.19) has a different ordering of map elements, the same ordering can be selected from within (17.21). If correspondence information is provided, then (17.22) can be expressed as:

$$p\left(\mathbf{x}_{0:k}, (\mathbf{m}_k^1, \mathbf{m}_k^2, \dots, \mathbf{m}_k^m)\middle|\mathcal{Z}_{1:k-1}, \mathbf{u}_{0:k}, \theta_{1:k-1}\right), \tag{17.23}$$

this is being identical to the joint likelihood of the vehicle pose and the RV map in (17.19).

The significance of Theorem 17.1 lies in its demonstration that the RV framework necessitates a deterministic map size and selects a specific ordering of map elements from all possible permutations during the estimation process. Additionally, it highlights the requirement for correspondence selection by a DA routine.

17.4.1 Notes on RV and RFS Measurement Likelihoods

In Section 17.3.2, a significant issue was highlighted regarding (17.5). It was noted that treating θ_k as a state variable necessitates a specific ordering of the measurements/detections $\mathbf{z}_k^1, \mathbf{z}_k^2, \dots, \mathbf{z}_k^{n_k}$ in $\mathcal{Z}_k$, an ordering that does not inherently exist. Mahler raised a pertinent question in [22] (p. 341):

"In choosing such an ordering, are we introducing extraneous information—and thus also a statistical bias?"

This suggests that accurately estimating the correct association hypothesis θ_k may not be feasible without imposing a priori order on the measurements/detections that have yet to be recorded.

To address this issue, it was demonstrated that a solution is necessary to derive a statistically unbiased multi-target measurement model [22]. The RFS measurement likelihood in (17.17), if based on an individual association hypothesis θ_k, would be expressed as $p\left(\mathcal{Z}_k^{\theta_k}\middle|\mathbf{x}_{0:k}, \mathcal{M}_k^{\theta_k}, \theta_k\right)$. This formulation introduces the problem of an association-dependent state in both (17.14) and (17.15). The solution involves replacing the likelihood functions $p\left(\mathcal{Z}_k^{\theta_k}\middle|\mathbf{x}_{0:k}, \mathcal{M}_k^{\theta_k}, \theta_k\right)$ with their average value with respect to θ_k, resulting in the unbiased RFS measurement likelihood $p\left(\mathcal{Z}_k\middle|\mathbf{x}_{0:k}, \mathcal{M}_k\right)$:

$$p\left(\mathcal{Z}_k\middle|\mathbf{x}_{0:k}, \mathcal{M}_k\right) \equiv \frac{1}{N_{\mathcal{Z}_k}} \sum_{\theta_k} p\left(\mathcal{Z}_k^{\theta_k}\middle|\mathbf{x}_{0:k}, \mathcal{M}_k^{\theta_k}, \theta_k\right). \tag{17.24}$$

$N_{\mathcal{Z}_k}$ serves as a normalization constant, ensuring that (17.24) integrates to unity. With this definition, the association variable θ_k is effectively marginalized out, meaning it does not influence the RFS measurement likelihood $p\left(\mathcal{Z}_k\middle|\mathbf{x}_{0:k}, \mathcal{M}_k\right)$.

In the RFS multi-object framework, the measurement likelihood can be expressed more generally in terms of the joint state's probability of detection $P_D(\mathbf{x}_k, \mathbf{m})$, the probability of misdetection $1 - P_D(\mathbf{x}_k, \mathbf{m})$, and the likelihood of the measurements $\mathcal{Z}_k$ being false positives, denoted as $p(\mathcal{Z}_k)$.

When the measurements are modeled using a multi-Bernoulli distribution, the RFS measurement likelihood function is [22]:

$$p\left(\mathcal{Z}_k \middle| \mathcal{M}_k, \mathbf{x}_{0:k}\right) = \sum_{\theta_k} \left(\underbrace{p\left(\mathcal{Z}_k^{\theta_k} \middle| \mathcal{M}_k^{\theta_k}, \mathbf{x}_{0,k}, \theta_k\right)}_{\substack{\text{Spatial measurement} \\ \text{likelihood for state –} \\ \text{measurement pairs} \\ \text{under association } \theta_k}} \underbrace{\left[\prod_{\mathbf{m} \in \mathcal{M}_k^{\theta_k}} P_D(\mathbf{x}_k, \mathbf{m}) \right]}_{\substack{\text{For all associated} \\ \text{state elements:} \\ \text{Product of their} \\ \text{detection probabilities}}} \underbrace{\left[\prod_{\mathbf{m} \in \overline{\mathcal{M}_k^{\theta_k}}} (1 - P_D(\mathbf{x}_k, \mathbf{m})) \right]}_{\substack{\text{For all unassociated} \\ \text{state elements: Product} \\ \text{of their misdetection} \\ \text{probabilities}}} \underbrace{p_\kappa\left(\overline{\mathcal{Z}_k^{\theta_k}}\right)}_{\substack{\text{Likelihood of} \\ \text{non – associated} \\ \text{measurements} \\ \text{being clutter}}} \right) \tag{17.25}$$

As indicated in (17.24), the RFS measurement likelihood is derived by averaging over all DA hypotheses θ_k, treating each hypothesis equally. Consequently, this measurement likelihood comprises additive terms corresponding to each measurement's association with the state hypothesis θ_k.

It is usually assumed that measurements are conditionally independent, when conditioned on the underlying state, which is also the case here, implying that detections are independent. For example, the detections of two objects, in which one occludes the other, are considered independent when conditioned on the states of both objects.

Theorem 17.2 The following conditions ensure the equivalence of the RFS measurement likelihood[7] to the conventional RV measurement likelihood:

1. The correspondence between measurements and map elements is explicitly defined.
2. Only the associated map elements are assumed to be detected with absolute certainty.
3. Any measurements that do not correspond to mapped elements are entirely attributed to false positives.

Proof: Based on the correspondence θ_k, define the following sets to represent the associated and nonassociated measurements, as well as the associated and nonassociated map elements, respectively:

$$\mathcal{Z}_k^{\theta_k} \equiv \left\{ \mathbf{z}_k^j \in \mathcal{Z}_k | \theta^j \neq 0 \right\}, \qquad \overline{\mathcal{Z}_k^{\theta_k}} \equiv \mathcal{Z}_k \setminus \mathcal{Z}_k^{\theta_k}, \tag{17.26}$$

$$\mathcal{M}_k^{\theta_k} \equiv \left\{ \mathbf{m}_k^i \in \mathcal{M}_k | \exists j, \theta^j = i \right\}, \quad \overline{\mathcal{M}_k^{\theta_k}} \equiv \mathcal{M}_k \setminus \mathcal{M}_k^{\theta_k}. \tag{17.27}$$

7 Commonly referred to as the multi-target likelihood function.

Based on the definitions of the measurement and map sets in (17.26) and (17.27) and given the correspondence θ_k, the RV measurement likelihood term that appears in both the Bayesian update (17.8) and the batch estimation formulation (17.9), can be expressed as follows:

$$p\left(\mathcal{Z}_k \middle| \mathbf{x}_{0:k}, \mathbf{M}_k, \theta_k\right) = p\left(\mathcal{Z}_k^{\theta_k} \middle| \mathbf{x}_{0:k}, \mathcal{M}_k^{\theta_k}, \theta_k\right). \tag{17.28}$$

The RFS measurement likelihood of (17.25) involves a summation over all possible configurations of the correspondence θ_k. This summation accounts for every feasible permutation of DA, where individual measurements can either be linked to a map element or classified as a false positive. Similarly, map elements can either be associated with a measurement or considered undetected (misdetection). The unassociated measurements contribute to the clutter (false positive) term, represented as $p_\kappa\left(\overline{\mathcal{Z}_k^{\theta_k}}\right)$, while the unassociated map elements contribute to the misdetection probability, denoted as $(1 - P_D())$. For associated pairs, where $\theta^j = i \neq 0$, the probabilities of detection, $P_D()$, and the measurement likelihood factor corresponding to the given DA hypothesis, $p\left(\mathcal{Z}_k^{\theta_k} \middle| \mathbf{x}_{0,k}, \mathcal{M}_k^{\theta_k}, \theta_k\right)$ result.

For an assumed set of correspondences θ^*:

$$p\left(\mathcal{Z}_k^{\theta} \middle| \mathcal{M}_k^{\theta}, \mathbf{x}_{0,k}, \theta_k\right) = 0 \quad \text{if } \theta \neq \theta^*. \tag{17.29}$$

Also, it is necessary to assume that only the associated map elements comprise the detections with complete certainty:

$$P_D(\mathbf{x}_k, \mathbf{M}) = \begin{cases} 1 & \text{if } \mathbf{m} \in \mathcal{M}_k^{\theta^*} \\ 0 & \text{otherwise} \end{cases}. \tag{17.30}$$

Assumptions (17.29) and (17.30) have significant implications for the behavior and performance of RV SLAM filters. Their impact will be discussed in Section 17.4.2.

Finally, the assumption that nonassociated measurements are false positives is necessary:

$$p_\kappa\left(\overline{\mathcal{Z}_k^{\theta^*}}\right) = 1. \tag{17.31}$$

The substitution of (17.29), (17.30), and (17.31) into (17.25) results in:

$$p\left(\mathcal{Z}_k \middle| \mathcal{M}_k, \mathbf{x}_{0:k}\right), \quad \text{which given } \theta^* = p\left(\mathcal{Z}_k^{\theta^*} \middle| \mathcal{M}_k^{\theta^*}, \mathbf{x}_{0,k}, \theta^*\right), \tag{17.32}$$

which is identical to the RV likelihood (17.28).

Theorem 17.2 is significant because it reveals the underlying assumptions regarding detection statistics inherent in the RV framework. Specifically, it assumes that outlier rejection mechanisms are employed to remove measurements that are not considered to originate from the mapped environment. Subsequently, DA is presumed to establish correct correspondences between the remaining measurements and map elements. However, it is important to note that this presumed DA is not always flawless. Incorrect associations can and regularly do occur, potentially causing the state estimates to diverge.

To summarize, the joint distribution of the vehicle's state and the map, along with the corresponding measurement likelihoods, are represented as conditional PDFs within both the RFS and RV frameworks of the Bayes filter and batch estimation methodologies. Theorems 17.1 and 17.2 demonstrate that, under specific conditions, these conditional PDFs, and consequently the estimation challenges of SLAM, are equivalent. These specific conditions are collectively referred to as the *ideal detection* conditions, defined as follows:

1. *Deterministic Map Size*: The size of the map is known and remains fixed.
2. *Assumed DA*.
3. *Perfect Detection of Associated Map Elements*: If a map element is associated with a measurement, it is detected with absolute certainty ($P_D = 1$). Conversely, if it is not associated, $P_D = 0$.
4. *Absolute False Positive Identification for Nonassociated Measurements*: Any measurement that is not associated with a map element is considered to be a false positive with unity probability.

17.4.2 Paradox of the "Ideal Detection Conditions"

To derive current RV Bayesian and ML SLAM estimators from the general RFS measurement likelihood described in (17.25), it is essential to apply the assumptions outlined in (17.29) to (17.31). A closer look at these assumptions reveals a fundamental issue within the conceptual framework of these estimators. As demonstrated in Section 17.4.1, the RFS measurement likelihood is constructed by averaging over all possible association hypotheses, treating each one with equal importance. However, to establish the foundation for the RV SLAM formulations presented in Section 17.4.1 and to connect the principles of current RV nonlinear, least-squares (batch) SLAM in (17.9) with the RFS measurement likelihood from (17.25), it becomes necessary to select a single association, specifically $\theta_k = \theta^*$. This choice leads to (17.29), which may inadvertently introduce statistical bias into RV SLAM estimators, as discussed in [22].

An additional issue arises in (17.30), where the detection probability is strictly defined as either 0 or 1 based on the specified association θ_k. Notably, the detection probability $P_D(\mathbf{x}_k, \mathbf{m})$ in (17.25) is solely dependent on the joint state $(\mathbf{x}_k, \mathbf{m})$. However, when (17.30) is applied, the same map feature and robot pose pair $(\mathbf{x}_k, \mathbf{m})$ can exhibit varying detection probabilities depending on the chosen association hypothesis. This is a clear inconsistency, which may introduce additional, unwarranted information into the state, potentially leading to bias.

The findings presented in [16] demonstrate that when clutter is minimal and map features are well-separated relative to the robot's path, the strict assumptions outlined in (17.29) to (17.31) in RV SLAM solutions are largely met, leading to favorable outcomes. However, when these "ideal detection conditions" are not maintained, a common scenario in real-world applications, the statistical biases introduced by these assumptions become evident, causing RV-based SLAM algorithms to diverge.

17.5 Batch RFS-based SLAM Solutions

Concurrently with the development of RFS-based filters in the tracking community, the robotics SLAM community has moved away from filtering-based solutions to the batch estimation approach, which uses nonlinear optimization methods to obtain either an ML or a Maximum a posteriori (MAP) solution [9, 27]. Such methods include GraphSLAM [6], iSAM [8] and iSAM2 [9]. These algorithms, which are usually based on nonlinear least squares optimization, provide more accurate solutions over larger data sets than their filtering counterparts.

This provides a compelling reason to derive batch RFS-based SLAM algorithms, which should result in a superior performance both for large data sets and under high clutter and circumvent the need for map management and DA heuristics.

SLAM solutions using the batch estimation approach are usually ML strategies in which the likelihood function,

$$p\left(\mathcal{Z}_{1:k}, \mathbf{u}_{0:k-1} \middle| \mathbf{x}_{0:k}, \mathcal{M}\right), \tag{17.33}$$

is maximized over all possible trajectories, $\mathbf{x}_{0:k}$, and maps, $\mathcal{M}$. Assuming that the number of landmarks is known, and using a known DA hypothesis, this likelihood can be expressed as:

$$\prod_{i=1}^{k} \mathbf{g}(\mathbf{x}_i | \mathbf{x}_{i-1}, \mathbf{u}_{i-1}) \prod_{j}^{|\mathcal{Z}_i|} \mathbf{h}(\mathbf{z}_i^j | \mathbf{x}_i, \mathbf{m}^{a_i(j)}), \tag{17.34}$$

where the function $a_i(j)$ represents the DA of measurement $\mathbf{z}_i^j$.

When the number of landmarks and DAs are unknown, the objective function of (17.33) can be formulated using the RFS framework, by using the set-based measurement and motion models:

$$p\left(\mathcal{Z}_{1:k}, \mathbf{u}_{0:k-1} \middle| \mathbf{x}_{0:k}, \mathcal{M}\right) = \prod_{i=1}^{k} \mathbf{g}(\mathbf{x}_i | \mathbf{x}_{i-1}, \mathbf{u}_{i-1}) p\left(\mathcal{Z}_i \middle| \mathbf{x}_k, \mathcal{M}\right). \tag{17.35}$$

Taking logs throughout (17.35) gives

$$\begin{aligned} \log\left(p\left(\mathcal{Z}_{1:k}, \mathbf{u}_{0:k-1} \middle| \mathbf{x}_{0:k}, \mathcal{M}\right)\right) &= \sum_{i=1}^{k}\left[\log(\mathbf{g}(\mathbf{x}_i | \mathbf{x}_{i-1}, \mathbf{u}_{i-1})) + \log\left(p\left(\mathcal{Z}_i \middle| \mathbf{x}_k, \mathcal{M}\right)\right)\right] \\ &= \sum_{i=1}^{k}\left[\log(\mathbf{g}(\mathbf{x}_i | \mathbf{x}_{i-1}, \mathbf{u}_{i-1})) + \log\left(\sum_{\theta_i} p\left(\mathcal{Z}_i \middle| \mathbf{x}_k, \mathcal{M}, \theta_i\right)\right)\right]. \end{aligned} \tag{17.36}$$

The maximization of (17.35) can be carried out using the nonlinear optimization of (17.36). This requires the substitution of the general RFS-based measurement likelihood function from (17.25), and should be performed over all possible map sizes $|\mathcal{M}|$.

In Section 17.5.1, the RFS framework will be combined with graph optimization approaches currently used in RV-based SLAM solutions. The resulting algorithm will exploit the correlation between SLAM state components, and the sparse nature of the SLAM state itself, via the use of the sparse information matrix.

17.5.1 JVS-SLAM

With the objective of using the specialized algorithms that are used in the GraphSLAM optimization approach (e.g., Levenberg Marquardt) to exploit the correlation between the states being estimated and their sparse nature (i.e., the sparse information matrix), this section will introduce an RFS-based formulation which includes these correlations.

In this section, the SLAM state is represented as a joint distribution that encapsulates both the vector-valued trajectory and the set-valued map, collectively termed the JVS distribution. This distribution comprises components that can be efficiently computed using state-of-the-art techniques such as the g2o solver [7]. By employing this model, it remains possible to apply an optimization-based approach to maximize the likelihood (17.36). The concept of the JVS distribution will be introduced and subsequently utilized to develop a SLAM formulation in the upcoming subsections, along with convincing performance comparisons of JVS-SLAM against state-of-the-art solutions.

17.5.1.1 A JVS Bernoulli Distribution

To develop a SLAM solution that harnesses the benefits of both batch processing and RFS-based approaches, we begin by introducing the simplest form of set representation: A random set that contains at most one element. In this context, a JVS random variable, denoted as $\mathbb{X}_k$, is defined according to a JVS Bernoulli distribution $\pi(\chi_K)$. This distribution is expressed over the tuple $\chi_K = (\mathbf{x}_{0:K}, \mathcal{M})$, where:

$$p(\mathbb{X}_k = \chi_k) = \pi(\chi_K) = \begin{cases} (1-r)\pi_0(\mathbf{x}_{0:K}) & \chi_K = (\mathbf{x}_{0:K}, \emptyset), \\ r\pi_1(\mathbf{x}_{0:K}, \mathbf{m}) & \chi_K = (\mathbf{x}_{0:K}, \{\mathbf{m}\}), \\ 0 & \text{else,} \end{cases} \tag{17.37}$$

and where r represents the probability of the feature $\mathbf{m}$ existing. Here, $\pi_0(\mathbf{x}_{0:K})$ is a vector distribution over the vehicle's trajectory, assuming the map is empty, while $\pi_1(\mathbf{x}_{0:K}, \mathbf{m})$ is a joint vector distribution over the vehicle's trajectory and the single map element $\mathbf{m}$ when the map is not empty.

It is important to clarify that $\mathbb{X}_k$ is a random variable, whereas $\chi_K = (\mathbf{x}_{0:K}, \mathcal{M})$ signifies a specific instance of the state $\mathbb{X}$. The vector state $\mathbf{x}_{0:K}$ evolves over time, whereas the set $\mathcal{M}$ is assumed to be static, simplifying the overall model. Nevertheless, extending the formulation to include a dynamic transition model for the map state $\mathcal{M}$ would not require significant algorithmic changes.

Additionally, the vector state $\mathbf{x}_{0:K}$ and the set element $\mathbf{m}$ can generally exhibit linear or nonlinear correlations. The proof that the JVS Bernoulli distribution defined in (17.37) is indeed a valid distribution is provided in [20].

17.5.1.2 The JVS-SLAM Distribution

The aforementioned distribution can be generalized to accommodate arbitrary set sizes by building on the work of [28], extending the JVS framework to multiple map elements. This generalization results in a distribution akin to the generalized labeled multi-Bernoulli (GLMB) distribution. Specifically, the JVS-SLAM distribution is expressed as follows:

$$p(\mathbb{X}_k = \chi_k) = \pi(\chi_K) = \Delta(\mathcal{M}) \sum_{c \in \mathcal{C}} w^c(\mathcal{L}(\mathcal{M}))\pi^c(\mathbf{x}_{0:K}, \vec{v}(\mathcal{M})), \tag{17.38}$$

where $\Delta(\mathcal{M})$ corresponds to a distinct label indicator:

$$\Delta(\mathcal{M}) = \delta_{|\mathcal{M}|}(|\mathcal{L}(\mathcal{M})|) = \begin{cases} 1 & \text{if there are no repeated labels in } \mathcal{M} \\ 0 & \text{otherwise.} \end{cases} \tag{17.39}$$

This indicator is equal to unity if all labels in the set $\mathcal{M}$ are unique, and zero if any labels are duplicated. In the context of feature-based SLAM, these "labels" represent unique features or landmark identities, which are crucial for accurate feature identification and hence mapping.

In (17.38), $\pi^c(\cdot)$ is a vector distribution capable of including correlations between the features in $\mathcal{M}$ and the vector state $\mathbf{x}_{0:k}$. Additionally, the operation:

$$\vec{v}(\mathcal{M}) = \mathbf{M} \tag{17.40}$$

defines a "vectorization" function, which converts the labeled RFS $\mathcal{M}$ into a vector representation $\mathbf{M}$, based on the order of its labels (or a fixed permutation of them).

It is important to note that in conventional RFS filter formulations, the elements within the set are generally treated as independent. Consequently, prior RFS-SLAM methods have been limited to approaches based on Rao-Blackwellized particle filters, such as PHD-SLAM, which assume conditional independence. In contrast, JVS-SLAM is a joint batch estimation method that eliminates this requirement.

To guarantee the normalization of (17.38), the following condition must hold:

$$\sum_{L\subseteq\mathbb{L}}\sum_{c\in C} w^c(L) \equiv 1, \tag{17.41}$$

as proven in [20]. Here, C represents a finite index set encompassing all possible DA hypotheses between the observed measurements and the map elements $\mathbf{m}$ in $\mathcal{M}$.

In the GLMB filter implementation [28], which applies Bayes' theorem for updating the belief state, each index c corresponds to a specific DA hypothesis. This hypothesis links a subset of the available measurements to the elements $\mathbf{m}$ in the map $\mathcal{M}$, while the remaining measurements are presumed to be false positives.

Equation (17.38) can be reformulated to match the δ-GLMB structure described in [28]:

$$w^c(\mathcal{L}(\mathcal{M})) = \sum_{I\in\mathcal{F}(\mathbb{L})} w^{(c,I)}\delta_I(\mathcal{L}(\mathcal{M})), \tag{17.42}$$

where $\delta_I(\mathcal{L}(\mathcal{M}))$ is the set-valued Kronecker delta function. This function equals unity if the set of labels in $\mathcal{M}$ matches the subset I exactly, and 0 otherwise. Here, $w^{(c,I)}$ is a weight associated with each hypothesis and $\mathcal{F}(\mathbb{L})$ represents the collection of all possible subsets of the label space $\mathbb{L}$.

Substituting (17.42) back into (17.38) results in:

$$\pi(\mathcal{X}_k) = \Delta(\mathcal{M})\sum_{c\in C}\sum_{I\in\mathcal{F}(\mathbb{L})} w^{(c,I)}\delta_I(\mathcal{L}(\mathcal{M}))\pi^c(\mathbf{x}_{0:k}, \vec{v}(\mathcal{M})). \tag{17.43}$$

Since only a finite number of weights $w^{(c,I)}$ are nonzero for any given c, the notation can be further simplified. This is achieved by introducing the index set C_k, which indexes the pairs (c_k, I). Here, the set of existing labels I is associated with the index c and is now denoted as I^c.

With this refinement, the JVS-SLAM distribution can be rewritten as follows:

$$\pi(\mathcal{X}_k) = \Delta(\mathcal{M})\sum_{c_k\in C_k} w^{c_k}\delta_{I^{c_k}}(\mathcal{L}(\mathcal{M}))\pi^{c_k}(\mathbf{x}_{0:k}, \vec{v}(\mathcal{M})). \tag{17.44}$$

This form represents a structured and compact representation of the distribution, where the indexing naturally captures the association between measurement hypotheses and labeled map elements.

To formulate the SLAM solution, Bayes' theorem can be applied to the JVS-SLAM distribution of (17.44), to derive the complete posterior distribution. The prediction phase can be expressed as follows:

$$p\left(\mathcal{X}_{k+1}\middle|\mathcal{Z}_{0:k}, \mathbf{u}_{1:k+1}\right) = p\left(\mathcal{X}_k\middle|\mathcal{Z}_{0:k}, \mathbf{u}_{1:k}\right)p\left(\mathbf{x}_{k+1}\middle|\mathbf{x}_k, \mathbf{u}_{k+1}\right), \tag{17.45}$$

with update step:

$$p\left(\mathcal{X}_{k+1}\middle|\mathcal{Z}_{0:k+1}, \mathbf{u}_{1:k+1}\right) = \frac{p\left(\mathcal{X}_k\middle|\mathcal{Z}_{0:k}, \mathbf{u}_{1:k}\right)p\left(\mathbf{x}_{k+1}\middle|\mathbf{x}_k, \mathbf{u}_{k+1}\right)p\left(\mathcal{Z}_{k+1}\middle|\mathbf{x}_{k+1}, \mathcal{M}\right)}{p(\mathcal{Z}_{k+1})}, \tag{17.46}$$

with $p(\mathcal{Z}_{k+1})$ serving as a normalization constant that does not require explicit computation.

It is important to highlight that, like other Bayesian approaches, the SLAM solution utilizes prediction and update steps to compute the posterior distribution. However, JVS-SLAM uniquely approximates the full posterior distribution of the joint trajectory-map state without resorting to typical approximations, such as linearization or Gaussian assumptions. Consequently, JVS-SLAM operates as a batch estimation method.

The substitution of the measurement likelihood expression from (17.25) and the JVS-SLAM distribution (17.44) into the Bayes update (17.46), while noting that $\pi(\mathcal{X}_k) = p\left(\mathcal{X}_k \middle| \mathcal{Z}_{0:k}, \mathbf{u}_{1:k}\right)$ results in the following expression:

$$p\left(\mathcal{X}_{k+1} \middle| \mathcal{Z}_{1:k+1}, \mathbf{u}_{1:k+1}\right) \propto \Delta(\mathcal{M}) \sum_{c_k \in \mathcal{C}_k} \sum_{\theta_{k+1}} \Bigg[w^{c_k} \delta_{I^{c_k}}(\mathcal{L}(\mathcal{M})) \pi^{c_k}(\mathbf{x}_{0:k}, \vec{v}(\mathcal{M})) \\ p\left(\mathbf{x}_{k+1} \middle| \mathbf{x}_k, \mathbf{u}_{k+1}\right) p_\kappa\left(\mathcal{Z}_{k+1}^{\bar{A}_{\theta_{k+1}}}\right) \prod_{\mathbf{m}^j \in \mathcal{M}^{\bar{A}_{\theta_{k+1}}}} (1 - P_\mathrm{D}(\mathbf{m}^j | \mathbf{x}_{k+1})) \\ \prod_{\mathbf{z}_{k+1}^i \in \mathcal{Z}_{k+1}^{A_{\theta_{k+1}}}} P_\mathrm{D}\left(\mathbf{m}^{a_{\theta_{k+1}}(i)} | \mathbf{x}_{k+1}\right) p\left(\mathbf{z}_{k+1}^i \middle| \mathbf{m}^{a_{\theta_{k+1}}(i)}, \mathbf{x}_{k+1}\right) \Bigg]. \tag{17.47}$$

where $\mathcal{Z}_{k+1}^{A_{\theta_{k+1}}}$ is the subset of measurements from $\mathcal{Z}_{k+1}$ that are matched to landmarks in $\mathcal{M}$ according to θ_{k+1} [16]:

$$\mathcal{Z}_{k+1}^{A_{\theta_{k+1}}} = \{\mathbf{z}^i \mid \mathbf{z}^i \in \mathcal{Z}_{k+1}, \exists (i, j) \in \theta_{k+1}\}, \tag{17.48}$$

and $\mathcal{M}^{\bar{A}_{\theta_k}}$ is the subset of landmarks from $\mathcal{M}$ that are not associated with any measurements in $\mathcal{Z}_k$:

$$\mathcal{M}^{\bar{A}_{\theta_k}} = \{\mathbf{m}^j \in \mathcal{M} \mid \nexists (i, j) \in \theta_k\}. \tag{17.49}$$

This results in a new form of Joint Vector-δ-GLMB distribution:

$$\pi(\mathcal{X}_{k+1}) = \Delta(\mathcal{M}) \sum_{c_{k+1} \in \mathcal{C}_{k+1}} w^{c_{k+1}} \delta_{I^{c_{k+1}}}(\mathcal{L}(\mathcal{M})) \pi^{c_{k+1}}(\mathbf{x}_{0:k}, \vec{v}(\mathcal{M})), \tag{17.50}$$

where the updated index c_{k+1} is extended with all possible DAs θ_{k+1}.

The new index set $\mathcal{C}_{k+1}$ is defined as the Cartesian product of the previous index set $\mathcal{C}_k$ and the set of all possible DAs Θ_{k+1}, such that:

$$\mathcal{C}_{k+1} = \mathcal{C}_k \times \Theta_{k+1}, \tag{17.51a}$$

$$c_{k+1} = (c_k, \theta_{k+1}), \tag{17.51b}$$

$$w^{c_{k+1}} = w^{c_k} \int \pi^{c_k}(\mathbf{x}_{0:k}, \vec{v}(\mathcal{M})) p\left(\mathbf{x}_{k+1} \middle| \mathbf{x}_k, \mathbf{u}_{k+1}\right) p_\kappa\left(\mathcal{Z}_{k+1}^{\bar{A}_{\theta_{k+1}}}\right) \\ \prod_{\mathbf{m}^j \in \mathcal{M}^{\bar{A}_{\theta_{k+1}}}} (1 - P_\mathrm{D}(\mathbf{m}^j | \mathbf{x}_{k+1})) \prod_{\mathbf{z}_{k+1}^i \in \mathcal{Z}_{k+1}^{A_{\theta_{k+1}}}} P_\mathrm{D}\left(\mathbf{m}^{a_{\theta_{k+1}}(i)} | \mathbf{x}_{k+1}\right) p\left(\mathbf{z}_{k+1}^i \middle| \mathbf{m}^{a_{\theta_{k+1}}(i)}, \mathbf{x}_{k+1}\right) d\mathcal{X}_{k+1}. \tag{17.51c}$$

The corresponding updated component distribution is given by:

$$\pi^{c_{k+1}}(\mathbf{x}_{0:k+1}, \vec{v}(\mathcal{M})) = \frac{w^{c_k}}{w^{c_{k+1}}} \pi^{c_k}(\mathbf{x}_{0:k}, \vec{v}(\mathcal{M})) p\left(\mathbf{x}_{k+1} \middle| \mathbf{x}_k, \mathbf{u}_k\right) p_\kappa\left(\mathcal{Z}_{k+1}^{\bar{A}_{\theta_k}}\right) \times \\ \prod_{\mathbf{m}^j \in \mathcal{M}^{\bar{A}_{\theta_{k+1}}}} (1 - P_\mathrm{D}(\mathbf{m}^j | \mathbf{x}_{k+1})) \prod_{\mathbf{z}_{k+1}^i \in \mathcal{Z}_{k+1}^{A_{\theta_{k+1}}}} P_\mathrm{D}\left(\mathbf{m}^{a_{\theta_{k+1}}(i)} | \mathbf{x}_{k+1}\right) p\left(\mathbf{z}_{k+1}^i \middle| \mathbf{m}^{a_{\theta_{k+1}}(i)}, \mathbf{x}_{k+1}\right). \tag{17.51d}$$

The updated expressions represent a refined JVS-SLAM distribution meaning that it is possible to start with a JVS-SLAM prior and recursively calculate the full posterior of the JVS-SLAM distribution.

17.5.1.2.1 Normal Distribution JVS-SLAM Approximation

To implement the JVS-SLAM solution, we use Eqs. (17.51a) through (17.51d) and approximate the distribution $\pi^{c_k}(\cdot)$ using a Gaussian model:

$$\pi^{c_k}(\mathbf{x}_{0:k}, \vec{v}(\mathcal{M})) = \mathcal{N}(\mathbf{x}_{0:k}, \vec{v}(\mathcal{M}); \mu^{c_k}, \Sigma^{c_k}), \tag{17.52}$$

where μ^{c_k} and Σ^{c_k} represent the mean and covariance matrix of the Gaussian distribution for the component c_k, respectively. This Gaussian approximation enables the application of graph-based optimization techniques as follows.

Substituting the Gaussian approximation from (17.52) into the expression for the JVS-SLAM distribution (17.38) gives:

$$\pi(\chi_k) = \Delta(\mathcal{M}) \sum_{c_k \in \mathcal{C}_k} w^{c_k} \delta_{I^{c_k}}(\mathcal{L}(\mathcal{M})) \mathcal{N}(\mathbf{x}_{0:k}, \vec{v}(\mathcal{M}); \mu^{c_k}, \Sigma^{c_k}). \tag{17.53}$$

This form represents the posterior as a weighted sum of Gaussian components, each associated with a specific hypothesis indexed by c_k.

The update of the posterior at time step $k+1$ involves recalculating the components $\mathcal{C}_{k+1}$, c_{k+1}, $w^{c_{k+1}}$ and $\pi^{c_{k+1}}(\mathbf{x}_{0:k}, \vec{v}(\mathcal{M}))$, which is achieved by approximating $\pi^{c_k}(\mathbf{x}_{0:k}, \vec{v}(\mathcal{M}))$ with the Gaussian distribution $\mathcal{N}(\mathbf{x}_{0:k}, \vec{v}(\mathcal{M}); \mu^{c_k}, \Sigma^{c_k})$ in (17.51). The updated equations are then:

$$\mathcal{C}_{k+1} = \mathcal{C}_k \times \Theta_{k+1}, \tag{17.54a}$$

$$c_{k+1} = (c_k, \theta_{k+1}), \tag{17.54b}$$

$$w^{c_{k+1}} = w^{c_k} \int \mathcal{N}(\mathbf{x}_{0:k}, \vec{v}(\mathcal{M}); \mu^{c_k}, \Sigma^{c_k}) p\left(\mathbf{x}_{k+1} \middle| \mathbf{x}_k, \mathbf{u}_{k+1}\right) p_\kappa\left(\mathcal{Z}_{k+1}^{\bar{A}_{\theta_{k+1}}}\right)$$

$$\prod_{\mathbf{m}^j \in \mathcal{M}^{\bar{A}_{\theta_{k+1}}}} (1 - P_D(\mathbf{m}^j | \mathbf{x}_{k+1})) \prod_{\mathbf{z}_{k+1}^i \in \mathcal{Z}_{k+1}^{A_{\theta_{k+1}}}} P_D\left(\mathbf{m}^{a_{\theta_{k+1}}(i)} | \mathbf{x}_{k+1}\right) p\left(\mathbf{z}_{k+1}^i \middle| \mathbf{m}^{a_{\theta_{k+1}}(i)}, \mathbf{x}_{k+1}\right) d\chi_{k+1}, \tag{17.54c}$$

$$\pi^{c_{k+1}}(\mathbf{x}_{0:k}, \vec{v}(\mathcal{M})) = \frac{w^{c_k}}{w^{c_{k+1}}} \mathcal{N}(\mathbf{x}_{0:k}, \vec{v}(\mathcal{M}); \mu^{c_k}, \Sigma^{c_k}) p\left(\mathbf{x}_{k+1} \middle| \mathbf{x}_k, \mathbf{u}_{k+1}\right) p_\kappa\left(\mathcal{Z}_{k+1}^{\bar{A}_{\theta_{k+1}}}\right)$$

$$\prod_{\mathbf{m}^j \in \mathcal{M}^{\bar{A}_{\theta_{k+1}}}} (1 - P_D(\mathbf{m}^j | \mathbf{x}_{k+1})) \prod_{\mathbf{z}_{k+1}^i \in \mathcal{Z}_{k+1}^{A_{\theta_{k+1}}}} P_D\left(\mathbf{m}^{a_{\theta_{k+1}}(i)} | \mathbf{x}_{k+1}\right) p\left(\mathbf{z}_{k+1}^i \middle| \mathbf{m}^{a_{\theta_{k+1}}(i)}, \mathbf{x}_{k+1}\right). \tag{17.54d}$$

Detection Probability Approximation: To simplify the update, it is assumed that the detection probability $P_D(\mathbf{m}^j | \mathbf{x}_{k+1})$ does not vary significantly with both the robot pose $\mathbf{x}_{k+1}$ and the landmark location $\mathbf{m}^j$. This is generally valid except near the boundaries of the sensor's FoV, where detection rates may fluctuate more. Accounting for this could be an avenue for future research. Therefore, $P_D(\mathbf{m}^j | \mathbf{x}_{k+1})$ is calculated at the mean μ^{c_k} of the Gaussian distribution. This approximation simplifies the integration step in the update equations, allowing it to be treated as a constant $P_D^{(j,k+1)}$.

Equations (17.54a) to (17.54d) express the posterior distribution after K sequential updates, as detailed in [20].

$$\mathcal{C}_K = \mathcal{C}_\emptyset \times \Theta_0 \times \Theta_1 \times \cdots \times \Theta_K, \tag{17.55a}$$

$$c_K = (c_\emptyset, \theta_0, \cdots, \theta_K), \tag{17.55b}$$

$$w^{c_K} = w^{c_\varnothing} \prod_{k=0}^{K} p_\kappa\left(\mathcal{Z}_k^{\bar{A}_{\theta_k}}\right) \prod_{\mathbf{m}^j \in \mathcal{M}^{\bar{A}_{\theta_k}}} \left(1 - P_{\mathrm{D}}^{(j,i)}\right) \prod_{\mathbf{z}_k^i \in \mathcal{Z}_k^{A_{\theta_k}}} P_{\mathrm{D}}^{(a_{\theta_k}(i),k)}$$

$$\int \mathcal{N}(\mathbf{x}_0, \vec{\upsilon}(\mathcal{M}); \mu^{c_\varnothing}, \Sigma^{c_\varnothing}) \prod_{k=1}^{K} p\left(\mathbf{x}_k \middle| \mathbf{x}_{k-1}, \mathbf{u}_k\right) \prod_{k=0}^{K} \prod_{\mathbf{z}_k^i \in \mathcal{Z}_k^{A_{\theta_k}}} p\left(\mathbf{z}_k^i \middle| \mathbf{m}^{a_{\theta_k}(i)}, \mathbf{x}_k\right) d\chi_K, \tag{17.55c}$$

$$\pi^{c_K}(\mathbf{x}_{0:K}, \vec{\upsilon}(\mathcal{M})) \propto \mathcal{N}(\mathbf{x}_0, \vec{\upsilon}(\mathcal{M}); \mu^{c_\varnothing}, \Sigma^{c_\varnothing}) \prod_{k=1}^{K} p\left(\mathbf{x}_k \middle| \mathbf{x}_{k-1}, \mathbf{u}_k\right) \prod_{k=0}^{K} \prod_{\mathbf{z}_k^i \in \mathcal{Z}_k^{A_{\theta_k}}} p\left(\mathbf{z}_k^i \middle| \mathbf{m}^{a_{\theta_k}(i)}, \mathbf{x}_k\right), \tag{17.55d}$$

where (17.55d) has been stated as a proportionality for brevity.

Equation (17.55b) implies that each posterior component index c_K is uniquely determined by the initial component $c_\varnothing$ and the series of DAs $(\theta_0, \cdots, \theta_K)$. This direct mapping signifies that the evolution of each hypothesis through time is fully described by its association history. Notably, the initial component set $C_\varnothing$ can be arbitrarily defined, representing different initial beliefs.

17.5.1.3 Calculation of the Component Weights

Equations (17.55a) to (17.55d) represent the complete JVS-SLAM posterior. To solve for the spatial posterior component $\pi^{c_K}(\mathbf{x}_{0:K}, \vec{\upsilon}(\mathcal{M}))$ efficiently, a nonlinear least squares optimization approach is applied, leveraging the Gaussian structure of the posterior distribution.

An efficient method for calculating the component weight w^{c_K} for each DA hypothesis is still required. The spatial posterior (17.55d) in its full form is:

$$\pi^{c_K}(\mathbf{x}_{0:K}, \vec{\upsilon}(\mathcal{M})) = \frac{\mathcal{N}(\mathbf{x}_0, \vec{\upsilon}(\mathcal{M}); \mu^{c_\varnothing}, \Sigma^{c_\varnothing}) \prod_{k=1}^{K} p\left(\mathbf{x}_k \middle| \mathbf{x}_{k-1}, \mathbf{u}_k\right) \prod_{i=0}^{K} \prod_{\mathbf{z}_k^i \in \mathcal{Z}_k^{A_{\theta_k}}} p\left(\mathbf{z}_k^i \middle| \mathbf{m}^{a_{\theta_k}(i)}, \mathbf{x}_k\right)}{q(c_k, \mathcal{Z}_{1:K}, \mathbf{u}_{1:K})}, \tag{17.56}$$

where the denominator

$$q(c_k, \mathcal{Z}_{0:K}, \mathbf{u}_{1:K}) = \int \mathcal{N}(\mathbf{x}_0, \vec{\upsilon}(\mathcal{M}); \mu^{c_\varnothing}, \Sigma^{c_\varnothing}) \prod_{k=1}^{K} p\left(\mathbf{x}_k \middle| \mathbf{x}_{k-1}, \mathbf{u}_k\right) \prod_{k=0}^{K} \prod_{\mathbf{z}_k^i \in \mathcal{Z}_k^{A_{\theta_k}}} p\left(\mathbf{z}_k^i \middle| \mathbf{m}^{a_{\theta_k}(i)}, \mathbf{x}_k\right) d\chi_K, \tag{17.57}$$

is the normalization term.

The motion and measurement models are approximated by nonlinear functions with additive Gaussian noise:

$$p\left(\mathbf{x}_k \middle| \mathbf{x}_{k-1}, \mathbf{u}_k\right) \approx \mathcal{N}(\mathbf{x}_k; f_u(\mathbf{x}_{k-1}), \Sigma^{odo}) \tag{17.58}$$

$$p\left(\mathbf{z}_k^i \middle| \mathbf{m}^{a_{\theta_k}(i)}, \mathbf{x}_k\right) \approx \mathcal{N}\left(\mathbf{z}_k^i; f_z(\mathbf{m}^{a_{\theta_k}(i)}, \mathbf{x}_k), \Sigma^z\right), \tag{17.59}$$

where $f_u(\mathbf{x}_{k-1})$ is the nonlinear motion model, $f_z(\mathbf{m}^{a_{\theta_k}(i)}, \mathbf{x}_k)$ is the nonlinear measurement model. Σ^{odo} and Σ^z are their associated noise covariances.

Given the nonlinear models, the spatial posterior $\pi^{c_K}(\mathbf{x}_{0:K}, \vec{v}(\mathcal{M}))$ can be approximated by a Gaussian distribution:

$$\pi^{c_K}(\mathbf{x}_{0:K}, \vec{v}(\mathcal{M})) = \mathcal{N}\left(\mathbf{x}_{0:K}, \vec{v}(\mathcal{M}); \mu^{\text{MAP}}, \Sigma^{\text{MAP}}\right), \tag{17.60}$$

where μ^{MAP} and Σ^{MAP} are obtained through the nonlinear least squares optimization. Then the normalization factor is:

$$q(c_K, \mathcal{Z}_{0:K}, \mathbf{u}_{1:K}) = \frac{\mathcal{N}(\mathbf{x}_0, \vec{v}(\mathcal{M}); \mu^{c_\varnothing}, \Sigma^{c_\varnothing}) \prod_{k=1}^{K} p\left(\mathbf{x}_k \middle| \mathbf{x}_{k-1}, \mathbf{u}_k\right) \prod_{k=0}^{K} \prod_{\mathbf{z}_k^i \in \mathcal{Z}_k^{A_{\theta_k}}} p\left(\mathbf{z}_k^i \middle| \mathbf{m}^{a_{\theta_k}(i)}, \mathbf{x}_k\right)}{\mathcal{N}(\mathbf{x}_{0:K}, \vec{v}(\mathcal{M}); \mu^{MAP}, \Sigma^{MAP})}, \tag{17.61}$$

where $\mu^{c_\varnothing}$ and $\Sigma^{c_\varnothing}$ are the mean and covariance of the initial spatial estimate of the prior distribution. Taking logs throughout (17.61) gives

$$\begin{aligned}\log(q(c_K, \mathcal{Z}_{0:K}, \mathbf{u}_{1:K})) &= \log(\mathcal{N}(\mathbf{x}_0, \vec{v}(\mathcal{M}); \mu^{c_\varnothing}, \Sigma^{c_\varnothing})) + \sum_{k=1}^{K} l\left(\mathbf{x}_k \middle| \mathbf{x}_{k-1}, \mathbf{u}_k\right) \\ &+ \sum_{k=0}^{K} \sum_{\mathbf{z}_k^i \in \mathcal{Z}_k^{A_{\theta_k}}} l\left(\mathbf{z}_k^i \middle| \mathbf{m}^{a_{\theta_k}(i)}, \mathbf{x}_k\right) - \log(\mathcal{N}(\mathbf{x}_{0:K}, \vec{v}(\mathcal{M}); \mu^{MAP}, \Sigma^{MAP})),\end{aligned} \tag{17.62}$$

where $l(\cdot) = \log(p(\cdot))$.

By expanding the Gaussian expressions and noting that $|\chi_K| = |\chi_0| + K|\mathbf{x}_k|$, the log-transformed normalization factor simplifies into a sum of Mahalanobis distances and log determinants:

$$\begin{aligned}\log(q(c_k, \mathcal{Z}_{0:K}, \mathbf{u}_{1:K})) = -\frac{1}{2}\Bigg[&(\chi_0 - \mu^{c_\varnothing})^{\mathsf{T}} (\Sigma^{c_\varnothing})^{-1} (\chi_0 - \mu^{c_\varnothing}) + \log(\det(\Sigma^{c_\varnothing})) \\ &+ \sum_{k=1}^{K} (\mathbf{x}_k - f_u(\mathbf{x}_{k-1}, \mathbf{u}_k))^{\mathsf{T}} (\Sigma^{odo})^{-1} (\mathbf{x}_k - f_u(\mathbf{x}_{k-1}, \mathbf{u}_k)) + \log(\det(\Sigma^{odo})) \\ &+ \sum_{k=0}^{K} \sum_{\mathbf{z}_k^i \in \mathcal{Z}_k^{A_{\theta_k}}} |\mathbf{z}_k^i| \log(2\pi) + (\mathbf{z}_k^i - f_z(\mathbf{m}^{a_{\theta_k}(i)}, \mathbf{x}_k))^{\mathsf{T}} (\Sigma^z)^{-1} (\mathbf{z}_k^i - f_z(\mathbf{m}^{a_{\theta_k}(i)}, \mathbf{x}_k)) \\ &+ \log(\det(\Sigma^z)) - \left(\log(\det(\Sigma_{MAP})) + (\chi_K - \mu^{MAP})^{\mathsf{T}} \Sigma_{MAP}^{-1} (\chi_K - \mu^{MAP})\right)\Bigg].\end{aligned} \tag{17.63}$$

Near the SLAM solution, the nonlinear models are approximated using a second-order Taylor series expansion:

$$\begin{aligned}
&(\chi_0 - \mu^{c\emptyset})^{\mathsf{T}} (\Sigma^{c\emptyset})^{-1} (\chi_0 - \mu^{c\emptyset}) + \sum_{k=1}^{K} (\mathbf{x}_k - f_u(\mathbf{x}_{k-1}, \mathbf{u}_k))^{\mathsf{T}} (\Sigma^{odo})^{-1} (\mathbf{x}_k - f_u(\mathbf{x}_{k-1}, \mathbf{u}_k)) \\
&+ \sum_{k=0}^{K} \sum_{\mathbf{z}_k^i \in \mathcal{Z}_k^{A_{\theta_k}}} (\mathbf{z}_k^i - f_z(\mathbf{m}^{a_{\theta_k}(i)}, \mathbf{x}_k))^{\mathsf{T}} (\Sigma^z)^{-1} (\mathbf{z}_k^i - f_z(\mathbf{m}^{a_{\theta_k}(i)}, \mathbf{x}_k)) \\
&\approx \sum_s e_s^{\mathsf{T}} \Omega_s e_s + \Delta\chi_K{}^{\mathsf{T}} J_s^{\mathsf{T}} \Omega_s J_s \Delta\chi_K,
\end{aligned} \tag{17.64}$$

where $e_s^{\mathsf{T}} \Omega_s e_s$ is the error of each constraint evaluated at the spatial estimate μ^{MAP}, and

$$e_s = \begin{cases} \chi_0(\mu^{MAP}) - \mu^{c\emptyset} & \text{if } s = 0 \\ \mathbf{x}_k(\mu^{MAP}) - f_u(\mathbf{x}_{k-1}(\mu^{MAP}), \mathbf{u}_k) & \text{if } 1 \le s \le K \\ \mathbf{z}_k^i - f_z(\mathbf{m}^{a_{\theta_k}(i)}(\mu^{MAP}), \mathbf{x}_k(\mu^{MAP})) & \text{else} \end{cases} \tag{17.65a}$$

$$\Omega_s = \begin{cases} (\Sigma^{c\emptyset})^{-1} & \text{if } s = 0 \\ (\Sigma^{odo})^{-1} & \text{if } 1 \le s \le K \\ (\Sigma^z)^{-1} & \text{else} \end{cases} \tag{17.65b}$$

$$J_s = \begin{cases} \dfrac{\delta \chi_0}{\delta \chi_K}(\mu^{MAP}) & \text{if } s = 0 \\ \dfrac{\delta f_u(\mathbf{x}_{k-1}, \mathbf{u}_k)}{\delta \chi_K}(\mu^{MAP}) & \text{if } 1 \le s \le K, \\ \dfrac{\delta f_z(\mathbf{m}^{a_{\theta_k}(i)}, \mathbf{x}_k)}{\delta \chi_K}(\mu^{MAP}) & \text{else} \end{cases} \tag{17.65c}$$

where the arbitrary index s is the index of the constraint, corresponding to either the prior, the odometry or the measurement constraint, and in (17.65a) the functions $\chi_0(\mu^{MAP})$, $\mathbf{x}_k(\mu^{MAP})$, and $\mathbf{m}^{a_{\theta_k}(i)}(\mu^{MAP})$ give the state variables at the spatial estimate μ^{MAP}.

Then the weight is given by

$$\begin{aligned}
\log(q(c_K, \mathcal{Z}_{1:K}, \mathbf{u}_{0:K})) \approx -\frac{1}{2}\Bigg[&\sum_s e_s^{\mathsf{T}} \Omega_s e_s + \log\left(\det\left(\sum_s J_s^{\mathsf{T}} \Omega_s J_s\right)\right) - \sum_s \log(\det(\Omega_s)) \\
&+ \sum_{k=0}^{K} \sum_{\mathbf{z}_k^i \in \mathcal{Z}_k^{A_{\theta_k}}} |\mathbf{z}_k^i| \log(2\pi) \Bigg].
\end{aligned} \tag{17.66}$$

17.5.1.3.1 Linear Complexity Optimization

A core observation is that the entire weight computation process maintains linear complexity in the number of landmarks and measurements except for the second term within the square brackets in (17.66). To efficiently solve this, methods including iSAM and the g2o library apply Cholesky decomposition, which decomposes the matrix $\sum_s J_s^\top \Omega_s J_s = LL^*$, where L is a lower triangular matrix, and L^* is its conjugate transpose. The determinant of a triangular matrix is simply the product of its diagonal elements keeping the complexity linear in the number of measurements. Therefore, the log-determinant simplifies to:

$$\log\left(\det\left(\sum_s J_s^\top \Omega_s J_s\right)\right) = 2\sum_i \log(L_{i,i}). \tag{17.67}$$

This approach dramatically reduces computational complexity, allowing the determinant to be evaluated in linear time.

By taking the log of both sides of the weight expression (17.55c) and substituting the log determinant from Cholesky decomposition (17.67) and the log of the normalization factor from the Taylor expansion approximation (17.66), the full expression for the log-weight of hypothesis c_K results:

$$\begin{aligned}\log(w^{c_K}) = {} & \log(w^{c_\emptyset}) + \sum_{k=0}^{K} \log\left(p_\kappa\left(\mathcal{Z}_k^{\bar{A}_{\theta_k}}\right)\right) + \sum_{\mathbf{m}^j \in \mathcal{M}^{\bar{A}_{\theta_i}}} \log\left(1 - P_{\mathrm{D}}^{(j,i)}\right) + \sum_{\mathbf{z}_k^i \in \mathcal{Z}_k^{A_{\theta_k}}} \log\left(P_{\mathrm{D}}^{(a_{\theta_k}(i),k)}\right) \\ & -\frac{1}{2}\left[\sum_s e_s^\top \Omega_s e_s + 2\sum_i \log(L_{i,i}) - \sum_s \log(\det(\Omega_s)) + \sum_{i=0}^{k} \sum_{\mathbf{z}_k^i \in \mathcal{Z}_k^{A_{\theta_k}}} |\mathbf{z}_k^i| \log(2\pi)\right],\end{aligned} \tag{17.68}$$

which is used in the implementation of JVS-SLAM.

The final step to complete the formulation is the inclusion of a reasonable prior:

$$\{w^{c_\emptyset}, \pi^{c_\emptyset}(\mathbf{x}_0, \vec{v}(\mathcal{M}))\}_{c_\emptyset \in \mathcal{C}_\emptyset}. \tag{17.69}$$

The robot prior is modeled as a known initial position for the robot, $\mathbf{x}_0^*$, represented as a Dirac delta:

$$p(\mathbf{x}_0) = \delta(\mathbf{x}_0 - \mathbf{x}_0^*). \tag{17.70}$$

The map prior is a multi-Bernoulli distribution to represent the initial map hypothesis $\mathcal{M}_0$, encapsulating the initial set of possible landmarks.

The expressions presented represent the complete posterior distribution for the JVS-SLAM formulation, now incorporating a detailed prior model for the robot's initial state and the map's landmarks.

The prior distribution for the robot and map is defined as follows:

$$\pi^{c_\emptyset}(\mathbf{x}_0, \mathcal{M}) = \delta(\mathbf{x}_0 - \mathbf{x}_0^*)\mathcal{N}(\vec{v}(\mathcal{M}); \mu^{c_\emptyset}, \Sigma^{c_\emptyset}), \tag{17.71}$$

$$w^{c_\emptyset} = \prod_{\mathbf{m}^j \in \mathcal{M}} \frac{P_E(j)}{1 - P_E(j)} \prod_{\mathbf{m}^j \in \mathcal{M}_0} 1 - P_E(j), \tag{17.72}$$

$$\mu^{c_\emptyset} = \begin{bmatrix} \mu_0 \\ \cdots \\ \mu_j \\ \cdots \\ \mu_{|\mathcal{M}|} \end{bmatrix}_{\mathbf{m}^j \in \mathcal{M}}, \quad \Omega^{c_\emptyset} = \begin{bmatrix} \Omega_0^{c_\emptyset} & 0 & 0 & 0 & 0 \\ 0 & \cdots & 0 & 0 & 0 \\ 0 & 0 & \Omega_j^{c_\emptyset} & 0 & 0 \\ 0 & 0 & 0 & \cdots & 0 \\ 0 & 0 & 0 & 0 & \Omega_{|\mathcal{M}|}^{c_\emptyset} \end{bmatrix}_{\mathbf{m}^j \in \mathcal{M}}, \tag{17.73}$$

where each element j of $\mathcal{M}_0$ can exist with probability $P_E(j)$ (set to a constant value according to the expected density of landmarks) and if it exists, it has a normal spatial distribution with mean μ_j and information matrix $\Omega_j^{c\emptyset}$.

Based on this prior, the complete posterior distribution after K time steps is represented as:

$$\mathcal{C}_K = \mathcal{C}_\emptyset \times \Theta_0 \times \Theta_1 \times \cdots \times \Theta_K, \tag{17.74a}$$

$$c_K = (c_\emptyset, \theta_0, \cdots, \theta_K), \tag{17.74b}$$

$$\begin{aligned}\log(w^{c_K}) = &\sum_{\mathbf{m}^j \in \mathcal{M}} \log(P_E(j)) - \log(1 - P_E(j)) + \sum_{\mathbf{m}^j \in \mathcal{M}_0} \log(1 - P_E(j)) \\ &+ \sum_{k=0}^{K} \log\left(p_\kappa\left(\mathcal{Z}_k^{\bar{A}_{\theta_k}}\right)\right) + \sum_{\mathbf{m}^j \in \mathcal{M}^{\bar{A}_{\theta_k}}} \log\left(\left(1 - P_{\mathrm{D}}^{(j,k)}\right)\right) + \sum_{\mathbf{z}_k^i \in \mathcal{Z}_k^{A_{\theta_k}}} \log(P_{\mathrm{D}}^{(a_{\theta_k}(i),k)}) \\ &- \frac{1}{2}\left[\sum_s e_s^\top \Omega_s e_s + 2\sum_i \log(L_{i,i}) - \sum_s \log(\det(\Omega_s)) + \sum_{k=0}^{K} \sum_{\mathbf{z}_k^i \in \mathcal{Z}_k^{A_{\theta_k}}} |\mathbf{z}_k^i| \log(2\pi)\right],\end{aligned} \tag{17.74c}$$

and

$$\begin{aligned}\pi^{c_K}(\mathbf{x}_{0:K}, \vec{v}(\mathcal{M})) \propto\ &\delta(\mathbf{x}_0 - \mathbf{x}_0^*)\mathcal{N}(\vec{v}(\mathcal{M}); \mu^{c\emptyset}, \Sigma^{c\emptyset}) \prod_{k=1}^{K} p\left(\mathbf{x}_k \middle| \mathbf{x}_{k-1}, \mathbf{u}_k\right) \\ &\times \prod_{k=0}^{K} \prod_{\mathbf{z}_k^i \in \mathcal{Z}_k^{A_{\theta_k}}} p\left(\mathbf{z}_k^i \middle| \mathbf{m}^{a_{\theta_k}(i)}, \mathbf{x}_k\right),\end{aligned} \tag{17.74d}$$

with

$$\mu^{c\emptyset} = \begin{bmatrix} \mu_0 \\ \cdots \\ \mu_j \\ \cdots \\ \mu_{|\mathcal{M}|} \end{bmatrix}_{\mathbf{m}^j \in \mathcal{M}}, \quad \Omega^{c\emptyset} = \begin{bmatrix} \Omega_0^{c\emptyset} & 0 & 0 & 0 & 0 \\ 0 & \dots & 0 & 0 & 0 \\ 0 & 0 & \Omega_j^{c\emptyset} & 0 & 0 \\ 0 & 0 & 0 & \dots & 0 \\ 0 & 0 & 0 & 0 & \Omega_{|\mathcal{M}|}^{c\emptyset} \end{bmatrix}_{\mathbf{m}^j \in \mathcal{M}}. \tag{17.74e}$$

The full posterior distribution is now complete and importantly:

- A structured prior is established for both robot and landmarks.
- The weight computation is efficient due to Cholesky decomposition.
- The spatial posterior is represented in a multi-hypothesis Gaussian form.

17.5.1.4 JVS-SLAM Algorithmic Implementation

The complete Bayes posterior, as outlined in Eqs. (17.74a) to (17.74e), encompasses all potential measurements to establish feature state DA hypotheses. However, computing each one is impractical for anything beyond the simplest scenarios. Consequently, we must approximate the posterior using a limited set of DA hypotheses.

In SLAM filtering methods, the spatial estimates remain unchanged when different DA hypotheses are selected. In contrast, this approach necessitates an optimization process to adjust the spatial estimates each time a new DA hypothesis is chosen, prior to calculating the new measurement likelihood. This means that it is not feasible to select an association based on weight before executing the optimization defined by the association hypothesis, nor is it possible to iterate through the associations in order of decreasing weight.

Nonetheless, adding each new component to the estimate reduces the error between the estimate and the true Bayesian posterior, as described in (17.74), which includes all possible components at time K in $\mathcal{C}_K$. This has been demonstrated for the GLMB distribution using the $L1$ distance in [29]. Therefore, it is not essential to incorporate components into the approximated posterior in strictly decreasing order of weight, even though doing so may accelerate error reduction. Importantly, it should be noted that:

Even a two component DA hypotheses JVS-SLAM estimate will have lower $L1$ error, relative to the true posterior, than a JVS-SLAM estimate based on any of its single component DA hypotheses.

In this step, the proposed algorithm could incorporate external information to enhance the exploration of DA hypotheses. For instance, integrating additional potential DA hypotheses from an external "loop closure" algorithm can be beneficial, as these solutions would naturally be reflected in the Bayes posterior.

Pseudo-code for the proposed algorithm, along with further implementation details are given in [20].

17.5.1.5 JVS-SLAM Computational Complexity

The JVS-SLAM algorithm operates by generating a specified number, H_{DA} of DA hypotheses. For each hypothesis, the algorithm optimizes the associated vehicle trajectories and map estimates, subsequently calculating their weights. The parameter H_{DA} defines how many hypotheses are created and evaluated.

These H_{DA} hypotheses are processed concurrently across multiple computer threads. Each thread has access to the complete set of available data $\mathcal{Z}_{0:K}, \mathbf{u}_{0:K}$ and the previous estimate $\pi(\chi_{K-1})$. Importantly, if two or more threads produce identical hypotheses, that hypothesis only contributes once to the posterior distribution.

Within each thread, the Gibbs sampler is used to generate a new DA hypothesis $(\theta_0, \cdots, \theta_K)$. This hypothesis is then processed through the g2o solver, which optimizes the spatial estimate. Finally, based on the optimization results, a weight is assigned to the hypothesis, reflecting its contribution to the overall map and trajectory estimate.

The computational complexity of the JVS-SLAM algorithm, in $O(\cdot)$ notation, is

$$O(\text{JVS-SLAM}) = O(H_{\text{DA}}(O(\text{Gibbs sampler}) + O(\text{Optimizer}) + O(\text{CalculateWeight}))) \tag{17.75}$$

$$O(\text{JVS-SLAM}) = O(H_{\text{DA}}(O(\text{CalcDAProbs} + N_{\text{Gibbs}}|\mathcal{Z}_{0:K}|) + O(\text{Optimizer}) + |\mathcal{Z}_{0:K}| + K)) \tag{17.76}$$

$$O(\text{JVS-SLAM}) = H_{\text{DA}}(|\mathcal{Z}_{0:K}||\mathcal{M}| + N_{\text{Gibbs}}|\mathcal{Z}_{0:K}| + O(\text{Optimizer}) + |\mathcal{Z}_{0:K}| + K) \tag{17.77}$$

$$O(\text{JVS-SLAM}) = H_{\text{DA}}(|\mathcal{Z}_{0:K}|(|\mathcal{M}| + N_{\text{Gibbs}}) + O(\text{Optimizer}) + K). \tag{17.78}$$

In the complexity analysis, N_{Gibbs} represents the number of iterations executed by the Gibbs sampler. It is important to note that the batch optimizer (e.g., the g2o solver) must be run H_{DA} times, once for each generated hypothesis. This results in a higher computational complexity compared to traditional SLAM algorithms that rely on a single, heuristically determined DA.

Table 17.1 Parameters H_{DA} (the number of times g2o is called per time step) and N_{Gibbs} for the real EuRoC experiments.

Parameter	EuRoC visual data set
H_{DA}	26
N_{Gibbs}	10

The number of hypotheses, H_{DA}, necessary to achieve reliable results is influenced by the efficiency of the sampling method used to generate candidate association hypotheses. If the Gibbs sampler samples too many unlikely associations, the computational load increases unnecessarily. Enhancing or replacing the current Gibbs sampling mechanism with a more effective method could reduce H_{DA} by focusing on more probable associations, thereby accelerating convergence. In particular, a linear complexity Gibbs sampling approach has been applied to the GLMB filter in [30] to truncate its final number of Bernoulli components. An investigation into its application to JVS-SLAM could result in computational savings.

The specific values for H_{DA} and N_{Gibbs} utilized in the experiments in Section 17.5.1.6 are provided in Table 17.1.

One potential approach to reduce the computational complexity of the algorithm is to integrate the hypo-tree structure as described in [31]. This structure allows for the efficient representation and storage of the spatial estimates across multiple JVS-SLAM components, minimizing redundancy and optimizing memory usage. Although the spatial estimates are compactly stored within the hypo-tree, the individual weights of each hypothesis would still need to be maintained separately. Implementing this strategy could substantially decrease the cost associated with each optimization step.

The implementation of these concepts is available as open-source code in this gitrepository.[8]

17.5.1.6 Visual JVS-SLAM Experiments

To demonstrate its robustness, the proposed JVS-SLAM algorithm was adapted for visual stereo data, specifically from the challenging EuRoC data set [32]. This visual-inertial data set was collected on board a micro aerial vehicle (MAV) and consists of stereo images with synchronized inertial measurement unit (IMU) measurements. For the experiments here, only the sequential images from one stereo camera pair were used to perform SLAM, and the IMU data was ignored. Figure 17.2a shows the drone used in the experiments, together with the multiple sensor systems. Visual data was chosen for the JVS-SLAM experiments because of the availability of keypoint detector descriptor pairs, enabling a feature-based SLAM formulation.

Therefore, the state-of-the-art visual SLAM algorithm, oriented FAST rotated brief SLAM—version 3 (ORBSLAM3) [13], was chosen as inspiration for the development of JVS-SLAM due to its performance, robustness and open-source code-base. As in all three ORBSLAM algorithms, the ORB keypoint detector and descriptor [11] are used by JVS-SLAM as its measurements, with the same parameters as in the ORBSLAM3 algorithm. The ORBSLAM3 algorithm will be one of the benchmark algorithms used in the performance comparisons.

To evaluate the performance of JVS-SLAM, and the chosen state-of-the-art SLAM competitors, MAV trajectory ground truth is available with this data set, based on a Vicon motion capture system

8 https://github.com/finostro/orb_jvs_slam

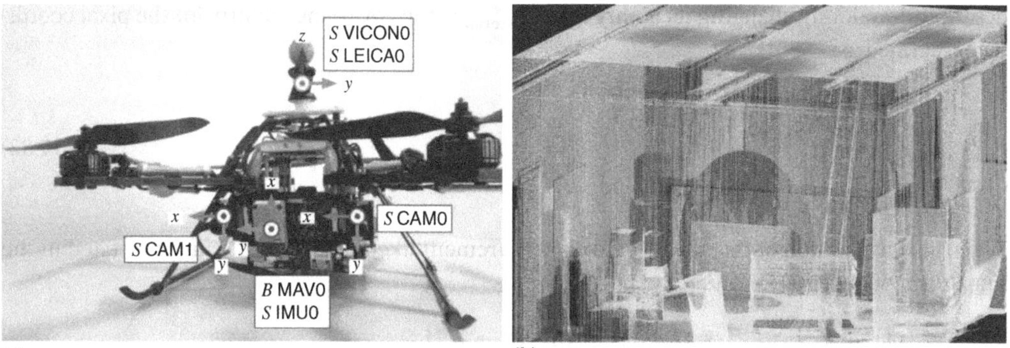

(a) (b)

Figure 17.2 (a) The Asctec Firefly hex-rotor helicopter MAV used to collect data for the EuRoC data set and (b) a ground truth 3D scan of the environment in which the JVS-SLAM experiments were conducted. *Source*: [3] / with permission of Sage Publications.

and Leica MS50 laser tracker. Figure 17.2b shows a ground truth 3D scan of the indoor environment in which the data sets were recorded.

17.5.1.6.1 Visual Measurement Model

Stereo measurements are represented as follows:

$$\mathbf{z}_k^i = [u_l\ ,\ v_l\ ,\ u_r\ ,\ \mathrm{orb}_z]^\mathsf{T}, \tag{17.79}$$

where $\mathbf{z}_k^i$ denotes an individual detection from the set $\mathcal{Z}_k$. In this representation, u_l and v_l correspond to the pixel coordinates of the detection in the left image, u_r is the x-coordinate of the detection in the right image, and orb_z represents the ORB descriptor of the measurement, which is composed of 256 binary values $\mathbf{z}_k^i$ [11]. Notably, the y-coordinate of the detection in the right image, v_r, contributes no further information, since its value is inherently determined by the stereo epipolar constraint.

The map elements of $\mathcal{M}$ are defined as:

$$\mathbf{m}^j = [x\ ,\ y\ ,\ z\ ,\ \mathrm{orb}_m]^\mathsf{T}, \tag{17.80}$$

where x, y, and z indicate the 3D coordinates of the keypoint and orb_m is its associated ORB descriptor.

The single feature ORB measurement likelihood is expressed as follows:

$$\begin{aligned} p\left(\mathbf{z}_k^i \middle| \mathbf{m}^{a_{\theta_k}(i)}, \mathbf{x}_k\right) &= \mathcal{N}(u_l, v_l, u_r | \Gamma(\mathbf{m}^j, \mathbf{x}_k), \Sigma_{\mathbf{stereo}}) \left(p^f_{\mathrm{orb}}\right)^{d_h(\mathrm{orb}_z, \mathrm{orb}_m)} \\ &\times \left(1 - p^f_{\mathrm{orb}}\right)^{(256 - d_h(\mathrm{orb}_z, \mathrm{orb}_m))}, \end{aligned} \tag{17.81}$$

where $\Gamma(\mathbf{m}^j, \mathbf{x}_k)$ is the transformation function, which projects feature $\mathbf{m}^j$, detected at robot pose $\mathbf{x}_k$, into camera pixel coordinates. $d_h(\cdot)$ is the Hamming distance between the ORB descriptors of the map point and the measurement and p^f_{orb} is the probability of each bit in the ORB descriptor randomly changing its binary state. $\Sigma_{\mathbf{stereo}}$ is the covariance matrix for the pixel coordinates, defined as:

$$\Sigma_{\mathbf{stereo}} = \begin{bmatrix} \sigma^2_{\text{pix}} & 0 & 0 \\ 0 & \sigma^2_{\text{pix}} & 0 \\ 0 & 0 & \sigma^2_{\text{pix}} \end{bmatrix}. \tag{17.82}$$

With this formulation of the single feature measurement likelihood, (17.81) can be restated in the same form of (17.2) as follows:

$$\mathbf{z}^i_k \equiv \begin{bmatrix} u_l \\ v_l \\ u_r \\ \text{orb}_z \end{bmatrix} = \mathbf{h}(\mathbf{x}_k, \mathbf{m}^j_k, \epsilon_k) \equiv \begin{bmatrix} \Gamma_{u_l}(\mathbf{m}^j, \mathbf{x}_k) + \epsilon_{u_l} \\ \Gamma_{v_l}(\mathbf{m}^j, \mathbf{x}_k) + \epsilon_{v_l} \\ \Gamma_{u_r}(\mathbf{m}^j, \mathbf{x}_k) + \epsilon_{u_r} \\ \text{orb}_m \oplus \epsilon_{\text{orb}} \end{bmatrix}, \tag{17.83}$$

where ϵ_{u_l}, ϵ_{v_l}, and ϵ_{u_r} represent normally distributed pixel noise random variables. ϵ_{orb} is the ORB noise, which is a string of bits of length 256 where every bit has a probability p^f_{orb} of being unity, and $\oplus$ is the bitwise exclusive or (XOR) operator.[9]

An essential aspect of the measurement model is the probability of detection, which is determined based on the structure of the ORBSLAM algorithms as follows:

$$P^{(j,k)}_{\text{D}} = P_{\text{D}}(\mathbf{m}^j, \mathbf{x}_k) = P_{\text{D}} \cdot \text{isInFrustum}(\mathbf{m}^j, \mathbf{x}_k) \cdot \text{depthCheck}(\mathbf{m}^j, \mathbf{x}_k) \cdot \text{angleCheck}(\mathbf{m}^j, \mathbf{x}_k), \tag{17.85}$$

where P_{D} is a predefined constant parameter.

The functions: isInFrustum$(\mathbf{m}^j, \mathbf{x}_k)$, depthCheck$(\mathbf{m}^j, \mathbf{x}_k)$ and angleCheck$(\mathbf{m}^j, \mathbf{x}_k)$ are key components of the ORBSLAM3 algorithm and explained in [13], with their implementation within the JVS-SLAM algorithm being discussed in [20].

17.5.1.6.2 MAV Motion Model

The drone's motion is represented as pure Brownian motion, which is the most basic form of motion modeling. This is implemented using the GTSAM *BetweenFactor* model [33]. BetweenFactor evaluates its error based solely on the difference between two consecutive poses, $\mathbf{x}_k$ and $\mathbf{x}_{k+1}$. The state at time k is composed of the position and orientation of the drone in 3D, i.e., an element in SE3:

$$\mathbf{x}_k = [x_k, y_k, z_k, \mathbf{q}_k]^\top, \tag{17.86}$$

---

9 Originally, this likelihood over the descriptor match was included heuristically, using an exponential distribution:

$$p\left(\mathbf{z}^i_k \middle| \mathbf{m}^{a_{\theta_k}(i)}, \mathbf{x}_k\right) \propto \mathcal{N}(u_l, v_l, u_r \mid \Gamma(\mathbf{m}^j, \mathbf{x}_k), \Sigma_{\mathbf{stereo}})\lambda_{\text{orb}} \exp(-\lambda_{\text{orb}} d_h(\text{orb}_z, \text{orb}_m)), \tag{17.84}$$

where parameter λ_{orb} was used to model the distribution over the Hamming distance, reflecting the quality of the descriptor match. This alternate form is proportional to (17.81) if $\lambda_{\text{orb}} = \log\left(\frac{1 - p^f_{\text{orb}}}{p^f_{\text{orb}}}\right)$ and therefore results in identical SLAM estimates as measurement likelihood (17.81).

with $\mathbf{q}_k$ being the orientation quaternion at time k. The "Between factor" within the *BetweenFactor* model of [33], can then be written in the form of (17.1) as:

$$\mathbf{x}_k = \mathbf{g}(\mathbf{x}_{k-1}, \mathbf{u}_k, \delta_k) = \mathrm{Exp}(\mathrm{Log}(\mathbf{x}_{k-1}) + \delta), \tag{17.87}$$

where Exp and Log are the Exponential and Logarithmic mappings inherent to SE3 as defined in [34]. δ is a random noise term, assumed normally distributed in $\mathcal{R}^6$ with zero mean and covariance

$$\Sigma_{odo} = \mathrm{diag}([\sigma_{wx}^2, \sigma_{wy}^2, \sigma_{wz}^2, \sigma_x^2, \sigma_y^2, \sigma_z^2]). \tag{17.88}$$

Therefore, Gaussian noise is added in the SE3 tangent space.

Notably, this factor only considers these two poses, which are part of the vector component in the JVS-SLAM model, and does not involve any set state variables. As a result, upgrading this factor to a more sophisticated motion model, such as one that incorporates IMU measurements, would not alter the underlying theoretical formulation of the model.

17.5.1.6.3 EuRoC Data Set JVS-SLAM Results

The complete algorithm is executed using ORB stereo detections on the EuRoC Vicon Room 2 (V02) data set (Figure 17.2b) [32], which is divided into three distinct trajectories: V201, V202, and V203. These trajectories represent different levels of motion complexity, referred to as "easy," "medium," and "difficult," respectively, due to varying speeds and accelerations. The source code for these results is publicly available at gitrepository.[10]

The parameters used for the algorithm are given in Table 17.2.

Table 17.2 JVS-SLAM parameters used for the EuRoC dataset.

Parameter	Value
σ_{pix}^2	1.0
λ_{orb}	−2.0
p_{orb}^{f} [11]	0.11920
P_{D}	0.6
$p_{\mathbf{x}}(\mathcal{Z})$	$(e^{-150})^{\|\mathcal{Z}\|}$
σ_{wx}^2	$1 \cdot 10^{-3}$
σ_{wy}^2	$1 \cdot 10^{-3}$
σ_{wz}^2	$1 \cdot 10^{-3}$
σ_x^2	$0.2 \cdot 10^{-3} [\mathrm{m}^2]$
σ_y^2	$0.2 \cdot 10^{-3} [\mathrm{m}^2]$
σ_z^2	$0.2 \cdot 10^{-3} [\mathrm{m}^2]$
P_E	$1 \cdot 10^{-2}$

10 https://github.com/finostro/orb_jvs_slam
11 This is the equivalent value calculated from λ_{orb} which is actually used in the code.

Table 17.3 RMS-ATE([m]) for the EuRoC, Vicon Room 2 data set.

Algorithm	V201	V202	V203
ORBSLAM2	0.037	0.035	–
ORBSLAM3	0.041	0.028	0.521
Vins-Fusion	0.230	0.200	–
SVO	0.050	0.090	0.790
JVS-SLAM	0.038	0.046	–

The algorithm's performance is evaluated against state-of-the-art SLAM methods using the absolute trajectory error (ATE) metric.[12] Although JVS-SLAM estimates the complete pose of the MAV, the following results omit the orientation components, since in SLAM usually only robot position is of interest. In particular, other notable state-of-the-art SLAM algorithms, such as those used for comparison in Table 17.3, only publish positional and not orientation errors. Therefore, the results focus on positional errors, for the sake of comparison purposes.

The root-mean-square (RMS) error values of the ATE[13] for JVS-SLAM, ORBSLAM2, ORBSLAM3, Vins-Fusion, and SVO [35] are given in Table 17.3.

The JVS-SLAM ATE values were calculated using the evo Python package [36], while the results for the other algorithms were sourced from [13].

Although JVS-SLAM does not achieve the lowest ATE in each category, it remains highly competitive. Its main advantage lies in its purely Bayesian framework, which eliminates the need for traditional DA heuristics (i.e., a SLAM front-end) that other algorithms rely on. This is achieved by considering multiple DA hypotheses simultaneously, unlike ORBSLAM3 and other SLAM algorithms, which evaluate only a single, externally determined hypothesis.

Entries marked with a "–" in Table 17.3 indicate cases where the respective SLAM algorithm failed to estimate the entire trajectory. For those that succeeded, the ATE values were unacceptably high. Additionally, all compared algorithms use robust cost functions to handle outliers effectively.

Figures 17.3 and 17.4 illustrate the detailed trajectories estimated by JVS-SLAM for the three distinct trajectories in the EuRoC Vicon Room 2 data set. The ground truth trajectory is depicted as a gray dashed line, while the estimated trajectory is represented by shaded lines, with lighter shades corresponding to larger errors. The trajectory's starting point is marked with a circle, and the endpoint is indicated with a cross. The shade of the estimated line corresponds to the instantaneous position error in meters, as shown in the associated bar.

In Figure 17.3a, which corresponds to the "easy" data set, the estimated trajectory closely follows the ground truth, with the largest deviation occurring toward the end. This increased error is attributed to the drone's landing phase, where it experiences its highest decelerations. Simultaneously, the stereo camera is primarily facing the floor, which has low texture and therefore produces fewer ORB detections. Figure 17.4a provides a top-down view of this trajectory.

12 The ATE metric provides a single, interpretable metric of the systems global consistency and drift. It quantifies the difference between the estimated and actual robot poses at corresponding points in time, after alignment into a common reference frame. Therefore, the ATE executes *alignment* between the estimated and ground truth coordinate systems and *error calculation* in the form of Euclidean distances between corresponding poses at each time stamp. This is typically followed by *summarization*, where the individual errors are summarized via the RMS error formula, to provide a single, average value [34]. The ATE is the de facto standard in the robotics community and was used to quantify the performance of the competing SLAM algorithms in Table 17.3. Therefore, it is used here also.

13 The RMS value of the ATE is averaged over the trajectory, not over Monte Carlo runs.

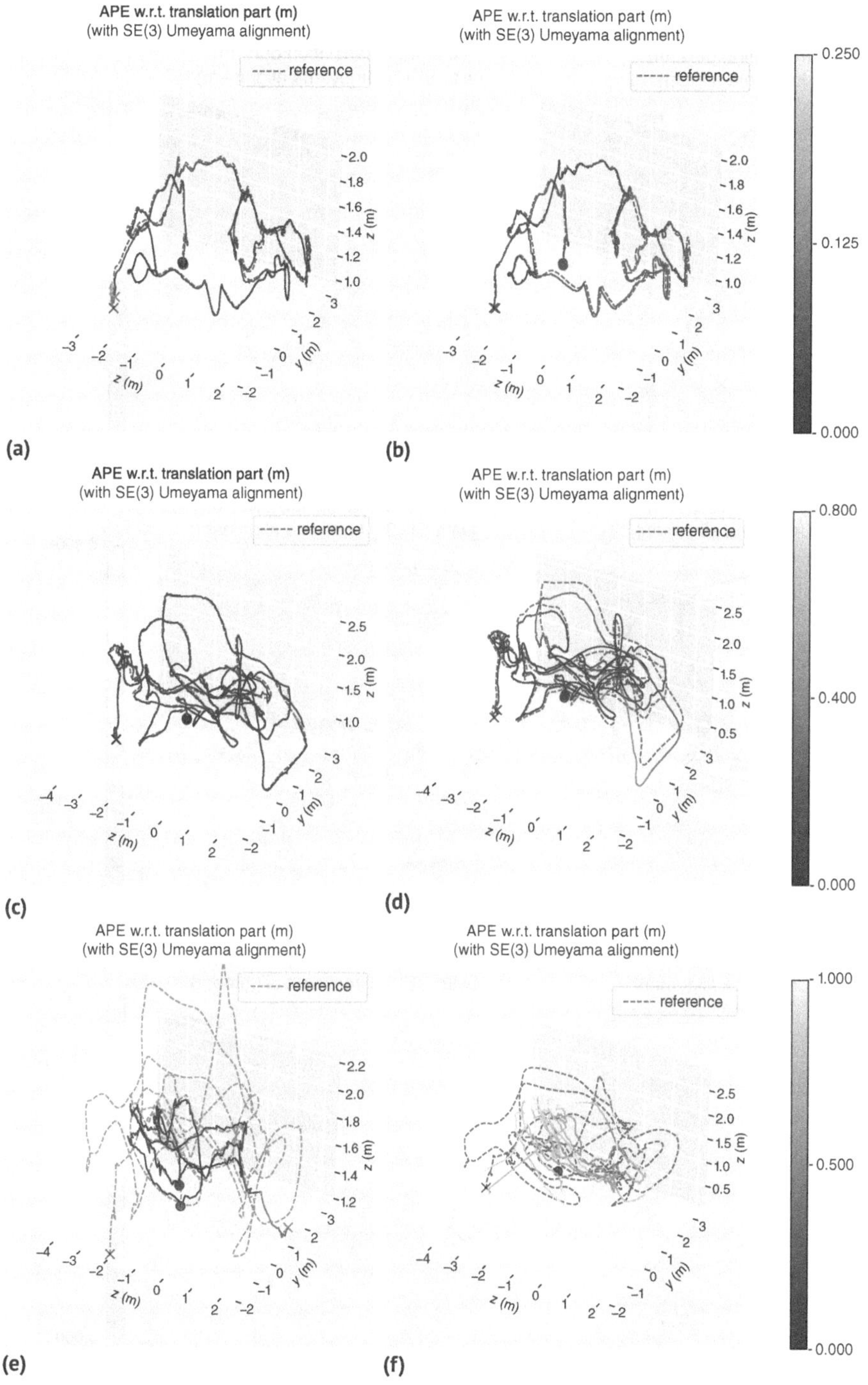

Figure 17.3 JVS-SLAM (left plots) and ORBSLAM3 (right plots) results on the EuRoC data set V201, V202, and V203 trajectories, 3D views. The trajectories are shaded according to the instantaneous error, with lighter shades corresponding to larger errors. The ground truth trajectory is shown as a gray dashed line for reference. (a) V201 "easy" dataset, JVS-SLAM, (b) V201 "easy" dataset, ORB-SLAM3, (c) V202 "medium" dataset, JVS-SLAM, (d) V202 "medium" dataset, ORB-SLAM3, (e) V203 "difficult" dataset, JVS-SLAM, and (f) V203 "difficult" dataset, ORB-SLAM3.

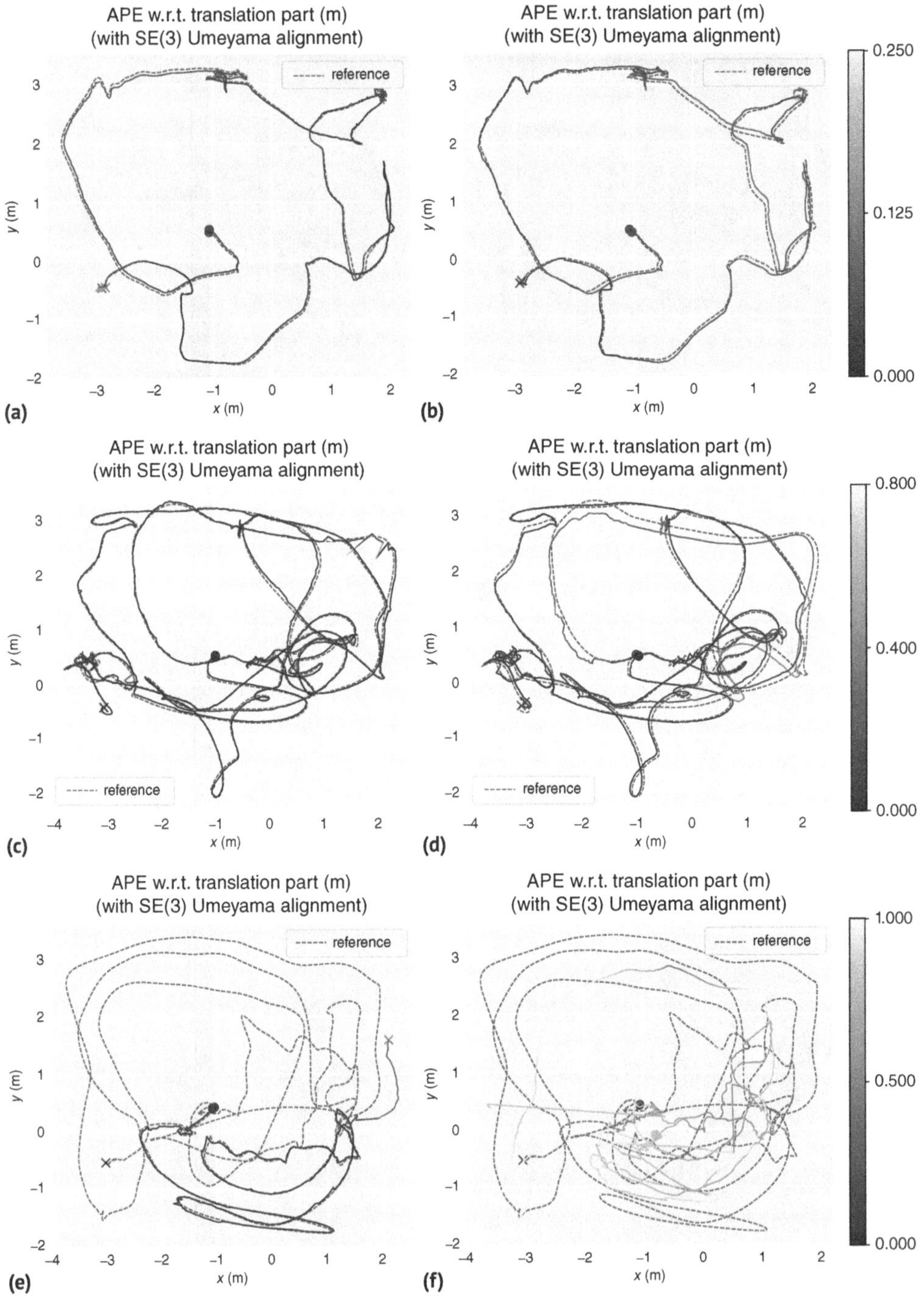

Figure 17.4 JVS-SLAM and ORBSLAM3 EuRoC data set results: V201, V202, and V203 trajectories, plan views. (a) V201 "easy" dataset, JVS-SLAM, (b) V201 "easy" dataset, ORB-SLAM3, (c) V202 "medium" dataset, JVS-SLAM, (d) V202 "medium" dataset, ORB-SLAM3, (e) V203 "difficult" dataset, JVS-SLAM, and (f) V203 "difficult" dataset, ORB-SLAM3.

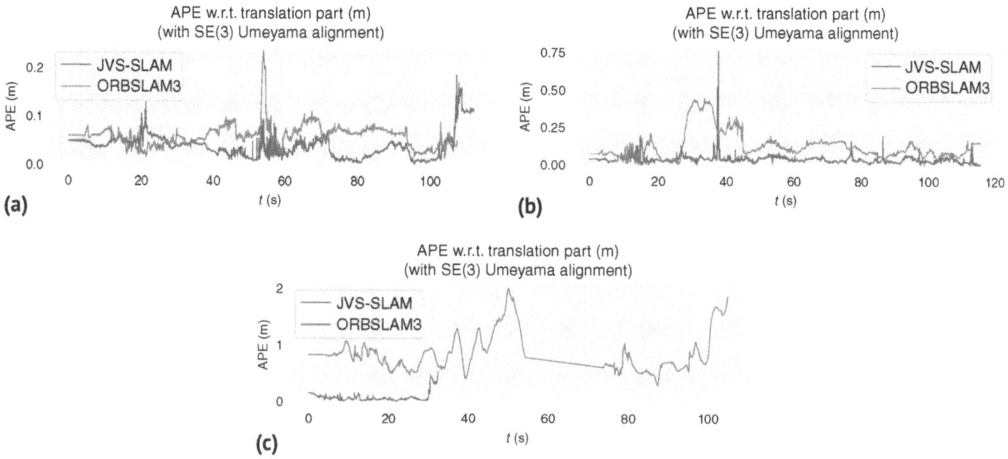

Figure 17.5 JVS-SLAM (lower curves) and ORBSLAM3 (upper curves) position error over time for the EuRoC data set V201, V202, and V203 trajectories. It can be seen that for trajectories V201 and V202, JVS-SLAM has lower position errors than ORBSLAM3, while for trajectory V203 neither algorithm is able to provide a full trajectory estimate. (a) V201 "easy" data set, (b) V202 "medium" data set, and (c) V203 "difficult" data set.

For comparison, Figure 17.3b shows the trajectory estimated using the latest publicly available version of ORBSLAM3.[14] Notably, the trajectory accuracy is lower than the values reported in Table 17.3. This discrepancy might be due to differences in ORBSLAM3 versions or variations in the computational power available.

Figure 17.5a presents the instantaneous position error for both JVS-SLAM (lower curve) and ORBSLAM3 (upper curve), which was used to compute the ATE metric. The plot indicates that JVS-SLAM generally maintains a lower error throughout the trajectory. At the end of the trajectory, ORBSLAM3 fails to produce an estimate, entering a lost state. This is consistent with the same challenging conditions where JVS-SLAM experiences its highest error, specifically, during the drone's landing with low-texture ground. It is worth noting that arguably, ORBSLAM3 could potentially recover and close the loop if the trajectory were to extend further.

Trajectory V202 demonstrated similar overall performance to the previous case, as illustrated in Figures 17.3c and 17.4c. However, this trajectory revealed some artifacts, represented by isolated poses with noticeably high instantaneous errors. This effect is particularly evident in Figure 17.4c near the coordinates $(x, y) = (2.0\text{ m}, 2.5\text{ m})$.

The trajectory estimate produced by the publicly available ORBSLAM3 implementation is shown in Figures 17.3d and 17.4d. To compare the two approaches, Figure 17.5b presents the instantaneous position error for both algorithms. The plot highlights that JVS-SLAM consistently achieves a lower error overall, although it experiences occasional spikes of high instantaneous error.

Interestingly, as observed in Figure 17.5b, the current version of ORBSLAM3 underperforms compared to JVS-SLAM for this trajectory. This outcome suggests that JVS-SLAM's Bayesian approach, which inherently incorporates multiple DA hypotheses, provides enhanced robustness in challenging sections where ORBSLAM3 fails.

14 Data taken from the official GitHub repository https://github.com/UZ-SLAMLab/ORB_SLAM3, latest commit: 4452a3c4ab75b1cde34e5505a36ec3f9edcdc4c4 as of December 2023.

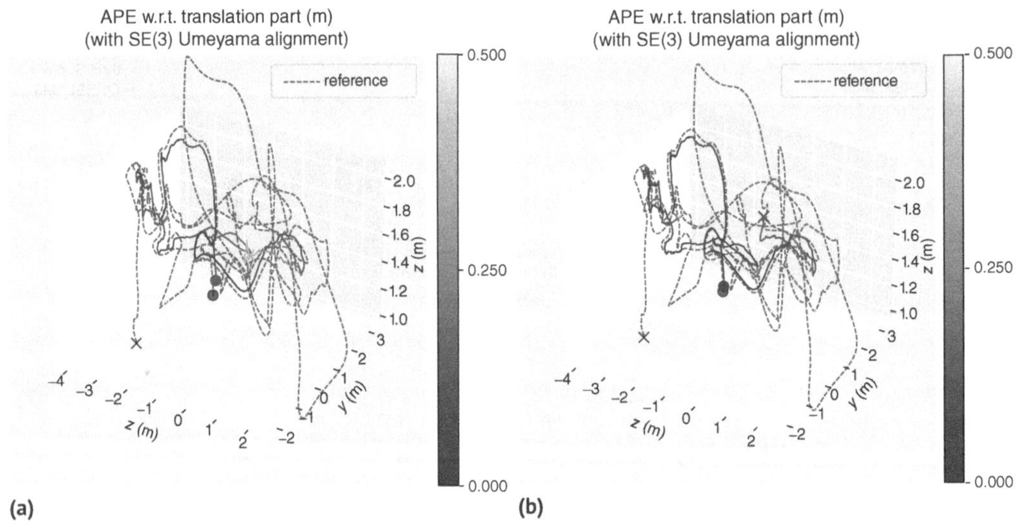

Figure 17.6 Intermediate JVS-SLAM results with the EuRoC V202 "medium" data set. 3D views just before and after a sharp turn. After the sharp turn, JVS-SLAM is able to correct a large accumulated error. (a) Just before sharp turn and (b) just after sharp turn.

In the "difficult" data set, characterized by the highest speeds and accelerations, JVS-SLAM is unable to generate a complete trajectory estimate. This failure occurs when the Gibbs sampler is unable to produce DA hypotheses that yield JVS-SLAM components with sufficiently high weights. As a result, the algorithm effectively stalls in its estimation process. The incomplete trajectory can be observed in Figures 17.3e and 17.4e.

In comparison, the publicly available version of ORBSLAM3 also suffers with this data set, displaying significant trajectory errors as shown in Figures 17.3f and 17.4f. The instantaneous position error for the estimated portion of the trajectory is illustrated in Figure 17.5c. The plot highlights that JVS-SLAM's trajectory error peaks just before it stops generating high-weight DA hypotheses, signaling the end of its estimate.

For ORBSLAM3, the figures reveal repeated instances of the algorithm entering a lost state. While it manages to partially recover from some of these states, it ultimately fails to recover from the final one, resulting in an incomplete trajectory estimate.

Videos demonstrating the ORB-based JVS-SLAM algorithm in action are available on YouTube for trajectories V201[15] and V202.[16]

Finally, Figure 17.6a,b displays notable intermediate results for JVS-SLAM on the V202 data set. These snapshots capture the state of the algorithm just before and just after the camera executes a sharp turn and reobserves keypoints that were originally detected near the start of the trajectory, effectively "closing a loop."

As shown in Figure 17.6b, the detection of previously visited locations significantly reduces the trajectory error and corrects the estimated path retroactively by more than 100 poses. This substantial correction demonstrates the strength of JVS-SLAM's "loop closure" capability, which is driven by its Bayesian multi-hypothesis framework.

15 https://youtu.be/WBJ66AwyfSs
16 https://youtu.be/DIFUC3DOneY

17.6 Conclusions

This chapter has demonstrated that the "meeting" between RFS's and SLAM offers significant benefits over state-of-the-art RV-SLAM solutions. In particular:

1. The SLAM problem can be expressed within the RFS framework and linked to the RV formulation, but only under ideal detection conditions. These include the absence of false positives, a single correct DA permutation, and a known map size.
2. In contrast, the RFS approach inherently considers all possible DA permutations, integrates detection statistics, and encompasses uncertainty in map cardinality. This principle applies not just to Bayesian filtering but also to ML batch estimation.
3. Leveraging the principles of RFSs, the application of Bayes' theorem led to the development of the batch-based, ML-JVS-SLAM algorithm. This approach integrates map management and measurement-to-landmark DA directly into the Bayesian JVS-SLAM estimator, unifying the SLAM back and front ends within a single Bayesian framework. It employs the well-established sparse graph optimization techniques such as the algorithms available within the g2o framework to estimate both the map and the robot's trajectory for each JVS-SLAM component.
4. A stereo vision-based JVS-SLAM algorithm was developed utilizing the ORB keypoint detector and descriptor pair, achieving performance comparable, and in many instances superior, to the state-of-the-art ORBSLAM3 algorithm, without the need for separate DA heuristics.

This chapter has uncovered various future research opportunities. The current implementation of the JVS-SLAM algorithm achieves joint estimation with a computational complexity comparable to multi-hypothesis SLAM methods. Since potential associations must be selected before the costly computation of the component weight w^{c_K}, the sequence in which these hypotheses are sampled significantly impacts the algorithm's performance.

One potential improvement could be improving the initial estimate provided to the Gibbs sampler. At present, the Gibbs sampler relies on the previously estimated pose $\mathbf{x}_k$ to approximate the pose after the graph optimization step. An alternative strategy would be to randomly sample the minimum number of associations needed to estimate the pose $\mathbf{x}_k$, similar to the RANSAC method. This approach could then initialize the Gibbs sampler with a more refined estimate of $\mathbf{x}_k$, potentially improving its convergence and accuracy.

Acknowledgments

This work was supported by ANID (National Research Agency, Chile) PIA Project AFB180004 and ANID FONDECYT project 1231658.

References

1 M. W. M. G. Dissanayake, P. Newman, S. Clark, H.F. Durrant-Whyte, and M. Csorba. A solution to the simultaneous localization and map building (SLAM) problem. *IEEE Transactions on, Robotics and Automation,* 17(3):229–241, 2001.

2 H. Durrant-Whyte and T. Bailey. Simultaneous localization and mapping: Part I. *IEEE Robotics & Automation Magazine,* 13(2):99–110, 2006.

3 J. Aulinas, Y. Petillot, J. Salvi, and X. Llado. The SLAM problem: A survey. In *Frontiers in Artificial Intelligence and Applications*, volume 184, 363–371, 2008.

4 T. Bailey and H. Durrant-Whyte. Simultaneous localization and mapping (SLAM): Part II. *IEEE Robotics & Automation Magazine*, 13(3):108–117, 2006.

5 I. A. Kazerouni, L. Fitzgerald, G. Dooly, and D. Toal. A survey of state-of-the-art on visual SLAM. *Expert Systems with Applications*, 205:117734, 2022.

6 S. Thrun and M. Montemerlo. The GraphSLAM algorithm with applications to large-scale mapping of urban structures. *International Journal of Robotics Research*, 25(5–6):403–429, 2006.

7 R. Kümmerle, G. Grisetti, H. Strasdat, K. Konolige, and W. Burgard. G2o: A general framework for graph optimization. In *2011 IEEE International Conference on Robotics and Automation*, pages 3607–3613, 2011.

8 M. Kaess, A. Ranganathan, and F. Dellaert. ISAM: Incremental smoothing and mapping. *IEEE Transactions on Robotics*, 24(6):1365–1378, 2008.

9 M. Kaess, H. Johannsson, R. Roberts, V. Ila, J. Leonard, and F. Dellaert. iSAM2: Incremental smoothing and mapping using the bayes tree. *International Journal of Robotics Research*, 31(2):216–235, 2012.

10 R. Mur-Artal, J. M. M. Montiel, and J. D. Tardós. ORB-SLAM: A versatile and accurate monocular slam system. *IEEE Transactions on Robotics*, 31(5):1147–1163, 2015.

11 E. Rublee, V. Rabaud, K. Konolige, and G. Bradski. ORB: An efficient alternative to SIFT or SURF. In *International Conference on Computer Vision*, pages 2564–2571, 2011.

12 R. Mur-Artal and J. D. Tardós. ORB-SLAM2: An open-source slam system for monocular, stereo, and rgb-d cameras. *IEEE Transactions on Robotics*, 33(5):1255–1262, 2017.

13 C. Campos, R. Elvira, J. J. Gómez Rodríguez, J. M. M. Montiel, and J. D. Tardós. Orb-slam3: An accurate open-source library for visual, visual–inertial, and multimap slam. *IEEE Transactions on Robotics*, 37(6):1874–1890, 2021.

14 J. Mullane, B.-N. Vo, and M. D. Adams. Rao-Blackwellised PHD SLAM. In *Proceedings of IEEE International Conference on Robotics and Automation (ICRA)*, pages 5410–5416, 2010.

15 J. Mullane, B.-N. Vo, M. D. Adams, and B.-T. Vo. A random-finite-set approach to Bayesian SLAM. *IEEE Transactions on Robotics*, 27(2):268–282, 2011.

16 K. Leung, F. Inostroza, and M. Adams. Relating random vector and random finite set estimation in navigation, mapping, and tracking. *IEEE Transactions on Signal Processing*, 65(17):4609–4623, 2017.

17 H. Deusch, S. Reuter, and K. Dietmayer. The labeled multi-Bernoulli SLAM filter. *IEEE Signal Processing Letters*, 22(10):1561–1565, 2015.

18 D. Moratuwage, M. Adams, and F. Inostroza. δ-Generalised labelled multi-Bernoulli simultaneous localisation and mapping. In *International Conference on Control, Automation and Information Sciences (ICCAIS)*, pages 175–182, 2018.

19 L. Gao, G. Battistelli, and L. Chisci. Phd-slam 2.0: Efficient slam in the presence of missdetections and clutter. *IEEE Transactions on Robotics*, 37(5):1834–1843, 2021.

20 F. Inostroza and M. Adams. Combining the SLAM back and front ends with a joint vector-set distribution. *The International Journal of Robotics Research*, 44(7), 1231–1254, 2025.

21 S. Thrun, W. Burgard, and D. Fox. *Probabilistic Robotics*. MIT press Cambridge, 2005.

22 R. Mahler. *Statistical Multisource-multitarget Information Fusion*. Artech House Norwood, MA, USA, 2007.

23 I. J. Cox. A review of statistical data association techniques for motion correspondence. *International Journal of Computer Vision*, 10(1) 53–66, 1993.

24 J. Neira and J. D. Tardós. Data association in stochastic mapping using the joint compatibility test. *IEEE Transactions on Robotics & Automation*, 17(6):890–897, 2001.

25 R. Mahler. *Advances in Statistical Multisource-multitarget Information Fusion*. Artech House, Norwood, MA, USA, 2014.

26 B-N. Vo, S. Singh, and A. Doucet. Sequential monte carlo implementation of the PHD filter for multi-target tracking. In *Proceedings of the International Conference on Information Fusion*, pages 1561–1565, 2003.

27 H. Strasdat, J. Montiel, and A. Davison. Visual SLAM: Why filter? *Image and Vision Computing*, 30(2):65–77, 2012.

28 B-T. Vo and B-N. Vo. Labeled random finite sets and multi-object conjugate priors. *IEEE Transactions on Signal Processing*, 61(13): 3460–3475, 2013.

29 B-N. Vo, B-T. Vo, and D. Phung. Labeled random finite sets and the Bayes multi-target tracking filter. *IEEE Transactions on Signal Processing*, 62(24):6554–6567, 2014.

30 C. Shim, B.-T. Vo, B.-N. Vo, J. Ong, and D. Moratuwage. Linear complexity Gibbs sampling for generalized labeled multi-Bernoulli filtering. *IEEE Transactions on Signal Processing*, 71:1981–1994, 2023.

31 M. Hsiao and M. Kaess. MH-iSAM2: Multi-hypothesis iSAM using Bayes tree and hypo-tree. In *2019 International Conference on Robotics and Automation (ICRA)*, pages 1274–1280, 2019.

32 M. Burri, J. Nikolic, P. Gohl, T. Schneider, J. Rehder, S. Omari, M. Achtelik, and R. Siegwart. The EuRoC micro aerial vehicle datasets. *International Journal of Robotics Research*, 35:1157–1163, 2016.

33 F. Dellaert and Contributors. 2022. URL https://github.com/borglab/gtsam).

34 L. Carlone, K. Ayoung, T. Barfoot, D. Cremers, and F. Dellaert, editors. *SLAM Handbook. From Localization and Mapping to Spatial Intelligence*. Cambridge University Press, 2026.

35 C. Forster, Z. Zhang, M. Gassner, M. Werlberger, and D. Scaramuzza. SVO: Semi-direct visual odometry for monocular and multicamera systems. *IEEE Transactions on Robotics*, 33(2):249–265, 2017.

36 M. Grupp. Evo: Python package for the evaluation of odometry and slam. 2017.https://github.com/MichaelGrupp/evo

Part III

Distributed Fusion

18

Centralized and Distributed Multiple-hypothesis Tracking

Stefano Coraluppi

Systems and Technology Research, Woburn, MA, USA

18.1 Introduction

Multi-target tracking (MTT) addresses a complex statistical estimation problem for which a variety of solution paradigms have been developed [1, 2]. A key challenge is data association, as the time-varying number of targets and measurement provenance (from the targets or due to clutter) is unknown. Traditionally, particularly with solution approaches that include *multiple-hypothesis tracking* (MHT), a principal focus is how to manage the complexity associated with high-rate high-volume measurement inputs. More recently, context-aided techniques have gained prominence in settings where data is scarce, and the key challenge is rather how to mitigate the growth in target state uncertainty [3].

In this chapter, we first introduce salient elements of MHT. Next, we discuss how performance and robustness of centralized MHT can often be improved upon via principled multistage distributed MHT solutions. As instances of first-stage processing, we give particular attention to *angle-of-arrival* (AOA) and *time-of-arrival* (TOA) measurement processing. Further aspects of advanced tracking discussed here include the merits of decoupled data association and track management, performance modeling, and performance evaluation.

18.2 Multiple-hypothesis Tracking

MHT is well-established as a leading paradigm for MTT [4]. It was first introduced in hypothesis-oriented form in [5] and made viable with the track-oriented formulation that avoids global-hypothesis enumeration via hypothesis score factorization [6]. Extensions to account for merged measurements, undetected birth, and multiple measurements include [7–9]. MHT is in fielded use in many operational domains.

MHT seeks the *maximum a posteriori* (MAP) global data association hypothesis. In fact, strictly speaking, MHT seeks the MAP hypothesis including both data association and target existence, in the sense that the global hypothesis identifies the time of birth and death of each target. Generally, there are many global hypotheses that are indistinguishable from a data-association perspective, and in some settings, one may perform better (with respect to operational performance metrics) by seeking the MAP *set* of identical (or nearly identical) global data association hypotheses.

Arguably, MHT is a bit of a misnomer. What makes MHT tractable is that one identifies (generally, within a temporal sliding window) a single data association solution, discarding competing hypotheses. This *hard data association* exhibits track repulsion that may induce suboptimal tracking performance; the effect is complementary to the track coalescence often seen in *soft data association* methods [10].

Derivation of the track-oriented MHT recursion depends crucially on the assumptions of a Poisson birth-death process and Poisson distributed clutter. We assume continuous-time target birth rate λ_b and target death rate λ_χ, leading to the following discrete-time expected number of births and probability of death over an interval of duration Δt:

$$\mu_b = \frac{\lambda_b}{\lambda_\chi}(1 - e^{-\lambda_\chi \Delta t}) \tag{18.1}$$

$$p_\chi = 1 - e^{-\lambda_\chi \Delta t} \tag{18.2}$$

For each sensor scan, we assume target detection probability p_d and false alarm rate Λ_{fa}. We generally assume target measurements to be a (possibly nonlinear) function of the target state with additive (possibly Gaussian) noise. The structure of the measurement equation will lead to more or less complex recursive filtering but is not crucial to our discussion here. Both Kalman-based and particle filtering methods are readily accommodated.

The track-oriented MHT recursion expresses the probability of the child global hypothesis $p(q|Z,\bar{Z})$ as a function of the probability of the parent global hypothesis $p(\bar{q}|\bar{Z})$ is given in (18.3–18.4), where $Z = \{z_j, j = 1, \ldots, r\}$ is the set of measurements in the current scan, r is the number of measurements in the current scan, and $\bar{Z}$ denotes all prior data.

In the recursion, τ is the number of targets present at the previous scan, χ is the number of targets that do not survive to the current scan, d is the number of target detections in the current scan, b is the number of target births in the current scan, J_d is the set of indices of target update measurements, J_{fa} is the set of indices of false alarms, J_b is the set of indices of target birth measurements, $f(\cdot|\bar{Z},q)$ is the conditional probability distribution of a target (update) measurement, $f(\cdot)$ is the probability distribution of a target birth measurement, and $f_{\text{fa}}(\cdot)$ is the probability distribution of false alarms in measurement space.

$$\begin{aligned} p(q|Z,\bar{Z}) &= C p(\bar{q}|\bar{Z}) p_\chi^\chi ((1-p_\chi)(1-p_d))^{\tau-\chi-d} \\ &\cdot \prod_{j \in J_d} \frac{(1-p_\chi) p_d f(z_j|\bar{Z},q)}{\Lambda_{\text{fa}} f_{\text{fa}}(z_j)} \prod_{j \in J_b} \frac{p_d \mu_b f(z_j)}{\Lambda_{\text{fa}} f_{\text{fa}}(z_j)} \end{aligned} \tag{18.3}$$

$$C = \frac{\left\{\frac{\exp(-\mu_b - \Lambda_{\text{fa}})\Lambda_{\text{fa}}^r}{r!}\right\} \prod_{z_j \in Z} f_{\text{fa}}(z_j)}{p(Z)} \tag{18.4}$$

The factor C in (18.3), given by (18.4), is a factor common to all global hypotheses, and thus need not be computed in practice. It may be understood as the probability of the null hypothesis in which there are no targets to account for from the previous scan, no target births, and all measurements are false alarms.

The MHT formalism prescribes batch data association and track extraction with the MAP global hypothesis. Owing to the hypothesis score factorization, this can be cast as a (simpler) integer programming problem. Practical MHT solutions rely on sliding-window data association and track extraction to address both computational and real-time surveillance requirements.

18.3 Distributed MHT

Despite efforts to extend MHT theory to address various challenges, often the most effective means to contend with complex sensor phenomena, disparate multi-sensor update rates and information content, and challenging multi-target ambiguities (due to closely spaced targets, large sensor measurement errors, Doppler ambiguity, and high sensor clutter rates) is to perform distributed (or multistage) MHT. This challenges the intuition afforded by the distributed detection and estimation literature, whereby one often seeks to approach centralized performance via effective distributed processing. In the MTT setting, we seek rather to outperform necessarily suboptimal centralized MHT via an effectively designed distributed MHT solution.

There is amble empirical evidence of the benefits of distributed MHT; see [4] and references therein. As a single-sensor example, a first stage of cautious tracking (low target process noise, low track coast times) followed by a second stage of more aggressive tracking (higher target process noise, larger track coast times) is effective to track maneuvering targets in heavy clutter (see Figure 18.1). As a (passive) multi-sensor example, it is beneficial to perform single-sensor measurement-space tracking upstream, with downstream multi-sensor (Cartesian) tracking (see Figure 18.2). The second stage has lower-error cross-sensor data association that mitigates ghost tracks.

In most of the distributed MHT architectures that we consider, measurement information is passed to downstream stages of processing. The motivation for this is to avoid the suboptimality associated with distributed estimation contending with correlated state estimation errors [1]. In this view, tracks are sequences of associated measurements, as this is more informative than the corresponding sequences of state estimates.

Note that the storage and transmission requirements associated with measurements (and associated measurement error covariances) is comparable to that of state estimates (and associated state covariances). The primary and drastic reduction in data flow is associated with false measurement

Figure 18.1 An example of distributed single-sensor tracking.

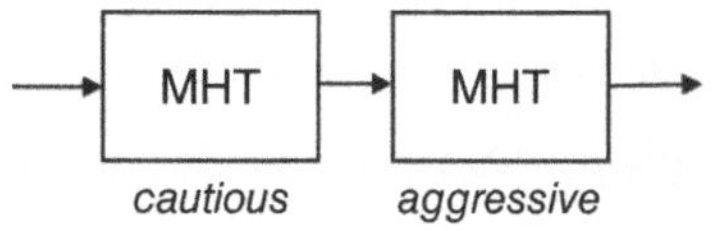

Figure 18.2 An example of distributed multi-sensor tracking.

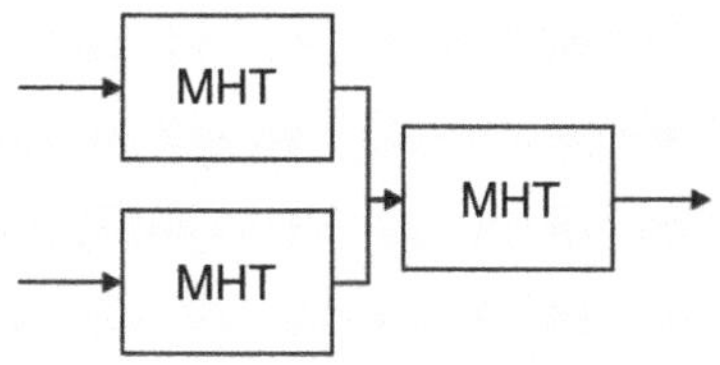

suppression in first-stage tracking. Hence, it is a good design choice to send track-relevant measurements to downstream processing.

Despite the proven effectiveness of distributed MHT in many single-sensor and multi-sensor settings, relying on upstream association decisions in downstream processing can be problematic. Corrective downstream logic to contend with necessarily imperfect upstream processing can be cumbersome, highly complex, and not sufficiently robust. Here, we identify through a series of vignettes several examples of difficulties that may emerge from distributed MHT.

18.3.1 Issue 1: Upstream Tracking Errors

Consider the case where the distributed architecture in Figure 18.2 is used, and each of the upstream trackers produces two tracks. In one solution, the tracks IA and IB cross, while in the second solution, the tracks IIA and IIB bounce. This is illustrated in Figure 18.3. The downstream MHT may associate IA and IIA as well as IB and IIB, only to be confronted with diverging input tracks with which to update fused tracks.

An example of simple corrective logic is to terminate the fused track IIIA and IIIB, and thereafter to treat the input tracks as new tracks amenable to new fusion decisions. This would induce fused tracks IIIC and IIID. More complex corrective logic to avoid the fragmented fused-tracking solution may require rollback logic and suitable fragmenting of upstream track feeds.

18.3.2 Issue 2: Cautious Upstream Processing

In some settings, one may need to contend with highly disparate sensors. For instance, we may have high-rate kinematic data from one sensor, and low-rate, highly informative identity measurements from a second sensor. Single-stage MHT exploiting identity information is difficult, as the required hypothesis tree depth is prohibitive.

A reasonable processing architecture is shown in Figure 18.4, whereby a first stage of cautious processing is employed to form kinematic track fragments that are stitched in a second stage of processing with the aid of the identity measurements [11]. Due to the drastic reduction kinematic track objects after the first stage of processing, the second stage MHT may rely on a large hypothesis depth to enable correct association decisions. A notional illustration is in Figure 18.5.

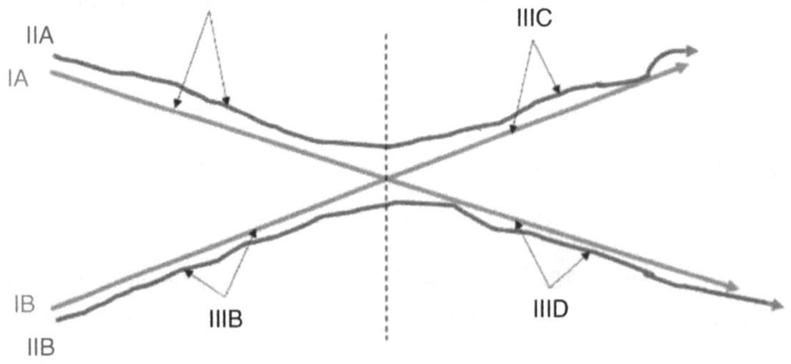

Figure 18.3 Fragmentation induced by upstream tracking errors.

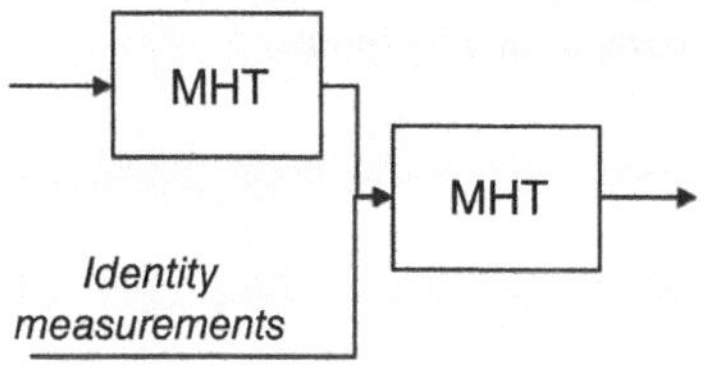

Figure 18.4 Multi-INT tracking architecture.

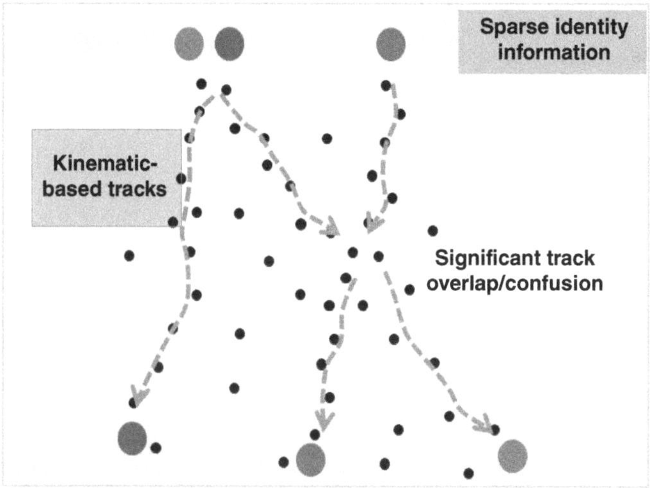

Figure 18.5 Multi-INT tracking problem illustration.

The success of this approach relies on the ability to ensure cautious processing that fragments first-stage tracks when association ambiguity is high. This is extremely difficult to achieve as tracking solutions are based on global hypothesis scores in both hypothesis-oriented and modern track-oriented MHT solutions.

18.3.3 Issue 3: Variable-quality Sensors

Downstream track-fusion processing generally assumes a negligible false track rate from upstream sensors. As such, downstream processing performs track association decisions, without discarding any unassociated tracks. Relaxing the *no false track* assumption while contending with unequal residual false track rates across sensors is nontrivial. In a sense, we face a generalization to the classic distributed detection problem of designing optimal fusion rules [12]. In distributed tracking, if we are to maintain upstream measurement association constraints, sequences of associated measurements are to be maintained or discarded consistently. The generalization to simple $M\text{-}of\text{-}N$ type rules requires further extension to contend with associated measurement sequences.

As an illustration (Figure 18.6), consider the case where System I produces fragmented tracks but with essentially no false tracks, while System II produces higher-continuity tracks but with nontrivial numbers of false tracks as well. Design of systematic and robust track-fusion logic (extensible to the case of N upstream systems) is lacking and appears to be highly complex.

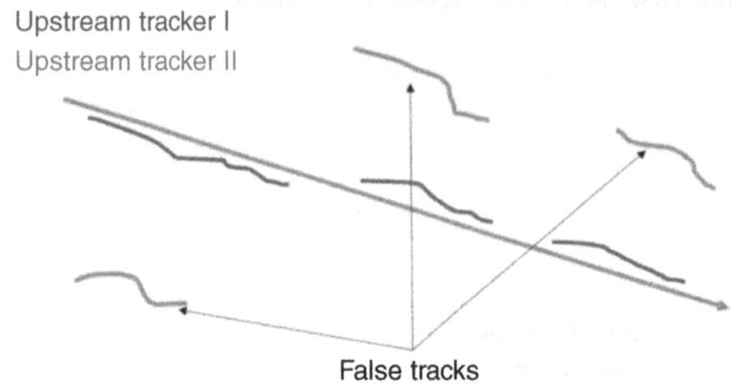

Figure 18.6 Track fusion with variable-quality inputs.

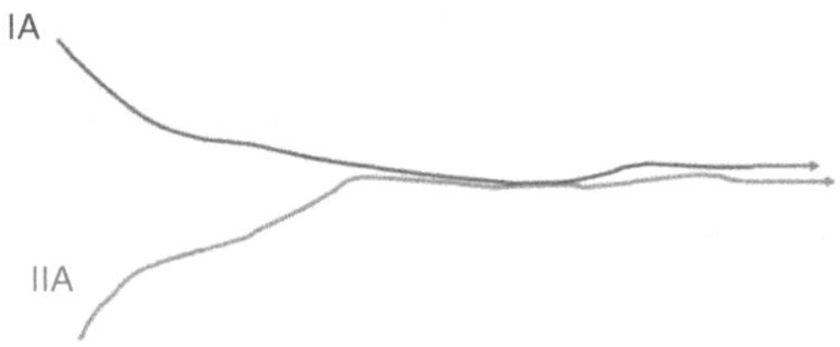

Figure 18.7 The case for revisiting a non-fusion decision.

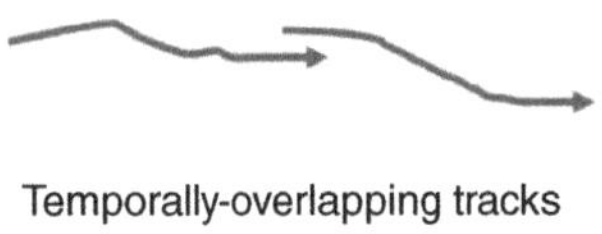

Figure 18.8 A challenging instance of first-stage track fragmentation.

18.3.4 Issue 4: Non-fusion Decisions

Consider again the multi-sensor case where track IA and track IIA from distinct upstream trackers are not fused downstream; see Figure 18.7. This decision may be due to initial poor track localization, even though the tracks may subsequently align well. Nonetheless, once it is determined not to fuse these tracks, subsequent track updates do not impact this decision. Modifying real-time track fusion logic to revisit earlier non-fusion decisions significantly complicates reasoning logic, potentially requiring that we disregard upstream single-sensor associations.

18.3.5 Issue 5: Upstream Fragmentation

A related instance of an upstream non-fusion decision is best explained in the single-sensor case, with the architecture in Figure 18.1. Refer to Figure 18.8. Consider the case where the first stage produces two track fragments. In many cases, the two tracks are temporally nonoverlapping and hence amenable to second-stage fusion.

In some cases, the fragmentation is such that the second track starts before the second is terminated. If we are fortunate, there will not be a sensor frame in which both tracks have a measurement update. If the situation occurs with both track having an update (as in the case where one track correctly makes use of a target-induced measurement while the other makes use of a spurious false alarm), we cannot fuse the tracks in downstream processing without violating the point-target assumption. This makes for a non-robust processing scheme in which the slightest data clash invalidates successful track fusion.

18.3.6 Issue 6: Wide-area Tracking

In wide-area settings, an effective distributed processing solution is to partition the field of regard and to apply a bank of trackers, each devoted to a subregion. Downstream fusion combines single-sensor tracks into an overall wide-area surveillance solution.

Solution robustness for targets that are near the boundary between subregions is achieved by introducing a small overlap between subregions. Refer to Figure 18.9. This, however, requires appropriate adjustments in downstream processing to avoid double counting of those measurements that contribute to more than one track.

18.3.7 Issue 7: Distributed Sensor Networks

Perhaps the most complex setting illustrating the difficulty of accounting for hard upstream data association decisions in downstream processing is in the context of distributed sensor networks with arbitrary topology. Here, we have sensors with associated processing that will perform data association decisions with local detection data as well as with associated measurement sequences received from neighbors.

In principle, maintaining pedigree (provenance) information for all detections prevents double counting, when the same data arrives to a sensor via multiple channels. On the other hand, how to reconcile conflicting fusion decisions involving the same measurement that arrives via multiple channels?

Referring to Figure 18.10, Track IA may be fused with Track IIA and, separately, with Track IIIA, with IIA and IIIA incompatible for fusion. How should System IV contend with Track IA received via two channels? A similar difficulty exists when a track produced by System IV flows through the network and returns to System IV, having undergone distinct (and potentially incompatible) associations along the way.

While the distributed tracking paradigm whereby tracks are sequences of associated measurements (rather than sequences of state estimates) is helpful to avoid distributed estimation suboptimality, the examples illustrated here serve to identify some of the difficulties with hard upstream data association decisions. In a sense, the same difficulty would exist if we were to require that the

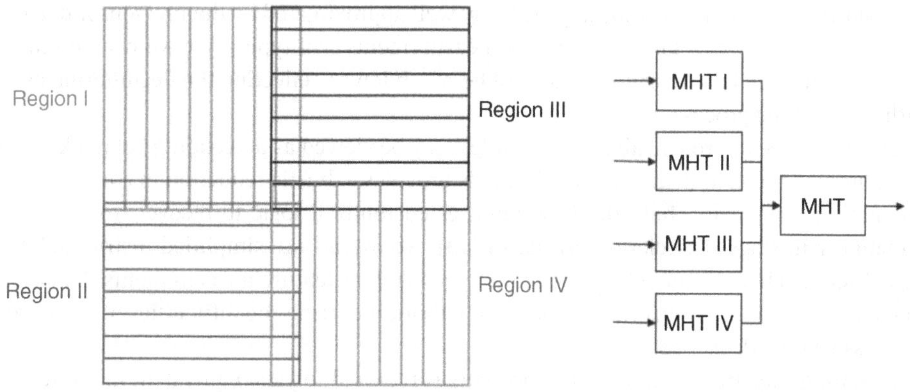

Figure 18.9 Wide-area distributed tracking.

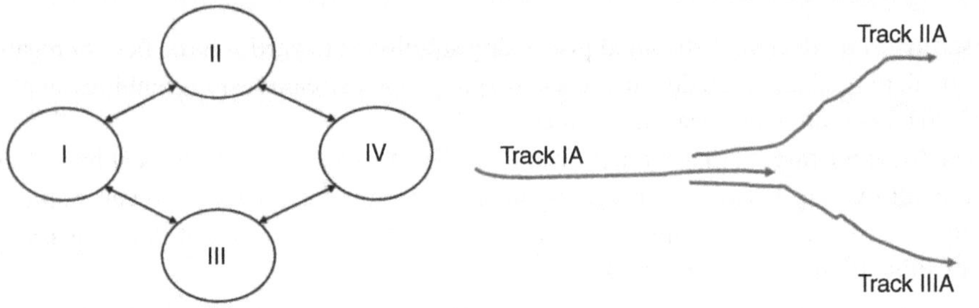

Figure 18.10 Tracking in a sensor network.

tracker not discard measurements deemed to be detections in upstream processing. No false alarms could be discarded by the tracker!

A more helpful paradigm is to view all tracking processes as (generalized) filters that discard spurious measurements and retain those deemed to be target originated. The association decisions established as part of this filtering process is an ancillary product that need not impact subsequent processing. Stated another way, we may simply ignore all upstream track labels and reason over the available detection data.

A reasonable objection might be offered, in that by ignoring previously established association decisions, inefficiency is introduced, and more complex tracking is needed. Referring for instance to Figure 18.4, the downstream fuser will need to consider a larger set of track hypotheses if measurement sequences associated with first-stage kinematic tracks need to be reassessed. On the other hand, most upstream association decisions incur little ambiguity once the confounding clutter has been removed. On balance, the modest increase in downstream complexity is easily offset by no longer having to contend with the many difficulties and instances of solution non-robustness identified above.

Accordingly, all MHT modules in distributed architectures may be viewed as *detection-level trackers*. As such, the classical track oriented MHT recursion (as found, e.g., in [9]) is directly applicable, with suitably defined target and sensor statistics. In particular, the sensor detection probability p_d following an initial tracking stage is likely well approximated by the raw sensor detection probability, as relatively few target-originated measurements will typically have been discarded. On the other hand, the false alarm rate Λ_{fa} will be much lower, reflecting the decluttering associated with the tracking process.

With our simplified approach, all MHT modules may be viewed as detection-level trackers with suitably calibrated target and sensor statistics. This provides a drastic reduction in conceptual and implementation complexity that otherwise requires customized logic to recover from a myriad of upstream (or in-stage) association errors. In [13], we extend the simplified distributed MHT paradigm discussed here by modeling upstream track labels as feature measurements. This allows for enhanced exploitation of upstream processes without incurring the difficulties of hard exogenous data association decisions.

In some settings, upstream estimation processing is required to yield equivalent measurements amenable to downstream tracking, as with AOA or TOA sensors. With multi-sensor AOA measurements, we require an equivalent measurement for track initialization; subsequently, track refinement may be achieved directly with AOA measurements. With TOA measurements, we require equivalent measurements both for track initialization and track update. As such, tracking with

AOA and TOA sensors necessarily requires decoupled estimation and tracking stages: a form of distributed tracking. In the next two sections, we discuss methods for target localization with AOA and TOA sensors.

18.4 Target Localization with Angle-only Measurements

We describe here the 3D track localization solution whereby only positional initialization is performed. Full 6D state initialization when angle rates are available as well is described in [14].

Figure 18.11 captures our basic approach. We seek to initialize the track at a point that minimizes the sum of the squared distances to the projections on the rays defined by the sensor data. In 2D, measurement lines intersect; this is not so in 3D.

We considering a *least squares* (LS) solution approach that does not depend on the accuracy of sensor data, though the corresponding solution uncertainty accounts for sensor error statistics. An alternative to our approach is to seek the *maximum likelihood* (ML) solution, requiring numerical optimization [15].

The objective J to be minimized is based on the Euclidean norm. We denote by $\widetilde{X}_i$ the location discrepancy with respect to sensor i, and by $\widehat{X}$ the location that achieves the minimum.

$$J = \sum_{i=1,2} \|\widetilde{X}_i\|^2 = \sum_{i=1,2} \|X - X_i\|^2 \tag{18.5}$$

Let $u_i = \begin{bmatrix} \sin\theta_i \cos\phi_i \\ \cos\theta_i \cos\phi_i \\ \sin\phi_i \end{bmatrix}$, where θ_i and ϕ_i are, respectively, the bearing and elevation measurements from sensor i, with $i = 1, 2$. First, we express X_i as a function of X and $\bar{X}_i$.

$$X_i = \bar{X}_i + (u_i \cdot (X - \bar{X}_i))u_i = \bar{X}_i + u_i u_i^T (X - \bar{X}_i) \tag{18.6}$$

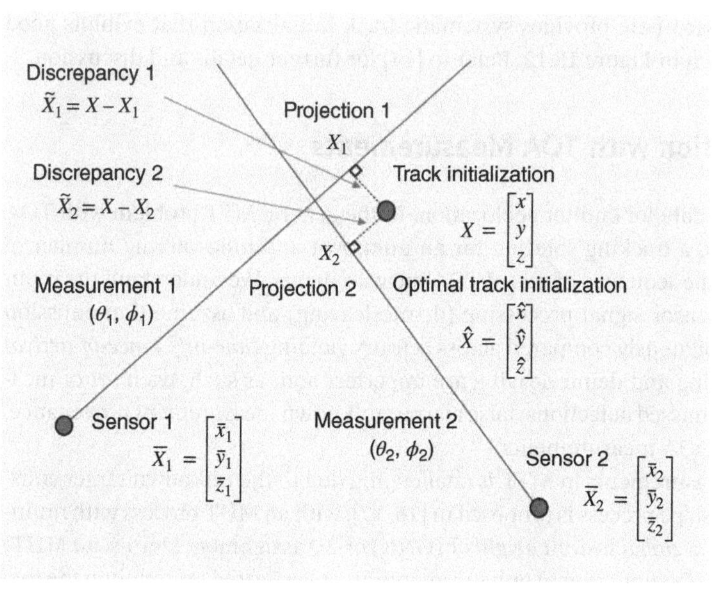

Figure 18.11 Illustration of geometry for track initialization.

Next, we express sensor discrepancy as a function of X and $\overline{X}_i$.

$$\widetilde{X}_i = X - X_i = (I - u_i u_i^T)\left(X - \overline{X}_i\right) \tag{18.7}$$

Thus, we can express the objective as follows:

$$J = \sum_{i=1,2} \left\| (I - u_i u_i^T)\left(X - \overline{X}_i\right) \right\|^2 \tag{18.8}$$

Minimization is straightforward.

$$\begin{aligned}
\frac{\partial J}{\partial X} &= \frac{\partial}{\partial X} \sum_{i=1,2} \left(X - \overline{X}_i\right)^T (I - u_i u_i^T)^2 \left(X - \overline{X}_i\right) \\
&= \sum_{i=1,2} 2\left(X - \overline{X}_i\right)^T (I - u_i u_i^T)^2 \\
&= \sum_{i=1,2} 2\left(X - \overline{X}_i\right)^T (I - u_i u_i^T).
\end{aligned}$$

$$\frac{\partial J}{\partial X} = 0$$

$$\Longrightarrow (I - u_1 u_1^T)\left(X - \overline{X}_1\right) + (I - u_2 u_2^T)\left(X - \overline{X}_2\right) = 0$$

$$\Longrightarrow ((I - u_1 u_1^T) + (I - u_2 u_2^T))X = (I - u_1 u_1^T)\overline{X}_1 + (I - u_2 u_2^T)\overline{X}_2$$

Track positional initialization is thus given by the following:

$$\widehat{X} = ((I - u_1 u_1^T) + (I - u_2 u_2^T))^{-1}\left((I - u_1 u_1^T)\overline{X}_1 + (I - u_2 u_2^T)\overline{X}_2\right). \tag{18.9}$$

Note that the matrix inversion requires $u_1 \neq u_2$. If equal, there is no viable initialization: this case can be avoided via measurement gating when the vectors are sufficiently similar.

An advantage of the approach described here is that it directly supports uncertainty quantification required for subsequent track processing. Hence, the LS approach to localization and uncertainty estimation discussed here provides systematic track initialization that exhibits good statistical consistency, as shown in Figure 18.12. Refer to [14] for further details and discussion.

18.5 Target Localization with TOA Measurements

TOA sensors provide relevant data for emitter geolocation. In the general MTT problem with TOA sensors, we seek to determine a tracking solution for an unknown and time-varying number of objects, based on a discrete-time sequence of sets of TOA measurements. We understand the input to be downstream of single-sensor signal processing (deinterleaving) and assume that emission time sequences can be unambiguously compared across sensors yielding *time-difference-of-arrival* (TDOA) measurements. Sensing and deinterleaving are imperfect and, as such, each set of measurements is characterized by missed detections, false alarms, unknown measurement provenance, and noisy target-originated TDOA measurements.

The direct use of TDOA measurements in MTT is challenging due to the unknown target emission time. Accordingly, a two-stage process is proposed in [16, 17], with an MHT process with multi-target geolocation followed by a *global nearest neighbor* (GNN) or 2D assignment (zero-scan MHT) approach to dynamic tracking. Computational enhancements are investigated by coupling the two

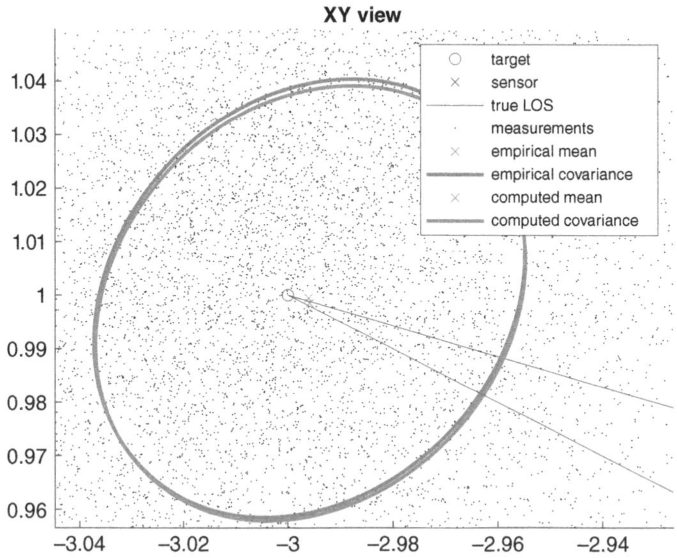

Figure 18.12 Numerical validation of 3D analytical LS localization and uncertainty estimation.

processing via efficient gating. Non-MHT solution approaches have been proposed as well; of particular interest is a recently introduced approach via particle flows [18]. The single-target geolocation problem with TDOA measurements has a long history. We highlight the analytical solution approach in [19], and subsequent refinements in [20, 21].

We discuss here a two-stage MHT approach that builds on these earlier efforts. Specific advances include the following. First, we derive explicit scoring of track hypotheses for the first-stage geolocation MHT based on target birth-death statistics, sensor characteristics, and target geolocation uncertainties. Crucially, we rely on the generalized track-oriented MHT recursion introduced in [8] that allows for target missed detection birth hypotheses. Second, we derive an explicit analytical covariance associated with target geolocation based on small-error analysis. This is crucial for association scoring as well as to provide statistically consistent positional measurements for second-stage dynamic tracking. Third, we derive the detection statistics associated with the geolocation MHT that is also required for downstream processing. Fourth, we consider a nonzero depth, second-stage dynamic MHT for improved data-association performance.

Further extensions may be investigated. These include exploiting *pulse descriptor word* (PDW) feature data associated with TOA measurements (here we consider the kinematic-only tracking problem), including additional sensors providing detections to second-stage tracking (and hence accounting for a mix of emitter-level and platform-level measurements), and allowing for limited-communication sensing networks (and hence with temporal integration needed for emitter geolocation). A moving reference sensor invalidates analytical geolocation methods, as discussed in [22].

18.5.1 Problem Formulation and Solution Framework

We consider a set of $N+1$ sensors, with $N \geq 3$, each reporting at the sequence of times $t_1, t_2, \ldots$ For simplicity and at each sensor, we assume target-independent detection probability p_D. The TOA measurement at sensor i takes the following form, with t_i the *true* signal reception time.

$$Y_i = t_i + v,\ v \sim N\left(0, \sigma_i^2\right) \tag{18.10}$$

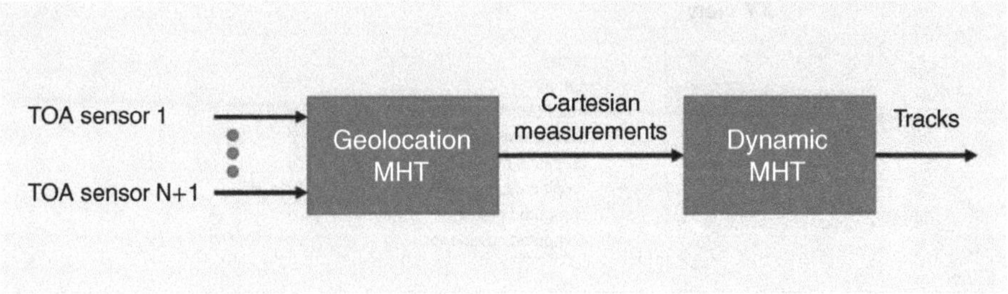

Figure 18.13 Two-stage MHT architecture for MTT with TOA sensors.

For a set of potentially associated measurements from $M + 1$ sensors, one sensor is chosen as the reference and M TDOA measurements can be computed. Measurement noises are assumed to be uncorrelated across sensors. We assume TDOA surveillance volume V and, in addition to target-originated measurements, we assume a Poisson distributed number of TOA clutter of measurements with rate Λ, with uncorrelated and uniformly distributed TDOA measurements in the surveillance volume.

We emphasize that each TOA measurement is in fact the time of one of a *series* of emitted pulses associated in upstream deinterleaving; we assume that this time can be compared unambiguously to one of a series of emitted pulses, detected and associated at another sensor, to produce a candidate TDOA measurement. That said, we do not know which measurements received at each sensor at each time are target originated, and how these are to be associated with measurements at another sensor.

Since target measurements are not easily related to target states, we adopt a two-stage MTT approach as illustrated in Figure 18.13, with distinct geolocation and dynamic tracking stages. Before describing each of these processing stages, we discuss geolocation for a hypothesized set of $M + 1$ associated measurements, with $M \geq 3$.

18.5.2 Geolocation

We first provide the explicit analytical geolocation solution that is derived in [19], while adopting the same notation. In particular, d_i is the difference in the distance of the emitter to the ith sensor and the distance of the emitter to the zeroth (reference) sensor, the rows of S encode the locations of the sensors (with the reference sensor assumed to be at the origin), R_{is} is the distance between the emitter and the ith sensor, and R_s is the distance between the emitter and the reference sensor. In (10.17), we take the real positive solution.

$$x = a - bR_s \tag{18.11}$$

$$a = \left(S^T S\right)^{-1} S^T z \tag{18.12}$$

$$b = \left(S^T S\right)^{-1} S^T d \tag{18.13}$$

$$z = \frac{1}{2}\begin{bmatrix} R_{1s}^2 \\ \vdots \\ R_{Ns}^2 \end{bmatrix} - \frac{1}{2}\begin{bmatrix} d_1^2 \\ \vdots \\ d_N^2 \end{bmatrix} \tag{18.14}$$

$$d = \begin{bmatrix} d_1 \\ \vdots \\ d_N \end{bmatrix} \tag{18.15}$$

$$d_i = c(t_i - t_0), i = 1, \dots, N \tag{18.16}$$

$$R_s = \frac{d^T W z \pm \sqrt{(d^T W z)^2 + z^T W z(1 - d^T W d)}}{d^T W d - 1} \tag{18.17}$$

$$W = S(S^T S)^{-2} S^T \tag{18.18}$$

Let us denote by Y the measurement vector.

$$Y = \begin{bmatrix} t_0 \\ \vdots \\ t_N \end{bmatrix} + v, v \sim N(0_{(N+1)\times 1}, \Sigma_v) \quad \Sigma_v = \text{diag}(\sigma_{t_0}^2, \dots, \sigma_{t_N}^2) \tag{18.19}$$

We denote by $\hat{x}$ our localization estimate; this is obtained using the equations above with the measurement Y replacing the true (unknown) values of $\begin{bmatrix} t_0 \\ \vdots \\ t_N \end{bmatrix}$. We may compute the covariance Σ_x associated with the localization solution $\hat{x}$ via small-error analysis. It is possible also to determine the bias associated with the geolocation solution by including a second-order term in the small-error expansion. See [23] for details, including the statistical consistency of the error statistics. An illustration of the consistency achieved is in Figure 18.14.

18.5.3 MHT: First Stage

The first-stage MHT performs distinct, batch track-oriented MHT processing for each set of $N + 1$ synchronous sensor frames for each time t_k. Hypothesis gating is beneficial for computational efficiency but is not described here. Each track hypothesis consisting of fewer than $M + 1$ associated measurements, with $M \geq 3$, is not of interest. For those that are of interest, we rely on the geolocation solution consistent with the discussion above.

The $N + 1$ sensor frames are ordered arbitrarily for MHT hypothesis generation. Unlike conventional MHT, we consider track hypotheses that may not initiate with a detection. This is illustrated in Figure 18.15. This shows hypothesis generation for three scans of data; the modification appears minor, and indeed has no impact on hypothesis-space complexity. Crucially, we allow target birth on a missed-detection event. Otherwise, for a set of synchronous sensor frames, birth hypotheses on all but the first frame would have a track log-likelihood score of $-\infty$; a similar problem would exist for any track terminations prior to the end of a set of synchronous frames.

Principled hypothesis scoring relies on the development in [8]. Here, we need not contend with the impact of statistically nonstationary target motion statistics as all frames are synchronous. Based on the generalized track-oriented MHT recursion, each track hypothesis will have a log-likelihood score that we denote as c_i for track hypothesis i; this is part of the overall cost vector $\bar{c}$.

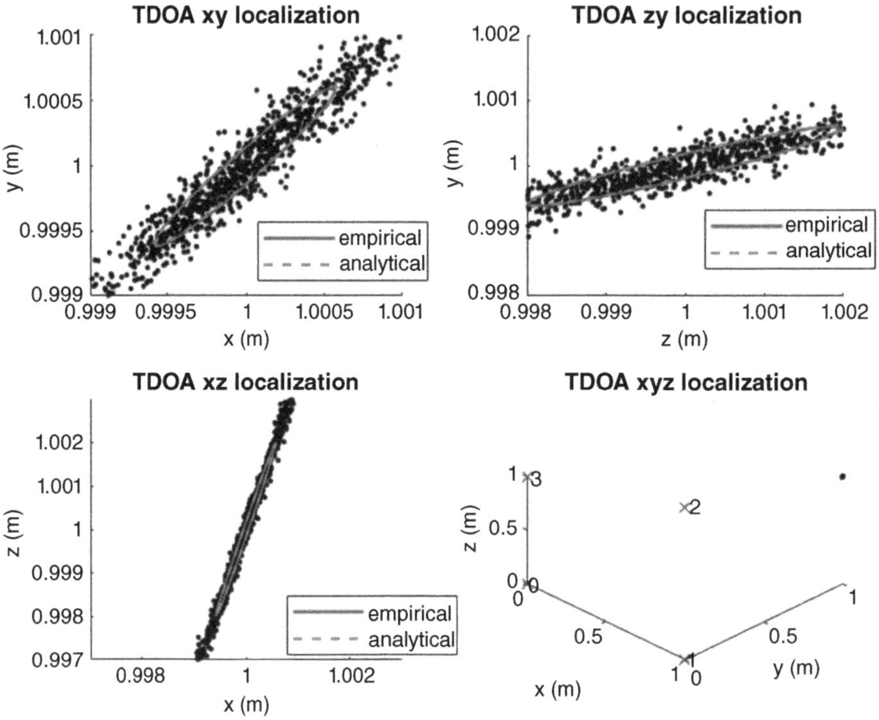

Figure 18.14 Agreement between empirical and analytical errors.

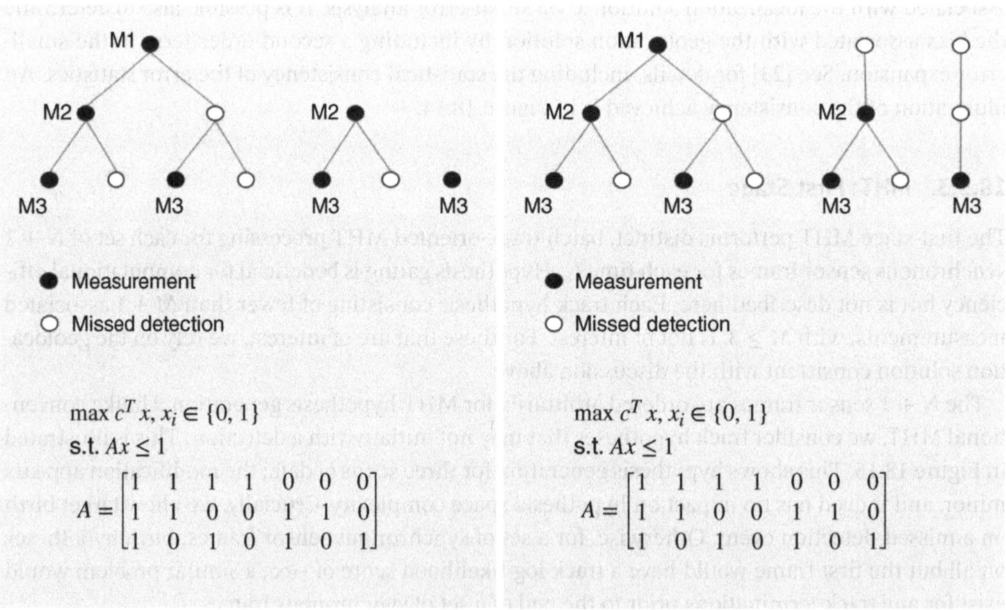

Figure 18.15 Track hypotheses and corresponding optimization problem for classical MHT (left) and enhanced MHT (right).

With $M + 1$ measurements, $M \geq 3$, we denote by Z the TDOA measurement vector resulting from the TOA measurement vector Y. Further, we have the corresponding geolocation solution $\hat{x}$.

$$Z = CY \tag{18.20}$$

$$C = \begin{bmatrix} -1 & 1 & 0 & 0 \\ \vdots & 0 & \ddots & 0 \\ -1 & 0 & 0 & 1 \end{bmatrix} \tag{18.21}$$

$$f(Z|\hat{x}) = N\left(Z; \frac{1}{c} \begin{bmatrix} \vdots \\ \|x_i - \hat{x}\| - \|x_0 - \hat{x}\| \\ \vdots \end{bmatrix}, C\Sigma_v C^T \right) \tag{18.22}$$

$$c_i = \begin{cases} \log\left(\dfrac{\overline{\mu}_b (1 - p_D)^{N-M} p_D^{M+1} f(Z|\hat{x}) V^M}{\Lambda^{M+1}} \right), & M \geq 3 \\ -\infty, M < 3 \end{cases} \tag{18.23}$$

Equation (18.23) includes the ratio of target-originated and false alarm TDOA measurement densities, given by $f(Z|\hat{x})$ and $\frac{1}{V^M}$, respectively. Note that [16, 17] do not include all relevant target and sensor statistics in the track score and appear erroneously to include a nondimensionless measurement likelihood ratio. In particular, (22–23) in [16] have different dimensionality ($S - 1$ and S, respectively) for the likelihood ratio in (24) of [16].

The track-oriented MHT *integer linear program* (ILP) of interest takes the following form. Its solution yields a set of emitter locations with uncertainties, amenable to downstream kinematic tracking.

$$\max \overline{c}^T x, x_i \in \{0, 1\}, i = 1, \ldots \tag{18.24}$$

$$Ax \leq 1 \tag{18.25}$$

18.5.4 MHT: Second Stage

In order to make proper use of the sequence of sets of emitter geolocation measurements and uncertainties from first-stage MHT, it is necessary to quantify the corresponding emitter detection and false alarm statistics. Neglecting incorrect data association decisions, detection performance is given by (18.26). Note that for 3D tracking we require at least four TOA detections to achieve a geolocated detection.

$$P_D = \sum_{i=4}^{N+1} \binom{N+1}{i} p_D^i (1 - p_D)^{N+1-i} \tag{18.26}$$

Precise false alarm statistics for emitter geolocation are more complex to obtain. The false alarm rate will be small, and we might reasonably assume that these measurements are uniformly distributed in the spatial surveillance volume.

The classical track-oriented MHT recursion applies directly to the second-stage MHT. There are real-time surveillance and computational benefits to temporal sliding-window processing. Further, as we discuss next, there are empirical beneficials to decoupling data association and track management functions.

18.6 Decoupled Data Association and Track Management

As discussed in [24], there is empirical evidence that decoupling data association and track extraction yields good performance. That is, owing to the need for sliding-window processing for computational and real-time requirements, it is generally the case that data association must operate on a short time window (for computational expediency) while track confirmation and deletion may operate on a longer time window (for improved performance. A limiting case of this approach is to consider a generalized GNN for data association, followed by (optimal) batch extraction based on track log-likelihood score. This offers a simple but good performing MHT solution. We discuss some elements of this approach here. Application to unframed measurement data may be found in [25].

Applying the MHT recursion with no hypothesis depth amounts to resolving association hypotheses immediately, while still relying on correct birth and coast scores consistent with the MHT recursion. It may be viewed as a generalized GNN tracking approach. One point of caution with immediate hypothesis resolution is that we risk indefinite track coasting. We address this by including as part of the track state a counter on the number of missed detections since the last detection. Correspondingly, we may consider a near-lossless track coast procedure whereby we terminate a track when it is nearly impossible for the track score of two associated track segments to exceed the sum of the fragmented track scores.

Given M tracks at the previous scan and N measurements in the current scan, we consider the following generalized assignment matrix of dimensions $(M+N)$ by $(M+N)$. The structure of the score matrix J is as follows:

$$J = \begin{bmatrix} \begin{pmatrix} \log \dfrac{(1-p_\chi)p_d f(z_1|X_1,\Sigma_1)}{\Lambda_{fa}/V} & \cdots & \log \dfrac{(1-p_\chi)p_d f(z_N|X_1,\Sigma_1)}{\Lambda_{fa}/V} \\ \vdots & \ddots & \vdots \\ \log \dfrac{(1-p_\chi)p_d f(z_1|X_M,\Sigma_M)}{\Lambda_{fa}/V} & \cdots & \log \dfrac{(1-p_\chi)p_d f(z_N|X_M,\Sigma_M)}{\Lambda_{fa}/V} \end{pmatrix} & \begin{pmatrix} \log((1-p_\chi)(1-p_d)) & \cdots & -\infty \\ \vdots & \ddots & \vdots \\ -\infty & \cdots & \log(p_\chi) \end{pmatrix} \\ \begin{pmatrix} \log \dfrac{p_d \mu_b}{\Lambda_{fa}} & \cdots & -\infty \\ \vdots & \ddots & \vdots \\ -\infty & \cdots & \log \dfrac{p_d \mu_b}{\Lambda_{fa}} \end{pmatrix} & \begin{pmatrix} 0 & \cdots & 0 \\ \vdots & \ddots & \vdots \\ 0 & \cdots & 0 \end{pmatrix} \end{bmatrix} \tag{18.27}$$

In (18.27), the top-left block has dimensions M by N, with each element corresponding to a track update hypothesis. We denote by $f(z_i|X_j,\Sigma_j)$ the filter residual associated with measurement z_i of a target with state estimate X_j and state covariance Σ_j. The top-right block has dimensions M by M, with each element corresponding either to a track coast or to a track death hypothesis; which of these applies depends on whether the maximum number of allowed missed detections K has been reached for the given track. (In the equation, we illustrate one of each of these two cases.) The bottom-left block is of dimension N by N, with each element corresponding to a track birth hypothesis. Finally, the bottom-right block is set to zero and corresponds to slack variables to allow for a nonzero number of track updates.

Optimal assignment results from the selection of $M + N$ elements with exactly one element from each row and from each column while maximizing the sum of the element scores. This is the classical 2D assignment problem for which many solution approaches exist; we opt for use of the `assignmunkres` function implementation in MATLAB.

The second element of our simplified approach is to perform batch extraction of tracks based on track log-likelihood scores. All tracks with a positive score contribute to the global hypothesis score. Further improvements can be achieved by removal of terminal coasts and, potentially, by considering extraction of only *portions* of full tracks, by identifying the largest log-likelihood ratio over the course of track evolution and only extracting the relevant track portion.

It is worth noting that, while suboptimal, our simple MHT does have the salient characteristics of mature MHT solutions. Most notably, practical MHT implementations generally decouple data association form track extraction, with the former operating on a shorter timeline (typically, a few scans) than the latter. The short time window for data association is done to mitigate computational explosion as well as due to the diminishing returns in deep hypothesis trees in many domains. The longer timeline for track extraction is used to enable accrual of positive log-likelihood ratios. For instance, on initiation, one typically starts with a very negative track score, as $\log \frac{(1 - p_\chi) p_d f(z_1|X_1, \Sigma_1)}{\Lambda_{\text{fa}}/V} \ll 0$ in most practical settings. Hence, track extraction on short timelines risks suppressing all sensor data and failing track formation.

Our simple MHT solution corresponds to the limiting case of no temporal depth on data association and arbitrarily long temporal depth on track extraction. The key advantage of this approach is its simplicity, as there are no tracker initiation, gating, or maintenance parameters. The algorithm is determined entirely by target and sensor statistical models. The only tracker setting is the number of allowed missed detections, which is not an exogenous parameter as it can be set as a function of target and sensor parameters.

The primary advantage of the simple MHT solution described here is that it enables principled analysis of performance as a function of other system settings. The "only" suboptimality associated with our MHT is its immediate, sequential hypothesis resolution.

18.7 Tracker Performance Modeling

Model-based *probability of correct association* (PCA) correlates strongly with empirical tracker performance. Model-based PCA is given in [26]; we describe it as the *Mori Chang Chong* (MCC) model. For empirical tracker performance, we will consider the *probability of correct decision* (PCD) or *weighted PC* (WPC) described in the next section; these quantify both false alarm suppression and correct measurement association.

The MCC model for PCA is as follows; refer (7.25) in [26]. Here, C_m is a constant that depends on measurement-space dimension m, β is the clutter density, and σ is the standard deviation (in each dimension) for the filter innovation.

$$P_{\text{CA}} = \exp(-C_m \beta \sigma^m) \tag{18.28}$$

We wish to express P_{CA} in terms of target and sensor statistical parameters. These include target birth rate λ_b, target death rate λ_χ, interframe time t, target detection probability p_d, false alarm rate Λ_{fa}, and measurement volume V. Assuming for simplicity uncorrelated target dynamics and measurements across dimensions, *nearly constant position* (NCP) motion, and positional measurements, we have process noise parameter q and measurement variance r.

With these modeling assumptions, we may express β and σ as follows. For the clutter rate, note that the expected number of (confuser) target detections is $\frac{p_d\lambda_b}{\lambda_\chi}$; this increases the effective clutter rate. For the filter innovation standard deviation, we use steady-state statistics via solution p of the (scalar) discrete algebraic Riccati equation, whose matrix form is given by (7) with solution P. Note that for the scalar NCP model with positional measurements, we have $A = C = 1$, $L = \frac{p}{p+r}$, and $Q = qt$. The (positive) solution is given explicitly in (18.32).

$$\beta = \frac{1}{V}\left(\frac{p_d\lambda_b}{\lambda_\chi} + \Lambda_{fa}\right) \tag{18.29}$$

$$\sigma^2 = p + r \tag{18.30}$$

$$P = A(I - LC)PA^T + Q \tag{18.31}$$

$$p = \frac{r}{p+r}p + qt \Longrightarrow p = \frac{qt + \sqrt{q^2t^2 + 4rqt}}{2} \tag{18.32}$$

We have:

$$P_{\text{CA}} = \exp\left(-C_m\frac{1}{V}\left(\frac{p_d\lambda_b}{\lambda_\chi} + \Lambda_{fa}\right)\left(\frac{qt + \sqrt{q^2t^2 + 4rqt}}{2} + r\right)^{m/2}\right) \tag{18.33}$$

Our primary interest is to have a model-based performance metric that is predictive of empirical performance. By this we mean that settings for which (model-based) P_{CA} is higher should yield higher (empirical) WPC. Note that P_{CA} and WPC values are not directly comparable. Thus, we may simplify (18.33) and consider the figure of merit J, which we hypothesize will be inversely correlated with WPC.

$$J = \frac{1}{V}\left(\frac{p_d\lambda_b}{\lambda_\chi} + \Lambda_{\text{fa}}\right)\left(\frac{qt + \sqrt{q^2t^2 + 4rqt}}{2} + r\right)^{m/2} \tag{18.34}$$

When $qt \gg r$, we have:

$$J \approx \frac{1}{V}\left(\frac{p_d\lambda_b}{\lambda_\chi} + \Lambda_{\text{fa}}\right)(qt + r)^{m/2} \approx \frac{1}{V}\left(\frac{p_d\lambda_b}{\lambda_\chi} + \Lambda_{\text{fa}}\right)(qt)^{m/2} \tag{18.35}$$

Likewise, when $r \gg qt$, we have:

$$J \approx \frac{1}{V}\left(\frac{p_d\lambda_b}{\lambda_\chi} + \Lambda_{\text{fa}}\right)r^{m/2} \tag{18.36}$$

Accordingly, we adopt the following simplified expression in lieu of (18.34), where β is as in (18.29) and s is the approximate filter innovation.

$$\tilde{J} = \frac{1}{V}\beta s^{m/2} \tag{18.37}$$

$$\beta = \frac{1}{V}\left(\frac{p_d \lambda_b}{\lambda_\chi} + \Lambda_{\text{fa}}\right) \tag{18.38}$$

$$s = qt + r \tag{18.39}$$

It is of interest to consider whether changes in the clutter rate β and filter innovation s yielding the same model-based performance $\tilde{J}$ yield similar empirical performance as measured by WPC. Changes for which $\tilde{J}$ increases (decreases) should yield a reduced (increased) WPC.

It is perhaps nonintuitive for the tracking system designers to relate the process noise parameter q to deterministic parameters such as maximum target acceleration $a_{\max}$; it can be shown that a good choice is given by (18.40) [27]. Alternatively, one may set q via numerical calibration; adaptive filtering methods are needed when ground truth calibration data is not available.

$$q = a_{\max}^2 T \tag{18.40}$$

As a back-of-the envelope approach to understanding (18.40), consider that for the 1D NCV model, we have Eq. (18.41), with T the discretization time interval.

$$\dot{x}_{k+1} - \dot{x}_k = w_k, w_k \sim N(0, qT) \tag{18.41}$$

From this, we have:

$$E\left[\left(\frac{\dot{x}_{k+1} - \dot{x}_k}{T}\right)^2\right] = \frac{q}{T} \tag{18.42}$$

Now consider replacing the Gaussian variable in the expectation, $\frac{\dot{x}_{k+1} - \dot{x}_k}{T}$, with a uniformly distributed variable on $[-a_{\max}, a_{\max}]$. For such a variable, the variance will be $\frac{a_{\max}^2}{3}$. Thus, we arrive at (18.43), which is roughly comparable to Eq. (18.40). We achieve precise agreement with (18.40) if, rather than the uniform, we use a distribution with a pair of Dirac delta functions at $\pm a_{\max}$.

$$q = \frac{a_{\max}^2 T}{3} \tag{18.43}$$

18.8 Tracker Performance Metrics

When speaking of optimal tracking, it is important to clarify what we intend. Optimal MHT identifies the MAP global hypothesis. Doing so does not necessarily lead to optimal performance with respect to operational performance metrics of interest. Further, there is no guarantee of optimality with respect to proposed scalar performance metrics that address multi-target detection and localization performance [28, 29], and the same is true when considering proposed track-level generalizations of these metrics.

Here, we consider the *probability of error* (PE) that focuses on data association performance without considering state estimation errors. Minimizing PE is similar to, but not the same as, maximizing the global hypothesis likelihood. Indeed, consider that the clairvoyant tracker that makes correct use of all sensor data is not guaranteed in any data realization to match the score of the MAP global hypothesis.

Strictly speaking, the PE assesses the probability that all decisions regarding measurements to classify as false, as well as association decisions on target-originated measurements, be correct. Here, we define the PE in a manner that allows for fractional correctness. Specifically, we consider a binary adjacency matrix associated with a dataset. If the ith measurement is a false alarm, all elements in the row and ith column are zero. If the ith measurement is target originated, element (i,i) is one. Further, if measurement i and measurement j originate form the same target, element (i,j) and (j,i) are one.

Equation (18.44) illustrates an adjacency matrix A that consists of six measurements, with three originating from one target, one from a second target, and two false alarms. The ordering of the measurements is arbitrary. For a particular data realization and tracker output, PE is computed as the fraction of elements that differ between the true adjacency matrix A and estimated adjacency matrix $\widehat{A}$. See the example in (18.44–18.46); the tracker output corresponding to adjacency matrix $\widehat{A}$ includes three tracks with a total of five measurements.

$$A = \begin{bmatrix} 1 & 1 & \mathbf{1} & 0 & 0 & 0 \\ 1 & 1 & \mathbf{1} & 0 & 0 & 0 \\ \mathbf{1} & \mathbf{1} & 1 & \mathbf{0} & 0 & 0 \\ 0 & 0 & \mathbf{0} & 1 & 0 & 0 \\ 0 & 0 & 0 & 0 & \mathbf{0} & 0 \\ 0 & 0 & 0 & 0 & 0 & 0 \end{bmatrix} \tag{18.44}$$

$$\widehat{A} = \begin{bmatrix} 1 & 1 & \mathbf{0} & 0 & 0 & 0 \\ 1 & 1 & \mathbf{0} & 0 & 0 & 0 \\ \mathbf{0} & \mathbf{0} & 1 & \mathbf{1} & 0 & 0 \\ 0 & 0 & \mathbf{1} & 1 & 0 & 0 \\ 0 & 0 & 0 & 0 & \mathbf{1} & 0 \\ 0 & 0 & 0 & 0 & 0 & 0 \end{bmatrix} \tag{18.45}$$

$$PE = \frac{1}{n^2} \sum_{\substack{i=1,\ldots,n \\ j=1,\ldots,n}} \left| A_{ij} - \widehat{A}_{ij} \right| = \frac{7}{36} \tag{18.46}$$

It is interesting to compare use of the PE scalar metric with classical detection statistics. One could imagine classifying each decision in the $\widehat{A}$ matrix as one of *true positive* (TP), *false negative* (FN), *false positive* (FP), and *true negative* (TN). Our PE metric and classical *precision* (P), *recall* (R), and F1 statistics are derived from these quantities as follows:

$$\mathrm{PE} = 1 - \frac{\mathrm{TP} + \mathrm{TN}}{\mathrm{TP} + \mathrm{FN} + \mathrm{FP} + \mathrm{TN}} \tag{18.47}$$

$$\mathrm{P} = \frac{\mathrm{TP}}{\mathrm{TP} + \mathrm{FP}} \tag{18.48}$$

$$R = \frac{TP}{TP + FN} \tag{18.49}$$

$$F1 = \frac{2 \cdot P \cdot R}{P + R} \tag{18.50}$$

Arguably, evaluation of tracking performance in terms of PE is more representative of overall algorithmic effectiveness, as, unlike the F1 score, it reflects the amount of false alarm suppression (i.e., TN).

It is worth noting that PE as computed here considers the full adjacency matrix. Due to symmetry and to avoid double counting, it would perhaps be more appropriate to consider only the upper triangular adjacency matrix (with diagonal included). That is, for an adjacency matrix of n^2 elements, there are only $\frac{n(n+1)}{2}$ independent decisions. For simplicity, the latter number still allows for any symmetric binary adjacency matrix solution $\widehat{A}$, even though, strictly speaking, some solutions may not be self-consistent. For example, we allow a solution to have element (i,j) and element (j,k) set to one, with $i \neq j \neq k$, but with element (i,k) set to zero. Likewise, we allow a solution to have element (i,i) set to 0, without having all elements of the ith row and ith column set to zero. By allowing for solutions that are not self-consistent, we always have $\frac{n(n+1)}{2}$ decisions to consider.

Rather than PE, we might equivalently consider the PCD that is given as follows:

$$PCD = 1 - PE = \frac{TP + TN}{TP + FN + FP + TN} \tag{18.51}$$

Further, we may wish to balance the weight given to TP and TN decisions. One way to achieve this is via the *weighted PCD* (WPCD) defined as follows:

$$WPCD = \frac{1}{2}\left(\frac{TP}{TP + FN} + \frac{TN}{FP + TN}\right) \tag{18.52}$$

The development of tracker performance metrics remains an ongoing area of research and analysis in the tracking community, and it is important when considering the relative merits of centralized and distributed MHT as well as of other MTT paradigms. Recent work in this direction may be found in [30, 31] and references therein.

18.9 Conclusions

High-quality MTT algorithms are essential in multi-sensor surveillance. MHT is arguably the leading operational paradigm and can be improved upon with judicious distributed processing. Generally, computational and real-time considerations dictate small hypothesis depth for data association and longer reasoning windows for track management. Research on evaluation metrics is ongoing. This chapter addresses these topics and provides references for further investigation.

References

1 Y. Bar-Shalom, P. Willett, and X. Tian. *Tracking and Data Fusion: A Handbook of Algorithms*. YBS Publishing, 2011.

2 B.-N. Vo, M. Mallick, Y. Bar-Shalom, S. Coraluppi, R. Osborne, R. Mahler, and B.-T. Vo. Multitarget tracking. In J. G. Webster, editor, *Wiley Encyclopedia of Electrical and Electronics Engineering*. Wiley-IEEE, 2015.

3 S. Coraluppi. Multi-target surveillance: Distributed multiple-hypothesis tracking, graph-based methods, and context exploitation. In *NATO STO SET-290 Lecture Series on AI for Military Multisensor Fusion Engines*. NATO Collaboration and Support Office, 2022.

4 C.-Y. Chong, S. Mori, and D. Reid. Forty years of multiple hypothesis tracking. *ISIF Journal of Advances in Information Fusion*, 14(2):131–151, 2019.

5 D. Reid. An algorithm for tracking multiple targets. *IEEE Transactions on Automatic Control*, 24(6):843–854, 1979.

6 T. Kurien. Issues in the design of practical multitarget tracking algorithms, Chapter 3. In Y. Bar-Shalom, editors, *Multitarget-multisensor Tracking: Advanced Applications*. Artech House, 1990.

7 W. Koch and G. van Keuk. Multiple hypothesis track maintenance with possibly unresolved measurements. *IEEE Transactions on Aerospace and Electronic Systems*, 33(3):883–892, 1997.

8 S. Coraluppi and C. Carthel. If a tree falls in the woods, it does make a sound: Multiple-hypothesis tracking with undetected target births. *IEEE Transactions on Aerospace and Electronic Systems*, 50(3):2379–2388, 2014.

9 S. Coraluppi and C. Carthel. Multiple-hypothesis tracking for targets producing multiple measurements. *IEEE Transactions on Aerospace and Electronic Systems*, 54(3):1485–1498, 2018.

10 T. Kropfreiter, F. Meyer, D. F. Crouse, S. Coraluppi, F. Hlawatsch, and P. Willett. Track coalescence and repulsion in multitarget tracking: An analysis of MHT, JPDA, and belief propagation methods. *IEEE Open Journal of Signal Processing*, 5:1–17, 2024.

11 S. Coraluppi, C. Carthel, and A. Willsky. Multiple-hypothesis tracking and graph-based tracking extensions. *ISIF Journal of Advances in Information Fusion*, 14(2):152–166, 2019.

12 P. Varshney. *Distributed Detection and Data Fusion*. Springer, 1997.

13 S. Coraluppi, N. Jenkins, and M. Lexa. Simplified distributed tracking. In *Proceedings of the 27th International Conference on Information Fusion*. Venice, Italy, 2024.

14 S. Coraluppi, C. Rago, C. Carthel, and B. Bale. Distributed MHT with passive sensors. In *Proceedings of the 24th International Conference on Information Fusion*. Sun City, South Africa, 2021.

15 R. Osborne and Y. Bar-Shalom. Statistical efficiency of composite position measurements from passive sensors. *IEEE Transactions on Aerospace and Electronic Systems*, 49(4):2799–2806, 2013.

16 T. Sathyan, A. Sinha, and T. Kirubarajan. Passive geolocation and tracking of an unknown number of emitters. *IEEE Transactions on Aerospace and Electronic Systems*, 42(2):740–750, 2006.

17 T. Sathyan, A. Sinha, T. Kirubarajan, M. McDonald, and T. Lang. MDA-based data association with prior track information for passive multitarget tracking. *IEEE Transactions on Aerospace and Electronic Systems*, 47(1):539–556, 2011.

18 W. Zhang, M. Khojasteh, and F. Meyer. Particle flows for localizing sources in 3-D based on TDOA measurements. In *Proceedings of the IEEE Aerospace Conference*. Big Sky MT, USA, 2024.

19 G. Mellen, M. Pachter, and J. Raquet. Closed-form solution for determining emitter location using time difference of arrival measurements. *IEEE Transactions on Aerospace and Electronic Systems*, 39(3):1056–1058, 2003.

20 R. Amiri, F. Behnia, and A. Noroozi. An efficient estimator for TDOA-based source localization with minimum number of sensors. *IEEE Communications Letters*, 22(12):2499–2502, 2018.

21 L. Zhang, T. Zhang, and H.-S. Shin. An efficient constrained weighted least squares method with bias reduction for TDOA-based localization. *IEEE Sensors Journal*, 21(8):10122–10131, 2021.

22 J. Dunham, J. Simmons, and S. Shapero. Analytic emitter geolocation and filtering via time difference of arrival. In *Proceedings of the IEEE Aerospace Conference*. Big Sky MT, USA, 2019.

23 S. Coraluppi. Nonlinear multiple-hypothesis tracking. In *NATO STO SET-336 Lecture Series on Artificially Intelligent Military Situational Awareness*. NATO Collaboration and Support Office, 2025.

24 S. Coraluppi and C. Carthel. Track management in multiple-hypothesis tracking. In *Proceedings of the 10th IEEE Sensor Array and Multichannel Signal Processing Workshop*. Sheffield, United Kingdom, 2018.

25 S. Coraluppi and A. Hunter. Multiple-hypothesis tracking with unframed sensor measurements. In *Proceedings of the IEEE Aerospace Conference*. Big Sky MT, USA, 2025.

26 S. Mori, K.-C. Chang, and C.-Y. Chong. Performance analysis of optimal data association with application to multiple target tracking. In Y. Bar-Shalom, editor, *Multitarget-multisensor Tracking: Applications and Advances Vol. II*. Artech House, 1992.

27 W. Dale Blair and Y. Bar-Shalom. MSE design of nearly constant velocity Kalman filters for tracking targets with deterministic maneuvers. *IEEE Transactions on Aerospace and Electronic Systems*, 59(4):4180–4191, 2023.

28 D. Schuhmacher, B.-T. Vo, and B.-N. Vo. A consistent metric for performance evaluation of multi-object filters. *IEEE Transactions on Signal Processing*, 56(8):3447–3457, 2008.

29 A. Rahmathullah, Á. García-Fernández, and L. Svensson. Generalized optimal sub-pattern assignment metric. In *Proceedings of the 20th International Conference on Information Fusion*. Xi'an, China, 2017.

30 Á. García-Fernández, A. Rahmathullah, and L. Svensson. A metric on the space of finite sets of trajectories for evaluation of multi-target tracking algorithms. *IEEE Transactions on Signal Processing*, 68:3917–3928, 2020.

31 S. Coraluppi. Advances in multi-target tracking performance evaluation. In *Proceedings of the 26th International Conference on Information Fusion*. Charleston SC, USA, 2023.

19

Distributed State Estimation on Sensor Networks

Giorgio Battistelli, Luigi Chisci, and Nicola Forti

Dipartimento di Ingegneria dell'Informazione (DINFO), Università degli Studi di Firenze, Firenze, Italy

19.1 Introduction

Multiagent systems—i.e., networks consisting of multiple nodes (agents) with sensing, processing, communication and, sometimes, actuation capabilities—are nowadays finding pervasive application in a variety of distributed control and monitoring tasks (e.g., surveillance, navigation, environmental, and health monitoring) [1–5]. Such systems pose a number of difficult and interesting challenges that have attracted great research efforts over the last decades.

This chapter is specifically devoted to multiagent state estimation [6–12] wherein the agents aim to cooperatively estimate the state of a dynamical system. Our treatment will follow a Bayesian approach wherein:

- available information on the state to be estimated is represented by a *probability density function* (PDF);
- each agent—sensing data from the environment, locally processing, and exchanging them with neighboring agents—propagates in time a PDF;
- PDFs of all agents are consistently fused so as to combine all available information while avoiding double counting of information;
- fusion is performed in a distributed and scalable fashion over a peer-to-peer multiagent network by exploiting the consensus protocol.

Consistent with the above-outlined Bayesian framework, the rest of this chapter will be organized as follows. Section 19.2 will provide notation as well as the necessary background on network modeling and recursive Bayesian estimation. Then, Section 19.3 will overview fusion of PDFs from an optimization-based perspective. Sections 19.4 and 19.5 will delve into specific optimization-based fusion rules relying on the *Kullback–Leibler divergence* (KLD) as a measure of discrepancy between PDFs; in particular, it will be shown how the exchange of arguments in the KLD leads to completely different rules consisting of geometric or, respectively, arithmetic averages of the PDFs. Next, Section 19.7 will address consensus for distributed fusion over a multiagent network. Section 19.8 will overview consensus-based algorithms for *distributed state estimation* (DSE) and their stability properties. Section 19.9 will present simulation results concerning DSE in a target tracking scenario. Finally, Section 19.10 will summarize the chapter and provide perspectives for future work.

19.2 Background

19.2.1 Notation

Throughout this chapter, the following notation is used: $\mathbb{R}$ is the set of real numbers; the cardinality of a finite set $\mathcal{N}$ is denoted by $|\mathcal{N}|$; the all-one column vector of any dimension is denoted by $\mathbb{1}$; the trace of a square matrix A is denoted by $\mathrm{tr}(A)$ and its determinant by $\det(A)$; $\delta(x - x_0)$ is the Dirac delta centered at x_0; $w \sim p_w(\cdot)$ indicates that the random variable w is distributed according to the PDF $p_w(\cdot)$; $\mathbb{E}_f[\cdot]$ denotes expectation with respect to the PDF f; $\mathcal{P} = \{f : \mathcal{X} \to \mathbb{R} \mid f(x) \geq 0 \ \forall x \in \mathcal{X} \text{ and } \int_{\mathcal{X}} f(x)\, dx = 1\}$ denotes the set of all admissible PDFs over the state space $\mathcal{X}$; $\mathcal{G}(\cdot; m, \Sigma)$ denotes the Gaussian PDF with mean m and covariance Σ; given a set of matrices $\{M_i\}_{i\in\mathcal{N}}$ having the same number of columns/rows, $\mathrm{col}\{M_i\}_{i\in\mathcal{N}}$ / $\mathrm{row}\{M_i\}_{i\in\mathcal{N}}$ denotes the collective matrix obtained by stacking M_i on top/right of each other; for a symmetric matrix M, $\mathrm{vec}(M) = [m_{11}, m_{12}, m_{22}, \dots]^{\mathsf{T}}$ denotes column-wise vectorization of its upper triangle.

19.2.2 Peer-to-peer Network Model

In this chapter, it is assumed that multiple agents aiming to cooperatively estimate the state of a dynamical system are connected to form a network (graph) $\mathfrak{G} = (\mathcal{N}, \mathcal{A})$, where $\mathcal{N}$ is the set of agents and $\mathcal{A} \subseteq \mathcal{N} \times \mathcal{N}$ the set of arcs, with $(j, i) \in \mathcal{A}$ indicating that agent i can receive data from agent j. The attention focuses on a peer-to-peer network model (see Figure 19.1) wherein:

- there is no *fusion center* accessing data from all agents, i.e., all agents are peers;
- each agent is unaware of the overall network topology;
- each agent (peer) can take the following actions: get measurements of the target system, receive data from in-neighbors, broadcast data to out-neighbors, process data.

For the subsequent developments on DSE, it is convenient to introduce $|\mathcal{N}|$ to denote the number of agents (cardinality of $\mathcal{N}$) and, for each agent i, its *in-neighborhood* $\mathcal{N}_i = \{j \in \mathcal{N} : (j, i) \in \mathcal{A}\}$ representing the subset of agents (including i itself) from which it can get data.

19.2.3 Recursive Bayesian Estimation

The state estimation problem of interest consists of estimating the state x_k of the discrete-time dynamical system

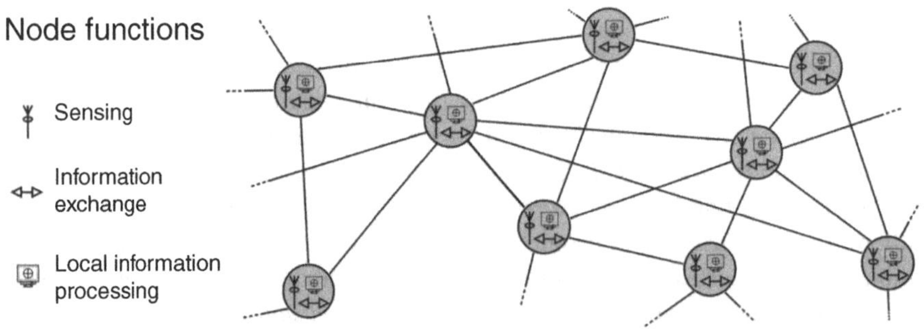

Figure 19.1 Peer-to-peer agent network.

$$\begin{cases} x_{k+1} = f_k(x_k, w_k) \\ w_k \sim p_{w_k}(\cdot) \end{cases} \tag{19.1}$$

by recursively processing the measurements

$$\begin{cases} y_{k,i} = h_{k,i}(x_k, v_{k,i}) \\ v_{k,i} \sim p_{v_{k,i}}(\cdot) \end{cases} \tag{19.2}$$

sensed by the agents $i \in \mathcal{N}$ for $k \geq 1$. It is assumed that w_k and $\{v_{k,i}\}_{i\in\mathcal{N}}$ are mutually independent white noise sequences with possibly time-varying PDFs $p_{w_k}(\cdot)$ and $\{p_{v_{k,i}}(\cdot)\}_{i\in\mathcal{N}}$, also independent of the initial state $x_1 \sim p_{1|0}(\cdot)$. In a Bayesian framework, x_k is regarded as a random variable and its estimation aims to find the PDF $p_{k|k}(x)$ of x_k conditional to all measurements up to time k, i.e., $p_{k|k}(x) = p(x_k|y_{1:k})$ where $y_t \stackrel{\triangle}{=} \text{col}\{y_{t,i}\}_{i\in\mathcal{N}}$ and $y_{1:k} \stackrel{\triangle}{=} \{y_1, \dots, y_k\}$. The state-space model (19.1)–(19.2) can be recast in Bayesian form as

$$\begin{cases} x_{k+1} \sim \varphi_{k+1|k}(\cdot|x_k) \\ y_{k,i} \sim l_{k,i}(\cdot|x_k) \end{cases} \tag{19.3}$$

where, assuming invertibility of the functions $f_k(\cdot,\cdot)$ and $h_{k,i}(\cdot,\cdot)$ with respect to their second argument, the *Markov transition PDF* $\varphi_{k+1|k}$ and *likelihoods* $l_{k,i}$ are given by

$$\begin{aligned} \varphi_{k+1|k}(x|\xi) &= \left[\frac{p_{w_k}(w)}{\left|\det \frac{\partial f_k}{\partial w}(\xi, w)\right|} \right]_{w=f_k^{-1}(\xi,x)} \\ l_{k,i}(y|x) &= \left[\frac{p_{v_{k,i}}(v)}{\left|\det \frac{\partial h_{k,i}}{\partial v}(x, v)\right|} \right]_{v=h_{k,i}^{-1}(x,y)} . \end{aligned} \tag{19.4}$$

In a centralized setting, i.e., when there is an agent having access to all measurements $y_{k,i}$, $i \in \mathcal{N}$, the solution to the state estimation problem is given by the Bayesian recursion:

$$p_{k|k}(x) = \frac{l_k(y_k|x)\, p_{k|k-1}(x)}{\int l_k(y_k|\xi)\, p_{k|k-1}(\xi)\, d\xi} \tag{19.5}$$

$$p_{k+1|k}(x) = \int \varphi_{k+1|k}(x|\xi)\, p_{k|k}(\xi)\, d\xi \tag{19.6}$$

initialized from $p_{1|0}(x)$, where: $p_{t|k}(\cdot)$ is the PDF of x_t given $y_{1:k}$ and, due to independence of $v_{k,i}$, the collective likelihood is given by

$$l_k(y_k|x) = \prod_{i\in\mathcal{N}} l_{k,i}(y_{k,i}|x). \tag{19.7}$$

Conversely, in the distributed setting of this work, each agent propagates in time its own PDF $p_{k|k,i}(\cdot)$ by fusing the available information over then subnetwork $\mathcal{N}_i$. The next section will address the fundamental task of PDF fusion [13].

19.3 Fusion of PDFs

Let us consider that each agent $i \in \mathcal{N}$ has a local PDF $f_i(x)$ available, summarizing its information about a common variable of interest x; e.g., referring to the previous section, $f_i(x) = p_{k|k,i}(x)$ represents the PDF at time k and agent i for the system state x_k.Then, the aim of PDF fusion (see Figure 19.2) is to combine all local (fusing) PDFs $\{f_i\}_{i\in\mathcal{N}}$ into a global (fused) PDF $\bar{f}$ that summarize, in the best possible way, both common and complementary information from all agents. This task is also referred to, in the statistical community, as *probabilistic opinion pooling*.

19.3.1 Optimization-based Fusion

The fusion/pooling map $\{f_i\}_{i\in\mathcal{N}} \to \bar{f}$ can be interpreted as a sort of *averaging* among PDFs. In this respect, taking inspiration from averaging of vectors $x_i \in \mathbb{R}^n$, it is easy to see that

$$\bar{x} \stackrel{\triangle}{=} \sum_i \pi_i\, x_i = \arg\min_x \sum_i \pi_i\, ||x - x_i||^2 = \arg\min_x \sum_i \pi_i\, d(x, x_i)$$

for any set of weights $\pi_i \geq 0$ such that $\sum_i \pi_i = 1$, reflecting the reliability/credibility of agent i. In words, the weighted arithmetic average $\bar{x}$ can be regarded as solution of an optimization problem consisting of the minimization, with respect to x, of the weighted sum of squared Euclidean distances $d(x, x_i) = ||x - x_i||^2$ between x and x_i. In a similar way, the above optimization-based interpretation can be applied to PDF fusion by defining the fused PDF as

$$\bar{f} = \arg\min_{f\in\mathcal{P}} \sum_i \pi_i\, d(f, f_i) \tag{19.8}$$

for a suitable measure of discrepancy (dissimilarity) $d(f, f_i)$ between PDFs f and f_i, and for suitably chosen *fusion weights*

$$\pi_i \geq 0 \text{ such that } \sum_i \pi_i = 1. \tag{19.9}$$

19.3.2 Conjugacy Guarantee

In recursive Bayesian filtering, it is desirable that the propagated PDFs all belong to a common parametric family $\mathcal{F} \stackrel{\triangle}{=} \{f(\cdot;\theta), \theta \in \Theta \subseteq \mathbb{R}^d\}$ where θ is some finite-dimensional parameter vector belonging to the parameter set Θ. A typical example is the Gaussian family where $f(\cdot;\theta) = \mathcal{G}(\cdot;\bar{x}, P)$ for some mean vector $\bar{x}$ and (symmetric positive definite) covariance matrix $P = P^\top > 0$.

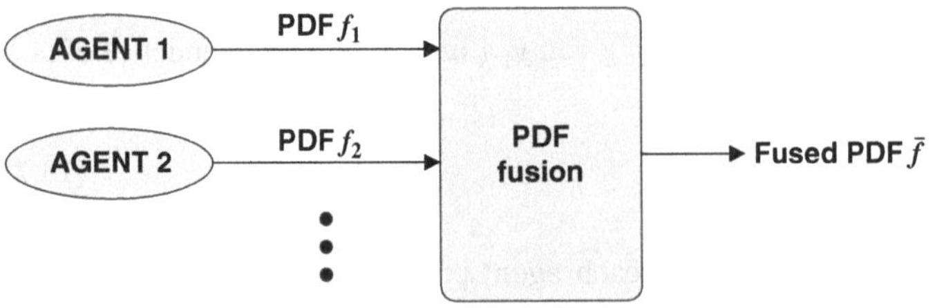

Figure 19.2 Fusion of PDFs.

To this end, the chosen parametric family $\mathcal{F}$ must be closed under both Bayesian correction (19.5) and prediction (19.6), in that: if $p_{k|k-1}(x) = f(x; \theta_{k|k-1})$ with $\theta_{k|k-1} \in \Theta$, then $p_{k|k}(x) = f(x; \theta_{k|k})$ and $p_{k+1|k}(x) = f(x; \theta_{k+1|k})$ for some $\theta_{k|k}, \theta_{k+1|k} \in \Theta$. It is also said that the parametric family $\mathcal{F}$ must be *conjugate (prior)* [14] under both the Bayes equation (19.5) and the Chapman–Kolmogorov equation (19.6). Since the fused PDF $\bar{f}$ must undergo the Bayesian recursion, it is also desirable that $\mathcal{F}$ be conjugate under the fusion rule (19.8), i.e.,

$$f_i(x) = f(x; \theta_i) \in \mathcal{F}, \quad \forall i \in \mathcal{N} \Rightarrow \bar{f}(x) = f(x; \bar{\theta}) \in \mathcal{F} \tag{19.10}$$

for some $\bar{\theta}$ depending on $\{\theta_i\}_{i\in\mathcal{N}}$ and the fusion weights $\pi = \text{col}\{\pi_i\}_{i\in\mathcal{N}}$. If the unconstrained fusion criterion (19.8) does unfortunately not satisfy the conjugacy property (19.10), one can either approximate the PDF $\bar{f} \notin \mathcal{F}$ resulting from (19.8) with some $\hat{f} \in \mathcal{F}$, i.e., $\bar{f} \approx \hat{f}$, or modify (19.8) into the constrained fusion criterion

$$\bar{f} = \arg\min_{f\in\mathcal{F}} \sum_i \pi_i \, d(f, f_i). \tag{19.11}$$

19.3.3 Discrepancy Measures Among PDFs

Different choices of the PDF discrepancy measure $d(f, f_i)$ in (19.8) imply different fusion criteria. A comprehensive overview of discrepancy measures between PDFs can be found in [15]. Any discrepancy measure $d(f, g)$ quantifies how far apart (dissimilar) the PDFs f and g are. It must satisfy the *positivity* property

$$d(f, g) \geq 0, \quad \forall f, g \text{ and } d(f, g) = 0 \Leftrightarrow f = g \tag{19.12}$$

but, in general, the discrepancy measure most commonly employed for the design of the fusion criterion is the *KLD:*

$$\mathrm{D_{KL}}(f \parallel g) = \int f(x) \log \frac{f(x)}{g(x)} \, dx = \mathbb{E}_f\left[\log \frac{f(x)}{g(x)}\right] \tag{19.13}$$

defined as the expectation of the logarithmic difference $\log f(x) - \log g(x)$ according to the probability distribution with PDF f, which is referred to as KLD from g to f. The KLD is not symmetric in that $\mathrm{D_{KL}}(f \parallel g) \neq \mathrm{D_{KL}}(g \parallel f)$; in fact, it will be shown in the next two sections how selecting the KLD as discrepancy measure in (19.8) and reversing the two arguments of the KLD leads to completely different fusion rules with complementary features.

It is worth to point out that discrepancy measures, besides their use in the design of the fusion rule, have other valuable applications in the context of DSE. Specifically, they can be used

- for *event-triggered transmission* [16–18] aiming at bandwidth reduction and energy saving, to decide whether or not transmit the current local PDF according to its discrepancy with respect to a reference PDF obtained from the most recently transmitted PDF via prediction only;
- for inter-agent registration [19] purposes, to estimate the registration parameters between two agents (e.g., relative position and/or orientation, time-offset, etc.) so as to minimize the discrepancy between the PDFs of the agents requiring registration;
- for fusion resilient to cyberattacks and/or sensor faults [20], to exclude from fusion potentially corrupted PDFs whose discrepancy with respect to a reference PDF, deemed to be reliable, exceeds a given threshold.

19.4 Left Kullback–Leibler Fusion

19.4.1 General Case

In this section, we consider the optimization-based fusion (19.8) when the discrepancy measure is chosen as

$$d(f, f_i) = D_{KL}(f \parallel f_i) = \int f(x) \log \frac{f(x)}{f_i(x)}\, dx \tag{19.14}$$

which, in Bayesian inference, can be regarded as the *information gain* obtained when passing from a prior PDF $f_i(\cdot)$ to a posterior PDF $f(\cdot)$. Accordingly, the resulting average PDF $\bar{f}(\cdot)$—referred to as *left Kullback–Leibler average* (LKLA) as it appears as the left argument of the KLD—minimizes the weighted sum of information gains from the local PDFs. Thus, LKLA is coherent with the *principle of minimum discrimination information* (PMDI) according to which the PDF that best represents the current state of knowledge is the one which provides an information gain as small as possible [21, 22]. The following fundamental result on the LKLA holds [6, 23].

Theorem 19.1 (LKLA as normalized weighted geometric average)—Let us consider the fusion criterion (19.8) with the discrepancy measure (19.14). Then,

$$\bar{f}(x) = \frac{\prod_i f_i(x)^{\pi_i}}{\int \prod_i f_i(x)^{\pi_i}\, dx}. \tag{19.15}$$

Proof: see the Appendix. ■

The above result states that the LKLA $\bar{f}$ is nothing but the normalized weighted geometric average of the PDFs f_i. The same result also holds for discrete probability distributions, i.e., for *probability mass functions* (PMFs), and in Chapter 20 will be extended to multi-object densities. LKLA is also referred to as log-linear or logarithmic pooling as the unnormalized LKLA $\tilde{f}(x) = \prod_i f_i(x)^{\pi_i}$ satisfies

$$\log\ \tilde{f}(x) = \sum_i \pi_i\ \log\ f_i(x),$$

i.e., its logarithm is the weighted arithmetic average of the logarithms of local PDFs f_i.

19.4.2 LKLA on Exponential Densities

In this section, the attention is devoted to exponential families (or classes) of PDFs. An exponential family is a parametrized set of PDFs of the form

$$f(x, \theta) = \exp\left[\theta^{\top} t(x) - F(\theta) + k(x)\right] \tag{19.16}$$

where: θ is the *natural parameter* vector belonging to an open convex set Θ; $t(x)$ is the *sufficient statistics*; $F(\theta)$ the *log normalizer*; $k(x)$ the *carrier measure*. It is worth to point out that exponential families include most probability distributions of interest like, e.g., *Gaussian, binomial, negative binomial, multinomial, gamma, beta, chi-squared, Dirichlet, Bernoulli, exponential, Poisson, Wishart, inverse Wishart, Rayleigh, categorical, geometric, Weibull* and others [24]. Remarkable examples of non-exponential-family distributions are *uniform, mixture, Student's t*.

The next result concerns conjugacy of exponential families under LKLA [16].

Theorem 19.2 (Conjugacy of exponential families under LKLA)—Let $f_i(x) = f(x, \theta_i), \forall i \in \mathcal{N}$. Then, the LKLA in (19.15) is given by

$$\bar{f}(x) = f(x, \bar{\theta}), \quad \bar{\theta} = \sum_{i \in \mathcal{N}} \pi_i \theta_i. \tag{19.17}$$

Proof: see the Appendix. ■

In words, the above result states that fusing, according to the LKLA criterion, PDFs of the same exponential family amounts to a weighted arithmetic average of their natural parameters. A case of great practical interest is when all local PDFs to be fused belong to the Gaussian family, which is a special case of exponential family. In fact, a Gaussian PDF $f(\cdot) = \mathcal{G}(\cdot; m, \Sigma)$ can be expressed in the form (19.16) with

$$\theta = \begin{bmatrix} \Sigma^{-1} m \\ \\ \mathrm{vec}(\Sigma^{-1}) \end{bmatrix}, \; t(x) = \begin{bmatrix} x \\ -\frac{1}{2}\,\mathrm{vec}(xx^{\top}) \end{bmatrix}, \; F(\theta) = \frac{1}{2} m^{\top} \Sigma^{-1} m + \frac{1}{2} \log\det(2\pi\Sigma), \tag{19.18}$$
$$k(x) = 0, \; \Theta = \{\theta : \Sigma = \Sigma^{\top} > 0\}.$$

Then, according to Theorem 19.2, LKLA of Gaussian PDFs $f_i(x) = \mathcal{G}(x; m_i, \Sigma_i)$ yields the Gaussian PDF $\bar{f}(x) = \mathcal{G}(x; \bar{m}, \bar{\Sigma})$ with

$$\begin{cases} \overline{\Sigma} = \left(\sum_i \pi_i \Sigma_i^{-1} \right)^{-1} \\ \overline{m} = \overline{\Sigma} \left(\sum_i \pi_i \, \Sigma_i^{-1} m_i \right). \end{cases} \tag{19.19}$$

Introducing the *information* (inverse covariance) matrix $\Omega \stackrel{\triangle}{=} \Sigma^{-1}$ and information vector $q \stackrel{\triangle}{=} \Sigma^{-1} m$ in place of mean m and covariance Σ, LKLA of Gaussian PDFs is more conveniently expressed in terms of the information pair (q, Ω) as

$$\begin{cases} \overline{q} = \sum_i \pi_i \, q_i \\ \overline{\Omega} = \sum_i \pi_i \, \Omega_i. \end{cases} \tag{19.20}$$

In words, normalized weighted geometric average (LKLA) of Gaussian PDFs corresponds to weighted arithmetic average of their information pairs. Notice that LKLA fusion of Gaussian PDFs reduces to the well-known *covariance intersection* [25] rule (19.19). In this respect, LKLA fusion can be regarded as a generalization of covariance intersection to generic (non-Gaussian) PDFs and, for this reason, is also referred to as *generalized covariance intersection* (GCI) [26].

It is common practice in nonlinear and/or non-Gaussian state estimation problems to represent the state PDF either in *Gaussian-mixture* (GM) or *delta-mixture* (also called *particle*) form. In fact, any (unimodal or multimodal) PDF can be approximated, to any desired degree of accuracy, via a

Gaussian/delta mixture with sufficiently large number of *Gaussian components/particles*. Unfortunately, however, both mixture representations are not closed under LKLA and, therefore, suitable approximations of the geometric averages of Gaussian/delta mixtures need to be found.

19.4.3 LKLA on Gaussian Mixtures

Let us consider, just for the sake of simplicity, only two agents ($i = 1, 2$) with GM PDFs

$$f_i(x) = \sum_{j=1}^{J_i} \alpha_{i,j}\, \mathcal{G}(x; m_{i,j}, \Sigma_{i,j}) \tag{19.21}$$

to be fused. It is natural to wonder whether the fused PDF

$$\bar{f}(x) = \frac{f_1(x)^{\pi}\, f_2(x)^{1-\pi}}{\int f_1(x)^{\pi}\, f_2(x)^{1-\pi}\, dx} \tag{19.22}$$

is a GM. Notice that (19.22) involves exponentiation and multiplication of GMs. To this end, the following properties concerning elementary operations on Gaussian components and mixtures hold.

- The power of a Gaussian is Gaussian, i.e.,

$$\begin{aligned} [\alpha\, \mathcal{G}(x; m, \Sigma)]^{\omega} &= \alpha^{\omega}\, \kappa(\omega, \Sigma)\, \mathcal{G}\left(x; m, \frac{\Sigma}{\omega}\right) \\ \kappa(\omega, \Sigma) &\triangleq \frac{\left[\det(2\pi\Sigma\omega^{-1})\right]^{1/2}}{[\det(2\pi\Sigma)]^{\omega/2}}. \end{aligned} \tag{19.23}$$

- The product of Gaussians is Gaussian, i.e.,

$$\begin{aligned} &\alpha_1\, \mathcal{G}(x; m_1, \Sigma_1) \cdot \alpha_2\, \mathcal{G}(x; m_2, \Sigma_2) = \alpha_{12}\, \mathcal{G}(x; m_{12}, \Sigma_{12}) \\ &\Sigma_{12} = \left(\Sigma_1^{-1} + \Sigma_2^{-1}\right)^{-1} \\ &m_{12} = \Sigma_{12}\left(\Sigma_1^{-1} m_1 + \Sigma_2^{-1} m_2\right) \\ &\alpha_{12} = \alpha_1\, \alpha_2\, \mathcal{G}(m_1 - m_2; 0, \Sigma_1 + \Sigma_2). \end{aligned} \tag{19.24}$$

- By the distributive property, the product of two GMs of J_1 and, respectively, J_2 components is a GM of $J_1 J_2$ components.
- Exponentiation of a GM is not, in general, a GM but, as suggested in [27], can be reasonably approximated as a GM by:

$$\left[\sum_{j=1}^{J} \alpha_j\, \mathcal{G}(x; m_j, \Sigma_j)\right]^{\omega} \approx \sum_{j=1}^{J} [\alpha_j\, \mathcal{G}(x; m_j, \Sigma_j)]^{\omega} = \sum_{j=1}^{J} \alpha_j^{\omega} \kappa(\omega, \Sigma_j)\, \mathcal{G}\left(x; m_j, \frac{\Sigma_j}{\omega}\right) \tag{19.25}$$

provided that the means (centers) m_i and m_j, $i \neq j$, of the Gaussian components are well-separated, as measured by the respective covariances Σ_i and Σ_j, i.e.,

$$(m_i - m_j)^{\top} \Sigma_i^{-1} (m_i - m_j) \gg 1 \text{ and } (m_i - m_j)^{\top} \Sigma_j^{-1} (m_i - m_j) \gg 1.$$

Summarizing, the fused PDF in (19.22) can be approximated by the following GM

$$
\begin{aligned}
\bar{f}(x) &\approx \sum_{j=1}^{J_1} \sum_{h=1}^{J_2} \alpha_{12,jh}\, \mathcal{G}(x; m_{12,jh}, \Sigma_{12,jh}) \\
\alpha_{12,jh} &= \alpha_{1,j}^{\omega}\, \alpha_{2h}^{1-\omega}\, \kappa(\omega, \Sigma_{1,j})\, \kappa(1-\omega, \Sigma_{2,h})\, \mathcal{G}\left(m_{1,j} - m_{2,h}; 0, \frac{\Sigma_{1,j}}{\omega} + \frac{\Sigma_{2,h}}{1-\omega}\right) \\
\Sigma_{12,jh} &= \left[\omega \Sigma_{1,j}^{-1} + (1-\omega)\Sigma_{2,h}^{-1}\right]^{-1} \\
m_{12,jh} &= \Sigma_{12,jh}\left[\omega \Sigma_{1,j}^{-1} m_{1,j} + (1-\omega)\Sigma_{2,h}^{-1} m_{2,h}\right].
\end{aligned}
\tag{19.26}
$$

An improved approximation better handling the case of closely-spaced Gaussian components can be found in [28].

Extension to LKLA of $|\mathcal{N}| > 2$ agents can be obtained by applying the pairwise fusion procedure (19.26) $|\mathcal{N}| - 1$ times. Notice that, since each pairwise fusion between GMs involves some approximation, their ordering affects the final fusion result while, in the case of exponential-family PDFs where LKLA fusion can be performed exactly, the order does not matter in view of the commutative and associative properties of LKLA.

19.4.4 LKLA on Particle PDFs

Another commonly employed representation of PDFs for nonlinear filtering is the so called *δ-mixture* or *particle* form

$$
f_i(x) = \sum_{j=1}^{N_i} w_{i,j}\, \delta(x - x_{i,j}). \tag{19.27}
$$

If the particles $\{(w_{i,j}, x_{i,j})\}_{j=1}^{N_i}$ are independently generated by each agent $i \in \mathcal{N}$, LKLA fusion of $\{f_i\}_{i \in \mathcal{N}}$ will produce, in view of its multiplicative nature, $\bar{f}(x)$ zero almost everywhere thus implying complete loss of information. To remedy this inconvenience, it is possible to proceed in the following way:

1. Approximate each local δ-mixture PDF $f_i(x)$ in (19.27) by a GM PDF $\hat{f}_i(x)$ of the form (19.21);
2. Fuse, via LKLA, the approximate GM PDFs $\hat{f}_i(x)$ into a GM PDF $\hat{f}(x)$ exploiting one of the methods discussed in the previous subsection;
3. Transform the GM PDF $\hat{f}(x)$ back to δ-mixture form $\bar{f}(x)$.

Step 3 trivially involves sampling of a GM PDF while Step 1 can be accomplished, e.g., by the *expectation-maximization* (EM) algorithm [29].

19.5 Right Kullback–Leibler Fusion

In this section, we consider the optimization-based fusion (19.8) by reversing the two arguments of the KLD, i.e., when the discrepancy measure is chosen as

$$
d(f, f_i) = \mathrm{D_{KL}}(f_i \parallel f) = \int f_i(x) \log \frac{f_i(x)}{f(x)}\, dx \tag{19.28}
$$

which, in Bayesian inference, can be regarded as the *information loss* incurred when replacing the local PDF $f_i(\cdot)$ by the fused one $f(\cdot)$. In this case, the resulting fused PDF $\bar{f}(\cdot)$ is referred to as *right*

Kullback–Leibler average (RKLA) as it appears as the right argument of the KLD. Since the RKLA minimizes the weighted sum of information losses from the local PDFs, it has also been named as *minimum information loss* (MIL) fusion [30, 31].

19.5.1 Arithmetic Average and MIL Fusion

The following fundamental result on the RKLA holds [30, 31].

Theorem 19.3 (RKLA as weighted arithmetic average)—Let us consider the fusion criterion (19.8) with the discrepancy measure (19.28). Then,

$$\bar{f}(x) = \sum_i \pi_i f_i(x). \tag{19.29}$$

Proof : see the Appendix. ■

The above result states that the RKLA $\bar{f}$ turns out to be the weighted arithmetic average of the PDFs f_i.

Unlike LKLA, RKLA of PDFs of an exponential family does not belong to the same exponential family. For instance, RKLA of Gaussian PDFs $f_i(x) = \mathcal{G}(x; m_i, \Sigma_i)$ provides the Gaussian mixture $\bar{f}(x) = \sum_i \pi_i \mathcal{G}(x; m_i, \Sigma_i)$ which is not Gaussian. Hence, to fuse estimates–covariances from Kalman-like filtering agents while preserving Gaussianity, the following constrained optimization problem can be formulated:

$$\overline{f}_{\text{MIL}} = \arg \min_{f \in \mathcal{P}_G} \sum_i \pi_i \, \mathrm{D}_{\text{KL}}(f_i \| f) \tag{19.30}$$

where $\mathcal{P}_G \stackrel{\triangle}{=} \{f \in \mathcal{P} : f(x) = \mathcal{G}(x; m, C)\}$ is the family of Gaussian PDFs over $\mathbb{R}^{\dim x}$. The PDF $\bar{f}_{\text{MIL}}$ defined in (19.30) is the best, in MIL sense, among all Gaussian PDFs. The following result holds.

Theorem 19.4 (Gaussian MIL fusion)—Let $f_i(\cdot) = \mathcal{G}(\cdot; m_i, \Sigma_i)$. Then,

$$\overline{f}_{\text{MIL}}(\cdot) = \arg \min_{f \in \mathcal{P}_G} \sum_i \pi_i \, \mathrm{D}_{\text{KL}}(f_i \| f) = \mathcal{G}(\cdot; \overline{m}, \overline{\Sigma}) \tag{19.31}$$

with

$$\begin{cases} \overline{m} = \sum_i \pi_i m_i \\ \overline{\Sigma} = \sum_i \pi_i \left[\Sigma_i + (m_i - \overline{m})(m_i - \overline{m})^{\top}\right]. \end{cases} \tag{19.32}$$

Proof : see the Appendix. ■

In words, the best, in the MIL sense, Gaussian PDF is the one with mean $\bar{m}$ and covariance $\bar{\Sigma}$ matching those of the Gaussian mixture $\sum_i \pi_i \mathcal{G}(\cdot; m_i, \Sigma_i)$. Notice from (19.32) that the fused covariance $\bar{\Sigma}$ also depends on the means (estimates). Hence, if MIL fusion is adopted in the context of multiagent Kalman filtering, covariances become measurement-dependent, thus making the multiagent Kalman filter nonlinear and hence complicating its stability analysis unlike in single-agent Kalman filtering or when LKLA (covariance intersection) fusion is adopted.

19.5.2 Implementation Issues

The RKLA of GMs $f_i(x)$ in (19.21) is trivially a GM, precisely

$$\begin{aligned}\bar{f}(x) &= \sum_i \pi_i \left[\sum_{j=1}^{J_i} \alpha_{i,j}\, \mathcal{G}(x; m_{ii,j}, \Sigma_{i,j}) \right] \\ &= \sum_i \sum_{j=1}^{J_i} \pi_i\, \alpha_{i,j}\, \mathcal{G}(x; m_{i,j} \Sigma_{i,j}) \\ &= \sum_{h=1}^{J} \bar{\alpha}_h\, \mathcal{G}(x; \bar{m}_h, \bar{\Sigma}_h)\end{aligned}$$

is a GM with $J = \sum_i J_i$ Gaussian components. Recall that, on the other hand, approximate LKLA of GMs yields a GM with $J = \prod_i J_i$ components. In any case, LKLA/RKLA fusion causes an increase of the number of Gaussian components which, however, needs to be kept limited, i.e., $J \leq \bar{J}$. To this end, a suitable *pruning and merging* procedure [32] can be applied to the GM.

19.6 Design of the Fusion Weights

The fusion weights π_i represent, along with the discrepancy measure $d(f, f_i)$, degrees of freedom to be selected, subject to the constraints (19.9), in the fusion criterion (19.8). This section provides considerations about the design of such fusion weights.

19.6.1 Uniform Weights

The simplest choice is to take uniform weights, i.e., the same weight $\pi_i = 1/|\mathcal{N}|$ for each agent $i \in \mathcal{N}$. This amounts to regarding all agents as equally reliable which, certainly, represents a natural choice whenever there is no available information about the reliability of each agent.

19.6.2 Quality-dependent Weights

Provided that each agent i has available some local measure $r_i > 0$ of its quality or accuracy or reliability such that the higher is r_i better is the agent quality, a possible choice is to take *quality-dependent* fusion weights [13, 33]

$$\pi_i = \frac{r_i}{\sum_j r_j}. \tag{19.33}$$

For instance, $r_i = \mathrm{tr}(\Omega_i)$ or $r_i = \det(\Omega_i)$, where Ω_i is the locally computed information matrix of agent i, can be considered as quality measure.

19.6.3 Optimization-based Weight Design

Several authors propose optimization-based fusion weights, i.e., to select $\pi = \mathrm{col}\{\pi_i\}_{i\in\mathcal{N}}$ that minimize some criterion, i.e.,

$$\pi^* = \arg \min_{\pi \in \mathbb{S}_{|\mathcal{N}|}} J(\pi) \tag{19.34}$$

where $\mathbb{S}_N = \{\pi \in [0,1]^N : \mathbb{1}^\top \pi = 1\}$ is the probability simplex on $[0,1]^N$. Possible choices for the criterion $J(\pi)$ are:

- trace or determinant of the fused covariance matrix for the fusion of Gaussian PDFs [23];
- average discrepancy between agent PDFs and fused one [34], i.e.,

$$J(\pi) = \frac{1}{|\mathcal{N}|} \sum_{i \in \mathcal{N}} d(\bar{f}, f_i)$$

where the fused PDF $\bar{f}$ clearly depends on π and $d(\cdot,\cdot)$ is any PDF discrepancy measure (not necessarily but preferably the one adopted in the fusion criterion).

Moreover, [35] proposed a max-min optimization approach to the joint design of fusion weights π and fused PDF $\bar{f}$, i.e.,

$$(\pi^*, \bar{f}) = \arg\ \max_{\pi \in \mathbb{S}_{|\mathcal{N}|}}\ \min_{f \in \mathcal{P}} \sum_{i \in \mathcal{N}} \pi_i\, d(f, f_i).$$

Some remarks on optimization-based fusion weight design are in order.

Remarks

- If $\bar{f} \propto \prod_i f_i^{\pi_i}$ is obtained by LKLA and $J(\pi) = |\mathcal{N}|^{-1} \sum_i D_{KL}(\bar{f} \parallel f_i)$, then the optimal weights are the uniform ones, i.e., $\pi^* = |\mathcal{N}|^{-1}\, \mathbb{1}$. Hence, when no basis is available to measure agent reliability, uniform fusion weighting seems to be a reasonable choice at least for LKLA fusion.
- Weight optimization has $|\mathcal{N}| - 1$ degrees-of-freedom due to the normalization constraint (19.9). It is, therefore, computationally cumbersome unless $|\mathcal{N}| = 2$ (pairwise fusion) for which optimization becomes one-dimensional (there is only one scalar weight $\pi \in [0,1]$ to be chosen, the other being $1 - \pi$). To simplify weight optimization, it is possible to adopt a sequential approach by splitting fusion over N agents into $N-1$ pairwise fusions; in this way, however, the outcome of weight design and, hence, of fusion is severely affected by the considered ordering of agents.
- The adoption of optimization-based fusion weights seems to be impractical in situations where fusion must be performed at a high rate (e.g., in multiagent state estimation), and the extra computational burden for weight optimization might not necessarily be remunerated in terms of performance improvement.

19.7 Scalable Distributed Fusion via Consensus

Fusion has been regarded so far as a global task involving all agents $i \in \mathcal{N}$. According to the network model in Subsection 19.2.2, however, each agent i has direct access only to the subnetwork $\mathcal{N}_i$ which, for large and sparse networks, typically represents a much smaller portion of the overall network, i.e., $|\mathcal{N}_i| \ll |\mathcal{N}|$. Hence, to perform fusion in a scalable and distributed way, a suitable *distributed information processing* protocol—like, e.g., *diffusion* [9], *gossiping* [36], *flooding* [12], or *consensus* [4]—needs to be adopted. This chapter specifically exploits the consensus protocol to spread PDF information throughout the network.

19.7.1 Consensus on the Average

Consensus is a general method that allows each agent i of a distributed network to perform some global task (e.g., averaging, min-max computation, optimization) over the whole network $\mathcal{N}$, by iteratively performing the same task over the subnetwork $\mathcal{N}_i$. In view of the interpretation of fusion as averaging on PDFs, we are interested to *consensus on the average* and, thanks to the conjugacy of exponential families as well as to the measurement-independent nature of covariances, we focus on the left Kullback–Leibler (geometric) average criterion.

For LKLA fusion/consensus, it is convenient to introduce the operators $\oplus$ and $\odot$ defined as follows:

$$\begin{aligned} f \oplus g\,(x) &\stackrel{\triangle}{=} \frac{f(x)g(x)}{\int f(x)g(x)dx} \\ \pi \odot f(x) &\stackrel{\triangle}{=} \frac{f(x)^{\pi}}{\int f(x)^{\pi}\,dx} \end{aligned} \tag{19.35}$$

for PDFs $f(\cdot)$, $g(\cdot)$ and $\pi \in (0, 1]$. It is easy to check, from their definitions, that such operators satisfy the following properties:

$$\begin{aligned} (f \oplus g) \oplus h &= f \oplus (g \oplus h) = f \oplus g \oplus h \\ f \oplus g &= g \oplus f \\ (\alpha\,\beta) \odot f &= \alpha \odot (\beta \odot f) \\ 1 \odot f &= f \\ \alpha \odot (f \oplus g) &= (\alpha \odot f) \oplus (\alpha \odot g) \\ (\alpha + \beta) \odot f &= (\alpha \odot f) \oplus (\beta \odot f) \end{aligned} \tag{19.36}$$

for PDFs f, g, h and positive scalars α, β. Further, the LKLA (19.15) can be rewritten as:

$$\bar{f} = \bigoplus_{j \in \mathcal{N}} (\pi_j \odot f_j). \tag{19.37}$$

Then, consensus on LKLA for agent i operates as follows:

$$\begin{cases} \text{initialize} & f_{i,0} = f_i \\ \text{for } \ell = 1, \dots, L & \\ \quad \text{receive} & f_{j,\ell-1}, \forall j \in \mathcal{N}_i \\ \quad \text{compute} & f_{i,\ell} = \bigoplus_{j \in \mathcal{N}_i} (\pi_{i,j} \odot f_{j,\ell-1}) \\ \quad \text{broadcast} & f_{i,\ell} \\ \text{endfor} & \end{cases} \tag{19.38}$$

where $\pi_{i,j} \geq 0$ are consensus weights such that

$$\sum_{j \in \mathcal{N}_i} \pi_{i,j} = 1, \quad \forall i \in \mathcal{N}. \tag{19.39}$$

A snapshot of the above consensus procedure is depicted in Figure 19.3.

Let us introduce the consensus matrix Π whose generic (i, j) entry coincides with the consensus weight $\pi_{i,j}$ if $j \in \mathcal{N}_i$ and is zero otherwise. Clearly, by (19.39), Π turns out to be row-stochastic. Thanks to the properties (19.36), it can be seen that

$$f_{i,\ell} = \bigoplus_{j \in \mathcal{N}} \left(\pi_{i,j}^{\ell} \odot f_j\right) \tag{19.40}$$

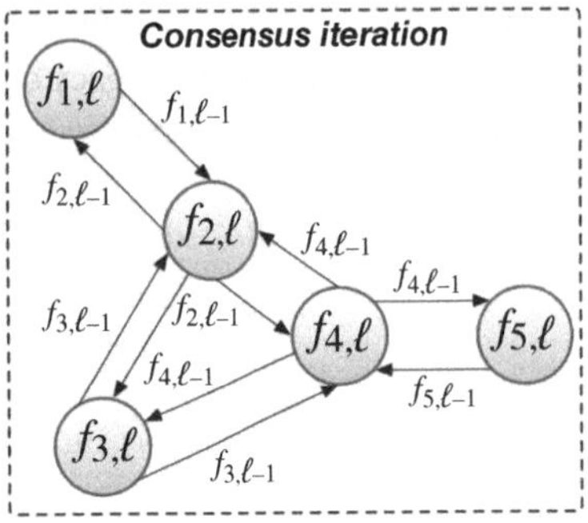

Figure 19.3 Consensus snapshot.

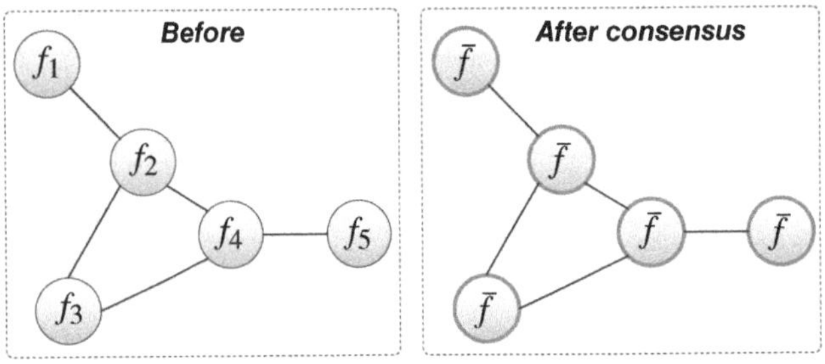

Figure 19.4 Situation at the beginning of consensus (on the left) and after consensus convergence (on the right).

where $\pi_{i,j}^{\ell}$ is the (i, j) entry of Π^{ℓ}. Next subsection will analyze the convergence of consensus which, by (19.40), is related to the convergence of Π^{ℓ} for $\ell \to \infty$.

19.7.2 Convergence Issues

Recall that a nonnegative matrix Π is said to be *primitive* if there exists an integer h such that all entries of Π^h are strictly positive. The following consensus convergence result holds.

Theorem 19.5 (Convergence of consensus)—Let the consensus weights $\pi_{i,j}$ be chosen such that Π be primitive, then all agents asymptotically achieve consensus on a weighted average, i.e.,

$$\lim_{\ell \to \infty} f_{i,\ell} = \bigoplus_{j \in \mathcal{N}} (\alpha_j \odot f_j), \quad \forall i \in \mathcal{N} \tag{19.41}$$

for suitable weights $\alpha_j > 0$, such that $\sum_j \alpha_j = 1$, depending on Π. If Π is also column-stochastic (hence doubly-stochastic), then consensus is on the unweighted average, i.e.,

$$\lim_{\ell \to \infty} f_{i,\ell} = \bigoplus_{j \in \mathcal{N}} \left(\frac{1}{|\mathcal{N}|} \odot f_j \right), \quad \forall i \in \mathcal{N}. \tag{19.42}$$

Proof: see the Appendix. ■

The above result states that, under primitiveness of the consensus matrix, all agents reach asymptotic consensus on a suitably weighted LKLA of the local PDFs (see Figure 19.4).

Primitiveness of Π requires that the network $\mathfrak{G}$ be *strongly connected*, i.e., for any pair of agents $i, j \in \mathcal{N}$ there must be a path from i to j and vice versa, and in such a case is guaranteed provided that $\pi_{i,j} > 0$ for all $j \in \mathcal{N}_i$. If the consensus matrix is also column-stochastic, then asymptotic consensus is on the unweighted LKLA. To this end, when the network $\mathfrak{G}$ is undirected, a possible choice of the consensus weights ensuring that Π is doubly stochastic is given by the so-called *Metropolis* weights [8, 12]

$$\pi_{i,j} = \begin{cases} \dfrac{1}{\max(|\mathcal{N}_i|, |\mathcal{N}_j|)}, & j \in \mathcal{N}_i \setminus \{i\} \\ 1 - \sum\limits_{h \in \mathcal{N}_i \setminus \{i\}} \pi_{i,h}, & j = i \\ 0, & j \notin \mathcal{N}_i . \end{cases} \tag{19.43}$$

The consensus algorithm (19.38) allows, under primitiveness and double-stochasticity of Π, to asymptotically compute in a distributed fashion the global LKLA (19.37) with uniform fusion weights $\pi_i = 1/|\mathcal{N}|$ for all $i \in \mathcal{N}$.

If quality-dependent fusion weights (19.33) are adopted, the consensus algorithm can be modified as follows:

$$\begin{cases} \text{initialize} \quad & f_{i,0} = r_i \odot f_i, \quad r_{i,0} = r_i \\ \text{for } \ell = 1, \dots, L & \\ \quad \text{receive} & f_{j,\ell-1} \text{ and } r_{j,\ell-1}, \forall j \in \mathcal{N}_i \\ \quad \text{compute} & f_{i,\ell} = \bigoplus\limits_{j \in \mathcal{N}_i} (\pi_{i,j} \odot f_{j,\ell-1}), \\ & r_{i,\ell} = \sum\limits_{j \in \mathcal{N}_i} \pi_{i,j} \, r_{j,\ell-1} \\ \quad \text{broadcast} & f_{i,\ell} \text{ and } f_{i,\ell} \\ \text{endfor} & \\ & \bar{f}_i = \dfrac{1}{r_{i,L}} \odot f_{i,L} \end{cases} \tag{19.44}$$

In fact, thanks to the properties (19.36) of the operators $\oplus$ and $\odot$, it can be checked that the PDF resulting from (19.44) turns out to be

$$\bar{f}_i = \bigoplus_{j \in \mathcal{N}} \left[\frac{\pi_{i,j}^L r_j}{\sum_{h \in \mathcal{N}} \pi_{i,h}^L \, r_h} \odot f_j \right]. \tag{19.45}$$

Then, under primitiveness and double-stochasticity of Π, $\pi_{i,j}^L \to 1 \setminus |\mathcal{N}|$ as $L \to \infty$ which, in turn, implies that, as $L \to \infty$, $\bar{f}_i$ converges to the LKLA (19.37) with quality-dependent fusion weights given by (19.33) as desired.

19.7.3 Immunity to Double Counting

The aim of this section is to show how LKLA is immune to *double counting of information*. To this end, let $\mathcal{F}(\mathcal{N})$ denote the power set of $\mathcal{N}$, i.e., the set of all possible subsets of agents. Then, for each $\mathcal{S} \in \mathcal{F}(\mathcal{N})$, let $Y_{\mathcal{S}}$ be the information set which is available to all, and only, the agents in $\mathcal{S}$.

For instance, $Y_{\{i\}}$ is the information set available exclusively to agent i, while $Y_{\mathcal{N}}$ represents the information set common to all agents. Finally, let $f(x|Y_{\mathcal{S}})$ be the PDF of x given the information set $Y_{\mathcal{S}}$. Notice that, by construction, the information sets are taken mutually independent in that $Y_{\mathcal{S}} \cap Y_{\mathcal{S}'} = \emptyset$ for any $\mathcal{S}, \mathcal{S}' \in \mathcal{F}(\mathcal{N})$ such that $\mathcal{S} \neq \mathcal{S}'$. In other words, the information sets $Y_{\mathcal{S}}$ provide a (nonoverlapping) partition of the collective information set. Hence, for each agent $i \in \mathcal{N}$, the PDF $f_i(x)$ can be factorized as

$$f_i(x) = \bigoplus_{\mathcal{S} \in \mathcal{F}(\mathcal{N}): i \in \mathcal{S}} f(x|Y_{\mathcal{S}}) \tag{19.46}$$

while the optimal fusion of all agent PDFs $\{f_i(x)\}_{i \in \mathcal{N}}$ could, in principle, be obtained as

$$f^*(x) = \bigoplus_{\mathcal{S} \in \mathcal{F}(\mathcal{N})} f(x|Y_{\mathcal{S}}) \tag{19.47}$$

where each information set $Y_{\mathcal{S}}$ is counted only once with unit weight. In practice, however, the above optimal fusion cannot be performed on a peer-to-peer network with loops due to impossibility to single out common information. Hence, some robust suboptimal fusion strategy has to be adopted. In this regard, LKLA provides an effective approach to PDF fusion in view of the following result [37].

Theorem 19.6 (Immunity to double counting of LKLA)—Let all the agent PDFs f_i be factorized as in (19.46). Then, the LKLA in (19.37) turns out to be given by

$$\bar{f}(x) = \bigoplus_{\mathcal{S} \in \mathcal{F}(\mathcal{N})} \left[\left(\sum_{i \in \mathcal{S}} \pi_i \right) \odot f(x|Y_{\mathcal{S}}) \right]. \tag{19.48}$$

Proof : see the Appendix. ∎

From (19.48), it is clear that each information set $Y_{\mathcal{S}}$ is counted only once with weight $\sum_{i \in \mathcal{S}} \pi_i \leq 1$, thus preventing double counting of common information. The price to be paid for such robustness to data incest is a conservative flattening of independent information which occurs whenever $\sum_{i \in \mathcal{S}} \pi_i < 1$.

When LKLA is computed in a distributed way via consensus, for each agent i and consensus step ℓ, the PDF $f_{i,\ell}$ turns out to be given by

$$f_{i,\ell}(x) = \bigoplus_{\mathcal{S} \in \mathcal{F}(\mathcal{N})} \left[\left(\sum_{j \in \mathcal{S}} \pi_{i,j}^{\ell} \right) \odot f(x|Y_{\mathcal{S}}) \right]. \tag{19.49}$$

Since $\sum_{j \in \mathcal{S}} \pi_{i,j}^{\ell} \leq 1$ for all i and ℓ, consensus produces PDFs that are immune to double counting for any agent i and consensus step ℓ.

19.8 Distributed State Estimation

This section exploits the previously introduced tools of PDF fusion and consensus on the average in order to develop scalable DSE algorithms for the peer-to-peer agent network of Subsection 19.2.2. Specifically, the aim is to tackle, in a distributed and scalable fashion, the solution of the Bayesian filtering problem (19.5)–(19.7). In this respect, the following general approach is pursued in this chapter:

- Each agent i runs a local Bayesian filter exploiting available measurements $y_{k,i}$ and interacting, via LKLA consensus iterations on PDFs, with the in-neighborhood $\mathcal{N}_i$;
- Consensus can be performed on the *likelihoods*, on the *posterior* PDFs or, in parallel, on both *likelihoods and prior PDFs*.

The choice of the LKLA criterion for PDF fusion is mainly for two reasons:

1. Exponential families are closed under LKLA;
2. Covariances are measurement-independent under LKLA of Gaussian PDFs, thus preserving linearity of the distributed multiagent filter in the linear Gaussian case.

19.8.1 Consensus on Likelihoods

Using the operators $\oplus$ and $\odot$ and removing for simplicity the likelihood arguments, the multiagent likelihood (19.7) can be expressed as

$$l_k = \bigoplus_{j \in \mathcal{N}} (1 \odot l_{k,j}) \tag{19.50}$$

which can be obtained by each agent i as outcome of the following consensus algorithm

$$\begin{cases} l_{k,i,0}(x) = l_{k,i}(y_{k,i}|x) \\ \text{for } \ell = 1, \dots , L \\ \quad l_{k,i,\ell}(x) = \bigoplus_{j \in \mathcal{N}_i} (\pi_{i,j} \odot l_{k,j,\ell-1}(x)) \\ \text{endfor.} \end{cases} \tag{19.51}$$

In fact, for a suitable choice of the consensus weights,

$$\lim_{\ell \to \infty} l_{k,i,\ell} = \bigoplus_{j \in \mathcal{N}} \left(\frac{1}{|\mathcal{N}|} \odot l_{k,j} \right) \stackrel{\triangle}{=} l_{k,i,\infty}$$

from which the global likelihood l_k can be recovered as $l_k = |\mathcal{N}| \odot l_{k,i,\infty}$. In practice, however, a finite (possibly small) number L of consensus steps is performed so that the actual outcome of the consensus on likelihoods (CL) (19.51) is

$$l_{k,i,L} = \bigoplus_{j \in \mathcal{N}} \left(\pi^L_{i,j} \odot l_{k,j} \right)$$

which certainly avoids double counting of information but, compared to (19.50), underweights the likelihood of agent j whenever $\pi^L_{i,j} < 1$. To compensate such underweighting, a suitable factor $\rho_{k,i} \geq 1$ can be introduced to replace l_k by $\rho_{k,i} \odot l_{k,i,L}$ in the Bayesian correction step performed by agent i. The resulting CL approach to DSE is summarized in Algorithm 1.

The choice of factor $\rho_{k,i}$ is a key issue to avoid underweighting/overweighting of the likelihoods. To this end, notice that Algorithm 1 updates the prior with the likelihood $\bigoplus_{j \in \mathcal{N}} \left(\rho_{k,i} \pi^L_{i,j} \odot l_{k,j} \right)$ so that, to recover the optimal Bayesian correction via (19.50), agent i should select $\rho_{k,i}$, if possible, so that $\rho_{k,i} \pi^L_{i,j} = 1$ for any $j \in \mathcal{N}$. Hence, the following rules for the choice of $\rho_{k,i}$ can be suggested.

1. Since $\lim_{L \to \infty} \pi^L_{i,j} = |\mathcal{N}|^{-1}$, for $L \gg 1$ a reasonable choice is

   $$\rho_{k,i} = |\mathcal{N}| \tag{19.52}$$

   even if, for finite L, this might cause overweighting of some local likelihood $l_{k,j}$.

Algorithm 1: CL (time k, agent i)

Inputs: $p_{k|k-1,i}(\cdot), l_{k,i}(\cdot|\cdot), \varphi_{k+1|k}(\cdot|\cdot), y_{k,i}, \rho_{k,i}, \{\pi_{i,j}\}_{j\in\mathcal{N}_i}$;

1: Perform consensus on likelihoods according to (19.51) ▷ Consensus on likelihoods

2: $p_{k|k,i}(x) = p_{k|k-1,i}(x) \oplus (\rho_{k,i} \odot l_{k,i,L}(x))$ ▷ Correction

3: $p_{k+1|k,i}(x) = \int \varphi_{k+1|k}(x|\xi)\, p_{k|k,i}(\xi)\, d\xi$ ▷ Prediction

Outputs: $p_{k|k,i}(\cdot), p_{k+1|k,i}(\cdot)$

2. For $L = 1$, since $\pi^L_{i,j} = \pi_{i,j}$ are known, a nonconservative choice that avoids overweighting is given by

$$\rho_{k,i} = \min_{j\in\mathcal{N}_i}\left(\frac{1}{\pi_{i,j}}\right). \tag{19.53}$$

Unfortunately, however, this choice cannot be easily extended to $L > 1$ as it requires $\pi^L_{i,j}$ which, in general, is not locally computable.

3. A choice that ensures $\rho_{k,i}\pi^L_{i,j} \leq 1$ for any $L \geq 1$ is to exploit consensus to estimate, at time k, the fraction of agents actually contributing to the Bayesian correction. To this end, the following consensus algorithm can be used:

$$\begin{cases} b_{k,i,0} = \begin{cases} 1, & \text{if a measurement } y_{k,i} \text{ is available} \\ 0, & \text{otherwise} \end{cases} \\ \text{for } \ell = 1, \dots, L \\ \quad b_{k,i.\ell} = \sum_{j\in\mathcal{N}_i} \pi_{i,j}\, b_{k,j,\ell-1} \\ \text{endfor} \\ \rho_{k,i} = \begin{cases} \dfrac{1}{b_{k,i,L}}, & \text{if } b_{k,i,L} \neq 0 \\ 1, & \text{otherwise} \end{cases} \end{cases} \tag{19.54}$$

which guarantees that $\rho_{k,i}\pi^L_{i,j} \leq 1$ without requiring a priori knowledge on the network.

19.8.2 Consensus on Posteriors

As an alternative to CL described above, another possible approach is for each agent $i \in \mathcal{N}$ to: (1) first perform correction with the local likelihood, i.e.,

$$p_{k|k,i} = p_{k|k-1,i} \oplus l_{k,i} \tag{19.55}$$

to get the local posterior $p_{k,k,i}$; (2) then carry out consensus with the local posteriors $p_{k|k,j}$ from in-neighbors $j \in \mathcal{N}_i$; finally apply prediction to the PDF. This approach, referred to as *consensus on the posteriors* (CP), is summarized in Algorithm 2.

19.8.3 Parallel Consensus on Priors and Likelihoods

The CL and CP approaches of the previous subsections have complementary positive and negative features. For the sake of comparison, the expressions for the local posterior $p_{k|k,i}$, resulting after

Algorithm 2: CP (time k, agent i)

Inputs: $p_{k|k-1,i}(\cdot)$, $l_{k,i}(\cdot|\cdot)$, $\varphi_{k+1|k}(\cdot|\cdot)$, $y_{k,i}$, $\{\pi_{i,j}\}_{j\in\mathcal{N}_i}$;

1: $p_{k|k,i,0}(x) = p_{k|k-1,i}(x) \oplus l_{k,i}(y_{k,i}|x)$ ▷ Correction
2: **for** $\ell = 1$ to L **do** ▷ Consensus on posteriors
3: $\quad p_{k|k,i,\ell}(x) = \bigoplus_{j\in\mathcal{N}_i} \pi_{i,j} \odot p_{k|k,j,\ell-1}(x)$
4: **end for**
5: $p_{k|k,i}(x) = p_{k|k,i,L}(x)$
6: $p_{k+1|k,i}(x) = \int \varphi_{k+1|k}(x|\xi)\, p_{k|k,i}(\xi)\, d\xi$ ▷ Prediction

Outputs: $p_{k|k,i}(\cdot)$, $p_{k+1|k,i}(\cdot)$

correction of the considered CL and CP filters, are reported in Table 19.1. Notice that CL involves consensus only on the novel information while CP involves consensus also on the local priors. As a consequence, the CL filter of agent i can only gather information from $\mathcal{N}_i^L \triangleq \{j \in \mathcal{N} : \pi_{i,j}^L > 0\}$ which is the subset of agents from which i can be reached in at most L hops. This means that the CL filter cannot guarantee boundedness of the estimation error unless observability (or, at least, detectability) from the subset of agents $\mathcal{N}_i^L$ holds which, in turn, requires a sufficiently high number L of consensus steps. On the other hand, CP unavoidably underweights the novel information since $\pi_{i,j}^L \le 1$, while CL can compensate this underweighting via multiplication by the factor $\rho_{k,i}$. In particular, for $L \to \infty$ and $\rho_{k,i} = |\mathcal{N}|$, CL actually tends to the optimal Bayesian filter. In order to combine benefits and counteract drawbacks of both CL and CP, the two approaches can be merged into a distributed filter that performs in parallel *CL and consensus on priors*, referred to hereafter as CLCP and detailed in Algorithm 3.

Algorithm 3: CLCP (time k, agent i)

Inputs: $p_{k|k-1,i}(\cdot)$, $l_{k,i}(\cdot|\cdot)$, $\varphi_{k+1|k}(\cdot|\cdot)$, $y_{k,i}$, $\rho_{k,i}$, $\{\pi_{i,j}\}_{j\in\mathcal{N}_i}$;

1: $p_{k|k-1,i,0}(x) = p_{k|k-1}(x)$; $l_{k,i,0}(x) = l_{k,i}(y_{k,i}|x)$; ▷ Parallel consensus on priors and likelihoods
2: **for** $\ell = 1$ to L **do**
3: $\quad p_{k|k-1,i,\ell}(x) = \bigoplus_{j\in\mathcal{N}_i} \pi_{i,j} \odot p_{k|k-1,j,\ell-1}(x)$;
4: $\quad l_{k,i,\ell}(x) = \bigoplus_{j\in\mathcal{N}_i} \pi_{i,j} \odot l_{k,j,\ell-1}(x)$;
5: **end for**
6: $p_{k|k,i}(x) = p_{k|k-1,i,L}(x) \oplus \left(\rho_{k,i} \odot l_{k,i,L}(x)\right)$; ▷ Correction
7: $p_{k+1|k,i}(x) = \int \varphi_{k+1|k}(x|\xi)\, p_{k|k,i}(\xi)\, d\xi$ ▷ Prediction

Outputs: $p_{k|k,i}(\cdot)$, $p_{k+1|k,i}(\cdot)$

Table 19.1 Posterior after correction in the considered consensus-based Bayesian filters.

Algorithm	$p_{k\|k,i}$
CL	$p_{k\|k-1,i} \oplus \left(\bigoplus_{j\in\mathcal{N}} \rho_{k,i}\pi_{i,j}^L \odot l_{k,j}\right)$
CP	$\left(\bigoplus_{j\in\mathcal{N}} \pi_{i,j}^L \odot p_{k\|k-1,j}\right) \oplus \left(\bigoplus_{j\in\mathcal{N}} \pi_{i,j}^L \odot l_{k,j}\right)$
CLCP	$\left(\bigoplus_{j\in\mathcal{N}} \pi_{i,j}^L \odot p_{k\|k-1,j}\right) \oplus \left(\bigoplus_{j\in\mathcal{N}} \rho_{k,i}\pi_{i,j}^L \odot l_{k,j}\right)$

19.8.4 Consensus-based Kalman-like Filters

The CL, CP, and CLCP filters of Algorithms 1–3 do not admit exact finite-dimensional implementation with the notable exception of the linear Gaussian case. Henceforth, Gaussian noises will be assumed, i.e., $p_{w_k}(\cdot) = \mathcal{G}(\cdot; 0, Q_k)$ and $p_{v_{k,i}}(\cdot) = \mathcal{G}(\cdot; 0, R_{k,i})$, while the functions $f_k(\cdot, \cdot)$ and $h_{k,i}(\cdot, \cdot)$ are possibly nonlinear. Exploiting linearization, according to the *extended Kalman filter* (EKF) paradigm, the measurement equation of agent i can be approximated as

$$
\begin{aligned}
y_{k,i} &\approx h_{k,i}(\hat{x}_{k|k-1,i}, 0) \; + \; C_{k,i}(x_k - \hat{x}_{k|k-1,i}) + E_{k,i} v_{k,i} \\
C_{k,i} &= \frac{\partial h_{k,i}}{\partial x}(\hat{x}_{k|k-1,i}, 0), \quad E_{k,i} = \frac{\partial h_{k,i}}{\partial v}(\hat{x}_{k|k-1,i}, 0)
\end{aligned}
\tag{19.56}
$$

from which

$$
v_{k,i} \approx E_{k,i}^{-1}(\bar{y}_{k,i} - C_{k,i} x_k)
$$

where, by invertibility of $h_{k,i}(x, v)$ with respect to v, $E_{k,i}$ is nonsingular and

$$
\bar{y}_{k,i} = y_{k,i} - h_{k,i}(\hat{x}_{k|k-1,i}, 0) + C_{k,i}\hat{x}_{k|k-1,i}. \tag{19.57}
$$

Hence, the local likelihood of agent i can be approximated by

$$
\begin{aligned}
l_{k,i}(x) &= p_{v_{k,i}}\left(E_{k,i}^{-1}(\bar{y}_{k,i} - C_{k,i}x)\right) = \mathcal{G}\left(E_{k,i}^{-1}(\bar{y}_{k,i} - C_{k,i}x); 0, R_{k,i}\right) \\
&\propto \exp\left[-\frac{1}{2}\left(\bar{y}_{k,i} - C_{k,i}x\right)^{\top}\left(E_{k,i}R_{k,i}E_{k,i}^{\top}\right)^{-1}\left(\bar{y}_{k,i} - C_{k,i}x\right)\right] \\
&\propto \exp\left[-\frac{1}{2}(x^{\top}\,\delta\Omega_{k,i}\,x - 2x^{\top}\,\delta q_{k,i})\right]
\end{aligned}
\tag{19.58}
$$

with

$$
\delta\Omega_{k,i} = C_{k,i}\left(E_{k,i}R_{k,i}E_{k,i}^{\top}\right)^{-1}C_{k,i}^{\top}, \quad \delta q_{k,i} = C_{k,i}\left(E_{k,i}R_{k,i}E_{k,i}^{\top}\right)^{-1}\bar{y}_{k,i}. \tag{19.59}
$$

From (19.58) and in view of Theorem 19.2, CL amount to consensus on the (arithmetic) average of the natural parameters $\delta q_{k,i}$ and $\delta\Omega_{k,i}$. In a similar way, Gaussian priors and posteriors propagated by EKF, or *unscented Kalman filter* (UKF), take the form:

$$
\begin{aligned}
p_{k|k-1,i}(x) &= \mathcal{G}(x; \hat{x}_{k|k-1,i}, P_{k|k-1,i}) \\
&\propto \exp\left[-\frac{1}{2}(x^{\top}\Omega_{k|k-1,i}x - 2x^{\top}q_{k|k-1,i})\right] \\
p_{k|k,i}(x) &\propto \exp\left[-\frac{1}{2}(x^{\top}\Omega_{k|k,i}x - 2x^{\top}q_{k|k,i})\right]
\end{aligned}
\tag{19.60}
$$

so that consensus on the Gaussian priors/posteriors reduces to a weighted arithmetic average of the information pairs $(q_{k|k-1,j}, \Omega_{k|k-1,j})$ / $(q_{k|k,j}, \Omega_{k|k,j})$ over $\mathcal{N}_i$.

Based on the above developments, finite-dimensional Kalman-like versions of the Bayesian CL, CP, and CLCP distributed filters can be derived; as in [7], they are referred to as *consensus on measurements* (CM), *consensus on information* (CI) and *hybrid consensus on measurements and information* (HCMCI) respectively. To save space, only the pseudo-code of the HCMCI filter is reported in Algorithm 4; pseudo-codes for CM and CI filters could be obtained in the same way, mutatis mutandis. Notice that all the considered (CM, CI, HCMCI) distributed filters involve computational complexity and communication cost per agent and filtering cycle that are cubic and, respectively, quadratic with respect to the state dimension.

It is worth to point out that the proposed consensus-based Kalman-like state estimators are in *information filter* (IF) form (propagating the information pair instead of the estimate–covariance pair) which, in the distributed case, is preferable to the covariance filter form for two reasons: (1) in IF form, the multiagent correction can be split into parallel single-agent corrections; (2) LKLA fusion of Gaussian PDFs simply reduces to arithmetic average of the information pairs.

Algorithm 4: HCMCI(time k, agent i)

Inputs: $q_{k|k-1,i}, \Omega_{k|k-1,i}, y_{k,i}, f_k(\cdot,\cdot), h_{k,i}(\cdot,\cdot), Q_k, R_{k,i}, y_{k,i}, \rho_{k,i}, \{\pi_{i,j}\}_{j\in\mathcal{N}_i}$.

Compute $C_{k,i}$ and $E_{k.i}$ via (19.56), then compute $\bar{y}_{k,i}$ via (19.57).

Compute $\delta q_{k,i}$ and $\delta\Omega_{k,i}$ via (19.59).

$\delta q_{k,i,0} = \delta q_{k,i}$; $\delta\Omega_{k,i,0} = \delta\Omega_{k,i}$; ▷ Parallel consensus on measurements and on information

$q_{k,i,0} = q_{k|k-1,i}$; $\Omega_{k,i,0} = \Omega_{k|k-1,i}$;

for $\ell = 1$ to L **do**

  $\left[\delta q_{k,i,\ell}, \delta\Omega_{k,i,\ell}\right] = \sum_{j\in\mathcal{N}_i} \pi_{i,j} \left[\delta q_{k,j,\ell-1}, \delta\Omega_{k,j,\ell-1}\right]$;

  $\left[q_{k,i,\ell}, \Omega_{k,i,\ell}\right] = \sum_{j\in\mathcal{N}_i} \pi_{i,j} \left[q_{k,j,\ell-1}, \Omega_{k,j,\ell-1}\right]$;

end for

$\left[q_{k|k,i}, \Omega_{k|k,i}\right] = \left[q_{k,i,L}, \Omega_{k,i,L}\right] + \rho_{k,i}\left[\delta q_{k,i,L}, \delta\Omega_{k,i,L}\right]$; ▷ Correction

$P_{k|k,i} = \Omega_{k|k,i}^{-1}$; $\hat{x}_{k|k,i} = P_{k|k,i}\, q_{k|k,i}$;

$\hat{x}_{k+1|k,i} = f_k(\hat{x}_{k|k,i}, 0)$; $A_{k,i} = \dfrac{\partial f_k}{\partial x}(\hat{x}_{k|k,i}, 0)$; $D_{k,i} = \dfrac{\partial f_k}{\partial w}(\hat{x}_{k|k,i}, 0)$; ▷ Prediction

$W_{k,i} = \left(D_{k,i}Q_k D_{k,i}^{\top}\right)^{-1}$; $\Omega_{k+1|k,i} = W_{k,i} - W_{k,i}A_{k,i}\left(\Omega_{k|k,i} + A_{k,i}^{\top}W_{k,i}A_{k,i}\right)^{-1} A_{k,i}^{\top}W_{k,i}$;

$q_{k+1|k,i} = \Omega_{k+1|k,i}\,\hat{x}_{k+1|k,i}$;

Outputs: $\hat{x}_{k|k,i}, P_{k|k,i}$.

19.8.5 Stability Analysis

To address stability, we consider the standard discrete-time linear state-space model

$$x_{k+1} = Ax_k + w_k, \tag{19.61}$$

$$y_{k,i} = C_i x_k + v_{k,i}, \quad i \in \mathcal{N} \tag{19.62}$$

and we assume that the process disturbance w_k and the measurement noises $v_{k,i}, i \in \mathcal{N}$ are bounded sequences. We focus on Algorithm 4, which performs consensus separately on prior and novel information, and then fuses the results locally at each node.

Theorem 19.7 (Stability of HCMCI) Let the HCMCI algorithm be applied to the system (19.61)–(19.62) with any initial estimate $\hat{x}_{1|0,i}$ and error covariance $P_{1|0,i} = P_{1|0,i}^{\top} > 0$ at each node $i \in \mathcal{N}$. Suppose that:

A1 - the matrix A is invertible;

A2 - the communication graph is quasi-strongly connected, i.e., for any two agents $i, j \in \mathcal{N}$ there is an agent $h \in \mathcal{N}$ from which both i and j are reachable;

A3 - the system is collectively observable, i.e., the pair (A, C) with $C = \mathrm{col}\{C_i\}_{i\in\mathcal{N}}$ is observable.

Then, the estimation error at each node remains uniformly bounded, i.e., there exists a constant $M > 0$ such that

$$\sup_k \| x_k - \hat{x}_{k|k,i} \|^2 \leq M, \quad \forall i \in \mathcal{N}.$$

Furthermore, in the noise-free case, i.e., if $w_k \equiv 0$ and $v_{k,i} \equiv 0$ for all k and i, the estimation error converges exponentially to zero in all network nodes.

Some comments on assumptions A1–A3 are in order.

- A1, which is just a technical assumption exploited to facilitate the proof, is anyway automatically satisfied when the discrete-time model is obtained via uniform sampling of a continuous-time linear system.
- A2 is a network connectivity condition ensuring that there are no isolated subnetworks so that information from all agents can successfully spread throughout the network.
- A3 requires that the system be observable from the whole set of agents though might be unobservable from any proper subset of agents.

The proof of Theorem 19.7 [7] relies on the construction of a collective Lyapunov function for the collective estimation error dynamics. In a probabilistic sense, a similar result holds. Specifically, if the process and measurement noise are zero-mean with bounded second moments, the estimation error remains uniformly bounded in the mean-square sense. Assumption A2 can be relaxed to time-varying communication graphs [38], provided they are *uniformly connected*, i.e., the union graph over bounded time windows is quasi-strongly connected. Further, assumption A3 can be weakened to detectability, by exploiting an observability decomposition and restricting estimation to the detectable subspace.

The stability guarantees can be extended to more general scenarios. For instance, they remain valid under *event-triggered communication* schemes [17], where each node transmits its information only when sufficiently different from the last transmitted message, thus allowing for bandwidth and energy savings without compromising stability and performance. Stability can also be preserved in the presence of *denial-of-service (DoS) attacks*, as long as the overall communication frequency is above a threshold that depends on the duration and frequency of DoS intervals [39]. In the nonlinear setting, the generalization of the HCMCI algorithm based on the EKF paradigm has been shown [40] to enjoy *local exponential stability*, provided that the system is locally uniformly observable and the Jacobians are bounded. In this case, Lyapunov analysis of the linearized error dynamics ensures that the estimation error remains bounded for sufficiently accurate initialization.

19.9 Numerical Simulation Examples

To evaluate the performance of the proposed DSE algorithms, we consider a nonlinear target tracking scenario. The target follows a nearly constant velocity motion model, with the state vector defined as $x_k = [\xi_k, \dot{\xi}_k, \eta_k, \dot{\eta}_k]^\top$, representing the Cartesian position and velocity. The sampling interval is set to $T = 1$ s, and the process noise is modeled as zero-mean Gaussian with intensity $\sigma_w = \sqrt{0.6}\,\mathrm{m}^2/\mathrm{s}^3$.

The surveillance region is a square area of size 5000 m × 5000 m. The sensor network comprises 50 nodes deployed over this region, as shown in Figure 19.5. Within this network, 10 nodes are designated as active sensors, while the remaining nodes act purely as *communication relays* (COM). Specifically, the sensing nodes include five *direction of arrival* (DOA) sensors (depicted

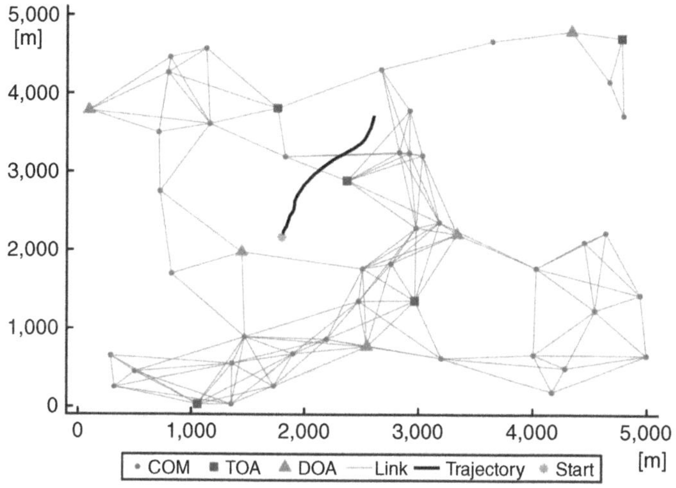

Figure 19.5 Simulation scenario: sensor network with 50 nodes (40 COM nodes, 5 TOA sensors, 5 DOA sensors), communication links, and an example target trajectory.

as triangles) and five *time of arrival* (TOA) sensors (squares). DOA sensors measure the bearing angle to the target with measurement noise standard deviation $\sigma_\theta = 1°$, while TOA sensors measure range with noise standard deviation $\sigma_r = 5$ m. For a sensor i located at (ξ_i, η_i), the DOA and TOA measurement functions are respectively:

$$h_{k,i}(x) = h_i(x) = \begin{cases} \text{atan2}(\eta - \eta_i, \xi - \xi_i), & \text{if } i \text{ is a DOA sensor,} \\ \sqrt{(\xi - \xi_i)^2 + (\eta - \eta_i)^2}, & \text{if } i \text{ is a TOA sensor.} \end{cases} \tag{19.63}$$

A representative target trajectory is illustrated in black, starting from the asterisk. Node communications use Metropolis weights (19.43) for consensus, and all estimation algorithms are initialized with the same prior mean $\hat{x}_{1|0} = [2000, 10, 2000, 5]^\top$ and covariance $P_{1|0} = \text{diag}([1500^2, 20^2, 1500^2, 20^2])$.

The performance of several distributed filters is compared: CI, two variants of *consensus on measurements* (CM-1 with $\rho_{k,i} = 1/b_{k,i,L}$ and CM-2 with $\rho_{k,i} = |\mathcal{N}|$), and two hybrid filters combining CM and CI (HCMCI-1 and HCMCI-2 using the same respective weights). These are benchmarked against the centralized EKF. The primary performance metric is the *position root mean square error* (PRMSE), averaged over 100 Monte Carlo trials of 150 time steps each. Figure 19.6 plots PRMSE versus the number of consensus steps L.

Several insights emerge from the results in Figure 19.6. The centralized EKF achieves a baseline PRMSE of approximately 4 m. Among the distributed approaches, the CI filter remains stable for all $L \geq 1$ but shows higher estimation error, starting from 29 m at $L = 1$ and gradually decreasing to 8.5 m at $L = 15$. In contrast, the CM filters (CM-1 and CM-2) show instability for a limited number of consensus steps. Specifically, CM-1 becomes stable for $L \geq 7$, while CM-2 requires $L \geq 8$ for stability. For $L \geq 8$, CM-1 consistently outperforms CM-2, achieving a PRMSE of 6 m at $L = 15$. The HCMCI filters provide the most favorable trade-off between stability and accuracy. Both HCMCI-1 and HCMCI-2 are stable for all values of $L \geq 1$. HCMCI-2, which uses $\rho_{k,i} = |\mathcal{N}|$, shows rapid improvement, reducing PRMSE from 29 m at $L = 1$ to under by $L = 3$, and nearing EKF performance (4 m) for $L \geq 13$. HCMCI-1 performs slightly better than HCMCI-2 at very low L (e.g., $L = 2$), but HCMCI-2 consistently dominates as L increases.

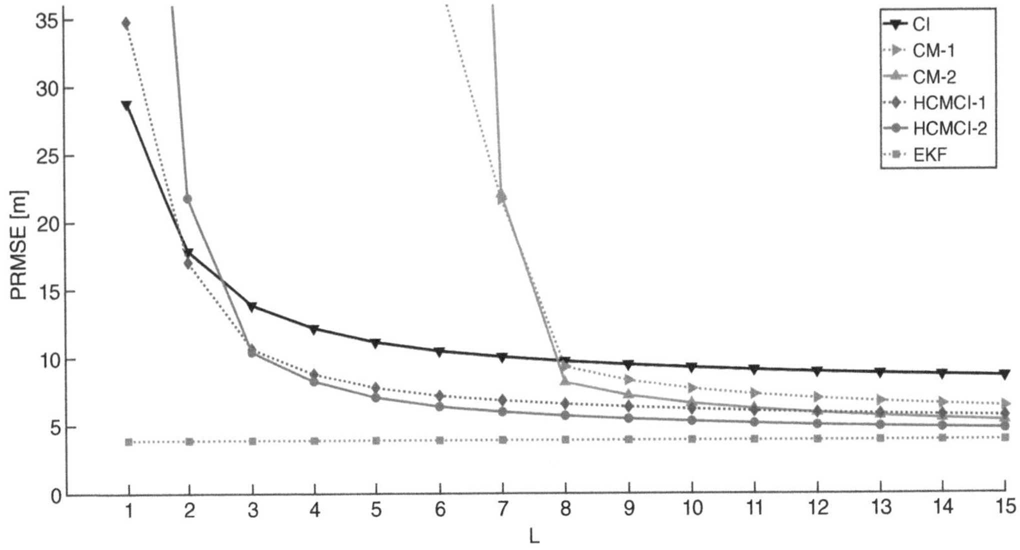

Figure 19.6 PRMSE for different consensus filters in the nonlinear scenario versus number L of consensus steps.

These simulation results underscore the benefits of the hybrid approach. Both HCMCI variants ensure stability even with minimal communication ($L = 1$), a critical advantage over CM. While HCMCI-1 can offer a slight edge for very few consensus steps (e.g., $L = 2$ in this specific setup), HCMCI-2 demonstrates superior convergence toward the centralized EKF performance as L increases, making it a compelling choice when a moderate number of consensus iterations are feasible. The choice of weighting $\rho_{k,i}$ in HCMCI thus presents a trade-off dependent on the communication budget L.

19.10 Conclusions

This chapter has provided an overview of the work on DSE carried out by the authors over the last decade. The goal was to allow multiple communicating, sensing, and processing agents to cooperatively estimate the state of a dynamical system in a distributed and scalable way, trying to get performance as close as possible to the one achievable by a centralized estimator having access to all sensor measurements, while counteracting the deleterious effects of data incest. The developed work concerns a multiagent networked system wherein:

- each agent is unaware of the overall network topology;
- there is no fusion center gathering data from all agents;
- all agents operate as peers running synchronously the same algorithm with computational and memory complexities independent of the network size.

The proposed solution to DSE follows a Bayesian approach wherein each agent propagates a PDF of the state to be estimated. Fusion among agent PDFs is cast as an optimization problem consisting of minimization of a weighted sum of discrepancy measures between the agent PDFs and the fused one. Special attention is devoted to the adoption of the KLD as discrepancy measure in the fusion criterion; in particular, it is shown how the fused PDF turns out to be the (weighted) geometric or arithmetic average of the agent PDFs according to whether such PDF is taken as left or, respectively, right argument of the KLD. It is also shown that exponential families are closed under geometric average, aka *left Kullback–Leibler average* (LKLA). Further, the consensus protocol is exploited to perform LKLA fusion in a distributed way over the multiagent network ensuring immunity to double counting of information. Finally, DSE algorithms based on three different types of consensus—specifically *consensus on measurements* (CM), *consensus on information* (CI) and *hybrid consensus on measurements and on information* (HCMCI)—are presented, their stability properties are discussed and a comparative evaluation via simulations is provided.

Beyond what presented in this chapter, research on DSE has proceeded in several directions, including: DSE with event-triggered communication [16–18] to save communication bandwidth and energy consumption; joint DSE and sensor registration [41]; DSE resilient to cyberattacks and sensor faults [20, 42]; moving-horizon DSE [43]; multiple-model DSE [44]; distributed field estimation [45]; adaptive DSE [46, 47] to jointly estimate state and unknown noise covariances; outlier mitigation in DSE [48]; stability of DSE under denial of service [39]; DSE over asynchronous multiagent networks [49]; distributed fusion with global and local variables [50]; machine learning-based DSE [51].

19.A Proof of Theorem 19.1

Let $\tilde{f} = c\overline{f}$ with $\tilde{f} = \prod_i f_i^{\pi_i}$ and $c = \int \prod_i f_i(x)^{\pi_i} dx$. Then,

$$
\begin{aligned}
&\sum_i \pi_i D_{KL}(f \parallel f_i) = \sum_i \pi_i \int f(x) \log \frac{f(x)}{f_i(x)} dx \\
=& \int f(x) \sum_i \pi_i \log \frac{f(x)}{f_i(x)} dx = \int f(x) \sum_i \log \frac{f(x)^{\pi_i}}{f_i(x)^{\pi_i}} dx \\
=& \int f(x) \log \frac{\prod_i f(x)^{\pi_i}}{\prod_i f_i(x)^{\pi_i}} dx = \int f(x) \log \frac{f(x)}{\tilde{f}(x)} dx = \int f(x) \log \frac{f(x)}{c\overline{f}(x)} dx \\
=& \int f(x) \log \frac{f(x)}{\overline{f}(x)} dx - \log c \int f(x) dx = D_{KL}(f \parallel \overline{f}) - \log c
\end{aligned}
$$

is minimized by taking $f = \overline{f}$ as stated. ■

19.B Proof of Theorem 19.2

First notice from (19.16) that the normalization condition $\int f(x, \theta)\, dx = 1$ implies

$$
\exp F(\theta) = \int \exp[\theta^\top t(x) + k(x)]\, dx. \tag{19.64}
$$

Next, the numerator of (19.15) turns out to be

$$
\begin{aligned}
\tilde{f}(x) &= \prod_i f(x, \theta_i)^{\pi_i} = \prod_i \{\exp[\theta_i^\top t(x) - F(\theta_i) + k(x)]\}^{\pi_i} \\
&= \exp\left[\left(\sum_i \pi_i \theta_i\right)^\top t(x)\right] \exp\left[-\sum_i \pi_i F(\theta_i)\right] \exp\,[k(x)]^{\sum_i \pi_i} \\
&= \exp[\bar{\theta}^\top t(x)]\, \exp\left[-\sum_i \pi_i F(\theta_i)\right] \exp[k(x)]
\end{aligned} \tag{19.65}
$$

where we have defined $\bar{\theta}$ as in (19.17) and we have exploited (19.9). Then, from (19.65), the normalization constant in the denominator of (19.15) is given by

$$c = \int \tilde{f}(x)\,dx = \exp\left[-\sum_i \pi_i F(\theta_i)\right] \int \exp[\bar{\theta}^\top t(x) + k(x)]\,dx. \tag{19.66}$$

Combining (19.65) and (19.66) yields

$$\bar{f}(x) = \frac{\tilde{f}(x)}{c} = \frac{\exp[\bar{\theta}^\top t(x) + k(x)]}{\int \exp[\bar{\theta}^\top t(x) + k(x)]\,dx} = \exp[\bar{\theta}^\top t(x) - F(\bar{\theta}) + k(x)] \triangleq f(x, \bar{\theta}) \tag{19.67}$$

where we have used (19.64) for $\theta = \bar{\theta}$. ■

19.C Proof of Theorem 19.3

Let $\bar{f} = \sum_i \pi_i f_i$, then

$$\sum_i \pi_i D_{KL}(f_i \parallel f) = \sum_i \pi_i \int f_i \log \frac{f_i}{f} dx$$
$$= \sum_i \pi_i \int f_i \log f_i dx - \sum_i \pi_i \int f_i \log f dx$$
$$+ \int (\sum_i \pi_i f_i) \log\left(\sum_i \pi_i f_i\right) dx - \int \left(\sum_i \pi_i f_i\right) \log\left(\sum_i \pi_i f_i\right) dx$$
$$= \sum_i \pi_i \int f_i \log f_i\,dx - \int \left(\sum_i \pi_i f_i\right) \log f\,dx + \int \overline{f} \log \overline{f}\,dx$$
$$- \sum_i \pi_i \int f_i \log \overline{f}\,dx$$
$$= \sum_i \pi_i \int f_i \log \frac{f_i}{\overline{f}}\,dx - \int \overline{f} \log f dx + \int \overline{f} \log \overline{f}\,dx$$
$$= \sum_i \pi_i D_{KL}(f_i \parallel \overline{f}) + D_{KL}(\overline{f} \parallel f)$$

is minimized by taking $f = \overline{f}$ as stated. ■

19.D Proof of Theorem 19.4

From the analytical formula for the KLD of Gaussian PDFs
with , the function to be minimized with respect to m and Σ is:

$$J(m, \Sigma) = \sum_i \pi_i \left[\text{tr}(\Sigma^{-1}\Sigma_i) + (m - m_i)^\top \Sigma^{-1}(m - m_i) + \log\ \det\ \Sigma\right].$$

Then, $\frac{\partial J}{\partial m} = 2\,\Sigma^{-1} \sum_i \pi_i (m - m_i) = 0$ implies $\overline{m} = \sum_i \pi_i m_i$. Further,

$$J(m, \Sigma) = \log\det\Sigma + \text{tr}\left\{\Sigma^{-1} \sum_i \pi_i \left[\Sigma_i + (m - m_i)(m - m_i)^\top\right]\right\}.$$

Using the scalar-by-matrix derivatives (see Wikipedia—Matrix calculus) $\dfrac{\partial \log \det \Sigma}{\partial \Sigma} = \Sigma^{-1}$ and $\dfrac{\partial \operatorname{tr}(\Sigma^{-1}A)}{\partial \Sigma} = -\Sigma^{-1}A\Sigma^{-1}$, we have

$$
\begin{aligned}
\frac{\partial J}{\partial \Sigma} &= \Sigma^{-1} - \Sigma^{-1}\left\{\sum_i \pi_i\left[\Sigma_i + (m-m_i)(m-m_i)^{\top}\right]\right\}\Sigma^{-1} \\
&= \Sigma^{-1}\left\{\Sigma - \sum_i \pi_i\left[\Sigma_i + (m-m_i)(m-m_i)^{\top}\right]\right\}\Sigma^{-1} = 0 \\
&\Rightarrow \overline{\Sigma} = \sum_i \pi_i\left[\Sigma_i + (m_i-\overline{m})(m_i-\overline{m})^{\top}\right]
\end{aligned}
$$

thus concluding the proof. ■

19.E Proof of Theorem 19.5

First notice that, by (19.39), Π is row-stochastic, i.e., its rows sum up to 1. Then, for a row-stochastic primitive matrix Π, the Perron–Frobenius theorem implies that:

1. Π has a unit eigenvalue $\lambda_1 = 1$ with multiplicity 1;
2. the all-one vector $\mathbb{1}$ is an eigenvector of Π relative to $\lambda_1 = 1$, i.e., $\Pi\,\mathbb{1} = \mathbb{1}$;
3. all other eigenvalues λ_i of Π are such that $|\lambda_i| < 1$.

Then, Π can be decomposed as

$$
\Pi = \underbrace{\left[v_1\ \widetilde{V}\right]}_{V} \begin{bmatrix} 1 & 0 \\ 0 & J \end{bmatrix} \underbrace{\begin{bmatrix} w_1 \\ \widetilde{W} \end{bmatrix}}_{W=V^{-1}} \tag{19.68}
$$

where: $v_1 = \alpha\,\mathbb{1}$ is the (right) eigenvector relative to the eigenvalue 1 and w_1 the corresponding left eigenvector; J is a Jordán matrix relative to the other eigenvalues λ_i, $i > 1$, such that $|\lambda_i| < 1$; $\widetilde{V}$ is a matrix of generalized eigenvectors spanning the eigenspace relative to all other eigenvalues but $\lambda_1 = 1$. Hence,

$$
\begin{aligned}
\Pi^{\ell} &= \left(V\begin{bmatrix} 1 & 0 \\ 0 & J \end{bmatrix}V^{-1}\right)^{\ell} = V\begin{bmatrix} 1 & 0 \\ 0 & J^{\ell} \end{bmatrix}V^{-1} \\
&= \left[v_1\ \widetilde{V}\right]\begin{bmatrix} 1 & 0 \\ 0 & J^{\ell} \end{bmatrix}\begin{bmatrix} w_1 \\ \widetilde{W} \end{bmatrix} \\
&= v_1 w_1 + \widetilde{V} J^{\ell}\,\widetilde{W}
\end{aligned}
$$

from which, since $J^{\ell} \to 0$ as $\ell \to \infty$, it turns out that

$$
\lim_{\ell\to\infty} \Pi^{\ell} = v_1\, w_1 = \alpha\,\mathbb{1}\, w_1 = \begin{bmatrix} \bar{w}_1 \\ \vdots \\ \bar{w}_1 \end{bmatrix} \stackrel{\triangle}{=} \Pi_{\infty} \tag{19.69}
$$

with $\bar{\omega}_1 = \alpha \omega_1$. Since Π is row-stochastic, Π^ℓ is row-stochastic for any $\ell \geq 1$ and, therefore, $\Pi_\infty = \lim_{\ell\to\infty} \Pi^\ell$ is row-stochastic, thus implying that $\bar{\omega}_1 = \omega_1(\omega_1 \, \mathbb{1})^{-1}$ is the normalized left eigenvector relative to the unit eigenvalue of Π. Let $\bar{\omega}_1 = \text{row}\{\alpha_j\}_{j\in\mathcal{N}}$, then

$$\lim_{\ell\to\infty} \pi_{i,j}^\ell = \alpha_j, \quad \forall i,j \in \mathcal{N}$$

which, applying $\lim_{\ell\to\infty}$ to both sides of (19.40), yields (19.41).

If the consensus matrix Π is doubly stochastic, then $\omega_1 = \beta \, \mathbb{1}^\top$ for some β. Then, $\Pi_\infty = \alpha\beta \, \mathbb{1}\mathbb{1}^\top$ which, by row-stochasticity of Π_∞, implies $\alpha\beta|\mathcal{N}| = 1$ which, in turn, yields

$$\alpha\beta = \frac{1}{|\mathcal{N}|} \Rightarrow \Pi_\infty = \frac{1}{|\mathcal{N}|} \mathbb{1}\mathbb{1}^\top \Rightarrow \lim_{\ell\to\infty} \pi_{i,j}^\ell = \frac{1}{|\mathcal{N}|}, \quad \forall i,j \in \mathcal{N}.$$

In this case, applying $\lim_{\ell\to\infty}$ to both sides of (19.40), one gets (19.42). ■

19.F Proof of Theorem 19.6

Exploiting the properties (19.36), (19.37) yields

$$\begin{aligned}
\bar{f}(x) &= \bigoplus_{i\in\mathcal{N}} [\pi_i \odot f_i(x)] = \bigoplus_{i\in\mathcal{N}} \left[\pi_i \odot \left(\bigoplus_{\mathcal{S}\in\mathcal{F}(\mathcal{N}):i\in\mathcal{S}} f(x|Y_\mathcal{S}) \right) \right] \\
&= \bigoplus_{i\in\mathcal{N}} \left[\bigoplus_{\mathcal{S}\in\mathcal{F}(\mathcal{N}):i\in\mathcal{S}} (\pi_i \odot f(x|Y_\mathcal{S})) \right] = \bigoplus_{\mathcal{S}\in\mathcal{F}(\mathcal{N})} \left[\bigoplus_{i\in\mathcal{S}} (\pi_i \odot f(x|Y_\mathcal{S})) \right] \\
&= \bigoplus_{\mathcal{S}\in\mathcal{F}(\mathcal{N})} \left[\left(\sum_{i\in\mathcal{S}} \pi_i \right) \odot f(x|Y_\mathcal{S}) \right]
\end{aligned}$$

thus proving (19.48). ■

References

1 J. Cortes, S. Martinez, T. Karatas, and F. Bullo. Coverage control for mobile sensing networks. *IEEE Transactions on Robotics and Automation*, 20(2):243–255, 2004.

2 M. Dunbabin and L. Marques. Robots for environmental monitoring: Significant advancements and applications. *IEEE Robotics & Automation Magazine*, 19(1):24–39, 2012.

3 C. Kwon and I. Hwang. Sensing-based distributed state estimation for cooperative multiagent systems. *IEEE Transactions on Automatic Control*, 64(6):2368–2382, 2019.

4 R. Olfati-Saber, J. A. Fax, and R. M. Murray. Consensus and cooperation in networked multi-agent systems. *Proceedings of the IEEE*, 95(1):215–233, 2007.

5 W. Ren, R. W. Beard, and E. M. Atkins. Information consensus in multivehicle cooperative control. *IEEE Control Systems Magazine*, 27(2):71–82, 2007.

6 G. Battistelli and L. Chisci. Kullback–Leibler average, consensus on probability densities, and distributed state estimation with guaranteed stability. *Automatica*, 50(3):707–718, 2014.

7 G. Battistelli, L. Chisci, G. Mugnai, A. Farina, and A. Graziano. Consensus-based linear and nonlinear filtering. *IEEE Transactions on Automatic Control*, 60(5):1410–1415, 2015.

8 G. C. Calafiore and F. Abrate. Distributed linear estimation over sensor networks. *International Journal of Control*, 82(5):868–882, 2009.

9 F. S. Cattivelli and A. H. Sayed. Diffusion strategies for distributed Kalman filtering and smoothing. *IEEE Transactions on Automatic Control*, 55(9):2069–2084, 2010.

10 A. T. Kamal, J. A. Farrell, and A. K. Roy-Chowdhury. Information weighted consensus filters and their application in distributed camera networks. *IEEE Transactions on Automatic Control*, 58(12):3112–3125, 2013.

11 S. S. Stanković, M. S. Stanković, and D. M. Stipanović. Consensus based overlapping decentralized estimation with missing observations and communication faults. *Automatica*, 45(6):1397–1406, 2009.

12 L. Xiao, S. Boyd, and S. Lall. A scheme for robust distributed sensor fusion based on average consensus. In *Proceedings of the 4th International Symposium on Information Processing in Sensor Networks*, pages 63–70, 2005.

13 G. Koliander, Y. El-Laham, P. M. Djurić, and F. Hlawatsch. Fusion of probability density functions. *Proceedings of the IEEE*, 110(4):404–453, 2022.

14 B.-T. Vo and B.-N. Vo. Labeled random finite sets and multi-object conjugate priors. *IEEE Transactions on Signal Processing*, 61(13):3460–3475, 2013.

15 S.-H. Cha. Comprehensive survey on distance/similarity measures between probability density functions. *International Journal of Mathematical Models and Methods in Applied Sciences*, 1(4):300–307, 2007.

16 G. Battistelli, L. Chisci, and D. Selvi. Distributed averaging of exponential-class densities with discrete-time event-triggered consensus. *IEEE Transactions on Control of Network Systems*, 5(1):359–369, 2018.

17 G. Battistelli, L. Chisci, and D. Selvi. A distributed Kalman filter with event-triggered communication and guaranteed stability. *Automatica*, 93:75–82, 2018.

18 X. Dong, G. Battistelli, L. Chisci, and Y. Cai. An event-triggered hybrid consensus filter for distributed sensor network. *IEEE Signal Processing Letters*, 29:1472–1476, 2022.

19 L. Gao, G. Battistelli, L. Chisci, and P. Wei. Distributed joint sensor registration and multitarget tracking via sensor network. *IEEE Transactions on Aerospace and Electronic Systems*, 56(2):1301–1317, 2020.

20 Y. Zhou, G. Battistelli, L. Chisci, L. Gao, G. Li, and P. Wei. Median-based resilient multi-object fusion with application to LMB densities. *IEEE Transactions on Signal and Information Processing over Networks*, 10:473–486, 2024.

21 H. Akaike. Information theory and an extension of the maximum likelihood principle. In *Proceedings of the Second International Symposium on Information Theory*, pages 267–281, 1973.

22 L. L. Campbell. Equivalence of Gauss's principle and minimum discrimination information estimation of probabilities. *The Annals of Mathematical Statistics*, 41(3):1011–1015, 1970.

23 M. B. Hurley. An information theoretic justification for covariance intersection and its generalization. In *Proceedings of the Fifth International Conference on Information Fusion*, 1:505–511, 2002.

24 F. Nielsen and V. Garcia. Statistical exponential families: A digest with flash cards. *arXiv preprint arXiv:0911.4863*, 2011.

25 S. J. Julier and J. K. Uhlmann. A non-divergent estimation algorithm in the presence of unknown correlations. In *Proceedings of the 1997 American Control Conference*, volume 4, pages 2369–2373, 1997.

26 R. P. S. Mahler. Optimal/robust distributed data fusion: A unified approach. In *Signal Processing, Sensor Fusion, and Target Recognition IX*, volume 4052, International Society for Optics and Photonics, SPIE, pages 128–138, 2000.

27 G. Battistelli, L. Chisci, C. Fantacci, A. Farina, and A. Graziano. Consensus CPHD filter for distributed multitarget tracking. *IEEE Journal of Selected Topics in Signal Processing*, 7(3):508–520, 2013.

28 M. Gunay, U. Orguner, and M. Demirekler. Chernoff fusion of Gaussian mixtures based on sigma-point approximation. *IEEE Transactions on Aerospace and Electronic Systems*, 52(6):2732–2746, 2016.

29 A. P. Dempster, N. M. Laird, and D. B. Rubin. Maximum likelihood from incomplete data via the EM algorithm. *Journal of the Royal Statistical Society*, 39(1):1–38, 1977.

30 L. Gao, G. Battistelli, and L. Chisci. Fusion of labeled RFS densities with minimum information loss. *IEEE Transactions on Signal Processing*, 68:5855–5868, 2020.

31 L. Gao, G. Battistelli, and L. Chisci. Multiobject fusion with minimum information loss. *IEEE Signal Processing Letters*, 27:201–205, 2020.

32 B.-N. Vo and W.-K. Ma. The Gaussian mixture probability hypothesis density filter. *IEEE Transactions on Signal Processing*, 54(11):4091–4104, 2006.

33 L. Zheng, G. Battistelli, L. Chisci, F. Yang, and L. Shi. Distributed state estimation for heterogeneous sensor networks. *Automatica*, 169:111839, 2024.

34 L. M. De Carvalho, D. A. M. Villela, F. C. Coelho, and L. S. Bastos. Combining probability distributions: Extending the logarithmic pooling approach. *arXiv preprint arXiv:1502.04206*, 2020.

35 M. Üney, J. Houssineau, E. Delande, S. J. Julier, and D. E. Clark. Fusion of finite-set distributions: Pointwise consistency and global cardinality. *IEEE Transactions on Aerospace and Electronic Systems*, 55(6):2759–2773, 2019.

36 S. Kar and J. M. F. Moura. Gossip and distributed Kalman filtering: Weak consensus under weak detectability. *IEEE Transactions on Signal Processing*, 59(4):1766–1784, 2011.

37 G. Battistelli, L. Chisci, C. Fantacci, A. Farina, and R. P. S. Mahler. Distributed fusion of multitarget densities and consensus PHD/CPHD filters. In *Signal Processing, Sensor/Information Fusion, and Target Recognition XXIV*, volume 9474, International Society for Optics and Photonics, SPIE, page 94740E, 2015.

38 S. Wang and W. Ren. On the convergence conditions of distributed dynamic state estimation using sensor networks: A unified framework. *IEEE Transactions on Control Systems Technology*, 26(4):1300–1316, 2018.

39 G. Battistelli, L. Chisci, D. Selvi, and P. Tesi. Stability of consensus-based distributed estimation under denial of service. *IEEE Transactions on Automatic Control*, 69(3):2036–2043, 2024.

40 G. Battistelli and L. Chisci. Stability of consensus extended Kalman filter for distributed state estimation. *Automatica*, 68:169–178, 2016.

41 L. Gao, G. Battistelli, L. Chisci, and P. Wei. Distributed joint sensor registration and target tracking via sensor network. *Information Fusion*, 46:218–230, 2019.

42 N. Forti, G. Battistelli, L. Chisci, S. Li, B. Wang, and B. Sinopoli. Distributed joint attack detection and secure state estimation. *IEEE Transactions on Signal and Information Processing over Networks*, 4(1):96–110, 2018.

43 G. Battistelli. Distributed moving-horizon estimation with arrival-cost consensus. *IEEE Transactions on Automatic Control*, 64(8):3316–3323, 2019.

44 G. Battistelli, L. Chisci, C. Fantacci, A. Farina, and A. Graziano. Consensus-based multiple-model Bayesian filtering for distributed tracking. *IET Radar, Sonar & Navigation*, 9(4):401–410, 2015.

45 G. Battistelli, L. Chisci, N. Forti, G. Pelosi, and S. Selleri. Distributed finite-element Kalman filter for field estimation. *IEEE Transactions on Automatic Control*, 62(7):3309–3322, 2017.

46 X. Dong, G. Battistelli, L. Chisci, and Y. Cai. An adaptive consensus filter for distributed state estimation with unknown noise statistics. *IEEE Signal Processing Letters*, 28:1595–1599, 2021.

47 X. Dong, G. Battistelli, L. Chisci, and Y. Cai. Consensus variational Bayesian moving horizon estimation for distributed sensor networks with unknown noise covariances. *Signal Processing*, 198:108571, 2022.

48 Z. Jiang, G. Battistelli, L. Chisci, N. Forti, and W. Zhou. Outlier-robust centralized and distributed variational Bayesian moving horizon estimation. *IEEE Transactions on Signal Processing*, 73:2527–2541, 2025.

49 G. Li, G.Battistelli, and L. Chisci. Target tracking in asynchronous sensor networks under temporal misalignment. *IEEE Transactions on Signal Processing*, 74:1064–1078, 2020.

50 G. Battistelli, L. Chisci, and N. Forti. Distributed fusion with global and local variables. *IEEE Transactions on Control of Network Systems*, 2026. doi: 10.1109/TCNS.2026.3691472

51 F. M. Zegers, R. Sun, G. Chowdhary, and W. E. Dixon. Distributed state estimation with deep neural networks for uncertain nonlinear systems under event-triggered communication. *IEEE Transactions on Automatic Control*, 68(5):3107–3114, 2023.

20

Fusion of Multiobject Densities

Giorgio Battistelli[1], *Luigi Chisci*[1], *Lin Gao*[2], *Amirali K. Gostar*[3], *and Reza Hoseinnezhad*[3]

[1]*Dipartimento di Ingegneria dell'Informazione (DINFO), Università degli Studi di Firenze, Firenze, Italy*
[2]*School of Information and Communication Engineering University of Electronic Science and Technology of China, Chengdu, China*
[3]*School of Engineering, RMIT University, Carlton, Victoria, Australia*

20.1 Introduction

In several practical contexts (e.g., autonomous driving, air, or maritime surveillance), the need arises to detect the presence of multiple objects in a given monitored region and estimate the state of each detected object. This task, referred to in this chapter as *multiobject estimation*, has several relevant applications like, e.g., *multitarget tracking* [1–3], *simultaneous localization and mapping* (SLAM) [4], *multisource estimation*. Notice that in multiobject estimation, both the number of objects and the state of each object need to be estimated. In this respect, while single-object estimation addressed in Chapter 19 concerns estimation of a random vector, multiobject estimation can be regarded as estimation of a *random finite set* (RFS) [5] in the state space, where both cardinality and elements (states) of the set need to be inferred from the available observations.

Advances in communication technologies are nowadays promoting the ever-growing diffusion of spatially distributed systems wherein multiple cooperating agents equipped with sensors are spread over the area of interest, thus allowing to expand system coverage and, thanks to the availability of different multiple views of the scene, also to improve situational awareness. Adopting a Bayesian approach to estimation and the RFS paradigm for multiobject representation, each agent summarizes its knowledge about the monitored scene in terms of a *multiobject density* (MOD) which characterizes the probability distribution of the multiobject RFS. Hence, a task of fundamental importance in multiagent multiobject estimation is fusion of MODs from the agents.

This chapter aims to provide an overview of effective criteria for MOD fusion. Particular attention will be devoted to the criterion of selecting the fused MOD as the one that minimizes a weighted average of *Kullback–Leibler divergences* (KLDs) with respect to the local MODs. In this respect, the reversal of the two arguments (local and fused MOD) in the unsymmetrical KLD yields two completely different fusion rules, a multiplicative geometric average or an additive arithmetic average according to whether the fused MOD is the left or right argument. As a result, the two fusion rules have complementary features. On one hand, the geometric average: (1) ensures conjugacy for the most commonly used (unlabeled and labeled) MODs; (2) is resilient to false information of some agents; (3) is highly sensitive to missing information in some agents. On the other hand, the arithmetic average: (1) does not ensure conjugacy, which must be enforced at the price of suboptimality; (2) is sensitive to false information; (3) is inclusive of complementary information from all agents, though with some weight reduction. The chapter will also focus on another fusion strategy, referred to as *complementary fusion*, which, thanks to its *union-centric* nature, is especially well suited to aggregate information from agents with limited and possibly nonoverlapping fields-of-view (FoVs). The focus in this chapter is on multiobject fusion criteria, which are introduced based on centralized fusion scenarios for ease of presentation. Extensions to distributed fusion scenarios can be easily accomplished exploiting, e.g., the consensus strategy as shown in Chapter 19 and [6, 7].

Consistent with the above work plan, the rest of this chapter is organized as follows. Section 21.2 provides some background on MODs with special emphasis on some RFS models (Poisson point process (PPP), i.i.d. cluster process (IIDCP), labeled multi-Bernoulli (LMB), δ-generalized labeled multi-Bernoulli, marginalized δ-generalized labeled multi-Bernoulli) that are commonly employed in multiobject estimation. Next, Section 21.3 introduces a general information-theoretic approach to MOD fusion. Sections 21.4 and 21.5 deal with two specific information-theoretic fusion rules based on the KLD, providing fusion formulas for all the considered MOD types. Section 21.6 provides a comparative evaluation of the two rules in multitarget tracking with different types of multiobject filters. Section 21.7 presents a complementary fusion approach devised for agents with limited and nonoverlapping FoVs. Finally, Section 21.8 summarizes the chapter and also outlines further work on the fusion of MODs.

20.2 Background on MODs

In this chapter, we denote by:

$$X \triangleq \{x_1, \dots , x_N\}$$

an (unlabeled) RFS. Similarly, we denote by:

$$\mathbf{X} \triangleq \{(x_1, l_1), \dots , (x_N, l_N)\}$$

a labeled RFS (LRFS) [8], wherein each element is attached to a unique label. Here we also define the label set of $\mathbf{X}$ as $L \triangleq \{l_1, \dots , l_N\}$ and the mapping from $\mathbf{X}$ to L as $\mathcal{L}(\mathbf{X}) = L$.

An MOD $f(\cdot)$ completely characterizes the probability distribution of a (labeled or unlabeled) RFS. Basic mathematical requirements for $f(X)$ are that $f(X) \geq 0$ for any X and $\int f(X)\delta X = 1$, where $\int \cdot \delta X$ denotes the set integral [9]; such constraints must also hold for an LRFS.

Hereafter, the attention focuses on some special types of RFS models commonly used for multiobject estimation. Specifically, the PPP and IIDCP) [10] models—on which PHD (probability hypothesis density) and, respectively, CPHD (cardinalized PHD) filters rely—are considered for unlabeled RFSs. Conversely, for LRFSs, we focus on LMB [8, 11] and *δ-generalized LMB* (δ-GLMB) [8, 12] models on which the respective LMB/δ-GLMB filters are based.

20.2.1 Poisson Point Process

A PPP X models a finite set of independent objects with a Poisson distributed number of objects. Its MOD is accordingly given by:

$$f(X) = \exp(-\lambda) \prod_{x \in X} \lambda p(x), \tag{20.1}$$

where λ denotes the intensity (i.e., expected object number) and $p(\cdot)$ the spatial PDF defined over the single-object state space. Resorting to the definition of PHD $D(x) = \lambda\, p(x)$, the MOD can be equivalently reexpressed in terms of $D(\cdot)$ only as:

$$f(X) = \exp\left[-\int D(x)dx\right] \prod_{x \in X} D(x). \tag{20.2}$$

20.2.2 I.I.D. Cluster Process

An IIDCP X models a finite set of independent objects with generic cardinality distribution, i.e., its MOD is given by:

$$f(X) = |X|!\, \rho(|X|) \prod_{x \in X} p(x), \tag{20.3}$$

where $\rho(\cdot)$ is the *probability mass function* (PMF) of the cardinality of X. Clearly, when $\rho(\cdot)$ is specified by the Poisson distribution, the IIDCP reduces to a PPP.

20.2.3 Labeled Multi-Bernoulli RFS

An LMB $\mathbf{X}$ models a set of independent labeled objects, each existing or not according to some existence probability (EP). The corresponding LMB MOD takes the form:

$$f(\mathbf{X}) = \prod_{l \in \mathbb{L}} (1 - r_l) \prod_{l \in \mathcal{L}(\mathbf{X})} \frac{r_l p_l(x)}{1 - r_l} \tag{20.4}$$

where: $\mathbb{L}$ denotes the label space; r_l and p_l denote the EP and *spatial PDF* of Bernoulli component with label l. Consequently, an LMB density is uniquely characterized by the EPs and spatial PDFs of all Bernoulli components. Hence, an LMB density is also denoted as:

$$f(\mathbf{X}) = \{r_l, p_l\}_{l \in \mathbb{L}}. \tag{20.5}$$

20.2.4 δ-GLMB and Marginalized δ-GLMB RFSs

A δ-GLMB RFS $\mathbf{X}$ models a multiobject as a set of weighted hypotheses, each consisting of a subset of labels and corresponding spatial PDFs. Accordingly, its MOD is specified by:

$$f(\mathbf{X}) = \Delta_{|\mathcal{L}(\mathbf{X})|}(|\mathbf{X}|) \sum_{(I,\xi) \in \mathcal{F}(\mathbb{L}) \times \Xi} w_{I,\xi}\, \Delta_I(\mathcal{L}(\mathbf{X})) \prod_{(x,l) \in \mathbf{X}} p_{l,I,\xi}(x), \tag{20.6}$$

where: $\Delta_X(\dot{X})$ is such that $\Delta_X(\dot{X}) = 1$ only if $X = \dot{X}$ and $\Delta_X(\dot{X}) = 0$ otherwise; Ξ is a discrete index set (whose elements represent object-to-measurement association hypotheses in the context of multitarget tracking); $\mathcal{F}(\cdot)$ denotes the power set; $w_{I,\xi}$ denotes the weight of label set I under hypothesis ξ satisfying $\sum_{I \subseteq \mathbb{L}} \sum_{\xi \in \Xi} w_{I,\xi} = 1$; $p_{l,I,\xi}$ represents the spatial PDF of object l conditional on I and ξ.

The marginalized δ-GLMB (Mδ-GLMB) density [13, 14] is obtained by performing marginalization of (20.6) with respect to the discrete index set Ξ. The resulting MOD takes the form:

$$f(\mathbf{X}) = \Delta_{|L(\mathbf{X})|}(|\mathbf{X}|)\, w_I\, \Delta_I(L(\mathbf{X})) \prod_{(x,l) \in \mathbf{X}} p_{l,I}(x), \tag{20.7}$$

where

$$w_I = \sum_{(I,\xi)\in\mathcal{F}(\mathbb{L})\times\Xi} w_{I,\xi}, \tag{20.8}$$

$$p_{l,I}(x) = \sum_{(I,\xi)\in\mathcal{F}(\mathbb{L})\times\Xi} p_{l,I,\xi}(x). \tag{20.9}$$

Notice that the Mδ-GLMB density is uniquely characterized by the weights w_I and spatial PDFs $p_{l,I}$ of all hypotheses. Hence, we also denote an Mδ-GLMB density as:

$$f(\mathbf{X}) = \{w_I, p_{l,I}\}_{I\subseteq\mathbb{L}, l\in I}. \tag{20.10}$$

20.3 Information-theoretic Criteria for MOD Fusion

This section addresses the fusion of MODs from an information-theoretic perspective. Let $\mathcal{N}$ denote the set of agents, each providing a posterior $f_i(\cdot)$ by a suitable local multiobject filter. For instance, a PHD filter [15] is used to produce $f_i(X)$ if the object RFS is modeled as unlabeled PPP, or an LMB filter [11] is used to produce $f_i(\mathbf{X})$ if the object set is modeled as LMB.

20.3.1 Optimization-based Fusion

According to the *principle of minimum discrimination information* (PMDI), fusion of densities can be obtained by solving the following optimization problem:

$$\overline{f} = \arg\min_f \sum_{i\in\mathcal{N}} \pi_i\, d(f, f_i), \tag{20.11}$$

where: $\pi_i \geq 0$ are fusion weights satisfying $\sum_{i\in\mathcal{N}} \pi_i = 1$; $d(f, g)$ represents a suitable measure of discrepancy between f and g, e.g., the KLD or *Cauchy–Schwarz divergence* (CSD), where each specific measure of discrepancy leads to a different fusion criterion.

20.3.2 Conjugacy Guarantee

An important requirement in multiobject fusion is that the fused MOD $\overline{f}$ be of the same type of the local ones f_i ($i \in \mathcal{N}$), so that such fused MOD can be fed back to the agents and employed as prior in the subsequent filter recursion. To this end, (20.11) should be modified into the following constrained optimization:

$$\begin{aligned} \overline{f} &= \arg\min_f \sum_{i\in\mathcal{N}} \pi_i\, d(f, f_i), \\ \text{such that} \quad & f \text{ is of the same type of } \{f_i\}_{i\in\mathcal{N}}. \end{aligned} \tag{20.12}$$

A more convenient way to express the constraint is to directly impose the required form of the fused MOD. For instance, if local densities are modeled as PPP, we have:

$$\begin{aligned} \overline{f} &= \arg\min_f \sum_{i\in\mathcal{N}} \pi_i\, d(f, f_i), \\ \text{such that} \quad & f(X) = \exp(-\lambda) \prod_{x\in X} \lambda p(x). \end{aligned} \tag{20.13}$$

20.3.3 Measure of Discrepancy Between MODs

The measure of discrepancy between MODs plays an essential role in deriving the fusion criterion. As a matter of fact, there are many candidates that could be adopted to serve the purpose, such as, for instance, the KLD $D_{KL}(f||g)$ or CSD $D_{CS}(f||g)$ [16] defined as follows:

$$D_{KL}(f||g) = \int f(X) \log \frac{f(X)}{g(X)} \delta X, \tag{20.14}$$

$$D_{CS}(f, g) = -\log \frac{\int f(X)\, g(X)\, \delta X}{\sqrt{\int f^2(X)\, \delta X}\sqrt{\int g^2(X) \delta X}}. \tag{20.15}$$

As it can be immediately seen, the KLD is not symmetric, i.e., $D_{KL}(f||g) \neq D_{KL}(g||f)$, while the CSD is symmetric. The unsymmetrical nature of the KLD results, therefore, in different criteria, i.e., *left or right Kullback–Leibler* (LKL or RKL) fusion, according to whether $d(f, f_i) = D_{KL}(f \parallel f_i)$ or $d(f, f_i) = D_{KL}(f_i \parallel f)$ in (20.11). Clearly such differences lead to different characteristics in fusing MODs, which will be detailed later.

20.4 Left Kullback–Leibler Fusion

This section introduces the *left Kullback–Leibler* (LKL) fusion criterion.

20.4.1 General Case

Setting $d(f, f_i) = D_{KL}(f \parallel f_i)$ in (20.11) yields:

$$\overline{f} = \arg\min_{f} \sum_{i \in \mathcal{N}} \pi_i\, D_{KL}(f||f_i), \tag{20.16}$$

whose solution turns out to be [6]:

$$\overline{f}(X) = \frac{\prod_{i \in \mathcal{N}} f_i(X)^{\pi_i}}{\int \prod_{i \in \mathcal{N}} f_i(X')^{\pi_i}\, \delta X'}, \tag{20.17}$$

i.e., $\overline{f}$ is the normalized weighted geometric average of the local MODs $\{f_i\}_{i \in \mathcal{N}}$. Notice that (20.17) is the MOD counterpart of Theorem 19.1 stated for PDF fusion, and that its proof can be obtained in a similar way, replacing ordinary vector integrals with set integrals. As it will be shown later, for most MOD densities (20.16) ensure conjugacy, so that the constraint in (20.12) can be removed in LKL fusion.

20.4.2 Poisson MODs

Let each local MOD f_i, $i \in \mathcal{N}$, be of Poisson type, i.e.,

$$f_i(X) = \exp(-\lambda_i) \prod_{x \in X} \lambda_i p_i(x) = \exp\left[-\int D_i(x) dx\right] \prod_{x \in X} D_i(x) \tag{20.18}$$

with PHD $D_i(x) = \lambda_i p_i(x)$. Substituting (20.18) into (20.17), the resulting fused density turns out to be also of Poisson type [6], i.e.,

$$\overline{f}(X) = \exp(-\overline{\lambda}) \prod_{x \in X} \overline{\lambda}\overline{p}(x) = \exp\left[-\int \overline{D}(x)dx\right] \prod_{x \in X} \overline{D}(x) \tag{20.19}$$

with

$$\overline{D}(x) = \prod_{i \in \mathcal{N}} D_i(x)^{\pi_i}, \quad \overline{\lambda} = \int \overline{D}(x)dx, \quad \overline{p}(x) = \frac{1}{\overline{\lambda}} \overline{D}(x). \tag{20.20}$$

In words, the fused PHD turns out to be the weighted geometric average of the local ones.

20.4.3 IID-cluster MODs

In case all local MODS f_i, $i \in \mathcal{N}$, are IID-cluster, i.e.,

$$f_i(X) = |X|!\, \rho_i(|X|) \prod_{x \in X} p_i(x), \tag{20.21}$$

then also the fused density according to the LKL criterion turns out to be IID-cluster [6], i.e.,

$$\overline{f}(X) = |X|!\, \overline{\rho}(|X|) \prod_{x \in X} \overline{p}(x), \tag{20.22}$$

with

$$\overline{p}(x) = \frac{\prod_{i \in \mathcal{N}} p_i(x)^{\pi_i}}{\int \prod_{i \in \mathcal{N}} p_i(x')^{\pi_i}\, dx'}, \tag{20.23}$$

$$\overline{\rho}(n) = \frac{\prod_{i \in \mathcal{N}} \rho_i(n)^{\pi_i} \left\{ \int \prod_{i \in \mathcal{N}} p_i(x)^{\pi_i} dx \right\}^n}{\sum_{n'=0}^{\infty} \prod_{i \in \mathcal{N}} \rho_i(n')^{\pi_i} \left\{ \int \prod_{i \in \mathcal{N}} p_i(x)^{\pi_i} dx \right\}^{n'}}. \tag{20.24}$$

Notice that the fused spatial PDF is the normalized weighted geometric average of the local ones. Clearly, to make the computation of the cardinality PMF $\overline{\rho}(\cdot)$ practically feasible, an upper bound $n_{\max}$ on the number of objects needs to be imposed.

20.4.4 Labeled (LMB and Mδ-GLMB) MODs

When considering fusion of labeled MODs, it must be taken into account that, in most practical cases, label generation is independently carried out by each agent $i \in \mathcal{N}$. This means that, in such cases, *label matching* among the MODs to be fused must be solved upstream of fusion. Label matching has received considerable attention in the literature on multiagent labeled multiobject estimation [12] and is not addressed in this chapter, where it is assumed that labels among all local MODs have been matched prior to fusion.

Let us first consider fusion of Mδ-GLMB MODs $f_i(\mathbf{X}) = \{w_{i,I}, p_{i,l,I}\}_{I \subseteq \mathbb{L}, l \in I}$. It has been shown [17] that the fused MOD $\overline{f} = \{\overline{w}_I, \overline{p}_{l,I}\}_{I \subseteq \mathbb{L}, l \in I}$, according to the LKL criterion, is also Mδ-GLMB with parameters:

$$\overline{w}_I = \frac{\prod_{i \in \mathcal{N}} w_{i}, I^{\pi_i} \left[\int \prod_{i \in \mathcal{N}} \prod_{l \in I} p_{i,l,I}(x)^{\pi_i} dx \right]}{\sum_{I' \subseteq \mathbb{L}} \prod_{i \in \mathcal{N}} w_{i}, I'^{\pi_i} \left[\int \prod_{i \in \mathcal{N}} \prod_{l \in I'} p_{i,l,I'}(x)^{\pi_i} dx \right]}, \tag{20.25}$$

$$\overline{p}_{i,l,I}(x) = \frac{\prod_{i \in \mathcal{N}} p_{i,l,I}(x)^{\pi_i}}{\int \prod_{i \in \mathcal{N}} p_{i,l,I}(x')^{\pi_i} dx'}. \tag{20.26}$$

Then, when the LKL fusion is applied to LMB densities $f_i(\mathbf{X}) = \{r_{i,l}, p_{i,l}\}_{l \in \mathbb{L}}$, the resulting fused MOD $\overline{f} = \{\overline{r}_l, \overline{p}_l\}_{l \in \mathbb{L}}$ also turns out to be LMB with EP $\overline{r}_l$ and spatial PDF $\overline{p}_l$ of each label $l \in \mathbb{L}$ given by:

$$\overline{r}_l = \frac{\int \prod_{i \in \mathcal{N}} (r_{i,l} p_{i,l}(x))^{\pi_i} dx}{\prod_{i \in \mathcal{N}} (1 - r_{i,l})^{\pi_i} + \int \prod_{i \in \mathcal{N}} (r_{i,l} p_{i,l}(x))^{\pi_i} dx}, \tag{20.27}$$

$$\overline{p}_l(x) = \frac{\prod_{i \in \mathcal{N}} [p_{i,l}(x)]^{\pi_i}}{\int \prod_{i \in \mathcal{N}} [p_{i,l}(x')]^{\pi_i} dx'}. \tag{20.28}$$

20.4.5 Implementation Issues

In practice, MODs are usually represented in *Gaussian mixture* (GM) or in *particle* form. These representations are adopted for the spatial PDFs of MODs: p_i of (20.18) and (20.3) for PPP and IID-cluster MODs, respectively; $p_{i,l,I}$ of (20.6) for Mδ-GLMB MODs; $p_{i,l}$ of (20.4) for LMB MODs. As it can be seen from (20.20), (20.23), (20.26), and (20.28), implementation of LKL fusion involves normalized weighted geometric averaging of spatial PDFs (denoted by p_i), i.e.,

$$\overline{p}(x) = \frac{\prod_{i \in \mathcal{N}} p_i(x)^{\pi_i}}{\int \prod_{i \in \mathcal{N}} p_i(x')^{\pi_i} dx'}. \tag{20.29}$$

Adopting GM representation, i.e.,

$$p_i(x) = \sum_{j=1}^{J_i} \omega_{i,j} \, \mathcal{G}(x; \mu_{i,j}, P_{i,j}), \tag{20.30}$$

where $\mathcal{G}(x; \mu, P)$ denotes the Gaussian PDF with mean μ and covariance P, the exponentiation of p_i can be approximately computed as [6]:

$$p_i(x)^{\pi_i} = \sum_{j=1}^{J_i} \tilde{\omega}_{i,j} \, \mathcal{G}\left(x; \mu_{i,j}, \frac{P_{i,j}}{\pi_i}\right), \tag{20.31}$$

where

$$\tilde{\omega}_{i,j} = \omega_{i,j}^{\pi_i} \left[\frac{\det[2\pi P_{i,j} \pi_i^{-1}]}{\det(2\pi P_{i,j})} \right]^{\frac{1}{2}}. \tag{20.32}$$

Next, the following product needs to be computed:

$$\prod_{i \in \mathcal{N}} p_i(x)^{\pi_i} = \sum_{j_1=1}^{J_1} \cdots \sum_{j_{|\mathcal{N}|}=1}^{J_{|\mathcal{N}|}} \left[\prod_{i \in \mathcal{N}} \tilde{\omega}_{i,j_i} \, \mathcal{G}\left(x; \mu_{i,j_i}, \frac{P_{i,j_i}}{\pi_i} \right) \right] \tag{20.33}$$

which, in turn, involves repeated appeal to the multiplication of two Gaussians by:

$$\mathcal{G}(x; \mu_1, P_1) \mathcal{G}(x; \mu_2, P_2) = \mathcal{G}(\mu_1 - \mu_2; 0, P_1 + P_2) \, \mathcal{G}\left(x; \overline{\mu}, \overline{P}\right), \tag{20.34}$$

with

$$\overline{P} = \left(P_1^{-1} + P_2^{-1}\right)^{-1}, \quad \overline{\mu} = \overline{P}(P_1^{-1}\mu_1 + P_2^{-1}\mu_2). \tag{20.35}$$

Please notice that the number of Gaussian components increases to $\prod_{i \in \mathcal{N}} J_i$, thus requiring the resort to a *pruning and merging* procedure to limit the number of Gaussian components.

In case the agent spatial PDFs are represented in *particle*, also called *δ-mixture*, form:

$$p_i(x) = \sum_{j=1}^{J_i} \omega_{i,j} \, \delta(x - x_{i,j}), \tag{20.36}$$

with state samples $x_{i,j}$ and corresponding importance weights $\omega_{i,j}$, LKL fusion cannot be directly implemented as applying (20.29) directly to (20.36) would yield $\overline{p}(x)$ zero almost everywhere. A possible implementation strategy is to first convert the particle representation (20.36) to GM form (20.30) via, e.g., the method in [18], then to get $\overline{p}$ in GM form, and finally to get back to particle form by sampling the GM.

20.5 RKL Fusion

This section introduces the RKL fusion criterion.

20.5.1 Arithmetic Average and MIL Fusion

The RKL fusion criterion modifies (20.16) by interchanging the order of the two arguments f and f_i in the KLD, i.e.,

$$\overline{f} = \arg\min_{f} \sum_{i \in \mathcal{N}} \pi_i \, D_{\mathrm{KL}}(f_i || f), \tag{20.37}$$

whose solution [19] turns out to be given by the weighted arithmetic average:

$$\overline{f}(X) = \sum_{i \in \mathcal{N}} \pi_i \, f_i(X). \tag{20.38}$$

Nevertheless, the resulting fused MOD $\overline{f}(X)$ is no longer conjugate to the local ones. For instance, if the local MODs are of PPP type, the resulting fused MOD $\overline{f}(X)$ turns out to be a "PPP mixture." Hence, constraints should be imposed to (20.37), i.e.,

$$\begin{aligned} \overline{f} &= \arg\min_{f} \sum_{i\in\mathcal{N}} \pi_i \, D_{\mathrm{KL}}(f_i || f), \\ \text{such that} \quad & f \text{ is of the same type of} \{f_i\}_{i\in\mathcal{N}}. \end{aligned} \tag{20.39}$$

Before providing results on RKL fusion for specific types of MODs, we would like to remind the interpretation of (20.37) from an information-theoretic viewpoint, specifically that the KLD $D_{\mathrm{KL}}(f_i || f)$ represents the information loss from f_i to f; as a result, solution of (20.39) yields the fused MOD $\overline{f}$ of the desired type leading to *minimum information loss* (MIL) [20].

20.5.2 Poisson MODs

In case the local MODs are of PPP type, i.e.,

$$f_i(X) = \exp(-\lambda_i) \prod_{x\in X} \lambda_i p_i(x), \tag{20.40}$$

the best, in MIL sense, PPP MOD is obtained by solving the following optimization problem:

$$\begin{aligned} \overline{f} &= \arg\min_{f} \sum_{i\in\mathcal{N}} \pi_i D_{\mathrm{KL}}(f_i || f), \\ \text{subject to} \quad & f(X) = \exp(-\lambda) \prod_{x\in X} \lambda p(x). \end{aligned} \tag{20.41}$$

Then, the resulting fused density turns out to be $\overline{f}(x) = \exp(-\overline{\lambda}) \prod_{x\in X} \overline{\lambda}\,\overline{p}(x)$, with

$$\overline{\lambda} = \sum_{i\in\mathcal{N}} \pi_i \lambda_i, \tag{20.42}$$

$$\overline{p}(x) = \frac{\sum_{i\in\mathcal{N}} \pi_i \, \lambda_i \, p_i(x)}{\sum_{j\in\mathcal{N}} \pi_j \lambda_j}. \tag{20.43}$$

Equivalently, the fused PHD $\overline{D}$ turns out to be:

$$\overline{D}(x) = \sum_{i\in\mathcal{N}} \pi_i \, D_i(x), \tag{20.44}$$

i.e., the weighted arithmetic average of the local PHDs.

20.5.3 IID-cluster MODs

For RKL fusion of IID-cluster MODs:

$$f_i(X) = |X|! \, \rho_i(|X|) \prod_{x\in X} p_i(x), \tag{20.45}$$

the best, in MIL sense, IID-cluster MOD is obtained by solving:

$$\begin{aligned} \overline{f} &= \arg\min_{f} \sum_{i\in\mathcal{N}} \pi_i \, D_{\mathrm{KL}}(f_i || f), \\ \text{subject to} \quad & f(X) = |X|! \, \rho(|X|) \prod_{x\in X} p(x). \end{aligned} \tag{20.46}$$

Then, the fused IID-cluster MOD turns out to be $\overline{f}(X) = |X|!\,\overline{\rho}(|X|)\,\prod_{x\in X}\overline{p}(x)$, with:

$$\overline{\rho}(n) = \sum_{i\in\mathcal{N}} \pi_i\,\rho_i(n), \tag{20.47}$$

$$\overline{p}(x) = \frac{\sum_{i\in\mathcal{N}} \pi_i\,\widehat{N}_i\,p_i(x)}{\sum_{j\in\mathcal{N}} \pi_j \widehat{N}_j}, \tag{20.48}$$

$\widehat{N}_i = \sum_{n=0}^{\infty} n\,\rho_i(n)$ being the expected number of objects for agent i.

20.5.4 Labeled MODs

Let us now consider RKL fusion of labeled MODs, assuming that labels among different agents have been matched. In case the MODs are Mδ-GLMB, i.e., $f_i(\mathbf{X}) = \{w_{i,I},\, p_{i,l,I}\}_{I\subseteq\mathbb{L},l\in I}$, we have the following optimization problem:

$$\begin{aligned} \overline{f} &= \arg\min_{f} \sum_{i\in\mathcal{N}} \pi_i\,D_{\mathrm{KL}}(f_i||f), \\ \text{subject to} \quad & f = \{w_I,\, p_{l,I}\}_{I\subseteq\mathbb{L},l\in I} \end{aligned} \tag{20.49}$$

whose solution yields:

$$\overline{f} = \{\overline{w}_I,\, \overline{p}_{l,I}\}_{I\subseteq\mathbb{L},l\in I}, \tag{20.50}$$

with

$$\overline{w}_I = \sum_{i\in\mathcal{N}} \pi_i\,w_{i,I}, \tag{20.51}$$

$$\overline{p}_{l,I}(x) = \sum_{i\in\mathcal{N}} \widetilde{\pi}_{i,I}\,p_{i,l,I}(x), \tag{20.52}$$

$$\widetilde{\pi}_{i,I} = \frac{\pi_i\,w_{i,I}}{\sum_{j\in\mathcal{N}} \pi_j\,w_{j,I}}. \tag{20.53}$$

It is also worth to point out that RKL fusion can also be applied to δ-GLMB MODs which, on the other hand, cannot be fused via the LKL criterion; see details in [7].

Finally, we consider RKL fusion of LMB MODs $f_i(\mathbf{X}) = \{r_{i,l},\, p_{i,l}\}_{l\in\mathbb{L}}$ which amounts to solving:

$$\begin{aligned} \overline{f} &= \arg\min_{f} \sum_{i\in\mathcal{N}} \pi_i D_{\mathrm{KL}}(f_i||f) \\ \text{subject to} \quad & f = \{r_l,\, p_l\}_{l\in\mathbb{L}}. \end{aligned} \tag{20.54}$$

The resulting fused LMB MOD $\overline{f} = \{\overline{r}_l,\, \overline{p}_l\}_{l\in\mathbb{L}}$ leading to MIL is then given by:

$$\overline{r}_l = \sum_{i\in\mathcal{N}} \pi_i\,r_{i,l}, \tag{20.55}$$

$$\overline{p}_l(x) = \sum_{i\in N} \widetilde{\pi}_{i,l}\,p_{i,l}(x), \tag{20.56}$$

$$\widetilde{\pi}_{i,l} = \frac{\pi_i\,r_{i,l}}{\sum_{j\in\mathcal{N}} \pi_j\,r_{j,l}}. \tag{20.57}$$

20.5.5 Implementation Issues

Implementation of RKL fusion for MODs requires, as key step, to compute the weighted average of spatial PDFs:

$$\overline{p}(x) = \sum_{i\in\mathcal{N}} \pi_i \, p_i(x). \tag{20.58}$$

Such operation can be performed in a straightforward way for both GM and particle representations. In fact, for spatial PDFs in GM form (20.30), (20.58) yields the GM:

$$\overline{p}(x) = \sum_{i\in\mathcal{N}} \sum_{j=1}^{J_i} \pi_i \, \omega_{i,j} \, \mathcal{G}(x; \mu_{i,j}, P_{i,j}). \tag{20.59}$$

Similarly, for spatial PDFs in δ-mixture form (20.36), we get the δ-mixture:

$$\overline{p}(x) = \sum_{i\in\mathcal{N}} \sum_{j=1}^{J_i} \pi_i \, \omega_{i,j} \, \delta(x - x_{i,j}). \tag{20.60}$$

Hence, the weighted arithmetic average preserves all mixture forms though increasing the number of components. To avoid unlimited growth of the number of components, a pruning and merging procedure can be applied to GMs, while resampling can be applied to δ-mixtures.

20.6 Numerical Simulation Examples

In this section, we are going to evaluate performance of the (LKL and RKL) fusion rules introduced in Sections 20.4 and 20.5 by considering the multiagent multitarget tracking system depicted in Figure 20.1. Specifically, each agent gets measurements of the multitarget state, based on which local filtering (prediction and update) is carried out to obtain a local posterior MOD. Local posteriors are then transmitted to the fusion center, wherein MOD fusion takes place thus producing a global posterior MOD which is then sent back to each agent to provide the prior MOD for the next local filtering recursion.

In our simulations, the following types of MOD are considered to model the multitarget state: PPP, IIDCP, LMB, Mδ-GLMB for which PHD, CPHD, LMB, Mδ-GLMB filters are respectively adopted for local filtering. Fusion of the aforementioned MODs is carried out based on both LKL and RKL fusion criteria. In all cases, MODs are represented in GM form.

20.6.1 Scenario

In the simulations, the single target state at discrete time k is denoted as:

$$x_k = [\xi_k \; \dot{\xi}_k \; \eta_k \; \dot{\eta}_k]^\top,$$

where $[\xi_k \; \eta_k]^\top$ and $[\dot{\xi}_k \; \dot{\eta}_k]^\top$ are respectively position and velocity in Cartesian coordinates ξ and η. It is supposed that the target motion is described by the following *nearly constant velocity* (NCV) model [21]:

$$x_k = Ax_{k-1} + w_k, \tag{20.61}$$

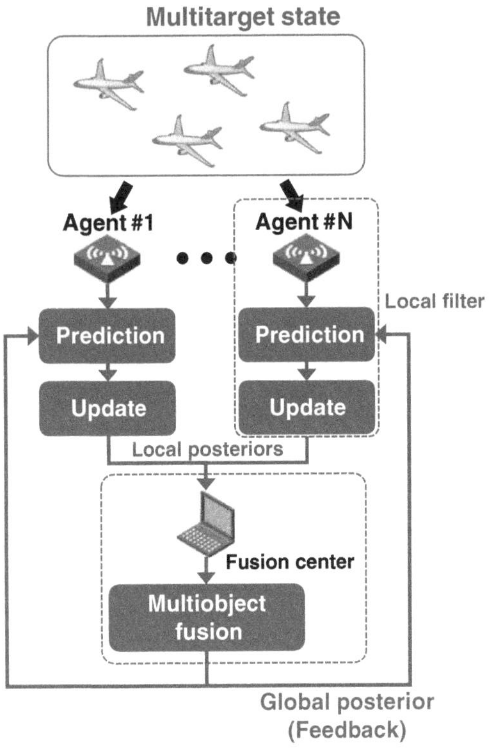

Figure 20.1 Scheme of multiagent multitarget tracking system.

where w_k represents additive zero-mean white Gaussian process noise with covariance:

$$Q = \begin{bmatrix} \frac{1}{3}T^3 & \frac{1}{2}T^2 & 0 & 0 \\ \frac{1}{2}T^2 & T & 0 & 0 \\ 0 & 0 & \frac{1}{3}T^3 & \frac{1}{2}T^2 \\ 0 & 0 & \frac{1}{2}T^2 & T \end{bmatrix}, \tag{20.62}$$

and

$$A = \begin{bmatrix} 1 & T & 0 & 0 \\ 0 & 1 & 0 & 0 \\ 0 & 0 & 1 & T \\ 0 & 0 & 0 & 1 \end{bmatrix}, \tag{20.63}$$

$T = 1$ s being the sampling interval. In the simulated scenario, it is supposed that five targets subsequently enter and then move inside a $5,000 \times 5,000$ m^2 square surveillance region. The considered multiagent system consists of $|\mathcal{N}| = 5$ agents deployed at known locations $[\xi_i \; \eta_i]^\top$ for each $i \in \mathcal{N}$. The considered scenario is illustrated in Figure 20.2.

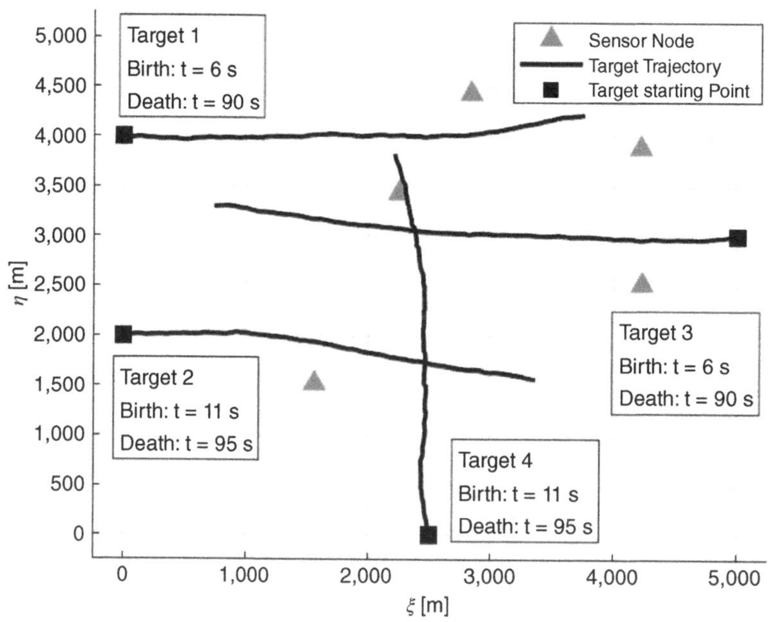

Figure 20.2 Tracking scenario with agents (sensors) sharing a common field-of-view.

Further, it is assumed that each agent is able to provide both range-of-arrival and direction-of-arrival target measurements. Thus, the measurement $z_{k,i}$ generated by a target with state x_k, at time k and agent $i \in \mathcal{N}$, is modeled as:

$$z_{k,i} = h_i(x_k) + v_{k,i}, \tag{20.64}$$

where $v_{k,i}$ is a zero-mean Gaussian measurement noise with covariance matrix $R_i =$ diag(400 [m^2], 1 [$^{\circ 2}$]) and

$$h_i(x) = \begin{bmatrix} \sqrt{(\xi - \xi_i)^2 + (\eta - \eta_i)^2} \\ \text{atan2}(\eta - \eta_i, \xi - \xi_i) \end{bmatrix}, \tag{20.65}$$

atan2 denoting the four-quadrant inverse tangent. Clutter at each sensor node is modeled with Poisson-distributed cardinality and expected number λ_c of clutter detections per time step.

The parameters of the local filters are set as follows:

- the probability of target survival is set to $P_s = 0.95$ for all agents;
- the intensity (for PHD and CPHD filters) and EP (for LMB and Mδ-GLMB filters) of new-born targets is set to 0.01, and the spatial PDF is taken Gaussian, where the position components of the mean vector are obtained by remapping measurements back to target state space, while the velocity components are set to zero, and the covariance matrix is set to diag(1,600 m^2, 400 m^2/s^2, 1,600 m^2, 400 m^2/s^2);
- pruning and merging thresholds for GMs are set respectively to 10^{-8} and 10;
- for fusion of MODs, we adopt uniform fusion weights, i.e., $\pi_i = \dfrac{1}{|\mathcal{N}|}$ for any $i \in \mathcal{N}$;
- the label matching strategy in [20] is leveraged for fusing LMB and Mδ-GLMB densities.

In our simulations, we adopt the *optimal sub-pattern assignment* (OSPA) distance (with order $p = 2$ and cutoff $c = 50$ m) as performance indicator. All results have been averaged over 100 independent Monte Carlo trials. To better illustrate performance of multitarget fusion, various detection probabilities and clutter rates are considered in the simulation setup.

20.6.2 Results

20.6.2.1 General Results

Let us first set the detection probability to $P_d = 0.98$ and average clutter rate to $\lambda_c = 5$. For better readability, we separately show the results of (unlabeled) RFSs and LRFSs in Figures 20.3 and 20.4, respectively. As it can be seen, when sensors are able to detect targets with high probability and compress clutter, performance obtained with LKL and RKL fusion criteria is similar for the same type of MOD.

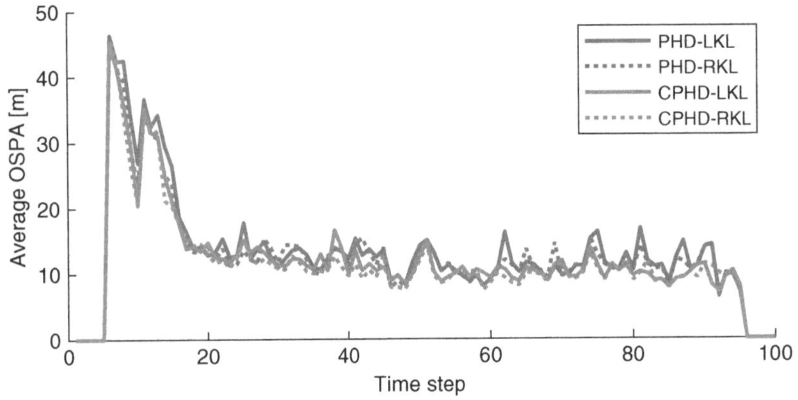

Figure 20.3 Tracking performance based on (unlabeled) RFS MODs and different fusion criteria.

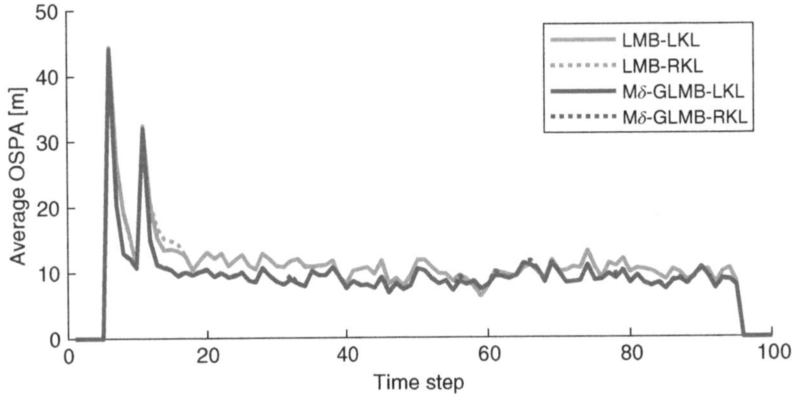

Figure 20.4 Tracking performance based on (labeled) RFS MODs and different fusion criteria.

20.6.2.2 Performance Under Different Detection Probabilities

Next, we fix clutter rate to $\lambda_c = 5$ and tune the detection probabilities of sensors. We first show the results of (unlabeled) RFS and LRFS MODs under low detection probability $P_d = 0.5$ in Figures 20.5 and 20.6, respectively.

It can be observed that the RKL fusion criterion (arithmetic average) has much better performance compared to LKL (geometric average). The reason of such result is the multiplicative nature of LKL in estimating the target intensity/EP, i.e., the fused intensity/EP tends to be extremely small when a target is miss-detected by an agent. In case a target is not detected by all agents (i.e., its intensity/EP equals zero for some agent), then the fused intensity/EP will always be zero.

Under a fixed detection probability ($P_d < 1$), the probability that a target is missed by at least one agent, i.e., $1 - P_d^{|\mathcal{N}|}$, increases with the number of agents. As a reasonable deduction, LKL fusion becomes increasingly sensitive to the number of participating agents and is, particularly, vulnerable in centralized fusion setups involving many agents.

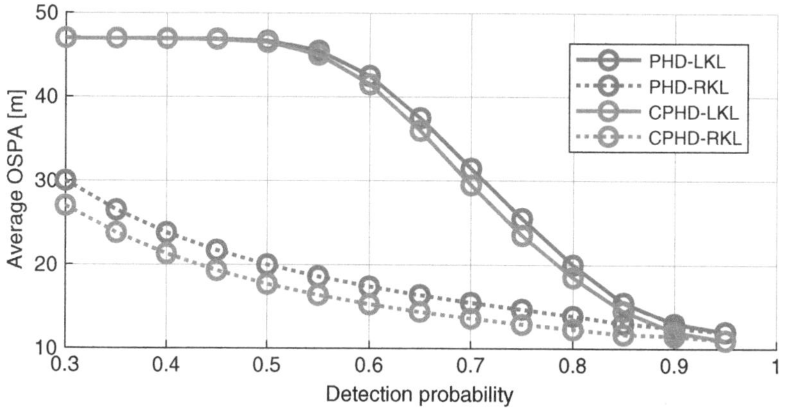

Figure 20.5 Average OSPA versus detection probability under different fusion criteria based on (unlabeled) RFS MODs.

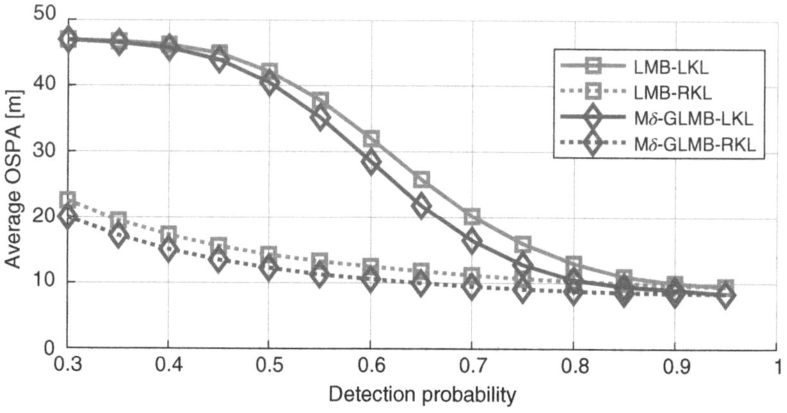

Figure 20.6 Average OSPA versus detection probability under different fusion criteria based on (labeled) RFS MODs.

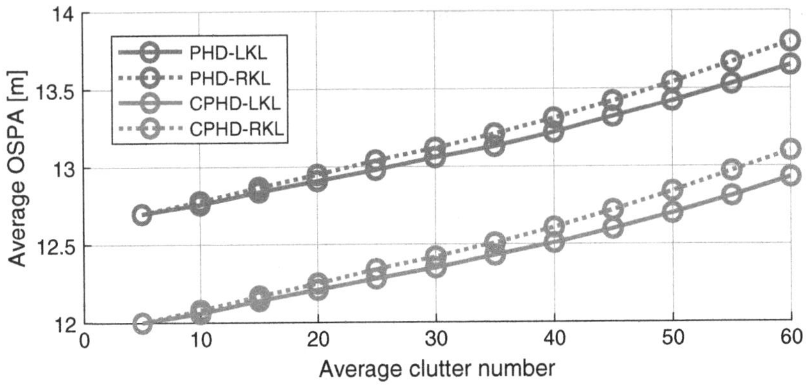

Figure 20.7 Average OSPA versus average clutter number under different fusion criteria based on (unlabeled) RFS MODs.

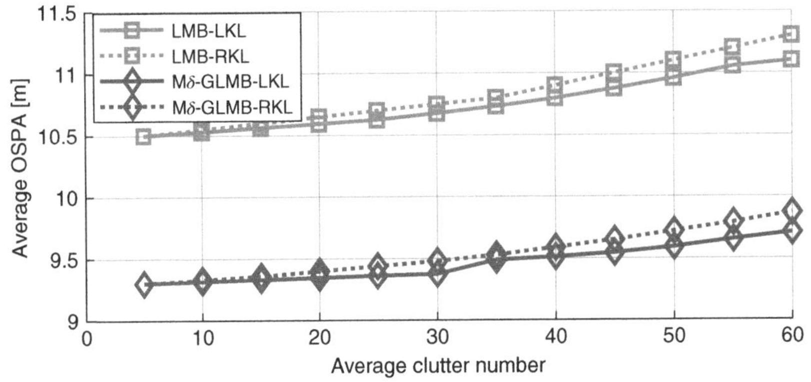

Figure 20.8 Average OSPA versus average clutter number under different fusion criteria based on (labeled) RFS MODs.

20.6.2.3 Performance Under Different False Alarm Rates

Let us now examine the performance of LKL and RKL fusion criteria versus clutter rate. In this case, we fix the detection probability to $P_d = 0.98$. The results are illustrated in Figures 20.7 and 20.8 for (unlabeled) RFS and LRFS MODs, respectively. It can be seen that the performance of LKL fusion is almost the same as RKL fusion under a low clutter rate. On the other hand, for higher clutter rates, LKL fusion performs better than its RKL counterpart. These results are due to the fact that RKL fusion tends to preserve all information, including the false one provided by clutter.

20.7 Complementary Fusion for Limited Fields-of-view

Fusing MODs is critical in multitarget tracking systems, as it enhances estimation accuracy, robustness, and overall situational awareness by integrating observations from multiple sensing nodes. Conventional fusion methods, such as LKL and RKL, considered in Sections 21.4 and 21.5, were originally devised under the assumption of overlapping or redundant FoVs. However, in many practical scenarios—including urban surveillance, intelligent transportation, and autonomous vehicle applications—sensor FoVs are inherently limited and often nonoverlapping. As a result:

- LKL fusion, due to its multiplicative nature, cannot directly deal with sensors having different FoVs, however countermeasures can be taken by considering suitable decompositions [22–26].
- RKL fusion, due to its additive nature, is more capable of naturally handling sensors with different FoVs but anyway there is the tendency to reduce the weights related to targets detected by only a subset of sensors, then also in this case, suitable countermeasures must be adopted to avoid track losses [27–30].

As an alternative to LKL and RKL fusion, this section presents a complementary fusion framework explicitly tailored to dynamic sensor networks with limited and partially disjoint sensor coverage. Unlike consensus-based approaches that focus on intersecting sensor information (common observations), complementary fusion aims to form a fused posterior density representing the *union* of observations from all agents. This strategy enables each sensor's unique contributions to be fully retained, preventing suppression of targets not observed by multiple agents.

The approach is developed within the RFS framework, employing LMB densities to represent each agent's local posterior. By utilizing an information-theoretic formulation based on minimizing the KLD, this method approximates the union of corresponding labeled Bernoulli components across agents. As a result, the final fused density encapsulates the complete set of available sensor observations, regardless of observation overlap. Empirical evaluations in subsequent subsections confirm that the complementary fusion approach outperforms traditional consensus-based methods in fragmented sensing scenarios. Specifically, the proposed fusion technique consistently and accurately estimates target states and cardinalities, even when individual sensors observe targets intermittently or uniquely.

20.7.1 Context and Challenges of Fusion with Limited FoVs

In multiobject tracking scenarios, it is common for sensors to operate with constrained and partially nonoverlapping FoVs. This limitation frequently arises in practical applications such as wide-area surveillance, intelligent transportation systems, autonomous vehicles, and urban monitoring environments, where physical obstructions or spatial distribution restrict each sensor's observation capability. Consequently, sensors typically provide only partial and distinct views of the scene, resulting in complementary rather than redundant observations. Conventional multiagent fusion methods, including multiplicative LKL fusion, predominantly rely on intersection-based strategies designed under the assumption of overlapping FoVs. These methods emphasize agreement among agents, making them effective when multiple agents observe common targets. However, in scenarios with limited or disjoint FoVs, intersection-based fusion tends to neglect or suppress target information observed uniquely by individual agents. Such suppression can degrade tracking performance, manifesting as missed detections and reduced accuracy in estimating object cardinality and states.

To address this fundamental challenge, a complementary fusion framework has been developed, focusing explicitly on aggregating sensor information characterized by limited and nonoverlapping coverage. Instead of striving for consensus, complementary fusion seeks to approximate the *union* of individual agent posteriors. This strategy ensures that unique observations from each sensor are preserved within the fused posterior. This approach represents local agent posteriors as LMB distributions, providing structured sets of hypothesized targets and corresponding labels. The fusion operation thus carefully combines local LMB densities, maintaining label consistency and rigorously avoiding double-counting of target information. The resulting fusion rule employs an information-theoretic optimization, specifically minimizing a weighted sum of KLDs, providing a robust and computationally efficient solution suitable for dynamic and spatially fragmented sensor network deployments.

20.7.2 Fusion Objective: Union versus Intersection

The LKL (aka GCI) fusion method aggregates sensor information through an intersection-based strategy, retaining only those hypotheses jointly supported across multiple agents. Formally, given local posteriors $\{f_i\}_{i\in\mathcal{N}}$, each represented as LMB density defined over a shared label space $\mathbb{L}$, LKL yields a fused density of the form:

$$\overline{f}_{\text{LKL}}(\mathbf{X}) \propto \prod_{i\in\mathcal{N}} f_i(\mathbf{X})^{\pi_i}, \tag{20.66}$$

where $\pi_i \in (0,1)$ and $\sum_{i\in\mathcal{N}} \pi_i = 1$. This formulation avoids double-counting but inherently filters out hypotheses observed by only a subset of the network. In scenarios with partially or fully disjoint sensor coverage, such as urban or wide-area surveillance, intersection-based fusion may miss valid targets, leading to poor tracking performance.

In contrast, complementary (union-based) fusion aims to approximate the *union* of local multiobject posteriors, preserving all target hypotheses—including those unique to individual agents. For each label $l \in \mathbb{L}$, the associated Bernoulli components $\{(r_{i,l}, p_{i,l}(x))\}_{i\in\mathcal{N}}$ are fused into a single component $(\overline{r}_l, \overline{p}_l(x))$. This is achieved by minimizing the weighted sum of KLDs between the fused and local components, a formulation detailed in Section 20.7.4. This approach inherently assigns greater importance to high-confidence sensor reports, integrates unique and shared observations, and maintains statistical consistency without enforcing consensus. It is particularly suited to networks with heterogeneous sensing capabilities and fragmented visibility, where consensus may be either suboptimal or infeasible.

20.7.3 Modeling Assumptions and Labeled Multi-Bernoulli Representation

The complementary fusion framework assumes a centralized multiagent architecture in which each agent independently performs multitarget tracking using the LMB filter. A key modeling assumption is the existence of a globally consistent label space, denoted by $\mathbb{L}$, which is shared and synchronized across all sensor nodes. This assumption allows for straightforward matching of target labels during the fusion process, eliminating the need for explicit label alignment or reconciliation.

At each time step, the local posterior at agent i is an LMB MOD over the labeled state space $\mathbb{X} \times \mathbb{L}$, denoted by f_i, and parameterized by a collection of Bernoulli components $\{(r_{i,l}, p_{i,l}(x)) : l \in \mathbb{L}_i\}$, where $\mathbb{L}_i \subseteq \mathbb{L}$ denotes the subset of labels observed by agent i. Each component encodes the probability of existence $r_{i,l}$ and the spatial PDF $p_{i,l}(x)$ associated with label l. The general form of the LMB density is:

$$f_i(\mathbf{X}) = \Delta(\mathbf{X})\, w_i(\mathcal{L}(\mathbf{X})) \prod_{(x,\ell)\in\mathbf{X}} p_{i,l}(x), \tag{20.67}$$

where $\Delta(\mathbf{X})$ is the distinct label indicator function ensuring that all labels in $\mathbf{X}$ are unique, and $\mathcal{L}(\mathbf{X})$ denotes the set of labels in the multiobject state $\mathbf{X}$. The label-dependent weight function is given by:

$$w_i(L) = \prod_{l\in L} r_{i,l} \prod_{l\in\mathbb{L}_i\setminus L} (1 - r_{i,l}), \tag{20.68}$$

for any label subset $L \subseteq \mathbb{L}_i$.

Due to limited and heterogeneous FoVs, each agent may observe only a partial view of the scene, thus resulting in unique and potentially disjoint sets of Bernoulli components across agents. The fusion objective is to approximate, for each label $l \in \bigcup_i \mathbb{L}_i$, the collection of local Bernoulli components $\{(r_{i,l}, p_{i,l}(x))\}$ with a single fused Bernoulli component $(\overline{r}_l, \overline{p}_l(x))$. The resulting fused components collectively define an LMB distribution over the global label space $\mathbb{L}$, preserving the independence structure between labels and supporting computationally efficient updates. This representation is compatible with standard LMB filtering cycles and enables robust multitarget tracking in environments with fragmented sensor coverage.

20.7.4 Formulation of the Fusion Rule

Assume that the local posterior formed by every agent i is an LMB MOD with parameters $\{(r_{i,l}, p_{i,l}(\cdot))\}_{l\in\mathbb{L}}$. The RFS at each agent i is itself a union of single-Bernoulli random sets:

$$\boldsymbol{X}_i = \bigcup_{l\in\mathbb{L}} \boldsymbol{X}_{i,l}, \tag{20.69}$$

where each $\boldsymbol{X}_{i,l}$ is a single-Bernoulli RFS associated with label l and characterized by $(r_{i,l}, p_{i,l}(\cdot))$. The union across all agents is:

$$\boldsymbol{X} = \bigcup_{i\in\mathcal{N}} \left[\bigcup_{l\in\mathbb{L}} \boldsymbol{X}_{i,l} \right] = \bigcup_{l\in\mathbb{L}} \left[\bigcup_{i\in\mathcal{N}} \boldsymbol{X}_{i,l} \right] = \bigcup_{l\in\mathbb{L}} Y_l. \tag{20.70}$$

Since each Y_l can contain up to $|\mathcal{N}|$ elements with the same label, we approximate Y_l with a single-Bernoulli $\overline{Y}_l \sim \{(\overline{r}_l, \overline{p}_l(\cdot))\}$. The corresponding density is:

$$\overline{f}_l(X) = \begin{cases} 1 - \overline{r}_l & \text{if } X = \emptyset \\ \overline{r}\, \overline{p}_l(x) & \text{if } X = \{x\} \\ 0 & \text{otherwise.} \end{cases} \tag{20.71}$$

To determine $\overline{r}_l$ and $\overline{p}_l(x)$, we model the elements $X_{i,l}$ as independent and define:

$$P_0 = f_l(\emptyset) \approx \prod_{i\in\mathcal{N}} (1 - r_{i,l}) \tag{20.72}$$

$$P_1(x) = f_l(\{x\}) \approx P_0 \sum_{i\in\mathcal{N}} \frac{r_{i,l}\, p_{i,l}(x)}{1 - r_{i,l}}. \tag{20.73}$$

We seek the single-Bernoulli density $\overline{f}_l$ that minimizes the KL divergence from f_l, i.e.,

$$\overline{f}_l = \arg\min_{f\in\mathcal{B}} D_{\mathrm{KL}}(f \parallel f_l), \tag{20.74}$$

where $\mathcal{B}$ is the set of all single-Bernoulli densities. The resulting parameters are given by:

$$\overline{r}_l = \frac{\int P_1(x)\, dx}{\int P_1(x)\, dx + P_0}, \tag{20.75}$$

$$\overline{p}_l(x) = \frac{P_1(x)}{\int P_1(x)\, dx}. \tag{20.76}$$

Using (20.72) and (20.73), the final simplified expressions become:

$$\overline{r}_l = \frac{\sum_{i\in\mathcal{N}} \frac{r_{i,l}}{1-r_{i,l}}}{1+\sum_{i\in\mathcal{N}} \frac{r_{i,l}}{1-r_{i,l}}}, \tag{20.77}$$

$$\overline{p}_{i,l}(x) = \frac{\sum_{i\in\mathcal{N}} \frac{r_{i,l}\, p_{i,l}(x)}{1-r_{i,l}}}{\sum_{i\in\mathcal{N}} \frac{r_{i,l}}{1-r_{i,l}}}. \tag{20.78}$$

20.7.5 Relation to Existing Fusion Methods

The fusion strategy developed in this section directly addresses a key limitation in classical multiobject fusion techniques: their inability to effectively exploit information from nonoverlapping sensor FoVs. Traditional fusion rules are typically designed with the assumption of significant overlap between sensor observations, often discarding target hypotheses that appear in only one local posterior. In contrast, the proposed method is explicitly constructed to preserve the *union* of sensor information, enabling robust tracking in spatially fragmented environments.

20.7.5.1 LKL Fusion (Intersection-based)

LKL fusion, aka GCI, constructs the fused MOD as a normalized weighted geometric mean of local densities according to (20.66), with nonnegative fusion weights π_i that sum up to one. This formulation minimizes the weighted sum of KLDs from the fused posterior to each local posterior and provides a consistent solution under unknown correlation assumptions. However, its intersection-based nature results in the suppression of object hypotheses not supported by all agents, leading to potential under-detection in networks with disjoint or partially overlapping FoVs.

20.7.5.2 MIL Fusion (Inclusive but Structurally Inexact)

MIL fusion aims to minimize the aggregate divergence from each agent's distribution to the fused result, often using arithmetic averaging. This inclusive strategy preserves all sensor-reported targets but does not maintain the structural form of distributions like LMBs. As a result, MIL fusion typically yields a fused density that is no longer of the original type, thus requiring projection or approximation to remain compatible with the recursive filtering framework.

20.7.5.3 CSD-based Fusion (Soft Consensus)

Fusion methods based on the CSD modulate the influence of each sensor's information per target label, resulting in a *soft-consensus* behavior. Targets with widespread support across agents are reinforced, while those observed by only a few agents are preserved with diminished confidence. This offers improved flexibility over LKL, especially in cases with limited FoV overlap, but still retains a bias toward consensus and often requires tractable approximations.

20.7.5.4 KL-based Union Fusion (FoV-aware Union)

The method proposed in this section takes a *union-centric* approach that is explicitly designed for dynamic sensor networks with limited and nonoverlapping FoVs. For each label $l \in \mathbb{L}$, local Bernoulli components $\{(r_{i,l}, p_{i,l}(x))\}$ are approximated by a single fused Bernoulli component

$(\overline{r}_l, \overline{p}_{i,l}(x))$, minimizing the KL divergence from the local posteriors to the fused MOD. The closed-form solutions for the fused EP and spatial density are given by (20.77)–(20.78). This method retains all plausible target hypotheses, even when observed by a single sensor, while maintaining label integrity and mitigating double-counting. The result is a fused LMB that accurately reflects the collective information in the network, supporting improved cardinality estimation and robustness to observation fragmentation.

20.7.5.5 Recent Developments

A promising alternative to the MOD fusion framework has recently emerged in the form of Fréchet mean fusion of trajectories and sets of trajectories [31]. This approach defines the mean trajectory (and multi-trajectory set) using the OSPA(2) metric and provides tractable algorithms—such as greedy and Gibbs sampling—for estimating this mean. Unlike MOD fusion, which operates on MODs and is rooted in information-theoretic divergence minimization (e.g., KLD), the Fréchet mean method directly minimizes geometric distance between finite sets of trajectories. This makes it particularly attractive for large-scale problems involving a high number of agents or objects, where computational efficiency and label ambiguity are significant concerns. However, the MOD fusion approach remains advantageous when preserving probabilistic information is critical, especially in Bayesian estimation frameworks where uncertainty quantification and recursive updates are required. A systematic comparison of these complementary approaches—particularly in distributed tracking settings—presents an interesting avenue for future research.

20.7.6 Implementation Strategies

The fusion framework admits practical implementation through either *sequential Monte Carlo* (SMC) or GM representations of the single-Bernoulli components.

20.7.6.1 SMC-based Implementation

In the particle filter (SMC) case, each local spatial PDF $p_{i,l}(x)$ is approximated using a common set of particles as:

$$p_{i,l}(x) \approx \sum_{j=1}^{J_l} \omega_{i,l,j}\, \delta(x - x_{l,j}), \quad \sum_{j=1}^{J_l} \omega_{i,l,j} = 1. \tag{20.79}$$

Since the prediction step is performed centrally, the particle locations $x_{l,j}$ are identical across agents, and only the weights $\omega_{i,l,j}$ vary. The fused density $\overline{p}_l(x)$ retains the common particle support and updates the weights via:

$$\overline{\omega}_{l,j} = \frac{\sum_{i\in\mathcal{N}} \dfrac{r_{i,l}}{1 - r_{i,l}}\, \omega_{i,l,j}}{\sum_{i\in\mathcal{N}} \dfrac{r_{i,l}}{1 - r_{i,l}}}. \tag{20.80}$$

20.7.6.2 GM-based Implementation

For the GM model, each local spatial PDF is represented as:

$$p_{i,l}(x) \approx \sum_{j=1}^{J_{i,l}} \omega_{i,l,j}\, \mathcal{G}(x; \mu_{i,l,j}, P_{i,l,j}), \quad \sum_{j=1}^{J_{i,l}} \omega_{i,l,j} = 1. \tag{20.81}$$

In this case, the prediction step need not be centralized, provided that label consistency is maintained. The fused density $\overline{p}_l(x)$ becomes a mixture of all J_l Gaussian components from the contributing agents, with

$$J_l = \sum_{i \in \mathcal{N}} J_{i,l}, \tag{20.82}$$

and the weight of each component updated via (20.80).

20.7.6.3 Post-fusion Processing

Due to the accumulation of similar components, the resulting GM may contain redundancies. To ensure numerical tractability, components with negligible weights should be pruned, and closely spaced components may be merged using standard mixture reduction techniques.

20.7.7 Simulation Scenarios and Evaluation

To validate the proposed complementary fusion approach, two simulation studies were conducted using LMB filters running locally at each agent. Both considered scenarios, described in the next subsections, emulate realistic multitarget tracking settings with limited field-of-view (FoV) sensing and varying degrees of sensor overlap.

20.7.7.1 Scenario 1: Comparison with Consensus-based Fusion

This first scenario investigates how the proposed KL-union fusion performs in contrast to traditional techniques, specifically, LKL and CSD-based fusion. A network of three static sensors observes up to five maneuvering targets. Each target follows a coordinated turn model with state vector: $x_k = \begin{bmatrix} \xi_k & \dot{\xi}_k & \eta_k & \dot{\eta}_k & \omega_k \end{bmatrix}^\top$ and transitions governed by $f_{k|k-1}(x_k|x_{k-1}) = \mathcal{G}(x_k; A(\omega_{k-1})x_{k-1}, Q)$, where $A(\omega)$ is the coordinated-turn transition matrix and Q the covariance of process noise. Sensors provide range-bearing measurements with additive Gaussian noise. Two experiments are carried out:

- **Experiment 1 (Wide FoV,** 175°**):** All sensors maintain visibility of all targets throughout the simulation. All fusion methods—including KL-union, LKL, and CSD—perform comparably as expected.
- **Experiment 2 (Limited FoV,** 110°**):** Sensors observe disjoint subsets of the environment. Many targets are only seen by one sensor at any given time. In this case, traditional LKL and CSD-based methods miss many targets or prematurely terminate tracks, whereas the proposed union-based method successfully tracks all targets using complementary observations.

A quantitative comparison is provided via the OSPA metric [32]. Figure 20.9a, b plot OSPA errors averaged over 100 Monte Carlo runs.

Remark: While the computational cost of KL-union fusion is higher than GCI and CSD due to the decomposition and recomposition of Bernoulli components, it enables robust tracking in fragmented sensing conditions.

20.7.7.2 Scenario 2: Robustness Evaluation

A second experimental scenario was designed to evaluate the robustness of the proposed fusion approach in dynamic and fragmented sensing environments, as illustrated in Figure 20.10. This

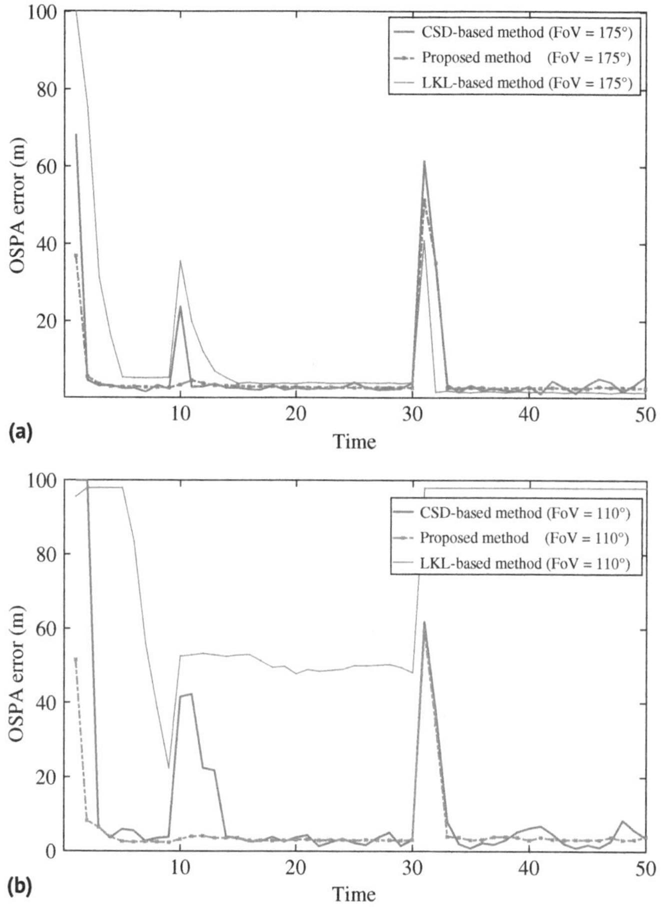

Figure 20.9 Scenario 1 performance: trajectory estimates and OSPA comparison across field-of-view conditions. (a) OSPA errors for wide FoV (175°) and (b) OSPA errors for narrow FoV (110°)

scenario involved eight targets with asynchronous birth and death times, reflecting more realistic multiobject dynamics. A range of simulation configurations was explored, varying the number of sensors, their FoVs, clutter rates, and detection probabilities. Each sensor independently executed an LMB filter to process local measurements, while a centralized node performed fusion of the local posteriors. The following sensitivity analyses were conducted to assess the method's performance under diverse conditions:

20.7.7.2.1 Field-of-view

The performance of the proposed fusion method was evaluated under varying sensor FoVs, ranging from narrow to unlimited angular coverage. As shown in Figure 20.11, a slight increase in localization error is observed as the FoV becomes wider, likely due to increased blind zones at the periphery. Nevertheless, the fusion framework preserves high accuracy in localization and cardinality estimation across all FoV settings, demonstrating robustness to fragmented visibility and limited overlap.

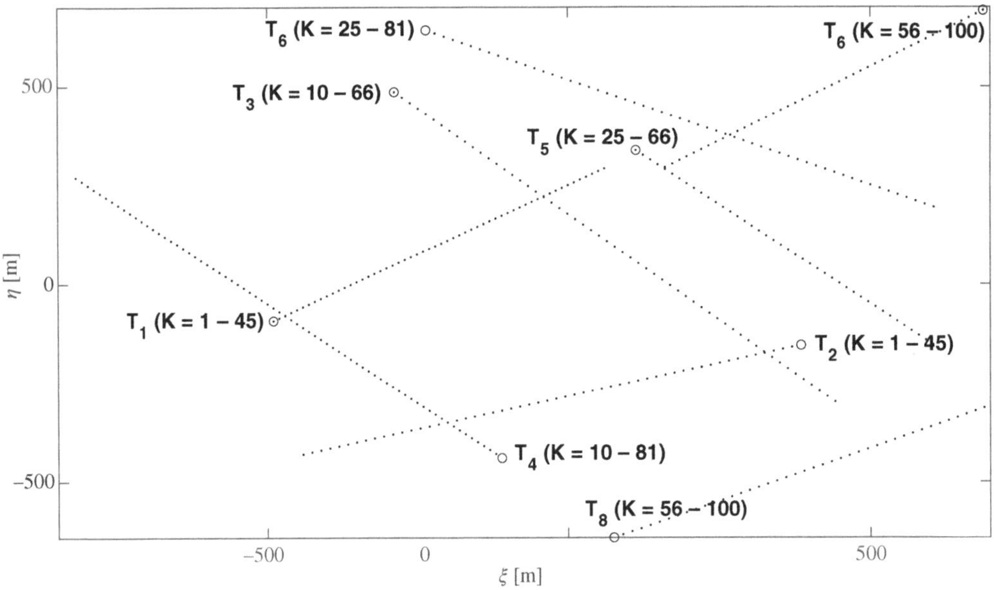

Figure 20.10 Ground truth trajectories for Scenario 2. Each target moves under constant velocity dynamics.

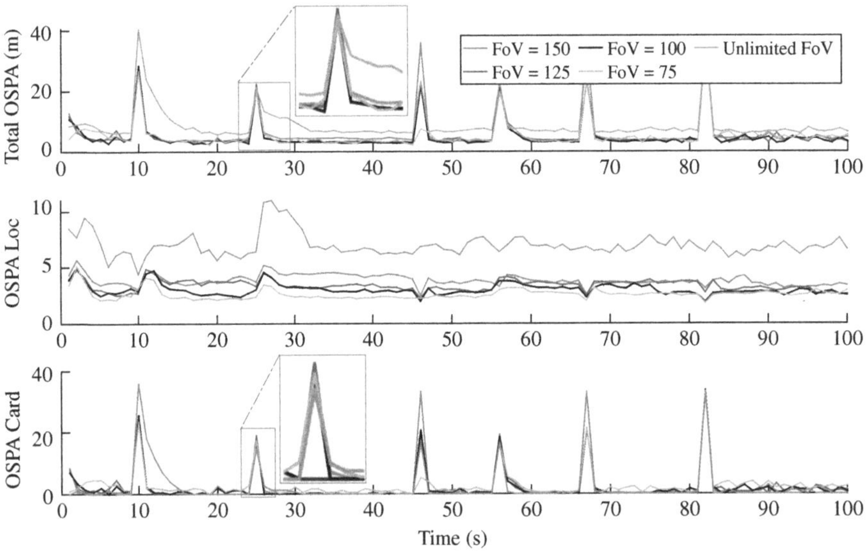

Figure 20.11 OSPA error versus FoV for proposed method. Top: total, middle: localization, bottom: cardinality.

20.7.7.2.2 Sensor Count

To assess scalability, the number of sensors in the network was varied. As shown in Figure 20.12, while increasing the number of nodes improves spatial coverage, it also leads to greater observation fragmentation. The proposed method effectively accommodates this fragmentation, maintaining stable multitarget tracking accuracy and consistent cardinality estimation, even as the sensor count increases substantially.

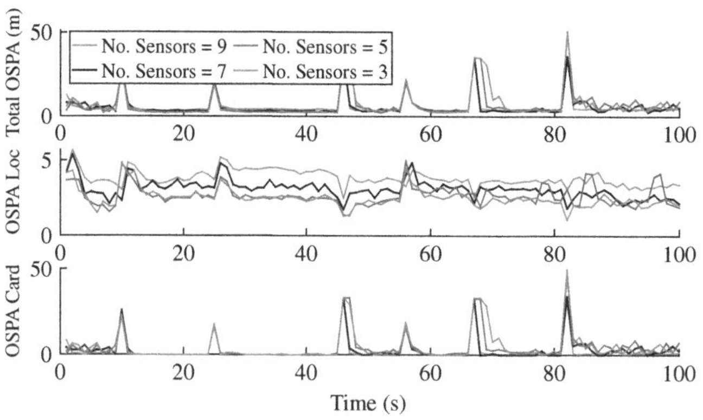

Figure 20.12 OSPA error versus sensor count. Top: total, middle: localization, bottom: cardinality.

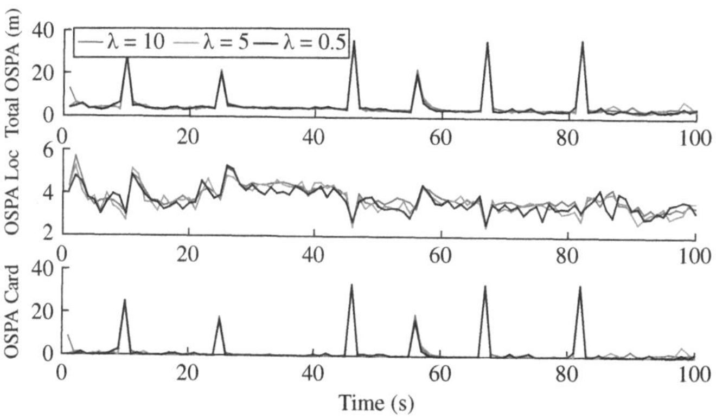

Figure 20.13 OSPA error versus clutter rate. The proposed method maintains robustness across all tested levels.

20.7.7.2.3 Clutter Rate

Robustness to false alarms was evaluated by varying the clutter rate across a wide range of values. As illustrated in Figure 20.13, the fusion approach consistently maintained low tracking error and reliable object count estimates, showing strong resilience to elevated clutter conditions.

20.7.7.2.4 Detection Probability

When the sensor detection probability was reduced, the fusion algorithm continued to deliver accurate estimates, with only a marginal rise in tracking error, as shown in Figure 20.14. This indicates the method's tolerance to occasional missed detections, a common challenge in real-world sensing environments.

20.7.7.3 Summary

The experimental results confirm that KL-union fusion matches or exceeds the performance of consensus-based methods under overlapping sensing conditions and significantly outperforms them when observations are fragmented due to limited or disjoint FoVs. The method maintains accurate cardinality estimates and robust tracking across a wide range of sensing parameters.

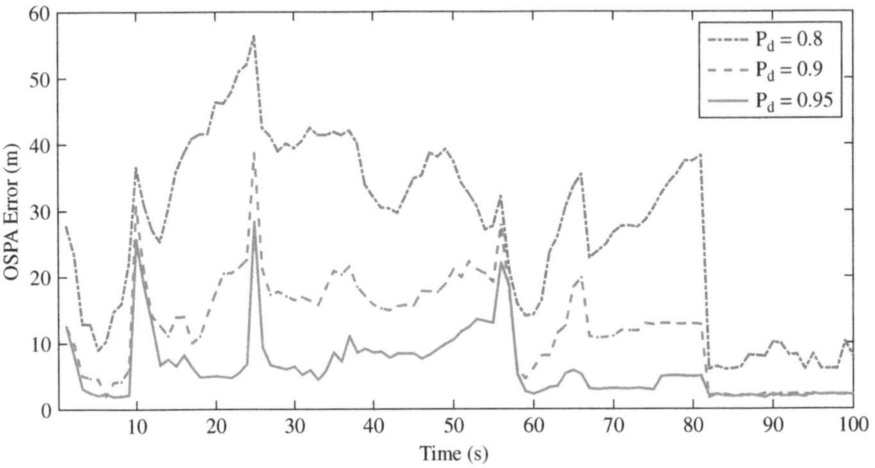

Figure 20.14 Effect of varying detection probability on OSPA error.

20.8 Conclusions

This chapter has provided an overview of fusion methods for MODs to be used to aggregate information from multiple agents in the context of multiagent multiobject estimation. In particular, two information-theoretic fusion criteria based on the unsymmetrical KLD, referred to as left (LKL) or right (RKL) Kullback–Leibler fusion, have been considered, minimizing either the information gain (LKL) or the information loss (RKL) and accordingly leading to either a geometric or, respectively, arithmetic average of the MODs. While LKL fusion ensures conjugacy, i.e., produces a fused MOD of the same type of the fusing ones, RKL fusion does not enjoy this property, which can, however, be enforced via a suitable constraint added to information loss minimization. Explicit formulas for both LKL and RKL fusion of different types of labeled and unlabeled MODs are provided. Via simulation experiments in a multitarget tracking case study, it is shown how LKL fusion, in view of its multiplicative nature, is robust to false detections but sensitive to missed detections, while the converse holds for the additive RKL fusion. Moreover, an alternative *complementary fusion* paradigm has been presented to address the practically relevant situation of agents with limited and nonoverlapping FoVs, and its effectiveness in incorporating complementary information from such agents has also been demonstrated via simulation experiments in multitarget tracking.

Beyond what is presented in this chapter, research on MOD fusion has proceeded in several directions, including: Kullback–Leibler fusion of MODs with different sensor FoVs [30]; leveraging event-triggered communication [33–35] to save communication bandwidth and energy consumption; resilient fusion of MODs [36–38]; distributed multi-robot SLAM [23, 39]; joint MOD fusion and sensor registration [40–42]; distributed *joint detection, tracking, and classification* (JDTC) of multiple objects [43, 44]. Furthermore, the idea of MOD fusion can also be applied to multiobject filtering scenarios sharing similar mechanisms to MOD fusion, e.g., multidetection multitarget tracking [45, 46].

References

1 A. Farina and F. Studer. *Radar Data Processing*. 1985.
2 S. Blackman and R. Popoli. *Design and Analysis of Modern Tracking Systems*. 1999.
3 Y. Bar-Shalom, P. Willett, and X. Tian. *Tracking and data fusion*, volume 11. YBS publishing Storrs, CT, USA, 2011.
4 S. Thrun. Probabilistic robotics. *Communications of the ACM*, 45(3):52–57, 2002.
5 B.-N. Vo, B.-T. Vo, T. Nguyen, and C. Shim. An overview of multi-object estimation via labeled random finite set. *IEEE Transactions on Signal Processing*, 72:4888–4917, 2024.
6 G. Battistelli, L. Chisci, C. Fantacci, A. Farina, and A. Graziano. Consensus CPHD filter for distributed multitarget tracking. *IEEE Journal of Selected Topics in Signal Processing*, 7(3):508–520, 2013.
7 L. Gao, G. Battistelli, and L. Chisci. Fusion of labeled RFS densities with minimum information loss. *IEEE Transactions on Signal Processing*, 68:5855–5868, 2020.
8 B.-T. Vo and B.-N. Vo. Labeled random finite sets and multi-object conjugate priors. *IEEE Transactions on Signal Processing*, 61(13):3460–3475, 2013.
9 R. Mahler. *Advances in Statistical Multisource-multitarget Information Fusion*. Artech House, 2014.
10 R. Mahler. *Statistical Multisource-multitarget Information Fusion*. Artech House, 2007.
11 S. Reuter, B.-T. Vo, B.-N. Vo, and K. Dietmayer. The labeled multi-Bernoulli filter. *IEEE Transactions on Signal Processing*, 62(12):3246–3260, 2014.
12 B.-N. Vo, B.-T. Vo, and D. Phung. Labeled random finite sets and the Bayes multi-target tracking filter. *IEEE Transactions on Signal Processing*, 62(24):6554–6567, 2014.
13 C. Fantacci, B.-T. Vo, F. Papi, and B.-N. Vo. The marginalized δ-GLMB filter. *arXiv preprint arXiv:1501.00926*, 2015.
14 C. Fantacci and F. Papi. Scalable multisensor multitarget tracking using the marginalized δ-GLMB density. *IEEE Signal Processing Letters*, 23(6):863–867, 2016.
15 R. Mahler. Multitarget Bayes filtering via first-order multitarget moments. *IEEE Transactions on Aerospace and Electronic Systems*, 39(4):1152–1178, 2004.
16 H. Hoang, B.-N. Vo, B.-T. Vo, and R. Mahler. The Cauchy–Schwarz divergence for Poisson point processes. *IEEE Transactions on Information Theory*, 61(8):4475–4485, 2015.
17 C. Fantacci, B.-N. Vo, B.-T. Vo, G. Battistelli, and L. Chisci. Robust fusion for multisensor multiobject tracking. *IEEE Signal Processing Letters*, 25(5):640–644, 2018.
18 J. Li and A. Nehorai. Distributed particle filtering via optimal fusion of Gaussian mixtures. *IEEE Transactions on Signal and Information Processing over Networks*, 4(2):280–292, 2017.
19 A. Abbas. A Kullback-Leibler view of linear and log-linear pools. *Decision Analysis*, 6(1):25–37, 2009.
20 L. Gao, G. Battistelli, and L. Chisci. Multiobject fusion with minimum information loss. *IEEE Signal Processing Letters*, 27:201–205, 2020.
21 Y. Bar-Shalom, X.-R. Li, and T. Kirubarajan. *Estimation with Applications to Tracking and Navigation: Theory Algorithms and Software*. John Wiley & Sons, 2001.
22 S. Li, G. Battistelli, L. Chisci, W. Yi, B. Wang, and L. Kong. Multi-sensor multi-object tracking with different fields-of-view using the LMB filter. In *2018 21st International Conference on Information Fusion (FUSION)*, IEEE, pages 1201–1208, 2018.
23 L. Gao, G. Battistelli, and L. Chisci. Random-finite-set-based distributed multirobot SLAM. *IEEE Transactions on Robotics*, 36(6):1758–1777, 2020.
24 G. Li, G. Battistelli, W. Yi, and L. Kong. Distributed multi-sensor multi-view fusion based on generalized covariance intersection. *Signal Processing*, 166:107246, 2020.

25 G. Li, G. Battistelli, L. Chisci, W. Yi, and L. Kong. Distributed multi-view multi-target tracking based on CPHD filtering. *Signal Processing*, 188:108210, 2021.

26 B. Wang, S. Li, G. Battistelli, L. Chisci, and W. Yi. Multi-agent fusion with different limited fields-of-view. *IEEE Transactions on Signal Processing*, 70:1560–1575, 2022.

27 T. Li, V. Elvira, H. Fan, and J. Corchado. Local-diffusion-based distributed SMC-PHD filtering using sensors with limited sensing range. *IEEE Sensors Journal*, 19(4):1580–1589, 2018.

28 W. Yi, G. Li, and G. Battistelli. Distributed multi-sensor fusion of PHD filters with different sensor fields of view. *IEEE Transactions on Signal Processing*, 68:5204–5218, 2020.

29 K. Da, T. Li, Y. Zhu, and Q. Fu. Gaussian mixture particle jump-Markov-CPHD fusion for multitarget tracking using sensors with limited views. *IEEE Transactions on Signal and Information Processing over Networks*, 6:605–616, 2020.

30 L. Gao, G. Battistelli, and L. Chisci. Fusion of labeled RFS densities with different fields of view. *IEEE Transactions on Aerospace and Electronic Systems*, 58(6):5908–5924, 2022.

31 T. Nguyen, B. T. Vo, B.-N. Vo, H. V. Nguyen, and C. Shim. The mean of multi-object trajectories, 2025. https://arxiv.org/abs/2504.20391.

32 D. Schuhmacher, B.-T. Vo, and B.-N. Vo. A consistent metric for performance evaluation of multi-object filters. *IEEE Transactions on Signal Processing*, 56(8):3447–3457, 2008. https://doi.org/10.1109/TSP.2008.920469.

33 L. Gao, G. Battistelli, L. Chisci, and P. Wei. Event-triggered consensus Bernoulli filtering. In *2018 21st International Conference on Information Fusion (FUSION)*, IEEE, pages 84–91, 2018.

34 L. Gao, G. Battistelli, and L. Chisci. Event-triggered distributed multitarget tracking. *IEEE Transactions on Signal and Information Processing over Networks*, 5(3):570–584, 2019.

35 G. Li, G. Battistelli, L. Chisci, L. Gao, and P. Wei. Event-triggered consensus LMB filter for distributed multitarget tracking. *IEEE Transactions on Aerospace and Electronic Systems*, 59(1):712–719, 2022.

36 L. Gao, G. Battistelli, and L. Chisci. Resilience of multi-object density fusion against cyber-attacks. In *2022 11th International Conference on Control, Automation and Information Sciences (ICCAIS)*, IEEE, pages 7–12, 2022.

37 L. Gao, G. Battistelli, and L. Chisci. Resilient labeled multi-Bernoulli fusion with peer-to-peer sensor network. *Information Fusion*, 100:101965, 2023.

38 Y. Zhou, G. Battistelli, L. Chisci, L. Gao, G. Li, and P. Wei. Median-based resilient multi-object fusion with application to LMB densities. *IEEE Transactions on Signal and Information Processing over Networks*, 10:473–486, 2024.

39 H. Kim, K. Granström, L. Gao, G. Battistelli, S. Kim, and H. Wymeersch. 5G mmWave cooperative positioning and mapping using multi-model PHD filter and map fusion. *IEEE Transactions on Wireless Communications*, 19(6):3782–3795, 2020.

40 L. Gao, G. Battistelli, L. Chisci, A. Gostar, and R. Hoseinnezhad. Distributed joint mapping and registration with limited fields-of-view. In *2019 International Conference on Control, Automation and Information Sciences (ICCAIS)*, IEEE, pages 1–6, 2019.

41 L. Gao, G. Battistelli, L. Chisci, and P. Wei. Distributed joint sensor registration and multitarget tracking via sensor network. *IEEE Transactions on Aerospace and Electronic Systems*, 56(2):1301–1317, 2019.

42 L. Gao, G. Battistelli, L. Chisci, and P. Wei. Distributed joint sensor registration and target tracking via sensor network. *Information Fusion*, 46:218–230, 2019.

43 G. Li, G. Battistelli, L. Chisci, L. Gao, and P. Wei. Distributed joint detection, tracking, and classification via labeled multi-Bernoulli filtering. *IEEE Transactions on Cybernetics*, 54(3):1429–1441, 2022.

44 G. Li, P. Wei, G. Battistelli, L. Chisci, L. Gao, and A. Farina. Distributed joint target detection, tracking and classification via Bernoulli filter. *IET Radar, Sonar & Navigation*, 16(6):1000–1013, 2022.

45 L. Gao, G. Battistelli, L. Chisci, and A. Farina. GCI fusion based multi-detection multitarget tracking. In *2020 IEEE Radar Conference (RadarConf20)*, IEEE, pages 1–6, 2020.

46 L. Gao, G. Battistelli, L. Chisci, and A. Farina. Fusion-based multidetection multitarget tracking with random finite sets. *IEEE Transactions on Aerospace and Electronic Systems*, 57(4):2438–2458, 2021.

Part IV

Sensor Management

21

Agile and Non-agile Multi-satellite Scheduling

Ratnasingham Tharmarasa[1], *Abhijit Chatterjee*[2], *and Aranee Balachandran*[1]

[1] *Department of Electrical and Computer Engineering, McMaster University, Hamilton, ON, Canada*
[2] *Department of Computer Science, University of Exeter, Exeter, UK*

21.1 Introduction

Earth-observing satellites (EOSs) provide a unique perspective from space orbits for information acquisition required in surveillance and tracking missions, which is often beyond the capability of sensors placed on the Earth's surface. Due to the advantage of the high altitude of EOS, they have better access to information arising from a much larger area of interest (AOI) and difficult to reach AOIs due to hostile or remote environmental conditions. These enable EOSs to perform real-world applications, including change detection and tracking land and marine traffic, often required in defense applications.

The number of satellites available to observe the Earth is much less than the number of AOIs needed to be serviced in a given time window. Multiple visits by the same satellite or different satellites may be needed to cover a whole AOI. The time a satellite can spend on imaging during an orbit is also limited by the available resources, such as energy and memory, and the assigned nonimaging tasks. Hence, it is necessary to have an algorithm that can provide the optimal task-to-satellite assignment over time.

The workflow of EOS systems at the ground stations generally consists of collecting observation tasks from the users, planning the optimal task schedule of the EOSs, uplinking the schedule to the EOSs, downlinking the obtained image data from the EOSs, and processing the information before sending it to the users. The focus of this chapter lies in analyzing and modeling the EOS scheduling problem for the user-requested observation tasks, handling uncertainties, identifying appropriate methods for scheduling the tasks and assigning the EOSs to the tasks, and considering variants in task specifications while satisfying the necessary resource constraints of the EOSs. This chapter is based on the works presented in [1–3].

In the literature, several open-loop algorithms are proposed for the multi-satellite scheduling problems [4, 5]. To the best knowledge of the authors, a very few publications apply the Markov-decision process (MDP) for satellite scheduling. For many reasons, there is always a chance for a scan to fail. Hence, the failure probability must be incorporated in the ground planning, and decisions may be revised based on the outcome (success or failure) of each scan. In addition to failures, new tasks can appear in the already planned horizon. Hence, replanning is necessary to get optimal performance. If we ignore the new tasks, an MDP-based closed-loop planning can be used to avoid replanning. In [6], the feasibility of using an MDP-based optimal closed-loop solution for small and large-scale problems is analyzed. As shown in [6], handling a large-scale problem is hard in an optimal MDP framework due to the exponential growth of the number of MDP states. Hence, replanning is inevitable to get optimal performance for large-scale problems. In this chapter, adaptive replanning at the ground and MDP-based satellite local planning are presented to make use of the benefits of open- and closed-loop planning in Section 21.3.

To extend the above work to Agile EOSs (AEOSs), a realistic AEOS scheduling problem, where satellite resource specifications like energy and memory are taken into consideration, is studied in this chapter [3]. Scheduling AEOSs becomes significantly more difficult due to a larger task allocation search space. First, an open-loop solution technique is presented to handle multiple small regions in Section 21.4, and then that work is extended to handle larger regions in Section 21.5.

21.2 Satellite Scheduling: An Overview

21.2.1 Agile Satellite versus Non-agile Satellite

Non-agile EOSs (N-AEOSs) like SPOT systems of EOS[1] use a single degree of freedom for sensors to maneuver in the *roll* axis for image acquisition. The roll axis is parallel to the orbital trajectory of the satellite. The N-AEOSs can only perform imaging tasks when they are positioned exactly above the AOIs. This makes the visible time window (VTW) of an AOI extremely limited. In conflicting situations arising from overlapping VTWs, the problem reduces to a selection problem rather than a true scheduling problem [7].

On the contrary, AEOSs like *GeoEye-1*[2] have three degrees of freedom in *roll*, *pitch* and *yaw* axes [8]. The additional agility of AEOS provides significantly more opportunities for AOI imaging. The maneuverability in pitch and yaw directions omits the restrictions posed in N-AEOS systems.

Figures 21.1 and 21.2 illustrate the difference in the VTWs arising from the varied degrees of freedom in N-AEOS and AEOS. Since the starting time of image acquisition for AEOS can be more flexible than N-AEOS, it provides infinitely more imaging possibilities to the AEOS. In situations where the N-AEOSs fail to perform imaging tasks due to the overlapping of AOIs along the timeline, the AEOSs can perform those imaging tasks by using the advantage of the higher degrees of freedom. However, this makes the scheduling problem for AEOSs significantly harder due to the substantially larger search space.

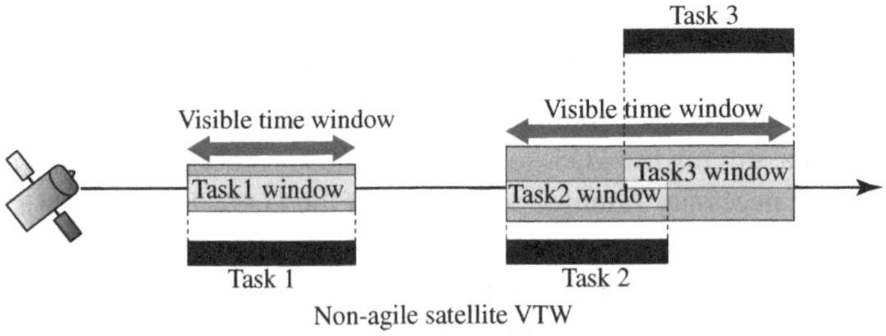

Figure 21.1 Visible time window for non-agile satellites.

1 https://spot.cnes.fr/en/SPOT/index.htm
2 https://earth.esa.int/eogateway/missions/geoeye-1

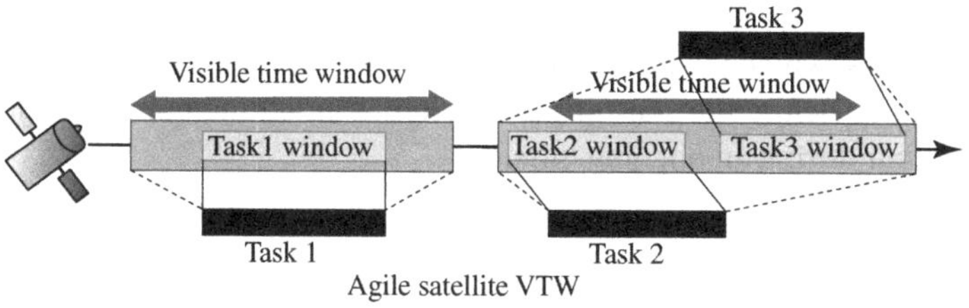

Figure 21.2 Visible time window for agile satellites.

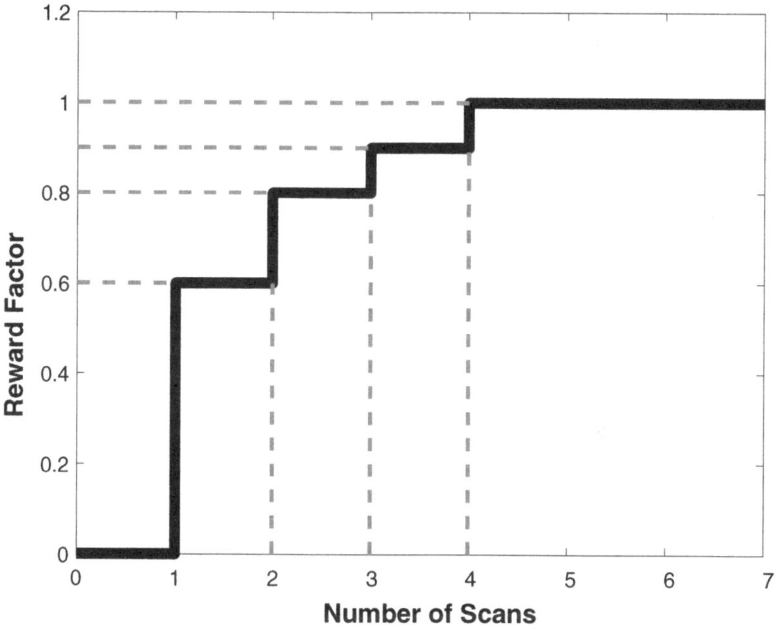

Figure 21.3 Reward factor versus completed scans.

21.2.2 What Is a Task?

An AOI defined by the user for satellite-sensor-based imaging is considered a task. AOIs might be spread across the Earth's surface in a scattered manner. The task AOIs for the AEOSs are usually classified as spot and polygon targets. Spot targets or polygon targets with smaller AOIs have a limited area to scan, which can be completed by one pass according to the observation scope of the AEOS sensor, while larger polygon targets may need to be captured by multiple passes of multiple AEOSs in a cooperative manner. When a request to scan a large region is received, the request either involves the region to be scanned in a geometrically consistent manner or involves scanning the whole region within a specified time window.

The reward after the successful completion of a task is also provided by the user. Depending on the nature of a task, single or multiple scans might be needed to complete that task. A reward function could be based on the quality of the image [9]. A step-function-based reward factor can also be used to deal with the multiple scan requirement for a task [1], see Figure 21.3 for a sample reward factor.

The user-requested tasks can be further classified based on the time of task arrival to the scheduler. Some tasks are daily imaging tasks, which are known to the scheduler prior to the start of the mission. These tasks can be scheduled before the start of the mission as an offline optimization problem. Some task arrivals can occur during an ongoing mission.

21.2.3 Resource Constraints

The EOSs do not require any energy to revolve around the Earth. The equilibrium of their centrifugal force and gravity helps the satellites to revolve. However, these satellites require energy to image using the sensors, transition from one task to another while orbiting by rotating the sensors along with the roll and pitch angles, and transfer data to and from the ground stations. The amount of energy needed for each action is different; thus, the discharging rate varies for each action. The satellites have solar sensors that automatically direct the solar panels toward the Sun so that the satellite's onboard batteries can charge at the maximum possible charging rate. Charging and discharging constraints have been used for AEOS scheduling [10, 11], N-AEOS scheduling [12], and nano-satellite swarms [13].

The task allocation and schedule of the tasks are uplinked to the satellites by ground stations. After completing the tasks, the satellites downlink the data to ground stations. In order to store the data until it is downlinked, a limited resource of onboard memory is used. Hence, periodically, the satellites need to transfer the collected data to the ground stations in order to free up the memory storage.

21.2.4 Uncertainty in the Satellite Scheduling

An assigned task may not be served successfully by a satellite due to possible cloud coverage, sensor and equipment malfunctions, and other factors. Some of the failures can be predicted to a certain accuracy with additional information such as weather forecast, while some are hard to predict. A better scheduling algorithm should factor in possible failures during the planning and revise actions based on failure reports.

21.2.5 Notations

Notations used for parameters related to satellites and tasks are summarized in Tables 21.1 and 21.2, respectively. The notations specific to the algorithms are described in corresponding sections.

Table 21.1 Parameters of satellites.

Notation	Meaning
N^s	Number of satellites in the mission
N^{os}	Number of revolutions in the trajectory of satellite s in the given time horizon
$T^{V_{st}}(s, o, g)$	Start time of the VTW for satellite s in revolution o to transfer data to ground station g
$T^{V_{end}}(s, o, g)$	End-time of the VTW for satellite s in revolution o to transfer data to ground station g
e_s^{roll}	Energy required by satellite s to roll a unit angle
e_s^{pitch}	Energy required by satellite s to pitch a unit angle
τ_s^{roll}	Time required by satellite s to roll a unit angle

Notation	Meaning
τ_s^{pitch}	Time required by satellite s to pitch a unit angle
e_s^{dl}	Average energy required for a downlink by satellite s
e_s^{ul}	Average energy required for an uplink by satellite s
$m_s^{\max}$	Maximum memory capacity of satellite s
m_s^{unit}	Memory used to scan per unit time step by satellite s
τ_s^{mem}	Time needed to transfer data between satellite and ground station
ch_s	Charging rate for satellite s
$\Theta_s^{\max}$	Maximum roll angles for satellite s
$\Phi_s^{\max}$	Maximum pitch angles for satellite s
ψ_s	Swath width angle of satellite s

Table 21.2 Parameters of tasks.

Notation	Meaning
N^t	Number of tasks to complete
$D_t^{\max}$	Maximum number of successful scans needed to complete task t
$\tau(s, o, t)$	Time needed by satellite s for scanning task t in revolution o
$T^{V_{\text{st}}}(s, o, t)$	Start time of the VTW for satellite s in revolution o for scanning task t
$T^{V_{\text{end}}}(s, o, t)$	End-time of the VTW for satellite s in revolution o for scanning task t
$R^f(t, d)$	Reward factor for completing d^{th} scan of task t
$R(s, \alpha, t)$	Reward for scanning task t by satellite s with observation angle α (includes roll and pitch angles)
$P(s, o, t)$	Probability of success for a particular scan of task t by satellite s in revolution o
$e(s, o, t)$	Energy required by satellite s to scan task t in revolution o
$m(s, t)$	Memory usage of satellite s to scan task t
A_t	Total area of task t
α_t	Earliness penalty cost for task t per unit time step
β_t	Tardiness penalty cost for task t per unit time step
$R_t^{\max}$	Maximum reward for the task t provided by the user
$\mathbb{R}_t^{\text{unit}}$	Reward for scanning unit area of task t
$\zeta(t, \theta, \phi)$	Reward factor for the AEOSs' roll and pitch angles during the scan

21.3 Non-agile Satellite Scheduling: Mixed Open-and Closed-loop

In this section, a mixed open-and closed-loop-based satellite scheduling for non-agile satellites is analyzed. As discussed in the introduction, a closed-loop planning approach will help avoid replanning to achieve an optimal solution when a failure occurs in a scan. However, managing large-scale problems involving hundreds of tasks is challenging within an optimal MDP framework, primarily because the number of MDP states grows exponentially [6]. This section introduces adaptive ground-level replanning and MDP-based satellite local planning to capitalize on the strengths of both open- and closed-loop planning methods.

To simplify the analysis, in this section, it is assumed that AOIs are already partitioned into multiple parallel strips [14], which are called tasks, so that each strip can be fully covered by a satellite during an orbit. The strip size depends on the satellite altitude, the sensor beam width and the satellite ground track trajectory. We may not be able to use the same strips for a non-trailing satellite constellation.

For a non-agile satellite, the start and end times of each task for a given orbit of the satellite are known and fixed. Hence, decision variables do not need to include the start and end times of the tasks for the satellite assignments. Instead, we need to decide if a task should be scanned or not during an orbit.

Another problem to solve is finding out the communication (download and upload) times for all the satellites based on the possible satellite-ground station coverages, remaining satellite resources, tasks serviced after the last communication time, etc.

The problem is to decide when to service, which task to service, and when to upload the instructions and download the images for each satellite in order to maximize the overall reward over the given horizon.

Ground stations can find the optimal task-to-satellite assignment over the given horizon by considering all the satellites together. Mostly, the satellites have limited onboard processing capacity. Hence, after the ground station assigns the tasks to a satellite for an intercommunication interval, a satellite may pick the best task based on the latest weather and cloud conditions obtained by the onboard sensors.

21.3.1 Assumptions

The following assumptions are made for this mixed open-and closed-loop formulation: (1) Satellites have onboard processing capabilities. They can update the cloud coverage information so that the action could be revised based on the coverage feedback; (2) If a task is started to service, it cannot be interrupted; (3) Each satellite can service only a task at a time; (4) Satellites' onboard memory will be cleared after each downlink; (5) Satellites' energy will be recharged after each orbit; (6) Ground stations are interconnected, so all the ground stations will have the same global view. Hence, instead of considering multiple ground stations in the formulation, all the possible ground stations are marked as the ground station, and a single ground planner is considered.

21.3.2 Framework

The flowchart of the considered mixed open-and closed-loop framework is shown in Figure 21.4.

In this approach, the ground station runs an open-loop with feedback solution while satellites run closed-loop solutions for the intercommunication interval, i.e., the interval between satellite's two consecutive communications with ground stations. The ground open loop solution should incorporate that the satellites have onboard processing capability and will use a closed-loop solution. If the satellites do not have onboard processing, the ground station can assign a series of tasks to service. However, when a satellite has onboard capability, it can service a different task that can have a time overlap with a preferred task (rather than waiting for the next task in the queue) if a preferred task cannot be serviced for any reason. In this case, the ground station should also assign alternatives.

The above problem can be divided into the following four subproblems: (1) Down/uplink time planning; (2) Ground open-loop planning; (3) Ground open-loop replanning; (4) Satellite closed-loop planning.

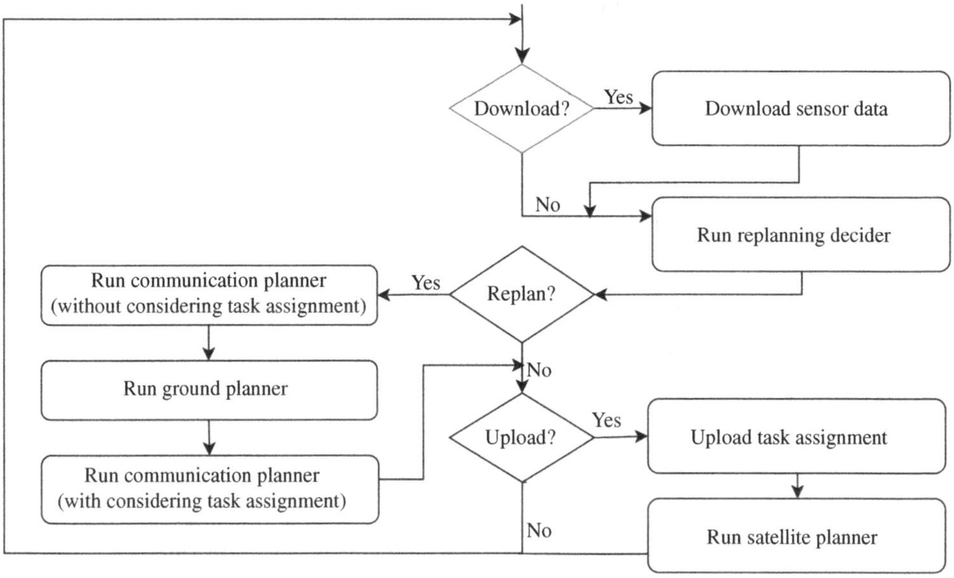

Figure 21.4 Flowchart of a mixed open-and closed-loop framework.

21.3.3 Down/Uplink Time Planning

We can find the predicted down/uplink times as an open-loop problem for the whole horizon. These predicted times can be used by the ground planner and the satellite planner. To simplify the problem, these predicted times can be calculated independently of assigned tasks. Even though this is a suboptimal approach, it will simplify all other planning.

To reduce the computational complexity, this planning can be performed using only the satellite-to-ground station coverage and the satellite's energy and memory limitations without considering the tasks to service. This planning can be performed for each satellite independently.

The information that can be used for the planning is the start and end times of satellite-ground station coverages, average duration of downlink/uplink, memory capacity of each satellite, average memory usage for a unit time, energy capacity of each satellite, and average energy required for downlink/uplink.

In most cases, we will have more than one satellite-to-ground station coverage intervals for an orbit. Hence, the outcomes (success and failure of assigned tasks) of a satellite after a downlink may not affect the task assignment for the same satellite for the next intercommunication interval, since the possible serviceable tasks will be different from the ones in the last intercommunication interval. If the downlink duration is significantly large, it is better to perform uplink (i.e., provide the next task assignment) before downlink. In most cases, the uplink (sending tasks assignment) duration is smaller than the downlink (receiving sensor raw data) duration.

A simple approach is to limit one uplink and one downlink for a ground-satellite interval with a maximum wait time.

21.3.4 Ground Open-loop Planning

A detailed open-loop planning will be discussed in Section 21.4 for agile satellite. A simplified version of agile-satellite planning can be used for non-agile satellites.

21.3.5 Ground Open-loop Replanning

Once we have a solution to the ground-open loop planning, we may need to rerun the planner (with the new world state, which affects only the variables, e.g., number of tasks, not the formulation) whenever a significant change occurs.

We may need to replan, if a new task appears, priorities of the unserviced tasks change, the weather forecast changes, a satellite fails to scan a task, a satellite's operation is interrupted for any reason, or the current time exceeds a predefined duration from the last replanning time.

21.3.6 Satellite Closed-loop Planning

An MDP-based closed-loop planning is performed at the satellites [15].

21.3.6.1 Inputs to the Satellite

In the ground open loop planning, only one task at a time is assigned to the satellites. However, we can send multiple tasks with different priorities for each overlapping duration since satellites use local MDPs to decide which tasks to service based on the latest weather forecasts.

Once we find the optimal assignment $a(s, o, t)$, assignment of task t to satellite s for orbit o ($\in \{0, 1\}$), from the ground open-loop planner, tasks that are within the coverage of the satellite during the next satellite planner horizon (intercommunication interval) are assigned to the satellite with the reward:

$$\bar{R}(t) = \lambda_t R^f(t, N_t^c + 1) \tag{21.1}$$

where N_t^c is the number of already completed scans for task t by any satellite at the time of satellite-planning, $R^f(t, N_t^c + 1)$ is the reward factor for $N_t^c + 1$-th scan and λ_t is the scaling factor for task t. If we are planning for satellite $\bar{s}$ for orbit $\bar{o}$, the scaling factor λ_t can be selected as

$$\lambda_t = \begin{cases} 1 & \text{if } a(\bar{s}, \bar{o}, t) = 1 \\ \bar{\alpha} b_t & \text{if } a(\bar{s}, \bar{o}, t) = 0 \end{cases} \tag{21.2}$$

where $\bar{\alpha}$ is the maximum factor for a second choice, e.g., 0.1, and $b_t \in [0\ 1]$ is the benefit of assigning task t as the second choice. The value of $\bar{\alpha}$ should be much smaller than 1.0 in order to avoid a second choice with a larger (unmodified) reward to be picked instead of a first choice (i.e., $a(\bar{s}, \bar{o}, t) = 1$) task with a lower reward.

The b_t can depend on several factors, including the next assignment time whether to the same or different satellite, the probability of success of that assignment, the probability of missing that task considering all the future assignments, the probability of having a new task closer to this task (a serving opportunity could be used for the new task), etc. A possible b_t could be

$$b_t = \beta_m p_m + \beta_n p_t^{\text{new}} + \beta_u \min\left(1.0, \frac{\nu_t}{\nu^{\max}}\right) + \beta_l \min\left(1.0, \frac{l_t^n}{L_t^{\max}}\right) \tag{21.3}$$

with

$$\beta_m + \beta_n + \beta_u + \beta_l = 1.0 \tag{21.4}$$

where, p_m—the probability of miss detection (zero success), β_m—weight for miss detection, β_n—weight for new target occurrence, β_u—weight for the number of overlapping unserviced tasks, β_l—weight for remaining time to try again after the next assignment, p_t^{new}—probability of getting a new task in a unit time (unit time can be decided based on horizon duration, etc.) near task t, v_t—number of overlapping unserviced tasks, v^{max}—maximum threshold for v_t, l_t^n—time duration between the current time and the next assignment time, and L_t^{max} is the maximum threshold for l_t^n.

Instead of finding the modified rewards at the ground station, all the satellite-to-task assignments could be sent to the satellite to find the MDP solution. However, any operations at the satellite level could be done at the ground station level in order to reduce the communication and computational loads of the satellite.

21.3.6.2 Local MDP

A local MDP at a satellite does not need to handle possible new tasks since the local MDP consider the actions from one communication time to the next communication time, during which new tasks that appear at the ground station level will not be available to the satellite. Hence, new tasks are handled by the ground station. This problem can be formulated as the following MDP problem. The state contains:

- ct—current task;
- sp—status of the previous task (0: failed or skipped, 1: completed);
- r—remaining resources;
- $\{c_i\}$—coverage of future task i (1: task i can be serviced, 0: otherwise (e.g., cloud presence preventing task i coverage);
  - All future tasks, including the current task, should be considered. For example, if we are at task 5, and ten more tasks could be in the coverage, then 2^{10} additional states are needed for the state with the current task 5.

The possible actions are:

- a_s—skip the current task;
- a_p—service the task.

The outcomes are:

- o_s—success;
- o_f—failure.

The transition probabilities are:

- p_s—probability of success of the last task (depends on the sensor);
- p_{c_i}—probability of coverage of future task i (depends on cloud coverage).

The rewards are:

- task's reward, $\bar{R}(t)$, (if completed);
- zero (if skipped or failed);
- $-\infty$ (if remaining resources become less than zero).

In order to handle a larger number of tasks, we may need to have an approximation to limit the number of states. A positive note is that even if we have many tasks in total, only a few may be within the coverage of the current satellite during the intercommunication interval.

An approach to reduce the number of states is to consider only a k (e.g., 5) significant future tasks that can change the serving decision of the given task. While deciding the k-significant tasks, we may need to consider future tasks with

- overlapping coverage duration (i.e., $t_c^s \leq t_a^s \leq t_c^e \; \forall \; a$, where t_c^s and t_c^e are the start and end times of the current task, respectively, and t_a^s is the start-time of task a).
- better rewards than the current reward. In some cases, saving energy and memory by skipping the current task to increase the possibility of servicing a future task may provide a better overall reward. If there is no energy or memory limitation, we do not need to consider this.

The value iteration dynamic programming algorithm can be used to solve the above MDP problem [16].

21.3.7 Simulation Results

In this simulation, a large-scale problem with 440 tasks is considered with the following settings.

- Lengths of tasks are uniformly distributed between 30 km and 350 km.
- Rewards are uniformly distributed between 10 and 100.
- Start-time is 0 for 20% of the tasks and uniformly distributed between 0 h and 5 h for the remaining 80%.
- Maximum allowed latency (task time window) is uniformly distributed between 3 h and 10 h.
- Reward factor 1.0 (i.e., one scan is enough) for 75% of the tasks and [0.8 1.0] (i.e., two scans are required to get the full reward, and 0.8 of the full reward will be given after one scan) for the remaining 25%.

Sample trajectories of the satellites are shown in Figure 21.5a. The locations of the tasks are shown in Figure 21.5b.

The completed tasks and the collected rewards at the end of the simulation are shown in Table 21.3. As expected, the mixed open-and closed-loop planning algorithm gives a significant improvement over an open-loop solution. The addition of satellite planning to the open-loop solution provides more improvement than the open-loop with replanning. Note that only 20% of the tasks, which were available at the start of the mission, are considered in the "open-loop" and the

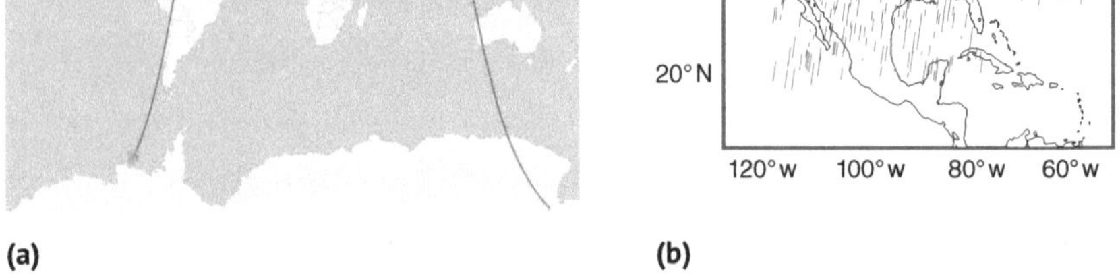

Figure 21.5 (a) Satellite trajectories and (b) task distribution.

Table 21.3 Performance comparison for a large-scale problem with three satellites.

Algorithm	Completed tasks (out of 440)	Partially completed tasks (out of 440)	Collected reward (out of 8,872.3)
Open-loop	40.19	9.31	2,819.13
Open-loop with replanning	53.36	11.81	4,608.99
Open-loop with satellite planning	46.77	10.18	3,084.39
Mixed open-and closed-loop	79.7	16.78	6,055.56

"open-loop with satellite local planning" algorithms, while all the tasks are considered in the other two algorithms.

The peak computation times over all the Monte Carlo runs for ground open-loop planning and satellite local planning were 0.888 seconds and 4.715 seconds, respectively. The average peak computation times of all the Monte Carlo runs were 0.184 seconds for ground open-loop planning and 3.859 seconds for satellite local planning.

21.4 Multi-agile Satellites Scheduling: Small Tasks

In this section, we consider that satellites need to perform some predefined observation tasks within the given time horizon. The total reward is calculated by accumulating all the rewards gained by the satellites at the end of the mission horizon. Satellites can scan the whole region of the task in one visit. The objective is to schedule observation tasks optimally by allocating the available resources of agile satellites.

In order to scan a particular task, the position of the satellites in their trajectory needs to be inside the VTW of the corresponding task. The VTW of a task for a satellite is calculated beforehand based on the geographic location of the task region, the orbit of the satellite and the maximum roll and pitch angles of the satellite, see Figure 21.6. Each possible start time of a scan for the task within the VTW will have a different roll and pitch angle combination for every satellite on every revolution. The reward accumulated by a satellite after scanning a task may depend on the corresponding roll and pitch angles. As a result, it is important to schedule the scans in an optimal way to maximize the reward with the above-mentioned factors.

The amount of energy needed for each action is different; thus, the discharging rate varies for each action. Figure 21.7 shows an example of satellite charging and discharging over multiple revolutions.

21.4.1 Problem Formulation

In this formulation, the start time of each operation is considered as a decision variable, which is continuous in nature. The operations of a satellite include performing the scans, and data transfer to and from the ground stations. A binary decision variable is also defined to determine whether a satellite is performing a particular scan on a specific revolution. The variables used for the formulation are described below.

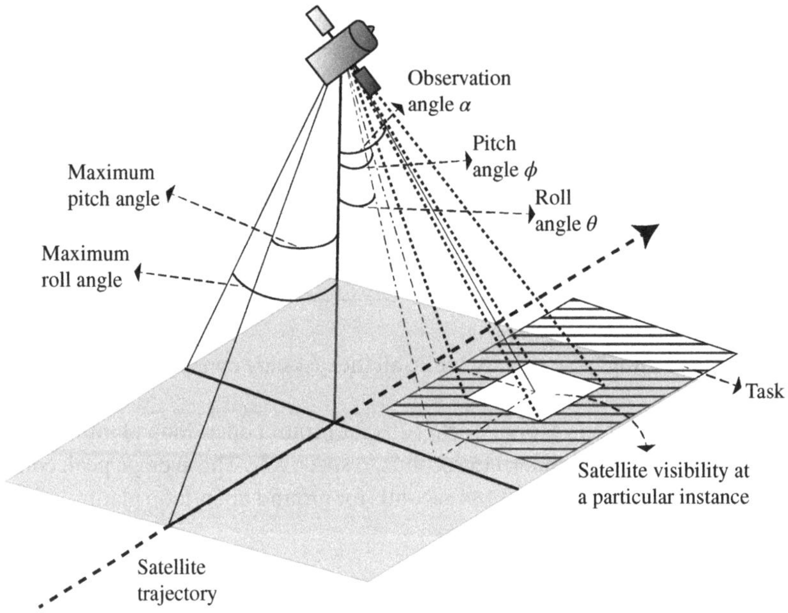

Figure 21.6 Observation angle of a satellite.

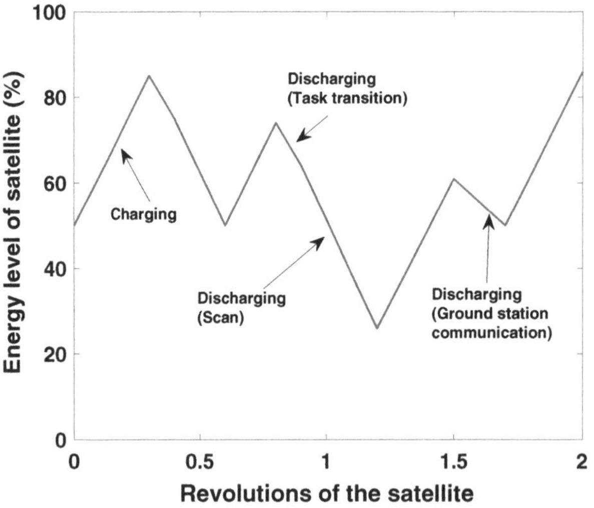

Figure 21.7 Charging and discharging of energy for satellites in each revolution.

Decision Variables:

- $a(s, o, t)$: Binary variable that indicates if satellite s performs task t in revolution o
- $T^{st}(s, o, t)$: Start time of scanning for task t by satellite s in revolution o
- $T^{sg}(s, o, g)$: Start time of data transfer from satellite s in revolution o to ground station g

Other Variables:

- $P^+(s, t, d)$: Probability of getting d successful scans for task t by satellite s
- $P^-(t, (d-1), s, o)$: Probability of having $d-1$ successful scans before the start time ($T^{st}(s, o, t)$) of the scan by satellite s in revolution o
- $N(s, g, \hat{g})$: Number of scans performed between communication of satellite s to ground station g and the next ground station $\hat{g}$
- $\theta_{t,\hat{t}}(s, o)$: Roll angle needed for satellite s in revolution o to move from task t to task $\hat{t}$
- $\phi_{t,\hat{t}}(s, o)$: Pitch angle needed for satellite s in revolution o to move from task t to task $\hat{t}$
- $C(s, \gamma)$: Available charge of satellite s at time γ

21.4.1.1 Objective Function

The objective function involves optimizing the reward accumulated by all the satellites during the mission by performing several successful scans of the tasks. The objective function is defined as follows:

$$\max \sum_{t=1}^{N^t} \sum_{d=1}^{D_t^{\max}} R^f(t, d)\rho(t, d) \tag{21.5}$$

$$\text{where,} \quad \rho(t, d) = \sum_{s=1}^{N^S} \sum_{o=1}^{N^{os}} R(s, \alpha, t)P^+(t, d, s, o) \tag{21.6}$$

Here, the reward factor $R^f(t, d)$ is included in the objective function to calculate reward for different scans of the same task and $\rho(t, d)$ calculates the reward for dth successful scan of task t. The reward $R(s, \alpha, t)$ achieved by scanning might depend on the observation angle α of the task t from the satellite s. The relationship of reward with the observation angle is specific to the task and defined by the user. In general, the reward is inversely proportional to the observation angle. The angle α is calculated as a combination of the roll and pitch angle movements of the satellite, depending on the position of the task. For example, the relationship of the observation angle to the reward can be calculated as:

$$R(s, \alpha, t) = R_t^{\max} - \left(\frac{R_t^{\max}}{2} \times \frac{\theta + \phi}{\Theta_s^{\max} + \Phi_s^{\max}}\right) \tag{21.7}$$

where $R_t^{\max}$ is the maximum reward possible to accumulate for task t, θ and ϕ are the roll and pitch angle for the scan, respectively, and $\Theta_s^{\max}$ and $\Phi_s^{\max}$ are the maximum allowable roll and pitch angle for the satellite, respectively. The above model for the reward calculation is used in Section 21.4.3 for all the simulations.

A scan can be unsuccessful due to several factors, such as weather influences and sensor malfunctions. $P^+(t, d, s, o)$ is defined as the probability of having successful scan for task t by satellite s in revolution o. The probability of successful scan is defined as:

$$P^+(t, d, s, o) = a(s, o, t)P(s, o, t)P^-(t, (d-1), s, o) \tag{21.8}$$

where, $P^-(t, (d-1), s, o)$ is the probability of having $d-1$ successful scans before the start time ($T^{st}(s, o, t)$) of the scan by satellite s in revolution o.

If the external factors are not taken into consideration, i.e., $P(s, o, t) = 1$ for all (s, o, t), then the probability of dth successful scan depends only on the decision variable $a(s, o, t)$ and the possible values are $\{0, 1\}$.

21.4.1.2 Energy Constraints

Satellites have a finite amount of energy and recharge using solar panels. It is important to ensure that the satellite will have enough energy to complete an operation. It is assumed that the satellite must have $\kappa(\geq 0)$ energy remaining after the completion of each operation.

The available charge at the starting time for a scan of task t by satellite s is denoted as $C(s, T^{st}(s, o, t))$. The energy $e(s, o, t)$ needed to scan task t in revolution o should always be lesser than or equal to the available charge at the starting time $C(s, T^{st}(s, o, t))$ plus the gained energy $ch_s \times \tau(s, o, t)$ during the scan. Therefore, the energy constraint for scanning a task can be written as,

$$\left[C(s, T^{st}(s, o, t)) + ch_s \times \tau(s, o, t) - e(s, o, t)\right] \geq \kappa \qquad \forall t, s, o \text{ when } a(s, o, t) = 1 \tag{21.9}$$

Another energy constraint is needed to deal with the resources required for the rotation of the satellite. When task $\hat{t}$ is followed by task t, the satellite needs energy to rotate along the roll and pitch angles. It is important to ensure that the satellite has enough energy to rotate. The total available energy at the start of the rotation is $C(s, T^{rot}(s, o, t, \hat{t}))$. The energy gained by the satellite during the roll and the pitch is $ch_s(\tau_s^{roll}\theta_{t,\hat{t}}(s, o))$ and $ch_s(\tau_s^{pitch}\phi_{t,\hat{t}}(s, o))$ respectively, where $\theta_{t,\hat{t}}$ is the roll angle change needed and $\phi_{t,\hat{t}}$ is the pitch angle change needed while transitioning from task t to task $\hat{t}$. Equation (21.10) describes the energy requirement constraint for the transition of the satellite from one task to another.

$$\begin{aligned} C(s, T^{rot}(s, o, t, \hat{t})) + ch_s\left(\tau_s^{roll}\theta_{t,\hat{t}}(s, o) + \tau_s^{pitch}\phi_{t,\hat{t}}(s, o)\right) \\ -\left(\theta_{t,\hat{t}}(s, o)e_s^{roll} + \phi_{t,\hat{t}}(s, o)e_s^{pitch}\right) \geq \kappa \quad \forall\, t, \hat{t} \neq t, s, o \end{aligned}$$

Another energy constraint is needed to handle the energy requirement for the communication between the satellite and the ground stations. The energy available at the start of a data transfer is $C(s, T^{sg}(s, o, g))$. The energy gained during the data transfer is $ch_s\tau_s^{mem}$. The required energy for the communication is $(e_s^{dl} + e_s^{ul})$. Hence, the constraint for uploading and downloading the data is defined as follows:

$$C(s, T^{sg}(s, o, g)) + ch_s\tau_s^{mem} - (e_s^{dl} + e_s^{ul}) \geq \kappa \quad \forall\, s, o, g \tag{21.10}$$

where, e_s^{dl} and e_s^{ul} are considered as the average energy required by a satellite to downlink and uplink the data, respectively, to the ground station.

21.4.1.3 Memory Constraints

Satellites have a finite resource of memory. While scanning, the memory of the satellite fills up and during the communication with the ground station, it resets the memory. The satellite s can only perform a specific number of scans between the communication to the ground station g and the next available ground station $\hat{g}$. Hence, the memory constraint is defined as,

$$\sum_{N(s,g,\hat{g})} m(s, t) \leq m_s^{max} \quad \forall\, s, g, \hat{g} \neq g \tag{21.11}$$

The value of $N(s, g, \hat{g})$ is calculated as the sum of the decision variable $a(s, o, t)$ in the interval between two communications with the ground station.

21.4.1.4 Constraints Related to the Start Time of an Operation

If the data transfer between the ground station and the satellite occurs before the scan of task t, the start time of the scan should be after it has completed the data transfer. So, if a download occurs and the satellite takes τ_s^{mem} time to transfer the data, then

$$T^{st}(s,o,t) > \tau_s^{\text{mem}} + T^{sg}(s,o,g) \tag{21.12}$$

Similarly, if the data transfer is done after the scan of task t, the data transfer should start after the scan is complete.

$$T^{sg}(s,o,g) > T^{st}(s,o,t) + \tau(s,o,t) \tag{21.13}$$

It is important to ensure that the scanning of the tasks or data transfer operations occurs when the satellites are in the VTW of the task and the ground station, respectively. The transfer should occur inside the VTW of the ground station, and the scan for the task should always start within the VTW of the tasks, i.e.,

$$\begin{aligned} T^{V_{st}}(s,o,g) \le T^{st}(s,o,g) \le T^{V_{\text{end}}}(s,o,g) \\ T^{V_{st}}(s,o,t) \le T^{st}(s,o,t) \le T^{V_{\text{end}}}(s,o,t) \end{aligned} \tag{21.14}$$

While transitioning from one task to another, before starting the second task, the satellite needs to ensure that it has completed the scan for the first task and has enough time to rotate. Hence, to ensure the minimum time interval between start time for scan of task t and start time for scan of $\hat{t}$, the constraint is defined as:

$$T^{st}(s,o,\hat{t}) \ge T^{st}(s,o,t) + \tau(s,o,t) + \tau_s^{\text{roll}}\theta_{t,\hat{t}}(s,o) + \tau_s^{\text{pitch}}\phi_{t,\hat{t}}(s,o) \tag{21.15}$$

21.4.2 Solution Technique

In this section, a heuristic search technique has been provided to find a near-optimal solution to the nonlinear mixed-integer programming problem formulated in Section 21.4.1. The solution methodology is based on a combination of local search and genetic algorithm with an elitist selection operator.

An elitist mixed coded genetic algorithm-based satellite scheduling (EMCGA-SS), which is based on a modified elitist genetic algorithm [17] with binary and continuous variables together in the structure of the chromosome, is used. The structure of the chromosome, having both binary and continuous variables, is specifically designed to solve the formulated problem in a simpler manner. Instead of having different chromosomes for the different types of variables, the combined chromosome helps to reduce the total population size of the solver. EMCGA-SS considers an initial population of candidate solutions, which are random schedules of all the satellites performing the tasks across the whole mission horizon. Our task is to find a near-optimal schedule, that is, near-optimal chromosome from the population. The algorithm uses crossover and mutation of schedules to create offspring, and an elitist selection operator is used to select from a pool of offspring schedules. The chromosomes are sorted based on the reward accumulated for the corresponding schedule. In each generation, the best half of the population is chosen to create the next generation with the help of mutation, crossover and elitist selection operators. The structure of the EMCGA-SS is described in Algorithm 21.1, where λ is the population size defined by the user based on the complexity of the problem.

Algorithm 21.1: Elitist Mixed Coded Genetic Algorithm-based Satellite Scheduling

Result: Optimal schedule for all satellites;
Set population of schedules, $\mathcal{P}_0 = \Phi$;
while *population size,* $|\mathcal{P}_0| \leq \lambda$ **do**
  Allocate a task randomly to each satellite in each revolution to generate a random chromosome p;
  $\mathcal{P}_0 = \mathcal{P}_0 \cup p$;
end
$Gen = 0$;
while *Stopping criteria reached* **do**
  Sort population $\mathcal{P}_{\text{Gen}}$ by reward accumulated (Eq. 21.5);
  Select $\lambda/2$ best chromosomes from population, $\mathcal{P}_{\text{Gen}}^{\lambda/2}$;
  for $|\mathcal{P}_{Gen+1}| \leq \lambda$ **do**
    Select 2 parents p_1, p_2 randomly from $\mathcal{P}_{\text{Gen}}^{\lambda/2}$;
    Obtain children c_1, c_2 by crossover on p_1, p_2;
    Mutate c_1, c_2 to obtain cm_1, cm_2;
    Check feasibility of c_1, c_2, cm_1, cm_2 with Eqs. (21.9)–(21.15);
    Add best feasible individual to $\mathcal{P}_{\text{Gen+1}}$;
  end
  $\text{Gen} = \text{Gen} + 1$;
end

Structure of Chromosome: Each chromosome in the population represents the task assignment and start time of the operations of the satellites. The respective schedules of N^s satellites are juxtaposed to create the chromosome. Since the exposed revolutions for each satellite corresponding to each task calculated in the preprocessing stage are not equal, the number of bits corresponding to the task allocation of each satellite is different. A schedule of a satellite s is designed as a combination of bits containing binary and continuous numbers. The possibility of occurrence of the scan for a task on an exposed revolution $a(s, o, t)$ is the binary bits stored at the beginning of the chromosome. The length of the binary bits is the same as the total number of exposed revolutions for all the tasks. The start time of each corresponding scan $T^{st}(s, o, t)$ is stored in the continuous bits, following the binary bits. The total length of the bits containing start time is the same as the length of the binary bits. The bits containing the start time of the communication to the ground station in each exposure $T^{sg}(s, o, g)$, are placed at the end of the schedule of the satellite s. The structure of the chromosome is illustrated in Figure 21.8.

Detailed explanation of the operations of this genetic algorithm is provided in [3].

21.4.3 Simulation Results

For a large-scale scenario, a significantly larger task allocation problem with 400 tasks is considered. These tasks have rewards uniformly distributed between 10 and 30. The lengths of these tasks are also uniformly distributed between 30 km and 350 km. For 75% of the considered tasks, it is assumed that one scan is enough to get the full reward, and for the remaining 25% of the tasks, two scans are required to get the full reward with the reward factor of (0.8, 0.2).

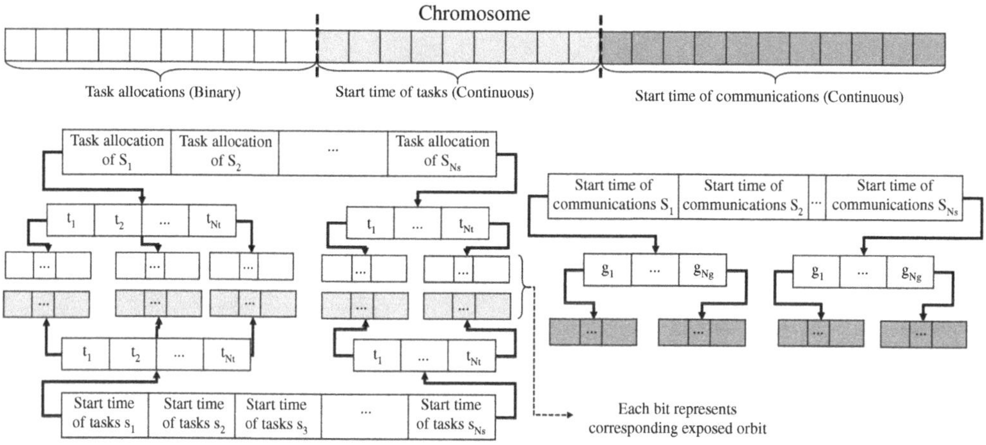

Figure 21.8 Structure of chromosome.

Table 21.4 Performance comparison between AEOS and N-AEOS satellites in large scale.

Satellite type	Number of scans	Total reward
AEOS	984	4,527
N-AEOS	731	2,886

Table 21.5 Performance comparison of different algorithms.

Algorithms considered	Reward	Time needed (s)
EMCGA-SS	**4,867**	1,309
Tabu search	3,240	10,500
Simulated annealing	2,969	9,644

The mission horizon is set to 12 hours. Five identical agile satellites with different trajectories have been considered for this simulation. The satellites periodically communicate with 6 ground stations.

Table 21.4 shows the comparison between the performance of AEOS and N-AEOS. As expected, AEOS performs a higher number of scans during the mission horizon and accumulates a larger reward than N-AEOS.

In Table 21.5, the performance of the EMCGA-SS algorithm is compared with Tabu-Search and Simulated Annealing for five AEOS satellites. The EMCGA-SS provides a better schedule in terms of total reward accumulated and lesser computation time.

A comparison among three, five, and eight satellites is also performed for 100 Monte Carlo runs, and the results are shown in Figure 21.9. The figure with box plots shows that if the number of satellites increases, the reward accumulation increases as well in the large-scale scenario.

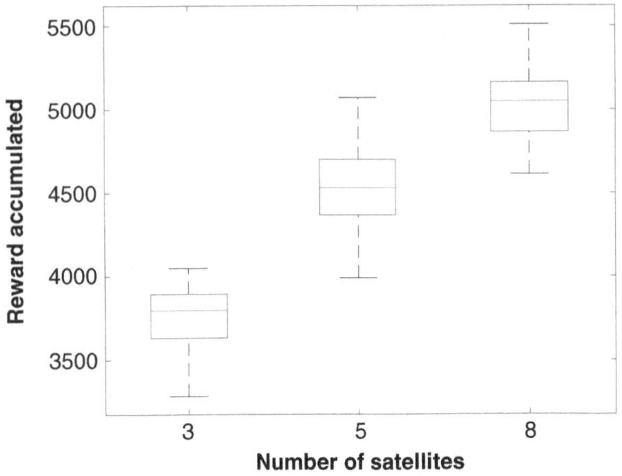

Figure 21.9 Reward accumulated with respect to the number of satellites for large-scale scenario.

21.5 Multi-agile Satellites Scheduling: Large Tasks

In this section, it is assumed that the task AOIs can be significantly larger than the visibility range of any AEOS while passing over it. These large tasks would need multiple *scans* of smaller regions that could be completed by a single AEOS visit. These user-requested tasks are known prior to the start of the mission. The reward for a scan is calculated by the amount of area covered by an AEOS in a single visit. As multiple scans are needed to cover the whole region of the task, the total reward accumulated for each task is the sum of the reward accumulated during each individual scan. The reward for each scan is assigned when the AEOS completes a scan successfully.

21.5.1 Problem Formulation

This formulation is also valid if some of the task regions are smaller than the scanning capacity of an AEOS.

The variables for the optimization problem are defined as follows:

- $x(s, o, t) \in \{0, 1\}$—decision variable denoting the assignment of AEOS s to task t in revolution o. Here 0 means that there are no assignments and 1 means at least one assignment.
- $T^{st}(s, o, t)$—a vector decision variable containing start time for all the assigned scans of AEOS s in revolution o to task t. If there are continuous scans of task t with different roll and pitch angles, those are considered discrete elements, and the start times are noted.
- $t^{st}(s, o, t, i)$—decision variable denoting ith element of $T^{st}(s, o, t)$.
- $\Theta(s, o, t)$—vector decision variable denoting the roll angle for the scan of task t by AEOS s during revolution o. The size of the vector would be the same as $T^{st}(s, o, t)$.
- $\Phi(s, o, t)$—vector decision variable denoting the pitch angle for the scan of task t by AEOS s during revolution o. The size of the vector would be the same as $T^{st}(s, o, t)$.
- θ_t^i—decision variable denoting roll angle for ith scan of task t.
- ϕ_t^i—decision variable denoting pitch angle for ith scan of task t.
- $D(s, o, t)$—a vector decision variable containing duration for all the assigned scans of AEOS s in revolution o to task t. The size of the vector would be the same as $T^{st}(s, o, t)$.

- $d(s, o, t, i)$—decision variable denoting ith element for $D(s, o, t)$, which is the duration of the ith scan for task t. It can take any value from 0 to the maximum duration required for a scan using satellite s.
- $\tilde{A}(s, o, t, i)$—area of task t scanned by AEOS s in revolution o in the ith scan.
- $\tilde{R}(s, o, t, i)$—the reward obtained for successful completion of ith scan by AEOS s for task t in revolution o.
- $\mathbb{R}^{unit}(s, o, t, i)$—unit reward for ith scan by AEOS s in revolution o for task t, including the penalty.
- $m(s, o, t, i)$—memory used for ith scan of task t by AEOS s in revolution o.
- $T_g(s, o)$—communication start time to ground station g for satellite s in revolution o.
- $d_g(s, o)$—communication duration to ground station g for satellite s in revolution o.
- $\tilde{B}_t$—Set of overlapped regions for task t due to multiple scans.
- $\mathbb{B}_t^n$—Reward accumulated for the nth overlapped scan for task t.

When the AEOSs perform scanning tasks, some scans need to be initiated within a specified time window defined by the user. Let the scanning of a task t have a requirement to be started within a particular time window, defined as $\left[\tau_t^l, \tau_t^u\right]$ where τ_t^l and τ_t^u denote the start time and the end time of the time window, respectively.

If a scanning is done before or after the specified time window, an earliness $\mathcal{E}(s, o, t, i)$ or a tardiness $\mathcal{L}(s, o, t, i)$ penalty factors are applied as follows,

$$\mathcal{E}(s, o, t, i) = \max\left(\tau_t^l - t^{st}(s, o, t, i), 0\right) \tag{21.16}$$

$$\mathcal{L}(s, o, t, i) = \max\left(t^{st}(s, o, t, i) - \tau_t^u, 0\right) \tag{21.17}$$

The reward achieved by scanning the area $\tilde{A}(s, o, t, i)$ in the ith scan will be penalized if it does not occur within the specific time window $\left[\tau_t^l, \tau_t^u\right]$. This penalty would be effective only on the area scanned by the AEOS $\tilde{A}(s, o, t, i)$.

The reward achieved by scanning might depend on the observation angle, which is the combination of roll (θ) and pitch (ϕ) angles for the position of task t from the AEOS s. A reward factor corresponding to the roll and pitch angles, $\zeta(t, \theta_t^i, \phi_t^i)$, is multiplied in the reward calculation to address this issue. If the reward does not depend on the angle of the scan, $\zeta(t, \theta_t^i, \phi_t^i)$ can be considered as 1. Considering the above factors, the reward accumulated by AEOS s in the revolution o for ith scan for task t can be given as,

$$\begin{aligned}\tilde{R}(s, o, t, i) &= \mathbb{R}_t^{\text{unit}}\tilde{A}(s, o, t, i)\zeta(t, \theta_t^i, \phi_t^i) - \left[\frac{\tilde{A}(s, o, t, i)}{A_t}(\alpha_t\mathcal{E}(s, o, t, i) + \beta_t\mathcal{L}(s, o, t, i))\right] \\ &= \tilde{A}(s, o, t, i)\mathbb{R}^{\text{unit}}(s, o, t, i)\end{aligned} \tag{21.18}$$

where, $\mathbb{R}^{\text{unit}}(s, o, t, i)$ is the unit reward for ith scan by AEOS s in revolution o for task t, which includes the penalty.

Since the tasks may need multiple scans to be completed successfully, there might be overlaps of areas during the scans. The reward calculation for a task comprising all scans must compute rewards for each overlapping area only once. Let $\tilde{B}_t$ be the n^t overlapped regions in the scans for task t.

$$\tilde{B}_t = \{\tilde{B}_t^1, \tilde{B}_t^2, \dots \tilde{B}_t^{n^t}\} = \bigcap_{(s,o,i)} \tilde{A}(s, o, t, i) \tag{21.19}$$

Since earliness or tardiness penalties are associated with the rewards based on the time of the scan, each scan might collect different rewards for the overlapped areas. The reward accumulated by the nth overlap by AEOS s in revolution o for ith scan can be given by

$$\mathbb{B}_t^n = \tilde{B}_t^n \mathbb{R}^{\text{unit}}(s, o, t, i) \tag{21.20}$$

To maximize the reward for a particular task, only the scan that collects the maximum reward for the overlapped areas will be considered. The reward to subtract from (21.20) is defined as

$$\sum_{k=1}^{n} \mathbb{B}_t^k - \max(\mathbb{B}_t) \quad \text{where } \mathbb{B}_t = \{\mathbb{B}_t^1, \mathbb{B}_t^2, \ldots, \mathbb{B}_t^n\} \tag{21.21}$$

Thus, the total reward for task t can be calculated as

$$\sum_s \sum_o \sum_i \tilde{R}(s, o, t, i) = \sum_s \sum_o \sum_i \tilde{A}(s, o, t, i) \mathbb{R}^{\text{unit}}(s, o, t, i) - \left(\sum_{k=1}^{n} \mathbb{B}_t^k - \max(\mathbb{B}_t) \right) \tag{21.22}$$

$$\text{for } s \in [1, N^s], o \in [1, N_{os}] i \in [i_1, i_2, \ldots, i_I]$$

where i_I is the total number of scans by AEOS s in revolution o for task t. The objective function of the considered optimization problem needs to be an aggregation of the accumulated rewards from all the user-specified tasks assigned to all the AEOSs in the mission and is defined as follows

$$\begin{aligned} &\underset{\mathcal{D}}{\text{maximize}} \sum_t \sum_s \sum_o x(s, o, t) \sum_i \tilde{R}(s, o, t, i) \\ &\text{where } \mathcal{D} \in \{x(s, o, t), T^{st}(s, o, t), D(s, o, t), \Theta(s, o, t), \Phi(s, o, t)\} \\ &\qquad \forall s \in [1, N^s], \forall o \in [1, N_{os}] \forall t \in [1, N^t] \end{aligned} \tag{21.23}$$

In order to deal with the complexity and size of the optimization problem defined above, a *multi-stage* optimization framework has been used. A flowchart of the described multistage optimization problem framework is illustrated in Figure 21.10.

In the *first* stage, the AEOSs are assigned to the tasks with the consideration of the user-specified time windows. At this stage, the assignment process would involve determining the decision variables associated with the assignment of tasks, their starting times, and their duration, i.e., $x(s, o, t)$, $T^{st}(s, o, t)$ and $D(s, o, t)$, respectively. The assignment of tasks having massive AOIs would involve processing on the basis of the information available prior to the start of the mission, i.e., the AEOS trajectories and the task requests with the corresponding coordinates of the AOIs.

Then, in the *second* stage, corresponding to the AEOS assignments in the first stage, each AOI is considered individually. Since these tasks are massive, each assigned scan to a single AOI can only scan a portion of the required AOI. The optimal roll and pitch angles of the AEOSs needed to perform the scanning tasks for each assigned scan are calculated at this stage. This means that in the second stage, the optimal roll and pitch angles, $\Theta(s, o, t)$ and $\Phi(s, o, t)$, respectively, are to be determined to cover the maximum area of the specified AOI by the assigned AEOSs. However, the determination of the optimal schedule would involve optimizing the first and the second stages hand-in-hand for several iterations.

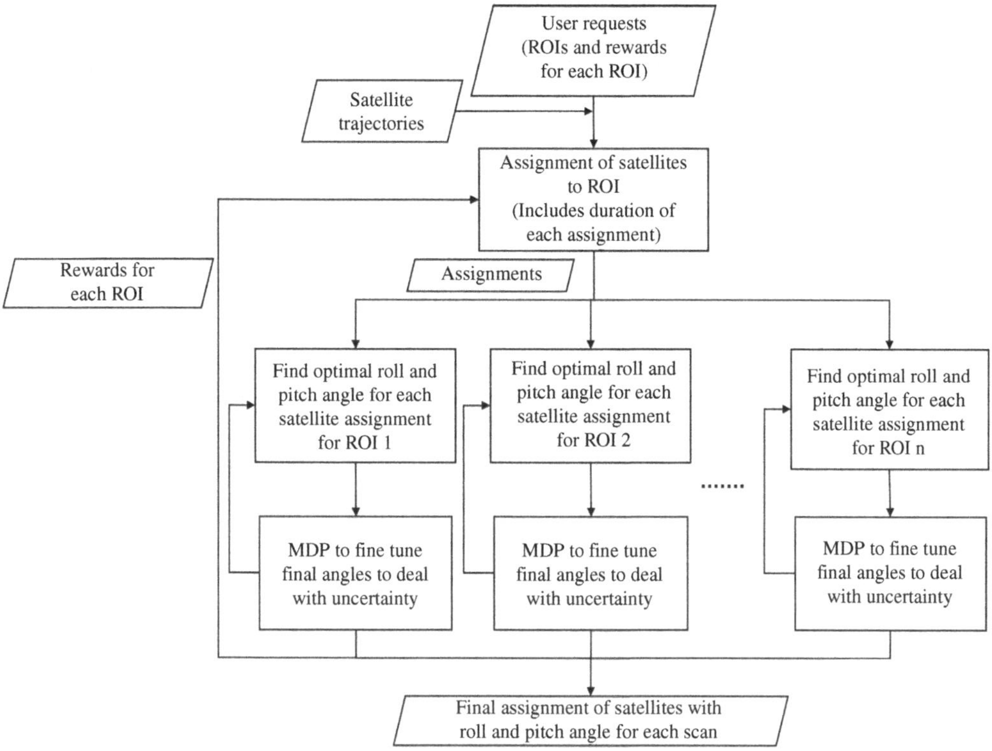

Figure 21.10 Flowchart of the multistage optimization problem framework.

As shown in Figure 21.10, at the beginning of the optimization process, the first stage finds a suboptimal assignment of the satellites, and then the second stage finds the roll and pitch angle corresponding to each assignment. The process goes back to the first stage to find a better solution to the assignment problem, and the cycle continues for several iterations. The optimization of $\Theta(s, o, t)$ and $\Phi(s, o, t)$ in the second stage will determine the scanned areas corresponding to the decision variables $x(s, o, t)$, $T^{st}(s, o, t)$ and $D(s, o, t)$ identified in the first stage. During a scan of a task, these roll and pitch angles will be fixed. This would consequently allow to calculate the reward obtained for the scans corresponding to this combination of decision variables. As these tasks are independent of each other, the second stage of optimization for all the tasks can be done with parallel processing to reduce computational time.

The *third* stage deals with the uncertainty originating from the failure of scans. A Markov decision process-based intermediate stage is introduced to fine-tune the roll and pitch angles for the consecutive assignments if any previous assignment has failed. Additionally, the role of the third stage will be effective in the identification of roll and pitch angles should any failure of scans occur in an online optimization setting.

The above-mentioned multistage optimization framework is illustrated in detail in the following sections.

21.5.2 Stage 1: The Optimal Assignment of Tasks to the AEOSs

The assignment of the AEOSs to the tasks in the first stage is optimized by maximizing the reward obtained by N^s AEOSs scanning N^t tasks. In the simplest form, the reward accumulation may be framed in the form of the following objective function,

$$\underset{\mathcal{D}_1}{\text{maximize}} \sum_{t=1}^{N^t} \sum_{s=1}^{N^S} \sum_{o=1}^{N_{OS}} x(s,o,t) \left(\sum_{i} \left(\prod_{j=1}^{i} C(j) \right) \tilde{R}(s,o,t,i) \right) \tag{21.24}$$
$$\text{where } \mathcal{D}_1 \in \{x(s,o,t), T^{st}(s,o,t), D(s,o,t)\}$$

Here, $x(s, o, t)$ is a binary decision variable that takes the value 1 if AEOS s is assigned to task t in revolution o; otherwise, it takes the value 0. Once an AEOS s has been assigned to a particular task t in revolution o, multiple scans might occur as discussed above. To monitor the number of scans in a particular revolution o by AEOS s for task t, a new binary decision variable $C(j)$ with values of 0 and 1 is introduced, where $j \in [1, i]$. The constraints of the objective function at this stage are defined as follows in Eqs. (21.25) and (21.27). The maximization of the reward $\tilde{R}(s, o, t, i)$ is performed with respect to determining the optimal decision variables $\mathcal{D}_1$.

21.5.2.1 Time Constraint

The schedule needs to consider that a particular AEOS has enough time to finish an assigned scan i before proceeding to the next assignment j. This is incorporated with respect to the user-specified time windows, in terms of the start time and duration of the consecutive scans as follows,

$$t^{st}(s,o,t,j) > t^{st}(s,o,t,i) + d(s,o,t,i) \quad \forall\, t, i \neq j \tag{21.25}$$

Hence, the constraint (21.25) ensures that any new assignment j for AEOS s in revolution o on task t occurs after the assigned duration for assignment i has passed.

21.5.2.2 Memory Constraint

The memory required for ith scan for task t by satellite s is defined as

$$m(s,o,t,i) = m_s^{\text{unit}} \times d(s,o,t,i) \tag{21.26}$$

When the AEOS passes over the VTW of the ground station, it transmits the data from the onboard memory. After each communication with the ground station, the onboard memory is reset. The scans can be performed by the AEOS until the maximum capacity of the onboard memory, is reached. The satellite s can only perform a specific number of scans in between the communication to the ground station g and the next available ground station $\hat{g}$, which is defined as,

$$\sum_{N_{(s,g,\hat{g})}} m(s,o,t,i) \leq m_s^{\max} \quad \forall\, s, g, \hat{g} \neq g \tag{21.27}$$

The value of $N(s, g, \hat{g})$ is calculated from the decision variable $T^{st}(s, o, t)$, which denotes the start time of the scan in the interval between two consecutive communications with the ground station g and $\hat{g}$ for AEOS s.

$$N(s,g,\hat{g}) = |T^{st}(s,o,t)| \text{ where } (T_g(s,o) + d_g(s,o)) < t^{st}(s,o,t,i) < T_{\hat{g}}(s,o) \tag{21.28}$$

The values of $T_g(s, o)$ and $d_g(s, o)$, which is a function of $m_s^{\max}$, are predefined.

21.5.3 Stage 2: The Optimal Selection of Roll and Pitch Angles

Now, in the next stage, the optimal roll and pitch angles for each assignment need to be determined in order to cover the AOI corresponding to the task maximally. In the first stage, the AEOS assignment to each task is set and passed on to the second stage of the optimization problem. In the context of each task, the combination of roll and pitch angles for each assignment decides the area that will be scanned. The area scanned for each unit of time is calculated with the help of trigonometric relationships between the position of the AEOS in its trajectory and the roll and pitch angle of the sensors, as illustrated in Figure 21.6.

The AOI corresponding to a single task is represented as the area $PQRS$ in Figure 21.11 on the Earth's surface. The dark-shaded region $DEGH$ inside $PQRS$ is the area covered in a single scan for a unit time stamp. Let the AEOS be at position B, and the projection of the AEOS's position on the Earth's surface be A. The coordinates of B, say, (x, y) are known prior to the mission from the trajectory of the AEOS, which is the same as A. Now, let (θ, ϕ) be the roll and pitch angles of the scan. In this figure, the pitch angle ϕ is considered zero, and the trajectory of the AEOS is parallel to the longitude lines of the Earth's surface.

Then, with the rules of trigonometry, the coordinates of the points C and F can be calculated as $(x, y + h\tan(\theta))$ and $(x, y + h\tan(\theta + \psi_s))$, where h is the height of the AEOS and ψ_s is the swath width of the AEOS. Since C is the mid-point between the line segment joining H and D and the swath width of the AEOS is known, the coordinates of H and D can be calculated. Since the length of the line segment HD is equal to CF, the coordinates of H and D can be calculated as $\left(\left(x + \frac{h}{2}(\tan(\theta + \psi_s) - \tan\theta)\right), (y + h\tan(\theta))\right)$ and $\left(\left(x - \frac{h}{2}(\tan(\theta + \psi_s) - \tan\theta)\right), (y + h\tan(\theta))\right)$ respectively. In a similar manner, the coordinates of G and E are $\left(x + \frac{h}{2}[\tan(\theta + \psi_s) - \tan\theta], y + h\tan(\theta + \psi_s)\right)$, and $\left(\left(x - \frac{h}{2}(\tan(\theta + \psi_s) - \tan\theta)\right), (y + h\tan(\theta + \psi_s))\right)$, respectively.

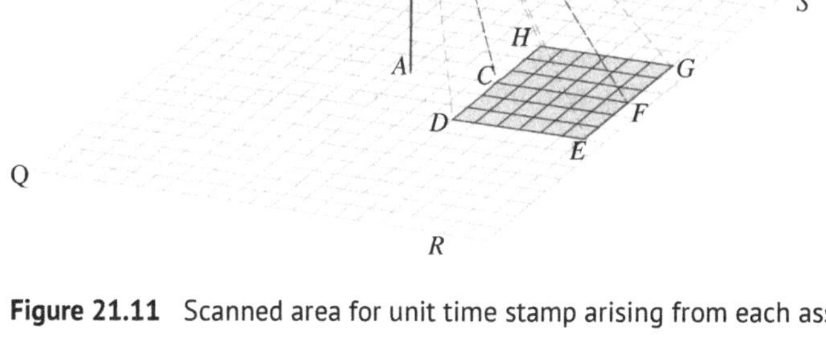

Figure 21.11 Scanned area for unit time stamp arising from each assignment.

Then, the area covered by a unit timestamp $DEGH$ can be determined. This illustrates an instance of the determination of the decision variable $\tilde{A}(s,o,t,i)$ corresponding to the area covered in the ith scan of task t by AEOS s with revolution o. At this stage, the following objective function maximizes the total area covered for task t:

$$\underset{\mathcal{D}_2}{\text{maximize}} \bigcup_{s,o,i} \tilde{A}(s,o,t,i) \quad \text{where } \mathcal{D}_2 \in \{\Theta(s,o,t), \Phi(s,o,t)\} \tag{21.29}$$

21.5.3.1 Roll and Pitch Angles Constraints

According to the specifications of the AEOS, the maximum possible roll and pitch angles for each scan are defined. During the scan, these decision variables need to follow the constraints stated below,

$$\theta_t^i \leq \Theta_s^{\max} \tag{21.30}$$

$$\phi_t^i \leq \Phi_s^{\max} \tag{21.31}$$

The above-defined objective function and constraints in the second stage allow for finding the maximum area coverage by optimizing the roll and pitch angles. As discussed in Eqs. (21.19) to (21.22), the rewards accumulated during the overlapped region are discarded. Since the first and second stage of optimization are performed simultaneously for multiple iterations, the reward obtained by scanning the region $\tilde{A}(s,o,t,i)$ is given as,

$$\tilde{R}(s,o,t,i) = \tilde{A}(s,o,t,i)\mathbb{R}^{\text{unit}}(s,o,t,i) - \left(\sum_{k=1}^{n} \mathbb{B}_t^k - \max(\mathbb{B}_t)\right) \tag{21.32}$$

It should be noted that the reward $R(s,o,t,i)$ is the same as in Eq. (21.24) in Stage 1 of the optimization framework. Since each task is independent of the other, maximizing the reward corresponding to each task can be calculated with parallel processing.

21.5.4 Stage 3: Markov Decision Process to Handle Uncertainty

Due to unpredictable cloud coverage, sensor and equipment malfunctions, and other factors, failure of scans might occur. To handle the uncertainties arising due to these unprecedented reasons, an MDP-based approach is used as an additional step in the multistage optimization framework. Since these tasks correspond to massive AOIs, multiple scans are needed to cover the whole region. Failure of a single scan would impact the coverage of the whole task corresponding to the existing assignment. A reallocation of scanning areas for the upcoming assignments in the mission horizon might be helpful to append some portions of the missed region of the failed scan. This is achieved through the determination of an updated optimal choice of roll and pitch angles of the consequent assignment to ensure maximum coverage, including portions of missed regions, and hence maximize the reward.

21.5.4.1 State Space ($\mathcal{S}$)

The sample state space for the MDP is defined as follows:

- ν—Area to be scanned in the next assignment;
- w—Weather prediction for the next scan.

Here,

$$\nu \in \mathcal{V}_t, \text{ where } \mathcal{V}_t = \bigcup_{\forall s, o, i} \tilde{A}(s, o, t, i) \tag{21.33}$$

The set of assignments corresponding to the areas $\tilde{A}(s, o, t, i)$ for task t is defined as $\mathcal{V}_t$. During the mission horizon, the AEOSs get updated real-time information about the weather. Based on the current weather, the state variable w is defined as 0 or 1, where 0 indicates bad weather conditions and 1 indicates good weather conditions. A multi-tier weather condition state variable can also be used instead of the two-tier (0, 1), but that will increase the number of states and the complexity.

21.5.4.2 Action Space ($\mathcal{A}$)

The possible actions for the MDP, depending upon the current state, are given as follows:

- a_s—Skip the next scan;
- a_p—Proceed to the next scan.

By the action a_s, the AEOS will skip the next scheduled scan and preserve the resources. The action a_p will perform the scan and depreciate resources irrespective of the outcome.

21.5.4.3 Transition Probability ($\mathcal{T}$)

The actions' transition probabilities depend on weather conditions, sensor failure, and other factors. When the weather condition is favorable ($w = 1$), the probability of success (ξ_g) for a scan is higher than when the weather is unfavorable, ($w = 0$). The probability of success when the weather condition is not favorable, is defined as ξ_b. Sensor failures and other factors may still result in a scan failure even if the weather is in favor. The probability condition is reversed when the weather condition is adverse ($w = 0$). The transition probability for having a successful or failed scan can be defined as:

$$\begin{aligned}
&\Pr(\text{Success}|a_p, w = 1) = \xi_g \\
&\Pr(\text{Failure}|a_p, w = 1) = (1 - \xi_g) \\
&\Pr(\text{Success}|a_p, w = 0) = \xi_b \\
&\Pr(\text{Failure}|a_p, w = 0) = (1 - \xi_b) \\
&\Pr(\text{Failure}|a_s, w = *) = 1
\end{aligned} \tag{21.34}$$

The probability of the weather prediction is defined as:

$$\begin{aligned}
&\Pr(w = 1|a_*, w = *) = w_p \\
&\Pr(w = 0|a_*, w = *) = (1 - w_p)
\end{aligned} \tag{21.35}$$

where $0 \leq \xi_g, \xi_b \leq 1$ is the success rate of a scan and $0 \leq w_p \leq 1$ is the weather prediction at each state, a_* denotes either of the action and $w = *$ denotes either state condition for w. The final transitional probability is calculated by the product of (21.34) and (21.35).

21.5.4.4 Reward Function ($\mathcal{R}$)

The reward function for the MDP includes the following factors:

- reward for completing the current scan;
- no reward for skipping the scan;
- penalty for utilizing the resource.

The penalty for using the resources is applied every time a scan is performed, irrespective of the success or failure of the scan.

$$R^t = \begin{cases} \tilde{R}(s,o,t,i) - p \times m(s,o,t,i) & \text{if scan is successful} \\ -p \times m(s,o,t,i) & \text{if scan is unsuccessful} \\ 0 & \text{if scan is skipped} \end{cases} \tag{21.36}$$

where R^t is the reward for the state transition. Here, p is the penalty for utilizing the unit resources. The penalty is chosen by the user. The AEOSs can conserve resources by skipping a scan.

A sample state space with actions for the example from Figure 21.12 for the above MDP model is shown in Figure 21.13. The Figure 21.12a illustrates an instance of area distribution for scans obtained by optimization of roll and pitch angles in the second stage. It is shown that there are four scans to be performed in the considered task AOI. The areas are named 1, 2, 3, and 4, corresponding to the order of the scan in the mission horizon. This area distribution will yield the maximum reward if all these four scans are successful as calculated in (21.24) and (21.32). However, as illustrated in Figure 21.12b, if scan 1 is unsuccessful, the second stage of the optimization is called again, and the roll and pitch angles are recalculated for the remaining three assignments to maximize the area coverage. In Figure 21.12b, the change in the area distribution after the recalculation and the updated numbering of the areas are illustrated. The adopted rule of numbering the reallocated areas includes the index of the unsuccessful scan, followed by the order of the scan according to the mission horizon. Since the scan corresponding to the area 11 is successful, no further reallocation of the area is needed for the remaining scans 12 and 13, as shown in Figure 21.12c.

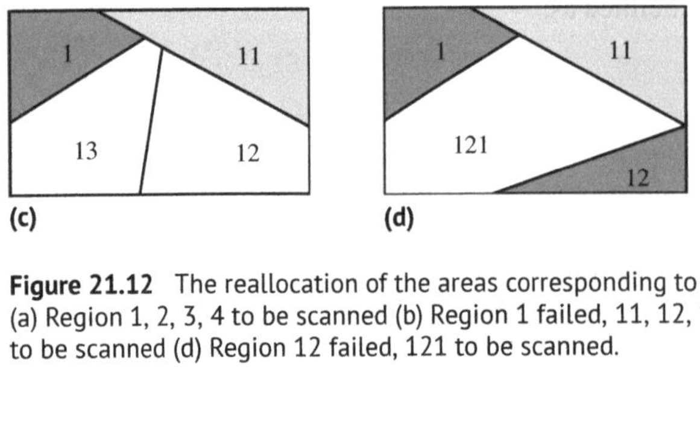

Figure 21.12 The reallocation of the areas corresponding to assignments arising from unsuccessful scans. (a) Region 1, 2, 3, 4 to be scanned (b) Region 1 failed, 11, 12, 13 to be scanned (c) Region 11 success, 12, 13 to be scanned (d) Region 12 failed, 121 to be scanned.

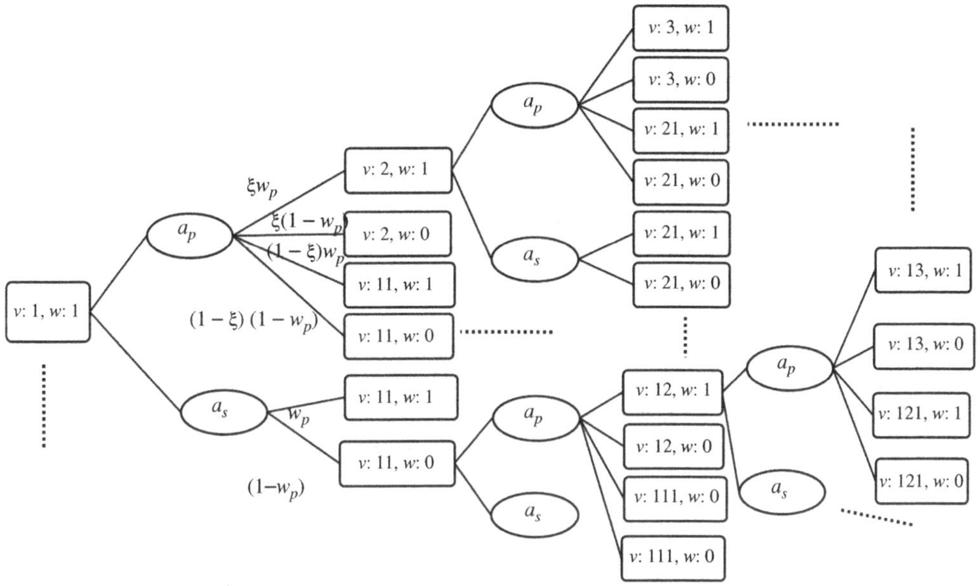

Figure 21.13 Tree diagram for the states of the MDP for the problem in Figure 21.12.

Further in the mission horizon, when the scan corresponding to area 12 is unsuccessful, the second stage is called again to find an optimal roll and pitch to maximize the area corresponding to scan 121, illustrated by Figure 21.12d. Figure 21.13 shows some of the states for the example above.

The following section illustrates the solution methodology for the problem formulated in this section.

21.5.5 Solution Methodology

As discussed in Section 21.5.1, a three-stage optimization model has been developed with the help of a randomized heuristic using an elitist genetic algorithm technique [17] (in the first stage), constrained nonlinear optimization solver (in the second stage) coupled with MDP solution strategy (in the third stage). Detailed explanation of the genetic algorithm used for the first stage can be found in [2]. For the second stage, any constrained nonlinear optimization algorithm will be suitable, however, for this chapter, the interior point method has been used in the simulations [18]. The third stage of the problem is solved using the value iteration technique [6]. Prior to the mission, the third stage fine-tunes the schedule from the first and second stages. In order to reduce the computational load, the third stage can be started after a few initial generations of the first and second stages.

21.5.6 Simulation Results

A scenario with ten user-requested tasks is considered. By considering the maximum roll and pitch angles, the exposure of each task to a particular AEOS on its revolution is calculated. Prior to the multistage optimization, a preprocessing of the simulated data is done to eliminate redundant revolutions in which no tasks are exposed to the AEOS.

In this simulation, the AEOSs periodically communicate with six ground stations. Prior to the mission, the number of possible communications of the AEOSs with the ground stations based on the VTW of the ground stations is calculated in a similar manner to the task exposures preprocessing.

Table 21.6 shows the number of times the AEOSs are exposed to a particular task AOI. This table further shows how many times each task has been scanned and the reward collected by all the AEOSs for each task. It is noticeable from Table 21.6 that for tasks 6–10 the number of scans is higher, and AEOSs have collected the full reward from them. For tasks 1–5 because of their close positioning (shown in Figure 21.14), the number of scans by all the AEOSs is less, and they have not received the full reward.

Table 21.7 demonstrates the effects of redistributing the roll and pitch angle by using the MDP when a scan is unsuccessful. For task 1, all the scans are successful, so the MDP did not affect the total reward accumulated. For task 2, only 2 out of the 3 scans were successful. The MDP redistributed the area, and it yielded more rewards. A similar scenario is observed for tasks 6 and 8.

Table 21.6 Total number of scans performed by all the AEOSs.

	Number of exposure	Scans done	Max reward	Reward collected
T1	6	3	10	8.3
T2	6	3	10	6.65
T3	6	1	10	4.15
T4	5	2	10	7.38
T5	5	2	25	9.36
T6	5	4	10	10
T7	8	5	17	17
T8	16	6	11	11
T9	16	6	15	15
T10	20	7	20	20

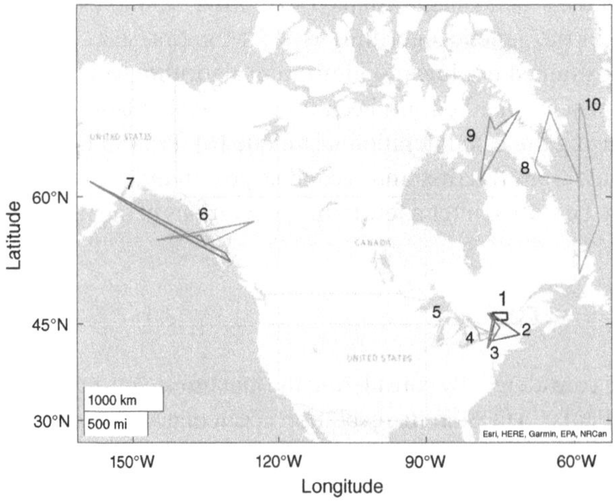

Figure 21.14 Locations of the tasks.

Table 21.7 Online performance with MDP.

	Scans assigned	Successful scans	Max reward	Reward collected with MDP	Reward collected without MDP
T1	3	3	10	8.3	8.3
T2	3	2	10	6.65	5.72
T3	1	1	10	4.15	4.15
T4	2	2	10	7.38	7.38
T5	2	2	25	9.36	9.36
T6	4	3	10	8.54	7.86
T7	5	5	17	17	17
T8	6	5	11	9.67	8.65
T9	6	6	15	15	15
T10	7	7	20	20	20
Total			**138**	**106.05**	**103.42**

Table 21.8 Performance comparison between the strip method and the discussed method.

	Max reward	Reward collected with strips method	Reward collected with discussed method	Memory consumed with strips method	Memory consumed with discussed method
T1	10	6.39	8.3	59	48
T2	10	5.48	6.65	46	35
T3	10	3.26	4.15	31	20
T4	10	6.11	7.38	55	47
T5	25	8.23	9.36	87	72
T6	10	6.92	8.54	57	50
T7	17	17	17	139	117
T8	11	8.32	9.67	84	74
T9	15	15	15	126	112
T10	20	20	20	157	129
Total	**138**	**96.71**	**106.05**	**841**	**704**

Table 21.8 compares the reward collected by the discussed method in this chapter and the conventional strip method. From the table, it can be observed that in most of the tasks, the discussed method of assigning roll and pitch angles to create strips that are not parallel to the trajectory of the AEOSs collects more reward than the conventional strip method [19]. It is also shown in Table 21.8, that the memory consumed by the strip-based method is more than the discussed method. Figure 21.15a shows that in the conventional strip method, the strips are formed parallel to the trajectory of the AEOSs, which are marked as the dotted line. The dart shape polygon is the area of task 4. Figure 21.15b shows which strips are actually scanned during the mission with the strip method. The duration of a scan will depend on the task size, satellite trajectory, and the swath

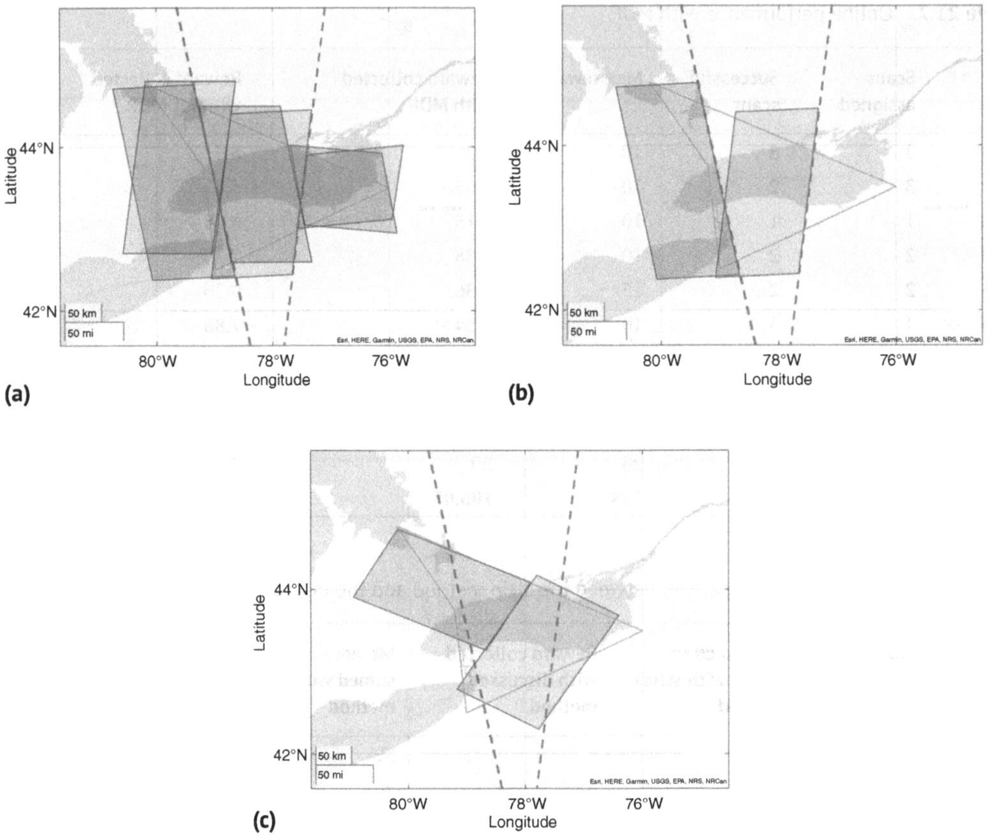

Figure 21.15 Comparison of strip method versus discussed method for task 4. (a) All possible strips, (b) scans with strip method [19], and (c) scans with the discussed approach.

width of the satellite. Figure 21.15c shows that with the methodology considered in the chapter, the area scanned by the two AEOSs is not parallel to the trajectory, and they cover more area than the strip method, as shown in Figure 21.15b.

21.6 Conclusions

In this chapter, agile and non-agile multi-satellite scheduling problems with realistic operational constraints and task specifications are studied. A complete closed-loop solution may not be feasible for large-scale scenarios. First, a mixed open-and closed-loop solution has been discussed for non-agile satellite scheduling to handle large-scale problems in real time. In order to handle the problem in real time, the problem has been decomposed into four subproblems, and an approximate formulation has been used to solve the ground open-loop planning. Then, a detailed analysis of an agile satellite scheduling algorithm for small tasks, or a large region already split into small sub-regions is provided. The problem is formulated as a mixed-integer nonlinear optimization formulation that finds the optimal schedule by maximizing the reward with realistic satellite resources as energy and memory constraints. A genetic algorithm-based methodology has been used to solve this scheduling model. A three-stage AEOS scheduling model has been used to find the optimal task assignment when the AOIs are too large to complete in a single scan.

References

1 R. Tharmarasa, T. Kirubarajan, J. Berger, and M. Florea. Mixed open-and-closed loop satellite task planning. In *2019 22th International Conference on Information Fusion (FUSION)*, IEEE, pages 1–8, 2019.

2 A. Chatterjee and R. Tharmarasa. Multi-stage optimization framework of satellite scheduling for large areas of interest. *Advances in Space Research*, 73(3):2024–2039, 2024.

3 A. Chatterjee and R. Tharmarasa. Reward factor-based multiple agile satellites scheduling with energy and memory constraints. *IEEE Transactions on Aerospace and Electronic Systems*, 58(4):3090–3103, 2022.

4 J. Berger, N. Lo, and M. Barkaoui. Quest - a new quadratic decision model for the multi-satellite scheduling problem. *Computers & Operations Research*, 115:104822, 2020.

5 X. Chen, G. Reinelt, G. Dai, and A. Spitz. A mixed integer linear programming model for multi-satellite scheduling. *European Journal of Operational Research*, 275(2):694–707, 2019.

6 R. Tharmarasa, A. Chatterjee, Y. Wang, T. Kirubarajan, J. Berger, and M. Florea. Closed-loop multi-satellite scheduling based on hierarchical MDP. In *2019 22th International Conference on Information Fusion (FUSION)*, IEEE, pages 1–7, 2019.

7 M. Lemaître, G. Verfaillie, F. Jouhaud, J. Lachiver, and N. Bataille. Selecting and scheduling observations of agile satellites. *Aerospace Science and Technology*, 6(5):367–381, 2002.

8 M. Lemaître, G. Verfaillie, F. Jouhaud, J. Lachiver, and N. Bataille. How to manage the new generation of agile earth observation satellites. In *Proceedings of the International Symposium on Artificial Intelligence, Robotics and Automation in Space*, Citeseer, pages 1–10, 2000.

9 W. Wolfe and S. Sorensen. Three scheduling algorithms applied to the earth observing systems domain. *Management Science*, 46(1):148–166, 2000.

10 X. Liu, G. Laporte, Y. Chen, and R. He. An adaptive large neighborhood search metaheuristic for agile satellite scheduling with time-dependent transition time. *Computers & Operations Research*, 86:41–53, 2017.

11 C. Han, Y. Gu, G. Wu, and X. Wang. Simulated annealing-based heuristic for multiple agile satellites scheduling under cloud coverage uncertainty. *IEEE Transactions on Systems*, 53(5):2863–2874, 2022.

12 S. Baek, S. Han, K. Cho, D. Lee, J. Yang, P. Bainum, and H. Kim. Development of a scheduling algorithm and GUI for autonomous satellite missions. *Acta Astronautica*, 68(7–8):1396–1402, 2011.

13 C. Pang, A. Kumar, C. Goh, and C. Le. Nano-satellite swarm for SAR applications: Design and robust scheduling. *IEEE Transactions on Aerospace and Electronic Systems*, 51(2):853–865, 2015.

14 J. Wang, E. Demeulemeester, D. Qiu, and J. Liu. Exact and inexact scheduling algorithms for multiple earth observation satellites under uncertainties of clouds. *Available at SSRN 2634934*, 2015.

15 M. L. Puterman. Markov decision processes: Discrete stochastic dynamic programming. In *Wiley-Interscience*, 2008.

16 R. Howard. Dynamic programming and Markov processes. In *Massachusetts*, MIT Press, 1960.

17 J. Kim and Ralph D. Ellis J. Permutation-based elitist genetic algorithm for optimization of large-sized resource-constrained project scheduling. *Journal of Construction Engineering and Management*, 134(11):904–913, 2008.

18 R. Byrd, M. Hribar, and J. Nocedal. An interior point algorithm for large-scale nonlinear programming. *SIAM Journal on Optimization*, 9(4):877–900, 1999.

19 Y. Xu, X. Liu, R. He, Y. Chen, and Y. Chen. Multi-satellite scheduling framework and algorithm for very large area observation. In *2018 IEEE Congress on Evolutionary Computation (CEC)*, IEEE, pages 1–8, 2018.

22

Sensor Control for Multi-target Tracking: A Review

Amirali K. Gostar[1], Aidan Blair[2], Branko Ristic[1], and Reza Hoseinnezhad[1]

[1] *School of Engineering, RMIT University, Melbourne, Victoria, Australia*
[2] *School of Computing Technologies, RMIT University, Melbourne, Victoria, Australia*

22.1 Introduction

In the era of intelligent sensing systems and autonomous platforms, the ability to dynamically control sensors with the aim of maximizing tracking performance has become a cornerstone of modern surveillance, navigation, and situational awareness applications. Sensor control in the context of multi-target tracking (MTT) refers to the process of adaptively managing sensor parameters, such as position, orientation, and sensing mode, to optimize tracking objectives under uncertainty. This chapter examines the principles, methods, and challenges associated with sensor control in multi-target environments. By leveraging models that account for target dynamics, sensor limitations, and environmental variability, sensor control algorithms aim to improve the accuracy, reliability, and efficiency of tracking systems. The field has witnessed significant advancements over the past two decades, driven by developments in Bayesian estimation, information theory, and decision-making under uncertainty. Readers interested in complementary topics—such as optimal sensor placement for localization [1], robust power allocation in MIMO radar systems [2], and multisensor resource allocation [3]—are encouraged to consult Chapters 24–26 of this volume for a broader view of advanced sensor management strategies.

22.1.1 Overview of Sensor Control in Multi-object Tracking

MTT is a fundamental task in many sensing and perception systems, wherein the objective is to simultaneously estimate the number of targets and their dynamic states (e.g., position, velocity) over time. This task is especially complex in real-world scenarios due to factors such as measurement noise, clutter (false alarms), missed detections, and the unknown or time-varying number of objects in the scene. While traditional tracking algorithms focus on state estimation using available sensor data, sensor control introduces a higher level of autonomy by actively influencing how and where the data is collected.

Sensor control in the context of MTT refers to the dynamic adjustment of sensor parameters, such as position, orientation, field of view, resolution, or sensing modality, to optimize tracking performance according to predefined objectives. The current state estimates and the prior knowledge of target dynamic models and sensor characteristics typically inform this adaptation. Unlike static sensing, where the sensor configuration is fixed, active sensor control closes the loop with an autonomous decision on the best course of sensing action, guided by the anticipated value of information to be gained.

The essence of sensor control lies in its ability to maximize the utility of collected data. In multi-target scenarios, this is especially valuable, as targets may be sparsely distributed, exhibit complex motion patterns, or be partially observable due to occlusions or environmental constraints. A well-designed sensor control strategy can prioritize regions with higher uncertainty, resolve ambiguities between closely spaced targets, and ensure robust coverage in large or cluttered environments.

Sensor control can be implemented using various paradigms, including information-theoretic, task-driven, or learning-based frameworks. These approaches often rely on probabilistic representations of target states, such as multi-Bernoulli or Poisson random finite sets (RFSs), and incorporate control decisions that aim to reduce uncertainty, enhance detection probability, or optimize downstream performance metrics, including classification accuracy and resource efficiency. Moreover, as modern systems often involve mobile sensors (e.g., UAVs, autonomous vehicles, robotic platforms), the integration of motion planning and sensor control becomes critical. This coupling introduces additional constraints related to sensor dynamics, field-of-view limitations, and operational safety, making sensor control a rich and interdisciplinary field of research.

22.1.2 Importance and Challenges in Modern Applications

Sensor control in MTT has become indispensable in a wide array of modern applications, ranging from defense systems to autonomous platforms and environmental monitoring. Dynamically adjusting sensor parameters—such as orientation, coverage area, and measurement mode—enhances the ability to track multiple moving objects in uncertain and cluttered environments. Adaptive sensor control plays a vital role across a wide range of modern applications where the ability to dynamically manage sensing resources can significantly improve system performance, resilience, and decision-making.

In *defense and security* contexts, sensor control is indispensable for detecting and tracking hostile targets, such as drones, submarines, or missiles, in complex, cluttered environments. For example, in anti-submarine warfare (ASW), effective management of sonar and surveillance assets is viewed as a critical capability, where the ability to coordinate sensor actions in real time can determine operational success [4]. Similarly, ground- and air-based defense platforms rely on active sensor control to maintain situational awareness and respond swiftly to evolving threats.

In *autonomous systems*, including self-driving cars and unmanned aerial vehicles (UAVs), real-time perception is essential for tasks like navigation, obstacle avoidance, and tracking of nearby entities. These systems often operate under energy and bandwidth constraints and must make rapid decisions using noisy, partial sensor data. Adaptive control of sensors—such as adjusting fields of view, scanning patterns, or sensing modes—enables better prioritization of relevant environmental cues, ultimately supporting safer and more efficient autonomy.

In *environmental monitoring*, sensor control enables efficient tracking of dynamic or sparsely distributed phenomena such as wildlife, pollutants, or fire fronts. UAV-based sensing platforms, for instance, can adjust their flight paths to maximize coverage of affected areas or to focus on zones of uncertainty. In large-scale deployments, coordinated control across sensor networks enhances spatial and temporal resolution while conserving energy and extending operational lifespan, especially in remote or hazardous environments [5–7].

22.1.2.1 Challenges in Modern Applications

Sensor control in MTT must navigate a range of complex challenges inherent to real-world applications. These challenges are particularly pronounced in dynamic, uncertain, and resource-constrained environments.

Balancing Exploration and Exploitation: A central challenge in sensor control is the trade-off between exploring the environment to detect new or hidden targets and exploiting the sensing systems to maintain accurate tracking of known targets. Optimally managing this balance in a dynamic setting is challenging, especially when sensor agility or sensing range is limited [8].

Decision-making Under Uncertainty: Sensor control decisions must be made using partially observed and noisy target states. The sensor platform often lacks complete knowledge of the number, type, or exact location of targets. Approaches such as partially observable markov decision processes (POMDPs) and RFS-based objective functions have been explored; however, solving these models remains computationally intensive and often requires approximations [9].

Multi-sensor Coordination: In multi-sensor systems, coordinating sensor actions to avoid redundant coverage while maximizing overall information gain is complex. Ensuring consistent data fusion and avoiding conflicting sensing objectives are ongoing concerns in distributed architectures [10].

Real-time Constraints and Computational Load: Effective sensor control requires real-time decision-making, which can be hindered by the computational complexity of optimizing sensor trajectories or orientations based on information-theoretic or task-based cost functions [9]. The challenge is further compounded when the control space is large or continuous.

Dynamic Constraints and Platform Limitations: Sensor control must account for physical constraints such as motion limits, field-of-view (FoV), and energy or bandwidth constraints. In UAV-based or mobile platforms, these restrictions can significantly limit feasible sensing strategies [11].

Scalability and Robustness: As the number of sensors and targets increases, sensor control strategies must remain scalable and robust. This includes adapting to sensor failures, communication delays, and the possibility of adversarial interference [10].

These challenges highlight the need for adaptive, scalable, and computationally efficient sensor control strategies that can maintain high tracking performance across diverse operating conditions.

22.1.3 Chapter Roadmap

This chapter is organized as a structured survey of sensor control for MTT. The survey is anchored in the taxonomy introduced in *Sensor Control Categorization* (objective function: information-driven vs. task-driven; purpose: nonselective vs. selective; architecture: single-sensor vs. multi-sensor vs. distributed).

- **Understanding the Sensor Control Problem** (Section 22.2) formulates sensor control for MTT within a Bayesian/POMDP framework and describes the closed-loop control pipeline, including practical implementation concepts such as predicted densities and PIMS-based pseudo-updates.
- **Sensor Control Categorization** (Section 22.3) defines the chapter's organizing taxonomy across three axes: (1) the nature of the underlying tracking method (conventional vs. RFS-based), (2) the objective function class (information-driven vs. task-driven), and (3) the tracking purpose (nonselective vs. selective).

- **In-depth Analysis of Sensor Control Techniques** (Section 22.4) surveys representative methods under each category, detailing widely used objective functions (e.g., Rényi, Cauchy–Schwarz, and KL divergences; PENT; cardinality variance; and PEECS) and summarizing key literature in Table 22.1.
- **Selective-Sensor Control** reviews label-aware control strategies for prioritizing targets of interest (ToIs), covering both the *Concept and Implementation* (Section 22.5.1) and *Case Studies and Applications* (Section 22.5.2).
- **Extension to Multi-sensor Control** generalizes the discussion to cooperative sensing in sensor networks, highlighting coordination challenges and reviewing recent multi-sensor control techniques and representative case-study scenarios.
- **Distributed Multi-sensor Control** discusses scalable networked architectures and distributed optimization mechanisms for multi-sensor control, with emphasis on communication constraints, neighbor-based decision-making, and consensus-style approaches.
- **Comparative Analysis and Simulation Results** presents the experimental setup and comparative results to illustrate accuracy–computation trade-offs across representative control strategies in multi-sensor and distributed settings.
- **Conclusions** summarizes the main findings and highlights practical considerations and future directions for robust, scalable sensor control in modern MTT applications.

22.2 Understanding the Sensor Control Problem

Based on the motivation and applications discussed earlier, we now focus on the technical formulation and solution strategies for the sensor control problem within the context of multi-object tracking. Multi-object sensor control represents a nonlinear stochastic control problem, with the primary objective of dynamically assigning appropriate sensor actions to optimize the estimation of multiple object states over time [12]. The complexity of this problem stems from multiple sources of uncertainty, including inaccuracies in object state estimates, false alarms, missed detections, and the unknown appearance or disappearance of objects within the sensing environment. Sensor control decisions are typically made under these uncertainties, often leveraging the latest available measurement history to guide the next control action.

Traditionally, MTT approaches—such as joint probabilistic data association (JPDA) and multiple hypothesis tracking (MHT)—model individual target states separately and maintain explicit data association hypotheses. While effective in structured environments, these methods face substantial challenges when the number of objects varies dynamically or when clutter and missed detections become significant, leading to prohibitive computational complexity and degraded performance. To address these challenges, the random finite set (RFS) framework provides a powerful alternative by modeling the entire collection of object states as a single random set. This unified approach naturally handles uncertainties in target number, states, false alarms, and missed detections without requiring explicit data association. Within this framework, the evolution of the multi-object state can be modeled as a Markov process, making the sensor control problem naturally formulated within the partially observable markov decision process (POMDP) paradigm [13]. Inference proceeds via sequential Bayesian estimation, where the goal is to estimate the posterior multi-target probability density $\pi_{k|k}(\cdot)$ conditioned on all observations up to time k. Sensor control strategies are then designed to optimize an objective function that quantifies the expected utility, such as information gain or reduction in estimation error, resulting from a given control action.

Table 22.1 Classification of sensor control techniques based on objective and selectivity.

Reference	Method description	Year	Task-driven	Info-driven	RFS-based	Selective
Kershaw and Evans [37]	Optimal waveform selection for tracking systems	1994	✓			
Mahler [30]	Csiszár divergence approach	1998		✓	✓	
Krishnamurthy and Evans [38]	Beam scheduling using HMM multiarmed bandits	2001	✓			
Kreucher et al. [39]	Alpha divergence via JMPD	2003		✓		
Mahler et al. [19]	PENT (works with PHD/MHT)	2004	✓		✓	
Kreucher et al. [18]	Sensor control using mutual information	2005		✓		
Ristic et al. [25]	Rényi alpha divergence	2010		✓	✓	
Ristic et al. [26]	Rényi divergence-based method	2011		✓	✓	
Hoang et al. [31]	Rényi divergence and MAP on cardinality	2014		✓	✓	
Gostar et al. [40]	Cardinality variance (mean)	2013	✓		✓	
Gostar et al. [22]	PEECS metric (estimation error)	2015	✓		✓	
Beard et al. [28]	Closed-form CS divergence (GLMB)	2015		✓	✓	
Gostar et al. [41]	OSPA-based cost function	2015	✓		✓	
Gostar et al. [42]	CS divergence for MB filters	2016		✓	✓	
Jiang et al. [43]	CS divergence for JDM/IDM on Labeled RFS	2016		✓	✓	
Beard et al. [44]	CS divergence + void function (GLMB)	2017		✓	✓	
Gostar et al. [45]	CS + void function (LMB)	2017		✓	✓	
Wang et al. [32]	Multi-sensor control with GCI/CDM (PEECS)	2018	✓		✓	
Panicker et al. [46]	Max-confidence + selective PEECS	2018	✓		✓	✓
Panicker et al. [47]	ARAPP method (adaptive resource allocation)	2018	✓		✓	✓
Li & Han [9]	Dual control using predicted detection probability and distance-based metrics	2018	✓		✓	
Lian *et al.* [48]	Sensor selection in large-scale networks using GLMB filtering	2018	✓		✓	
Panicker *et al.* [49]	Selective tracking of ToIs using LMB	2020	✓		✓	✓
Ding *et al.* [50]	Dwell-time and bandwidth optimization under CRLB constraint	2020	✓			
Zhu *et al.* [51]	Receiver selection via cardinality variance minimization (LMB)	2021	✓		✓	
Liang *et al.* [52]	Multi-objective sensor selection with posterior uncertainty minimization	2022	✓		✓	
Liu *et al.* [53]	Cost function combining KL divergence, detection probability, and collision avoidance	2023	✓	✓	✓	

A general multi-object tracking system receives sensor data and outputs state estimates over time. Sensor control algorithms interface with this system by generating commands that influence sensor behavior, such as positioning, orientation, or activation, prior to the next data acquisition step, creating a closed feedback loop. Figure 22.1 illustrates a high-level overview of the sensor control feedback loop within a multi-object tracking framework, highlighting the interplay between measurement acquisition, tracking, and control.

In Bayesian multi-object tracking, the system typically proceeds through prediction, update, and estimation phases. However, only the prediction step is available before new measurements are collected. Consequently, the control module must rely on the predicted multi-object density, denoted as $\pi_{k|k-1}(\cdot)$, to make decisions. Let $\mathbb{U} = \{u_1, \dots, u_m\}$ represent the discrete and finite set of admissible control actions. For each control command $u_i \in \mathbb{U}$, a predicted ideal measurement set (PIMS) [14] is generated. The PIMS approximates the measurements that would be obtained under ideal conditions—assuming perfect detections, no clutter, and noise-free observations—given the predicted object states. While PIMS is widely used in practical implementations due to its computational efficiency, it is worth noting that it does not fully account for the expected distribution of the predicted measurements. Ideally, the expected utility of a control command should be computed by integrating over all possible measurement sets, weighted by their likelihood under the predicted multi-object density and sensor model. That is, the expected reward (or cost) should involve an expectation operator over the measurement space:

$$\mathbb{E}_{Z \sim g(Z|\pi_{k|k-1}, u)}[R(Z, u)]. \tag{22.1}$$

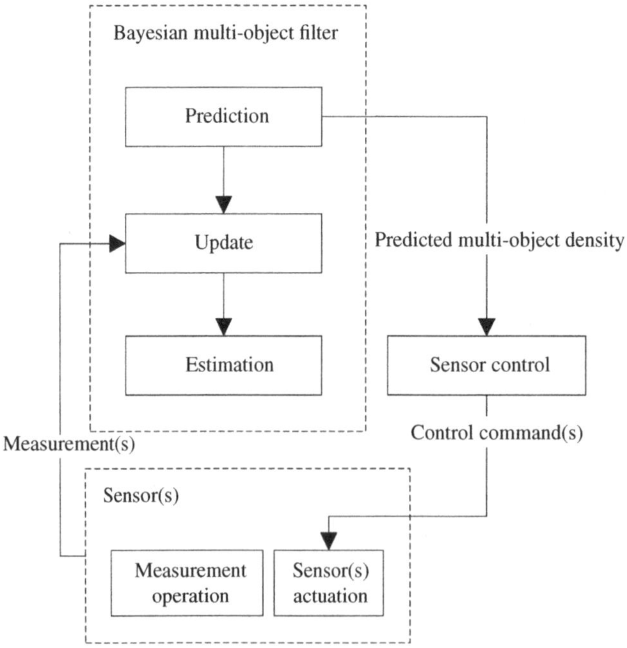

Figure 22.1 A framework incorporating Bayesian multi-object tracking. Here, the sensor control module operates based on the predicted multi-object density, issuing control commands that influence the measurement operation.

However, performing this integration is often computationally prohibitive. As a practical alternative, PIMS approximates the most likely measurement set that would be observed for a given control command under ideal conditions (i.e., no noise or clutter). The PIMS corresponding to control command u_i is constructed as follows:

$$\tilde{Z}(u_i) = \{\tilde{z}(\hat{x}; u_i)\}_{\hat{x} \in \hat{X}_{k|k-1}}, \tag{22.2}$$

where $\hat{X}_{k|k-1}$ is the predicted set of object states and $\tilde{z}(\hat{x}; u_i)$ is given by:

$$\tilde{z}(\hat{x}; u_i) = \arg\max_{z} g(z \mid \hat{x}; u_i), \tag{22.3}$$

with $g(z \mid \hat{x}; u_i)$ being the single-object likelihood function conditioned on the control command. Figure 22.2 illustrates the internal workflow of a typical sensor control module used in multi-object tracking systems. This process begins with generating a PIMS based on the predicted multi-object density and a candidate control command. The PIMS is then used to perform a pseudo-update, which approximates the effect of taking a given control action u on the resulting posterior distribution. This yields a pseudo-posterior density $\tilde{\pi}_k(\cdot; u)$, which is evaluated using an objective function to quantify the quality of the control action. Depending on the formulation, this objective can be expressed either as a reward function $R(u)$, which is to be maximized, or a cost function $C(u)$, which is to be minimized. These functions capture metrics such as estimation accuracy, information

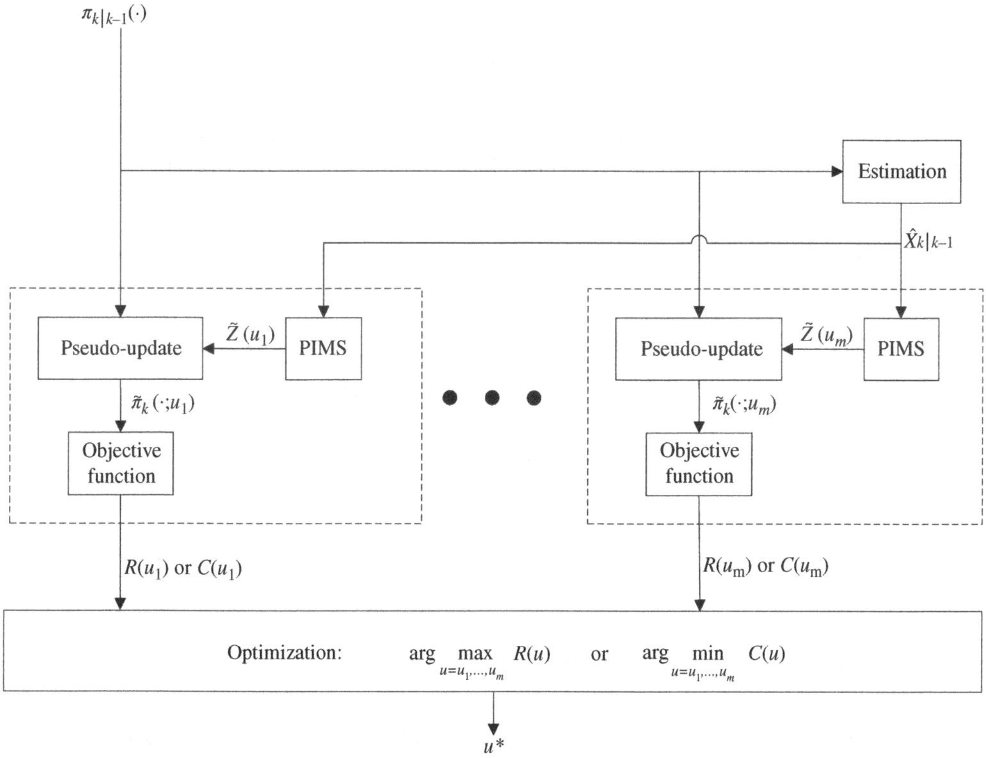

Figure 22.2 Internal architecture of a typical sensor control algorithm.

gain, or confidence in the existence of an object. As illustrated in Figure 22.2, the optimal control command u^* is obtained by executing an optimization procedure over all admissible actions in $\mathbb{U}$.

22.3 Sensor Control Categorization

To systematically understand the wide range of methods used in sensor control for MTT, it is useful to organize them into clear categories. In this section, we categorize sensor control strategies based on three primary aspects:

- The nature of the multi-object tracking system,
- The type of objective function guiding the control decisions, and
- The overall purpose of the tracking task.

This classification is illustrated in Figure 22.3. Each category and its subcategories are described in the following subsections, providing a structured foundation for understanding the landscape of sensor control techniques.

22.3.1 Nature of Multi-target Tracking

Sensor control methods are closely tied to the design and behavior of the underlying multi-object tracking system. In most advanced applications, multi-object estimation is performed using Bayesian multi-object filters. These filters produce a posterior multi-object probability density function (PDF) over the space of finite sets, capturing both the uncertainty in the number of objects and their states. In frameworks based on labeled random finite sets (RFSs) [15], each object is assigned a unique label, enabling the filter to maintain consistent identities over time. In this case, the multi-object Bayesian filter not only estimates the multi-object state but also supports trajectory tracking, functioning effectively as a Multi-target tracker. Based on the underlying tracking methodology, we classify sensor control methods into two major types:

1. **Conventional Filters**: These include well-known approaches such as global nearest neighbor (GNN), joint probabilistic data association (JPDA), and multiple hypothesis tracking (MHT) [16]. These trackers use explicit data association strategies to match measurements to targets. While they are effective, they face challenges related to data association ambiguity and computational complexity, especially as the number of targets increases.

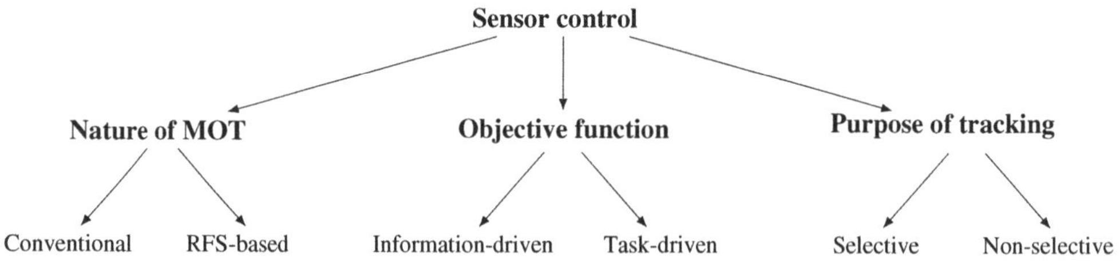

Figure 22.3 Hierarchical categorization of sensor control strategies used in multi-object tracking systems. The classification is based on three key aspects: (1) the nature of the underlying tracking method (e.g., conventional vs. RFS-based), (2) the type of objective function used for decision-making (task-driven or information-driven), and (3) the overall tracking goal (selective vs. nonselective). This structure provides a comprehensive overview of the diverse approaches found in recent literature.

2. **Random Finite Set (RFS)-based Filters**: These approaches model the entire multi-target state as an RFS [12], using the finite set statistics (FISST) framework [17]. RFS-based methods elegantly account for uncertainties in object number and data association in a unified probabilistic framework. This makes them highly suitable for sensor control in complex environments, where missed detections, false alarms, and varying target cardinality are common.

22.3.2 Objective Function

The core of any sensor control algorithm is its decision-making mechanism, which determines the most beneficial sensor action to execute at each time step. This decision is guided by an *objective function*—a mathematical criterion used to evaluate and compare the utility or cost associated with different sensing actions. The objective function encapsulates what the sensor control algorithm aims to optimize, whether that is tracking accuracy, information gain, robustness, or a task-specific performance metric. Broadly, sensor control strategies can be categorized into two principal classes based on the nature of their objective functions: *task-driven* and *information-driven* methods [18].

- **Task-driven Methods**: These approaches explicitly define the objective function in terms of tracking performance criteria. That is, the cost function directly quantifies how well the sensor system performs a specific tracking task, such as minimizing the mean squared error (MSE) of state estimates, reducing the variance of the target cardinality estimate, or maximizing detection rates. A well-known example in this category is the *Posterior Expected Number of Targets (PENT)* [19], where sensor actions are chosen to maximize the expected number of targets detected. Another line of work employs the *Posterior Cramér–Rao Bound (PCRB)* [20], which serves as a theoretical lower bound on estimation error. In this context, the sensor control problem is framed as minimization of the trace or determinant of the PCRB matrix, which indirectly ensures optimal estimation accuracy over future states [21]. Recent contributions like PEECS (Posterior Expected Error of Cardinality and States) [22] integrate both state and cardinality errors into a unified cost function. Task-driven methods are well-suited when the end-user's priorities, such as tracking accuracy or energy consumption, are clearly defined and directly measurable.
- **Information-driven Methods**: Instead of targeting specific task outcomes, these methods aim to reduce uncertainty about the multi-object state by maximizing the *expected information gain* derived from new measurements. The objective function here is typically a reward function based on information-theoretic divergences between prior and posterior probability distributions. Pioneering work in this domain includes the use of Shannon entropy [23], Kullback–Leibler (KL) divergence [18, 24], Rényi divergence [25, 26], and Cauchy–Schwarz divergence [27, 28]. These measures quantify how much a particular sensor action is expected to improve the knowledge about the targets, irrespective of any predefined task objective. Information-driven methods are particularly attractive in multi-object settings with high uncertainty or when multiple competing objectives (e.g., localization and classification) must be balanced without explicitly prioritizing one over the other [29].

The distinction between task-driven and information-driven paradigms provides crucial insight into the philosophy and design choices of sensor control algorithms. Task-driven strategies offer precise control over mission-specific goals. In contrast, information-driven approaches are more flexible, offering a general-purpose framework for improving target state knowledge through principled uncertainty reduction. Hybrid methods have also emerged, combining the benefits of both by linking information gain to task-relevant performance metrics [18].

22.3.3 Purpose of Tracking

Finally, sensor control methods can also be classified based on the specific purpose they serve in the tracking task. This dimension has become increasingly relevant with the development of selective tracking strategies. Accordingly, sensor control techniques are divided into two groups:

22.3.3.1 Non-selective Sensor Control

Traditional sensor control methods fall under this category. Their objective is to improve tracking performance for all targets in the surveillance region. For static sensors, this often involves adjusting sensor orientations to align with target directions. For mobile sensors, a common heuristic is to navigate toward the centroid of the estimated target positions. This strategy is particularly effective when targets are clustered or move cohesively, and it is widely used in early sensor control literature. However, it may be suboptimal in scenarios involving dispersed, maneuvering, or adversarial targets, where more sophisticated, model-based approaches are required to account for future target motion and sensor constraints [12–14, 19, 25, 30–32].

22.3.3.2 Selective Sensor Control

Sensor control strategies can differ based on whether they aim to improve tracking performance for *all* targets in the surveillance region or for a *specific subset* of them. The latter is referred to as selective sensor control, where the system allocates sensing resources to track only a designated subset of objects, commonly called *Targets of Interest (ToIs)*. These may be high-priority assets, anomalous entities, or targets satisfying mission-specific criteria. Importantly, nonselective sensor control can be viewed as a special case of selective control, in which the ToI set includes all known or detectable targets. This inclusive formulation highlights that selective control generalizes the traditional sensor control paradigm. The concept of focusing on specific targets is not new and has existed since early developments in MTT. Classical algorithms such as multiple hypothesis tracking (MHT) and joint probabilistic data association (JPDA) have long been used to maintain target identity over time, supporting decisions that emphasize particular tracks.

More recently, the development of *labeled Random Finite Set (RFS)* frameworks has provided a principled mathematical foundation for incorporating target identities directly into the Bayesian multi-target state space. Filters such as the *Labeled Multi-bernoulli (LMB)* and *Generalized Labeled Multi-bernoulli (GLMB)* return labeled multi-object posteriors, making it easier to formulate control strategies explicitly targeted at selected identities [15, 33–36]. Selective sensor control is especially important in *mission-critical applications*, such as VIP protection, anomaly detection, or tracking high-value military assets, where not all objects are of equal interest. It enables the system to prioritize sensing actions that yield the most significant benefit for ToIs while potentially ignoring or deprioritizing others.

22.4 In-depth Analysis of Sensor Control Techniques

This section presents a detailed examination of key sensor control strategies developed for single- and multi-sensor MTT, specifically within the stochastic framework of POMDPs. These methods are primarily designed to work in conjunction with RFS-based MTT filters. Table 22.1 provides a comprehensive summary of notable techniques introduced in recent years, classifying them based on the type of objective function (task-driven or information-driven), their reliance on RFS-based filtering, and their applicability to selective sensor control. The survey highlights that while the majority of recent advancements are grounded in the RFS paradigm, particularly in the context of multi-target filtering, there also exist methods that maintain compatibility with both traditional tracking algorithms and RFS-based filters. It is important to note the distinction: *multi-target filters* produce multi-object posterior densities and are primarily concerned with probabilistic state estimation, whereas *multi-target trackers (MTTs)* seek to extract and maintain individual target tracks (trajectories) over time, often incorporating track continuity and identity management. For instance, Mahler's posterior expected number of targets (PENT) approach [19] can be seamlessly integrated with RFS-based filters such as the probability hypothesis density (PHD) filter, as well as with classical data association-based MTT algorithms like multiple hypothesis tracking (MHT). This underscores the versatility of certain foundational sensor control strategies that bridge the conceptual gap between filtering and tracking methodologies. By organizing the literature along these methodological dimensions, we aim to clarify the evolution and taxonomy of sensor control approaches. This structure also helps contextualize the trajectory from early information-theoretic objectives to more recent task-specific and selective sensor-control techniques. Ultimately, this synthesis provides a comprehensive foundation for designing future sensor management systems, accounting for both practical considerations and theoretical advances.

22.4.1 Information-driven Sensor Control Methods

Information-driven sensor control aims to optimize the use of sensor resources by guiding their actions to maximize information acquisition about the target environment. The core principle is to reduce uncertainty in the multi-target pdf $\pi_{k|k}$ through active interaction with the environment. In this context, each sensor control decision is directed toward improving knowledge of the number and states of the objects being tracked, enhancing either the accuracy of state estimates or the confidence in the existence of targets.

The foundation of information-driven control lies in quantifying the information gain resulting from a sensor action. This gain is typically measured by the reduction in uncertainty, most commonly expressed as the change in entropy between the predicted (prior) and updated (posterior) probability densities. One of the earliest formal applications of this concept was introduced by Hintz and McVey [54], who used Shannon entropy in conjunction with Kalman filtering to quantify information gain for target tracking.

Following this pioneering work, numerous information-theoretic divergence measures have been proposed to serve as objective functions for sensor control. Among the most prominent are the Kullback–Leibler (KL) divergence [24], the R'enyi divergence [55], and the Cauchy–Schwarz (CS) divergence [56], all of which assess the difference between prior and posterior target state distributions. These divergences are used to evaluate the expected information gain associated with a candidate control action, thereby enabling the selection of the most informative sensor command.

In addition to these, a range of alternative statistical distance measures have been explored for sensor control, including total variation distance, Bhattacharyya distance, Hellinger-Matusita distance, and Wasserstein distance. These metrics capture the dissimilarity between probability distributions and are useful for comparing random variables in tracking contexts. However, most of these measures lack closed-form expressions and require computationally intensive approximations, such as Monte Carlo sampling, to be evaluated in practical scenarios.

Despite this computational burden, information-driven methods remain widely used due to their general applicability and strong theoretical foundations. The remainder of this section provides an overview of some of the most influential techniques in this category, emphasizing their formulation, implementation, and practical considerations in both single-sensor and multi-sensor control settings.

Table 22.2 summarizes the main objective functions used in the literature and highlights computational trade-offs and typical use cases.

22.4.1.1 Rényi Divergence-based Sensor Control

Rényi divergence, also known as alpha divergence, is a widely adopted objective function in information-driven sensor control. It quantifies the information gain achieved when transitioning from a prior to a posterior multi-target density. Referring to Figure 22.2, consider that at time k, a specific sensor control command u is selected. The predicted multi-object density $\pi_{k|k-1}(\cdot)$ is then transformed into a pseudo-posterior $\tilde{\pi}_k(\cdot; u)$ via a pseudo-update step, using the PIMS $\tilde{Z}(u)$. The information gained through this transformation is measured using the Rényi divergence between these two densities, given by:

$$\mathcal{R}(u) = \frac{1}{\alpha - 1} \log\left(\int [\tilde{\pi}_k(X; u)]^{\alpha} [\pi_{k|k-1}(X)]^{1-\alpha} \delta X \right) \tag{22.4}$$

Here, $\alpha \in (0, 1)$ is a tunable parameter. As special cases, the Rényi divergence approaches the Kullback–Leibler divergence as $\alpha \to 1$, and it corresponds to the Hellinger affinity when $\alpha = 0.5$. The set integral in the equation above is defined as [34]:

$$\begin{aligned} \int_S f(X)\delta X = f(\emptyset) + \int_S f(x)dx + \frac{1}{2!} \int_{S\times S} f(x_1, x_2)dx_1dx_2 + \frac{1}{3!} \\ \int_{S\times S\times S} f(x_1, x_2, x_3)dx_1dx_2dx_3 + \cdots . \end{aligned} \tag{22.5}$$

Ristic and Vo [25] introduced an information-driven sensor control method based on Rényi divergence using a particle implementation of the Bayes multi-object filter within the RFS framework. They employed the Rényi divergence as the reward function. Since a closed-form solution is generally unavailable for Rényi divergence, the authors proposed a numerical approximation via Sequential Monte Carlo (SMC) methods. Assuming that the prior density is approximated by N particles:

$$\pi_{k|k-1}(X) \approxeq \sum_{i=1}^{N} w_i \delta(X - X_i), \tag{22.6}$$

Table 22.2 Objective functions commonly used in sensor control for MTT: interpretation, tractability, and typical use.

Objective	**What it encourages**	**Tractability**	**When it is preferable/caveats**
Rényi divergence	Information gain (prior → posterior)	Often Monte Carlo/ particles	Flexible; can be costly; sensitive to PIMS/measurement approximation
CS divergence	Symmetric distribution separation	Closed-form for key RFS families (in some cases)	Efficient/stable; may be less sensitive than KL in some regimes
KL divergence	Relative entropy/ information gain	Often approximated	Sensitive; can be expensive; decompositions may bias emphasis
PENT	Coverage/expected posterior cardinality	Often closed-form	Strong for FoV-limited sensing; may neglect localization quality
Cardinality variance	Reduce uncertainty in target count	Closed-form for MB/LMB	Stabilizes cardinality; does not directly optimize localization
PEECS	Joint cardinality + localization error	Closed-form/ particle-based	Task-aligned; needs careful weighting; relies on state-error proxy
Selective variants	Focus on ToIs only	Same as base objective	Prioritizes ToIs; risk of deprioritizing scene coverage if not managed

where each (X_i, w_i) pair denotes a multi-object set particle and its corresponding weight. Using Bayes' rule, which relates the posterior to the prior via the multi-object likelihood function, the Rényi divergence can be approximated as:

$$\mathcal{R}(u) \approxeq \frac{1}{\alpha - 1} \log \frac{\sum_{i=1}^{N} w_i [g(\tilde{Z}(u) \mid X_i, u)]^{\alpha}}{\left[\sum_{i=1}^{N} w_i g(\tilde{Z}(u) \mid X_i, u)\right]^{\alpha}}, \tag{22.7}$$

where $g(Z \mid X, u)$ represents the multi-object likelihood function (sensor model). The optimal control action is selected by maximizing the divergence:

$$u^* = \arg \max_{u \in \mathbb{U}} [\mathcal{R}(u)]. \tag{22.8}$$

22.4.1.2 Cauchy–Schwarz Divergence-based Sensor Control

An alternative information-theoretic criterion that has gained traction in recent years is the *Cauchy–Schwarz (CS) divergence*. This measure is derived from the Cauchy–Schwarz inequality for inner products and provides a symmetric, nonnegative quantification of the dissimilarity between probability densities. Given two probability density functions f and g, the CS divergence is formally defined as [27]:

$$D_{\mathrm{CS}}(f, g) = -\log \frac{\langle f, g \rangle}{\| f \| \| g \|}. \tag{22.9}$$

In this expression, the inner product $\langle f, g \rangle$ is computed as:

$$\langle f, g \rangle = \int f(x)\, g(x)\, dx, \tag{22.10}$$

and the norm is defined as:

$$\| f \| = \sqrt{\langle f, f \rangle}. \tag{22.11}$$

The CS divergence takes on a value of zero when the two densities are identical ($f = g$), and is positive otherwise. Its symmetry ensures that $D_{\mathrm{CS}}(f, g) = D_{\mathrm{CS}}(g, f)$, distinguishing it from some other divergence measures such as the Kullback–Leibler divergence.

From a geometric standpoint, the argument of the logarithm in Eq. (22.9) represents the cosine of the angle between the two functions in a Hilbert space of square-integrable functions. This provides an intuitive interpretation: the more "orthogonal" or dissimilar the densities, the smaller the cosine term, and thus the greater the CS divergence. Importantly, since the integrands are nonnegative (as valid probability densities), the argument of the logarithm remains within the interval [0, 1], ensuring the divergence is well-defined.

In the context of sensor control, the CS divergence can serve as an effective reward function for quantifying the information gain between a predicted multi-object density and its corresponding pseudo-posterior. Notably, the CS divergence can also be interpreted as a tractable approximation to the Kullback–Leibler divergence [24], making it particularly appealing in scenarios where computational resources are limited or where analytic forms of the Kullback–Leibler divergence are unavailable.

Cauchy–Schwarz divergence has emerged as a valuable information-theoretic reward function for sensor control, particularly within the RFS framework. Initially, Hoang et al. [57] introduced a closed-form CS divergence formulation for Poisson RFS densities and applied it to the Probability Hypothesis Density (PHD) filter for sensor control. Although this approach is computationally efficient, its limitations include poor cardinality estimation and the absence of target tracking capabilities. To overcome these issues, Gostar et al. [42] extended the CS divergence to multi-Bernoulli filters by approximating them with Poisson RFS densities, enabling sensor control within a cardinality-balanced multi-Bernoulli (CB-MeMBer) framework. This methodology was later advanced to labeled multi-bernoulli (LMB) filters [45], which support target labeling and enable more effective selective tracking. The CS divergence was further extended by Beard et al.[44] to generalized labeled multi-bernoulli (GLMB) filters, providing an exact closed-form solution that accounts for label consistency and target identities. Moreover, both Beard et al. [44] and Gostar et al. [45] incorporated void probability constraints into the sensor control decision, ensuring that the sensor remains outside sensitive or restricted zones. These advancements establish CS divergence as a robust and versatile tool for information-driven sensor control in modern multi-object tracking systems.

22.4.1.3 Kullback–Leibler Divergence-based Sensor Control

Another information-theoretic criterion that has been used is the *Kullback–Leibler (KL) Divergence*. Given two probability density functions f and g, the KL divergence between them is defined as:

$$D_{\mathrm{KL}}(f \,||\, g) = \int f(x) \log\left(\frac{f(x)}{g(x)}\right) dx \tag{22.12}$$

The KL divergence between two RFS distributions, P and Q, is defined as:

$$D_{\mathrm{KL}}(P \,||\, Q) = \int_{\mathbb{X}} P(X) \log\left(\frac{P(X)}{Q(X)}\right) \delta X \tag{22.13}$$

Similarly to the CS divergence, the KL divergence is zero when the two densities are identical, and it is positive otherwise; however, it is not symmetric, i.e., $D_{\mathrm{KL}}(P \,||\, Q) \neq D_{\mathrm{KL}}(Q \,||\, P)$, and is therefore not a *metric*, but a *divergence*. KL divergence can be used as an effective reward function that measures the relative entropy between a predicted multi-object density and its corresponding pseudo-posterior and is more sensitive to small differences in the distributions than CS divergence. KL divergence is more computationally expensive than CS divergence, and is typically approximated for practical use in sensor control applications.

For example, for RFS densities, the KL divergence can be decomposed into two components, one for the cardinalities and the other for the target states. While the KL divergence between two cardinality distributions is simple to calculate, the KL divergence between two target state densities may not be computationally tractable. Therefore, in scenarios where coverage (ensuring the sensor observes all targets) is considered more important than localization (ensuring that the estimated target locations are highly accurate), the significance of information gain from the state component is negligible compared to the information gain from the cardinality component. The state component can be dropped for a computationally tractable approximation.

The Kullback–Leibler divergence has been successfully used as an information-theoretic reward function for sensor control within the RFS framework. Blair et al. [58] introduced using an approximation of the KL divergence without the state component to focus on maximum coverage and speed up computation. Additional modifications that have been used include an additional penalty term to punish dropped targets [59] and using constraints on actions that move sensors too close to targets using void probability, actions that avoid sensor collision, and actions that move sensors out of communication range [60].

22.4.2 Task-driven Sensor Control Methods

The development of RFS-based sensor control methods, particularly those following a task-driven approach, can be traced back to the seminal works of Mahler in 2003 [12] and 2004 [19]. These foundational studies laid the groundwork for formulating sensor management strategies within the framework of RFSs. In the task-driven paradigm, the objective function is a cost function that reflects the performance of the tracking system, typically measured by metrics such as state estimation error and cardinality variance.

22.4.2.1 Posterior Expected Number of Targets (PENT)

In the early 2000s, Mahler [12] demonstrated that using Csiszár divergence and related geometric functionals as objective functions can yield tractable solutions to the sensor control problem, especially when combined with multi-hypothesis correlator (MHC) filters. Building on this foundation, Mahler *et al.* [19] introduced a novel and probabilistically grounded objective function for sensor control: the *Posterior Expected Number of Targets* (PENT).

PENT was explicitly proposed for scenarios involving sensors with a finite field-of-view (FoV), where the sensor control strategy aims to select the control action that maximizes the expected number of detected objects—formally, the expected posterior cardinality—after updating the multi-object density. This approach fits naturally within the sensor control framework depicted in Figure 22.2 and leads to the following reward function:

$$\mathcal{R}(u) = \mathbb{E}_{\tilde{\pi}_k(\cdot;u)}\{|\tilde{X}_k|\} = \int |\tilde{X}_k|\tilde{\pi}_k(X;u)\delta X. \tag{22.14}$$

The expectation term $\mathbb{E}_{\tilde{\pi}_k(\cdot;u)}\{|\tilde{X}_k|\}$ represents the *expected a posteriori* (EAP) estimate of the cardinality. In practice, this estimate can be replaced by the *maximum a posteriori* (MAP) estimate to simplify decision-making, yielding a general control policy expressed as:

$$u^* = \arg\max_{u\in\mathbb{U}} \left|\widehat{X}_k(u)\right|. \tag{22.15}$$

Although PENT was initially derived in the context of PHD and MHC filters [12, 19], this general formulation is broadly applicable to a wide range of multi-object tracking filters, including JPDA, MHT, MB, LMB, and GLMB filters.

For instance, when the underlying tracking system employs a PHD filter implemented via Sequential Monte Carlo (SMC), the control decision simplifies to:

$$u^* = \arg\max_{u\in\mathbb{U}} \int_{\mathbb{X}} \tilde{v}_k(x;u)dx \tag{22.16}$$

$$= \arg\max_{u\in\mathbb{U}} \sum_{i=1}^{N} \widetilde{w}_{k,i}(u), \tag{22.17}$$

where $\tilde{v}_k(x;u)$ denotes the updated intensity function and $\tilde{w}_{k,i}(u)$ are the particle weights after the pseudo-update.

In the case of a multi-Bernoulli (MB) or Labeled Multi-Bernoulli (LMB) filter with SMC implementation, the optimal control action becomes:

$$u^* = \arg\max_{u\in\mathbb{U}} \sum_{i=1}^{M} \tilde{r}_k^{(i)}(u), \tag{22.18}$$

or equivalently,

$$u^* = \arg\max_{u \in \mathbb{U}} \sum_{\ell \in \mathbb{L}} \tilde{r}_k^{(\ell)}(u), \tag{22.19}$$

where $\tilde{r}_k^{(i)}(u)$ or $\tilde{r}_k^{(\ell)}(u)$ denote the pseudo-updated existence probability of object i or label ℓ under the control action u.

PENT thus offers a computationally efficient and intuitive criterion for control, especially in systems where maximizing the number of correctly detected and tracked objects is paramount.

22.4.2.2 Cardinality Variance-based Sensor Control

While the PENT approach is well-suited for scenarios with limited sensor field-of-view (FoV), an alternative strategy prioritizes minimizing the uncertainty in the estimated number of targets. This task-driven method focuses on reducing the variance of the posterior cardinality estimate, formulated as the following cost function:

$$\mathcal{C}(u) = \sigma^2_{\tilde{\pi}_k(\cdot;u)}\{|\tilde{X}_k|\} = \sum_{m=1}^{M} \left(m - \mathbb{E}_{\tilde{\pi}_k(\cdot;u)}\{|\tilde{X}_k|\}\right)^2 \Pr(|\tilde{X}_k| = m). \tag{22.20}$$

Gostar *et al.* [40] proposed a closed-form expression for this cost when the underlying multi-object tracker is a multi-Bernoulli filter:

$$\mathcal{C}(u) = \sum_{i=1}^{M} \tilde{r}_k^{(i)}(u)\left[1 - \tilde{r}_k^{(i)}(u)\right]. \tag{22.21}$$

A similar expression holds when using a Labeled Multi-Bernoulli (LMB) filter:

$$\mathcal{C}(u) = \sum_{\ell \in \mathbb{L}} \tilde{r}_k^{(\ell)}(u)\left[1 - \tilde{r}_k^{(\ell)}(u)\right]. \tag{22.22}$$

Hoang *et al.* [31] introduced an alternative approach that utilizes the maximum a posteriori (MAP) estimate of cardinality. The cost function in this case becomes:

$$\mathcal{C}(u) = \sum_{m=1}^{M} \left(m - \widehat{|X|}_{\text{MAP}}\right)^2 \Pr(|\tilde{X}_k| = m), \tag{22.23}$$

where the MAP cardinality estimate is defined as:

$$\widehat{|X|}_{\text{MAP}} = \arg\max_{m=1:M} \Pr(|\tilde{X}_k| = m). \tag{22.24}$$

For the LMB filter, the cost becomes:

$$\mathcal{C}(u) = \sum_{i=1}^{M} \tilde{r}_k^{(i)}(u)\left(1 - \tilde{r}_k^{(i)}(u)\right) + \left[\widehat{|X|}_{\text{MAP}} - \sum_{i=1}^{M} \tilde{r}_k^{(i)}(u)\right]^2, \tag{22.25}$$

where the MAP estimate is computed as:

$$\widehat{|X|}_{\text{MAP}} = \arg\max_{m=1:M} \left[\prod_{i=1}^{M}\left(1 - \tilde{r}_k^{(i)}(u)\right) \sum_{1 \le i_1 \ne \cdots \ne i_m \le M} \prod_{j=1}^{m} \frac{\tilde{r}_k^{(i_j)}(u)}{1 - \tilde{r}_k^{(i_j)}(u)}\right]. \tag{22.26}$$

22.4.2.3 Posterior Expected Error of Cardinality and States (PEECS)

To jointly account for errors in both cardinality and target state estimates, Gostar *et al.* [22] introduced the PEECS framework. This cost function incorporates a weighted combination of normalized cardinality and state estimation errors:

$$\mathcal{C}(u) = \eta\, \xi^2_{|X|}(u) + (1-\eta)\, \xi^2_{X}(u), \tag{22.27}$$

where $\eta \in [0, 1]$ controls the emphasis on cardinality versus state accuracy. The normalized cardinality variance is defined by:

$$\xi^2_{|X|}(u) = \frac{4}{M}\sum_{i=1}^{M} \tilde{r}_k^{(i)}(u)\left[1 - \tilde{r}_k^{(i)}(u)\right]. \tag{22.28}$$

For state error estimation, assume that object location in 2D is of primary concern, and the multi-object system is a multi-Bernoulli filter with particles $\left\{\left(\tilde{w}_{k,j}^{(i)}(u), x_{k,j}^{(i)}\right)\right\}_{j=1}^{J^{(i)}}$. Let $p_{x,k,j}^{(i)}$ and $p_{y,k,j}^{(i)}$ be the x and y components of the state vector. Then, the variance terms are:

$$\sigma^2_{p_x^{(i)}} = \sum_{j=1}^{J^{(i)}} \tilde{w}_{k,j}^{(i)}(u)[p_{x,k,j}^{(i)}]^2 - \left(\sum_{j=1}^{J^{(i)}} \tilde{w}_{k,j}^{(i)}(u) p_{x,k,j}^{(i)}\right)^2, \tag{22.29}$$

$$\sigma^2_{p_y^{(i)}} = \sum_{j=1}^{J^{(i)}} \tilde{w}_{k,j}^{(i)}(u)[p_{y,k,j}^{(i)}]^2 - \left(\sum_{j=1}^{J^{(i)}} \tilde{w}_{k,j}^{(i)}(u) p_{y,k,j}^{(i)}\right)^2. \tag{22.30}$$

Maximum variances occur when all particle weights are equal, yielding:

$$\max(\sigma^2_{p_x^{(i)}}) = \frac{1}{J^{(i)}}\left(1 - \frac{1}{J^{(i)}}\right)\sum_{j=1}^{J^{(i)}} [p_{x,k,j}^{(i)}]^2, \tag{22.31}$$

$$\max(\sigma^2_{p_y^{(i)}}) = \frac{1}{J^{(i)}}\left(1 - \frac{1}{J^{(i)}}\right)\sum_{j=1}^{J^{(i)}} [p_{y,k,j}^{(i)}]^2. \tag{22.32}$$

Normalized state error variance for the i-th component is:

$$\xi^2_{x^{(i)}} = \frac{\sigma^2_{p_x^{(i)}}\sigma^2_{p_y^{(i)}}}{\max(\sigma^2_{p_x^{(i)}})\max(\sigma^2_{p_y^{(i)}})}. \tag{22.33}$$

The total normalized state error variance is computed by:

$$\xi^2_X = \frac{\sum_{i=1}^{M} \tilde{r}_k^{(i)}(u)\xi^2_{x^{(i)}}}{\sum_{i=1}^{M} \tilde{r}_k^{(i)}(u)}. \tag{22.34}$$

22.5 Selective Sensor Control

22.5.1 Concept and Implementation

Conventional sensor control strategies are typically designed to manage sensor resources to collectively track all targets within a scene. However, in practical scenarios such as anomaly monitoring, VIP protection, or asset tracking, only a specific subset of targets—referred to as ToIs—require high-fidelity tracking. The emergence of labeled RFS filters, including the Labeled Multi-Bernoulli (LMB) and Generalized Labeled Multi-Bernoulli (GLMB) filters, has enabled the development of selective sensor control strategies that cater to such scenarios.

Selective sensor control methods exploit target-label information to steer sensing resources toward ToIs. These methods differ from traditional approaches in that they evaluate sensor actions solely based on their impact on estimation accuracy for ToIs, rather than for all detected targets.

22.5.1.1 Maximum Confidence in Existence

Panicker *et al.* [46] proposed a task-driven selective sensor control method that focuses on maximizing the confidence in the existence of ToIs. Let L_{ToI} denote the set of labels for the ToIs. After performing a pseudo-update on the LMB density, the confidence in the cardinality estimate of the ToIs is inversely proportional to its variance. Hence, the control decision aims to minimize the following cost function:

$$\mathcal{C}(u) = \sum_{\ell \in L_{\text{ToI}}} \tilde{r}_k^{(\ell)}(u)\left(1 - \tilde{r}_k^{(\ell)}(u)\right), \tag{22.35}$$

where $\tilde{r}_k^{(\ell)}(u)$ is the pseudo-updated probability of existence for label ℓ under control command u.

22.5.1.2 Selective-PEECS

An alternative strategy, known as Selective-PEECS [46], extends the Posterior Expected Error of Cardinality and States (PEECS) approach [22] by restricting calculations to the ToIs. The cost function integrates both the normalized error in cardinality and in state estimates:

$$\mathcal{C}(u) = \eta \xi_{|X|}^2(u; L_{\text{ToI}}) + (1 - \eta)\xi_X^2(u; L_{\text{ToI}}), \tag{22.36}$$

where $\eta \in [0, 1]$ balances the emphasis on cardinality vs. state estimation accuracy. The normalized cardinality variance is given by:

$$\xi_{|X|}^2(u; L_{\text{ToI}}) = \frac{4}{|L_{\text{ToI}}|} \sum_{\ell \in L_{\text{ToI}}} \tilde{r}k^{(\ell)}(u)[1 - \tilde{r}k^{(\ell)}(u)], \tag{22.37}$$

and the normalized state variance is:

$$\xi_X^2(u; L_{\text{ToI}}) = \frac{\sum_{\ell \in L_{\text{ToI}}} \tilde{r}k^{(\ell)}(u)\xi x^{(\ell)^2}}{\sum_{\ell \in L_{\text{ToI}}} \tilde{r}_k^{(\ell)}(u)}. \tag{22.38}$$

22.5.2 Case Studies and Applications

Selective sensor control has significant practical relevance in applications where not all detected targets are equally important. The ability to allocate sensing resources preferentially to a subset of targets leads to improved performance under resource constraints. In the study by Panicker *et al.* [46], the maximum confidence and Selective-PEECS methods were evaluated in scenarios involving multiple targets and occlusions, where only a few targets required continuous and accurate tracking. The selective control strategies were shown to maintain higher fidelity in estimating ToIs than nonselective methods, which dilute resources across all targets. Such methods have applications in:

- Security and surveillance, where certain individuals or vehicles need to be closely monitored in a crowded environment.
- Search-and-rescue operations, where tracking known survivors or critical equipment is prioritized.
- Wildlife tracking, where researchers may focus on endangered species within a broader animal population.

These case studies demonstrate the flexibility and efficiency of selective sensor control strategies, particularly when combined with modern labeled RFS filters.

22.6 Extension to Multi-sensor Control

22.6.1 Challenges and Solutions in Multi-sensor Environments

When the scenario shifts from having a single sensor to having a network of multiple sensors $\mathbb{S}$, the problem shifts from single-sensor control to multi-sensor control. Multi-sensor control aims to select the optimal control command $u_s^* \in \mathbb{U}_s$ for each sensor $s \in \mathbb{S}$. This can be represented by a multi-sensor control command $\mathbf{u}^* = \{u_1^*, u_2^*, \dots, u_{\mathcal{S}}^*\} \in \mathbb{U}^{\mathcal{S}}$, where $\mathcal{S} = |\mathbb{S}|$ is the number of sensors in the network. Each sensor will have a local RFS density $\pi_k^{(s)}$ that undergoes prediction and update at each time step. Additionally, at each time step, information fusion must be performed, allowing for information on the states of targets observed by an individual sensor to be broadcast throughout the sensor network and influence each sensor's local density. This becomes critical when each sensor has a limited FoV and there are many targets to track, because a single sensor cannot observe all targets in the scene.

Multi-sensor control brings unique challenges compared to single-sensor control. A crucial challenge is sensor coordination, where each sensor in the network should be controlled to maximize the sensor network's collective performance while minimizing redundancy. This necessitates that individual sensors consider the actions the other sensors in the network will take, choosing locally suboptimal control commands that achieve a globally optimal result.

Achieving sensor coordination, or cooperative multi-sensor control, requires robust communication between the sensors in the network, which presents practical problems, including limited bandwidth, communication delays, and dropout, that can hinder performance. Additionally, achieving consensus among sensors, such that each sensor's selected control command is aligned with the strategy of the rest of the network, is complicated by asynchronous communication and the need for scalability with respect to the size of the network, as the space of possible multi-sensor control commands increases exponentially with the number of sensors.

22.6.2 Review of Recent Advances and Techniques

Recent works have introduced multi-sensor control for labeled RFS filters. For instance, Jiang et al. [43] proposed a CS divergence-based objective function combined with generalized labeled multi-Bernoulli (GLMB) filters. Another example is the multi-sensor control solution, in which a task-driven cost (a mixture of expected localization and cardinality errors) was chosen to be minimized [32]. In this work, an LMB filter was assumed to be running on board each sensor node. Panicker et al. [46] developed a multi-sensor control solution that focuses on ToIs, again with LMB filters running on each node. An information-theoretic objective function for RFS-based multi-sensor control was also proposed [61]. Rathnayake et al [62] developed a novel track management scheme that uses label recovery to handle occlusions in visual tracking.

Akselrod & Kirubarajan [63] proposed a distributed control algorithm for MTT by a swarm of UAVs, where local measurements from each UAV are broadcast throughout the network and fused together. However, each UAV chooses an action to take without considering the actions of the other UAVs. To address this issue, Fu & Yang [64] proposed a control framework aimed at maximizing tracking accuracy while guaranteeing tracking coverage. Each sensor can be allocated to track a particular target or group of targets, and sensor positions and allocations are distributed throughout the network to jointly optimize tracking accuracy while ensuring that at least one sensor is allocated to every target. Li et al. [65] recently proposed a fully distributed multi-sensor control approach that uses auctioned partially observable Markov decision processes (POMDPs) to determine each sensor's action. While this approach is distributed and computationally efficient, it has the limitation that each action is taken by at most one sensor at a time.

Coordinate Descent is a family of optimization techniques in which each dimension (or parameter) is optimized sequentially while the others are held fixed, thereby approximately optimizing along coordinate directions. An overview of various coordinate descent algorithms is provided by Wright [66]. Panicker et al. [49] extended their previous work [46] on multi-sensor control for tracking ToIs by adopting coordinate descent optimization, significantly reducing computation time.

Genetic algorithms (GAs) are stochastic population-based optimization methods based on the principles of natural selection [67]. They are generic, robust, and applicable to a wide range of optimization problems. GAs have recently begun to be used in various MTT applications, including data association [68], data fusion [69], and single-sensor control of UAVs [70, 71]. Most recently, the application of a GA to enhance multi-sensor control was demonstrated in the context of a centralized sensor network. This advanced approach was integrated explicitly into an RFS-based filter, designed for the concurrent tracking and searching of multiple targets. This development marks a significant stride in the field of sensor network optimization [72].

22.6.3 Multi-sensor Control Case Study

Figure 22.4a illustrates a basic multi-sensor control scenario where sensor coordination is required to achieve optimal MTT performance. At time step $k-1$, all three targets are visible to both sensors; however, the top and bottom targets are moving outside of the sensors' current FoVs, therefore the sensors will need to move and therefore change their FoV so that at time step k all of the targets are still observed.

In this scenario, the sensors can take any of 3 possible actions: remain stationary, rotate 30° clockwise, and rotate 30° anticlockwise. From the perspective of either sensor individually, remaining stationary will result in observing only the center target, whereas taking either of the other two actions will result in observing two targets. From a single-sensor control perspective, rotating clockwise and rotating anticlockwise have an equal objective score and are therefore equally valid. This can result in one sensor rotating clockwise and the other rotating anticlockwise. Hence, both sensors observe the same two targets, and the third target remains unobserved, which is clearly not optimal. Therefore, a multi-sensor control approach where each sensor considers the information gathered from the other sensors in the network and the actions they will take is essential to achieve the optimal result shown in Figure 22.4b.

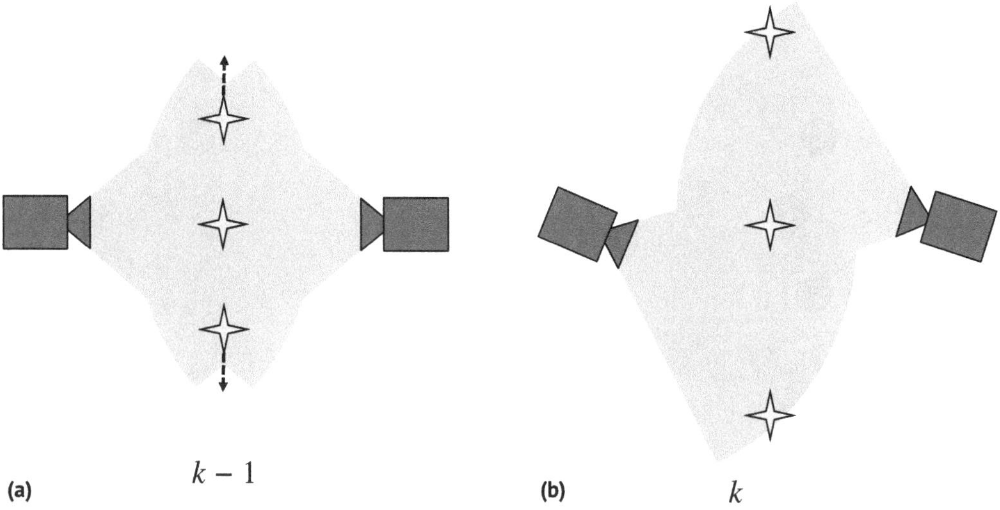

Figure 22.4 An example of sensor coordination: To ensure that all targets are observed, the sensors must consider the actions that the other sensors in the network will take.

22.7 Distributed Multi-sensor Control

22.7.1 Formulation

Typically, multi-sensor control is performed in a *centralized* manner, where each sensor's action is computed at a single computing node and then broadcast to each sensor. While conceptually simple, unfortunately, centralized approaches suffer from a scaling problem. A linear increase in the number of sensors results in an exponential growth in the number of possible multi-sensor actions, increasing the computation time required to compute the selected multi-sensor action. This necessitates efficient optimization algorithms to find a near-optimal or optimal multi-sensor action. However, even with an efficient optimization algorithm, the computation time will inevitably grow with the number of sensors, resulting in a practical limit on the size of the sensor network.

One approach to alleviate this problem is to perform multi-sensor control in a *distributed* manner. In a distributed implementation of multi-sensor control, rather than all sensors broadcasting their information to a single computing node that calculates the actions for all of the sensors in the network, instead each sensor communicates only with its neighbors and calculates its own action, as shown in Figure 22.5. This has the advantage of being able to scale to significantly larger sensor networks than the centralized approach. However, distributed multi-sensor control implementations are significantly more complex, as no individual sensor node has complete knowledge of the state of the rest of the network.

Recently, two types of distributed approaches for multi-sensor control have been proposed. The first type is where each sensor node calculates its optimal action based on its local neighborhood, and the second type is a truly distributed approach where there is consensus across the network on the optimal action for each sensor. Both approaches are detailed in the following sections.

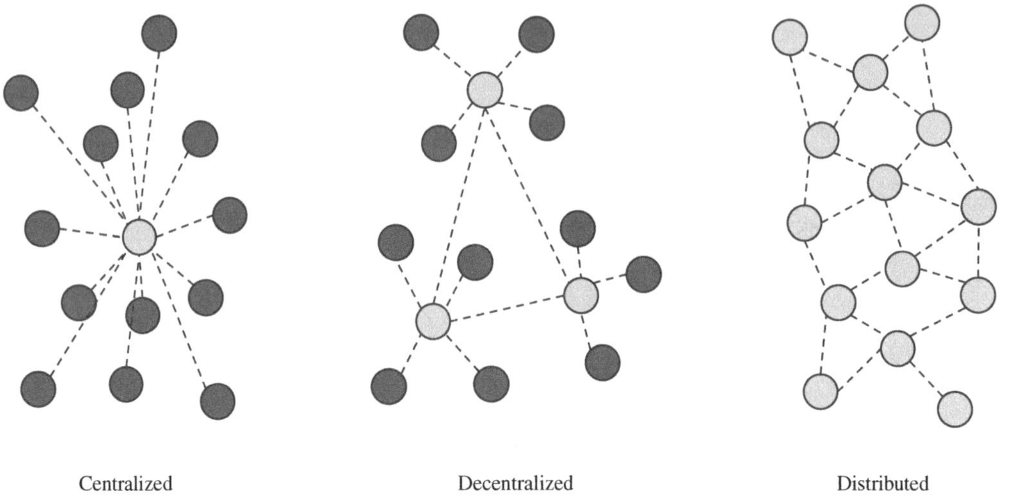

Figure 22.5 Centralized, decentralized, and distributed network architectures. Highlighted nodes perform decision-making.

22.7.2 Local Neighborhood Method

The first distributed multi-sensor control approach is based on the principle that each local sensor node s will broadcast and receive its information with just its local neighbors $N^{(s)}$ (i.e., the set of sensor nodes that can directly communicate with the node). Then each sensor node will calculate an optimal multi-sensor control command for itself and its local neighbors, $\mathbf{u}^{(s)*} = \{u_n^{(s)*}\}_{n\in N^{(s)}\cup\{s\}}$. Each sensor will send the locally calculated optimal single-sensor control command for each neighboring node, i.e., $u_n^{(s)*}$, and the associated multi-sensor control command's objective score $C^{(s)*}$ to each respective neighboring node $n \in N^{(s)}$. Each sensor will then have a collection of single-sensor control commands $\{u_s^{(n)*}\}_{n\in N^{(s)}\cup\{s\}}$ and their associated objective scores $\{C^{(n)*}\}_{n\in N^{(s)}\cup\{s\}}$.

The choice of which final single-sensor control command to select from the collection at each local sensor is nontrivial. Several possibilities were considered, including averaging the rewards of multiple instances of the same control command and selecting the control command with the highest average reward, summing multiple instances of the same control command before selecting the highest overall reward, and weighting each reward by the number of neighbors the sensor that generated the command has. The chosen method was to select the control command with the highest associated reward in the set. This is a simple way for an individual sensor's selected multi-sensor control command, if it has a high associated reward, to propagate into the network and affect the actions of neighboring sensors.

Two optimization methods have been used for this distributed multi-sensor control approach: a distributed implementation of coordinate descent [58] and a variety of genetic algorithm implementations [59].

22.7.3 Distributed Method

The second distributed multi-sensor control approach is truly distributed, meaning that each sensor in the network is in consensus on the optimal multi-sensor control command for the entire network. This is achieved through a multi-agent, fully distributed coordinate descent optimization method (FDCD-SC) [60].

In FDCD-SC, each sensor node iteratively updates its own selected action, taking into account the actions that every other sensor has selected. In a distributed network, two sensors cannot know each other's selected action in real-time. But they can use the flooding method of inter-node communication to *flood-in* the information related to the action most recently selected by others, and *flood-out* their own selection to others for future use in their own asynchronous time.

At each sensor node, an iterative operation occurs, comprising flooding-in, optimization, and flooding-out steps. Let us denote the iteration counter for this process by t, in contrast to filtering iterations denoted by k. In iteration t, the sensor node s first receives all currently selected control commands $u_{s'}^{(t-1)}$ from other sensors $s' \in \mathbb{S} \setminus \{s\}$ via a series of flooding steps. That is the realization of the flood-in step. Note that communicating only the selected control commands from other nodes typically does not place a heavy load on communication. Hence, we can practically assume that after a number of flooding steps, depending on the network size, each sensor node will have communicated with the entire network, or, in the case of an extremely large network, with a large subset of it.

The optimization step of iteration t involves an exhaustive search over the space of single-sensor commands as follows:

$$u_s^{(t)} = \arg\max_{\boldsymbol{u} \in \mathbb{U}} \nu\left(u_{1:(s-1)}^{(t-1)}, \boldsymbol{u}, u_{(s+1):\mathcal{S}}^{(t-1)}\right) \tag{22.39}$$

where

$$\begin{aligned} u_{1:(s-1)}^{(t-1)} &\triangleq \left(u_1^{(t-1)}, u_2^{(t-1)}, \dots, u_{s-1}^{(t-1)}\right) \\ u_{(s+1):\mathcal{S}}^{(t-1)} &\triangleq \left(u_{s+1}^{(t-1)}, u_{s+2}^{(t-1)}, \dots, u_{\mathcal{S}}^{(t-1)}\right). \end{aligned} \tag{22.40}$$

At the conclusion of iteration t at node s, the locally decided optimal command for the sensor is flooded-out to all the other sensor nodes for use in their own optimization step.

An important note to make is that solving the optimization problem in (22.39) by exhaustive search is not computationally prohibitive. The main reason is that it involves computing the reward for just a finite number of possible single-sensor commands. In addition, note that when computing the reward for each possible sensor command, the PIMS and pseudo-update calculations need to be performed only for the actual node s, not for the others.

The proposed method has a well-defined stopping criterion, where convergence of the multi-sensor control command is mathematically guaranteed when it occurs [60].

22.8 Comparative Analysis and Simulation Results

22.8.1 Methodology and Setup

The probability of detection of an object with state x is denoted by $p_{D,k}^{(s)}(x)$, implying that it may differ from one sensor node to another, may vary with time and may be dependent on the object's state. The dependence on the object's state is particularly of interest and significance. The *limits* of the sensor's field of view is mathematically formulated by incorporating such a dependency. An example is shown in Figure 22.6. In a 2D tracking application, the sensor is oriented downward

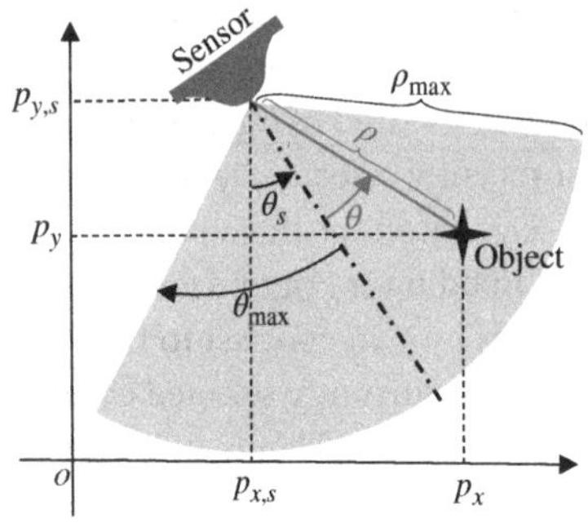

Figure 22.6 Schematic diagram of a sensor with limited FoV, and the notation used for the formulation of the probability of detection.

at an angle θ_s. Its field of view is limited in range (can only detect objects at distances between ρ_{min} and ρ_{max}) and angle (up to θ_{max} radians away from the sensor's tilted axis). Having the sensor located at $(p_{x,s}, p_{y,s})$, the probability of detection of an object located at (p_x, p_y) can be given by:

$$p_{D,k}^{(s)}(\rho, \theta) = \begin{cases} \dfrac{p_{D,\max} \tanh\left(\frac{|\rho_{max}-\rho|}{\lambda}\right)}{\tanh\left(\frac{|\rho_{max}-\rho_{min}|}{\lambda}\right)} & \text{if } |\theta| \leq \theta_{max}, \rho_{min} \geq \rho \leq \rho_{max} \\ 0 & \text{else} \end{cases} \tag{22.41}$$

where

$$\rho = \sqrt{(p_x - p_{x,s})^2 + (p_y - p_{y,s})^2}$$

$$\theta = \text{atan2}(p_x - p_{x,s}, p_y - p_{y,s}) + \frac{\pi}{2} - \theta_s,$$

where atan2 means the four-quadrant inverse tangent function.

There is distance-dependent measurement noise on each observation. Measurements are returned as a vector with two elements: the distances from the sensor to the measurement along the x- and y-axes, d_x and d_y, respectively. The measurement noise along each axis is normally distributed, with mean 0 and variance $1 + 0.00001 \times d_x^2$ along the x-axis and $1 + 0.00001 \times d_y^2$ along the y-axis.

22.8.2 Performance Comparison and Discussion

22.8.2.1 Scenarios

Multiple challenging multi-object tracking scenarios were developed to evaluate the performance of the FDCD-SC method. Figure 22.7 illustrates the first scenario, with 6 sensors tracking 11 targets for a total of 50 time steps with a period of 1 second. The targets are distributed within a 1000 m × 2000 m area and follow linear trajectories. All of the targets are present at the first time step and are observed by at least one sensor; however, as the scenario progresses, the targets will move away from the center of the scenario, following a linear motion model, and therefore out of the initial sensor FoVs. This dynamic nature of the multi-target tracking problem is handled by the complementary information fusion between sensors, so that targets aren't dropped when they move out of one sensor's FoV but remain in other sensors' FoVs, and they aren't double-counted when multiple sensors observe the same target. Additionally, two of the targets will die prematurely.

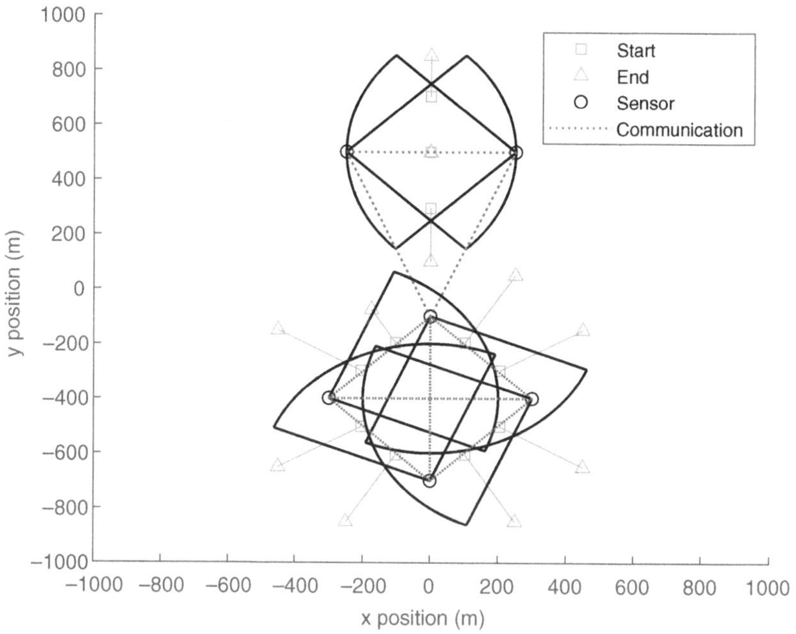

Figure 22.7 Overview of the six sensor scenario: Initial sensor positions are represented by small circles, the initial FoV of each sensor is shown by the sectors, a dotted line indicates the presence of communication between a pair of sensors, and the target trajectories are denoted by dashed lines.

The detection profile of each sensor is modeled by (22.41), in the form depicted in Figure 22.6, with $\theta_{\max} = 45°(90°$ field of view), a maximum detection range of $\rho_{\max} = 500$ m, and constants $p_{D,\max} = 0.99$, $k_\rho = 0.5$, and $k_\theta = 20$. The maximum communication range between sensors is simulated to be 800 m. Each sensor has 3 possible actions: remain stationary, rotate 22.5° clockwise, and rotate 22.5° anticlockwise.

Figure 22.8 illustrates the second, more challenging, scenario, with 8 sensors tracking 20 targets in a 800 m × 800 m area for a total of 100 time steps with a period of 1 second. In this scenario, there are two circular groups of targets that are each following a linear motion model, with trajectories such that the two target groups are initially separate, with four sensors tracking each group, followed by a period with significant overlap between the two groups, with a final period of separation again. The maximum detection range of the sensors is set to $\rho_{\max} = 100$ m, with no angular FoV restrictions, $p_{D,\max} = 0.99$, $k_\rho = 0.5$, and the maximum communication range between sensors is set at 300 m. In the second scenario, each sensor has 9 possible actions: remain stationary, move 15 m east, move 15 m north-east, move 15 m north, move 15 m north-west, move 15 m west, move 15 m south-west, move 15 m south, and move 15 m south-east.

22.8.2.2 Methods

The FDCD-SC method for multi-sensor control is evaluated and compared against several other approaches. Sensors with fixed positions and orientations are examined to establish baseline results. An individual sensor control (I-SC) method, in which each sensor independently finds the optimal action for itself without considering the other sensors' actions, is evaluated to assess the

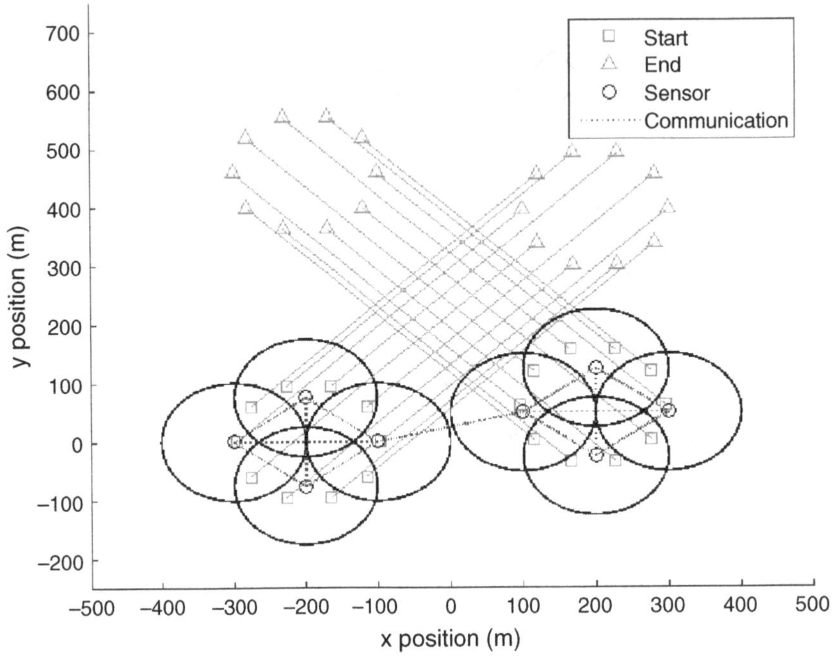

Figure 22.8 Overview of the 20 sensor scenario.

benefits of cooperative multi-sensor control. For the I-SC approach, the fused pseudo-updated posterior in the objective function is replaced with the local pseudo-updated posterior, and each sensor independently performs an exhaustive search over the possible single-sensor control commands.

Finally, the recently proposed Distributed Coordinate Descent approach (DCD-SC) is also tested to assess its comparative multi-object tracking accuracy and computational time. Wang et al. [32] describe the relationship between the number of coordinate descent runs required to reach a likelihood of finding the global optimum. Given $\mathfrak{M}$ local optima and the probability of finding the global optimum P_{success}, the number of required runs is given by (22.42).

$$m = \left\lceil \frac{\log(1 - P_{\text{success}})}{\log(1 - \frac{1}{\mathfrak{M}})} \right\rceil \tag{22.42}$$

where $\lceil \cdot \rceil$ means "rounded up to the next integer".

Assuming that $\mathfrak{M} = 2|N_{\max}|$, where $N_{\max} = \max_{s \in \mathbb{S}} |N^{(s)}|$, then to achieve a 95% probability of finding the global optimum in the first scenario, the sensor with the most neighbors requires 35 runs, and in the second scenario, the sensor with the most neighbors requires 47 runs. This is an impractical number of runs given the high computational cost; therefore, we aim for a sub-optimal but computationally tractable solution. To compare the computation time and accuracy trade-off, several different numbers of runs are tested.

22.8.2.3 Results

For the first scenario, Figure 22.9 shows the average estimated target cardinality (number of estimated targets) across the 30 Monte Carlo runs at each time step, for each of the tested methods. The closer the thin lines (estimated target cardinality) are to the thick black line (true target car-

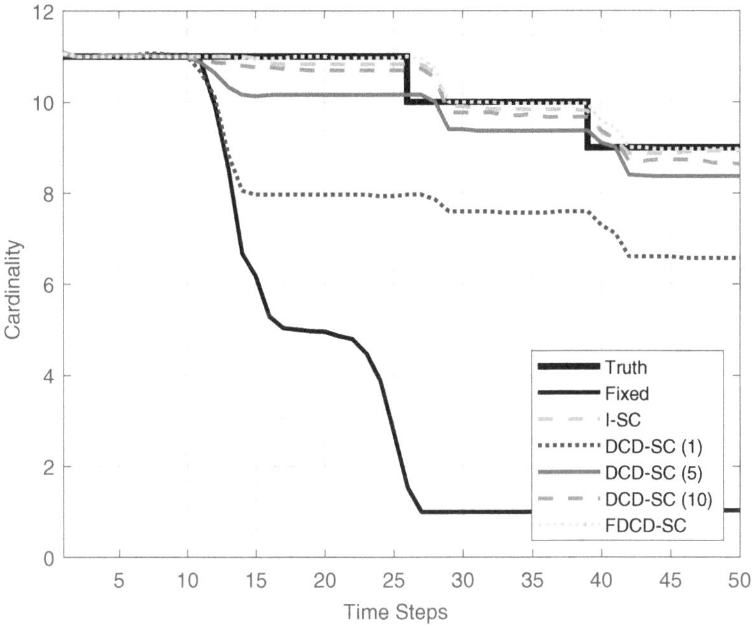

Figure 22.9 The average cardinality error of the fixed sensors, I-SC, DCD-SC, and FDCD-SC methods for the first scenario. The true number of targets is indicated by the thick black line, and the average estimated numbers of targets are shown by the other lines.

dinality), the more accurate the estimate of the number of targets and, therefore, the multi-target tracking accuracy. Table 22.3 details the average OSPA and OSPA$^{(2)}$ distances across the 30 Monte Carlo runs for each method, as well as the average computation time per sensor for each method.

For the second more challenging scenario, Figure 22.10 shows the average estimated target cardinality (number of estimated targets) across the 30 Monte Carlo runs at each time step, for each of the tested methods. The closer the thin lines (estimated target cardinality) are to the thick black line (true target cardinality), the more accurate the estimate of the number of targets and, therefore, the MTT accuracy. Table 22.4 details the average OSPA and OSPA$^{(2)}$ distances across the 30 Monte Carlo runs for each method, as well as the average computation time per sensor for each method.

22.8.2.4 Discussion

As discussed previously, the primary source of multi-object tracking error in a scenario with a large number of mobile, dynamic targets is when targets are not observed by any sensor, i.e., cardinality error. For the first scenario, Figure 22.9 shows how, when the sensors are fixed, initially the targets are all within the FoV of the sensor network; when the targets begin to move outside of the network's FoV, the targets are rapidly lost, leading to a high cardinality error and poor multi-object tracking performance. In comparison, when multiple-sensor control methods are used, the sensors can follow the targets' trajectories, resulting in a much lower cardinality error. The proposed FDCD-SC method has the best cardinality error of all of the evaluated methods, rarely dropping targets due to subsequent missed detections, with the majority of the cardinality inaccuracies occurring at target deaths. The I-SC method performs similarly to FDCD-SC, with only a slightly worse cardinality error. DCD-SC with $m = 1$ performs poorly, however, as m increases, the cardinality error decreases, approaching the errors of I-SC and FDCD-SC at $m = 10$. This suggests that

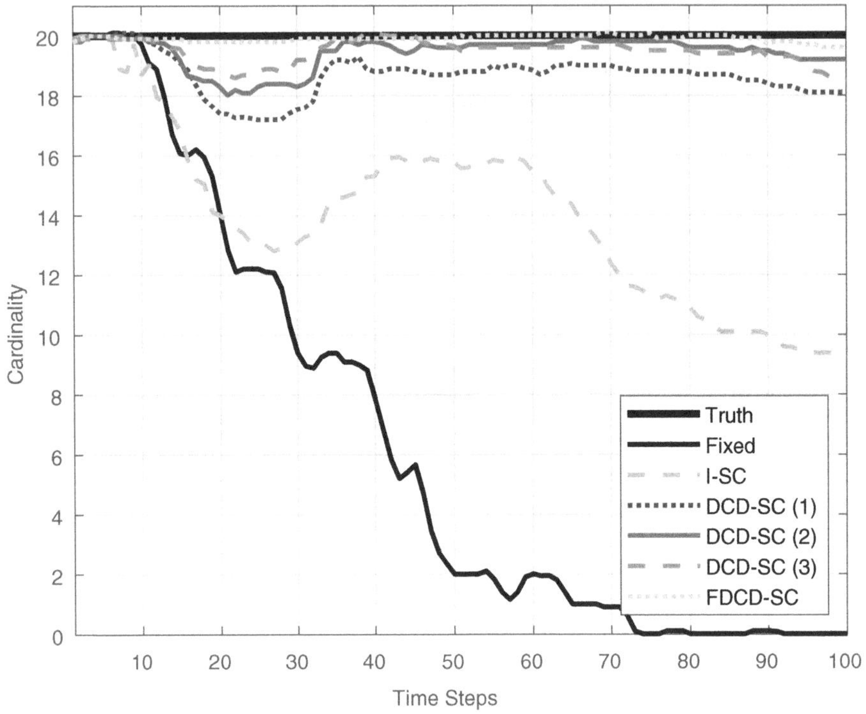

Figure 22.10 The average cardinality error of the fixed sensors, I-SC, DCD-SC, and FDCD-SC methods for the second scenario. The true number of targets is indicated by the thick line, and the average estimated numbers of targets are shown by the other lines.

Table 22.3 Average multi-object tracking performance results over 30 Monte Carlo runs of fixed sensors, I-SC, DCD-SC, and FDCD-SC.

Control method	OSPA (m)	OSPA$^{(2)}$ (m)	Comp. time (s)
Fixed Sensors	60.36	48.97	N/A
I-SC	3.94	5.35	0.06
DCD-SC (1 run)	22.39	21.34	0.10
DCD-SC (5 runs)	8.36	8.65	0.43
DCD-SC (10 runs)	5.54	7.34	0.83
FDCD-SC	**3.39**	**4.86**	**0.74**

a large portion of the cardinality error at lower m values is due to poor optimization of the local multi-sensor control commands, while the smaller portion that remains at any m is due to the lack of global coordination between the sensors, resulting in sensors taking actions that drop targets.

Table 22.3 highlights the significant improvement that multi-sensor control has over fixed, stationary sensors when tracking multiple dynamic targets, as well as the considerable reduction in both the OSPA and OSPA$^{(2)}$ distances in FDCD-SC compared to I-SC and particularly DCD-SC.I The lower multi-object tracking errors of FDCD-SC compared to I-SC come at the cost of a significantly longer computation time than I-SC (~12 times longer). By the time $m = 10$, the computation time of DCD-SC has already surpassed FDCD-SC, suggesting that there is no accuracy-computation tradeoff between the two methods.

Table 22.4 Average multi-object tracking performance results over 30 Monte Carlo runs of fixed sensors, I-SC, DCD-SC, and FDCD-SC.

Control method	OSPA (m)	OSPA$^{(2)}$ (m)	Comp. time (s)
Fixed Sensors	68.58	70.76	N/A
I-SC	31.82	46.44	0.54
DCD-SC (1 run)	8.13	23.40	3.31
DCD-SC (2 runs)	5.12	20.85	7.44
DCD-SC (3 runs)	4.16	13.35	13.75
FDCD-SC	**1.73**	**5.87**	**12.80**

Figure 22.10 again shows that the proposed FDCD-SC has significantly less cardinality error than both fixed cameras and I-SC, further indicating that the cooperative nature of FDCD-SC allows for better coverage than individual sensor control in complex MTT scenarios. Table 22.4 highlights that FDCD-SC has the lowest OSPA and OSPA$^{(2)}$ errors of all the methods tested. The computation time for FDCD-SC is again significantly longer in comparison to I-SC (~24 times longer) and performs better than DCD-SC at a comparable computation time ($m = 3$).

The following presents a comparison of the computational costs of the tested methods. The computational complexity of I-SC is $\mathcal{O}(|\mathbb{U}|)$. The computational complexity of DCD-SC is $\mathcal{O}(m \times n \times N_{\max} \times |\mathbb{U}|)$, where m represents the number of runs, n is the number of iterations per run, $N_{\max}$ is the maximum size of a sensor's neighborhood, and $|\mathbb{U}|$ is the number of possible control commands. The computational complexity of FDCD-SC is expressed as $\mathcal{O}(n \times \mathcal{S} \times |U|)$.

For the same scenario, FDCD-SC is $I\mathcal{S}$ times more expensive than I-SC and $m\dfrac{N_{\max}}{\mathcal{S}} > 1$ times more expensive than DCD-SC. If $m\dfrac{N_{\max}}{\mathcal{S}} > 1$, which is more likely in densely connected networks, then FDCD-SC will be computationally faster than DCD-SC. Conversely, if $m\dfrac{N_{\max}}{\mathcal{S}} < 1$, which is more likely in sparsely connected networks, then FDCD-SC will be computationally slower than DCD-SC. Interestingly, neither method's computational cost depends on the number of possible actions, $|U|$.

22.9 Limitations and Practical Considerations

This section summarizes practical limitations and trade-offs that commonly arise in sensor control for MTT, complementing the method survey in Section 22.4 and the multi-sensor/distributed results presented earlier. The intent is to clarify when certain design choices are likely to succeed, when they may fail, and what practitioners should check in implementation.

22.9.1 Modeling Assumptions and Sensitivity

Most sensor control objectives—whether information-driven (e.g., KL/CS/Rényi divergences) or task-driven (e.g., PENT, cardinality variance, PEECS)—depend critically on the fidelity of the underlying models used in prediction and pseudo-update (Section 22.2). In practice, performance can degrade substantially under:

- **Detection and clutter mismatch:** Incorrect assumptions about $p_D(x)$ clutter rate/intensity, or gating can bias the expected utility of actions, causing systematic preference for suboptimal viewpoints or regions.
- **Measurement noise mismatch:** Under- or over-estimating range/angle noise (or distance-dependent noise) can distort information gain estimates and localization-driven objectives.
- **Motion model mismatch:** When target maneuvers differ from the assumed dynamic model, control can over-commit to predicted motion and under-explore alternative regions, particularly in sparse-observation regimes.

A practical implication is that objective functions should be interpreted as *model-conditioned* measures of value. When model uncertainty is significant, robustness mechanisms (e.g., conservative p_D models, regularization, or multi-model filtering) become important.

22.9.2 Limitations of PIMS and Pseudo-update Approximations

To make online control tractable, many implementations replace the expectation over all possible measurement sets with a PIMS and a pseudo-update (Section 22.2). While effective in many regimes, this approximation introduces limitations:

- **Low p_D and heavy clutter:** PIMS implicitly reflects an idealized, high-confidence update. When missed detections and false alarms dominate, PIMS may overestimate the benefit of aggressive actions.
- **Strong nonlinearities:** For highly nonlinear sensors or observation geometries, a single idealized measurement set may not represent the true spread of informative outcomes.
- **Risk of over-exploitation:** PIMS-based objectives can favor actions that look optimal under idealized updates, potentially reducing exploration and increasing the chance of track loss.

Where computationally feasible, one mitigation is to approximate the expected utility using a small number of sampled measurement sets (rather than a single PIMS), trading runtime for reduced bias.

22.9.3 Computational Scaling and Optimization Constraints

A central practical issue is the computational cost of evaluating objectives across the action space:

- **Action-space growth:** Exhaustive search scales with $|\mathbb{U}|$. Continuous actions (e.g., heading/position) require discretization, sampling, or derivative-free optimization, often without global optimality guarantees.
- **Objective evaluation cost:** Information-driven criteria may require Monte Carlo approximations or expensive set-integral surrogates, whereas certain task-driven objectives admit closed-form expressions under specific RFS families.
- **Multi-sensor combinatorics:** Centralized multi-sensor control can scale exponentially in the number of sensors if joint actions are enumerated, motivating coordinate descent, greedy heuristics, or distributed optimization.

Accordingly, objective selection is often inseparable from the available compute budget: closed-form or low-variance surrogates can outperform theoretically richer criteria when runtime constraints prevent adequate approximation accuracy.

22.9.4 Multi-sensor Coordination and Communication Limitations

In multi-sensor and distributed settings, coordination gains must be weighed against communication and fusion realities:

- **Delays and dropout:** Asynchronous communication, packet loss, and limited bandwidth can lead to inconsistent beliefs across nodes, degrading cooperative decision-making.
- **Fusion-consistency effects:** In distributed architectures, approximate fusion and local neighborhood exchanges can introduce information overlap or inconsistency, impacting both filtering accuracy and the control objective values.
- **Consensus vs agility:** Methods that pursue network-wide consensus can be robust, but may respond more slowly to rapid scene changes; local methods can be agile but risk redundant sensing or temporary coverage gaps.

22.9.5 Selective Control: Benefits and Failure Modes

Selective sensor control improves fidelity for ToIs, but introduces its own trade-offs:

- **Coverage risk:** Over-focusing on ToIs can reduce surveillance coverage and increase the probability of missing new births or losing non-ToI tracks that later become relevant.
- **Label/identity fragility:** Selective policies depend on reliable label management; identity swaps, coalescence, and prolonged occlusion can reduce the effectiveness of ToI-focused objectives.

In practice, many systems benefit from hybrid policies that enforce minimum coverage or maintain a small exploration budget while prioritizing ToIs.

22.9.6 Selection Guidance

The following heuristics are often useful when selecting a control objective and architecture:

- **Coverage-dominated regimes (FoV-limited sensing):** Prefer cardinality-focused objectives (e.g., PENT, cardinality variance) or add explicit penalties for dropped tracks.
- **Localization-dominated regimes:** Prefer joint objectives (e.g., PEECS-style) provided state-error proxies are reliable and appropriately weighted.
- **Tight compute budgets:** Prefer objectives with closed-form evaluation (when available) or low-variance approximations.
- **Large sensor networks:** Prefer scalable optimization (coordinate descent, greedy, decomposed methods, or distributed schemes) and explicitly account for communication constraints.

22.9.7 Open Problems

Despite extensive progress, several research challenges remain open and practically important:

- **Continuous control with guarantees:** Designing sensor control for continuous action spaces (e.g., trajectories) with stability/performance guarantees under realistic constraints.
- **Learning-based policies with robustness:** Integrating reinforcement learning or imitation learning while preserving safety, interpretability, and robustness to model mismatch.

- **Adversarial and contested environments:** Control under deception, jamming, spoofing, or strategic target behavior, including robust objective design and resilient distributed coordination.
- **Scalable distributed consensus:** Achieving reliable cooperation under limited bandwidth, asynchronous updates, and partial connectivity without sacrificing responsiveness.
- **Principled evaluation:** Establishing benchmarking protocols that jointly capture track continuity, cardinality stability, communication load, and compute cost across representative scenarios.

22.10 Conclusions

This chapter presents a comprehensive examination of sensor control strategies in the domain of MTT, with a particular emphasis on integrating Bayesian estimation frameworks and RFS theory. By systematically exploring task-driven and information-theoretic approaches, we have highlighted how modern sensor control methodologies address the inherent challenges of uncertainty, limited field of view, resource constraints, and dynamic target behaviors. The chapter has also underscored the role of objective function design—whether based on estimation accuracy, information gain, or mission-specific priorities—in shaping control decisions across diverse application contexts.

A salient development in recent years is the shift toward selective sensor control, enabled by advances in labeled RFS filtering. These methods enable prioritization of high-value targets in cluttered environments and improve tracking performance under operational constraints. Moreover, we have examined the transition from single-sensor to multi-sensor control frameworks, illustrating how coordinated and distributed control architectures can significantly enhance system-level performance. Techniques such as flooding-based optimization, coordinate descent, and genetic algorithms have demonstrated scalability and robustness in distributed sensor networks, supporting real-time tracking in complex environments.

The synthesis presented in this chapter lays a foundation for future research directions aimed at developing adaptive, efficient, and context-aware sensor control systems. Open problems include integrating learning-based policies, handling adversarial conditions, and designing distributed architectures that balance local autonomy with global coordination. As sensing systems become increasingly autonomous and interconnected, sensor control will remain a cornerstone of robust MTT and situational awareness.

References

1 K. Dogancay and H. Hmam. Optimal sensor placement for aoa localization and tracking. In M. Mallick, B.-N. Vo, and R. Tharmarasa, editors, *Advanced Techniques in Tracking and Sensor Management: Theory and Applications*, chapter 24. Wiley/IEEE, 2025.

2 Y. Yuan, W. Yi, R. Hoseinnezhad, and P. K. Varshney. Robust power allocation for multi-target tracking in colocated mimo radars with minimized resource consumption. In M. Mallick, B.-N. Vo, and R. Tharmarasa, editors, *Advanced Techniques in Tracking and Sensor Management: Theory and Applications*, chapter 25. Wiley/IEEE, 2025.

3 W. Pu, J. Yan, P. Zhang, H. Jiao, and H. Liu. Multisensor resource allocation for multitarget tracking. In M. Mallick, B.-N. Vo, and R. Tharmarasa, editors, *Advanced Techniques in Tracking and Sensor Management: Theory and Applications*, chapter 26. Wiley/IEEE, 2025.

4 X. Wang, J. Zhang, H. Wang, and J. Zhang. Sensor control in anti-submarine warfare—a digital twin and random finite sets based approach. *Sensors*, 19(24):5439, 2019. https://doi.org/10.3390/s19245439.

5 Y. Ko, T. Kim, D. Lee, and J. Lee. Mutual information-based path planning for multi-target tracking using uavs. *Sensors*, 22(1):346, 2022. https://doi.org/10.3390/s22010346.

6 X. Guo, S. L. Smith, and M. E. Broucke. Multi-agent informative path planning with energy awareness for persistent environmental monitoring. *IEEE Transactions on Control of Network Systems*, 8(2):690–702, 2021. https://doi.org/10.1109/TCNS.2020.2989252.

7 J. Na, T. Kim, H. Kim, and D. Lee. Autonomous sensing for multi-target tracking using uavs with particle phd filter. *IEEE Transactions on Aerospace and Electronic Systems*, 57(6):4181–4195, 2021. https://doi.org/10.1109/TAES.2021.3076813.

8 S. Panicker, S. Arulampalam, and A. Sinha. Recent advances in stochastic sensor control for multi-object tracking. *Sensors*, 19(17):3692, 2019. https://doi.org/10.3390/s19173692.

9 W. Li and C. Han. Dual sensor control scheme for multi-target tracking. *Sensors*, 18(5):1653, 2018. https://doi.org/10.3390/s18051653. ISSN 1424-8220.

10 Q. Li, R. Gan, and S. Godsill. Distributed multi-sensor multi-target tracking with fault detection and exclusion. *Information Fusion*, 97:101948, 2024. https://doi.org/10.1016/j.inffus.2023.101948.

11 S. A. Fard, R. Hoseinnezhad, and B.-N. Vo. Adaptive non-myopic sensor control with guaranteed bounded suboptimality for multi-object tracking. *IEEE Transactions on Aerospace and Electronic Systems*, 59(6):5029–5045, 2023. https://doi.org/10.1109/TAES.2023.3265197.

12 R. P. S. Mahler. Multisensor-multitarget sensor management: a unified bayesian approach. In *Proceedings of the SPIE Acquisition, Tracking, and Pointing XII*, volume 5096,pages 222–234, 2003.

13 D. A. Castanõn and L. Carin. Stochastic control theory for sensor management. In A. O. Hero III, D. A. Castanõn, D. Cochran, and K. Kastella, editors, *Foundations and Applications of Sensor Management*, 7–32. Springer, Boston, MA, 2008.

14 R. P. S. Mahler, D. Grundel, R. Murphey, and P. M. Pardalos. Multitarget sensor management of dispersed mobile sensors. In *Theory and Algorithms for Cooperative Systems*, World Scientific, Singapore, 239–310, 2004.

15 B. T. Vo and B. N. Vo. Labeled random finite sets and multi-object conjugate priors. *IEEE Transactions on Signal Processing*, 61(13):3460–3475, 2013.

16 Y. Bar-Shalom and T. E. Fortmann. *Tracking and Data Association*. Academic Press, Boston, 1988.

17 R. Mahler. *Statistical Multisource-multitarget Information Fusion*. Artech House, 2007.

18 C. Kreucher and A. O. Hero. A comparison of task driven and information driven sensor management for target tracking. In *Proceedings of the IEEE Conference on Decision and Control (CDC-ECC)*, pages 4004–4009, 2005.

19 R. P. S. Mahler and T. R. Zajic. Probabilistic objective functions for sensor management. In *Proceedings of the SPIE Signal Processing, Sensor Fusion, and Target Recognition XIII*, volume 5429, pages 233–244, 2004.

20 P. Tichavsky, C. H. Muravchik, and A. Nehorai. Posterior Cramér–Rao bounds for discrete-time nonlinear filtering. *IEEE Transactions on Signal Processing*, 46(5):1386–1396, 1998.

21 Y. Liu, L. Zhou, Q. Wei, and B. Zhao. Sensor management based on convex optimisation via pcrlb and joint interception probability. In *2022 IEEE Sensors*, pages 1–4, 2022. https://doi.org/10.1109/SENSORS52175.2022.9967316.

22 A. K. Gostar, R. Hoseinnezhad, and A. Bab-Hadiashar. Multi-bernoulli sensor control via minimization of expected estimation errors. *IEEE Transactions on Aerospace and Electronic Systems*, 51(3):1762–1773, 2015.

23 C. E. Shannon. A mathematical theory of communication. *Bell System Technical Journal*, 27(2):379–423, 1948.

24 S. Kullback and R. A. Leibler. On information and sufficiency. *The Annals of Mathematical Statistics*, 22:79–86, 1951.

25 B. Ristic and B.-N. Vo. Sensor control for multi-object state-space estimation using random finite sets. *Automatica*, 46:1812–1818, 2010.

26 B. Ristic, B.-N. Vo, and D. Clark. A note on the reward function for phd filters with sensor control. *IEEE Transactions on Aerospace and Electronic Systems*, 47:1521–1529, 2011.

27 K. Kampa, E. Hasanbelliu, and J. C. Principe. Closed-form Cauchy-Schwarz pdf divergence for mixture of gaussians. In *Proceedings of the 2011 International Joint Conference on Neural Networks (IJCNN)*, pages 2578–2585, 2011.

28 M. Beard, B. T. Vo, B. N. Vo, and S. Arulampalam. Sensor control for multi-target tracking using Cauchy-Schwarz divergence. In *Proceedings of the 18th International Conference on Information Fusion (FUSION)*, pages 937–944, 2015.

29 A. O. Hero and D. Cochran. Sensor management: Past, present, and future. *IEEE Sensors Journal*, 11(12):3064–3075, 2011.

30 R. P. S. Mahler. Global posterior densities for sensor management. In *SPIE Conference on Acquisition, Tracking, and Pointing XII*, volume 3365, pages 252–263, 1998.

31 H. G. Hoang and B. T. Vo. Sensor management for multi-target tracking via multi-bernoulli filtering. *Automatica*, 50(4):1135–1142, 2014.

32 X. Wang, R. Hoseinnezhad, A. K. Gostar, T. Rathnayake, B. Xu, and A. Bab-Hadiashar. Multi-sensor control for multi-object bayes filters. *Signal Processing*, 142:260–270, 2018.

33 B. N. Vo, B. T. Vo, and D. Phung. Labeled random finite sets and the bayes multi-target tracking filter. *IEEE Transactions on Signal Processing*, 62(24):6554–6567, 2014.

34 R. P. S. Mahler. *Advances in Statistical Multisource-multitarget Information Fusion*. Artech House, 2014.

35 S. Reuter, B. T. Vo, B. N. Vo, and K. Dietmayer. The labeled multi-bernoulli filter. *IEEE Transactions on Signal Processing*, 62(12):3246–3260, 2014.

36 B. N. Vo, B. T. Vo, and H. G. Hoang. An efficient implementation of the generalized labeled multi-bernoulli filter. *IEEE Transactions on Signal Processing*, 65(8):1975–1987, 2016.

37 D. J. Kershaw and R. J. Evans. Optimal waveform selection for tracking systems. *IEEE Transactions on Information Theory*, 40(5):1536–1550, 1994. https://doi.org/10.1109/18.312154.

38 V. Krishnamurthy and R. J. Evans. Hidden markov model multiarm bandits: A methodology for beam scheduling in multitarget tracking. *IEEE Transactions on Signal Processing*, 49(12):2893–2908, 2001. https://doi.org/10.1109/78.968685.

39 C. Kreucher, K. Kastella, and A. O. Hero. Multi-target sensor management using alpha-divergence measures. In F. Zhao and L. Guibas, editors, *Information Processing in Sensor Networks*, Springer, Berlin, Heidelberg, 209–222, 2003.

40 A. K. Gostar, R. Hoseinnezhad, and A. Bab-Hadiashar. Robust multi-bernoulli sensor selection for multi-target tracking in sensor networks. *IEEE Signal Processing Letters*, 20(12):1167–1170, 2013.

41 A. K. Gostar, R. Hoseinnezhad, A. Bab-Hadiashar, and F. Papi. Ospa-based sensor control. In *IEEE Conference on Control Automation and Information Sciences (ICCAIS)*, pages 214–218, 2015.

42 A. K. Gostar, R. Hoseinnezhad, and A. Bab-Hadiashar. Multi-bernoulli sensor control using Cauchy-Schwarz divergence. In *Proceedings of the 19th International Conference on Information Fusion (FUSION)*, pages 651–657, 2016.

43 M. Jiang, W. Yi, and L. Kong. Multi-sensor control for multi-target tracking using Cauchy-Schwarz divergence. In *Proceedings of the 19th International Conference on Information Fusion (FUSION)*, pages 2059–2066, 2016.

44 M. Beard, B. T. Vo, B. N. Vo, and S. Arulampalam. Void probabilities and Cauchy–Schwarz divergence for generalized labeled multi-bernoulli models. *IEEE Transactions on Signal Processing*, 65(19):5047–5061, 2017.

45 A. K. Gostar, R. Hoseinnezhad, T. Rathnayake, X. Wang, and A. Bab-Hadiashar. Constrained sensor control for labeled multi-bernoulli filter using Cauchy-Schwarz divergence. *IEEE Signal Processing Letters*, 24(9):1313–1317, 2017.

46 S. Panicker, A. K. Gostar, A. Bab-Hadiashar, and R. Hoseinnezhad. Sensor control for selective object tracking using labeled multi-bernoulli filter. In *Proceedings of the 21st International Conference on Information Fusion (FUSION)*, pages 2218–2224, 2018.

47 S. Panicker, A.K. Gostar, A. Bab-Hadiashar, and R. Hoseinnezhad. Accelerated multi-sensor control for selective multi-object tracking. In *Proceedings of the International Conference on Control, Automation and Information Sciences (ICCAIS)*, pages 183–188, 2018.

48 F. Lian, L. Hou, B. Wei, and C. Han. Sensor selection for decentralized large-scale multi-target tracking network. *Sensors*, 18(12):4115, 2018. https://doi.org/10.3390/s18124115. ISSN 1424-8220.

49 S. Panicker, A. K. Gostar, A. Bab-Hadiashar, and R. Hoseinnezhad. Tracking of targets of interest using labeled multi-bernoulli filter with multi-sensor control. *Signal Processing*, 171:107451, 2020. https://doi.org/10.1016/j.sigpro.2019.107451. ISSN 0165-1684.

50 L. Ding, C. Shi, W. Qiu, and J. Zhou. Joint dwell time and bandwidth optimization for multi-target tracking in radar network based on low probability of intercept. *Sensors*, 20(5):1269, 2020. https://doi.org/10.3390/s20051269. ISSN 1424-8220.

51 Y. Zhu, L. Zhao, Y. Zhang, and X. Wu. Receiver selection for multi-target tracking in multi-static doppler radar systems. *EURASIP Journal on Advances in Signal Processing*, 2021(1):118, 2021. https://doi.org/10.1186/s13634-021-00826-3.ISSN 1687-6180.

52 S. Liang, Y. Zhu, H. Li, and J. Yan. Evolutionary computational intelligence-based multi-objective sensor management for multi-target tracking. *Remote Sensing*, 14(15):3624, 2022. https://doi.org/10.3390/rs14153624.ISSN 2072-4292.

53 Q. Liu, W. Liu, Z. Zhao, and Z. Peng. Multisensor multitarget tracking and collision avoidance based on the labeled rfs void probability and kl-pac cost function. *IEEE Sensors Journal*, 23(9):9770–9785, 2023. https://doi.org/10.1109/JSEN.2023.3263289. ISSN 2379-9153.

54 K. J. Hintz and E. S. McVey. Multi-process constrained estimation. In *Multitarget-multisensor Tracking: Advanced Applications*, pages 237–244. Artech House, 1991.

55 A. Rényi. On measures of entropy and information. In *Proceedings of the Fourth Berkeley Symposium on Mathematical Statistics and Probability*, volume 1, University of California Press, Berkeley, California, 547–561, 1961.

56 J. Principe and J. Iii. Information-theoretic learning. In *Advances in Unsupervised Adaptive Filtering*, 2000.

57 H. G. Hoang, B. N. Vo, B. T. Vo, and R. Mahler. The Cauchy–Schwarz divergence for poisson point processes. *IEEE Transactions on Information Theory*, 61(8):4475–4485, 2015.

58 A. Blair, A. K. Gostar, R. Tennakoon, A. Bab-Hadiashar, X. Li, J. Palmer, and R. Hoseinnezhad. Distributed multi-sensor control for multi-target tracking. In *International Conference on Control, Automation and Information Sciences (ICCAIS 2022)*, page 72, 2022. https://doi.org/10.1117/12.2637606.

59 A. Blair, A. K. Gostar, X. Li, A. Bab-Hadiashar, and R. Hoseinnezhad. Evaluation of genetic algorithms in multi-vehicle control for multi-target tracking applications. In *Proceedings of the Genetic and Evolutionary Computation Conference Companion, GECCO '24 Companion.* Association for Computing Machinery, New York, NY, USA, pages 631–634, 2024. https://doi.org/10.1145/3638530.3654261. ISBN 9798400704956.

60 A. Blair, A. K. Gostar, A. Bab-Hadiashar, and R. Hoseinnezhad. Distributed multi-sensor control for multi-target tracking using adaptive complementary fusion for LMB densities. Preprint on webpage athttps://arxiv.org/abs/2604.19160.

61 N. Ishtiaq, S. Panicker, A. K. Gostar, A. Bab-Hadiashar, and R. Hoseinnezhad. *Selective Sensor Control via Cauchy Schwarz Divergence*, Springer, pages 113–124. 2021. https://doi.org/10.1007/978-981-15-5224-3-11. ISBN 978-981-15-5223-6.

62 T. Rathnayake, A. K. Gostar, R. Hoseinnezhad, R. Tennakoon, and A. Bab-Hadiashar. On-line visual tracking with occlusion handling. *Sensors*, 20(3), 929, 2020. https://doi.org/10.3390/s20030929. ISSN 1424-8220.

63 D. Akselrod and T. Kirubarajan. Collaborative distributed sensor management and information exchange flow control for multitarget tracking using Markov decision processes. In I. Kadar, editor, *Signal Processing, Sensor Fusion, and Target Recognition XVII*, 6968, 165–175. International Society for Optics and Photonics, SPIE, 2008. https://doi.org/10.1117/12.779348.

64 Y. Fu and L. Yang. Sensor mobility control for multitarget tracking in mobile sensor networks. *International Journal of Distributed Sensor Networks*, 10(3):278179, 2014. https://doi.org/10.1155/2014/278179.

65 Z. Li, Y. Cai, and H. Leung. Distributed multi-sensor control for multi-target tracking with a sparsity-promoting objective function. *IEEE Signal Processing Letters*, 31:621–625, 2024. https://doi.org/10.1109/LSP.2024.3362189.

66 S. J. Wright. Coordinate Descent Algorithms, 2015.https://arxiv.org/abs/1502.04759.

67 D. E. Goldberg. *Genetic Algorithms in Search, Optimization, and Machine Learning.* Addison-Wesley, New York, 1989.

68 G. Chen and L. Hong. A genetic algorithm based multi-dimensional data association algorithm for multi-sensor—multi-target tracking. *Mathematical and Computer Modelling*, 26(4):57–69,1997. https://doi.org/10.1016/S0895-7177(97)00144-1. ISSN 0895-7177.

69 G. Sun, Z. Zhang, B. Zheng, and Y. Li. Multi-sensor data fusion algorithm based on trust degree and improved genetics. *Sensors*, 19(9):2139, 2019. https://doi.org/10.3390/s19092139. ISSN 1424-8220.

70 L. Hua, J. Zhang, D. Li, X. Xi, and M. Shah. Sensor fault diagnosis and fault tolerant control of quadrotor uav based on genetic algorithm. *Journal of Sensors*, 2022:1–8, 2022. https://doi.org/10.1155/2022/8626722.

71 V. D'antuono, G. D. Matteis, D. Trotta, and A. Zavoli. Optimization of uav robust control using genetic algorithm. *IEEE Access*, 11:122252–122272, 2023. https://doi.org/10.1109/ACCESS.2023.3325845.

72 S. Papaioannou, P. Kolios, T. Theocharides, C. G. Panayiotou, and M. M. Polycarpou. Jointly-optimized searching and tracking with random finite sets. *IEEE Transactions on Mobile Computing*, 19(10):2374–2391, 2020. https://doi.org/10.1109/TMC.2019.2922133.

23

Optimal Sensor Placement for AOA Localization and Tracking

Kutluyil Dogancay[1] and Hatem Hmam[2]

[1] *UniSA STEM, University of South Australia, Mawson Lakes, SA, Australia*
[2] *Defence Science & Technology Group, Sensors and Effectors Division, Edinburgh, SA, Australia*

23.1 Introduction

This chapter addresses the problem of optimal sensor placement for angle-of-arrival (AOA) or bearings-only target localization and tracking in the 2D plane. Optimal sensor placement is a fundamental aspect of sensor management, directly influencing the performance of target localization systems. The relationship between relative sensor-target geometry and target localization performance is well established. Several optimal sensor placement strategies have been proposed to exploit this relationship [1–9]. Optimal sensor placement has wide-ranging applications, including surveillance and monitoring of underwater, sea-surface, and aerial objects, search and rescue operations, and mobile user localization in wireless communications.

Broadly speaking, optimal AOA sensor placement algorithms can be classified as either non-Bayesian or Bayesian. The former treats the localization problem as a deterministic parameter estimation problem and the latter as a Bayesian parameter estimation problem with a Gaussian prior representing probabilistic knowledge of the target location [10–12]. Target localization and tracking problems are inherently nonlinear estimation problems whereby the unknown parameters are related to the sensor measurements through a nonlinear function. Working out a closed-form expression for the covariance of a nonlinear estimator is a difficult task. It is therefore often the case that the performance of a target localization/tracking algorithm is approximated by a lower bound on the covariance of estimates. The lower bound of choice is usually the Cramér–Rao lower bound (CRLB) for non-Bayesian (deterministic) estimators and the Bayesian CRLB (BCRLB) for Bayesian estimators (BCRLB also known as the posterior CRLB) [13, 14]. If the AOA measurement noise is Gaussian as commonly assumed in practice, the CRLB has a simple form that is easy to compute. Since the CRLB is a matrix, various optimality criteria have been proposed with scalar performance measures that can be adopted for optimal sensor placement [15–17]. The CRLB requires knowledge of the true target location, which is not available. This limitation is addressed by substituting the unknown parameters with their estimates.

The chapter begins with a concise overview of linear parameter estimation for deterministic, random and time-varying random parameters, establishing a foundation for the subsequent sections. It then introduces the Bayesian AOA target localization problem and presents the Bayesian Fisher information matrix (BFIM), which is the inverse of the BCRLB. Since the BFIM for target location estimates is difficult to compute directly, an approximate form is proposed by substituting the unknown random target location vector with the mean of the Gaussian prior. This approximation is shown to be valid under mild conditions. Building on the approximate BFIM, the chapter discusses the A-optimality and D-optimality criteria for sensor placement. The optimal sensor placement problem is formulated as maximizing the determinant of the approximate BFIM (D-optimality), which is demonstrated to be equivalent to the A-optimality criterion. Target localization geometries are categorized into far-field and near-field scenarios, and their implications for sensor placement are examined, highlighting the relative simplicity of optimal placement in the far-field case. Results for varying sensor configurations are analyzed, followed by an extension of the framework to optimal sensor trajectories for maneuvering target tracking. An efficient approximation of the recursive BFIM is derived for extended Kalman filter (EKF) state estimates, leveraging readily available EKF outputs. Finally, A-optimal and D-optimal waypoint optimization algorithms are developed using the recursive BFIM in its approximate form, enabling practical solutions for dynamic sensor deployment in target tracking applications.

Section 23.2 presents an overview of Bayesian estimation fundamentals. Section 23.3 discusses the BCRLB and BFIM for AOA target localization and the optimal sensor placement criteria. Section 23.4 defines the optimization problem for sensor placement. Optimal sensor placement for the special cases of far-field and near-field targets is addressed in Section 23.5. Section 23.6 presents optimal sensor placement results for sensor-target geometries with varying numbers of sensors. Section 23.7 discusses optimal sensor trajectories for target tracking. Concluding remarks are made in Section 23.8.

23.2 Bayesian Estimation Fundamentals

23.2.1 Deterministic Parameter Estimation

Consider the linear matrix equation

$$\boldsymbol{y} = \boldsymbol{A}\boldsymbol{x} + \boldsymbol{w}, \tag{23.1}$$

where $\boldsymbol{y}$ is an $N \times 1$ measurement vector, $\boldsymbol{A}$ is a full-rank $N \times M$ deterministic matrix with $N \geq M$, $\boldsymbol{x}$ is an $M \times 1$ unknown *deterministic (nonrandom)* parameter vector and $\boldsymbol{w}$ is an $N \times 1$ noise vector, which is Gaussian distributed with zero mean and covariance $\boldsymbol{R}$ (i.e., $\boldsymbol{w} \sim \mathcal{N}(\boldsymbol{0}, \boldsymbol{R})$).

The *weighted least squares* (WLS) estimator of the unknown vector $\boldsymbol{x}$ is

$$\hat{\boldsymbol{x}} = (\boldsymbol{A}^T\boldsymbol{R}^{-1}\boldsymbol{A})^{-1}\boldsymbol{A}^T\boldsymbol{R}^{-1}\boldsymbol{y}, \tag{23.2}$$

which is the *best linear unbiased estimator* (BLUE) of $\boldsymbol{x}$ [18]. The WLS solves the optimization problem:

$$\hat{\boldsymbol{x}} = \arg\min_{\boldsymbol{x}\in\mathbb{R}^M} (\boldsymbol{y} - \boldsymbol{A}\boldsymbol{x})^T\boldsymbol{R}^{-1}(\boldsymbol{y} - \boldsymbol{A}\boldsymbol{x}) \tag{23.3}$$

where $\mathbb{R}^M$ denotes the set of $M \times 1$ real-valued vectors. The WLS is not only unbiased ($E\{\hat{\boldsymbol{x}}\} = \boldsymbol{x}$), but also efficient; i.e., its covariance is equal to the CRLB:

$$E\{(\hat{\boldsymbol{x}} - \boldsymbol{x})(\hat{\boldsymbol{x}} - \boldsymbol{x})^T\} = \boldsymbol{\Sigma}_0, \tag{23.4}$$

where

$$\Sigma_0 = (A^T R^{-1} A)^{-1} \tag{23.5}$$

is the CRLB, which is the lower bound on the covariance of any unbiased estimator of x [14].

23.2.2 Random Parameter Estimation

Suppose that in (23.1) the unknown vector to be estimated, x, is random and Gaussian distributed with mean x_0 and covariance P_0. It is assumed that both x_0 and P_0 are known, forming the *Gaussian prior*, and x and w are statistically independent.

The *maximum a posteriori* (MAP) estimator of x is given by

$$\hat{x}_{\text{MAP}} = \arg \max_{x \in \mathbb{R}^M} \mathcal{L}(x|y), \tag{23.6}$$

where $\mathcal{L}(x|y)$ is the Bayesian log-likelihood function. Applying the Bayesian theorem, the conditional joint probability density function (pdf) of x (posterior pdf) can be written as

$$f(x|y) = \frac{f(y|x)f(x)}{f(y)}. \tag{23.7}$$

Taking the logarithm of both sides of (23.7) gives the Bayesian log-likelihood function:

$$\mathcal{L}(x|y) = \ln f(x|y) \tag{23.8}$$

$$= \ln f(y|x) + \ln f(x) - \ln f(y) \tag{23.9}$$

$$= -\frac{1}{2}(y - Ax)^T R^{-1}(y - Ax) - \frac{1}{2}(x - x_0)^T P_0^{-1}(x - x_0) + \kappa \tag{23.10}$$

where κ denotes the terms that are independent of x. A comparison with (23.3) reveals that the Bayesian log-likelihood function is made up of weighted contributions from the measurements and Gaussian prior.

To solve (23.6) for $\hat{x}_{\text{MAP}}$, take the gradient of $\mathcal{L}(x|y)$ with respect x and set it equal to zero:

$$P_0^{-1}x_0 + A^T R^{-1} y - (P_0^{-1} + A^T R^{-1} A)x = 0. \tag{23.11}$$

Solving the above equation for $\hat{x}_{\text{MAP}}$ yields

$$\hat{x}_{\text{MAP}} = (P_0^{-1} + A^T R^{-1} A)^{-1}(P_0^{-1}x_0 + A^T R^{-1} y) \tag{23.12a}$$

$$= x_0 + (P_0^{-1} + A^T R^{-1} A)^{-1} A^T R^{-1}(y - Ax_0) \tag{23.12b}$$

$$= x_0 + P_0 A^T (A P_0 A^T + R)^{-1}(y - Ax_0), \tag{23.12c}$$

where the matrix inversion lemma

$$(P_0^{-1} + A^T R^{-1} A)^{-1} A^T R^{-1} = P_0 A^T (A P_0 A^T + R) \tag{23.13}$$

was used to rewrite (23.12b) as (23.12c).

The MAP estimator is Bayesian efficient, which means that its covariance is identical to the BCRLB; that is,

$$E\{(\hat{x}_{\text{MAP}} - x)(\hat{x}_{\text{MAP}} - x)^T\} = \Sigma_0, \tag{23.14}$$

where $\boldsymbol{\Sigma}_0$ is the BCRLB [14]:

$$\boldsymbol{\Sigma}_0 = (\boldsymbol{P}_0^{-1} + \boldsymbol{A}^T\boldsymbol{R}^{-1}\boldsymbol{A})^{-1} \tag{23.15a}$$

$$= \left(\boldsymbol{I} - \boldsymbol{P}_0\boldsymbol{A}^T(\boldsymbol{A}\boldsymbol{P}_0\boldsymbol{A}^T + \boldsymbol{R})^{-1}\boldsymbol{A}\right)\boldsymbol{P}_0. \tag{23.15b}$$

The matrix inversion lemma was employed to arrive at (23.15b). The MAP estimator is also a minimum mean-square error (MMSE) estimator given by the conditional mean of $\boldsymbol{x}$:

$$\hat{\boldsymbol{x}}_{\text{MAP}} = E\{\boldsymbol{x}\,|\,\boldsymbol{y}\}. \tag{23.16}$$

23.2.3 Time-varying Random Parameter Estimation

Suppose that the unknown random vector $\boldsymbol{x}$ evolves over time according to the Gauss-Markov model

$$\boldsymbol{x}_{k+1} = \boldsymbol{F}\boldsymbol{x}_k + \boldsymbol{n}_k, \quad k = 0, 1, 2, \ldots \tag{23.17a}$$

with the measurements given by

$$\boldsymbol{z}_k = \boldsymbol{H}_k\boldsymbol{x}_k + \boldsymbol{w}_k \tag{23.17b}$$

where $\boldsymbol{F}$ and $\boldsymbol{H}_k$ are deterministic, $\boldsymbol{n}_k \sim \mathcal{N}(\boldsymbol{0}, \boldsymbol{Q}_k)$, $\boldsymbol{w}_k \sim \mathcal{N}(\boldsymbol{0}, \boldsymbol{R}_k)$, and $\boldsymbol{n}_k$ and $\boldsymbol{w}_k$ are statistically independent. Eq. (23.17) defines a linear *state-space* model with $\boldsymbol{x}_k$ being the state vector and $\boldsymbol{z}_k$ the measurement vector. The state vector $\boldsymbol{x}_k$ is a Gauss-Markov process because $f(\boldsymbol{x}_k|\boldsymbol{x}_{k-1}, \boldsymbol{x}_{k-2}, \ldots, \boldsymbol{x}_0) = f(\boldsymbol{x}_k|\boldsymbol{x}_{k-1})$ [18]. The objective is to estimate $\boldsymbol{x}_k$ from the measurements $\boldsymbol{z}_k, \boldsymbol{z}_{k-1}, \ldots, \boldsymbol{z}_0$. The Kalman filter [14, 19] recursively computes the conditional mean $E\{\boldsymbol{x}_k|\boldsymbol{z}_k, \boldsymbol{z}_{k-1}, \boldsymbol{z}_0\}$, which is the MMSE estimator of $\boldsymbol{x}_k$. For nonlinear and non-Gaussian state-space models, various modifications of the Kalman filter have been proposed, such as the extended Kalman filter (EKF) [20], the unscented Kalman filter [21], the cubature Kalman filter [22] and the particle filter [23].

23.3 Optimality Criteria for Sensor Placement

23.3.1 Bayesian AOA Target Localization

In stationary target localization, the objective is to geolocate an emitter from signals intercepted at multiple sensors. This has applications, for example, in emergency beacon localization in search and rescue missions and user localization in wireless communication systems.

Consider a stationary target in the 2D plane whose location is a random vector with a known Gaussian prior:

$$\boldsymbol{x} = [x, y]^T \sim \mathcal{N}(\boldsymbol{x}_0, \boldsymbol{P}_0). \tag{23.18}$$

A set of N stationary sensors are located at $\boldsymbol{s}_i = [s_{x,i}, s_{y,i}]^T$, $i = 1, 2, \ldots, N$, and collect AOA measurements from the target:

$$\hat{\theta}_i = \theta_i(\boldsymbol{x}) + w_i, \quad -\pi < \hat{\theta}_i(\boldsymbol{x}) \leq \pi, \tag{23.19}$$

where w_i is the AOA measurement noise, which is Gaussian distributed with zero mean and variance σ_i^2 ($w_i \sim \mathcal{N}(0, \sigma_i^2)$). The w_i is assumed to be independent. The true AOA is

$$\theta_i(\boldsymbol{x}) = \tan^{-1}\frac{y - s_{y,i}}{x - s_{x,i}}, \tag{23.20}$$

where $\tan^{-1}(\cdot)$ is the four-quadrant arctangent function.

In Bayesian AOA target localization, the task is to estimate the unknown target location $\boldsymbol{x}$ from noisy AOA measurements $\hat{\theta}_i$ taken at multiple sensors, combined with the *a priori* knowledge of the target location expressed as a Gaussian prior with mean $\boldsymbol{x}_0$ and covariance $\boldsymbol{P}_0$ (see Figure 23.1). As the AOA measurements are related to the unknown target location $\boldsymbol{x}$ in a nonlinear way (refer to (23.19) and (23.20)), this is a *nonlinear* Bayesian estimation problem with the Bayesian log-likelihood function

$$\mathcal{L}(\boldsymbol{x}|\hat{\boldsymbol{\theta}}) = -\frac{1}{2}(\hat{\boldsymbol{\theta}} - \boldsymbol{\theta}(\boldsymbol{x}))^T \boldsymbol{R}^{-1}(\hat{\boldsymbol{\theta}} - \boldsymbol{\theta}(\boldsymbol{x})) - \frac{1}{2}(\boldsymbol{x} - \boldsymbol{x}_0)^T \boldsymbol{P}_0^{-1}(\boldsymbol{x} - \boldsymbol{x}_0) + \kappa, \tag{23.21}$$

where

$$\hat{\boldsymbol{\theta}} = [\hat{\theta}_1, \hat{\theta}_2, \ldots, \hat{\theta}_N]^T$$

$$\boldsymbol{\theta}(\boldsymbol{x}) = [\theta_1(\boldsymbol{x}), \theta_2(\boldsymbol{x}), \ldots, \theta_N(\boldsymbol{x})]^T$$

$$\boldsymbol{R} = \begin{bmatrix} \sigma_1^2 & & & \boldsymbol{0} \\ & \sigma_2^2 & & \\ & & \ddots & \\ \boldsymbol{0} & & & \sigma_N^2 \end{bmatrix}.$$

The MAP estimator of the target location $\boldsymbol{x}$ is given by

$$\hat{\boldsymbol{x}}_{\text{MAP}} = \arg\min_{\boldsymbol{x}\in\mathbb{R}^2}(\hat{\boldsymbol{\theta}} - \boldsymbol{\theta}(\boldsymbol{x}))^T \boldsymbol{R}^{-1}(\hat{\boldsymbol{\theta}} - \boldsymbol{\theta}(\boldsymbol{x})) + (\boldsymbol{x} - \boldsymbol{x}_0)^T \boldsymbol{P}_0^{-1}(\boldsymbol{x} - \boldsymbol{x}_0), \tag{23.22}$$

which does not have a closed-form solution, but can be numerically solved using an iterative search algorithm such as the Gauss-Newton method [24].

If no Gaussian prior is available, the unknown vector $\boldsymbol{x}$ is assumed to be deterministic, reducing the MAP estimator to the maximum likelihood estimator (MLE) with the log-likelihood function

$$\mathcal{L}(\boldsymbol{x}|\hat{\boldsymbol{\theta}}) = -\frac{1}{2}(\hat{\boldsymbol{\theta}} - \boldsymbol{\theta}(\boldsymbol{x}))^T \boldsymbol{R}^{-1}(\hat{\boldsymbol{\theta}} - \boldsymbol{\theta}(\boldsymbol{x})) + \kappa. \tag{23.23}$$

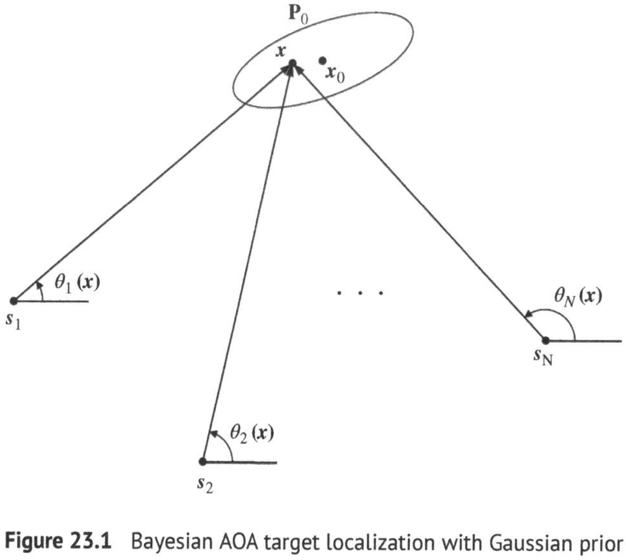

Figure 23.1 Bayesian AOA target localization with Gaussian prior $\boldsymbol{x} \sim \mathcal{N}(\boldsymbol{x}_0, \boldsymbol{P}_0)$.

The MLE of $\boldsymbol{x}$ is a nonlinear weighted least squares estimator given by

$$\hat{\boldsymbol{x}}_{\text{MLE}} = \arg\min_{\boldsymbol{x}\in\mathbb{R}^2} (\hat{\theta} - \theta(\boldsymbol{x}))^T \boldsymbol{R}^{-1} (\hat{\theta} - \theta(\boldsymbol{x})). \tag{23.24}$$

In general, the MLE does not have a closed-form solution, therefore necessitating the use of a numerical search algorithm.

Recognizing that the MAP estimator is a generalization of the MLE, this chapter examines Bayesian estimation problems and the formulation of optimal Bayesian sensor–target geometries with Gaussian priors.

23.3.2 Performance Bounds

In general, finding an optimal sensor placement is an intractable problem. This is mainly because of the lack of exact analytical expressions for the covariance of nonlinear estimators as a function of the sensor-target geometry and sensor measurement noise variances. A common way to address this issue is to use a lower bound on the covariance of unbiased estimators and assume that the estimator approximately achieves it for sufficiently small measurement noise. The CRLB is often the bound of choice in deterministic parameter estimation problems as it easily lends itself to analysis for Gaussian noise. The CRLB tends to predict the estimation covariance well if the estimator is operating away from the noise threshold region where the estimation performance deteriorates sharply. By definition, the CRLB is given by the inverse of the Fisher information matrix (FIM), which measures the information contained in an estimate. In good target localization geometries, the FIM is relatively "large," making the CRLB relatively "small." Several optimization criteria have been developed to derive scalar measures from the FIM/CRLB for assessing sensor placement quality [17].

In Bayesian estimation problems, the CRLB is replaced with the BCRLB and the FIM with the BFIM. For the Bayesian AOA target localization problem, the BFIM is

$$\boldsymbol{\Phi} = \boldsymbol{P}_0^{-1} + E_{\boldsymbol{x}}\left\{\sum_{i=1}^{N} \frac{1}{\sigma_i^2 d_i^2(\boldsymbol{x})} \boldsymbol{u}_i(\boldsymbol{x}) \boldsymbol{u}_i^T(\boldsymbol{x})\right\}, \tag{23.25}$$

which is comprised of contributions from the Gaussian prior $\boldsymbol{P}_0^{-1}$ and the AOA measurements (the second term with expectation over $\boldsymbol{x} \sim \mathcal{N}(\boldsymbol{x}_0, \boldsymbol{P}_0)$). In (23.25), $d_i(\boldsymbol{x})$ is the distance between the target and the ith sensor and $\boldsymbol{u}_i(\boldsymbol{x})$ is the unit vector orthogonal to the range vector $\boldsymbol{x} - \boldsymbol{s}_i$:

$$d_i(\boldsymbol{x}) = \| \boldsymbol{x} - \boldsymbol{s}_i \| \tag{23.26a}$$

$$\boldsymbol{u}_i(\boldsymbol{x}) = \begin{bmatrix} -\sin\theta_i(\boldsymbol{x}) \\ \cos\theta_i(\boldsymbol{x}) \end{bmatrix}. \tag{23.26b}$$

Taking the expectation of the second term on the right-hand side of (23.25) is challenging as it requires computationally expensive Monte Carlo simulations. However, it can be approximated by

$$E_{\boldsymbol{x}}\left\{\sum_{i=1}^{N} \frac{1}{\sigma_i^2 d_i^2(\boldsymbol{x})} \boldsymbol{u}_i(\boldsymbol{x}) \boldsymbol{u}_i^T(\boldsymbol{x})\right\} \approx \sum_{i=1}^{N} \frac{1}{\sigma_i^2 d_i^2(\boldsymbol{x}_0)} \boldsymbol{u}_i(\boldsymbol{x}_0) \boldsymbol{u}_i^T(\boldsymbol{x}_0), \tag{23.27}$$

where the random vector $\boldsymbol{x}$ is replaced by its mean $\boldsymbol{x}_0$, making the summation terms deterministic and allowing the expectation to be stripped off. Thus, using (23.27), the BFIM in (23.25) is approximately given by

$$\tilde{\boldsymbol{\Phi}} = \boldsymbol{K}_0 + \sum_{i=1}^{N} \frac{1}{\sigma_i^2 d_i^2(\boldsymbol{x}_0)} \boldsymbol{u}_i(\boldsymbol{x}_0)\boldsymbol{u}_i^T(\boldsymbol{x}_0) \tag{23.28}$$

where $\boldsymbol{K}_0 = \boldsymbol{P}_0^{-1}$. Unlike the exact BFIM expression, the approximate BFIM is easy to compute, thereby simplifying the optimal sensor placement analysis. Therefore, (23.28) will be adopted as the BFIM expression for the Bayesian AOA target localization problem. Appendix 23.A provides a justification for why (23.28) is a good approximation.

If $\boldsymbol{x}$ is deterministic with no Gaussian prior, the BFIM in (23.25) is reduced to

$$\boldsymbol{\Phi} = \sum_{i=1}^{N} \frac{1}{\sigma_i^2 d_i^2(\boldsymbol{x})} \boldsymbol{u}_i(\boldsymbol{x})\boldsymbol{u}_i^T(\boldsymbol{x}), \tag{23.29}$$

which is the FIM and can be computed directly unlike the BFIM.

23.3.3 Optimality Criteria

The optimal sensor placement problem is solved using a scalar performance measure based on the BFIM, which is referred to as the optimality criterion. Several scalar functions of BFIM have been proposed as optimality criteria [15–17]. Among them, the A-optimality and D-optimality criteria have been particularly popular.

The objective function of the A-optimality criterion is the trace of the BCRLB:

$$J_A = \operatorname{tr} \boldsymbol{\Phi}^{-1}. \tag{23.30}$$

As $\boldsymbol{\Phi}^{-1}$ is the BCRLB, J_A gives the optimal mean-square error (MSE) of the parameter estimator. The objective function of the D-optimality criterion is, on the other hand, defined as the determinant of the BFIM:

$$J_D = |\boldsymbol{\Phi}|, \tag{23.31}$$

which is inversely proportional to the area of the error ellipse in 2D localization problems (or the volume of the error ellipsoid in 3D localization).

Both J_A and J_D are scalar functions of the BFIM, which makes it easy to formulate the optimal sensor placement problem as an optimization problem over sensor locations:

$$\min_{s_1, s_2, \ldots, s_N} J_A \tag{23.32a}$$

$$\max_{s_1, s_2, \ldots, s_N} J_D. \tag{23.32b}$$

As illustrated in Figure 23.2, the A-optimality and D-optimality criteria seek to minimize the size of the inverse BFIM error ellipse by minimizing the sum of its squared semiaxis lengths and the product of its squared semiaxis lengths (i.e., the area of the error ellipse), respectively.

For a 2×1 Gaussian random vector, the error ellipse defines a region centered at its mean capturing a certain percentage of the random vectors (see Figure 23.3). An error ellipse at m-σ confidence

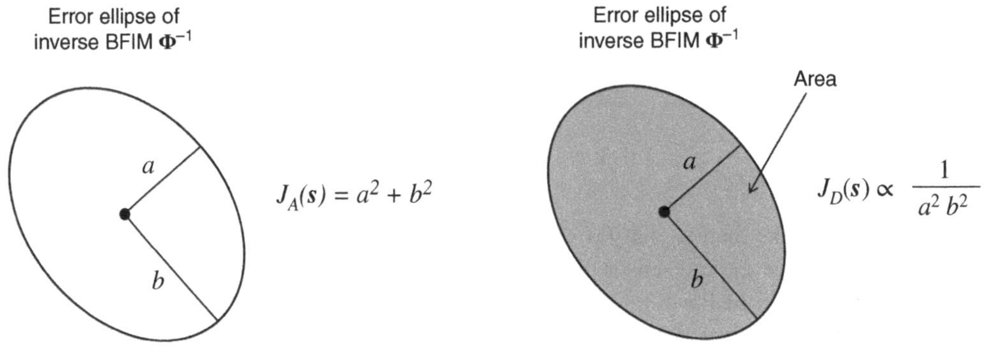

Figure 23.2 Comparison of A-optimality and D-optimality criteria in terms of inverse BFIM error ellipse.

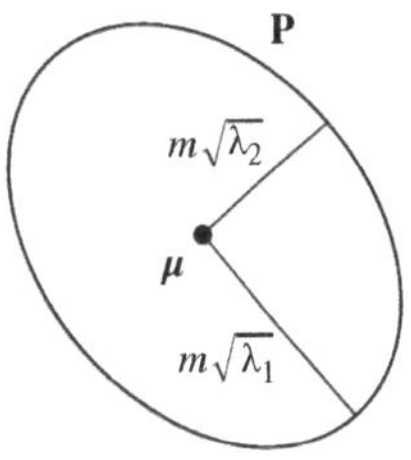

Figure 23.3 Illustration of m-σ error ellipse for a Gaussian random vector $[x, y]^T \sim \mathcal{N}(\boldsymbol{\mu}, \boldsymbol{P})$. $\boldsymbol{P}$ has eigenvalues $\lambda_1 \geq \lambda_2$ with major and minor axis lengths $m\sqrt{\lambda_1}$ and $m\sqrt{\lambda_2}$, respectively.

level (e.g., 1-σ, 2-σ, 3-σ, etc.) is defined by a collection of points $[x, y]^T$ with mean $\boldsymbol{\mu}$ and covariance $\boldsymbol{P}$, satisfying

$$([x, y]^T - \boldsymbol{\mu})^T \boldsymbol{P}^{-1}([x, y]^T - \boldsymbol{\mu}) < m^2. \tag{23.33}$$

The probability of a point being within the m-σ error ellipse is given by the chi-square cumulative distribution function for two degrees of freedom:

$$\Pr\{([x, y]^T - \boldsymbol{\mu})^T \boldsymbol{P}^{-1}([x, y]^T - \boldsymbol{\mu}) < m^2\} = \Pr\{\chi_2^2 < m^2\} \tag{23.34a}$$

$$= 1 - \exp\left(-\frac{m^2}{2}\right). \tag{23.34b}$$

For example, for $m = 2$, $\Pr(\chi_2^2 < m^2) = 0.8647$; hence, the 2-$\sigma$ error ellipse defines a 86.5% confidence region.

23.4 Optimization Problem for Sensor Placement

In keeping with Bayesian estimation, a Gaussian prior for the target location is assumed to be available with a known mean $\boldsymbol{x}_0$ and covariance $\boldsymbol{P}_0$. In optimal sensor placement, the sensors are assumed to be at fixed distances from $\boldsymbol{x}_0$. This means that they can be anywhere on circles of radii $d_i(\boldsymbol{x}_0)$, $i = 1, \dots, N$, centered at $\boldsymbol{x}_0$ with all the $d_i(\boldsymbol{x}_0)$ fixed and given (see Figure 23.4). The AOA noise covariance $\boldsymbol{R}$ is also assumed to be known. The objective of optimal sensor placement is to find specific sensor locations on these circles or, equivalently, the sensor angles from the mean of the Gaussian prior, $\theta_i(\boldsymbol{x}_0)$, that solve (23.32) to achieve the best localization performance.

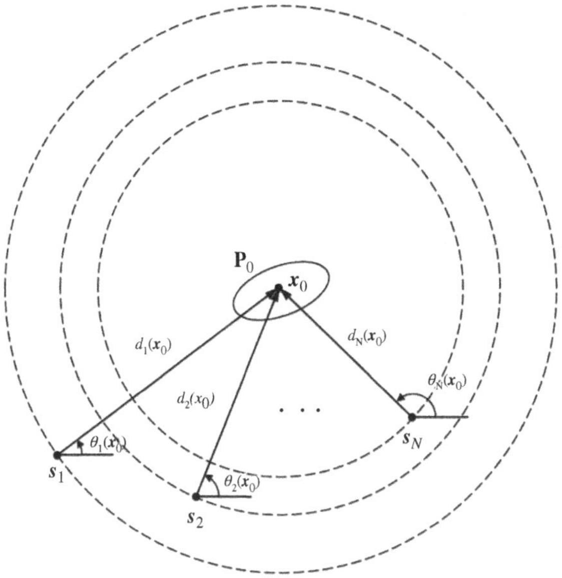

Figure 23.4 Optimal sensor placement problem. Given the Gaussian prior for target location, the objective is to find optimal sensor angles $\theta_i(\boldsymbol{x}_0)$ at fixed distances from $\boldsymbol{x}_0$, $d_i(\boldsymbol{x}_0)$, to achieve best localization performance.

Using the simplified approximation of BFIM in (23.28) and letting

$$\tilde{\boldsymbol{\Phi}} = \begin{bmatrix} \tilde{\phi}_{11} & \tilde{\phi}_{12} \\ \tilde{\phi}_{21} & \tilde{\phi}_{22} \end{bmatrix}, \tag{23.35}$$

where $\tilde{\phi}_{12} = \tilde{\phi}_{21}$, the objective function of the A-optimality criterion J_A can be written as

$$\operatorname{tr}\tilde{\boldsymbol{\Phi}}^{-1} = \frac{1}{|\tilde{\boldsymbol{\Phi}}|}\operatorname{tr}\begin{bmatrix} \tilde{\phi}_{22} & -\tilde{\phi}_{12} \\ -\tilde{\phi}_{21} & \tilde{\phi}_{11} \end{bmatrix} \tag{23.36a}$$

$$= \frac{1}{|\tilde{\boldsymbol{\Phi}}|}\operatorname{tr}\tilde{\boldsymbol{\Phi}} \tag{23.36b}$$

$$= \frac{1}{|\tilde{\boldsymbol{\Phi}}|}\left(\operatorname{tr}\boldsymbol{K}_0 + \sum_{i=1}^{N}\frac{1}{\sigma_i^2}\left(\frac{\sin^2\theta_i(\boldsymbol{x}_0)}{d_i^2(\boldsymbol{x}_0)} + \frac{\cos^2\theta_i(\boldsymbol{x}_0)}{d_i^2(\boldsymbol{x}_0)}\right)\right) \tag{23.36c}$$

$$= \frac{1}{|\tilde{\boldsymbol{\Phi}}|}\left(\operatorname{tr}\boldsymbol{K}_0 + \sum_{i=1}^{N}\frac{1}{\sigma_i^2 d_i^2(\boldsymbol{x}_0)}\right). \tag{23.36d}$$

As $\boldsymbol{K}_0$, σ_i^2 and $d_i(\boldsymbol{x}_0)$ are fixed and given, minimizing the trace of $\tilde{\boldsymbol{\Phi}}^{-1}$ is equivalent to maximizing the determinant of $\tilde{\boldsymbol{\Phi}}$. In other words, the A-optimality and D-optimality criteria yield identical optimal sensor placement results for the Bayesian AOA target localization problem described in Section 23.3.1. The equivalence of the A-optimality and D-optimality criteria for non-Bayesian AOA target localization was also proved in [7]. As a consequence, we opt to use the D-optimality criterion to solve the optimal sensor placement problem described in Figure 23.4.

For fixed σ_i^2 and $d_i(\boldsymbol{x}_0)$, the optimization problem in (23.32b) takes the form

$$\underset{\theta_1(\boldsymbol{x}_0),\ldots,\theta_N(\boldsymbol{x}_0)}{\arg\max} \ |\tilde{\boldsymbol{\Phi}}|. \tag{23.37}$$

Using the following result on the determinant of matrix sums [25]

$$|\boldsymbol{A} + \boldsymbol{U}\boldsymbol{V}^T| = |\boldsymbol{A}||\boldsymbol{I} + \boldsymbol{V}^T\boldsymbol{A}^{-1}\boldsymbol{U}|, \quad \boldsymbol{A} \text{ is non-singular} \tag{23.38}$$

the determinant of $\tilde{\boldsymbol{\Phi}}$ can be expressed more conveniently as

$$|\tilde{\boldsymbol{\Phi}}| = \left|\boldsymbol{K}_0 + \sum_{i=1}^{N} \alpha_i(\boldsymbol{x}_0)\boldsymbol{u}_i(\boldsymbol{x}_0)\boldsymbol{u}_i^T(\boldsymbol{x}_0)\right| \tag{23.39a}$$

$$= \left|\boldsymbol{K}_0 + \boldsymbol{U}\boldsymbol{V}^T\right| \tag{23.39b}$$

$$= |\boldsymbol{K}_0|\,|\boldsymbol{I} + \boldsymbol{V}^T\boldsymbol{P}_0\boldsymbol{U}|, \tag{23.39c}$$

where

$$\alpha_i(\boldsymbol{x}_0) = \frac{1}{\sigma_i^2 d_i^2(\boldsymbol{x}_0)} \tag{23.40a}$$

$$\boldsymbol{U} = \left[\boldsymbol{u}_1(\boldsymbol{x}_0),\ \boldsymbol{u}_2(\boldsymbol{x}_0),\ \cdots,\ \boldsymbol{u}_N(\boldsymbol{x}_0)\right] \tag{23.40b}$$

$$\boldsymbol{V} = \boldsymbol{U}\begin{bmatrix} \alpha_1(\boldsymbol{x}_0) & & & \boldsymbol{0} \\ & \alpha_2(\boldsymbol{x}_0) & & \\ & & \ddots & \\ \boldsymbol{0} & & & \alpha_N(\boldsymbol{x}_0) \end{bmatrix}. \tag{23.40c}$$

As $|\boldsymbol{K}_0|$ does not change with sensor placement, using (23.39c), the optimization problem in (23.37) can be written as

$$\max_{\theta_1(\boldsymbol{x}_0),\ldots,\theta_N(\boldsymbol{x}_0)} |\boldsymbol{I} + \boldsymbol{V}^T\boldsymbol{P}_0\boldsymbol{U}|, \tag{23.41}$$

where

$$\boldsymbol{I} + \boldsymbol{V}^T\boldsymbol{P}_0\boldsymbol{U} = \boldsymbol{I} + \begin{bmatrix} \alpha_1(\boldsymbol{x}_0) & & & \boldsymbol{0} \\ & \alpha_2(\boldsymbol{x}_0) & & \\ & & \ddots & \\ \boldsymbol{0} & & & \alpha_N(\boldsymbol{x}_0) \end{bmatrix}\boldsymbol{U}^T\boldsymbol{P}_0\boldsymbol{U} \tag{23.42a}$$

$$= \begin{bmatrix} 1 + \alpha_1\boldsymbol{u}_1^T\boldsymbol{P}_0\boldsymbol{u}_1 & \alpha_1\boldsymbol{u}_1^T\boldsymbol{P}_0\boldsymbol{u}_2 & \cdots & \alpha_1\boldsymbol{u}_1^T\boldsymbol{P}_0\boldsymbol{u}_N \\ \alpha_2\boldsymbol{u}_2^T\boldsymbol{P}_0\boldsymbol{u}_1 & 1 + \alpha_2\boldsymbol{u}_2^T\boldsymbol{P}_0\boldsymbol{u}_2 & \cdots & \alpha_2\boldsymbol{u}_2^T\boldsymbol{P}_0\boldsymbol{u}_N \\ \vdots & & \ddots & \vdots \\ \alpha_N\boldsymbol{u}_N^T\boldsymbol{P}_0\boldsymbol{u}_1 & & \cdots & 1 + \alpha_N\boldsymbol{u}_N^T\boldsymbol{P}_0\boldsymbol{u}_N \end{bmatrix}_{N\times N}. \tag{23.42b}$$

Here the notation for $\alpha_i(\boldsymbol{x}_0)$ and $\boldsymbol{u}_i(\boldsymbol{x}_0)$ has been simplified by suppressing their dependence on $\boldsymbol{x}_0$.

23.5 Far-field vs Near-field Target

Simple optimal sensor placement results become available if the sensor-target geometry obeys certain requirements in concert with the assumption in (23.102). This section discusses these requirements in relation to the target range from the sensors.

To begin with, consider the entries of the $N \times N$ matrix $\boldsymbol{I} + \boldsymbol{V}^T \boldsymbol{P}_0 \boldsymbol{U}$ in (23.42b). The quadratic forms, $\boldsymbol{u}_i^T \boldsymbol{P}_0 \boldsymbol{u}_i$, $i = 1, 2, \dots, N$, are maximized if the sensor angles $\theta_i(\boldsymbol{x}_0)$ are selected to align the unit vectors $\boldsymbol{u}_i$ with the major axis of the error ellipse of $\boldsymbol{P}_0$. This is easy to show by noting that

$$\max_{\theta_i(\boldsymbol{x}_0)} \boldsymbol{u}_i^T \boldsymbol{P}_0 \boldsymbol{u}_i \quad \text{or} \quad \max_{\boldsymbol{u}_i} \boldsymbol{u}_i^T \boldsymbol{P}_0 \boldsymbol{u}_i \tag{23.43}$$

is a problem of quadratic form maximization over the unit circle with solution given by [12]

$$\lambda_1 = \max_{\boldsymbol{u}_i} \boldsymbol{u}_i^T \boldsymbol{P}_0 \boldsymbol{u}_i \quad \text{and} \quad \pm \boldsymbol{q}_1 = \arg \max_{\boldsymbol{u}_i} \boldsymbol{u}_i^T \boldsymbol{P}_0 \boldsymbol{u}_i \tag{23.44}$$

where $\boldsymbol{q}_1$ is the orthonormal eigenvector of $\boldsymbol{P}_0$ associated with λ_1 (the maximum eigenvalue of $\boldsymbol{P}_0$):

$$\boldsymbol{P}_0 = \begin{bmatrix} \boldsymbol{q}_1 & \boldsymbol{q}_2 \end{bmatrix} \begin{bmatrix} \lambda_1 & 0 \\ 0 & \lambda_2 \end{bmatrix} \begin{bmatrix} \boldsymbol{q}_1^T \\ \boldsymbol{q}_2^T \end{bmatrix}, \quad \lambda_1 \geq \lambda_2, \quad \boldsymbol{q}_i^T \boldsymbol{q}_j = \begin{cases} 1 & \text{if } i = j \\ 0 & \text{otherwise.} \end{cases} \tag{23.45}$$

Suppose that the target is far from the sensors so that

$$\max_{\boldsymbol{u}_i} \alpha_i \boldsymbol{u}_i^T \boldsymbol{P}_0 \boldsymbol{u}_i = \alpha_i \lambda_1 = \frac{\lambda_1}{\sigma_i^2 d_i^2(\boldsymbol{x}_0)} \tag{23.46}$$

is very small for all $i = 1, 2 \dots, N$. Let us refer to this situation as a "far-field" target. Asymptotically as $\alpha_i \to 0$, $i = 1, 2, \dots, N$, (23.42b) can be seen to become a diagonally dominant matrix:

$$\boldsymbol{I} + \boldsymbol{V}^T \boldsymbol{P}_0 \boldsymbol{U} \approx \begin{bmatrix} 1 + \alpha_1 \boldsymbol{u}_1^T \boldsymbol{P}_0 \boldsymbol{u}_1 & & & \boldsymbol{0} \\ & 1 + \alpha_2 \boldsymbol{u}_2^T \boldsymbol{P}_0 \boldsymbol{u}_2 & & \\ & & \ddots & \\ \boldsymbol{0} & & & 1 + \alpha_N \boldsymbol{u}_N^T \boldsymbol{P}_0 \boldsymbol{u}_N \end{bmatrix}. \tag{23.47}$$

This means that in a far-field target situation, the determinant of $\boldsymbol{I} + \boldsymbol{V}^T \boldsymbol{P}_0 \boldsymbol{U}$ is approximately given by the product of its diagonal entries:

$$|\boldsymbol{I} + \boldsymbol{V}^T \boldsymbol{P}_0 \boldsymbol{U}| \approx \prod_{i=1}^{N} \left(1 + \alpha_i \boldsymbol{u}_i^T \boldsymbol{P}_0 \boldsymbol{u}_i\right), \tag{23.48}$$

which simplifies the optimization problem in (23.41) to

$$\max_{\theta_1(\boldsymbol{x}_0), \dots, \theta_N(\boldsymbol{x}_0)} \prod_{i=1}^{N} \left(1 + \alpha_i \boldsymbol{u}_i^T \boldsymbol{P}_0 \boldsymbol{u}_i\right). \tag{23.49}$$

Using (23.44), it is easy to see that each product term, $1 + \alpha_i \boldsymbol{u}_i^T \boldsymbol{P}_0 \boldsymbol{u}_i$, is maximized if $\boldsymbol{u}_i = \pm \boldsymbol{q}_1$; i.e., the $\boldsymbol{u}_i$, $i = 1, 2, \dots, N$, are aligned with the major axis of the error ellipse of $\boldsymbol{P}_0$, yielding

$$\max_{\theta_1(\boldsymbol{x}_0), \dots, \theta_N(\boldsymbol{x}_0)} \prod_{i=1}^{N} \left(1 + \alpha_i \boldsymbol{u}_i^T \boldsymbol{P}_0 \boldsymbol{u}_i\right) = \prod_{i=1}^{N} \left(1 + \alpha_i \lambda_1\right) \tag{23.50}$$

regardless of the number of sensors and specific sensor noise variances. Since the $\boldsymbol{u}_i$ are orthogonal to $\boldsymbol{x}_0 - \boldsymbol{s}_i$, $i = 1, 2, \dots, N$, the optimality condition $\boldsymbol{u}_i = \pm\boldsymbol{q}_1$, $i = 1, 2, \dots, N$, implies that an optimal sensor-target geometry for a far-field target is one in which all range vectors $\boldsymbol{x}_0 - \boldsymbol{s}_i$, $i = 1, 2, \dots, N$, are aligned with the minor axis of the error ellipse of $\boldsymbol{P}_0$, producing a collinear sensor placement along the minor axis of the error ellipse as depicted in Figure 23.5. As changing the sign of $\boldsymbol{u}_i$ does not affect the value of the quadratic form $\boldsymbol{u}_i^T\boldsymbol{P}_0\boldsymbol{u}_i$, flipping the sensor locations about $\boldsymbol{x}_0$ does not affect the optimality of the sensor-target geometry.

It is worth noting that, for fixed α_i, $\prod_{i=1}^{N}\left(1 + \alpha_i\boldsymbol{u}_i^T\boldsymbol{P}_0\boldsymbol{u}_i\right)$ does not vary greatly with $\boldsymbol{u}_i$ as it is bounded by

$$\prod_{i=1}^{N}\left(1 + \alpha_i\lambda_2\right) \le \prod_{i=1}^{N}\left(1 + \alpha_i\boldsymbol{u}_i^T\boldsymbol{P}_0\boldsymbol{u}_i\right) \le \prod_{i=1}^{N}\left(1 + \alpha_i\lambda_1\right), \tag{23.51}$$

where λ_2 is the minimum eigenvalue of $\boldsymbol{P}_0$ and $\boldsymbol{u}_i^T\boldsymbol{P}_0\boldsymbol{u}_i$ is minimized at λ_2 when $\boldsymbol{u}_i = \pm\boldsymbol{q}_2$. Therefore, the lower bound in (23.51) is attained if all the $\boldsymbol{u}_i$ are aligned with the minor axis of the error

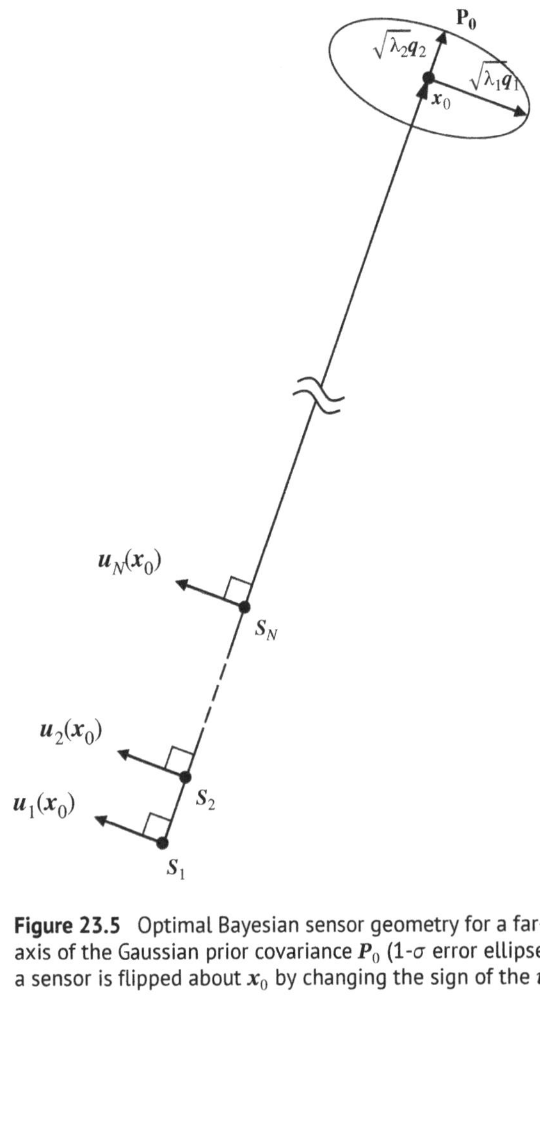

Figure 23.5 Optimal Bayesian sensor geometry for a far-field target. All sensors are aligned with the minor axis of the Gaussian prior covariance $\boldsymbol{P}_0$ (1-σ error ellipse of $\boldsymbol{P}_0$ is shown). The geometry remains optimal if a sensor is flipped about $\boldsymbol{x}_0$ by changing the sign of the $\boldsymbol{u}_i$ vector.

ellipse of $\boldsymbol{P}_0$, making all the sensors aligned with the major axis. As $\alpha_i \to 0$, the difference between the upper and lower bounds in (23.51) becomes very small, implying that the performance loss for nonoptimal geometries, whereby the sensors are not strictly aligned with the minor axis, may not be significant.

An exception occurs to the optimal sensor placement for a far-field target if the covariance of the Gaussian prior happens to be of the form $\boldsymbol{P}_0 = \lambda \boldsymbol{I}$ (i.e., a circular error ellipse). In this case, the diagonal entries of (23.47) become $1 + \alpha_i \lambda$, $i = 1, 2, \dots, N$, which are not dependent on the $\boldsymbol{u}_i$. This suggests that to maximize $|\boldsymbol{I} + \boldsymbol{V}^T \boldsymbol{P}_0 \boldsymbol{U}|$ the focus should be on minimizing the off-diagonal entries by making $\boldsymbol{u}_i^T \boldsymbol{u}_j$, $i \neq j$, as small as possible. The resulting optimal sensor-target geometry will not necessarily have a collinear sensor placement. This kind of Gaussian prior is rare and therefore will not be considered any further in this chapter.

A "near-field" target is characterized by $\alpha_i \gg 0$ for all $i = 1, 2, \dots, N$. This is usually encountered in applications where the AOA sensor noise is very small, as is, for example, the case with high-resolution cameras, and the target range is not significantly large relative to the "size" of the error ellipse of $\boldsymbol{P}_0$. For a near-field target, the terms in the expanded expression for $|\boldsymbol{I} + \boldsymbol{V}^T \boldsymbol{P}_0 \boldsymbol{U}|$ that are multiplied by $\alpha_i \alpha_j$, $i \neq j$, will make the most significant contribution to the determinant of BFIM and therefore should be maximized in the first place to establish sensor placement constraints. This will be discussed in more detail in the next section.

23.6 Optimal Sensor Placement Results

23.6.1 Single-sensor Geometry

For $N = 1$, the optimization problem in (23.41) takes the form

$$\max_{\theta_1(\boldsymbol{x}_0)} |1 + \alpha_1 \boldsymbol{u}_1^T \boldsymbol{P}_0 \boldsymbol{u}_1|, \tag{23.52}$$

which is reduced to

$$\max_{\boldsymbol{u}_1} \boldsymbol{u}_1^T(\boldsymbol{x}_0) \boldsymbol{P}_0 \boldsymbol{u}_1(\boldsymbol{x}_0). \tag{23.53}$$

Referring to (23.44), an optimal sensor-target geometry for a single AOA sensor is obtained by aligning the range vector $\boldsymbol{x}_0 - \boldsymbol{s}_1$ with the minor axis of the error ellipse of $\boldsymbol{P}_0$ (see Figure 23.6). This result holds true regardless of the target range and AOA noise. It is also consistent with the optimal sensor placement result for a far-field target in Section 23.5. Flipping the sensor about $\boldsymbol{x}_0$ does not affect the optimality of the geometry.

23.6.2 Two-sensor Geometry

In 2D AOA localization problems, geometric rotations do not change the values of J_A and J_D. In order to simplify the optimal geometry analysis for two or more sensors, the sensor-target geometry will be rotated around $\boldsymbol{x}_0$ to align the major and minor axes of the error ellipse of $\boldsymbol{P}_0$ with the 2D Cartesian coordinates. To this end, consider the eigen-decomposition of $\boldsymbol{P}_0$ in (23.45), which is reproduced here for convenience:

$$\boldsymbol{P}_0 = [\boldsymbol{q}_1, \boldsymbol{q}_2] \boldsymbol{\Lambda}_0 [\boldsymbol{q}_1, \boldsymbol{q}_2]^T. \tag{23.54}$$

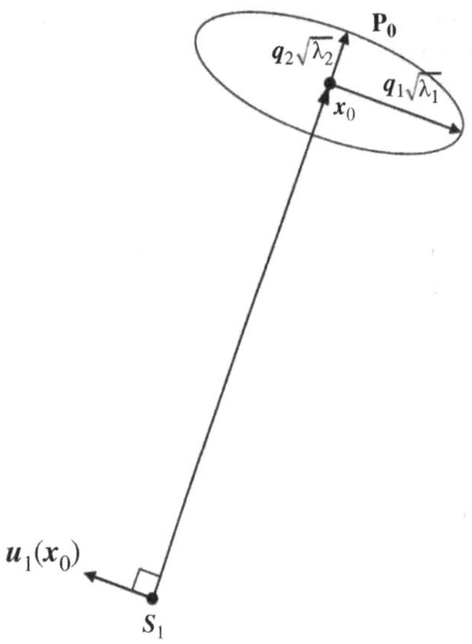

Figure 23.6 Optimal Bayesian sensor geometry for a single AOA sensor. The sensor is aligned with the minor axis of the Gaussian prior covariance $\boldsymbol{P}_0$.

Here $\boldsymbol{\Lambda}_0 = \text{diag}(\lambda_1, \lambda_2)$ is the diagonal matrix of eigenvalues of $\boldsymbol{P}_0$ with $\lambda_1 \geq \lambda_2$. Using the eigen-decomposition of $\boldsymbol{P}_0$, the matrix entries in (23.42b) can be rewritten as

$$\alpha_i \boldsymbol{u}_i^T \boldsymbol{P}_0 \boldsymbol{u}_j = \alpha_i v_i^T \boldsymbol{\Lambda}_0 v_j, \tag{23.55}$$

where

$$v_i = \begin{bmatrix} \cos\vartheta_i \\ \sin\vartheta_i \end{bmatrix} = [\boldsymbol{q}_1, \boldsymbol{q}_2]^T \boldsymbol{u}_i \tag{23.56}$$

is the rotated unit vector $\boldsymbol{u}_i$ and the diagonalized Gaussian prior covariance $\boldsymbol{\Lambda}_0$ has major and minor axes given by $[\sqrt{\lambda_1}, 0]^T$ and $[0, \sqrt{\lambda_2}]^T$, respectively, which are aligned with the 2D Cartesian coordinates as desired. The angles of the major and minor axes are 0 and $\pi/2$ rad, respectively.

For $N = 2$, using the diagonalized Gaussian prior covariance $\boldsymbol{\Lambda}_0$, (23.42b) takes the form

$$\boldsymbol{I} + \boldsymbol{V}^T \boldsymbol{P}_0 \boldsymbol{U} = \begin{bmatrix} 1 + \alpha_1 v_1^T \boldsymbol{\Lambda}_0 v_1 & \alpha_1 v_1^T \boldsymbol{\Lambda}_0 v_2 \\ \alpha_2 v_2^T \boldsymbol{\Lambda}_0 v_1 & 1 + \alpha_2 v_2^T \boldsymbol{\Lambda}_0 v_2 \end{bmatrix} \tag{23.57}$$

with its determinant given by

$$|\boldsymbol{I} + \boldsymbol{V}^T \boldsymbol{P}_0 \boldsymbol{U}| = 1 + \alpha_1 v_1^T \boldsymbol{\Lambda}_0 v_1 + \alpha_2 v_2^T \boldsymbol{\Lambda}_0 v_2 + \alpha_1\alpha_2 \left(v_1^T \boldsymbol{\Lambda}_0 v_1 v_2^T \boldsymbol{\Lambda}_0 v_2 - \left(v_1^T \boldsymbol{\Lambda}_0 v_2\right)^2 \right) \tag{23.58a}$$

$$= 1 + \frac{1}{2}(\alpha_1 + \alpha_2)(\lambda_1 + \lambda_2) + \frac{1}{2}(\lambda_1 - \lambda_2)(\alpha_1 \cos 2\vartheta_1 + \alpha_2 \cos 2\vartheta_2) + \alpha_1\alpha_2\lambda_1\lambda_2 \sin^2(\vartheta_1 - \vartheta_2). \tag{23.58b}$$

For a far-field target ($\alpha_i \to 0$, $i = 1, 2$), the term

$$\frac{1}{2}(\lambda_1 - \lambda_2)(\alpha_1 \cos 2\vartheta_1 + \alpha_2 \cos 2\vartheta_2) \tag{23.59}$$

dominates (23.58b), which is maximized at $\frac{1}{2}(\lambda_1 - \lambda_2)(\alpha_1 + \alpha_2)$ if $\vartheta_1 = \vartheta_1 = 0$; i.e., $\boldsymbol{v}_1$ and $\boldsymbol{v}_2$ are both aligned with the major axis of the error ellipse of $\boldsymbol{\Lambda}_0$. This implies that an optimal sensor placement maximizing (23.58) is given by both sensors being aligned with the minor axis of the error ellipse of $\boldsymbol{\Lambda}_0$ in the rotated geometry or $\boldsymbol{P}_0$ in the original geometry. Specifically, the condition for (23.59) to be the dominant term is

$$\frac{\alpha_1 + \alpha_2}{\alpha_1 \alpha_2} \gg 2\frac{\lambda_1 \lambda_2}{\lambda_1 - \lambda_2}, \tag{23.60}$$

which captures the interplay between the eigenvalues λ_1 and λ_2, and α_1 and α_2 for a far-field target.

In the near-field target case, which is characterized by $\alpha_i \gg 0$, $i = 1, 2$, the dominant term in (23.58b) is

$$\alpha_1 \alpha_2 \lambda_1 \lambda_2 \sin^2(\vartheta_1 - \vartheta_2) \tag{23.61}$$

if the following condition is satisfied

$$\frac{\alpha_1 + \alpha_2}{\alpha_1 \alpha_2} \ll 2\frac{\lambda_1 \lambda_2}{\lambda_1 - \lambda_2}. \tag{23.62}$$

Equation (23.61) is maximized if $|\vartheta_1 - \vartheta_2| = \pi/2$ rad. This means that the condition to maximize (23.61) is to have $\boldsymbol{v}_1^T \boldsymbol{v}_2 = 0$ or $\boldsymbol{u}_i^T [\boldsymbol{q}_1, \boldsymbol{q}_2][\boldsymbol{q}_1, \boldsymbol{q}_2]^T \boldsymbol{u}_i = \boldsymbol{u}_i^T \boldsymbol{u}_i = 0$; i.e., $\boldsymbol{v}_1$ and $\boldsymbol{v}_2$ or, equivalently, $\boldsymbol{u}_1$ and $\boldsymbol{u}_2$ should be orthogonal. In Eq. (23.58b) the only other term that depends on ϑ_1 and ϑ_2 is $\frac{1}{2}(\lambda_1 - \lambda_2)$ $(\alpha_1 \cos 2\vartheta_1 + \alpha_2 \cos 2\vartheta_2)$. Maximizing it over ϑ_1 and ϑ_2 subject to the constraint $|\vartheta_1 - \vartheta_2| = \pi/2$ rad gives

$$\max_{\vartheta_1} \frac{1}{2}(\lambda_1 - \lambda_2)(\alpha_1 \cos 2\vartheta_1 + \alpha_2 \cos 2\vartheta_2), \quad \vartheta_1 - \vartheta_2 = \pm\pi/2 \tag{23.63a}$$

$$\max_{\vartheta_1}(\lambda_1 - \lambda_2)(\alpha_1 - \alpha_2) \cos 2\vartheta_1. \tag{23.63b}$$

Thus, the optimal ϑ_1 is

$$\vartheta_1 = \begin{cases} 0 & \text{if } \alpha_1 > \alpha_2 \text{ or } \sigma_2 d_2(\boldsymbol{x}_0) > \sigma_1 d_1(\boldsymbol{x}_0) \\ \pm\pi/2 & \text{if } \alpha_2 > \alpha_1 \text{ or } \sigma_1 d_1(\boldsymbol{x}_0) > \sigma_2 d_2(\boldsymbol{x}_0) \\ \text{not unique} & \text{if } \alpha_1 = \alpha_2 \text{ or } \lambda_1 = \lambda_2. \end{cases} \tag{23.64}$$

Assuming identical AOA noise for both sensors ($\sigma_1 = \sigma_2$) and recalling that the $\boldsymbol{v}_i$ are orthogonal to the range vectors from the sensors to $\boldsymbol{x}_0$, the result in (23.64) means that in an optimal two-sensor near-field target geometry

1. the sensor closest to the mean of the Gaussian prior $\boldsymbol{x}_0$ (i.e., the sensor with larger α_i) will align with the minor axis of the error ellipse of $\boldsymbol{\Lambda}_0$, or $\boldsymbol{P}_0$ in the unrotated geometry, and the other sensor with the major axis,
2. the sensors can be placed at an arbitrary angle as long as they are separated by $\pi/2$ rad if both sensors have the same distance from $\boldsymbol{x}_0$ (i.e., $\alpha = \alpha_1 = \alpha_2$).

In scenarios where neither a far-field nor near-field target assumption can be applied, (23.58b) will need to be maximized over the angles $\{\vartheta_1, \vartheta_2\}$ by analytical or numerical means.

23.6.2.1 Example

Consider a Gaussian prior for the target location with

$$\boldsymbol{x}_0 = \begin{bmatrix} 0 \\ 0 \end{bmatrix}, \quad \boldsymbol{P}_0 = \begin{bmatrix} 15 & 6 \\ 6 & 10 \end{bmatrix}. \tag{23.65}$$

The eigenvalues of $\boldsymbol{P}_0$ are $\lambda_1 = 19$ and $\lambda_2 = 6$. The major and minor axes of the error ellipse of $\boldsymbol{P}_0$ make the angles of 33.7° and 123.7° with the positive x-axis, respectively. The sensors have the same AOA noise standard deviation $\sigma_1 = \sigma_2 = 1°$. Three scenarios are simulated corresponding to a near-field target, a target at an intermediate range (neither near-field nor far-field) and a far-field target. The simulation parameters selected for these scenarios are listed in Table 23.1. In the near-field target case, $\frac{\alpha_1+\alpha_2}{\alpha_1\alpha_2}$ is much smaller than $2\frac{\lambda_1\lambda_2}{\lambda_1-\lambda_2}$ as per (23.62). Likewise, $\frac{\alpha_1+\alpha_2}{\alpha_1\alpha_2}$ is much larger than $2\frac{\lambda_1\lambda_2}{\lambda_1-\lambda_2}$ for the far-field target in agreement with (23.60). In the case of intermediate-range target, $\frac{\alpha_1+\alpha_2}{\alpha_1\alpha_2}$ and $2\frac{\lambda_1\lambda_2}{\lambda_1-\lambda_2}$ are comparable.

Numerical search was used to determine the sensor angles θ_1 and θ_2 that maximize $|\tilde{\boldsymbol{\Phi}}|$ in the angle range $0 \le \theta_i < \pi$ rad, $i = 1, 2$. The computed values of $|\tilde{\boldsymbol{\Phi}}|$ versus θ_1 and θ_2 are depicted in Figure 23.7. The optimal sensor angles are also shown. For the near-field target case, the optimal angle for $\boldsymbol{s}_2$, the sensor closer to $\boldsymbol{x}_0$, is $\theta_2 = 123.7°$, which is the same as the angle of the minor axis of the error ellipse of $\boldsymbol{P}_0$, and the optimal angle of $\boldsymbol{s}_1$ is $\theta_1 = 33.7°$, which is the angle of the major axis. In other words, $\boldsymbol{s}_1$ is aligned with the major axis of $\boldsymbol{P}_0$ and $\boldsymbol{s}_2$ with the minor axis. For the target at intermediate range, the optimal sensor angles are $\theta_1 = 171.0°$ and $\theta_2 = 102.4°$, which are not aligned with the major or minor axis. The optimal sensor angles for the far-field target are $\theta_1 = \theta_2 = 123.7°$, which make the sensors collinear with the minor axis of the error ellipse of $\boldsymbol{P}_0$. These numerical results confirm the analytical findings for two-sensor optimal geometries.

Figure 23.8 shows the optimal sensor-target geometries for the optimal sensor angles in Figure 23.7. The unique features of optimal geometries for near-field and far-field targets are evident.

23.6.3 Three-sensor Geometry

Using the diagonalized Gaussian prior covariance $\boldsymbol{\Lambda}_0$, for a three-sensor geometry ($N = 3$) (23.42b) becomes

$$\boldsymbol{I} + \boldsymbol{V}^T\boldsymbol{P}_0\boldsymbol{U} = \begin{bmatrix} 1+\alpha_1 v_1^T\boldsymbol{\Lambda}_0 v_1 & \alpha_1 v_1^T\boldsymbol{\Lambda}_0 v_2 & \alpha_1 v_1^T\boldsymbol{\Lambda}_0 v_3 \\ \alpha_2 v_2^T\boldsymbol{\Lambda}_0 v_1 & 1+\alpha_2 v_2^T\boldsymbol{\Lambda}_0 v_2 & \alpha_2 v_2^T\boldsymbol{\Lambda}_0 v_3 \\ \alpha_3 v_3^T\boldsymbol{\Lambda}_0 v_1 & \alpha_3 v_3^T\boldsymbol{\Lambda}_0 v_2 & 1+\alpha_3 v_3^T\boldsymbol{\Lambda}_0 v_3 \end{bmatrix}. \tag{23.66}$$

Table 23.1 Simulation parameters ($N = 2$).

Scenario	$d_1(\boldsymbol{x}_0)$	$d_2(\boldsymbol{x}_0)$	$\frac{\alpha_1+\alpha_2}{\alpha_1\alpha_2}$	$2\frac{\lambda_1\lambda_2}{\lambda_1-\lambda_2}$
Near-field target	70 km	40 km	1.98	17.54
Intermediate range	170 km	140 km	14.77	17.54
Far-field target	320 km	290 km	56.81	17.54

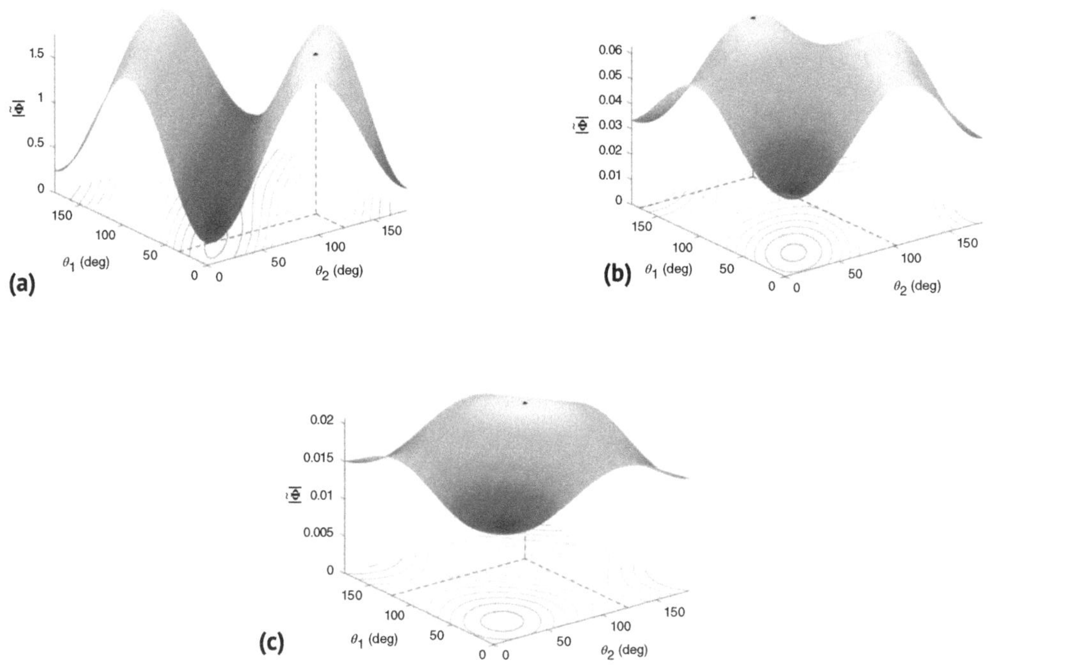

Figure 23.7 Determinant of BFIM $|\tilde{\Phi}|$ versus sensor angles $\theta_1(\boldsymbol{x}_0)$ and $\theta_2(\boldsymbol{x}_0)$. The optimal angles maximizing $|\tilde{\Phi}|$ are (a) $\theta_1 = 33.7°$ and $\theta_2 = 123.7°$ (near-field target), (b) $\theta_1 = 171°$ and $\theta_2 = 102.4°$ (intermediate-range target), (c) $\theta_1 = \theta_2 = 123.7°$ (far-field target).

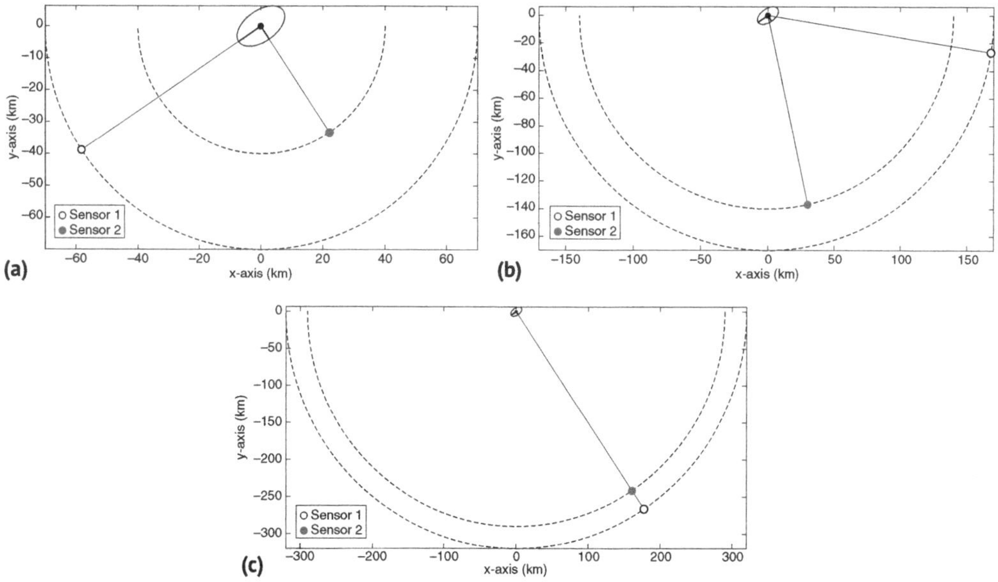

Figure 23.8 Optimal Bayesian sensor geometries for two sensors ($N = 2$): (a) near-field target with $\alpha_1 = 0.67$ and $\alpha_2 = 2.05$, (b) target at intermediate range with $\alpha_1 = 0.11$ and $\alpha_2 = 0.17$, and (c) far-field target with $\alpha_1 = 0.03$ and $\alpha_2 = 0.04$. The 2-σ error ellipse of $\boldsymbol{P}_0$ is shown.

The determinant of $\boldsymbol{I} + \boldsymbol{V}^T\boldsymbol{P}_0\boldsymbol{U}$ is

$$|\boldsymbol{I} + \boldsymbol{V}^T\boldsymbol{P}_0\boldsymbol{U}| = 1 + \frac{1}{2}(\lambda_1 + \lambda_2)(\alpha_1 + \alpha_2 + \alpha_3) + \frac{1}{2}(\lambda_1 - \lambda_2)(\alpha_1 \cos 2\vartheta_1 + \alpha_2 \cos 2\vartheta_2 + \alpha_3 \cos 2\vartheta_3) + \lambda_1\lambda_2\left(\alpha_1\alpha_2 \sin^2(\vartheta_1 - \vartheta_2) + \alpha_1\alpha_3 \sin^2(\vartheta_1 - \vartheta_3) + \alpha_2\alpha_3 \sin^2(\vartheta_2 - \vartheta_3)\right). \quad (23.67)$$

If the far-field target condition

$$\frac{\alpha_1 + \alpha_2 + \alpha_3}{\alpha_1\alpha_2 + \alpha_1\alpha_3 + \alpha_2\alpha_3} \gg 2\frac{\lambda_1\lambda_2}{\lambda_1 - \lambda_2} \quad (23.68)$$

is satisfied, the term

$$\frac{1}{2}(\lambda_1 - \lambda_2)(\alpha_1 \cos 2\vartheta_1 + \alpha_2 \cos 2\vartheta_2 + \alpha_3 \cos 2\vartheta_3) \quad (23.69)$$

becomes dominant in (23.67) and is maximized by $\vartheta_1 = \vartheta_2 = \vartheta_3 = 0$; that is, all the sensors are collinear with the minor axis of the error ellipse of $\boldsymbol{\Lambda}_0$ in the unrotated geometry or $\boldsymbol{P}_0$ in the original geometry.

On the other hand, if

$$\frac{\alpha_1 + \alpha_2 + \alpha_3}{\alpha_1\alpha_2 + \alpha_1\alpha_3 + \alpha_2\alpha_3} \ll 2\frac{\lambda_1\lambda_2}{\lambda_1 - \lambda_2} \quad (23.70)$$

we have a near-field target, which makes the following term dominant in (23.67):

$$\lambda_1\lambda_2\left(\alpha_1\alpha_2\sin^2(\vartheta_1 - \vartheta_2) + \alpha_1\alpha_3\sin^2(\vartheta_1 - \vartheta_3) + \alpha_2\alpha_3\sin^2(\vartheta_2 - \vartheta_3)\right). \quad (23.71)$$

Maximizing it over ϑ_1, ϑ_2, and ϑ_3 gives the optimal angle constraints between the sensors. Then maximizing the second term on the right-hand side of (23.67) subject to these constraints gives the specific orthogonal unit vector angles for the optimal near-field target geometry.

If $\alpha_1 = \alpha_2 = \alpha_3 = \alpha$, the near-field target condition in (23.70) becomes

$$\frac{1}{\alpha} \ll 2\frac{\lambda_1\lambda_2}{\lambda_1 - \lambda_2}, \quad (23.72)$$

and it is easy to show that (23.71) is maximized by $\{\vartheta_1, \vartheta_2, \vartheta_3\} = \{0, \pm\pi/3, \mp\pi/3\} + \varphi$ where φ is an arbitrary angle rotation. For these optimal angle constraints, the second term on the right-hand side of (23.67) is zero regardless of φ. Therefore, for a near-field target geometry with $\alpha_1 = \alpha_2 = \alpha_3 = \alpha$, the optimal sensor-target geometry is defined by $\{\vartheta_1, \vartheta_2, \vartheta_3\} = \{0, \pm\pi/3, \mp\pi/3\} + \varphi$. A special case for $\alpha_1 = \alpha_2 = \alpha_3 = \alpha$ is a sensor-target geometry where each sensor has the same AOA noise $\sigma_1 = \sigma_2 = \sigma_3$ and the sensor distances from $\boldsymbol{x}_0$ are equal ($d_1(\boldsymbol{x}_0) = d_2(\boldsymbol{x}_0) = d_3(\boldsymbol{x}_0)$). Recalling that the sensors can be flipped about $\boldsymbol{x}_0$ without affecting the optimality, equal angular spacing between the sensors, e.g., $\vartheta_2 - \vartheta_1 = \vartheta_3 - \vartheta_2 = 2\pi/3$ rad, is also an optimal geometry.

It is difficult to offer general guidelines for optimal sensor placement in intermediate-range target scenarios and near-field target scenarios where the α_i take on different values. In such scenarios, (23.67) will have to maximized numerically to find the optimal sensor angles.

23.6.3.1 Example

The Gaussian prior for the target location is the same as in Section 23.6.2.1. The sensors have the same AOA noise $\sigma_1 = \sigma_2 = \sigma_3 = 1°$. Four scenarios are considered corresponding to near-field and far-field target situations, compliant with the conditions in (23.68) and (23.70). Table 23.2 lists the simulation parameters for these scenarios.

The optimal sensor angles θ_1, θ_2 and θ_3 were determined using exhaustive numerical search to maximize $|\tilde{\boldsymbol{\Phi}}|$. The resulting optimal sensor-target geometries are illustrated in Figure 23.9. For near-field target i, $\alpha_1 = \alpha_2 = \alpha_3 = \alpha$ and the optimal sensor-target geometry is defined by $\{\vartheta_1, \vartheta_2, \vartheta_3\} = \{0, \pm\pi/3, \mp\pi/3\} + \varphi$ (φ is arbitrary), which is depicted in Figure 23.9a. For near-field targets ii and iii, the α_i are not the same and the optimal geometries have different configurations, depending on the relative difference between the α_i. If one of the α_i is relatively small (see Figure 23.9c), the optimal geometry resembles that for a two-sensor geometry (cf. Figure 23.8a), resulting in the sensor with the largest α_i to be aligned with the minor axis of the error ellipse of $\boldsymbol{P}_0$ and the remaining sensors to be aligned with the major axis. For the far-field target, the optimal sensor locations are collinear with the minor axis of the error ellipse of $\boldsymbol{P}_0$ as shown in Figure 23.9d.

Table 23.2 Simulation parameters ($N = 3$).

Scenario	$d_1(\boldsymbol{x}_0)$	$d_2(\boldsymbol{x}_0)$	$d_3(\boldsymbol{x}_0)$	$\frac{\alpha_1+\alpha_2+\alpha_3}{\alpha_1\alpha_2+\alpha_1\alpha_3+\alpha_2\alpha_3}$	$2\frac{\lambda_1\lambda_2}{\lambda_1-\lambda_2}$
Near-field target i	25 km	25 km	25 km	0.19	17.54
Near-field target ii	25 km	30 km	35 km	0.27	17.54
Near-field target iii	20 km	25 km	50 km	0.24	17.54
Far-field target	380 km	400 km	420 km	48.66	17.54

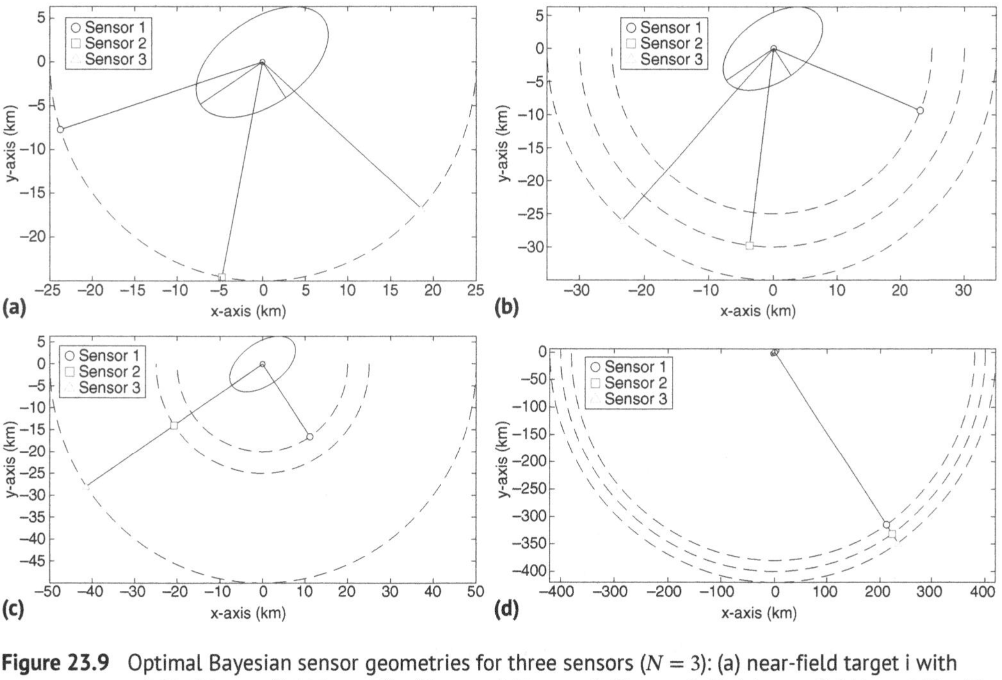

Figure 23.9 Optimal Bayesian sensor geometries for three sensors ($N = 3$): (a) near-field target i with $\alpha_1 = \alpha_2 = \alpha_3 = 5.25$, (b) near-field target ii with $\alpha_1 = 5.25$, $\alpha_2 = 3.65$, $\alpha_3 = 2.68$, (c) near-field target iii with $\alpha_1 = 8.21$, $\alpha_2 = 5.25$, $\alpha_3 = 1.31$, and (d) far-field target. The 2-σ error ellipse of $\boldsymbol{P}_0$ is shown.

23.6.4 Generalization to More than Three Sensors

For $N > 3$, the general expression for $|\boldsymbol{I} + \boldsymbol{V}^T\boldsymbol{P}_0\boldsymbol{U}|$ is

$$
\begin{aligned}
|\boldsymbol{I} + \boldsymbol{V}^T\boldsymbol{P}_0\boldsymbol{U}| = 1 + \frac{1}{2}(\lambda_1 + \lambda_2)\sum_{i=1}^{N}\alpha_i + \frac{1}{2}(\lambda_1 - \lambda_2)\sum_{i=1}^{N}\alpha_i\cos 2\vartheta_i \\
+\lambda_1\lambda_2\sum_{i=1}^{N}\sum_{j=i+1}^{N}\alpha_i\alpha_j\sin^2(\vartheta_i - \vartheta_j).
\end{aligned}
\tag{23.73}
$$

A far-field target case prevails if

$$
\frac{\sum_{i=1}^{N}\alpha_i}{\sum_{i=1}^{N}\sum_{j=i+1}^{N}\alpha_i\alpha_j} \gg 2\frac{\lambda_1\lambda_2}{\lambda_1 - \lambda_2}. \tag{23.74}
$$

Reversing the inequality in (23.74) gives the condition for a near-field target:

$$
\frac{\sum_{i=1}^{N}\alpha_i}{\sum_{i=1}^{N}\sum_{j=i+1}^{N}\alpha_i\alpha_j} \ll 2\frac{\lambda_1\lambda_2}{\lambda_1 - \lambda_2}. \tag{23.75}
$$

For a near-field target, the general approach to finding optimal sensor placement proceeds as follows:

1. Solve the optimization problem

$$
\max_{\vartheta_1,\ldots,\vartheta_N}\sum_{i=1}^{N}\sum_{j=i+1}^{N}\alpha_i\alpha_j\sin^2(\vartheta_i - \vartheta_j) \tag{23.76}
$$

   to find optimal sensor angle separations

$$
\{\vartheta_1, \ldots, \vartheta_N\} = \{0, \pm\varphi_1, \ldots, \pm\varphi_{N-1}\} + \varphi, \tag{23.77}
$$

   where φ is an arbitrary rotation angle.
2. Solve the optimization problem

$$
\{\vartheta_1^{\text{opt}}, \ldots, \vartheta_N^{\text{opt}}\} = \arg\max_{\vartheta_1,\ldots,\vartheta_N}\sum_{i=1}^{N}\alpha_i\cos 2\vartheta_i \tag{23.78}
$$

   subject to the constraints in (23.77), to determine the specific optimal angles for the orthogonal unit vectors in the rotated geometry.
3. Rotate ϑ_i^{opt} to find the optimal sensor angles in the original unrotated geometry:

$$
\theta_i^{\text{opt}} = \pm\left(\angle\left([\boldsymbol{q}_1, \boldsymbol{q}_2]\begin{bmatrix}\cos\vartheta_i^{\text{opt}} \\ \sin\vartheta_i^{\text{opt}}\end{bmatrix}\right) + \pi/2\right), \quad i = 1, \ldots, N \tag{23.79}
$$

   where $\angle(\cdot)$ denotes the angle of its vector argument.

If $\alpha_1 = \ldots = \alpha_N = \alpha$, (23.75) becomes

$$
\frac{2}{(N-1)\alpha} \ll 2\frac{\lambda_1\lambda_2}{\lambda_1 - \lambda_2}. \tag{23.80}
$$

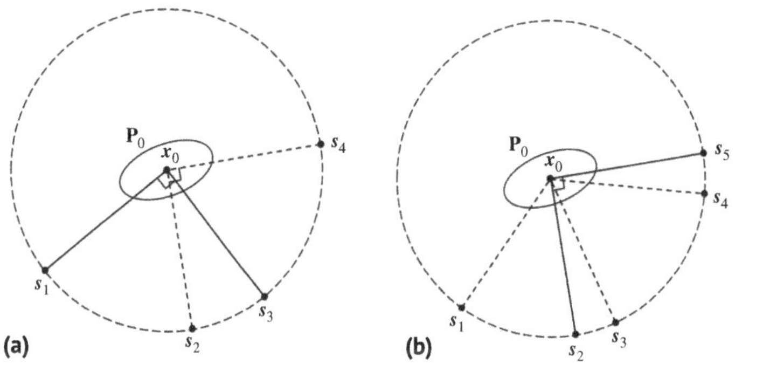

Figure 23.10 Optimal sensor-target geometries for a near-field target with $\alpha_1 = ... = \alpha_N$ and $\sigma_1 = ... = \sigma_N$: (a) a four-sensor geometry where two-sensor partitions $\{\boldsymbol{s}_1, \boldsymbol{s}_3\}$ and $\{\boldsymbol{s}_2, \boldsymbol{s}_4\}$ have sensors separated by $\pi/2$ rad, and (b) a five-sensor geometry where the three-sensor partition $\{\boldsymbol{s}_1, \boldsymbol{s}_3, \boldsymbol{s}_4\}$ has sensors with $\pi/3$ rad separation while the two-sensor partition $\{\boldsymbol{s}_2, \boldsymbol{s}_5\}$ has $\pi/2$ rad separation. Rotating individual sensor partitions around $\boldsymbol{x}_0$ does not alter optimality.

In a near-field target scenario with $\alpha_1 = ... = \alpha_N = \alpha$, an optimal sensor-target geometry is obtained by partitioning the sensors into subsets of two and/or three sensors where each subset has optimal sensor angle separation $|\vartheta_i - \vartheta_j| = \pi/2$ rad for a two-sensor partition, and $\{\vartheta_i, \vartheta_j, \vartheta_m\} = \{0, \pm\pi/3, \mp\pi/3\} + \varphi$ for a three-sensor partition (φ is an arbitrary rotation). In addition, equal angular spacing between the sensors (e.g., $\vartheta_{i+1} - \vartheta_i = \pi/N$, $i = 1, ..., N-1$) will also produce an optimal geometry. These optimal sensor-target geometries solve both (23.76) and (23.78) as desired. New optimal geometries can be generated by flipping the sensors about $\boldsymbol{x}_0$ and/or rotating individual sensor partitions around $\boldsymbol{x}_0$. Examples of optimal geometries for $N = 4$ and $N = 5$ are illustrated in Figure 23.10.

For a far-field target satisfying (23.74), the optimal sensor placement is collinear with the minor axis of the error ellipse of $\boldsymbol{P}_0$. For all other cases, $|\boldsymbol{I} + \boldsymbol{V}^T \boldsymbol{P}_0 \boldsymbol{U}|$ will need to be maximized numerically to find an optimal sensor placement, which can be computationally demanding if N is large.

23.7 Optimal Sensor Trajectories for Target Tracking

23.7.1 Embedding Optimal Bayesian Sensor Placement into the Kalman Filter

The Kalman filter is the algorithm of choice in AOA target tracking problems. In multi-sensor single-target tracking, the Kalman filter computes the target location and velocity estimates, using the prior information (state prediction) available from the previous recursion and the current AOA measurements at each recursion. Each Kalman filter recursion effectively computes the MAP estimate of the state vector consisting of target location and velocity. The state transition equation models the evolution of the time-varying state vector (target maneuvers). Optimal sensor placement is determined from the state prediction (Gaussian prior) and other statistical parameters either assumed to be known or readily available from the Kalman filter. The AOA measurements are taken at the optimal sensor locations, which are subject to constraints due to sensor dynamics, environment, and time interval between Kalman filter recursions.

For illustration purposes, a nearly constant-velocity motion model is assumed for the maneuvering target:

$$\boldsymbol{x}_{k+1} = \boldsymbol{F}\boldsymbol{x}_k + \boldsymbol{n}_k, \quad k = 0, 1, 2, ... \tag{23.81}$$

where $\boldsymbol{F}$ is the state transition matrix

$$\boldsymbol{F} = \begin{bmatrix} \boldsymbol{A} & \boldsymbol{0} \\ \boldsymbol{0} & \boldsymbol{A} \end{bmatrix}, \quad \boldsymbol{A} = \begin{bmatrix} 1 & T \\ 0 & 1 \end{bmatrix} \tag{23.82}$$

with T denoting the time interval between discrete-time instants k, $\boldsymbol{x}_k$ is the state vector comprised of target location $[x_k, y_k]^T$ and velocity $[\dot{x}_k, \dot{y}_k]^T$ at time k:

$$\boldsymbol{x}_k = [x_k, \dot{x}_k, y_k, \dot{y}_k]^T \tag{23.83}$$

and $\boldsymbol{n}_k \sim \mathcal{N}(\boldsymbol{0}, \boldsymbol{Q})$ is the process noise that accounts for unknown target maneuvers. The process noise covariance is

$$\boldsymbol{Q} = \begin{bmatrix} q_x\boldsymbol{B} & \boldsymbol{0} \\ \boldsymbol{0} & q_y\boldsymbol{B} \end{bmatrix}, \quad \boldsymbol{B} = \begin{bmatrix} T^4/4 & T^3/2 \\ T^3/2 & T^2 \end{bmatrix} \tag{23.84}$$

where q_x and q_y are related to maximum target acceleration [26].

The AOA measurements taken by the moving sensors at $\boldsymbol{s}_{i,k} = [s_{x,i,k}, s_{y,i,k}]^T$, $i = 1, \dots, N$, are a *nonlinear* function of the state vector:

$$\boldsymbol{z}_k = \boldsymbol{\theta}(\boldsymbol{x}_k) + \boldsymbol{w}_k \tag{23.85}$$

where

$$\boldsymbol{\theta}(\boldsymbol{x}_k) = \begin{bmatrix} \theta_1(\boldsymbol{x}_k) \\ \vdots \\ \theta_N(k) \end{bmatrix}, \quad \theta_i(\boldsymbol{x}_k) = \tan^{-1}\frac{y_k - s_{y,i,k}}{x_k - s_{x,i,k}} \tag{23.86}$$

and $\boldsymbol{w}_k \sim \mathcal{N}(\boldsymbol{0}, \boldsymbol{R}_k)$ is the AOA measurement noise with covariance

$$\boldsymbol{R}_k = \begin{bmatrix} \sigma_{1,k}^2 & & \boldsymbol{0} \\ & \ddots & \\ \boldsymbol{0} & & \sigma_{N,k}^2 \end{bmatrix}. \tag{23.87}$$

Suppose that the EKF is used to estimate the state vector $\boldsymbol{x}_k$ from the AOA measurements $\boldsymbol{z}_k$. This entails linearizing (23.85) as follows:

$$\boldsymbol{z}_k = \boldsymbol{H}_k\boldsymbol{x}_k + \boldsymbol{w}_k \tag{23.88}$$

where

$$\boldsymbol{H}_k = \left.\frac{\partial\boldsymbol{\theta}(\boldsymbol{x}_k)}{\partial\boldsymbol{x}_k}\right|_{\boldsymbol{x}_k=\boldsymbol{x}_{k|k-1}} \tag{23.89a}$$

$$= \begin{bmatrix} -\dfrac{\sin\theta_1(\boldsymbol{x}_{k|k-1})}{d_1(\boldsymbol{x}_{k|k-1})} & 0 & \dfrac{\cos\theta_1(\boldsymbol{x}_{k|k-1})}{d_1(\boldsymbol{x}_{k|k-1})} & 0 \\ -\dfrac{\sin\theta_2(\boldsymbol{x}_{k|k-1})}{d_2(\boldsymbol{x}_{k|k-1})} & 0 & \dfrac{\cos\theta_2(\boldsymbol{x}_{k|k-1})}{d_2(\boldsymbol{x}_{k|k-1})} & 0 \\ \vdots & \vdots & \vdots & \vdots \\ -\dfrac{\sin\theta_N(\boldsymbol{x}_{k|k-1})}{d_N(\boldsymbol{x}_{k|k-1})} & 0 & \dfrac{\cos\theta_N(\boldsymbol{x}_{k|k-1})}{d_N(\boldsymbol{x}_{k|k-1})} & 0 \end{bmatrix}_{N\times 4}. \tag{23.89b}$$

Here $d_i(\boldsymbol{x}_{k|k-1}) = \| \boldsymbol{\mu}_{k|k-1} - \boldsymbol{s}_{i,k} \|$, $i = 1, \dots, N$, is the target range estimate from the sensors using the predicted target location $\boldsymbol{\mu}_{k|k-1} = [x_{k|k-1}, y_{k|k-1}]^T$. The innovation equation of the EKF is

$$\tilde{\boldsymbol{z}}_k = \boldsymbol{z}_k - \theta(\boldsymbol{x}_{k|k-1}). \tag{23.90}$$

A lower bound on the covariance of state estimates of the EKF is given by the recursive BFIM [14]:

$$\boldsymbol{\Phi}_k = (\boldsymbol{F}\boldsymbol{\Phi}_{k-1}^{-1}\boldsymbol{F}^T + \boldsymbol{Q})^{-1} + E_{\boldsymbol{x}_k}\left\{\bar{\boldsymbol{H}}_k^T \boldsymbol{R}_k^{-1} \bar{\boldsymbol{H}}_k\right\} \tag{23.91}$$

where $\bar{\boldsymbol{H}}_k$ is the Jacobian matrix in (23.89b) evaluated at the true target state vector $\boldsymbol{x}_k$. Similarly to (23.28), (23.91) can be simplified by stripping off the expectation and replacing $\boldsymbol{x}_k$ with the mean of the Gaussian prior $\boldsymbol{x}_{k|k-1}$ in the second term on the right-hand side of (23.91), and by substituting the covariance of the Gaussian prior $\boldsymbol{P}_{k|k-1}$ for $\boldsymbol{F}\boldsymbol{\Phi}_{k-1}^{-1}\boldsymbol{F}^T + \boldsymbol{Q}$. This results in an approximate expression for the recursive BFIM, $\tilde{\boldsymbol{\Phi}}_k$, which is simply equal to the inverse of the EKF state estimate covariance:

$$\tilde{\boldsymbol{\Phi}}_k \approx \boldsymbol{\Phi}_k \tag{23.92a}$$

$$= \boldsymbol{P}_{k|k-1}^{-1} + \boldsymbol{H}_k^T \boldsymbol{R}_k^{-1} \boldsymbol{H}_k \tag{23.92b}$$

$$= \boldsymbol{P}_{k|k}^{-1}. \tag{23.92c}$$

The trajectory optimization for moving sensors is constrained by several factors including sensor platform speeds, turn-rates, and distances between sensor waypoints. It is also important to observe that the optimal sensor placement results that were derived in the previous section are not directly applicable to dynamically changing sensor-target geometries as the assumption of fixed target-sensor distances is no longer valid and the change in these distances should also be accounted for. Figure 23.11 shows the effect of speed and turn-rate constraints on the sensor motion. The permissible sensor waypoints at time k are given by the set $\mathcal{S}_{i,k}$ with cardinality of M, which contains uniformly spaced candidate waypoints compliant with the sensor motion constraints. The distance between the permissible waypoints at time k and the sensor location at $k-1$, $\boldsymbol{s}_{i,k-1}$, is given by $s = \| \boldsymbol{s}_{i,k} - \boldsymbol{s}_{i,k-1} \| = vT$, $i = 1, \dots, N$, where v is the sensor speed, which is assumed constant with no loss of generality. The maximum turn-rate angle $\vartheta_{\max}$ reflects the maneuvering capability of the sensors.

A block diagram of the EKF, incorporating sensor trajectory optimization, is shown in Figure 23.12. At each EKF recursion, the Gaussian prior available from the previous recursion $\mathcal{N}(\boldsymbol{x}_{k|k-1}, \boldsymbol{P}_{k|k-1})$ is utilized to determine optimal sensor waypoints subject to the constraints $\mathcal{S}_{i,k}$, $i = 1, 2, \dots, N$.

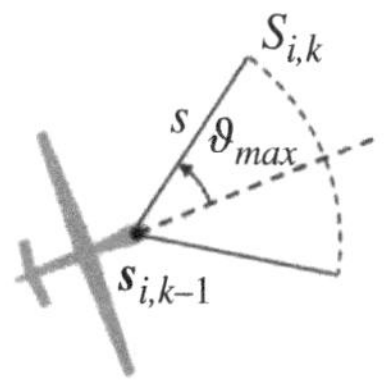

Figure 23.11 Sensor motion constraints where $\vartheta_{\max}$ is the maximum turnrate and s is the distance between waypoints.

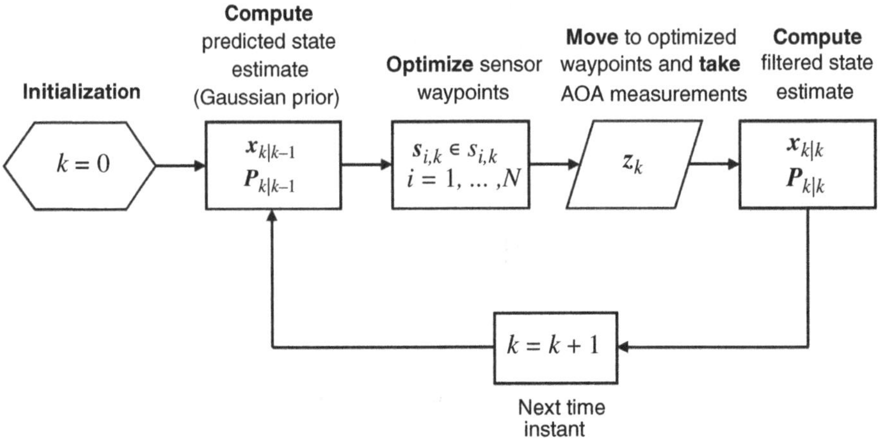

Figure 23.12 Bayesian waypoint optimization embedded into EKF.

23.7.2 Optimal Sensor Control

Although the A-optimality and D-optimality criteria yield identical optimal sensor placement results for 2D AOA localization as established in Section 23.4, they do not generate identical sensor trajectories for moving sensors when motion constraints are imposed.

As the approximate BFIM at time k is $\tilde{\boldsymbol{\Phi}}_k = \boldsymbol{P}_{k|k}^{-1}$ (see (23.92)), $\boldsymbol{P}_{k|k}$ gives the approximate BCRLB for the state vector estimate $\boldsymbol{x}_{k|k}$ at time k. Remaining focused on the optimization of target localization performance, the BCRLB for the target location estimate $\boldsymbol{\mu}_{k|k} = [x_{k|k}, y_{k|k}]^T$ is approximately given by

$$\boldsymbol{P}_{l,k|k} = \begin{bmatrix} p_{k|k}(1,1) & p_{k|k}(1,3) \\ p_{k|k}(3,1) & p_{k|k}(3,3) \end{bmatrix} \tag{23.93}$$

where $p_{k|k}(i, j)$ is the (i, j)th entry of $\boldsymbol{P}_{k|k}$. It is now straightforward to devise optimal sensor control algorithms based on the A- and D-optimality criteria in (23.32). A-optimal sensor waypoints are obtained from

$$\min_{\substack{\boldsymbol{s}_{i,k} \in \mathcal{S}_{i,k},\\ i=1,\dots,N}} \; p_{k|k}(1,1) + p_{k|k}(3,3), \quad k = 0, 1, \dots \tag{23.94}$$

where $p_{k|k}(1,1) + p_{k|k}(3,3) = \mathrm{tr}\boldsymbol{P}_{l,k|k}$, and D-optimal sensor waypoints are

$$\min_{\substack{\boldsymbol{s}_{i,k} \in \mathcal{S}_{i,k},\\ i=1,\dots,N}} \; p_{k|k}(1,1)p_{k|k}(3,3) - p_{k|k}^2(1,3), \quad k = 0, 1, \dots \tag{23.95}$$

where $p_{k|k}(1,1)p_{k|k}(3,3) - p_{k|k}^2(1,3) = |\boldsymbol{P}_{l,k|k}|$.

The optimization problems (23.94) and (23.95) can become computationally demanding or even prohibitive if the size of the search space for optimal sensor waypoints, M^N, is large. Alternative optimization methods based on approximate gradient-descent optimization using first-order finite-difference approximation of gradients may help alleviate the computationally complexity.

In single-sensor target tracking applications, projection methods can be devised to simplify the search for optimal sensor waypoints by leveraging the fact that sensor placement closer to the minor axis of the Gaussian prior yields better localization performance [12]. Sensor trajectory constraints, such as no-fly zones, maximum inter-sensor distances and minimum target clearance, are easily implemented by modifying the objective function [9]. Non-myopic (multi-step) trajectory optimization methods have been proposed in [27, 28]. Optimal sensor control for multitarget tracking is addressed in [29].

23.7.2.1 Example

A nearly constant-velocity target is considered with acceleration noise parameters $q_x = q_y = 10^{-10}$ km^2/s^4. The initial Gaussian prior for the target location at $k = 0$ is $[x_{0|-1}, y_{0|-1}]^T \sim \mathcal{N}(\boldsymbol{\mu}_{0|-1}, \boldsymbol{P}_{l,0|-1})$ with

$$\boldsymbol{\mu}_{0|-1} = \begin{bmatrix} 0 \\ 0 \end{bmatrix} \text{km} \quad \text{and} \quad \boldsymbol{P}_{l,0|-1} = \begin{bmatrix} 20 & 9 \\ 9 & 10 \end{bmatrix} \text{km}^2. \tag{23.96}$$

The initial Gaussian prior for the target velocity is $[\dot{x}_{0|-1}, \dot{y}_{0|-1}]^T \sim \mathcal{N}(\boldsymbol{v}_{0|-1}, \boldsymbol{P}_{v,0|-1})$ with

$$\boldsymbol{v}_{0|-1} = \begin{bmatrix} 0.0111 \\ 0 \end{bmatrix} \text{km/s} \quad \text{and} \quad \boldsymbol{P}_{v,0|-1} = \begin{bmatrix} 10^{-6} & 0 \\ 0 & 10^{-6} \end{bmatrix} \text{km}^2/\text{s}^2 \tag{23.97}$$

which corresponds to an average initial target speed of $\| \boldsymbol{v}_{0|-1} \| = 40$ km/h. Thus, the EKF is initialized by

$$\boldsymbol{x}_{0|-1} = \begin{bmatrix} 0 \\ 0.0111 \\ 0 \\ 0 \end{bmatrix}, \quad \boldsymbol{P}_{0|-1} = \begin{bmatrix} 20 & 0 & 9 & 0 \\ 0 & 10^{-6} & 0 & 0 \\ 9 & 0 & 10 & 0 \\ 0 & 0 & 0 & 10^{-6} \end{bmatrix}. \tag{23.98}$$

First, the target is assumed to be tracked by a single sensor ($N = 1$). The sensor moves with a constant speed of $v = 90$ km/h and has the maximum turnrate of 2 deg/s. The initial sensor location is $\boldsymbol{s}_{1,0} = [-22.98, -19.28]^T$ km, which is 30 km away from $\boldsymbol{\mu}_{0|-1}$, and the initial sensor velocity is $[0.025, 0]^T$ km/s. The time interval between EKF state updates is $T = 10$ s. The sensor waypoints are therefore separated by $s = vT = 0.025 \times 10 = 0.25$ km with $\vartheta_{\max} = 2 \times 10 = 20$ deg. The set of candidate waypoints $\mathcal{S}_{1,k}$ is comprised of $M = 10$ uniformly spaced sensor locations spanning the maximum turnrate around the current sensor heading. The AOA measurement noise is $\sigma_{1,k} = 2°$.

The optimal sensor trajectories computed using the A-optimality and D-optimality criteria (refer to (23.94) and (23.95)) and the average root-mean-square error (RMSE) for target location estimates are shown in Figure 23.13. Figure 23.13a and b shows the sensor and target trajectories for a single noise realization, and Figure 23.13c presents an RMSE comparison of target location estimates obtained from 1000 Monte Carlo simulation runs. The D-optimality criterion is seen to produce a better target tracking performance.

Next, a two-sensor target tracking problem ($N = 2$) is considered. The initial sensor locations are $\boldsymbol{s}_{1,0} = [-93.97, -34.20]^T$ and $\boldsymbol{s}_{2,0} = [-93.97, 34.20]^T$ km, which are 100 km away from $\boldsymbol{\mu}_{0|-1}$.

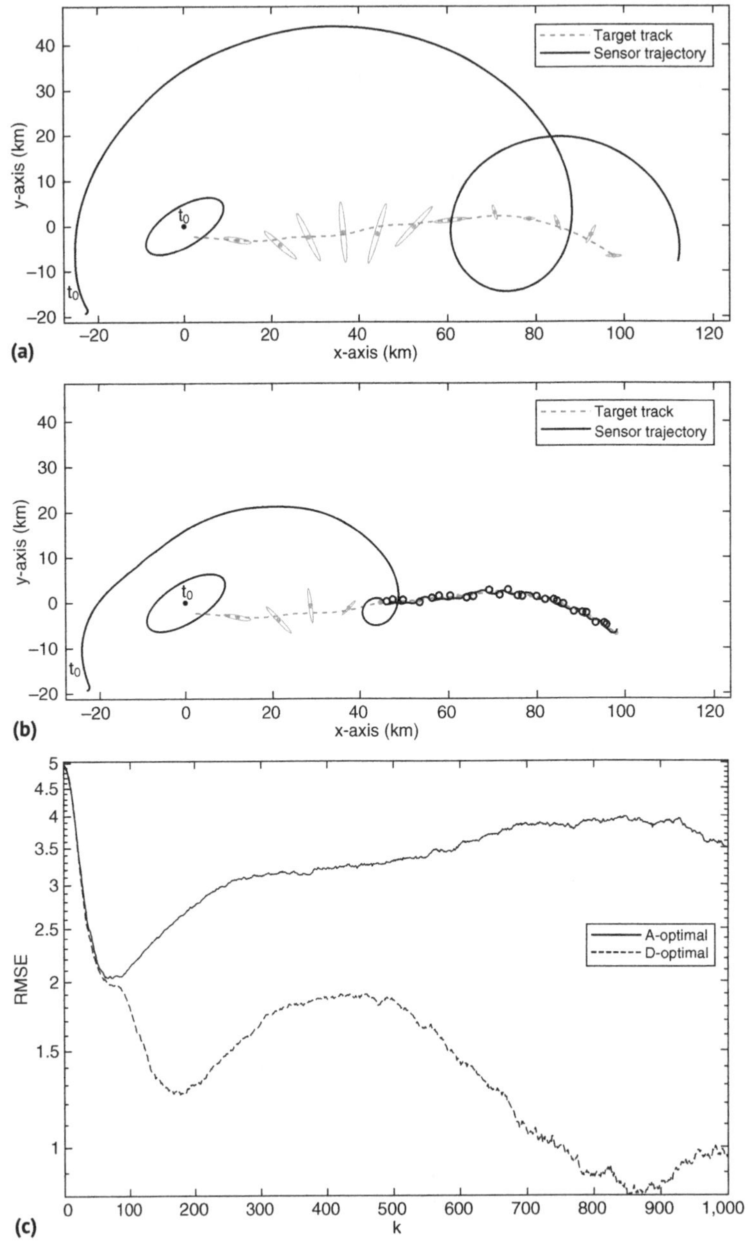

Figure 23.13 Optimal sensor trajectories for AOA target tracking ($N = 1$): (a) A-optimal sensor trajectory, (b) D-optimal sensor trajectory, and (c) average RMSE for target location estimates. Initial target and sensor locations are labelled t_0. 2-σ error ellipses at initialization (black ellipse) and every 100 recursions afterwards (gray ellipses) are also shown.

The initial sensor velocities are $[0.025, 0]^T$ km/s. Both sensors have the same AOA noise $\sigma_{1,k} = \sigma_{2,k} = 2°$. Figure 23.14 shows the optimal sensor trajectories and a comparison of average RMSE for target location estimates. The D-optimality criterion produces a better target tracking

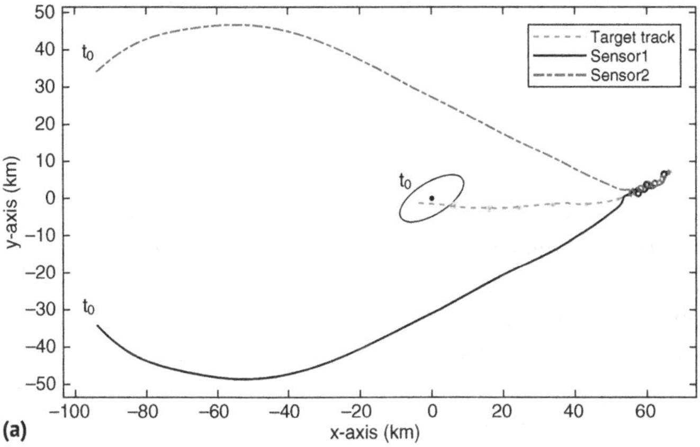

(a)

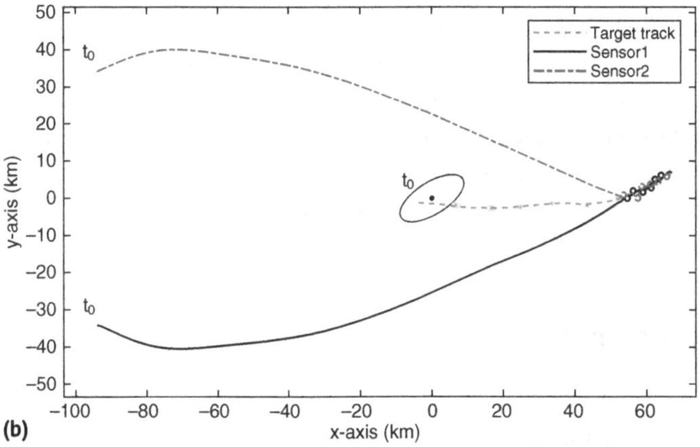

(b)

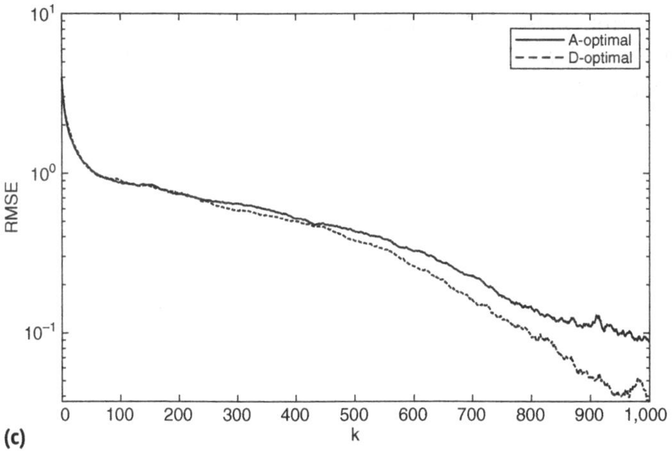

(c)

Figure 23.14 Optimal sensor trajectories for AOA target tracking ($N = 2$): (a) A-optimal sensor trajectories, (b) D-optimal sensor trajectories, and (c) average RMSE for target location estimates. Initial target and sensor locations are labelled t_0. 2-σ error ellipses at initialization (black ellipse) and every 100 recursions afterwards (gray ellipses) are also shown.

performance for $N = 2$ as well. Overall, the RMSE performance is much better with two sensors than a single sensor, as expected.

23.8 Conclusions

This chapter has presented a comprehensive study of optimal sensor placement for two-dimensional (2D) AOA target localization and tracking with Gaussian priors. The A-optimality and D-optimality criteria were considered to establish information-theoretic optimal sensor placement results based on the BFIM. An approximate expression for the BFIM was adopted to simplify its computation. For the 2D AOA target localization problem, the A-optimality and D-optimality criteria were shown to be equivalent. AOA localization scenarios were classified into three cases; viz., far-field target, near-field target, and intermediate-range target. The conditions under which these cases prevail were established with reference to the eigenvalues of the Gaussian prior covariance, target ranges and measurement noise variances. For the far-field target case, the optimal sensor placement was simply shown to be a collinear arrangement for all the sensors with the minor axis of the error ellipse of the Gaussian prior covariance. Optimal sensor trajectories for multi-sensor maneuvering target tracking were derived and illustrated using the A-optimality and D-optimality criteria. The recursive BFIM for EKF state estimates was approximated to facilitate the development of A-optimal and D-optimal sensor waypoint optimization algorithms. All the analytical results were verified with numerical examples.

23.A Validity of BFIM Approximation

A natural question to ask is "when is (23.27) a good approximation?" Figure 23.15 illustrates how approximating the true target location $\boldsymbol{x}$ with the mean of the Gaussian prior $\boldsymbol{x}_0$ affects the AOA angle. Intuitively, if the size of the error ellipse of $\boldsymbol{P}_0$ is small compared with the distance $d_i(\boldsymbol{x}_0)$, $\theta_i(\boldsymbol{x}_0)$ will not significantly deviate from $\theta_i(\boldsymbol{x})$, therefore justifying (23.27) [12]. To show this more rigorously, rewrite the left-hand side of (23.27) as

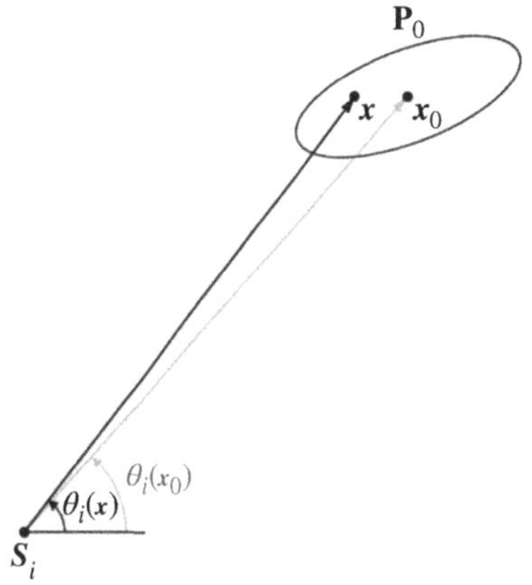

Figure 23.15 Approximating the random target location $\boldsymbol{x}$ with its mean $\boldsymbol{x}_0$.

$$E_{\boldsymbol{x}}\left\{\sum_{i=1}^{N}\frac{1}{\sigma_i^2 d_i^2(\boldsymbol{x})}\boldsymbol{u}_i(\boldsymbol{x})\boldsymbol{u}_i^T(\boldsymbol{x})\right\}=\sum_{i=1}^{N}\frac{1}{\sigma_i^2}\begin{bmatrix} E\left\{\frac{\sin^2\theta_i(\boldsymbol{x})}{d_i^2(\boldsymbol{x})}\right\} & -\frac{1}{2}E\left\{\frac{\sin 2\theta_i(\boldsymbol{x})}{d_i^2(\boldsymbol{x})}\right\} \\ -\frac{1}{2}E\left\{\frac{\sin 2\theta_i(\boldsymbol{x})}{d_i^2(\boldsymbol{x})}\right\} & E\left\{\frac{\cos^2\theta_i(\boldsymbol{x})}{d_i^2(\boldsymbol{x})}\right\} \end{bmatrix}. \tag{23.99}$$

Letting $\boldsymbol{x} = \boldsymbol{x}_0 + \boldsymbol{\delta}$ where $\boldsymbol{\delta} = [\delta_x, \delta_y]^T \sim \mathcal{N}(\boldsymbol{0}, \boldsymbol{P}_0)$ and $\boldsymbol{x}_0 = [x_0, y_0]^T$, and assuming a sufficiently small error ellipse for $\boldsymbol{P}_0$ compared with the target range $d_i(\boldsymbol{x})$, the entries of the matrix on the right-hand side of (23.99) can be approximated as

$$E\left\{\frac{\sin^2\theta_i(\boldsymbol{x})}{d_i^2(\boldsymbol{x})}\right\} \approx \frac{E\{\sin^2\theta_i(\boldsymbol{x})\}}{E\{d_i^2(\boldsymbol{x})\}} \tag{23.100a}$$

$$E\left\{\frac{\cos^2\theta_i(\boldsymbol{x})}{d_i^2(\boldsymbol{x})}\right\} \approx \frac{E\{\cos^2\theta_i(\boldsymbol{x})\}}{E\{d_i^2(\boldsymbol{x})\}} \tag{23.100b}$$

$$E\left\{\frac{\sin 2\theta_i(\boldsymbol{x})}{d_i^2(\boldsymbol{x})}\right\} \approx \frac{E\{\sin 2\theta_i(\boldsymbol{x})\}}{E\{d_i^2(\boldsymbol{x})\}}. \tag{23.100c}$$

On the right-hand side of (23.100a), $E\{\sin^2\theta_i(\boldsymbol{x})\}$ is approximately given by

$$E\{\sin^2\theta_i(\boldsymbol{x})\} = E\left\{\frac{(y_0+\delta_y-s_{y,i})^2}{(x_0+\delta_x-s_{x,i})^2+(y_0+\delta_y-s_{y,i})^2}\right\} \tag{23.101a}$$

$$\approx \frac{E\{(y_0+\delta_y-s_{y,i})^2\}}{E\{(x_0+\delta_x-s_{x,i})^2+(y_0+\delta_y-s_{y,i})^2\}} \tag{23.101b}$$

$$\approx \frac{(y_0-s_{y,i})^2+E\{\delta_y^2\}}{d_i^2(\boldsymbol{x}_0)+\operatorname{tr}\boldsymbol{P}_0} \tag{23.101c}$$

$$\approx \frac{(y_0-s_{y,i})^2}{d_i^2(\boldsymbol{x}_0)} \tag{23.101d}$$

$$\approx \sin^2\theta_i(\boldsymbol{x}_0), \tag{23.101e}$$

where the following assumption is made

$$d_i^2(\boldsymbol{x}_0) \gg \operatorname{tr}\boldsymbol{P}_0, \quad \operatorname{tr}\boldsymbol{P}_0 = E\{\delta_x^2\} + E\{\delta_y^2\} \tag{23.102}$$

that is, the Gaussian prior has a small error ellipse compared with the target range, which is the same assumption made in (23.100).

Furthermore, under the assumption of (23.102), in (23.100) we have

$$E\{\cos^2\theta_i(\boldsymbol{x})\} = E\left\{\frac{(x_0+\delta_x-s_{x,i})^2}{(x_0+\delta_x-s_{x,i})^2+(y_0+\delta_y-s_{y,i})^2}\right\} \tag{23.103a}$$

$$\approx \frac{(x_0-s_{x,i})^2}{d_i^2(\boldsymbol{x}_0)} \tag{23.103b}$$

$$\approx \cos^2\theta_i(\boldsymbol{x}_0) \tag{23.103c}$$

$$E\{\sin 2\theta_i(\boldsymbol{x})\} = 2E\left\{\frac{(x_0+\delta_x-s_{x,i})(y_0+\delta_y-s_{y,i})}{(x_0+\delta_x-s_{x,i})^2+(y_0+\delta_y-s_{y,i})^2}\right\} \tag{23.104a}$$

$$\approx 2\frac{(x_0-s_{x,i})(y_0-s_{y,i})}{d_i^2(\boldsymbol{x}_0)} \tag{23.104b}$$

$$\approx 2\sin\theta_i(\boldsymbol{x}_0)\cos\theta_i(\boldsymbol{x}_0) = \sin 2\theta_i(\boldsymbol{x}_0) \tag{23.104c}$$

$$\frac{1}{E\{d_i^2(\boldsymbol{x})\}} = \frac{1}{E\{(x_0+\delta_x-s_{x,i})^2+(y_0+\delta_y-s_{y,i})^2\}} \tag{23.105a}$$

$$= \frac{1}{d_i^2(\boldsymbol{x}_0)+\operatorname{tr}\boldsymbol{P}_0} \tag{23.105b}$$

$$\approx \frac{1}{d_i^2(\boldsymbol{x}_0)}. \tag{23.105c}$$

Substituting (23.101e), (23.103c), (23.104c) and (23.105c) into (23.100) results in the approximation in (23.27), which is valid if (23.102) is satisfied. In most practical target localization problems, (23.102) applies as the target is usually located within a small region away from the sensors.

References

1 S. C. Nardone and V. J. Aidala. Observability criteria for bearings-only target motion analysis. *IEEE Transactions on Aerospace and Electronic Systems*, 17(2):162–166, 1981. https://doi.org/10.1109/TAES.1981.309141.

2 S. C. Nardone, A. G. Lindgren, and K. F. Gong. Fundamental properties and performance of conventional bearings-only target motion analysis. *IEEE Transactions on Automatic Control*, 29(9):775–787, 1984. https://doi.org/10.1109/TAC.1984.1103664.

3 J. M. Passerieux and D. Van Cappel. Optimal observer maneuver for bearings-only tracking. *IEEE Transactions on Aerospace and Electronic Systems*, 34(3):777–788, 1998. https://doi.org/10.1109/7.705885.

4 J.-P. Le Cadre and S. Laurent-Michel. Optimizing the receiver maneuvers for bearings-only tracking. *Automatica*, 35(4):591–606, 1999. https://doi.org/10.1016/S0005-1098(98)00183-6.

5 Y. Oshman and P. Davidson. Optimization of observer trajectories for bearings-only target localization. *IEEE Transactions on Aerospace and Electronic Systems*, 35(3):892–902, 1999. https://doi.org/10.1109/7.784059.

6 A. Farina. Target tracking with bearings-only measurements. *Signal Processing*, 78(1):61–78, 1999. https://doi.org/10.1016/S0165-1684(99)00047-X.

7 K. Dogancay and H. Hmam. Optimal angular sensor separation for AOA localization. *Signal Processing*, 88(5):1248–1260, 2008. https://doi.org/10.1016/j.sigpro.2007.11.013.

8 A. N. Bishop, B. Fidan, B. D.O. Anderson, K. Dogancay, and P. N. Pathirana. Optimality analysis of sensor-target localization geometries. *Automatica*, 46(3):479–492, 2010. https://doi.org/10.1016/j.automatica.2009.12.003.

9 K. Dogancay. UAV path planning for passive emitter localization. *IEEE Transactions on Aerospace and Electronic Systems*, 48(2):1150–1166, 2012. https://doi.org/10.1109/TAES.2012.6178054.

10 C. Yang, L. Kaplan, E. Blasch, and M. Bakich. Optimal placement of heterogeneous sensors for targets with Gaussian priors. *IEEE Transactions on Aerospace and Electronic Systems*, 49(3):1637–1653, 2013. https://doi.org/10.1109/TAES.2013.6558009.

11 M. L. Hernandez, T. Kirubarajan, and Y. Bar-Shalom. Multisensor resource deployment using posterior Cramer-Rao bounds. *IEEE Transactions on Aerospace and Electronic Systems*, 40(2): 399–416, 2004. https://doi.org/10.1109/TAES.2004.1309993.

12 K. Dogancay. Optimal geometries for AOA localization in the Bayesian sense. *Sensors*, 22(24):9802, 2022. https://doi.org/10.3390/s22249802.

13 P. Tichavsky, C.H. Muravchik, and A. Nehorai. Posterior Cramer-Rao bounds for discrete-time nonlinear filtering. *IEEE Transactions on Signal Processing*, 46(5):1386–1396, 1998. https://doi.org/10.1109/78.668800.

14 H. L. van Trees, K. L. Bell, and Z. Tian. *Detection, Estimation, and Modulation Theory: Part I - Detection, Estimation, and Filtering Theory*. Wiley, Hoboken, NJ, 2nd edition, 2013.

15 V. V. Fedorov. *Theory of Optimal Experiments*. Academic Press, New York, 1972.

16 G. C. Goodwin and R. L. Payne. *Dynamic System Identification: Experiment Design and Data Analysis*. Academic Press, New York, 1977.

17 D. Ucinski. *Optimal Measurement Methods for Distributed Parameter System Identification*. CRC Press, Boca Raton, 2004.

18 J. M. Mendel. *Lessons in Estimation Theory for Signal Processing, Communications, and Control*. Prentice Hall, Englewood Cliffs, NJ, 1995.

19 Y. Bar-Shalom, X. R. Li, and T. Kirubarajan. *Estimation with Applications to Tracking and Navigation*. Wiley, New York, 2001.

20 B. D. O. Anderson and J. B. Moore. *Optimal Filtering*. Prentice-Hall, Englewood Cliffs, NJ, 1979.

21 S. J. Julier and J. K. Uhlmann. Unscented filtering and nonlinear estimation. *Proceedings of the IEEE*, 92(3):401–422, 2004. https://doi.org/10.1109/JPROC.2003.823141.

22 I. Arasaratnam and S. Haykin. Cubature Kalman filters. *IEEE Transactions on Automatic Control*, 54(6):1254–1269, 2009. https://doi.org/10.1109/TAC.2009.2019800.

23 N. J. Gordon, D. J. Salmond, and A. F. M. Smith. Novel approach to nonlinear/non-Gaussian Bayesian state estimation. *IEE Proceedings-F*, 140(2):107–113, 1993.

24 S. M. Kay. *Fundamentals of Statistical Signal Processing: Estimation Theory*. Prentice-Hall, Upper Saddle River, NJ, 1993.

25 P. Balestra. Determinant and inverse of a sum of matrices with applications in economics and statistics. Technical Report hal-01527161, Laboratoire d'analyse et de techniques économiques (LATEC), 1978. https://ideas.repec.org/p/hal/wpaper/hal-01527161.html.

26 Y. Bar-Shalom and W. D. Blair, editors. *Multitarget-multisensor Tracking: Applications and Advances*, volume III, Artech, Norwood, MA, 2000.

27 M. L Hernandez. Optimal sensor trajectories in bearings-only tracking. In *Proceedings of the Seventh International Conference on Information Fusion*, volume 2, Stockholm, Sweden, pages 893–900, 2004.

28 M. Hernandez. Performance bounds for Target Tracking: Computationally efficient formulations and associated applications. In *Integrated Tracking, Classification, and Sensor Management*, Wiley/IEEE, Hoboken, New Jersey, 2013.

29 A. K. Gostar, A. Blair, B. Ristic, and R. Hoseinnezhad. Sensor control for multitarget tracking. In *Advanced Techniques in Tracking and Sensor Management: Theory and Applications*, Wiley/IEEE, 2025

24

Robust Power Allocation for Multitarget Tracking in Colocated MIMO Radars With Minimized Resource Consumption

Ye Yuan[1], Wei Yi[1], Reza Hoseinnezhad[2], and Pramod K. Varshney[3]

[1] *School of Electronic Engineering, University of Electronic Science and Technology of China, Chengdu, Sichuan, China*
[2] *School of Engineering, RMIT University, Melbourne, Victoria, Australia*
[3] *Department of Electrical Engineering & Computer Science, Syracuse University, Syracuse, NY, USA*

24.1 Introduction

Sensor management has emerged as a critical area of research, particularly for systems that demand efficient allocation of limited resources to fulfill multiple objectives. Among these systems, colocated multiple-input multiple-output (C-MIMO) radar has emerged as a powerful technology, providing key benefits in target detection, tracking, and resource optimization [1–4]. By leveraging colocated, i.e., closely spaced antennas and waveform diversity, a C-MIMO radar enhances target identifiability, tracking accuracy, and overall system flexibility [2, 5]. A C-MIMO radar generates multiple transmitting beams capable of performing independent tasks simultaneously, such as multitarget tracking (MTT) [6]. This multifunction capability makes it an ideal candidate for adaptive sensor management.

To efficiently execute multiple tasks, optimal sharing of finite transmit resources is required. C-MIMO radar's high degrees of freedom (DoF) in beam synthesis enable flexible resource management, optimizing task performance across multiple objectives [6]. The problem of radar resource management has been extensively studied in the literature [7–12]. A variety of schemes/solutions have been proposed based on the quality of service-based resource allocation model (Q-RAM) [10, 11, 13]. Q-RAM formulates resource allocation as an optimization problem, where each task's performance is quantified using a utility function [13]. A global objective function, representing overall mission effectiveness, combines these utility functions. By optimizing the objective function under resource constraints, efficient resource allocation solutions can be achieved. Herein, the design of task utility functions is inherently tied to specific task objectives and optimization criteria [14].

For radar resource management in multitask scenarios, two commonly pursued objectives [15] are: maximizing task performance with all available resources and minimizing resource consumption while meeting minimum performance requirements. Based on the first objective, several criteria have been proposed. For example, in MTT applications, a classic criterion is to minimize the target state estimation error associated with the worst predicted tracking performance, which is known as the MinMax criterion [6, 16, 17]. Alternatively, a scaled accuracy-based criterion based on Q-RAM has been proposed to optimize resource allocation by minimizing global cost or maximizing global task utility [14, 18].

The second objective seeks to minimize resource consumption while meeting minimum task performance requirements [7, 15, 19]. This approach is crucial for applications prioritizing resource conservation over achieving high-performance levels. Examples include radar networks powered by non-rechargeable batteries, where reducing transmitted power extends network lifetime, or in complex electromagnetic combat environments, where minimizing radiated energy ensures a low probability of intercept (LPI) [8]. This enhances the radar system's survivability. Considering this objective in the multitask case, a criterion called direct resource minimization (DRM) was proposed to directly minimize the sum of resource consumptions, such as the total radiated power of radar systems, while ensuring that the tolerable maximum tracking error (TMTE) requirements for each task are not exceeded [15]. Building on this, robust chance-constrained power allocation schemes have been proposed for multitarget localization [19] and MTT tasks using Bayesian Cramér–Rao lower bound (BCRLB) metrics [20]. These DRM-based approaches often fail when resources are insufficient to satisfy strict constraints, leading to infeasible solutions.

To combat the drawbacks of DRM, a robust power allocation (RPA) approach is proposed in this chapter to address the MTT power allocation problem in a C-MIMO radar system. Preliminary results were published in [21]; this chapter elaborates further on some aspects of our prior work in [21]. The major contributions of this work are summarized as follows:

1) *Robust Power Allocation for MTT with Minimized Resource Consumption:* The proposed RPA approach applies the Q-RAM framework to MTT power allocation. A weighted sum of piecewise task utility functions serves as the objective function that incorporates individual MTT goals to be achieved under given power constraints. The approach incorporates tight performance constraints into the utility functions, offering flexibility in resource allocation for both sufficient and insufficient power scenarios. When resources are ample, RPA minimizes total transmit energy while meeting individual target performance objectives. When resources are scarce, it maximizes task utility functions to ensure robust system performance.
2) *General Constrained Polynomial Approximation to Task Utility Functions:* To improve numerical tractability, a constrained polynomial method is proposed to approximate task utility functions, which are often non-differentiable. Approximation of these functions with differentiable polynomials enables the use of standard optimization solvers. This method is applied to RPA, converting the original piecewise utility functions into differentiable approximations.
3) *Theoretical Proof of BCRLB Monotonicity:* While previous works assume that increasing transmit power reduces the trace of BCRLB, this work provides a rigorous theoretical proof of this monotonic relationship. This result strengthens the theoretical foundation for power allocation problems and aids in designing task utility functions.
4) *Iterative Parallel Search Algorithm for RPA:* The polynomially approximated RPA problem is proven to be non-convex, but the monotonicity of BCRLB ensures that task utility functions are unimodal. Based on this property, an iterative parallel search algorithm with time complexity $\mathcal{O}(n \log_2 n)$ is proposed to optimize power allocation. This algorithm prioritizes targets with the lowest utility values, ensuring robust solutions under limited resources and efficient termination when resources are sufficient.

The rest of this chapter is organized as follows: The system model of the C-MIMO radar along with the MTT model are introduced in Section 24.2. Cognitive MTT, which is a sensing mechanism that continuously perceives the environment, learns from incoming measurements, and dynamically adjusts its resource allocation strategies to optimize tracking performance, along with its closed-form data processing approach, are developed in Section 24.3. The Q-RAM and its solutions in the context of minimized resource consumption are introduced in Section 24.4. Some experimental results are presented in Section 24.5. Section 24.6 provides some concluding remarks.

24.2 System Model for C-MIMO Radar with MTT

24.2.1 System Configuration and Capability Constraints

Consider a C-MIMO radar positioned at coordinates $(x_\mathrm{s}, y_\mathrm{s})$, tasked with simultaneously tracking Q point-like targets labeled as $1, 2, \cdots, Q$. The q-th target is initially located at (x_0^q, y_0^q) moving with velocity $(\dot{x}_0^q, \dot{y}_0^q)$. The observation interval between two contiguous tracking frames is denoted by T. At time k (equivalent to continuous time kT), target q is located at (x_k^q, y_k^q) with the velocity $(\dot{x}_k^q, \dot{y}_k^q)$. We further define the state vector of the q-th target as

$$\mathbf{x}_k^q = \left[x_k^q \dot{x}_k^q y_k^q \dot{y}_k^q\right]^\mathrm{T}, \tag{24.1}$$

where $[\cdot]^\mathrm{T}$ is the transpose operation. During each tracking frame, the C-MIMO radar operates in multibeam mode, generating Q focused beams to track Q targets. In this mode, both the total radiated power $P_{\mathtt{total}}$ and each beam's transmit power P_k^q are constrained. In radar power allocation problems, usually the upper and lower bounds for P_k^q are specified, i.e.,

$$P_{\min} \leq P_k^q \leq P_{\max}, \tag{24.2}$$

where $P_{\max}$ and $P_{\min}$ represent the upper and lower bounds on the allocated power, respectively. To simplify the problem, the following assumptions are made:

Assumption 1: The tracked targets have already been initialized and the number of targets is known in advance. This assumption can be realized by available approaches such as maximum likelihood probabilistic data association (ML-PDA) [22] or multiframe detection [23] before target tracking.

Assumption 2: The targets are well-separated and each of them moves independently in the surveillance region. Hence, there is no measurement origin uncertainty (MOU) regarding each tracked object.

24.2.2 Target Motion Model

Each target is assumed to follow a constant velocity (CV) model, with the dynamic model of the q-th target at the k-th tracking frame described by [24]

$$\mathbf{x}_k^q = \mathbf{F}\mathbf{x}_{k-1}^q + \mathbf{n}_k^q, \tag{24.3}$$

where $\mathbf{F}$ denotes the state transition matrix expressed as

$$\mathbf{F} = \mathbf{I}_2 \otimes \begin{bmatrix} 1 & T \\ 0 & 1 \end{bmatrix}, \tag{24.4}$$

in which $\otimes$ denotes the Kronecker product and $\mathbf{I}_2$ represents the identity matrix of order 2. In (24.3), $\mathbf{n}_k^q$ denotes the state noise, which is assumed to be a zero-mean Gaussian process with the covariance $\mathbf{Q}_T$ given by

$$\mathbf{Q}_T = q_t \mathbf{I}_2 \otimes \begin{bmatrix} T^3/3 & T^2/2 \\ T^2/2 & T \end{bmatrix}, \tag{24.5}$$

where q_t is used to control the amount of process noise.

24.2.3 Radar Signal and Measurement Models

24.2.3.1 Signal Model

At frame k, the baseband echo from the q-th target is represented as [6]

$$r^q(t) = h_k^q \sqrt{\alpha_k^q P_k^q} s(t - \tau_k^q) e^{-j2\pi f_k^q(\mathbf{x}_k^q)t} + \omega(t), \tag{24.6}$$

where $\omega(t)$ represents zero-mean complex white Gaussian noise in the received signal. Here, h_k^q represents the q-th target's reflectivity, and $s(t)$ is the radar's transmit signal waveform. Also, $\alpha_k^q \propto (r_k^q(\mathbf{x}_k^q))^{-4}$ and $\tau_k^q = 2r_k^q(\mathbf{x}_k^q)/c$ represent the attenuation of the signal strength due to the path loss and the time delay, respectively, in which c denotes the speed of light and

$$r_k^q(\mathbf{x}_k^q) = \sqrt{(x_s - x_k^q)^2 + (y_s - y_k^q)^2} \tag{24.7}$$

denotes the range from the radar to the q-th target. In (24.6), $f_k^q(\mathbf{x}_k^q)$ is the Doppler shift caused by target motion. It is given by

$$f_k^q(\mathbf{x}_k^q) = -\frac{2[\dot{x}_k^q(x_k^q - x_s) + \dot{y}_k^q(y_k^q - y_s)]}{\lambda_f r_k^q(\mathbf{x}_k^q)}, \tag{24.8}$$

where λ_f denotes the carrier wavelength.

24.2.3.2 Measurement Model

At each tracking frame, the C-MIMO radar observes the targets and acquires measurement data corresponding to returns for subsequent processing, wherein the measurement model is comprised of a nonlinear mapping of the state vector $\mathbf{x}_k^q$ in the presence of additive Gaussian noise. For the q-th target, the measurement is modeled as a vector given by

$$\mathbf{z}_k^q = \mathbf{h}(\mathbf{x}_k^q) + \mathbf{v}_k^q, \tag{24.9}$$

where $\mathbf{h}(\mathbf{x}_k^q)$ denotes the nonlinear mapping from the state space to the measurement space that includes range, bearing and Doppler shift, denoted by [14]

$$\mathbf{h}(\mathbf{x}_k^q) = [r_k^q(\mathbf{x}_k^q) \theta_k^q(\mathbf{x}_k^q) f_k^q(\mathbf{x}_k^q)]^{\mathrm{T}}, \tag{24.10}$$

in which $\theta_k^q(\mathbf{x}_k^q)$ is the bearing angle of the q-th target:

$$\theta_k^q(\mathbf{x}_k^q) = \text{atan2}(y_k^q - y_s, x_k^q - x_s), \tag{24.11}$$

and atan2(Y, X) denotes the four-quadrant inverse tangent function of Y and X. In (24.9), the measurement noise $\mathbf{v}_k^q$ is zero-mean Gaussian with a diagonal covariance matrix given by

$$\mathbf{R}_k^q(P_k^q) = \text{diag}\left(\sigma^2_{r_k^q}(P_k^q), \sigma^2_{\theta_k^q}(P_k^q), \sigma^2_{f_k^q}(P_k^q)\right), \tag{24.12}$$

where $\sigma^2_{r^q_k}(P^q_k)$, $\sigma^2_{\theta^q_k}(P^q_k)$ and $\sigma^2_{f^q_k}(P^q_k)$ respectively, denote the variances of range, bearing and Doppler estimates, which are negatively correlated with the power P^q_k [16].

24.3 Closed-form Data Processing for Cognitive MTT

Target tracking is a discrete-time Bayesian filtering problem that aims to sequentially estimate the state parameters $\mathbf{x}^q_k$ using historical target information and current measurement data. In this process, inaccuracies arise from noise affecting the target state and radar measurements. Uncertainty can be reduced by the mechanism of cognitive tracking.

Cognitive tracking integrates tracking data processing with resource allocation approaches. We explore the details in the following. We first outline the basic Bayesian filtering process for target tracking and then introduce the tracking performance metric. Finally, the overall operating mechanism of the cognitive MTT is presented based on the data processing and resource allocation models.

Assuming targets are well-separated and the radar operates in simultaneous multibeam mode, MTT is divided into single-target tracking sub-tasks [16]. The extended Kalman filter (EKF) [25] is used to estimate the states of tracked objects in the nonlinear Gaussian model formulated in (24.3) and (24.9). Then, the prior probability density function (PDF)$p(\mathbf{x}^q_k \mid \mathbf{z}^q_{1:k-1})$ and posterior PDF$p(\mathbf{x}^q_k \mid \mathbf{z}^q_{1:k})$ can all be characterized by Gaussian distributions:

$$\begin{aligned} p(\mathbf{x}^q_k \mid \mathbf{z}^q_{1:k-1}) &\sim \mathcal{N}\left(\mathbf{x}^q_k; \hat{\mathbf{x}}^q_{k|k-1}, \hat{\Sigma}^q_{k|k-1}\right), \\ p(\mathbf{x}^q_k \mid \mathbf{z}^q_{1:k}) &\sim \mathcal{N}(\mathbf{x}^q_k; \hat{\mathbf{x}}^q_k, \hat{\Sigma}^q_k), \end{aligned} \tag{24.13}$$

where $\mathcal{N}(\mathbf{x}; \hat{\mathbf{x}}, \hat{\Sigma})$ denotes the Gaussian distribution with argument $\mathbf{x}$, mean $\hat{\mathbf{x}}$ and covariance $\hat{\Sigma}$. For the employed EKF, the mean and covariance of the above densities can be written as

$$\hat{\mathbf{x}}^q_{k|k-1} = \mathbf{F}\hat{\mathbf{x}}^q_{k-1}, \qquad \hat{\Sigma}^q_{k|k-1} = \mathbf{Q}_T + \mathbf{F}\hat{\Sigma}^q_{k-1}\mathbf{F}^\top, \tag{24.14}$$

and

$$\begin{aligned} \hat{\mathbf{x}}^q_k &= \hat{\mathbf{x}}^q_{k|k-1} + \mathbf{K}_k(P^q_k)\left(\mathbf{z}^q_k - \mathbf{h}\left(\hat{\mathbf{x}}^q_{k|k-1}\right)\right) \triangleq \hat{\mathbf{x}}^q_k(P^q_k) \\ \hat{\Sigma}^q_k &= \hat{\Sigma}^q_{k|k-1} - \mathbf{K}_k(P^q_k)\mathbf{H}^q_k\hat{\Sigma}^q_{k|k-1} \triangleq \hat{\Sigma}^q_k(P^q_k), \end{aligned} \tag{24.15}$$

with $\mathbf{K}_k(P^q_k) = \hat{\Sigma}^q_{k|k-1}\mathbf{H}^q_k[\mathbf{S}_k(P^q_k)]^{-1}$ being the Kalman gain, and

$$\mathbf{S}_k(P^q_k) = \mathbf{H}^q_k\hat{\Sigma}^q_{k|k-1}(\mathbf{H}^q_k)^\top + \mathbf{R}^q_k(P^q_k) \tag{24.16}$$

representing the covariance of the innovation. Also, $\mathbf{H}^q_k$ denotes the local linearization of the measurement function calculated by the state prediction, i.e., $\mathbf{H}^q_k = \nabla_{\mathbf{x}^q_k}\mathbf{h}(\mathbf{x}^q_k)|_{\mathbf{x}^q_k=\hat{\mathbf{x}}^q_{k|k-1}}$, in which $\nabla_{\mathbf{x}^q_k}[\cdot]$ denotes the first-order gradient of $[\cdot]$.

The above steps can be independently implemented for the Q targets given the assumption that all targets are well-separated. For the more complicated problem of near-target tracking in the clutter environment, additional data association processing needs to be introduced, and interested readers can refer to [24] for more details.

24.3.1 Performance Bound for Single Beam Target Tracking

Since the BCRLB represents an intuitive surrogate for the estimation error and is widely adopted by current works [6, 14, 16, 26], it is utilized as the performance metric for each tracked target in this work. In general, the mean-square-error (MSE) of any unbiased estimate [27] is bounded by the BCRLB, i.e.,

$$E\left\{[\mathbf{x}_k^q - \hat{\mathbf{x}}_k^q(P_k^q)][\mathbf{x}_k^q - \hat{\mathbf{x}}_k^q(P_k^q)]^{\mathsf{T}}\right\} \geq \left[\mathbf{J}_{\mathbf{x}_k^q}(P_k^q)\right]^{-1}, \tag{24.17}$$

where $E\{\cdot\}$ is the mathematical expectation with respect to the joint density function of the states and measurements up to time k. Also,

$$\mathbf{J}_{\mathbf{x}_k^q}(P_k^q) = E\{-\Delta_{\mathbf{x}_k^q}^{\mathbf{x}_k^q} \log p(\mathbf{x}_k^q, \mathbf{z}_{1:k}^q)\}, \tag{24.18}$$

denotes the Fisher information matrix (FIM), and $\Delta_{\mathbf{x}_k^q}^{\mathbf{x}_k^q}[\cdot]$ is the second-order gradient of $[\cdot]$. For the aforementioned Gaussian models, the FIM can be written as the sum of the following two components:

$$\mathbf{J}_{\mathbf{x}_k^q}(P_k^q) \triangleq \mathbf{J}_{P,k} + \mathbf{J}_Z(P_k^q, \mathbf{x}_k^q), \tag{24.19}$$

where $\mathbf{J}_{P,k}$ corresponds to the prior information regarding the target state:

$$\mathbf{J}_{P,k} = \left[\mathbf{Q}_T + \mathbf{F}\mathbf{J}_{\mathbf{x}_{k-1}^q}^{-1}(P_{k-1}^q)\mathbf{F}^{\mathsf{T}}\right]^{-1}, \tag{24.20}$$

and $\mathbf{J}_Z(P_k^q, \mathbf{x}_k^q)$ is the component arising from the measurements:

$$\mathbf{J}_Z(P_k^q, \mathbf{x}_k^q) = E\left\{\nabla_{\mathbf{x}_k^q}^{\mathsf{T}} \mathbf{h}(\mathbf{x}_k^q)[\mathbf{R}_k^q(P_k^q)]^{-1} \nabla_{\mathbf{x}_k^q} \mathbf{h}(\mathbf{x}_k^q)\right\}. \tag{24.21}$$

In order to simplify the calculations, the statistical mean in (24.21) is usually approximated by its predicted value, i.e., by substituting $\mathbf{x}_k^q$ with $\hat{\mathbf{x}}_{k|k-1}^q$, which leads to

$$\mathbf{J}_{\hat{\mathbf{x}}_{k|k-1}^q}(P_k^q) = \mathbf{J}_{P,k} + (\mathbf{H}_k^q)^{\mathsf{T}}[\mathbf{R}_k^q(P_k^q)]^{-1}\mathbf{H}_k^q. \tag{24.22}$$

Specifically, the trace of BCRLB is adopted as the metric of tracking performance, i.e.,

$$\mathcal{C}_{\hat{\mathbf{x}}_{k|k-1}^q}(P_k^q) = \mathrm{Tr}\left\{\mathbf{J}_{\hat{\mathbf{x}}_{k|k-1}^q}^{-1}(P_k^q)\right\}. \tag{24.23}$$

24.3.2 Closed-loop Framework for Multibeam Data Processing and Resource Allocation

Examination of Eq. (24.15) reveals that both the mean and covariance of the posterior depend explicitly on the transmit power. This dependency highlights the direct influence of transmit power on the state estimation quality in MTT. Leveraging this property enables the dynamic scheduling of radar power to control and enhance MTT accuracy over time. The cognitive MTT framework for C-MIMO radar capitalizes on this principle by optimally allocating limited system resources during each frame. This strategy enhances target measurement quality and state filtering, improving the overall task utility for MTT.

Moreover, as illustrated in Figure 24.1, the cognitive MTT process operates as a closed-loop system, using posterior densities from previous frames as priors for state predictions and adaptive resource allocation, thereby ensuring robust and efficient tracking performance.

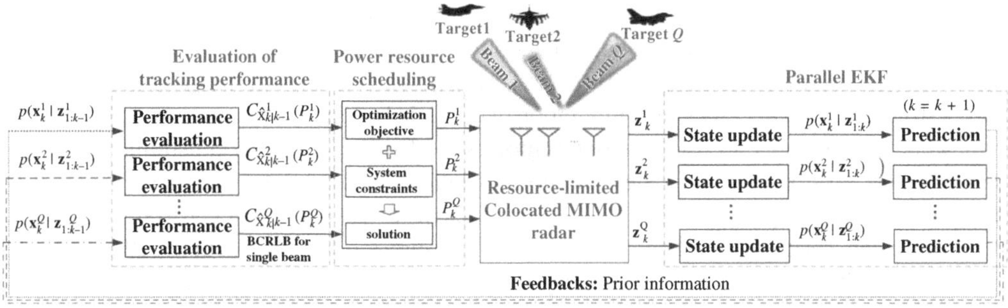

Figure 24.1 Closed-form data processing and resource allocation flow for cognitive MTT

24.3.3 Resource Allocation Solutions

In the above closed-loop framework, a key challenge lies in devising strategies for efficient and flexible resource allocation that align with specific task objectives. This has driven extensive research in radar resource allocation, with the goal of optimizing radar system performance under constrained resources. Two prominent paradigms for radar resource allocation are:

1) *Optimizing task performance by utilizing all available transmit resources.* This approach emphasizes maximizing tracking accuracy and other metrics without explicit regard for resource efficiency.
2) *Minimizing the consumed resources while ensuring minimum acceptable task performance.* This strategy focuses on conserving system resources, such as power, while meeting predefined performance thresholds.

This work adopts the second criterion and introduces two approaches to achieve resource-efficient radar operation, aligning with the goals of cognitive radar systems. In existing resource allocation methodologies based on the second criterion, a widely adopted approach is the DRM. This approach aims to minimize the total resource consumption while ensuring that the performance requirements for each task are met [15]. To formalize this, we denote the TMTE of the q-th task at the k-th frame by η_k^q. With respect to power allocation for MTT, the DRM can then be formulated as:

$$
\begin{aligned}
&\arg\min_{\mathbf{P}_k} \quad \mathbf{1}^{\mathrm{T}}\mathbf{P}_k \\
&\text{s.t.} \quad \begin{cases} \mathcal{C}_{\hat{\mathbf{x}}^q_{k|k-1}}(P_k^q) \le \eta_k^q \\ \mathbf{1}^{\mathrm{T}}\mathbf{P}_k \le P_{\mathtt{total}} \\ P_{\min} \le P_k^q \le P_{\max} \\ q = 1, 2, \cdots, Q, \end{cases}
\end{aligned}
\tag{24.24}
$$

where $\mathbf{1}$ denotes the vector whose elements are all equal to unity, and $\mathbf{P}_k = [P_k^1, \cdots, P_k^Q]^{\mathrm{T}}$. Also, $P_{\mathtt{total}}$ is the total radiated power of the radar system. Since the trace of BCRLB$\mathcal{C}_{\hat{\mathbf{x}}^q_{k|k-1}}(P_k^q)$ represents the lower bound on the estimation error for the target to be tracked, the first constraint $\mathcal{C}_{\hat{\mathbf{x}}^q_{k|k-1}}(P_k^q) \le \eta_k^q$ ensures that the tracking error for each target is not expected to exceed TMTE.

The problem outlined in Eq. (24.24) is convex, as demonstrated in [28], and it can be effectively solved using established convex optimization methods such as the gradient projection (GP) method or the interior point method (IPM) [29]. These methods take advantage of the problem's convex structure to ensure efficient computation of the global optimal solution. However, the DRM framework, as formulated in (24.24), strictly enforces the TMTEη_k^q as a hard constraint. While this ensures robust performance under ideal conditions where sufficient power is available, it lacks flexibility. When the total power budget $P_{\mathtt{total}}$ is insufficient to satisfy the first constraint of (24.24), the formulation becomes infeasible, resulting in a failure to allocate resources effectively. To address this limitation, we propose leveraging the Q-RAM framework for resource allocation in the next section.

24.4 Robust Resource Allocation for C-MIMO Radar

Unlike the traditional DRM approach, Q-RAM offers flexibility in resource allocation by incorporating utility-based performance metrics. This approach minimizes resource consumption while adapting to varying resource availability, ensuring performance under constraints. Q-RAM enhances resource allocation strategies' robustness and applicability across diverse scenarios.

24.4.1 QoS Model-based Resource Allocation

By applying Q-RAM to the resource minimization criterion, the strict requirement of TMTE is represented as a weighted sum of individual task utility functions, preventing the formulated problem from failing due to performance constraints that remain unsatisfied in case of insufficient power.

24.4.1.1 Problem Formulation

The Q-RAM for MTT power allocation is mathematically formulated as [30]

$$\begin{aligned} \arg\max_{\mathbf{P}_k} \quad & U(\mathbf{P}_k) = \sum_{q=1}^{Q} \omega^q \times u^q(\mathcal{T}^q(P_k^q \mid \mathbf{e}_k^q)) \\ \text{s.t.} \quad & \begin{cases} \mathbf{1}^{\mathrm{T}}\mathbf{P}_k \le P_{\mathtt{total}} \\ P_{\min} \le P_k^q \le P_{\max} \\ q = 1, 2, \cdots, Q, \end{cases} \end{aligned} \tag{24.25}$$

where ω^q denotes the task importance of the q-th task and is normalized as $\sum_{q=1}^{Q} \omega^q = 1$. The term $\mathcal{T}^q(P_k^q \mid \mathbf{e}_k^q)$ is the task quality function for the q-th task, which is a mapping from the allocated power P_k^q to tracking quality with an environment parameter $\mathbf{e}_k^q$, which represents a set of unconstrained/uncontrolled factors that affect the tracking utility, e.g., the state of targets to be tracked and the tracking accuracy requirement. Task QoS is derived from the task quality function. Also, $u^q(\mathcal{T}^q(P_k^q \mid \mathbf{e}_k^q))$ denotes the task utility function, which is used to describe the satisfaction associated with the achieved task quality. In (24.25), $U(\mathbf{P}_k)$ is the mission effectiveness and represents the ability of the radar system to satisfy the mission specific task quality requirements. It is defined as the weighted sum of the target task utility functions.

The primary insight from (24.25) is that the mission effectiveness is expressed as a fixed weighted summation of task utility functions, where the structure and design of the task quality function are heavily influenced by the task utility function. This dependency places the design of task utility functions at the core of the Q-RAM framework. However, designing these functions remains an open challenge [18]. Typically, such designs are guided by two critical requirements:

1) *Flexibility in Task Satisfaction:* The utility function must be highly responsive to variations in allocated power and should adapt automatically to meet task objectives. For example, it should incorporate penalty mechanisms that prioritize resource allocation to tasks where the BCRLB exceeds the TMTE.
2) *Solvability of the Optimization Problem:* The utility function must adhere to the mathematical and computational requirements of optimization solvers. For instance, gradient-based algorithms like GP and IPM require differentiable utility functions to determine a feasible search direction at each iteration.

Balancing these two requirements poses a significant challenge. Flexibility often demands composite or non-differentiable forms, whereas solvability favors simpler, differentiable formulations. Historically, task utility functions have utilized elementary functions (e.g., sums, products, trigonometric, and exponential functions) that balance these competing needs [31].

To address this trade-off, a constrained polynomial approximation approach is proposed to enhance flexibility by accommodating non-differentiable task utilities, while maintaining the solvability of optimization problems. This method approximates complex, non-differentiable task utility functions with differentiable polynomials, enabling smoother optimization while preserving adaptability.

Remark 1: Note that as pointed out in [14], determining ω^q remains an open problem. These weights are intended to quantify the relative importance of individual tasks among all tasks currently executed by the radar. Such importance can be evaluated based on multiple factors, including the threat level posed by the target, target type, and target state. The value of ω^q can be derived using methods such as analytic hierarchy processes or fuzzy logic. Since the focus of this paper is not on developing novel methods for determining task-importance weights, but rather on designing a dynamic and robust radar resource allocation scheme given such weights, we present only comparative results under several representative configurations in the subsequent simulation section.

24.4.1.2 Constrained Polynomial-based Approximation to Utility Function

The essence of this method lies in approximating non-differentiable task utility functions, $u^q(\mathcal{T}^q(P_k^q \mid \mathbf{e}_k^q))$, with a weighted sum of differentiable polynomials. This approach minimizes the MSE between the task utility function and its polynomial approximation, subject to optional constraints on the weights. The approximate task utility function is expressed as:

$$\bar{u}^q(\mathcal{T}^q(P_k^q \mid \mathbf{e}_k^q); \mathbf{a}^q) = \sum_{i=1}^{N} a_i^q \cdot f(\mathcal{T}^q(P_k^q \mid \mathbf{e}_k^q), i), \tag{24.26}$$

where N is the number of polynomials, $\mathbf{a}^q = \left[a_1^q, \cdots, a_N^q\right]^{\mathrm{T}}$ is the weight vector, and $f(\cdot, i)$ is the i-th polynomial, chosen from elementary functions to ensure continuity and differentiability within a specified range.

The approximation error is defined as:

$$\tilde{u}^q(\mathcal{T}^q(P_k^q \mid \mathbf{e}_k^q); \mathbf{a}^q) = u^q(\mathcal{T}^q(P_k^q \mid \mathbf{e}_k^q)) - \bar{u}^q(\mathcal{T}^q(P_k^q \mid \mathbf{e}_k^q); \mathbf{a}^q). \tag{24.27}$$

To optimize the approximation, the MSE is minimized within a defined range $[-K, K]$, leading to the following constrained optimization problem:

$$\mathbf{a}^{q*} = \arg\min_{\mathbf{a}^q} \int_{-K}^{K} \left[\tilde{u}^q(\mathcal{T}^q(P_k^q \mid \mathbf{e}_k^q); \mathbf{a}^q)\right]^2 d\mathcal{T}^q(P_k^q \mid \mathbf{e}_k^q) \tag{24.28}$$

subject to $g(\mathbf{a}^q) \leq 0$, where $g(\mathbf{a}^q)$ imposes constraints on the weights, and K determines the approximation range. Larger K values improve accuracy across the entire range but may reduce precision for specific points. By solving the optimization problem, the optimal weights $\mathbf{a}^{q^*}$ are obtained. Substituting these into the polynomial representation yields the desired approximate task utility function.

This method allows flexible modeling of non-differentiable utilities while preserving optimization solvability, providing a robust tool for cognitive resource allocation. Now, the key challenge is applying constrained polynomial approximation to the resource allocation problem to achieve high flexibility in utility design while maintaining problem solvability. To address this, leveraging the Q-RAM framework, we design an exact task utility function that minimizes power consumption while ensuring the desired tracking performance.

24.4.2 Application of Constrained Polynomial Approximation to Minimize Resource Consumption

24.4.2.1 Non-differentiable Task Utility Function

An exponential function is selected as the primary building block for the task utility function. For target q, it is defined using the predicted BCRLB as follows:

$$u^q(\mathcal{T}^q(P_k^q \mid \mathbf{e}_k^q)) = \begin{cases} 1 - e^{\mathcal{T}^q(P_k^q \mid \mathbf{e}_k^q)}, & \mathcal{T}^q(P_k^q \mid \mathbf{e}_k^q) > 0 \\ 1 - e^{-\mathcal{T}^q(P_k^q \mid \mathbf{e}_k^q)}, & \mathcal{T}^q(P_k^q \mid \mathbf{e}_k^q) \leq 0 \end{cases}, \tag{24.29}$$

where the task quality function is:

$$\mathcal{T}^q(P_k^q \mid \mathbf{e}_k^q) \triangleq \frac{\mathcal{C}_{\hat{\mathbf{x}}_{k|k-1}^q}(P_k^q)}{\eta_k^q} - 1, \tag{24.30}$$

and $\mathbf{e}_k^q \triangleq \{\hat{\mathbf{x}}_{k|k-1}^q, \eta_k^q\}$ contains the predicted state and the required TMTE for the BCRLB, respectively. The task utility function is symmetric about $\mathcal{T}^q(P_k^q \mid \mathbf{e}_k^q) = 0$, reaching its maximum when $\mathcal{C}_{\hat{\mathbf{x}}_{k|k-1}^q}(P_k^q) = \eta_k^q$. This ensures that any deviation in allocated power from the optimal value reduces utility. However, the function is non-differentiable at $\mathcal{T}^q(P_k^q \mid \mathbf{e}_k^q) = 0$, complicating optimization. To address this, constrained polynomial approximation is proposed.

Remark 2: Prior studies [10, 32, 33] have demonstrated the effectiveness of exponential functions in capturing the relationship between tracking performance and system resources. Additionally, the superiority of exponential functions for task utility fitting was validated in [14] in the context of maximizing MTT performance. Consequently, exponential functions are adopted here for quantifying task quality.

24.4.2.2 Constrained Polynomial-based Approximation

To approximate the non-differentiable task utility function in (24.29), we employ differentiable even power functions:

$$f(\mathcal{T}^q(P_k^q \mid \mathbf{e}_k^q), i) \triangleq -[\mathcal{T}^q(P_k^q \mid \mathbf{e}_k^q)]^{2i}. \tag{24.31}$$

Summing these components with appropriate weights gives the approximate utility function:

$$\bar{u}^q(\mathcal{T}^q(P_k^q \mid \mathbf{e}_k^q); \mathbf{a}^q) \triangleq -\sum_{i=1}^{N} a_i^q \cdot [\mathcal{T}^q(P_k^q \mid \mathbf{e}_k^q)]^{2i}. \tag{24.32}$$

where a_i^q is constrained to be nonnegative to ensure unimodality. Substituting this approximation into the optimization problem results in:

$$\begin{aligned} \mathbf{a}^{q,*} &= \arg\min_{\mathbf{a}} && \int_{-K}^{K} [\tilde{u}^q(\mathcal{T}^q(P_k^q \mid \mathbf{e}_k^q); \mathbf{a}^q)]^2 d\mathcal{T}^q(P_k^q \mid \mathbf{e}_k^q) \\ & s.t. && -\mathbf{a}^q \preceq \mathbf{0}, \end{aligned} \tag{24.33}$$

where the constraint indicates that all elements in $\mathbf{a}^q$ are nonnegative, and

$$\tilde{u}^q(\mathcal{T}^q(P_k^q \mid \mathbf{e}_k^q); \mathbf{a}^q) \triangleq 1 - e^{\mathcal{T}^q(P_k^q \mid \mathbf{e}_k^q)} + \sum_{i=1}^{N} a_i^q \cdot [\mathcal{T}^q(P_k^q \mid \mathbf{e}_k^q)]^{2i}. \tag{24.34}$$

Since the even power functions are all symmetric about $\mathcal{T}^q(P_k^q \mid \mathbf{e}_k^q) = 0$, their linear combination is also symmetric. Therefore, only considering the right half part when computing the integral in (24.33) will not affect the differentiability and the accuracy of the approximated task utility function. Therefore, we only consider $(0, K)$ when performing the integration to simplify computation. The following proposition is presented to show the convexity of (24.33):

Proposition 1: The problem in (24.33) can be reformulated as the following quadratic programming (QP) problem:

$$\begin{aligned} &\arg\min_{\mathbf{a}} && (\mathbf{a}^q)^\mathsf{T}\mathbf{A}\mathbf{a}^q + 2(\mathbf{a}^q)^\mathsf{T}(\mathbf{b} + \mathbf{c}) \\ &\text{s.t.} && \mathbf{a}^q \succeq \mathbf{0}, \end{aligned} \tag{24.35}$$

where $\mathbf{A} = [a_{ij}]_{N\times N}$, $\mathbf{b} = [b_i]_{N\times 1}$ and $\mathbf{c} = [c_i]_{N\times 1}$, in which

$$a_{ij} = \frac{K^{2i+2j+1}}{2i + 2j + 1}, \quad b_i = \frac{K^{2i+1}}{2i + 1}, \quad c_i = -e^K \sum_{j=0}^{2i} \left\{(-1)^{2i-j} \frac{(2i)!}{j!} K^j\right\} + (2i)!. \tag{24.36}$$

Proof: See Section 24.A.

Remark 3: The problem in (24.35) remains computationally efficient due to its convexity, enabling real-time parameter adjustments in dynamic scenarios. Moreover, this framework allows flexibility in choosing alternate polynomial forms if they fit the task utility function's trend.

Remark 4: Besides the proposed constrained polynomial method, subgradient methods offer an alternative for handling non-differentiable functions but often suffer from higher computational complexity and feasibility issues, especially near concave local maxima [34].

24.4.3 Mathematical Structure of the RPA Problem

When applying the approximate task utility function defined in (24.32) to Q-RAM, the RPA problem can be formulated as the problem in (24.37), in which the strict constraints are only applied to power budgets, and the constraint on the TMTE is incorporated in the objective function. As such, the failure to reach a feasible solution caused by unsatisfied performance constraints does not result in an exit from the optimization algorithm while solving (24.37), thus achieving a robust power allocation.

$$\begin{aligned} &\arg\max_{\mathbf{P}_k} \quad \bar{U}(\mathbf{P}_k) = \sum_{q=1}^{Q} \omega^q \cdot \bar{u}^q(\mathcal{T}^q(P_k^q \mid \mathbf{e}_k^q); \mathbf{a}^q) \\ &\quad s.t. \quad \begin{cases} \mathbf{1}^{\mathsf{T}}\mathbf{P}_k \leq P_{\mathtt{total}} \\ P_{\min} \leq P_k^q \leq P_{\max}, \end{cases} \end{aligned} \tag{24.37}$$

where $\bar{U}(\mathbf{P}_k)$ can be expressed in the following form by substituting (24.32) into the objective function in (24.25):

$$\bar{U}(\mathbf{P}_k) = -\sum_{q=1}^{Q}\sum_{i=1}^{N} \omega^q \cdot a_i^q \cdot \left[\frac{C_{\hat{\mathbf{x}}_{k|k-1}^q}(P_k^q)}{\eta_k^q} - 1\right]^{2i}. \tag{24.38}$$

We then define the weighted task utility function specified for target q as follows:

$$U(\omega^q, P_k^q) = \omega^q \cdot \bar{u}^q(\mathcal{T}^q(P_k^q \mid \mathbf{e}_k^q); \mathbf{a}^q). \tag{24.39}$$

The optimization problem in (24.37) is non-convex, the corresponding proof is given in Section 24.B. Based on it, the optimality of the solution for (24.37) by convex solvers cannot be guaranteed. It is well-known that the monotonicity of a function enables us to determine the minimum/maximum value. In the following, we study the optimality of the proposed RPA by discussing the monotonicity of its objective function.

24.4.3.1 Monotonicity of the RPA's Objective Function

For a function that is only related to a one-dimensional variable, its monotonicity can be determined by the sign of the first-order derivative. Here, since the objective function of RPA in (24.37) is the linear combination of Q task utility functions, it can be divided into Q independent components when calculating derivatives with respect to each corresponding transmit power. Therefore, we can discuss the monotonicity of each independent component first. To do that, we give the expression of the derivative with respect to the q-th power variable as

$$\begin{aligned} \frac{\partial \bar{U}(\mathbf{P}_k)}{\partial P_k^q} &= \frac{\partial U(\omega^q, P_k^q)}{\partial P_k^q} \\ &= -\omega^q \frac{\partial C_{\hat{\mathbf{x}}_{k|k-1}^q}(P_k^q)}{\partial P_k^q} \sum_{i=1}^{N} \frac{2ia_i^q}{\eta_k^q} [\mathcal{T}^q(P_k^q \mid \mathbf{e}_k^q)]^{2i-1}, \end{aligned} \tag{24.40}$$

where ω^q and η_k^q are fixed and non-negative. Also, due to the positive definiteness of the BCRLB matrix, the BCRLB trace is non-negative [28]. Meanwhile, in the proposed constrained polynomial method, a_i^q is also constrained to be nonnegative. Consequently, the sign of (24.40) is dependent on

$$\begin{cases} \dfrac{\partial C_{\hat{\mathbf{x}}^q_{k|k-1}}(P^q_k)}{\partial P^q_k} \\ \left[\mathcal{T}^q(P^q_k \mid \mathbf{e}^q_k)\right]^{2i-1} = \left[\dfrac{C_{\hat{\mathbf{x}}^q_{k|k-1}}(P^q_k)}{\eta^q_k} - 1\right]^{2i-1} \end{cases} \tag{24.41}$$

on the right-hand side of (24.40). It can be observed that both of these two terms are related to BCRLB, which requires us to discuss the values and the first-order derivative of BCRLB to further determine the sign of (24.40). For the BCRLB, it is intuitive that an increase of allocated resources improves tracking accuracy, and previous works have assumed that the BCRLB is a monotonically decreasing function without a theoretical proof [14, 16, 28]. In view of that, we first give the following proposition to theoretically show the monotonicity of BCRLB:

Proposition 2: The BCRLB is a monotonically decreasing function of P^q_k. That is, the first order derivative of the BCRLB is less than 0, i.e.,

$$\frac{\partial C_{\hat{\mathbf{x}}^q_{k|k-1}}(P^q_k)}{\partial P^q_k} < 0, (P^q_k > 0). \tag{24.42}$$

Proof: See Section 24.C.

According to Proposition 2, the sign of the first term in (24.41) can be determined. We then give the following proposition to determine the sign of the other term in (24.41) and further study the trend of the weighted task utility function given in (24.40):

Proposition 3: $U(\omega^q, P^q_k)$ is a unimodal function related to P^q_k, in which the only local maximum occurs at the point where P^q_k satisfies either of the two conditions $P^q_k = 0$ or $C_{\hat{\mathbf{x}}^q_{k|k-1}}(P^q_k) = \eta^q_k$.

Proof: See Section 24.D.

By combining the results of (24.40) and Propositions 2 and 3, we sketch the weighted task utility function for the q-th target in Figure 24.2 to intuitively show where the maximum value is located for two different cases of the initial BCRLB. Figure 24.2a shows the case where the initial BCRLB is greater than its TMTEη^q_k. In this case, the function value increases first and then decreases with the improvement of P^q_k. Figure 24.2b represents the case where the initial BCRLB is already lower than η^q_k, which is monotone decreasing and concave. In this case, the optimal value lies at the point $P^q_k = 0$.

Based on the special structure of the task utility function, an iterative parallel search algorithm is proposed to find the solution of the non-convex but unimodal objective function in (24.37).

24.4.4 Iterative Parallel Search-based Fast Solution Algorithm

The basic idea of the proposed algorithm is to iteratively assign the power resource in the order of increasing utility values. Specifically, the system we consider consists of Q target tracking tasks denoted by $\mathcal{T} = \{T^1, \cdots, T^Q\}$. We then gather all $U(\omega^q, P^q_k)$ for $q = 1, \cdots, Q$ as

$$\mathbf{U}_k = \{U(\omega^1, P^1_k), \cdots, U(\omega^Q, P^Q_k)\}. \tag{24.43}$$

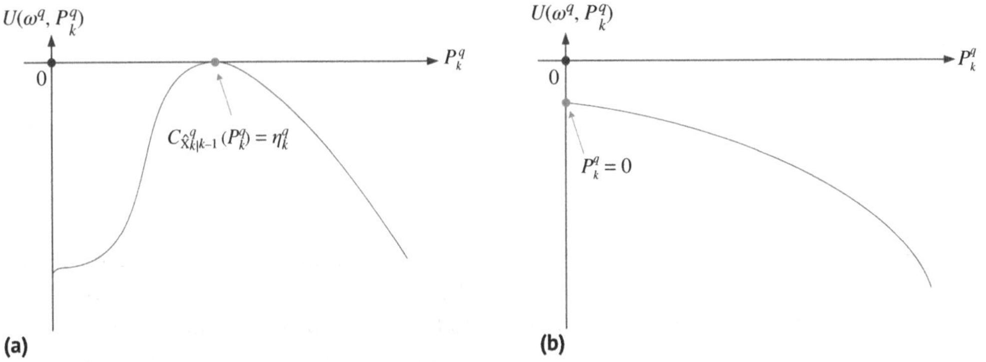

Figure 24.2 Sketches of the weighted task utility function for two different initial cases (a) when $C_{\hat{x}^q_{k|k-1}}(0) > \eta_k^q$ and (b) when $C_{\hat{x}^q_{k|k-1}}(0) \leq \eta_k^q$.

Also, the first-order derivative of $U\left(\omega^q, P_k^q\right)$ is computed by using (24.40) and the result is gathered as

$$\nabla \mathbf{U}_k = \left\{ \frac{\partial U\left(\omega^1, P_k^1\right)}{\partial P_k^1}, \cdots, \frac{\partial U\left(\omega^Q, P_k^Q\right)}{\partial P_k^Q} \right\}. \tag{24.44}$$

The detailed steps are given as follows:

Step 1 - *Select the targets with the smallest task utility:* At the i-th ($i = 1, 2, \cdots$) iteration, targets are ranked in ascending order of utility function values, and the targets with the smallest utility function value (note that the number of such targets can be greater than or equal to one) are selected from $\mathcal{T}$, in which the corresponding smallest utility value is found from $\mathbf{U}_k$ and denoted by $\mathcal{U}_{\mathrm{S},(i)}^{(0)}$. We also find the second smallest value of $\mathbf{U}_k$ and denote it as $\mathcal{U}_{\mathrm{SS},(i)}$, which will be manually updated by using the following expression if $\mathcal{U}_{\mathrm{SS},(i)} = \mathcal{U}_{\mathrm{S},(i)}^{(0)} < 0$:

$$\mathcal{U}_{\mathrm{SS},(i)} = \min\{\mathcal{U}_{\mathrm{SS},(i)} + \varepsilon, 0\}, \tag{24.45}$$

where $\varepsilon > 0$ is a predetermined step size.

Step 2 - *Iterative parallel power allocation:* After selecting the targets with the smallest task utility function value, iterative parallel power allocation is implemented for them in an iterative manner. At the j-th ($j = 1, 2, \cdots$) iteration, the power is allocated to the selected targets in parallel until the task utility function value of each selected target is increased to

$$\mathcal{U}_{\mathrm{S},(i)}^{(j)} = \mathcal{U}_{\mathrm{S},(i)}^{(j-1)} + \min\left\{\varepsilon, \mathcal{U}_{\mathrm{SS},(i)} - \mathcal{U}_{\mathrm{S},(i)}^{(j-1)}\right\}. \tag{24.46}$$

Step 2 terminates when either of the following conditions is satisfied:

1) The utility function values of all these selected targets are equal to the second smallest utility function value, i.e., $\mathcal{U}_{\mathrm{S},(i)}^{(j)} = \mathcal{U}_{\mathrm{SS},(i)}$.
2) The overall power resource is exhausted.

Then, the power allocation result after Step 2 is recorded as $\mathbf{P}_{k,(i)} = [P_{k,(i)}^1, \cdots, P_{k,(i)}^Q]^{\mathrm{T}}$. Otherwise, the next iteration (the $j+1$-th) of Step 2 is executed.

Step 3 - *Update the set of targets to be allocated:* After obtaining $\mathbf{P}_{k,(i)}$, we can compute $\nabla \mathbf{U}_k$ at the i-th iteration by substituting $\mathbf{P}_{k,(i)}$ into (24.44). We then delete the targets that satisfy their own accuracy requirements from $\mathcal{T}$, i.e.,

$$\mathcal{T} = \mathcal{T} - \mathcal{T}_{(i)}, \tag{24.47}$$

where

$$\mathcal{T}_{(i)} = \left\{ \mathcal{T}^q \left| \frac{\partial U\left(\omega^q, P^q_{k,(i)}\right)}{\partial P^q_{k,(i)}} = 0 \right. \right\}. \tag{24.48}$$

Step 4 - *Termination of the iterative parallel search:* The algorithm terminates when either of the following conditions is satisfied: (1) The set $\mathcal{T}$ is empty. (2) The overall power resource is exhausted. Then, $\mathbf{P}_{k,(i)}$ is outputted as the finalized power allocations. Otherwise, go to the Step 1 and execute the next iteration. The pseudo-code of the proposed algorithm is presented in Algorithm 1.

Algorithm 1: Iterative Parallel Search Algorithm

Input: ε, n_1, n_2 and the residual power $P_{\mathtt{total},i}$,
Output: $\mathbf{P}_{k,\mathrm{opt}}$
Set the initial power allocation vector as a zero vector, i.e., $\mathbf{P}_{k,(0)} = \mathbf{0}$;
for $i = 1 : n_1$ **do**
    Find $\mathcal{U}^{(0)}_{\mathrm{S},(i)}$ and $\mathcal{U}_{\mathrm{SS},(i)}$ based on **Step 1**;
    for $j = 1 : n_2$ **do**
        Calculate $\mathcal{U}^{(j)}_{\mathrm{S},(i)}$ based on **Step 2**;
        if $\mathcal{U}^{(j)}_{\mathrm{S},(i)} = \mathcal{U}_{\mathrm{SS},(i)}$ *or* $P_{\mathtt{total},i} = 0$ **then**
            Record $\mathbf{P}_{k,(i)}$;
            Break;
        end
    end
    Update $\mathcal{T}$ based on **Step 3**;
    if $\mathcal{T} = \emptyset$ *or* $P_{\mathtt{total},i} = 0$ **then**
        Break;
    end
end
Set $\mathbf{P}_{k,\mathrm{opt}} = \mathbf{P}_{k,(i)}$

24.4.5 Optimality and Convergence Analysis for the Proposed Algorithm

In the proposed iterative parallel search algorithm, it is clear that the ε-optimal solution can be found in the case of sufficient power due to the unimodal structure of the task utility function. While in the case of insufficient power, it can be easily seen that the power resource will be preferentially allocated to the targets with smaller task utility function values, and the algorithm produces an acceptable solution in the sense of maximizing the objective function at this time.

As illustrated in Algorithm 1, the proposed algorithm will iterate at most $n_1 = \lceil 0 - \mathcal{U}^{(0)}_{\mathrm{S},(1)}/\varepsilon \rceil$ times to get the power allocation result, in which $\lceil \cdot \rceil$ denotes the ceiling operation. Also, during the execution of the proposed algorithm, Step 2, the iterative parallel power allocation actually performs a search in an ordered sequence (note that the utility function within this domain is monotonically increasing as illustrated in Figure 24.2(a)). If we employ the binary search method here, then it will be executed $n_2 = \lceil \log_2[(\mathcal{U}_{SS,(i)} - \mathcal{U}^{(0)}_{S,(i)})/\varepsilon] \rceil$ times to get the feasible power allocation. Hence, the computational complexity as well as the precision of the proposed algorithm are mainly dependent on the step-size and the interval between $\mathcal{U}^{(0)}_{S,(1)}$ and 0, and the overall time complexity of the proposed algorithm is polynomial, i.e., executing at most $\mathcal{O}(n_1 \times n_2) = \mathcal{O}(\lceil -\mathcal{U}^{(0)}_{S,(1)}/\varepsilon \rceil \times \lceil \log_2[(\mathcal{U}_{SS,(i)} - \mathcal{U}^{(0)}_{S,(i)})/\varepsilon] \rceil) = \mathcal{O}(1/\varepsilon \times \log_2 1/\varepsilon)$ times.

Remark 5: Note that the proposed RPA in (24.37) is a general framework that is applicable to numerous resource management problems. For example, although the radar revisit time of the phased array radar [35] can affect the statistical property of state noise $\mathbf{n}_k^q$ and further influence the information matrix $J_{P,k}$, the processing framework, the performance metric, and the polynomial approximation of the RPA can be extended to this case.

24.5 Numerical Simulations

Numerical simulation examples are conducted to evaluate the effectiveness of the proposed RPA strategy. The RPA approach is compared to the DRM method, which achieves optimal power allocation under sufficient resources, to validate its superiority. RPA's flexibility is demonstrated in low-power scenarios with varying task importance values.

24.5.1 Parameter Setting

As illustrated in Figure 24.3, the radar is positioned at coordinates (0, 0). To discriminate the desired accuracy requirement for targets in different range cells, the whole region of interest is divided into three parts, i.e., Areas 1, 2 and 3, depending on their distance from the radar. Their corresponding TMTEs are set as 150 m, 300 m, and 500 m, respectively.

A sequence of 40 frames with the time interval $T = 2$ s is observed in the simulations. The process noise parameter $q_t = 2.56 \times 10^4$, and the total transmit power $P_{\mathtt{total}} = 50$ dBw. The allocated power

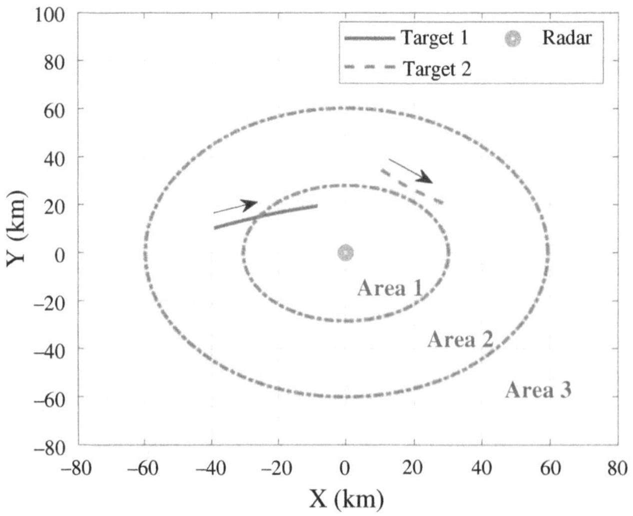

Figure 24.3 Scenario 1 with two targets to be tracked.

ranges from $P_{\min} = 10^{-3} \times P_{\texttt{total}}$ to $P_{\max} = P_{\texttt{total}}$. In the following, tracking root MSE (RMSE) is defined as the root of the MSEs after $M_c = 200$ Monte Carlo averaging given by

$$\texttt{RMSE}_k^q = \sqrt{\frac{1}{M_c}\sum_{m=1}^{M_c}\left((\hat{\mathbf{x}}_{k,m}^q - \mathbf{x}_k^q)^2 + (\hat{\mathbf{y}}_{k,m}^q - \mathbf{y}_k^q)^2\right)}, \tag{24.49}$$

where $(\hat{\mathbf{x}}_{k,m}^q, \hat{\mathbf{y}}_{k,m}^q)$ denotes the state estimation of target q at the m-th trial. Also, the corresponding BCRLB can be written as $\texttt{CRB}_k^q = \sqrt{C_{\hat{\mathbf{x}}_{k|k-1}^q}(P_k^q)}$. At time $k = 0$, the process noise covariance is defined as the inverse of the initial FIM, i.e., $\texttt{CRB}_0^q = \sqrt{\texttt{Tr}(\mathbf{Q}_T)} = 489$ m.

24.5.2 Approximation of the Task Utility Function

Figure 24.4 illustrates the exact task utility function and its polynomial approximations, computed using CVX tools [36] with parameters $K = 6, 11$ and $N = 1, 3, 5$. Figure 24.4 (a) and (b) show that increasing N improves approximation accuracy as fitted curves approach the exact curve. Figure 24.4 (c) illustrates the accumulated MSEs from 0 to K, and demonstrates that a larger value of K can worsen the accumulated MSE due to accumulation of errors. Increasing N mitigates the error accumulation associated with larger K.

The value of K reflects the deviation range between the BCRLB$C_{\hat{\mathbf{x}}_{k|k-1}^q}(P_k^q)$ and the TMTE η_k^q. After K is determined, the value of N is determined by the required fitting accuracy. The fitting MSEs for different values of K and N are given in Figure 24.4 (c), and users can choose the parameters accordingly.

For $K = 11$ and $N = 5$, the normalized MSE decreases to -25dB as shown in Figure 24.4(c). The corresponding fitting result is also shown as the dash-dotted line in Figure 24.4(b). It can be seen that the approximate task utility function keeps a high precision for fitting the exact function within $[-15, 15]$ even when we only constrain K to be equal to 11. This indicates that the fitting MSE is acceptable when the possible values of the BCRLB are within 16 times its required estimation error, or in other words, when the task quality function $\mathcal{T}^q(P_k^q \mid \mathbf{e}_k^q) = C_{\hat{\mathbf{x}}_{k|k-1}^q}(P_k^q)/\eta_k^q - 1 < 15$. This setting is generally sufficient to cover the possible range of $C_{\hat{\mathbf{x}}_{k|k-1}^q}(P_k^q)/\eta_k^q$ and satisfy the need for fitting accuracy. From these findings, K and N are set as 11 and 5, respectively, and the corresponding power values of the polynomials used are $[246810]^{\mathrm{T}}$. The resulting weights are given as follows by solving the optimization problem in (24.35):

$$\mathbf{a}^q = \left[3.3619\ 4.5094 \times 10^{-2}\ 6.3269 \times 10^{-4}\ 1.0622 \times 10^{-5}\ 2.0430 \times 10^{-6}\right]^{\mathrm{T}}. \tag{24.50}$$

24.5.3 Sufficient Power Scenario

In this scenario, we consider two targets to be tracked, named Targets 1 and 2, in which Target 1 moves from Area 2 to Area 1 at the 18-th frame while Target 2 moves in Area 2 for the entire period of 40 frames. The motion parameters of these two targets are given in Table 24.1. To ensure that the DRM is applicable, the transmit power budget is set large enough (50 dBw as previously mentioned) so that the total amount of allocated power would not all be used up. In addition, equal task importance is considered for these two targets $\mathbf{w} = [0.50.5]^{\mathrm{T}}$. The tracking deployment is shown in Figure 24.3.

The tracking BCRLBs and RMSEs of the two targets achieved by DRM and the proposed RPA are illustrated in Figure 24.5(a) and (b), respectively. The three gray dotted lines of Requirements

Table 24.1 Motion parameters of targets in Scenario 1.

	Initial locations (km)	Initial velocities (m/s)
Target 1	(−40, 10)	(380, 120)
Target 2	(10, 35)	(200, −200)

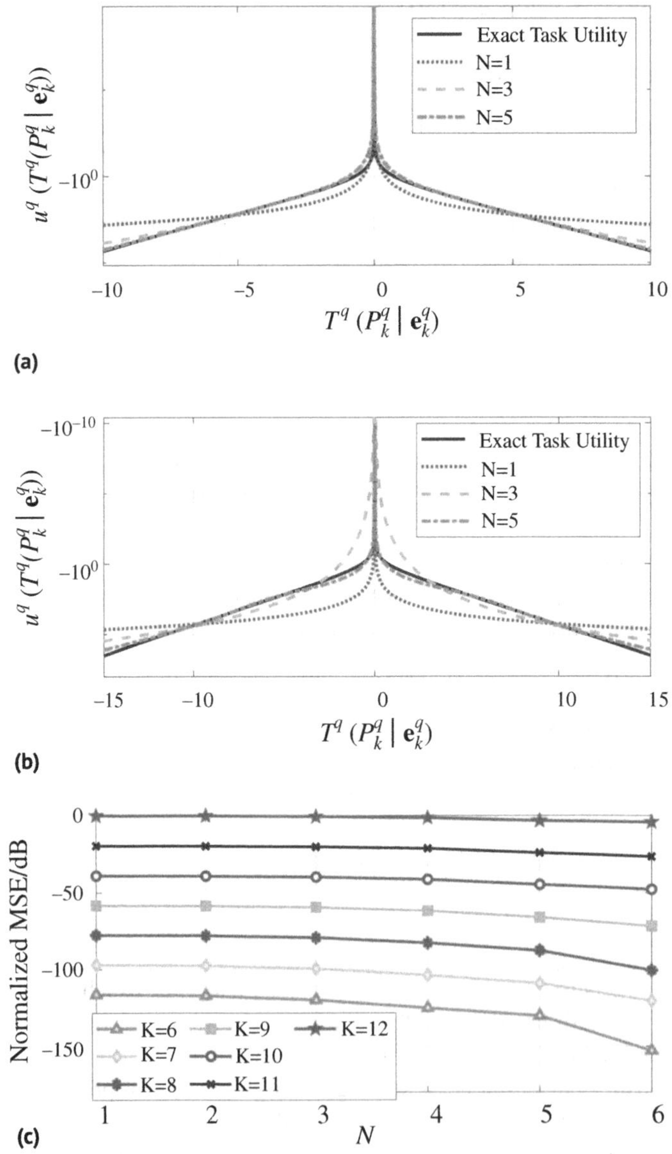

Figure 24.4 The plots of the exact task utility function and its approximate forms for different values of K and N. (a) K = 6. (b) K = 11. (c) MSE between exact task utility function and its approximations.

denote the TMTEs of 150 m, 300 m, and 500 m for the three areas. The results presented in Figure 24.5 (a) and (b) show that both with DRM and the proposed RPA method, both targets' RMSEs gradually converge to their corresponding BCRLBs. Furthermore, the targets' BCRLBs approach their respective TMTE requirements.

To compare the task utility of the proposed RPA with the DRM, the percentages of consumed power resource during the period of 40 frames are given in Figure 24.6. It is clear that both DRM and the RPA consume nearly the same amount of transmit power in each tracking frame. In addition, when Target 1 transits from Area 2 to Area 1 in the 18-th frame, a sudden jump occurs in both curves due to the increase of accuracy requirement for Target 1. The DRM and our proposed RPA strategy achieve almost the same tracking performance as well as consume the same amount of power resource when sufficient power resource is available.

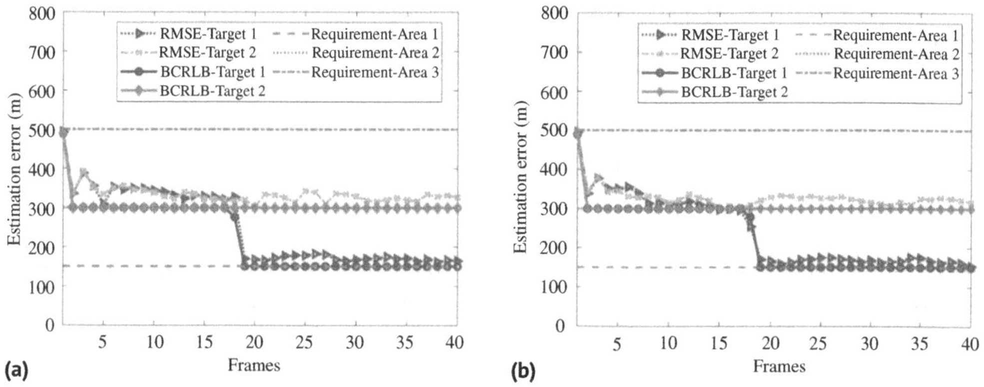

Figure 24.5 Tracking RMSEs and BCRLBs of the two targets to be tracked (Scenario 1). (a) DRM. (b) Proposed RPA.

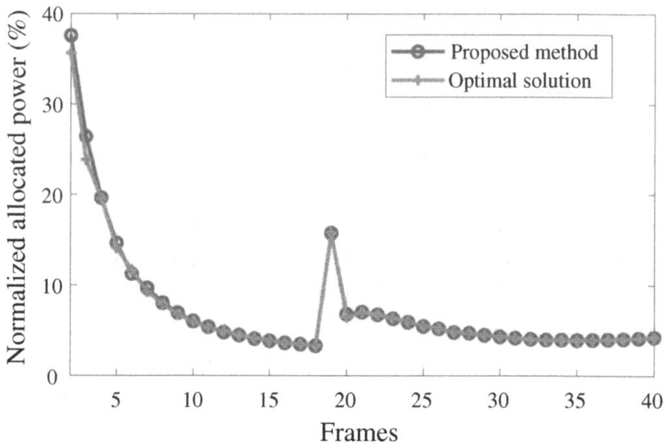

Figure 24.6 The percentages of allocated power consumed by two strategies (Scenario 1).

24.5.4 Insufficient Power Scenario

A scenario of insufficient power is considered to further verify the flexibility of the proposed RPA approach. To this end, two additional targets denoted as Targets 3 and 4 are assumed and their parameters are given in Table 24.2. The tracking deployment of Scenario 2 is illustrated in Figure 24.7. To demonstrate the practical use of the proposed RPA approach, the task importance for targets with different positions and motion parameters, i.e., two sets of task importances, namely, Case 1 ($\mathbf{w} = [0.250.250.250.25]^{\mathrm{T}}$) and Case 2 ($\mathbf{w} = [0.40.40.10.1]^{\mathrm{T}}$), are considered.

In this scenario, the DRM is not applicable. Here a 2-objective optimization scheme, which is to run an optimization problem with tunable parameters for the balance between the BCRLB and power consumption, is provided as the comparison of the proposed RPA approach. The basic idea of the 2-objective optimization scheme is to minimize the weighted sum of the normalized BCRLB and normalized power, i.e.,

$$\mathbf{P}_{k,\mathrm{opt}} = \arg\min_{\mathbf{P}_k} \quad \alpha_1 \times \left(\sum_{q=1}^{Q} \bar{C}_{\hat{\mathbf{x}}^q_{k|k-1}} (P^q_k) \right) + \alpha_2 \times \left(\mathbf{1}^{\mathrm{T}} \bar{\mathbf{P}}_k \right)$$
$$\text{s.t.} \quad \begin{cases} \mathbf{1}^{\mathrm{T}} \mathbf{P}_k \leq P_{\mathtt{total},i} \\ P_{\min} \leq P^q_k \leq P_{\max} \\ \alpha_1 + \alpha_2 = 1, \end{cases} \tag{24.51}$$

where α_1 and α_2 represent the weights of the BCRLB component and power component, respectively. Also, $\bar{C}_{\hat{\mathbf{x}}^q_{k|k-1}}(P^q_k)$ is a normalized BCRLB associated with its TMTE requirement given by

$$\bar{C}_{\hat{\mathbf{x}}^q_{k|k-1}} (P^q_k) = \frac{\eta_{\min}}{\eta^q_k} \times \frac{C_{\hat{\mathbf{x}}^q_{k|k-1}} (P^q_k)}{C_{\hat{\mathbf{x}}^q_{k|k-1}} (0)} \in (0, 1) \tag{24.52}$$

and $\eta_{\min} = \min\{\eta^1_k, \cdots, \eta^Q_k\}$. In (24.51), $\bar{\mathbf{P}}_k = [\bar{P}^1_k, \cdots, \bar{P}^Q_k]^{\mathrm{T}}$ is the normalized power vector, in which the q-th element is

$$\bar{P}^q_k = \frac{P^q_k}{P_{\max}} \in (0, 1]. \tag{24.53}$$

It can be easily seen that the problem in (24.51) is convex, thus, the CVX tools are recommended to solve it.

Figure 24.8 shows the tracking BCRLBs of these four tracked targets, in which Figure 24.8 (a)–(d) represent the results achieved by RPA with Case 1, RPA with Case 2 and 2-objective optimization with two different weights, respectively. For 2-objective optimization, the two weights correspond to different design philosophies: one balancing tracking performance and resource conservation, and the other emphasizing resource saving. Note that RMSE curves are omitted for the sake of clarity of figures. As observed in Figure 24.8 (a), none of the BCRLBs of the four targets in RPA with

Table 24.2 Motion parameters of additional targets in Scenario 2.

	Initial locations (km)	Initial velocities (m/s)
Target 3	(65, 30)	(−250, 250)
Target 4	(−60, −40)	(200, −250)

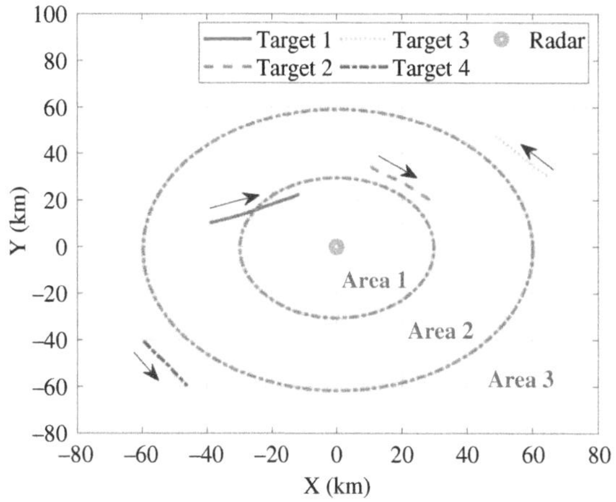

Figure 24.7 Scenario 2 with two additional targets adding to Scenario 1.

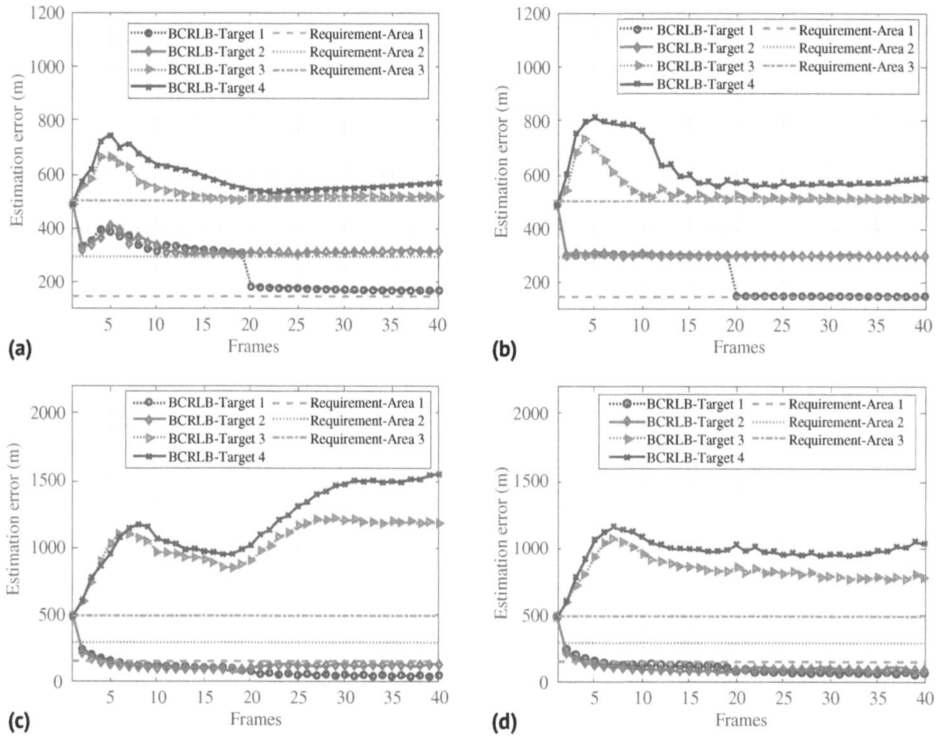

Figure 24.8 Tracking BCRLBs of the RPA and 2-objective optimization in Scenario 2. (a) RPA with Case 1. (b) RPA with Case 2. (c) 2-objective opt. ($\alpha_1 = 0.5$ and $\alpha_2 = 0.5$). (d) 2-objective opt. ($\alpha_1 = 0.1$ and $\alpha_2 = 0.9$).

Case 1 satisfy their TMTE requirements in the initial frames due to the lack of power resource. As the tracking progresses, the BCRLBs of Targets 1, 2 and 3 begin to satisfy the TMTE requirement while Target 4 never satisfies the requirement within the whole 40 frames. It should be noted that after increasing the task importance for Targets 1 and 2 in Case 2, as illustrated in Figure 24.8 (b),

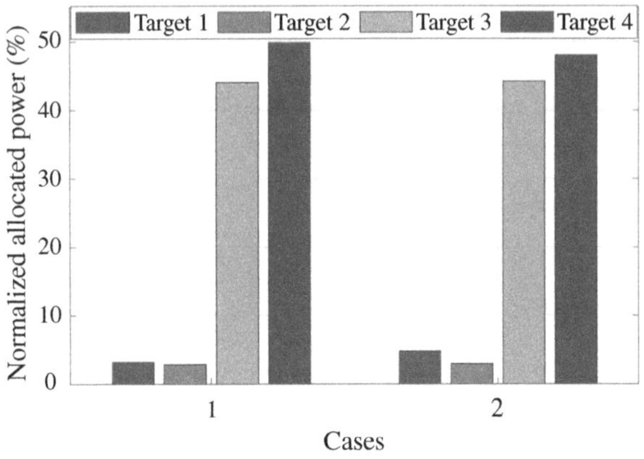

Figure 24.9 The amount of allocated power under two cases of RPA in Scenario 2.

the BCRLB of Targets 1 and 2 tightly stick to their TMTE requirements in the whole tracking process due to their higher task importance. It can be seen from Figure 24.8 (c) and (d) that by using the 2-objective optimization scheme, the tracking accuracy of the targets will be affected by the desired TMTE requirement and the values of weights α_1 and α_2, but the BCRLB cannot tightly converge to the desired TMTE requirement. The reason is that the optimal value of the 2-objective optimization's objective function is not strictly located at the point where the BCRLB just equals to the requirement. It should also be noted that the results achieved by 2-objective optimization scheme are mainly dependent on the desired TMTE requirements, the weights and the distance between the target and radar. Compared to 2-objective optimization, the proposed RPA can provide one more degree of freedom for the control of tracking performance by adjusting the task importance.

To further discuss the reason for achieving different tracking performance for Targets 1 and 2 in the two cases of RPA, Figure 24.9 illustrates the amount of allocated power for each target. It demonstrates that as the task importance of Targets 1 and 2 increases in case 2, the power previously allocated to the far-field Target 4, is released and then allocated to Targets 1 and 2, resulting in a better performance of Targets 1 and 2 in the presence of insufficient power.

24.6 Conclusions

This work proposed a robust power allocation (RPA) approach for solving the multitarget tracking (MTT) problem using a C-MIMO radar. Task utility functions were developed to quantify the ability to satisfy tolerable maximum tracking error (TMTE) requirements for multiple targets. The sum of these task utility functions was used to formulate the RPA problem as an optimization task. The formulated problem's objective function, though non-convex, was shown to exhibit unimodal characteristics. To address this, an efficient iterative algorithm was proposed to solve the RPA. Simulations showed that the proposed RPA achieves task utility comparable to the optimal DRM approach in scenarios with sufficient power. The RPA approach also demonstrated superior flexibility and performance in low-power scenarios where DRM is ineffective.

24.A Proof of Proposition 1

Concentrating on the integrand of the objective function in (24.33), we have:

$$\begin{aligned}\text{Int} &= e^{2\mathcal{T}^q(P_k^q|\mathbf{e}_k^q)} + 1 - 2e^{(\mathcal{T}^q(P_k^q|\mathbf{e}_k^q)} + \left\{\sum_{i=1}^{N} a_i^q[\mathcal{T}^q(P_k^q \mid \mathbf{e}_k^q)]^{2i}\right\}^2 \\ &+2\sum_{i=1}^{N} a_i^q[\mathcal{T}^q(P_k^q \mid \mathbf{e}_k^q)]^{2i} - 2e^{\mathcal{T}^q(P_k^q|\mathbf{e}_k^q)} \cdot \sum_{i=1}^{N} a_i^q \cdot [\mathcal{T}^q(P_k^q \mid \mathbf{e}_k^q)]^{2i}.\end{aligned} \tag{24.54}$$

Since the first three terms are not dependent on the coefficients $\mathbf{a}^q$, we can discard them and rewrite the objective function in (24.33) as follows:

$$\bar{\mathbf{a}}^{q,*} = \arg\min_{\mathbf{a}^q} \int_0^K g(\mathcal{T}^q(P_k^q \mid \mathbf{e}_k^q); \mathbf{a}^q) d\mathcal{T}^q(P_k^q \mid \mathbf{e}_k^q), \tag{24.55}$$

where $g(\mathcal{T}^q(P_k^q \mid \mathbf{e}_k^q); \mathbf{a})$ is the sum of the last three terms of (24.54). The integral of each component of it can be further calculated as

$$\int_0^K \left\{\sum_{i=1}^{N} a_i^q[\mathcal{T}^q(P_k^q \mid \mathbf{e}_k^q)]^{2i}\right\}^2 d\mathcal{T}^q(P_k^q \mid \mathbf{e}_k^q) = (\mathbf{a}^q)^{\mathsf{T}}\mathbf{A}\mathbf{a}^q, \tag{24.56}$$

$$\int_0^K \sum_{i=1}^{N} a_i^q[\mathcal{T}^q(P_k^q \mid \mathbf{e}_k^q)]^{2i} d\mathcal{T}^q(P_k^q \mid \mathbf{e}_k^q) = (\mathbf{a}^q)^{\mathsf{T}}\mathbf{b}, \tag{24.57}$$

and

$$\int_0^K \sum_{i=1}^{N} a_i^q[\mathcal{T}^q(P_k^q \mid \mathbf{e}_k^q)]^{2i} e^{\mathcal{T}^q(P_k^q|\mathbf{e}_k^q)} \, d\mathcal{T}^q(P_k^q \mid \mathbf{e}_k^q) = -(\mathbf{a}^q)^{\mathsf{T}}\mathbf{c}, \tag{24.58}$$

where $\mathbf{A}$, $\mathbf{b}$ and $\mathbf{c}$ are defined in (24.36). Then, by substituting (24.56)–(24.58) into (24.55) and considering the constraint of $\mathbf{a}^q$, the problem can be reconstructed as the form of (24.35), where the constraint is linear, and $\mathbf{A}$ represents a positive semi-definite Hankel matrix [37].

As such, the objective function combines a positive semi-definite quadratic term and a linear term of $\mathbf{a}^q$. As defined in [29], the optimization problem in (24.35) qualifies as a convex QP problem.

24.B Proof of the Non-convexity of (24.37)

Since the constraints in (24.37) are all linear, (24.37) is convex if the second-order gradient of the objective function is positive definite. Because the objective function is a linear combination of task utility functions, the second-order gradient of it can be written as the following diagonal matrix:

$$\Delta_{\mathbf{P}_k}^{\mathbf{P}_k} \bar{U}(\mathbf{P}_k) = \text{diag}\left(\frac{\partial^2 U(\omega^1, P_k^1)}{\partial P_k^1 \partial P_k^1}, \cdots, \frac{\partial^2 U(\omega^Q, P_k^Q)}{\partial P_k^Q \partial P_k^Q}\right), \tag{24.59}$$

where $\dfrac{\partial^2[\cdot]}{\partial P_k^q \partial P_k^q} = \dfrac{\partial}{\partial P_k^q}\left(\dfrac{\partial[\cdot]}{\partial P_k^q}\right)$ is the second-order derivative operation. It is known that (24.59) is positive definite as long as its diagonal elements are all positive in the entirety of their own defined domains [29]. The q-th diagonal element of (24.59) is expressed as

$$\frac{\partial^2 U(\omega^q, P_k^q)}{\partial P_k^q \partial P_k^q} = -\frac{2\omega^q}{\eta_k^q}\sum_{i=1}^{N} i a_i^q [\mathcal{T}^q(P_k^q \mid \mathbf{e}_k^q)]^{2i-2} \cdot \left\{ \frac{2i-1}{\eta_k^q}\left[\frac{\partial \mathcal{C}_{\hat{\mathbf{x}}_{k|k-1}^q}(P_k^q)}{\partial P_k^q}\right]^2 + \mathcal{T}^q(P_k^q \mid \mathbf{e}_k^q)\frac{\partial^2 \mathcal{C}_{\hat{\mathbf{x}}_{k|k-1}^q}(P_k^q)}{\partial P_k^q \partial P_k^q} \right\}. \tag{24.60}$$

In (24.60), since the negative of the term on the left side of the multiplication and the first term in the braces are nonnegative, the sign of (24.60) is dependent on the second term in the braces. Wherein, the sign of $\mathcal{T}^q(P_k^q \mid \mathbf{e}_k^q)$ defined in (24.30) is not fixed in the whole domain, and we cannot guarantee that (24.60)> 0 always holds. Therefore, the indeterminate positive definiteness of $\Delta_{\mathbf{P}_k}^{\mathbf{P}_k} \bar{U}(\mathbf{P}_k)$ confirms that (24.37) is non-convex.

24.C Proof of BCRLB Monotonicity

BCRLB's decreasing property is proven by showing its negative first-order derivative. Since P_k^q is a scalar, it is easy to know that the derivative of BCRLB's trace related to P_k^q can be written as the trace of BCRLB's first order derivative. By using (24.22) and (24.23), we have

$$\begin{aligned}\frac{\partial \mathcal{C}_{\hat{\mathbf{x}}_{k|k-1}^q}(P_k^q)}{\partial P_k^q} &= \mathrm{Tr}\left\{\frac{\partial\left[\mathbf{J}_{P,k} + \mathbf{J}_Z\left(P_k^q, \hat{\mathbf{x}}_{k|k-1}^q\right)\right]^{-1}}{\partial P_k^q}\right\} \\ &= -\mathrm{Tr}\left\{\mathbf{J}_{\hat{\mathbf{x}}_{k|k-1}^q}^{-1}(P_k^q)\left[\mathbf{J}_{\hat{\mathbf{x}}_{k|k-1}^q}^{-1}(P_k^q)\right]^{\mathrm{T}}\frac{\partial \mathbf{J}_Z\left(P_k^q, \hat{\mathbf{x}}_{k|k-1}^q\right)}{\partial P_k^q}\right\}.\end{aligned} \tag{24.61}$$

In (24.61), $\mathbf{J}_{P,k}$ and $\mathbf{J}_Z(P_k^q, \hat{\mathbf{x}}_{k|k-1}^q)$ are shown to be symmetric positive definite [6]. To further confirm the sign of (24.61), we first give the following lemma [38]:

Lemma 1: For two positive definite matrices **A** and **B** with the same dimension, **AB** is positive definite only if **AB** = **BA** holds.

According to Lemma 1, we can confirm that the product of the first two terms in the brace of (24.61)

$$\mathbf{J}_{\hat{\mathbf{x}}_{k|k-1}^q}^{-1}(P_k^q)\left[\mathbf{J}_{\hat{\mathbf{x}}_{k|k-1}^q}^{-1}(P_k^q)\right]^{\mathrm{T}} \triangleq \mathbf{J}_{\mathrm{inv}}^2(P_k^q) \tag{24.62}$$

is a symmetric positive definite matrix. Then we need to discuss the sign of the third term in the brace of (24.61), wherein $\mathbf{J}_Z(P_k^q, \hat{\mathbf{x}}_{k|k-1}^q)$ can be written as a linear function of P_k^q given by [14]

$$\mathbf{J}_Z\left(P_k^q, \hat{\mathbf{x}}_{k|k-1}^q\right) = P_k^q \times \mathbf{Y}\left(\hat{\mathbf{x}}_{k|k-1}^q\right), \tag{24.63}$$

where

$$\mathbf{Y}\left(\hat{\mathbf{x}}_{k|k-1}^q\right) = (\mathbf{H}_k^q)^{\mathrm{T}}\mathbf{W}_k^q\mathbf{H}_k^q. \tag{24.64}$$

Then, by comparing (24.63) and (24.64) with (24.22), we have

$$P_k^q \times \mathbf{W}_k^q = [\mathbf{R}_k^q(P_k^q)]^{-1}, \tag{24.65}$$

since the diagonal elements of $[\mathbf{R}_k^q(P_k^q)]^{-1}$ are all positive and linear functions of P_k^q ($P_k^q > 0$), $\mathbf{W}_k^q$ is also a diagonal matrix given by

$$\mathbf{W}_k^q = \mathrm{diag}\left(\gamma_{r_k^q}, \gamma_{\theta_k^q}, \gamma_{f_k^q}\right), \tag{24.66}$$

where the three diagonal elements $\gamma_{r_k^q} = (\sigma^2_{r_k^q}(P_k^q))^{-1}/P_k^q$, $\gamma_{\theta_k^q} = (\sigma^2_{\theta_k^q}(P_k^q))^{-1}/P_k^q$ and $\gamma_{f_k^q} = (\sigma^2_{f_k^q}(P_k^q))^{-1}/P_k^q$ are all positive. As such, $\mathbf{W}_k^q$ is symmetric positive definite, thus according to (24.63) and (24.64), we can know that $\frac{\partial \mathbf{J}_Z\left(P_k^q, \hat{\mathbf{x}}_{k|k-1}^q\right)}{\partial P_k^q} = \mathbf{Y}(\hat{\mathbf{x}}_{k|k-1}^q)$ is a quadratic form of $\mathbf{W}_k^q$, hence also symmetric positive definite. By substituting the above expression and Eqs. (24.62) into (24.61), the term in the brace of (24.61) can be expressed as the product of two symmetric positive definite matrices, hence its trace is positive. By these findings, we can obtain that the first order derivative of $\mathcal{C}_{\hat{\mathbf{x}}_{k|k-1}^q}(P_k^q)$ is less than 0, i.e.,

$$\frac{\partial \mathcal{C}_{\hat{\mathbf{x}}_{k|k-1}^q}(P_k^q)}{\partial P_k^q} = -\mathrm{Tr}\left\{\mathbf{J}^2_{\mathrm{inv}}(P_k^q)\mathbf{Y}\left(\hat{\mathbf{x}}_{k|k-1}^q\right)\right\} < 0. \tag{24.67}$$

According to the definition of function monotonicity, the BCRLB is monotonically decreasing with respect to P_k^q.

24.D Proof of Weighted Task Utility Function Unimodality

We first define the initial BCRLB without allocating any power resource as $\mathcal{C}_{\hat{\mathbf{x}}_{k|k-1}^q}(0)$. Then, the proof is respectively implemented for two different initial BCRLBs. In the first case, where the initial BCRLB exceeds its TMTE, i.e., $\mathcal{C}_{\hat{\mathbf{x}}_{k|k-1}^q}(0) > \eta_k^q$, we can find only one point $P_k^q = P_k^{q,*}$ satisfying

$$\mathcal{C}_{\hat{\mathbf{x}}_{k|k-1}^q}(P_k^{q,*}) = \eta_k^q. \tag{24.68}$$

Then, we prove that this point is the only one local minimizer in this case. According to the expression of (24.40) and Proposition 2, the first order derivative of the weighted task utility function has the following property:

$$\frac{\partial U(\omega^q, P_k^q)}{\partial P_k^q} \begin{cases} \leq 0 & P_k^q \geq P_k^{q,*} \\ > 0 & P_k^q < P_k^{q,*}, \end{cases} \tag{24.69}$$

where the equal sign is taken at the point of $P_k^q = P_k^{q,*}$. By using (24.69), for any two of the variables P_1 and P_2 that satisfied $P_1 > P_2$, we have:

$$\begin{cases} U\left(\omega^q, P_k^{q,*} + P_1\right) < U\left(\omega^q, P_k^{q,*} + P_2\right), & P_1 > P_2 > 0, \\ U\left(\omega^q, P_k^{q,*} + P_1\right) > U\left(\omega^q, P_k^{q,*} + P_2\right), & 0 > P_1 > P_2, \end{cases} \tag{24.70}$$

due to the arbitrariness of P_1 and P_2, we can obtain that $U\left(\omega^q, P_k^q\right)$ decreases to the right of $P_k^{q,*}$ and increases to its left. Based on these findings, it can be obtained that $P_k^q = P_k^{q,*}$ is the only one local maximizer of $U\left(\omega^q, P_k^q\right)$. While for the second case where the initial BCRLB is lower than the

TMTE, i.e., $\mathcal{C}_{\hat{\mathbf{x}}^q_{k|k-1}}(0) \leq \eta^q_k$, based on the expression of (24.40) and Proposition 2, the derivative of the weighted task utility function is constantly lower than 0 for any $P^q_k > 0$. Therefore, the utility function is monotonically decreasing in this case, hence the global maximizer is taken at $P^q_k = 0$.

These findings confirm the utility function's unimodality, whose global maximizer is taken at the point indicated in (24.68) or $P^q_k = 0$.

References

1 J. Li and P. Stoica. MIMO radar with colocated antennas. *IEEE Signal Processing Magazine*, 24(5):106–114, 2007.

2 A. A. Gorji, R. Tharmarasa, W. D. Blair, and T. Kirubarajan. Multiple unresolved target localization and tracking using colocated MIMO radars. *IEEE Transactions on Aerospace and Electronic Systems*, 48(3):2498–2517, 2012.

3 D. S. Kalogerias and A. P. Petropulu. Matrix completion in colocated MIMO radar: Recoverability, bounds & theoretical guarantees. *IEEE Transactions on Signal Processing*, 62(2):309–321, 2013.

4 P. Chen, L. Zheng, X. Wang, H. Li, and L. Wu. Moving target detection using colocated MIMO radar on multiple distributed moving platforms. *IEEE Transactions on Signal Processing*, 65(17):4670–4683, 2017.

5 J. Li and P. Stoica. *MIMO Radar Signal Processing*. Wiley-IEEE Press, Hoboken, NJ, USA, 2009.

6 J. Yan, H. Liu, B. Jiu, B. Chen, Z. Liu, and Z. Bao. Simultaneous multibeam resource allocation scheme for multiple target tracking. *IEEE Transactions on Signal Processing*, 63(12):3110–3122, 2015.

7 N. Garcia, A. M. Haimovich, M. Coulon, and M. Lops. Resource allocation in MIMO radar with multiple targets for non-coherent localization. *IEEE Transactions on Signal Processing*, 62(10): 2656–2666, 2014.

8 C. Shi, F. Wang, M. Sellathurai, J. Zhou, and S. Salous. Power minimization-based robust OFDM radar waveform design for radar and communication systems in coexistence. *IEEE Transactions on Signal Processing*, 66(5):1316–1330, 2017.

9 J. Chen, C. Wen, and K. Wong. An efficient sensor-node selection algorithm for sidelobe control in collaborative beamforming. *IEEE Transactions on Vehicular Technology*, 65(8):5984–5994, 2016.

10 A. Irci, A. Saranli, and B. Baykal. Study on Q-RAM and feasible directions based methods for resource management in phased array radar systems. *IEEE Transactions on Aerospace and Electronic Systems*, 46(4):1848–1864, 2010.

11 A. Fallahi and E. Hossain. A dynamic programming approach for QoS-aware power management in wireless video sensor networks. *IEEE Transactions on Vehicular Technology*, 58(2):843–854, 2009.

12 X. Shen and P. K. Varshney. Sensor selection based on generalized information gain for target tracking in large sensor networks. *IEEE Transactions on Signal Processing*, 62(2):363–375, 2013.

13 A. Charlish and R. Nadjiasngar. Quality of service management for a multi-mission radar network. In *2015 IEEE 6th International Workshop on Computational Advances in Multi-Sensor Adaptive Processing (CAMSAP)*, IEEE, pages 289–292, 2015.

14 Y. Yuan, W. Yi, T. Kirubarajan, and L. Kong. Scaled accuracy based power allocation for multi-target tracking with colocated MIMO radars. *Signal Processing*, 158:227–240, 2019.

15 H. Godrich, A. P. Petropulu, and H. Vincent Poor. Power allocation strategies for target localization in distributed multiple-radar architectures. *IEEE Transactions on Signal Processing*, 59(7):3226–3240, 2011.

16 M. Xie, W. Yi, T. Kirubajaran, and L. Kong. Joint node selection and power allocation strategy for multitarget tracking in decentralized radar networks. *IEEE Transactions on Signal Processing*, 66(3):729–743, 2018.

17 M. Xie, W. Yi, L. Kong, and T. Kirubarajan. Receive-beam resource allocation for multiple target tracking with distributed MIMO radars. *IEEE Transactions on Aerospace and Electronic Systems*, 54(5):2421–2436, 2018.

18 W. Yi, Y. Yuan, R. Hoseinnezhad, and L. Kong. Resource scheduling for distributed multi-target tracking in netted colocated MIMO radar systems. *IEEE Transactions on Signal Processing*, 68:1602–1617, 2020.

19 J. Yan, W. Pu, H. Liu, B. Jiu, and Z. Bao. Robust chance constrained power allocation scheme for multiple target localization in colocated MIMO radar system. *IEEE Transactions on Signal Processing*, 66(15):3946–3957, 2018.

20 X. Wang, W. Yi, M. Xie, and L. Kong. A joint beam and dwell time allocation strategy for multiple target tracking based on phase array radar system. In *2017 20th International Conference on Information Fusion (Fusion),* IEEE, pages 1–5, 2017.

21 Y. Yuan, W. Yi, R. Hoseinnezhad, and P. K. Varshney. Robust power allocation for resource-aware multi-target tracking with colocated MIMO radars. *IEEE Transactions on Signal Processing*, 69:443–458, 2020.

22 C. Jauffret and Y. Bar-Shalom. Track formation with bearing and frequency measurements in clutter. *IEEE Transactions on Aerospace and Electronic Systems*, 26(6):999–1010, 1990.

23 W. Yi, H. Jiang, T. Kirubarajan, L. Kong, and X. Yang. Track-before-detect strategies for radar detection in G0-distributed clutter. *IEEE Transactions on Aerospace and Electronic Systems*, 53:2516–2533, 2017.

24 Y. Bar-Shalom, P. K. Willett, and X. Tian. *Tracking and Data Fusion*. YBS publishing Storrs, CT, USA, 2011.

25 M. S. Arulampalam, S. Maskell, N. Gordon, and T. Clapp. A tutorial on particle filters for online nonlinear/non-Gaussian Bayesian tracking. *IEEE Transactions on Signal Processing*, 50(2):174–188, 2002.

26 L. Zuo, R. Niu, and P. K. Varshney. Conditional posterior Cramér–Rao lower bounds for nonlinear sequential Bayesian estimation. *IEEE Transactions on Signal Processing*, 59(1):1–14, 2010.

27 P. Tichavsky, C. H. Muravchik, and A. Nehorai. Posterior Cramér-Rao bounds for discrete-time nonlinear filtering. *IEEE Transactions on Signal Processing*, 46(5):1386–1396, 1998.

28 J. Yan, H. Liu, J. Bo, and Z. Bao. Power allocation algorithm for target tracking in unmodulated continuous wave radar network. *IEEE Sensors Journal*, 15(2):1098–1108, 2014

29 S. Boyd and L. Vandenberghe. *Convex Optimization*. Cambridge University Press, Cambridge, U.K., 2013.

30 A. Charlish and F. Hoffmann. Cognitive radar management. In *Novel Radar Techniques and Applications: Waveform Diversity and Cognitive Radar and Target Tracking and Data Fusion,* 2:157–193, 2017.

31 D. Shanks. *Solved and Unsolved Problems in Number Theory*, volume 297. American Mathematical Soc., 2001.

32 C. Lee, C. Shih, and L. Sha. Online QoS optimization using service classes in surveillance radar systems. *Real-Time Systems*, 28(1):5–37, 2004.

33 D. Seto, J. P. Lehoczky, L. Sha, and K. G. Shin. On task schedulability in real-time control systems. In *17th IEEE Real-time Systems Symposium*, IEEE, pages 13–21, 1996.

34 S. Boyd, L. Xiao, and A. Mutapcic. *Subgradient Methods. Lecture Notes of EE392o, Stanford University, Autumn Quarter*, 2004–2005, 2004.

35 X. Wang, W. Yi, M. Xie, B. Zhai, and L. Kong. Time management for target tracking based on the predicted Bayesian Cramér-Rao lower bound in phase array radar system. In *International Conference on Information Fusion,* pages 1–5, 2017.

36 Inc. CVX Research. CVX: Matlab software for disciplined convex programming, version 2.0. 2012. http://cvxr.com/cvx.

37 J. R. Partington. *An Introduction to Hankel Operators,* volume 13. Cambridge University Press, 1988.

38 R. A. Horn and C. R. Johnson. *Matrix Analysis.* Cambridge University Press, 2012.

25

Multiple Radar Resource Allocation for Multitarget Tracking

Wenqiang Pu, Junkun Yan, Peng Zhang, Hao Jiao, and Hongwei Liu

National Key Laboratory of Radar Signal Processing, Xidian University, Xi'an, China

25.1 Introduction

Due to inherent limitations in radiated power, antenna aperture, and signal processing capabilities, a single radar platform often struggles to achieve both continuous wide-area surveillance and high-precision target tracking. In contrast, a radar sensor network (RSN) can acquire comprehensive target information from multiple nodes with diverse viewpoints, thereby overcoming many of the constraints faced by monostatic radar systems. Numerous studies have demonstrated the potential advantages of RSNs over traditional monostatic radars, making them suitable for target detection and tracking in both military and civilian applications. The effectiveness of target tracking within an RSN is influenced by several factors, including the accuracy of target motion models, the quality of radar measurements, and the geometric configuration between the targets and radar sensors. Notably, the measurement quality of each radar sensor is intrinsically dependent on the resource allocated to tracking a specific target. Ideally, to maximize measurement quality, each node would allocate the maximum possible resource to each target. However, in the context of multiple target tracking (MTT), the total resource required to track all targets at a desired performance level may exceed the resource constraints of individual radar nodes. Therefore, it is essential to account for the resource limitations of each node when designing effective resource allocation (RA) strategies. While previous studies have provided initial surveys of RA algorithms for single radar platforms, this chapter offers a focused introduction of RA techniques within RSN for MTT applications.

Conventional MTT techniques in RSN typically adopt a segmented, open-loop processing paradigm. In this framework, multiple radar nodes independently detect or track a target and subsequently transmit either raw data or processed measurements to a central fusion center for integrated analysis. This approach does not support the feedback of prior information from the fusion center to individual nodes for RA. As a result, radar sensors operate in a predetermined manner, allocating their limited resource among multiple targets without accounting for target priorities, geometrical relationships, or specific performance requirements. To allocate resource effectively, exploiting the tracking information is essential, as it provides critical information about the targets needed for resource distribution. The RA scheme necessitates that the tracker generates predicted states to form a closed-loop system. As depicted in Figure 25.1, the dashed line illustrates the data flow of heterogeneous radar measurements transmitted to the fusion center, where they undergo further processing by target-specific trackers to generate state estimates. These target state

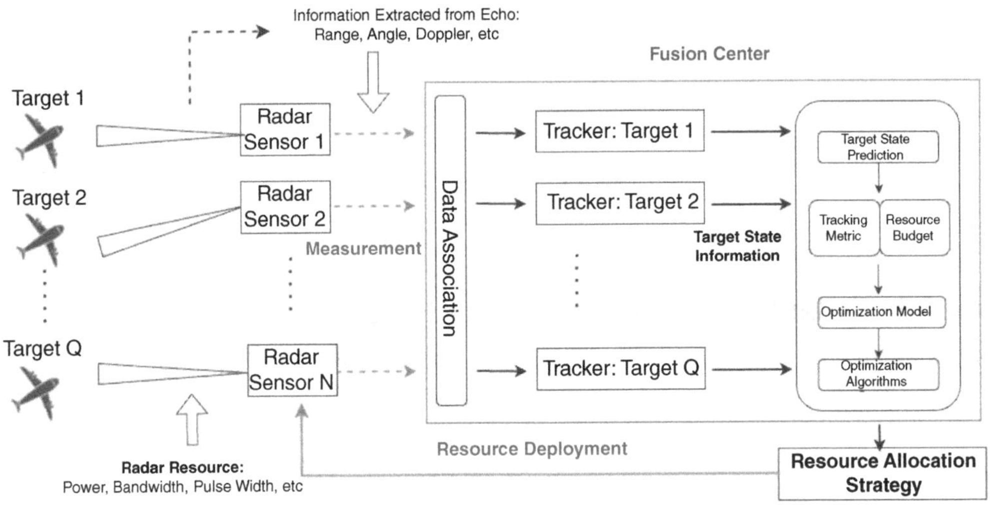

Figure 25.1 Diagram of resource allocation for MTT.

estimates are subsequently utilized to formulate an appropriate resource allocation optimization model based on predefined performance criteria. The solution to this optimization problem yields a resource allocation strategy that is deployed to each radar sensor, as indicated by the solid line. The integration of these two information pathways establishes a closed-loop system for adaptive resource allocation in MTT applications.

Organization: In this chapter, we begin by introducing basic mechanism for resource allocation in Section 25.2, including the basic target motion models, measurement models, and the underlying resource allocation mechanism. Section 25.3 provides an overview of various resource types commonly encountered in RSN applications, and Section 25.4 presents different tracking performance metrics. In Section 25.5, existing RA works are divided into two types of resource allocation models and are subsequently discussed, accompanied by illustrative examples. Finally, Section 25.6 concludes this chapter with a summary of key findings and future research directions.

25.2 Resource Allocation Mechanism

25.2.1 Target Motion and Radar Measurement Models

To begin with, a network of N radars is deployed to track Q spatially distributed and independent point targets. Multiple radar sensors transmit signals with different carrier frequencies and thus operate in a monostatic way. Let k be the time instance index and denote $\mathbf{x}_k^q \in \mathbb{R}^d$ as the qth target state in the reference coordinate system of dimension d, e.g., Cartesian coordinate, then its motion dynamics is characterized as

$$\mathbf{x}_k^q = \mathbf{f}(\mathbf{x}_{k-1}^q) + \mathbf{u}_{k-1}^q. \tag{25.1}$$

Here, $\mathbf{f}(\cdot) : \mathbb{R}^d \mapsto \mathbb{R}^d$ is the base dynamic function of target motion, $\mathbf{u}_{k-1}^q \in \mathbb{R}^d$ represents the motion process noise, typically modeled as zero-mean Gaussian noise with covariance $\mathbf{Q}_{k-1}^q$. Remarkably, the dynamics of the target in (25.1) are solely dependent on the target and cannot be influenced by the radar. Instead, radars optimize sensing resource to improve measurement quality, thereby enhancing target state estimation accuracy.

Let $s_k^{i,q} \in \mathbb{R}^{\ell}$ represents the resource vector associated with radar sensor i to target q, where ℓ is the dimension of the vector. The measurement of target q is affected by the resource of all sensors, denoted as $\mathbf{s}_k^q = (s_k^{1,q}, s_k^{2,q}, \dots, s_k^{N,q}) \in \mathbb{R}^{N\ell}$. Then the measurement can be modeled as

$$\mathbf{z}_k^{i,q} = \mathbf{h}_i(\mathbf{x}_{k-1}^q) + \mathbf{w}_k^{i,q}(s_k^{i,q}) \quad \text{or empty.} \tag{25.2}$$

Here, $\mathbf{h}_i(\cdot) : \mathbb{R}^d \mapsto \mathbb{R}$ is the measurement function of radar sensor i, which maps the target state into the radar sensor's local measurement space, e.g., range, azimuth, Doppler, etc. The measurement noise $\mathbf{w}_k^{i,q}(\cdot)$ is zero mean with covariance $\boldsymbol{\Sigma}_k^{i,q}(s_k^{i,q})$, which is influenced by the allocated sensing resource $s_k^{i,q}$, and the case 'empty' stands for no measurement, which may be caused by missed detection or no resource allocated to target q.

In this chapter, the tracker is modeled as an oracle that processes measurements $\{\mathbf{z}_k^{i,q}\}_{i,q}$ and outputs the target state estimate, along with additional information such as the posterior covariance. With such information from trackers, the Bayesian information matrix (BIM) [1] $\mathbf{J}(\mathbf{x}_k^q) \in \mathbb{R}^{d\times d}$ can be recursively determined at each time index through an elegant form expressed as[1]

$$\mathbf{J}(\mathbf{x}_k^q, \mathbf{s}_k^q) = \underbrace{\left[\mathbf{Q}_{k-1}^q + \mathbf{F}^q \mathbf{J}^{-1}(\mathbf{x}_{k-1}^q)(\mathbf{F}^q)^T\right]^{-1}}_{\mathbf{J}_\mathrm{P}(\mathbf{x}_k^q)} + \underbrace{\sum_{j=1}^{N} \mathbb{E}_{\mathbf{x}_k^q}\left\{\mathbf{H}_{j,q,k}^T \left[\boldsymbol{\Sigma}_k^{j,q}(s_k^{j,q})\right]^{-1} \mathbf{H}_{j,q,k}\right\}}_{\mathbf{J}_\mathrm{Z}(\mathbf{x}_k^q, \mathbf{s}_k^q)}, \tag{25.3}$$

where $\mathbf{F}^q \in \mathbb{R}^{\ell\times d}$ is the state transition matrix, $\mathbf{H}_{j,q,k} \triangleq \left[\Delta_{\mathbf{x}_k^q} \mathbf{h}_j^T(\mathbf{x}_k^q)\right]^T \in \mathbb{R}^{d\times\ell}$ is the Jacobian matrix of the measurement function of radar j with respect to the target state. $\mathbf{J}_\mathrm{P}(\mathbf{x}_k^q)$ and $\mathbf{J}_\mathrm{Z}(\mathbf{x}_k^q, \mathbf{s}_k^q)$ denote the prior information matrix and the data information matrix, respectively. Note that (25.3) indicates that resource controls $\mathbf{J}_\mathrm{Z}(\cdot)$ and thereby affects tracking performance. Next, we will illustrate how this control mechanism operates.

25.2.2 Effect of Resource on BIM

Consider radar i transmitting a single-pulse narrowband signal, each target's attenuation response can be represented by a complex-valued scalar. For spatially separated targets, the echo signal $e_{i,q}(t)$ from the qth target is

$$e_{i,q}(t) = \sqrt{P_{i,q}}\alpha_{i,q} s_i(t - \tau_{i,q}) \exp(-j2\pi f_{i,q}^d t) + w_i(t) = \tilde{e}_{i,q}(t) + w_i(t). \tag{25.4}$$

In (25.4), $\tau_{i,q}$ and $f_{i,q}^d$ denote the time delay and Doppler shift of target q relative to radar i, respectively. $P_{i,q}$ represents the transmit power of radar i for target q. The attenuation coefficient $\alpha_{i,q}$ is inversely proportional to the fourth power of the distance, and the term $w_i(t)$ denotes additive white Gaussian noise with zero mean and variance σ_w^2. Based on this signal model, several parameters—including transmit power $P_{i,q}$, coefficient $\alpha_{i,q}$, waveform $s_i(t)$, etc., have a direct influence on the echo signal, thereby determining the quality of measurement given in (25.2). Next, we introduce how they affect the measurement.

Through advanced signal processing techniques, target parameters such as range, azimuth angle, and Doppler shift can be extracted from the received echo signals. Typically, the maximum likelihood (ML) criterion is adopted for parameter estimation, which is asymptotically unbiased under regular conditions. To objectively quantify the impact of resource allocation on parameter

---

1 For simplicity, we consider the target motion function takes linear form as $\mathbf{f}(\mathbf{x}_{k-1}^q) = \mathbf{F}^q \mathbf{x}_{k-1}^q$.

estimation performance without being influenced by the choice of estimation algorithm, the Cramér–Rao Lower Bound (CRLB) is commonly used, which provides a theoretical lower bound on the variance of all unbiased estimators. The CRLB for key parameters is given as follows.

- Range $r_{i,q}$: Let $\sigma^2_{\tau,i,q}$ be the variance of time delay $\tau_{i,q}$ estimation, which is lower bounded by its CRLB expressed as [2]

$$\sigma^2_{\tau,i,q} \geq \frac{\sigma^2_w}{\mathrm{Re}\left\{2\int\left|\frac{\partial e_{i,q}(t)}{\partial \tau_{i,q}}\right|^2 d\tau_{i,q}\right\}} = \frac{\sigma^2_w}{8\pi^2\alpha_{i,q}P_{i,q}|h_{i,q}|^2\beta^2_{i,q}},$$

  where β_i is the bandwidth of $s_i(t)$. Denote $r_{i,q}$ as the range between target q and radar i, by relation $r_{i,q} = c\tau_{i,q}/2$ (c is the speed of light). Let $\sigma^2_{r,i,q}$ be the variance of range estimation, the CRLB of $r_{i,q}$ is

$$\sigma^2_{r,i,q} \geq \frac{c^2\sigma^2_w}{32\pi^2\alpha_{i,q}P_{i,q}|h_{i,q}|^2\beta^2_{i,q}}. \tag{25.5}$$

- Doppler $f^d_{i,q}$: Suppose there are $N_{d,q}$ consecutive echo signals from target q, which can be expressed as [3]

$$e^n_{i,q}(t) = \tilde{e}_{i,q}(t)\exp(-j2\pi f^d_{i,q}(n-1)T^r_{i,q}) + w_{i,n}(t),\ n = 1,\cdots,N_{d,q}$$

  where $T^r_{i,q}$ represents the pulse repetition time (PRT) of radar i for target q. Let $\sigma^2_{f,i,q}$ be the estimation variance of $f^d_{i,q}$, then the CRLB of Doppler estimation is

$$\sigma^2_{f,i,q} \geq \frac{\sigma^2_w}{2\mathrm{Re}\left\{\sum_{n=1}^{N_{d,q}}\left[\frac{\partial e^n_{i,q}(t)}{\partial f^d_{i,q}}\right]^*\left[\frac{\partial e^n_{i,q}(t)}{\partial f^d_{i,q}}\right]\right\}} \approx \frac{6\sigma^2_w}{8\pi^2\alpha_{i,q}P_{i,q}|h_{i,q}|^2(T^c_{i,q})^2N_{d,q}}, \tag{25.6}$$

  where $T^c_{i,q} = N_{d,q}T^r_{i,q}$ is the effective time width.
- Angle $\theta_{i,q}$: Considering a uniform linear array of size N_r, denote $\theta_{i,q}$ as the relative azimuth angle between radar i and target q, then the received signals of each array element can be expressed as

$$e^n_{i,q}(t) = \tilde{e}_{i,q}(t)\exp(-j2\pi(n-1)d\sin(\theta_{i,q})/\lambda) + w^n_{i,q}(t),\ n = 1,\cdots,N_r$$

  where d is the array element spacing. Let $\sigma^2_{\theta,i,q}$ be the variance of angle estimation, then its CRLB can be expressed as

$$\begin{aligned}\sigma^2_{\theta,i,q} &\geq \frac{\sigma^2_w}{2\mathrm{Re}\left\{\sum_{n=1}^{N_{r,q}}\left[\frac{\partial e^n_{i,q}(t)}{\partial \theta_{i,q}}\right]^*\left[\frac{\partial e^n_{i,q}(t)}{\partial \theta_{i,q}}\right]\right\}} \\ &= \frac{6\sigma^2_w\lambda^2}{8\pi^2\alpha_{i,q}P_{i,q}|h_{i,q}|^2d^2\cos^2(\theta_{i,q})N_{r,q}(N_{r,q}-1)(2N_{r,q}-1)},\end{aligned} \tag{25.7}$$

  where λ is the wavelength, $\theta_{3dB,i,q} = k\lambda/(N_{r,q}d\cos\theta_{i,q})$ is the 3dB beam width, and k is the beam width factor. For N_r being large enough, (25.7) can be approximated as [3]

$$\sigma^2_{\theta_{i,q}} \geq \frac{6\sigma^2_w\theta^2_{3dB,i,q}}{8\pi^2\alpha_{i,q}P_{i,q}|h_{i,q}|^2N_{r,q}} \tag{25.8}$$

The CRLBs in (25.5), (25.6), and (25.8) establish the mathematical foundation for resource allocation in RSN, as they establish the relationship between radar resource and the quality of measurements in (25.2). For example, consider function $\mathbf{h}_i(\cdot)$ measuring the range, angle, and Doppler, $s_k^{i,q}$ represents the allocation of power, bandwidth, and the effective time width, i.e., $s_k^{i,q} = (P_{i,q}, \beta_{i,q}, T_{i,q}^c)$. In this case, taking these CRLBs as the metric, the covariance matrix thus is modeled as

$$\boldsymbol{\Sigma}_k^{i,q}(s_k^{i,q}) = \mathrm{diag}(\sigma_{\tau,i,q}^2, \sigma_{\theta,i,q}^2, \sigma_{f,i,q}^2)$$

where

$$\sigma_{\tau,i,q}^2 \propto \left(P_{i,q}\beta_{i,q}^2\right)^{-1}, \sigma_{\theta,i,q}^2 \propto P_{i,q}^{-1}, \sigma_{f,i,q}^2 \propto \left(P_{i,q}(T_{i,q}^c)^2\right)^{-1}.$$

This formulation demonstrates that the resource vector $s_k^{i,q}$ affects measurement quality in different ways; for example, signal bandwidth primarily influences range estimation, while transmit power impacts all parameter estimates. This principle provides a foundational guideline for resource allocation in RSNs, that resource should be allocated in a manner that minimizes measurement uncertainty, thereby enhancing overall tracking performance.

25.3 Radar Resource

In practical applications, an RSN may comprise multiple types of radar sensors, each operating in distinct modes and thus possessing unique resource characteristics. For example, a collocated MIMO radar can employ multi-beam operation to track multiple targets [4, 5]. In this configuration, the transmit power allocated to each beam is individually adjustable, subject to a total power constraint across all beams [6, 7]. Comparatively, phased array radar leverages beam agility, enabling rapid and adaptive adjustments of dwell times and revisit intervals for multiple targets. In this chapter, radar resource required for effective target tracking are systematically categorized into five primary regulation domains, include power resource, time domain resource, spatial domain resource, frequency domain resource, and coding domain resource. A short summary of different resources is presented in Table 25.1.

Table 25.1 Summary of performance metrics.

Resource type	Primary effect	Key formula
Power domain (Section 25.3.1)	Parameter estimation accuracy	(25.9)
Time domain (Section 25.3.2)	Parameter estimation accuracy fisher information matrix (tracking)	(25.10) (25.11)
Spatial domain (Section 25.3.3)	Parameter estimation accuracy fisher information matrix (Tracking)	(25.14) (25.15)
Frequency domain (Section 25.3.4)	Parameter estimation accuracy fisher information matrix (Tracking)	(25.17) (25.19)
Code domain (Section 25.3.5)	Parameter estimation accuracy	(25.20)

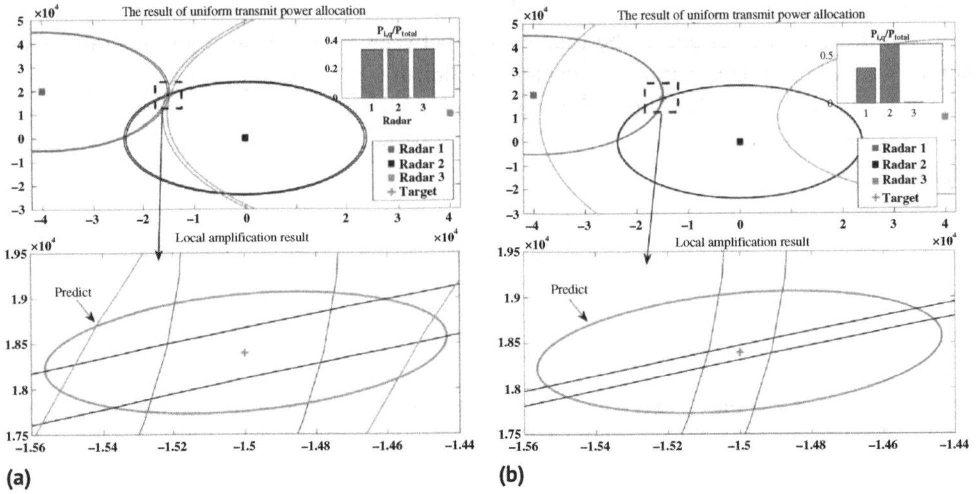

Figure 25.2 The transmit power allocation schematic diagram of physical meaning. (a) Result of uniform transmit power allocation. (b) Result of optimal transmit power allocation.

25.3.1 Power Resource

Transmit power is one fundamental resource in RSN, which represents the intensity of the radar signal emission toward targets. Due to variations in the quality of prior information of targets, e.g., relative positions, target-specific characteristics, etc., identical transmission power can yield different tracking performances for different targets. For instance, by adaptively allocating less power to targets with larger radar cross-section (RCS) values and more power to those with smaller RCS values, the overall tracking performance can be significantly enhanced [8–10]. The transmit power directly impacts measurement quality, which can be described by the following relationship:

$$\sigma^2_{\tau,i,q} \propto (P_{i,q})^{-1}, \sigma^2_{\theta,i,q} \propto (P_{i,q})^{-1}, \sigma^2_{f,i,q} \propto (P_{i,q})^{-1}. \tag{25.9}$$

In Figure 25.2, the localization CRLB of three radars is indicated by several circles with different colors. The predicted ellipse characterizes the predicted track information of the target, and its size relates to target's motion model. In such a case, the area where the ellipse intersects with multiple concentric circles indicates the tracking accuracy of the target. Since the size of circles can be controlled by adjusting the power, tracking accuracy can be optimized through proper power allocation. The results show that limited resource are preferentially allocated to Radar 1 and Radar 2, making the intersection area smaller.

25.3.2 Time Domain Resource

Time-domain resource in radar system determine the temporal characteristics of sensing operations and significantly influence key radar performance metrics, such as target illumination duration, Doppler resolution, maximum unambiguous range, and maximum unambiguous velocity [11]. A schematic diagram of a typical time-resource relationship is shown in Figure 25.3, and representative time-domain resource include [11]:

- Pulse Width $T^{\varepsilon}_{i,q}$: Determines range resolution and energy delivered on target, directly influencing detection sensitivity and resolution capabilities.

- Pulse Repetition Time (PRT) $T^p_{i,q}$: Defines the interval between consecutive pulses, affecting Doppler resolution and maximum unambiguous range.
- Dwell Time $T^d_{i,q}$: Indicates radar observation duration on a particular target, crucial for reliable detection, tracking, and measurement accuracy.
- Revisit Interval $T^r_{i,q}$: Specifies the elapsed time between successive observations of the same target, impacting continuous monitoring and tracking performance.

Similar to Section 25.2.2, time resource effects measurement quality as [3]

$$\sigma^2_{\tau,i,q} \propto (T^{\varepsilon}_{i,q} T^d_{i,q} / T^p_{i,q})^{-1}, \sigma^2_{\theta,i,q} \propto (T^{\varepsilon}_{i,q} T^d_{i,q} / T^p_{i,q})^{-1}, \sigma^2_{f,i,q} \propto (T^{\varepsilon}_{i,q} T^d_{i,q} / T^p_{i,q})^{-1}. \tag{25.10}$$

Reasonable allocation of time-domain resource can make the radar have better measurement information matrix $\mathbf{J}_Z(\mathbf{x}^q_k, \mathbf{s}^q_k)$ and thus effectively improve the target tracking performance. Note that sometimes the time resource does not have the monotonic relation as shown above. For high-speed moving targets, too large a value of $T^d_{i,q}$ may cause range migration and Doppler migration, resulting in a waste of resource [12].

In addition, different time resource may have coupled effects on the quality of measurement. For example, when the revisit interval $T^r_{i,q}$ is less than the processing interval T_i (as shown in Figure 25.3), the target state must be predicted using measurements obtained. In such scenarios, the BFIM for tracking performance is expressed as

$$\mathbf{J}(\mathbf{x}^q_k) = \mathbf{J}_P(\mathbf{x}^q_k) + u_t(T_i - T^r_{i,q})\mathbf{J}_Z(\mathbf{x}^q_k), \tag{25.11}$$

where $u_t(T_i - T^r_{i,q}) \in (0, 1]$ represents the reduction factor for predicted measurement information [13]. When considering the impact of signal processing procedures, such as moving target indicator (MTI) and space-time adaptive processing (STAP) techniques, the detection gains vary significantly across different Doppler channels (as shown in Figure 25.4). Under these conditions, the relationship between the target measurement variance matrix and $T^p_{i,q}$ can be expressed as

$$\bar{\boldsymbol{\Sigma}}^{i,q}_k(s^{i,q}_k) = \boldsymbol{\Sigma}^{i,q}_k(s^{i,q}_k)/\mu(\, mod\,(f^d_{i,q} T^p_{i,q}, 1)) \tag{25.12}$$

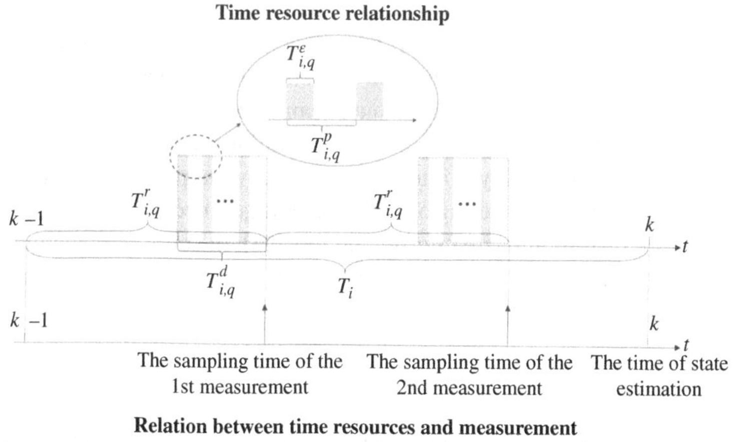

Figure 25.3 Illustration of typical time resource in radar system.

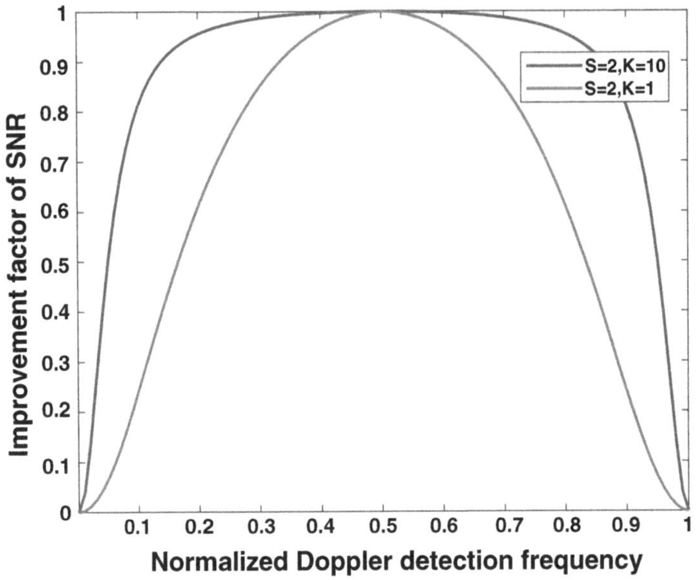

Figure 25.4 The improvement factor function of SNR after MTI filtering (S is the canceller order and K is the noise bandwidth coefficient).

where $\mu(\cdot)$ is the improvement factor function of SNR [14, 15] and $mod\,(a, b)$ is the modulo operation with b being the divisor. As derived above, optimizing $T^p_{i,q}$ can steer target echoes into high-gain Doppler channels, thereby enhancing measurement quality.

Additionally, other critical time-domain resource include data rate and processing intervals [16]. The data rate governs the data rate provided for processing, directly impacting system responsiveness and adaptability in rapidly evolving scenarios [17]. Processing intervals define the allocated duration for computational tasks within the radar system, influencing real-time performance and efficient data analysis capabilities [18].

The effective allocation of time resource has been a prominent focus in the literature, highlighting its critical role in improving tracking accuracy [19], reducing mutual interference among radar nodes [20], and enhancing transmission efficiency [21]. The efficiency and constraints associated with time-domain resource allocation vary depending on operational scenarios and task priorities. In tracking applications, allocating greater time resource can significantly enhance tracking performance [22]. Conversely, in anti-reconnaissance and anti-jamming scenarios, excessive time allocation may increase the radar's vulnerability to interception, thereby compromising system survivability [23]. Furthermore, tracking maneuvering or high-speed targets necessitates higher data rates and shorter processing intervals to mitigate information loss [24]. In contrast, detecting low, small, and slow targets demands longer pulse width, higher PRF, and extended dwell times to ensure an adequate probability of detection [25].

25.3.3 Spatial Domain Resource

Spatial resource define the geometric relationships between radar nodes and targets, influencing both the relative positioning of targets and the pointing direction of the radar main lobe. In RSNs, spatial resource include node placement, beam direction, and beam width. A representative

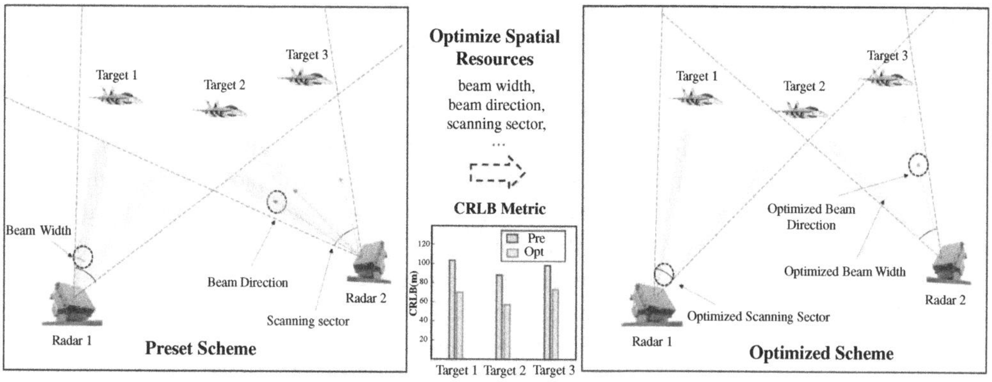

Figure 25.5 Illustration of typical spatial resource in radar system.

diagram illustrating typical spatial resource optimization is provided in Figure 25.5. Key spatial resource parameters include:

- Beam direction $\vartheta_{i,q}$: Indicates the specific spatial angle at which the radar signal is directed into the electromagnetic environment.
- Beam width $\theta_{3dB,i,q}$: Specifies the azimuth/elevation span between the half-power points of the antenna main lobe, which directly affects target angular resolution and energy coverage efficiency.
- Scanning sector $\Omega_i^s = \{(\theta, \phi) \in \mathbb{R}^2 | \theta_{\min} \leq \theta \leq \theta_{\max}, \phi_{\min} \leq \phi \leq \phi_{\max}\}$: Refers to the angular surveillance coverage, determining whether targets can be effectively detected and tracked, where θ and ϕ are azimuth and elevation, respectively.

The beam direction and beam width jointly determine the antenna transmission gain to target q, when a Gaussian function is used to approximate the transmission gain of a one-way antenna, $G_{i,q}(\vartheta_{i,q}, \theta_{3dB,i,q})$ can be expressed as [26]

$$G_{i,q}(\vartheta_{i,q}, \theta_{3dB,i,q}) \approx \exp\left(-\frac{2.776(\vartheta_{i,q} - \theta_{i,q})^2}{\theta_{3dB,i,q}^2}\right), \tag{25.13}$$

where $\theta_{i,q}$ is the azimuth of the target q relative to the radar i. These spatial resource effects measurement quality with the following relation [3]

$$\sigma_{\tau,i,q}^2 \propto G_{i,q}^{-1}(\vartheta_{i,q}, \theta_{3dB,i,q}), \sigma_{\theta,i,q}^2 \propto \theta_{3dB,i,q}^2 G_{i,q}^{-1}(\vartheta_{i,q}, \theta_{3dB,i,q}), \sigma_{f,i,q}^2 \propto G_{i,q}^{-1}(\vartheta_{i,q}, \theta_{3dB,i,q}). \tag{25.14}$$

For target tracking, the scanning sector Ω_i^s directly determines the radar's illumination decision on targets. The quantitative relationship between this resource allocation and tracking performance can be expressed as

$$\mathbf{J}\left(\mathbf{x}_k^q\right) = \mathbf{J}_P(\mathbf{x}_k^q) + u_s(\mathbf{x}_k^q, \Omega_i^s)\mathbf{J}_Z(\mathbf{x}_k^q, \mathbf{s}_k^q), \tag{25.15}$$

where $u_s(\mathbf{x}_k^q, \Omega_i^s)$ represents the resource-to-illumination mapping function. It captures the coupling between spatial resource configuration and target illumination strategy given as

$$u_s(\mathbf{x}_k^q, \Omega_i^s) = \begin{cases} 1, (\theta_{i,q}, \phi_{i,q}) \in \Omega_i^s \\ 0, else \end{cases}, \tag{25.16}$$

where $(\theta_{i,q}, \phi_{i,q})$ is the azimuth and elevation angle of target q w.r.t radar i.

The positions of radar nodes and the target jointly determine the Jacobian matrix $\mathbf{H}_{i,q,k}$ within $\mathbf{J}_Z(\mathbf{x}_k^q, \mathbf{s}_k^q)$, representing the partial derivatives of radar measurements with respect to the target states. The Jacobian matrix depends on the geometric configuration between the radar and the target. Such a mathematical relationship fundamentally accounts for the spatial diversity gain: under identical energy constraints, different spatial configurations result in distinct information matrices. For instance, consider a scenario in which two radars are deployed for target localization. The radar bandwidth is set to 300 MHz, the beam width is 3°, and the target is positioned at (300, 300)m. Radar 1 is fixed at (300, 100)m, while the position of radar 2 is variable but always maintains a constant distance from the target, as shown in Figure 25.6a. Under these conditions, the CRLB for target localization is illustrated in Figure 25.6b. Optimal positioning accuracy occurs when the angle formed between the two radars and the target approaches 90°. This phenomenon arises from the geometric characteristics of error ellipses, that the distance error contours of the two radars become nearly orthogonal, thereby minimizing the area of the position uncertainty region.

Moreover, other spatial resource encompass the maneuvering capabilities of radar nodes, including acceleration and angular velocity [27]. These maneuvering capabilities determine the radar's motion state, thereby affecting not only the quality of acquired radar measurement information but also the nonlinear mapping relationship between measured data and the target state estimates.

Efficient management and allocation of spatial resource in RSN plays a crucial role in optimizing system coverage and target tracking capabilities [28–30]. For instance, array optimization in RSN involves determining the positions and numbers of radar nodes [28]. By conducting a reasonable array optimization, the distance between radar nodes can be appropriately maintained, ensuring good coverage and target detection capabilities [31]. For search/tracking missions, the dynamic allocation of airspace resource can enhance the diversity of observation data in RSN [29]. This allows the fusion center to make better utilization of the spatial diversity information provided by radar nodes, leading to improved target detection and tracking performance.

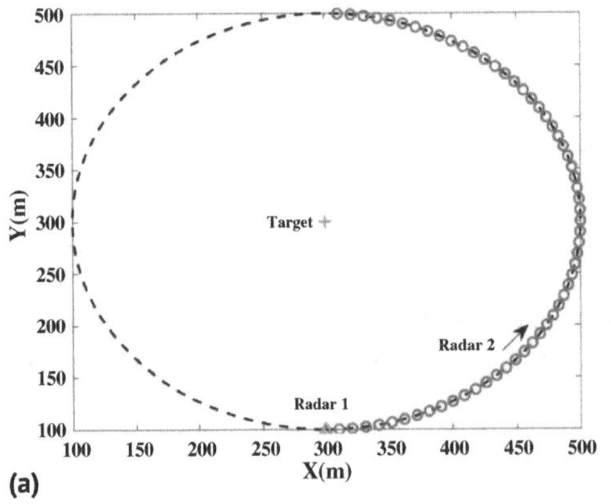

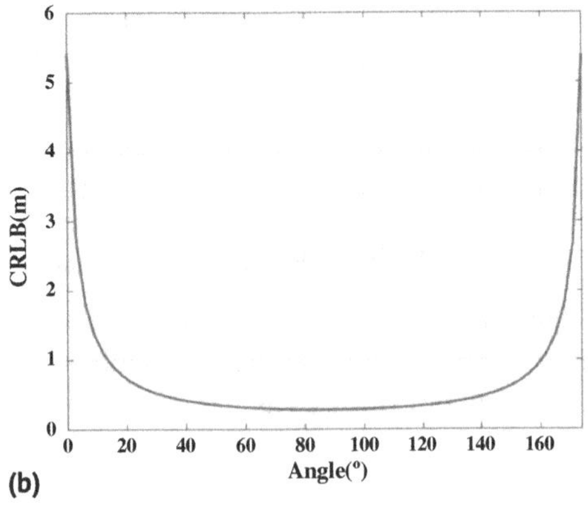

Figure 25.6 Relationship between radar position and CRLB. (a) Relationship between target and radar. (b) The target location CRLB.

25.3.4 Frequency Domain Resource

Frequency domain resource determine the response characteristics of targets and their ability to resist interference. These resource include carrier frequency, bandwidth, spectrum planning, communication bandwidth, etc. Representative frequency-domain resource include:

- Carrier frequency $f^c_{i,q}$: It directly impacts key performance indicators such as detection range, resolution, tracking accuracy, and anti-jamming capability.
- Bandwidth $\beta_{i,q}$: Bandwidth is the frequency range covered by the transmission signal, determining the radar's ability for distance resolution and ranging performance.

For target detection, the effective receiving area of the radar receiving antenna can be expressed as $A_{r,i,q} = G_{r,i}\lambda^2_{i,q}/4\pi$, where $G_{r,i}$ is the antenna gain of the receiving antenna, and $\lambda_{i,q} = c/f^c_{i,q}$ is the wavelength of the transmitted signal. These frequency resource effects measure quality with the following relation [3]

$$\sigma^2_{\tau,i,q} \propto (f^c_{i,q})^2/\beta^2_{i,q}, \sigma^2_{f,i,q} \propto (f^c_{i,q})^2. \tag{25.17}$$

For the range, velocity, and angular resolution of target, carrier frequency effects can be expressed as

$$\Delta_{r,i,q} = \frac{c}{2\beta_{i,q}}, \Delta_{\theta_{i,q}} \approx \frac{\lambda_{i,q}}{N_{r,q}d}k, \Delta_{v_{i,q}} = \frac{\lambda_{i,q}}{2T^c_{i,q}} \tag{25.18}$$

It is worth noting that to avoid the distortion of the transmitted signal, the instantaneous bandwidth usually does not exceed 10% of the carrier frequency, i.e., $\beta_{i,q} \leq 0.1 f^c_{i,q}$. For target tracking, frequency resource allocation can determine the detection ability of the system under interference conditions. The quantitative relationship between this resource allocation and tracking performance can be expressed as

$$\mathbf{J}(\mathbf{x}^q_k) = \mathbf{J}_P(\mathbf{x}^q_k) + u_f(f^c_{i,q}, \beta_{i,q})\mathbf{J}_Z(\mathbf{x}^q_k, \mathbf{s}^q_k), \tag{25.19}$$

where $u_f(f^c_{i,q}, \beta_{i,q})$ represents the attenuation factor of measurement information under interference conditions [32]. Note that the choice of carrier frequencies not only counts the above effects but also depends on scenarios. For instance, low-frequency radars exhibit longer wavelengths experience lower atmospheric attenuation, making them suitable for long-range early warning [33]. High-frequency radars have shorter detection ranges but offer higher resolution and greater spectral flexibility [34]. Furthermore, frequency domain resource encompass the radar's transmission communication bandwidth and spectrum planning schemes [20, 35]. Optimizing communication bandwidth ensures seamless exchange and synchronization of information between nodes [36]. Through spectrum planning techniques like frequency agility, the anti-jamming capability of networked radars can be effectively enhanced [32].

Reasonable dynamic allocation of frequency domain resource can effectively improve target detection and tracking performance [37, 38]. For instance, by effectively optimizing frequency domain resource, RSNs can enable efficient data sharing and communication among networked nodes [39]. By utilizing multi-carrier frequencies, radar can increase maximum detectable unambiguous velocity [40]. Through environment-aware techniques, intelligent dynamic adjustment of frequency domain resource, including carrier frequency and bandwidth can enhance the radar system's resistance to jamming [41, 42].

25.3.5 Coding Domain Resource

Coding domain resource primarily involve the encoding style of radar transmission waveforms. It determines radar transmission pattern, and can enhance detection, tracking, and anti-jamming capabilities. Coding domain resource refer to the waveform coding strategies that govern signal structure through $s_i(t) = \sum_{n=0}^{N-1} c_{i,n}p(t - nT_c)$, where $\mathbf{c}_i = [c_{i,0}, \cdots, c_{i,N-1}]$ denotes the phase-coded sequence of radar i, $c_{i,n}$ is the nth coding phase, $p(t)$ is the sub-pulse waveform, and T_c the chip duration. Key code-domain metric is the orthogonality, defined as $\rho_{ij} = \frac{|\langle \mathbf{c}_i, \mathbf{c}_j \rangle|}{\| \mathbf{c}_i \| \| \mathbf{c}_j \|}$. It affects the measurement quality with the following relation [2]:

$$\sigma^2_{\tau,i,q} \propto (1 - \sum_{i \neq j} \rho_{ij})^{-1}, \sigma^2_{\theta,i,q} \propto (1 - \sum_{i \neq j} \rho_{ij})^{-1}, \sigma^2_{f,i,q} \propto (1 - \sum_{i \neq j} \rho_{ij})^{-1}. \tag{25.20}$$

Optimization of code domain resource can effectively improve the performance of target detection and tracking. For example, optimizing the design of radar transmission waveforms with low side-lobes can improve radar detection and tracking of targets [43, 44]. Designing multiple waveform encoding styles with different gain main lobes can enable differential detection and tracking of multiple targets [45]. In the presence of prior interference information, optimizing the encoding style to design nulls at specific locations can effectively reduce the impact of interference signals [46, 47].

25.4 Tracking Performance Metrics

Performance evaluation of target tracking is of paramount importance, not only for the refinement of tracking algorithms, but also for the development of robust and realistic resource allocation schemes. A variety of quantitative metrics have been established in the literature. The most commonly employed metrics include signal-to-noise ratio (SNR), signal-to-interference-and-noise ratio (SINR), tracking position error, number of tracks, mutual information, etc. A short summary of different metric is presented in Table 25.2.

25.4.1 Signal Quality

SNR is a fundamental metric that quantifies the relative strength of the desired signal to background noise, while the SINR extends this concept by accounting for both interference and noise. Higher values of SNR or SINR generally correspond to improved detection reliability and tracking

Table 25.2 Summary of performance metrics.

Performance metric	Description	Key formula
Signal quality (Section 25.4.1)	SNR/SINR	(25.21)
Tracking position error (Section 25.4.2)	Predicted BCRLB predicted conditional BCRLB PC-BCRLB under clutters	(25.22) (25.23) (25.24)
Number of tracks (Section 25.4.3)	Number of tracks with satisfied quality	(25.25)
Mutual information (Section 25.4.4)	Information gain about targets	(25.26)
Others (Section 25.4.5)	Low probability of intercept area coverage, false track rate, etc.	–

accuracy, making these metrics essential indicators of system performance, particularly in cluttered or contested environments. In the context of RSN, the SNR for radar i observing target q can be expressed as [48]:

$$\mathrm{SNR}_{i,q} = \frac{P_{i,q} G_{\mathrm{T},i,q} G_{\mathrm{R},i,q} \sigma_q \lambda_i^2}{(4\pi)^3 k T B_i F_i R_{i,q}^4 L_i} \propto \frac{P_{i,q} \sigma_q \lambda_i^2}{R_{i,q}^4}, \tag{25.21}$$

where $P_{i,q}$ is the transmit power of radar i for target q, σ_q denotes the RCS of target q, λ_i is the transmit wavelength, $R_{i,q}$ is the distance between radar i and target q, $G_{\mathrm{T},i,q}$ and $G_{\mathrm{R},i,q}$ are the transmit and receive antenna gains, k is Boltzmann's constant, T is the noise temperature, B_i is the receiver bandwidth, F_i is the receiver noise figure, and L_i represents system losses. From this formula, it is evident that several resource parameters can be adjusted to improve SNR, such as increasing transmit power $P_{i,q}$, selecting a carrier frequency and radar system with larger $\sigma_q \lambda_i^2$, deploying or selecting radars at shorter distances to the target $R_{i,q}$, optimizing antenna gains and minimizing system losses. For SINR, the formulation is similar but incorporates the power of interference sources in the denominator [49].

Both SNR and SINR have a direct and significant impact on the quality of the received echo signal and, consequently, on the performance of detection and tracking algorithms. Owing to their critical role in characterizing signal quality, these metrics are widely adopted in the literature [50–52] as key performance indicators for resource allocation in RSNs.

25.4.2 Tracking Position Error

In the field of target tracking, target position error is the most important evaluation metric. Generally speaking, it measures the deviation between the estimated target position and the true (ground truth) position. As a direct indicator, the Bayesian Cramér–Rao lower bound (BCRLB) provides a lower bound for the mean square error (MSE) of target tracking problems, and it is related to the Bayesian information matrix (BIM) in (25.3) and the BCRLB is the inverse of the BIM [53].

However, the BCRLB is an offline bound for static systems, and it cannot be directly applied to resource management, which needs to be predictive and allocate resource online. Thus, variants of BCRLB have been developed for resource management. Common criteria include predicted BCRLB [7], predicted conditional Cramér–Rao lower bound (PC-CRLB) [54] and PC-CRLB with measurement origin uncertainty (MOU):

- Predicted BCRLB: In general, predicted BCRLB is the simplification of BCRLB [7], which removes the expectation calculation, given as

$$\mathbf{J}\left(\mathbf{x}_k^q, \mathbf{s}_k^q\right) = \left[\mathbf{Q}_{k-1}^q + \mathbf{F}^q \mathbf{J}^{-1}(\mathbf{x}_{k-1}^q)(\mathbf{F}^q)^T\right]^{-1} + \sum_{i=1}^{N} \left\{\mathbf{H}_{i,q,k}^T \left[\boldsymbol{\Sigma}_k^{i,q}(\mathbf{s}_k^q)\right]^{-1} \mathbf{H}_{i,q,k}\right\}. \tag{25.22}$$

  As the earliest widely adopted online evaluation criteria, the effectiveness of predicted BCRLB has been demonstrated in numerous studies [19, 55, 56].
- Predicted Conditional CRLB (PC-CRLB): Predicted conditional CRLB (PC-CRLB) characterizes performance conditioned on the actual measurements that have been received and do not need to be calculated recursively like the predicted BCRLB. It is more suitable for online resource management. Specially, the PC-FIM can still be expressed as [54]:

$$\mathbf{J}\left(\mathbf{x}_k^q, \mathbf{s}_k^q\right) \approx (\mathbf{P}_{k|k-1}^q)^{-1} + \sum_{i=1}^{N} \mathbf{H}_{i,q,k}^T \left[\boldsymbol{\Sigma}_k^{i,q}(\mathbf{s}_k^q)\right]^{-1} \mathbf{H}_{i,q,k}. \tag{25.23}$$

where $\mathbf{P}^q_{k|k-1}$ is the estimates for the state covariance matrix of target q from tracker. Finally, the PC-CRLB is the inverse of the PC-FIM. Since the implementation of PC-CRLB depends on recently acquired measurements, this performance criterion offers a tighter bound—particularly beneficial for managing resource when tracking targets with complex motion patterns [57, 58].

- PC-CRLB under the Condition of Clutter (also see details in Chapter 10) [59] For target tracking in clutter, multiple measurements may be obtained at each scan, and it is not known which one of them comes from the target. This is termed MOU. Similar to the performance criterion in non-clutter environment, the PC-FIM with MOU is calculated recursively without manipulating the large matrices at each time index [6]:

$$\mathbf{J}\left(\mathbf{x}_k^q, \mathbf{s}_k^q\right) = (\mathbf{P}^q_{k|k-1})^{-1} + \sum_{i=1}^{N}\left[T\left(P_{\mathrm{d}}^{i,q,k}, \mathbf{\Sigma}_k^{i,q}(\mathbf{s}_k^q)\right)\mathbf{H}^T_{i,q,k}\left[\mathbf{\Sigma}_k^{i,q}(\mathbf{s}_k^q)\right]^{-1}\mathbf{H}_{i,q,k}\right], \tag{25.24}$$

where the scalar $T\left(P_{\mathrm{d}}^{i,q,k}, \mathbf{\Sigma}_k^{i,q}(\mathbf{s}_k^q)\right)$ is the information reduction factor (IRF), which is related to the probability of detection (PD) $P_{\mathrm{d}}^{i,q,k}$ and measurement covariance matrix $\mathbf{\Sigma}_k^{i,q}$ [60]. Finally, the PC-CRLB with MOU is the inverse of the PC-FIM.

The key to constructing a performance criterion with MOU lies in introducing the IRF to quantify the degradation of the data information matrix under imperfect detection conditions. Its effectiveness in resource management has also been well demonstrated [6, 53, 61, 62].

25.4.3 Number of Tracks

Based on the aforementioned performance metric related to target tracking, the number of tracks reflects the capability of RSN, which is particularly relevant for evaluating scalability and robustness. Specifically, its primary objective is to maximize the number of targets tracked to requirements of RSN within a fixed resource budget, while also promoting efficient resource usage. Accordingly, the performance criterion can be expressed as:

$$f(\mathbf{s}_k^1, \dots, \mathbf{s}_k^Q) = \sum_{q=1}^{Q}\Gamma(\eta_{q,k} - \mathbb{F}_q(\mathbf{s}_k^q)), \tag{25.25}$$

where $\eta_{q,k}$ denotes the requirements of RSN for target q, $\Gamma(m)$ is the indicator function, i.e., it equals to 1 for $m \geq 0$ and otherwise is 0, and $\mathbb{F}_q(\mathbf{s}_k^q)$ is the tracking performance of target q, such as SNR, SINR and PC-CRLB. Essentially, this metric aims to increase the number of tracks that satisfy system requirements [5, 63].

25.4.4 Mutual Information

This information-theoretic metric quantifies the reduction in uncertainty about the target state obtained and thus offers a principled means for evaluating and optimizing sensor resource allocation. Suppose that the RSN has M radar sensor as transmitters and N radar sensor as receivers, denote X as the set of transmission sequences, Y as the sampled backscatter signals for all radars, K as the discrete time index, and G as the target reflection gain matrix between the different targets, then the definition of mutual information (MI) can be expressed as [50]:

$$I(Y, G|X) = \sum_{i=1}^{M}\sum_{j=1}^{N}\ln\left(\sigma_\theta^2 + P_i\sigma_g^2 p_{ij}^2\right) - NK\ln\sigma_\theta^2, \tag{25.26}$$

where P_i is the power, σ_g^2 is the covariance of target reflection coefficient, σ_θ^2 is covariance of the Gaussian white noise, $p_{i,j}$ represents the variation in the signal strength due to antenna properties and propagation losses.

Intuitively, maximizing mutual information (MI) often entails increasing transmit power, which can make the radar network more susceptible to detection. Therefore, MI is commonly used as a constraint in radar resource management problem [51, 64]. In particular, [65] adopts MI as the objective function, which aims to maximize the MI between transmit and received signals to enhance information acquisition from targets, while simultaneously minimizing the MI between echo signals at adjacent time instances to reduce redundant information.

25.4.5 Other Metrics

According to the specific requirements and operating environment of the tracking application, there are many other task-specific metrics considered in the literature. For example, the search radar equation is used to design a resource management model to improve the radar search performance [66]. The probability of intercept has been widely employed to evaluate the low probability of intercept (LPI) performance for RSN [67]. The area coverage emphasizes the coverage of the radar perception area, which is used to evaluate the search performance under a given radar configuration scheme [28]. The track confirmation time is defined as the time from the initial detection of the target to the confirmation of the target through continuous observation [68]. In the target tracking aspect, the false track rate is also a major metric to be considered in resource management, which aims to control the number of ghost tracks [69]. Besides, the optimal subpattern assignment metric is characterized by its unique capability to simultaneously account for cardinality errors and localization errors. It is adopted in the literature [70–72] due to its tunable balance between these two aspects through adjustable parameters.

25.5 Resource Allocation Models

Overall, existing RA work for MTT can be divided into two types of optimization models, tracking quality constrained resource allocation (TQC-RA) and performance-driven resource allocation (PD-RA) models, each designed to enhance resource utilization efficiency in RSNs, but prioritizing differently between resource consumption and tracking performance based on the application context. The TQC-RA model focuses on minimizing system resource consumption while meeting specified tracking performance requirements. This optimization approach allows the saved resource to be redirected toward other radar functions, such as surveillance or fire control. For the PD-RA model, it aims to maximize overall MTT performance within a predefined radar resource budget. This model optimizes a specific system-level utility function, subject to defined resource constraints.

It is important to recognize that the resource allocation problem is inherently dynamic, as resource must be reallocated at each tracking frame. Both the availability of resource and the geometric relationships between targets and radar sensors can vary over time. Consequently, it is necessary to periodically reconstruct the optimization models based on the current resource availability and to recalculate the corresponding solutions for each time frame. Besides, an essential aspect of RA is the inherent trade-offs between different types of resources and performance metrics. For instance, while increasing transmit power can improve the signal-to-noise ratio (SNR) and overall tracking accuracy, it may lead to higher energy consumption and reduced system stealth.

Similarly, allocating longer dwell times can enhance Doppler resolution but may result in fewer opportunities to revisit other targets, thereby impacting coverage.

Next, to better illustrate the resource allocation strategy, transmit power is selected as the representative resource. For notational simplicity, we formulate generalized RA models and omit the time index k in this section.

25.5.1 Tracking Quality Constrained Resource Allocation Model

Literature: The TQC-RA model has been extensively studied across various scenarios, focusing on optimizing resource efficiency while meeting tracking performance constraints. For example, references [73–75] aim to minimize transmit power under detection or tracking constraints, while [51] employs the SINR metric for target detection and optimizes power spectral density to reduce energy consumption. Reference [76] introduces an RA scheme for MTT localization, minimizing power while ensuring localization accuracy, and [77] extends this by formulating the localization task as a bi-objective constrained optimization problem. Other works, such as [64] and [78], incorporate the CRLB, communication data rate, and mutual information as performance constraints, while reference [79] focuses on optimizing dwell time resource using the BCRLB. Hybrid RA methods have also been explored in heterogeneous RSN environments, as shown in [19] and [22], where generalized TQC-RA models aim to minimize resource consumption while ensuring tracking performance. In addition to transmit resource optimization, RA studies for receiving roles in RSN have examined sensor selection and joint transmit-receive strategies. References [80, 81] address sensor selection to meet tracking accuracy demands, while joint optimization of transmitting and receiving resource is explored in [4, 67, 82–85]. These works propose various radar selection and power allocation schemes tailored for localization and tracking. For example, [84] incorporates deep learning for joint radar selection, while [67] optimizes radar node selection, power, and bandwidth to enhance LPI performance. Reference [85] uses SINR to design a joint power allocation and sensor selection scheme. For more details on the literature, please refer to the discussion in [8].

Next, we give the general form of TQC-RA model. The objective is to minimize resource consumption while ensuring guaranteed performance [76], which can be formulated as

$$\min_{\mathbf{s}\in\mathcal{S}} \sum_{i=1}^{N} \mathbf{1}^T \cdot \sum_{q=1}^{Q} \mathbf{g}_{i,q}(\mathbf{s}_q) \quad \text{s.t.} \quad B_q(\mathbf{s}_q) - \eta_q \leq 0, \forall q, \tag{25.27a}$$

$$\mathbf{c}_q(\mathbf{s}_q) - \boldsymbol{\kappa}_q \leq \mathbf{0}, \forall q, \tag{25.27b}$$

$$\sum_{q=1}^{Q} \mathbf{g}_i(\mathbf{s}_q) \leq \mathbf{r}_i, \forall i. \tag{25.27c}$$

The detailed explanation of this formulation is as follows.

- Optimization Variable: The variable $\mathbf{s} = (\mathbf{s}_1, \cdots, \mathbf{s}_Q) \in \mathcal{S} \subset \mathbb{R}^{QMN}$ represents the resource allocation vector for Q targets. For each target q, $\mathbf{s}_q \in \mathbb{R}^{MN}$ contains the allocation of M types of resource across N radar sensors. The feasible set $\mathcal{S}$ encodes basic resource constraints due to physical limitations, such as nonnegativity of power or other resource bounds.
- Objective Function: The objective is to minimize the resource consumption where $\mathbf{g}_i(\cdot) : \mathbb{R}^{MN} \mapsto \mathbb{R}^M$ is the normalized resource consumption function. It ensures that different types of resource (e.g., power, bandwidth, etc.) are adapted to a common scale despite being measured in different units.

- Constraints: The constraints in the problem are structured as below.
  - performance constraints (25.27a): The performance for tracking target q is denoted by $B_q(\mathbf{s}_q) : \mathbb{R}^{MN} \mapsto \mathbb{R}$, it must meet the required threshold $\eta_q \in \mathbb{R}$.
  - operational constraints (25.27b): Additional operational function is expressed as $\mathbf{c}_q(\cdot) : \mathbb{R}^{MN} \mapsto \mathbb{R}^L$, representing requirements such as target SNR or MI, where $\boldsymbol{\kappa}_q \in \mathbb{R}^L$ is the vector of threshold values for the operational requirements.
  - resource constraints (25.27c): For each radar sensor i, the total allocated resource across all targets must not exceed the sensor's maximum available resource $\mathbf{r}_i \in \mathbb{R}^M$.

The complexity of solving problem (25.27) depends on the properties of the mapping functions $\mathbf{g}_i(\cdot)$, $B_q(\cdot)$, and $\mathbf{c}_q(\cdot)$. If these functions are convex, the problem becomes a convex optimization problem that can be efficiently solved using methods like the interior point method, with many toolboxes available. However, in practical applications, the problem often involves non-convex or discrete structures, necessitating the development of specialized algorithms that leverage the problem's unique features. A detailed discussion on algorithm development can be found in [8] and is not elaborated upon here.

25.5.1.1 TQC-RA Example

To illustrate the application of the TQC-RA scheme, we consider an example where the RSN consists of $N = 3$ radar sensors. The positions of these radar sensors are outlined in Table 25.3. The RSN is capable of acquiring both distance and angle information for multiple targets, enabling target tracking at each frame. The effective signal bandwidth and null-to-null beamwidth for each radar sensor are set as 1 MHz and 1°, respectively. Additionally, the available power budget for each radar sensor is denoted as $s_{i,\text{total}}$ ($i = 1, 2, 3$. Different radar sensors may have different transmit resource budgets, and the budget model is set to $s_{1,\text{total}} = s_{2,\text{total}} = s_{\text{total}}$ and $s_{3,\text{total}} = 7.5s_{\text{total}}$. In this scenario, there are $Q = 4$ targets present within the surveillance area, each moving with nearly constant velocities. The RCS of all targets is assumed to be 1 m^2. For a reference target located at a distance of 30 km, the SNR is set to 10 dB when the transmit power is allocated as $s_{i,\text{total}}/Q$. Using the radar equation, the SNR for the remaining targets can be calculated accordingly. The additional parameters for all targets are detailed in Table 25.4, while Figure 25.7 illustrates the spatial distribution of the RSN in relation to these targets.

The objective of the TQC-RA scheme is to minimize resource consumption while ensuring guaranteed performance. In this example, the normalized resource consumption function is expressed as $\mathbf{g}_i(s_{i,q}) = \dfrac{s_{i,q}}{s_{i,\text{total}}}$. The constraints primarily consist of performance requirements and resource

Table 25.3 Parameters setting.

Radar index	1	2	3
Position(km)	(−40, 0)	(0, 0)	(40, 0))

Table 25.4 Parameters setting.

Target index	1	2	3	4
Position(km)	(−60, 20)	(−20, 20)	(20, 20)	(60, 20)
Speed(m/s)	(0, −80)	(0, −80)	(0, −80)	(0, −80)
η_q(m)	100	20	20	120

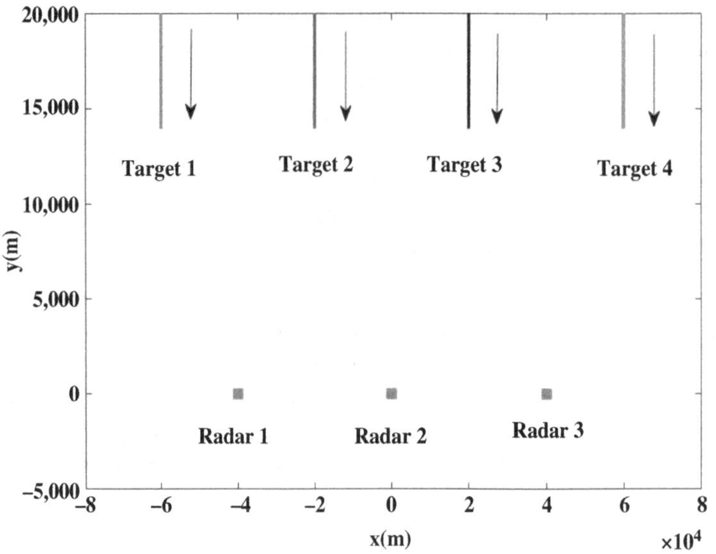

Figure 25.7 Deployment of targets with respect to the RSN.

limitations. Specifically, the predicted BCRLB is used to constrain the target tracking performance. Consequently, $B_q(\mathbf{s}_q)$ is defined as:

$$B_q(\mathbf{s}_q) = \sqrt{\mathrm{Trace}\left(\mathbf{J}^{-1}\left(\mathbf{x}_k^q, \mathbf{s}_q\right)\right)} \leq \eta_q,$$

where η_q represents the required target tracking accuracy. The resource constraints ensure that the allocated resource of each radar node do not exceed their total available resource. This constraint can be formulated as $\sum_{q=1}^{Q} \mathbf{g}_i(\mathbf{s}_q) \leq 1, \quad \forall i$. Combining the objective function and the constraints, the optimization model for the TQC-RA scheme is given by:

$$\min_{\mathbf{s} \in \mathcal{S}} \sum_{i=1}^{N} \sum_{q=1}^{Q} \frac{s_{i,q}}{s_{i,\mathrm{total}}} \quad \text{s.t.} \quad B_q(\mathbf{s}_q) - \eta_q \leq 0, \forall q, \qquad \sum_{q=1}^{Q} s_{i,q} \leq s_{i,\mathrm{total}}, \forall i. \tag{25.28}$$

Here, $\mathcal{S}$ represents the nonnegative orthogonal space, implying that all power allocations must be nonnegative. Problem (25.28) is a nonlinear constrained convex optimization problem. The Karush-Kuhn-Tucker (KKT) conditions provide the necessary and sufficient conditions for achieving optimality [22]. The detailed solution process is omitted here for brevity. In the following simulation, the benchmark denotes the case that the limited transmit resource of each radar sensor is uniformly allocated to multiple targets, the accuracy requirement of the RSN for each target is different and total 20 frames are used. RSN initializes the target state within the first three frames to ensure stable tracking of each target. The TQC-RA scheme is then executed from the frame 4.

It can be observed from Figure 25.8 that TQC-RA framework can ensure that the BCRLB of RSN for each target is capable of meeting the requirements $\eta_{i,k}$ for tracking accuracy. Here, we only present the BCRLBs w.r.t. multiple targets without giving their root mean quare errors (RMSEs) for the following reasons: 1) The RMSE is shown to be asymptotically tight to the BCRLB in some

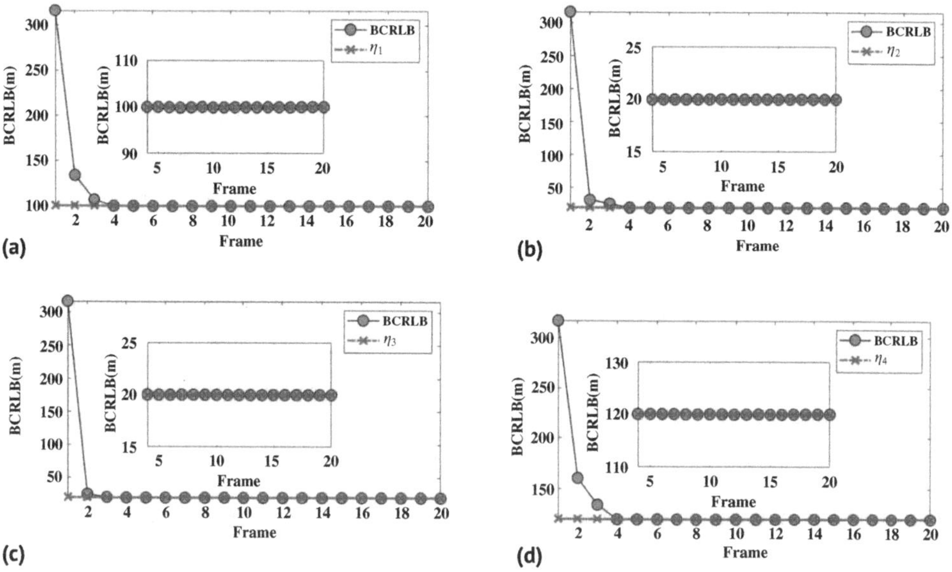

Figure 25.8 Tracking BCRLB of each target in the TQC-RA scheme. (a) Target 1. (b) Target 2. (c) Target 3. (d) Target 4.

works [29, 55, 56]; 2) The RMSE may fluctuate from time to time due to noise uncertainty, which may prevent us from verifying optimality condition explicitly.

In addition to the tracking accuracy requirements, another objective of resource management is to reduce the power resource consumption of the RSN. In Figure 25.9, the resource saving ratios w.r.t. the TQC-RA scheme as well as the benchmark are presented for given MTT accuracy requirements. Here, the resource saving ratio is denoted as:

$$\rho_k = \left(1 - \frac{\sum_{i=1}^{N} \sum_{q=1}^{Q} s_{i,q,k}^{\text{opt}} / s_{i,\text{total}}}{\sum_{i=1}^{N} \sum_{q=1}^{Q} s_{i,q,k}^{\text{uni}} / s_{i,\text{total}}}\right)$$

where $s_{i,q,k}^{\text{opt}}$ and $s_{i,q,k}^{\text{uni}}$ denote the resource allocation results w.r.t. the TQC-RA scheme and the benchmark. Figure 25.9a shows TQC-RA scheme can reduce power consumption by more than 40% when compared with the benchmark. Specially, there exists no prior information for RA, and the limited resource of each radar sensor is uniformly allocated to multiple targets at frame $k = 1, 2, 3$, and thus $\rho_1 = 0$.

The normalized resource allocation results $\delta_{i,q} = s_{i,q}/s_{i,\text{total}}, \forall i, q$ of each radar sensor w.r.t. each target are shown separately in Figure 25.9b. The results show that the TQC-RA scheme can automatically select radar sensors to illuminate each target while optimizing the transmit resource of those selected radar sensors for given tracking accuracy thresholds. Generally, the following types of radar sensors are more likely to be selected for a specified target. The radar sensor with closer distance and better angular spread. For example, radar sensor 2 is designated to work with radar sensor 1 and radar sensor 3 to track target 2 and target 3, respectively. The radar sensor with relatively larger available resource budget. For instance, radar sensor 3 cooperates with radar sensor 1 to track target 1 due to its larger resource budget and better angular spread compared with other radar sensors.

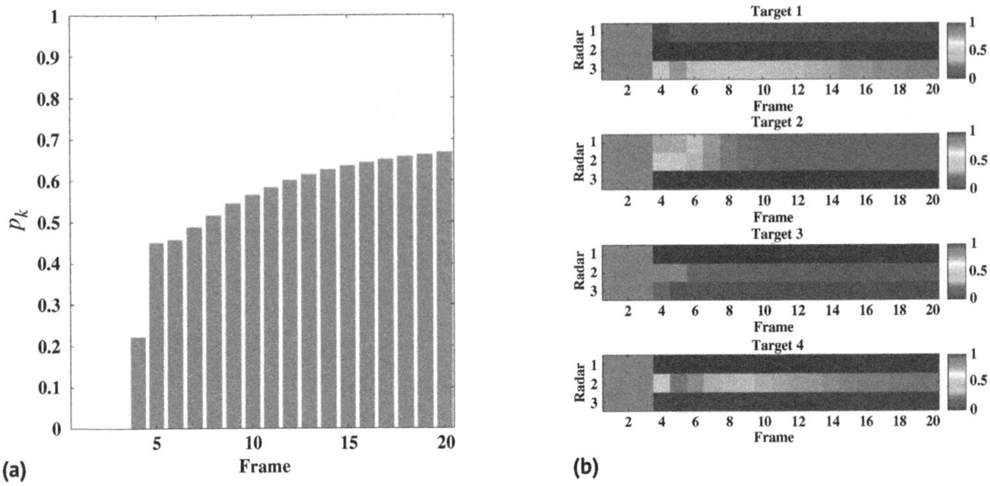

Figure 25.9 Resource utilization of TQC-RA scheme. (a) Resource saving ratio. (b) Normalized resource allocation results.

25.5.2 Performance-driven Resource Allocation Model

Literature: Overall, the PD-RA model has received more extensive examination than the TQC-RA model in the literature. The PD-RA model typically utilizes a function related to each target's tracking performance as the optimization criterion, which is the core in many RSN applications. Numerous studies, including [57, 86–88], have employed the BCRLB to formulate power allocation schemes aimed at minimizing tracking mean squared error (MSE), while others, such as [56, 89, 90], extend this approach to dwell time allocation. Further advancements address cluttered environments [62], multifunction radar platforms [91, 92], and asynchronous heterogeneous RSNs [19]. Low probability of intercept (LPI) objectives [52], target capacity [63], and finite set statistics theory [93, 94] have also driven the development of hybrid allocation schemes optimizing transmit power, dwell time, signal bandwidth, and receiving resource [95]. Combinatorial optimization frameworks [80, 96] have been applied to radar subset selection to maximize target state estimation, while joint transmit-receive RA schemes optimize multitarget tracking (MTT) performance under resource constraints [25, 55]. Maneuvering RA, focusing on trajectory planning for tasks such as surveillance and attack, optimizes resource like acceleration and speed to achieve feasible paths or minimize mission time [97–100]. Studies also explore task-oriented maneuvering RA to enhance tracking performance by fully utilizing transmit and maneuvering resource [101, 102], with online path planning schemes improving detection and tracking of multiple targets [103]. Collectively, these efforts underscore the crucial role of RA in advancing RSN capabilities across diverse operational scenarios.

The PD-RA model aims to maximize the overall tracking performance for given radar resource budgets. The overall target tracking performance achieved by the RSN depends on its operational requirements. For different applications, different utility functions will be defined, which can be viewed as a mapping from individual target tracking performance to an overall tracking performance

$$u(\mathbf{s}) = f_u\big(B_1(\mathbf{s}), \cdots, B_Q(\mathbf{s})\big), \tag{25.29}$$

where $B_q(\cdot)$ represents the tracking performance of the target q and $f_u(\cdot)$ is a mapping function from single target tracking performance to overall tracking performance. Then, the RA strategy can be formulated as a problem of optimizing a certain system-level utility function subject to some resource constraints

$$\min_{\mathbf{s}\in\mathcal{S}} u(\mathbf{s}) \quad \text{s.t.} \quad \sum_{q=1}^{Q} \mathbf{g}_i(s_{i,q}) - \mathbf{r}_i \le \mathbf{0}, \forall i, \\ \mathbf{c}_q(\mathbf{s}_q) - \kappa_q \le \mathbf{0}, \forall q, \tag{25.30}$$

where set $\mathcal{S}$, $\mathbf{g}_i(\cdot)$, and $\mathbf{c}_q(\cdot)$ are the same as used in TQC-RA model.

Generally, the challenges in solving the PD-RA problem stem from the continuity and smoothness of the functions involved in the objective and constraints. When the PD-RA model is not a continuous or smooth optimization problem—for instance, when it includes binary variables for sensor selection—various approximation or relaxation techniques are employed. Similar to the TQC-RA problem, solving different PD-RA problems often necessitates the development of specialized algorithms that exploit the unique characteristics of each problem. A discussion on algorithm development is provided in [8], and thus, it is not elaborated upon here.

25.5.2.1 PD-RA Example

We consider a PD-RA example with a RSN consisting of N=3 radar sensors. The basic settings are the same as the TQC-RA example except the location of radars and trajects of targets, as shown in Figure. 25.10. The initial locations and speeds of each target are given in Table 25.5. In this example, the predicted BCRLB $\mathbf{J}^{-1}(\mathbf{x}_k^q)$ in (25.22) is used to evaluate the state estimation performance for each target. As BCRLB is a matrix related to the radar resource vector, and its diagonal elements

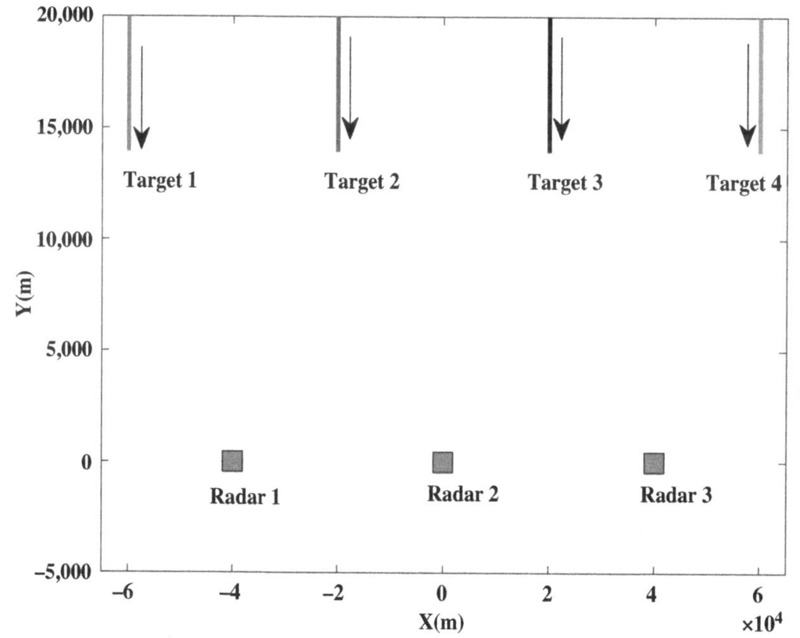

Figure 25.10 Deployment of targets with respect to RSN.

Table 25.5 Parameters setting.

Target index	1	2	3	4
Position (km)	(−60, 20)	(−20, 20)	(20, 20)	(60, 20)
Speed (m/s)	(0, −80)	(0, −80)	(0, −80)	(0, −80)

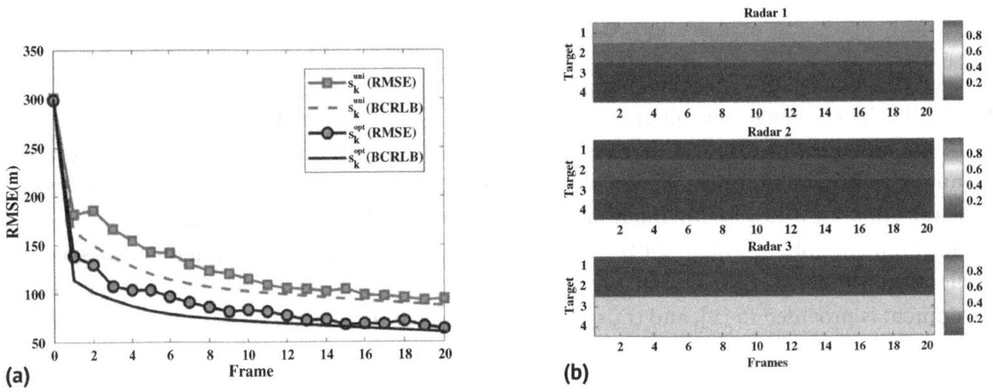

Figure 25.11 Performance comparison of PD-RA. (a) RMSE comparison. (b) The allocation results.

have different units. Therefore, by standardizing these elements, the target tracking performance metric can be defined as

$$\mathbf{B}_q(\mathbf{s}) = \mathrm{Tr}\left[\Lambda \cdot \mathbf{J}^{-1}\left(\mathbf{x}_k^q\right) \cdot \Lambda^T\right]$$

where Λ is the standardization matrix $\Lambda = \mathbf{I}_2 \otimes \begin{bmatrix} 1 & 0 \\ 0 & 0 \end{bmatrix}$ and $\mathrm{Tr}(\mathbf{A})$ returns the sum of the diagonal elements of matrix $\mathbf{A}$. The objective function for the PD-RA problem can be modeled as

$$u(\mathbf{s}) = \frac{1}{Q} \sum_{q=1}^{Q} \mathbf{B}_q(\mathbf{s})$$

Consequently, the optimization problem can be formulated as

$$\min_{\mathbf{s}} u(\mathbf{s}) \qquad \text{s.t.} \sum\nolimits_{q=1}^{Q} s_{i,q} = s_{i,\text{total}}, \forall i, \tag{25.31}$$

This optimization problem is to achieve optimal MTT performance by allocating power under total power limits. It is convex and can be solved for its optimal solution using methods such as the interior-point method or gradient descent.

To better examine the optimality of the PD-RA strategy, we compare its tracking accuracy, in terms of the RMSE and $u(\mathbf{s}_k)$, with the uniform resource allocation scheme. In our simulation, resource consumptions are always normalized into consumption ratios (by dividing $s_{i,\text{total}}$). The performance results of the PD-RA scheme are shown in Figure 25.11a. Note that the resource constraint is not to limit the total transmit resource of multiple radars, but to limit the transmit resource of each radar in this simulation.

The tracking results in Figure 25.11a show that the state estimation accuracy is very close to the BCRLB. Generally, the benchmark (uni, in Figure 25.11a) exhibits higher MTT BCRLB compared with an optimized system resource. The ratio of the transmit resource in this scheme is shown in

Figure 25.11b. The results in Figure 25.11b suggest that each radar needs to allocate its predetermined resource to multiple targets, and the resource are more likely to be distributed to its preferred targets, in the PD-RA scheme. When the radar resource is abundant, it may also distribute its redundant resource to other less preferred targets. Physically speaking, a radar is said to prefer target q, if it can provide a larger accuracy growth rate for this target than for any other. For example, radar 3 has sufficient resource, and the relationship between the improvement of target tracking performance and the occupied resource is not linear, so radar 3 is more inclined to allocate resource to target 3 and target 4 with higher improvement performance.

25.6 Conclusions

This chapter presents RA strategies for MTT in RSNs, which in principle, is the allocation of resource to control the measurement quality. Various radar resource types, performance metrics, and their impacts on RA were discussed to provide understanding of the current solutions in RSNs. Both TQC-RA and PD-RA optimization models were introduced and demonstrated with examples, showcasing their efficiency in optimizing resource for tracking. Given that RA in RSNs continues to be a vibrant research field, future investigations could focus on developing real-time RA strategies and their adaptation to dynamic and complex environments, leveraging online learning techniques to propel the field forward.

References

1 P. Tichavsky, C. H. Muravchik, and A. Nehorai. Posterior Cramér-Rao bounds for discrete-time nonlinear filtering. *IEEE Transactions on Signal Processing*, 46(5):1386–1396, 1998.

2 S. M. Kay. *Fundamentals of Statistical Signal Processing: Estimation Theory*. Prentice-Hall, Inc., 1993.

3 M. A. Richards. *Fundamentals of Radar Signal Processing*, Mcgraw-hill, 2nd edition, 2005.

4 Y. Lu, C. Han, Z. He, S. Liu, and Y. Wang. Adaptive jspa in distributed colocated mimo radar network for multiple targets tracking. *IET Radar, Sonar & Navigation*, 13(3):410–419, 2019.

5 J. Yan, P. Zhang, J. Dai, and H. Liu. Target capacity based simultaneous multibeam power allocation scheme for multiple target tracking application. *Signal Processing*, 178:107794, 2021.

6 J. Yan, B. Jiu, H. Liu, B. Chen, and Z. Bao. Prior knowledge-based simultaneous multibeam power allocation algorithm for cognitive multiple targets tracking in clutter. *IEEE Transactions on Signal Processing*, 63(2):512–527, 2014.

7 J. Yan, H. Liu, B. Jiu, B. Chen, Z. Liu, and Z. Bao. Simultaneous multibeam resource allocation scheme for multiple target tracking. *IEEE Transactions on Signal Processing*, 63(12):3110–3122, 2015.

8 J. Yan, H. Jiao, W. Pu, C. Shi, J. Dai, and H. Liu. Radar sensor network resource allocation for fused target tracking: A brief review. *Information Fusion*, 86:104–115, 2022.

9 C. Shi, F. Wang, S. Salous, and J. Zhou. Joint subcarrier assignment and power allocation strategy for integrated radar and communications system based on power minimization. *IEEE Sensors Journal*, 19(23):11167–11179, 2019.

10 H. Jiao, J. Yan, W. Pu, Y. Chen, H. Liu, and M. S. Greco. Wideband sensor resource allocation for extended target tracking and classification. *IEEE Transactions on Signal Processing*, 73:55–66, 2024.

11 N. Levanon. *Radar Principles*. Wiley-Interscience, New York, 1988.

12 P. C. Suo, S. Tao, R. Tao, and Z. Nan. Detection of high-speed and accelerated target based on the linear frequency modulation radar. *IET Radar, Sonar & Navigation*, 8(1):37–47, 2014.

13 X. Yang, W.-A. Zhang, M. Chen, and L. Yu. Hybrid sequential fusion estimation for asynchronous sensor network-based target tracking. *IEEE Transactions on Control Systems Technology*, 25(2):669–676, 2016.

14 A. Ludloff and M. Minker. Reliability of velocity measurement by mtd radar. *IEEE Transactions on Aerospace and Electronic Systems*, AES-21(4):522–528, 2007.

15 C. Liu, T. Wang, S. Zhang, and B. Ren. A clutter suppression algorithm via weighted $l_2 - norm$ penalty for airborne radar. *IEEE Signal Processing Letters*, 29:1522–1525, 2022.

16 H. You, X. Jianjuan, and G. Xin. *Radar Data Processing with Applications*. John Wiley & Sons, 2016.

17 V. Lakshmanan. Overview of radar data compression. In *Satellite Data Compression, Communications, and Archiving III*, volume 6683, SPIE, pages 75–79, 2007.

18 R. J. Purdy, P. E. Blankenship, C. E. Muehe, C. M. Rader, E. Stern, and R. C. Williamson. Radar signal processing. *Lincoln Laboratory Journal*, 12(2):297–320, 2000.

19 J. Yan, W. Pu, S. Zhou, H. Liu, and M. S. Greco. Optimal resource allocation for asynchronous multiple targets tracking in heterogeneous radar networks. *IEEE Transactions on Signal Processing*, 68:4055–4068, 2020.

20 N. C. Luong, X. Lu, D. T. Hoang, D. Niyato, and D. I. Kim. Radio resource management in joint radar and communication: A comprehensive survey. *IEEE Communications Surveys & Tutorials*, 23(2):780–814, 2021.

21 P. Moo and Z. Ding. *Adaptive Radar Resource Management*. Academic Press, 2015.

22 J. Yan, J. Dai, W. Pu, S. Zhou, H. Liu, and Z. Bao. Quality of service constrained-resource allocation scheme for multiple target tracking in radar sensor network. *IEEE Systems Journal*, 15(1):771–779, 2020.

23 C. Shi, J. Zhou, and F. Wang. Adaptive resource management algorithm for target tracking in radar network based on low probability of intercept. *Multidimensional Systems and Signal Processing*, 29:1203–1226, 2018.

24 J. L. Williams. Search theory approaches to radar resource allocation. In *Proceedings of the 7th US/Australia Joint Workshop on Defense Applications of Signal Processing (DASP), Coolum, Australia*, 2011.

25 P. Chavali and A. Nehorai. Scheduling and power allocation in a cognitive radar network for multiple-target tracking. *IEEE Transactions on Signal Processing*, 60(2):715–729, 2011.

26 B. R. Mahafza. *Radar Systems Analysis and Design using MATLAB*. Chapman and Hall/CRC, 2005

27 J. Yan, T. He, L. Ma, W. Pu, H. Liu, and M. S. Greco. Maneuvering resource allocation for coordinated target tracking in airborne radar network. *IEEE Transactions on Signal Processing*, 71:1563–1573, 2023.

28 J. Yan, T. Zhang, L. Ma, W. Pu, and H. Liu. Deployment optimization for integrated search and tracking tasks in netted radar system based on pareto theory. *IEEE Transactions on Aerospace and Electronic Systems*, 60(3):3664–3672, 2024.

29 M. Xie, W. Yi, T. Kirubarajan, and L. Kong. Joint node selection and power allocation strategy for multitarget tracking in decentralized radar networks. *IEEE Transactions on Signal Processing*, 66(3):729–743, 2017.

30 C. Shi, L. Ding, F. Wang, J. Zhou, and J. Yan. Collaborative radar node selection and transmitter resource allocation algorithm for target tracking applications in multiple radar architectures. *Digital Signal Processing*, 121:103318, 2022.

31 J. Sun, W. Yi, P. K Varshney, and L. Kong. Resource scheduling for multi-target tracking in multi-radar systems with imperfect detection. *IEEE Transactions on Signal Processing*, 70:3878–3893, 2022.

32 K. Dabcevic. Intelligent Jamming and Anti-jamming Techniques using Cognitive Radios. PhD Programme in Computational Intelligence University of Genoa, 2015

33 E. F. Knott, J. F. Schaeffer, and M. T. Tulley. *Radar Cross Section*. SciTech Publishing, 2004.

34 J. M. Headrick and J. F. Thomason. Applications of high-frequency radar. *Radio Science*, 33(4):1045–1054, 1998.

35 S. Sahin and T. Girici. Resource allocation in networked joint radar and communications. *IEEE Transactions on Aerospace and Electronic Systems*, 61(2):4697–4706, 2024.

36 C. Bell, D. Bonachea, R. Nishtala, and K. Yelick. Optimizing bandwidth limited problems using one-sided communication and overlap. In *Proceedings 20th IEEE International Parallel & Distributed Processing Symposium*, IEEE, page 10, 2006.

37 D.-W. Yun and W.-C. Lee. Intelligent dynamic spectrum resource management based on sensing data in space-time and frequency domain. *Sensors*, 21(16):5261, 2021.

38 C. Shi, Z. Tang, L. Ding, and J. Yan. Multidomain resource allocation for asynchronous target tracking in heterogeneous multiple radar networks with nonideal detection. *IEEE Transactions on Aerospace and Electronic Systems*, 60(2):2016–2033, 2023.

39 A. Ahmad, S. Ahmad, M. H. Rehmani, and N. U. Hassan. A survey on radio resource allocation in cognitive radio sensor networks. *IEEE Communications Surveys & Tutorials*, 17(2):888–917, 2015.

40 M. Bică and V. Koivunen. Generalized multicarrier radar: Models and performance. *IEEE Transactions on Signal Processing*, 64(17):4389–4402, 2016

41 Y. Gao, Y. Yuan, H. Li, and W. Yi. Reinforcement learning-based anti-jamming strategy for self-defense jammer-aided radar systems. *IEEE Transactions on Aerospace and Electronic Systems*, 61(2):3852–3867, 2024.

42 W. Zhang, T. Zhao, Z. Zhao, D. Ma, and F. Liu. Performance analysis of deep reinforcement learning-based intelligent cooperative jamming method confronting multi-functional networked radar. *Signal Processing*, 207:108965, 2023.

43 K. Roussel, G. Babur, and F. Le Chevalier. Optimization of low sidelobes radar waveforms: Circulating codes. In *2014 International Radar Conference*, IEEE, pages 1–6, 2014.

44 J. S. Kulpa, Ł. Maślikowski, and M. Malanowski. Filter-based design of noise radar waveform with reduced sidelobes. *IEEE Transactions on Aerospace and Electronic Systems*, 53(2):816–825, 2017.

45 G. Galati, G. Pavan, Waveform design and related processing for multiple target detection and resolution. In G. Weinberg, editor, *Topics in Radar Signal Processing*, IntechOpen, 3–24, 2018.

46 N. Levanon and E. Mozeson. *Radar Signals*. John Wiley & Sons, 2004.

47 M. R. Bell. Information theory and radar waveform design. *IEEE transactions on Information Theory*, 39(5):1578–1597, 2002

48 M. I. Skolnik. Radar Handbook. 1970.

49 P. Zhang, J. Yan, W. Pu, H. Liu, and M. S. Greco. Multi-dimensional resource management scheme for multiple target tracking under dynamic electromagnetic environment. *IEEE Transactions on Signal Processing*, 722377–2393, 2024.

50 C. Shi, J. Zhou, F. Wang, and J. Chen. Mutual information–based lpi optimisation for radar network. *International Journal of Electronics*, 102(7):1114–1131, 2015.

51 C. Shi, F. Wang, M. Sellathurai, J. Zhou, and H. Zhang. Robust transmission waveform design for distributed multiple-radar systems based on low probability of intercept. *ETRI Journal*, 38(1):70–80, 2016.

52 F. Wang, H. Li, and M. A. Govoni. Power allocation and co-design of multicarrier communication and radar systems for spectral coexistence. *IEEE Transactions on Signal Processing*, 67(14):3818–3831, 2019.

53 H. Zhang, W. Liu, J. Shi, T. Fei, and B. Zong. Joint detection threshold optimization and illumination time allocation strategy for cognitive tracking in a networked radar system. *IEEE Transactions on Signal Processing*, 70:5833–5847, 2022.

54 K. L. Bell, C. J. Baker, G. E. Smith, J. T. Johnson, and M. Rangaswamy. Cognitive radar framework for target detection and tracking. *IEEE Journal of Selected Topics in Signal Processing*, 9(8):1427–1439, 2015.

55 J. Yan, H. Liu, W. Pu, S. Zhou, Z. Liu, and Z. Bao. Joint beam selection and power allocation for multiple target tracking in netted colocated mimo radar system. *IEEE Transactions on Signal Processing*, 64(24):6417–6427, 2016

56 J. Yan, W. Pu, H. Liu, S. Zhou, and Z. Bao. Cooperative target assignment and dwell allocation for multiple target tracking in phased array radar network. *Signal Processing*, 141:74–83, 2017.

57 H. Zhang, W. Liu, B. Zong, J. Shi, and J. Xie. An efficient power allocation strategy for maneuvering target tracking in cognitive mimo radar. *IEEE Transactions on Signal Processing*, 69:1591–1602, 2021.

58 J. Dai, J. Yan, W. Pu, H. Liu, and M. S. Greco. Adaptive channel assignment for maneuvering target tracking in multistatic passive radar. *IEEE Transactions on Aerospace and Electronic Systems*, 59(3):2780–2793, 2022.

59 M. Hernandez and A. Farina. Posterior Cramér-Rao Bounds in Cluttered Environments with Measurement Origin and Accuracy Uncertainties. In M. Mallick, B.-N. Vo, and R. Tharmarasa, editors, *Advanced Techniques in Tracking and Sensor Management: Theory and Applications*, Wiley/IEEE, 10, 2025.

60 X. Zhang, P. Willett, and Y. Bar-Shalom. Dynamic Cramér-Rao bound for target tracking in clutter. *IEEE Transactions on Aerospace and Electronic Systems*, 41(4):1154–1167, 2005.

61 J. Yan, H. Liu, W. Pu, H. Liu, Z. Liu, and Z. Bao. Joint threshold adjustment and power allocation for cognitive target tracking in asynchronous radar network. *IEEE Transactions on Signal Processing*, 65(12):3094–3106, 2017

62 J. Yan, H. Liu, and Z. Bao. Power allocation scheme for target tracking in clutter with multiple radar system. *Signal Processing*, 144:453–458, 2018.

63 J. Yan, J. Dai, W. Pu, H. Liu, and M. Greco. Target capacity based resource optimization for multiple target tracking in radar network. *IEEE Transactions on Signal Processing*, 69:2410–2421, 2021.

64 C. Shi, F. Wang, M. Sellathurai, J. Zhou, and S. Salous. Power minimization-based robust ofdm radar waveform design for radar and communication systems in coexistence. *IEEE Transactions on Signal Processing*, 66(5):1316–1330, 2017.

65 X. Fan, P. Bai, H. Wang, J. Zhang, and H. Li. Joint optimization method of airborne mimo radar track and radiated power based on mutual information. *IEEE Access*, 8:136278–136295, 2020.

66 J. Yan, W. Pu, J. Dai, H. Liu, and Z. Bao. Resource allocation for search and track application in phased array radar based on pareto bi-objective optimization. *IEEE Transactions on Vehicular Technology*, 68(4):3487–3499, 2019.

67 C. Shi, L. Ding, F. Wang, S. Salous, and J. Zhou. Low probability of intercept-based collaborative power and bandwidth allocation strategy for multi-target tracking in distributed radar network system. *IEEE Sensors Journal*, 20(12):6367–6377, 2020.

68 M. Malanowski and K. Kulpa. Optimization of confirmation time of bistatic tracks in passive radar. *IEEE Transactions on Aerospace and Electronic Systems*, 47(2):1060–1072, 2011.

69 J. Yan, W. Pu, S. Zhou, H. Liu, and Z. Bao. Collaborative detection and power allocation framework for target tracking in multiple radar system. *Information Fusion*, 55:173–183, 2020.

70 D. Schuhmacher, B.-T. Vo, and B.-N. Vo. A consistent metric for performance evaluation of multi-object filters. *IEEE Transactions on Signal Processing*, 56(8):3447–3457, 2008.

71 A. K. Gostar, R. Hoseinnezhad, T. Rathnayake, X. Wang, and A. Bab-Hadiashar. Constrained sensor control for labeled multi-bernoulli filter using Cauchy-Schwarz divergence. *IEEE Signal Processing Letters*, 24(9):1313–1317, 2017.

72 Y. Zhu, J. Wang, and S. Liang. Multi-objective optimization based multi-bernoulli sensor selection for multi-target tracking. *Sensors*, 19(4):980, 2019.

73 A. Deligiannis, A. Panoui, S. Lambotharan, and J. A. Chambers. Game-theoretic power allocation and the nash equilibrium analysis for a multistatic mimo radar network. *IEEE Transactions on Signal Processing*, 65(24):6397–6408, 2017.

74 C. Shi, F. Wang, M. Sellathurai, and J. Zhou. Non-cooperative game theoretic power allocation strategy for distributed multiple-radar architecture in a spectrum sharing environment. *IEEE Access*, 6:17787–17800, 2018.

75 Y. Yuan, W. Yi, R. Hoseinnezhad, and P. K. Varshney. Robust power allocation for resource-aware multi-target tracking with colocated mimo radars. *IEEE Transactions on Signal Processing*, 69:443–458, 2020.

76 H. Godrich, A. P. Petropulu, and H. V. Poor. Power allocation strategies for target localization in distributed multiple-radar architectures. *IEEE Transactions on Signal Processing*, 59(7):3226–3240, 2011.

77 H.-Z. Feng, H.-W. Liu, J.-K. Yan, F.-Z. Dai, and M. Fang. A fast efficient power allocation algorithm for target localization in cognitive distributed multiple radar systems. *Signal Processing*, 127:100–116, 2016.

78 C. Shi, Y. Wang, F. Wang, S. Salous, and J. Zhou. Power resource allocation scheme for distributed mimo dual-function radar-communication system based on low probability of intercept. *Digital Signal Processing*, 106:102850, 2020.

79 X. Liu, Z. Xu, W. Dong, L. Wang, and X. Li. Cognitive resource allocation for target tracking in location-aware radar networks. *IEEE Signal Processing Letters*, 27:650–654, 2020.

80 H. Godrich, A. P. Petropulu, and H. V. Poor. Sensor selection in distributed multiple-radar architectures for localization: A knapsack problem formulation. *IEEE Transactions on Signal Processing*, 60(1):247–260, 2011.

81 H. Godrich, A. Petropulu, and H. V. Poor. Cluster allocation schemes for target tracking in multiple radar architecture. In *2011 Conference Record of the Forty Fifth Asilomar Conference on Signals, Systems and Computers (ASILOMAR)*, IEEE, pages 863–867, 2011.

82 H. Godrich, A. Petropulu, and H. V. Poort. Estimation performance and resource savings: tradeoffs in multiple radars systems. In *2012 IEEE International Conference on Acoustics, Speech and Signal Processing (ICASSP)*, IEEE, pages 5205–5208, 2012.

83 Y. Zhang, M. Pan, and Q. Han. Joint sensor selection and power allocation algorithm for multiple-target tracking of unmanned cluster based on fuzzy logic reasoning. *Sensors*, 20(5):1371, 2020.

84 Y. Shi, B. Jiu, J. Yan, and H. Liu. Data-driven radar selection and power allocation method for target tracking in multiple radar system. *IEEE Sensors Journal*, 21(17):19296–19306, 2021.

85 A. Deligiannis, S. Lambotharan, and J. A. Chambers. Game theoretic analysis for mimo radars with multiple targets. *IEEE Transactions on Aerospace and Electronic Systems*, 52(6):2760–2774, 2016.

86 J. Yan, H. Liu, B. Jiu, and Z. Bao. Power allocation algorithm for target tracking in unmodulated continuous wave radar network. *IEEE Sensors Journal*, 15(2):1098–1108, 2014.

87 G. Alirezaei, M. Reyer, and R. Mathar. Optimum power allocation in sensor networks for passive radar applications. *IEEE Transactions on Wireless Communications*, 13(6):3222–3231, 2014.

88 G. Alirezaei, O. Taghizadeh, and R. Mathar. Optimum power allocation in sensor networks for active radar applications. *IEEE Transactions on Wireless Communications*, 14(5):2854–2867, 2015.

89 X. Liu, Z.-H. Xu, L. Wang, W. Dong, S. Xiao. Cognitive dwell time allocation for distributed radar sensor networks tracking via cone programming. *IEEE Sensors Journal*, 20(10):5092–5101, 2020

90 H. Zhang, W. Liu, J. Xie, J. Shi, Z. Zhang, and W. Lu. Space-time allocation for transmit beams in collocated mimo radar. *Signal Processing*, 164:151–162, 2019.

91 H. Zhang, B. Zong, and J. Xie. Power and bandwidth allocation for multi-target tracking in collocated mimo radar. *IEEE Transactions on Vehicular Technology*, 69(9):9795–9806, 2020.

92 N. Garcia, A. M. Haimovich, M. Coulon, and M. Lops. Resource allocation in mimo radar with multiple targets for non-coherent localization. *IEEE Transactions on Signal Processing*, 62(10):2656–2666, 2014.

93 H. G. Hoang and B. T. Vo. Sensor management for multi-target tracking via multi-bernoulli filtering. *Automatica*, 50(4):1135–1142, 2014.

94 A. K. Gostar, R. Hoseinnezhad, and A. Bab-Hadiashar. Multi-bernoulli sensor control for multi-target tracking. In *2013 IEEE Eighth International Conference on Intelligent Sensors, Sensor Networks and Information Processing*, IEEE, pages 312–317, 2013.

95 M. Xie, W. Yi, L. Kong, and T. Kirubarajan. Receive-beam resource allocation for multiple target tracking with distributed mimo radars. *IEEE Transactions on Aerospace and Electronic Systems*, 54(5):2421–2436, 2018.

96 X. Li, T. Zhang, W. Yi, L. Kong, and X. Yang. Radar selection based on the measurement information and the measurement compensation for target tracking in radar network. *IEEE Sensors Journal*, 19(18):7923–7935, 2019.

97 H. Duan, J. Zhao, Y. Deng, Y. Shi, and X. Ding. Dynamic discrete pigeon-inspired optimization for multi-uav cooperative search-attack mission planning. *IEEE Transactions on Aerospace and Electronic Systems*, 57(1):706–720, 2020.

98 C. Zhao, J. Liu, M. Sheng, W. Teng, Y. Zheng, and J. Li. Multi-uav trajectory planning for energy-efficient content coverage: A decentralized learning-based approach. *IEEE Journal on Selected Areas in Communications*, 39(10):3193–3207, 2021.

99 S. Perez-Carabaza, E. Besada-Portas, J. A. Lopez-Orozco, and J. M. de la Cruz. Ant colony optimization for multi-uav minimum time search in uncertain domains. *Applied Soft Computing*, 62:789–806, 2018.

100 J. Sun, J. Tang, and S. Lao. Collision avoidance for cooperative uavs with optimized artificial potential field algorithm. *IEEE Access*, 5:18382–18390, 2017

101 T. He, X. Mo, P. Zhang, H. Jiao, J. Yan, and H. Liu. Joint trajectory design and power allocation for target tracking in airborne radar network. In *2021 CIE International Conference on Radar (Radar)*, IEEE, pages 1165–1168, 2021.

102 C. Shi, X. Dai, Y. Wang, J. Zhou, and S. Salous. Joint route optimization and multidimensional resource management scheme for airborne radar network in target tracking application. *IEEE Systems Journal*, 16(4):6669–6680, 2021.

103 H. V. Nguyen, H. Rezatofighi, B.-N. Vo, and D. C. Ranasinghe. Online uav path planning for joint detection and tracking of multiple radio-tagged objects. *IEEE Transactions on Signal Processing*, 67(20):5365–5379, 2019.

Index

j

k

l

m

n

o

Printed in the USA/Agawam, MA
September 17, 2026

910660.016